Iron Ore and Open Pit Operators Conference 2026

23–25 June 2026
Perth, Australia

The Australasian Institute of Mining and Metallurgy
Publication Series No 2/2026

AusIMM

Published by:
The Australasian Institute of Mining and Metallurgy
Ground Floor, 204 Lygon Street, Carlton Victoria 3053, Australia

All papers published in this volume were peer reviewed before publication.

ISBN 978-1-922395-57-3

Advisory Committee

IRON ORE

Keith Vining
Conference Advisory Committee Chair

Dr John Clout
FAusIMM

Amir Cooper
GAusIMM

James Crump

Peter Fisher
FAusIMM

Meg Fletcher

Michelle Helm

Ralph Holmes
HonFAusIMM(CP)

Tom Honeyands
MAusIMM

Jillian James
MAusIMM

Theresa Nguyen

Dr Johan Utomo
MAusIMM(CP)

OPEN PIT

Daniel Goldstein
MAusIMM(CP)
Conference Advisory Committee Chair

Chris Carr
FAusIMM(CP)

Carol De Moliner
MAusIMM(CP)

Sjoerd Duim
MAusIMM

Peter Fisher
FAusIMM

Darren Head
FAusIMM

Paul Hughes
MAusIMM

Jacinta Ireland
FAusIMM

Taylor McDonald
AAusIMM

Adrian Pratt
FAusIMM (CP)

AusIMM

Julie Allen
General Manager, Events

Raha Karimi
Event Manager, Events

Sasha Pell
Senior Manager, Conference Growth, Events

Sharon Lim
Program Manager, Events

Sioban Rocks
Marketing Manager, Events

Reviewers

We would like to thank the following people for their contribution towards enhancing the quality of the papers included in this volume:

IRON ORE

Oluwatosin Aladejebi

Maryam Barmi

Nathan Barrett

Chunlin Chen

Henry Connor

Mingyong Du

Fang Huang

Vlad Jokovic

Liming Lu

Angus McFarlane

Alex Minns

Alan Ooi

Kwaku Owusu

Will Patton

Michael Peterson

Mark Pownceby

Clayton Simpson

Tejbir Singh

Duncan Street

Adel Vatandoost

Rou Wang

Wendy Wang

OPEN PIT

Tony Deng

Megan Gaida

John Mulcahy

Jeffrey Telfer

Mathew Widelski

Reviewers

Foreword

It is with great pleasure that we present the proceedings of the Iron Ore 2026 Conference, bringing together a global community of industry leaders, technical experts, researchers, and practitioners dedicated to advancing the iron ore sector. As one of the cornerstone commodities underpinning modern infrastructure and economic development, iron ore continues to play a critical role in shaping the future of our society.

Iron Ore 2026 occurs against a backdrop of resilient export volumes but mounting structural pressure. Shipments are expected to remain near record levels, underpinned by Australia's low-cost position and established supply chains, although growth will be constrained as new global supply enters the market. At the same time, demand is becoming less dependent on China—where steel production has likely peaked and is expected to moderate—and increasingly supported by emerging markets. However, a softer and more volatile price environment is anticipated as supply growth outpaces demand, placing greater emphasis on cost competitiveness and, critically, ore quality. In this evolving landscape, Australia's long-term value proposition will depend not only on scale and reliability, but on its ability to respond to shifting market preferences for higher-grade, lower-emissions feedstocks.

Simultaneously, the technological landscape is undergoing rapid transformation, with digitalisation, automation, advanced analytics, and artificial intelligence redefining how ore is discovered, mined, processed, and delivered to market. The rapid integration of these emerging technologies into an operational environment is critical for maintaining the industry's existing competitive advantages.

A central theme throughout these proceedings is sustainability. The imperative to reduce carbon emissions, minimise environmental impact, and operate responsibly is no longer aspirational—it is fundamental. Contributions within this volume highlight innovative approaches in ore beneficiation, energy efficiency, water management, tailings stewardship, and mine rehabilitation. Importantly, many papers also reflect the growing integration of sustainability metrics into operational decision-making and long-term strategic planning. In particular, the Federal Government's Future Made in Australia policy framework is driving emergence of green iron production projects and demonstrators in the Mid-West region of WA to exploit future markets for premium green products.

Safety and people remain at the heart of our industry. The continued focus on zero harm, combined with the changing nature of work in increasingly automated and remote operations, presents both challenges and opportunities. These proceedings showcases advance in safety systems, workforce development, and organisational culture, underscoring the ongoing commitment to protecting and empowering the people who drive the industry forward.

Innovation is another defining thread of Iron Ore 2026. The proceedings capture leading-edge developments across the value chain—from exploration techniques and geometallurgy to comminution, materials handling, and logistics optimisation. The integration of real-time data, predictive modelling, and system-wide optimisation is enabling more resilient and efficient operations, even in the face of variability in orebodies and market conditions.

AusIMM and CSIRO extend our sincere appreciation to all sponsors, authors, reviewers, session chairs, and organising committee members whose dedication and expertise made this conference possible. Your contributions ensure that these proceedings serve not only as a record of current practice but also as an enduring resource for future advancements.

We trust that readers will find this volume both informative and inspiring, and that it will contribute meaningfully to the continued evolution of a safe, sustainable, and innovative iron ore industry.

Yours faithfully,

Keith Vining

Iron Ore 2026 Conference Advisory Committee Chair

Foreword

On behalf of the Conference Organising Committee, I am delighted to welcome you to the AusIMM Open Pit Operators Conference 2026.

Australia remains a global leader in mining, underpinning the supply of iron ore, gold and increasingly critical minerals essential to the energy transition. This conference provides an important platform to bring together industry leaders, practitioners, researchers and technology partners who are shaping the future of open pit mining.

The 2026 program showcases the breadth and depth of innovation transforming our industry. Across the plenary sessions and technical streams, delegates will engage with leading-edge advancements in automation and autonomous systems, data-driven optimisation, and strategic mine planning. Strong focus is also placed on geotechnical excellence and slope stability, alongside advances in drill and blast practices at increasingly larger and more complex scales. Importantly, the program reflects the growing emphasis on integrating ESG considerations, risk-informed decision-making, and sustainable, lower-carbon operations into everyday mining practice.

A defining feature of this year's conference is the integration of digital technologies and real-world application, from AI-enabled operational optimisation and autonomous haulage to advanced geotechnical monitoring and analytics. Equally, we recognise that operational success is grounded in people and systems, with sessions exploring capability development, workforce resilience and psychosocial safety in modern, open pit mining operations.

The technical program draws on high-quality contributions from across the industry, with practical case studies and lessons learned that will support better decision-making and improved operational outcomes.

I would like to thank our presenters and authors for their valuable contributions, as well as the Conference Organising Committee and our dedicated paper reviewers for their efforts in delivering a strong and relevant program. I also acknowledge our sponsors and exhibitors, BHP, Dredge Robotics, Sandvik, Micromine, SGS, RPMGlobal, RioTinto, ALS, AMC Consultants, MX3, Elastomers Australia, Fortescue, Hatch and McLanahan, whose ongoing support is essential to the success of this event.

Finally, thank you to our delegates. Your participation and engagement are what make this conference a vibrant and valuable forum for learning, collaboration and advancing our industry.

I look forward to a highly productive and insightful conference.

Yours faithfully,

Daniel Goldstein

MAusIMM(CP)

Open Pit Operators 2026 Conference Advisory Committee Chair

Sponsors

Major Conference Sponsor

BHP

Platinum Sponsors

Dredge Robotics

Sandvik

Gold Sponsors

Micromine

SGS

Silver Sponsor

RPMGlobal

Lanyard Sponsor

Rio Tinto

Coffee Cart Sponsor

ALS

Conference App Sponsor

AMC Consultants

Hydration Sponsor

MX3

Conference Proceedings Sponsor

elastomers AUSTRALIA

Welcome Reception Sponsor

Fortescue

Notepad Sponsor

HATCH

Technical Session Sponsor

McLanahan

elastomers AUSTRALIA
easa SCREEN MEDIA SOLUTIONS
lastoEasi-Lok™
SAFER FIXING SYSTEM BUILT
OR MODERN WORKSITES
REVOLUTIONISING SAFETY IN SCREENING OPERATIONS.
ElastoEasi-Lok™ represents the next step in our modular fixing journey, evolving over 40 years from ElastoLok® to ElastoKey-Lok® and now to ElastoEasi-Lok™.
Developed with refined fastening technology, ElastoEasi-Lok™ is designed to significantly reduce hammer use during panel and locking strip installation, supporting reduced line of fire risks and labour intensity.
KEY
BENEFITS
afety above shortcuts
astoEasi-Lok™ prioritises safety over shortcuts. By iminating repetitive impact tasks, it slashes the risk of cidents and costly Lost Time Injuries (LTIs) – which irrently average $40,000 per event.
educed hammer use during installation
anels and locking strips are secured using a plug stening system creating a significant reduction in ammer use.
ustomisable and retrofittable
astoEasi-Lok™ fits most screening equipment and can e seamlessly integrated with existing ElastoKey-Lok® anels and side slats, allowing sites to upgrade ogressively without full system replacement and inimised costs.
ow risk
astoEasi-Lok can be integrated in stages with the oility to seamlessly lock into other Elastomers Australia odular systems, minimising risk to your screening ficiency and plant reliability.
trength without compromise
erformance is benchmarked to our proven odular fixing systems including ElastoLok® and astoKey-Lok®, ensuring secure long-term stability nd fastening integrity.
RESULTS
GUARANTEED
ElastoEasi-Lok™ is a safer, more efficient screen media system that supports your people, your plant, and your long-term operational goals.
SEE THE
DIFFERENCE
Whether you're upgrading or starting fresh, ElastoEasi-Lok™ delivers safe long-term performance you can rely on.
• Express your interest via the contact details below
• Seek a product demonstration
• Request a tailored trial plan to suit your operations
Elastomers Australia
(03) 9335 7999
sales@elastomersaustralia.com.au
www.elastomersaustralia.com.au

Contents

Iron Ore

Iron Ore: Digital transformation and AI in mining

Iron Ore: Exploration and geology

Iron Ore: Health, safety, environment and community

Iron Ore: **Iron and steelmaking**

Iron Ore: **Mining and processing**

Iron Ore: **Ore characterisation and geometallurgy**

Contents

Open Pit Operators

Open Pit: AI, automation and autonomous systems

Open Pit: Capability development

Open Pit: Critical minerals

Open Pit: Drill and blast – design, execution and scale

Open Pit: Mine planning

Open Pit: Mine planning and strategic optimisation

Open Pit: Operational data, analytics and optimisation

Open Pit: People, capability and operational systems

Open Pit: **Risk management**

Open Pit: **Risk, ESG and system-level decision making**

Open Pit: **Slope stability and geotechnical analysis**

Iron Ore

Digital transformation and AI in mining

AI-driven operational excellence for iron ore processing

J Erome-Utunedi[1], T Carricajo[2] and J Sepulveda[3]

1. Manager, Automation and Digitalisation, ANDRITZ Pty Ltd, Carrum Downs Vic 3201.
   Email: john.erome-utunedi@andritz.com
2. Product Manager Digital Solution Mining, ANDRITZ Pty Ltd, Carrum Downs Vic 3201.
   Email: tomas.carricajo@andritz.com
3. Regional Manager Australia and New Zealand, ANDRITZ Pty Ltd, Carrum Downs Vic 3201.
   Email: juan.sepulveda@andritz.com

INTRODUCTION

Advanced Process Control (APC) has historically provided substantial benefits in iron ore operations by stabilising circuits and reducing process variability. Despite these advantages, many sites continue to encounter persistent challenges related to extreme ore variability, rapidly shifting operational constraints, and competing production targets. These dynamic variables severely limit the ability of static control logic to consistently achieve optimal economic performance.

To bridge this gap, this paper outlines a practical method for enhancing existing APC infrastructure through AI-driven optimisation, emphasising Reinforcement Learning (RL) as a supervisory layer. Rather than replacing established controls, this approach enables operators to maintain baseline stability while AI continuously adjusts operational targets in response to actual plant conditions.

OPTIMISATION – BREAKING THE CONTROL CEILING WITH MACHINE LEARNING

Traditional control technologies excel at managing predictable, linear environments. Systems like PID loops and standard APC rely heavily on three fundamental assumptions to function optimally: a consistent cause-and-effect relationship, predictable process gain, and measurable, continuous feedback.

In the reality of iron ore processing, these assumptions frequently fail. Fluctuating ore hardness, sensor degradation, and mechanical wear cause traditional control models to drift, leading to performance plateaus or erratic system behaviour as controllers apply outdated logic to new physical realities.

Machine Learning, particularly RL, thrives in these complex environments where conventional control methods fall short. By acting as a supervisory layer, RL replaces rigid, pre-programmed rules with adaptive intelligence, learning underlying process dynamics directly from historical and real-time data.

For instance, the Shuttle AI project deployed RL to resolve material handling bottlenecks in a Pilbara iron ore screening plant (Versiani *et al*, 2021). As process conditions evolved, the RL agent continuously self-corrected. By analysing real-time shuttle coordinates, payloads, and downstream bin capacities, the system executed dynamic 'bin jumping' to navigate material flows. This implementation adapted to feed surges seamlessly without manual intervention, reducing operational instability and production losses by 17 per cent while virtually eliminating high-level alarms. Furthermore, the dynamic control strategy delivered a 15 per cent increase in ramp-up speed and achieved a 40 per cent increase in production during ore break operations.

SIMULATION – THE PROVING GROUND FOR INDUSTRIAL AI

The transition to AI-driven control requires robust training data, but training an AI agent directly on a live iron ore circuit introduces unacceptable operational and safety risks. Consequently, high-fidelity simulation environments serve as the foundational proving ground for industrial AI.

Before deployment, RL agents are trained within digital replicas of the plant. During the development of the Shuttle AI solution, the system executed millions of virtual cycles, condensing years of operational scenarios, including rare edge cases and severe upsets, into a highly accelerated time frame. This generated the massive volumes of synthetic data required for the agent to test optimal routing strategies. By bypassing exploratory errors in a completely risk-free environment, the AI converged on a stable control policy prior to live deployment. Once integrated with the existing PLC architecture, this seamless sim-to-real transfer ensured the AI arrived fully prepared, applying proven strategies from day one.

SMART MONITORING – CLOSING THE DECISION GAP

While autonomous control optimises the physical process, sustaining operational excellence requires empowering the engineering and reliability teams who manage the plant's assets. Historically, a severe 'Decision Gap' has existed in the critical middle of the organisation. While operators rely on real-time DCS screens and management tracks high-level KPIs, process engineers are frequently left troubleshooting complex root causes using disconnected spreadsheets and ad hoc tools.

AI-driven Smart Monitoring platforms and generative AI assistants shift these teams from manual data gatherers to strategic decision-makers. By achieving cross-domain data fusion, integrating machine condition data with process data, AI engines provide a unified, real-time view of asset health. For instance, the ANDRITZ Copilot, developed in partnership with Microsoft, allowed engineers from industrial plants to rapidly diagnose anomalies using natural language queries (Fuentealba, Torres and Olivares, 2025). Instead of hunting through isolated sensor trends, the AI fused live process data with digitised technical manuals to identify the root cause and recommend specific counteractions. This capability accelerates troubleshooting and facilitates scalable knowledge transfer, standardising workflows and reducing equipment downtime.

CONCLUSION – THE PRAGMATIC PATH FORWARD

The integration of Artificial Intelligence on top of Advanced Process Control represents the next mandatory evolution in mineral processing. True, unassisted 'lights-out' autonomy remains a future horizon; however, the strategic deployment of AI technologies today brings operations significantly closer to that goal. By applying Machine Learning to optimise complex constraints, utilising Simulation for risk-free training, and deploying Smart Monitoring to close the engineering decision gap, operations can achieve adaptive, data-driven optimisation without requiring disruptive modifications to their existing infrastructure.

REFERENCES

Versiani, B, Gooch, A, Horeth, A and Duggan, E, 2021. Application of a reinforcement learning AI solution for optimisation of screening shuttles control, in *Proceedings of the Iron Ore Conference 2021*, pp 237–248 (The Australasian Institute of Mining and Metallurgy: Melbourne).

Fuentealba, A, Torres, G and Olivares, B, 2025. Artificial Intelligence Agent for decision support in CMP Tailings Filtration Process, presented at the 10th International Conference on Operational Excellence in Mining, Minexcellence 2025 (Gecamin: Santiago).

'DrillTime' orebody definition via blasthole logging

J Market[1] and M Mayne[2]

1. Chief Scientist, Wireline Services Group, Osborne Park WA 6017. Email: jennifermarket@wirelineservices.com.au
2. Managing Director, Wireline Services Group, Osborne Park WA 6017. Email: mattmayne@wirelineservices.com.au

ABSTRACT

As iron ore deposits become deeper, lower grade, and more geologically complex, rapid and accurate orebody characterisation is essential. Borehole geophysical logs—such as density, porosity, gamma ray, resistivity, televiewers (borehole images), magnetic susceptibility, sonic velocity, nuclear magnetic resonance and spectroscopy—are routinely used to construct and refine 3D mine models. However, conventional data acquisition and processing often take days or weeks, limiting their use in production, where rapid decisions and automated workflows are required.

To fully exploit geophysical data for mine productivity, two advances are critical: (1) developing a geophysics-based 3D mine model that is continuously refined as new data are acquired, and (2) integrating 'drilltime' data acquisition to enable near-real-time model updates. The workflow begins with an initial model built from exploration logs, refined as laboratory and geological data become available, and continually updated throughout production. Semi-autonomous calibration of geophysical relationships ensures that the model remains accurate even as the mine expands into new geological domains.

Recent developments in robotic logging platforms now make near-real-time geophysical integration feasible. These systems can automatically navigate between holes, acquire multiple geophysical measurements, process data, and upload results directly to the mine database to update the 3D model within minutes of drilling. The novelty of this approach is that a robotic logging platform removes the risk of personnel on the ground in the blast pattern – something which has limited the applicability of blasthole geophysical logging in the past due to safety concerns. Now it is possible to safely and efficiently collect a wide variety of geophysical logs in the pit.

Real-time geophysical logging shortens the discovery-to-recovery timeline by providing continuously updated, high-resolution orebody information. The combination of automated data collection and dynamic 3D modelling enhances mine safety, precision, and productivity, enabling data-driven, optimised resource recovery throughout the mine life cycle.

This paper is presented as a technology introduction of 'drilltime' Orebody definition. Field trials of the systems described are underway in late 2025 to mid 2026 in the Pilbara and limited examples are presented at this time, with further publications expected in late 2026 once full scale (1000+ hole) trials are completed.

INTRODUCTION

As iron ore mining progresses toward greater depths, lower grades, and increasingly complex geological settings, the need for rapid and accurate orebody characterisation has become more critical than ever. Traditional exploration strategies, while effective during early project stages, are often too slow and episodic to support the operational decision-making required during active production. In this environment, the ability to continuously refine geological understanding directly influences mine safety, productivity, and resource recovery. While geophysical logging data are not fully available in real-time (eg logging while drilling), they can be available in 'drilltime' – within minutes after drilling.

Borehole geophysical logging plays a central role in modern orebody characterisation and 3D mine model construction. Measurements such as density, porosity, gamma ray, resistivity, borehole imaging (televiewer), magnetic susceptibility, sonic velocity, nuclear magnetic resonance, and spectroscopy provide high-resolution information that complements geological logging and laboratory analyses. These data sets are routinely used to build and refine 3D mine models that guide planning, grade control, and production optimisation. However, conventional logging workflows typically involve delayed data acquisition, off-site processing, and manual interpretation, resulting in turnaround times of days to weeks. Such delays limit the effective use of geophysical data in production environments, where rapid decisions, automated workflows, and near-real-time feedback are critical to optimise productivity.

To fully exploit borehole geophysics as an operational tool rather than a post-hoc interpretive aid, two key advances are required. First, mine models must evolve from static to geophysics-driven 3D frameworks that are continuously updated as new data become available. Second, geophysical data acquisition must be integrated directly into drilling operations, enabling model updates in drilltime. In this paradigm, an initial model constructed from exploration data is progressively refined using laboratory measurements, geological observations, and ongoing production logging. Semi-autonomous calibration of geophysical–geological relationships is essential to maintain model accuracy as mining advances into new lithological and structural domains.

Recent developments in robotic borehole logging platforms provide a practical pathway toward this vision. These systems are capable of autonomously navigating between drill holes, acquiring multiple geophysical measurements, processing data on-site, and uploading results directly to mine databases. By eliminating the need for personnel within active blast patterns, robotic platforms address longstanding safety constraints that have historically limited the application of blasthole geophysical logging. As a result, it is now feasible to safely and efficiently collect a wide range of geophysical logs in open pit production settings and to integrate them into 3D mine models within minutes of drilling.

GEOPHYSICAL LOGGING MEASUREMENTS IN IRON ORE

Borehole geophysics logging via wireline has long been standard in the resource development stage of iron ore mines (Jones, Walraven and Knott, 1973; Fullagar, Zhou and Fallon, 1999; Wanstedt, 1982). Essential measurements commonly used in iron ore include the following.

The standard iron ore suite, usually run on nearly every resource development hole and used to create the mine resource map:

- gyro/directional survey.
- natural gamma ray – lithology/depth correlation.
- resistivity/conductivity/SP/induced polarisation – electrical rock properties/water table location.
- magnetic susceptibility – rock magnetic properties.
- gamma-gamma density – rock density, porosity indicator, tonnage.
- calliper – hole size, indicating breakouts/enlargements/collapse.

Figure 1 shows an example of a conventional iron ore geophysical logging suite.

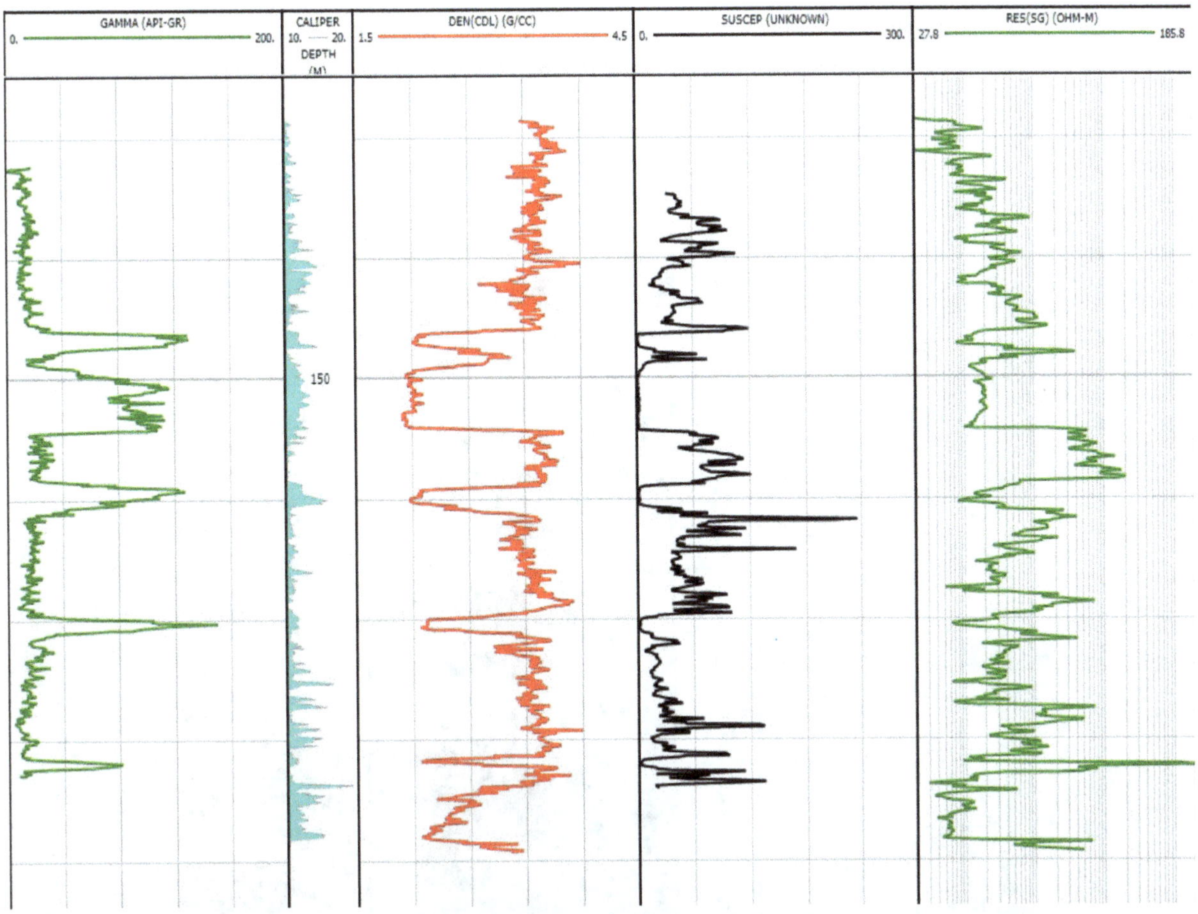

FIG 1 – Conventional iron ore geophysical log suite, including natural gamma, calliper, magnetic susceptibility, density, and resistivity (directional data also collected but not shown).

Geotechnical suite: In addition to the classic suite, televiewer imaging logs provide 360° images of the borehole to characterise structure such as stratigraphy, faults, fractures, and near-borehole voids. They are typically run on a subset of resource definition holes to aid in mapping the orebody structure and in geotechnical modelling.

- Optical televiewers (OPTV): 3D images from a high-resolution downhole camera. OPTV imaging works both above and below the water table, can be affected by 'dirty' mud as it is hard to see through murky water.
- Acoustic televiewers (ATV/BHTV): 3D images created by ultrasonic transducers which provide 360° maps of the borehole shape (form the ultrasonic travel time) and acoustic impedance (from the ultrasonic reflected amplitude). BHTV imaging only works below the water table but is much less affected by water/mud quality than the OPTV.

Figure 2 shows an example of high-resolution stratigraphy in an iron ore resource definition hole (Hawke, 2015).

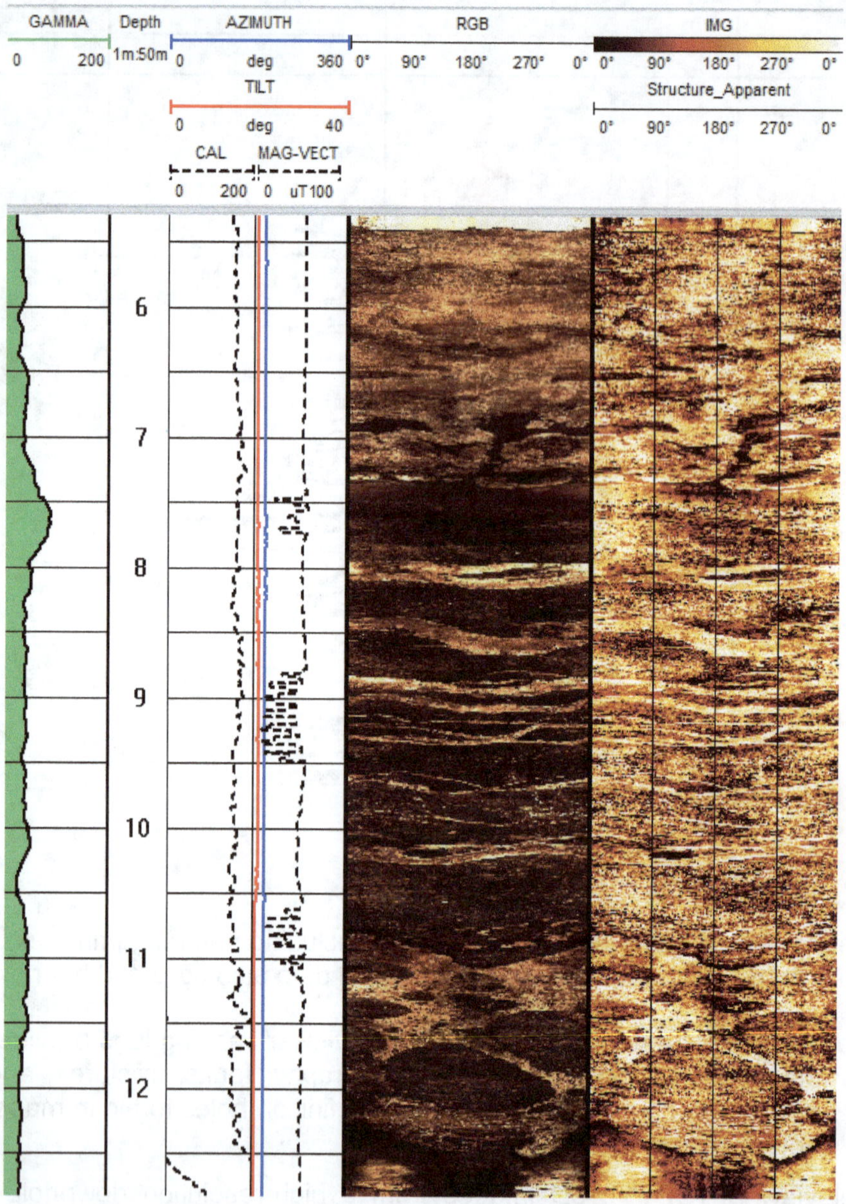

FIG 2 – Televiewer images showing high resolution stratigraphy.

'Specialty' logging suite: additional measurements are also less commonly acquired in iron ore mines and include such measurements as:

- Full waveform sonic: compressional and shear velocities which can be used to determine porosity, correlate with seismic, and calculate rock mechanics properties such as unconfined compressive strength (UCS), bulk modulus, Young's modulus and shear modulus (Elkington, Stouthamer and Brown, 1982). It can also be combined with density measurements to estimate dry bulk density. Sonic logs only work below the water table.
- Vertical seismic profile (VSP): used to measure compressional and shear velocities using a surface source and downhole receivers. It is clamped to the borehole wall and thus works both above and below the water table. It is not yet widely used as there are challenges in both logistics and capabilities.
- Neutron porosity – measures/estimates porosity using a geochemical neutron source/neutron detector. Works above and below the water table if appropriate environmental corrections are applied. Is not as widely used as the traditional suite as the source is significantly stronger than the density tool and the operating cost is higher.
- Nuclear Magnetic Resonance (NMR/BMR): uses nuclear magnetic resonance to determine porosity and permeability. It can also be combined with density measurements to estimate dry

bulk density. It works best in non-magnetic environments. It is not widely used in iron ore due the interference from magnetic rocks as well as the fact that it is generally a more expensive tool to operate.

- Elemental Spectroscopy: uses a neutron source and gamma detector to measure the gamma spectra of the rock which can be converted into elemental analysis or minerology. The tools measure significantly deeper (~30 cm) into the surrounding formation and work above/below the water table. Elemental spectroscopy tools are relatively recent introductions to iron ore and work is still underway in the industry to understand how best to use the measurements (eg as 'downhole assays', minerology indicators etc) (Chi *et al*, 2017).
- Borehole Radar: uses downhole radar sources and receivers to map electrical reflections in the surrounding rock. It maps near- and far- borehole structures and can be used for mapping voids tens of metres from the wellbore. These measurements are not yet widely used in iron ore as the capabilities of the measurements in the conductive rock and the optimal data collection scenarios are still being tested.

Borehole geophysical measurements are used in mapping:

- ore/waste
- grade
- geometry
- tonnage
- geotechnical structure
- mineralogy.

Many established mine sites have 'pre-set' models which input measurements such as gamma, density, and magnetic susceptibility to create the orebody map. However, particularly as orebodies become more heterogeneous, it is increasingly important to have a good understanding of the more complex minerology. To enable this high-resolution mapping, combinations of the traditional iron ore borehole logs can be integrated to map geological and mineral boundaries, while adding additional measurements such as sonic, porosity, and elemental spectroscopy can create insightful mine maps that surpass traditional resolutions and can reduce the reliance on physical assays for orebody definition.

The large amounts of borehole geophysical data collected in iron ore are ideal for applying Artificial Intelligence (AI) techniques to determine lithology (Kitzig, Kepic and Grant, 2018). Figure 3 shows an example where a combination of gamma ray, magnetic susceptibility, resistivity, and induced polarisation are used to compute geological bodies. Using a combination of geophysics and machine learning, accurate models can be created to compute geology and mineralogy from a combination of downhole measurements. This example was a simple but effective case. Machine learning was applied to a training set of holes to determine the best combination of measurements that would reliably map the geology. Then an AI model was developed that uses the pertinent logs to determine lithology/mineralogy zones. While some of these zones mapped directly to manually-picked geologies, the method revealed that the model was more accurate at identifying distinct geological zones. The geologist was still very much part of the solution as local expertise was used to refine the labels of the zones identified by the automation.

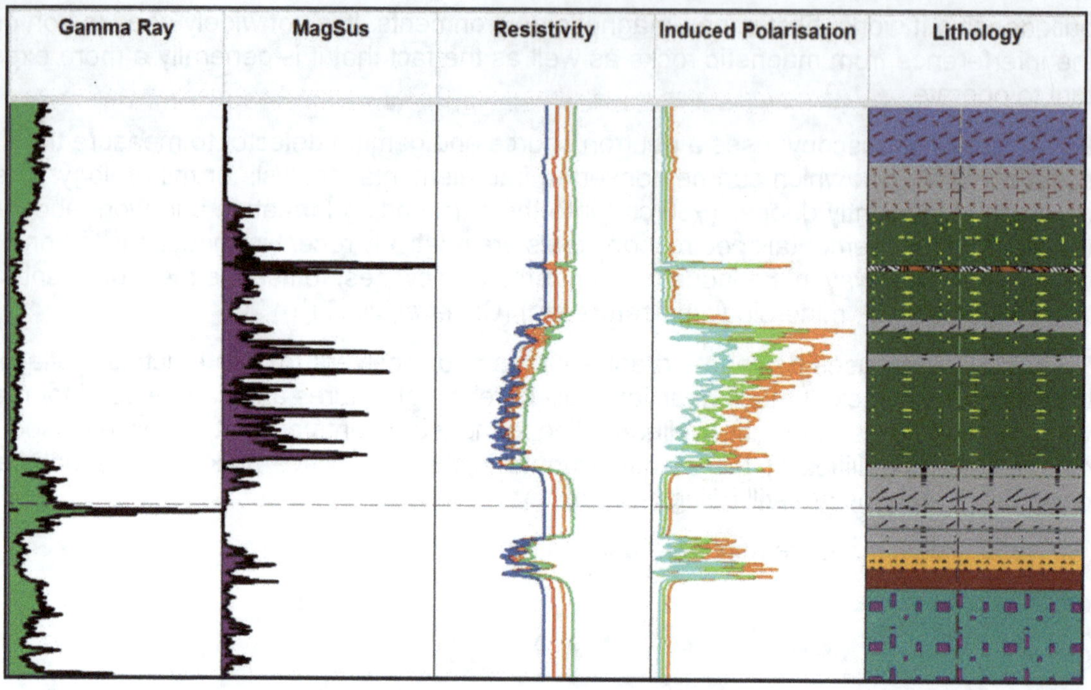

FIG 3 – Example of combining multiple geophysical measurements to determine lithology.

GEOPHYSICAL DATA COLLECTION AND DELIVERY

To collect the conventional borehole geophysical data, manned wireline logging vehicles are generally operated by one person who drives to each resource development hole (RC/diamond/water bore), gets out of the vehicle and uses a winch to lower the wireline logging vehicles down the hole to acquire measurements of the formation. Depending upon the application, several types of measurements might be made on a single sonde (such as gamma + density + resistivity), several sondes might be screwed together to make many measurements at once (time-saving, but suffers from the fact that all the measurements do not reach the bottom of the hole), or multiple logging runs made to collect a full suite of data over the maximum range of hole depths. While data is sometimes at the end of the shift, it is much more common that the data is sent to processing centres in town and processed/reviewed before delivery, which is typically in a few days.

Figure 4 shows an example of a conventional iron ore logging vehicle.

FIG 4 – Conventional wireline logging truck with an operator using a winch to lower a logging tool into the hole.

'DRILLTIME' OREBODY DEFINITION IN REAL-TIME

Traditionally, the borehole geophysical data collected in the resource definition stage are used to build the initial orebody models, alongside assay measurements and (limited) core data sets, then blasthole assays are collected manually in the pit to refine/update the orebody map. Blasthole assay collection is far from ideal:

- There are safety issues around having personnel working in the pit to physically collect the assays.
- Blasthole assays are not ideal for defining heterogenous areas – one assay is used to represent an entire 10+ metre bench.
- Often fines are lost in the drilling process and the blasthole assays miss essential composition of the hole.
- Handling the physical samples and processing them in the lab can be challenging and costly in both logistics and lab availability (McArthur, Jones and Murphy, 2010; Pitard, 2008; Simpson and Market, 2021).

There is great interest in being able to refine the orebody model during production (eg 'in the pit'). In some iron ore mines, elemental spectroscopy is being used to replace/reduce the need for blasthole assays, though elemental spectroscopy alone is not ideal for every mine. Combining multiple measurements can be a better solution to defining the orebody and reducing dependence on blasthole assay. Some of the more promising combinations are gamma, density, magnetic susceptibility and elemental spectroscopy.

Interest is increasing in in-pit logging, particularly in RCGC (reverse circulation grade control) campaigns. One of the keys to success is to be able to collect a broader suite of geophysical data in the pit and be able to deliver it in near-real-time so that the production models can be updated in 'drilltime'. Not only are measurements such as gyros, gamma, density, and magnetic susceptibility well suited in the blasthole (and very familiar from resource definition mapping), but televiewers are particularly insightful and enable increased safety and wall stability by confirming or refining the geotechnical risks in the pit.

To make 'drilltime' geophysics effective, robotic logging vehicles have been introduced which can:

- operate in the pit without the need for the driver to leave the vehicle to use the logging tools.
- run multiple logging tools consecutively within minutes of drilling.
- use the mine site connectivity to process and deliver data before the end of shift.

At the time of publication, the full robotic logging system was in field trials in the Pilbara region. Operational statistics are not yet available for publication as the data set is small (less than 10 000 holes).

Figure 5 shows an example of a robotic logging vehicle logging in the Pilbara.

FIG 5 – A robotic logging truck is shown on the left. Unlike the conventional logging vehicle on the right, the operator never has to leave the vehicle, and a suite of four logging tools are run in sequence on each hole.

NEXT STEPS

The methods described, both in automated geophysics solutions and robotic geophysical data collection, were in field trials at the time of publication. Ongoing work includes collecting a minimum of 1000 holes in an active pit in the Pilbara to fully characterise the operational efficiency and the reliability of using automated geophysics interpretations in the mine model. Results are expected to be presented in late 2026.

CONCLUSIONS

Near-real-time geophysical logging and dynamic model updating have the potential to significantly shorten the discovery-to-recovery timeline by providing continuously refreshed, high-resolution orebody information. The combination of automated data acquisition, robotic deployment, and adaptive 3D modelling represents a step change in how geophysical data are used throughout the mine life cycle. A wide suite of borehole geophysical measurements can now be acquired in blastholes safely, and the data delivered in near-real-time. The next stage is to collect and evaluate 10 000 m of data to evaluate the potential of the method for more precise grade control, improved operational safety, and data-driven optimisation of resource recovery.

ACKNOWLEDGEMENTS

The authors wish to thank WSG and its clients for collaboration in developing this work and permission to publish.

REFERENCES

Chi, B, Maddever, R A M, Mahanta, A and Dominguez, O, 2017. Near real-time assay with downhole assay tool, in *Proceedings of the Eighth World Conference on Sampling and Blending*, pp 137–144 (The Australasian Institute of Mining and Metallurgy: Melbourne).

Elkington, P, Stouthamer, P and Brown, J, 1982. Rock strength predictions from wireline logs, *Int J Rock Mech Min Sci Geomech Abstr,* 19:91–97.

Fullagar, P K, Zhou, B and Fallon, G N, 1999. Automated interpretation of geophysical borehole logs for orebody delineation and grade estimation, *Miner Resour Eng,* 8:269–284.

Hawke, P, 2015. Application of wireline logging in iron ore evaluation, Australian Society of Exploration Geophysicists (ASEG) Workshop. Available from: <https://www.aseg.org.au/resources/workshop-proceedings/2015-aseg-borehole-logging-workshop/>

Jones, H, Walraven, F and Knott, G, 1973. Natural gamma radiation as an aid to iron ore exploration in the Pilbara region of Western Australia, in Proceedings of the Australasian Institute of Mining and Metallurgy Conference 1973.

Kitzig, M C, Kepic, A and Grant, A, 2018. Near Real-Time Classification of Iron Ore Lithology by Applying Fuzzy Inference Systems to Petrophysical Downhole Data, *Minerals*, 8(7):276. https://doi.org/10.3390/min8070276

McArthur, G, Jones, C and Murphy, M, 2010. Blast hole cone sampling experiments for Iron ore flitch mining, in *Proceedings of the Sampling Conference 2010*, pp 732–741 (Australasian Institute of Mining and Metallurgy: Melbourne).

Pitard, F, 2008. Blast hole sampling for grade control – the many problems and solutions, in *Proceedings of the Sampling Conference 2008*, pp 15–21 (Australasian Institute of Mining and Metallurgy: Melbourne).

Simpson, C and Market, J, 2021. Geophysical blast hole sampling, in *Proceedings of the Iron Ore Conference 2021*, pp 176–192 (Australasian Institute of Mining and Metallurgy: Melbourne).

Wanstedt, S, 1982. Geophysical logging applied to ore characterization in the Zinkgruvan mine, Sweden, *Explor Geophys,* 23:401–406.

Leveraging Model-Based Design for advanced control system development of long overland conveyor belts

S Oliver[1] and D Lloyd[2]

1. Principal Technical Consultant, MathWorks, Chatswood NSW 2067. Email: soliver@mathworks.com
2. Director, DJL Automation, Newcastle NSW 2302. Email: daniel@djlautomation.com.au

ABSTRACT

The development and commissioning of control systems for long overland conveyor belts present significant engineering challenges, particularly in mining operations where conveyors must transport material over extended distances and varying elevations. Traditional approaches to control system design often struggle to address the complex dynamics introduced by variable loading, gradient changes, and the need for precise belt tension management. This paper presents a comprehensive methodology that leverages Model-Based Design to support the development, evaluation, and deployment of advanced process controllers (APCs) for overland conveyor systems.

A detailed dynamic model of the conveyor is developed, incorporating critical subsystems such as belt tension regulation, pulley slip dynamics and variable material loading. The model includes typical head-drive and tripper configurations, capturing the interaction between multiple electric drives and the conveyor belt. By simulating these elements, the model provides a high-fidelity platform for understanding system behaviour under a wide range of operating conditions, including disturbances caused by fluctuating loads, gradient changes, and mechanical wear.

Model-Based Design enables iterative development and rigorous testing of control algorithms in a virtual environment prior to hardware implementation. Control strategies can be evaluated for robustness and reliability using the validated conveyor model, significantly reducing the risk of software-induced fault and downtime during commissioning. Feedforward and model predictive control concepts can be applied efficiently by using reduced-order models tailored to the conveyor dynamics. The methodology supports automatic code generation, allowing the tested control logic to be deployed to production-grade programmable logic controllers (PLCs). This workflow ensures that controller software is robust and maintainable, with traceability from system requirements through to deployed code.

Validation of the conveyor model is performed using field data, ensuring that simulated responses accurately reflect real-world system dynamics. This validation step is critical for building confidence in the model's predictive capability and for supporting the development of advanced control strategies tailored to site-specific requirements. The proposed workflow also establishes a regression testing environment, enabling continuous verification of controller performance in desktop simulations and on the deployed PLC hardware. This systematic approach reduces development time and cost by identifying and resolving issues early in the design process, while improving software quality and reliability through comprehensive simulation and testing.

In summary, the paper demonstrates the substantial benefits of applying Model-Based Design to the development of control systems for long overland conveyor belts. The approach facilitates integration of multidisciplinary engineering teams, accelerates the adoption of advanced control technologies, and ultimately delivers more reliable and efficient conveyor operations. The methodology and results presented are broadly applicable to similar industrial automation challenges, offering a scalable framework for future APC development projects.

INTRODUCTION

Long overland conveyor belts in mining operations are required to transport ore and waste over many kilometres, often with complex vertical and horizontal profiles, multiple drive stations, and a mix of uphill and downhill sections (Lodewijks, 1999, 2002; Nordell, 2003). These systems often operate as production bottlenecks, where instability during start-up, shutdown, or emergency stopping can cause significant downtime, belt damage, and safety risks. Traditional control system design for such conveyors relies heavily on conservative rules-of-thumb, steady-state calculations, and incremental tuning while running in production, limiting the scope for innovation and systematic optimisation.

Dynamic design methods for belt conveyors emerged to address problems such as excessive transient belt tensions, slack belt, and pulley slippage that are not captured by quasi-static models (Lodewijks, 1996, 2002). Pioneering work by Lodewijks (2002) and those cited therein, showed that the belt must be treated as an elastic, distributed system and that start-up, shutdown, and load disturbances must be analysed using time-domain simulations, often based on finite-element or lumped-mass models. However, while these models were originally developed to support component sizing and drive selection, they also provide an ideal foundation for modern control system development workflows using Model-Based Design, where the same model underpins control design, verification, and implementation.

This paper presents a methodology where a dynamic conveyor model is developed in Simscape™ within the Simulink® environment and used as the central artefact throughout the control system life cycle. The focus is a long overland conveyor with head and tripper drives, where both belt speed and belt tension must be tightly controlled under variable loading. The key contributions are:

- A practical lumped-element conveyor model tailored to control design and implemented using Simscape™ belts and pulleys libraries (The MathWorks, Inc , 2018).
- A control architecture that separates speed and tension objectives across coordinated drives and is compatible with industrial PLC implementations.
- A development workflow using Model-Based Design that supports virtual commissioning, regression testing, and automatic code generation for production controllers.

The remainder of the paper is structured as follows. Firstly, it introduces the concept of Model-Based Design and motivates its use for overland conveyor applications. Followed by a review of conveyor dynamic modelling, which led to the choice of the lumped-element model for this study. Using this lump architecture, it then describes the Simscape™ implementation of the conveyor model. Leveraging this model, it then presents the controller design, comparing a baseline and a proposed advanced controller. Finally, it discusses the results and concludes with lessons learned and directions for future work.

MODEL-BASED DESIGN FOR OVERLAND CONVEYORS

Concept and workflow

Model-Based Design is a model-centric approach in which an executable model becomes the reference for system requirements, design, implementation, and verification. In the context of industrial control, Model-Based Design typically uses block-diagram, state machines, and physical-modelling formalisms (for example, Simulink®, Stateflow®, and Simscape™) to represent plant dynamics, control algorithms, and supervisory logic. These models support time-domain simulation, automatic test execution, and automatic code generation for targets such as PLCs and embedded controllers (The MathWorks, Inc, 2018).

Using Model-Based Design for a conveyor control system typically comprises of the following steps:

1. **Plant modelling** – develop and calibrate a dynamic model of the conveyor, including belt dynamics, drives, take-up, and loading (Lodewijks, 1996, 2002).
2. **Control design** – implement candidate control algorithms, often starting with classical structures and progressing to more advanced schemes (for example, model-based and predictive control) as needed (Chen and Li, 2016).
3. **Simulation and test** – execute scripted scenarios such as start-up, shutdown, load changes, and faults and assess performance against requirements.
4. **Code generation** – generate controller code automatically for PLC or drive-embedded platforms, preserving the algorithm structure proven in simulation.

5. **Hardware-in-the-loop (HIL) and regression testing** – connect the plant model to real controller hardware and integrate automated test suites to guard against regressions as the design evolves.

This approach contrasts with traditional methods, where the plant model is implicit in the designer's calculations, and the first integrated test of the final control software occurs on the real conveyor during commissioning. For long, high-power mining conveyors, performing extensive trial-and-error tuning on the actual plant is undesirable due to production loss and risk of equipment damage (Lodewijks, 1996, 2002; He, Pang and Lodewijks, 2016).

Motivation in the conveyor context

Early conveyor dynamics work already highlighted the need for a 'testing environment' where start-up and shutdown strategies could be evaluated without risking the real equipment (Lodewijks, 1996, 2002). In practice, many operations still rely on conservative ramp profiles and limited use of advanced control options in variable-speed drives, because the consequences of more aggressive tuning are uncertain and difficult to test online.

Model-Based Design directly addresses this barrier by:

- Allowing engineers to experiment with alternative ramp shapes, drive coordination schemes, and tension control strategies in simulation.
- Supporting systematic 'what-if' studies, including emergency stop behaviour and failure of a drive or braking element.
- Providing a common artefact that mechanical, electrical, and control engineers can review and validate together, improving cross-disciplinary understanding.

For overland conveyors with multiple drives, regenerative sections, and tight production constraints, this capability is essential to fully exploit modern APC techniques while maintaining safety and availability (Feng *et al,* 2020; Chen and Li, 2016; Cornet, 2002; Kruse and Lemmon, 2024).

CONVEYOR MODELLING OVERVIEW

Evolution of dynamic models

Quasi-static models, still widely used for sizing, treat the belt as a rigid element and compute steady-state tensions and power from distributed resistances and lift. These methods cannot predict transient phenomena such as tension waves and local slack, which have been shown to be critical for long overland conveyors and high-speed belts (Lodewijks, 1995, 1999; Nordell, 2003).

Dynamic models treat the belt as an elastic medium. Lodewijks (1996, 2002) and others describe an evolution from:

- Simple two-mass representations for carry and return strands, linked by springs and dashpots.
- Through multi-mass, lumped-element models representing the belt as a chain of segments.
- To finite-element and multibody models that capture longitudinal and transverse dynamic behaviour (Lodewijks, 1996, 2002).

State-of-the-art models treat the belt as a 2D or 3D viscoelastic body and use wave-propagation theory to capture complex vibration phenomena, particularly in high-speed or very long conveyors (Lodewijks, 1999, 2002). These models are indispensable for detailed mechanical design but can be computationally intensive for routine control design and PLC-oriented simulation.

Rationale for a lumped axial model

For the control design tasks addressed in this paper, a 1D lumped-element model that captures axial vibration and longitudinal motion provides an effective compromise between fidelity and complexity. Similar reduced-order models have been used successfully for speed-control studies and control design (Chen and Li, 2016; Feng *et al,* 2020; Pihnastyi, Kozhevnikov and Cherniavska, 2021). They are particularly well suited to:

- Capturing propagation of tension waves during start-up and shutdown.
- Representing the effect of variable loading and drive torque changes on belt speed and tension.
- Running fast enough to support extensive simulation campaigns and HIL set-ups.

The conceptual model follows Lodewijks' head–tail conveyor representation, extended in this work to include an additional drive (tripper) on the carry strand (Lodewijks, 1996, 2002).

CONVEYOR MODEL IMPLEMENTATION

Simscape™ provides a physical-modelling environment in which mechanical components, powertrains and control systems are represented by interconnected blocks with energy-based ports. For the conveyor model in this study (The MathWorks, Inc, 2018), the top-level Simscape™ conveyor model is shown in Figure 1.

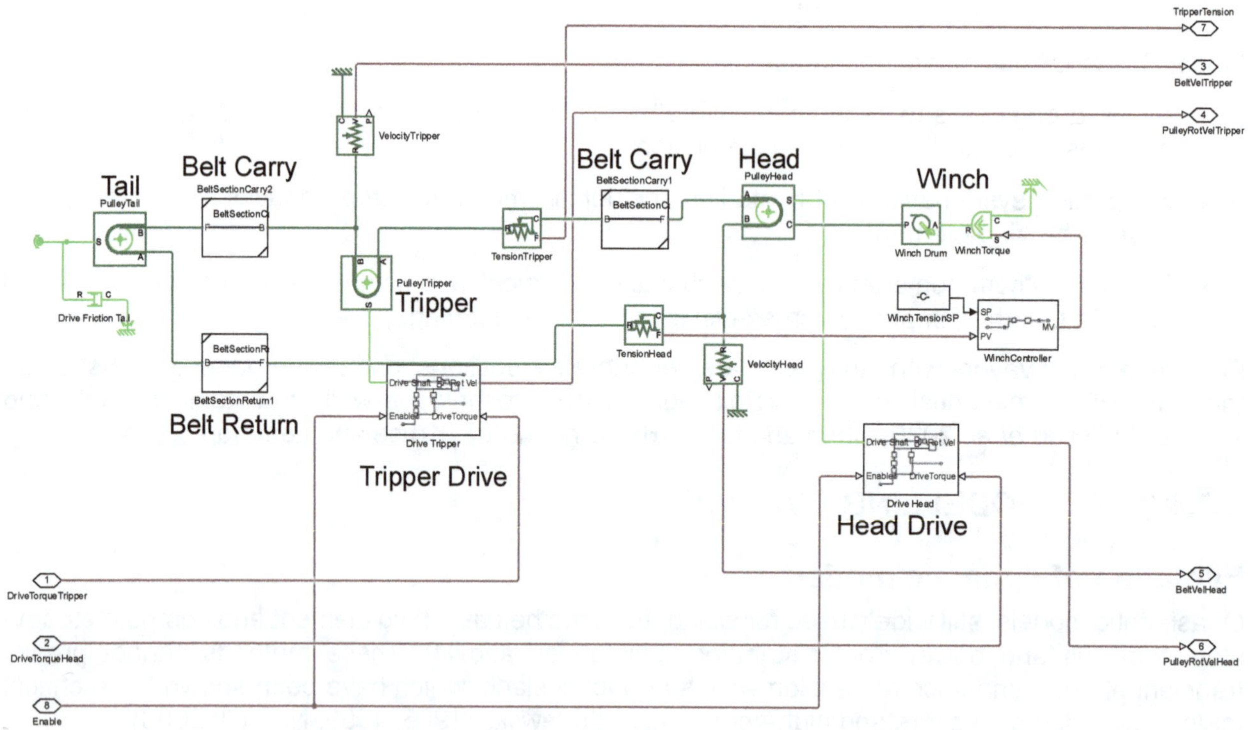

FIG 1 – Simscape™ model of the conveyor system.

The conveyor considered comprises:

- a head pulley with an electric drive at the discharge end
- a tail pulley at the feed end
- an additional pulley and drive pair on the carry side acting as a tripper or booster drive, primarily used to control belt tension in a critical section
- a winch that controls the belt tension at the head.

In this model, following the conceptual model of Lodewijks (1996, 2002), the head drive, adjacent pulley and nearby winch are aggregated into a single dynamic element because their separation is small relative to the total conveyor length. The head, tail, and tripper pulleys are connected by belt sections: two sections on the carry side connecting tail, tripper, and head pulleys, and a single belt section on the unloaded return side that connects the head and tail pulleys.

Each belt segment is divided into a series of belt elements. The Simscape™ belt element is shown in Figure 2 and includes:

- Lumped belt mass with gravitational effects to represent the carried material mass.

- Axial stiffness and damping representing the combined elasticity of belt carcass and material; this element provides resistance while the belt is in tension.
- Friction representing internal losses and rolling resistance effects in a simplified form.

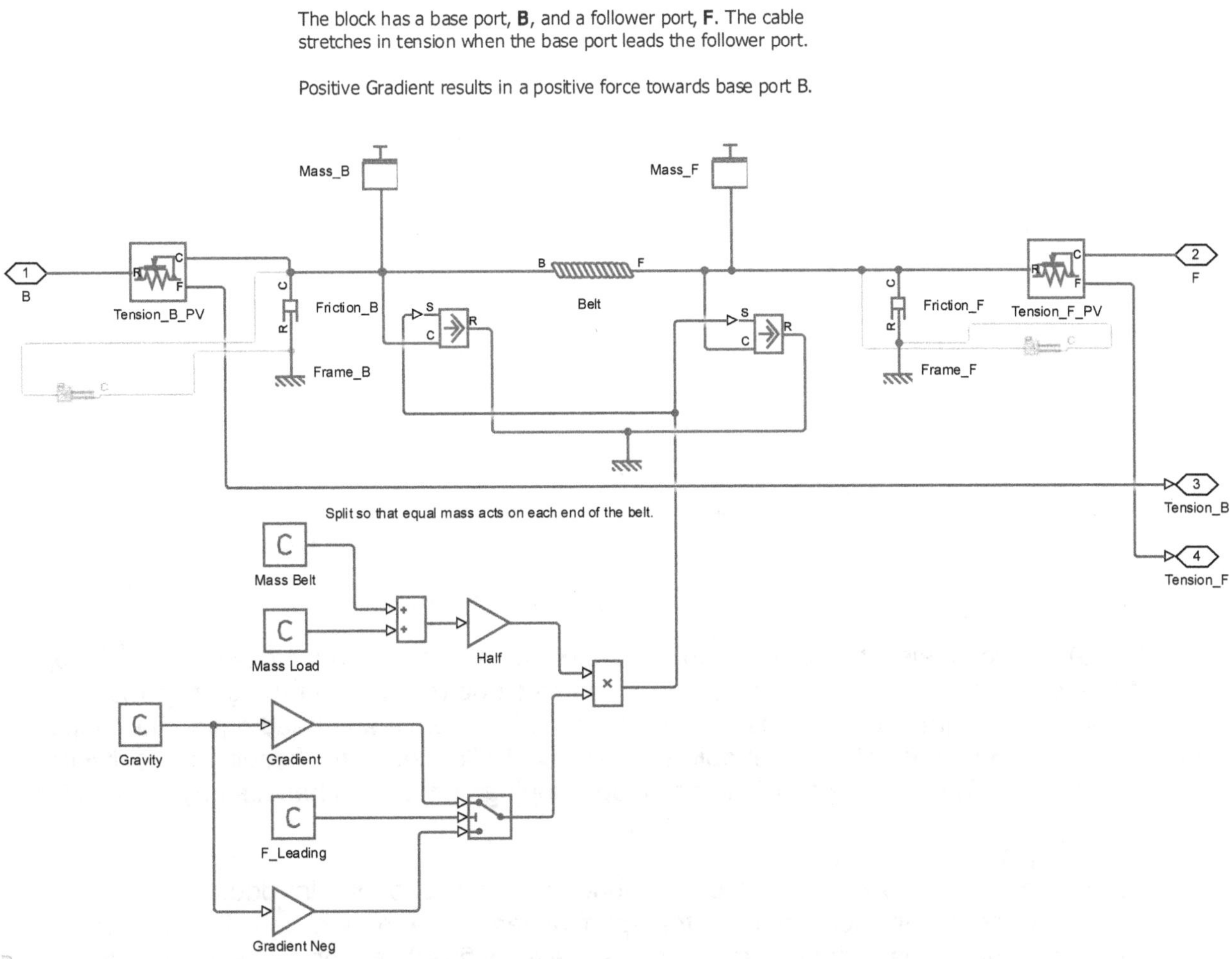

FIG 2 – Simscape™ model of the belt element.

The resulting chain of segments provides a spatial discretisation of longitudinal belt motion suitable for time-domain simulation of tension waves and representing tension around the belt loop.

A MATLAB®-based script is used to customise the belt sections to match the topography of a specific conveyor. This modular approach allows the same belt-segment, pulley, and drive subsystems to be reused across different conveyor configurations. Alternative drive layouts or take-up strategies can be evaluated by rewiring existing modules rather than rebuilding the model from scratch.

The topography for the conveyor under study is shown in Figure 3a, which is 4800 m long and 70 m total elevation change.

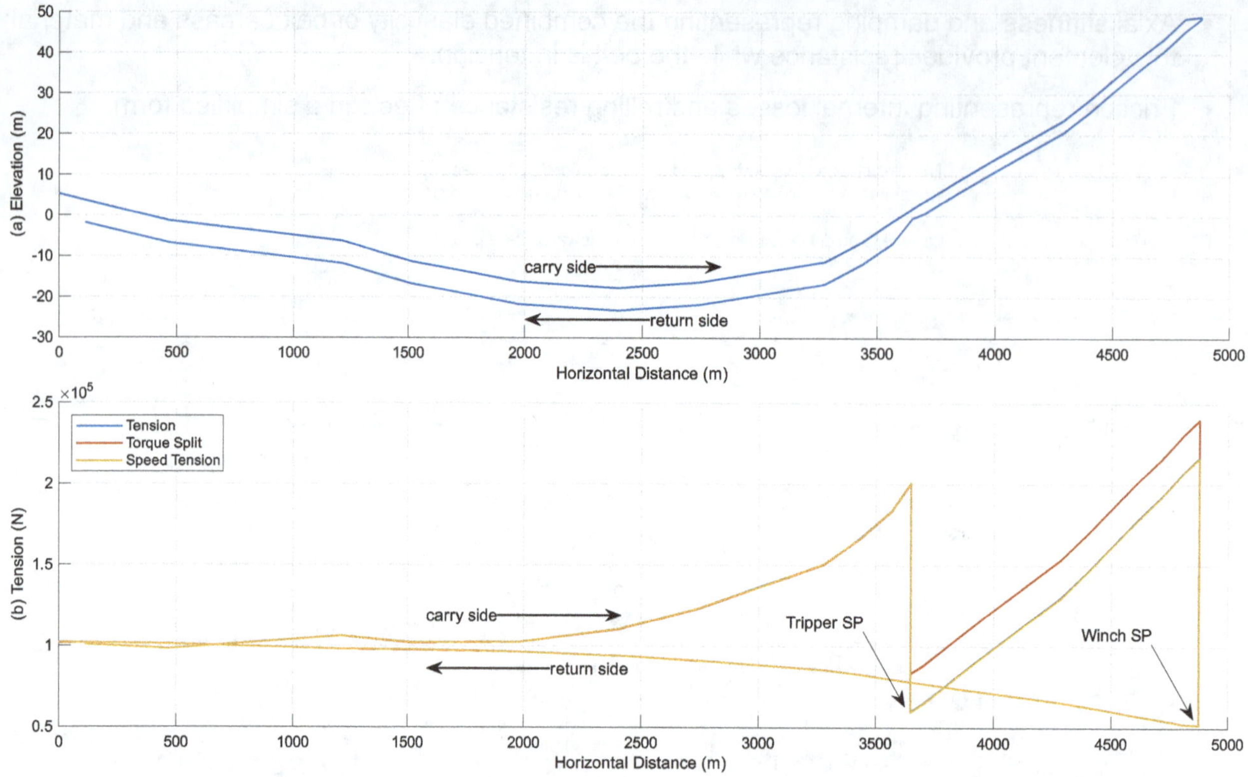

FIG 3 – (a) Conveyor elevation profile showing both the carry and return sides of the belt loop; (b) belt tension loop. The material carry side of the belt loop (running left to right) shows large tension variations with step changes across both the tripper drive and head drive. The tripper (tripper SP) and winch (winch SP) set points are the controller set points keeping the belt within operational limits. The return side of the belt loop (running right to left) has a smoother tension gradient.

This reduced-order model does not represent vibration and oscillations introduced by transverse motion. It also does not include material transport dynamics. Although both these phenomena represent logical extensions to the model, the current level of fidelity supports the immediate control development objects of this study.

Model calibration and validation

Calibration relies on readily available design and operational data:

- Belt properties (mass per unit length, elastic modulus, cross-sectional area).
- Conveyor geometry (length, vertical profile, pulley locations).
- Drive characteristics (rated torque, inertia, torque–speed envelope).
- Measured start-up and shutdown traces where available (belt speed, drive torque, tension at take-up or head).

Model parameters such as belt–pulley friction, drive friction, and belt rolling resistance determine the drive torque–speed characteristics. Other model parameters such as belt effective stiffness and damping coefficients determine the frequency response. Continuous-dynamics modelling work on intermediate-drive conveyors shows that such calibration enables accurate prediction of dynamic tension distributions under varying load conditions (Feng *et al,* 2020; Nordell, 2003).

Parameters were tuned by comparing the model's response to operational data and minimising the difference between simulated and measured responses. Optimisation-based tools automated this workflow by tuning the parameters with the objective of minimising the discrepancy between model and operational data.

The conveyor's open-loop drive-torque-to-belt-speed behaviour is shown in Figure 4. In this case, while the steady-state speed response matches, there are deviations. Fitting the parameters for the belt loaded and unloaded cases shows that there is a doubling of the load-induced roller friction. Initially, there is a difference in the step response: in the test, = a pulley slip event occurs, whereas the model continues to ramp and subsequently overshoots due to timing effects. The other deviation occurs in the steady-state speed relationship as the test continues, likely due to viscoelastic effects in the belt as it heats up or similar effects in the roller friction. Deviations such as these are due to tuning and unmodelled phenomena, and continued evaluation of the model response against reality allows the modeller to continue the cycle of question, explore, and learn to understand what drives the system behaviour.

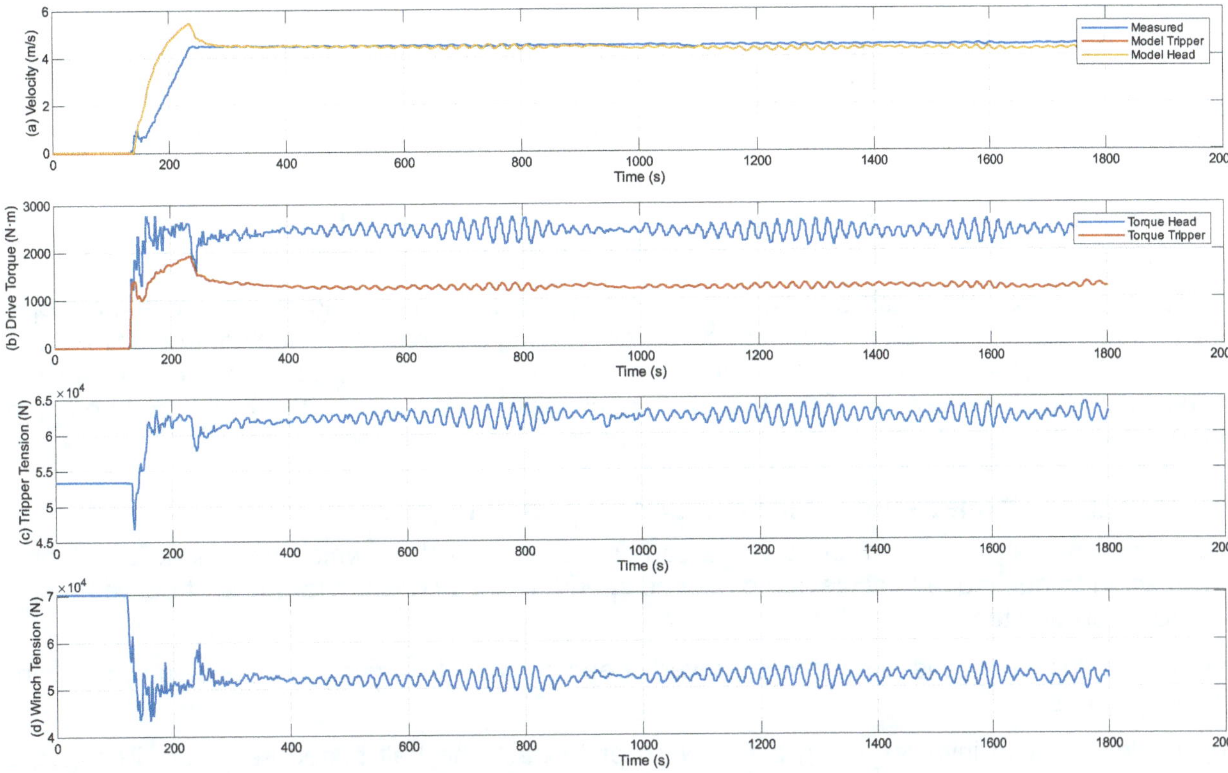

FIG 4 – Conveyor field measured data in no-load case using the proposed Torque Speed controller. (a) Measured velocity (blue) and model response at the tripper (red) and head (yellow) drives to the measured open loop torque inputs. (b) Head (blue) and tripper (red) drive torque. (c) Belt tension at the tripper. (d) Belt tension at head/winch.

CONTROLLER DESIGN

Control objectives and constraints

The primary control objectives for the overland conveyor are to:

- Maintain the belt at a commanded speed trajectory during start-up, normal operation, and shutdown.
- Regulate belt tension within safe limits, avoiding both excessive tension (risk of belt or splice failure) and slack (risk of lift-off or slip). Figure 3a shows the tripper tension set point. The winch set point, which is also shown in this figure, is related but controlled separately by the winch controller.

Operational constraints include:

- Limiting drive torque and torque rate to avoid exciting harmful tension waves.
- Coordinating head and tripper drives so that neither is overloaded, and the resulting tension distribution remains within design limits.

- Keeping controller structures sufficiently simple so that maintenance engineers and operators can understand and tune them in the field.
- Distribute starting torque between the head and tripper drives with a minimum drive torque of 10 per cent as they are operating without encoders.

These constraints favour relatively simple PI-type loops and feed-forward structures, rather than highly complex multivariable controllers (Chen and Li, 2016), while still benefiting from tuning and validation directly in design models developed using Model-Based Design.

Baseline controller – separated speed and tension control

The baseline controller structure reflects common practice in multi-drive conveyors:

- The head drive operates in speed-control mode, following a predefined speed ramp profile during start-up and shutdown.
- The tripper drive primarily regulates belt tension at a chosen location, using feedback from a load cell.
- A passive speed ramp is used to mitigate start-up transients, with conservative torque limits applied at both drives.

In this configuration, the head drive is responsible for tracking the speed profile, while the tripper adjusts its torque to maintain tension, effectively decoupling speed and tension control to first order. However, this separation is imperfect and poor coordination can still lead to transient tension overshoots or under-utilisation of the tripper drive capacity (Lodewijks, 2002; Kruse and Lemmon, 2024).

Proposed controller – coordinated speed control

A proposed new controller adopts a more coordinated approach in which both head and tripper drives act in torque mode to achieve the desired speed, while a constant torque split between them is used to target a tension.

The goal of this strategy is to provide improved speed control while setting the tension by selecting a fixed torque split.

The implementation involves using a PI controller to track the belt speed set point. The torque request is split with a fixed ratio across the two drives. This ratio and the PI gain are tuned in the model through simulation tests.

Proposed controller – coordinated speed and tension control

The coordinated speed-and-tension controller extends the speed-control concept—where both head and tripper drives act in torque mode to achieve the desired speed—by actively controlling the torque split between them to target a tension at the tripper.

The goal of this strategy is to provide improved speed control while maintaining target tension at the tripper.

The implementation involves extending the speed control by adding a second PI loop that tracks the tripper tension target by adjusting the torque-split command between the head and tripper drives.

Testing and controller implementation

The validated conveyor model provides a safe environment to exercise controller concepts before making any changes to the production system. Candidate algorithms are implemented directly in the model, where their responses to representative start-up, shutdown, and disturbance scenarios can be evaluated and compared with the existing baseline controller. This enables many ideas to be trialled quickly, with only the most promising structures progressing towards implementation.

A library of scripted test cases was developed, including cold and warm start-ups, partially loaded conditions, emergency stops, and step changes in loading or rolling resistance. Each controller concept is subjected to the same test suite, allowing objective comparison of speed tracking, tension envelopes, and drive torque limits using common performance metrics. Once a controller structure and tuning are accepted, the control logic can be exported as structured-text code suitable for industrial PLCs, preserving the algorithm hierarchy defined in the modelling environment. The Simscape™ conveyor model is then linked to the production PLC, forming a hardware-in-the-loop set-up where the same automated test scenarios are executed against the real control hardware. This step confirms that implementation details—such as data types, execution order, and scaling—do not introduce unintended behaviour and establishes a regression test environment for future software changes.

Overall, this test-early, test-often workflow increases confidence in the final controller, shortens on-site integration time, and reduces the risk of unplanned downtime when the controller is deployed to the live conveyor. By the time commissioning begins, the controller has already been exercised extensively in both purely virtual and hardware-in-the-loop configurations, with known performance envelopes for the key operating scenarios.

RESULTS

Dynamic simulations were conducted using the calibrated lumped-element conveyor model for a standard cold start-up scenario, comparing three controller configurations: the baseline separated-role controller, a coordinated-speed controller with fixed torque split, and a coordinated speed-and-tension controller with active torque-split control. A summary of control algorithm performance is included in Table 1. Figure 5 presents the resulting belt speed, drive torques, and belt tensions at the tripper and head/winch for the three cases.

TABLE 1

Control algorithm capability summary.

Control algorithm	Speed set point tracking	Tension set point tracking	Speed and tension limits
Separated speed and tension	Slow	Slow	Implicit
Coordinated speed	Fast	Constant (no tracking)	Explicit
Coordinated speed and tension	Fast	Fast	Explicit

Results taken from simulation testing using a constant ramp for start-up are shown in Figure 5.

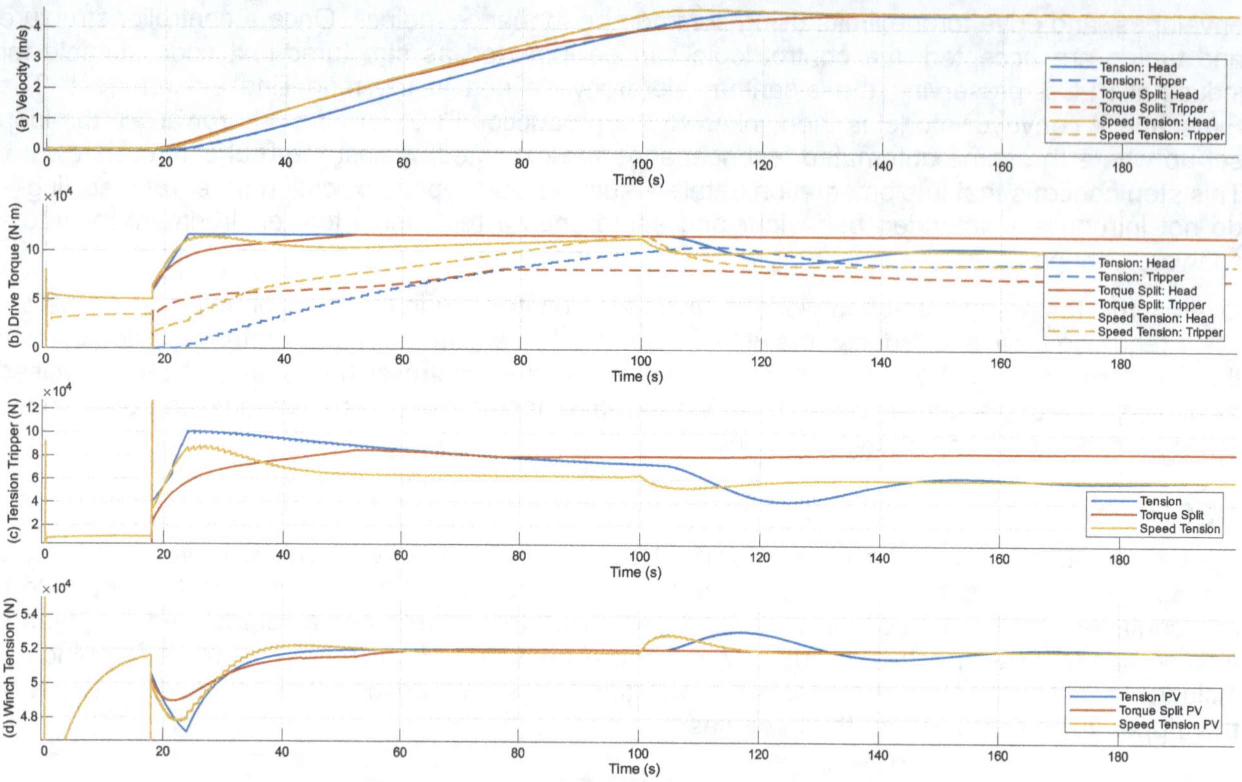

FIG 5 – Conveyor model results comparing performance of the Tension (blue), Torque Split (red), and Speed Tension (yellow) controller strategies. (a) Conveyor belt velocity at the head (solid line) and tripper (dashed line) drives. (b) Conveyor drive torque velocity at the head (solid line) and tripper (dashed line) drives. (c) Belt tension at the tripper. (d) Belt tension at the head/winch.

In all cases the belt accelerates to the commanded speed within the required start-up time, but the detailed responses differ. The baseline controller shows small speed-tracking errors during the ramp and a slightly slower approach to steady-state, reflecting its conservative torque limits and lack of explicit coordination between drives. The coordinated-speed controller improves speed tracking and reduces overshoot by allowing both drives to contribute actively according to a fixed torque split, while remaining within drive torque constraints.

The largest performance differences appear in the tension responses. Under the baseline controller, peak drive-side tensions at the head and tripper are higher, and the decay of the start-up tension transient is slower, indicating less effective damping of the tension wave launched during acceleration. The coordinated-speed controller reduces these peaks modestly by sharing torque more evenly, but residual oscillations remain because the torque split is fixed and does not respond to measured tension.

The coordinated speed-and-tension controller achieves the lowest peak tensions and the smoothest tension envelopes at both the tripper and head. By adjusting the torque split in response to measured tripper tension, the controller actively shapes the tension profile along the belt, suppressing overshoot and reducing the risk of slack-belt or excessive-tension conditions. The additional PI loop on tension does not materially degrade speed tracking; instead, the belt speed follows the reference closely while the tension loop mitigates disturbances and accelerates the decay of residual oscillations.

Across the simulated scenarios, the proposed coordinated speed-and-tension controller therefore offers the best overall trade-off between speed tracking, tension control and drive utilisation. The ability to compare controllers side-by-side in the same model and dashboard accelerates design iteration and provides quantitative evidence to support the selection of the final control strategy.

The coordinated speed-and-tension controller has since been deployed to the production conveyor. Initial operating data, shown in Figure 4, confirm the improved speed-tracking and tension-tracking performance anticipated from the simulation study. The measured belt speed closely follows the commanded ramp, while both tripper and head-winch tensions remain within their target envelopes during start-up and steady operation. The production traces also reveal a low-frequency oscillation in both torque and tension. This underlines the need to refine the model structure to include appropriate measurement delays and belt element properties—such as string rates and damping—to replicate the behaviour. Subsequently then correct in the simulation and then deploy the updates to the production controller.

CONCLUSION AND FUTURE WORK

This paper has demonstrated how Model-Based Design can be applied to the development of advanced control systems for long overland conveyor belts in mining applications. A lumped-element axial model of the conveyor—implemented using Simscape™—provides a sufficiently detailed yet tractable representation of belt dynamics, including the interaction between head and tripper drives and the influence of variable loading. This model forms the backbone of a workflow that supports controller design, virtual commissioning, and regression testing before deployment to PLC hardware.

The case study illustrates how the modelling exercise itself yields insight into fundamental system behaviour, such as the interplay between speed and tension control and the importance of drive coordination. These insights inform controller-structure decisions, such as the separation of speed and tension objectives and the use of coordinated torque control to improve transient performance. Automatic code-generation tools allow the validated controller logic to be transferred to production controllers with high fidelity and traceability, reducing the risk of introducing errors during manual implementation.

Future work will focus on extending the control architecture to include explicit slip control, using estimates of belt–pulley traction margins to adapt torque limits in real time and mitigate slip or drive-failure conditions. Additional developments include exploring model predictive control strategies derived from reduced-order conveyor models and integrating energy-optimised speed control to reduce operating costs while respecting mechanical constraints. Further model extensions are planned to support variable mass representations and material transport tracking along the belt, enabling spatially resolved loading profiles, localised disturbances, and material-flow dynamics to be captured explicitly in simulation. These enhancements will allow more operating scenarios, including complex loading patterns and time-varying friction conditions, to be modelled and assessed within the same Model-Based Design framework.

ACKNOWLEDGEMENTS

The authors would like to acknowledge the support and technical expertise from DJL Automation during the development work. Also, would like to thank Kestrel Coal Resources for supporting the project and for sharing the results.

REFERENCES

Chen, W and Li, X, 2016. Model predictive control based on reduced order models applied to belt conveyor system, *ISA Transactions*, 65:350–360.

Cornet, J-L, 2002. Head and tail controls in long overland conveyors, SME Annual Conference and Exhibit, USA.

Feng, Y, Zhang, M, Li, G and Meng, G, 2020. Dynamic characteristic analysis and startup optimization design of an intermediate drive belt conveyor with non-uniform load, *Science Progress*, 103(1):0036850419881089.

He, D, Pang, Y and Lodewijks, G, 2016. Belt conveyor dynamics in transient operation for speed control, *International Journal of Civil, Environmental, Structural, Construction and Architectural Engineering*, 10(7).

Kruse, D and Lemmon, R, 2024. Case study: advancements in overland conveyor technologies, Proceedings of Beltcon 20, South Africa.

Lodewijks, G, 1995. The rolling resistance of conveyor belts, *Bulk Solids Handling*, 15:15–22.

Lodewijks, G, 1996. Dynamics of belt systems, PhD thesis, Delft University of Technology, The Netherlands.

Lodewijks, G, 1999. The design of high-speed belt conveyors, Proceedings of Beltcon 10, South Africa.

Lodewijks, G, 2002. Two decades dynamics of belt conveyor systems, *Bulk Solids Handling*, 22(2):124–132.

The MathWorks, Inc, 2018. Simscape Multibody belts and pulleys examples [online], MathWorks File Exchange, The MathWorks, Inc. Available from: <https://www.mathworks.com/matlabcentral/fileexchange/66519-simscape-multibody-belts-and-pulleys-examples> [Accessed: 6 February 2026].

Nordell, L, 2003. Overland conveyors designed for efficient cost and performance, Proceedings of Beltcon 12, South Africa.

Pihnastyi, O, Kozhevnikov, G and Cherniavska, S, 2021. Conveyor belt speed control efficiency using the energy management methodology, in P*roceedings of the 2021 IEEE 8th International Conference on Problems of Info Communications, Science and Technology (PIC S&T)*, Ukraine, pp 357–362 (IEEE: New York).

Solving iron ore dust challenges with DEM and CFD

L Stone[1], S Lord[2], W Taylor[3] and G Wenham[4]

1. Bulk Materials Handling Lead AUA, Hatch, Wollongong NSW 2500.
   Email: luke.stone@hatch.com
2. Mechanical Engineer, Hatch, Wollongong NSW 2500. Email: samuel.lord@hatch.com
3. Regional Lead Specialised Engineering, Hatch, Brisbane Qld 4000.
   Email: wes.taylor@hatch.com
4. Manager Sustaining Capital – Iron Ore, Hatch, Perth WA 6000.
   Email: garry.wenham@hatch.com

ABSTRACT

Dust abatement is a common challenge faced by most processing industries, wherever product is degraded through processing and transported; dust is involved. The iron ore industry is no exception to this with dust becoming a key focus for many sites as it represents either lost product revenue, environmental repercussions due to uncontrolled emission or increased downtime due to unplanned maintenance activities. In order to solve the challenges surrounding dust, the need for more novel and complex analysis has never been greater. This paper provides a decision framework for selecting CFD–DEM (computational fluid dynamics – discrete element method) or CFD–DDPM (dense discrete phase method) approaches for iron ore dust problems, supported by laboratory calibration and industrial-scale case studies.

This builds on the work previously completed by Stone *et al* (2016, 2019, 2021), who focused specifically on biomass and alumina dust modelling in enclosed vessels, as well as developing a calibration methodology. To extend on this previous work, a literature survey of recent advances is presented both with respect to dust generally, the iron ore industry, and DEM coupled with CFD. The findings from this survey, as well as industry-based experience is then used to develop a practical guide and methodology to model dust such that it can be used to solve real world problems. Finally, a case study is presented that outlines how the developed methodology was implemented, and the outcomes form undertaking a complex DEM-CFD coupled analysis and design approach.

INTRODUCTION

Dust generation is a persistent challenge in bulk materials handling, particularly for iron ore operations where fines are liberated at transfer points, stockpiles, and during conveying. Uncontrolled emissions lead to environmental compliance risks, health and safety concerns, equipment bogging, and direct product loss. While empirical rules of thumb can guide early design and analysis, physics-based simulations have become necessary to reliably predict both dust generation and airborne transport in complex operational scenarios such as flow over stockpiles (Schulz *et al,* 2021) and enclosed systems such as cyclones (Chu *et al,* 2022).

This paper develops a practical modelling methodology for iron ore dust in enclosed systems and directly compares two workflows: a coupled CFD–DEM (computational fluid dynamics – discrete element method) approach using Ansys Fluent and Rocky DEM, and a CFD–DDPM (dense discrete phase method) approach using Fluent's in-built Discrete Phase Model.

A discussion is presented on when each method is most appropriate, the level of fidelity achievable, computational requirements, and how to deploy them efficiently within engineering project timelines.

The work builds on prior studies by Stone *et al* (2016, 2019, 2021), extending earlier validation and review efforts beyond alumina and biomass to iron ore and providing a workflow guide to industry. What existing papers do not guide engineers on (model selection logic, calibration workflow, compute trade-offs) is discussed in this paper, as well as the challenge of iron ore re angularity of particles, broad PSD, and high throughput rates.

LITERATURE SURVEY

A literature survey has been completed as part of the development of the methodology; in order to determine how others have been leveraging CFD and DEM to model dust flows. A key finding from this review is that the scale of research is quite broad with authors such as Chu *et al* (2022) modelling 1 to 200 micron particles in a 200 mm cyclone, while others such as Toraño *et al* (2011) modelled a 5 m wide roadway with CFD only. This demonstrates the challenge presented whereby, scale of domain, and particle size become crucial factors in determining which simulation technique, model and package should be used to capture the phenomenon of interest.

Dust generation mechanisms

It is well understood that dust in iron ore handling facilities originates from degradation of material, breakage on impact, and shear during flow as well as through active processing steps ie intentional crushing and grinding.

What is difficult to quantify is the level of dust that will be liberated from particles during the impacts experienced throughout the handling sequence. An interesting approach that has been presented by Schulz *et al* (2022) is to model individual particles coated with fine dust particles and measure detachment upon impact. The goal was to close the gap on dust dispersion based on the coarse bulk solid contacts using the Johnson, Kendall, Roberts (JKR) and Derjaguin, Muller, Toporov (DMT) theories of adhesion on DEM particles. Although beneficially in understanding how dust may be liberated, it is computationally intensive to model each collision, Figure 1 depicts the scale of the model with the main particle diameter = 1.5 mm.

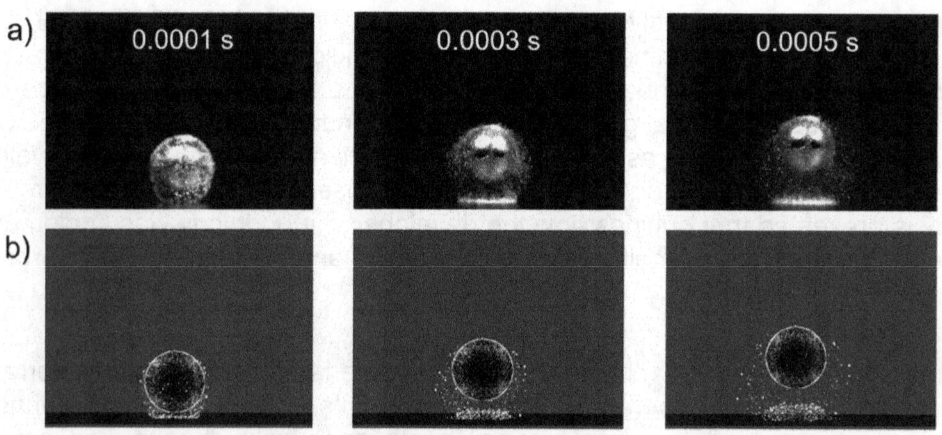

FIG 1 – Qualitative comparison of: (a) a single drop test recorded with a high-speed camera; and (b) a dust-resolved DEM-JKR simulation with γ_{12} = 0.0372J/m^2 and θ_{crit} = 0.5 for a drop height of 30 cm. Source Schulz *et al* (2022).

This supports the argument for a calibrated model whereby larger particles represent fine material behaviour. This is further supported by the work of Shi *et al* (2023) who looked at the effect of modifying bulk flow behaviour due to dust content ie the changing the internal friction due to fine particles. They leveraged DEM to model the coarse particles and modified friction values to capture the 'lubrication' effects of dust based on experimental data, refer to the schematic (Figure 2). Similar to Schulz *et al* (2022) the main problem faced is the shear time required to simulate this phenomenon as well as the hardware required is not practical for design cases.

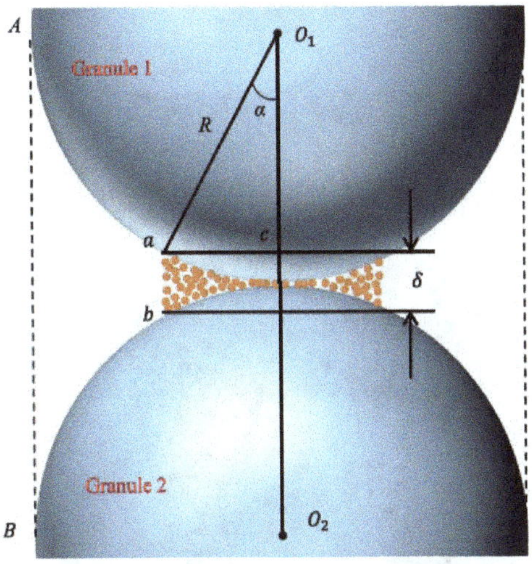

FIG 2 – Schematic diagram of dusty angle, the blue ball represents granules, the orange portion represents the dust, α is the dusty angle of the granule, O_1;O_2 is the centre of the two granules. Source Shi *et al* (2023).

Modelling approaches

Modelling strategies range from empirical correlations to continuum two-fluid (Eulerian–Eulerian) methods and Lagrangian particle tracking. The Discrete Element Method (DEM) resolves particle to particle and particle to wall contacts and thus can represent mechanisms that generate dust from the bulk stream. When DEM is coupled with CFD, momentum exchange between particles and air is accounted for, enabling prediction of dust liberation and the modelling of physical dust particle movements in the air stream as described by Woschny *et al* (2021).

Two approaches can be leveraged here to improve simulation progression rates:

1. Particle scaling – density and drag coefficients are adjusted to account for the fact that the particle is physically larger.
2. Coarse graining – particles are representative of clusters of material.

These two approaches are required as it is not feasible to model the dust particles (or in many cases the granular material particle sizes) at real scale in a typical industrial domain as it would result in billions of particles.

Chu *et al* (2022) undertook an extensive modelling exercise using coarse graining to model a cyclone separator. A key takeaway from this work was that although the model showed promise with comparable results between the simulation and experiments; there were concerns with accuracy for more complex conditions with turbulent dispersion of particles. Refer to Figure 3, which depicts the experimental cyclone to a one way coupled simulation and a two-way coupled simulation.

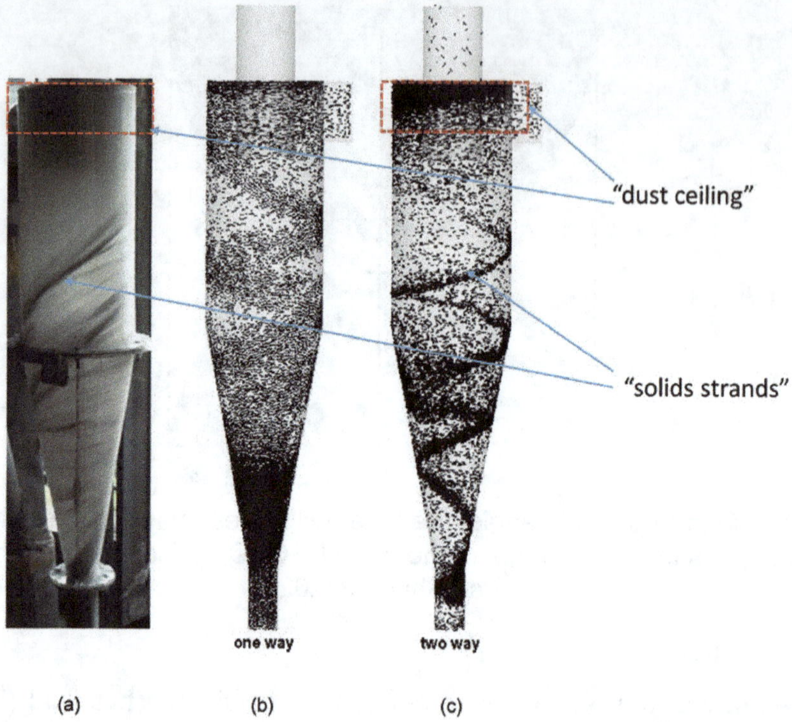

FIG 3 – Comparison of solids flow pattern between: (a) physical experiments; (b) one way; (c) two-way coupled simulations. Source Chu *et al* (2022).

In the CFD–DEM workflow, DEM resolves granular mechanics (collisions, friction, rolling, cohesion), while Fluent solves the transient airflow. Two-way coupling exchanges momentum source terms between phases, enabling local turbulence and drag feedback from particles to air. This allows 'physical' prediction of dust associated with impact energy, bed dilation, and shear at transfer points. A key aspect of particles influencing the air medium component as compared to a one-way resolved model where particles do not influence the air flow fields.

Strengths include realistic bulk flow, handling of broad particle size distributions (PSD) spanning lumps to fines, and the ability to diagnose generation hotspots. Limitations include higher computational cost, the need for material calibration (eg friction, restitution, cohesion), and stricter time-step control to maintain numerical stability in the coupled solution.

Conversely, the Dense Discrete Phase Model (DDPM) in Fluent tracks Lagrangian particles through a pre solved or concurrently solved airflow field but does not resolve inter-particle collisions. DDPM is efficient and valuable for modelling transport and capture of already airborne dust; however, it requires assumptions about where, how much, and with what size distribution, dust is released into the air and is more aligned to coarse graining of DEM particles.

An example of implementation of the CFD-DDPM model is presented by Ren, Wang and Cooper (2014) where they were able to use the CFD-DDPM to represent respirable dust in an underground roadway. Refer to Figure 4 which depicts the results of the model. Ren, Wang and Cooper (2014) were able to use this model successfully to reduce the air borne dust through a suppression system.

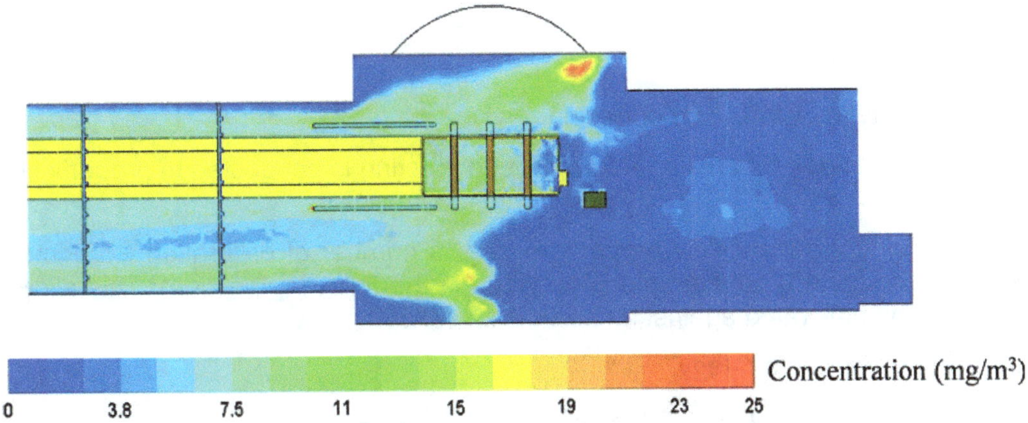

FIG 4 – Respirable dust concentration distribution at 1.7 m above floor. Source Ren *et al* (2014).

Fluent's DDPM tracks representative particles within a turbulent airflow field using drag and turbulence dispersion models. Because particle to particle interactions are not resolved, DDPM is most suitable for dilute airborne dust where the bulk solids flow is known or modelled separately.

Strengths include fast turnaround time, simpler set-up, and the ability to explore many ventilation scenarios. Limitations are the absence of explicit dust generation mechanisms and the need to prescribe injection locations, rates, and PSD based on measurements or upstream modelling.

Summary of fundamental differences

Table 1 outlines a high-level comparison between CFD-DEM and CFD-DDPM methods.

TABLE 1

Comparison of CFD-DEM and CFD-DDPM.

Feature	CFD–DEM	CFD–DDPM
Particle collisions	Resolved (typically with coarse graining or particle scaling)	Not resolved
Dust liberation prediction	Supported via contact mechanics and coupling	Not inherent; requires assumed sources
PSD handling	Broad PSD (lumps to fines)	Airborne fines only
Computational cost	High	Low
Calibration required	Yes (material parameters)	Minimal (air/particle properties)
Best suited for	Transfer points, loading zones, equipment where interactions dominate	Airborne transport, hood/duct optimisation

Gaps identified

Despite many studies on general dust control, few publications quantify the practical differences between CFD-DEM and CFD-DDPM. Industry case studies often report outcomes but omit the modelling decision logic, calibration steps, and computational trade-offs. The findings from the literature review show that successful implementation of CFD with DEM or DDPM models is possible, however, guidance is required with respect to where, when, and how each should be leveraged.

METHODOLOGY

A methodology has been developed based on the literature reviewed, the work completed by Stone (2021) and experience from industry to help provide direction to designers of systems where dust is involved:

1. Define the operational scenario and objectives of the study:

- o Dust liberation, flow patterns, build-up or dispersion.

2. Characterise the material:
   - o PSD, density, moisture, and cohesion.
   - o It is recommended that material flow property testing is conducted to properly characterise the material.
   - o In addition to standard flow property testing, a number of calibration experiments should also be conducted (an example is provided in the following section). This allows the dust model to be calibrated against a known data set using a feedback loop, refer to Figure 5.
3. Choose the modelling strategy:
   - o CFD–DEM for generation mechanisms ie bulk flow behaviour:
     - It is recommended, that this approach is adopted where there is interest in understanding the air flow fields generated by a bulk stream, and where particle deposition/loading will drastically affect the result.
   - o CFD–DDPM for airborne transport only:
     - It is recommended that this approach is adopted where the flow fields are already established and the material loading is low, such that it does not drastically affect the result.
4. Build and calibrate a DEM model for the bulk flow (for CFD–DEM) using angle-of-repose, shear, and impact tests:
   - o Guidance can be drawn from white papers such as that developed by (Katterfeld *et al*, 2019).
5. Develop the CFD model:
   - o Geometry simplification, mesh strategy, turbulence model, and boundary conditions.
6. Execute transient simulations and perform sensitivity studies (mesh, timestep, turbulence, PSD).
7. Validate against site measurements and iterate the design.

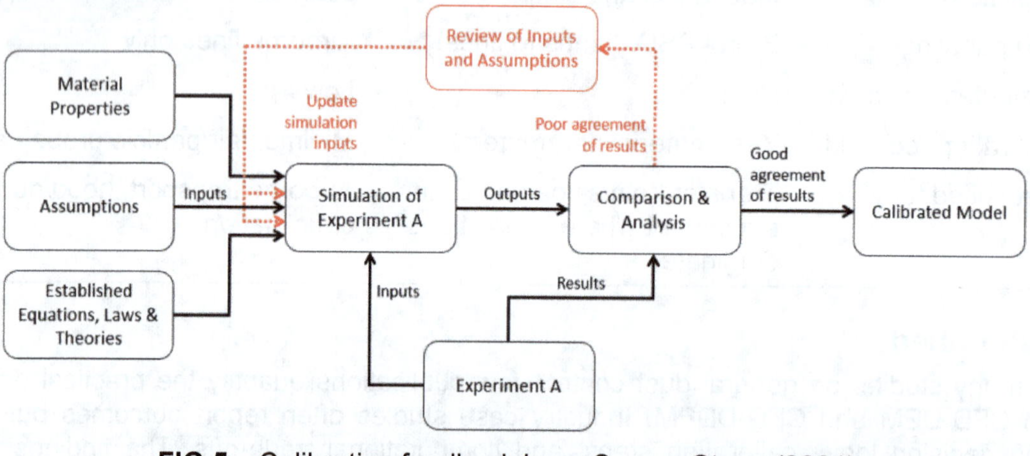

FIG 5 – Calibration feedback loop. Source Stone (2021).

It is important to note the effect of particle shape on the dust model, as outlined previously, experiments should be undertaken to quantify a known scenario and particle behaviour. Refer to Figure 6 which depicts a comparison made by Stone (2021) of three dissimilar materials. Iron ore dust is quite angular due to the mechanisms that generate the particles; however, this should be verified for each site.

FIG 6 – Scanning electron microscope images of dust particles from various products. Source Stone (2021).

The experiment proposed and examined by Stone (2021), leverages a wind tunnel with collection trays for calibration of dust deposition behaviour. Figure 7 outlines one of the test apparatuses used for the calibration process, with resulting iron ore dust distributed after an experiment shown in Figure 8. This particular wind tunnel type approach links the particle shape factor, and deposition directly to an air velocity which can then be leverage easily in a simulation.

FIG 7 – Schematic of test apparatus for calibration of dust. Source Stone (2021).

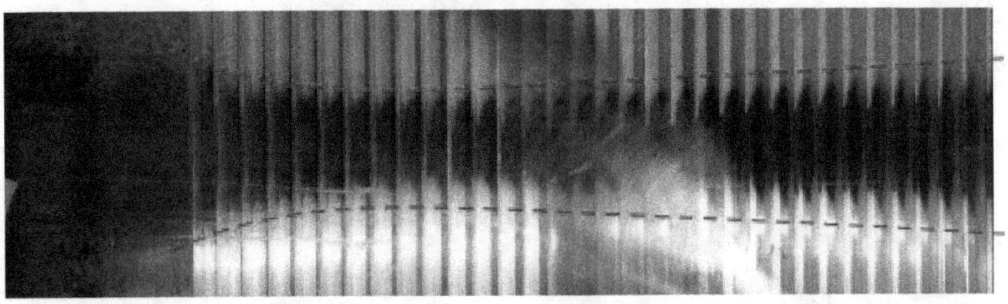

FIG 8 – Dust deposition pattern for iron ore dust. Source Stone (2021).

Once enough runs are completed and data gathered the database can be used for calibration purposes. It is recommended that at least three air velocities are examined around the target/expected on-site ie one lower, one higher and one close to the velocity of interest. The selection of air velocity could be determined from DEM bulk flow simulations ie the air flows in a transfer chute.

An example is shown in Figure 9 for the effect of shape factor on deposition distances based on the experimental test outlined in Figures 7 and 8. It is important to note that there is a particle size that is sufficiently small as compared to the transport velocity whereby the shape factor no longer plays a significant role in transport distance, and equally that at larger particle sizes the sensitivity to shape factor is more significant. This becomes clearer when considering the terminal velocity for a particle, as outlined in Figure 10, as the particle size decreases, the shape factor effects reduce.

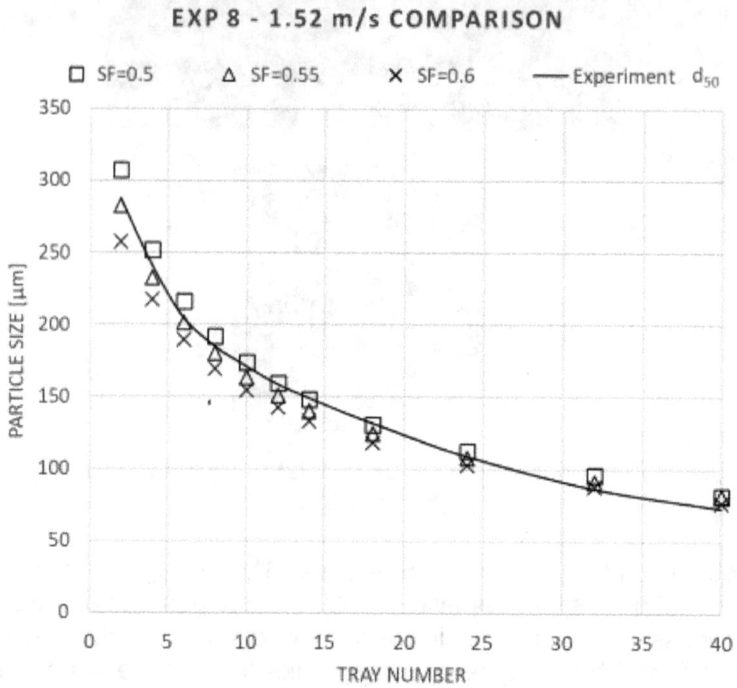

FIG 9 – Shape factor effect on transport distance for iron ore dust. Source Stone (2021).

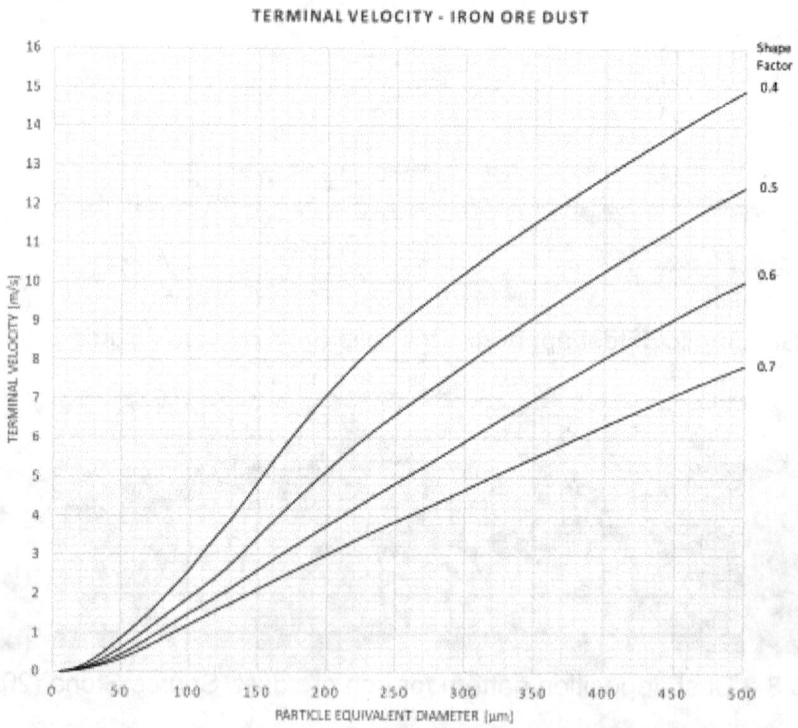

FIG 10 – Terminal velocity plot for various shape factors. Source Stone (2021).

Another helpful plot that can be generated from the wind tunnel testing, is a velocity versus d_{50} particle travel distance comparison. Figure 11 shows the trend of deposition for a number of experiments, at various air velocities. This plot can be used to understand what will happen to dust particles under different design solutions, ie reduction of air velocities, introduction of collection hoods, or applying misting sprays.

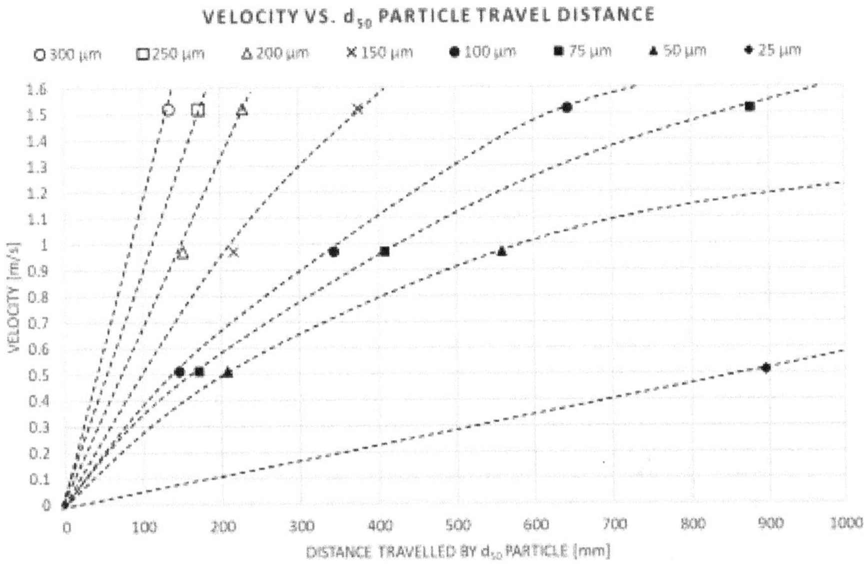

FIG 11 – Experimental results for particle 'drop out' at different velocities. Source Stone (2021).

In summary, the methodology outlined at the beginning of this section can be applied to most scenarios and with lab based experimental data as well as material flow property data, a sound and data back design can be developed.

CASE STUDY – EVALUATION: CFD–DEM VERSUS CFD–DDPM

This section outlines an example case study, focusing on the impact of modelling method selection on the deposition results. Limited data was available, and as such assumptions were made surrounding shape factor, particle distributions, and operating conditions of the system. In this example, the system is comprised of a series of five hoppers under a baghouse used for collecting iron ore dust.

In following the established methodology, a DEM and CFD model were developed and calibrated based on-site data available. Three models were then analysed, a DDPM coupled model, a one-way CFD and DEM model and finally a two-way CFD and DEM model. Table 2 outlines the distribution of mass between the hoppers for each simulation, and Table 3 outlines the errors.

TABLE 2

Comparison of normalised CFD-DEM and CFD-DDPM deposition mass data.

Percentage Fill	1	2	3	4	5	Total
Site Data	53.1%	10.8%	9.0%	11.5%	15.6%	100.0%
CFD-DDPM	54.6%	11.5%	15.6%	9.6%	8.7%	100.0%
CFD-DEM (one-way)	50.2%	17.0%	10.0%	9.4%	13.3%	100.0%
CFD-DEM (two-way)	37.7%	7.1%	19.6%	28.3%	7.3%	100.0%

TABLE 3

Comparison of normalised CFD-DEM and CFD-DDPM errors.

Percentage Fill	1	2	3	4	5	Sum
CFD-DDPM	2.8%	6.5%	72.8%	-15.9%	-44.2%	30.7%
CFD-DEM (one-way)	-5.4%	57.7%	11.3%	-17.8%	-14.8%	22.4%
CFD-DEM (two-way)	-29.0%	-33.6%	116.7%	146.8%	-53.1%	70.7%

The CFD-DDPM and CFD-DEM (one-way) model agree for initial deposition, however the error increases further downstream. The CFD-DEM (two-way) model performed the worst out of the three examined for this particular system. In part this may be due to assumptions made on particle shape factor, or scaling effects implemented. Further investigation is required to understand what mechanism is leading to pockets of under prediction and over prediction of deposition mass. Visual representations of the particles suspended in the baghouse are depicted in Figures 12, 13 and 14.

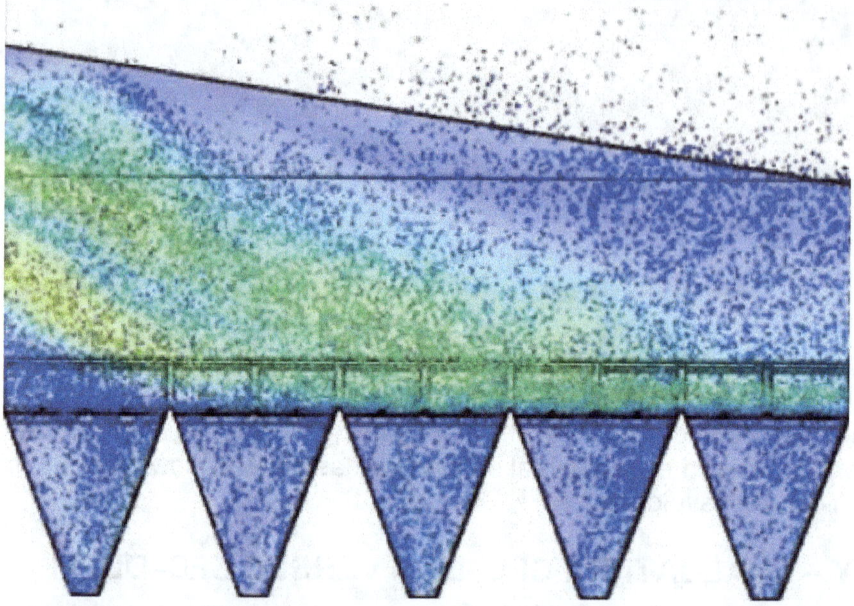

FIG 12 – CFD-DDPM simulation snapshot of suspended particles.

FIG 13 – CFD-DEM (one-way) simulation snapshot of suspended particles.

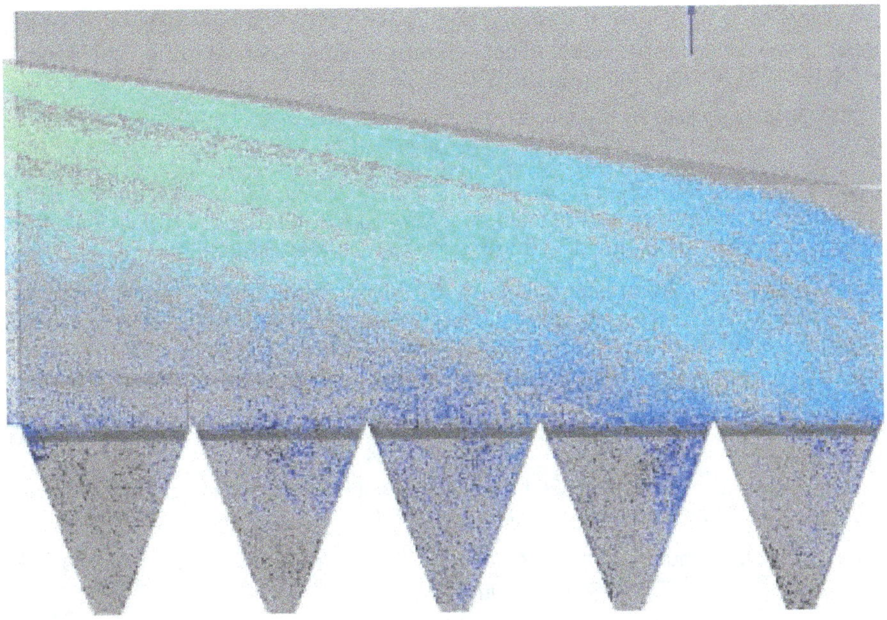

FIG 14 – CFD-DEM (two-way) simulation snapshot of suspended particles.

CONCLUDING REMARKS

This paper has investigated some of the latest developments on modelling dust in a CFD-DEM environment, as well as presented a step-by-step methodology for simulating dust. A crucial component is the need to undertake experiments in a laboratory setting, which can then be used to calibrate an accurate model. The case study presented outlines the variability of results when there is a lack of data available to calibrate a model. However, it also demonstrated, that there is not a single method that is going to give 100 per cent accurate results. Namely due to the limitations of modelling such small particles, in typically, significantly larger domains.

CFD–DEM is essential when dust generation depends on particle interactions: drop points, loading zones, and equipment where impact energy, shear, and bed dilation control fines liberation. It provides spatially resolved insight into where dust originates and how it couples into the airflow.

CFD–DDPM is sufficient when the primary question concerns transport and capture of an already airborne dust cloud ie evaluating hood placement, extraction flow rate, and enclosure design. Its speed enables rapid design iteration across many scenarios.

Computationally, CFD–DEM requires high particle counts and small time-steps, often necessitating HPC resources, whereas CFD–DDPM can be executed on a standard engineering workstation for large domains. Selection should therefore balance fidelity needs against project timelines and available compute.

REFERENCES

Chu, K, Chen, Y, Ji, L, Zhou, Z, Yu, A and Chen, J, 2022. Coarse-grained CFD-DEM study of Gas-solid flow in gas cyclone, *Chemical Engineering Science*, 260.

Katterfeld, A, Coetzee, C J, Donohue, T J, Fottner, J, Grima, A, Ramírez-Gómez, A, Ilic, D, Kačianauskas, R, Necas, J, Schott, D, Williams, K C and Zegzulka, J, 2019. Calibration of DEM Parameters for Cohesionless Bulk Materials under Rapid Flow Conditions and Low Consolidation, *Research Gate.* https://doi.org/10.13140/RG.2.2.26318.31048/1

Ren, T, Wang, Z and Cooper, G, 2014. CFD modelling of ventilation and dust flow behaviour above an underground bin and the design of an innovative dust mitigation system, *Tunnelling and Underground Space Technology,* 41:241–254.

Schulz, D, Schwindt, N, Schmidt, E and Kruggel-Emden, H, 2021. Modelling of dust emissions induced by flow over stockpiles and through packed beds, *Particuology*, pp 55–63.

Schulz, D, Woschny, N, Schmidt, E and Kruggel-Emden, H, 2022. Modelling of the detachment of adhesive dust particles during bulk solid particle impact to enhance dust detachment functions, *Powder Technology,* 400.

Shi, S, Wu, P, Fu, H, Jiang, M, Li, L, Zhang, S, Dong, C and Wang, L, 2023. A new method for DEM simulation of dust-bearing granular flow – Modeling the effect of dust on the change in granule friction coefficient, *Advanced Powder Technology*, 34(10):104156. https://doi.org/10.1016/j.apt.2023.104156

Stone, L E, 2021. CFD-DEM Modelling of Powder Flows and Dust Generation Mechanisms, PhD thesis, University of Wollongong.

Stone, L E, Hastie, D B, Wypych, P W, Roberts, J and Zigan, S, 2019. Verification of dust particle behaviour in 2D and 3D symmetrical simulations, in *Proceedings of the 13th International Conference on Bulk Materials Storage, Handling and Transportation (ICBMH 2019)*, pp 807–816 (Engineers Australia: Barton).

Stone, L E, Wypych, P W, Hastie, D B and Zigan, S, 2016. CFD-DEM Modelling of Powder Flows and Dust Generation Mechanisms – A review, in *Proceedings of the 12th International Conference on Bulk Materials Storage, Handling and Transportation*, pp 417–426.

Stone, L E, Zigan, S, Lulbadda Waduge, L L and Hastie, D B, 2021. A Modelling and Validation Approach for Predicting Particle Concentration of Airborne Dust during the Filling Process of Cylindrical Silos, *Applied Science,* 11(1794):1–19.

Toraño, J, Torno, S, Menéndez, M and Gent, M, 2011. Auxiliary ventilation in mining roadways driven with roadheaders: Validated CFD modelling of dust behaviour, *Tunnelling and Underground Space Technology,* 26:201–210.

Woschny, N, Schulz, D, Kruggel-Emden, H and Schmidt, E, 2021. Predicting dust emissions – Experimental study compared to coupled DEM/CFD simulations using a reference test bulk material, *Advanced Powder Technology,* 32:3189–3206.

Iron Ore

Exploration and geology

Dolerite dyke geochemical discrimination using routine assay data for hydrogeological modelling

A Garay[1]

1. Geoscience Specialist, Rio Tinto, Perth WA 6000. Email: amos.garay@riotinto.com

ABSTRACT

Western Australia's geological record contains multiple swarms of mafic dykes that significantly influence hydrogeological compartmentalisation within potential dewatering footprints of mining operations. These dykes can impact groundwater flow and mine water management. Accurate characterisation is therefore critical for hydrogeological modelling and mine planning.

However, dyke identification is challenging. Their sub-horizontal, tabular geometry complicates drill hole interception, while extensive regolith and detrital cover limit conventional mapping. Weathering further obscures lithological features in drill core, reducing the reliability of visual logging. These constraints highlight the need for alternative approaches that utilise existing data sets without incurring additional costs.

This study presents a litho-geochemical method that leverages immobile and semi-immobile elements commonly included in standard and historical assay suites, avoiding reliance on specialised or expensive analyses. Immobile elements are less affected by alteration and weathering, making them reliable indicators of igneous lithologies. A ranking system was developed to classify elements by mobility, enabling dolerite identification through immobile or less-mobile elements. Specifically, the TiO_2/Cr ratio combined with vanadium (V) content forms the basis of a discriminant plot that separates dolerite from other lithologies in the Pilbara Province such as banded iron formation (BIF), shale, dolomite, and igneous rocks of the Fortescue and Weeli Wolli formations.

The generation of the discriminant plot employed a bootstrap approach using interpreted stratigraphy as a guiding framework. However, samples plotting within the 'geochemical dolerite' field on the V versus TiO_2/Cr diagram were identified as dolerites, independent of logging. Spatial correlation between these geochemically classified samples and modelled dyke intrusions shows strong alignment with geological and geophysical interpretations, confirming the robustness of the method.

This approach offers two key advantages: cost-effectiveness, by extracting additional value from existing data sets, and resilience to weathering, mitigating the impact of mineralogical changes on lithological identification. The results provide a validation tool for lithological logging alongside traditional mapping, structural interpretation, and geophysical modelling. Ultimately, this method improves understanding of dyke distribution and continuity, supporting hydrogeological modelling and mine planning.

INTRODUCTION

Efficient dewatering is critical to safe, economic, socially responsible, and sustainable iron ore mining in the Pilbara, where highly permeable units and major structures can transmit large groundwater volumes. Mafic dykes locally compartmentalise aquifers, altering water-flow dynamics and potentially affecting significant cultural and environmental sites. However, dyke identification is often uncertain because tabular geometries are difficult to intersect with drilling, regolith and detrital cover obscure mapping, and weathering hampers visual recognition in reverse circulation drill chips or diamond drilling core.

This paper summarises a simple litho-geochemical workflow that uses elements already present in routine four-acid inductively coupled plasma (ICP) assay suites to discriminate dolerite dykes from common Pilbara lithologies. By extracting geochemical signals from existing data sets, the approach provides a rapid, low-cost line of evidence that complements mapping and geophysical interpretation, supporting more reliable data and boundaries in hydrogeological models.

Disclaimer: The methods and interpretations presented reflect a case study application and are not intended to represent formal Rio Tinto standards or prescribed workflows.

REGIONAL CONTEXT – PILBARA DYKE SWARMS AND LIP FRAMEWORK

Multiple dyke and sill suites occur across the Pilbara–Yilgarn region and relate either to large igneous province (LIP) events or to local extensional tectonic settings that typically develop following orogenic processes (Wingate, Pirajno and Morris, 2004). In the Hamersley basin, the Moodogara dyke suit (2208 Ma) and the Panhandle dyke suit (2008) are associated with the Ophthalmia Orogeny and the Panhandle event, respectively. The Warakurna LIP (~1078–1070 Ma) emplaced voluminous mafic dykes, sills, and layered intrusions over a broad area within a relatively short period. The Mundine Well dyke suit (755) intruded in the Channar area (Wingate and Martin, 2017). All these dyke swarms established long-lived structural fabrics and rheological contrasts that can influence present-day groundwater partitioning in the Pilbara.

APPLICATION TO HYDROGEOLOGICAL MODELLING AND MINE PLANNING

Dykes can act as hydraulic barriers when fracture networks are sealed and mafic mineralogies remain intact; conversely, weathered or brecciated margins may become local conduits. Accurate discrimination and mapping of dyke networks therefore influence boundary conditions, anisotropy, and connectivity assumptions in numerical groundwater models. By identifying probable dyke intervals from assays—particularly where textures are obscured by weathering and/or faulting—the method presented here improves confidence in the position, continuity, and extent of barrier features.

DATA AND METHOD

To reduce weathering bias, elements were ranked by expected mobility and prioritised according to their immobile or semi-immobile behaviour—particularly Ti, Zr, V, and Cr—consistent with established exploration geochemistry practice (Halley, 2020). This approach ensures that the elements characteristic of dolerite dykes remain in place despite leaching during weathering. A discriminant diagram using vanadium (V, ppm) versus the titanium-to-chromium ratio (TiO_2/Cr) was found to be effective in isolating a geochemical population that correlates well with logged dolerite dykes while excluding other igneous units present in the Jeerinah Formation (part of the Fortescue Group) and in the Weeli Wolli Formation (Figure 1).

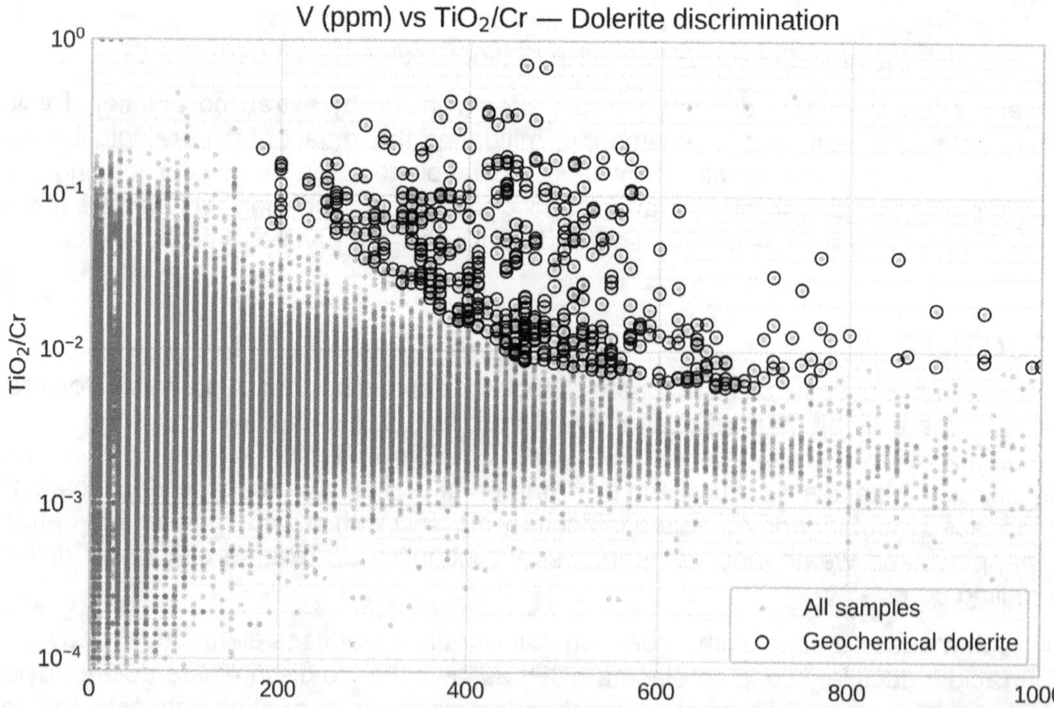

FIG 1 – V (ppm) versus TiO_2/Cr discriminant diagram: geochemical dolerite (open circles) relative to the full data set (grey dots).

Accordingly, the discriminant fields were constructed using a bootstrap approach, initially guided by interpreted stratigraphic logs and subsequently refined based on independent geochemical behaviour. Samples plotting within the 'geochemical dolerite' field were then classified as dolerite regardless of logged lithology. This procedure is robust in identifying dolerite dykes and can be implemented entirely on standard and historical assay data sets without requiring specialised analyses.

CASE STUDY RESULTS AND VALIDATION

Applying the V versus TiO_2/Cr diagram to Pilbara data sets reliably separates dolerite dykes from banded iron formation (BIF), shale, dolomite, and other mafic igneous units. When the geochemical dolerite population is reviewed in its geographic context, the points consistently confirm, infill and extend mapped or interpreted dyke corridors in plan view and honour existing interpretations in cross-section (Figure 2).

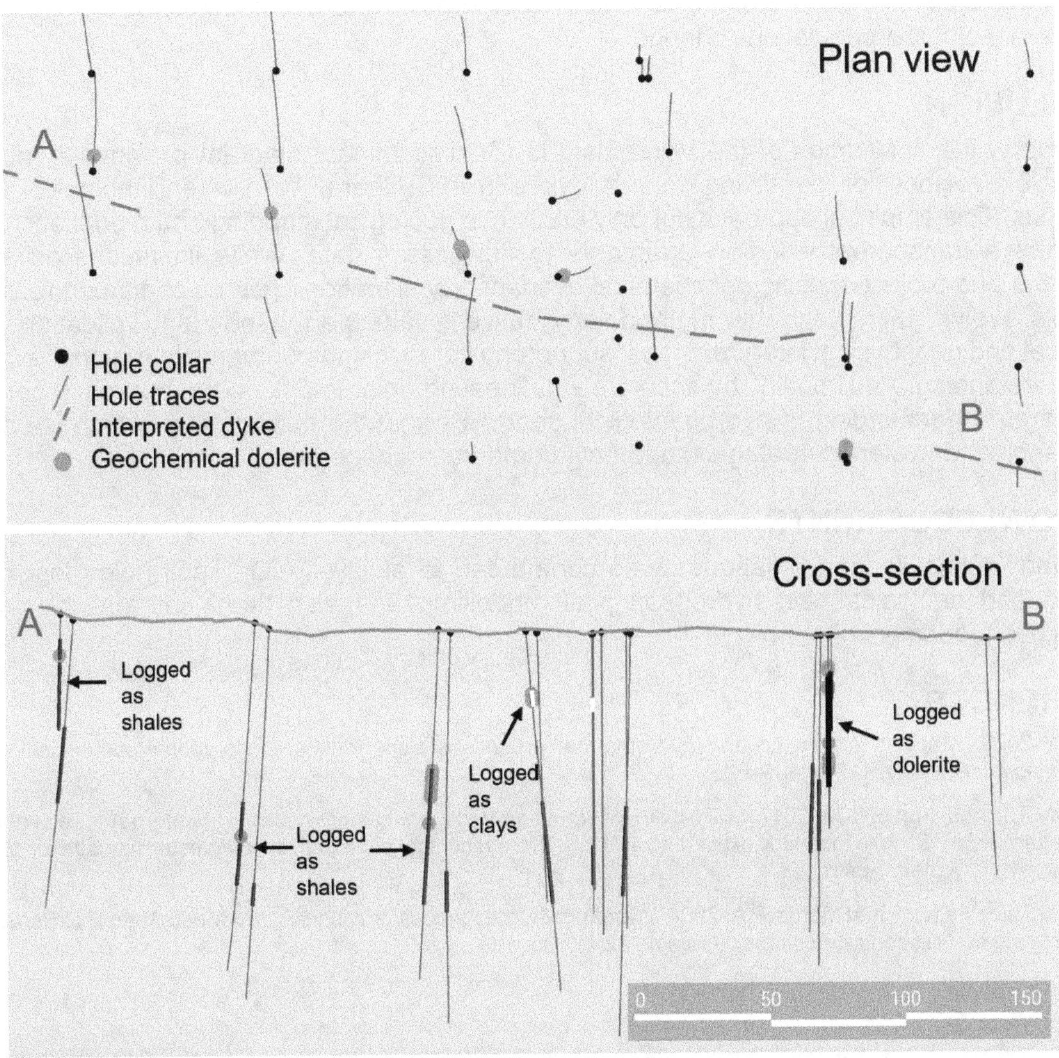

FIG 2 – Plan view and cross-section of a drilled area. In plan view, 'geochemical dolerites' match the interpreted dyke from magnetic survey data. In cross-section, only one hole logs dolerite; the other 'geochemical dolerites' outline the dyke in areas logged as shales or clays under conditions of intense weathering. *Coordinates and location references have been intentionally removed.*

In the example illustrated in Figure 2, the 'geochemical dolerites' identified using the described methodology align with an interpreted dyke as delineated by magnetic survey data in plan view. In the corresponding cross-section, dolerite has been logged in only one drill hole, while the remaining 'geochemical dolerites' outline the dyke trend at depth, even though the samples were identified as weathered shales or clays rather than dolerite. This logging interpretation does not reflect poorly on the geologists involved; it is entirely understandable to confuse a weathered dolerite dyke with a weathered shale. Additionally, dykes and faults are often spatially associated, and the reactivation of faults previously sealed by intrusions can introduce intense fracturing and, consequently, result in deeper weathering within the dykes.

Spatial comparisons show that points classified as 'geochemical dolerite' confirm previous geological and geophysical dyke interpretations, providing an independent check on lithological logging and structural interpretation, while also supplying additional data. Because the workflow is deterministic and transparent, results can be reproduced rapidly as new assays become available, allowing iterative updates to geological models.

LIMITATIONS AND FUTURE WORK

Although Ti, V, and Cr are comparatively immobile, severe weathering, metasomatism, or sampling issues can blur class boundaries. The approach distinguishes dolerite from sedimentary units effectively, but overlaps can occur with the geochemically similar igneous rocks mentioned above. Because the discriminants are empirical and project-specific, the envelopes should be recalibrated whenever lithological populations change.

CONCLUSION

In summary, the application of the V versus TiO_2/Cr discriminant diagram provides a robust and reproducible method for identifying dolerite dykes in the Pilbara, even when lithological logs are ambiguous. This empirical approach not only enhances geological mapping and model accuracy but also offers a transparent workflow adaptable to new assay data. While limitations exist due to weathering and project-specific geochemical overlaps, recalibration ensures continued reliability as data sets evolve. Ultimately, this methodology delivers valuable independent validation for both geological and geophysical interpretations, supporting iterative improvements in regional exploration and understanding. Additionally, by accurately delineating dolerite dykes, the approach contributes to a clearer understanding of hydrogeological controls within the mining area, which is crucial for mine planning, dewatering strategies, and environmental management.

ACKNOWLEDGEMENTS

The author acknowledges colleagues who contributed to assay QA/QC, drill holes logging, field mapping and geological and hydrogeological modelling, and also thank internal reviewers for constructive feedback.

REFERENCES

Halley, S, 2020. Mapping magmatic and hydrothermal processes from routine exploration geochemical analyses, *Economic Geology*, 115(3):489–503.

Wingate, M T D and Martin, D M, 2017. Mafic dyke swarms and large igneous provinces in Western Australia get a digital makeover, in GSWA Extended Abstracts 2017 – Promoting the prospectivity of Western Australia, Geological Survey of Western Australia.

Wingate, M T D, Pirajno, F and Morris, P A, 2004. Warakurna large igneous province: A new Mesoproterozoic large igneous province in west-central Australia, *Geology*, 32(2):105–108.

Structural controls on the metallogeny and distribution of the Mary River Fe-deposits, Canada

B M Saumur[1], S T Johnston[2], B Dyck[3], K P Larson[4], J Hey[5] and M Robatian[6]

1. Professor, Université du Québec à Montréal and GEOTOP, Montréal QC H2X 3Y7, Canada. Email: saumur.benoit@uqam.ca
2. Professor, University of Alberta, Edmonton AB T6G 2E3, Canada. Email: stjohnst@ualberta.ca
3. Associate Professor, University of British Columbia – Okanagan, Kelowna BC V1V 1V7, Canada. Email: brendan.dyck@ubc.ca
4. Professor, University of British Columbia – Okanagan, Kelowna BC V1V 1V7, Canada. Email: kyle.larson@ubc.ca
5. Senior Director, Exploration and Strategic Planning, Baffinland Iron Mines Ltd., Oakville Corporate Office, Oakville ON L6H 6K8, Canada. Email: jon.hey@baffinland.com
6. Manager, Ferrous Exploration, Baffinland Iron Mines Ltd., Oakville Corporate Office, Oakville ON L6H 6K8, Canada. Email: massoud.robatian@baffinland.com

INTRODUCTION

The Mary River Group (MRG) refers a suite of Neoarchean supracrustal rocks within greenstone belts of the Rae Craton of northern Baffin Island (Saumur *et al,* 2025). The MRG hosts the high-grade, high-tonnage Mary River Fe-ore deposits, which were discovered in the early 1960s. Renewed exploration efforts, feasibility studies and public consultations in the mid-2000s led to open pit mining ore production at Deposit 1 starting in 2014 (Iannelli, Dammeier and Samiei, 2016). Iron-ore grades that average 63–65 wt per cent Fe allow for direct shipping to market without concentration or processing (Baffinland Iron Mines, 2026). This, as well as the sheer size of the deposits (roughly 1 billion tonnes (Bt) of Measured, Indicated and Inferred Mineral Resources; ArcelorMittal, 2024), renders Mary River economic despite challenging logistics associated with remote Arctic operations. As of 2026, Mary River's Deposit 1 is the site of Canada's northernmost active mine, and accounts for 9 per cent of Canada's total Fe production (NRCan, 2025) and 4.2 million tonnes (Mt) of shipped ore per annum.

Recent field studies have elucidated some of the important geological ore-forming controls at play at Mary River. We highlight those herein, as well as new geochronological information about ore formation.

STRUCTURAL ELEMENTS

Updated mapping (Saumur *et al,* 2020; 2025) shows that the MRG forms discontinuous, 1–10 km scale, variably deformed and metamorphosed panels within orthogneiss and granitic rocks of the Rae Craton. In the Mary River area, economic to sub-economic Fe-ore deposits occur within three 10 km-scale synformal panels of MRG (Figure 1a) that open eastwards and plunge shallowly to moderately to the ESE. The folding and the associated metamorphism are linked to a regional D3 event attributed to the Palaeoproterozoic Trans-Hudson Orogeny. Despite the overall ESE plunge of the system, metamorphic grade (~upper greenschist) and the level of structural and stratigraphic exposure are broadly consistent across the region.

The Mary River orebodies occur along the footwall of the Central Borden Fault Zone (CBFZ), one of several important WNW-ESE brittle fault zones that affect the region. In addition, a minor set of NNW-SSE trending faults occur; incised valleys within the glaciated, boulder-strewn plateau reflect the surface expression of these faults (Figure 1a). Both fault sets control the geometry of Mesoproterozoic and younger basins occurring on Baffin Island (Jackson and Iannelli, 1981).

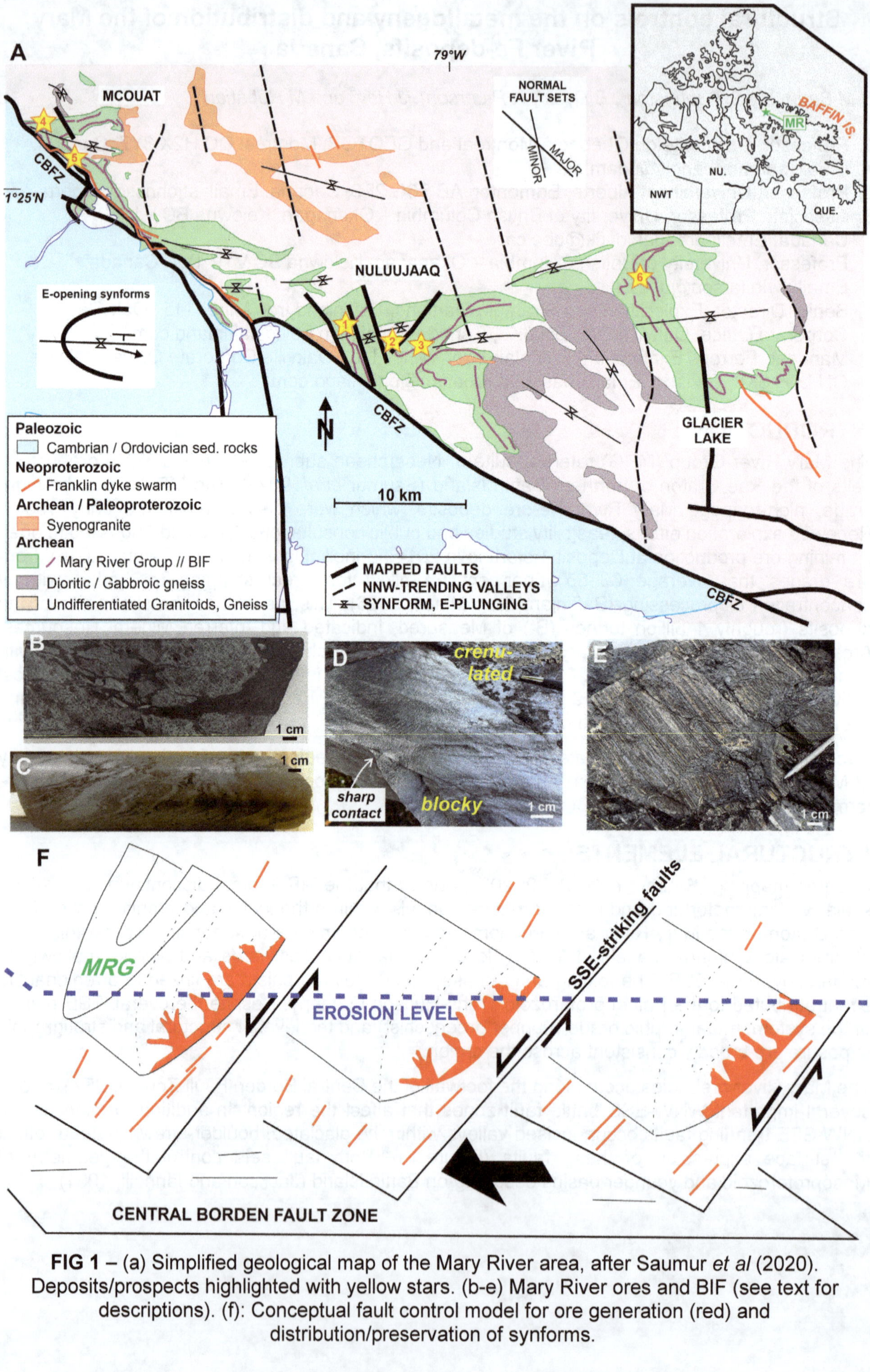

FIG 1 – (a) Simplified geological map of the Mary River area, after Saumur *et al* (2020). Deposits/prospects highlighted with yellow stars. (b-e) Mary River ores and BIF (see text for descriptions). (f): Conceptual fault control model for ore generation (red) and distribution/preservation of synforms.

ORE CHARACTERISTICS

The Mary River deposits are hypogene, Banded Iron Formation (BIF) associated Fe-ore deposits. Primary BIF has been recrystallised and metamorphosed to anhydrous mineral assemblages such as hematite-magnetite-quartz (ie itabirite). Metamorphosed BIF is associated with hematite-rich or magnetite-hematite ore, where significant quartz has been removed. Fe-ore can also occur in metamorphosed supracrustal rocks (eg chlorite schist) surrounding BIF, where infiltration of an Fe-rich front and metasomatic replacement of silicates by Fe-oxides are evident (Figure 1b). BIF occasionally exhibits vugs within quartz-bands indicative of silica dissolution (Figure 1c). The occurrence of this texture in folded BIF indicates that silica dissolution post-dates, at least in part, regional ductile deformation.

Fe-ores of the Mary River deposits, as best observed at Deposits 1, 2 and 3 (Figure 1a), can be described by two end-member textures: crenulated ore and blocky/massive ore (Figure 1d). Crenulated ore is characterised by specular hematite showing cm-scale tectonic crenulations and/or mm-scale tectonic lineations. In contrast, blocky ore, consisting of fine-grained magnetite and/or hematite, does not exhibit tectonic fabrics. Blocky ore locally overprints crenulated ore (Figure 1d), indicating that it represents a subsequent, post-tectonic hypogene mineralisation event. In addition, at Glacier Lake (Figure 1a), decimetre-scale injections of blocky iron ore occur along SSE-striking normal faults that cross-cut and offset primary BIF fabrics (Figure 1e); this relationship is indicative of a genetic link between late blocky ore and the NNW-SSE trending faults.

Blocky magnetite-rich ore associated with metasomatic replacement of chlorite schist (Figure 1b) contains accessory monazite suitable for geochronological studies. U-Pb monazite geochronology was performed at the Fipke Laboratory for Trace Element Research using an Agilent 8900 inductively coupled plasma mass spectrometer (UBC Okanagan; methods analogous to Larson *et al,* 2025). Monazite yields a $^{207}Pb/^{206}Pb$ weighted mean date of 1844 ± 13 Ma (Saumur *et al,* in review) that is interpreted to record the time of the late 'blocky-ore event' and is consistent with metasomatism being a product of Trans-Hudsonian orogenesis.

DISCUSSION

Since blocky-ore is genetically linked to NNW-SSE trending faults, the monazite geochronology of blocky-ore highlights that the regional fault system is at least Palaeoproterozoic in age, and has been subsequently sporadically reactivated since the Mesoproterozoic. The faults are genetically linked to the Trans-Hudson orogen and may reflect an expression of orogenic collapse.

Palaeoproterozoic folding and metamorphism structurally thickened the primary MRG banded iron formation (BIF), and resulted in the development of crenulated specular hematite ore, silica dissolution and metasomatic replacement of silicates by Fe-oxides. Subsequent fault activity during the Palaeoproterozoic is responsible for accentuated fluid flow leading to further metasomatic removal of silicates, Fe-recrystallisation and upgrading and the formation of blocky ore that locally overprints crenulated ore.

Later Mesoproterozoic activity along pre-existing faults, concentrated along SSE-striking listric normal faults, led to structural repetition of the MRG and its Fe-ore deposits by extensional bookshelf faulting (Figure 1f). The occurrence of a series of synforms without intervening antiforms is a result of fault repetition of a single synform, which similarly explains the consistent metamorphic grade across the Mary River area.

The Mary River system illustrates the importance of brittle faulting for the genesis and preservation of hypogene Fe-ore deposits. Moreover, brittle (or ductile) faults and shear zones represent favourable environments for ore concentration and should be considered as 'guides' for iron exploration in polydeformed terranes.

REFERENCES

ArcelorMittal, 2024. Annual Report 2024, ArcelorMittal Société Anonyme. Available from: <https://cdn.arcelormittal.com/media/hoclv2an/mt-31-12-2024-annual-report.pdf> [Accessed: January 2026].

Baffinland Iron Mines, 2026. Mary River Mine, Baffinland Iron Mines. Available from: <https://www.baffinland.com/operation/mary-river-mine/> [Accessed: January 2026].

Iannelli, T R, Dammeier, M and Samiei, A, 2016. Baffinland Iron Mines Corporation Mary River Project: Assessment Report: Rock Sampling, Prospecting, Mapping, and Geophysical Surveying, AANDC Assessment Report 086030. Available from: <nunavutgeoscience.ca> [Accessed: January 2026].

Jackson, G D and Iannelli, T R, 1981. Rift-related Cyclic Sedimentation in the Neohelikian Borden Basin, northern Baffin Island, in *Symposium on Proterozoic Basins of Canada, Geological Survey of Canada*, Paper 81–10, pp 269–302. https://doi.org/10.4095/109369

Larson, K, Dyck, B, Cottle, J, Canam, R and Milkowski, E, 2025. Complementary in situ garnet Lu-Hf geochronology and monazite U (-Th)-Pb petrochronology: an example from the Archean Slave craton, Northwest Territories, Canada, *Geological Magazine*, 162:e14. https://doi.org/10.1017/S001675682500007X

NRCan, 2025. Iron-Ore Facts, Natural Resources Canada, Government of Canada. Available from: <https://natural-resources.canada.ca/minerals-mining/mining-data-statistics-analysis/minerals-metals-facts/iron-ore-facts> [Accessed: January 2026].

Saumur, B M, Folkesson, C-P, Skipton, D R, Davies, J H F L, Wodicka, N and Johnston, S T, 2025. Precambrian Greenstone Belts of North Baffin Island, Canadian Arctic: Geochronological Constraints on Magmatism and Stratigraphy in the Rae Craton, *Geochemistry, Geophysics, Geosystems*, 26:e2025GC012206. https://doi.org/10.1029/2025GC012206

Saumur, B M, Johnston, S T, Dyck, B, Larson, K, Adibi, M and Hey, J, (in review). On the Genesis of the Mary River Fe-ore Deposits, Nunavut, Canada: A Tale of Four Synforms, *Economic Geology*.

Saumur, B M, Skipton, D R, St-Onge, M R, Jackson, G D, Wodicka, N, Bros, E R, Weller, O M and Johnston, S T, 2020. Bedrock geology, Angijurjuk-Mary River, Baffin Island, Nunavut, NTS 37-G west, Geological Survey of Canada, Canadian Geoscience Map 403, scale 1:100 000, 1 sheet.

The geology of the structurally complex Wyloo North Deposit

O P Shufeldt[1]

1. Principal Geologist, Fortescue, Perth WA 6000. Email: owen.shufeldt@fortescue.com

ABSTRACT

The Fortescue-owned Wyloo North iron ore deposit is located on the northern limb of the Wyloo Anticline in the south-western Hamersley Province, Western Australia. It forms part of a structurally-controlled iron ore system which, together with the Wyloo South and Rio Tinto's Metawandy deposits, contains hundreds of millions of tonnes of ore. This study integrates detailed >1:5000 scale geological mapping, structural analysis, drill core logging, and petrography to constrain the structural evolution, stratigraphic controls, and mineralisation history of the deposit. Wyloo North is dissected by at least 15 major faults over a ~23 km strike length, comprising multiple generations of predominantly normal faults with evidence of reactivation, back-thrusting, and minor strike-slip movement. Cross-cutting NW–SE faults, variably intruded by ~2008 Ma dolerite dykes, are interpreted to have acted as key permeability pathways for hydrothermal fluids and are spatially associated with high-grade martite–microplaty hematite mineralisation. Local stratigraphic variations within the Brockman Iron and Weeli Wolli Formations, including the local absence of the Paraburdoo Member and reduced sill development, further influence ore distribution and upgrading. Petrographic evidence indicates a multi-stage mineralising history involving Proterozoic hypogene martite–microplaty hematite alteration overprinted by younger supergene processes. Mineralisation is interpreted to reflect long-lived, structurally focused hydrothermal fluid flow with mode-switching behaviour within a damage zone bounded by major faults. These results refine the structural framework of the Wyloo Anticline and improve orebody knowledge.

INTRODUCTION

The Fortescue-owned Wyloo North deposit sits on the northern limb of the Wyloo Anticline (aka Wyloo Dome), located at 22°33'48.2"S 116°26'45.6"E (GDA94 Z50 44300E, 7504700) (Figure 1) in the south-western Hamersley Province. It is part of one of the largest undeveloped ore deposits that contains a component of hydrothermal martite microplaty-hematite (M-mpl H) in the western Hamersley Province. Together, the Wyloo South/North (Fortescue) and Rio Tinto's (RTIO's) Metawandy deposits host hundreds of millions of tonnes of iron ore and, at least in part, are structurally controlled. This paper will outline the complex geology of the Wyloo North deposit, including the following: local and regional structural evolution; the relationship between deformation and mineralisation; local stratigraphic variations; ore properties; petrography; relative timing of deformation and ore genesis.

Hosted on the northern limb of the Wyloo Anticline, the deposit is extensively faulted. Over its ~23 km strike length there are at least 15 major faults with displacements ranging from 50 m to 2.5 km. Most faults show normal displacement, although evidence exists for back-thrusting, reverse reactivation and minor strike-slip movement likely related to the regional Menindee Fault Zone (MFZ). At least three generations of major faults are recognised, some of which show evidence of multiple phases of movement and control on mineralisation. Timing relationships, and interpretations of mineralising fluid pathways will be discussed.

Stratigraphic variations within the Brockman Iron and Weeli Wolli Formations from typical Hamersley Group geology also influence ore distribution and prospectivity. A brief overview of this and its influence on the orebody will be provided.

The iterative nature of exploration in the Wyloo North area leading to the detailed approach to mapping and sampling over the last five years of exploration will be highlighted. Petrographic work from both surface samples and drill hole samples has further bolstered this work. This combination of methods has elucidated timing relationships between the different generations of folds, faults, dykes, veins and fluid flow and has been invaluable to unravelling the complex nature of the deposit leading to multiple exploration successes and vastly improved orebody knowledge.

REGIONAL GEOLOGY

The Wyloo North deposit sits on the northern limb of the regional Wyloo Anticline (prev. Wyloo Dome). The Metawandy deposit forms the hinge region of the anticline and Wyloo South, the southern limb. The mapped geology spans most of the Hamersley Group from the base of the Archean Marra Mamba Iron Formation (MMIF) through to the Proterozoic Weeli Wolli Formation. Bedding dominantly dips moderately to the north (~40–50°) with some minor folding.

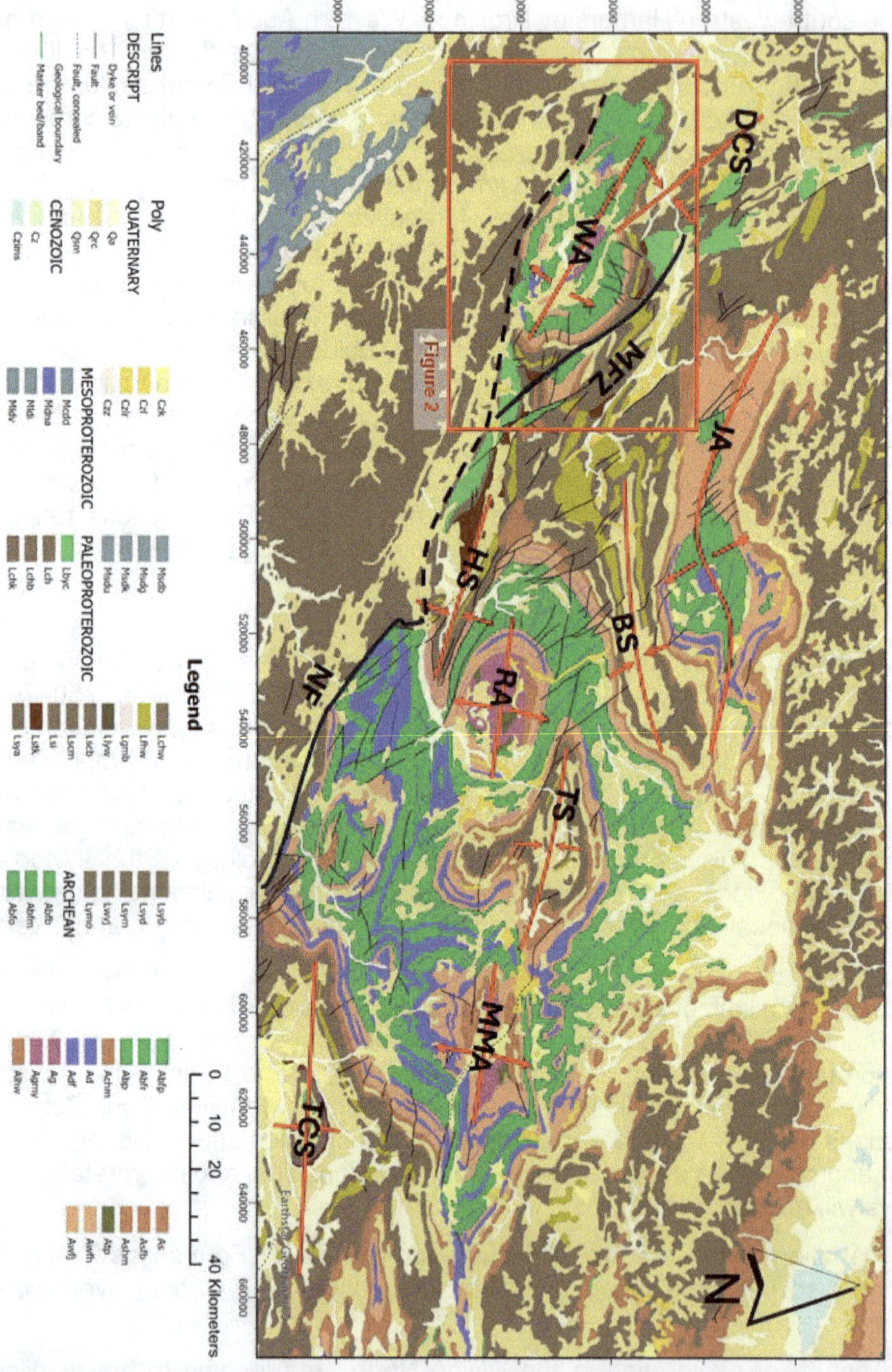

FIG 1 – Regional setting of Wyloo North (red box outlines area of Figure 2) in the south-western Hamersley Province. Abbreviations as follows: WA, Wyloo Anticline; DCS, Duck Creek Syncline; MFZ, Menindee Fault Zone; JA, Jeerinah Anticline; BS; Brockman Syncline; HS, Hardey Syncline; RA, Rocklea Anticline; TS, Turner Syncline; MMA, Milli Milli Anticline; NF, Nanjilgardy Fault; TCS, Turee Creek Syncline. Basemap and legend abbreviations from GSWA 1:1 000 000 Surface Geology Map.

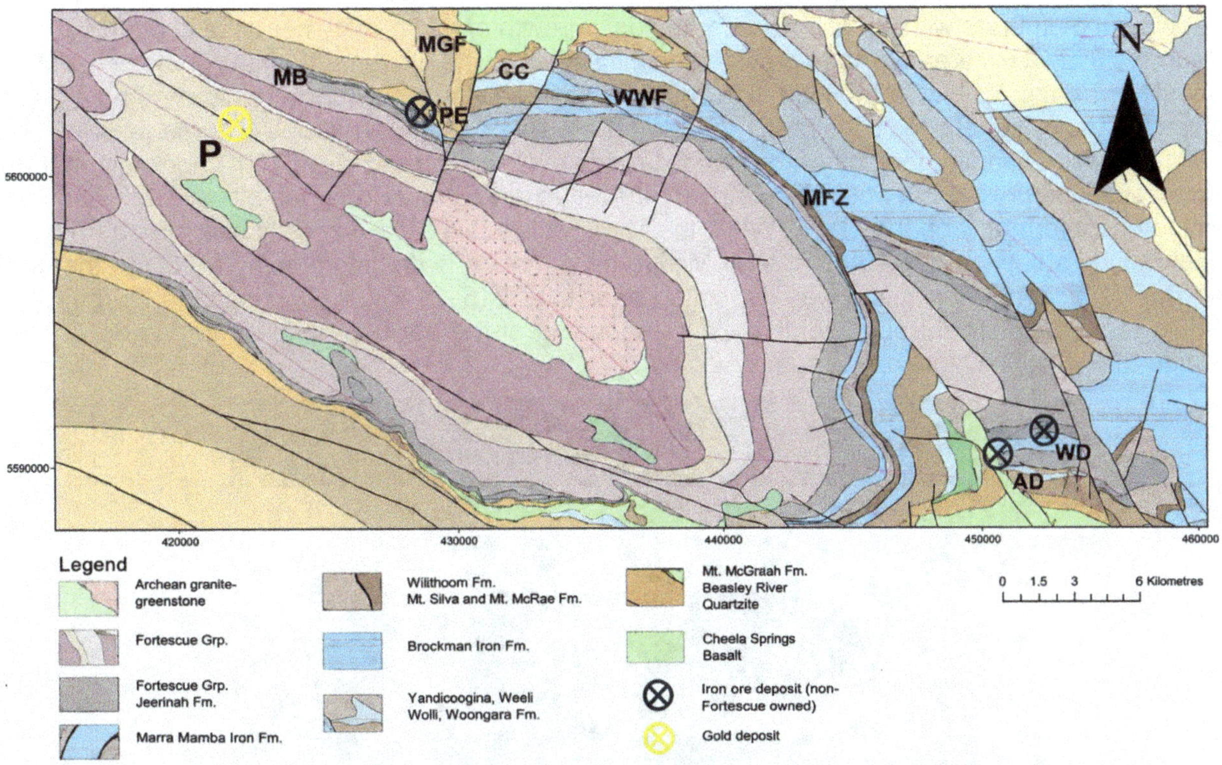

FIG 2 – Map of the Wyloo Anticline showing the structural complexity of the area. Abbreviations: P, Paulsen's Gold Deposit; MB, Madala Bore; PE, Paulsen's East; MGF, Mitala Gap Fault; CC, Centipede Creek; WWF, Worm World Fault; AD, Arochar Deposit; WD, Wellthandalthaluna Deposit. Wyloo North Deposit extents in red box. Basemap from GSWA 1:500 000 state interpreted bedrock geology.

The Wyloo Anticline has a core of Pilbara Supergroup rocks intruded by the Metawandy Granite (Seymour, Thorne and Blight, 1988) and unconformably overlain by the Fortescue Group with Hamersley, Shingle Creek and Wyloo Group sediments on the limbs of the anticline. It has long been recognised as a predominantly Ashburton structure with evidence for prior deformation (ie Ophthalmian folding) before deposition of the Shingle Creek and Wyloo Groups (Dalstra, 2014; Martin, 2020).

The Wyloo Anticline sits between two major fault zones: the Nanjilgardy Fault, which reaches lower crustal levels (Johnson *et al,* 2013); and the Menindee Fault Zone, a shallower set of faults. Early work done in the area proposed that the entire structural feature of the Wyloo Anticline itself had been 'translated 10–12 km to the north-west, along steeply dipping wrench faults which bound the south-western Hamersley Basin' (Thorne and Seymour, 1991). However, this was based on the interpretation that the complex fault system in the area was predominantly made up of strike-slip faults, which persisted into recent work descriptions of dominantly dextral strike-slip movement (Martin and Morris, 2010). Attempting to reconstruct the geology of the eastern side of the Wyloo Anticline including the Rio Tinto Metawandy, Arochar and Wellthandalthaluna deposit areas reveals that dip slip is dominant, as the hinges of the regional folds have all been displaced vertically, rather than along strike (Figures 2 and 12 of Dalstra, 2014). Additionally, for 'exclusively dextral movement' (Martin and Morris, 2010) to have been the case, attempting to reconstruct the area produces a result where the Arochar area would have originated as part of the Metawandy deposit. However, this contradicts the mapped geology where Arochar sits on the southern limb of the regional anticline (Wyloo-Mt Wall Anticline) and has been subsequently offset by normal faulting. This interpretation here agrees with work done in the region that also concluded most fault systems to be normal (Dalstra, 2014) with minor transpressional reactivation.

To the west of the deposit sits the unique 10 million tonne (Mt) Paulsen's East Deposit (aka Barret-Lennard). The host for this deposit is the Proterozoic Mt McGrath Formation (PWm), a pebble conglomerate containing mineralised BIF clasts. As is the case with most things in the area, there is some debate whether the Mt McGrath Formation has been mineralised/upgraded *in situ* or is an ore-pebble conglomerate with occasional post-tectonic growth of micro-platy hematite in matrix goethite (Morris, 2012). This area can also provide some insight into the movement on the limbs of the Wyloo Anticline, as the Mt McGrath Formation must have been deposited on a shallowly dipping slope given the depositional conditions required for a delta-plain or delta-front facies (Thorne and Seymour, 1991) but the open-file geology of the deposit reveals a moderate-to-steep (50°) north-dipping unit. Fortescue's tenure 1.5 km west along strike, referred to as Madala Bore, has Mt McGrath Formation sitting unconformably on the MMIF. While the MMIF at Madala Bore is mostly overturned (Figure 3) and dipping steeply to the south, the Mt McGrath Formation is dipping steeply (>70°) to the north. As the Mt McGrath Formation was deposited after the 2215–2145 Ma Ophthalmian Orogeny, this would require significant post-Ophthalmian steepening of the northern limb of the Wyloo Anticline during the ~1.8 Ga Capricorn Orogeny.

FIG 3 – South dipping Nammuldi Mbr. with clear axial planar cleavage (S_1) to bedding (S_0) relationship showing overturned bedding in the MMIF.

The Paulsen's East Deposit and continuation of the northern limb westward is separated from the Wyloo North project area by at least one significant fault, here referred to as the Mitala Gap Fault (MGF). These faults (or antithetic pair of faults) displace units up to the Duck Creek Dolomite of the Wyloo Group and place the units of the Shingle Creek and Wyloo Groups at a high angle to rocks of the Hamersley Group rocks. This suggests that the MGF could be a growth fault formed as the Turee Creek and Wyloo Basins were deepening. The steeply dipping units also bound the limbs of the Duck Creek Syncline near its hinge, possibly indicating the syncline at least in part is sagging on the growth faults and forming syn-sedimentarily with the upper units of the Turee Creek and the Shingle Creek Groups. The MGF is interpreted here as a significant domain boundary. Rocks to the west are majorly influenced by Capricorn and later deformation and to the east, heading into the Wyloo North deposit, are more influenced by the Ophthalmia Orogeny and post-Ophthalmian relaxation.

GEOLOGICAL MAPPING IN THE HAMERSLEY PROVINCE

The Wyloo North deposit has proven the value of a combination of detailed contact and exposure mapping (Coe, 2011) at >1:5000 scales of iron deposits in the Hamersley Province. In contrast to the traverse-style mapping traditionally taught in universities, the methodology relied on here emphasises direct tracking of stratigraphic contacts until termination by cover units or faults. The exposure of the different stratigraphic units in the Hamersley Province in general is excellent, and the sheer number of faults at Wyloo North requires careful boundary following to narrow down the location of structures.

The complexity of the deposit also required multiple revisits by the generative mapping team after successive phases of drilling. Faults are often poorly exposed, and the hard-capping on the mineralisation can often obscure boundaries and faults. As new faults were intersected during drill campaigns, the areas were revisited during the following mapping campaign and with the drill hole information the faults could be identified and properly mapped out. Faults often exposed as narrow (10–75 cm) wide zones of hematite, goethite, ferruginised dolerite or gravel which are easily missed without drill hole data (Figure 4). This iterative mapping and data review proved successful from an exploration perspective.

FIG 4 – Typical exposure of a dolerite-intruded fault (within white lines) that has been subsequently altered.

STRUCTURAL GEOLOGY OF WYLOO NORTH

The deposit is cross-cut by a complex series of SW-NE, NNW-SSE, NW-SE and some rarer E-W (or roughly strike-parallel) normal faults (Figure 5). The complexity is interpreted to be the result of a combination of the following: the long-lived folding of the Wyloo anticline, the core of which is relatively stiff Archean granite and Fortescue Group basalts acting as structural buttresses causing strain partitioning elsewhere; the proximity to the MFZ; and the NW-extent of the Nanjilgardy Fault. The relative age, orientation, and a description of the fault plane material for each fault (either through field observations or in drill hole intersections) is described below.

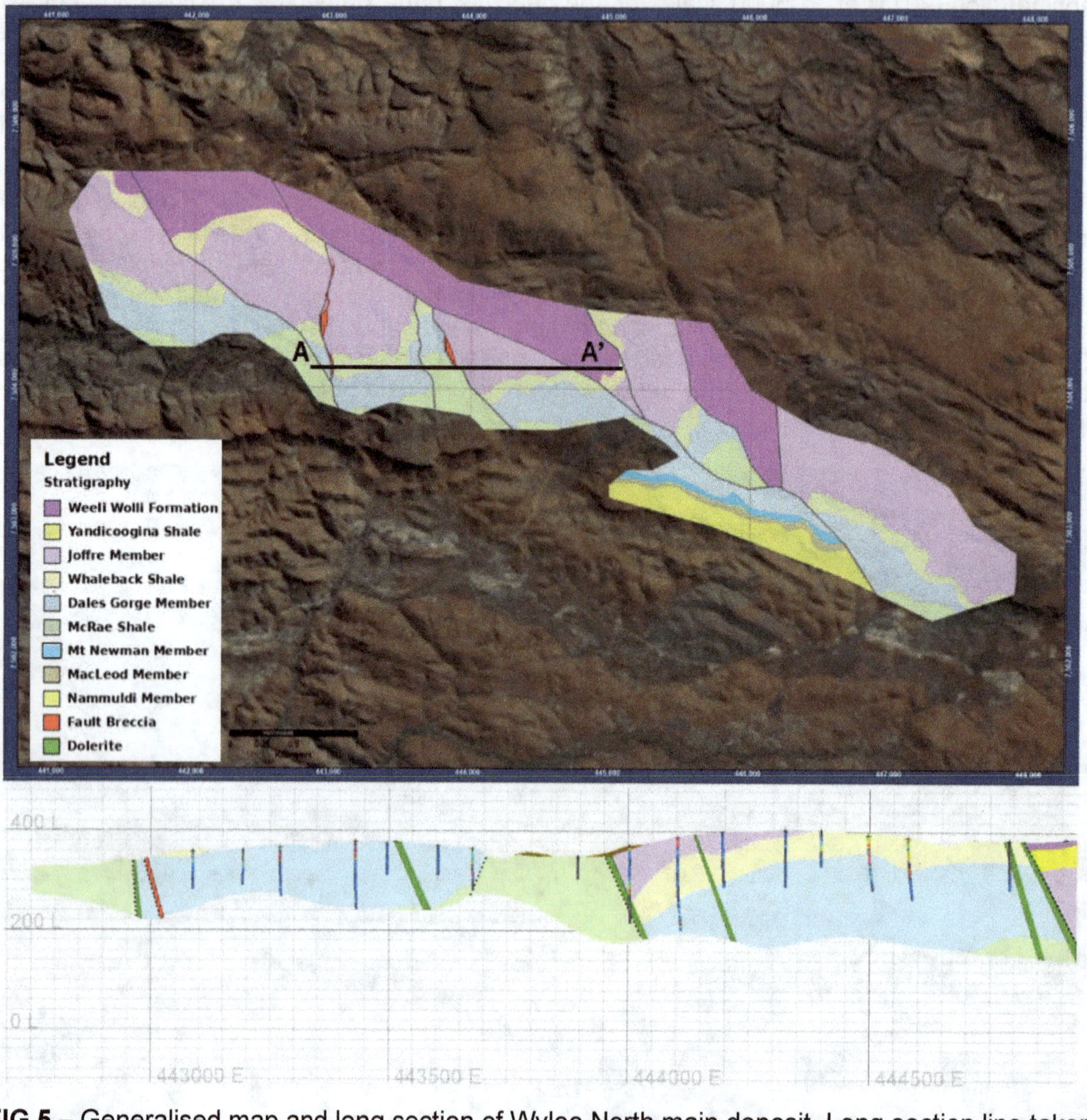

FIG 5 – Generalised map and long section of Wyloo North main deposit. Long section line taken from A-A' on map.

The oldest set of faults is interpreted to be striking roughly N-S and dipping steeply E or W (Figure 5). Displacement ranges from ~100–250 m and appears to be completely dip-slip. Diamond drilling through one of these faults has revealed that there have been multiple movements along this fault as evidenced by sub-angular to sub-rounded clasts of breccia with brecciated clasts within (Figures 6 and 7). These faults are localised, with only one or two across deposit confidently falling into this set and clearly bounded to the north and south by the through-going NW-SE trending faults.

FIG 6 – Picture of PQ sized drill core from hole WND0004 showing fault breccia at 105.9 m downhole.

FIG 7 – Zeiss Axioscan of thin section in reflected light (RL) showing sub-rounded to sub-angular brecciated clasts of altered BIF in an iron, goethite, clay and silica matrix. Sample taken from same drill hole as Figure 6 at 100.59 m downhole. Scale bar = 5 mm/5000 µm.

The next major set of structures trend NE-SW. There is no immediately apparent age relationship between the N-S trending faults and the major NE-SW trending fault that separates Wyloo North main deposit from the down-thrown block internally named as Centipede Creek (Figure 2). According to the GSWA 1:250 000 scale mapping over the Wyloo Sheet and geophysical evidence from the radiometric data, this fault (here named the Centipede Creek Fault) offsets the Beasley River Quartzite (2276 ± 41 Ma) but is truncated by the 2203 ± 4 Ma Cheela Springs Basalt (Krapež *et al,* 2017). There was a zone of weathered goethite and clay-rich gouge or siliceous and goethitic breccia along the fault, making any measurements of the plane difficult to obtain, but the fault appears to be steeply W-NW dipping with normal displacement on the order of 2.3–2.5 km with at least one synthetic splay with a displacement of 300 m.

The main orebody is separated from the eastern area by a complex set of anastomosing faults, one of which (here called the Worm World Fault) is relatively flat, ENE-trending and NW dipping at ~30–40° (no fault measurement could be taken, so this was calculated from 2 m contour data). Along the fault plane is a ferruginised, fine-grained clay or hard, remobilised hematite. The western portion of this fault system is cut by a NW-oriented fault, while the eastern portion of the flat fault is truncated by a steeper NE-trending fault, indicating the Worm World Fault could be an older structure. It could possibly be an antithetic pair to the south-dipping Mt Wall Fault of Dalstra (2014), although Capricorn folding likely has reoriented it, making it difficult to ascertain its original attributes and true age.

The most recent set of faults to majorly affect the deposit appear to be NW-SE trending. They run parallel to the MFZ and are interpreted to be reactivated during its development. These faults cannot have been developed as part of the MFZ however, because they are variably intruded by the 2008 ± 6 Ma Panhandle Dolerite dyke swarm (Martin, 2020; Müller *et al*, 2005), which pre-dates the MFZ. This suite also has an association with martite microplaty-hematite mineralisation, likely providing a pathway for hydrothermal fluid transport. Field relationships and deposit scale mapping reveals they cross-cut and truncate all other faults. The highest-grade and thickest mineralised intercepts occur where these faults cross-cut older faults, indicating that these zones provided increased permeability for mineralising fluids. This study interprets this set to be part of the Metawandy Fault system based on similar orientations and characteristics in proximal deposits (Dalstra, 2014; Martin, 2020). Where observed, the fault planes dip steeply (~70–85° to the NE) with thin 5–25 cm damage zones (Figure 8) forming a series of half grabens across the deposit. These faults are often intruded by dolerite (Figure 4) that is assumed to be part of the ~2008 Ma Panhandle dyke swarm (Martin, 2020; Müller *et al,* 2005).

FIG 8 – Clear fault in outcrop with mid-Joffre Mbr. downthrown against lower Dales Gorge Member of Brockman Iron Formation Fault was dipping at 71→ 015 (Dip→ Dip Direction).

LOCAL STRATIGRAPHY

While the Hamersley Province stratigraphy is notorious for its relatively uniform stratigraphy across the more than 60 000 km^2 area, there are some local variations at Wyloo North.

One of the more noticeable features of Wyloo North is that in the main orebody the dolomites of the Paraburdoo Mbr. (Wittenoom Formation) are locally missing. There are areas where this is interpreted to be faulted out, but there are other areas without a clear fault. This has been noted elsewhere in proximity to the hypogene Wellthandalthaluna and Arochar deposits which sit in a similar structural setting (Figure 2).

Also noteworthy is the lack of sills in the lower Weeli Wolli Formation. At about the same time as Fortescue was noting this in our mapping, the GSWA published an offhand remark about it as a revision to the Wyloo 1:250 000 map sheet (Martin, Allen and Blay, 2022). This difference in stratigraphy is important in that the distinctive 'razorback ridge' appearance of Weeli Wolli BIF separated by weathered dolerite sills is not common in the south-western Hamersley Province and instead there is a thick (65 m) sequence of BIFs and thin shales (Figure 9) above the Yandicoogina Shale Mbr. of the Brockman Iron Formation.

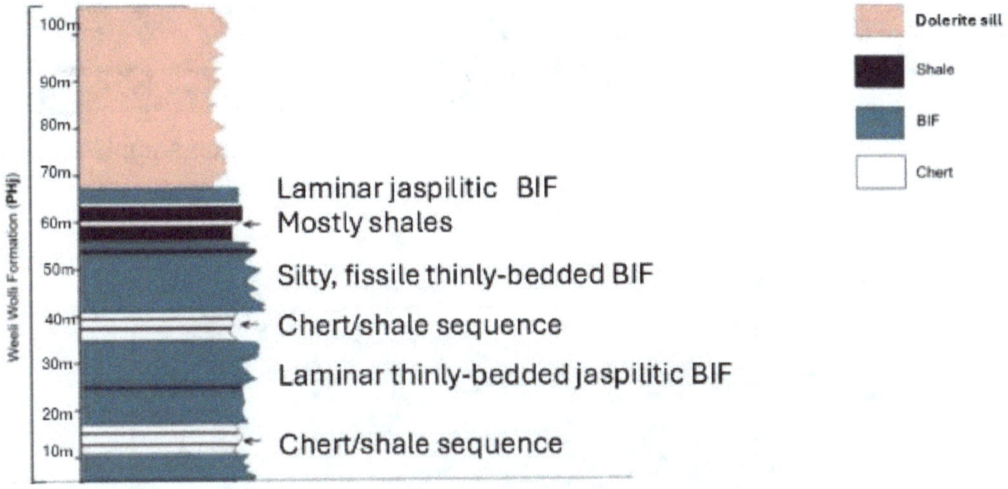

FIG 9 – Generalised stratigraphy of the Weeli Wolli Formation as mapped at Wyloo North. Modified from Tannock (2021).

PETROGRAPHY

Downhole and surface samples have been made into thin sections for analysis. Immediately apparent was the different types of hematite observed in samples including martite (hematite pseudomorph of magnetite), micro-platy hematite (mpl H) and anhedral hematite (Figure 10).

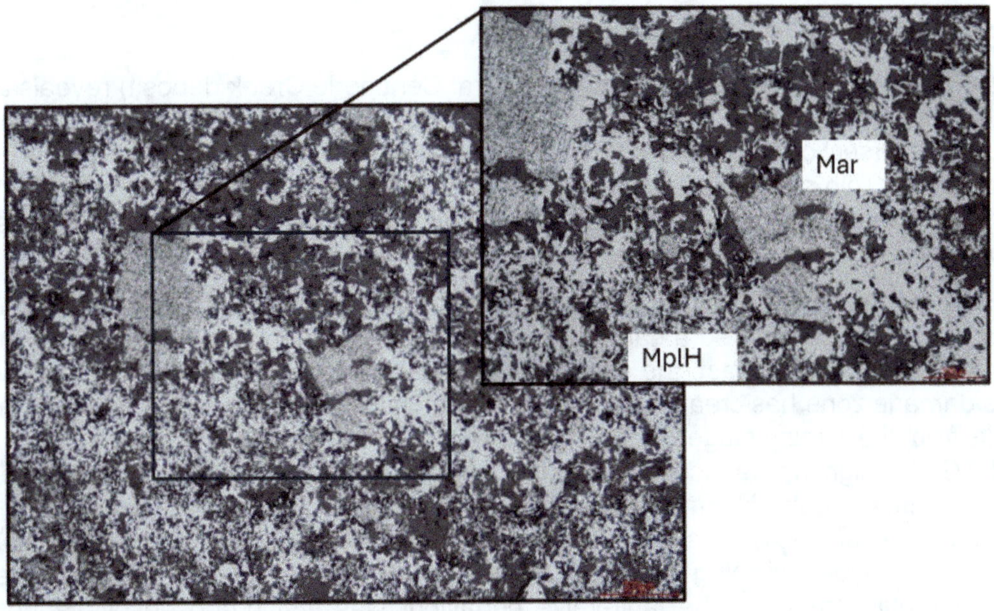

FIG 10 – Photomicrograph (RL) from 134 m downhole in one of the most mineralised drill holes in the project. Platy hematite nucleating off leached martite grain boundary.

Veins are common throughout the project. Veins contain quartz, goethite, minor specular hematite (mostly in Centipede Creek) and minor calcite (or dolomite; Figures 11 and 12).

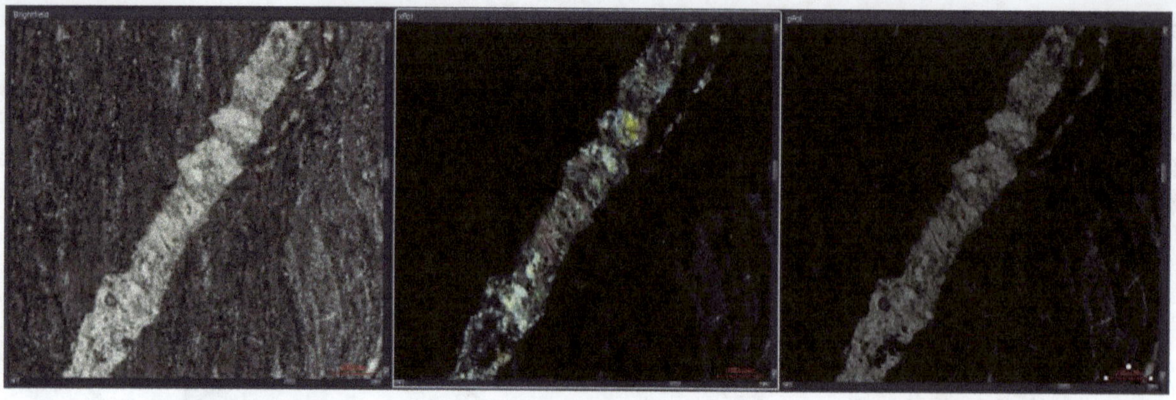

FIG 11 – Quartz + calcite (or dolomite) vein assemblage. From surface sample at Centipede Creek.

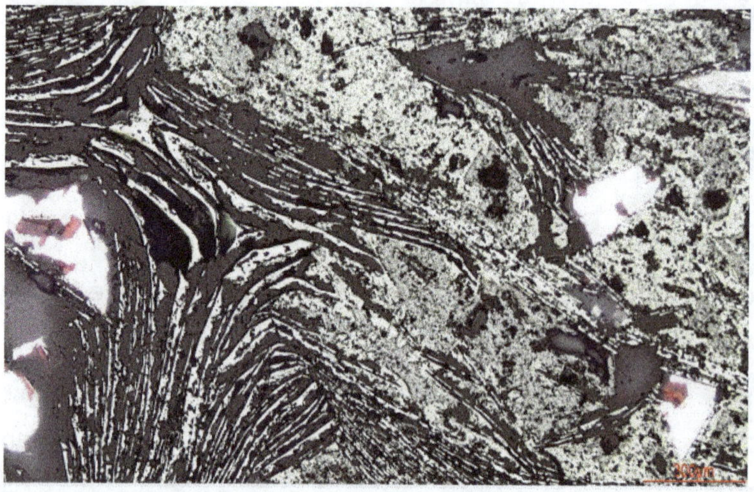

FIG 12 – Specular hematite on the margins of quartz veins in surface sample (RL).

ORE GENESIS

The study of the Wyloo North deposit (and peripheral Centipede Creek deposit) reveals a complex, multi-stage mineralising system. There is significant outcropping and near-surface mineralisation as well as >50 m intersections downhole of alteration and mineralisation below unmineralised zones proximal to faults. Thin section evidence combined with the structural setting indicates there is a component of Proterozoic hypogene-hydrothermal M-mpl H alteration as well as younger supergene martite-goethite (M-G) style of alteration.

As discussed, the deposit is heavily influenced by faults and dykes. On a broad scale, the whole of the deposit should be considered a damage zone between the Centipede Creek Fault in the west, the Worm World Fault in the east, the core of the Wyloo Anticline to the south and the MFZ to the north. This damage zone has created permeability pathways with evidence for multiple movements on the faults from the waning stages of the Ophthalmian Orogeny and post-compressional relaxation around ~2.2 Ga through the Capricorn Orogeny (~1.8 Ga), and, if the new ages of hematite formation apply, to the breakup of the Nuna supercontinent and/or amalgamation of Rodinia at ~1.4–1.1 Ga (Courtney-Davies *et al,* 2024). This is a possible setting whereby either: 1) fault-valve action (Sibson, 1990, 2020); or 2) mode-switching behaviour (Hobbs and Ord, 2023) played a role in generating the high-grade hematite orebody. The fault-valve behaviour was first proposed by Tannock (2021). Given the heterogeneity of BIFs interbedded with shales creating permeability gradients, episodic hydrothermal fluid flow demonstrated in the region (Courtney-Davies *et al,* 2024; Rasmussen *et al,* 2007), field evidence from faults and veins, and petrographic observations, the mode-switching model is preferred.

A modification of the compressional mode-switching model (Hobbs and Ord, 2023), exemplifies how this could work at Wyloo North where a vertical stress state created steep normal faults and permeability pathways along the fault planes (Figure 13). During a period of 'imposed fluid flux', likely during subsequent deformation events to the earliest normal faults, dissolution at grain boundaries initiates, driven further by induced stress. Soluble material, in this case SiO_2, is transported to 'dilatant sites,' ie along the fault or into extensional veins and leaching fluids move along BIF bedding planes, funnelled by bounding shales acting as aquicludes (Figure 14). This is proposed to form the thick intersections of mineralisation drilled beneath unmineralised material and adjacent to faults in multiple places in the deposit (Figure 15).

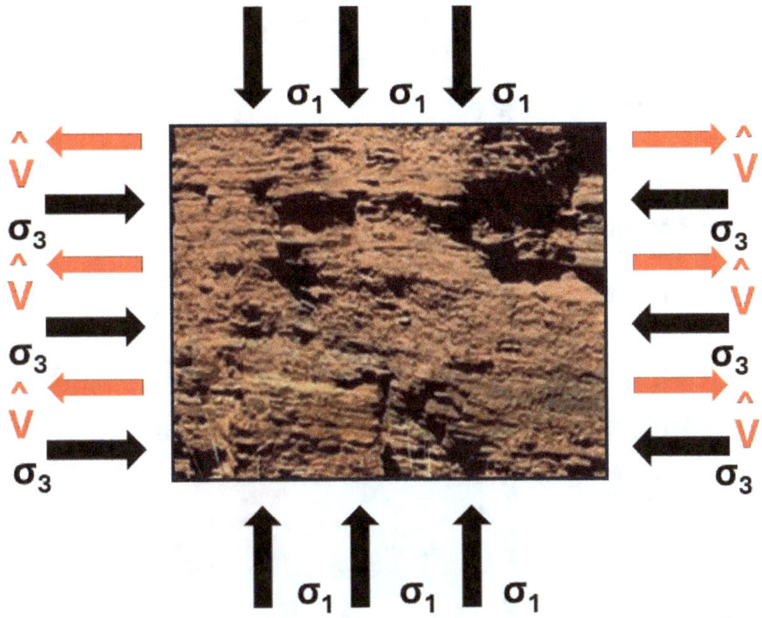

FIG 13 – Modified from Hobbs and Ord (2023), the initial state during early extension in which BIF is unaltered. Stress state, σ_3, and fluid flux, $\hat{v}$ marked by arrows.

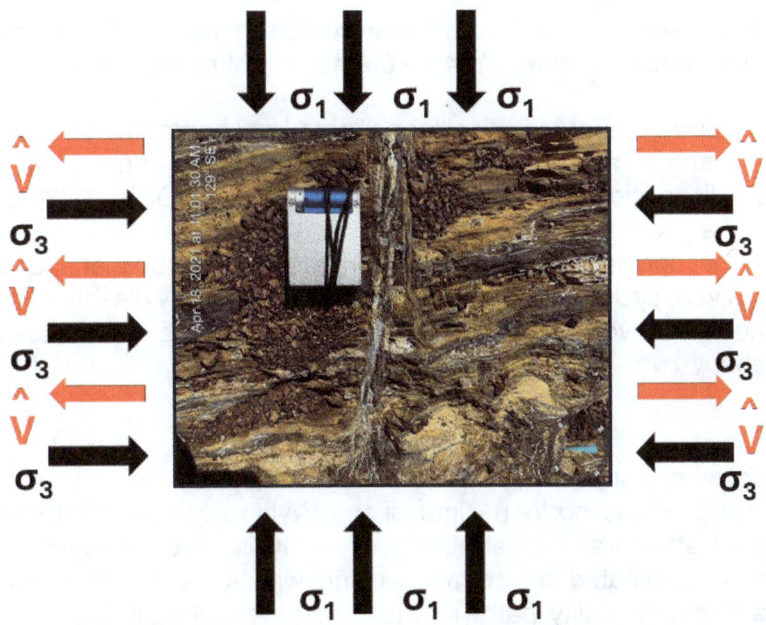

FIG 14 – Modified from Hobbs and Ord (2023), the final state (post- extension) in which BIF is altered, and a vein has formed with remobilised SiO_2 reprecipitated along bedding planes. Stress state, σ_3, and fluid flux, $\hat{v}$, marked by arrows. Picture taken at Centipede Creek.

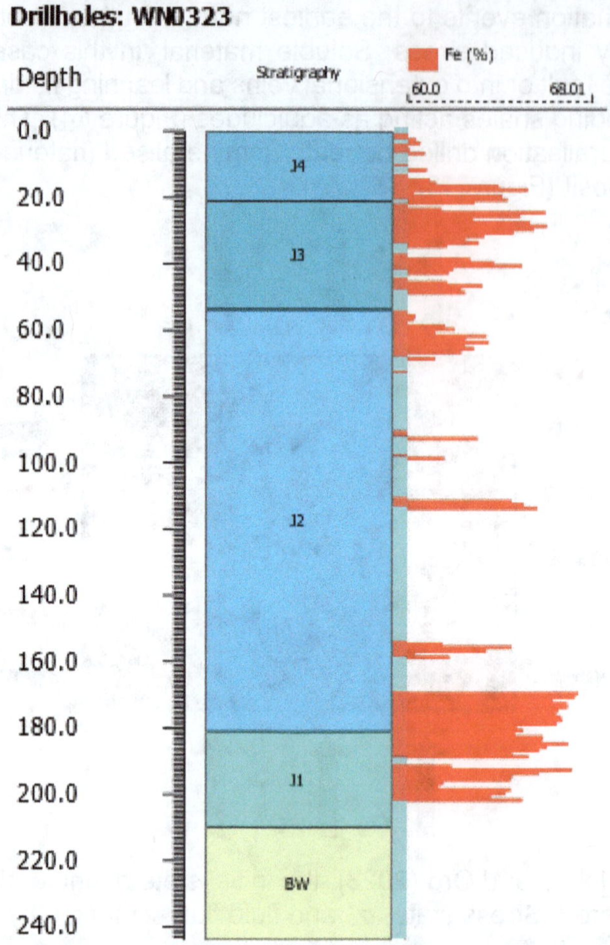

FIG 15 – Example of mineralisation at depth below surficial mineralisation (supergene) and a thick interval of unmineralised material. J1–4 = subunits of Joffre Mbr., BW = Whaleback Mbr.

A cycle 'whereby the rock mass automatically oscillates between a low-stress–high-fluid-pressure, low-permeability state and a high-stress–low-fluid-pressure, high-permeability state controlled by competition between dissolution and precipitation' (Hobbs and Ord, 2023) is set-up. At Wyloo, another variable that likely switches is between an extensional stress field and compressional stress field, in which case the original model shown in Figure 9 of Hobbs and Ord (2023) should be considered. This long lived, cyclical nature also may have completely dissolved the Paraburdoo Mbr., changing the fluid chemistry involved in the hydrothermal reactions and potentially resulting in further upgrading of the orebody.

CONCLUSIONS

The Wyloo North deposit is a structurally controlled, multi-stage iron ore system developed within a long-lived damage zone on the northern limb of the Wyloo Anticline. Detailed contact mapping, iterative drill-integrated structural and stratigraphic analysis, and petrography demonstrate that polyphase deformation dominated by normal faulting with subsequent reactivation created and repeatedly rejuvenated permeability pathways for hydrothermal fluid flow. Cross-cutting NW–SE faults, correlated with the broader Metawandy Fault system, are the principal controls on high-grade martite–microplaty hematite mineralisation, particularly where they intersect earlier fault sets.

Local stratigraphic variations, including the absence of the Paraburdoo Member and reduced sill development in the lower Weeli Wolli Formation, further influenced fluid chemistry, permeability contrasts, and ore upgrading. Petrographic relationships indicate a Proterozoic hypogene martite–microplaty hematite event overprinted by supergene martite–goethite alteration. Mineralisation is best explained by cyclic, structurally focused hydrothermal fluid flow consistent with stress-driven mode-switching behaviour operating over multiple tectonic events from the late Ophthalmian through Capricorn deformation and potentially into Mesoproterozoic tectonism.

This integrated structural–stratigraphic framework refines the tectonic evolution of the south-western Hamersley Province and improves orebody knowledge at Wyloo North.

ACKNOWLEDGEMENTS

The author would like to thank the Fortescue project generation team, as well as the greater exploration team for their contributions to the unravelling of the Wyloo North deposit. Special thanks to Alysia Kane for producing some of the figures used in the paper and Doug Kepert for helpful reviews and structural discussions. He would also like to recognise the company for allowing this paper to be published.

The field mapping methods discussed in the paper were heavily influenced by the mentorship of Dr. Andrew Duncan during his career training mapping geologists for Rio Tinto and Fortescue. Thank you for your many years of tromping around the Pilbara and passing your knowledge along.

REFERENCES

Coe, A L (ed), 2011. *Geological Field Techniques* (Wiley-Blackwell).

Courtney-Davies, L, Fiorentini, M, Dalstra, H, Hagemann, S, Ramanaidou, E, Danišik, M, Evans, N J, Rankenburg, K and McInnes, B I A, 2024. A billion-year shift in the formation of Earth's largest ore deposits, in *Proceedings of the National Academy of Sciences*, 121(31):e2405741121. https://doi.org/10.1073/pnas.2405741121

Dalstra, H J, 2014. Structural evolution of the Mount Wall region in the Hamersley province, Western Australia and its control on hydrothermal alteration and formation of high-grade iron deposits, *Journal of Structural Geology*, 67:268–292. https://doi.org/10.1016/j.jsg.2014.03.005

Hobbs, B E and Ord, A, 2023. An alternative to the fault-valve model, *Australian Journal of Earth Sciences*, 70(7):958–971. https://doi.org/10.1080/08120099.2023.2218452

Johnson, S P, Thorne, A M, Tyler, I M, Korsch, R J, Kennett, B L N, Cutten, H N, Goodwin, J, Blay, O, Blewett, R S, Joly, A, Dentith, M C, Aitken, A R A, Holzschuh, J, Salmon, M, Reading, A, Heinson, G, Boren, G, Ross, J, Costelloe, R D and Fomin, T, 2013. Crustal architecture of the Capricorn Orogen, Western Australia and associated metallogeny, *Australian Journal of Earth Sciences*, 60(6–7):681–705. https://doi.org/10.1080/08120099.2013.826735

Krapež, B, Müller, S G, Fletcher, I R and Rasmussen, B, 2017. A tale of two basins? Stratigraphy and detrital zircon provenance of the Palaeoproterozoic Turee Creek and Horseshoe basins of Western Australia, *Precambrian Research,* 294:67–90. https://doi.org/10.1016/j.precamres.2017.03.020

Martin, D M, 2020. Geology of the Hardey Syncline: The key to understanding the northern margin of the Capricorn Orogen: Report 203. Geological Survey of Western Australia.

Martin, D McB and Morris, P A, 2010. Tectonic setting and regional implications of ca 2.2 Ga mafic magmatism in the southern Hamersley Province, Western Australia, *Australian Journal of Earth Sciences*, 57(7):911–931. https://doi.org/10.1080/08120099.2010.510172

Martin, D, Allen, H and Blay, O, 2022. Stratigraphic and structural revisions on the Wyloo and Yarraloola 1:250 000 map sheets, Fortescue–Hamersley, East Perth, Western Australia.

Morris, R C, 2012. Microplaty hematite—Its varied nature and genesis, *Australian Journal of Earth Sciences*, 59(3):411–434. https://doi.org/10.1080/08120099.2011.626453

Müller, S G, Krape, B, Barley, M E and Fletcher, I R, 2005. Giant iron-ore deposits of the Hamersley province related to the breakup of Paleoproterozoic Australia: New insights from in situ SHRIMP dating of baddeleyite from mafic intrusions, *Geology*, 33(7):577. https://doi.org/10.1130/G21482.1

Rasmussen, B, Fletcher, I R, Muhling, J R, Thorne, W S and Broadbent, G C, 2007. Prolonged history of episodic fluid flow in giant hematite ore bodies: Evidence from in situ U–Pb geochronology of hydrothermal xenotime, *Earth and Planetary Science Letters*, 258(1–2):249–259. https://doi.org/10.1016/j.epsl.2007.03.033

Seymour, D B, Thorne, A M and Blight, D F, 1988. Wyloo, Western Australia, 2nd edn, Geological Survey of Western Australia.

Sibson, R H, 1990. Conditions for fault-valve behaviour, Geological Society, London, *Special Publications*, 54(1):15–28. https://doi.org/10.1144/GSLS.P.1990.054.01.02

Sibson, R H, 2020. Preparation zones for large crustal earthquakes consequent on fault-valve action, *Earth, Planets and Space*, 72(1):31. https://doi.org/10.1186/s40623-020-01153-x

Tannock, L, 2021. WylooNorth 2021 Mapping Report LTA (Full report), Internal Fortescue document.

Thorne, A M and Seymour, D B, 1991. Geology of the Ashburton Basin, Western Australia: Bulletin 139, Geology Survey of Western Australia.

ACKNOWLEDGEMENTS

Geological setting, structural evolution, and mineralisation of the Ouéléba Iron Ore Deposit, Simandou, Guinea – implications for Archean-Proterozoic crustal development and later supergene ore formation

M Zelic[1], *A A Diragitch*[2] *and M C Diallo*[3]

1. Principal Structural Geologist, Rio Tinto Copper, Perth WA 6000.
   Email: mario.zelic@riotinto.com
2. Principal Geologist, Rio Tinto Exploration, Belmont WA 6104.
   Email: alexandre.diragitch@riotinto.com
3. Senior Resource Geologist, Simfer, Conakry, 848, Republic of Guinea.
   Email: mamadoucellou.diallo@riotinto.com

INTRODUCTION

The Simandou Range is a 110 km-long chain of hills located in south-eastern Republic of Guinea, stretching across the Nzérékoré and Kankan regions within the country's forested Guinée Forestière zone. The range runs north–south and is known for its rugged, mountainous terrain. Its highest peak, Pic de Fon, rises to about 1656 m in the southern section. The Ouéléba deposit lies just north of the Pic de Fon deposit, with the two separated by approximately 4 km.

The Simandou range comprises a strongly deformed supracrustal depositional sequence, and locally, of a stack of tectonic units, deformed predominantly during Euburian event (2.0 to 2.25 Ga; Cope *et al,* 2008), forming a narrow N–S to NE-elongated structural keel within the eastern Kenema–Man Domain. The preserved stratigraphy, although discontinuous, records deposition within a tectonically active basin and includes Archean basement, Footwall Schist, Footwall BIF, Siliciclastic, Phyllite and Itabirite units. These units have been affected by progressive, polyphase deformation dominated by transpressive, largely sinistral shear, resulting in complex folding, transposition of earlier fabrics, and widespread development of shear zones. Marked structural asymmetry, the local omission of stratigraphic units, and strong lateral facies variations, all observed across the Oueleba Deposit, reflect the combined influence of depositional controls and subsequent high strain deformation. Supergene mineralisation and weathering have further altered the succession, leading to extensive carapace development, bauxite and iron enrichment of itabirite (silica removal).

Despite numerous geological investigations undertaken in the Simandou region, much of the previous work has focused on the Pic de Fon deposit. Consequently, several fundamental questions remain unresolved for Ouéléba deposit, which lies within the immediate mining area. Uncertainties persist regarding: (1) the detailed stratigraphy and geological architecture of Ouéléba, (2) the structural evolution and tectonic history, especially in the absence of geochronological constraints on deformation and mineralisation, and (3) the role and extent of supergene processes and weathering, for which no dedicated data sets currently exist.

This paper aims to address these gaps by integrating new structural, stratigraphic and mineralogical observations to refine the geological framework and evolutionary history of the Ouéléba segment of the Simandou range, particularly related to late Cretaceous to recent time events.

GEOLOGICAL FRAMEWORK

Simandou deposit is located in south-eastern Republic of Guinea, West Africa, within the eastern area of the Archean Kenema-Man Domain. Kenema-Man Domain is a part of Leo-Man Scheid of the West African Craton, and it is bounded to the west by Todi-Shear Zone, to the south by Trou Shear Zone, and to the east by the Sassandra-Trou Shear One (Cope *et al,* 2008). Majority of West African crust was already in existence 2.5 Ga ago. The West African crust has undergone steady growth since the Archean through repeated orogeny and lateral accretion, reworking older crust along subduction zones and adding metamorphosed and intrusive rocks. The eastern areas of the Western African Craton are particularly affected by the Palaeoproterozoic reactivations, particularly active during the Eburnean Orogeny (sometimes referred as Birimian, as the original type locality of these rocks, the Birim River valley of Ghana).

The geology of West Africa is generally characterised by basement rocks, elongated supracrustal belts and syn- to late-tectonic plutonic intrusions (Figure 1a). Basement rocks are characterised mainly by quartz-feldspar-biotite (±hornblende) gneiss and migmatite, ranging in composition from diorite through tonalite to granite, as well as lenses and intrusions of quartzite, marble and amphibolite, metamorphosed up to amphibolite facies metamorphic grade. Moreover, veins of coarse-grained pegmatite and fine-grained aplite have been detected within the basement. Basement rocks, subjected to multiple reactivation and thermo-tectonic events, are characterised by N-S to NE-SE steeply dipping foliation. There are three gneissic domains identified in Guinea (Wright *et al,* 1985): Mahana Gneiss east of the Simandou belt, Macenta Gneiss to the west, and the Guinea Gneiss to the north-west.

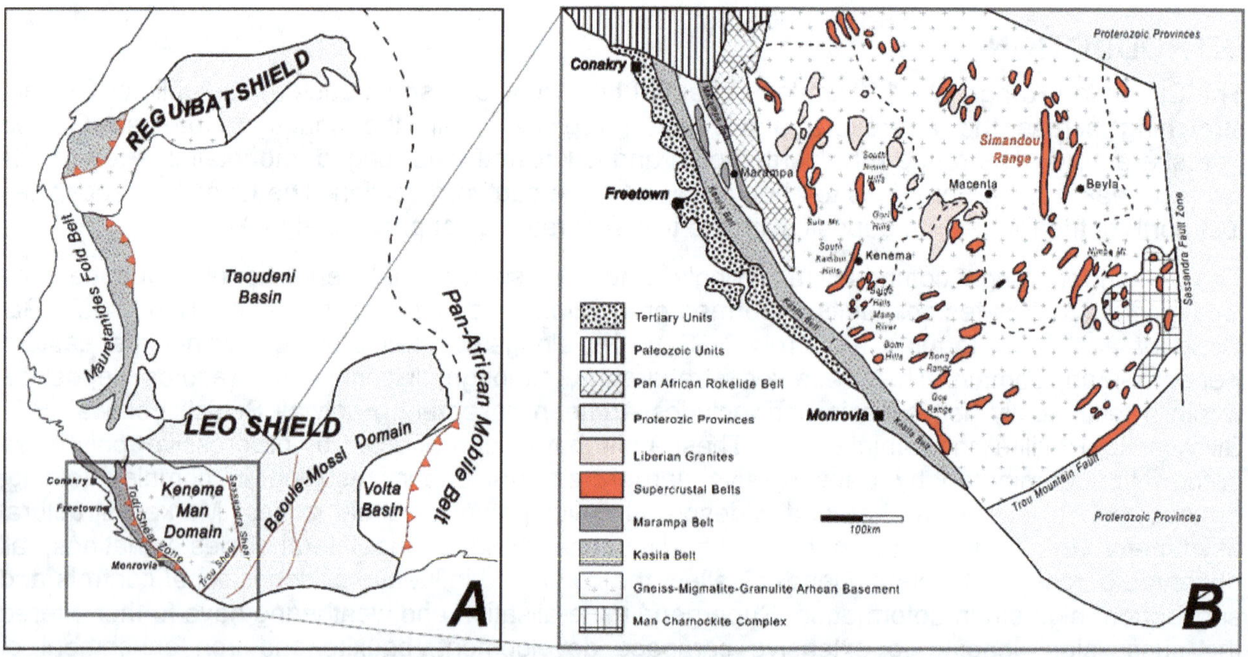

FIG 1 – (a) Geology sketch map of West Africa, from Cope *et al* (2008); modified after Rocci, Bronner and Deschamps (1991); (b) Schematic map of Archean age provinces of West Africa (from Wright *et al,* 1985).

Elongated supracrustal belts consist of phyllites, (mica-) schists and volcanic igneous rocks (greenstones), and subordinately, psammitic-composition rocks, quartzites and quartz-feldspathic units (from a quartz-sandstones, arkoses and greywackes protolith). The supracrustal belts are metamorphosed up to greenschist to lower amphibolite metamorphic facies, but it was noted that metamorphic grade generally increases towards the boundaries with the basement and large syn-tectonic granite bodies (Wright *et al,* 1985). Metavolcanic (ie metabasic rocks) component of the belts originated from basaltic lavas, ashes, pyroclastics and minor intrusive units. Quartzites are locally banded with magnetite or hematite (banded iron formation) or itabirite (Wright *et al,* 1985). The relative proportions of volcanic and sedimentary units within the supracrustal belts vary regionally. In general, greenstonetype volcanic sequences (ie volcanic igneous derived rocks) dominate the thicker western belts, interpreted as deeperwater, tectonically less stable settings underlain by oceanic crust. On the other hand, BIF and siliciclastic successions are more abundant in the thinner eastern belts, reflecting shallowerwater, more stable settings developed on continental crust (Wright *et al,* 1985; Cope *et al,* 2008). The metamorphic grade increases south-eastwards (Rocci, Bronner and Deschamps, 1991). However, a confident reconstruction of the original basin architecture is challenging due to pronounced lateral facies variations associated with tectonically controlled deposition, overprinted by subsequent deformation.

The supracrustal are tightly to isoclinally folded on all scales, showing a steep, N-S and NE-SW striking foliation. The contact with the basement rocks is mostly tectonic in nature, characterised by sheared fabrics, and showing characteristics of transpression/transtenssion evident in places (eg Simandou belt), interpreted to be of Eburnian event in age. Metamorphic grade increases from west to east, reaching around 600°C/5 kbar, and 750°C/7 kbar, respectively (Wright *et al,* 1985).

The range of tectonic settings (eg retro-arc/fore-arc during subduction, or foreland after retreating collisional setting; DeCelles, 2012) reflected on variation in the stratigraphic record. The thickness of supracrustal sequences, although uncertain due to intense folding, are estimated high as 10–12 km (Wright *et al,* 1985). The age of the supracrustal belt is speculated as Archaean (Wright, 1984) to Lower Proterozoic (Billa *et al,* 1999; Cope *et al,* 2008).

Syn-tectonic to late-tectonic plutonic intrusions, intruding both basement and supracrustal belts, are granite to granodiorite in composition, and subordinately diorite, gabbro, and syenite (Wright *et al,* 1985). When intruded into basement, granitic intrusions show mostly elongate and concordant geometries with the host rocks.

Archean Basement was affected by Liberian and Leonian thermo-tectonic event (Wright, 1984). Since the Liberian, the Archaean nucleus has been tectonically stable. Euburian event (ended 2 Ga ago; Cope *et al,* 2008) is characterised by many minor collisions involving the aggregation of continental blocks and the closure of numerous back-arc and inter-arc basins (Wright *et al,* 1985). Pan African event (750–550 Ma; Cope *et al,* 2008), regarded as the last reactivation event in the West Africa, is characterised by many low-grade supracrustal belts, tightly to isoclinally folded, whose size and general NNE-SSW trend is similar to that of supra-crustal in the cratonic nucleus (Wright *et al,* 1985).

GEOLOGY OF THE SIMANDOU AREA

Local stratigraphy

Ouéléba local stratigraphy is divided into Basement and the Simandou Units, where the latest is sub-divided into five sub-units that are field-mappable and constrained by mining propose. The contact of Basement and Simandou Group is tectonic in nature, resulting in a series of discordant faults/shear zones between basement rocks and any of the five units of the Simandou Group (Figure 2). As a general feature, the eastern portion of Ouéléba Deposit areas show a more preserved packages and all the five units are detected, while the western contact is more structured controlled and parts of the lower units are not outcropping, resulting in a series of structural contacts (Figure 3). The main Itabirite three Members are based on the work from Cope *et al* (2008) and Cope (2009).

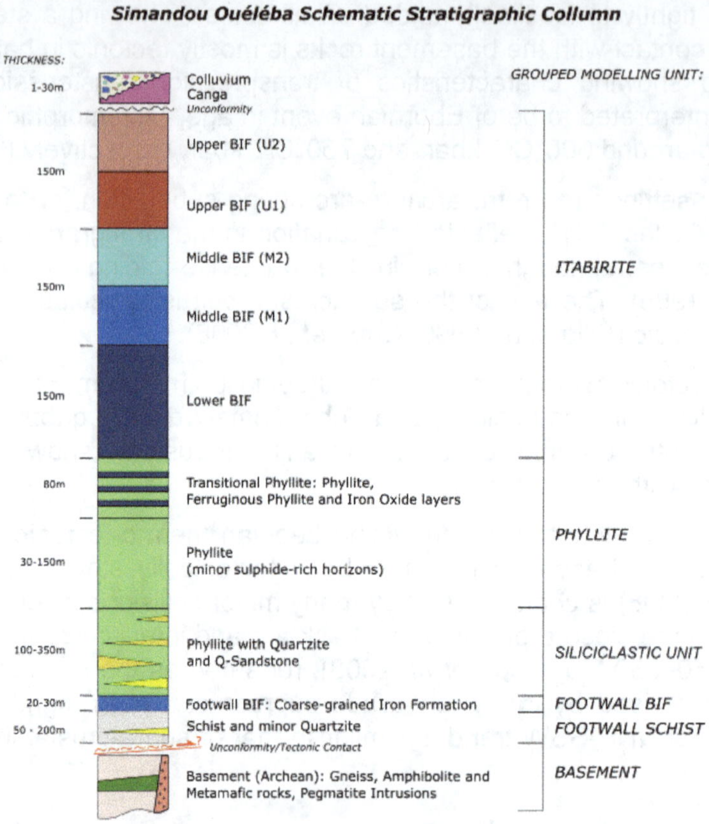

FIG 2 – Stratigraphy column for Simandou Project area based on field evidence and drilling data from Ouéléba Deposit.

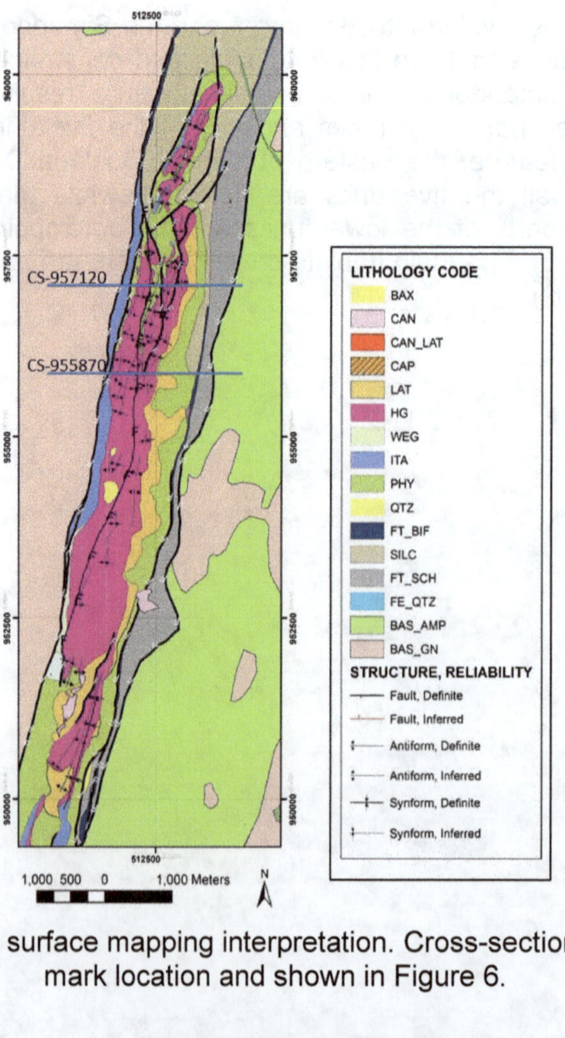

FIG 3 – Ouéléba Geology surface mapping interpretation. Cross-sections CS955870 CS957120 mark location and shown in Figure 6.

Basement

The basement is divided into two domains, a western domain composed mostly of quartz-biotite gneiss with minor quartzites, and quartz veins and an eastern domain composed mostly of the Amphibolite units (palaeo mafic) with minor granite-gneiss lenses.

Gneissic basement. This unit is detected on both sides of the Basement Unit. However, it is more pervasive along the western margin of Ouéléba deposit. The principal lithologies are felsic gneiss (quartz-biotite granite-paragneiss with occasional plagioclase and garnet porphyroblast) intercalated with quartzite. There is a strong fabric, and the main foliation (S1) is defined by biotite and amphibole mineral (hornblende). Foliation is on a general N30 strike direction, evident also by topographic features and geophysical signal in the basement. Rocks are generally strong to partially weathered and best examples are encountered in deep drilling, located in Ouéléba. Amphibolite Basement Unit is observed along the eastern margin. It is characterised by greenish-grey, medium-grained amphibole-mica-quartz schists grey, garnet-mica schists; and mica-quartz schists, sometimes with secondary pyrite along foliation near shear zones. Both basement units are intruded by later pegmatite dykes composed of quartz-tourmaline-muscovite-plagioclase.

Simandou Group

The Simandou Group sequence is mostly preserved along the eastern portion of the deposits, although still intensely faulted. Best outcrops/examples are only located in deep valleys, due to intense weathering near the Simandou Range (supported by the strong Iron Formations) and occasional drilling for geotechnical studies along the proposed mining infrastructure. The western portion of the range is extremely faulted, and full sequence is not preserved at surface.

Footwall 'Schists': Outcropping is fragmented on the eastern flank of the deposits only. Due to intense deformation and high strain, true thickness and facies variation is not defined, as it was not confirmed in areas below the hinge of Simandou Syncline. It is mostly composed of alternating mica schists and quartzites. In colour, the schists are lilac yellow, dark grey or greenish. They are often pyritised and sometimes contain garnet. The principal difference to the overlying phyllite is that individual mica grains are visible in the schist, stronger fabric. Disseminated magnetite or garnet may be present in discrete horizons. Grey garnet-biotite schists with sulfides are also present.

Footwall BIF: Exposure of Footwall BIF Unit are mostly detected along the eastern margin of the area, with some sporadic outcrop north-west of the Deposit. Its thickness varies from 10 to 30 m, which makes it technically very thin for a single entity, however, due to previous confusions of geologists on-site and surface mapping interpretation, it was deemed appropriate to show its clear location in the stratigraphy column. This clastic-BIF is coarse-grained and shows presence of stratigraphy textures, (cross bedding and fine lamination), suggesting it was deposited in a shallower environment, compared to the main Simandou Itabirite/BIF package. It is mineralised in places and always very goethite rich, suggesting presence of Fe-rich silicates.

Siliciclastic: Mostly quartzite with fine-grained granulometry, composition with minor intercalations of clean fine-grained quartz sandstone (locally variations to quartzite near base) with rare quartz-mica schist intercalated with minor thin bands of phyllite (more frequently towards the top). The sandstone facies are extremely variable in thickness, characterised by thickening towards north, with sandstone becoming the prevalent rock type along the northern third portion of the deposit. The quick facies variation suggests presence of structural control during deposition. A review of the drilling and historical maps (including Pic de Fon) shows the deepening of facies towards the south, where no major sandstone units have been observed south of Ouéléba.

Phyllite: Thick package of very fine laminated green to black pelitic material, although real rocks only identified in drilling, as it rarely outcrops as fresh, instead only presence of weathered yellow-red Fe-rich and white clay-rich soil. Locally small cm-scale thick quartz rich bands (fine siltstone). Noted presence of nodular sedimentary pyrite bands (evidence of reduced environment deposition). Pyrite can be re-mobilised along foliation in intense deformed areas, but drilling is rare to identify fresh material.

<u>Simandou Itabirite</u>: Outcropping Itabirite Unit were all grouped as one as the facies variant between BIF members are all linked to Al/Ti ratio and those cannot be detected by the field observation. Work from Cope *et al* (2008) best describes the Itabirite sequence, and it was not modified in this review, described as Lower itabirite: itabirite containing well-defined thin brown, grey and white bands (0.1–3 cm) of phyllite-goethite-silica. Middle itabirite: itabirite containing well-defined thick grey and white bands (0.6–5 cm) of hematite/magnetite-silica. Upper Itabirite: itabirite containing poorly-defined to diffuse grey and white thin bands (0.1–1 cm) of hematite/magnetite-silica. The itabirite units host several submetre to metre-scale phyllite horizons, which exhibit sheared contacts and appear to have been structurally emplaced. From exploration drilling and surface mapping, the overall thickness of the itabirite package is estimated to be not less than 250 m. To date, no carbonate-bearing lithologies have been identified in the Oueleba Deposit.

<u>Neogene Canga and Colluvium Covers</u>: At Ouéléba, most of the surface exposed material is covered by either Quaternary colluvium (1 to 30 m) and canga. Canga material is hard, irregular to sub-rounded fragments composed predominantly of massive to earthy hematite, cemented by iron oxides and hydroxides. They form part of a ferruginous duricrust developed by supergene weathering and induration. The blocks are typically highly dense, resistant, and porous to vuggy, reflecting dissolution and re-cementation during lateritic weathering.

STRUCTURAL GEOLOGY

The basement underlying the Simandou Group is Archean in age and occurs in a sheared contact with the overlying supracrustal succession. In several locations, this basal contact is intensely deformed and locally mylonitic (Figure 4). Deformation within the Simandou Group is dominated by two principal phases, D1 and D2, interpreted as part of a progressive deformation continuum developed between the late Archean and early Palaeoproterozoic. No clear evidence for an extensional interval separating D1 and D2 has been identified in the field, although such features may have been obliterated by the strong overprinting effects of D2 deformational phase.

FIG 4 – Meso-scale structures in the Oueleba deposit area. (a) Relationship between S0, S1 and S2 foliations. Diamond core of the drilled itabirite unit. (b) L2 crenulation lineation of the northern Oueleba area. (c) and (d) F1 folds, refolded into F2 folds during D2 deformation phase, characterised by different order of tight, isoclinal, curtain to sheath fold geometries. (e) The sub-horizontal to low-angle east- and west-dipping F3 fold axial planes developed in Itabirite Unit in the Oueleba area. (f) Shear zone developed at the contact of Simandou Group with the Basement rocks. (g) Shear folds within shear zone of the itabirite unit, west of Oueleba. (h) Thrust fault marked by a prominent shale band in the north Oueleba area (dashed line).

The finite structural architecture in the Ouéléba area is best interpreted as sinistral transpression, supported by geophysical magnetic data, regional geological mapping, and field observations. Fold vergence varies systematically along strike, with predominantly west-verging folds in the northern sector and east-verging folds in the south. Intense strain has produced highly non-cylindrical and locally sheath-like folds, characterised by strong variability in plunge from sub-horizontal to sub-vertical. Thrusts and steep reverse faults, inferred from core-logging and structural modelling, are consistent with the expected kinematics of a high-strain, transpressive regime.

In several locations, the deposit is overprinted by sub-horizontal folding, most likely reflecting late-stage orogenic and/or gravitational collapse. However, a component of N–S shortening cannot be fully excluded, and this late deformation is therefore referred to herein as D3. Field evidence at multiple outcrops also documents mineralisation-related dissolution collapse structures. Mineralisation spans pre-, syn- and post-tectonic stages: early stages are indicated by penetrative hematite-defined foliation within the ore (also noted by Marten, 2004), whereas many of the massive, high-grade orebodies are interpreted as predominantly supergene in origin, as shown by ore–host contacts that clearly cross-cut the earlier D1 and D2 structural fabrics.

The oldest structures recognised in the Simandou Group is the D1 deformational phase. D1 is characterised by S1 foliation, detected mostly in itabirite, as relicts in the microlithons of the S2 foliation in phyllite, re-folded F1 isoclinal folds of S0/S1 and intrafolial folding (Figures 4 and 5). S1 is mostly parallel to primary lamination and lithological layering. The meso-scale evidence of S1 in phyllites was defined as well-developed, continuous and transposive anisotropy (eg continuous or slaty cleavage; Passchier and Trouw, 2005), generally sub-parallel to the lithological boundaries. S1 foliation in itabirite is mainly parallel to primary bending and lamination, characterised by a strongly transposed bedding and disjunctive foliation (Passchier and Trouw, 1996). At the meso-scale F1 folds are detected as strongly non-cylindrical isoclinal to closed, recumbent, as well as sheath to curtain fold geometries. Folds in the high strain areas are disrupted by strong limb attenuation and detected mostly as re-folded isoclinal hinges. The limb attenuation process was recognised both on the outcrop scale, diamond core and during the synthetic modelling. The distribution of S0/S1 data collected during field mapping, diamond core logging and televiewer surveys forms a WNW-striking broad, elongate gridle consistent with F2 folding and associated structural transposition (Figure 5). The measured F1 fold axes plunge shallowly towards NNE-SSW and moderately steeply towards SW, also evident as sub-clusters in Figure 5.

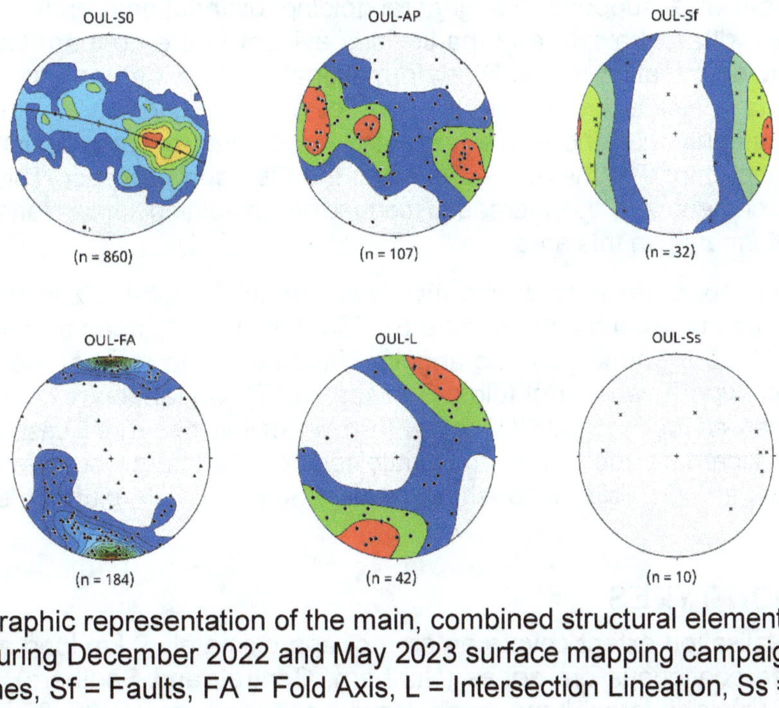

FIG 5 – Stereographic representation of the main, combined structural elements in the Oueleba area, collected during December 2022 and May 2023 surface mapping campaign (S0 = Bedding, AP = Axial Planes, Sf = Faults, FA = Fold Axis, L = Intersection Lineation, Ss = Shear Zones).

D2 structures mainly consist of open to isoclinal F2 folds, generally upright and symmetric in the axial areas of the ridge, and asymmetric, both upright and inclined, towards the contact with the basement, characterised by S and Z second-order fold geometries (Figure 5). F2 folds are characterised by subrounded to subangular fold hinges and thickened hinge zones (ie similar type of folds (Ramsay, 1967)). The fold limbs are often affected by shearing and shear fabric along the fold limbs are often recognised as well developed. F2 folds are mapped as non-cylindrical, characterised by a variation of fold plunge, as a result of non-coaxial deformation during D2 deformational phase (Figure 5). This produced a number of structural culminations and depression in the study area. Based on fold geometries, as well as some other evidence such as intersecting lineation F2 folds can be interpreted as a result of non-coaxial deformation during D2. D2 folding is associated with the development of an axial-planar crenulation cleavage (S2) locally. S2 foliation was mostly detected in itabirite, and based on the field evidence, can be classified as spaced foliation, ie gradational, zonal to discrete crenulation cleavage, showing rough cleavage domains with parallel relationships (Passchier and Trouw, 1996). In places S2 was recognised as shear fabric along the F2 fold limbs, parallel to S2 crenulation cleavage. Field mapping evidenced that the variation of the F2 plunge is not a result of later cross-folding or cross faulting but the non-coaxial deformation, dominant during D2 phase. This does not completely exclude the possibility of later re-folding overprint (post D2), as well as late faulting, cross-cutting the Simandou Group units. However, the collected structural data suggest that most of the present-day geometries are product of non-coaxial deformation and high strain regime during D1/D2 phase. L2 lineation have been recognised as intersection lineation and crenulation lineation, given the localised development of crenulation cleavage. In some cases, L2 has been detected as stretching lineation, represented by elongated mm-scale Fe-minerals or quartz grains. The data collected during mapping campaigns, diamond core logging and downhole geophysics indicate at the meso-scale SSW-NNE trend of the F2 axes with the shallow plunge and plunge direction predominantly towards SSW (Figures 4 and 5). The mapped axial surfaces of the F2 folds indicate predominately steep to moderately steep dips and NNE-SSW strike (Figures 4 and 5). Mapped fold axes data, as well as collected D2 fold axial planes are strongly concordant with the theoretical/statistical fold elements given by the S0/S1 distribution from the minimum eigenvectors as they plunge dominantly shallowly to the SSW, near the theoretical fold hinge. Televiewer S0 data distribution independently confirm this orientation, showing the same dominant shallow SSW plunge.

However, a sub-maximum is suggested as plunging rather steeply to the SW and WSW, suggesting an earlier folding event (eg F1), transposed within S2 and re-folded during D2. Most of the mapped faults show predominately steep dips and N-S strike. Moreover, mapped intersection lineation of Oueleba area supports the general folding distribution trends, but suggesting an angular discrepancy with fold axial trend, particularly evident in the northern Oueleba areas, as a result of non-coaxial deformation during D2 deformational phase. The analysis of surface mapping and televiewer data indicated moderate to low levels of cylindricity in the context of fold morphology. This means that there are expected moderate to high deviations of fold axes orientations along the plunge of the fold axes within the Oueleba prospect. This can be attributed to high strain environment and the increased complexity in folding characteristics due the long-lived polyphase deformation in this area.

D3 deformational phase is associated with the development of sub-horizontal F3 folds, showing sub-horizontal to horizonal axial planes (Figure 5). F3 folds are very open to gentle, showing inter-limb angles from 160°–80° and mostly sub-angular hinge zones. In most of the cases the F3 folds can be classified as Type 3 (ie parallel folds; Ramsay, 1967). Shear zones characterised by brittle cataclasites are often associated with F3 folds. In most of the observed cases, these faults and shear zones show kinematic indicators and brittle tectonic structures suggesting normal motion along D3 shear zones. D3 fold axes show mostly SSW to SW plunge, sub-parallel to D2 orientations.

MINERALISATION TYPES

The Ouéléba Mineralisation extent covers an area of approximately 9 km long and 1 km wide and is divided into three contiguous ore zones (Northern, Central, and Southern). The itabirite units (proto-ore) in the Ouéléba area have been locally enriched from ~30–35 per cent Fe (and >40 per cent SiO_2) to hematite- and hematite/goethite-rich rocks containing >65 per cent Fe with low deleterious elements. Most of the deposit is associated with the Lower and Middle Itabirite units (the Upper BIF is better preserved at Pic de Fon and is limited in the Ouéléba area), which have also been enriched within the ore zones. The resulting mineralisation includes hematite and goethite–hematite, with later predominantly associated with the Lower Itabirite units.

The enrichment of the itabirite units is transitional downward from hematite-rich material, through siliceous hematite rocks, to partially and non-enriched itabirite (<50 per cent Fe) and poor or fresh itabirite at the base of the mineralisation profile.

The enrichment of Fe from above the fresh/poor zone of this transitional enrichment zone is represented by itabirite rich friable (IRF—representing the transition between hard and friable poor Fe itabirite, and mineralised domains) is typically 20–50 m but may reach 300 m. Average grades in this zone are approximately 56 per cent Fe, 16 per cent SiO_2, and 2.22 per cent LOI (Figures 6a and 7).

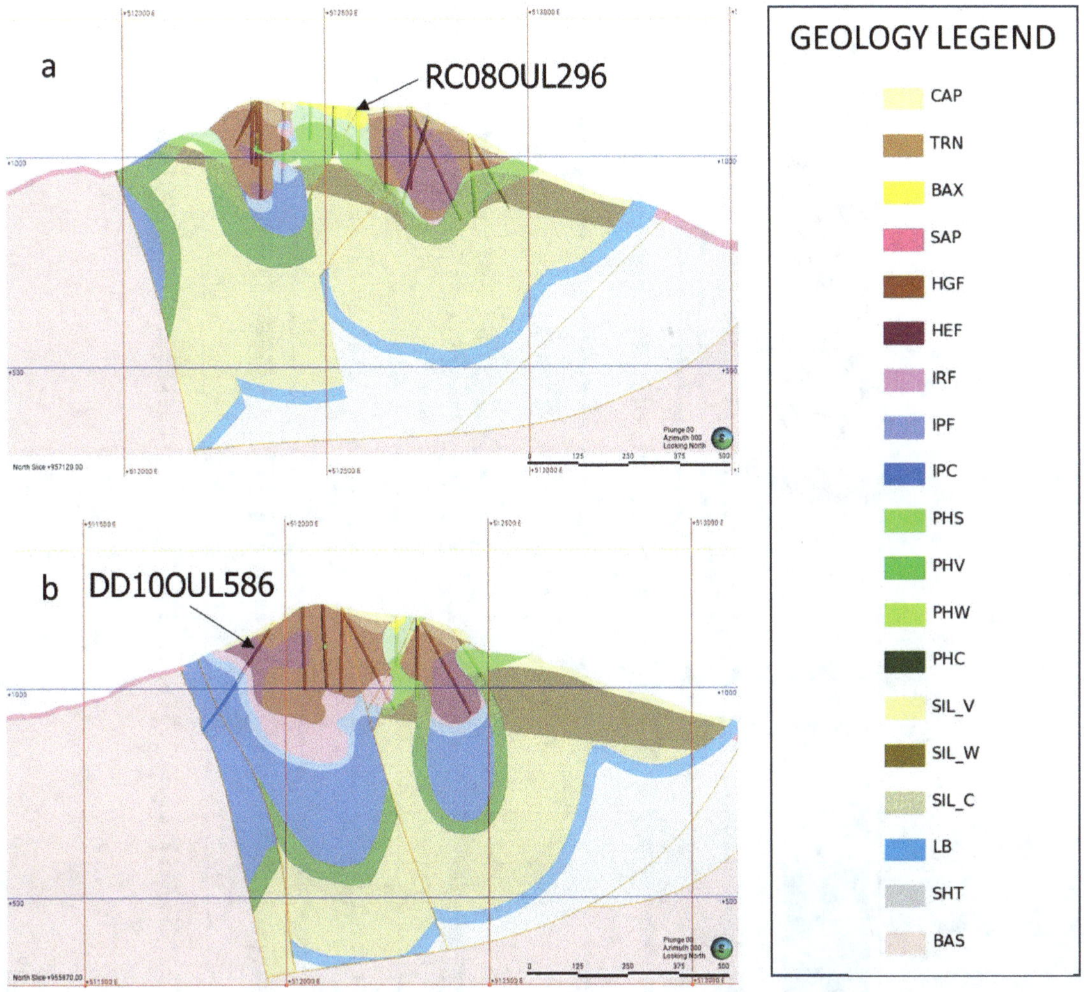

FIG 6 – Ouéléba Geology Cross-section looking north 957120 (a) and Cross-section 955870 (b), showing the geological model, which integrates stratigraphy, major faults, and weathering products. The geology legend includes: CAP (carapace of itabirite), TRN (transition between CAP and the main mineralisation), BAX (gibbsite-rich cap over phyllite), SAP (regolith over basement and lower schists), HGF (hematite/goethite friable), HEF (hematite friable), IRF (itabirite rich), IPF (itabirite poor friable), IPC (itabirite poor compact), PHS to PHC (degrees of weathering over phyllite, from soft and extremely weathered to compact, fresh/oxidised), SIL_V to SIL_C (degrees of weathering over the siliciclastic unit, from soft and extremely weathered to compact, fresh/oxidised), LB (lower clastic sedimented iron formation AKA footwall BIF), SHT (Simandou footwall schist), and BAS (undifferentiated basement comprising granite/gneiss and amphibolite with minor pegmatite dyke intrusions). Note: Mineral Resources compliance statement applies (see Important Information – Mineral Resources).

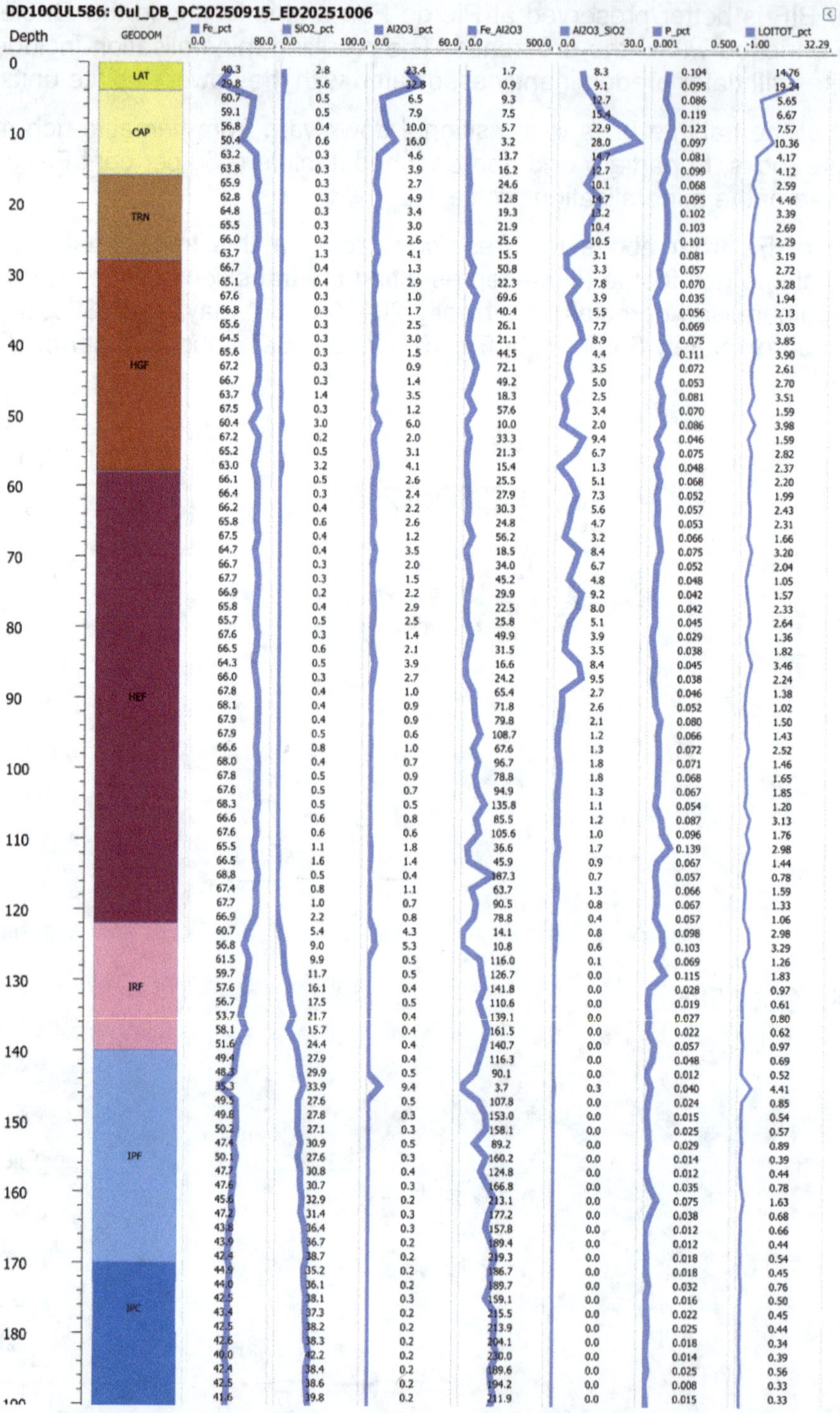

FIG 7 – Strip log from mineralised to fresh itabirite from DD10OUL586, showing major elements and the Fe/Al_2O_3 and Al_2O_3/SiO_2 ratios that define alteration contacts from CAP to compact and partially oxidised itabirite. DH from Section 955870 (Figure 6). Note only upper portion of IPC is shown in figure. Note: Mineral Resources compliance statement applies (see Important Information – Mineral Resources).

Above the transitional zone (IRF), the main mineralisation consists of two dominant material types:

1. A purer friable hematite zone (HEF), with average values of ~67 per cent Fe, 0.35–0.5 per cent SiO_2, 0.5–1 per cent Al_2O_3, 1–2 per cent LOI, and 0.05–0.06 per cent P. This unit is typically 70–100 m thick but can reach 400 m. It is dominantly controlled by and located within deep N–NE-trending D2 synforms positioned along the eastern side of the central and northern Ouéléba area, close to steep reverse faults. These structures have facilitated deep leaching, which also intensely weathers phyllites at surface into gibbsite-rich

bauxite up to 65–70 m thick. The cause of locally purer hematite zones is not fully understood, but these occur in fully developed, doubly plunging synclines, suggesting that aquifer pathways controlled by underlying phyllite and itabirite promote deeper leaching in these structures (Figure 6a and 7).

2. A mixed friable hematite/goethite zone (HGF), with average values of ~66 per cent Fe, 0.8–1 per cent SiO_2, 1.1–1.25 per cent Al_2O_3, 2.5–3.5 per cent LOI, and 0.075–0.09 per cent P. This unit is located mainly along the elevated areas of the deposit and near the surface plateau. It is typically 100–130 m thick but may also extend to 400 m (Figures 6a and 7).

At surface, a hard 2–10 m thick (with one location up to 60 m) weathered goethite-rich and secondary gibbsite-bearing carapace (CAP) has developed over outcropping mineralised iron formations. This carapace generally contains average values of ~59 per cent Fe, ~1 per cent SiO_2, 5.5–7 per cent Al_2O_3, 6–8 per cent LOI, and 0.125–0.15 per cent P. Beneath it lies a transition zone (TRN), 20–40 m thick, situated between the CAP and HEF/HGF zones, where local clay and limonite horizons have formed as products of advanced weathering. TRN material typically contains with average values of ~64 per cent Fe, 0.7–1 per cent SiO_2, 2–3 per cent Al_2O_3, 4–5 per cent LOI, and 0.09–0.125 per cent P. This later weathered zones (CAP + TRN) may extend deeper near fault zones.

A pronounced full lateritic weathering profile (CAP to IRF) is present at Ouéléba. In contrast, the Northern most part of the Deposit exhibits only a very restricted surface CAP, no TRN and limited HGF, suggesting deeper erosion post main mineralisation and CAP/TRN events at some parts of the deposit. This weathering profile reflects a lateritic weathering process affecting itabirite under wet tropical conditions, producing a well-developed weathering mantle composed of distinct horizons with differing chemical, mineralogical, and textural characteristics. This profile occasionally correlates laterally with canga surfaces located outside the deposit along the lower flanks of the ridgeline at lower contacts. The low alumina content of the itabirite, combined with intense silica leaching, inhibited the formation of clay minerals and resulted in Fe-oxide–rich material (limonite) with high LOI. Some alumina derived from weathered phyllite layers is incorporated into secondary minerals. Alternating dissolution and reprecipitation of Fe and Al oxides and oxyhydroxides control the lateritic weathering and in cases where phyllite is outcropping near top of ranges, a strong Al_2O_3 bauxite style mineralising occurs (Figures 6a and 8).

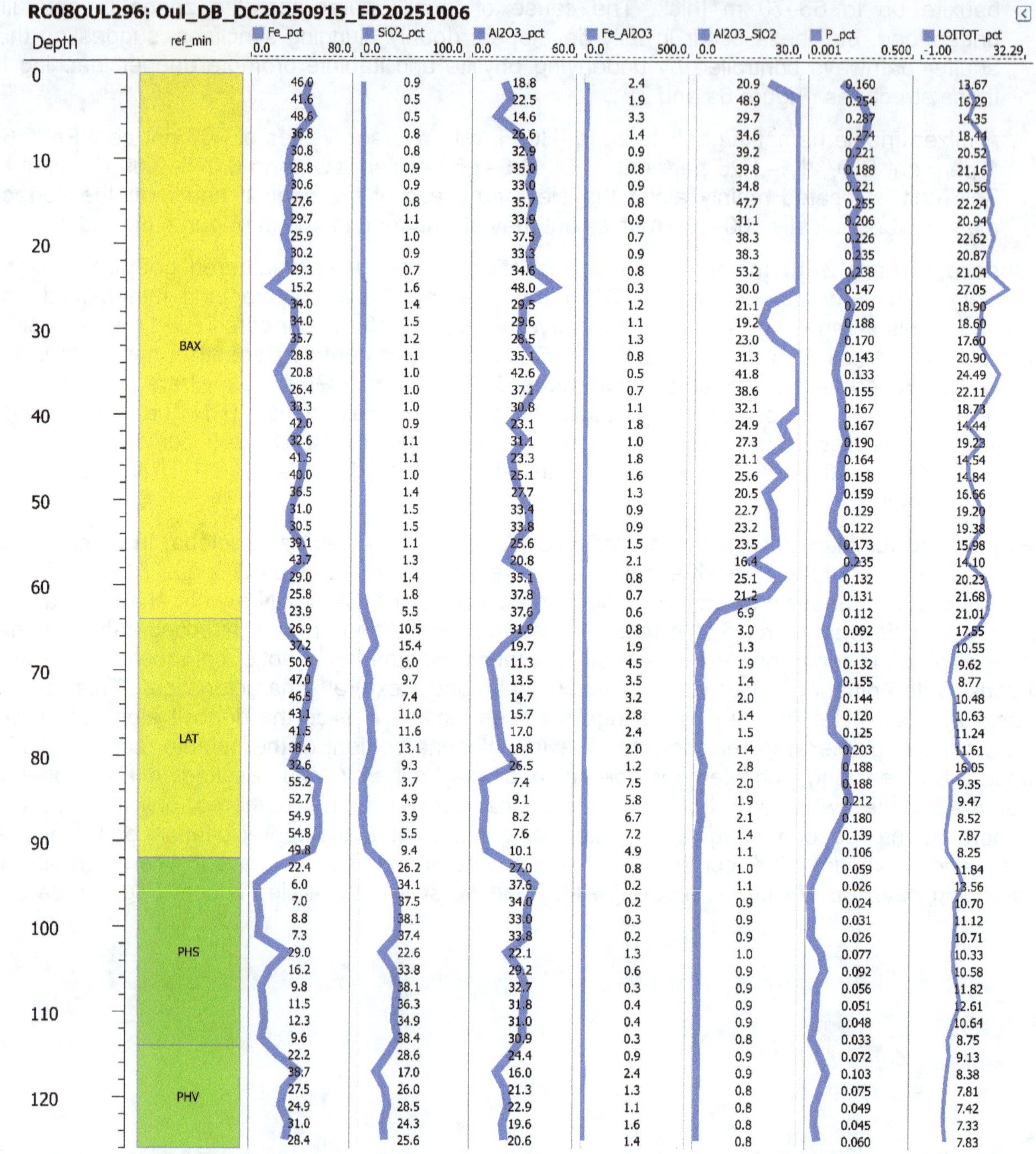

FIG 8 – Strip log from deeply weathered phyllite from RC08OUL296, showing major elements and the Fe/Al_2O_3 and Al_2O_3/SiO_2 ratios that define alteration contacts from BAX (gibbsite enriched material from weathered phyllite) and LAT is transition to soft weathered phyllite (PHS) to oxidised and weathered phyllite (PHV). DH from Section 957120 (Figure 6). Note: Mineral Resources compliance statement applies (see Important Information – Mineral Resources).

Processes including desilication, oxide dissolution, cementation, duricrust formation, and erosion collectively destroy or displace primary textures such as bedding and foliation. A summary of thickness and location of each weathering product over Iron Formation (Itabirite) is shown in Figure 9 and major element chemistry comparison (Fe (%), SiO_2 (%), Al_2O_3 (%), LOI (total,%), P (%) and Al/Ti molar ration) is shown in Figure 10.

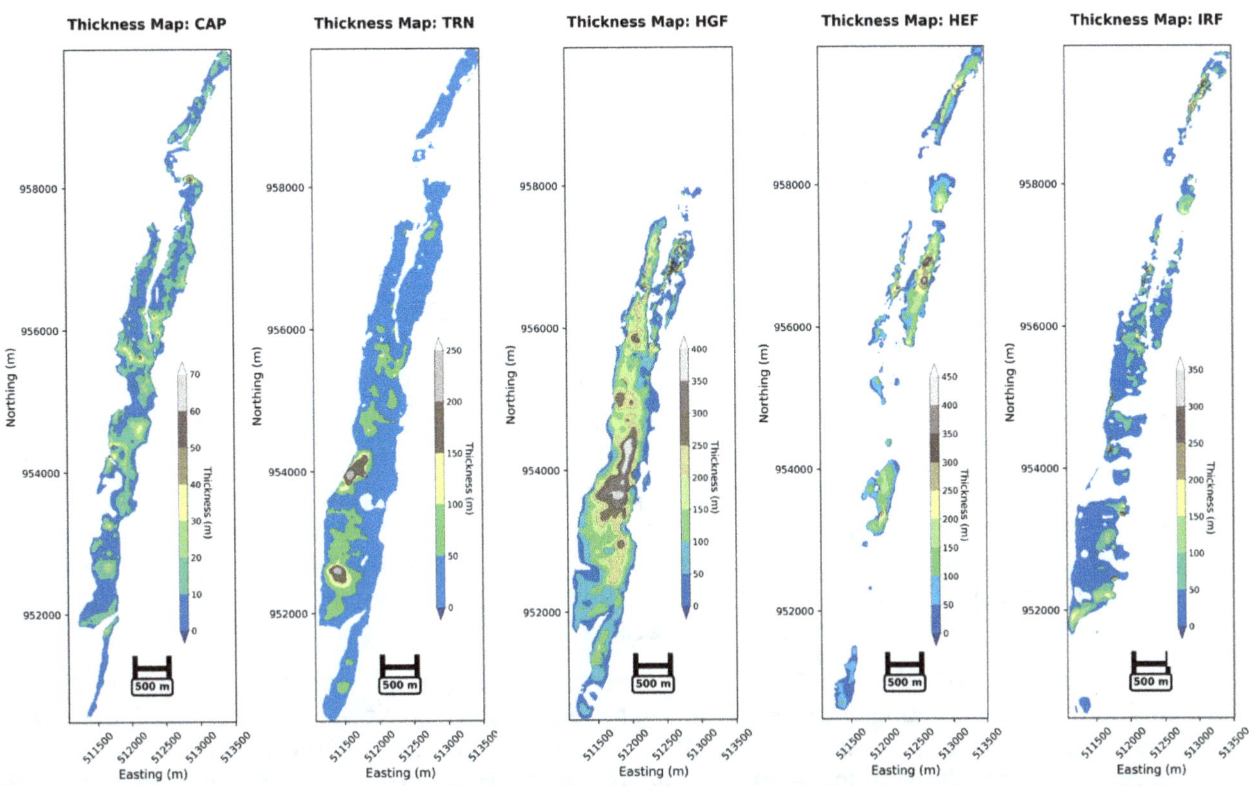

FIG 9 – Gridded thickness maps of weathering products of itabirite units. Note: Mineral Resources compliance statement applies (see Important Information – Mineral Resources).

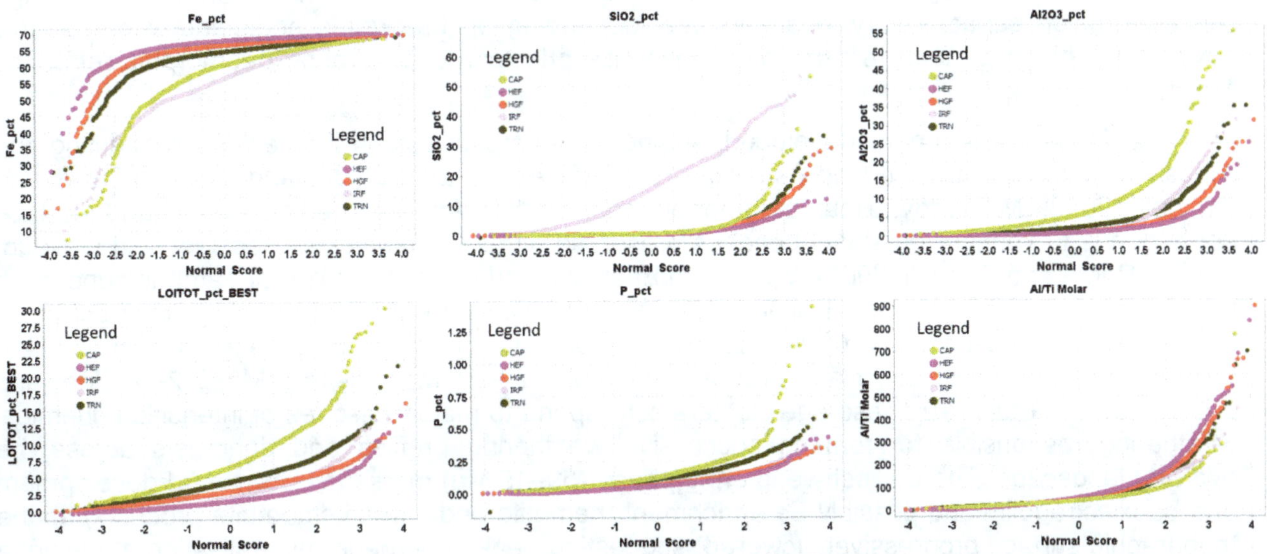

FIG 10 – Probability plots of major variables (legend shows weathering products of itabirite units). Note: Mineral Resources compliance statement applies (see Important Information – Mineral Resources).

GENETIC MODEL FOR IRON ENRICHMENT AT OUÉLÉBA DEPOSIT AND RELATIONSHIP TO LANDSCAPE EVOLUTION

Ouéléba Deposit mineralisation is linked to a long lived supergene system that started from events linked to the African Surface (described by Burke and Gunnell, 2008). The African Surface represents a vast composite erosion surface that developed across the Afro Arabian continent during a prolonged interval of relative tectonic quiescence between roughly 180 Ma and 30 Ma. Its formation integrates the effects of deep mantle structure, plume events, rifting, climate change, marine incursions, and extended episodes of intense tropical weathering. A defining control has been Africa's position above the large low shear wave velocity province (LLSVP) in the deep mantle, a longlived, compositionally distinct structure that has remained stable for at least 300 million years (Garnero, Lay and McNamara, 2007). Instead, Africa's remarkably stationary behaviour over this mantle domain, combined with episodic plume eruptions along LLSVP margins, governed its evolution during the Mesozoic and Cenozoic.

Over the past 200 million years, Africa underwent a tectonic history distinct from all other continents. New oceans formed around almost the entire continental margin in two major phases beginning near 180 Ma and 130 Ma, while Africa remained nearly fixed over the mantle due to the stabilising LLSVP beneath it. Additional tectonic influence came from plume driven upwellings at the margins of the LLSVP, which gave rise to Large Igneous Provinces such as those related to the Karroo and Tristan plumes (Burke, MacGregor and Cameron, 2003; Davaille *et al,* 2005). After these rifting episodes, Africa experienced a long period of minimal internal tectonism from approximately 130 Ma to 30 Ma, enabling prolonged denudation and the gradual planation of a continentwide erosion surface.

Prior to this plate arrest, a finite pole of rotation produced gentle counterclockwise rotation of Africa from 195 Ma to 30 Ma, shifting eastern regions farther latitudinally than western ones (Burke and Torsvik, 2004). This differential migration shaped climatic gradients, altering the distribution of humidity and the potential for deep weathering and bauxite development across the continent, particularly around West Africa (Burke and Torsvik, 2004).

Within this stable setting, the African Surface emerged as a low-lying, deeply weathered regional surface. Its history has long interested geomorphologists, who view it as the most extensive preserved erosion surface of Africa (King, 1962; Partridge and Maud, 1987). The surface is best defined as the composite landscape that dominated Afro Arabia around 30 Ma, formed during the 150 million year denudational interval beginning with early Pangean breakup near 180 Ma and reinforced by South Atlantic rifting at 125 Ma (Burke and Gunnell, 2008). Its upper age limit is set by the onset of swell uplift near 30 Ma, which deformed and elevated many remnants. Although Africa's erosion surfaces vary locally in age, depending on bracketing sedimentary or volcanic cover, the African Surface remains the continent's principal long duration geomorphic reference frame.

In West Africa, researchers developed an independent correlation framework connecting the African Surface with the *surface bauxitique* or *surface de Fantofa* (Boulangé, 1984; Burke and Gunnell, 2008; Mamedov, Bouféév and Nikitine, 2010). An intermediate level land surface corresponds to younger erosional benches preserved across swell flanks (Bocquier and Gavaud, 1964 in Burke and Gunnell, 2008). Work from Beauvais and Chardon (2013) at Tambao supergene manganese deposit in northern Burkina Faso is associated with a series of palaeo land surfaces. Radiometric $^{40}Ar/^{39}Ar$ dating of supergene cryptomelane from both surface and subsurface samples identifies five distinct age groups: 59–45 Ma, 29–24 Ma, 18–11.5 Ma, 7.2–5.8 Ma, and 3.4–2.9 Ma (Figure 11a). These age clusters correspond to major episodes of intensified chemical weathering responsible for forming successive weathering profiles and duricrusts across the evolving landscape. The earliest weathering phase (59–45 Ma) marks the downward development of a bauxitic regolith and possibly Fe enrichment (hematite and hematite/goethite products) as the topographic surface progressively lowered, suggesting peak supergene mineralisation at Ouéléba deposit (although requiring age dating to confirm). Correlating regional weathering to what is observed at Ouéléba deposit, it is possible to suggest that Fe enrichment (hematite and hematite/goethite mineralisation, occurred between 59 to 43 Ma and possible 29/24 Ma and later surfaces (TRN and CAP) 18 to 11 Ma and 7 to 6 Ma at Simandou.

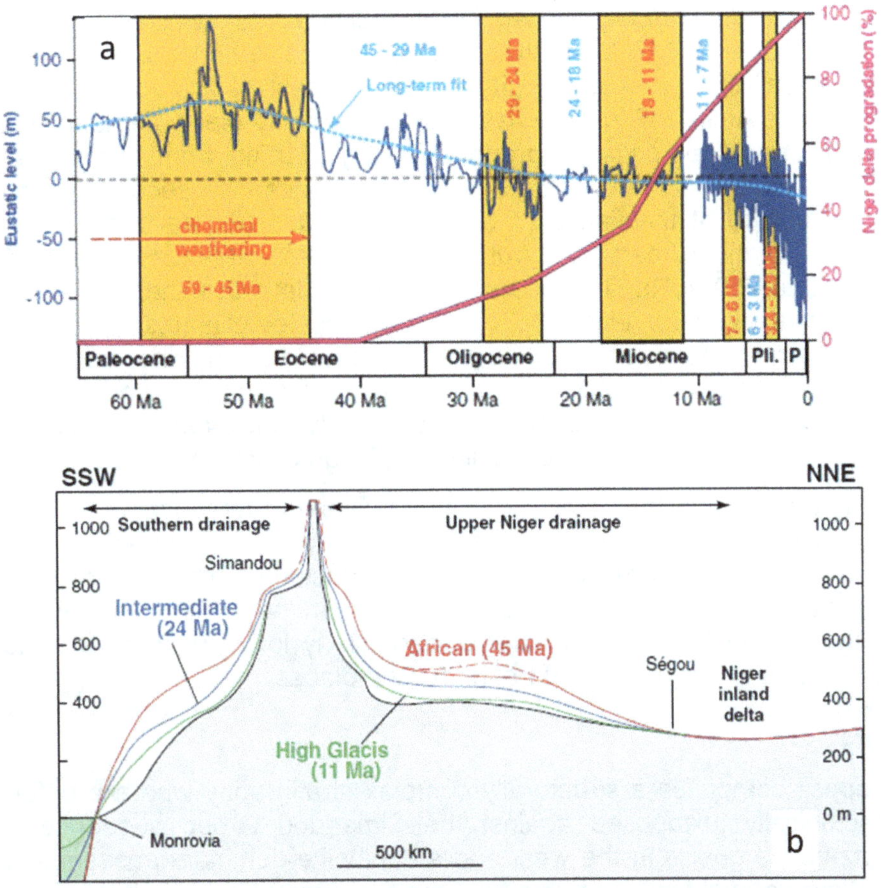

FIG 11 – (**a**) Data and image adapted from Beauvais and Chardon (2013) showing eustatic curve (adapted from Miller *et al,* 2005) and Niger delta progradation curve (Beauvais and Chardon, 2013) and yellow bars illustrate deeper chemical weathering events, which could correlate to Fe enrichment (hematite and hematite/goethite mineralisation, 59 to 43 Ma and possible 29/24 Ma) and later surfaces (TRN and CAP, 18 to 11 Ma and 7 to 6 Ma) at Simandou. (b) Synthetic cross-sections of the West African marginal upwarp showing the envelopes of the African (S1, also related to main hematite and hematite/goethite enrichment at Simando), Intermediate (S2, goethite/hematite and TRN), and high glacis (S3, CAP and cangas at Simandou) palaeo land surfaces, as well as regional current topography (in grey). Dashed lines represent reliefs abandoned above the regional base level at each denudation stage. (a) S1 results from differential fluvial erosion of an old, inherited upwarp now preserved as the Guinean rise (image from Beauvais and Chardon, 2013).

Climatic conditions during the mid-Cretaceous (~110–90 Ma) corresponded to global hothouse conditions with intensified humidity and reduced continental aridity (Fawcett and Barron, 1998). Deep weathering would have been intense in zones that lay within the equatorial belt, which migrated slowly as Africa rotated; maximum Cretaceous weathering likely occurred along a Conakry–Aden transect. The Palaeocene–Eocene Thermal Maximum around 53 Ma intensified lateritisation in West Africa, leaving isotopic signatures in Mn-rich laterites (Colin *et al,* 2005). Around 34 Ma, Southern Ocean gateway opening and Antarctic glaciation established a cold 'icehouse Earth' (Cande *et al,* 2000), shifting Hadley circulation northward and lowering global sea levels by ~50 m. These events terminated African Surface development and the last major bauxitic episode, also possibly linked to major deep Fe enrichment.

Beauvais and Chardon (2013) compare new denudation estimates with global Cenozoic data across cratonic and passive margin settings. Long-term denudation rates consistently fall within a narrow range (~2–20 m/Ma), regardless of timescale or setting, suggesting that cratonic regions experience low, spatially distributed denudation unaffected significantly by climate or mild tectonic uplift. Modern and intermediate term denudation rates generally match this long-term range, with higher values in areas influenced by strong topography or high run-off. Overall, denudation in cratons appears controlled primarily by regolith production and availability. In the tropics, thick inherited regolith is moderated by protective vegetation and duricrusts, whereas in midlatitudes, thinner regolith can be offset by higher run-off. The concept of a chrono sequence of stepped lateritic surfaces across Africa reflects the rise of swells after 30 Ma (note near Simandou area the erosion marked by the Niger delta progradation really starting from late Eocene (Figure 11a adapted from Beauvais and Chardon, 2013), which correlates with secondary iron mineralisation style, aka cangas located at the flanks of Simandou ranges (Figure 11b). It is also noted two distinct surfaces at Simandou, particularly at Ouéléba, where plateaus of Cangas occur at 950–1030 m RL and 650 to 750 m RL. The acceptance among laterite mineralogists that gibbsite and goethite are hydrated minerals related to humid and cooler climates, whereas boehmite and hematite are dehydrated minerals that are the product of warmer and drier tropical climates (Tardy, 1993 in Beauvais and Chardon, 2013), constitutes a helpful guide to understanding the lateritic formations types across the African continent, particularly link events at Simandou weathering profiles and links to regional features (Intermediate and high Glacis events).

After 2.8/2.9 Ma, the establishment of the Sahara Desert (deMenocal, 1995) and intensified glacial–interglacial cycling reshaped African denudation patterns. Pediment gravels on low slopes near the Gulf of Guinea record repeated southward expansion of desert conditions (Nichol, 1998). This is marked but deeper erosional features at the Simandou Ranges, resulting in localised removal of weathering blanket in deep valleys (particularly lower units of Simandou Group and Basement rocks), resulting in exposure of fresh rocks in places.

CONCLUSION

The Ouéléba deposit comprises a structurally complex stratigraphy where a deformed crystalline Basement is tectonically juxtaposed against the Simandou Group sequence. The Basement consists of quartz-biotite gneiss in the west and amphibolite-dominated assemblages in the east, both intruded by late pegmatites. Overlying this, the Simandou Group is best preserved along the eastern flank and includes, from base to top, schists, a thin but distinct Footwall BIF, variable siliciclastic quartzites and sandstones, phyllite, and a thick Itabirite package exceeding 250 m. Outcropping itabirite units were grouped into a single unit because facies variations between BIF members depend on Al/Ti ratios, which cannot be distinguished through field observation. The itabirite stratigraphy described by Cope *et al* (2008)—a ≥250 m-thick Lower, Middle, and Upper itabirite sequence with structurally emplaced phyllite horizons and no carbonate-bearing lithologies—remains valid for the Ouéléba deposit.

The Simandou Group of the Ouéléba deposit overlies Archean basement along a locally mylonitic, sheared contact and records a progressive D1–D2 history culminating in sinistral transpression at Oueleba, expressed by west- and east-verging F2 folds along strike, steep axial planes, non-cylindrical to sheath geometries with variable fold plunge, and associated thrusts/steep reverse faults consistent with a high-strain regime. A later D3 overprint comprises very open, sub-horizontal folding with local brittle normal reactivation (collapse-style), while mineralisation spans pre-/syn-tectonic stages (hematite-defined foliation) and post-tectonic supergene enrichment where ore contacts cross-cut D1–D2 fabrics.

Strong structural control, rapid facies transitions, and intense weathering have obscured much of the primary stratigraphy, particularly in the west. Neogene canga and colluvial covers further complicate surface mapping but also reflect long-term supergene processes active across the Ouéléba deposit.

Mineralisation at Ouéléba is the result of deep lateritic weathering and supergene enrichment of the Lower and Middle Itabirite units, producing laterally extensive hematite-rich ore zones with Fe grades exceeding 66 per cent. A vertical profile—from hard canga cap through clay- and limonite-bearing transition horizons to friable hematite and hematitegoethite ores—records progressive desilification, Fe-oxide precipitation, and structural focusing of leaching fluids within synformal traps and fault zones. Variations in the thickness and preservation of weathering horizons, particularly the reduced development of CAP/TRN in the northern deposit, reflect differential erosion following enrichment. Overall, the Ouéléba mineral system integrates stratigraphic architecture, deformation, and tropical weathering to create a robust high-grade hematite resource with predictable, zone controlled geometallurgical characteristics.

This deep weathering is linked to the Africa's geomorphic evolution reflects a single, long duration erosion cycle represented mostly by The African Surface, now partly elevated on swells and elsewhere buried beneath younger basins, records the integrated effects of tectonics, mantle dynamics, climate oscillations, and eustatic change over geological timescales (Burke and Gunnell, 2008). Laterites, supergene Iron Ore mineralisation (although still requires age dating to confirm Fe enrichment age) and bauxites formed extensively on this surface during the major global supergene ore systems forming between 70 and 40 Ma, although many younger laterites developed during the uplift phase post 30 Ma, possibly related to secondary cap and cementation and secondary mineralisation (cangas).

In synthesising, the main event controlling Mineralisation at Ouéléba is related to a deep Earth processes with geomorphic, climatic, and resource related outcomes, the African Surface illustrates how lithospheric plate behaviour, mantle plumes, atmospheric circulation, and global sea level changes combine to shape continental landscapes (Burke and Gunnell, 2008). This integrated perspective highlights Africa as a key example of Earth system interactions over geologic time, illuminating the distribution of resources such as, iron ore and also aluminium providing a framework for understanding continent scale surface evolution (Burke and Gunnell, 2008) and long-lived weathering system.

ACKNOWLEDGEMENTS

The authors would like to acknowledge all previous workers in the region, whose historical contributions have been fundamental to developing an understanding of the project area. We also acknowledge Rio Tinto (Simfer, Copper, and RTX divisions) for authorising the publication of this work. Special thanks are extended to Kaye Tindale and Matt Styles for their support and sponsorship. We further recognise Zeze Kalivogui, Abdoulaye Barry, Cece Lamah, Abdoulaye Bangoura, Mohamed Alpha, and Mamadoudjan Diallo for their invaluable assistance during field activities.

IMPORTANT INFORMATION – MINERAL RESOURCES

The information shows on the plans, cross-sections and graphs in this paper form part of the input data to the Simandou Mineral Resources which were reported in accordance with the JORC Code and the ASX Listing Rules in Rio Tinto's 2025 Annual Report released to the ASX on 19 February 2026. The Competent Person responsible for the information in the annual report that relates to Simandou Mineral Resources is Matthew Styles (MAusIMM).

Rio Tinto confirms that it is not aware of any new information or data that materially affects the information included in the annual report, that all material assumptions and technical parameters underpinning the estimates continue to apply and have not materially changed, and that the form and context in which the Competent Persons' findings are presented have not been materially modified.

REFERENCES

Beauvais, A and Chardon, D, 2013. Modes, tempo and spatial variability of Cenozoic cratonic denudation: The West African example, *Geochemistry, Geophysics, Geosystems*, 14(5):1590–1608. https://doi.org/10.1002/ggge.20093

Billa, M, Feybesse, J-L, Bronner, G, Lerouge, C, Milési, J-P, Traoré, S and Diaby, S, 1999. Les formations à quartzites rubanés ferrugineux des Monts Nimba et du Simandou: des unités empilées tectoniquement, sur un soubassementplutonique Archéen (craton de Kénéma-Man), lors de l'orogène Éburnéen [Banded ferruginous quartzite formations of the Nimba and Simandou ranges: tectonically stacked units on an Archean plutonic 'basement' (Kénéma-Man craton), during the Eburnean orogeny], *Comptes Rendus de l'Académie des Sciences - Series IIA - Earth and Planetary Science*, 329(4):287–294. https://doi.org/10.1016/S1251-8050(99)80248-1

Boulangé, B, 1984. Les formations bauxitiques et latéritiques de Côte d'Ivoire, Travaux et Documents 175. Paris: Éditions ORSTOM.

Burke, K and Gunnell, Y, 2008. The African Erosion Surface: A Continental Scale Synthesis of Geomorphology, Tectonics and Environmental Change over the Past 180 Million Years, Geological Society of America, Memoir 201.

Burke, K and Torsvik, T H, 2004. Derivation of large igneous provinces of the past 200 M.y. from long-term heterogeneities in the deep mantle, *Earth and Planetary Science Letters*, 227:531–538. https://doi.org/10.1016/j.epsl.2004.09.015

Burke, K, MacGregor, D and Cameron, N, 2003. African petroleum systems: Four tectonic 'aces' in the past 600 million years, in *Petroleum Geology of Africa: New Themes and Developing Technologies* (eds: T J Arthur, D S MacGregor and N R Cameron), 207:21–60 (Geological Society of London Special Publication).

Cande, S C, Stock, J M, Müller, R D and Ishihara, T, 2000. Cenozoic motion between East and West Antarctica, *Nature*, 404:145–150. https://doi.org/10.1038/3500450

Colin, F, Beauvais, A, Ruffet, G and Hénocque, O, 2005. First 40Ar/39Ar geochronology of lateritic manganiferous pisolites: Implications for the Paleogene history of a West African landscape, *Earth and Planetary Science Letters*, 238:172–188. https://doi.org/10.1016/j.epsl.2005.06.052

Cope, I L, 2009. Geological mapping at the Pic de Fon and Ouelaba deposits, Simandou Iron Ore Project: Q4 2008 – Q1 2009, Internal report.

Cope, I L, Wilkinson, J J, Boyce, A J, Chapman, J B, Herrington, R J and Harris, C J, 2008. Genesis of the Pic de Fon Iron Oxide Deposit, Simandou Range, Republic of Guinea, West Africa, in Banded Iron Formation Related High Grade Iron Ore, *Reviews in Economic Geology*, 15:339–360.

Davaille, A, Stutzmann, E, Silveira, G, Besse, J and Courtillot, V, 2005. Convective patterns under the Indo-Atlantic 'box, *Earth and Planetary Science Letters*, 239:233–252.

DeCelles, P G, 2012. Foreland basin systems revisited: Variations in response to tectonic settings, in *Tectonics of Sedimentary Basins* (eds: C Busby and A Azor), Recent advances, ch 20 (Wiley-Blackwell).

deMenocal, P B, 1995. Plio-Pleistocene African climate, *Science*, 270:53–59. https://doi.org/10.1126/science.270.5233.53

Fawcett, P J and Barron, E J, 1998. The role of geography and atmospheric CO_2 in long-term climate change: Results from model simulations for the Late Permian to the present, in *Tectonic Boundary Conditions for Climate Reconstructions* (eds: T J Crowley and K Burke), pp 21–36 (Cambridge: Cambridge University Press).

Garnero, E J, Lay, T and McNamara, A, 2007. Implications of lower-mantle structural heterogeneity for the existence and nature of whole-mantle plumes, in *Plates, Plumes and Planetary Processes* (eds: G R Foulger and D M Jurdy), Geological Society of America Special Paper, 430:79–101). https://doi.org/10.1130/2007.2430(05)

King, L C, 1962. *The Morphology of the Earth: A Study and Synthesis of World Scenery*, 699 p (London: Oliver and Boyd).

Mamedov, V I, Bouféév, Y V and Nikitine, Y A, 2010. Geologie de la République de Guinee, Ministère des Mines et de la Geologie, Université d'Etate de Moscou, pp 75–96.

Marten, B E, 2004. Simandou Iron Ore Project, Guinea, Report on Structural Interpretation, Rio Tinto Mining and Exploration Internal Report.

Miller, K G, Kominz, M A, Browning, J V, Wright, J D, Mountain, G S, Katz, M E, Sugaman, P J, Cramer, B S, Christie-Blick, N and Pekar, S F, 2005. The Phanerozoic record of global sea-level change, *Science*, 310:1293–1298.

Nichol, J E, 1998. Quaternary climate and landscape development in West Africa: Evidence from satellite images, *Zeitschrift für Geomorphologie*, 42:329–347.

Partridge, T C and Maud, R R, 1987. Geomorphic evolution of southern Africa since the Mesozoic, *South African Journal of Geology*, 90:179–208.

Passchier, C W and Trouw, R A J, 2005. *Microtectonics*, 366 p (Springer: Berlin).

Ramsay, J G, 1967. *Folding and Fracturing of Rocks*, 568 p (McGraw-Hill: New York).

Rocci, G, Bronner, G and Deschamps, M, 1991. Crystalline Basement of the West African Craton, in *The West African Orogens and CircumAtlantic Correlatives* (eds: R D Dallmeyer and J P Lécorché), pp 31–61) (SpringerVerlag). https://doi.org/10.1007/978-3-642-84153-8_3

Tardy, Y, 1993. *Pétrologie des latérites et des sols tropicaux*, 461 p (Paris: Masson).

Wright, J B, 1984. The Pan-African of West Africa — the eastern domain, *Geology and Mineral Resources of West Africa*, pp 56–69 (Springer).

Wright, J B, Hastings, D A, Jones, W B and Williams, H R, 1985. *Geology and Mineral Resources of West Africa*, 204 p (Springer).

Iron Ore

Health, safety, environment and community

Geomechanical stability of filtered tailings stacks founded on hydraulic tailings – critical state behaviour, pore-pressure dissipation, and deformation controls

D Eloi[1], L Cedro[2] and M Sanin[3]

1. Senior Geotechnical Consultant, WSP Canada, Vancouver BC V6Z2M1, Canada. Email: danilo.eloi@wsp.com
2. Geotechnical Mine Waste Consultant, WSP Canada, Vancouver BC V6Z2M1, Canada. Email: leticia.cedro@wsp.com
3. Senior Principal Geotechnical Engineer, WSP Australia, Perth WA 6000, Australia. Email: maria.sanin@wsp.com

ABSTRACT

The placement of filtered tailings over hydraulic deposits creates a geotechnical condition in which a stiff, partially unsaturated engineered fill is constructed over a soft, compressible, saturation-controlled foundation. Hydraulic tailings from past operations often exhibit loose structures, high void ratios, and stratified fabrics that may lie near or above their critical state line under modest loading. These characteristics make the foundation response highly sensitive to stress path, pore pressure generation, and strain accumulation during sequential stacking.

This study develops an integrated stability assessment focused on controlling critical state tendencies, tracking pore pressure evolution, and predicting deformations during embankment construction. Field and laboratory testing on representative hydraulic tailings indicate contractive behaviour within operational stress ranges, low initial stiffness, and high susceptibility to rapid pore pressure buildup under undrained loading. Critical state parameters were incorporated into numerical models to simulate stress paths during staged deposition.

Coupled stress deformation and transient seepage analyses indicate that, as filtered tailings are placed, excess pore pressures concentrate near depositional interfaces and within low-permeability laminae inherited from hydraulic sedimentation. Without drainage enhancement, these pressures reduce effective stress and may shift portions of the foundation toward the critical state surface, promoting localised softening or delayed settlement. Conversely, basal drains, relief wells, and engineered drainage corridors accelerate dissipation and maintain effective stress trajectories within stable domains.

Results also show that controlled compaction of filtered tailings improves load distribution and limits the propagation of shear strains into weaker hydraulic layers. Monitoring settlement and lateral displacement is essential to detect transitions toward the critical state. Seasonal infiltration modelling suggests that transient wetting may reduce suction in filtered layers but does not compromise stability when drainage pathways remain functional.

Overall, safe stacking on hydraulic foundations is feasible when design explicitly manages state parameter evolution, pore pressure dissipation, and deformation control, ensuring the long-term stability of the combined system.

INTRODUCTION

The placement of filtered tailings and dry stacking has become a technically robust and increasingly adopted alternative in the mining industry, driven by growing water availability constraints, closure performance requirements, and, above all, the need to reduce risks associated with conventional hydraulic deposition systems. This shift aligns with a broader change in the tailings management safety paradigm, which now emphasises governance, operational control, and performance demonstration throughout the full life cycle of tailings facilities, rather than isolated prescriptive checks (Morgenstern *et al,* 2016).

Despite the growing adoption of filtered tailings, many operations face area constraints imposed by legacy hydraulic deposits. In this context, studies involving the controlled placement of filtered tailings over hydraulic tailings structures can provide viable disposal solutions for mining operations with long depositional histories. Hydraulic tailings are characterised by loose initial states, high void ratios, and pronounced stratigraphic heterogeneity resulting from depositional processes, grain size segregation, and variable drainage and consolidation histories (Vick, 1990). Field and laboratory studies show that hydraulic tailings may exhibit mechanical behaviour highly sensitive to imposed stress paths, with significant pore pressure generation under relatively modest loading, particularly when subjected to rapid increments in vertical stress (Been, Jefferies and Hachey, 1991; Shuttle and Jefferies, 2016).

The interaction between filtered tailings stacks and foundations composed of hydraulic tailings creates an inherently layered and time-dependent geotechnical system in which stability is governed not only by drained peak strength but also by drainage evolution and stress path effects during staged construction. Anisotropic permeability, thin low-permeability laminae, and loose initial states can induce locally undrained responses, excess pore pressure buildup, and transient reductions in effective stress. From a critical state perspective, hydraulic tailings tend to contract under progressive vertical loading, potentially approaching the critical state surface and developing localised softening or delayed settlement when dissipation is limited. As a result, stability assessment requires an integrated framework that simultaneously considers strength, deformation, and pore pressure evolution, recognising that drained strength analyses alone may overlook key mechanisms of instability. Within this context, the present study investigates the geomechanical stability of filtered tailings stacks founded on hydraulic tailings using a framework that combines critical state concepts, pore pressure dissipation, and deformation control to support safe stacking practices on legacy hydraulic foundations.

BACKGROUND AND CONCEPTUAL FRAMEWORK

Hydraulic tailings as foundations

Hydraulic tailings deposits were historically developed as low-cost slurry-based disposal systems, producing stratified and spatially variable foundations due to grain-size segregation, fluctuating hydraulic regimes, and operational cycles (Vick, 1990). As a result, these deposits rarely form homogeneous masses and typically exhibit significant heterogeneity in density, void ratio, and effective stress state. Their sensitivity to additional loading is well documented: studies show strong stress-path dependence and substantial pore pressure generation even under modest vertical stress increments, particularly in the presence of layering and permeability anisotropy (Sharma and Fahey, 2003; Yin and Graham, 1996). These characteristics make hydraulic tailings critical foundation materials for new structures such as filtered-tailings stacks, which impose additional and often rapid loading stages.

Critical state framework and stress-path sensitivity

Critical State Soil Mechanics provides a unified framework for interpreting the behaviour of granular and fine materials under varying drainage and loading conditions, organising soil response around the Normal Consolidation Line (NCL) and Critical State Line (CSL) (Schofield and Wroth, 1968). The state parameter (ψ), defined as the difference between the current void ratio and that at the CSL for the same mean effective stress, quantifies the material's tendency toward contraction or dilation (Been and Jefferies, 1985). This concept remains one of the fundamental tools for interpreting state-dependent soil behaviour and liquefaction susceptibility within the Critical State framework (Been and Jefferies, 2015). Hydraulic tailings often lie in loose states with positive ψ, making them prone to contractive behaviour and pore-pressure buildup under vertical loading (Jefferies and Been, 2006). Because response is governed by stress path effects rather than drained peak strength alone, materials with similar strengths may behave very differently under partially or fully undrained conditions. This reinforces the need to evaluate stability through the evolution of effective stresses and deformations over time, a principle widely validated for tailings, including iron-ore tailings with well-defined CSLs and strong dependence on initial state (Sharma and Fahey, 2003; Boulanger and Idriss, 2014).

Pore pressure generation and dissipation in stratified deposits

The hydromechanical response of hydraulic tailings during filtered tailings stacking is strongly controlled by drainage conditions and the balance between pore pressure generation and dissipation. Ideally, slow loading and adequate drainage promote drained behaviour and progressive increases in effective stress. However, stratified deposits with low vertical permeability can exhibit locally undrained responses even under globally slow loading (Yin and Graham, 1996). Thin layers, depositional interfaces, and zones of segregation act as partial drainage barriers, concentrating excess pore pressures and influencing the location and magnitude of deformation. In staged construction, incomplete dissipation between lifts may lead to cumulative pore pressure buildup and altered stress path trajectories. Typical anisotropic permeability high horizontal and low vertical further promotes lateral dissipation while restricting vertical flow, generating steep hydraulic gradients and localised undrained zones (Vick, 1990). When combined with contractive tendencies, these conditions may induce localised softening, shear-strain concentration, and temporary reductions in effective stress, underscoring the need for integrated analyses that explicitly consider pore pressure generation, dissipation, stratigraphy, and hydraulic anisotropy in evaluating the stability of filtered tailings stacks founded on hydraulic tailings.

MATERIALS AND EXPERIMENTAL PROGRAM

The hydraulic tailings considered in this study originate from iron ore beneficiation and were historically deposited as slurry, forming a saturated foundation now intended to support the placement of filtered tailings. Deposition occurred over multiple operational phases, resulting in a heterogeneous stratigraphy typical of legacy hydraulic fills, with variations in grain size distribution, permeability, and initial state. The material is predominantly fine grained, with coarser fractions associated with hydraulic segregation. Characterisation tests indicate high initial void ratios and significant spatial variability in density, reflecting local differences in particle settling and consolidation history. Depositional interfaces linked to operational changes form potential hydraulic and mechanical discontinuities that influence pore pressure dissipation and deformation localisation. The deposit is fully saturated, with relatively low effective stresses at depth, particularly in less consolidated zones—conditions representative of legacy hydraulic foundations and critical for evaluating the response under the planned filtered tailings stack, which is expected to reach approximately 35 m in height. Although the focus of this study is the hydraulic foundation, the properties of the filtered tailings were considered when defining loading conditions, stress paths, and subsequent numerical analyses.

Field testing program

The field investigation comprised an extensive campaign of Cone Penetration Test with pore pressure measurement (CPTu) and Seismic CPTu (SCPTu) tests distributed across the study area, complemented by dissipation tests, boreholes, and hydrogeological measurements. These tests enabled the characterisation of stratigraphy, identification of contrasting layers, estimation of derived parameters including corrected cone resistance (qt), sleeve friction (fs), pore pressure measured at the shoulder (u_2), soil behaviour type index (Ic), normalised cone resistance (Qtn), pore pressure ratio (Bq), and undrained shear strength normalised by vertical effective stress (Su/σ'v) and assessment of spatial variability within the hydraulic deposit. The results revealed zones of low resistance, strongly contractive behaviour, and significant pore pressure generation, consistent with fine-grained, lightly consolidated hydraulic tailings. The integrated interpretation of the profiles allowed the delineation of representative regions for subsequent analyses, as well as the definition of hydraulic and mechanical parameters used in numerical modelling.

The combined interpretation of CPTu and geophysical data allowed the identification of zones with more contractive behaviour, which were explicitly represented in the numerical model through distinct initial state conditions. The interpreted zoning is illustrated in Figure 1.

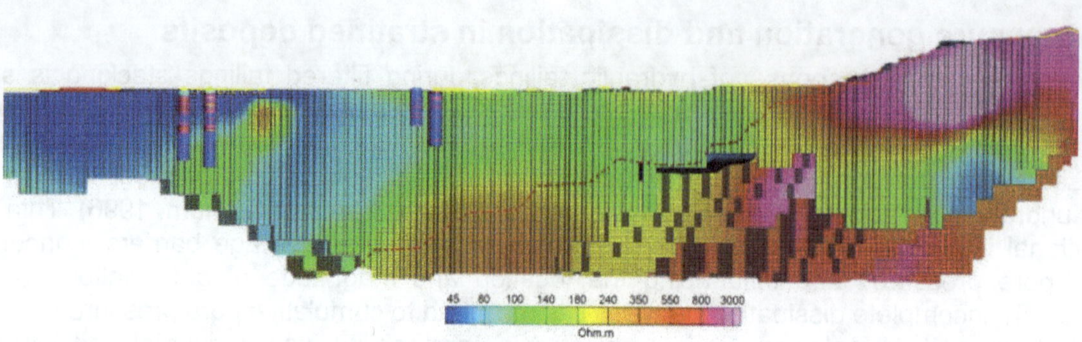

FIG 1 – Geotechnical zoning of the hydraulic tailings foundation based on electrical resistivity profiling and CPTu correlation, highlighting zones of lower resistivity associated with more contractive and weaker tailings.

Laboratory testing program

The laboratory program was designed to characterise the mechanical behaviour of the hydraulic tailings under field-representative stress conditions and included physical characterisation, one-dimensional consolidation tests, and consolidated drained (CD) and consolidated undrained (CU) triaxial tests on reconstituted samples covering the range of effective stresses expected during filtered tailings placement. The CU tests indicated predominantly contractive behaviour, with substantial excess pore pressure generation and strong dependence on the initial state. The Critical State Line (CSL) was defined from the convergence of drained and undrained final states, enabling consistent estimation of the compression index (λ), intercept of the CSL in e–ln p′ space (Γ), and critical state friction angle (φ′cs), which were used to calibrate constitutive models and interpret stress-path evolution. The combined results provide a robust basis for evaluating the response of the hydraulic foundation under progressive loading. The adopted critical state framework is illustrated in Figure 2, which presents the CSL defined from the laboratory results.

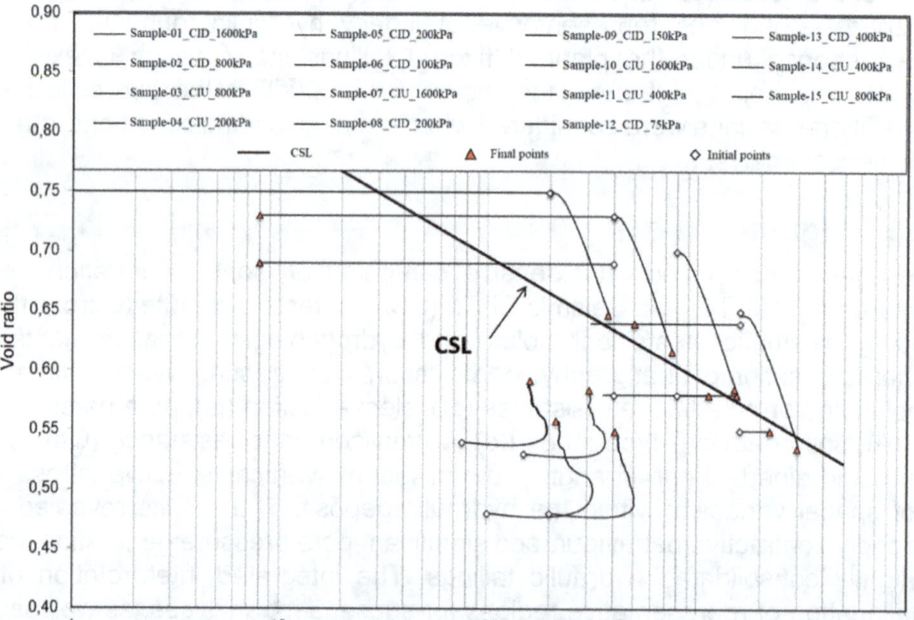

FIG 2 – CSL defined from triaxial test results on hydraulic tailings in e–ln p′ space.

NUMERICAL MODELLING FRAMEWORK

The numerical analyses were developed to capture the interaction between loading, pore pressure generation and dissipation, and deformation within a stratified hydraulic tailings foundation subjected to filtered tailings stacking. Particular emphasis was placed on representing the state dependent behaviour of the hydraulic tailings, especially the contractive response observed in laboratory testing and the stress-path sensitivity identified from CPTu and SCPTu interpretations.

The selection of the constitutive model was guided by experimental results, particularly the robust definition of the CSL and the consistent evidence of contractive behaviour under undrained loading. The adopted model incorporates an elasto-plastic formulation with stress dependent hardening, allowing representation of void ratio evolution, progressive strength mobilisation, and the tendency toward contraction or dilation as a function of the initial state relative to the CSL. This approach is consistent with critical state based frameworks widely used for granular soils and tailings, in which the state parameter governs the undrained response.

Constitutive parameters were derived directly from laboratory testing. The slopes of the Normal Consolidation Line and the CSL in e–ln p′ space were defined based on consolidation tests and the final states of drained and undrained triaxial tests. The critical state friction angle (φ′cs) was obtained from the critical state condition, while initial elastic parameters were estimated from the small-strain response observed during the early stages of loading. Where necessary, secondary adjustments were performed to ensure consistency between simulated and observed responses without compromising the physical meaning of the parameters.

The initial state of the hydraulic tailings in the numerical model explicitly incorporated the spatial variability observed in the field. Distinct initial states were assigned to the different stratigraphic zones, reflecting local variations in void ratio associated with hydraulic segregation, depositional processes, and consolidation history. Zones identified through CPTu and geophysical interpretation as more contractive and weaker were represented with higher state parameters and lower initial effective stresses.

The hydraulic foundation was assumed to be fully saturated, which is consistent with field investigations and the site-specific hydrogeological model. Initial stresses were defined from the self-weight of the deposit and the simulated phreatic surface, ensuring consistency between the initial mechanical and hydraulic conditions of the system.

The numerical analyses were conducted using a coupled stress–seepage formulation to simulate the hydromechanical response of the system during filtered tailings stacking. This approach allowed simultaneous evaluation of effective stress evolution, excess pore pressure generation, and dissipation over time, which are the key mechanisms controlling the stability of filtered stacks founded on hydraulic tailings.

The construction sequence was modelled incrementally, representing the progressive placement of filtered tailings up to the final height of approximately 35 m for the pilot stack. Each construction stage was simulated as a load increment applied at the surface of the hydraulic foundation, followed by a consolidation period compatible with the assumed construction rate. This strategy enabled evaluation of the influence of construction sequencing on stress path evolution and on localised undrained responses in the more contractive zones of the foundation.

Hydraulic boundary conditions were defined based on the site-specific hydrogeological model, including the initial phreatic surface, hydraulic gradients, and anisotropic permeability conditions. The vertical and horizontal permeabilities of the different layers were assigned consistently with the stratigraphic interpretation, CPTu-derived parameters, and the geophysical zoning of contractive and non-contractive regions.

Different drainage conditions were considered, including scenarios without drainage enhancement and scenarios incorporating the basal drainage system defined in the design. This system consists of a drainage blanket with granular filter transitions placed at the base of the filtered tailings stack. In the numerical model, the drainage system was represented through equivalent high-permeability layers, allowing assessment of its influence on pore pressure dissipation and stress path evolution.

The adopted stress–seepage coupling allows excess pore pressure generation to be directly controlled by the contractive constitutive response of the material, while dissipation occurs as a function of hydraulic properties, drainage elements, and boundary conditions. As a result, the model captures the possibility of locally undrained responses during loading increments, even when the overall system behaviour is predominantly drained. This effect is particularly relevant near depositional interfaces and low-permeability laminae identified in the hydraulic tailings foundation.

The set of analyses provides the basis for an integrated stability assessment, explicitly relating stress path evolution, pore pressure generation and dissipation, and deformation development during filtered tailings stacking.

To facilitate interpretation of the modelling approach and calibration process, the key material parameters adopted for both hydraulic and filtered tailings are summarised in Table 1. The parameters include strength, stiffness, and state-dependent variables derived from CPTu interpretation, laboratory testing, and constitutive model calibration.

TABLE 1

Summary of key material parameters adopted for deformation analyses and constitutive modelling.

Material	γ (kN/m^3)	φ' (°)	Su/σ'v (peak)	Su/σ'v (residual)	State parameter ψ	Behaviour	Basis
Hydraulic tailings (beach)	—	32	0.22–0.24	0.05–0.08	—	Contractive	CPTu interpretation
Hydraulic tailings (lake)	—	28	~0.22	0.04–0.06	—	Contractive	CPTu interpretation
Filtered tailings	21.5	34.5	—	—	-0.078	Dilatant (compacted)	Laboratory + calibration

The overall deformation pattern obtained from the coupled analysis is shown in Figure 3, illustrating the distribution of settlements within the hydraulic foundation under the 35 m filtered stack and the absence of a global failure mechanism.

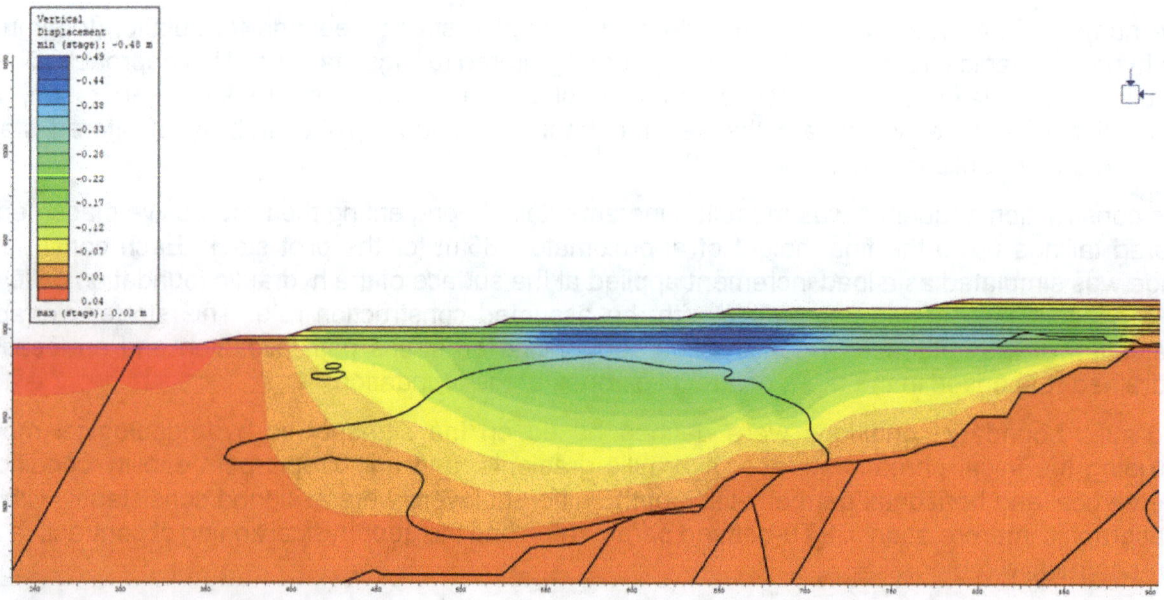

FIG 3 – Vertical displacement contours obtained from the coupled stress–seepage analysis, showing settlement concentrated within the hydraulic tailings foundation under the 35 m filtered stack, without development of a global instability mechanism.

RESULTS AND DISCUSSION

The coupled stress–seepage analyses provide an integrated view of the mechanical and hydraulic response of the hydraulic tailings foundation during staged filtered tailings stacking. The results indicate that the system response is governed by the interaction between the initial state of the hydraulic tailings, the loading rate, and the available drainage conditions.

The evolution of stress paths within the hydraulic foundation shows spatially variable responses associated with the stratified and state-dependent nature of the deposit. In the p'–q space, layers with looser initial states exhibit stress paths that progressively approach the CSL, particularly during intermediate construction stages, when excess pore pressures are only partially dissipated between load increments. In some zones, stress paths become tangential or slightly transgressive to the CSL, indicating conditions under which the material responds in a strongly contractive manner, accompanied by temporary reductions in effective stress. This behaviour is consistent with critical-state-based interpretations, in which materials with positive state parameters tend to generate excess pore pressures under partially drained or undrained loading.

The most critical responses are typically concentrated in intermediate layers of the hydraulic foundation, where three conditions coincide: (i) loose initial states associated with higher void ratios, (ii) reduced vertical permeability due to depositional stratification, and (iii) effective stress levels sufficient to mobilise nonlinear behaviour. In these regions, stress path evolution is strongly influenced by the loading rate and the efficiency of pore pressure dissipation, resulting in distinct responses even under similar load increments.

The computed pore pressure response during staged construction exhibits a strong dependence on time and stratigraphy. In scenarios without drainage enhancement, incremental loading results in the generation of excess pore pressures that do not fully dissipate between successive construction stages. This leads to progressive accumulation of pore pressures in specific layers, modifying effective stress paths and amplifying the contractive tendency of the material. Peak pore pressures typically occur near depositional interfaces and in zones of lower vertical permeability, confirming the role of these discontinuities as critical hydraulic control layers. This behaviour is consistent with observations reported in stratified natural and hydraulic deposits, in which anisotropy and layering govern pore pressure concentration.

In scenarios incorporating basal drainage, both the magnitude and persistence of excess pore pressures are significantly reduced. The drainage system promotes more efficient dissipation between construction stages, resulting in stress paths that remain farther from the CSL and in more stable effective stress conditions over time. Nevertheless, locally undrained responses may still occur in geometrically confined or low-permeability zones, indicating that drainage systems mitigate, but do not eliminate, the need for strict control of construction sequencing.

The construction sequence was not governed by a single prescriptive raising rate, but rather by the interaction between loading increments and the pore pressure dissipation capacity of the hydraulic foundation. The coupled stress–seepage analyses indicate that when load increments are applied at a rate compatible with the available drainage conditions, excess pore pressures are partially dissipated between stages, resulting in predominantly drained or transitional responses. Conversely, rapid construction rates lead to progressive pore pressure accumulation, particularly within low-permeability and stratified zones, shifting stress paths toward more contractive conditions. Therefore, the definition of staging intervals is intrinsically linked to the dissipation response of the foundation, and construction progression should be conditioned on the stabilisation of pore pressures and deformation trends rather than on a fixed nominal rate.

The deformation results indicate that the global response of the system is dominated by vertical settlements associated with consolidation of the hydraulic foundation. These settlements develop progressively during the construction stages and continue after load increments, reflecting the time-dependent dissipation of pore pressures. The pore pressure distribution at the end of stacking, considering the effect of the phreatic surface in the initial elastic state of the model, is presented in Figure 4, which highlight the role of the groundwater level in defining the stress field and effective loading conditions within the foundation.

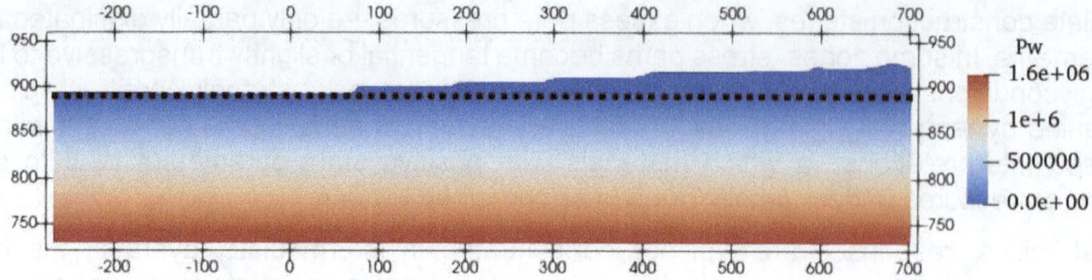

FIG 4 – Pore pressure distribution considering the phreatic surface in the initial elastic stress state of the model at the end of the filtered tailings stacking.

The vertical displacement contours show that deformations are concentrated within the hydraulic foundation, without the development of a global failure mechanism under the filtered stack with an approximate height of 35 m.

In addition to vertical settlements, the analyses reveal the formation of localised shear deformation zones, particularly in layers where the combination of contractive behaviour and limited pore pressure dissipation leads to temporary reductions in effective stress. These zones do not necessarily evolve into global failure surfaces but represent regions of strain concentration that may propagate progressively under unfavourable loading or drainage conditions. The displacement vector field at the end of construction, presented in Figure 5, illustrates this distributed deformation pattern within the hydraulic foundation.

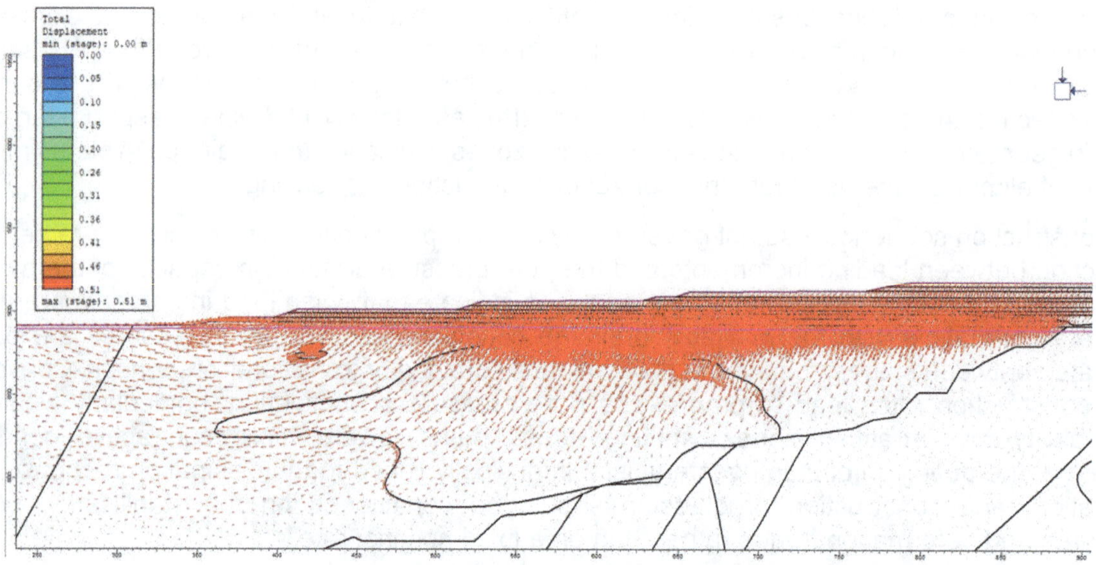

FIG 5 – Displacement vector field at the end of construction, showing the deformation pattern concentrated within the hydraulic tailings foundation beneath the filtered stack.

From a critical state perspective, operational stability is closely linked to maintaining stress paths that avoid strongly contractive conditions under limited drainage. An integrated interpretation of state, pore pressures, and deformation therefore provides a more realistic and defensible assessment of system performance than approaches based solely on ultimate strength.

The analyses also demonstrate the central role of drainage provisions and construction controls in the stability of filtered stacks over hydraulic foundations. Basal drainage systems significantly reduce the magnitude and persistence of excess pore pressures, promoting more favourable stress paths and more predictable mechanical responses. Controlled compaction of the filtered tailings also contributes to improved load distribution and reduced strain concentration in the underlying hydraulic foundation.

Construction sequencing emerges as one of the most critical elements of risk control. Load increments compatible with the dissipation capacity of the foundation lead to predominantly drained responses, whereas excessive construction rates promote locally undrained behaviour. These results are consistent with observations reported in case studies and with recent technical guidance, which emphasise the integration of construction rate, drainage, and monitoring in filtered tailings projects (Davies and Martin, 2009; Morgenstern *et al,* 2016; ANCOLD, 2024; ICOLD, 2023).

Practical monitoring requirements are directly linked to the governing mechanisms identified in the analyses. In particular, the control of pore pressure dissipation, vertical settlements, and deformation localisation within the hydraulic foundation is essential to ensure safe construction progression. This can be achieved through a combination of vibrating wire piezometers to monitor pore pressure response, settlement plates and surface monuments to track vertical displacements, and, where applicable, internal deformation measurements such as magnetic extensometers. In addition, operational controls such as periodic topographic surveys during stacking provide critical information on the effective raising rate and geometry evolution. As construction progresses, monitoring data should be continuously compared with model predictions to confirm expected behaviour or identify deviations, allowing for timely adjustment of construction rates and sequencing. This integrated approach ensures that field performance remains consistent with the assumptions adopted in the numerical analyses.

CONCLUSIONS

This study presented an integrated assessment of the geomechanical stability of filtered tailings stacks founded on hydraulic tailings, combining field geotechnical characterisation, laboratory testing within a critical state framework, and coupled stress–seepage numerical analyses. The results demonstrate that hydraulic tailings used as foundation materials typically exhibit loose initial states, pronounced stratigraphic heterogeneity, and predominantly contractive behaviour, making their response highly sensitive to the stress paths imposed during filtered tailings stacking.

The evolution of stress paths in p′–q space indicates that, during staged construction, portions of the foundation may approach or reach conditions near the critical state, particularly when pore pressure dissipation is limited by stratification and hydraulic anisotropy. In this context, pore pressure generation and dissipation play a central role in system stability, controlling the evolution of effective stresses and the localisation of deformations. Depositional interfaces and layers with low vertical permeability emerge as critical zones for pore pressure concentration and the development of localised shear deformation.

From a design perspective, the provision of adequate drainage systems is essential to reduce the persistence of excess pore pressures and to promote predominantly drained responses during stacking. Controlled compaction of the filtered tailings contributes to a more uniform stress distribution and reduces localised deformation concentrations within the hydraulic foundation. In addition, construction sequencing emerges as one of the most effective risk mitigation mechanisms and should be compatible with the pore pressure dissipation capacity of the deposit.

Monitoring strategies focused on pore pressures, settlements, and deformations are fundamental for performance management throughout the service life of the structure. The integration of detailed geotechnical characterisation, coupled numerical analyses, and continuous monitoring enables the adoption of an adaptive design and operational approach, consistent with contemporary tailings management practices.

Overall, the results indicate that filtered tailings stacking over hydraulic foundations is technically feasible, provided that design and operation explicitly recognise the state-dependent, time-dependent, and drainage-controlled nature of these systems, incorporating appropriate construction controls and monitoring strategies. The integration of critical state soil mechanics concepts, detailed characterisation, and coupled hydromechanical analyses contributes to reducing uncertainties, improving performance predictability, and aligning tailings engineering with modern safety and governance expectations throughout the life cycle of the facility.

REFERENCES

Australian National Committee on Large Dams (ANCOLD), 2024. Guidelines on Tailings Dams – Planning, Design, Construction, Operation and Closure, Australian National Committee on Large Dams (ANCOLD), Sydney.

Been, K and Jefferies, M G, 1985. A state parameter for sands, *Géotechnique*, 35(2):99–112.

Been, K and Jefferies, M G, 2015. *Soil Liquefaction: A Critical State Approach*, 2nd edn (CRC Press: Boca Raton).

Been, K, Jefferies, M G and Hachey, J, 1991. The critical state of sands, *Géotechnique*, 41(3):365–381.

Boulanger, R W and Idriss, I M, 2014. CPT and SPT Based Liquefaction Triggering Procedures, Report No, UCD/CGM-14/01, University of California, Davis.

Davies, M P and Martin, T E, 2009. Mining market cycles and tailings dam incidents, Proceedings of Tailings and Mine Waste Conference, Banff, Canada.

International Commission on Large Dams (ICOLD), 2023. *Tailings Dam Safety Guidelines,* International Commission on Large Dams (ICOLD), Bulletin.

Jefferies, M and Been, K, 2006. *Soil Liquefaction: A Critical State Approach* (Taylor and Francis: London).

Morgenstern, N R, Vick, S G, Viotti, C B and Watts, B D, 2016. *Fundão Tailings Dam Review Panel Report,* Report to Samarco Mineração S A.

Schofield, A N and Wroth, C P, 1968. *Critical State Soil Mechanics* (McGraw-Hill: London).

Sharma, J S and Fahey, M, 2003. Deformation and failure of tailings, *Canadian Geotechnical Journal*, 40(2):252–268.

Shuttle, D A and Jefferies, M G, 2016. State parameter interpretation of liquefaction triggering, *Canadian Geotechnical Journal*, 53(11):1808–1819.

Vick, S G, 1990. *Planning, Design and Analysis of Tailings Dams*, 2nd edn (BiTech Publishers: Vancouver).

Yin, J H and Graham, J, 1996. Elastic viscoplastic modelling of one-dimensional consolidation, *Géotechnique*, 46(3):515–527.

Evolving from engineered to geomorphic landform, incorporating traditional values

J MacRae[1]

1. Founder and Principal Mining Engineer, Landform Logic, Perth WA 6000.
   Email: jon@landformlogic.com

ABSTRACT

In the Pilbara region of Western Australia, mining companies and Traditional Owners hold different views on how final mine landforms should be designed. Mining engineers often develop closure designs that are efficient to model, construct, and execute within existing mine planning schedules and systems. These designs favour simple, planar slopes and trapezoidal profiles that are readily optimised using current mining software and equipment. The focus is on material movement, geotechnical stability, and cost-effectiveness. The primary design is the repose profile, and the final landform is derived from this structure. The result is a technically sound but visually artificial landform.

In contrast, Traditional Owners are requesting geomorphic landforms that reflect the natural country. Their vision includes convex and concave slope profiles, natural flow paths within catchments, and varied topography that reconnects surface water flow and supports ecological and cultural renewal. These landforms draw on the surrounding hills and valleys, aiming to restore the land's natural appearance and function.

The challenge for industry is to bridge these perspectives—merging technical stability with cultural authenticity to deliver landforms and landscapes that endure both physically and culturally. This raises an important question: who is truly designing the post-mining landscape? Engineers often treat the pit, the dump, and the haul road as separate design elements, whereas Traditional Owners see the entire Country as a connected system of land, water, and meaning. To move forward, mining engineers must look beyond the engineered footprint of the landform on the mining lease and design within the broader landscape.

INTRODUCTION

Waste rock dumps are among the most prominent and persistent features of post-mining landscapes (ICMM, 2025). Unlike pits, which may be backfilled or flooded, waste rock landforms often remain elevated and continue to influence drainage, ecology and land use for centuries (ICMM, 2025). Consequently, the design of these landforms has a critical influence on long-term closure outcomes.

In Australian mining practice, waste rock dumps are commonly designed as engineered structures defined by uniform batter angles and regular berm spacing (DMIRS, 2021). These geometries reflect the requirements of mine planning software, haulage efficiency and construction practicality (Taylor, Wrigth and Bettison, 2018). While such designs can satisfy geotechnical stability criteria and regulatory approvals, they frequently result in landforms that appear artificial (Hine and Erskine, 2016).

At the same time, Traditional Owners increasingly articulate expectations for post-mining landscapes that resemble natural Country (Reeves, Dowell and Welch, 2025). These expectations extend beyond visual amenity to include natural water movement, landscape diversity, ecological function and cultural connectivity (Reeves, Dowell and Welch, 2025). This divergence of views on engineered landform outcomes and Traditional Owner expectations presents a challenge in the approvals process (Malan and Murphy, 2024; Hine and Erskine, 2016).

This paper examines how current engineering approaches to waste rock dump design influence final landform outcomes and explores how these approaches can be adapted to support landforms that better align with both physical geomorphic and cultural objectives (Howard, Loch and Vacher, 2010; ICMM, 2025). The paper argues that geomorphic landform design is not a departure from engineering rigour, but an extension of it, requiring earlier integration of closure intent and closer collaboration with Traditional Owners (Leidich, Lacy and van Zyl, 2014; Hancock and Martin Duque, 2018; Reeves, Dowell and Welch, 2025).

WASTE ROCK DUMP DESIGN IN CURRENT PRACTICE

Waste rock dump design is typically embedded within life-of-mine planning processes and driven by operational priorities (Taylor, Wrigth and Bettison, 2018; ICMM, 2025). Dump locations, footprints and heights are selected to minimise haul distances, support production schedules whilst avoiding known environmental, heritage and tenure constraints (McCluskey and Barros, 2014). Waste placement is commonly undertaken in lifts on flat benches, with the batter slopes at or near the angle of repose. The incorporated berms are for access, safety, and rainfall and surface water control (DMIRS, 2021).

Geotechnical stability is a key consideration in design. Material strength, foundation conditions, and pore pressure responses are evaluated using limit equilibrium or numerical methods to ensure that safety factors are acceptable (Hawley and Cunning, 2017). Surface water management involves engineered controls, such as bunds and contour drains, on the waste rock dump during operations (Hawley and Cunning, 2017). For longer-term solutions, permanent measures like diversions and external bunds are used to direct water away from the landform (DMIRS, 2021).

Environmental considerations are generally incorporated through sediment control measures and progressive rehabilitation of completed slopes. Closure requirements are often treated as a later design phase, with reshaping, slope flattening and soil placement undertaken once operational dumping has ceased (Taylor, Wrigth and Bettison, 2018; ICMM, 2025). As a result, the final landform is typically a modified version of the operational dump rather than a purpose-designed landscape (Hine and Erskine, 2016).

This method depends on several implicit assumptions. It presumes that operational efficiency takes precedence over long-term landform impacts, that engineered drainage structures will function effectively throughout the closure period, and that reshaping at closure is both economically viable and practical (Hawley and Cunning, 2017; ICMM, 2025).

FINAL LANDFORM DESIGN AND GEOMORPHIC APPROACHES

Final landform design has traditionally focused on modifying engineered dumps to meet closure criteria rather than redefining landform intent (Kemp, Taylor and Hancock, 2014). Current practice commonly involves flattening outer batters, reducing slope lengths and installing surface drainage to limit erosion (DMIRS, 2021). While these measures can improve performance relative to unmodified dumps, they often reinforce the original engineered geometry and provide limited opportunity for natural landscape evolution (Hine and Erskine, 2016; Howard, Loch and Vacher, 2010).

Geomorphic landform design offers an alternative approach by focusing on long-term landscape processes rather than static form (Hancock and Martin Duque, 2018). Geomorphic designs draw on natural analogues observed in surrounding terrain, incorporating variable slope profiles, concave lower slopes, convex crests and integrated drainage networks aligned with catchment processes (Kemp, Taylor and Hancock, 2014). These forms reduce flow concentration, promote sediment deposition and improve long-term stability (Howard, Loch and Vacher, 2010; Hancock and Martin Duque, 2018).

The concave landform design combines traditional elements with a cascading effect, featuring gentler slope angles. This profile resembles the Messa formations found in the Pilbara region, with a stepped natural landform in which the upper surfaces transition into a reduced-gradient profile on the lower slopes (Kemp, Taylor and Hancock, 2014). In practice, concave and variable slope profiles can manage surface water run-off and reduce erosion potential, with the lower slope surfaces acting as an alluvial fan and promoting material deposition on the lower slope surface (Kemp, Taylor and Hancock, 2014; Howard, Loch and Vacher, 2010; Hancock and Martin Duque, 2018).

Waste placement strategies, lift sequencing and haul road alignment must be planned to progressively build the final landform (Taylor, Wrigth and Bettison, 2018; ICMM, 2025). While this adds complexity, it significantly reduces rehandling at closure and improves long-term performance of the landform (Leidich, Lacy and van Zyl, 2014; ICMM, 2025).

TRADITIONAL OWNER PERSPECTIVES ON LAND AND CLOSURE

Traditional Owners view land as Country, an interconnected system encompassing landforms, water, ecology, culture and law (Reeves, Dowell and Welch, 2025). From this perspective, landforms are not inert structures but living elements that carry cultural meaning and responsibility across generations (Reeves, Dowell and Welch, 2025).

Engineered landforms with uniform slopes and artificial drainage are often perceived as ongoing disturbances rather than rehabilitated landscapes. Such landforms can interrupt natural surface water movement, obscure culturally significant features and diminish the integrity of Country (Reeves, Dowell and Welch, 2025). Even where regulatory closure criteria are met, these outcomes may be unacceptable to Traditional Owners (Malan and Murphy, 2024).

Traditional Owner perspectives emphasise landforms that resemble natural systems, support the return of native flora and fauna, and enable the continuation of cultural practices. Visual alignment with surrounding landscapes is important, but it is inseparable from functional outcomes such as restored catchments and water flows (Hancock and Martin Duque, 2018). The failure to address these dimensions introduces significant social and closure risk (Malan and Murphy, 2024).

INTEGRATING TRADITIONAL OWNER EXPECTATIONS INTO LANDFORM DESIGN

Traditional Owners consistently seek early and meaningful involvement in landform design decisions (Reeves, Dowell and Welch, 2025). Rather than being consulted on predetermined designs, they seek to participate in defining closure objectives and landform concept and design. Desired outcomes include landforms that reconnect catchments, restore natural drainage patterns and support ongoing cultural use (Reeves, Dowell and Welch, 2025).

For engineers, this requires shifting from transactional consultation to a collaborative design process (ICMM, 2025). Knowledge of the cultural landscape can inform the selection of natural analogues, drainage alignments and landform geometry (Reeves, Dowell and Welch, 2025). Progressive rehabilitation offers opportunities to test designs, preform trails and build confidence through demonstrated outcomes and increased knowledge (ICMM, 2025). Clear communication of constraints and trade-offs is essential to maintaining trust and achieving practical solutions (Reeves, Dowell and Welch, 2025).

IMPLICATIONS FOR ENGINEERING PRACTICE

The transition from engineered to either concave or geomorphic landforms requires a shift in how closure is considered within mine planning (Taylor, Wrigth and Bettison, 2018; ICMM, 2025). Closure objectives must be embedded early and treated as primary design constraints rather than downstream modifications (ICMM, 2025). Engineers must work across disciplines to balance operational efficiency with long-term landscape performance (Leidich, Lacy and van Zyl, 2014; ICMM, 2025).

Immediate actions involve assessing current designs by comparing them with natural analogues, testing various linear and concave profiles, and experimenting with geomorphically styled landforms. When possible, update internal standards to align with evolving design and construction practices (Hancock and Martin Duque, 2018; Howard, Loch and Vacher, 2010). Gaps in data persist, particularly in long-term performance monitoring and establishing metrics that reflect cultural acceptance alongside physical stability (Malan and Murphy, 2024; ICMM, 2025). Closing these gaps will require continued collaboration and flexible management approaches.

LEARNING FOR ENGINEERS TO BE PART OF COUNTRY

Concave and Geomorphic landforms provide a practical means of reconciling engineering requirements with cultural expectations (Hancock and Martin Duque, 2018). When designed and implemented collaboratively, they offer the potential for landscapes that endure physically, function ecologically and are accepted culturally (Reeves, Dowell and Welch, 2025; ICMM, 2025).

CONCLUSIONS

Waste rock dumps were generally designed for operational efficiency and struggled to meet long-term closure objectives (Taylor, Wrigth and Bettison, 2018; Hine and Erskine, 2016). Engineers can reduce long-term risks by designing waste rock dumps as long-term landforms rather than temporary storage sites (ICMM, 2025; Leidich, Lacy and van Zyl, 2014). They can create landforms that mimic natural features, reduce erosion risks, and integrate more seamlessly into the landscape, whilst on Country (Kemp, Taylor and Hancock, 2014; Hancock and Martin Duque, 2018). This represents an opportunity to evolve current engineering practices (Howard, Loch and Vacher, 2010).

REFERENCES

Department of Mines, Industry Regulation and Safety (DMIRS), 2021. *Waste Rock Dump Design Guidelines*, Government of Western Australia.

Hancock, G R and Martin Duque, J F, 2018. Geomorphic design and landscape evolution modelling for best practice mine rehabilitation, in *Proceedings of the Life-of-Mine Conference 2018*, pp 119–121 (The Australasian Institute of Mining and Metallurgy: Melbourne).

Hawley, M and Cunning, J (eds), 2017. *Guidelines for Mine Waste Dump and Stockpile Design*, CSIRO Publishing, Melbourne.

Hine, A and Erskine, P, 2016. Recognising and integrating stakeholder landform expectations into *Life-of-Mine* planning, in *Proceedings of the Life-of-Mine Conference 2016*, pp 154–157 (The Australasian Institute of Mining and Metallurgy: Melbourne).

Howard, E J, Loch, R J and Vacher, C A, 2010. Evolution of landform design concepts, in *Proceedings of the First International Seminar on the Reduction of Risk in the Management of Tailings and Mine Waste, Mine Waste 2010* (eds: A B Fourie and R J Jewell), pp 83–90 (Australian Centre for Geomechanics: Perth). https://doi.org/10.36487/ACG_rep/1008_08_Howard

International Council on Mining and Metals Limited (ICMM), 2025. *Integrated Mine Closure; Good Practice Guide*, 3rd edn. Available from: <https://www.icmm.com/integrated-mine-closure> [Accessed: 7 March 2025].

Kemp, D, Taylor, I and Hancock, G, 2014. Landform design for Pilbara mine sites, in *Proceedings of the Life-of-Mine Conference 2014*, pp 47–54 (The Australasian Institute of Mining and Metallurgy: Melbourne).

Leidich, T, Lacy, H and van Zyl, M, 2014. Application of design concepts and engineering controls necessary to deliver well designed and rehabilitated landforms, in *Proceedings of the Life-of-Mine Conference 2014*, pp 55–66 (The Australasian Institute of Mining and Metallurgy: Melbourne).

Malan, S and Murphy, D, 2024. Navigating divergent expectations, in *Proceedings of the 17th International Conference on Mine Closure, Mine Closure 2024* (eds: A B Fourie, M Tibbett and G Boggs), pp 337–350 (Australian Centre for Geomechanics: Perth). https://doi.org/10.36487/ACG_repo/2415_24

McCluskey, J and Barros, N, 2014. Life of asset planning, uncertainty and value optimisation of the iron ore business – from exploration, to market, to closure, in *Proceedings of the Orebody Modelling and Strategic Mine Planning Symposium 2014*, pp 33–40 (The Australasian Institute of Mining and Metallurgy: Melbourne).

Reeves, D, Dowell, K and Welch, J, 2025. Argyle diamond mine: a model for First Nations engagement, in *Proceedings of the 18th International Conference on Mine Closure, Mine Closure 2025* (eds: S Knutsson, A B Fourie and M Tibbett), 15 p (Australian Centre for Geomechanics: Perth). https://doi.org/10.36487/ACG_repo/2515_42

Taylor, N, Wrigth, R and Bettison, M, 2018. Utilising mine planning to deliver closure landforms through productive movement, in *Proceedings of the Life-of-Mine Conference 2018*, pp 96–99 (The Australasian Institute of Mining and Metallurgy: Melbourne).

Integrating progressive backfill – Mining Area C's 'finish and fill' strategy

G Menaria[1] and J Frewen[2]

1. Superintendent Production Scheduling, Mining Area C BHP, Perth WA 6000. Email: gaurav.menaria@bhp.com
2. Manager Site Support, Mining Area C BHP, Perth WA 6000. Email: jeremiah.frewen@bhp.com

SUMMARY

The iron ore sector is increasingly focused on achieving safe, stable post-mining landscapes through innovative closure strategies. Mining Area C has pioneered the 'Finish and Fill' approach, which integrates progressive closure into operational planning by sequentially backfilling pits as mining advances. This method minimises reliance on external waste dumps, reduces closure costs, and delivers operational efficiencies such as shorter haul distances, lower fuel consumption, and reduced greenhouse gas emissions. By embedding closure into the mining sequence, the strategy transforms rehabilitation from an end-of-life activity into a continuous process, setting a benchmark for sustainable mine planning.

EXTENDED ABSTRACT

Increasing regulatory expectations, stakeholder scrutiny, and corporate sustainability commitments have placed progressive closure strategies at the forefront of contemporary mine planning. This paper introduces Mining Area C's 'Finish and Fill' approach which is an integrated and practical strategy designed to maximise opportunities for sequential pit backfilling while minimising long-term closure liabilities.

Mining Area C is establishing a benchmark in progressive closure by adopting a development sequence that enables waste material from active mining stages to directly backfill previously mined voids.

In FY24 and FY25, actual performance shows that 66 per cent and 68 per cent, respectively, of total waste movement was backfill. These results establish a strong operational baseline for the 'Finish and Fill' strategy.

From FY26 onward, the proportion of waste as backfill will increase materially, reaching 71 per cent in FY26 and peaking at 76 per cent in FY27 (+5 percentage points year-on-year). While a minor consolidation occurs in FY28 (75 per cent), the backfill share recovers to 76 per cent in FY29, indicating stabilisation at a materially higher level than early-period actuals.

Overall, the shift from 66 per cent (FY24 actual) to ~75–76 per cent (FY27–FY29) represents a ~10 percentage point uplift in the proportion of waste placed in-pit. This sustained increase demonstrates the progressive embedding of 'Finish and Fill' principles into operational mine planning, materially reducing reliance on external waste dumps and supporting progressive closure outcomes.

This approach significantly reduces the need for external waste storage facilities, which traditionally occupy large cleared areas and require extensive and complex rehabilitation works. By eliminating or substantially reducing overburden storage areas, the strategy minimises environmental disturbance and lowers the scope and complexity of post-mining landform re-profiling and rehabilitation.

A key feature of the 'Finish and Fill' strategy is its emphasis on pit sequencing that prioritises areas with lower strip ratios early in the mine life. This delivers two primary benefits: improved early cash flow and the timely creation of voids capable of accommodating waste from subsequent mining phases. The operational advantages of this sequencing are substantial. Shorter waste haul distances result in reduced fuel consumption, lower greenhouse gas emissions, and decreased equipment wear. These efficiencies translate into direct cost savings and align with both community expectations and corporate sustainability objectives, including emissions-reduction targets.

Embedding progressive closure into operational planning from the outset represents a paradigm shift in mine rehabilitation. Historically, closure activities have been deferred until the end of mine life, often resulting in significant financial, technical, and logistical challenges. In contrast, the 'Finish and Fill' approach integrates rehabilitation into daily operations, making closure an ongoing process rather than a terminal obligation. This proactive model reduces closure costs by minimising the need for extensive re-profiling, rock armouring, and large-scale rehabilitation works. Furthermore, it decreases post-closure monitoring and maintenance requirements, delivering long-term cost savings and reducing residual risk.

A critical enabler of this approach is the Backfill Business Case (BBC), which provides the formal assurance mechanism to confidently sterilise backfilled areas. The BBC delivers cross-functional endorsement across Resource Engineering, Production Scheduling and Geoscience that the backfill design and operational strategy have been fully assessed, and that the area will not be re-mined. This assurance allows backfilled areas to be treated as permanently rehabilitated for planning, closure and liability purposes.

The benefits of this strategy extend beyond cost efficiency. By reducing the footprint of external waste dumps, the approach mitigates visual and environmental impacts, supporting improved stakeholder relationships and facilitating regulatory compliance. It also enhances safety by reducing the number of large, unconsolidated waste landforms that require long-term stability management. Collectively, these outcomes contribute to the development of safer, more stable post-mining landscapes that align with industry best practice and community expectations.

From an operational perspective, the 'Finish and Fill' approach delivers measurable improvements in productivity. Reduced truck and operator hours, lower diesel consumption, and optimised haulage cycles all contribute to enhanced operational performance. These efficiencies are particularly valuable in remote mining regions such as the Pilbara, where fuel logistics, workforce availability, and equipment utilisation present ongoing challenges. By minimising waste haulage distances, the strategy also reduces exposure to supply chain disruptions and fuel cost volatility.

The practical application of this approach at Mining Area C demonstrates its potential for replication across the iron ore sector and beyond. While geological, spatial, and operational constraints will influence site-specific feasibility, The principles underpinning 'Finish and Fill': (1) sequential pit development, (2) integrated closure planning, and (3) progressive backfilling—are broadly applicable to open pit mining operations. Widespread adoption of similar strategies could deliver substantial environmental, operational, and economic benefits, supporting the industry's transition toward more sustainable mining practices.

In conclusion, Mining Area C's 'Finish and Fill' strategy represents a significant advancement in the integration of mine planning and closure. By embedding progressive rehabilitation into operational workflows, the approach reduces closure costs, enhances productivity, and minimises environmental impacts. This model sets a new benchmark for sustainable mining, providing a practical pathway to achieving safe, stable post-mining landscapes while delivering operational and financial value throughout the life-of-mine.

ACKNOWLEDGEMENTS

The authors acknowledge the Banjima people, the Traditional Custodians of the land on which this work is undertaken. We pay our respects to Banjima Elders past and present.

Australia's contribution to international standards in iron ore

B Russell[1]

1. Engagement Manager, Standards Australia, Sydney NSW 2000.
   Email: ben.russell@standards.org.au

ABSTRACT

Australia plays a pivotal role in shaping the global standards landscape for iron ore through its leadership of the ISO Technical Committee 102 (ISO/TC 102) Iron Ore, Subcommittee on Chemical Analysis (SC 2). This subcommittee, chaired and managed by Standards Australia, is responsible for the development and maintenance of internationally recognised standards that underpin the accurate and consistent chemical analysis of iron ores and direct reduced iron. These standards are critical to global trade, resource valuation, and metallurgical performance, and contribute directly to several UN Sustainable Development Goals, including responsible consumption and production, industry innovation, and economic growth.

Australia's stewardship of SC 2 has resulted in the publication of over 50 ISO standards and the ongoing development of new methods, including advanced techniques such as X-ray fluorescence (XRF), EDTA titrimetric analysis, and potentiometric titration. These standards ensure analytical consistency across laboratories worldwide, supporting equitable commercial transactions and robust quality assurance frameworks.

The standards developed by SC 2 directly support several UN Sustainable Development Goals, particularly those focused on responsible consumption and production (SDG 12), industry innovation and infrastructure (SDG 9), and economic growth (SDG 8). Reliable analytical methods reduce waste, improve resource utilisation, and support the development of new technologies in ore characterisation and processing.

The committee has published a wide suite of standards covering wet chemical methods, instrumental techniques, and sample preparation. Some notable examples include:

- ISO 9516 series – Methods for determining major and minor elements in iron ores using X-ray fluorescence (XRF).
- ISO 2597 series – Determination of total iron content using titrimetric methods, including the classic potassium dichromate titration.
- ISO 4687 series – Determination of phosphorus content by molybdenum blue spectrophotometry.
- ISO 10204 – Determination of calcium using EDTA titrimetric analysis.
- ISO 13930 – Determination of total iron in direct reduced iron by dichromate titration.
- ISO 3082 – Sampling and sample preparation procedures (developed by SC 1 but closely linked to SC 2's analytical work).

Australia has identically adopted many of these ISO standards as Australian Standards (AS), ensuring national alignment with global best practice. Examples include AS ISO 2597, AS ISO 9516, and AS ISO 3082, which are widely used across Australian laboratories, mining operations, and export terminals. Identical adoption means that Australian professionals operate under the same technical frameworks as their international counterparts, strengthening Australia's reputation for high-quality, reliable analytical results.

SC 2's work is not static. New analytical technologies and industry needs continually drive the development of updated or entirely new standards. Recent and ongoing work includes:

- Advanced XRF calibration and matrix correction techniques, improving accuracy for complex ore types such as magnetite concentrates and high-loss-on-ignition ores.

- Potentiometric titration methods for elements such as manganese and chromium, offering improved precision and automation potential.
- Refinements to EDTA titrimetric methods, which remain essential for determining calcium and magnesium in iron ores.
- Harmonisation of sample decomposition procedures, including fusion and acid digestion methods, to support compatibility across instrumental techniques.

These advancements reflect the evolving nature of iron ore products, including higher-grade magnetite concentrates, direct reduction feedstocks, and ores with increasingly complex gangue mineralogy.

Without standardisation in these crucial areas, the industry finds itself at risk of inconsistent sample data and methods. This can create challenges across the entire supply chain, from testing to commercial negotiations as the information coming through from these keys early stages would be unreliable. Australia's direct involvement and leadership in this space has helped advocate and protect our interests by pushing for the tolerance levels and product quality that our local industry demands from the rest of the international community.

ISO/TC 102 brings together experts from 18 participating and 22 observing member countries. This diversity ensures that standards reflect global needs rather than the preferences of any single nation. Australia's leadership role means it helps shape the agenda, coordinate working groups, and mediate technical discussions.

Standards Australia supports this work through its mirror committee MN-002, which includes representatives from mining companies, commercial laboratories, universities, research organisations, equipment manufacturers, and government agencies. MN-002 reviews draft ISO standards, develops Australian positions, nominates experts to international working groups, and identifies emerging areas where new standards may be needed.

AusIMM members, with their deep technical expertise and industry insight, are uniquely positioned to influence the future of iron ore standardisation. With professionals across the mining value chain—geologists, metallurgists, process engineers, laboratory managers, and researchers— they have a direct stake in the quality and relevance of iron ore standards. Their involvement ensures that standards reflect real-world operational challenges and emerging industry trends.

By participating in Standards Australia's mirror committees or contributing to public consultations on draft standards, members can help ensure that Australian innovation and operational excellence are reflected in global best practice. Opportunities also exist to join working groups, propose new work items, and provide feedback on emerging standards relevant to chemical analysis, sampling, and physical testing.

Use cases where AusIMM members and other mining industry stakeholders benefit from involvement:

- Resource evaluation and reporting: Geologists rely on consistent analytical methods to classify resources and reserves. Participation helps ensure standards reflect the needs of modern exploration and metallurgy.
- Process optimisation: Metallurgists depend on accurate chemical data to model beneficiation circuits, design processing plants, and troubleshoot operational issues. Their insights can guide improvements to analytical precision and method robustness.
- Quality assurance in laboratories: Laboratory professionals use ISO standards daily. Their practical experience is invaluable in identifying ambiguities, proposing method refinements, and ensuring standards are implementable.
- Commercial negotiations: Professionals involved in marketing and sales benefit from standards that reduce the risk of analytical disputes and support transparent pricing mechanisms.

- Research and innovation: Academics and researchers can influence the adoption of new technologies—such as laser-induced breakdown spectroscopy (LIBS) or automated mineralogy—by proposing new work items or contributing validation data.

Engagement with AusIMM members and other mining industry stakeholders can take several forms, each offering different levels of involvement:

- Join Standards Australia's MN-002 committee to participate directly in reviewing draft ISO standards and shaping Australia's national position.
- Contribute to public consultations, where draft standards are released for broader industry feedback.
- Nominate as an expert on an ISO working group, contributing technical knowledge to the development of new or revised standards.
- Propose new work items, particularly where emerging technologies or industry challenges are not yet addressed by existing standards.
- Participate in interlaboratory trials, which provide the data needed to validate new analytical methods.

This presentation will outline Australia's contributions to ISO/TC 102/SC 2, showcase recent advancements in chemical analysis standards, and provide practical guidance on how AusIMM members and other mining industry stakeholders can engage with the standards development process. By strengthening the link between professional practice and international standardisation, we can enhance Australia's influence, support global harmonisation, and drive continuous improvement across the iron ore value chain.

Spontaneous combustion hazard assessment of carbonaceous shales in Pilbara iron ore mining operations

J Theiler[1], B B Beamish[2], R Cooke[3], J Hurst[4] and S Lee[5]

1. Principal Mining Engineer, B3 Mining Services Pty Ltd, Brisbane Qld 4076. Email: jan@b3miningservices.com
2. Managing Director, B3 Mining Services Pty Ltd, Brisbane Qld 4076. Email: basil@b3miningservices.com
3. Lead Environmental Geologist, Fortescue Ltd, Perth WA 6000. Email: rachael.cooke@fortescue.com
4. Senior Environmental Geologist, Fortescue Ltd, Perth WA 6000. Email: josh.hurst@fortescue.com
5. Principal Environmental Geoscientist, Fortescue Ltd, Perth WA 6000. Email: steven.lee@fortescue.com

ABSTRACT

Carbonaceous shales contain organic carbon that can react with air when either exposed during the mining process or when buried in waste rock dumps. However, at ambient mine temperatures their intrinsic reactivity is extremely low due to the presence of a high mineral matter content (often silicate, carbonate and sulfate minerals) acting as a diluent and heat sink. However, if pyrite is present in these rocks, in a reactive form (size and morphology), it can act as an initiator of self-heating to raise the temperature of a broken rock pile to the point where the organic carbon takes over the self-heating process (generally > 120°C) leading to subsequent thermal runaway. A systematic approach has been used to assess this spontaneous combustion hazard across Fortescue iron ore operations in the Pilbara. A substantial database of adiabatic self-heating Incubation test results has enabled the development of a spontaneous combustion hazard potential chart based on the total sulfur and total organic carbon content of the shale. The chart can be used as a ready reckoner to classify whether thermal runaway or no thermal runaway is achievable assuming a critical pile thickness is present and from Mine Ambient conditions, which can then be used to make informed decisions on waste rock management to control the spontaneous combustion hazard. In addition, some fundamental differences between the carbonaceous shale units at the various mine sites has also been identified from the chart.

INTRODUCTION

The major overburden rock unit in the Pilbara iron ore operations that has been identified as a spontaneous combustion hazard is the Mt McRae Shale which is characterised by a high concentration of organic carbon (~2–8 per cent TOC) an abundance of disseminated pyrite (~1–10 per cent TS) and abundant pyrite nodules (~1–10 cm radius) (Kakegawa, Kawai and Ohmoto, 1999). Examples of the nodular pyrite can be seen in Figure 1. A systematic hazard assessment of this rock unit enables informed decision-making for waste rock management.

There are several spontaneous combustion tests documented in the literature (Nelson and Chen, 2007). Most of these tests produce a relative rating of hazard potential but they do not provide any indication of the time frame for spontaneous combustion issues to occur. They also do not provide a proper hazard evaluation of coal or non-coal (carbonaceous shale) self-heating that takes into consideration the effects of extrinsic factors (Beamish and Theiler, 2017a). Beamish and Beamish (2011, 2012) have shown how spontaneous combustion hazard assessment can be improved by benchmarking against known coal self-heating experience from operating mines and built stockpiles using an adiabatic incubation test method. Beamish and Theiler (2015, 2019) provide working examples of the success of this new technique. Earlier efforts to identify and quantify any accelerating effects from the presence of reactive pyrite in coal and non-coal strata were not successful (Miron, Lazzara and Smith, 1992), but the new incubation test method has overcome this problem (Beamish *et al*, 2018; Beamish and Theiler, 2017b; Beamish, Theiler and Ward, 2017). Incubation testing has been applied to samples of the Mt McRae Shale from two mines covering a range of TS and TOC contents to establish a site-specific Spontaneous Combustion Hazard Potential Classification Chart for Fortescue iron ore operations in the Pilbara.

FIG 1 – Photograph of Mt McRae Shale sample 0081.903 showing the presence of multiple pyrite bands and nodules.

INCUBATION TESTING AND ANALYSIS PROCEDURE

This incubation test is designed to replicate the true self-heating behaviour from low ambient temperature. As such, the normal in-mine temperature as defined for the mining situation is used as the starting point for the test. The nature of the test also assumes that in the real operational situation there is a critical pile thickness present that minimises any heat dissipation (represented by an adiabatic oven testing environment) and there is a sufficient supply of oxygen present to maintain the oxidation reaction. The sample either reaches thermal runaway or begins to lose heat due to insufficient intrinsic reactivity to overcome heat loss from moisture liberation/evaporation and/or heat sink effects from non-reactive mineral matter. As the sample is tested with its as-received moisture content, for some samples this may produce a prolonged temperature shoulder on the way to thermal runaway. For samples with a low self-heating rate, a proprietary step-heat procedure is used to establish the behaviour of the sample after incubating to thermal runaway. For carbonaceous shales where moisture is absent a small amount of water is added to quantify any self-heating acceleration from the presence of reactive pyrite.

The self-heating behaviour of coal or other carbonaceous strata recorded by the incubation test is compared against the documented self-heating performance of a range of samples with known spontaneous combustion histories (incubation periods). Due to the nature of the moisture evolution that occurs during self-heating, where some samples develop a prolonged temperature shoulder on the way to thermal runaway, the time taken for the sample temperature to reach 120°C is used as the key indicator of the sample to reach thermal runaway (t_{TR}). This time obtained in laboratory hours is converted to site days using the comparison with samples of known behaviour. The site days are expressed as a range rather than an exact value as it is impossible to project the outcome of a real event in exact terms. It should be noted that this analysis is for loosely piled coal or carbonaceous strata.

INCUBATION PERIOD ANALYSIS

Initial incubation testing has been conducted on Mt McRae Shale from Mine A. The carbon content and the sulfur content of the samples is contained in Table 1. Blending of samples 0081.902 and 0081.901 provided a larger range of sample geochemistries to be tested. The initial conditions for each incubation test and the inferred minimum incubation periods are contained in Table 2. Self-heating curves for the individual carbonaceous shale samples are shown in Figure 2. Three of the five samples tested have very short minimum incubation periods. The rapid self-heating associated with these samples is due to the presence of reactive pyrite that is predominantly in the form of nodules or bands. The best example of this is shown by Sample 0081.903 (Figure 1), which self-heats in the shortest period of time of all the samples in its as-received state. Retesting the sample with additional moisture added shows a longer incubation period, which is due to the extra evaporative heat loss from the excess moisture. The other two samples are effectively inert with no thermal runaway recorded. Blends of the inert 0081.901 sample and reactive 0081.902 sample show a successive dilution effect that increases the incubation period with increasing amounts of the inert 0081.901 added to the blend (Figure 3).

TABLE 1

Carbon and sulfur contents for Mt McRae Shale samples from Mine A (%, air-dried basis).

Sample ID	Total organic carbon	Total sulfur
0080.901	4.72	0.16
0080.902	2.80	17.76
0081.901	5.68	0.66
0081.902	3.06	17.43
0081.903	2.59	12.84
0081.904	3.33	3.02
0081.905	3.26	2.68
Sample mixtures of 0081.902 and 0081.901		
75/25	3.81	13.27
50/50	4.58	8.90
40/60	4.82	7.55
30/70	4.93	5.86
25/75	5.20	4.58

TABLE 2

Incubation test conditions and minimum incubation period for Mt McRae Shale samples from Mine A.

Sample	Initial temperature (°C)	Moisture content (%)	Minimum incubation period (days)
0080.902	34.9	2.0	6–7
0081.901	34.7	5.6	NTR
0081.902	35.0	3.2	4–5
0081.903	35.5 34.8	1.4 6.0	3–4 8–11
0081.905	35.0	3.6	NTR
Sample mixtures of 0081.902 and 0081.901			
75/25	35.1	4.2	9–12
50/50	35.1	4.6	19–24
40/60	35.1	4.6	25–31
30/70	34.8	4.8	49–60
25/75	34.8	4.8	106–125

NTR – No Thermal Runaway.

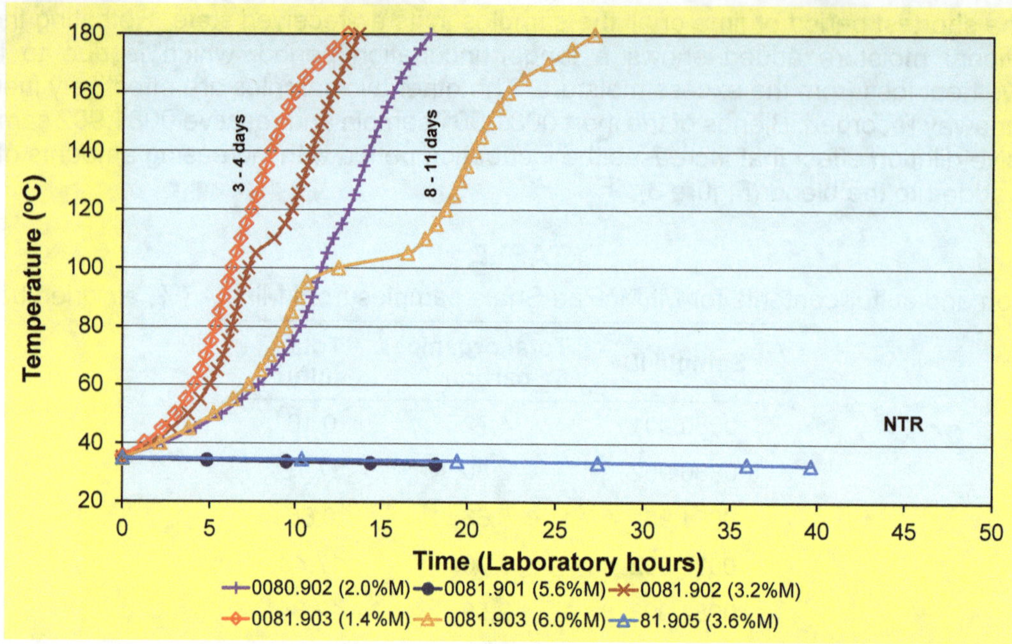

FIG 2 – Incubation test results for Mt McRae Shale samples from Mine A. Note the relative timescale comparing projected minimum incubation period in days for loosely piled material is also shown.

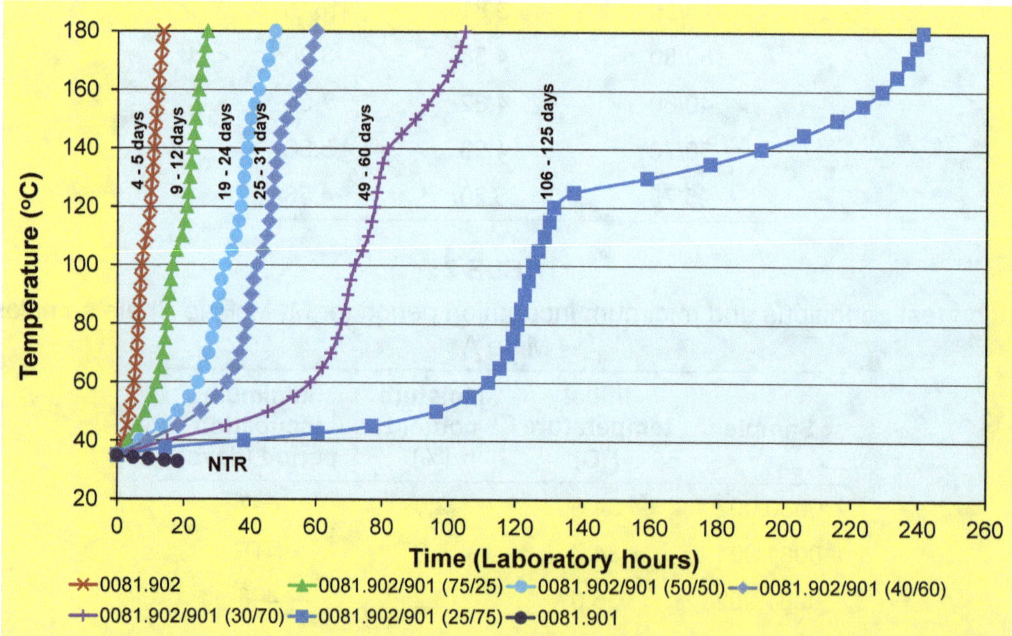

FIG 3 – Incubation test results for Mt McRae Shale sample blends of 0081.902 and 0081.901. Note the relative timescale comparing projected minimum incubation period in days for loosely piled material is also shown.

Additional incubation testing has been conducted on Mt McRae Shale from Mine B. The carbon content and the sulfur content of the samples is contained in Table 3. The initial conditions for each incubation test and the inferred minimum incubation periods are contained in Table 4. Self-heating curves for the individual carbonaceous shale samples and blended samples are shown in Figure 4.

TABLE 3

Carbon and sulfur contents for Mt McRae Shale samples from Mine B (%, air-dried basis).

Sample ID	Total organic carbon	Total sulfur
Composite 1	1.81	1.63
Composite 2	2.51	3.48
Composite 3	3.34	13.76
Composite 4	3.09	7.09
Composite 5	4.25	6.09
Composite 6	3.15	7.54
Composite 7	4.87	8.60
Composite 8	4.68	5.18
Composite 9	2.55	1.72
Composite 3_6 50/50	3.25	10.65
Composite 1_3 75/25	2.19	4.66
Composite 1_3 50/50	2.58	7.70

TABLE 4

Incubation test conditions and minimum incubation period for Mt McRae Shale samples from Mine B.

Sample	Initial temperature (°C)	Moisture content (%)	Minimum incubation period (days)
Composite 3	45.0	2.8	14–18
Composite 6	45.0	2.3	31–39
Composite 3_6 50/50	45.0	2.5	24–31
Composite 1_3 75/25	45.0	2.3	NTR
Composite 1_3 50/50	45.0	2.5	55–67

NTR – No Thermal Runaway.

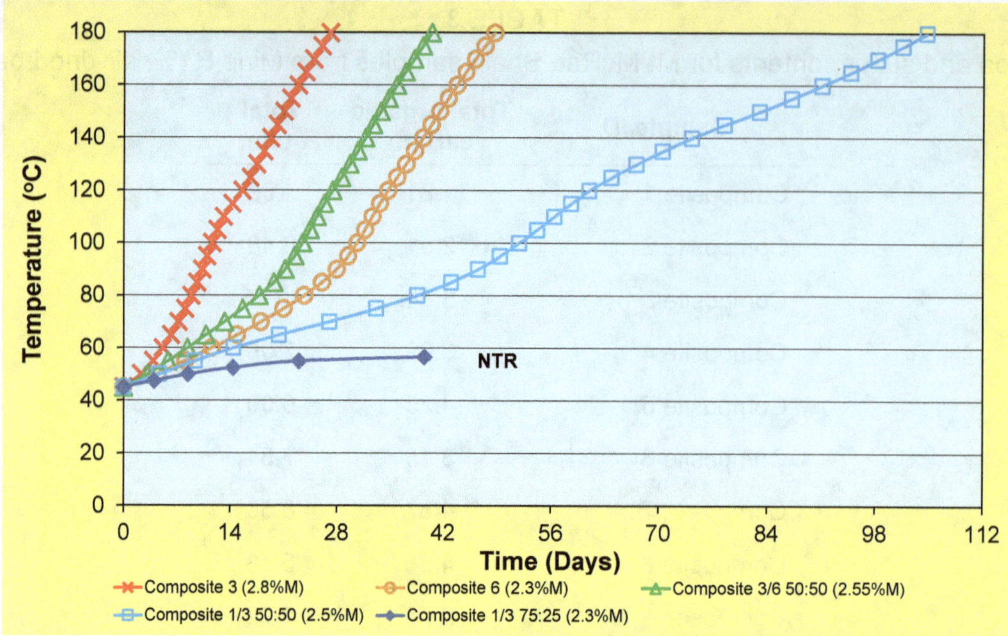

FIG 4 – Incubation test results for Mt McRae Shale samples from Mine B with incubation period shown for a loosely piled state under ideal conditions.

SPONTANEOUS COMBUSTION HAZARD POTENTIAL CLASSIFICATION CHART

The incubation test results show there is a general tendency for the potential of thermal runaway to increase as both the total sulfur and total organic carbon contents of a sample increase, particularly if the sulfur is associated with the presence of reactive pyrite (Beamish *et al,* 2018). Therefore, a plot of total sulfur and total organic carbon can be used as a rapid filter to assess the potential of samples to proceed to thermal runaway. This is shown in Figure 5 for all the samples from Mine A that have been subjected to adiabatic spontaneous combustion incubation testing, with the samples from Mine B added. It would appear that the Mt McRae Shale at Mine B is less reactive than Mine A. Figure 5 shows that for samples with total sulfur values less than 5 per cent no thermal runaway is recorded for the Mine B samples, although there is some elevation in temperature above ambient possible as shown in Figure 4 for the 75/25 composite 1/3 blend. The corresponding cut-off for Mine A is 3 per cent TS. This can be attributed to a fundamental difference in the pyrite morphology at the two sites. The pyrite present in the Mt McRae Shale at Mine A (Figure 6) shows a high proportion of spongy pyrite (indicated by a matt lustre and the absence of visible crystal faces). It is postulated that the pyrite from Mine B contains a lower proportion of reactive spongy pyrite and therefore a higher total concentration of pyrite is required to produce a comparable rate of heat production. The incubation periods shown in Figure 5 also provide an indication that a lower critical pile thickness is required to achieve thermal runaway for samples with short incubation periods.

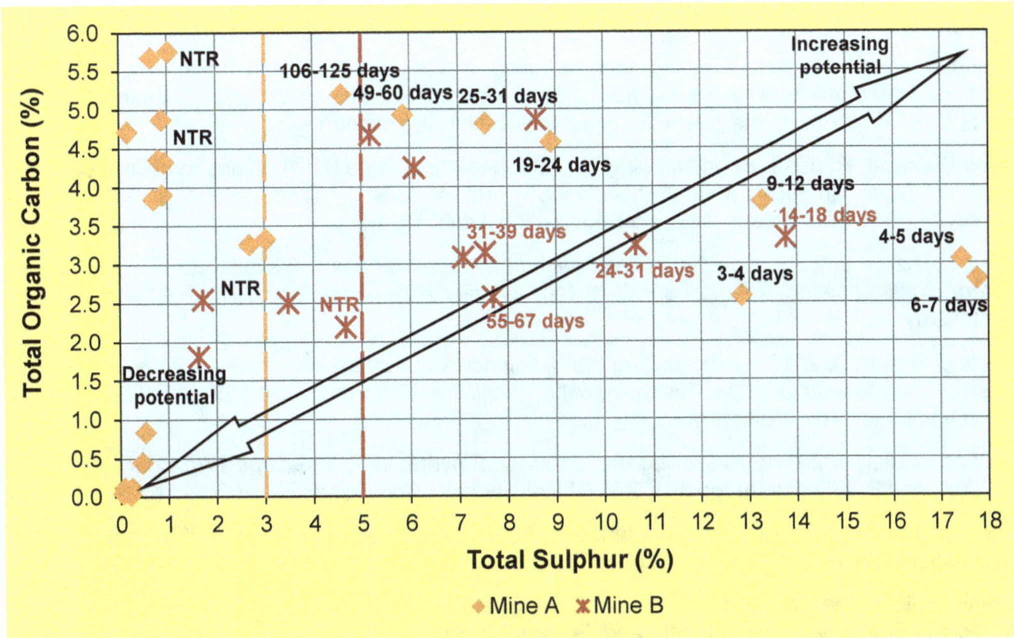

FIG 5 – Spontaneous combustion hazard potential classification chart based on total sulfur and total organic carbon content. TR – Thermal Runaway; NTR – No Thermal Runaway in a practical time frame without the presence of an additional heat source.

FIG 6 – Photographs of core samples from the Mt McRae Shale at Mine A showing lenses and nodules of spongy reactive pyrite oriented parallel to bedding.

CONCLUSIONS

The presence of reactive pyrite in the Mt McRae Shale acts as an initiator of self-heating which increases the temperature of the shale to a point where the organic Carbon that is present can proceed to thermal runaway and subsequent ignition. A systematic sampling and testing program has been conducted across iron ore operations within the Pilbara to characterise this spontaneous combustion hazard. The results from adiabatic incubation testing enables the development of a simple and effective spontaneous combustion hazard potential classification chart which plots the total Sulfur content of the shale against the total organic Carbon content of the shale. Each sample can then be classified as either having the potential to achieve thermal runaway or no thermal runaway is possible in a practical time frame without the presence of an additional heat source. If a sample is classified as able to achieve thermal runaway this can be verified with incubation testing using the mine site boundary conditions that are present, in particular the likely mine site temperature and moisture content of waste rock piles.

ACKNOWLEDGEMENTS

The authors would like to thank Fortescue Ltd and the respective iron ore operations for granting approval to publish these results.

REFERENCES

Beamish, B and Beamish, R, 2011. Testing and sampling requirements for input to spontaneous combustion risk assessment, in *Proceedings of the Australian Mine Ventilation Conference* (eds: B Beamish and D Chalmers), pp 15–21 (The Australasian Institute of Mining and Metallurgy: Melbourne).

Beamish, B and Beamish, R, 2012. Benchmarking coal self-heating using a moist adiabatic oven test, in *Proceedings of the 14th US/North American Mine Ventilation Symposium* (eds: F Calizaya and M G Nelson), pp 423–427 (University of Utah, Department of Mining Engineering, Utah, USA).

Beamish, B B and Theiler, J, 2015. Contrast in self-heating behaviour for coals of similar rank, in *Proceedings 15th Coal Operators' Conference* (ed: N Aziz), pp 300–304, (University of Wollongong and The Australasian Institute of Mining and Metallurgy).

Beamish, B B and Theiler, J, 2017a. Recognising the deficiencies of current spontaneous combustion propensity index parameters, in *Proceedings of the Australian Mine Ventilation Conference*, pp 113–117 (The Australasian Institute of Mining and Metallurgy: Melbourne).

Beamish, B B and Theiler, J, 2017b. Assessing the reactivity of pyrite, in *Proceedings 17th Coal Operators' Conference* (eds: N Aziz and B Kininmonth), pp 391–394 (University of Wollongong).

Beamish, B B and Theiler, J, 2019. Coal spontaneous combustion: Examples of the self-heating incubation process, *International Journal of Coal Geology*, 215:103297.

Beamish, B, Marianelli, P, Theiler, J and Spillman, A, 2018. Spontaneous combustion assessment of McArthur River Mine waste rock, in *Proceedings Mine Waste and Tailings 2018 Conference*, pp 422–426 (The Australasian Institute of Mining and Metallurgy: Melbourne).

Beamish, B, Theiler, J and Ward, C, 2017. Spontaneous combustion behaviour of coal from the Greta seam, in *Proceedings of the 40th Symposium on the Geology of the Sydney Basin*, pp 154–162 (Coalfield Geology Council of New South Wales).

Kakegawa, T, Kawai, H and Ohmoto, H, 1999. Origins of pyrites in the ~2.5 Ga Mt. McRae Shale, the Hammersley District, Western Australia, *Geochimica et Cosmochimica Acta*, 62(19/20):3205–3220.

Miron, Y, Lazzara, C P and Smith, A C, 1992. Cause of floor self-heatings in an underground coal mine, United States Bureau of Mines Report of Investigation RI9415.

Nelson, M I and Chen, X D, 2007. Survey of experimental work on the self-heating and spontaneous combustion of coal, *The Geological Society of America Reviews in Engineering Geology*, 18:31–83.

Optimising growth media ratios for open pit rehabilitation early successes from the Yandi Trial

E Trusz[1], E Forbes[2], B A Johnson[3], K Eldridge[4] and J Fisher[5]

1. Environmental Consultant, Okane Consultants, Brisbane Qld 4000. Email: etrusz@okaneconsultants.com
2. Senior Environmental Geologist, Okane Consultants, Osborne Park WA 6017. Email: eforbes@okaneconsultants.com
3. Senior Environmental Scientist, Okane Consultants New Zealand Office, Lower Hutt Wellington 5040, New Zealand. Email: bjohnson@okaneconsultants.com
4. Specialist Site Rehabilitation, BHP, Newman WA 6753. Email: kelsey.eldridge@bhp.com
5. Superintendent Closure Planning and Rehabilitation, BHP, Perth WA 6000. Email: jonathan.fisher@bhp.com

ABSTRACT

The requirement to achieve safe, stable, and self-sustaining post-mining landscapes drives innovation in the iron ore sector, particularly regarding rehabilitation and closure. This paper presents the initial findings from the Yandi Growth Media Field Trial, known by the project team as 'Project Grow'. This research initiative is designed by Okane Consultants, to inform closure planning at the BHP Western Australia Iron Ore (WAIO) Yandi operation. Limited availability of topsoil is a common issue for rehabilitation and mine closure and creates the need for alternative growth media options. Project Grow's overarching objective is to determine preferred growth media and rehabilitation methods for establishing vegetation communities at Yandi.

The Project Grow field trial is a controlled, replicated study using 12 plots, each measuring 38 m by 38 m. The trial is specifically designed to test the performance of alternative growth media options for post-mining rehabilitation. The trial employs a factorial design that investigates the combined effects of five growth media ratios, two distinct seed mixes, and variable fertiliser treatments. The trial design centres on investigating various combinations of topsoil and Channel Iron Deposit (CID) waste material to test growth media efficacy. The two seed mixes used locally sourced native species and comprised: Mix 1 Grassland species (*Triodia* wiseana and *T. pungens*) and Mix 2 Grassland and Shrubland species (the two Triodia species plus key woody perennial species including Acacia, Eucalyptus, Grevillea, Hakea and Senna). The 0.2 m deep growth media plots, which were established using the hand broadcasting seeding technique, provide a robust platform to directly compare the efficacy of different materials blends available on-site, against ecological completion criteria.

Initial monitoring was conducted four months (April 2025) and one year (December 2025) after seeding, confirmed excellent overall germination and robust establishment across the trial. The results underscore the critical role of topsoil in supporting early ecological recovery, with the beneficial effect clearly evident in the mixed material blends, although diminishing as the proportion of topsoil is reduced. Blends between 100 per cent and approximately 60 per cent topsoil showed acceptable initial vegetation establishment, indicating the CID material as a viable growth medium alternative up to a proportion of approximately 40 per cent. Assessments were made regarding fertiliser application, impact of a residual native seed bank from the topsoil, and observations of colonising fauna (birds, bees, and lizards), providing encouraging early indicators of restored ecological function.

The ongoing monitoring program for Project Grow will continue to track the long-term resilience and viability of these established vegetation communities. This research provides novel, evidence-based data directly quantifying the environmental and economic trade-offs of using alternative growth media, generating a scalable best-practice approach critical for sustainable closure of future open pit iron ore landforms.

INTRODUCTION

As the global footprint of open cut mining continues to expand, the inherent scarcity of high-quality topsoil has emerged as a limiting factor in achieving functional and self-sustaining post-mining landscapes (OQMRC, 2024). In many jurisdictions, the volume of salvaged topsoil is insufficient to cover the vast areas disturbed by extraction, particularly in arid regions like the Pilbara of Western Australia where soil horizons are naturally thin and nutrient deficient. This global topsoil deficit drives an urgent requirement for alternative growth media options that can supplement or substitute natural soil without compromising ecological outcomes. At the BHP Western Australia Iron Ore Yandi operation, this challenge is being addressed through the investigation of blended substrates that combine available topsoil with lithic materials. By identifying the optimal ratios of these materials, it is possible to engineer a growth medium that supports keystone species and restores ecosystem services, even when primary topsoil resources are severely constrained.

To address these challenges in practice, the Yandi Growth Media Field Trial—known as 'Project Grow'—was established as a series of replicated vegetation plots. This trial is specifically designed by Okane Consultants to evaluate the potential of alternative growth media to meet Yandi's overarching closure strategy: the creation of self-sustaining vegetation communities that are comparable to the surrounding undisturbed natural environment. By rigorously testing these growth media materials, the initiative aims to provide the evidence-based data required to inform long-term closure planning. Subsequently, the key objectives of Project Grow are as follows:

- Determine if Channel Iron Deposit (CID) waste rock material, combined with topsoil at varying ratios, can provide a growth medium capable of supporting successful and sustainable vegetation establishment, that is comparable to an undisturbed natural environment.
- Evaluate the viability and germinability of native vegetation keystone species.
- Provide an embedded research trial for growth media that will inform the Yandi closure landform design development.

METHOD

Trial construction

The field trial consists of 12 plots, with dimensions of 38 m by 38 m, at 0.2 m in height, and with an 8 m minimum buffer between each plot (Figure 1). All plots (#1 to #12) include specifications for a growth media and various treatment options. The trial tested three key variables:

1. Growth media (topsoil, topsoil mixed with CID material, and CID material alone).
2. Seed mix (Grassland and Shrubland).
3. Fertiliser application.

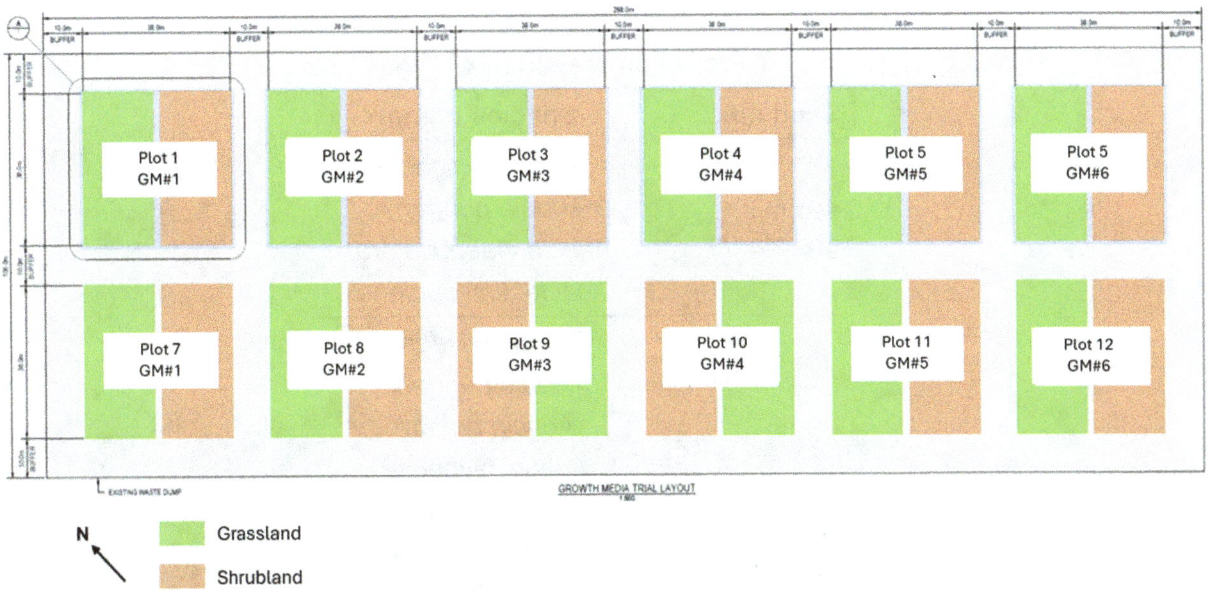

FIG 1 – Field trial plot layout (as-built).

Establishment of the trial plots took place over two phases: construction and seeding. Six growth media options were prepared for the field trial. The growth media trial location was constructed with a CID material base layer to replicate typical waste rock conditions of rehabilitation areas at the site. Grading was initially conducted to remove larger rocks from the plot surface and fill any large depressions. Growth media was prepared from topsoil and CID material stockpiles, with the desired ratios of topsoil and CID material achieved by mechanically mixing of the two material types using a dozer and an excavator. The growth media were spread on the demarcated growth plot areas, and the surface of each plot was levelled to a thickness of approximately 0.2 m above the existing ground level.

The growth media options were designed to allow for comparison between types and ratios of growth media material inputs in relation to the success of sustainable vegetation growth and viability of keystone plant species. The six growth media options are:

- Growth Media #1 – 100 per cent Topsoil.
- Growth Media #2 – 80 per cent Topsoil and 20 per cent CID material.
- Growth Media #3 – 60 per cent Topsoil and 40 per cent CID material.
- Growth Media #4 – 20 per cent Topsoil and 80 per cent CID material.
- Growth Media #5 – 100 per cent CID material.
- Growth Media #6 – 100 per cent CID material with fertiliser.

A diverse range of locally native and functionally diverse plant species were selected for the field trial. Two seed mixes, representative of the main vegetation communities of the Pilbara region, were prepared: 1) Grassland and 2) Grass and Shrubs (Shrubland) (Table 1). Each plot was divided into two halves, with the Grassland mix sown on the north-western half of the plot and the Shrubland mix sown on the south-eastern half (excluding Plots #9 and #10 which were sown on the opposite sides). To ensure even distribution across each plot half, the relatively small volume of seed required was amended with perlite as a bulking agent. This increased the total volume of the mixture, thereby facilitating a more uniform application of the diverse seed mixes. Hand broadcasting methods were used to disperse the seeds and plots were then scarified by a Polaris with harrows attached, preparing the ground and increasing soil contact with seeds.

TABLE 1

Project grow seed mix.

Seed mix	Species name
Seed mix 1 – grassland	*Triodia wiseana* *Triodia pungens* *Acacia inaequilatera* *Acacia dictyophleba*
Seed mix 2 – grass and shrubs (shrubland)	*Triodia wiseana* *Triodia pungens* *Acacia aneura* *Acacia bivenosa* *Acacia pyrifolia* *Eucalyptus gamophylla* *Eucalyptus leucophloia* *Grevillea wickhamii* *Hakea lorea* *Senna artemisioides* *Senna pleurocarpa*

Each growth media option was run in duplicate, resulting in a total of 12 plots. Among these 12 plots, fertiliser treatments were applied to Plots #6 and #12 following seeding. Two fertiliser products were used:

1. Soil stabilising agent (gypsum) – applied using a skid steer at a rate of 1 kg per square metre.
2. A mixture of NPK (nitrogen, phosphorous, and potassium) and micronutrient fertiliser (Osmocote All Purpose Controlled Release Fertiliser) – applied by hand broadcast method at a rate of 0.1 kg per square metre.

Material sampling and testing

Two samples of each growth media option trialled (ten samples in total) were collected and tested to determine the baseline geochemical and physical properties of the material used in the construction. Samples were taken directly from growth media stockpiles using hand and shovel method. Laboratory testing results showed the baseline conditions of the topsoil and CID material posed no immediate risk of acidity, salinity or elevated metal concentrations that would adversely affect plant growth and the success of the growth media trial plots.

Trial monitoring

A comprehensive monitoring program was implemented to evaluate trial performance, integrating qualitative observations with quantitative vegetation monitoring. This dual-approach methodology ensured the systematic collection of the following data:

- General vegetation condition and health.
- Erosional features and soil condition.
- Evidence of surface water ponding.
- Presence of coarse woody debris.
- Presence or evidence of fauna/fauna habitat.
- Presence of weed species.

- Percentage of vegetation cover.
- Percentage of dead leaf litter cover.
- Species richness – data was collected by counting the number of different vegetation species from each of the seed mixes at each trial plot.
- Species count – survey quadrat level.

Drone surveys were recommended to provide high resolution aerial imagery to support analysis and comparison of vegetation growth over the plot area over time.

RESULTS AND DISCUSSION

Vegetation monitoring was completed four months post seeding (April 2025) to capture vegetation trends following germination, and again 12 months following seeding (December 2025) to assess vegetation performance across the trial plots during summer months. To facilitate a direct temporal comparison with the 2025 data set, a continued monitoring program has been proposed. This program includes an assessment following the wet season in May 2026 to capture peak biomass and germination, and a subsequent survey in December 2026 to evaluate vegetation resilience during the summer months.

April 2025 monitoring

Vegetation cover

The vegetation monitoring undertaken on the field trial revealed clear differences in germination success across the various growth media substrates (Figure 2). Topsoil outperformed CID material as a growth media for vegetation germination and establishment. The highest vegetation cover (~20 per cent) was observed in the Grassland seed mix section of Plot 1, grown in 100 per cent topsoil, followed closely by other plots with either 100 per cent topsoil or higher proportions of topsoil mixed with smaller amounts of CID material (eg Plot 7, Plot 2, Plot 8). In contrast, vegetation cover decreased progressively as the proportion of CID material in growth media increased. Plots containing 100 per cent CID material or CID material with fertiliser (Plots 5, 11, 6, and 12) exhibited the lowest vegetation cover (~2 per cent or less). This trend was consistent for both the Grassland and Shrubland seed mixes.

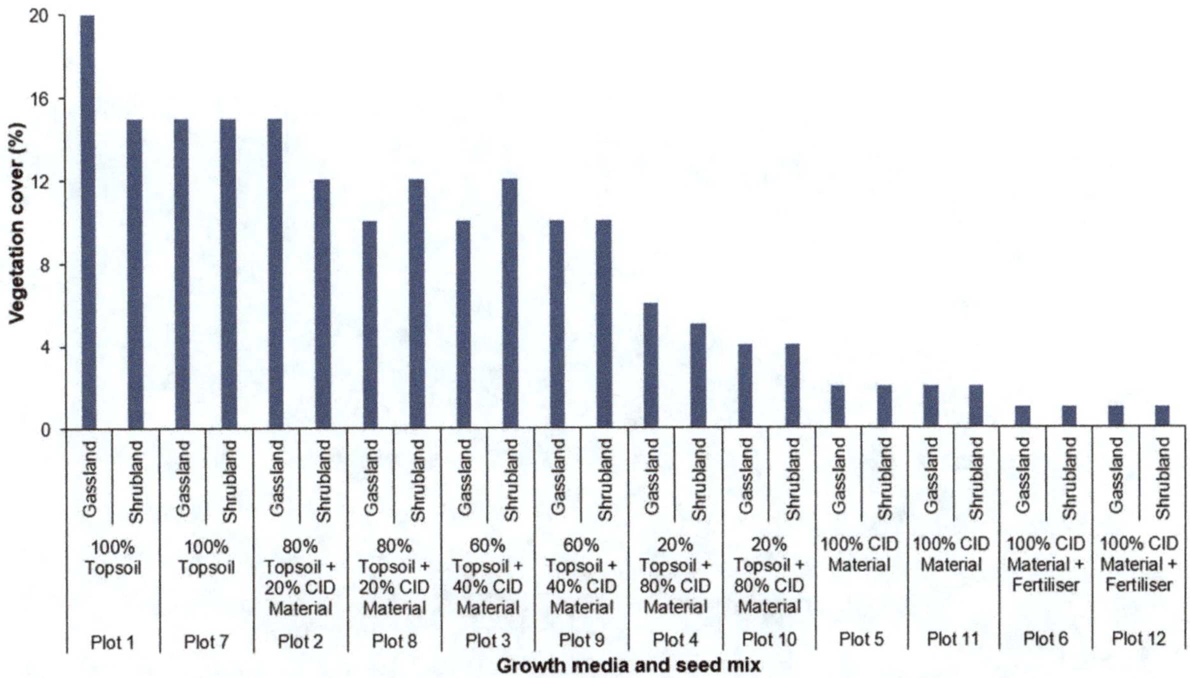

FIG 2 – Percentage of vegetation cover observed in each growth media trial plot during post germination monitoring. Data collected 15 and 16 April 2025.

The initial performance of the fertilised 100 per cent CID material plots provides a valuable opportunity to refine nutrient application strategies for alternative growth media. Observations of remnant fertiliser approximately four months after application suggest that the concentrated presence of gypsum and NPK products may have created a nutrient-dense environment during the sensitive germination phase. While intended to boost growth, these localised concentrations can influence seedling establishment through natural processes such as osmotic regulation or pH adjustment, particularly in the low-moisture conditions characteristic of the Pilbara (Haden, Xiang and Peng, 2011; Gashi *et al,* 2025). This finding is an encouraging indicator for future trials, suggesting that further optimising the integration, distribution and timing of soil amendments could unlock the full potential of CID material as a standalone growth medium.

A large number of plant species, not included in the original seed mix, were observed across the plots in both Grassland and Shrubland seed mix sections of the trial plots. These plants are believed to have dispersed and germinated either through natural wind dispersal or were present as dormant seeds within the topsoil. This unplanned recruitment suggests that topsoil not only supports seeded species but also facilitates the return of local biodiversity.

Aerial imagery showing a comparison of the trial plots is presented in Figures 3 and 4.

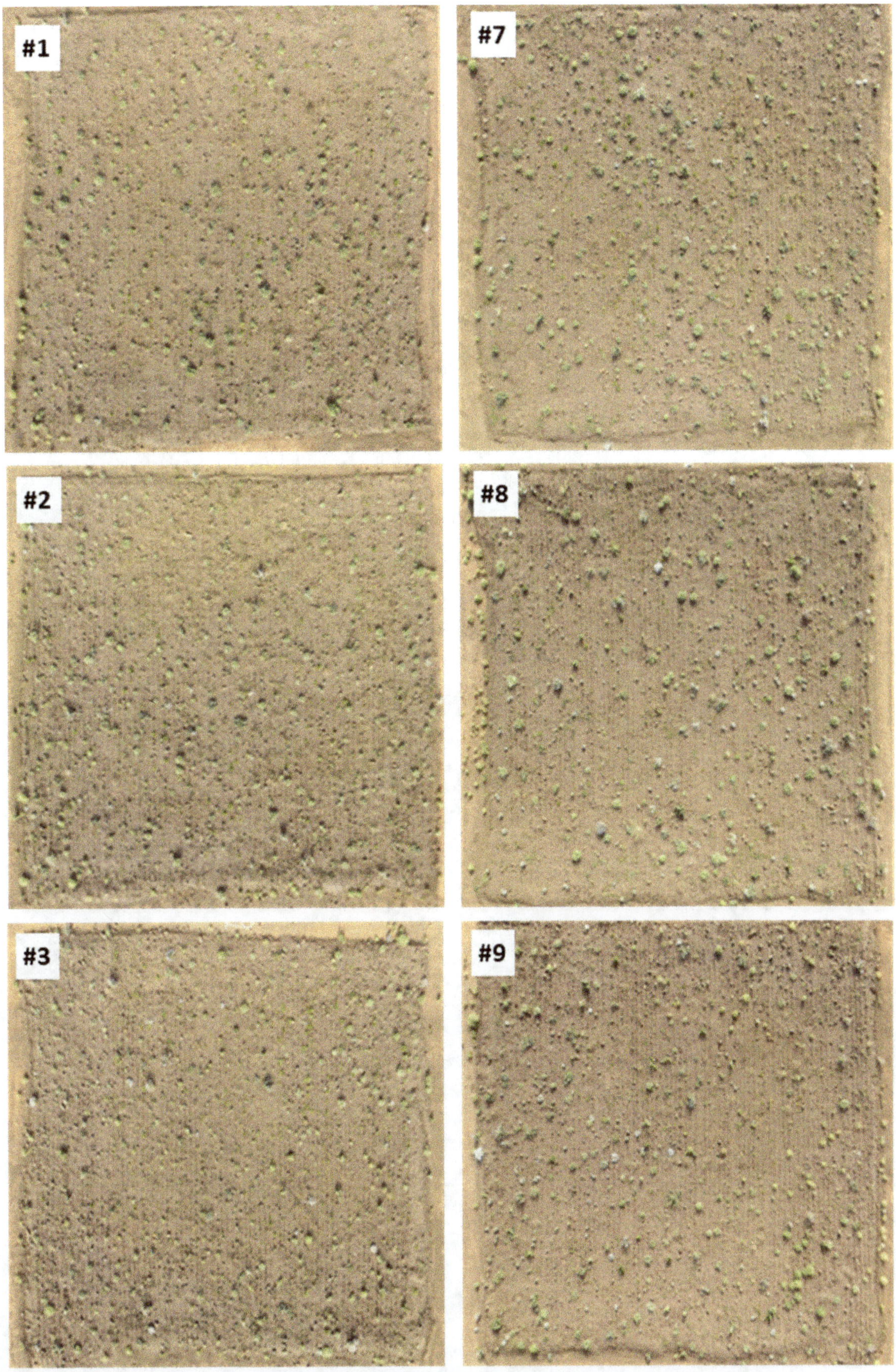

FIG 3 – Vegetation cover – Plots 1 and 7 (100 per cent Topsoil), 2 and 8 (80 per cent Topsoil; 20 per cent CID material), and 3 and 9 (60 per cent Topsoil; 40 per cent CID material). Aerial imagery collected April 2025.

FIG 4 – Vegetation cover – Plots 4 and 10 (20 per cent Topsoil; 80 per cent CID material), 5 and 1 (100 per cent CID material, and 6 and 12 (100 per cent CID material with fertiliser). Aerial imagery collected April 2025.

Vegetation species richness

Overall, the Shrubland seed mix section of the plots with higher proportions of topsoil than CID material exhibited the greatest vegetation species richness, with the exception of Plot 9 in which Grassland exhibited highest species richness (Figure 5). At this early stage of the trial, the specific drivers for this anomaly remain unconfirmed; however, it may reflect localised variations in micro-site conditions or water availability in Plot 9. Approximately 20 species were recorded in plots containing topsoil, even though only four species were seeded in the Grassland mix and 11 in the Shrubland mix. Among the species, *Triodia spp.*, *Acacia spp. Corchorus spp.*, *Ptilotus spp.* and *Senna spp.* were the most dominant across the trial plots.

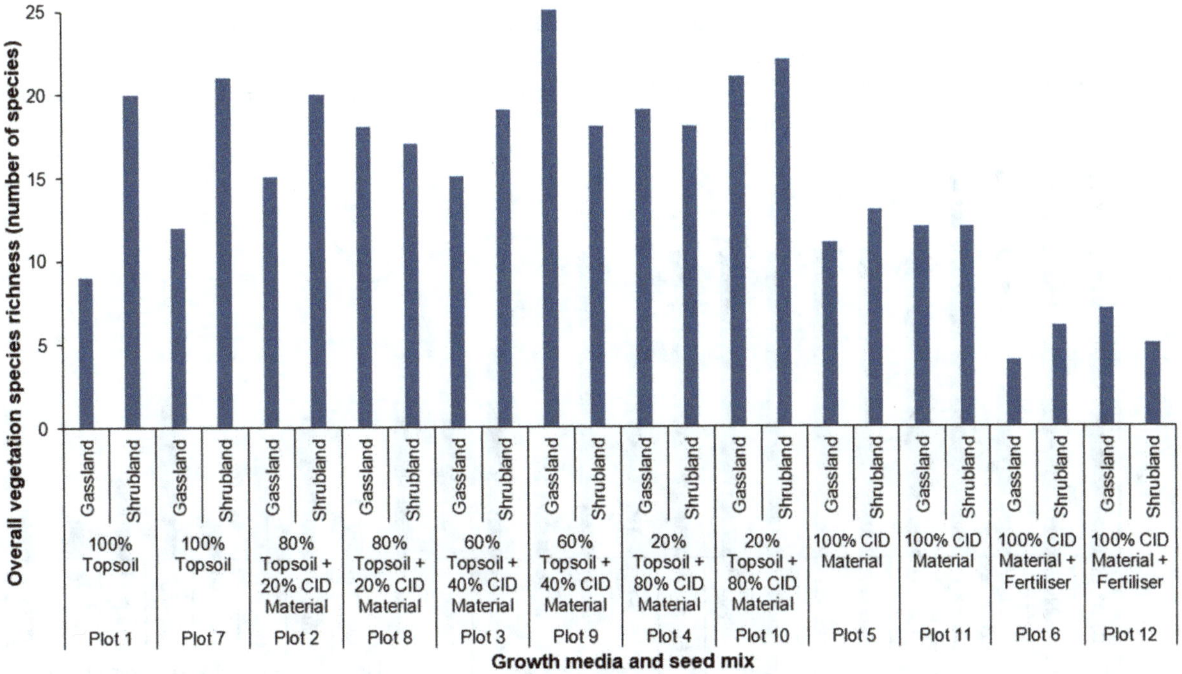

FIG 5 – Overall vegetation species richness observed in each growth media trial plot during post germination monitoring. Data collected 15 and 16 April 2025.

Plots which contained a mixed ratio of topsoil and CID material (Plot 2, 8, 3, 9, 4, and 10) still showed levels of species richness comparable to the plots containing only topsoil, suggesting that adequate richness could still be achievable even in a mixed material growth medium. In contrast, the lowest species richness was observed in plots consisting of 100 per cent CID material with added fertiliser (Plots 6 and 12).

Presence of weeds

Buffel grass (*Cenchrus ciliaris*) was the main weed species consistently observed across the different growth media plots, indicating its high adaptability and resilience under varying substrate conditions. Buffel grass was observed in Plots 2, 3, 4 and 7. Its presence suggests it may have been introduced through external means such as wind dispersal or was already present in the original topsoil seed bank.

Additional observations

In addition to vegetation establishment, early signs of fauna colonisation were observed. Native birds, bees, and lizards were seen utilising the trial plots, particularly those plots with more developed vegetation cover. The presence of fauna suggests that the trial areas are beginning to support broader ecosystem functions, such as habitat use and pollination.

December 2025 monitoring

Vegetation cover

Trends in the vegetation cover observed in December of 2025 are comparable to the April 2025 monitoring. Topsoil outperformed CID material as a growth media for vegetation cover (Figure 6). The highest vegetation cover (20 per cent) was seen in the plots containing 100 per cent topsoil (Plots 1 and 7). The vegetation cover decreased as the proportion of CID material increased. Plots containing 100 per cent CID material or CID material with fertiliser (Plots 5, 11, 6 and 12) showed the lowest vegetation coverage (<10 per cent). Remnant fertiliser was still clearly visible in trial plots during the survey, 12 months after application, suggesting that concentrated pockets may have inhibited seedling establishment and long-term vegetation growth. This trend was consistent across both Grassland and Shrubland seed mixes. Plots that contained some topsoil (60 per cent topsoil and 40 per cent CID material) still showed moderate levels of vegetation coverage (10–15 per cent).

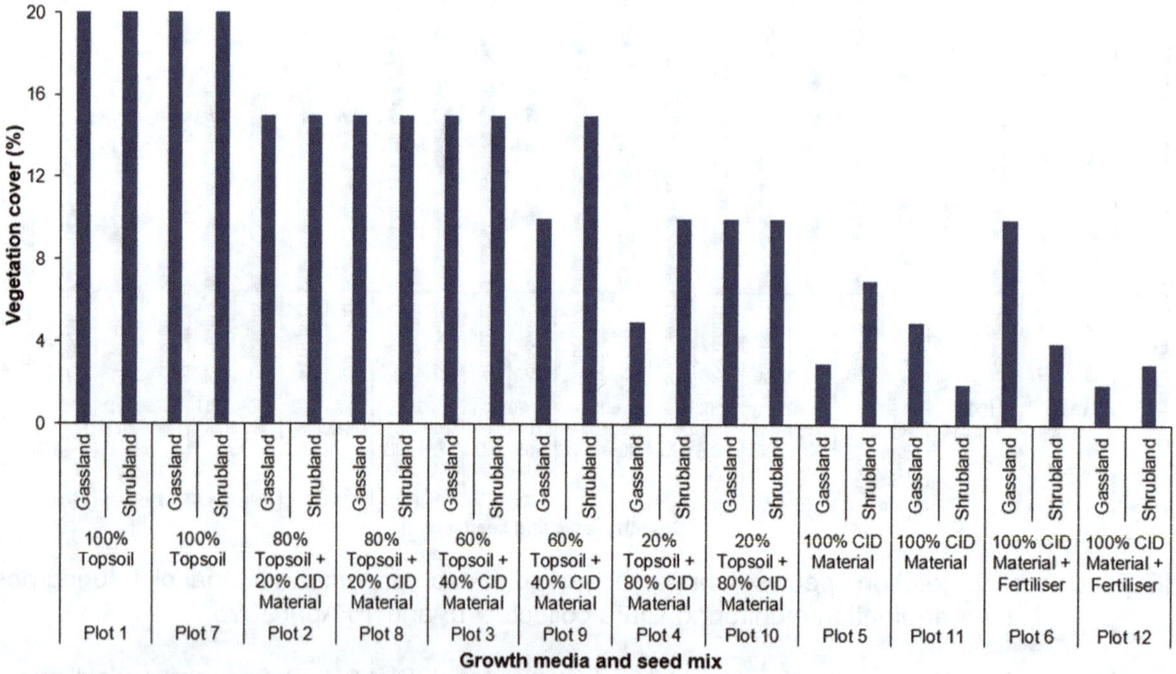

FIG 6 – Percentage of vegetation cover observed in each growth media trial plot during post germination monitoring. Data collected 15 and 16 December 2025.

Aerial imagery of the plots is shown in Figure 7. Hot and dry conditions during the summer are evident with reduced vegetation coverage across all plots.

FIG 7 – Vegetation cover overview – all trial plots. Aerial imagery collected December 2025.

The percentage of leaf litter cover in the topsoil plots was comparable to that of the CID material plots. The exception being Plot #7 (100 per cent topsoil, Shrubland seed mix section) and Plot #8 (80 per cent topsoil and 20 per cent CID material, Grass seed mix section), which showed the highest percentage of leaf litter coverage (15 per cent and 20 per cent, respectively). This is likely a result of a greater number of annual plant individuals within these plots dying back over the summer months.

Vegetation species richness

The vegetation monitoring conducted in December 2025 revealed that species richness remained highest in plots with a high proportion of topsoil, while also highlighting an interesting secondary peak in richness within high-ratio CID material blends (Figure 8).

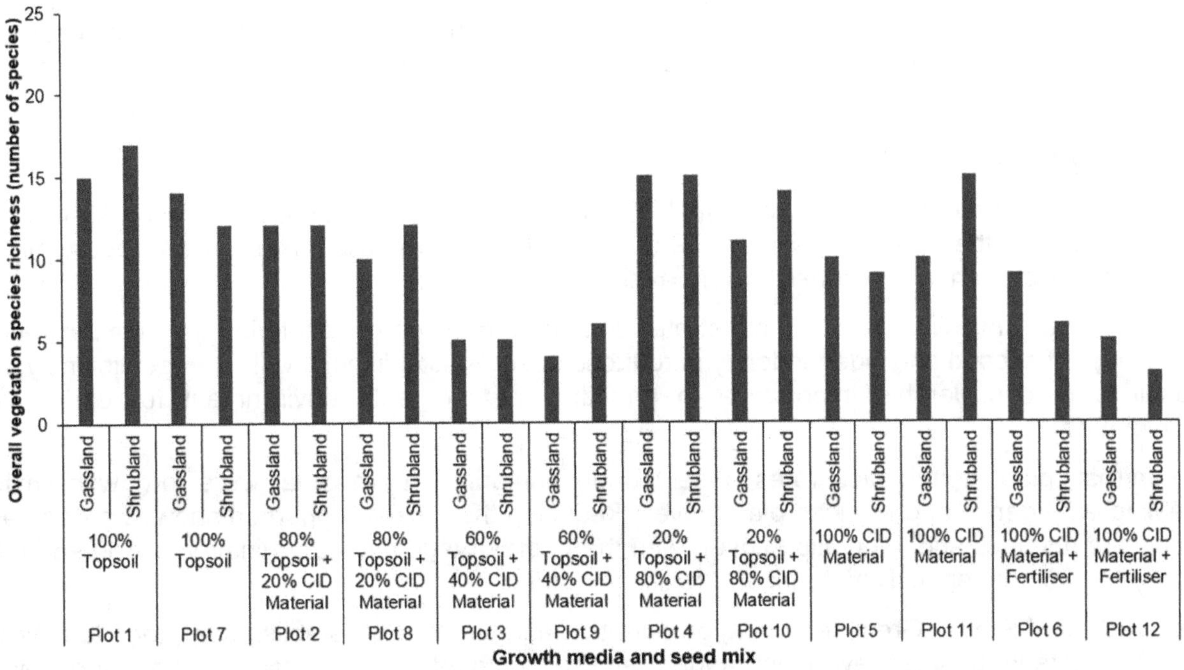

FIG 8 – Overall vegetation species richness observed in each growth media trial plot during post germination monitoring. Data collected 15 and 16 December 2025.

Plots 1 and 7, consisting of 100 per cent topsoil, supported the greatest species richness, with up to 17 species recorded in the Shrubland seed mix section. A general trend of declining richness was observed as topsoil content decreased from 100 per cent to 60 per cent. Notably, the 60 per cent topsoil and 40 per cent CID material blends (Plots 3 and 9) exhibited a distinct dip in richness, recording as few as four to six species.

Conversely, a resurgence in richness was observed in Plots 4 and 10, which contained only 20 per cent topsoil and 80 per cent CID material. These plots supported between 11 and 15 species, outperforming the 60:40 blends and reaching richness levels comparable to those of the 80 per cent and 100 per cent topsoil treatments.

In the plots lacking topsoil entirely, species richness was significantly lower. The 100 per cent CID material plots without fertiliser (Plots 5 and 11) recorded moderate species richness (between nine and 15 species), whereas the addition of fertiliser to the same substrate (Plots 6 and 12) resulted in the lowest species richness across the trial, with only three to nine species surviving by the December monitoring. Across nearly all growth media types, the Shrubland seed mix maintained higher species counts than the 'Grassland' only mix.

Photo-based assessments identified native tussock-form grasses and shrub or forb taxa consistent with the original Project Grow seed mix, including native grasses (Poaceae) and *Ptilotus* spp. Species-level identification was not always possible from photographs alone and was conservatively reported at genus or functional group level. Species-level identification will be completed as part of monitoring in mid-2026, following the wet season when seed heads and flowers are more present.

Presence of weeds

Buffel Grass (*Cenchrus ciliaris*) was observed across four of the plots (Plots 1, 2 6 and 12). The occurrence of Buffel Grass in 100 per cent CID material plots indicates that the topsoil seed bank is not the sole source. The presence of Buffel grass does not appear to correlate specifically with growth media substrate.

Additional observations

Similar to the April 2025 monitoring, various fauna species were observed using the plots. Ants, bees and lizards were observed, particularly in those plots with more vegetation cover (Plots 1, 2 7 and 8). There was also evidence of cattle (droppings) at Plot 11. No surface water ponding was observed on any of the trial plots. No evidence of major erosion features (gullies, channels) on any of the plots was observed.

CONCLUSION

Vegetation species richness was strongly linked to the presence of topsoil. However, trial results thus far support the use of topsoil and CID material growth media mixes as an opportunity to maximise the efficient use of topsoil where availability is limited.

The trial plots containing topsoil demonstrated a larger proportion of vegetation germination, with a wide range of seeded and independently introduced species establishing well. The results imply that topsoil is of considerable importance in providing a favourable environment for early plant development.

In contrast, plant germination rates observed on primarily CID material were low, with limited seedling emergence and reduced plant cover. Reducing the topsoil proportion showed a decrease in species richness and vegetation cover, with the poorest vegetation germination observed in the plots with 100 per cent CID material and fertilisers added.

Early results of Project Grow, regarding germination and vegetation establishment, tend to indicate that CID material by itself, even with added fertiliser, presents the poorest conditions for growth, particularly during the critical germination phase. However, some plots which contained a mixed ratio of topsoil and CID material exhibited a level of species richness comparable to the plots containing only topsoil, suggesting that adequate species richness and vegetation growth is achievable in a mixed growth media.

The presence of fauna suggests that the trial area is beginning to support broader ecosystem functions, such as pollination and habitat use. These early indicators of ecological recovery are encouraging for future rehabilitation planning and demonstrate the importance of suitable growth media in supporting both flora and fauna in post-mining landscapes.

Ultimately, the findings from Project Grow provide a critical evidence-based framework for addressing the global challenge of topsoil scarcity in large-scale mining. By demonstrating that specific blends of topsoil and CID material can support diverse vegetation communities and early faunal colonisation, this research indicates a scalable solution for creating self-sustaining landscapes that are comparable to the surrounding undisturbed environment. These results directly inform the long-term closure planning at Yandi, ensuring that the primary limiting factor of growth media availability is transformed into a manageable strategic asset. As the trial progresses through the 2026 monitoring cycle, it will continue to bridge the gap between theoretical rehabilitation goals and the practical delivery of resilient, functional ecosystems in the Pilbara.

ACKNOWLEDGEMENTS

The authors acknowledge the Banjima people, who are the traditional custodians of the land this research took place on. We would like to pay our respects to the Banjima Elders past and present. We thank the BHP WAIO Yandi Site Rehabilitation Team for their expertise and assistance in supporting this project. We extend our appreciation to all reviewers for their input into this manuscript.

REFERENCES

Gashi, N, Szőke, Z, Czakó, A, Fauszt, P, Dávid, P, Mikolás, M, Stündl, L, Gál, F, Remenyik, J, Sándor, Z and Paholcsek, M, 2025. Gypsum and Tillage Practices for Combating Soil Salinity and Enhancing Crop Productivity, *Agriculture*, 15(6):658. https://doi.org/10.3390/agriculture15060658

Haden, V R, Xiang, J and Peng, S, 2011. Relative effects of ammonia and nitrite on the germination and early growth of aerobic rice, *Journal of Plant Nutrition and Soil Science*, 174(2):292–300. https://doi.org/10.1002/jpln.201000222

Office of the Queensland Mine Rehabilitation Commissioner (OQMRC), 2024. Review of techniques to address topsoil deficit in open cut coal mines under rehabilitation in Queensland [online]. Available from <https://www.qmrc.qld.gov.au/__data/assets/pdf_file/0019/336043/review-techniques-address-topsoil-deficit-coal-mines-qld.pdf> [Accessed: 23 October 2025].

REFERENCES

[illegible]

[illegible]

[illegible]

Iron Ore

Iron and steel-making

Slag design principles for the NeoSmelt electric smelting furnace

S J Chew[1] *and R M Brell*[2]

1. Principal Technology and Development Engineer, BlueScope Steel, Port Kembla NSW 2505. Email: sheng.chew@bluescopesteel.com
2. Process and Technology Manager, NeoSmelt, Perth WA 6000. Email: ruwan.brell@bluescopesteel.com

ABSTRACT

Project NeoSmelt represents a major collaborative effort between BlueScope, BHP, Rio Tinto, Woodside Energy, and Mitsui Iron Ore Development to develop Australia's largest ironmaking Electric Smelting Furnace (ESF) pilot plant. The initiative aims to demonstrate the technical feasibility of a Direct Reduced Iron (DRI)–ESF process using predominantly Pilbara iron ores, offering a potential lower emissions alternative to traditional blast furnace routes. Supported by $19.8 million in Australian Renewable Energy Agency (ARENA) funding for the front-end engineering phase (FEED), the project is progressing through FEED, with a final investment decision anticipated in 2026 and pilot operations targeted for 2028 in Western Australia's Kwinana Industrial Area.

The NeoSmelt pilot plant is designed to produce 30 000–40 000 t of molten iron annually. Early operations will rely on natural gas for ore reduction, transitioning to lower carbon-intensity hydrogen as the process matures. Success would provide a significant pathway to decarbonising steelmaking while continuing to leverage Australia's substantial Pilbara ore resources, which contribute 38 per cent of global iron ore supply. For BlueScope, NeoSmelt aligns with long-term planning for Port Kembla Steelworks, where the reline of No. 6 blast furnace will maintain sovereign flat steel production while alternative low emissions technologies are developed and piloted.

A central technical challenge for ESF viability is slag design, which critically influences heat transfer, feed dissolution, reaction kinetics, refractory interaction, and overall furnace stability. Slag properties including electrical and thermal conductivity, density, viscosity, liquidus temperature, solidification behaviour, interfacial tension, and chemical activity interact in complex and often competing ways. For example, slag resistivity enhances electrical heating but increases viscosity; flux additions reduce viscosity but alter density and settling behaviour; higher operating temperatures improve smelting performance but elevate the risk of refractory wear. Understanding these trade-offs is essential for optimising both furnace performance and campaign life. Systematic technical work from laboratory to pilot scale is required to address both slag fundamentals for the Pilbara system and the realities of industrial slags containing minor elements, entrained solids, gases, and metal droplets.

INTRODUCTION

The successful deployment of Electric Smelting Furnace (ESF) technology within Project NeoSmelt depends critically on developing a robust and adaptable slag design that can accommodate the complexities of processing Pilbara-based direct reduced iron (DRI) feeds at industrial scale. As feed delivery and electrical heating input to the ESF occur entirely through the slag phase, the slag must simultaneously enable efficient heat transport, feed dissolution, chemical reaction, and protection of furnace refractories, while supporting operational objectives related to productivity, product quality, efficiency, and slag reuse, as represented in Figure 1. Achieving these outcomes requires navigating a series of interdependent and frequently competing material properties including electrical and thermal conductivity, viscosity, density, liquidus temperature and solidification behaviour, interfacial tension, and chemical activity, each of which strongly influences overall furnace performance. The interplay of these properties defines the operating window for the NeoSmelt system and underscores the need for a comprehensive, multi-disciplinary technical program to guide slag design from laboratory scale through to pilot operations.

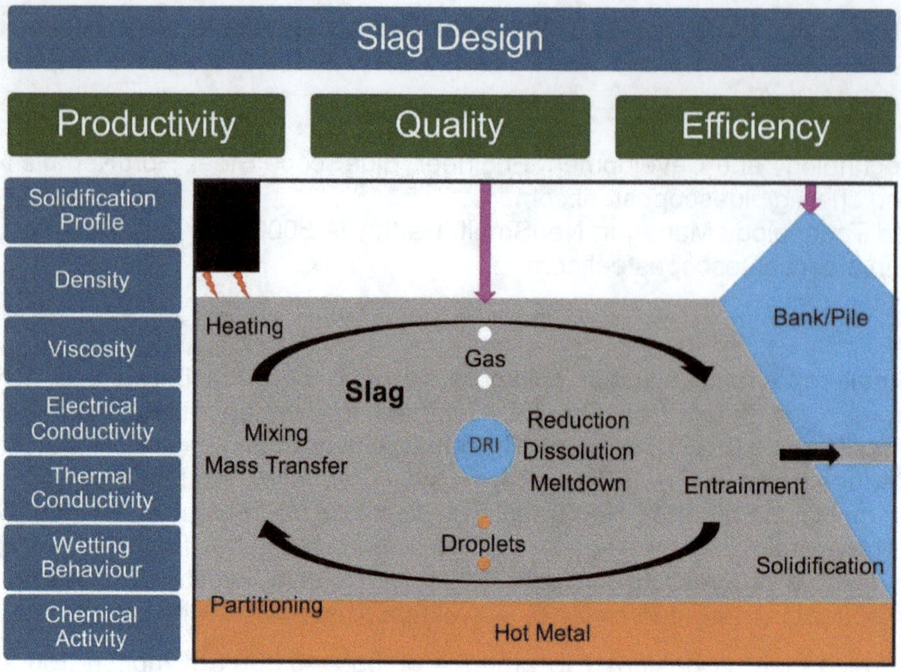

FIG 1 – Schematic of ESF slag design factors.

DISCUSSION

Productivity in the ESF is governed by the rate of hot metal generation for a given furnace geometry and power input, both of which are influenced by slag characteristics affecting heat distribution, feed meltdown, reaction kinetics, and slag–metal separation. Product quality, particularly with respect to hot metal carbon and silicon content and slag chemistry such as FeO levels, is similarly dependent on slag–metal equilibrium, mixing, reduction pathways, and the amount of entrained droplets or undissolved feed retained in the slag. Process efficiency adds further constraints, as the slag influences electrical energy consumption, carbon utilisation, flux requirements, electrode wear, and refractory campaign life. Consequently, slag design must strike a balance across these metrics, optimising the system holistically rather than for any single objective.

Feed and power input pathways illustrate the centrality of slag behaviour. Solid feed particles may enter the slag directly or accumulate as a pile on the slag surface, where local temperature, slag composition, and dissolution kinetics determine the rate of meltdown. Electrical power is delivered either by direct resistance heating within the slag or by arcing between the electrode and slag surface. The resulting convective flow transports heat, dissolved species, entrained gases and solids, and metal droplets between the electrode zone, the slag surface, near-wall region, and the slag–metal boundary. Heat may also be transferred via conduction through the slag, guided by its thermal conductivity and structural state. Reaction pathways including the carburisation of solid iron and hot metal droplets, dissolution of gangue, ash and flux, and reduction of FeO, Si, Mn, and P are strongly dependent on the interfacial area available between slag and feed materials as well as the local chemical environment, which may vary widely across the furnace.

Each major slag property exerts multiple and sometimes opposing influences on ESF performance. Electrical resistivity determines the maximum achievable heat input via immersed electrodes, making it a key design variable for maintaining sufficient energy flux. Resistivity is generally increased by acidic gangue components typical of Pilbara ores but reduced when fluxes are added to modify viscosity, chemical activity or solidification profile. Higher resistivity can enhance electrical heating but may correlate with increased viscosity, complicating effective heat distribution through mass transfer. Thermal conductivity typically increases alongside resistivity, adding another layer of interdependence that must be considered within the furnace energy balance.

Density and viscosity further shape feed behaviour, droplet separation, bath homogenisation, and overall reaction efficiency. Dilution of silica with fluxes commonly increases slag density and decreases viscosity. Higher density can improve feed buoyancy, potentially increasing residence time for meltdown and reaction, while lower viscosity improves mass transport, slag–metal separation, and homogenisation. However, excessively low viscosity may reduce entrainment of fine particles and accelerate settling, which can alter reduction pathways and change the distribution of reaction zones.

The slag's solidification profile and operating temperature window are essential for both furnace stability and metal–slag processing. Pilbara gangue compositions are highly refractory, meaning that slag formation requires the dissolution of acidic and basic oxides from below their melting points. Lower liquidus temperatures and higher operating temperatures minimise shell formation on feed particles, reduce effective viscosity, enhance smelting rates, aid separation of droplets and solids, and support granulation for supplementary cementitious material application. However, these conditions increase refractory wear risks, including dissolution, infiltration, and accelerated taphole degradation. In contrast, raising liquidus temperature may encourage freeze lining development, offering improved refractory protection but potentially reducing productivity and altering flow behaviour, heat transfer, and overall process stability.

Interfacial tension plays a key role in determining the wetting behaviour between molten slag and metal droplets as well as between slag and solid feed materials such as carbon sources or DRI pellets. These interactions influence carburisation mechanisms, metal droplet coalescence, residence time, and the extent to which solids are encapsulated or repelled by the slag. Chemical activity, governed by slag composition and temperature, further determines both the thermodynamic driving forces and kinetic viability of dissolution and reduction reactions. Flux adjustments can modify chemical activity, but slag volume and FeO must be minimised to maintain commercial viability through their influence on energy consumption and metal yield.

Given these interconnected challenges, a coordinated, multi-faceted research and development program is essential to establish a validated approach to slag design for NeoSmelt. Laboratory-scale work is needed to address gaps in published phase diagrams and thermodynamic models, which have been shown to inaccurately predict solidification behaviour for compositions relevant to the Pilbara ESF system. Experimental studies must also characterise viscosity, electrical conductivity, wetting behaviour, and carburisation mechanisms across candidate feed and slag compositions. These efforts will support the development of predictive models for slag liquidus, viscosity, electrical properties, and reaction behaviour, providing NeoSmelt engineers with practical tools for campaign planning and operational control.

Complementary pilot-scale test work and knowledge-sharing partnerships are equally important to bridge the gap between laboratory findings and industrially relevant behaviour. Industrial slags inevitably contain minor elements, entrained gases and solids, and metal droplets, and behaviour is shaped by furnace geometry, heat transfer constraints, and operational transients that cannot be fully replicated at bench scale. Pilot data will therefore be critical for validating models, understanding process sensitivities, refining flux and feed strategies, and informing the pathway toward larger-scale ESF deployment.

CONCLUSIONS

Collectively, this integrated technical approach will equip NeoSmelt with the necessary understanding to intentionally manipulate slag properties from the beginning of the pilot campaign. By resolving key uncertainties in slag behaviour and enabling accurate prediction tools, the program will support BlueScope and its partners in developing a viable low-emissions ironmaking pathway that leverages Australia's unique Pilbara ore resources while meeting the technical, economic, and environmental requirements of future steelmaking.

ACKNOWLEDGEMENTS

This Project received funding from the Australian Renewable Energy Agency (ARENA) Powering the Regions Industrial Transformation Stream.

The views expressed herein are not necessarily the views of the Australian Government, and the Australian Government does not accept responsibility for any information or advice contained herein.

Unlocking low-emission steelmaking pathways for Pilbara iron ore – a value chain perspective

S Ge[1], K Joubarani[2], J Zhang[3], J Els[4] and I Cameron[5]

1. Process Specialist, Hatch, Mississauga ON L5K2R7, Canada. Email: sa.ge@hatch.com
2. Process Engineer, Hatch, Mississauga ON L5K2R7, Canada.
   Email: kamal.joubarani@hatch.com
3. Process Engineer, Hatch, Mississauga ON L5K2R7, Canada. Email: jilai.zhang@hatch.com
4. Senior Consultant, Hatch, Brisbane Qld 4000. Email: jeanne.els@hatch.com
5. Principal Metallurgist, Hatch, Mississauga ON L5K2R7, Canada.
   Email: ian.cameron@hatch.com

ABSTRACT

The steel industry's decarbonisation imperative presents a unique challenge for Western Australian/ Pilbara hematite-goethite iron ore, whose mineralogical characteristics–such as chemically bound moisture and elevated alumina and phosphorus–complicate its use in conventional blast furnace and direct reduction routes. Despite these constraints, Pilbara iron ore remains a globally significant resource, and unlocking its potential through innovative processing pathways is essential for a sustainable steel future.

This paper explores the integration of electric smelting furnace (ESF) technology with direct reduction processes as a viable solution for the low emission, low waste, and high efficiency utilisation of Pilbara iron ore. ESFs enable complete metallisation of iron minerals and effective gangue removal, offering flexibility in feed preparation and compatibility with various reduction technologies. The analysis emphasises the broader and comparative business case for ESF-based pathways.

A comparative value chain assessment is presented, evaluating two ESF-integrated flow sheets– one based on shaft furnace reduction and the other on fluidised bed technology–against the conventional shaft furnace–electric arc furnace (EAF) route. Key metrics include iron yield, metallics quality, greenhouse gas emissions, and implications for capital investment, operational complexity, and market readiness.

The findings demonstrate that ESF-based pathways can mitigate Pilbara iron ore's processing challenges while delivering competitive environmental and economic performance. In particular, the production of low-emission pig iron offers strategic advantages in terms of product flexibility, supply chain resilience, and alignment with emerging green steel markets.

By reframing Pilbara iron ore's limitations as opportunities for innovation, this study highlights how targeted process integration and value chain optimisation can position Australian iron ore as a cornerstone of sustainable steelmaking. The insights presented will inform technology selection, investment decisions, and policy development as the industry transitions toward net-zero emissions.

INTRODUCTION

The decarbonisation of iron and steel production is reshaping the iron ore market by shifting value drivers away from solely blast furnace (BF) operability, and toward feed attributes that enable low-emissions processing at scale. The International Energy Agency (IEA) identifies electrification, hydrogen-based processes, resource efficiency, and carbon capture, utilisation and storage (CCUS) as core elements of deep emissions reduction pathways for the sector (IEA, 2020). At the same time, the scale of the transition is considerable: World Steel Association reports a global average CO_2 emissions intensity of approximately 1.92 t CO_2 per tonne of crude steel cast in 2023 (World Steel Association, 2024, 2025).

In the near- to mid-term, direct reduction coupled with electric arc furnace steelmaking (DR–EAF) is the principal route for reducing emissions where low-carbon electricity and suitable reductants are available (Ge and Cameron, 2025). A central raw material constraint is that conventional DR reactors are solid–gas reduction units and do not include a melting step; consequently, they cannot reject gangue as slag. Gangue is therefore retained in the ore-based metallics (OBM) and must be removed during downstream melting and refining. This is especially impactful for the EAF steelmaking process as due to its inherent melting mode and slag chemistry requirements, additional gangue input increases slag burden and imposes energy, yield, and productivity penalties when metallics purity deteriorates (IIMA, 2018). This sensitivity underpins the preference for DR-grade pellets or pellet feed and contributes to growing concerns over DR-suitable feed availability, which currently comprises a small portion of seaborne iron ore trade (Barrington, 2025; Ge, 2024; Nicholas and Basirat, 2022), as illustrated by the '>67 per cent' band in Figure 1.

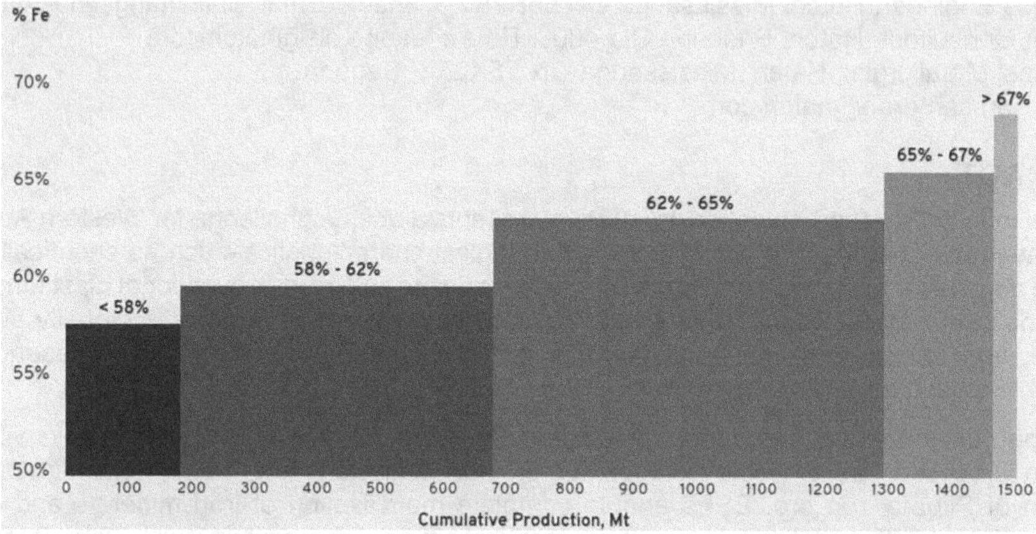

FIG 1 – Share of global seaborne iron ore by grade/source (MPP and NZSI, 2021).

A conventional response is deeper beneficiation followed by agglomeration to produce DR-grade pellets. However, intensive beneficiation can shift costs and environmental burdens upstream by increasing comminution and water intensity, enlarging tailings streams, and reducing iron-unit recovery across the mine-to-metal value chain. This reframes the challenge as an optimisation problem across the value chain: where should gangue be rejected to maximise ore utilisation efficiency and minimise total residues–during beneficiation as tailings, or later through pyrometallurgical separation as slag?

Electric smelting furnaces (ESFs) provide a pathway to relocate the task of gangue removal to a dedicated melting and final-reduction step (BHP, 2023; Cleveland-Cliffs Inc, 2024). In a DR–ESF configuration, the DR unit supplies partially reduced metallics to the ESF, which melts the metallic intermediate product and completes the reduction process to produce a molten iron product (pig iron or hot metal) together with a separable slag phase (Chomyn *et al*, 2023). Electric ironmaking draws on extensive industrial experience in electric smelting and on a documented history of electric ironmaking development, providing relevant foundations for equipment design, scale-up, and operational de-risking.

This paper evaluates whether ESF integration can expand low-emissions processing options for Pilbara DSO, which today is globally significant yet not naturally aligned with conventional DR–EAF feed requirements. Two ESF-integrated flow sheets are assessed–one based on a shaft-furnace DR and one based on a fluidised-bed DR process–against a conventional shaft-furnace DR–EAF route, using a value-chain framework focused on iron yield, metallics quality, greenhouse gas emissions within defined boundaries, and key cost and readiness considerations.

BACKGROUND AND CHALLENGES

Australia's iron ore export system has historically been anchored by high-volume production of hematite-goethite DSO, enabled by comparatively low processing intensity, proximity to Asian markets, and highly developed mine–rail–port infrastructure. A frequently cited advantage of Pilbara DSO is that it can be supplied as lump and fines products without the fine grinding and complex beneficiation commonly required for lower grade itabirites, taconites, and many magnetite deposits, lowering capital intensity and energy demand in ore preparation. As the steel industry decarbonises and shifts toward DR-based ironmaking, the implications of this legacy positioning change: constraints expand beyond BF operability and sinter plant performance to include gangue management across the value chain, performance in electric steelmaking, solid residue generation (eg tailings, dusts), and liquid steel cleanliness with respect to tramp elements.

Pilbara DSO is not a single quality class. Production and remaining resources span multiple ore types, commonly grouped into banded iron formation-derived ores (including Brockman and Marra Mamba variants) and channel iron deposits, with systematic differences in Fe grade, gangue (SiO_2 and Al_2O_3), LOI, phosphorus, and natural lump yield. Table 1 summarises indicative ranges for representative Pilbara DSO ore types and product attributes.

TABLE 1

Indicative Pilbara DSO ore types and chemistry ranges (Brent, 2021).

Typical properties	Banded Iron Formation				Channel iron deposits	
	Brockman (BKM)		Marra Mamba (MM)			
	Low P BKM	High P BKM	Competent dense MM	Friable, ocherous goethitic MM	Low Al/Si CID	High Al/Si CID
Fe (wt%)	60–67	58–65	58–63	56–62	55–59	55–57
P (wt%)	0.06–0.1	0.1–0.15	0.05–0.07	0.05–0.07	0.03–0.06	0.03–0.06
Al_2O_3 (wt%)	1.4–2.2	1.8–3.0	1.8–2.8	2–4	1–2	2–5
SiO_2 (wt%)	3.5–4.0	3.5–5.0	3–5	4–6	4–7	5–8
LOI (wt%)	2–3	4–5	4–6	6–8	9–10	9–10
Ease of beneficiation	High	High	Medium	Poor	Medium/ Poor	Medium/ Poor
Key Attribute	Premium ore, approaching depletion	High P ore, now a key component of blend with MM and a major portion of remaining Pilbara DSO	Now a major component of blended product with BKM to dilute P	Mid-range Fe, a key diluent for high P BKM for blended Pilbara DSO product	High calcined Fe, low Al and P, high LOI	High calcined Fe and Al, low P, high LOI

Across both incumbent BF and emerging DRI-based routes, the ore attributes that most strongly influence value-in-use are gangue load and chemistry (notably LOI, Al_2O_3, and phosphorus). In molten metallurgical processing, higher gangue increases the mass that must be heated, melted, and removed as slag, raising energy demand and increasing molten slag volumes as well as slag-to-metal ratio that can constrain throughput. In practice, this effect manifests as higher coke rate and greater liquid handling burden in the BF, and higher electrical energy demand and slag burden in electrically heated furnaces; depending on slag practice and oxidation potential, higher slag burdens can also increase iron losses to slag and reduce metallic yield. The sensitivity to gangue is therefore a general consequence of slag burden growth in molten processing, rather than a feature of a single furnace type.

The integrated steelmaking route is generally more accommodating of Pilbara DSO, consistent with the sustained prevalence of Pilbara fines and lump in BF ironmaking. The BF operates under strongly reducing conditions and is designed to reject essentially all gangue (from ore, coke ash, and fluxes) into an ironmaking slag phase. This slag is typically low in FeO and is widely valorised as a substitute for clinker in cement production, providing an established outlet that can improve value chain economics and sustainability. Impurities persisting beyond BF ironmaking are essentially only those dissolved in hot metal (eg Si, P, S), which are removed before or during steelmaking and refining. Alumina nevertheless remains an important constraint because excessively high Al_2O_3 increases BF slag viscosity and can impair drainage and permeability, reducing BF productivity; however, established BF slag-management practices provide a means to accommodate typical Pilbara DSO variability.

LOI and associated goethitic mineralogy introduce an additional constraint that depends strongly on the agglomeration route. For fines used in sintering, higher LOI affects thermal balance and bed permeability but is generally manageable through modified sintering practice developed over decades of Pilbara product use. In pelletising, however, induration performance and pellet quality are more sensitive to mineralogy and volatile matter released during firing; high-LOI feeds typically require moderation of temperature profiles and/or lower productivity to avoid pellet cracking, which in turn can translate to substantial productivity and capital penalties at scale. Pilbara ores can be incorporated into pellet production, but typically only in limited proportions and almost always in blends with other ores, with very limited application to pellets intended for DRI production under current commercial practice.

As the industry shifts toward DRI–EAF flow sheets to reduce GHG emissions, Pilbara DSO becomes structurally disadvantaged because the route removes two separation and refining 'buffers' that underpin Pilbara DSO's strong performance in BF-based routes: gangue rejection during ironmaking, and deep impurity removal during basic oxygen furnace (BOF)-based refining. This is consequential because EAF steelmaking typically relies on an oxidising slag practice and high basicity targets to support refining reactions. This requires high FeO slags and increases flux demand. Under such conditions, the combination of retained gangue in the metallics and oxidising, high-basicity slag practice increases total slag volume, raises energy consumption, and reduces iron yield through higher iron losses to slag, while tightening operating constraints associated with slag foaming, slag handling capacity, and tap-to-tap time.

Phosphorus further widens the gap between BF–BOF and DRI–EAF routes for Pilbara DSO. In integrated steelmaking, higher-P inputs can be managed through hot metal pretreatment and established BOF dephosphorization practices, including multi-slag or multi-stage approaches where required. In DRI–EAF there is no intermediate hot metal phase, so hot metal pretreatment steps (de-S, de-P, de-Si) are not possible. As such the EAF generally provides a narrower and less forgiving operating window for deep dephosphorization than the BOF. As a result, DRI–EAF operations place a strong premium on low-phosphorus ore-based metallics. Where higher-P DRI must be used, phosphorus control often depends more heavily on charge mix management and dilution (eg with lower-P scrap) than on a dedicated high-efficiency de-P stage. This is especially pertinent to EAF operators targeting higher-quality grades that require tighter control of residual elements in scrap (eg Cu, Ni), which increases the need for ore-based metallics for residual dilution. The resulting tension between impurity dilution drivers–greater ore-based metallics to manage scrap residuals, but greater scrap to dilute P (and S) in ore-based metallics–reduces the value-in-use of ore-based metallics derived from higher-P, higher-gangue feeds.

Feed form and agglomeration requirements compound these challenges. Most commercial DR plants are shaft-furnace-based and configured around iron ore pellets (and, in some cases, high-quality lump) with narrow specifications on chemistry and high-temperature performance. The difficulty of producing DR-quality pellets with high proportions of Pilbara ore–together with Fe-grade requirements–reduces the compatibility of Pilbara products with today's mainstream shaft DR–EAF configurations at scale. Alternative DR technologies that can directly accept fine ore, such as fluidised-bed reduction processes, can relax the pelletising constraint and provide a plausible entry point for Pilbara DSO; however, these routes do not eliminate the underlying penalties associated with gangue carryover into the metallics stream, and the downstream EAF must still accommodate that oxide load as slag under oxidising conditions.

In summary, (i) pellet dependence in shaft-furnace DR, (ii) gangue carryover that increases slag burden and iron losses under typical EAF slag practice, and (iii) limited dephosphorisation capability relative to BF–BOF create a systematic disadvantage for Pilbara DSO in conventional DRI–EAF decarbonisation pathways. The implication is not that Pilbara ores are unusable; rather, large-scale deployment in DRI–EAF value chains typically requires either substantial upstream upgrading and blending to meet DR-grade specifications, or an enabling downstream unit operation that can efficiently separate gangue and support impurity control without incurring excessive yield and energy penalties. In the absence of such measures, the expansion of DRI–EAF risks increasing dependence on a narrower subset of DR-suitable feed supply, with implications for cost, scalability, and the participation of ore supply systems historically optimised around BF–BOF routes.

ANALYSIS SET-UP AND METHODOLOGY

This study applies an integrated, first-principles modelling framework to evaluate low-emissions pathways for Pilbara DSO and related Western Australian hematite-goethite products. The assessment is built on custom, physics-based process models that solve mass and energy balances across the full value chain, which can accurately extrapolate for alternative production scenarios and conditions. The modelling scope follows a mine-to-metallics structure, shown in Figure 2, consistent with the reference scenario framework (Mineral Processing → Agglomeration → Direct Reduction → Smelting), enabling internally consistent tracking of iron units, gangue partitioning, energy demand, and residue generation across alternative flow sheets.

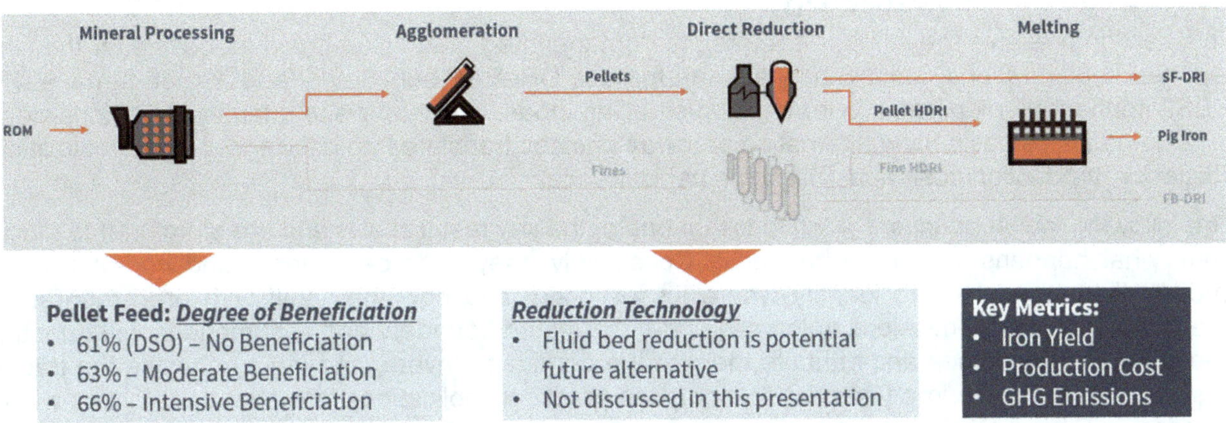

FIG 2 – System boundary and scenario set-up for the study.

A series of modelling scenarios were selected to accurately assess implications and opportunities for Pilbara iron ore. Modelled permutations span (i) different run-of-mine (ROM) feed bases representative of various Pilbara iron ores, (ii) alternative mineral processing intensities and circuit options (including 'no beneficiation' through intensive wet beneficiation with state-of-the-art processes), (iii) cases with and without agglomeration (including conventional pelletising and cold briquetting), (iv) multiple direct reduction technologies and reductant options (natural gas, hydrogen, and a blend thereof), and (v) alternative melting configurations, including cases with and without an ESF, to produce a variety of OBM (including hot- and cold-direct reduced iron (H/CDRI), hot-briquetted iron (HBI), hot metal (HM) and pig iron (PI)). In addition, the broader modelling platform used to develop and test these flow sheets has been applied to a wider set of iron ore types from multiple producing regions, and to downstream steelmaking configurations spanning both EAF- and BOF-centric routes, including variants that impose different steel product quality constraints and ferrous feed blend requirements (eg differing scrap reliance and ore-based metallics supplementation). This broader capability is used to ensure that pathway implications are interpreted within realistic downstream constraints, although the present paper reports only the subset of cases required to address the Pilbara DSO question posed in the abstract.

The published comparison includes two representative pathways selected to isolate the key trade-offs for Pilbara DSO: a conventional shaft-furnace DR–EAF benchmark, and an ESF-integrated pathways comprising shaft-furnace DR followed by ESF melting and separation. In the ESF-integrated cases, the ESF is treated explicitly as an electric ironmaking/smelting device (not in a steelmaking mode, which is also possible), producing a molten iron product together with a separable slag phase; published design and de-risking considerations for electric ironmaking furnaces provide supporting context for equipment-scale assumptions and integration concepts (Chomyn *et al*, 2024). Model development and parameterisation draw on engineering fundamentals, extensive project experience, and internal benchmarking, supplemented by proprietary and real-world operating data where available, implemented within a physics-consistent modelling structure.

Results are reported against three primary metrics aligned with iron ore producer decision drivers and downstream constraints: (i) iron yield (ore utilisation efficiency) across the value chain, (ii) ore-based metallics (OBM) production cost, and (iii) GHG emissions within explicitly stated boundaries. Iron yield is tracked by following iron units from ROM through each unit operation and accounting for losses to tailings (where applicable), dust/fines, and slag. OBM production cost is compiled from major cost drivers at each step (ore consumption as affected by yield, energy, reductants, fluxes, and key consumables). GHG emissions are compiled on an energy and material input basis consistent with the mine-to-metallics boundary, with electricity-related emissions treated via an explicit grid emissions factor within the analysis assumptions.

The broader modelling program described above includes additional ores, processing variants, steelmaking routes, and product/blend constraints beyond the scope of this paper; these form the basis for ongoing and future reporting, while the present manuscript focuses on the most relevant subset for evaluating low-emissions pathways for Pilbara DSO.

RESULTS AND DISCUSSION

This section presents the comparative results from the integrated value-chain modelling for the two reported configurations: a conventional shaft-furnace DR–EAF benchmark and a shaft-furnace DR + ESF configuration operated in ironmaking/smelting mode. The intent is to quantify, on a consistent mine-to-metallics basis, how alternative pathway choices translate into differences in ore utilisation efficiency, production cost, and emissions performance.

The following four subsections each focus on one of the key result charts and are structured to move from 'what happens' to 'why it happens'. Collectively, they (i) trace material and iron-unit flows through the value chain to identify where losses occur and how they shift with ore preparation intensity, (ii) quantify emissions outcomes under the stated boundary and assumptions, (iii) quantify operating cost outcomes and attribute them to the dominant drivers, and (iv) synthesise the trade-offs across the three primary metrics to clarify the practical implications for pathway selection when processing Pilbara DSO in low-emissions flow sheets.

Mine-to-metallics yield

Figure 3 summarises the mine-to-metallics mass balance for three representative configurations, constructed on a common starting basis of using Pilbara DSO/ROM. The cases differ only in where gangue is rejected and, by consequence, where iron units are lost. In the base case 'no beneficiation' configuration, gangue contained in the DSO is carried forward and rejected in the downstream melting step as ESF slag. In the intermediate configuration, moderate beneficiation is applied to produce BF-grade pellets; the remaining gangue is again rejected as slag in the ESF. In the 'fully beneficiated' configuration, intensive beneficiation is applied to produce DR-grade pellets suitable for producing a DRI/HBI product that can be directly made into steel without an ESF; the required gangue rejection occurs predominantly during beneficiation as tailings so that the DRI/HBI contains sufficiently low gangue for EAF steelmaking (Note that this is in fact a near-DR-grade pellet as producing a true DR-grade pellet, ie >68 per cent Fe and (Al_2O_3 + SiO_2) <2 per cent, from Pilbara DSO is very challenging, and therefore not used as a basis in the model). This methodology isolates a single controlling question: how placing the task of gangue rejection upstream (tailings) versus downstream (slag) affects overall iron-unit yield (Ge, 2025).

Two features of Figure 3 are important for interpreting process feasibility. First, incremental improvements in pellet Fe grade require disproportionately large gangue removal and can drive significant yield loss before ironmaking even begins. As illustrated in the reported cases, the yield loss associated with beneficiation is significant – upgrading from a DSO-equivalent pellet (~64 per cent Fe) to BF-grade pellet (~66 per cent Fe) requires removal of ~30 per cent of the total gangue during beneficiation and leads to ~14 per cent Fe yield loss. This is even more pronounced when comparing DSO to DR-grade pellets: beneficiating from ~61 per cent Fe to ~66 per cent Fe to reach DR-grade pellet feed specifications requires the rejection of ~60 per cent of gangue and leads to ~28 per cent Fe yield loss.

Note that these Pilbara DSO beneficiation results were based on in-house modelling of hypothetical/developmental circuits and represent optimistic assumptions and scenarios – including the assumed use of 100 per cent Pilbara DSO as a DR-grade pellet feed for DRI-EAF after intensive beneficiation.

More conservative/realistic assumptions in beneficiation and pelletising will further penalise the use of Pilbara DSO for DRI-EAF flow sheet.

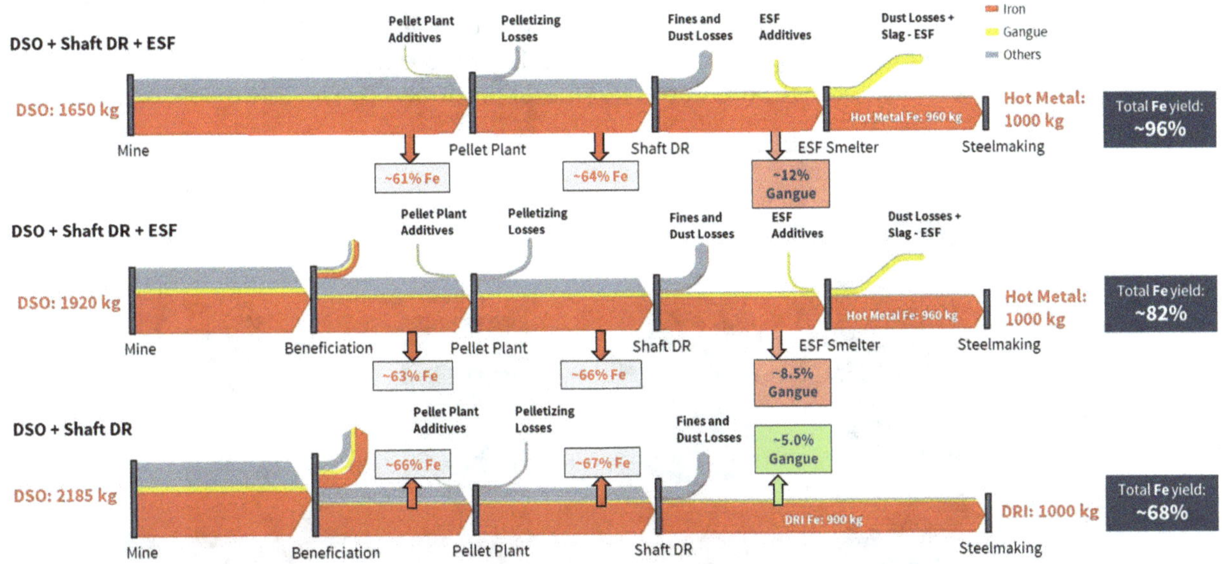

FIG 3 – Mine-to-metallics mass balance and total Fe yield for three flow sheets.

The yield penalty associated with beneficiation results in higher ROM demand and substantially larger residue streams to deliver the same metallic output to steelmaking. This effect is visible in the progressively higher DSO mass required across the three configurations and is a first-order driver of both cost and footprint (1.65 versus 1.92 versus 2.19 t-ROM per t-metallic).

A key distinction is therefore not only how much gangue is removed, but the form in which it exits the system. In the DR-grade route, gangue rejection occurs primarily as beneficiation tailings, which simultaneously (i) removes iron units as yield loss and (ii) creates a large tailings stream requiring long-term containment and management. By contrast, in the ESF-integrated configurations, gangue is rejected downstream as a separable, potentially sellable, ironmaking slag, avoiding the generation of tailings streams and their associated handling and risk profile.

Figure 3 also partially illustrates a product-quality and steelmaking-integration implication that is not captured by yield alone. Even with intensive beneficiation, DRI/HBI is not fully gangue-rejected; the residual oxide burden reports to EAF slag and therefore induces an energy/flux penalty relative to a fully gangue-rejected metallic such as pig iron. In addition, DRI/HBI is typically not fully metallised (often ~94–95 per cent; IIMA, 2018), so residual FeO must still be reduced during melting, consuming more energy and reductant during steelmaking. By contrast, pig iron (and especially hot metal integrated with steelmaking) delivers higher chemical, latent, and sensible energy to the meltshop via dissolved carbon and silicon, improving the energy balance compared with DRI/HBI, which generally carries lower carbon and essentially no silicon contribution (Chomyn *et al*, 2025). These differences also affect transition pathways: a DRI-ESF configuration can be flexibly arranged to feed either BOF-based steelmaking and enables brownfield BF replacement while retaining liquid-metal logistics, or produce hot metal or pig iron to feed collocated or remote EAF steelmaking operations, whereas DRI ironmaking without ESF more naturally aligns with an EAF-centric transition.

Emissions performance of DRI-ESF flow sheet

Figure 4 compares aggregate greenhouse gas (GHG) emissions for the shaft-furnace DRI–ESF pathway across three Pilbara ore scenarios (DSO, BF-grade beneficiation, and DR-grade beneficiation), illustrated here by nominal pellet feed grades of 61 per cent, 63 per cent, and 66 per cent Fe and expresses the total emissions relative to a 'typical BF' reference (normalised to 100 per cent). The results are calculated on a consistent basis using natural-gas-based direct reduction and a grid emissions factor of 0.51 kg-CO_2e/kWh for Scope 2 emissions and presented on a per-tonne hot metal basis. It is important to note that Scope 1/2/3 definitions and inclusion boundaries vary across organisations and reporting standards; here, the scope breakdown is from the ironmaking's perspective and used primarily to show how the distribution of emissions shifts between direct process sources and electricity- and supply-chain-related contributions under a single, internally consistent accounting framework. Mining, beneficiation and pellet production are classified as upstream Scope 3 emissions. Downstream Scope 3 emissions (including steelmaking) are excluded as they are indifferent to the same hot metal produced by the three flow sheets.

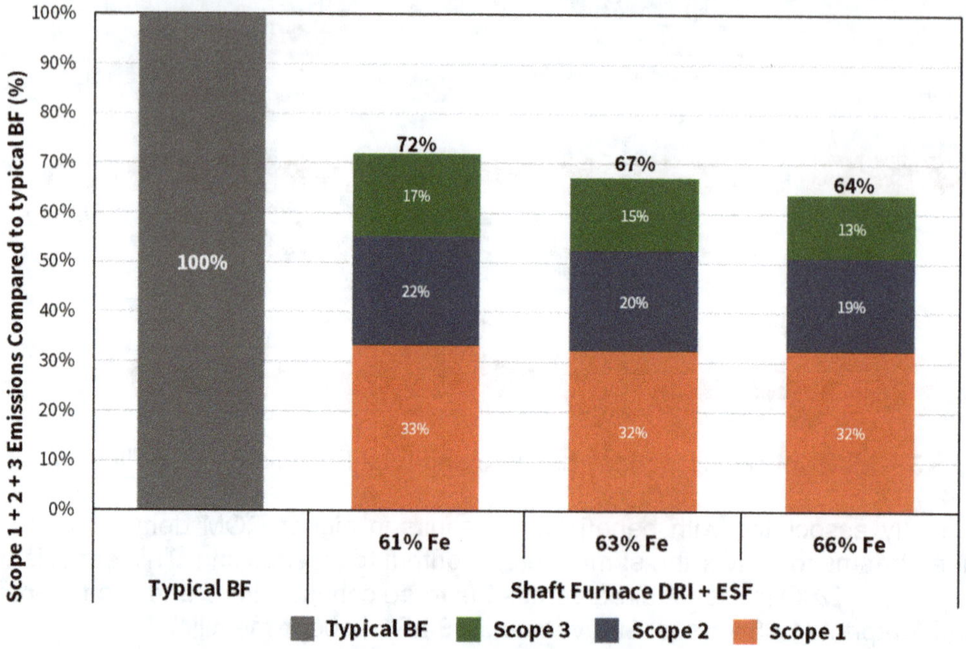

FIG 4 – Estimated emissions compared to a typical BF for DRI-ESF across ore grades.

Three salient trends are shown in Figure 4. First, Scope 1 emissions remain broadly comparable across the three ore grades. Since the dominant Scope 1 contributor is the direct reduction process, which scales primarily with the oxygen removal requirement per unit of metallic iron produced, and this requirement is approximately constant per tonne of metallic output across the compared ore grades, the Scope 1 component does not vary strongly with changes in ore grade.

Second, increasing ore grade produces only a modest reduction in Scope 2 emissions, reflecting reduced electricity demand in the ESF: higher-grade feed introduces less gangue to be melted in the ESF and therefore lowers the electricity requirement per tonne of hot metal under otherwise comparable operating assumptions. Notably, the change in Scope 2 between the DSO and DR-grade cases is modest (~15 per cent higher) relative to the change in gangue content (~2.5 times higher for DSO), indicating that the ESF can accommodate substantially higher gangue burdens without a proportional increase in power consumption. In contrast, upgrading ore through beneficiation to higher pellet feed grades typically requires increasingly intensive processing which incurs sharply rising yield penalties, with outcomes that are strongly mineralogy-dependent. This contrast underscores the ESF as a robust downstream separation step for gangue rejection in a value-chain context.

Third, there is also a modest reduction in Scope 3 emissions with increasing ore grade, driven mainly by reduced upstream processing emissions and consumables per tonne of hot metal (notably from the pelletising process and ESF flux usage), and reduced pellet consumption when higher-grade ore is used. Note that even though there are larger amounts of ROM that must be mined and beneficiated per unit metallic in higher-grade ore cases, mining and beneficiation's contribution to GHG emissions is much smaller compared to pelletising (due to pelletisations high direct emissions from induration and direct carbon usage), leading to lower upstream Scope 3 emissions in in 66 per cent Fe pellet feed scenario versus 61 per cent Fe, primarily driven by lower amount of pellets needed in the higher-grade pellet feed scenario – despite processing larger amounts of ROM.

A critical observation when interpreting these emissions comparisons is that the mining and beneficiation stages contribute relatively little to total emissions within the boundary used here, compared with agglomeration and ironmaking processes. As a result, the differences shown in Figure 4 are informative for GHG intensity under the stated boundary and do not fully reflect the system-level consequences and comprehensive ecological impact of higher ROM demand and lower iron-unit yield that accompany more intensive beneficiation (such as natural resource depletion, yield loss, waste generation, resource circularity, land and water use etc). In other words, intensive iron ore beneficiation can reduce total GHG emissions while shifting material-efficiency and residue-management burdens upstream.

A pathway can show modest improvement in the emissions breakdown while still imposing substantial upstream penalties in ROM demand and tailings. For this reason, the emissions results should be read alongside mine-to-metallics yield and residue outcomes: incremental GHG reductions achieved through deeper beneficiation can coincide with materially lower ore utilisation efficiency and greater upstream waste-management burden.

The analysis in Figure 4 is therefore best interpreted as a consistent snapshot under NG-DRI and the assumed (relatively high) grid emissions factor, where higher ore grades modestly reduce total emissions at the expense of other non-emissions ecological and environmental factors. Looking forward, the DRI-ESF configuration retains a clear pathway to near-net-zero: with low-emissions/green electricity, Scope 2 emissions can be largely eliminated, and with low-emissions hydrogen as the reductant, Scope 1 (and associated fuel-related Scope 3 contributions) can be reduced substantially as well. In that context, DRI-ESF remains competitive as both a gangue-rejection solution for Pilbara DSO and a future-ready low-emissions ironmaking route (BHP, 2024).

Operating costs

Figure 5 compares estimated operating cost (OPEX) for the shaft-furnace DRI + ESF pathway across three iron ore scenarios represented by nominal pellet feed grades of 61 per cent, 63 per cent, and 66 per cent Fe. Costs are normalised against the 61 per cent Fe case (100 per cent) and grouped according to major categories (iron ore feed, reductant, electricity, and other costs). OPEX is calculated from model-derived process inputs and a consistent set of commodity prices representative of 2023–2024. The effective cost of higher-grade feed reflects DSO/ROM input, beneficiation operating costs, and beneficiation yield loss; no additional price premium for DR-grade material is assumed.

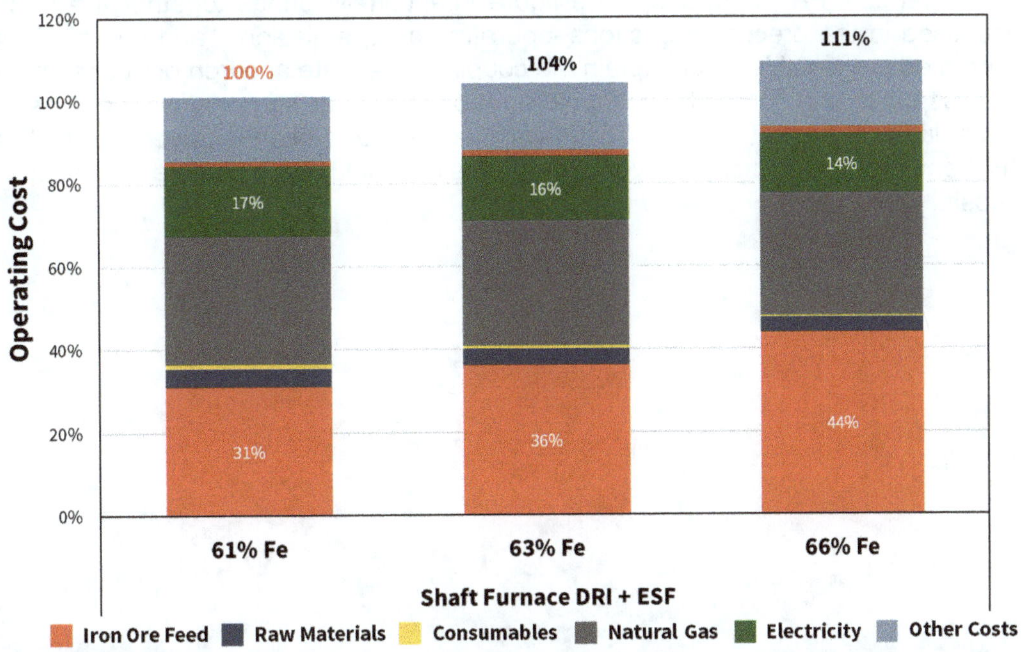

FIG 5 – Operating cost breakdown for shaft-furnace DRI–ESF (61/63/66 per cent Fe pellet feeds).

The reductant (natural gas) contribution remains broadly similar across the three cases, consistent with the expectation that natural gas consumption during direct reduction scales primarily with oxygen removal and total amount of reduction work, which is constant across the cases on per unit hot metal basis. Most remaining line items are comparatively stable or minor, leaving total OPEX sensitivity dominated by two opposing effects: electricity cost decreases with increasing ore grade, while total ore cost increases.

The key outcome is that the electricity benefit from upgrading is modest relative to the yield-driven ore penalty. Upgrading the feed from 61 per cent to 66 per cent Fe reduces ESF power consumption and electricity cost by roughly 3 per cent (via lower gangue input), but increases the iron ore feed cost component by ~13 per cent once beneficiation yield loss and the associated increase in ROM requirement are carried through the value chain. Total OPEX therefore increases from 100 per cent (61 per cent Fe) to ~104 per cent (63 per cent Fe) and ~111 per cent (66 per cent Fe). In this configuration and under the stated assumptions, upgrading toward DR-grade feed increases operating cost by ~11 per cent relative to direct DSO use.

This result is consistent with the yield behaviour discussed earlier: as pellet feed grade targets increase, the marginal gangue removed upstream is achieved with a disproportionate loss of iron units to tailings, which increases ROM throughput per tonne of hot metal and dominates total cost. In other words, within the DRI-ESF pathway, the cost leverage of reducing smelting electricity by lowering gangue input is much smaller than the cost impact of losing iron units during beneficiation, highlighting the ESF as a more cost-effective option for gangue rejection compared to intensive beneficiation for Pilbara iron ore.

There are two important exclusions to note. First, these OPEX estimates exclude tailings processing, transport, storage, and long-term management; inclusion would further penalise cases requiring more intensive beneficiation. Second, they exclude levelised capital costs for beneficiation and associated infrastructure, which are expected to be material and would increase with beneficiation intensity. Overall, Figure 5 indicates that, for the cases assessed, the operating-cost optimum for Pilbara DSO lies toward minimising or eliminating iron ore beneficiation and relying on the ESF for gangue rejection.

Holistic comparison

Figure 6 provides a consolidated comparison of the three Pilbara ore grades assessed within the shaft-furnace DRI + ESF pathway, using 61 per cent Fe pellet feed as the baseline (DSO/ROM basis) and showing the relative response when beneficiation is increased to 63 per cent Fe (BF-grade) and 66 per cent Fe (DR-grade). The figure intentionally brings together the three primary metrics discussed in the preceding sections–operating cost, emissions, and overall Fe yield–and adds two intermediate variables that explain the coupling: slag rate and iron ore consumption. Read as a system response, the figure shows that even though increasing beneficiation does deliver the expected metallurgical benefit in the ESF (lower slag rate and slightly lower electricity use) and modest GHG emissions reduction, it adversely impacts iron yield and overall resource utilisation and operating costs.

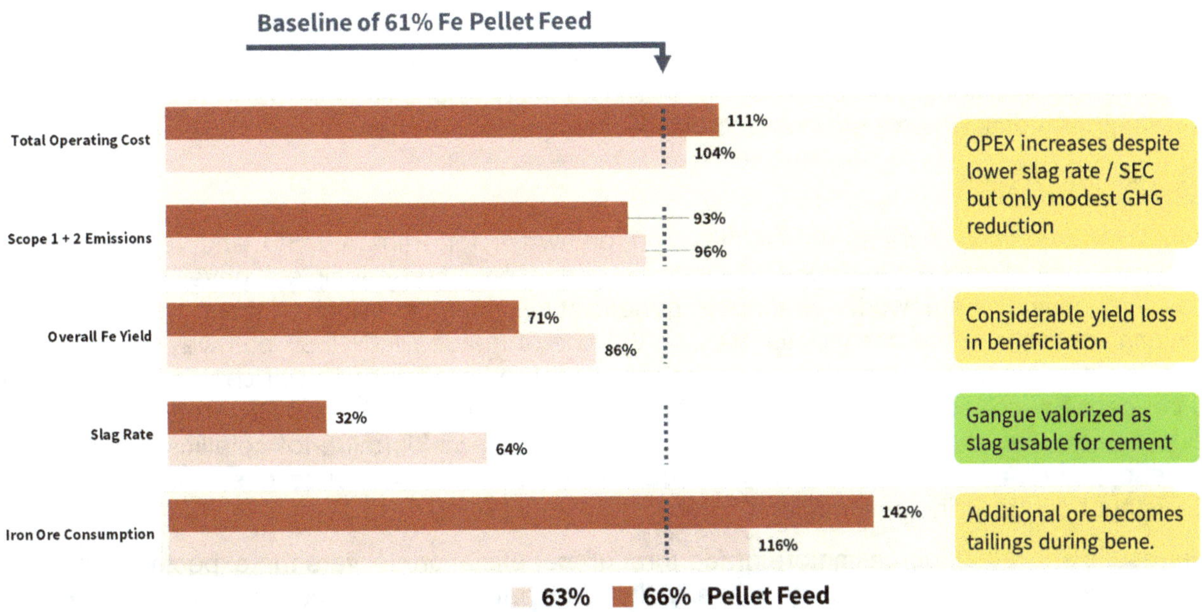

FIG 6 – Normalised trade-off summary for DRI + ESF across iron ore feed grades.

Two performance metrics are favoured as iron ore grade increases. First, slag rate decreases substantially (to ~64 per cent and ~32 per cent of baseline for the BF- and DR-grade cases, respectively), which reduces the ESF's melting burden and contributes to the modest reduction in emissions and electricity-related operating cost observed earlier. Second, Scope 1+2 emissions improve modestly with increasing ore grade, consistent with reduced electricity demand per tonne of hot metal when less gangue is melted in the ESF, while direct reduction emissions remain broadly similar on a hot-metal basis. These improvements are technically meaningful for the furnace operation and energy balance; however, the figure demonstrates that they do not dominate overall system performance at the value-chain level.

The limiting factor is the yield penalty embedded in beneficiation. Overall Fe yield decreases from the baseline toward ~86 per cent (63 per cent case) and ~71 per cent (66 per cent case), and the associated iron ore consumption increases to ~116 per cent and ~142 per cent of baseline, respectively. This significant increase in ore consumption and the additional 16 per cent and 42 per cent ore consumption primarily manifests as upstream solid waste produced by intensive beneficiation – ie tailings generation – required to achieve higher feed grades. In practical terms, the 'extra ore' mined and processed to make one tonne of hot metal is material that must end up in a residue stream. This leads to large quantities of tailings being generated at higher beneficiation intensity (order of hundreds of kilograms per tonne of hot metal, and potentially approaching ~750 kg/t hot metal for the most intensive case), consistent with the magnitude of yield loss shown in the figure and with typical beneficiation reject characteristics for Pilbara hematite-goethite circuits. This is the main reason why total OPEX increases to ~104 per cent and ~111 per cent despite lower slag rates: the effective ore feed cost increases because iron units are discarded upstream to reach tighter specifications, forcing higher ROM throughput per tonne of product.

This comparison also reframes 'waste' from a purely quantitative concept (mass of residues) into a qualitative one (type of residue and its value-chain implications). In the baseline configuration (DSO/ROM basis), gangue is rejected primarily as ESF slag. While lower slag volumes can improve furnace performance, appropriately designed ironmaking slags can be suitable for use as clinker substitutes, providing a potential by-product outlet and revenue stream. This outlet also has broader decarbonisation relevance because clinker substitution reduces the need for energy- and CO_2-intensive clinker production in the cement sector. In contrast, beneficiation rejects report as tailings, which generally have limited valorisation pathways and impose long-term handling, storage, and containment requirements.

Taken together, Figure 6 summarises the central conclusion emerging from the present study: within the DRI-ESF pathway, deeper beneficiation produces large reductions in slag rate but only modest improvements in emissions and electricity-related cost, while imposing large penalties in overall Fe yield and iron ore consumption that drive higher OPEX and substantially greater upstream residue generation. The key trade-off is therefore not simply 'energy versus grade', but slag versus tailings– ie whether gangue is rejected downstream as a potentially valorisable ironmaking slag, or upstream as tailings with associated yield loss and waste-management burden.

CONCLUSIONS

This work assessed low-emissions utilisation pathways for Pilbara DSO and related Western Australian hematite-goethite products using an integrated, first-principles, physics-based value-chain modelling framework spanning beneficiation, agglomeration, direct reduction, and smelting/separation. The analysis focused on the shaft-furnace DRI + ESF pathway and compared three ore-grade cases–represented by nominal pellet feed grades of 61 per cent, 63 per cent, and 66 per cent Fe–corresponding to DSO/ROM, BF-grade beneficiation, and DR-grade beneficiation. Performance was evaluated on the core metrics of iron yield (mine-to-metallics ore utilisation efficiency), ore-based metallics (OBM) production cost, and greenhouse gas emissions under a consistent boundary and assumptions.

The results show that increasing ore grade through beneficiation delivers the expected reduction in slag rate and a modest reduction in electricity consumption, with a corresponding reduction in emissions from electricity and pelletisation, but these benefits are offset by the upstream iron-unit losses required to achieve higher Fe grades. The yield penalty propagates through the value chain as significantly increased ROM demand per tonne of hot metal and, substantially greater tailings generation. Consequently, beneficiation of Pilbara iron ore toward DR-grade specifications is difficult to justify on resource-efficiency and operating cost grounds within the DRI-ESF context, particularly once tailings management costs, capital intensity, and other non-GHG-related environmental factors are considered.

Finally, the emissions analysis indicates a credible pathway to near-net-zero ironmaking for DRI–ESF as the energy system decarbonises: low-carbon electricity can largely eliminate Scope 2 contributions, and low-emissions hydrogen as the reductant can substantially reduce direct process emissions.

Beyond the select yield, cost, GHG results reported here, ESF integration offers additional advantages for Australian hematite ores that were not explicitly quantified in this paper but are relevant to pathway selection. First, the naturally elevated alumina of many Pilbara ores—typically treated as a penalty in BF—can be accommodated in ESF slag design and, within appropriate composition windows, can contribute to achieving target slag physical and electrical properties (eg viscosity, liquidus temperature, and resistivity) needed for stable smelting operation and may reduce fluxing costs. Second, producing a molten iron product (hot metal/pig iron) rather than a solid metallic (DRI/HBI) restores established hot metal refining options—most notably hot metal dephosphorization—providing an additional control point for higher-phosphorus feeds that is not available in DR- EAF flow sheets.

In conclusion, the findings position DRI–ESF as a competitive, decarbonisation-ready route that can utilise Pilbara DSO effectively while minimising value-chain yield loss and shifting gangue rejection toward a more manageable, and potentially valorisable, residue stream.

REFERENCES

Barrington, C, 2025. Iron ore for direct reduction: the challenge updated [online], Midrex Technologies Inc. Available from: <https://www.midrex.com/tech-article/iron-ore-for-direct-reduction-the-challenge-updated/>

BHP, 2023. Pathways to decarbonisation episode seven: the electric smelting furnace [online], BHP. Available from: <https://www.bhp.com/news/bhp-insights/2023/06/pathways-to-decarbonisation-episode-seven-the-electric-smelting-furnace>

BHP, 2024. BlueScope, BHP and Rio Tinto select WA for Australia's largest ironmaking electric smelting furnace [online], BHP media release, 17 December 2024. Available from: <https://www.bhp.com/news/media-centre/releases/2024/12/bluescope-bhp-and-rio-tinto-select-wa-for-australias-largest-ironmaking-electric-smelting-furnace>

Brent, A, 2021. Best in class: Australia's bulk commodity giants – Australian iron ore: when quality meets opportunity, Minerals Council of Australia, Canberra.

Chomyn, K, Ge, S, Koehler, T and Walker, C, 2023. Applications of advanced smelting furnace technology for sustainable steelmaking, AISTech 2023 Proceedings of the Iron and Steel Technology Conference, Michigan.

Chomyn, K, Ge, S, Koehler, T, Sweeten, N, Walker, C and Rudge, D, 2025. Electric ironmaking – history and design considerations, AISTech 2025 Proceedings of the Iron and Steel Technology Conference, Tennessee.

Chomyn, K, Ge, S, Koehler, T, Walker, C and Rudge, D, 2024. Electric smelting technology implementation road map, AISTech 2024 Proceedings of the Iron and Steel Technology Conference, Ohio.

Cleveland-Cliffs Inc, 2024. Cleveland-Cliffs reaffirms commitment to Middletown Works decarbonization project and ongoing partnership with the US Department of Energy [online], news release, 16 September 2024. Available from: <https://www.clevelandcliffs.com/news/news-releases/detail/652/cleveland-cliffs-reaffirms-commitment-to-middletown-works>

Ge, S and Cameron, I, 2025. Iron Ore for the Future of Steelmaking, presentation at 2025 SME Minnesota Conference / 10th North American Iron Ore Symposium, Minnesota.

Ge, S, 2024. DRI and the Future of Steelmaking, presentation at AIST DRI and HBI: Logistics, Production and Utilization Seminar, 3 October 2024, Ohio.

Ge, S, 2025. Processing Pilbara Iron Ore in Electric Smelting Furnaces, presentation at The Breakthrough Technology Conference 2025, Singapore.

International Energy Agency (IEA), 2020. Iron and steel technology roadmap: towards more sustainable steelmaking, IEA, Paris.

International Iron Metallics Association (IIMA), 2018. Use of direct reduced iron (DRI) in the electric arc furnace (EAF) for steelmaking (Fact Sheet No 11).

Mission Possible Partnership (MPP) and Net-Zero Steel Initiative (NZSI), 2021. Net-zero steel sector transition strategy, October 2021.

Nicholas, S and Basirat, S, 2022. Iron ore quality a potential headwind to green steelmaking: technology and mining options are available to hit net-zero steel targets, Institute for Energy Economics and Financial Analysis, June 2022.

World Steel Association, 2024. Sustainability indicators report 2024: sustainability performance of the steel industry 2003–2023, World Steel Association, Brussels.

World Steel Association, 2025. World steel in figures 2025, World Steel Association, Brussels.

Beyond charcoal – pathways to reduce CO_2 emissions in hematite pelletising

F A V Guimaraes[1], G Martins[2] and L Filippov[3]

1. Global R&D Mining and Resources Portfolio Leader, ArcelorMittal Global R&D, Maizières-les-Metz, France. Email: filipe.vasconcelos@arcelormittal.com
2. Research Engineer, ArcelorMittal Global R&D, Maizières-les-Metz, France. Email: giovanni.albarellomartins@arcelormittal.com
3. Professor, Université de Lorraine, CNRS, Georessources, Nancy, France. Email: lev.filippov@univ-lorraine.fr

ABSTRACT

The steel industry is undergoing a profound transformation driven by global decarbonisation targets and increasing social and regulatory pressure to reduce CO_2 emissions. As the primary raw material for steelmaking, iron ore accounts for approximately 5 per cent of the sector's total emissions, making its role in this transition critical. Within mining operations, the pelletising process, particularly for hematite-based pellets, represents the largest source of emissions, with fossil fuel combustion contributing up to 60 per cent of CO_2 output and fluxes might correspond up to 25 per cent. Efforts to replace coke breeze with charcoal have shown potential but face significant limitations: the charcoal industry cannot meet the required supply volumes while ensuring carbon neutrality, and pellet quality is often compromised due to differences in reactivity. Different challenges happen on the fluxes and additives, as little research has been done on this context. This study investigates alternative pathways that can be sourced in Europe and the Americas, evaluating their fixed carbon and ash content levels. It also takes into consideration the replacement of usual fluxes by iron carbonates, in attempt to reduce CO_2 emissions coming from this step. The analysis also examines their impact on pellet quality and induration performance. The findings aim to provide actionable pathways for reducing emissions in pelletising without sacrificing product integrity, supporting the iron ore industry's alignment with global sustainability goals and the steel sector's net-zero roadmap.

INTRODUCTION

The iron and steel industry remains a major source of global industrial CO_2 emissions, with the blast furnace (BF) route still dominating iron production. As ore grades decline, pelletising fine iron ores has become essential for ensuring consistent agglomeration, controlled mineralogy, and strong metallurgical performance in both BF and emerging direct reduction (DR) processes. Although indispensable, pelletising is energy-intensive, and much of its carbon footprint comes from the combustion of solid fuels during induration and from the calcination of carbonate fluxes.

Fluxes such as limestone and dolomite help control pellet basicity and slag formation but generate significant CO_2 through calcination. Research at ArcelorMittal has shown that replacing part of these fluxes with siderite-rich $FeCO_3$ can lower emissions while improving Fe grade and reducing CaO demand (Aguiar, 2018a, 2018b). Controlled $FeCO_3$ additions also influence microstructure development and reducibility, although high carbonate levels may reduce mechanical resistance (Penedo Lima, 2019). Moreover, the mechanism of thermal decomposition of siderite depends on its composition and on experimental conditions such as temperature, gas atmosphere, heating rate, and microstructure (Szubka *et al,* 2023).

The most impactful lever for decarbonising pelletising, however, lies in solid-fuel replacement. Traditional fuels such as coke breeze provide the heat required for pellet firing, reducing reliance on gaseous fuels in the furnace burners. However, they are entirely fossil derived and constitute a major source of emissions in the pelletising process.

The usage of biomass materials, such as biochar, torrefied biomass, or other lignocellulosic residues, release biogenic carbon, exhibit high reactivity, and devolatilise readily, enabling efficient heat generation during induration (Sefidari *et al,* 2020). When properly selected, these fuels can maintain or even enhance fired pellet strength, producing microstructures with stable slag bonding. Poorly reactive biomasses, however, lead to unmelted phases and weaker bonding, underscoring the importance of matching fuel reactivity and ash chemistry to process requirements (Martins, 2025). When combined with optimised carbonate blends, biomass co-firing significantly reduces fossil-carbon demand without compromising pellet integrity.

Together, biomass integration, flux optimisation, and microstructural control form a coherent strategy for low-carbon pelletising. These advances align with the industry's broader transition toward hydrogen-based direct reduction and net-zero ironmaking. In this context, engineered pellets, designed with the right flux chemistry, fuel mix, and thermal profile, become a central enabler of future low carbon steel production. Table 1 shows the CO_2 contribution from each component of hematite pellet most common recipes.

TABLE 1

CO_2 balance comparison.

Material	Main reaction/process	CO_2 released	Net CO_2 impact
Iron Carbonate ($FeCO_3$)	$3FeCO_3 \rightarrow Fe_3O_4 + CO + 2CO_2$	0.38–0.40 $kgCO_2/kg$	Moderate (partially offset)
Limestone ($CaCO_3$)	$CaCO_3 \rightarrow CaO + CO_2$	0.44 $kgCO_2/kg$	High (process CO_2)
Coke Breeze	$C + O_2 \rightarrow CO_2$	3.67 $kgCO_2/kg$	High (fossil CO_2)
Biochar	$C_{bio} + O_2 \rightarrow CO_2$	2.80–3.00 $kgCO_2/kg$	Low–Neutral

MATERIALS AND METHODS

Pellet mixtures were produced using industrial hematite concentrate processed to laboratory specifications. Fluxes (limestone, dolomite, and siderite) were ground to match the particle size distribution used in pellet plants. Their carbonate decomposition behaviour was central to the study due to its impact on CO_2 release and oxide-phase formation.

Solid fuel trials replaced fossil carbon with alternative fuels using a fixed carbon equivalence approach, ensuring comparable heat generation during firing. All ore, fluxes, binders, and fuels were weighed precisely and blended under controlled moisture to achieve consistent green pellet formation.

Pellets were prepared following standardised internal laboratory procedures. After dry mixing of all components, water was added gradually to reach the optimal moisture for balling. Green balls were produced on a laboratory disc pelletiser operating at controlled speed and angle, generating 8–12 mm pellets through layering and rolling. Pellets were screened for size uniformity and stored in sealed containers to maintain moisture before firing.

Thermal treatment simulated the drying, pre-heating, firing, and cooling stages of travelling-grate pelletising. Green pellets were heated in electric furnaces through a defined temperature program: drying to remove moisture, pre-heating for initial oxidation and carbonate breakdown, and firing at 1200 to 1350°C to promote slag formation and grain bonding.

Fired pellets underwent Cold Compression Strength (CCS) testing using calibrated universal testing machines. Density and porosity were measured using helium and sand pycnometer as well as optical microscopy analysis to assess how flux and fuel variations influenced pellet structure and void distribution.

Pellet cross-sections were prepared through resin impregnation and polishing. Optical microscopy was used to observe pore networks and bonding phases. Scanning electron microscope - energy dispersive X-ray spectroscopy (SEM-EDS) provided detailed characterisation of ferrite phases, slag chemistry, and reduction related features. X-ray diffraction (XRD) complemented imaging by identifying crystalline phases such as hematite, magnetite, and calcium or iron ferrites. Together, these techniques enabled correlation between thermal behaviour, phase evolution, and pellet strength.

FLUX REPLACEMENT

The decomposition of carbonate fluxes occurs during the induration phase (Borim, 2000). It is one of the most persistent sources of process CO_2 in hematite pelletising. Because $CaCO_3$ and $CaMg(CO_3)2$ release CO_2 as part of their fundamental calcination reactions, emissions occur regardless of furnace efficiency or fuel type. As global iron ore grades decline and flux demand increases, these calcination related emissions become even more significant.

Replacing traditional Ca based fluxes with siderite changes both the thermodynamics and the mineralogy of slag formation. Limestone supplies CaO needed for calcium-ferrite bonding phases, while siderite introduces FeO and Fe_3O_4, shifting the melt toward Fe-rich ferrites and reducing CaO content. At moderate substitution levels, this shift is manageable: oxidation and phase transformations occur earlier, total slag volume decreases, and CaO remains sufficient to maintain wetting and strong bonding typical of well-formed microstructures. Figure 1 shows the transition.

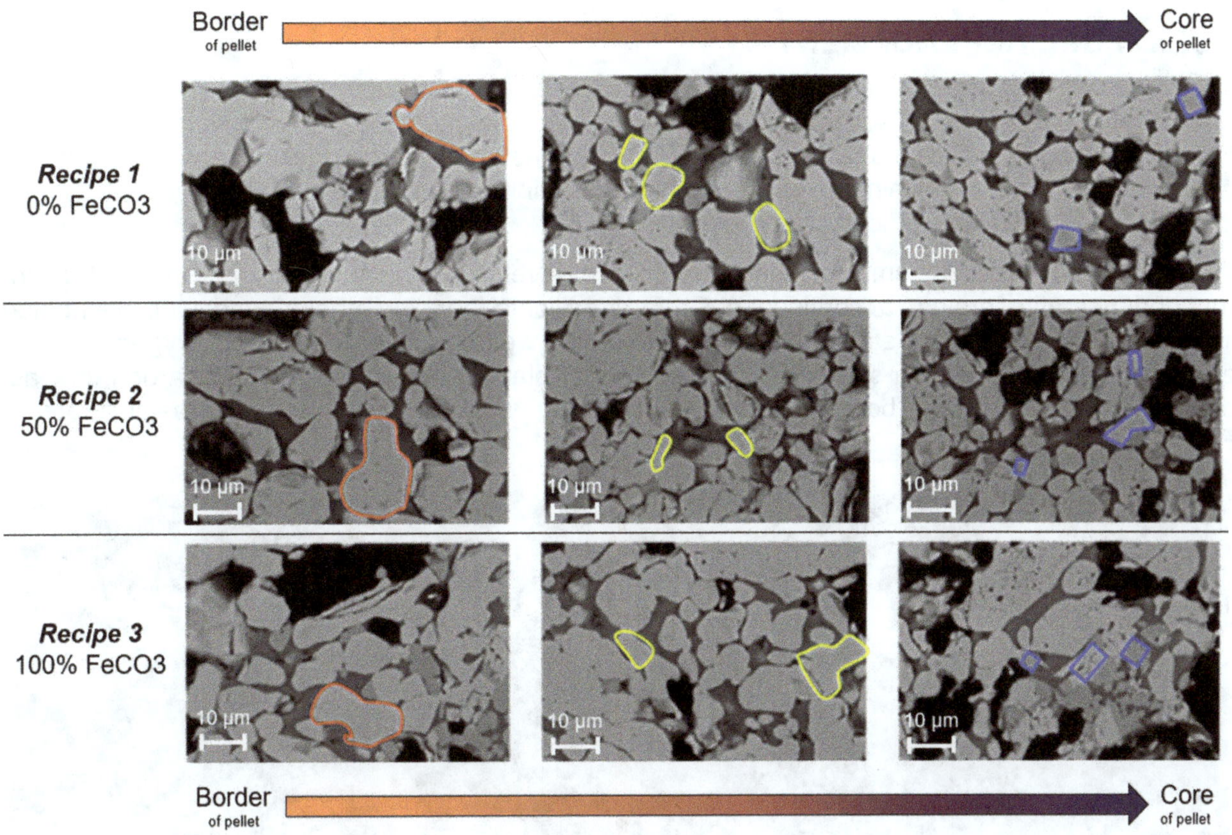

FIG 1 – Transition on bonding microstructures.

At higher substitution levels, however, falling CaO content weakens the calcium-ferrite network, Sslag formation becomes dominated by Fe-magnesioferrites with poorer wetting behaviour, while internal CO generation promotes premature reduction and metallic shell formation. These combined effects lead to over sintered microstructures, marked by coarser porosity, brittle slag films, and weaker bonding. Mechanical strength drops sharply once $FeCO_3$ replacement exceeds roughly 70–75 per cent, reflecting the loss of Ca-bearing liquid phases essential for high-strength sintering. The impact on CCS can be seen in Figure 2.

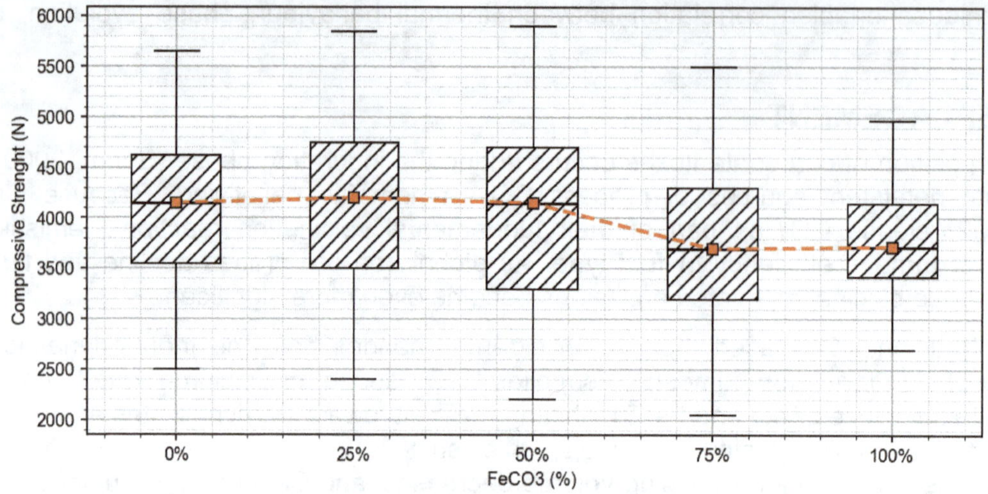

FIG 2 – CCS versus rate of replacement of $FeCO_3$.

SOLID FUEL REPLACEMENT

Decarbonising iron ore pelletising has become increasingly important as the steel industry shifts toward hydrogen-based, syngas-assisted, and biomass-integrated ironmaking. Among the remaining Scope 1 emissions, CO_2 released from carbonate flux decomposition during induration is particularly difficult to eliminate, as it is chemically inherent and cannot be avoided through improvements in furnace efficiency or fuel choice.

When internal reduction remains moderate, pellets develop a typical well-formed microstructure: rounded hematite grains, a continuous Ca-ferrite-silicate bonding phase, and well distributed, interconnected pores. This structure ensures good slag wetting, uniform heat and oxygen diffusion, and high cold compression strength. Biomass quality plays a key role high fixed carbon, low ash biomasses support proper bonding and result in strong, well consolidated pellets, as illustrated in Figure 3.

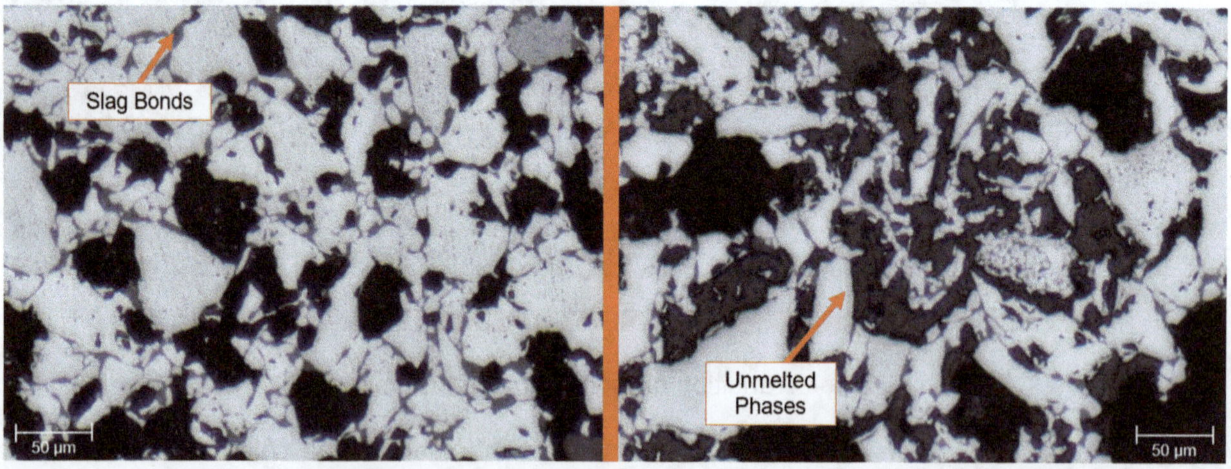

FIG 3 – Unformed slag bonds due high reactivity of low-quality biomass.

Microstructural analysis shows that slag bonds development depends mainly on the internal heat released by the fuel, while the pellet's chemical equilibrium is largely unaffected once the biomass burns off. High quality biomasses promote stable slag networks and improve overall pellet performance.

However, when reduction becomes too strong, due to highly volatile biomass, excessive reductant, or intense *in situ* CO/H_2 generation, the microstructure shifts toward over melted behaviour. Sun *et al* (2025) stated that this equilibrium is affected not by the type of biomass but the temperature. Premature metallic iron shells form around partially reduced grains, blocking re-oxidation and disrupting liquid phase redistribution. Instead of smooth bonding, coarse and brittle slag layers develop with magnetite or FeO-rich surfaces. Porosity becomes irregular, and mechanical strength typically falls by 5–15 per cent, reflecting weakened bonding and a more fragile structure.

CONCLUSIONS

This study demonstrates that meaningful CO_2 reductions in pelletising can only be achieved when flux chemistry, fuel selection, and induration conditions are optimised together. Replacing part of the conventional Ca-based fluxes with siderite lowers calcination-related CO_2 while still supporting the formation of strong bonding phases, provided CaO remains within the required range.

The choice of solid fuel plays an equally critical role. Controlled reduction promotes stable bonding phases and strong microstructures, while overly reducing conditions, driven by high-reactivity fuels or excess internal CO/H_2, produce poor slag distribution, and weaker structures.

Flux basicity, fuel reactivity, and thermal behaviour directly influence slag formation, grain evolution, and mechanical performance. When these parameters are jointly balanced, pellets achieve lower CO_2 emissions while maintaining the strength and metallurgical quality needed for both blast furnace and direct-reduction ironmaking.

ACKNOWLEDGEMENTS

The authors acknowledge Yolwin Perales for his contribution to the microstructural analyses conducted in this work. Appreciation is also extended to Paula Burgarelli and Wilker Penedo for their work on flux replacement studies. The authors further thank Tatiana Rozhkova, Rémi Belissont, and Marie Egert for coordinating and managing the experimental activities. Finally, the authors acknowledge ArcelorMittal S.A. for enabling and supporting the execution of this project.

REFERENCES

Aguiar, P, 2018a. Characterization of Pellets with Carbonate, Maizières lès Metz, internal report, ArcelorMittal Global R&D.

Aguiar, P, 2018b. Influence of Firing Temperature on the Mechanical Resistance of Pellets with Carbonate, Maizières lès Metz, internal report, ArcelorMittal Global R&D.

Borim, J, 2000. Modelagem e Controle de um processo de endurecimento de pelotas de minério de ferro (Modeling and Control of an Iron Ore Pellet Hardening Process), Master of Science Thesis.

Martins, G, 2025. Evaluation of Coke Breeze Substitution by Biomass in Iron Pellets, Maizières lès Metz, internal report, ArcelorMittal Global R&D.

Penedo Lima, W F, 2019. Replacement of Current Fluxes in Pelletising, Maizières lès Metz, internal report, ArcelorMittal Global R&D.

Sefidari, H, Lindblom, B, Nordin, L and Wiinikka, H, 2020. The feasibility of replacing coal with biomass in iron ore pelletizing plants with respect to melt induced slagging, *Energies*, 13(20):5386. https://doi.org/10.3390/en13205386

Sun, G, Guo, C, Ma, H, Xu, W and Wang, L, 2025. Chemical equilibrium and energy consumption analysis on biomass and iron oxides direct reduction ironmaking process, *Metals,* 15(1):57. https://doi.org/10.3390/met15010057

Szubka, M, Kądziołka Gaweł, M, Nowak, J, Klimontko, J and Wojtyniak, M, 2023. Thermal decomposition of siderite and characterization of the decomposition products under O_2 and CO_2 Atmospheres, *Minerals*, 13(8):1066. https://doi.org/10.3390/min13081066

Hydrogen reduction of iron ore – insights from 3D analysis

A Kiy[1], M Shadwell[2], T Evans[3] and M Knackstedt[4]

1. Postdoctoral Fellow, Department of Materials Physics, The Australian National University, Canberra ACT 2601. Email: alexander.kiy@anu.edu.au
2. Engineer Officer, Department of Materials Physics, The Australian National University, Canberra ACT 2601. Email: matt.shadwell@anu.edu.au
3. Chief Engineer R&D, Rio Tinto Iron Ore Technical Marketing and Steel Decarbonisation, Perth WA 6000. Email: tim.evans@riotinto.com
4. Professor, Department of Materials Physics, The Australian National University, Canberra ACT 2601. Email: mark.knackstedt@anu.edu.au

INTRODUCTION

Traditional iron production relies on carbon-intensive reductants like metallurgical coke, contributing significantly to greenhouse gas emissions. Low-carbon ironmaking technologies primarily use hydrogen-based reduction to produce direct reduced iron (DRI), leveraging green hydrogen from renewable electricity. Understanding iron ore behaviour during hydrogen reduction is essential for efficient and economical low-carbon processes.

This study uses a custom bench-top furnace for hydrogen reduction of iron ore lump, pellets, and fines. High-resolution micro-computed tomography (µCT) offers non-destructive 3D insights into microstructural evolution, including porosity development, crack formation, and phase-specific responses. Combined with scanning electron microscopy (SEM), X-ray diffraction (XRD), and X-ray fluorescence (XRF), these analyses reveal how structural differences affect reduction outcomes and disintegration. Adjusting parameters like temperature ramp rates and gas composition can mitigate these impacts, enhancing low-carbon ironmaking efficiency.

EXPERIMENTAL

Experiments were carried out in a custom bench-top furnace capable of simulating heating and reduction for iron ore products in industrial furnace conditions. Samples were limited to a maximum of 14 mm in diameter and 40 mm height. This allowed analysis of individual pellets (diameter <14 mm), up to 20 g of fines, or 12 mm cores drilled from +40 mm lump ore (yielding 3–4 cores per lump for comparisons or multi-technique analysis). The procedure involved three steps:

1. Heating the sample to 900°C at 20°C/min in argon (100 cm^3/min).
2. Introducing a reducing gas (90 cm^3/min hydrogen, 10 cm^3/min argon) and holding at 900°C for 2 hrs.
3. Natural cooling to room temperature in argon (100 cm^3/min, ~1 hr).

This ensured isothermal reduction at 900°C with controlled gas conditions.

Reductions took place in a custom holder with controlled gas atmosphere that allowed for seamless transfer between the µCT analysis and furnace equipment allowing for pre- and post-reduction comparisons. As the sample maintained its position before and after reduction, particle breakage behaviour could be investigated.

µCT was conducted at 6 µm voxel resolution. Porosity maps down to sub-micron resolution were further acquired via 'difference imaging': samples were imaged with and without an attenuating fluid, with the difference allowing for the quantification of open and closed porosity at sub-voxel resolutions (Latham, Varslot and Sheppard, 2008). Supplementary techniques used included SEM for 250 nm structural/mineral maps, XRD/XRF for mineral/elemental composition, and helium pycnometry for skeletal density.

RESULTS

High-resolution 3D µCT enables tailored analyses for industrial outcomes, such as minimising fine generation in pellet reduction influenced by composition (eg basicity) and structure (eg porosity).

Figure 1 shows μCT surface images of a pellet before and after reduction, revealing surface cracks post-reduction. Table 1 summarises porosity: total porosity rises from 23.0 per cent to 58.9 per cent (+156 per cent), dominated by micro-porosity (83 per cent and 88 per cent of the total porosity pre- and post-reduction, respectively). Figure 1c indicates homogeneous radial porosity distribution. An AI/ML model trained on 2D labelled data was applied to 3D data to distinguish cracks from porosity. Figure 1d shows homogeneous pre-reduction cracks shifting to surface dominated cracks post-reduction, potentially driving fines generation. Engineering pellet properties, including surface characteristics and composition, could mitigate fines generation and improve DRI performance.

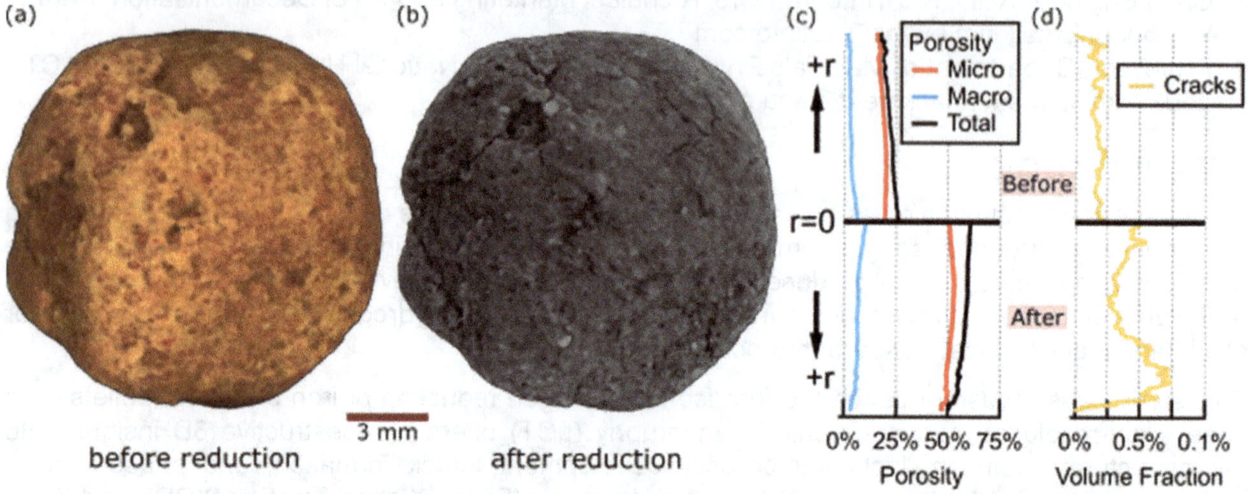

FIG 1 – μCT image of the surface of an iron ore pellet: (a) before reduction; (b) after reduction, with visible surface cracks; (c) radial porosity distribution pre- and post-reduction; (d) radial crack distribution pre- and post-reduction.

TABLE 1

Porosity of an iron ore pellet before and after reduction.

	Total porosity	Micro-porosity	Macro-porosity	Crack fraction
Before reduction	23.0%	19.2%	3.6%	0.2%
After reduction	58.9%	51.7%	6.8%	0.4%
Change	+156%	+169%	+89%	+100%

Another example of the utility of 3D μCT structural analysis is shown in Figure 2. μCT reveals a 12 mm diameter core taken from an iron ore lump exhibiting a layered structure (see Figure 2a). The top and bottom layer are dominated by Hematite (light grey), and the middle layer by Goethite (dark grey). Similarly to the pellet, the majority of the porosity (93 per cent) stems from micro-porosity (see Table 2). Once the lump is reduced (see Figure 2b), large cracks appear in the original Goethite layer, and at the interface between Goethite and Hematite rich regions (see Figure 2c). These originate from the high volume change that Goethite undergoes during reduction, due to the removal of water. Consequently, as the volume shrinks, stress and strain are induced on the original Goethite layer and interface, which are relieved by crack formation. As quantified in Figure 2d, cracks occupy as much as 25 per cent of the volume in the original Goethite layer but are absent in the original Hematite layer. This highly damaged region appears as a weak point in the reduced lump, which would contribute to a weak mechanical strength under load and would lead to high fine generation in industrial furnaces. Conversely, a lump that appears identical based on bulk properties such as mineral composition, porosity, and density, but with a different internal structure like finely inter-grained Hematite and Goethite would perform vastly different during reduction under load in a commercial plant.

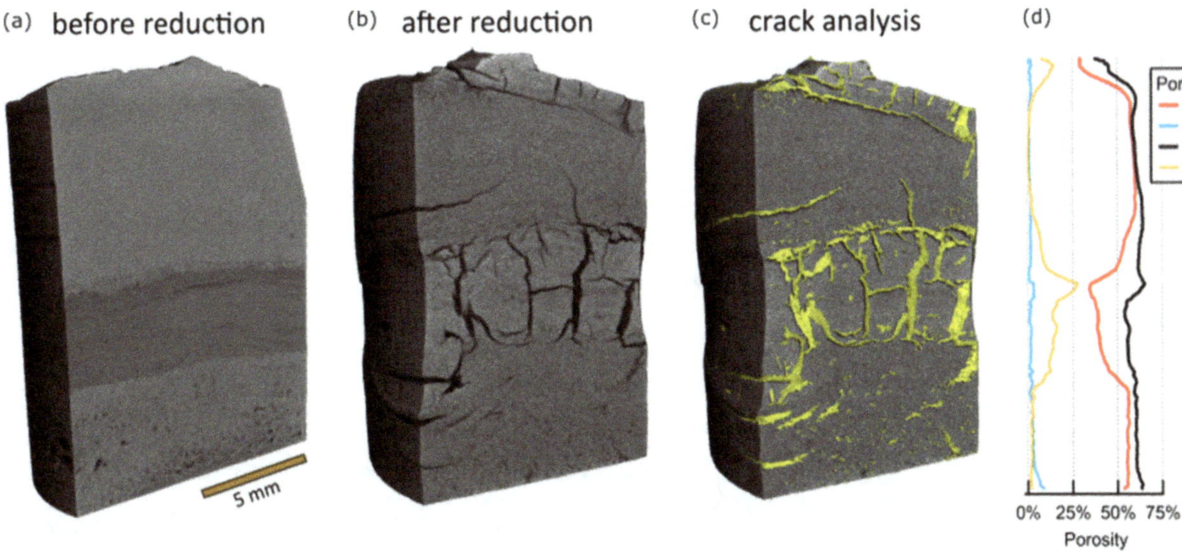

FIG 2 – µCT cross-section of an iron ore lump: (a) before reduction – the bright layers at the top and bottom mainly consist of Hematite, the dark central layer of Goethite; (b) after reduction; (c) cracks highlighted in yellow; (d) porosity and crack distribution along the z-axis after reduction.

TABLE 2

Porosity of an iron ore lump before and after reduction.

	Total porosity	Micro-porosity	Macro-porosity	Crack fraction
Before reduction	15.1%	14.0%	1.1%	~0%
After reduction	59.7%	51.4%	1.1%	7.2%
Change	+295%	+267%	±0%	N/A

CONCLUSION

This work underscores the value of high-resolution 3D µCT for understanding DRI processes across different iron ore products. The techniques provide insights for ore selection or process optimisation, advancing efficient low-carbon ironmaking and global decarbonisation.

REFERENCES

Latham, S, Varslot, T and Sheppard, A, 2008. Image registration: Enhancing and Calibrating X-ray Micro-CT Imaging, in *Proceedings of the International Symposium of the Society of Core Analysts 2008*, 12 p (Society of Core Analysts: New York).

Experimental study on the melting behaviour of different forms of H_2-DRI in electric smelting furnace-type slags

B Kumar[1], G Brooks[2], K Owusu[3], C Chen[4], A Guiraud[5], S Sabah[6], U Kudaru[7] and M A Rhamdhani[8]

1. Post Doctoral Fellow, CSIRO Clayton, Melbourne Vic 3168. Email: banty.kumar@csiro.au
2. Professor, Swinburne University of Technology, Hawthorn Vic 3122. Email: gbrooks@swin.edu.au
3. Post Doctoral Fellow, CSIRO Clayton, Melbourne Vic 3168. Email: kwaku.owusu@csiro.au
4. Research Scientist, CSIRO Clayton, Melbourne Vic 3168. Email: chunlin.chen@csiro.au
5. Principal Research Consultant, CSIRO Clayton, Melbourne Vic 3168. Email: adrien.guiraud@csiro.au
6. Postdoctoral Research Engineer, Swinburne University of Technology, Hawthorn Vic 3122. Email: ssabah@swin.edu.au
7. Research Fellow, Swinburne University of Technology, Hawthorn Vic 3122. Email: ukudaru@swin.edu.au
8. Professor, Swinburne University of Technology, Hawthorn Vic 3122. Email: arhamdhani@swin.edu.au

INTRODUCTION

Steel is one of the most widely used construction materials worldwide due to its exceptional durability, ease of fabrication, and cost-effectiveness, with applications spanning manufacturing, transportation, infrastructure, and agriculture. Global production of crude steel reached about 1.88 billion tonnes (Bt) in 2024. Steelmaking continues to be one of the most carbon-intensive industrial sectors, accounting for nearly 7–9 per cent of global CO_2 emissions, second only to cement production, despite its essential role in modern society (Shahabuddin, Brooks and Rhamdhani, 2022). The primary cause of this high carbon footprint is the predominance of the conventional coal-coke-based blast furnace–basic oxygen furnace (BF–BOF) route, which currently produces about 73 per cent of the world's steel and emits about 2 t of CO_2 per tonne of steel produced (Shahabuddin, Brooks and Rhamdhani, 2022).

To reach global net-zero emission goals, we need to switch to low-carbon steelmaking technologies. The hydrogen-based direct reduced iron (H_2-DRI) route, combined with an electric smelting furnace (ESF), has received a lot of attention as a possible way to reduce carbon emissions, especially for processing low-grade iron ores. This integrated approach significantly reduces greenhouse gas emissions by substituting hydrogen for carbon as the primary reducing agent and using renewable electricity for melting. Furthermore, the ESF route offers considerable operational flexibility, enabling the use of a range of feedstocks, such as DRI, hot briquetted iron (HBI), and fine iron-bearing residues. Additionally, the process improves material circularity and overall resource efficiency by producing value-added slags appropriate for cement applications.

METHODOLOGY

In this study, we investigated the melting behaviour of various forms of H_2-DRI, including H_2-reduced iron ore fines and tablet briquettes of different sizes produced from these fines. We used a synthetic slag composed of 39.4 per cent CaO, 36 per cent SiO_2, 14 per cent Al_2O_3, and 10.4 per cent MgO, which closely resembles typical blast furnace slag. The chemical composition of the H_2-DRI is given in detail in Table 1 and was characterised using X-ray fluorescence (XRF) at CSIRO Clayton, in Melbourne, Victoria.

Experiments were conducted at various temperatures and holding times to understand the interactions between the H_2-DRI samples and the slag during melting in a kilogram-scale induction furnace, simulating the bath conditions of an Electric Smelting Furnace (ESF). To determine the sinking time, the tablets were freely dropped from a height of approximately 200–250 mm above the slag surface. The time taken for each tablet to completely disappear or sink into the liquid slag was recorded. However, the exact melting time could not be clearly observed using this method. Therefore, to more accurately assess the melting behaviour, the tablets were held using a molybdenum wire for different durations, after which samples were collected. Based on these observations, the melting time was determined.

The results indicated distinct melting and sinking behaviours that depended on sample type, drop conditions, and size. The collected samples were further characterised using X-ray tomography analysis (XCT).

TABLE 1

Chemical Composition of H_2-DRI fines (wt %).

Fe(t)	Fe(m)	SiO_2	Al_2O_3	Na_2O	MgO	K_2O	CaO	P_2O_5	Mn_3O_4	TiO_2
77.6	51.8	8.2	4.3	0.008	0.107	0.037	0.063	0.16	1.52	0.311

RESULTS AND DISCUSSION

Sinking tests were conducted to evaluate the melting behaviour of H_2-DRI tablets (13 mm diameter, density ≈ 3.9 $g.cm^{-3}$) in molten slag mix of SiO_2–CaO–Al_2O_3–MgO at 1520±10°C and 1460±10°C. When tablets were freely dropped into the slag, they initially rebounded to the surface due to buoyancy and interfacial resistance before sinking.

At 1520±10°C, three experiments produced an average sinking time of approximately 26 sec. Rapid heat transfer upon contact with the molten slag led to the formation of a transient solidified slag shell, which affected both sinking and melting behaviour. No visible melting was observed up to 40 sec; partial melting occurred at 50 sec, and complete melting was achieved by 55 sec. The relatively short melting time at this temperature suggests favourable heat transfer conditions for rapid dissolution of H_2-DRI in high-temperature slag environments.

At 1460±10°C, the average sinking time increased significantly to approximately 42 sec. A more stable slag shell formed as a result of reduced heat transfer, which substantially delayed melting. In this case no visible melting was observed even after 100 sec of holding. The slag shell thickness initially increased up to 20 sec, followed by gradual thinning due to melting, with complete dissolution of the shell occurring after 45 sec. These results indicate a strong temperature dependence of slag shell stability and melting kinetics.

XCT was done to quantify the evolution of porosity with holding time. The results show a progressive increase in porosity with increasing holding time. The tablet recovered after holding for 50 sec at 1520°C exhibited a total porosity of approximately 30 per cent, as revealed by the 3D XCT reconstruction (Figure 1). The segmented porosity is widely distributed throughout the sample volume and forms a largely interconnected pore network, indicating extensive internal gas evolution and expansion of trapped gas inside the tablet during high-temperature exposure, which can enhance fragmentation and accelerate melting.

FIG 1 – 3D XCT reconstruction of the sample after holding at 1520°C for 50 sec, with coloured regions indicating porosity.

Additional experiments were performed using H_2-DRI fines, where 5 g of fines were directly introduced into molten slag. Upon contact with the slag, a rapid colour change from black to red was observed within a few seconds, indicating intense heat transfer to the fines. The recovered sample after 50 sec, shown in Figure 2, exhibited characteristics of 'Ferroberg' formation and remained afloat on the slag surface. This behaviour is attributed to rapid thermal arrest of the room-temperature fines, which promoted agglomeration and partial melting rather than complete dissolution into the slag. Due to the fine particle size and the transient nature of the melting process, the complete melting time could not be quantified under the present experimental conditions. Future studies will focus on employing modified experimental techniques to enable quantitative assessment of the melting kinetics of H_2-DRI fines.

FIG 2 – Photograph of the recovered sample after introducing 5 g of H_2-DRI fines into molten slag and holding for 50 sec at 1520°C.

SUMMARY

H_2-DRI tablet sinking and melting in molten slag are strongly influenced by temperature and slag shell formation. Higher temperatures (1520°C) led to faster sinking and complete melting, while lower temperatures (1460°C) slowed the process. Porosity increased with holding time (~30 per cent at 50 sec), and fines formed 'Ferroberg' on the slag surface.

ACKNOWLEDGEMENT

The authors gratefully acknowledge the support of HILT CRC, its industrial partners, and the Australian Government's Cooperative Research Centre Programme. We thank CSIRO colleagues Keith Vining, Chunlin Chen, Adrien Guiraud, Kwaku Owusu, Suneeti Purohit, and Tao Kan for technical discussions, experimental support, and access to facilities. We also acknowledge the Swinburne University of Technology team, including Prof. Geoff Brooks, Prof. Akbar Rhamdhani, Shabnam Sabah, and Ujwal Kudaru, for collaboration within the HILT CRC framework. Sherman Wong is acknowledged for assistance with XCT analysis.

REFERENCES

Shahabuddin, M, Brooks, G and Rhamdhani, M A, 2022. Decarbonisation and hydrogen integration of steel industries: Recent development, challenges and techno-economic analysis, *Journal of Cleaner Production*, 371:133611.

Metallurgical properties of pellets and lump ores under different temperatures and reducing atmosphere

C Liang[1], J Pan[2], D Q Zhu[3], Z Q Guo[4], C C Yang[5] and S W Li[6]

1. PhD, Central South University, Changsha Hunan 410083, China. Email: liangc@csu.edu.cn
2. Professor, Central South University, Changsha Hunan 410083, China. Email: pjcsu@csu.edu.cn
3. Professor, Central South University, Changsha Hunan 410083, China. Email: dqzhu@csu.edu.cn
4. Professor, Central South University, Changsha Hunan 410083, China. Email: guozqcsu@csu.edu.cn
5. Associate Professor, Central South University, Changsha Hunan 410083, China. Email: smartyoung@csu.edu.cn
6. Associate Researcher, Central South University, Changsha Hunan 410083, China. Email: swli@csu.edu.cn

ABSTRACT

Accordingly, this study first conducts a systematic comparative evaluation of the metallurgical behaviours of pellets and lump ore as standalone feed under a range of operating conditions. The study focus on critical factors presiding on the shaft furnace including reduction behaviour, sticking tendency, and low-temperature disintegration (LTD) characteristics in H_2-CO atmospheres, at various reduction temperatures. Under the reduction conditions predetermined for the individual lump and pellet samples, the metallurgical quality of pellet–lump composites was further interrogated. Implication to full scale industrial shaft furnace utilising a mixed burden of lump and pellet was assessed. The outcomes are expected to provide a scientific basis and engineering guidance for defining an appropriate blending ratio and optimising operating parameters in hydrogen-based reduction processes.

Pellets and lump ore with similar chemical compositions were samples for the investigation. The effects of reduction temperature (850, 900, 950, and 1000°C) and reducing atmosphere ($H_2/(H_2 + CO)$ = 0.60, 0.72, 0.84 and 1.00) on metallurgical performance were investigated. The results demonstrate that increasing the reduction temperature and enriching the hydrogen fraction both markedly enhance the reducibility of the two burden types. Specifically, when the reduction temperature increased from 850 to 1000°C, the reduction index (RI) of pellets increased from 7.33 to 8.30, whereas the RI of lump ore increased substantially from 3.67 to 7.66, indicating that higher temperature promotes the reduction reaction and that lump ore is more sensitive to temperature. In addition, as $H_2/(H_2 + CO)$ increased from 0.60 to 1.00, the RI of pellets increased dramatically from 3.37 to 15.6, and the RI of lump ore rose from 2.36 to 13.4. These results confirm that increasing the $H_2/(H_2 + CO)$ ratio significantly improves reducibility, and the enhancement is more pronounced for lump ore than for pellets.

To further clarify the influence of reduction temperature and gas composition on clustering tendency, the sticking index (SI) of pellets and lump ore was evaluated under the same operating conditions. The results show that the SI of both burden types increases significantly as the reduction temperature rises from 850 to 1000°C, suggesting that elevated temperature promotes inter-particle sticking during reduction. In contrast, increasing $H_2/(H_2 + CO)$ from 0.60 to 1.00 leads to a pronounced decrease in SI for both pellets and lump ore, indicating that a higher hydrogen fraction helps suppress sticking. Within the investigated conditions, the maximum SI reached approximately 90 per cent for pellets, whereas that for lump ore was only about 70 per cent, implying an overall stronger sticking tendency for pellets.

To elucidate the effects of reduction temperature and atmosphere composition on low-temperature disintegration (LTD), LTD indices were compared for pellets and lump ore under different conditions. When the reduction temperature was increased from 450 to 600°C, LTD + 6.3 mm and LTD + 3.15 mm for pellets decreased initially and then increased, while LTD - 0.5 mm exhibited the opposite trend (increasing first and then decreasing). In contrast, for lump ore, LTD + 6.3 mm and

LTD + 3.15 mm decreased continuously with increasing temperature, whereas LTD - 0.5 mm increased markedly. Moreover, as $H_2/(H_2 + CO)$ was increased from 0.60 to 1.00, LTD + 6.3 mm and LTD + 3.15 mm increased monotonically for both pellets and lump ore, while LTD - 0.5 mm decreased significantly. This indicates that increasing the hydrogen fraction suppresses the generation of fine particles (<0.5 mm) and improves resistance to LTD.

Based on the above comparative trends, to reduce dependence on pellet-only burdening and to assess the engineering applicability of a 'pellet–lump ore' composite burden under hydrogen-based reduction conditions, lump ore was progressively blended into pellets and the overall metallurgical performance of the resulting composite burden was evaluated.

The results show that the RI of the composite burden decreases overall with increasing lump-ore addition, which is consistent with the intrinsically lower reducibility of lump ore than pellets under comparable operating conditions. Nevertheless, when the lump-ore share reached 30 per cent, the RI of the composite burden still satisfied the HYL-III requirement (RI > 4), indicating that reducibility remains acceptable for practical operation at this blending level. Meanwhile, as the lump-ore proportion increased, LTD + 6.3 mm and LTD + 3.15 mm decreased gradually, whereas the fine fraction LTD - 0.5 mm increased continuously, suggesting that lump-ore addition tends to intensify LTD to some extent. Notably, even at a lump-ore share of 30 per cent, the composite burden retained good-sized stability (LTD + 6.3 mm > 80 per cent and LTD - 0.5 mm < 20 per cent), meeting the HYL-III constraints on reduction-disintegration performance. Overall, within the investigated range, the 'pellets + 30 per cent lump ore' composite burden simultaneously meets the process thresholds for RI and LTD, providing direct evidence for defining the blending window and optimising burden cost.

Characteristics of sinter return and its impact on sintering performance and decarbonisation

L Lu[1], T Evans[2] and K Takahara[3]

1. Chief Scientist, CSIRO Mineral Resources, Brisbane Qld 4000. Email: liming.lu@csiro.au
2. Principal Engineer, Rio Tinto Iron Ore, Perth WA 6000.
3. Senior Researcher, JFE Steel Research Centre, Fukuyama, Japan.

ABSTRACT

Sinter return is an undersized material, typically less than 5 mm, that is screened out either from the sintering machine after the cooler or just before being loaded into blast furnaces. It is generally recirculated back to the sintering machine and forms a significant portion of sinter blends. As it has undergone the sintering reactions and shows a considerably lower melting temperature compared with the natural ores in the sinter mixture, sinter return is believed to promote formation of sinter melt, the precursor of sinter bonding phase. As such use of sinter return during sintering has rarely been questioned despite its being a revert material from the process. However, with the global response to the threat of climate change, sinter plants are under increasing pressure to improve their efficiency and reduce their carbon footprint from both environmental and economic perspectives. Hence, it becomes increasingly important to re-visit and understand the role of the sinter return in the sinter mixture to optimise sintering performance and support future steel industry decarbonisation.

FORMATION MECHANISMS OF SINTER RETURN

Sintering is a raw material preparation process used worldwide in the production of iron and steel. The sintering process begins with preparation of a sinter mixture consisting of iron ores and recycled ferruginous materials, fuels and fluxes. After being mixed in a rotating drum, water is then added, often in conjunction with binders, to promote formation of green agglomerates of the components in the sinter mixture or granules (also called micro-pellets, quasiparticles, or pseudo-particles). Compared with the original sinter mixture, these granules are much coarser with virtually no ultrafine particles, assisting in achieving optimum permeability of the sinter bed during the sintering process. The green granules of the sinter mixture are then loaded to a depth of typically 0.5–1 m on a sinter strand, which is a continuous grate moving constantly at typically 2–3 m/min. After the granules are loaded and levelled on the sinter machine, the sinter bed passes a series of gas or oil burners, which heat up the granules and ignite the fuel (often coke) particles on the surface of sinter bed. Under the influence of suction, the heat generated from combustion of the coke particles on the surface is drawn down and continues to ignite the coke particles of successive layers of the sinter bed. The peak temperature of the burning coke layer (also called the flame front) reaches approximately 1350°C, generating a melt phase first from adhering fines and then by assimilation of coarse nucleus particles. On cooling the sinter melt solidifies into a sinter matrix that bonds the loose and not fully reacted iron ore particles into a cake of clinker-like material.

At the end of the sinter strand, the sinter cake falls off the grate and is crushed to a specified top size. The crushed sinter particles are then conveyed to a screening station and separated into product sinter (typically +5 mm), hearth material (typically -16+10 mm), and sinter return (typically -5 mm). The product sinter is then conveyed to the blast furnaces and screened again before being charged. Sinter return refers to the undersized materials screened out at the screening station and from the blast furnaces. The former is generated within a sinter plant and often called 'internal sinter return', while the latter is generated outside a sinter plant and is called 'external sinter return'. Both are too fine for use in blast furnaces and are reverted to a bin for recycling in the sintering process.

Figure 1 shows an optical micrograph of a typical pilot-scale pot-grate sinter. It consists of relict particles cemented in a porous matrix. Relict particles mostly originate from coarse dense iron ore particles in the mixture. Flux and coke particles are seldom observed as relict particles to be presented in the sinter structure. The coarse and white relict particles (labelled relict hem) in Figure 1 are all residues of iron ore particles. These iron ore particles are sufficiently stable and hence survived wholly or partially throughout the sintering process. As shown in Figure 1, the relict particles are quite dense or micro-porous depending on their precursors. Although they have not fully undergone the sintering reactions, these particles are fully dehydrated into hematite if they were not originally goethite particles. Therefore, the relict particles are often classified as 'primary hematite'. In contrast, the sinter matrix is more predominant in sinter structure and is significantly more porous both at macro and microscopic scales than the relict particles. Depending on the localised thermal and compositional conditions, the following mineral phases can be observed in the sinter matrix, silico-ferrite of calcium and aluminium (SFCA), magnetite, hematite, quartz, larnite and calcium disilicate. Unlike the relict particles, these phases are precipitated from the sinter melt and are classified as 'secondary phases'. They are often characterised of crystal defects and high impurities, such as SiO_2 and Al_2O_3, making them less competent than primary hematite.

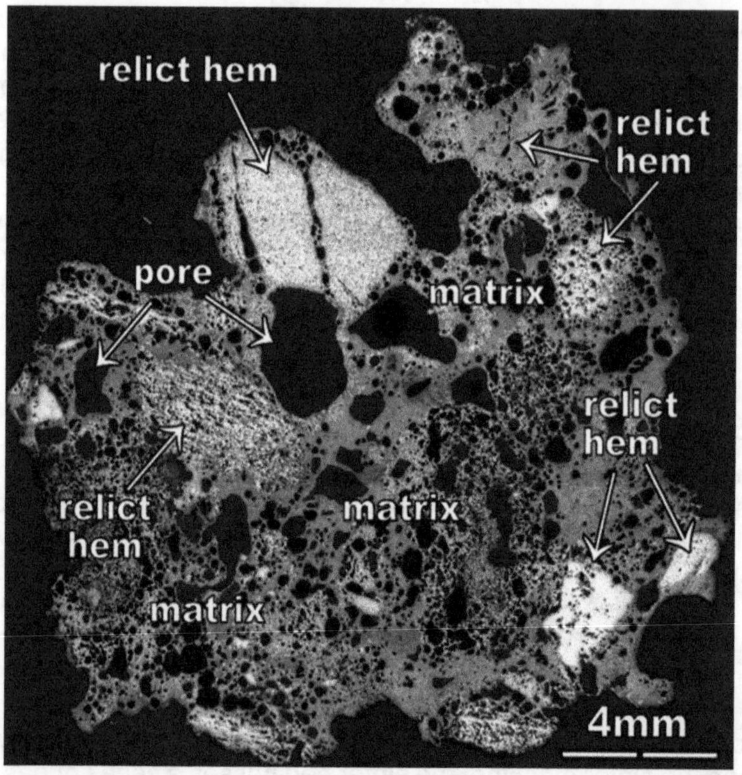

FIG 1 – Optical micrograph of a typical pilot scale pot-grate sinter displaying sinter nuclei (hematite) cemented by a continuous porous sinter matrix.

As shown in Figure 1, sinter is very heterogeneous in terms of its mineral phases, composition and pore structure, leading to the heterogeneous nature of its resistance to breakage under the influence of impact due to collision and attrition. As the sinter matrix contains considerable porosity and some of its phases are naturally brittle with defects, cracks are readily initiated and developed in the matrix. The sinter cake is thus broken into sinter particles of different sizes, inevitably some of these particles are below 5 mm, contributing to sinter return. Since sinter breakage takes place preferentially across the matrix, this will expose lots of matrix material on the surface of sinter particles. As the exposed matrix material is brittle and often angular, many small chips are expected to break off the sinter particles during transportation within sinter plants and to blast furnaces. The small chips will contribute to either internal sinter return or external sinter return depending on the location of their formation. In addition, some small sinter particles may fall below the blast furnace screen due to the size reduction taking place in transportation.

CHARACTERISTICS OF SINTER RETURN

Chemical characteristics

Table 1 shows the compositions of two sinter returns (SR1 and SR2) and three typical iron ore fines (Ores 1, 2 and 3). Both sinter returns have distinct chemistries from the natural ores. As the sinter returns have already been thermally processed, they contain no combined water. In contrast, the natural ores show a varying amount of hydrated iron bearing minerals. For the same reason, the sinter returns show a higher FeO content than the natural ores. As additional fluxes are added during the sintering process, sinter returns are also higher in SiO_2, CaO and MgO compared with the natural ores. Sinter returns are more complicated in chemistry and need be represented with quaternary Fe_xO-CaO-SiO_2-Al_2O_3 or higher system. In addition, the absence of hydrated iron bearing minerals in sinter returns will also have a profound impact on sintering performance and sinter quality.

TABLE 1

Chemical composition of sinter returns and typical iron ore fines (%).

	TFe	SiO_2	Al_2O_3	P	S	CaO	Mn	MgO	$LOI_{371^{\circ}C}$	LOI_{Total}
Ore 1	62.53	4.5	2.19	0.071	0.011	0.1	0.07	0.19	2.12	3.02
Ore 2	57.9	5.47	1.45	0.039	0.016	0.04	0.02	0.05	8.83	9.87
Ore 3	67.07	0.94	0.93	0.019	0.006	0.02	0.43	0.02	1.03	1.4
SR1	54.44	6.27	1.93	0.057	0.016	11.49	0.73	1.59	-0.09	-0.53
SR2	54.39	6.25	1.97	0.060	0.015	11.52	0.73	1.52	-0.06	-0.43

Physical characteristics

Figure 2 compares the size distribution of the same sinter returns and natural iron ores. Compared with natural iron ores, sinter returns are narrower in size distribution and lack both very coarse and adhering fines. Therefore, the return fines (RF) particles most likely serve as nucleus particles during granulation. This may explain why they are generally believed to be beneficial to the green bed permeability. However, RF has larger pores, is less hydrophilic and has lower moisture holding capacity than natural iron ore fines, which will adversely impact its granulation efficiency.

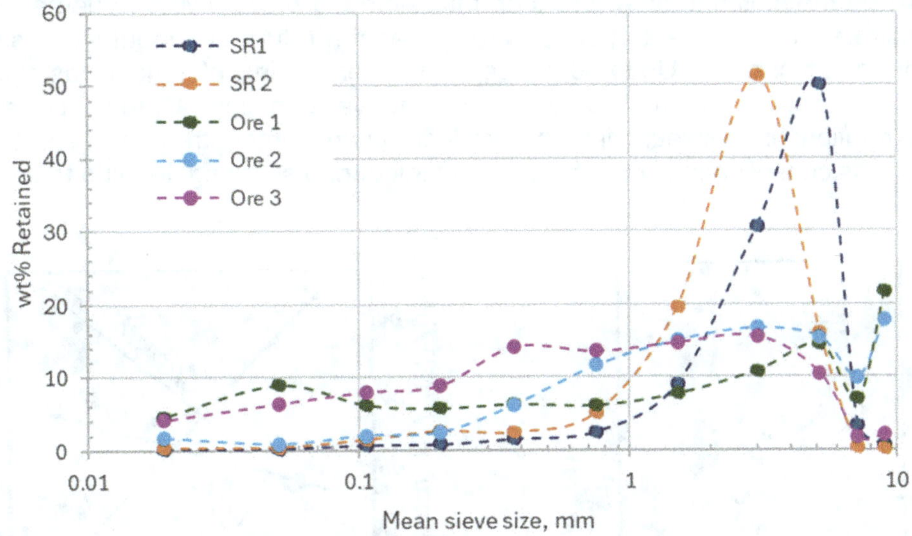

FIG 2 – Size distribution of sinter returns and typical natural iron ore fines.

Mineralogical characteristics

Based on the RF formation mechanisms presented above, it is not difficult to conclude that sinter returns have microstructure similar to the sinter matrix. In their study Wu *et al* (2013) classified the sinter return particles into four different types based on visual observations (A, B, C and D). Type A sinter return particles have no visible bonding phase on their surface, Type B sinter return particles are partially coated with bonding phase and Type C and D sinter return particles are predominantly coated by the bonding phase. Compared with Type C, Type D sinter return particles have a thick fracture interface. After sorting the sinter return particles in the size range of 3 to 5 mm based on the above classification, Wu *et al* (2013) found that 75 per cent of the return particles sorted fell in the category of Types C and D and a further 20 per cent in Type B. This work together with the difference observed in chemistry between sinter product and return further confirms the authors' conclusion on the formation of sinter return and mineralogical similarity of sinter returns and sinter matrix.

It is worth noting that the green granules on the sinter strand are loaded non-uniformly, either intentionally or un-intentionally, in particle size, mineral composition and fuel content. This is further complicated by the uneven air and heat flows resulting from the heterogeneous nature of bed porosity and varying thermal conditions around the sinter bed. As such, different parts of the sinter bed are expected to contribute to different amounts and types of sinter returns. Type A return particles most likely originate from the top of sinter bed, where heat is insufficient for melt formation due to significant heat loss from top and the strong cooling effect by cold incoming air. These sinter return particles are calcined iron ore particles of less than 5 mm. Due to a lack of matrix forming material, they are not bonded together and report to sinter return during screening. Type B return particles are formed in the subsurface area, side and sometimes bottom of the bed where sinter melt is formed but not sufficient to coat the whole particles. Types C and D return particles are believed to originate from the inner bed which accounts for most of the bed volume. This explains the dominance of these two types of return particles. It is possible that Types C and D return particles, in particular, Type C particles, have no relicts in them due to excessive heat and severe melting. Based on the discussion above, it is not difficult to conclude that more sinter returns are generated from the top, then side and bottom, finally core of the bed.

SINTER RETURN AND SINTERING PERFORMANCE

The role of sinter return with regards to the sintering performance has been rarely studied, particularly in recent years. This is largely since it is inevitable in the process and is often believed to be beneficial to the sintering process.

Two studies have been done in the 1960s in Japan to examine the effects of sinter return on sintering time, productivity and sinter strength. Figures 3 and 4 show the test results by Sanematsu *et al* (1960) and Shono, Ito and Oobuchi (1963), respectively. Both studies showed the beneficial impact of sinter return on the sintering time and therefore sintering productivity. Sanematsu *et al* (1960) further demonstrated improvement in sinter strength as the return fine input increased from 10 to 40 per cent (ore and flux base). Up to 60 per cent improvement in sinter yield was also reported by Shono, Ito and Oobuchi (1963). The findings reported above seem contrary to our current knowledge on the inverse relationship between sintering productivity and sinter strength. A short sintering time often indicates less consolidation time of solid and melt during sintering, leading to a weaker sinter.

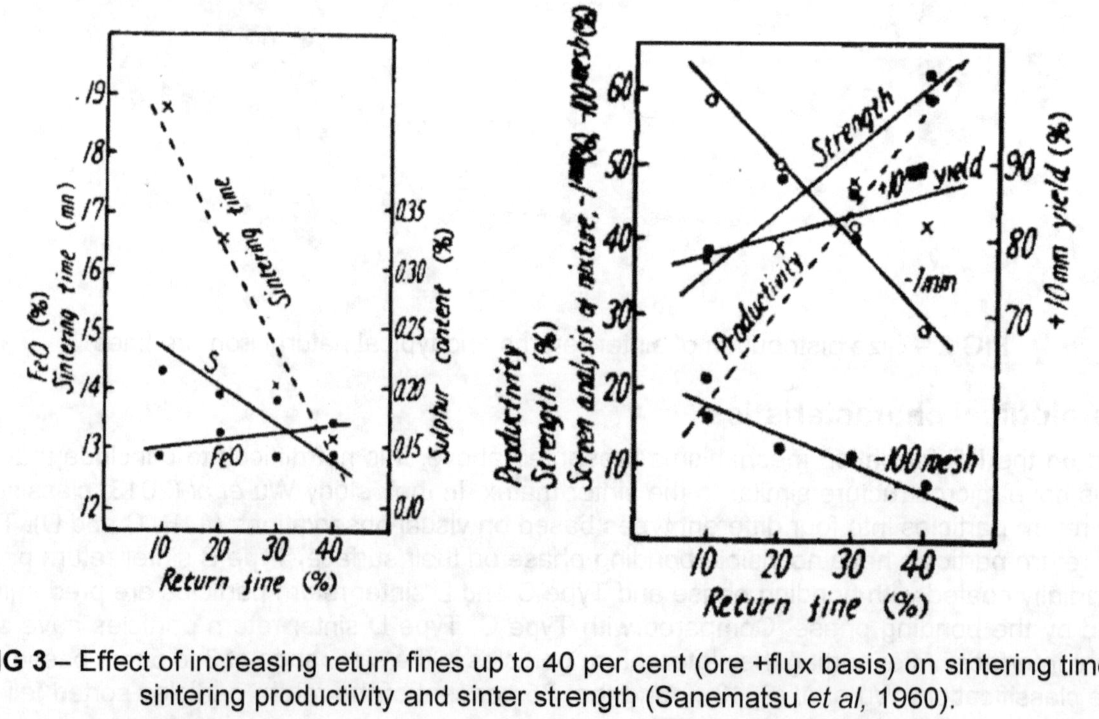

FIG 3 – Effect of increasing return fines up to 40 per cent (ore +flux basis) on sintering time, sintering productivity and sinter strength (Sanematsu *et al,* 1960).

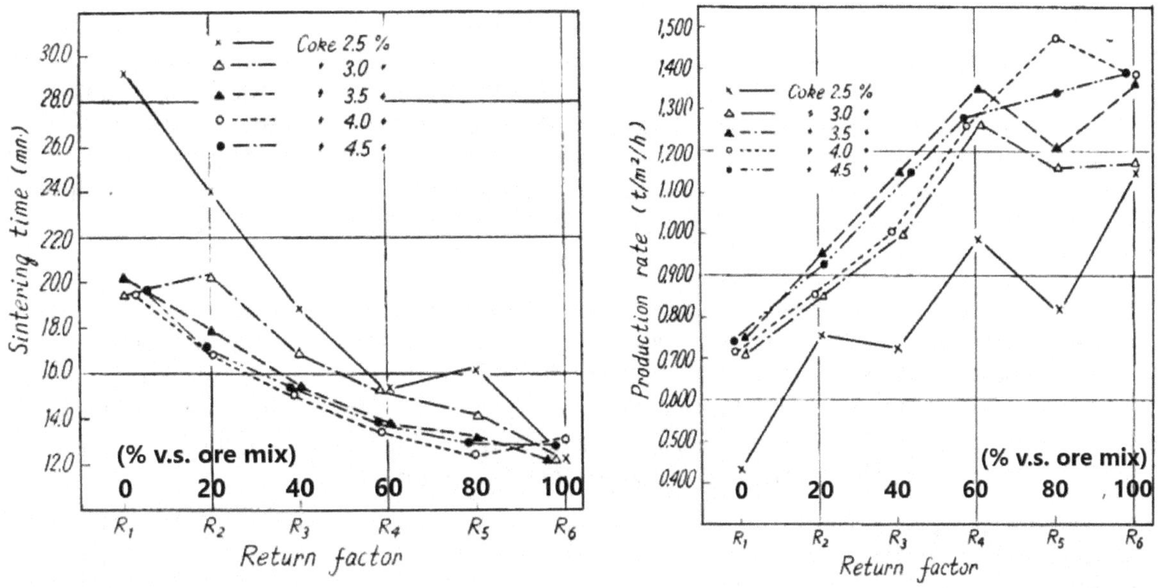

FIG 4 – Effect of increasing return fines up to 100 per cent (ore + flux basis) on sintering time and production rate (Shono, Ito and Oobuchi, 1963).

In addition, both studies attributed the observed improvement in sintering productivity to good green bed permeability of sinter mixtures at higher additions of sinter return. However, no permeability results of the sinter mixtures containing different proportions of RF were reported in the literature above. The effect of increasing sinter returns on the granulation efficiency of an ore blend was studied by Wu *et al* (2013). As shown in Figure 5, the granulation index of sinter mixtures was found to decrease with increasing addition of both types of sinter returns. However, since their granulation work was conducted at a constant water addition and there is a considerable difference in water holding capacity between sinter returns and natural ores, the sinter mixtures were not compared under the same saturation level. Therefore, the impact of sinter return on sintering performance is still debatable and further research is needed to clarify the test results in the literature.

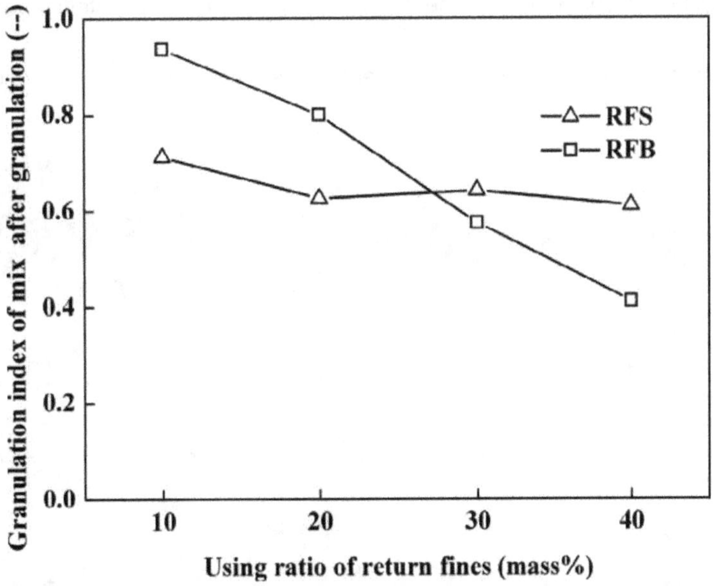

FIG 5 – Granulation index of granulated mixes at different ratios of sinter returns (Wu *et al,* 2013).

SINTER RETURN AND DECARBONISATION

As fossil fuels are key sources of heat for sintering reactions, sinter plants are a major point source of carbon emissions into the atmosphere, accounting for about 13–15 per cent of the CO_2 emitted from integrated steel mills. On the other hand, sinter return is inevitable and has shown some beneficial impact in the early literature on sintering productivity and sinter quality. Hence the application of sinter return has rarely been questioned and are unlikely optimised. Thus far, the proportion of sinter return in sinter mixtures has reached a high proportion accounting for more than one third of iron bearing materials in sinter mixtures. Furthermore, it has tended to trend up in recent years due to the deterioration of iron ore quality. Since sinter return is generally recirculated back to the sintering machine, the heat used for sintering this part of the material is largely wasted. It also limits the amount of virgin ores in the sinter mixture. Therefore, massive generation of sinter return should be avoided as it will affect sintering productivity and fuel consumption adversely, hence the carbon footprint of sinter plants.

Past efforts have been focused on improving sintering performance and reducing fuel consumption by optimising the sintering process and sinter blends without addressing sinter return. This has brought us to the point where further improvement in these areas becomes increasingly difficult. On the other hand, sinter return that makes up at least one third of the iron bearing materials has yet been sufficiently examined for the reasons discussed above. This will no doubt offer the opportunity for decarbonisation through optimising the usage of sinter return.

CLOSING REMARKS

Sinter return is largely formed due to breakage of the sinter matrix during crushing and transportation. Hence generation of sinter return is inevitable. Compared with natural iron ores, sinter return: 1) has a narrower size distribution and lacks very coarse and very fine adhering particles; 2) is high in FeO, SiO_2, CaO, MgO and hence lower in Fe; and 3) has mineralogy close to sinter matrix. As it contains phases of low melting point, sinter return is believed to promote melt formation, which is particularly important in achieving sinter strength as the sintering process evolves kinetically very fast. The role of sinter return to the sintering performance has been somewhat overlooked, particularly in recent years, and the results from limited research in the literature are debatable. Sinter return currently makes up at least one third of the iron bearing materials in sinter mixtures, optimising its usage will no doubt offer new opportunities for ironmaking decarbonisation.

REFERENCES

Sanematsu, T and Toyozawa, H, 1960. On the quantity of return fines in the sinter mixture, *Tetsu-to-Hagané*, 46(3):243–245. https://doi.org/10.2355/tetsutohagane1955.46.3_243

Shono, S, Ito, K and Oobuchi, S, 1963. Effects of the mixing rate of return ore on sintering properties, *Tetsu-to-Hagané*, 49(10):1264–1266.

Wu, S, Zhu, J, Fan, J, Zhang, G and Chen, S, 2013. Sintering Behaviour of Return Fines and their Effective Utilization Method, *ISIJ International*, 53(9):1561–1570.

Pelletising of Australian DSO sinter fines – challenges and opportunities

L Lu[1]

1. Chief Research Scientist, CSIRO Mineral Resources, Pullenvale Qld 4069.
   Email: liming.lu@csiro.au

ABSTRACT

This presentation will first discuss the challenges to make pellets from the existing Direct Ship Ore (DSO) sinter fines and then give a brief introduction of CSIRO's research and development in this area. The metallurgical performance of the pellets made from various DSO sinter fines will be presented, demonstrating the potential to produce pellets from this type of ores with a wide spectrum of goethite content.

INTRODUCTION

Due to their potential in 100 per cent hydrogen substitution for fossil carbon and together with their operation maturity and comparable output, the shaft furnace Direct Reduction (DR) processes are expected to gradually overtake the traditional blast furnace and other smelting processes to make steel. The current shaft furnace processes use predominantly premium grade iron ore pellets. Hematite–goethitic ore type represents by far the largest proportion of iron ores being produced in Australia. While this type of ores has been widely used as feed material for producing iron ore sinter fed to the conventional blast furnace process, unfortunately there has been no precedent success in making suitable pellets from this type of iron ores. It is hence important to develop agglomeration technologies for production of pellets from DSO sinter fines that can withstand the DR shaft furnace process. This would provide a commercially viable route for Australian ores into green steelmaking processes.

CHALLENGES TO PELLETISE DSO FINES

DSO fines defer from traditional pellet feeds in particle size distribution, mineralogy and chemistry. Many challenges need be overcome before their usage in pellet production. Compared with pellet feeds (usually <150 μm with a P_{80} of around 45 μm (Patra, Kumar and Venugopal, 2017; Abazarpoor *et al*, 2018), DSO fines are much coarser typically having a top size of 10 mm. The fines need first be ground to a size distribution and surface morphology suitable for pelletising. Typical DSO fines from Australia are dominated with goethite with a Loss On Ignition (LOI) content ranging from 3 to 10 per cent, this being an analytical measure for the amount of volatile matter contained in the sample. The unique mineralogical characteristics have impeded our past effort to make pellets from the type of fines. The volatile matter (often crystal water) they carry affect not only the formation of green balls but also their induration process, leading to the inferior quality of fired pellets. In addition, unlike pellet feeds which have been through multiple steps of grinding and concentration to achieve the size and chemical specifications required as pellet feeds, DSO fines are so called 'direct ship ore' and have undergone minimum upgrading. Therefore, DSO fines are considerably lower in Fe grade and higher in gangue minerals, particularly SiO_2, Al_2O_3, P. These minerals will affect not only the metallurgical performance of fired pellets but also their subsequent smelting processes.

CSIRO R&D DEVELOPMENT IN PELLETISING DSO FINES

Over the years, CSIRO has tested extensively on a variety of DSO fines from Australia and India. Figure 1 shows the upper and lower contents of SiO_2, Al_2O_3 and LOI contents of the DSO fines tested. The fines have been ground under various conditions using different milling methods to prepare feed materials suitable for producing green balls, followed by induration tests to establish the optimal firing conditions of the green pellets and metallurgical tests to evaluate the quality of the fried pellets.

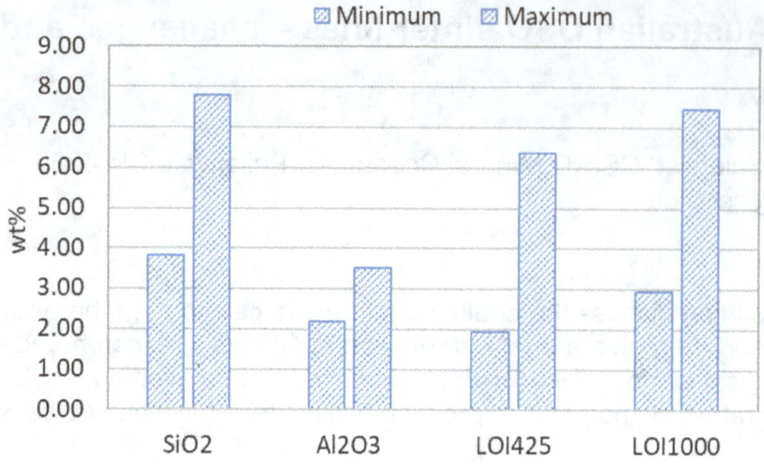

FIG 1 – SiO_2, Al_2O_3, and LOI ranges of DSO fines tested.

OPPORTUNITIES FOR PELLETISING DSO FINES

Figure 2 shows the quality of green balls obtained from various DSO fines. The green balls from the DSO fines tested achieved the required quality for subsequent handing and induration conditions during the pelletising process.

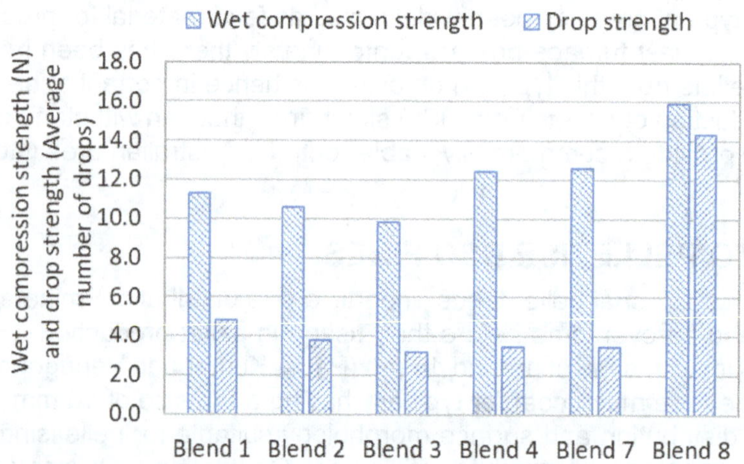

FIG 2 – Compression and drop strengths of green balls from various DSO fines.

Figure 3 shows the compression strength of fired pellets obtained from various DSO fines. All pellets except the pellets from Blend 2 achieved a compression strength of about 3000N or above. The pellets from Blend 2 were fired at a lower temperature compared with those from other DSO fines and hence showed a lower compression strength just under 2500N.

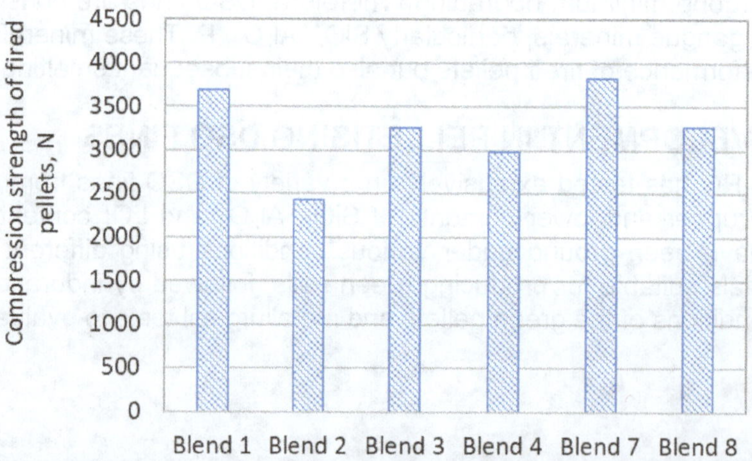

FIG 3 – Compression strength of fired pellets from various DSO fines.

Figure 4 shows the degree of reduction of pellets from Blend 1 as a function of reaction time at different temperatures in hydrogen and CO enriched gas atmospheres. Compared with CO, pellets were reduced considerably quicker in the hydrogen enriched atmosphere. The sample was fully reduced within 90 min at 950°C.

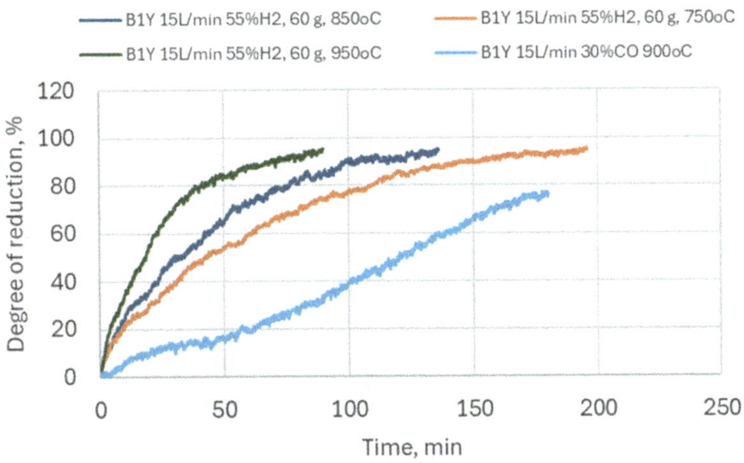

FIG 4 – Reduction – time curve of fired pellets at different temperatures.

Selected pellets were also tested for their resistance to low temperature reduction disintegration using the low temperature disintegration testing method by HYL Technologies. As shown in Figure 5, the pellets tested achieved a high % unbroken pellets of 88.0–94.4 per cent, excellent %+6.3 mm of 91.9–93.7 per cent and low %-3.15 mm of 6.0–6.8 per cent.

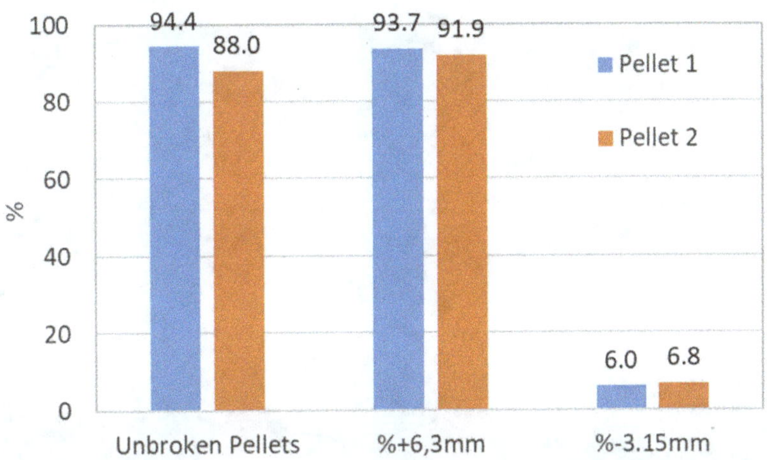

FIG 5 – %unbroken pellets, %+6.3 mm and %-3.15 mm of pellets resulting from HLY low temperature reduction disintegration tests.

CLOSING REMARKS

Test work on the Australian and Indian DSO fines so far confirmed that severe cracks can be avoided in the roasting process. From limited physical and reducibility tests done for the pellets from the DSO fines, the pellets meet the physical and metallurgical properties required for the DR shaft furnace process.

ACKNOWLEDGEMENTS

The author would like to thank Dr Han and Dr Aladejebi, for their contributions to prepare and characterise green and fired pellets and Dr Tang for her contribution to evaluate the reducibility of fired pellets.

REFERENCES

Abazarpoor, A, Halali, M, Hejazi, R and Saghaeian, M, 2018. HPGR effect on the particle size and shape of iron ore pellet feed using response surface methodology, *Mineral Processing and Extractive Metallurgy*, 127(1):40–48.

Patra, S, Kumar, A and Venugopal, R, 2017. The effect of particle size on green pellet properties of iron ore fines, *Journal of Mining and Metallurgy*, 53A(1):31–41.

Calix's zero emissions steel technology (ZESTY) – from proof of concept to demonstration plant

C Ormston[1], Y Xia[2], T Dufty[2], S Sabah[3], S Al-Assafi[2], A Okely[2], J Fu[2], S van Dorp[2], A R Rhamdani[3], G Brooks[4], M G Sceats[2] and M E Boot-Handford[2]

1. Calix Limited, Pymble NSW 2073. Email: cormston@calix.global
2. Calix Limited, Pymble NSW 2073.
3. FPD (Fluid and Process Dynamics) Group, Department of Mechanical and Product Design Engineering, Swinburne University of Technology, Hawthorn Vic 3122.
4. FPD (Fluid and Process Dynamics) Group, Department of Mechanical and Product Design Engineering, Swinburne University of Technology, Hawthorn Vic 3122. Email: gbrooks@swin.edu.au

INTRODUCTION

Decarbonising iron and steel production remains one of the most critical challenges for global industry, particularly for regions such as Australia where iron ore exports represent a cornerstone of economic activity. Hydrogen based direct reduction provides a pathway to near zero emissions ironmaking however, the technical hurdles associated with processing Australia's lower grade, fine grained hematite and goethite dominant ores, combined with operational issues such as sticking, heat transfer limits, and fines handling, have constrained the adoption of conventional shaft furnace or fluidised bed technologies. Calix's Zero Emissions Steel TechnologY (ZESTY) offers an alternative process route engineered to overcome these limitations. Built on Calix's patented indirectly heated Flash Calcination (CFC) technology, ZESTY enables flash hydrogen reduction of iron ore fines in a lean phase environment, with high metallisation, minimal hydrogen consumption, and lower processing temperatures than those of traditional ironmaking technologies.

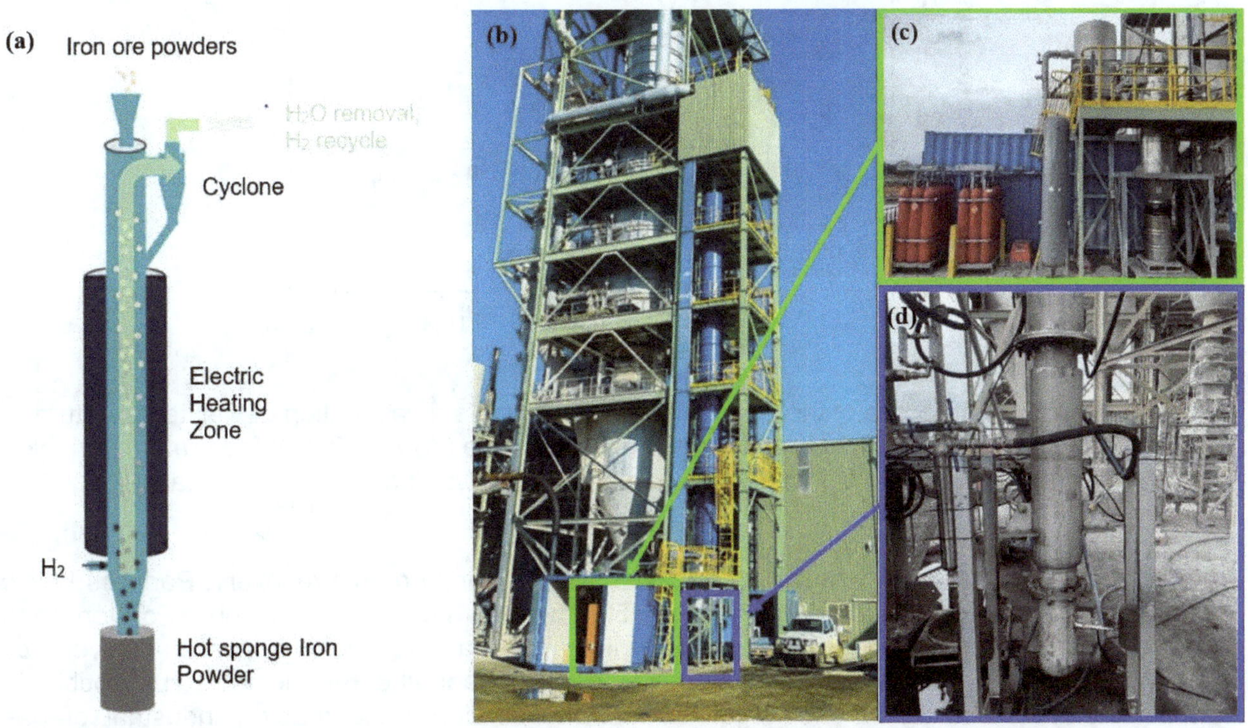

FIG 1 – (a) Schematic of ZESTY, (b) the pilot plant flash reduction reactor (blue reactor tube to the right of the Calix CFC1500 flash calcination demonstrator), (c) manifold and sealed bag house used and, (d) the modified powder/product collection set-up to allow safe collection and recovery of the fine DRI product.

Since 2021, Calix has completed over 130 pilot runs across nine hematite/goethite and magnetite ores. Test work at the Calix Technology Centre in Victoria, using the pilot electric CFC reactor modified for direct hydrogen reduction (Figure 1), generated critical process data to validate ore compatibility, optimise reduction parameters, informing the basis of design for the 30 000 t/a ZESTY H_2-DRI demonstration plant. Supported by a A$44.9M ARENA grant and more than A$35M in cash and in-kind contributions from Rio Tinto through a Joint Development Agreement, the demonstration plant will represent the world's first largescale implementation of an electrically heated, fines-based hydrogen reduction technology.

A key innovation of the ZESTY reactor is its indirect and electrically heated, vertical tube design. Ore fines, typically milled to <500 µm, are injected into the top of the reactor and descend through a precisely controlled hydrogen/steam atmosphere. Heat is supplied via external electrical furnace, which avoids hydrogen combustion and enables the process to approach the theoretical minimum hydrogen requirement. Short residence times of 40–120 seconds are sufficient to achieve high metallisation, as the combination of fine particle size, precise temperature control, and counter-flow hydrogen configuration ensures high reaction kinetics while mitigating the onset of sticking and accretion formation.

Pilot testing of hematite–goethite ores demonstrated clear temperature-dependent behaviour, with metallisation increasing sharply between 750°C and 950°C (Figure 2). Above approximately 950°C, metallisation gains diminished and sticking became more evident, particularly at 1050°C. As such, controlled operation within 950–1000°C was identified as the optimal regime for high metallisation while maintaining operational stability. All hematite–goethite ores processed at these conditions exceeded the 75 per cent metallisation threshold required for electric smelting furnace (ESF) entry.

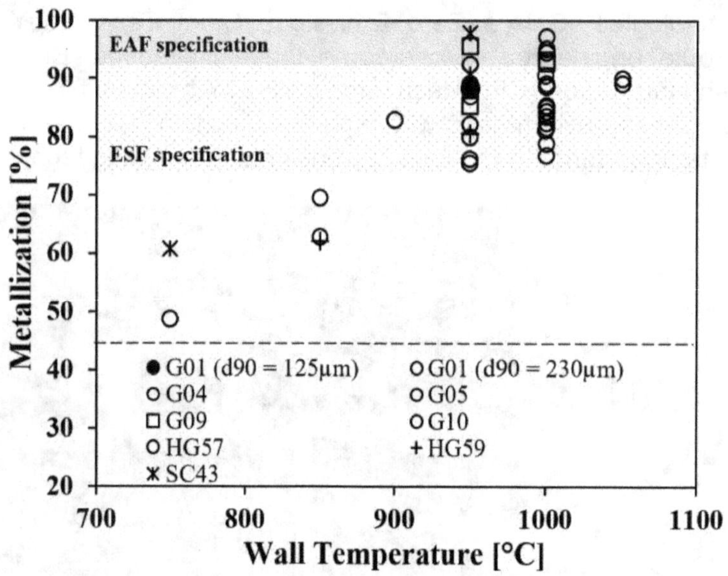

FIG 2 – Metallisation degree of the reduced iron ore products as a function of wall temperature. (Ore = various hematite-goethite, wall temperature = variable (750–1050°C); feed rate = variable (30 120 kg/h); H_2/O_{RED} stoichiometry = variable (target > 1.5); particle size = variable (d_{90} = < 300 µm).

Fine-particle behaviour proved critical for both metallisation and product recovery. Particles below ~300 µm reduced efficiently, while ultra-fines (<50 µm) were prone to elutriation. A cyclone and baghouse were introduced to capture and reinject fines at the pilot, improving recovery by ~13 per cent, though the small dimensions of the pilot made the recycle loop susceptible to accumulation and blockages. The larger 36 m demonstration scale reactor, industrial off-gas handling, and new fines-treatment strategies are being designed to support higher solids loadings and enable reliable ultra-fines processing.

PSD testing highlighted clear trade-offs between metallisation and yield. Increasing d_{90} from 125 µm to 230 µm lifted recovery from ~51 per cent to ~78 per cent with little impact on metallisation, while coarser PSDs (460–605 µm) achieved >95 per cent recovery but lower metallisation (66–69 per cent). Screening trials confirmed high metallisation for fractions <300 µm at pilot scale, with performance expected to extend beyond 500 µm in the demonstrator.

Beyond reduction testing, downstream qualification of ZESTY H_2DRI included ESF smelting and hot briquetting trials. Briquetting conducted at TU Bergakademie Freiberg showed that increased furnace temperature and compaction pressure improved briquette strength and density to >4.5 t/m^3, yielding products approaching IMO HBI shipping standards. ESF trials at the University of Newcastle demonstrated effective slag micro foaming, strong iron partitioning into the metal phase, and successful carburisation using reductant additions.

The ZESTY 30 000 kt/a demonstration project represents a major step toward TRL 7. The US Department of Energy (DOE) Technology Readiness Level (TRL) framework is applied to ZESTY project. TRL provides a consistent, externally recognised measure of technology maturity, spanning early concept development through to full commercial operation. TRL assessment at the program level, with supporting evidence generated through laboratory, pilot, demo and operational activities (see <www.energy.gov/eere/funding/technology-readiness-levels>). The FEED study, completed in early 2024, delivered a fully specified design envelope, including heat-transfer modelling, hydrogen-utilisation strategies, DRI/HBI product-handling systems, fines-recirculation concepts, and a comprehensive risk-mitigation framework. Considered in the design for the top of the reactor, feeding approach and fines recovery system to recycle particles back to the reactor. Particle size and milling circuit parameter selection to target optimised PSD to further limit extent of elutriation and losses. Building on this technical foundation, the project has now advanced to site-specific detailed engineering for a dedicated demonstration plant in Kwinana, Western Australia. The Kwinana Industrial Area offers established utilities, port access, proximity to major iron-ore producers and is minutes from the NeoSmelt Electric Smelting Furnace demonstration site. Key scale-up risks, which include ultra-fines management, heat-transfer limits, sticking avoidance, hydrogen utilisation and recycle, and safe hydrogen/DRI handling, remain central engineering priorities. Extended-duration campaigns at Kwinana will validate full-scale reactor performance, de-risk commercial deployment, and accelerate the transition from TRL 5 to TRL 7.

Preliminary technoeconomic analysis based on FEED outcomes forecasts commercial scale HBI production costs in the range of USD 410–520/t Fe, contingent on renewable electricity pricing and hydrogen sourcing strategies. Third-party hydrogen supply has emerged as the preferred configuration to minimise capital intensity and accelerate deployment.

Overall, the ZESTY pilot program has demonstrated that Australian fine hematite/goethite ores can be rapidly and efficiently reduced using an electrically heated flash reactor, achieving high metallisation with near theoretical hydrogen consumption. The pilot scale outcomes, combined with demonstrated briquetting and smelting performance, provide strong validation of the process fundamentals. The forthcoming 30 000 t/a demonstration plant will complete the critical scale-up step, resolve the remaining technical risks, and establish ZESTY as a viable, flexible, and renewably powered pathway for green iron and steel production.

ACKNOWLEDGEMENTS

The authors acknowledge the financial support of the Australian Renewable Energy Agency (ARENA) for the design and construction of the 30 000 t/a ZESTY Green Iron Demonstration Plant in Kwinana, Western Australia. The authors further acknowledge Rio Tinto for its recently announced support for the ZESTY demonstration project, including over A$35 million in combined cash and in-kind contributions, the provision of the demonstration plant site, engineering and technical support, product qualification pathways, and the supply of iron ore for commissioning and early operating campaigns. We also thank the Heavy Industry Low-carbon Transition Cooperative Research Center (HILT CRC) and its industry partners for funding fellowships at Swinburne University of Technology and the University of Adelaide, which supported pilot plant testing, process development, and techno-economic modelling of the ZESTY process. Additionally, we acknowledge the University of Newcastle for conducting electric smelting tests on flash-reduced DRI samples, and ITUN (TU Bergakademie Freiberg) and Maschinenfabrik Köppern GmbH and Co. for their work on HBI testing.

Thermodynamic assessment of gangue effect in hydrogen-direct reduced iron smelting in electric smelting furnaces

K Owusu[1], B Kumar[2], C Chen[3], A Guiraud[4], S Sabah[5], U Kudaru[6], M A Rhamdhani[7] and G Brooks[8]

1. Post Doctoral Fellow, CSIRO Clayton, Melbourne Vic 3168. Email: kwaku.owusu@csiro.au
2. Post Doctoral Fellow, CSIRO Clayton, Melbourne Vic 3168. Email: banty.kumar@csiro.au
3. Principal Research Scientist, CSIRO Clayton, Melbourne Vic 3168. Email: chunlin.chen@csiro.au
4. Minerals Processing Lead, CSIRO Clayton, Melbourne Vic 3168. Email: adrien.guiraud@csiro.au
5. Postdoctoral Research Engineer, Swinburne University of Technology, Hawthorn Vic 3122. Email: ssabah@swin.edu.au
6. Research Fellow, Swinburne University of Technology, Hawthorn Vic 3122. Email: ukudaru@swin.edu.au
7. Professor, Swinburne University of Technology, Hawthorn Vic 3122. Email: arhamdhani@swin.edu.au
8. Professor, Swinburne University of Technology, Hawthorn Vic 3122; CSIRO Clayton, Melbourne Vic 3168. Email: gbrooks@swin.edu.au

INTRODUCTION

Growing global concern over greenhouse gas emissions has intensified the need to decarbonise the iron and steel industry, a major contributor to global CO_2 emissions (7–9 per cent) (Brooks *et al,* 2024). However, current steelmaking pathways, particularly the traditional blast furnace and basic oxygen furnace route, which accounts for more than 72 per cent of global steel production, remain a major challenge due to their heavy reliance on metallurgical coke. Consequently, significant research efforts are underway to decarbonise the industry. Among emerging technologies, the electric smelting furnace (ESF) offers a promising low-emission pathway that can be deployed at large scale within a short time frame. The potential for emission reduction increases further as hydrogen becomes economically viable for direct reduced iron production. A key advantage of the ESF is its flexible implementation, as it can be integrated with BOF and EAF or retrofitted into existing steelmaking facilities. Moreover, ESF enables the processing of DRI derived from low- to medium-grade iron ores.

The application of the electric smelting furnace (ESF) is not novel, as the technology has been widely used in industry for processing high-slag titaniferous materials and for ferronickel production (Garlick *et al,* 2025). However, its application in iron and steelmaking remains limited. Currently, only three commercial-scale DRI–ESF operations are recognised worldwide, with historical deployment largely driven by process requirements rather than decarbonisation objectives. In recent years, interest in advancing ESF technology for decarbonising the industry has grown, driven by the need to reduce emissions and improve understanding for broader deployment. Despite this growing interest, limited data continues to constrain broader deployment. As technical uncertainties are reduced and system-level understanding improves, wider implementation of ESF is expected, particularly in the context of reducing the reliance on blast furnace-based steelmaking as green hydrogen becomes economically viable for DRI production.

Given Australia's position as the largest global supplier of iron ore and its substantial low- to medium-grade resources, this study examines the de-risking of ESF-based steelmaking using Australian iron ores from a thermodynamic perspective. Key process variables are evaluated, focusing on metal and slag chemistry and the potential use of slag in value-added applications such as cement production.

METHODOLOGY

In this study, three hydrogen-based DRIs produced from low-grade Australian iron ores were assessed using thermodynamic modelling with FactSage® (version 7.0). The modelling assumes an equilibrium/closed system and uses selected databases, including FToxid, FactPS, and FTmisc.

The effects of DRI gangue content (13–16 wt per cent) and key operating parameters on smelting performance were predicted, including hot metal and slag chemistry, as well as specific energy consumption. The base conditions were fixed at 100 per cent metallisation ratio (MR), a slag basicity (CaO/SiO_2) of 1.1, dissolved carbon in hot metal of 3.5 wt per cent, MgO flux as ~11 wt per cent, a smelting temperature of 1500°C and pressure of 1 atm, unless otherwise specified. Figure 1 presents a schematic representation of the ESF modelling framework used in this study.

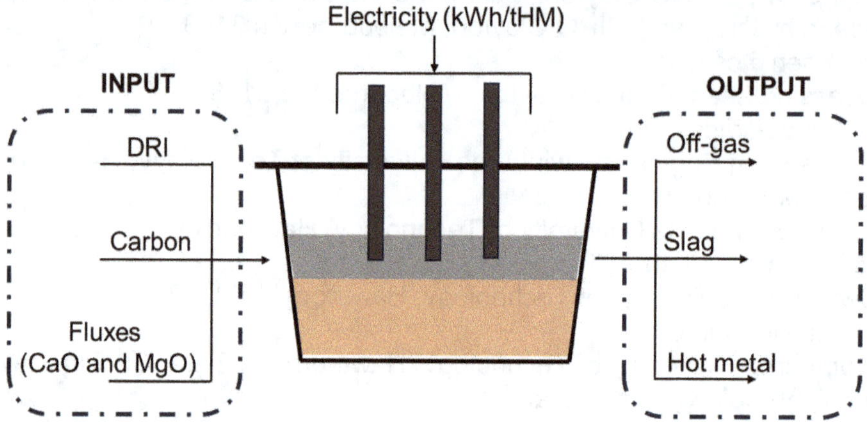

FIG 1 – Schematic overview of the thermodynamic modelling framework used in this study.

MAIN FINDINGS

Following the smelting of three DRI types under the fixed base conditions, the predicted hot metal and slag compositions reflected their respective gangue contents. With fixed carbon in hot metal of 3.5 wt per cent, the predicted hot metal consisted primarily of Fe (94–95 wt per cent, corresponding to >99 per cent Fe recovery) along with minor impurities. Most impurities, including detrimental elements such as sulfur and phosphorus, were within acceptable levels and therefore were not the focus of subsequent discussions. Greater emphasis was placed on the Si deportment to the hot metal, which ranged from 0.3 to 0.36 wt per cent. These levels could be ideally suited to BOF operation; however, excessive levels can affect mass or energy balance (Kexin *et al,* 2017; Hage, van der Stel and Yang*,* 2022; Madhavan *et al,* 2021a, 2021b). It is also worth noting that the issue with what is optimum Si level in an ESF hot metal is complex, as it depends greatly on the hot metal carbon level, the amount of scrap being added to the BOF and the control of product quality coming from the BOF (which in turn is a strong function of hot metal chemistry and tapping temperature) and beyond the scope of this study. The slag chemistry was comparable to that of blast furnace slag, with a viscosity of 0.25–0.27 Pa·s and an Al_2O_3 content of 15–17 wt per cent.

Varying the metallisation ratio (MR) from 80 to 100 per cent was predicted to result in only minor changes in hot metal and slag composition; however, a considerable reduction in specific energy consumption (SEC) was predicted. This reduction was primarily attributed to lower carbon requirements, as increasing MR reduces the presence of unreduced iron oxide (eg FeO). A similar trend was predicted with increasing scrap addition in the DRI–scrap blends. In contrast, increasing fayalite content led to a slight increase in SEC without significantly affecting metal or slag chemistry.

Temperature was identified as a key parameter influencing silicon levels in the hot metal, slag behaviour, and SEC. Another key parameter is the Al_2O_3/SiO_2 mass ratio, which affected both metal and slag chemistry, with spinel formation predicted when the ratio exceeded approximately 0.5, as shown in Figure 2. Also, Figure 2c predicted that the SEC decreased with increasing Al_2O_3/SiO_2 mass ratio, reflecting the corresponding reduction in slag mass generation. It should be noted that the Al_2O_3/SiO_2 mass ratio was derived by keeping the Al_2O_3 content in the DRI constant while varying the SiO_2 content. The DRI composition was then normalised to maintain 100 wt per cent.

Increasing dissolved carbon in hot metal increased silicon pickup and resulted in higher predicted SEC. The predicted off-gas was predominantly CO, confirming a strongly reducing smelting environment that supports efficient iron recovery and stable ESF operation.

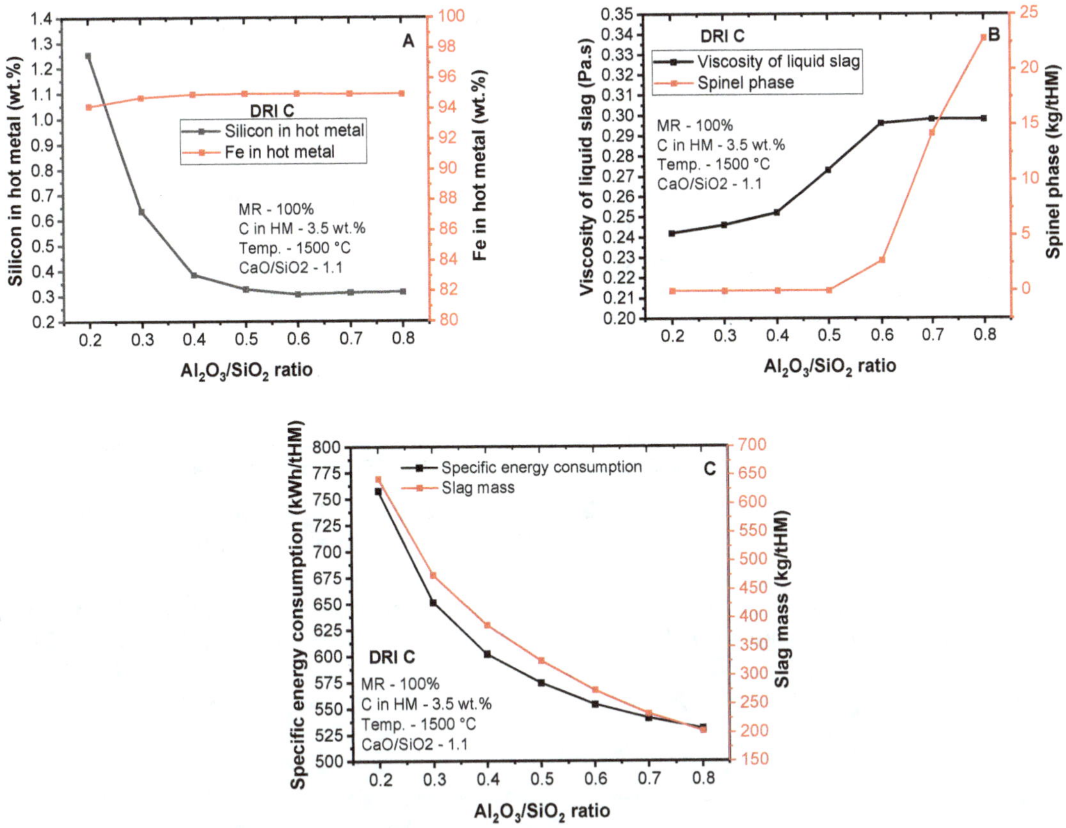

FIG 2 – Thermodynamic assessment of (a) hot metal chemistry, (b) slag viscosity, and (c) specific energy consumption and slag mass for different Al_2O_3/SiO_2 under fixed conditions of 100 per cent MR, slag basicity of 1.1, temperature of 1500°C, and pressure of 1 atm.

MODELLING LIMITATIONS

The main limitations of this study include:

- The use of thermodynamic equilibrium calculations only, without considering kinetics or mass transfer within the system.
- Limited validation with industrial data. Due to the scarcity of such data, validation was performed using a small industrial data set.

CONCLUSION

Using FactSage® thermodynamic modelling, this study assessed the smelting behaviour of three hydrogen-based DRIs produced from Australian iron ores with gangue contents of 13–16 wt per cent. The results predict high iron recovery (>99 wt per cent) and acceptable impurity levels, with silicon in hot metal generally within the range of 0.3–0.5 wt per cent, suitable for BOF operations. While the predicted slag chemistry and viscosity were broadly comparable to blast furnace slag, slag composition was sensitive to DRI feed characteristics and operating parameters. Maintaining an appropriate alumina-to-silica mass ratio in the DRI was found to minimise spinel formation in the slag while controlling silicon levels in the hot metal. Overall, the study defines an optimal operating window that supports efficient smelting, production of high-quality hot metal, and generation of slag suitable for value-added applications such as cement making. While this area is relatively new and has very limited data, it is important to note that the results of the study need to be validated against experimental or industrial data as it becomes available.

ACKNOWLEDGEMENTS

This research was supported by the Heavy Industry Low-carbon Transition Cooperative Research Centre (HILT CRC), which is funded through contributions from industry, research organisations, and government partners, together with the Australian Government's Cooperative Research

Centre Program (Project No. HILT.RP1.014). The support of the High-Temperature Chemistry Team (HTC) within the Mineral Resources unit of the Commonwealth Scientific and Industrial Research Organisation (CSIRO) is greatly acknowledged.

REFERENCES

Brooks, G, Jensen, P, Petrie, S and Webster, E, 2024. Can Australia Become a Green Iron Powerhouse?, *Australian Economic Review,* 57(4). https://doi.org/10.1111/1467-8462.12578

Garlick, C, Singh, T, Paymooni, K, O'Dea, D and Honeyands, T, 2025. Electrode smelting of hydrogen-reduced hematite–goethite at laboratory scale, *Journal of Sustainable Metallurgy*, 12:1114–1127. https://doi.org/10.1007/s40831-025-01281-2

Hage, J L T, van der Stel, J and Yang, Y, 2022. Silicon in hot metal from a blast furnace, the role of FeO, *Ironmaking and Steelmaking*, 49(6):581–587. https://doi.org/10.1080/03019233.2022.2027135

Kexin, J, Jianliang, Z, Chunlin, C, Zhengjian, L and Xi, J, 2017. Operation Characteristic of Super-Large Blast Furnace Slag in China, *ISIJ International*, 57(6):983–988. https://doi.org/10.2355/isijinternational.ISIJINT-2016-615

Madhavan, N, Brooks, G A, Rhamdhani, M A, Rout, B K and Overbosch, A, 2021b. General heat balance for oxygen steelmaking, *Journal of Iron and Steel Research International*, 28:538–551. https://doi.org/10.1007/s42243-020-00491-0

Madhavan, N, Brooks, G A, Rhamdhani, M A, Rout, B K, Schrama, F N H and Overbosch, A, 2021a. General mass balance for oxygen steelmaking, *Ironmak Steelmak*, 48:40–54. https://doi.org/10.1080/03019233.2020.1731252

Processing route for Australian low-grade iron ores

S Purohit[1], M Shaw[2], N A S Webster[3], M I Pownceby[4], G Brooks[5], A R Rhamdani[6] and K Vining[7]

1. Research Scientist, CSIRO Mineral Resources, Melbourne Vic 3168.
   Email: suneeti.purohit@csiro.au
2. Postdoctoral Researcher, Swinburne University of Technology, Melbourne Vic 3122.
   Email: mgshaw@swin.edu.au
3. Research Group Leader, Characterisation, CSIRO Mineral Resources, Melbourne Vic 3168.
   Email: nathan.webster@csiro.au
4. Senior Principal Research Scientist, CSIRO Mineral Resources, Melbourne Vic 3168.
   Email: mark.pownceby@csiro.au
5. Professor, Swinburne University of Technology, Melbourne Vic 3122.
   Email: gbrooks@swin.edu.au
6. Postdoctoral Researcher, Swinburne University of Technology, Melbourne Vic 3122.
   Email: drhamdani@swin.edu.au
7. Research Group Leader, Green Metals Production, CSIRO Mineral Resources, Brisbane Qld 4069. Email: keith.vining@csiro.au

INTRODUCTION

The ongoing decarbonisation efforts of the steelmaking industry, together with increasingly stringent climate policies, are driving a growing demand for high-grade ironmaking feedstocks. However, the availability of such materials remains limited. According to the 2020 IEEFA report, only around 3 per cent of the global seaborne iron ore supply qualifies as direct-reduction (DR) grade (Nicholas and Basirat, 2022). Projections by Bloomberg New Energy Finance further suggest that achieving a Net Zero pathway could require up to a tenfold increase in the supply of DR-grade pellets by 2050, unless hydrogen-based direct reduction (H_2-DRI) technologies evolve to accommodate lower-grade ores (Nicholas and Basirat, 2022). This anticipated supply–demand mismatch underscores the growing importance of strategies that enable the efficient utilisation of lower-grade iron ores, with significant implications for countries such as Australia, where iron ore exports are predominantly of lower grade.

Meanwhile, conventional blast furnace-basic oxygen furnace (BF-BOF) steelmaking continues to dominate global steel production, despite being the most carbon-intensive route due to its reliance on coke as both fuel and reductant. However, substantial effort is now being directed towards lower-emission alternatives like the hydrogen-based direct reduction pathways, DRI melter technologies for lower-grade ores, increased scrap utilisation in electric arc furnace (EAF) steelmaking, and the deployment of carbon capture technologies on existing BF assets. Recent industrial-scale implementations demonstrate the maturity and scalability of the DRI-EAF route; for example, Tosyali Algerie's MIDREX DRI operation achieved a world-record annual production of 2.43 Mt in 2025 and has been designed for future transition to 100 per cent hydrogen operation, illustrating that DRI-based steelmaking can deliver both high productivity and a credible pathway to emissions reduction (Midrex Technologies, 2026). Taken together, these trends highlight the growing need for lower carbon footprint iron ore agglomerates that can deliver immediate emissions reductions in existing BF-BOF operations, while also meeting the quality and performance requirements of future hydrogen-based DRI steelmaking routes.

This study proposes a novel, lower-temperature agglomeration route for Australia's low-grade goethite-rich hematite ores to produce a premium quality ironmaking feedstock. The novel approach in this project is referred to as the Lime Magnetite Pellet (LMP) process, which has previously been tested and validated at laboratory and kilogram-scales for magnetite concentrates (Purohit *et al,* 2020a; Purohit and Pownceby, 2022). Pilot-scale metallurgical testing of the magnetite concentrate-based LMPs with varying basicity ($CaO:SiO_2$ = 1.1, 1.5 and 2.3) demonstrated high reducibility (RI: 78 to 93 per cent) and low reduction degradation (RDI <5 per cent) (Purohit and Pownceby, 2022). Tumble strength (TI) was acceptable (89 to 90 per cent) at lower basicity but requires improvement at higher basicity. These results indicate the potential of LMPs as feedstocks for blast furnace ironmaking, while further evaluation is required for other shaft furnace ironmaking applications.

The novelty of the LMP process lies in its significantly lower induration temperature of 1050°C to 1100°C, about 300°C lower than that required for conventional sintering and pelletising, which when combined with a mildly reducing environment favours the formation of the $CaFe_3O_5$ (CWF) phase. Although originally developed for magnetite iron ores, this study employs a combined thermodynamic modelling and experimental approach to examine how hematite and goethite ore types behave in the LMP process, especially in terms of the formation of the CWF phase. The paper also analyses the effects of different operating conditions such as the oxygen partial pressure, CaO concentration, and induration time, using experimental findings and characterisation results. Results indicate that the $CaFe_3O_5$ phase can form in both hematite and goethite derived LMPs.

THERMODYNAMIC MODELLING

Thermodynamic modelling was conducted using the software package FactSage 8.3. The Equilibrium module was used in conjunction with the FactPS database for pure compounds and a private database for compounds and solutions within the Ca-Fe-O system (Hidayat *et al,* 2015). Modelling was aimed at evaluating the equilibrium phases in the three CaO-Fe oxide/oxyhydroxide systems (CaO-Fe_2O_3, CaO-FeO(OH) and CaO-Fe_3O_4) at various reducing conditions, represented by various CO_2/CO ratios. The equilibrium calculations were performed at three different $CaCO_3$ addition levels – the different CaO levels were selected based on their composition relative to the stoichiometric Ca/Fe ratio of 1:3 required for the formation of CWF phase in LMPs. This stoichiometric condition was implemented using 2 mol $CaCO_3$ per 6 mol Fe in the iron oxide feed, corresponding to 2 mol Fe_3O_4, 3 mole Fe_2O_3 or 6 mole FeO(OH). In addition, half- and quarter-stoichiometric calcium additions were examined and expressed as Ca/Fe ratios of ½:3 and ¼:3 (equivalent to 1:6 and 1:12), respectively. The selected $CaCO_3$ additions span a wide compositional range relevant to basicity variations in fluxed sinters and pellets, enabling evaluation of phase stability across low to high CaO levels. This approach allows systematic investigation of the evolution of different silicate phases to calcium ferrites with increasing CaO addition. These variations play a critical role in governing pellet strength and bonding, reducibility, and swelling behaviour, thereby helping to identify the optimal CaO level for balancing phase stability and overall metallurgical performance in the LMP system.

In addition to composition, the gas atmosphere was controlled to assess phase stability under relevant oxygen potentials. CO_2/CO ratios between 1 to 10 were selected to target oxygen partial pressures associated with wüstite and magnetite stability, while an induration temperature of 1050°C was selected based on prior optimisation (Purohit *et al*, 2020b, 2021).

The resulting equilibrium phase compositions for the three different stoichiometric calcium addition levels (stoichiometric, half-stoichiometric and quarter-stoichiometric) are shown in Figure 1. Under equilibrium conditions, the stable phase compositions are determined by oxygen partial pressure, temperature and are independent of the initial iron oxide phase. Hence, for all three iron oxides, identical stoichiometric, half-stoichiometric and quarter-stoichiometric trends are observed, as shown in Figure 1.

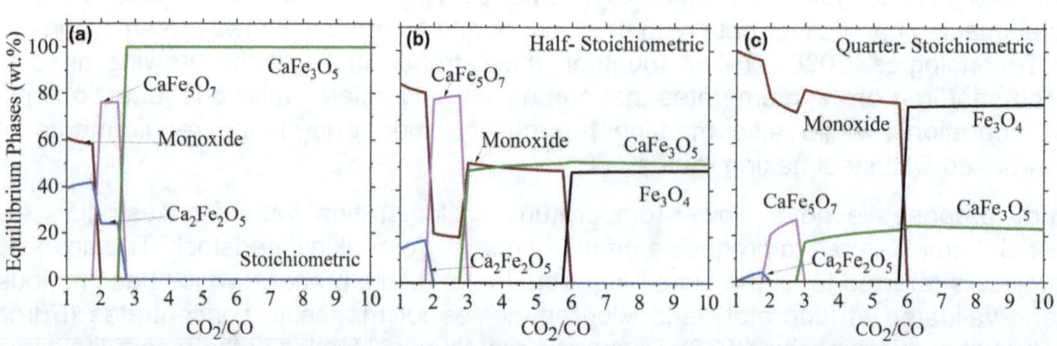

FIG 1 – Thermodynamic modelling results showing the equilibrium phases in; (a) stoichiometric, (b) half-stoichiometric and (c) quarter stoichiometric LMP compositions.

For all Ca/Fe ratios investigated, the CWF phase remains stable over a wide range of oxygen partial pressures. The fraction of the CWF phase decreases systematically with decreasing CaO addition, accounting for approximately 100 per cent of the solid phases at stoichiometric CaO addition, 51 per cent at half-stoichiometric, and 23 per cent at quarter-stoichiometric conditions. At CO_2/CO ratios below six, magnetite is reduced to wüstite (Figure 1b and 1c); however, the CWF phase, which contains iron in a similar oxidation-state environment to magnetite (Fe_3O_4) remains stable and is not reduced until the CO_2/CO ratio decreases to approximately 2.5. In addition to CWF, minor amounts of other calcium ferrite phases, including $Ca_2Fe_2O_5$ and $CaFe_5O_7$, are also observed to form.

EXPERIMENTAL STUDY

Following the thermodynamic modelling, experimental studies were carried out using high-purity (>99 per cent) hematite, goethite and limestone powders obtained from Sigma-Aldrich. Previously, detailed time-series experiments were performed using pure magnetite samples at stoichiometric and half-stoichiometric compositions (Purohit *et al,* 2020a). These experiments, conducted at 950°C under a fixed CO_2/CO ratio of four, showed that CWF formation was nearly complete after 60 min of reaction time. In the present study, this work was extended to hematite and goethite to investigate the influence of the initial iron oxide on CWF formation. Time-series induration experiments (5, 10, 20, 30 and 60 mins) were conducted at 1050°C and a CO_2/CO ratio of four using LMPs prepared at three different $CaCO_3$ additions. Experiments were conducted in a horizontal tube furnace under a nitrogen atmosphere while heating to the target temperature, followed by induration under the set oxygen partial pressure (~10^{-12} at CO_2/CO ratio of four). The indurated briquettes were then analysed using X-ray diffraction (XRD) and scanning electron microscopy (SEM) to evaluate the pellet phase composition and microstructure.

EXPERIMENTAL RESULTS

For the XRD analysis, all samples were first micronised to ensure compositional homogeneity and fine particle size before being presented to the X-ray beam for XRD data collection. XRD data were collected using a Panalytical Empyrean fitted with Co X-ray tube operated at 40 kV and 40 mA. Phase identification was performed using PANalytical HighScore Plus software in conjunction with the ICDD PDF5+ database. Quantitative phase analysis (QPA) was carried out by Rietveld refinement using TOPAS software (Rietveld, 1969) and the results are summarised in Figure 2.

Figure 2 shows that CWF is the predominant calcium ferrite phase formed during induration across all Ca/Fe ratios for both hematite and goethite based LMPs. However, both the kinetics of CWF formation and its final phase fraction are strongly dependent on the $CaCO_3$ level and the initial iron oxide. In stoichiometric LMPs, CWF becomes the dominant phase in both systems, whereas in half and quarter stoichiometric LMPs its formation is progressively limited by reduced CaO availability and other competing reactions. Figure 2 suggest CWF forms more rapidly in goethite based LMPs than in hematite based LMPs across all compositions. This is possible due to the formation of highly reactive intermediate hematite and the development of a porous microstructure during goethite dehydroxylation.

Both hematite and goethite based LMPs follow similar phase evolution pathways, involving intermediate calcium ferrite phases like $CaFe_2O_4$, $Ca_2Fe_2O_5$ and $Ca_3Fe_{14.8}O_{25}$ along with the progressive reduction of the initial hematite or goethite into magnetite and wüstite. Compared to these hematite and goethite based LMPs indurated at 1050°C, the previously reported magnetite-based LMPs showed only two calcium ferrite phases, CWF and $Ca_2Fe_2O_5$, at 950°C induration temperature. The simultaneous presence of magnetite and wüstite in both hematite and goethite based LMPs indicate that the system was not at equilibrium and the oxygen partial pressure was not uniform within the pellet structure. For the goethite based LMPs, the wüstite percentage is higher than that of hematite-based LMPs, indicating enhanced reduction due to higher porosity. Although the CWF percentage remains relatively unchanged with prolonged induration, the magnetite to wüstite reduction continues to happen till 60 mins induration.

Overall, the QPA results demonstrate that CWF is both the thermodynamically and kinetically favoured calcium ferrite phase under the investigated conditions. Its formation is kinetically promoted in goethite-based systems and with increasing CaO availability. It should be noted, however, that industrial iron ores can contain significant amounts of gangue components like SiO_2 and Al_2O_3, which will consume CaO and thereby reduce its availability for CWF formation. This is expected to result in different CWF phase fractions in practical ore systems and current is being expended to examine CWF formation in natural hematite, goethite and hematite-goethite ores.

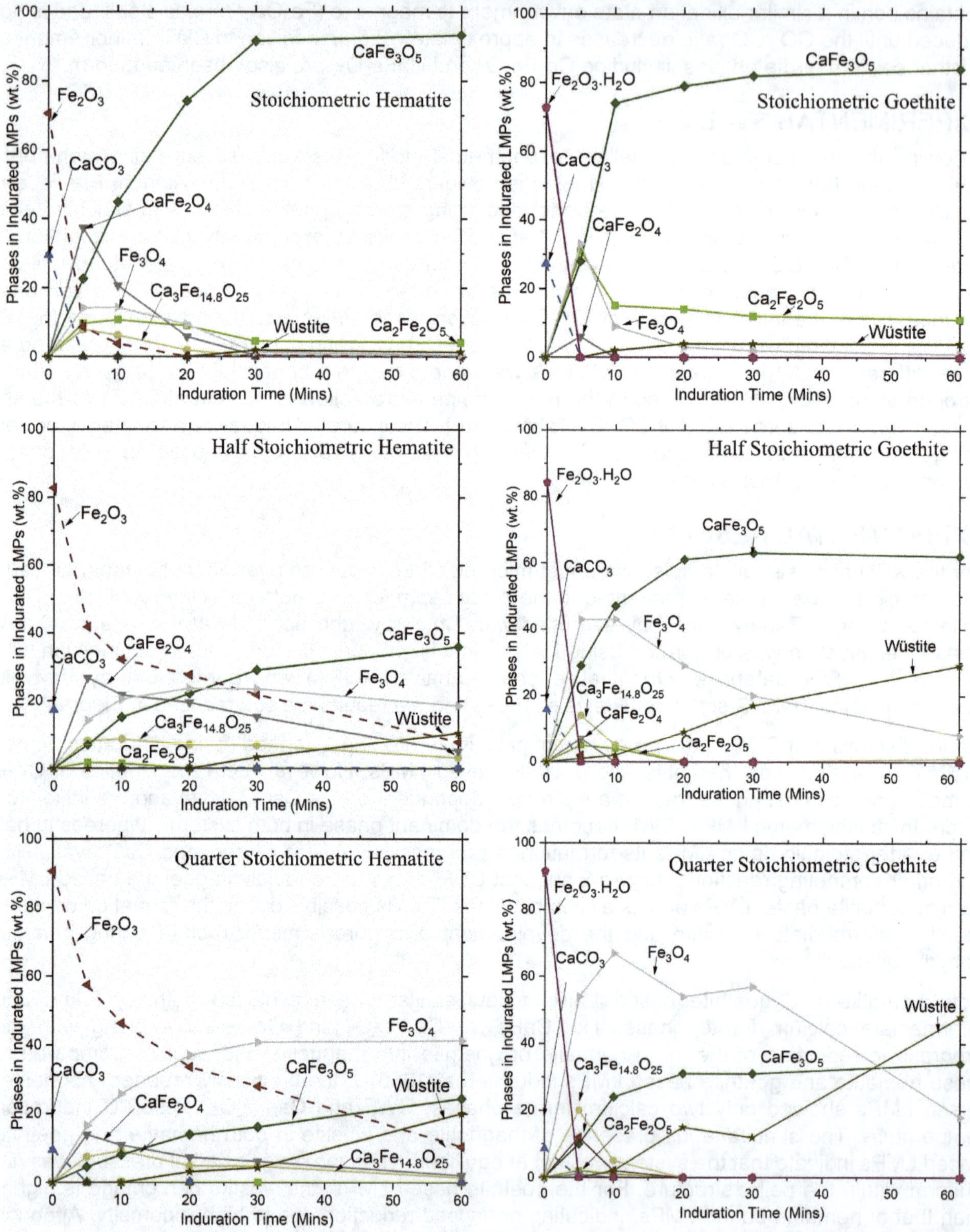

FIG 2 – XRD QPA results showing the product compositions of stoichiometric, half-stoichiometric and quarter stoichiometric LMPs made from pure hematite (LHS) and goethite (RHS), respectively.

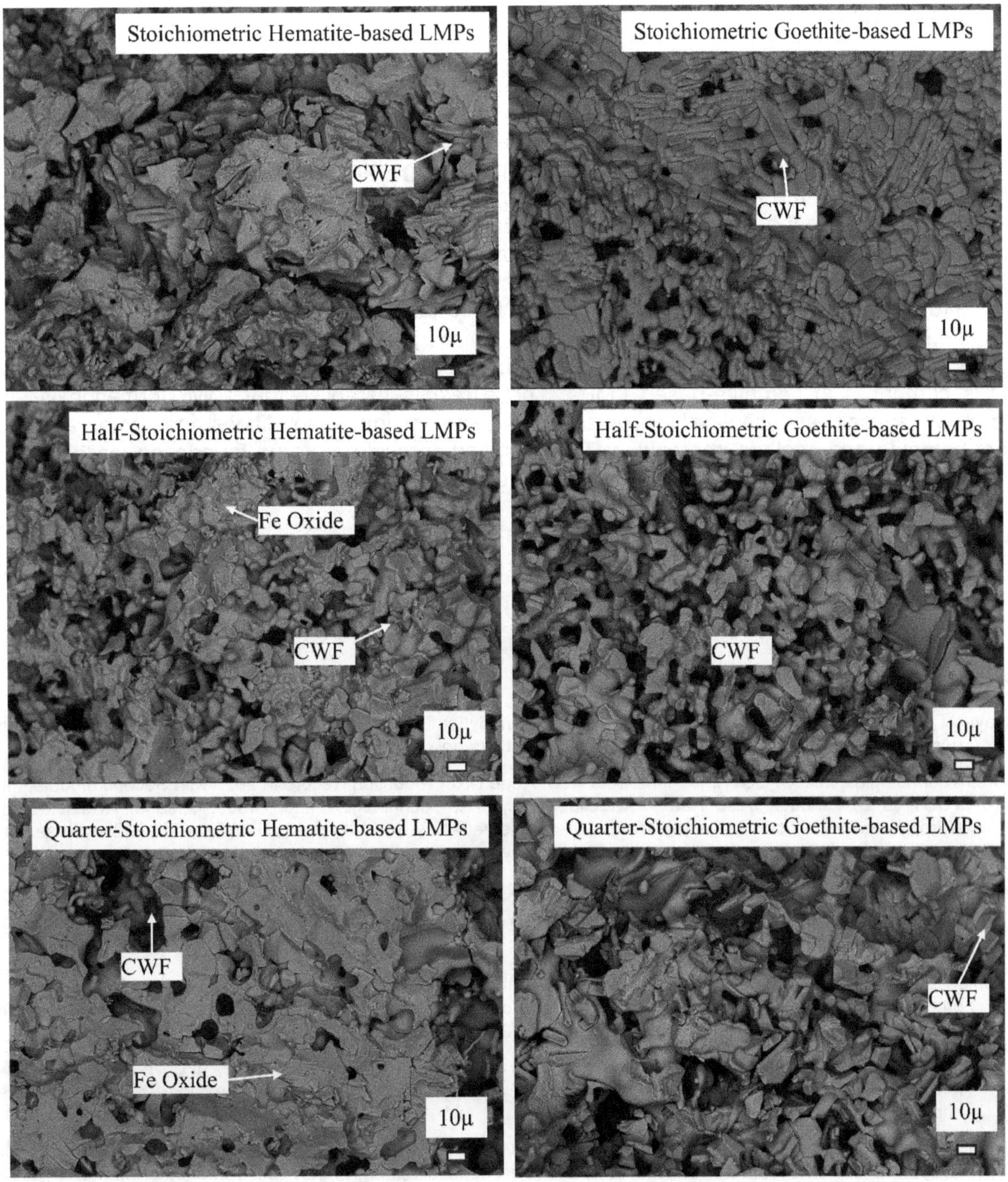

FIG 3 – SEM secondary electron images showing the typical microstructure of 60 min indurated stoichiometric, half-stoichiometric and quarter stoichiometric LMPs from hematite (LHS) and goethite (RHS), respectively.

Figure 3 shows clear differences in microstructure and porosity between hematite- and goethite-based LMPs. Goethite-based LMPs are consistently more porous than their hematite counterparts at the same Ca/Fe ratios. This trend is also reflected in the overall weight loss during induration, with hematite-based LMPs exhibiting a maximum weight loss of approximately 17 wt per cent compared to about 24 wt per cent for goethite-based LMPs. The additional weight loss, and the corresponding increase in porosity in goethite based LMPs, is attributed to goethite dehydroxylation during induration, which forms a highly porous microstructure and enhances gas permeability. In all samples, the dominant bonding phase appears as platelet-shaped features, which are identified as the CWF phase and can be observed across all compositions in both systems.

CONCLUSION

This paper investigates the viability of using the LMP process as a lower-temperature agglomeration route for low-grade goethite-rich iron ores. A combined thermodynamic modelling and experimental study was conducted for hematite and goethite based LMPs for three different Ca/Fe ratios (1:3, ½:3 and ¼:3). Thermodynamic modelling using FactSage predicts the stability of CWF over a wide range of CO_2/CO ratios (~3 to 10), with its equilibrium percentage strongly governed by CaO availability and independent of the initial iron oxide mineralogy. Experimental investigations confirmed these predictions, showing that CWF is the dominant calcium ferrite phase formed in both hematite and goethite derived LMPs across all three Ca/Fe ratios studied. While both systems follow similar phase evolution pathways involving intermediate calcium ferrite phases and progressive reduction of iron oxides to magnetite and wüstite, the kinetics of CWF formation are significantly enhanced in goethite-based LMPs due to goethite dehydroxylation, which generates a highly porous and reactive microstructure. SEM microstructural analysis revealed consistently higher porosity in goethite-based LMPs compared to hematite-based LMPs. These results establish that CWF formation is both thermodynamically and kinetically favoured under the investigated conditions and can be achieved from hematite and goethite iron oxides. These findings provide a strong foundation for the further development of LMPs using natural industrial lower-grade hematite-goethite iron ores.

ACKNOWLEDGEMENTS

The authors gratefully acknowledge the financial support provided by the Australian Renewable Energy Agency (ARENA) for this project. The authors also acknowledge the contributions of the research partner and industry collaborators for their valuable support and technical input.

REFERENCES

Hidayat, T, Shishin, D, Decterov, S A and Jak, E, 2015. Thermodynamic Optimization of the Ca-Fe-O System, *Metallurgical and Materials Transactions B,* 47:1:47(1):256–281. https://doi.org/10.1007/s11663-015-0501-0

Midrex Technologies, 2026. TOSYALI Algerie Sets New World DRI Production Record in 2025 with Second Module; First TOSYALI DRI Module Also Ranks 3rd Globally, Midrex Technologies, Inc, Press Release. Available from: <https://www.midrex.com/press-release/tosyali-algerie-sets-new-world-dri-production-record-in-2025-with-second-module-first-tosyali-dri-module-also-ranks-3rd-globally/>

Nicholas, S and Basirat, S, 2022. *Iron Ore Quality a Potential Headwind to Green Steelmaking,* IEEFA.org

Purohit, S and Pownceby, M, 2022. Metallurgical Properties of Lime Magnetite Pellets, The 1st International Symposium on Iron Ore Agglomerates.

Purohit, S, Brooks, G, Rhamdhani, M A, Pownceby, M I and Webster, N A S, 2020a. Alternative route for magnetite processing for lower carbon footprint iron-making through lime-magnetite pellets containing $CaFe_3O_5$, *Ironmaking and Steelmaking*, 47(6):674–685. https://doi.org/10.1080/03019233.2019.1569811

Purohit, S, Brooks, G, Rhamdhani, M A, Pownceby, M I and Webster, N A S, 2020b. Analyses of CWF ($CaFe_3O_5$):Phase formation in lime-magnetite pellets, *Ironmaking and Steelmaking*, 47(8):852–864. https://doi.org/10.1080/03019233.2019.1623594

Purohit, S, Pownceby, M I, Brooks, G and Rhamdhani, M A, 2021. Lime Magnetite Pellets-An Alternative Iron Ore Feedstock for Lower Carbon Footprint Ironmaking, in *Proceedings of the Iron Ore Conference*, pp 605–612 (The Australasian Institute of Mining and Metallurgy: Melbourne).

Rietveld, H M, 1969. A profile refinement method for nuclear and magnetic structures, *Journal of Applied Crystallography*, 2(2):65–71. https://doi.org/10.1107/s0021889869006558

Fayalite formation from high gangue ores in DRI production

A R Rhamdani[1], E Ansah[2], S Sabah[3], G Brooks[4] and M A Rhamdhani[5]

1. Postdoctoral Researcher, Swinburne University of Technology, Hawthorn Vic 3122. Email: drhamdani@swin.edu.au
2. PhD student, Swinburne University of Technology, Hawthorn Vic 3122. Email: eansah@swin.edu.au
3. Postdoctoral Researcher, Swinburne University of Technology, Hawthorn Vic 3122. Email: ssabah@swin.edu.au
4. Professor, Swinburne University of Technology, Hawthorn Vic 3122. Email: gbrooks@swin.edu.au
5. Professor, Swinburne University of Technology, Hawthorn Vic 3122. Email: arhamdhani@swin.edu.au

EXTENDED ABSTRACT

Steel is one of the world's most widely produced and consumed alloys, and its production has grown substantially to nearly 1.9 Bts, contributing about 7–9 per cent of global CO_2 emissions due to its heavy reliance on the blast furnace ironmaking process. There is growing pressure on steel producers to lower greenhouse gas emissions in the steelmaking processes. Achieving the net-zero emissions goal will require low-carbon or carbon-neutral steelmaking routes. To address this, the sector constantly explores direct reduced iron (DRI) as an alternative iron source for steelmaking. Natural gas or hydrogen is commonly used to produce DRI, which can make DRI more environmentally friendly than the conventional blast furnace route, which relies on coke. However, one of the challenges associated with this route is the formation of fayalite (Fe_2SiO_4). Fayalite forms when FeO reacts with SiO_2 during iron ore reduction. A review by Heidari *et al* (2021) stated that the formation of $Fe_xSi_yO_z$ is faster at high temperature. There were concerns about fayalite formation in the DRI making process, such as flash reduction, as it is operating at high temperature (eg Calix's ZESTY at 600 to 1050°C) when silica-rich ores are used. Understanding its formation mechanism is crucial, particularly for low-grade iron ores with high levels of SiO_2. In this study, both literature review and thermodynamic calculations were employed to understand the formation of fayalite during iron ore reduction.

The literature review shows that the presence of fayalite inhibits the iron ore reduction process. Studies show that fayalite formation decreases the sample porosity and the reduction degree when reduced by H_2. Fayalite could form a dense layer, which reduces the permeability of the sample, thus inhibiting the reduction gas to reach unreduced ore. Not to mention that the formation of fayalite itself is contributing to the iron loss, thus lowering the overall metallisation degree of the product. Kim *et al* (2012) show that the addition as small as 2.5 per cent of SiO_2 to FeO slows down the early stages of the reduction, while the addition of 10 per cent SiO_2 limits the reduction degree of the product. Kinetic study of fayalite formation by Minowa, Yamada and Torii (1968) shows that fayalite formation is relatively slow at 800°C, while it becomes quicker as the temperature increases.

Thermodynamic calculations were carried out using FactSage 8.2®. The databases used in this study are FactPS for representing compound phases, FSstel for iron phases and solution, and FToxid for oxide phases and solution. The thermodynamic calculation result (as shown in Figure 1) shows that FeO, an intermediate product of iron ore reduction, prefers to react with silica to form fayalite rather than being reduced by the assessed reductant. Once formed, fayalite is more difficult to reduce compared to iron oxides. Although its reduction is possible, it requires either higher temperatures or greater amounts of reductant. For example, reduction of fayalite at 1000°C requires at least 33 per cent higher H_2 partial pressure (pH_2) compared to reduction of FeO (0.6 versus 0.8, respectively). Similarly, in the case of CO as a reductant, the partial pressure of CO needed to reduce FeO is lower than that needed to reduce fayalite (0.7 versus 0.9, respectively). Hydrogen is the preferred reductant to reduce fayalite rather than CO, as it needed lower partial pressure for the reduction to proceed at temperatures above 827°C. Fayalite is predicted to be one of the last phases to be reduced in the reduction sequence, after the reduction of FeO, changing the reduction stages from $Fe_2O_3 \rightarrow Fe_3O_4 \rightarrow FeO \rightarrow Fe$ to $Fe_2O_3 \rightarrow Fe_3O_4 \rightarrow FeO \rightarrow Fe_2SiO_4 \rightarrow Fe$. The thermodynamic calculation will be discussed in detail in a separate paper (Rhamdani *et al*, in prep).

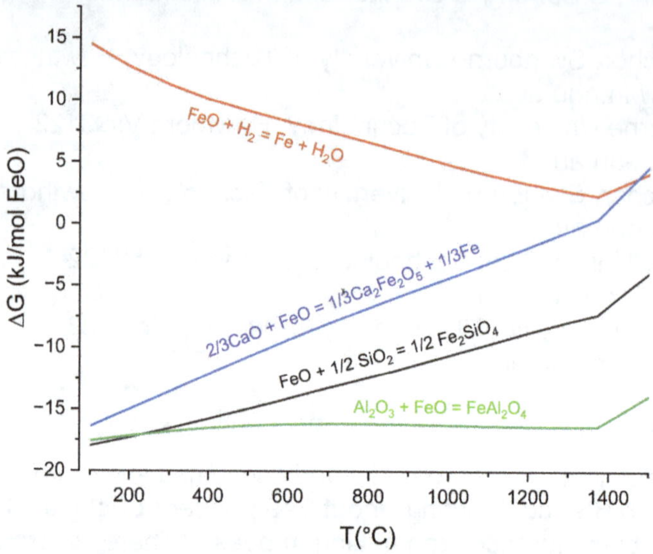

FIG 1 – Standard Gibbs free energy change ($\Delta G°$) of various FeO-based reactions as a function of temperature (Rhamdani *et al*, in prep).

Fayalite formation ($2Fe + SiO_2 \rightarrow Fe_2SiO_4$) is an exothermic reaction ($\Delta H°$= -44.7 kJ/mol at T = 900°C), while its reduction ($FeSiO_4 + 2H_2 \rightarrow 2Fe + SiO_2 + 2H_2O$) is a highly endothermic reaction $\Delta H°$= 76.5 kJ/mol at T = 900°C). Hydrogen reduction of Fe_2SiO_4 needs ~2.4 times more heat compared to the reduction of FeO at 900°C. This would mean that in a reactor where the DRI containing fayalite is formed, significantly more heat input is needed to improve the metallisation level from DRI containing iron and fayalite to DRI containing only iron and silica (all fayalite is reduced).

Thermodynamic calculations suggest several methods to reduce fayalite formation. One of the possible methods of reducing fayalite formation is adding another oxide phase. The idea is to bind SiO_2 by reacting it with these other oxide phases. Possible options for the oxide phases are CaO and MgO. Thermodynamic calculation shows that SiO_2 prefers to react with CaO or MgO to form $CaSiO_3$ or $MgSiO_3$, respectively, rather than to react with FeO to form fayalite. Thermodynamic calculations predicted that the addition of CaO to a simulated high gangue ore reduces the required H_2 to reach 100 per cent metallisation level, as shown in Figure 2. The addition of oxide to the iron ore will increase slag formation in the steelmaking process, although it is common to add oxides such as CaO to maintain good slag chemistry.

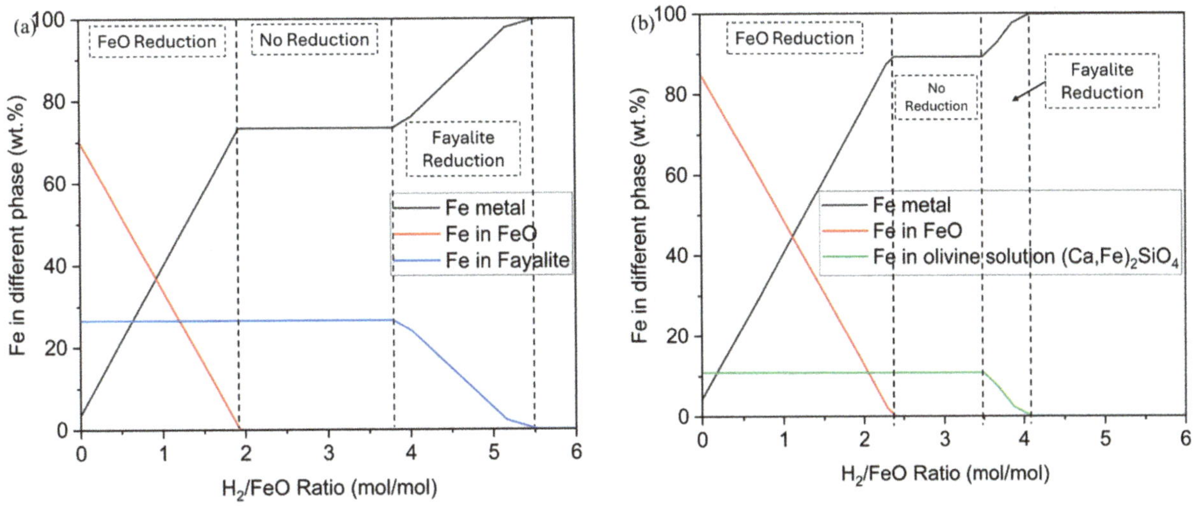

FIG 2 – Iron fraction in different phases as a function of H_2/FeO ratio for simulated ore (FeO-10 wt per cent SiO_2) (a); and 90 wt per cent Ore – 10 wt per cent CaO mixtures (b) (Rhamdani *et al*, in prep).

An alternative approach to reducing fayalite in the DRI is to improve the reduction power of the reduction gas. This approach involves adding CH_4 into the H_2 as the reducing gas (Park and Jung, 2015). Calculation predicted that CH_4 will react with H_2O, thus keeping the pH_2/pH_2O ratio remain high in the reduction gas. This higher pH_2/pH_2O ratio increases the reduction power of the gas and is able to reduce the fayalite. The findings of this study will benefit DRI producers, especially those utilising low-grade iron ores. These predictions of fayalite formation are consistent with what has been observed in Calix's ZESTY pilot plant (Sabah *et al*, in prep). The thermodynamic calculations are an equilibrium based and their results may vary depending on the selected database. One method of calculation which could represent closer to the actual conditions is thermodynamic modelling using effective equilibrium reaction zone (EERZ) which is currently an ongoing project in the team. It should be noted, even though thermodynamic calculations can offer solution to the fayalite issue, these results still need to be confirmed by experimental data, as factors such as kinetics, heat distribution, and transport phenomena are also influencing the reaction.

ACKNOWLEDGEMENTS

This work is funded by HILT CRC RP1.015 De-risking flash reduction of Australian iron ores.

REFERENCES

Heidari, A, Niknahad, N, Iljana, M and Fabritius, T, 2021. A Review on the Kinetics of Iron Ore Reduction by Hydrogen, *Materials*, 14(24):7540.

Kim, W-H, Lee, Y-S, Suh, I-K and Min, D-J, 2012. Influence of CaO and SiO_2 on the Reducibility of Wustite Using H_2 and CO Gas, *ISIJ International*, 52(8):1463–1471.

Minowa, S, Yamada, M and Torii, Y, 1968. A Study on the Formation and Reduction of Fayalite (Fe_2SiO_4), *Tetsu-to-Hagane*, 54(12):1203–1216.

Park, W-I and Jung, S-M, 2015. Effective Use of CH_4 Gas as a Reducing Agent in Suspension Reduction Process, *ISIJ International*, 55(1):166–174.

Rhamdani, A R, Sabah, S, Brooks, G and Rhamdhani, M A, in prep. Fayalite formation in hydrogen reduction of iron ore and strategies to reduce it from a thermodynamic perspective, *Journal of Sustainable Metallurgy*.

Sabah, S, Boot-Handford, M E, Brooks, G, Natha, M G, Triana, T, Rich, B, Rhamdhani, M A, Xia, Y, Dufty, T, Sceats, M G, Nathan, G and Chinnici, A, in prep. Pilot plant study of the ZESTY flash hydrogen direct reduction ironmaking process, *Journal of Sustainable Metallurgy*.

Impact of gangue in Australian iron ore on green steelmaking

S Sabah[1], M Shahabuddin[2], G Brooks[3], M A Rhamdhani[4], A Rahbari[5] and J Pye[6]

1. Postdoctoral Research Engineer, Swinburne University of Technology, Hawthorn Vic 3122.
   Email: ssabah@swin.edu.au
2. Senior Lecturer, University of Malaya, Malaysia. Email: shahabuddin@um.edu.my
3. Professor, Swinburne University of Technology, Hawthorn Vic 3122.
   Email: gbrooks@swin.edu.au
4. Professor, Swinburne University of Technology, Hawthorn Vic 3122.
   Email: arhamdhani@swin.edu.au
5. Senior Research Fellow, Australian National University, Canberra ACT 2601.
   Email: alireza.rahbari@anu.edu.au
6. Associate Professor, Australian National University, Canberra ACT 2601.
   Email: john.pye@anu.edu.au

ABSTRACT

The growing demand for decarbonisation and the declining iron ore quality present significant challenges for the global ironmaking and steelmaking industry. The Australian iron ore industry, as the leading iron ore producer, needs to understand the implications of high gangue on the green steelmaking routes. Green ironmaking pathways often involve the use of direct reduced iron (DRI). To prevent gangue materials from going to the downstream steelmaking (electric arc furnace (EAF) and basic oxygen furnace (BOF)), an intermediate smelting stage consisting of an electric smelting furnace (ESF) is currently under investigation.

Researchers from Swinburne University and the Australian National University have been looking into the effect of gangue-rich Australian iron ore on the overall process design, ore beneficiation, with a particular focus on the downstream steelmaking option: EAF and ESF integrating a basic oxygen furnace (ESF-BOF). This paper outlines the influence of gangue materials in DRI, ESF, EAF, and BOF processes.

The EAF model showed that with 100 wt per cent cold DRI, the specific electric energy requirement (SEC) ranged from 514 to 651 kWh/tls (tonne of liquid steel), yield decreased from 88.0 wt per cent to 75.5 wt per cent as gangue increased from 7.1 wt per cent to 15.8 wt per cent, and slag weight increased from about 200 to 630 kg/tls. On the other hand, high gangue increased the energy demand of the DRI reactor and SEC of the ESF. Every 1.0 wt per cent increase in gangue led to an increase in the slag production by 23 kg in the ESF. The ESF slag is expected to be used in cement production. The amount of required flux and slag produced did not change much in the BOF with changing gangue. The results indicated that ESF-BOF is more compatible to process lower grade ores than EAF, where the quantity of slag and loss of yield are very high with increasing gangue content.

INTRODUCTION

Australia is the world's largest producer of iron ore. According to the US Geological Survey, in 2024 Australia produced 37 per cent of the world's total iron ore (US Geological Survey, 2025). At present, Australian iron ore companies are navigating the challenges of decarbonisation and declining ore quality (Lu, 2015). Understanding the implications of these challenges is crucial for the Australian iron ore industry to remain competitive and sustainable. The green steelmaking pathway consists of producing direct reduced iron (DRI) using hydrogen, followed by downstream processing via the electric arc furnace (EAF) or basic oxygen furnace (BOF) pathway. Since the DRI process does not eliminate impurities such as silica and alumina from the iron ore, the use of electric smelting furnaces (Shahabuddin, 2024a; Wimmer, Rosner and Bohm, 2022) has been recommended for further processing of DRIs obtained from low-grade iron ores (as shown in Figure 1).

A collaborative project between Swinburne University of Technology and the Australian National University, which was funded by HILT CRC, developed an end-to-end model evaluating the process outputs and the techno-economic performance of the entire green ironmaking and steelmaking route.

The research project aimed to understand the effect of gangue in the iron ore on the green steelmaking pathways, specifically the H_2DRI-EAF pathway and the H_2DRI-ESF-BOF pathway. This paper highlights the impacts of utilising lower-grade iron ore in different processing pathways.

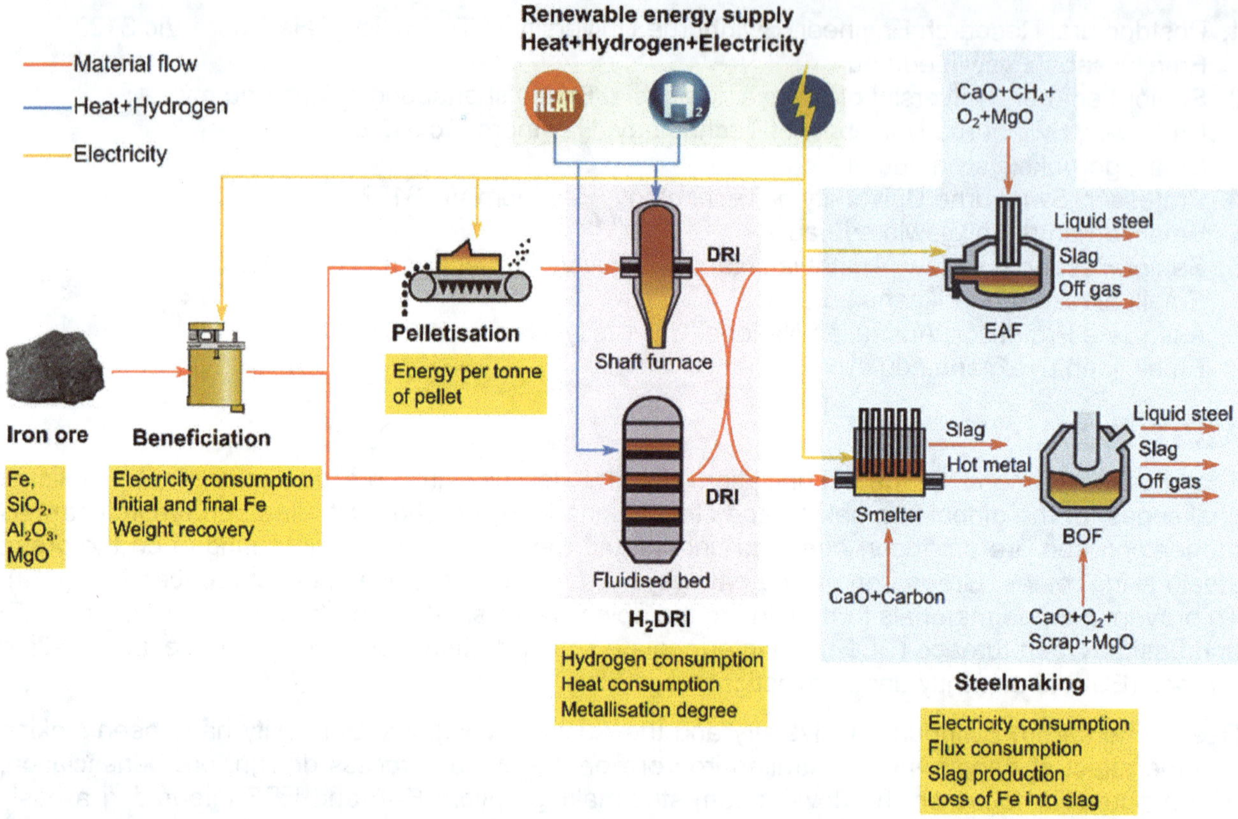

FIG 1 – Schematic of H_2DRI-EAF and H_2DRI-ESF-BOF pathways (Rahbari *et al*, 2025).

DESCRIPTION OF THE MODELS

Steady-state mass and energy balance models were developed for each stage of the processing route. The models were developed in HSC Chemistry 10.0 (Ronie and Kobylin, 2021) using the flow sheet simulation module coupled with the equilibrium module. The models calculated the output at equilibrium conditions using the Gibbs free energy minimisation method (Kho *et al,* 2010). HSC uses a simplified thermodynamic approach to analyse metallurgical reactions through mass and energy balances, without considering reaction kinetics. This approach allows relatively simple models of activity to be included for solution behaviour. It uses fixed activity coefficients, which means it is not intended to capture the detailed thermodynamics of complex reactions in steelmaking, such as dephosphorisation reactions. Five different iron ores – Ore A to Ore E (as shown in Table 1) with various levels of gangue content were processed via the H_2DRI-EAF and the H_2DRI-ESF-BOF route. Compositions of the DRI for processing in the electric smelting furnace are given in Table 2. The details of the models have been discussed in separate publications (Shahabuddin *et al,* 2024a, 2024b; Sabah *et al*, 2024). The outputs from the shaft DRI model have been used as the input of the EAF model and the ESF model. The outputs of the ESF have been further processed in a BOF model.

TABLE 1

Composition of the iron ores (Shahabuddin *et al,* 2024b).

Oxides of ore	A (wt.%)	B (wt.%)	C (wt.%)	D (wt.%)	E (wt.%)
Fe_2O_3	94.81	93.60	83.53	81.07	80.64
NiO	0.00	0.04	0.00	0.00	0.00
TiO_2	0.15	1.07	0.00	0.00	0.23
V_2O_5	0.00	0.68	0.00	0.00	0.00
Al_2O_3	2.50	0.30	2.57	3.28	3.01
SiO_2	2.50	1.94	5.97	6.56	6.71
MgO	0.00	2.17	0.00	0.00	0.07
CaO	0.00	0.12	0.00	0.00	0.04
S	0.01	0.06	0.03	0.03	0.04
P_2O_5	0.03	0.02	0.19	0.14	0.06
Mn	0.00	0.00	0.51	0.72	0.40
LOI	0.00	0.00	7.20	8.20	8.80
Total (%)	100	100	100	100	100
Fe (%)	69.0	67.5	58.0	57.5	56.5
Gangue (%)	5.20	6.40	16.47	18.93	19.35
Ore/tonne Fe (kg)	1506	1526	1712	1764	1765

TABLE 2

Composition of the DRI considering full metallisation (Shahabuddin *et al*, 2024a).

Ore	DRI (kg/h)	Fe	FeO	SiO_2	Al_2O_3	TiO_2	P_2O_5	S	CaO	MgO	Mn	Gangue
A	1076.97	92.85	0.00	3.44	3.44	0.21	0.05	0.02	0.00	0.00	0.00	7.15
B	1086.44	92.04	0.00	2.72	0.42	1.50	0.03	0.08	0.16	3.04	0.00	7.88
C	1159.78	86.22	0.00	8.87	3.82	0.00	0.28	0.05	0.00	0.00	0.76	13.78
D	1183.47	84.50	0.00	9.87	4.42	0.34	0.09	0.06	0.05	0.10	0.59	15.50
E	1189.83	84.05	0.00	9.76	4.88	0.00	0.21	0.04	0.00	0.00	1.07	15.95

DISCUSSION

Effect of gangue content on DRI production

Gangue content in the iron ore did not influence the level of metallisation of the DRI, as thermodynamic modelling assumes no interaction of gangue during reduction at such low temperatures. Kinetic modelling and plant data are required to understand the impact beyond thermodynamic calculations (Shahabuddin *et al,* 2024b). As expected, gangue content was found to significantly affect the energy consumption of the reactor (as shown in Figure 2). As the gangue content of the iron ore increased, reactor energy demand also increased. For example, at 900°C, as gangue content increased from 5.2 per cent to 19.4 per cent, shaft reactor energy demand increased by 27 per cent (Shahabuddin *et al,* 2024b). The energy required for pelletising is not included in this calculation.

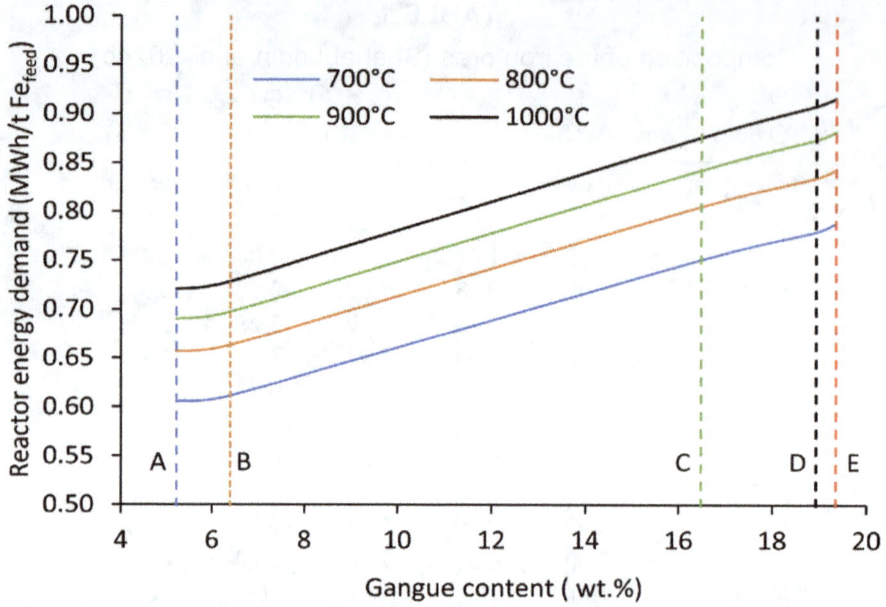

FIG 2 – Effect of gangue content on shaft reactor energy demand at different temperatures (Shahabuddin *et al*, 2024b).

Effect of gangue content on the performance of the ESF

Increasing the gangue content increased the specific energy consumption (SEC) of the ESF (as shown in Figure 3). This is due to the additional energy required to melt the gangue materials. At a 100 wt per cent metallisation level (ie metallisation rate, MR), as gangue content increased from 7 per cent to 16 per cent, SEC increased from 572 kWh/tHM (tonne of hot metal) to 684 kWh/tHM. Every 1.0 wt per cent increase in the gangue increased the SEC by 13 to 20 kWh/tHM, depending on the MR. Every 1.0 wt per cent increase in MR also decreased the SEC by an average of 12 kWh/tHM (Shahabuddin *et al*, 2024a). For lower MR, a higher specific energy is consumed to produce FeO.

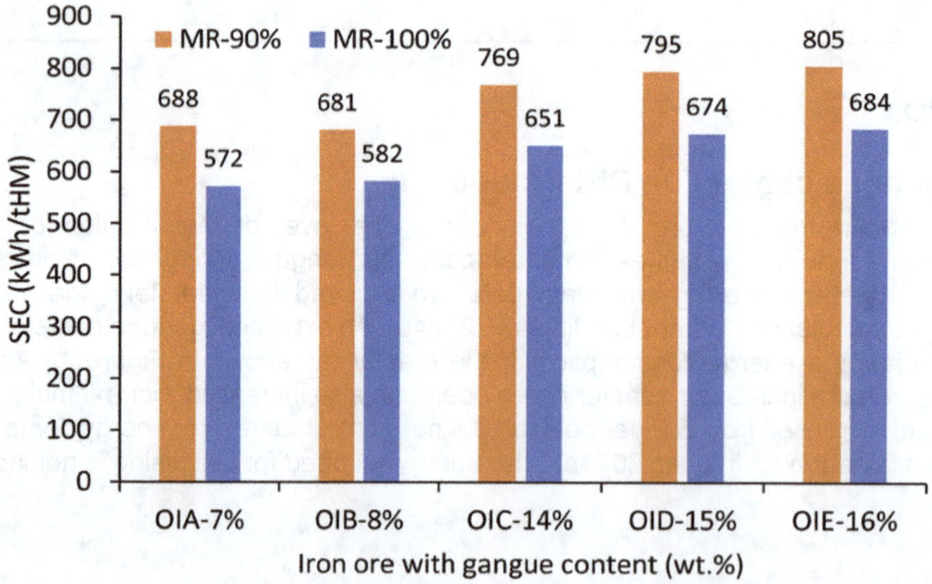

FIG 3 – Specific energy consumption (SEC) at different gangue content and MR (Shahabuddin *et al*, 2024a).

Increasing gangue content also increased the slag volume. For example, at 100 per cent MR with 7.0 wt per cent gangue, ESF produced 93 kg of slag. As gangue content increased to 16 wt per cent, slag weight increased to 302 kg. Every 1.0 wt per cent increase in gangue increased the slag production by 23 kg. The higher the MR, the lower the slag production. The slag produced from the

ESF is expected to be used in cement production, as the basicity of the ESF slag is similar to that of BF slag (Shahabuddin *et al*, 2024a).

Effect of gangue content on the performance of the EAF

The performance of the EAF was evaluated in terms of SEC, production yield, and slag generation. SEC requirement for an EAF using cold DRI, hot DRI, and 100 per cent scrap is presented in Figure 4. As the gangue content of the DRI increased, there was an increase in the SEC of the EAF process. When hot DRI (600°C) was used instead of cold DRI (25°C), there was 21 per cent reduction in the SEC. Using hot DRI is profitable in terms of saving the costs of electric energy in an EAF (Sabah *et al*, 2024).

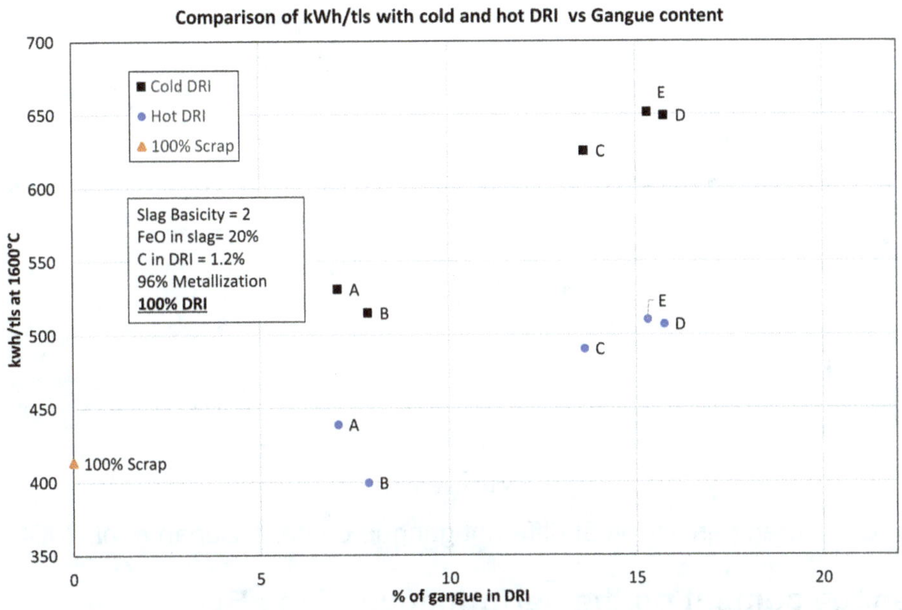

FIG 4 – SEC requirement of an EAF using cold DRI, hot DRI, and 100 per cent scrap (Sabah *et al*, 2024).

With increasing gangue content in the DRI, the yield (wt per cent) of the EAF process decreased. When gangue content increased from 7.1 wt per cent to 15.8 wt per cent, yield decreased from 88 wt per cent to 75.5 wt per cent. This represents a significant loss of materials when the gangue content of the DRI increases (as shown in Figure 5) (Sabah *et al*, 2024).

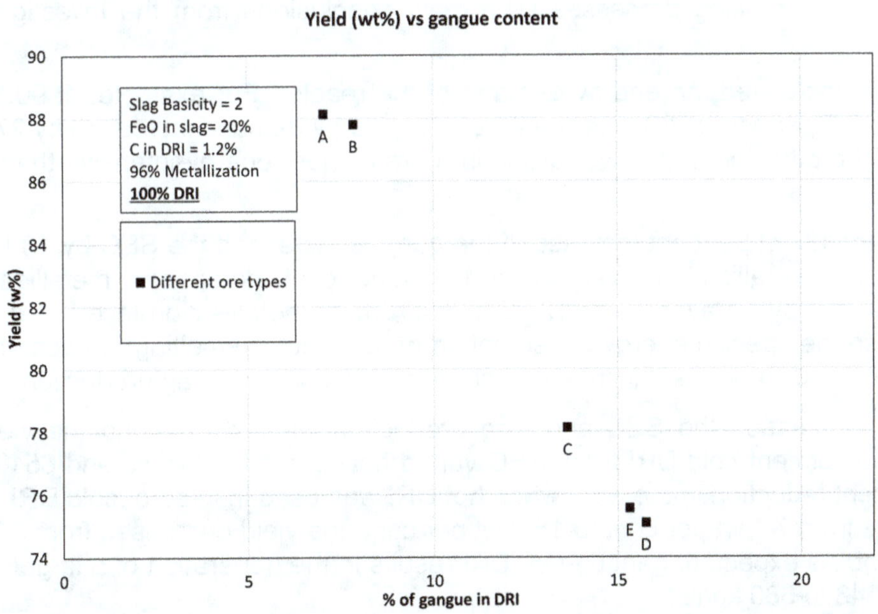

FIG 5 – Yield (%) of an EAF at different gangue content (Sabah *et al*, 2024).

A strong relationship was found between slag produced from the EAF and gangue content of the DRI. Figure 6 shows that with increasing gangue content, the weight of slag considerably increased. For higher grade DRI (A and B), about 200 kg/tls (tonne of liquid steel) slag weight was produced, whereas for the lower grade DRI obtained from lower grade ores (C, D, E), the amount of slag was found to be between 543 kg/tls and 630 kg/tls. Lower-grade ores contained higher silica levels, requiring greater CaO addition to maintain a slag basicity of 2. Consequently, the slag mass produced from lower-grade ores was significantly higher than that from higher-grade ore.

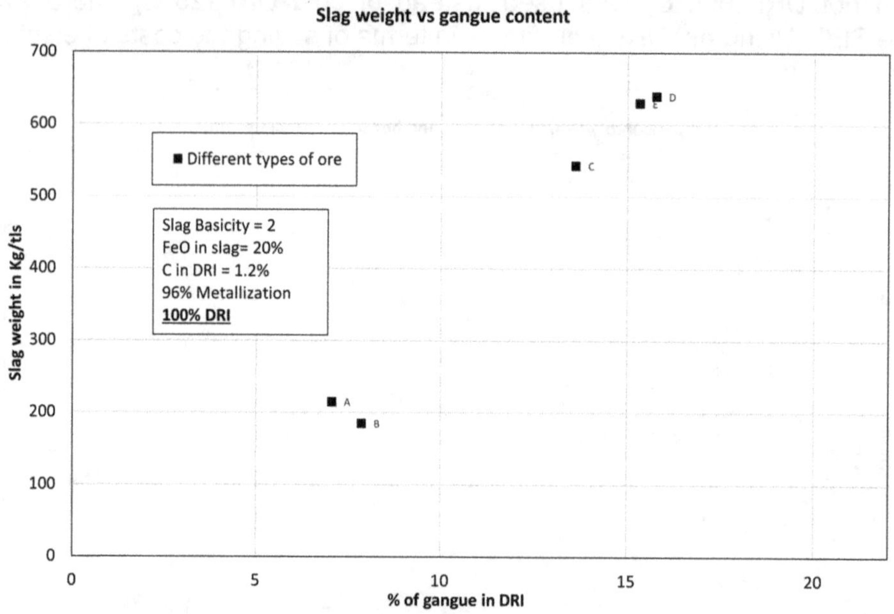

FIG 6 – Slag generation at different gangue content (Sabah *et al*, 2024).

Effect of gangue content on the performance of the BOF

Compared with slag produced in EAF processes, the slag quantity in a BOF showed little variation with changes in iron ore quality. This is due to the presence of the electric smelting furnace (ESF) between the DRI process and BOF, where most of the gangue was removed. The flux requirement showed very little variation across the different ore grades (Sabah *et al*, 2024).

CONCLUSIONS

This paper examines the impact of gangue content in iron ore across different stages of green ironmaking and steelmaking processes. The main conclusions from the investigation are given below.

Gangue content increased the energy demand of the reactor. For example, at 900°C, as gangue increased from 5.2 per cent to 19.4 per cent, reactor energy demand increased by 27 per cent. This implies that DRI production from lower-grade iron ore is more energy-intensive than that of higher-grade iron ore.

In an ESF, every 1.0 wt per cent increase in the gangue increased the SEC by 13 to 20 kWh/tHM, depending on the metallisation level. Every 1.0 wt per cent increase in metallisation level also decreased the SEC by an average of 12 kWh/tHM. Lower metallisation levels and higher gangue content increase the specific energy consumption of the electric smelting furnace. In terms of slag production, every 1.0 wt per cent increase in gangue increased the slag production by 23 kg.

Higher gangue increased the SEC and slag production while decreasing the yield of the EAF process. At 100 per cent cold DRI feed, SEC varied between 514 kWh/tls and 651 kWh/tls. There was a 21 per cent reduction in the SEC when hot DRI was used instead of cold DRI. Increasing the gangue content from 7.1 wt per cent to 15.8 wt per cent, the yield decreased from 88 wt per cent to 75.5 wt per cent. As expected, gangue-rich DRI results in the generation of a larger volume of slag (ranging from 543 to 630 kg/tls).

The amount of required flux and slag produced did not vary much in the BOF with changing gangue content, as most of the gangue materials were removed in the ESF process.

Processing lower-grade iron ore via DRI-EAF, compared to DRI-ESF-BOF, involves higher energy consumption, increased slag generation, and greater yield loss in the EAF process. Also, EAF slag is typically disposed of without any further applications. In contrast, the slag from ESF is predicted to be used for cement production (Shahabuddin *et al*, 2024a), and the process parameters of the BOF process are not affected by the gangue content, which makes the ESF-BOF route attractive for processing high-gangue iron ore. The viability of the electric smelter-BOF pathway depends on addressing challenges associated with smelter technology and maximising the utilisation of scrap in the BOF steelmaking.

ACKNOWLEDGEMENTS

The work has been supported by the Heavy Industry Low-carbon Transition Cooperative Research Centre (HILT CRC), whose activities are funded by its industry, research, and government Partners along with the Australian Government's Cooperative Research Centre Programme, with project number HILT RP 1.004/RP1.005.

REFERENCES

Kho, T S, Swinbourne, D, Blanpain, B, Arnout, S and Langberg, D, 2010. Understanding stainless steelmaking through computational thermodynamics Part 1: electric arc furnace melting, *Mineral Processing and Extractive Metallurgy*, 119(1):1–8.

Lu, L, 2015. *Iron ore: mineralogy, processing and environmental sustainability* (Elsevier).

Rahbari, A, Shahabuddin, M, Sabah, S, Brooks, G and Pye, J, 2025. Production of green steel from low-grade ores: An end-to-end techno-economic assessment, *Cell Reports Sustainability*, 2(1).

Ronie, A and Kobylin, P, 2021. HSC Chemistry User's Guide, Version 10, Metso: Outotec.

Sabah, S, Shahabuddin, M, Rahbari, A, Brooks, G, Pye, J and Rhamdhani, M A, 2024. Effect of gangue on CO2 emission for different decarbonisation pathways, *Ironmaking and Steelmaking*, 51(4):356–368.

Shahabuddin, M, Rahbari, A, Sabah, S, Brooks, G, Pye, J and Rhamdhani, M A, 2024a. The performance and charge behaviour in melter/smelter for the production of hot metal in hydrogen DRI-based steelmaking, *Ironmaking and Steelmaking*, 53(1):59–70.

Shahabuddin, M, Rahbari, A, Sabah, S, Brooks, G, Pye, J and Rhamdhani, M A, 2024b. Process modelling for the production of hydrogen-based direct reduced iron in shaft furnaces using different ore grades, *Ironmaking and Steelmaking,* 52(1).

US Geological Survey, 2025. Mineral Commodity Summaries 2025 [online], ver 1.2, March 2025. Available from: <https://pubs.usgs.gov/periodicals/mcs2025/mcs2025-iron-ore.pdf>

Wimmer, G, Rosner, J, Bohm, C, 2022. Smelter Technology-Transforming Integrated Steelmaking Toward Net Zero Carbon, in Proceedings of AISTech 2022.

Thermodynamic assessments of Australian iron ore DRI smelting in a laboratory-scale electric smelting furnace

A Taimullah[1], C Garlick[2], T Singh[3], T B T Nguyen[4], D O'Dea[5], A Rich[6] and T Honeyands[7]

1. PhD scholar, BHP Centre for Sustainable Steelmaking Research, School of Engineering, The University of Newcastle, Callaghan NSW 2308. Email: abrar.taimullah@uon.edu.au
2. Research Associate, BHP Centre for Sustainable Steelmaking Research, School of Engineering, The University of Newcastle, Callaghan NSW 2308. Email: craig.garlick@newcastle.edu.au
3. Research Associate, BHP Centre for Sustainable Steelmaking Research, School of Engineering, The University of Newcastle, Callaghan NSW 2308. Email: tejbir.singh@newcastle.edu.au
4. Lecturer, BHP Centre for Sustainable Steelmaking Research, School of Engineering, The University of Newcastle, Callaghan NSW 2308. Email: thibangtuyen.nguyen@newcastle.edu.au
5. BHP Marketing Sustainability, BHP, Brisbane Qld 4000. Email: damien.p.odea@bhp.com
6. Practice Leader Smelting, BHP, Perth WA 6000. Email: anthony.rich2@bhp.com
7. Associate Professor, BHP Centre for Sustainable Steelmaking Research, School of Engineering, The University of Newcastle, Callaghan NSW 2308. Email: tom.a.honeyands@newcastle.edu.au

ABSTRACT

Australia leads the global iron ore industry, supplying about 37 per cent of global production and over 52 per cent of total exports, playing a crucial role in supporting the world's iron and steel production. Most of the exports are medium grade ores, which are currently processed through the blast furnace-basic oxygen furnace (BF-BOF) route. However, the BF-BOF route emits around 2.3 t of CO_2 per tonne of steel. The direct reduced iron-electric smelting furnace (DRI-ESF) has emerged as a promising pathway to utilise medium-grade iron ore, particularly Australian hematite-goethite iron ores, in a low emission ironmaking route. A comprehensive understanding of thermodynamic principles is essential to evaluate the metallurgy and physicochemical reactions during DRI-ESF ironmaking as they provide insights into equilibrium of the reactions, phase stability and energy calculations. This study presents preliminary thermodynamic calculations of Australian hematite-goethite ore hydrogen-direct reduced iron (H-DRI) smelting in electric smelting furnace (ESF) using FactSage 8.3 software and compares the results with a lab-scale electrode smelting furnace experiment carried out with H-DRI produced from MAC iron ore fines. The calculations evaluated phase stability, hot metal and slag compositions as functions of temperature and charged carbon-to-iron ratio (C/Fe). The results showed that higher temperature promoted SiO_2 reduction, leading to higher silicon content in the hot metal. Increasing charged C/Fe increased both carbon and silicon contents in the hot metal. Comparison with the smelting experiment showed that the experimental carbon and silicon contents were lower than those predicted by FactSage, highlighting the limitation of single-zone equilibrium calculations, which do not account for kinetic effects, temperature gradients, and carbon losses in real smelting operation. Adjustment of the system pO_2 improved the agreement in carbon content between calculated and experimental result. However, silicon remained overpredicted. Overall, this study highlights the need for an advanced multi-zone thermodynamic model to better represent the actual furnace operation during H-DRI smelting in ESF conditions.

INTRODUCTION

Australia plays a critical role in the global iron ore supply chain, contributing around 37 per cent of global iron ore production and approximately 53 per cent of global iron ore exports in 2023 (World Steel Association, 2025). Around 70 per cent of Australia's iron ore resources are in the Pilbara region, which are characterised by hematite-goethite ores. These ores are considered medium-grade, with Fe contents of 58–62 per cent and medium levels of gangue and phosphorus (Wang *et al,* 2023). As a result, Australian iron ore is commonly processed via the blast furnace-basic oxygen furnace (BF-BOF) route, which currently produces an average emission intensity of around 2.3 t of CO_2 per tonne of crude steel (World Steel Association, 2025). Therefore, the development of alternative low-carbon ironmaking routes is essential, particularly for the processing of medium-grade iron ores (Brooks and Hornby, 2024).

The direct reduced iron-electric smelting furnace (DRI-ESF) has emerged as a promising pathway to utilise medium-grade iron ore in a low emission ironmaking route. In this process, natural gas or hydrogen-reduced DRI is smelted in the ESF to produce BF-like hot metal, which is then refined through a steelmaking process (Gadd *et al,* 2023). Based on current natural gas-based direct reduction processes, the DRI-ESF-BOF route is projected to emit around 1.3 t of CO_2 per tonne of crude steel, while utilising hydrogen-based DRI in combination with renewable electricity could reduce emissions to as low as 0.21 t CO_2 per tonne of crude steel (Paymooni *et al,* 2024).

So far, DRI smelting in the ESF has been investigated through both experiments and thermodynamic calculations; however, the number of published studies remains limited. Pfeiffer *et al* (2022) conducted simplified thermodynamic calculations using FactSage combined with empirical equations from literature to evaluate the suitability of medium-grade ores for the DRI-ESF-BOF routes. However, the results were not validated with experimental and/or operational data (Pfeiffer *et al,* 2022). Bae and Park (2025) combined lab-scale experiments with FactSage thermodynamic calculations to investigate hot metal production from synthetic slag, solid carbon, and electrolytic iron in a 600 g scale induction furnace at 1550°C. The study reported good agreement between measured and calculated carbon content in the metal, while significant discrepancies were observed for silicon content in metal, indicating that silica reduction did not attain equilibrium even under controlled conditions (Bae and Park, 2025). Nevertheless, this finding was limited by the use of an induction furnace, which does not represent the temperature gradients and reaction zones generated by arc heating in industrial ESF operation, as well as by the simplified raw materials used in the tests.

A more representative approach was demonstrated by Garlick *et al* (2025), who investigated hydrogen-DRI (H-DRI) produced from Australian hematite-goethite iron ore. The H-DRI, fluxes, and carburiser were smelted in a 2.5–6 kg lab-scale electrode smelting furnace. They found that this approach successfully produced temperature gradients inside the furnace, which better represents industrial ESF conditions. Their experiments successfully produced hot metal with 2.5–4.5 per cent C and 0.05–1.2 per cent Si. The hot metal chemistry was found to be influenced by several key operating parameters, including the charged carbon-to-iron ratio (C/Fe), oxygen-to-iron ratio (O/Fe), power density, and silica input (Garlick *et al,* 2025). However, a thermodynamic assessment of Australian iron ore H-DRI smelting using FactSage, and its comparison with lab-scale electrode smelting furnace results, has not yet been reported.

The present study addresses this knowledge gap by conducting thermodynamic equilibrium calculations using FactSage to evaluate the effects of temperature and charged C/Fe ratio on hot metal and slag chemistries. The calculation results were then compared with lab-scale electrode smelting furnace experimental results. In addition, the oxygen partial pressure (pO_2) and temperature were adjusted in FactSage calculations to match the experimental carbon and silicon contents.

METHOD

H-DRI smelting in the ESF was thermodynamically evaluated using FactSage 8.3 software, utilising the Equilib module based on FactPS, the UQPY (University of Queensland Pyrosearch) private database and FSstel. FSstel database was used for the solid and liquid metal phases, UQPY private database for liquid slag, solid oxide solutions, pure solids and gaseous phases. Meanwhile, FactPS covered gaseous and pure phases not covered by the UQPY database.

Input data for the calculations in this study were based on a smelting experiment of H-DRI fines produced from Australian hematite-goethite iron ore (MAC fines) using a lab-scale hydrogen fluidised bed reactor, with a metallisation of around 92 per cent. Smelting of H-DRI fines was conducted in a lab-scale electrode smelting furnace at a 2.5 kg scale by adding blast furnace slag and calcium oxide powder as fluxes, calcined anthracite as carburiser and 110 g scrap to initiate the arcing at the start of the experiment (more details are given in Garlick *et al,* 2025).

Detailed input materials, chemical compositions and material masses used in the thermodynamic calculations are summarised in Table 1. The base case key parameters included a charged C/Fe of

0.104 and charged O/Fe of 0.021. The total iron input to the system was 1590 g, and carbon was added in excess to the stoichiometric requirement to a target carbon content in hot metal, FeO reduction, oxides reduction and carbon losses. The slag basicity was set to 1.03 to closely represent a typical blast furnace slag.

TABLE 1

Input materials, chemical compositions and material masses used in this study.

DRI/Fluxes	**TFe (wt%)**	**SiO_2 (wt%)**	**Al_2O_3 (wt%)**	**CaO (wt%)**	**MgO (wt%)**	**P (wt%)**	**Mass (g)**
H-DRI (91% met)	86.75	6.10	3.30	0	1.40	0.10	1700
Blast furnace slag	0.97	35.30	13.40	41.40	5.75	-	390
Calcium oxide powder	-	1.50	-	97.00	-	1.50	67
Carburiser	**C (wt%)**	**SiO_2 (wt%)**	**Al_2O_3 (wt%)**	**CaO (wt%)**	**H_2O (wt%)**	**S (wt%)**	**Mass (g)**
Calcined anthracite	91.90	1.98	0.99	0.43	1.90	0.32	184

The thermodynamic calculations in this study included phase stability assessments of the charged materials, investigation of the influence of the temperature and charged C/Fe ratio on metal and slag chemical compositions, and comparison between experimental results and thermodynamic calculations.

RESULTS AND DISCUSSION

H-DRI smelting experimental results

The experimental results used in this study were obtained from the published work of Garlick *et al*, (2025), with additional calculations of carbon utilisation to estimate carbon consumption for reduction reactions and metal carburisation during smelting. H-DRI smelting was conducted for 40 mins with a total energy input of 6.6 kWh. Temperature measurements were taken at three positions (20, 5 and 0 mm from the sidewall) at a height of 12 mm from base, as presented in Figure 1. A clear temperature gradient was observed from the wall towards the electrode, with temperatures increasing closer to the electrode. The maximum temperatures recorded at 20 mm, 5 mm and 0 mm from the sidewall were 1431, 1291 and 1112°C, respectively. The temperature near the electrode was not measured, but it typically exceeded 2000°C.

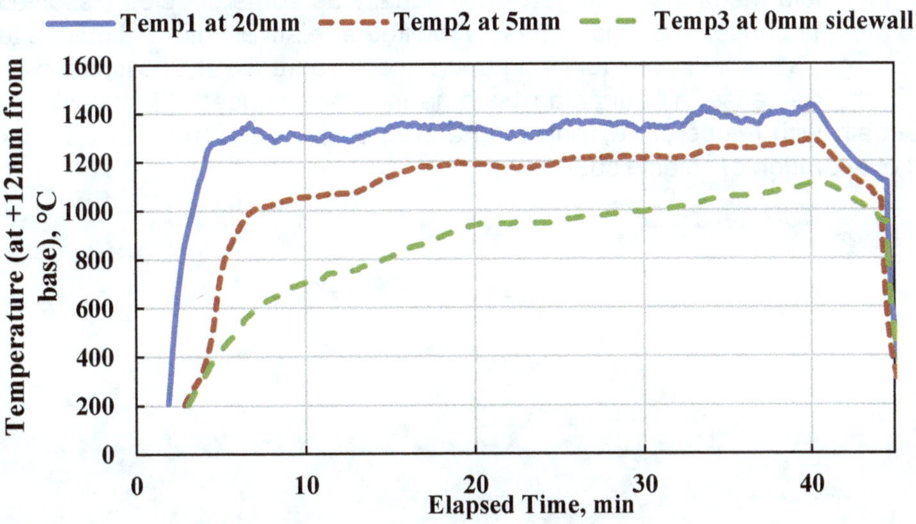

FIG 1 – Measured temperatures at 20, 5 and 0 mm from sidewall at +12 mm from base from the H-DRI smelting experiments (Garlick *et al,* 2025).

The masses and chemical compositions of the hot metal and slag produced from experiments are presented in Table 2. The test successfully produced 1500 g of hot metal with the desired carbon content of 3.74 per cent. However, the silicon content was relatively high for typical hot metal, reaching 0.67 per cent. The slag composition showed a basicity of 0.107 with a relatively high Al_2O_3 content of 24.62 per cent. Based on the charged mass balance calculation, the slag mass produced in the experiment was estimated to be around 629 g.

TABLE 2

Final hot metal and slag masses and compositions after 40 min of H-DRI smelting test.

Hot metal	Mass (g)	C (wt%)	Si (wt%)	Mn (wt%)	P (wt%)	S (wt%)
	1500	3.74	0.67	0.07	0.09	0.08
Slag	**Mass (g)**	**CaO (wt%)**	**SiO_2 (wt%)**	**Al_2O_3 (wt%)**	**MgO (wt%)**	**S (wt%)**
	629*	37.40	34.74	24.62	2.81	0.23

*Calculated value from mass balance calculation.

From the total carbon charged into the system, only a fraction was consumed by oxides reduction and hot metal carburisation. Based on the carbon utilisation calculation, 35.8 per cent of the charged carbon was used for hot metal carburisation, 15.6 per cent for FeO reduction, 5.6 per cent for SiO_2 reduction, and around 1 per cent for the reduction of other oxides. The remaining carbon loss may be attributed to oxidation by air ingress and carbon entrapment in slag/unmelted feed, resulting in an overall carbon utilisation of 58 per cent.

The electrode smelting furnace experiment for H-DRI smelting highlights several important factors that should be considered, including the presence of temperature gradient in the furnace, operating conditions that deviate from equilibrium and carbon losses. In contrast, single-zone thermodynamic calculations in FactSage assume equilibrium at a uniform temperature. Therefore, deviations between the experimental results and FactSage calculations are expected.

Phase stability of charged materials in H-DRI in smelting

The phase stability of charged materials mixture for H-DRI smelting is shown in Figure 2. Generally, the phases formed could be classified into solid oxides, liquid slag, solid metal, liquid metal and gas. The metal phase was completely melted at 1160°C, indicating a carbon-saturated iron composition for equilibrium calculations, with a liquid metal mass of 1064 g. The mass of liquid metal gradually increased to 1645 g at 1500°C and reached a maximum of 1689 g at 1700°C. At higher temperatures, the liquid metal mass decreased gradually as some species dissolved in the metal volatilised into the gas phase. The oxide phase exhibited a relatively higher liquidus temperature of 1460°C, with 565 g. The amount of liquid slag decreased progressively, reaching 545 g at 1500°C. Above 1700°C, the decrease in liquid slag became less pronounced. Meanwhile, the gas phase increased gradually with temperature, indicating a progressive gas formation from oxide reduction reactions and evaporation of volatile species.

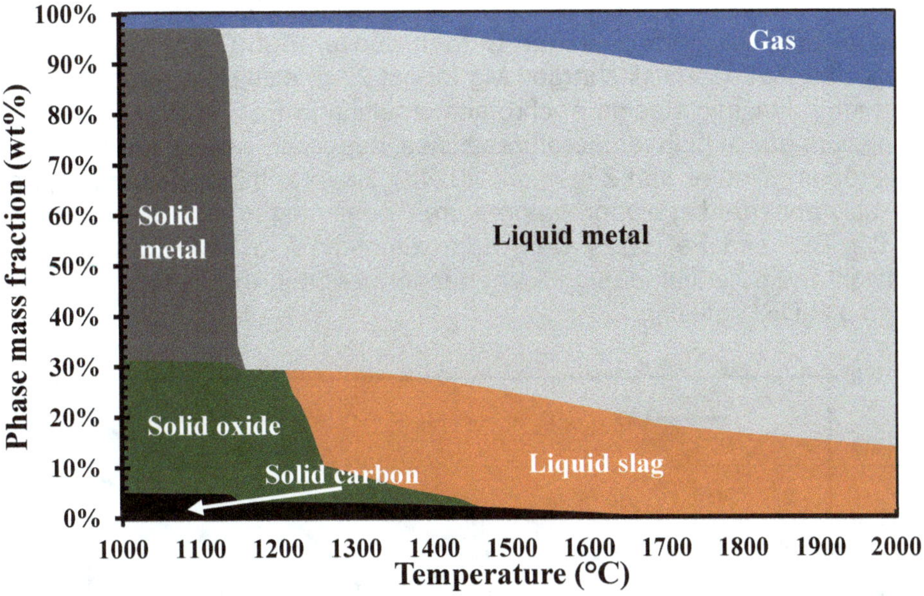

FIG 2 – Phase stability of charged materials in H-DRI smelting calculated using FactSage 8.3.

The details of the oxide phases formed, along with their respective masses at various temperatures, are presented in Figure 3. At temperatures below 1220°C, three stable oxide phases were predicted, consisting of melilite, feldspar and clinopyroxene, with masses of approximately 359 g, 189 g and 77 g, respectively. Clinopyroxene disappeared at 1220°C, while feldspar disappeared at 1260°C. The mass of melilite gradually decreased from 1220°C and fully melted at around 1460°C. This indicates that melilite ($Ca_2(Mg,Fe)Si_2O_7$) was the major precipitated oxide phase in this system, which is also a typical primary phase in industrial blast furnace slag. The presence of melilite as the primary precipitated phase in blast furnace slag has also been reported by Ren *et al* (2018) using FactSage calculations. Furthermore, experimental investigations have confirmed melilite as a major phase in blast furnace slag through XRD and SEM analyses (Kashiwaya *et al,* 2007; Qin *et al,* 2015; Ren *et al,* 2018).

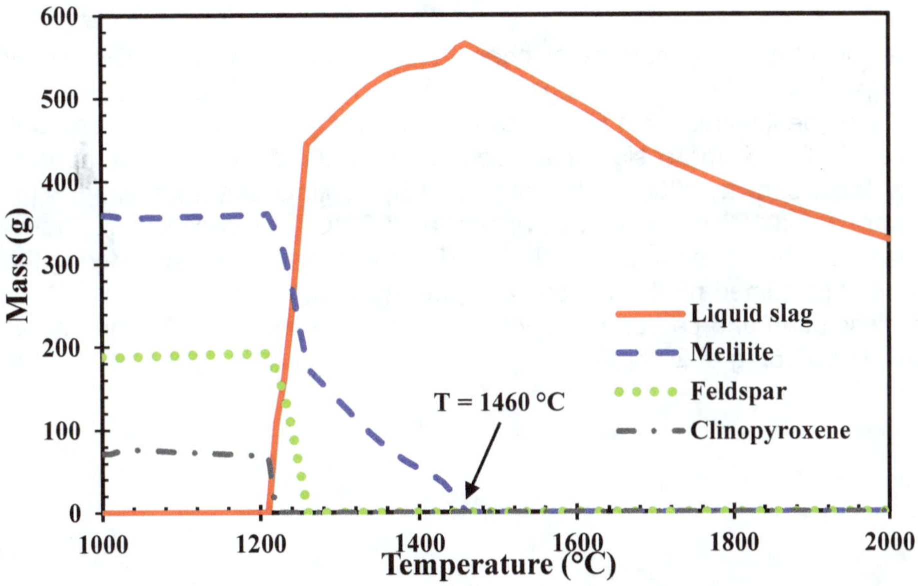

FIG 3 – Oxide phases formation in H-DRI smelting calculated using FactSage 8.3.

The calculated gaseous species in the range of 1000–2000°C are shown in Figure 4. CO gas was the dominant gas species, with a relatively constant mass of around 70 g below 1300°C. Above 1300°C, the CO mass increased significantly reaching approximately 321 g at 2000°C. In smelting of iron-bearing materials, CO generation is primarily attributed to the reduction of FeO, with minor contributions from the reduction of SiO_2, MnO and other oxides (Turkdogan, 1984). In addition, Mg was another major gaseous species, started to form above 1500°C and reached around 25 g at 2000°C. This indicates that nearly all charged Mg (about 27 g) existed in vapour form. This result is consistent with thermodynamic studies of chromite smelting in ESF systems, where MgO may be reduced to magnesium gas at high temperatures zones, especially near to the arc (Barcza, Curr and Jones, 1990; Oterdoom, Reuter and Zietsman, 2024a). Several minor gaseous species, including SiO, Fe and Ca, also appeared at higher temperatures, beginning to form above 1700°C. At 2000°C the masses of Ca, SiO and Fe gases were approximately 5, 3 and 1.5 g, respectively. These calculations highlight the potential for significant gas formation in the high temperature zones near the electrode during H-DRI smelting.

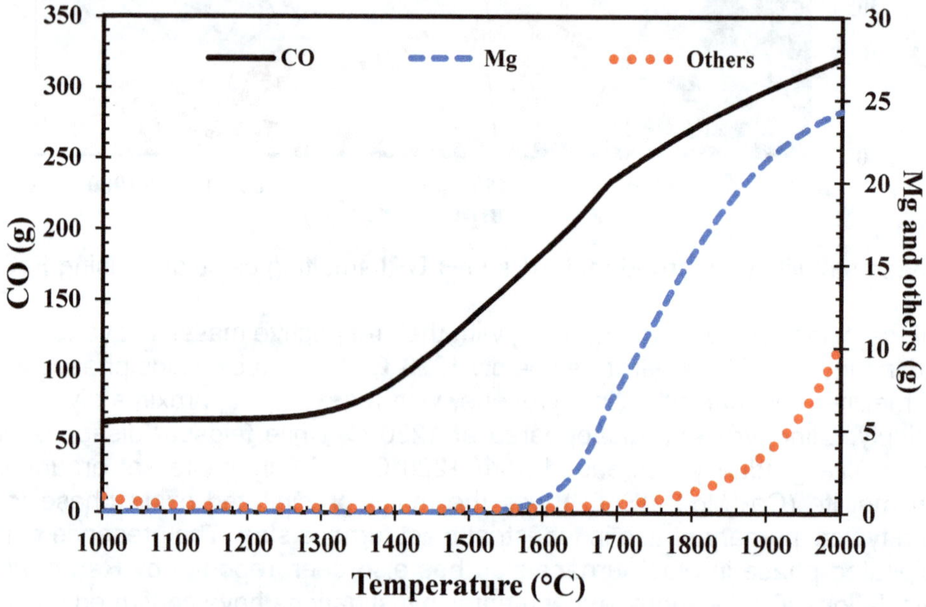

FIG 4 – Cumulative gas species formed in H-DRI smelting calculated using FactSage 8.3.

The influence of temperature on hot metal and slag composition in H-DRI smelting

Figure 5 shows the change in hot metal chemical composition from H-DRI smelting over the temperature range of 1200–2000°C at a charged C/Fe ratio of 0.1. The carbon content in hot metal ([C]) reached its highest value of 4.57 per cent at 1360°C, followed by a gradual decrease to 4.19 per cent at 1700°C. A more significant decrease occurred at higher temperatures, with [C] dropping below 2 per cent at 2000°C. In contrast, [Si] continuously increased with temperature, reaching 2.14 per cent at 1500°C and 5.92 per cent at 2000°C. The decline in [C] between 1360 and 1700°C is primarily attributed to the dissolution of silicon into the metal, considering that the mass of carbon in hot metal remained relatively constant (around 70 g). Above 1700°C, [C] decreased due to both silicon dissolution and carbon oxidation to CO gas, as indicated by the decrease in carbon mass in the metal to 33.53 g at 2000°C.

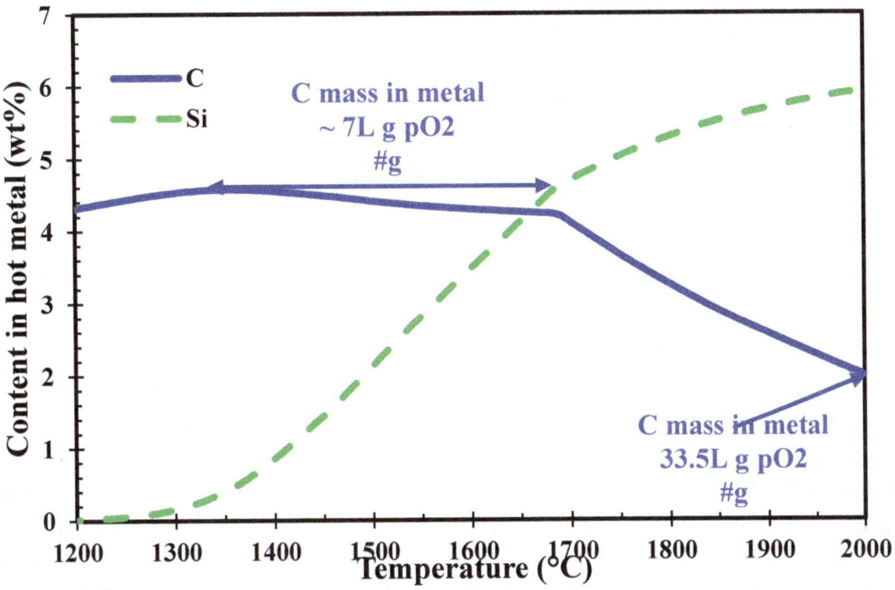

FIG 5 – The influence of temperature on hot metal composition calculated using FactSage 8.3.

Figure 6 presents the evolution of slag chemical composition in the temperature range of 1400–2000°C. A significant decrease in SiO_2 content was observed as temperature increased. This behaviour validated that the increase of [Si] in the hot metal was driven by the more extensive reduction of SiO_2 at higher temperatures. This is consistent with the findings of Meraikib (2000), who reported a strong temperature influence on silicon partitioning in blast furnace operation (Meraikib, 2000). The substantial reduction of SiO_2 resulted in an increase in CaO and Al_2O_3 content in the slag. Meanwhile, MgO content remained relatively stable up to 1650°C but decreased progressively at higher temperatures. MgO content dropped from 8.67 per cent at 1650°C to 1.41 per cent at 2000°C due to the reduction of MgO into gaseous magnesium at high temperatures.

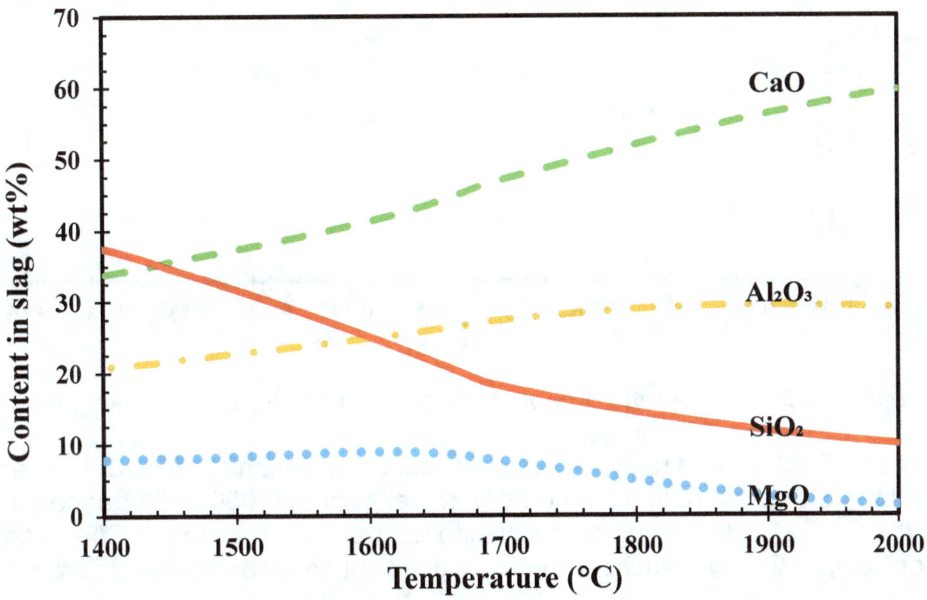

FIG 6 – The influence of temperature on slag composition calculated using FactSage 8.3.

These calculations provided insight into the application of H-DRI smelting in ESF, suggesting that excessively high temperatures should be avoided to prevent excessive silicon pickup into the hot metal. This observation is consistent with lab-scale H-DRI smelting experiments reported by Garlick *et al* (2025), where [Si] increased with rising average melt pool temperature. In practice, melt pool temperature can be controlled by adjusting the power density to avoid excessive reduction of silica (Garlick *et al*, 2025).

Moreover, at higher temperatures, magnesium volatilisation could cause significant MgO losses from the slag. This indicates that MgO reduction and Mg evaporation may occur in high-temperature zones, particularly near the electrode. Conversely, operating at lower temperatures could lead the formation of solid oxide phases, which may hinder furnace operation. However, it should be noted that single zone FactSage calculations only accounted for uniform temperature and equilibrium conditions, which do not fully represent the actual furnace operation (Oterdoom, Reuter and Zietsman, 2024b).

The influence of charged carbon-to-iron ratio (C/Fe) on hot metal and slag composition in H-DRI smelting

Figure 7 presents the influence of the charged C/Fe ratio on hot metal composition in H-DRI smelting at 1500°C. The results showed that increasing C/Fe ratio led to higher [C] and [Si]. However, both [C] and [Si] reached a plateau at C/Fe ratio above 0.082, approximately at 4.4 per cent and 2.2 per cent, respectively. This plateau indicates that both carbon and silicon reached their maximum solubility in hot metal at 1500°C at equilibrium conditions. Lab-scale H-DRI smelting experiments by Garlick *et al* (2025) also reported a positive correlation between [C] and [Si] and the charged C/Fe ratio. However, in their experiments, [C] and [Si] continued to increase over a C/Fe range of 0.107–0.171, with carbon utilisation was estimated to be only around 60 per cent in several tests (Garlick *et al,* 2025). This suggests that the discrepancy may be attributed to the limitations of FactSage calculations, which do not consider carbon losses due to mechanical entrapment in slag/unmelted feed and oxidation due to air ingress. Consequently, the charged C/Fe ratio may not fully represent the effective C/Fe ratio in real processes.

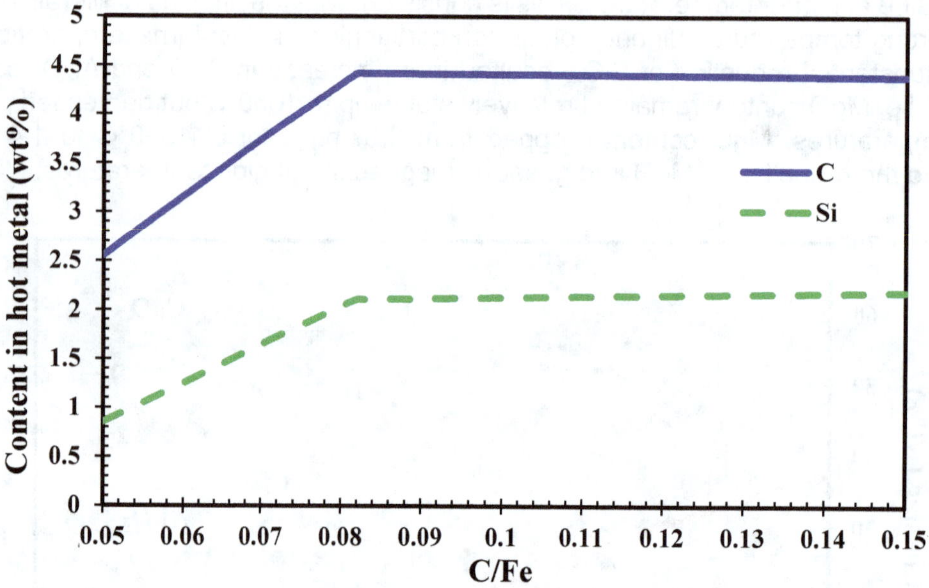

FIG 7 – The influence of C/Fe ratio on hot metal composition calculated using FactSage 8.3.

Figure 8 shows the effect of C/Fe ratio on slag chemical composition at 1500°C. Increasing the C/Fe ratio decreased the SiO_2 content in the slag from 36 per cent at 0.050 to 32 per cent at 0.082. This indicates that the SiO_2 reduction is more sensitive to temperature than to the C/Fe ratio. Meanwhile, the contents of CaO, Al_2O_3, and MgO showed no significant changes across different C/Fe ratios.

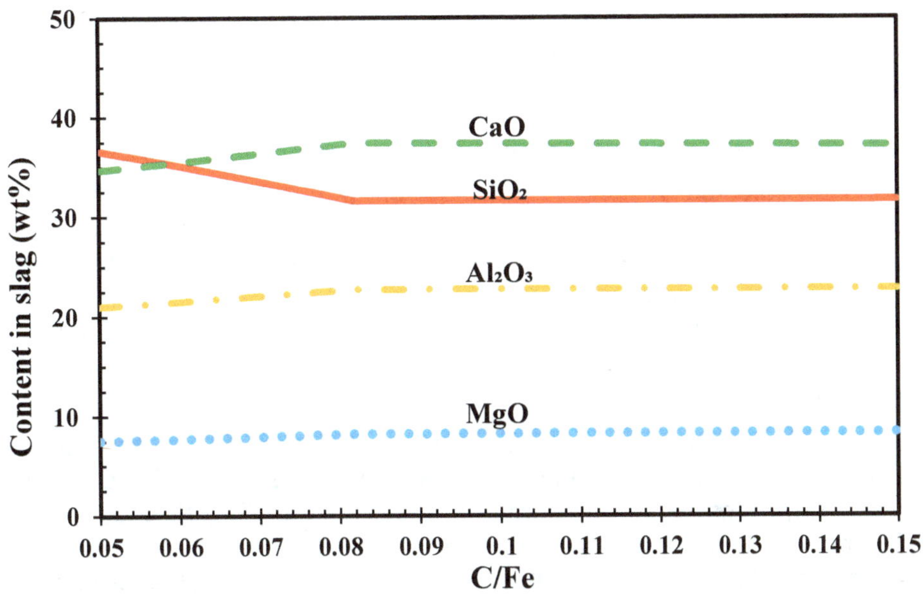

FIG 8 – The influence of C/Fe ratio on slag composition calculated using FactSage 8.3.

Comparison between thermodynamic calculations and experimental results

Base case calculation

Thermodynamic calculations were conducted to compare the results with those from the H-DRI smelting experiment. A reference temperature of 1500°C was selected for the base case of the thermodynamic calculations. Table 3 presents the comparison of hot metal and slag masses and compositions obtained from the experiment and thermodynamic calculation. Differences were observed in both metal and slag masses between the experiment and FactSage calculation. The metal mass produced in the experiment (1500 g) was lower than the calculated value (1645 g), whereas the slag mass showed the opposite trend. This difference may be attributed to incomplete participation of charged materials in the experiment, as indicated by splashing, sidewall accretion, and unmelted feed remaining at the sidewall and furnace base.

TABLE 3

Comparison between experimental result and FactSage calculation.

Hot metal	**Mass (g)**	**C (wt%)**	**Si (wt%)**	**Mn (wt%)**	**P (wt%)**	
Experiment	1500	3.74	0.67	0.07	0.09	
FactSage 8.3	1645	4.39	2.17	0.21	0.10	
Slag	**Mass (g)**	**CaO (wt%)**	**SiO_2 (wt%)**	**Al_2O_3 (wt%)**	**MgO (wt%)**	**Fe_{total} (wt%)**
Experiment	629*	37.40	34.74	24.62	2.81	3.90
FactSage 8.3	545	36.08	30.40	25.51	7.90	0.02

*Calculated value from mass balance calculation.

In more detail, the calculated [C] and [Si] were considerably higher than the experimental result. The experimental [C] and [Si] were 3.74 per cent and 0.67 per cent, while the calculated values were 4.39 per cent and 2.17 per cent, respectively. This again highlights the limitation of equilibrium FactSage calculations, which do not account for carbon losses through mechanical or chemical losses. From the slag composition, the SiO_2 content calculated by FactSage was relatively lower than the experimental result, which is consistent with the higher calculated [Si]. In addition, the notable difference was observed in MgO content, with calculated value was 7.90 per cent, while experimental result was 2.81 per cent. This suggested significant magnesium volatilisation occurred in the experiment, likely in high-temperature zones (as predicted in Figure 4 and Figure 6), which cannot be captured by single-temperature equilibrium calculations in FactSage.

Adjustment of partial pressure of oxygen (pO_2) and temperature

Carbon utilisation was adjusted by modifying the system pO_2 in the thermodynamic calculations to better represent the experimental conditions. Temperature was also adjusted, as the large temperature gradients in the furnace made it difficult to define an effective operating temperature. The higher pO_2 in the system (less reducing conditions) increases the carbon distribution in gas phase because under more oxidative conditions, more carbon can be oxidised into CO or CO_2. This can be represented as carbon utilisation, although in the experiment, carbon loss was not only through gas formation, but some was also found unreacted as the solid phase.

Figure 9 shows the effect of pO_2 and temperature on [C] and [Si] in the hot metal based on thermodynamic calculations. Two approaches were applied: targeting [C] to match experimental result and targeting [Si] to match the experimental result. When [C] was used as the reference for determining pO_2, the values ranged from $10^{-15.35}$ to $10^{-15.25}$. However, within this range, the [Si] was relatively higher compared to the experimental result. This may suggest that silica reduction reaction did not attain equilibrium, which is consistent with results reported in previous studies (Shankar, 2006; Bae and Park, 2025). On the other hand, when [Si] was used as the reference, the pO_2 values ranged from $10^{-14.80}$ to $10^{-13.20}$, however, this resulted in calculation of [C] contents that were too low. Nevertheless, this pO_2 determination was consistent with the study by Mitsufuji *et al* (1992), in which the pO_2 for silicon equilibrium was higher than carbon equilibrium in industrial blast furnace operation (Mitsufuji *et al,* 1992).

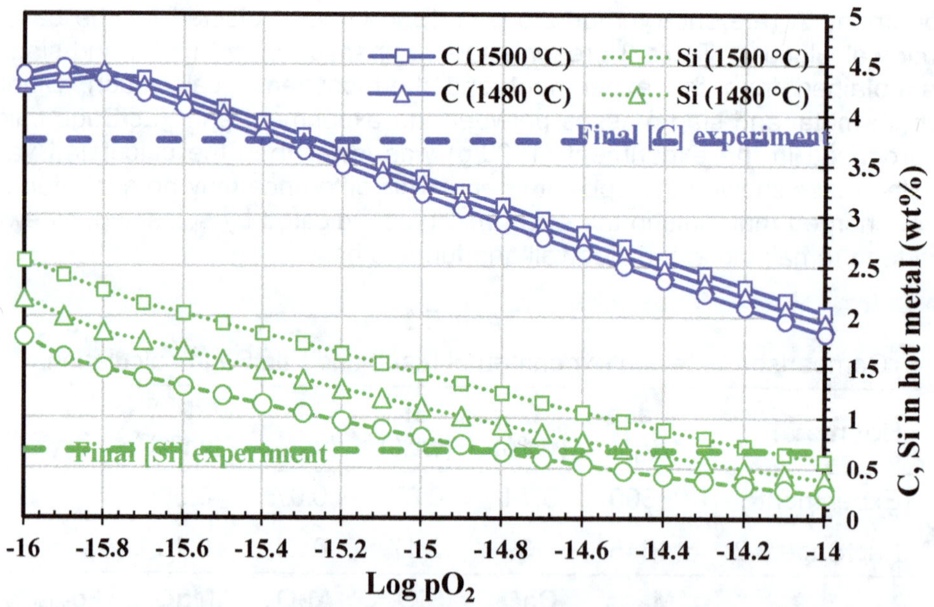

FIG 9 – The influence of pO_2 and temperature on carbon and silicon content in the hot metal calculated using FactSage 8.3.

The difficulty in predicting the experimental hot metal chemistry using single-temperature equilibrium calculations highlights the need for more advanced modelling approaches. One promising method is the effective equilibrium reaction zone (EERZ) concept, in which a complex process is simplified into several reaction zones and equilibrium is calculated for each zone (Jung and van Ende, 2020). EERZ has been applied in several smelting furnace studies, such as ferrochrome production (Nam, van Ende and Jung, 2022; Oterdoom, Reuter and Zietsman, 2024b) and alternative hot metal production in the New Zealand Steel (NZS) process (Büttner, 2024). However, an EERZ-based study for Australian iron ore-based DRI smelting has not yet been reported. Therefore, this represents an important opportunity for future work.

CONCLUSIONS

This study conducted a single-zone equilibrium thermodynamic assessment of Australian hematite-ore H-DRI smelting in an electric smelting furnace (ESF) using FactSage 8.3 and compared the results with a laboratory-scale electrode smelting furnace experiment. The main conclusions are:

- FactSage calculations showed that temperature strongly influenced hot metal chemistry. Higher temperatures promoted SiO_2 reduction and increased [Si]. This trend is consistent with laboratory-scale H-DRI smelting experiments, highlighting that power control is critical to avoid excessive silicon dissolution into the metal.
- The influence of charged C/Fe ratio on [C] and [Si] followed a similar trend to experimental results. However, discrepancies remained because equilibrium FactSage calculations do not account for carbon losses, indicating that the charged C/Fe ratio may not fully represent the effective C/Fe ratio in real operation.
- Comparison with the 40 min lab-scale smelting experiment showed that the [C] and [Si] were relatively lower than FactSage calculation. This is attributed to incomplete participation of charged materials due to the strong temperature gradient within the furnace and limited carbon utilisation.
- Adjusting pO_2 and temperature improved the agreement for [C], but [Si] remained overpredicted. This is because FactSage only calculates silicon under equilibrium conditions, whereas the experiment deviated from equilibrium, confirming the limitation of single-zone equilibrium calculations.

Overall, this work demonstrates the equilibrium thermodynamic calculations as a baseline for interpreting H-DRI smelting behaviour in ESF, while also highlighting the need for more advanced multi-zone approaches, such as EERZ, to better represent non-uniform temperature fields and reaction zoning in practical ESF operation.

ACKNOWLEDGEMENTS

The authors gratefully acknowledge BHP for their financial support for the Sustainable Steelmaking Research, The University of Newcastle, and permission to publish this work. This Project received funding from the Australian Renewable Energy Agency (ARENA) as part of ARENA's Transformative Research Accelerating Commercialisation Program. The authors also acknowledge the Pyrosearch Research Group at The University of Queensland for the development of the UQPY private database utilised in this study.

REFERENCES

Bae, S J and Park, J H, 2025. Sustainable hot metal production by solid and liquid carbon from H_2 direct reduced iron (H_2-DRI) in electric smelting furnace (ESF) conditions, *Journal of Cleaner Production*, 524.

Barcza, N A, Curr, T R and Jones, R T, 1990. Metallurgy of open-bath plasma processes, *Pure and Applied Chemistry*, 62(9):1761–1772.

Brooks, G and Hornby, S, 2024. Future Processing Options for Hydrogen DRI [online], MIDREX. Available from: <https://www.midrex.com/tech-article/future-processing-options-for-hydrogen-dri/> [Accessed: 15 January 2026].

Büttner, V, 2024. Zoning of an electric melting reactor for alternative pig iron production in reaction zones and their existing equilibria, Master thesis (unpublished), RWTH Aachen University.

Gadd, A, Tame, N, Lu, X and Dukino, R, 2023. Pathways to decarbonisation episode seven: the electric smelting furnace [online], BHP. Available from: <https://www.bhp.com/news/bhp-insights/2023/06/pathways-to-decarbonisation-episode-seven-the-electric-smelting-furnace> [Accessed: 10 January 2026].

Garlick, C, Singh, T, Paymooni, K, O'Dea, D and Honeyands, T, 2025. Electrode Smelting of Hydrogen-Reduced Hematite–Goethite at Laboratory Scale, *Journal of Sustainable Metallurgy*, 12(2026):1114–1127. https://doi.org/10.1007/s40831-025-01281-2

Jung, I H and van Ende, M-A, 2020. Computational Thermodynamic Calculations: FactSage from CALPHAD Thermodynamic Database to Virtual Process Simulation, *Metallurgical and Materials Transactions B,* 51:1851–1874.

Kashiwaya, Y, Nakauchi, T, Pham, K S and Akiyama, S, 2007. Crystallization Behaviors Concerned with TTT and CCT Diagrams of Blast Furnace Slag Using Hot Thermocouple Technique, *ISIJ International,* 47(1):44–52.

Meraikib, 2000. Silicon distribution between blast furnace slag and hot metal, *Ironmaking and Steelmaking*, 27(4):280–285.

Mitsufuji, H, Sakurai, M, Maki, A, Sumigama, T and Niwa, Y, 1992. Estimation of Silicon Transfer Based on Radially Sampled Slag and Metal at Tuyere Level, *Iron and Steel*, 78(7):1148–1155.

Nam, J, van Ende, M-A and Jung, I-H, 2022. Ferromanganese Production in a Submerged Arc Furnace: Thermodynamic and Energy Balance Analysis, *JOM*, 74:1624–1632.

Oterdoom, H, Reuter, M and Zietsman, J, 2024a. DC Ferrochrome Smelting: The Arcing Zone and Its Influence on Energy Transport and Exergy Dissipation, *Metallurgical and Materials Transactions B,* 56(1):890–912.

Oterdoom, H, Reuter, M and Zietsman, J, 2024b. The Importance of Understanding Mechanisms in Open-Bath (DC) Processes Related to Furnace Containment, in *Advances in Pyrometallurgy* (eds: G R F A Flores, C Fleuriault, D Gregurek, Q G Reynolds, H Joubert, S L Nicol, P J Mackey, J F White and I Nolet), pp 15–31 (Springer: Cham).

Paymooni, K, Garlick, C, O'Dea, D, Gadd, A and Honeyands, T, 2024. Emissions abatement potential of the DRI-ESF process route, in 2nd Symposium on Carbon Ultimate Utilisation Technologies for the Global Environment (CUUTE-2).

Pfeiffer, A, Wimmer, G, Wurm, J and Schenk, J, 2022. Processing low-grade Direct Reduced Iron – Definition and comparison of melting strategies [online]. Available from: <https://www.researchgate.net/publication/375519317_Processing_low_grade_Direct_Reduced_Iron-Definition_and_comparison_of_melting_strategies> [Accessed: 6 January 2026].

Qin, Y L, Lv, X W, Zhang, J, Hao, J L and Bai, C G, 2015. Determination of Optimum Blast Furnace Slag Cooling Rate for Slag Recycling in Cement Manufacture, *Ironmaking and Steelmaking*, 42(5):395–400.

Ren, Q Q, Zhang, Y Z, Long, Y, Zou, Z S, Chen, S S and Li, J, 2018. Investigation on the effect of MgO content on the crystallization behavior of synthetic BF slag, *Materials and Manufacturing Processes*, 33(6):1–7. https://doi.org/10.1080/10426914.2018.1424903

Shankar, A, 2006. Sulphur partition between hot metal and high alumina blast furnace slag, *Ironmaking and Steelmaking*, 33(5):413–418.

Turkdogan, E T, 1984. Physicochemical Aspects of Reactions in Ironmaking and Steelmaking Processes, *Transactions of the Iron and Steel Institute of Japan*, 24(8):591–611.

Wang, C, Walsh, S D C, Weng, Z, Haynes, M W, Summerfield, D and Feitz, A, 2023. Green steel: Synergies between the Australian iron ore industry and the production of green hydrogen, *International Journal of Hydrogen Energy*, 48(83):32277–32293.

World Steel Association (WSA), 2025. 2025 World Steel in Figures [online]. Available from: <https://worldsteel.org/data/world-steel-in-figures/world-steel-in-figures-2025/> [Accessed: 5 January 2026].

Metallurgical properties of pellets from a hematite–goethite ore

Y Tang[1], L Lu[2] and J Manuel[3]

1. Project Research Officer, Mineral Resources, CSIRO QCAT (Queensland Centre for Advanced Technologies), Pullenvale Qld 4069. Email: yajun.tang@csiro.au
2. Chief Research Consultant, Mineral Resources, CSIRO QCAT, Pullenvale Qld 4069. Email: liming.lu@csiro.au
3. Senior Experimental Scientist, Mineral Resources, CSIRO QCAT, Pullenvale Qld 4069. Email: james.manuel@csiro.au

ABSTRACT

Hydrogen-based direct reduction presents a promising pathway for decarbonising steel sector by producing water vapour instead of carbon dioxide during iron ore reduction compared with conventional fossil fuel-based blast furnace process. Iron ore pellets are a critical feedstock for direct reduced iron process. Magnetite concentrates are preferred for pelletising due to their favourable pelletisation characteristics and high quality after beneficiation. Australian iron ores are predominantly hematite-goethite, traditionally exported as sinter fines for blast furnace ironmaking. In particular, the hydrated nature of goethite presents challenges for pelletisation, often necessitating the use of selective binders or specialised processing routes to achieve sufficient pellet strength and quality. As a result, Australian hematite-goethite ores are more challenging to pelletise than magnetite concentrate. Understanding the metallurgical properties of pellets under hydrogen reduction conditions is therefore crucial not only for understanding their behaviour during the direct reduction process in shaft furnace but also for optimising the pelletising process.

In the present study, pellets produced from an Australian hematite–goethite ore at CSIRO's Queensland Centre for Advanced Technologies were evaluated in terms of their metallurgical properties. Two reducing gas atmospheres were employed to simulate operational environments: a reference gas mixture consisting of CO and N_2 representing of conventional reduction conditions and a target hydrogen-enriched reduction atmospheres applicable to the emerging low-carbon steel production technologies. This study systematically examined the reduction swelling behaviours at different reduction durations and assessed the effects of reaction temperature as well as low temperature reduction disintegration under the target gas composition on the metallurgical performance of the fired pellets. The microstructure and mineral phase of both raw and partially reduced pellet samples were examined using X-ray diffractometry (XRD) and optical microscopy to understand the structural evolution of pellets during reduction. The findings of this study demonstrated that pellets produced from Australian hematite–goethite ores exhibit measurable structural integrity and satisfactory reduction performance under hydrogen-enriched conditions.

INTRODUCTION

The iron and steel industry is one of the most energy and carbon intensive industrial sectors, accounting for approximately 7–9 per cent of global greenhouse gas emissions (World Steel Association, 2021). Most of these emissions arise from the conventional blast furnace-basic oxygen furnace route, in which carbon-based reductants such as coke and pulverised coal are used to reduce iron oxides, inherently generating large quantities of carbon dioxide. As global efforts to mitigate climate change intensifying, decarbonising steel production has become a critical priority for both policymakers and industry stakeholders.

Among the emerging low-carbon steelmaking pathways, the hydrogen-based direct reduction of iron process has attracted significant attention due to its strong potential to substantially reduce or even eliminate carbon emissions from the ironmaking stage (Vogl, Åhman and Nilsson, 2018). In this process, hydrogen replaces carbon monoxide as the primary reducing agent, producing water vapour instead of carbon dioxide as the main reaction by-product. When hydrogen is sourced from renewable electricity, the hydrogen direct reduced iron route offers a viable pathway toward near-zero-emission or green iron production (Wang *et al,* 2021). Consequently, several steel producers worldwide have announced plans to transition from traditional blast furnace operations to hydrogen-based direct reduction technologies (Shahabuddin, Brooks and Rhamdhani, 2023).

Iron ore pellets are essential for direct reduction processes, especially in shaft furnaces, due to their uniform size, high mechanical integrity and controlled chemical composition (Spreitzer and Schenk, 2019). The use of high-quality pellets promotes stable gas permeability within the reactor bed, enhances heat and mass transfer and enables predictable reduction kinetics during the gas–solid reactions governing the direct reduction process. Consequently, the pelletising characteristics of iron ores including mineralogical composition, gangue content, particle size distribution and surface chemistry play a decisive role in determining their suitability for hydrogen-based direct reduction technologies.

Australia, the world's largest iron ore exporter, has historically optimised its resources for blast furnace use, with most ores being hematite or goethite. While high-grade hematite can be used directly in steelmaking, medium to lower-grade or goethite-rich ores require beneficiation due to their complex mineral content. The hydrous nature of goethitic ores complicates pelletisation, often necessitating binders or special processing. As a result, Australian ores have been considered less suitable for pellet production compared to magnetite concentrates used in regions like Canada, Brazil and Sweden. However, the global shift to hydrogen-based direct reduction presents a challenge, as demand for sinter fines is expected to decline. This underscores the need to assess the pelletising behaviour and reduction performance of Australian ores, especially in the context of hydrogen-based technologies (International Energy Agency, 2021). This issue is critical not only for the long-term sustainability of Australia's iron ore exports, but also for supporting the global steel industry's transition toward low-carbon and environmentally responsible production routes.

A thorough understanding of the metallurgical behaviour of iron ore pellets under hydrogen reduction conditions is essential, as it governs their performance during direct reduction in shaft furnaces and provides a basis for optimising pelletising strategies. In this study, fired pellets prepared at CSIRO's Queensland Centre for Advanced Technology from an Australian hematite–goethite iron ore were systematically investigated with respect to their swelling behaviour, reducibility and low-temperature reduction disintegration. These properties are critical indicators of pellet mechanical integrity and reduction efficiency under hydrogen-based direct reduction conditions, thereby influencing overall efficiency and operational stability.

EXPERIMENTAL

Materials

The fired pellets and reducing gases were used in this study. The pellets contained 64.44 per cent total iron (TFe), 0.18 per cent FeO, and 4.14 per cent SiO_2, with a diameter of 10.0–12.5 mm. Prepared by CSIRO Mineral Resources, the fired pellets exhibited a cold compressive strength of 2373 N and a porosity of 25.0 per cent. TFe was determined by XRF (ISO 9516-1; ISO, 2003), FeO by volumetric analysis and compressive strength according to ISO 4700 (ISO, 2015). The reducing gases (N_2, CO, CO_2, and H_2) had purities above 99.9 per cent with H_2 exceeding 99.99 per cent.

Experimental method

Reduction swelling test

Each reduction swelling test used 18 fired pellets. As shown in Figure 1, the pellets were placed in a cylindrical retort (75 mm internal diameter) and arranged in three tiers, with three pellets per layer. A thermocouple was positioned in the middle tier to monitor temperature. Pellet volumes before and after reduction were measured to evaluate swelling behaviour. The reduction swelling index (RSI) was determined in accordance with ISO 4698 (ISO, 2022) and calculated using Equation 1:

$$RSI = (V_1 - V_0) / V_0 \times 100\% \qquad (1)$$

Where V_0 and V_1 are the volumes of the pellets before and after reduction, respectively

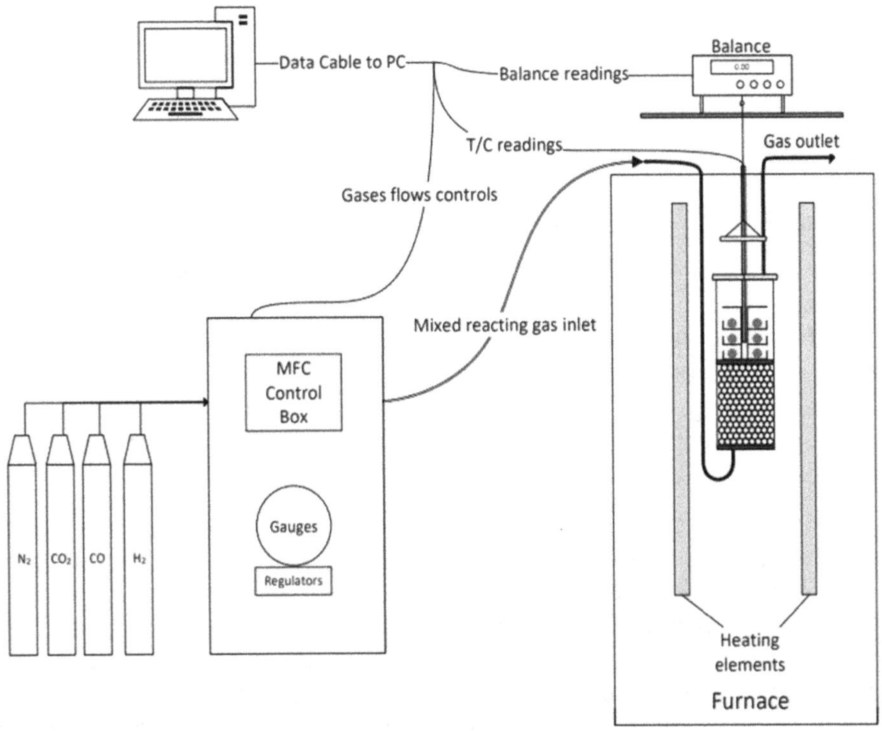

FIG 1 – Schematic diagram of reduction apparatus.

Reducibility

Two gas mixtures were investigated: a reference mixture of CO (30 per cent) – N_2 (70 per cent) and a hydrogen-enriched mixture of H_2 (55 per cent), CO (21 per cent), CO_2 (14 per cent) and N_2 (10 per cent). All tests were conducted at a total flow rate of 15 L/min, with additional hydrogen-enriched tests performed at varying temperatures and reaction times. Reduction progress was monitored by mass change and the reduction degree (R) was calculated using Equation 2.

$$R = (0.111W_1 / 0.430W_2 + \Delta m / m \times 0.430W_2) \times 100\% \quad (2)$$

Where:

m_0 is the mass of the fired pellets

Δm is the mass loss of the pellets during reduction

W_1 and W_2 are the FeO content and total Fe content of the pellets before the test, respectively

Low-temperature reduction disintegration tests

A total of 500 g of fired pellets was heated to 500°C under N_2, conditioned for 10 mins and reduced for 2 hours using a gas mixture of H_2 (55 per cent), CO (21 per cent), CO_2 (14 per cent) and N_2 (10 per cent) at 20 L/min. Low-temperature reduction disintegration was evaluated using a rotating drum at 18 rev/min for 50 mins, followed by sieving at 3.15 mm. The reduction disintegration index (RDI) was calculated using Equation 3, with $RDI_{-3.15}$ indicating the degree of disintegration.

$$RDI_{-3.15} = (m_1 - m_2) / m_1 \times 100\% \quad (3)$$

Where:

m_1 is the total mass of the reduced pellets

m_2 is the mass of reduced pellet fraction larger than 3.15 mm particle size

Characterisation

Phase compositions were determined by Rietveld refinement of XRD data collected using a PANalytical X'Pert Pro diffractometer. Phase identification was performed with PANalytical HighScore Plus©, and quantitative analysis with Bruker-AXS TOPAS. Polished pellets, before and after reduction, were imaged using a ZEISS Axio Imager.Z2m optical microscope with a motorised stage. The MosaiX function was used to acquire 4 × 4 image arrays at magnifications of 200× and 500×.

RESULTS AND DISCUSSION

Reduction swelling index at different reduction durations

Swelling represents a critical phenomenon that can constrain the large-scale industrial utilisation of pellets in shaft furnace operation. To ensure bed permeability and stable shaft furnace operation, the swelling index of iron ore pellets is generally limited to ≤20 per cent, as exceeding this threshold impairs gas flow and disrupts process stability (Purohit, Pownceby and Guiraud, 2025). Consequently, understanding and controlling swelling behaviour is essential for optimising the performance of oxidised pellets and ensuring their safe and efficient application in shaft furnace direct reduction processes.

The RSI of the pellets varies with reduction time. The RSI of the pellets were measured at 13.8 per cent after 10 mins, 14.9 per cent after 28 mins and 12.2 per cent after 60 mins of reduction. Correspondingly, the cold compressive strength of the pellets decreased from 2373 N before reduction to 655 N at 10 mins and 468 N at 28 mins but subsequently increased to 915 N at 60 mins as shown in Figure 2. The results demonstrate a clear correlation between pellet swelling behaviour and mechanical strength during reduction. The swelling behaviour is strongly governed by the evolution of mineral phases and the associated microstructural transformations occurring during reduction.

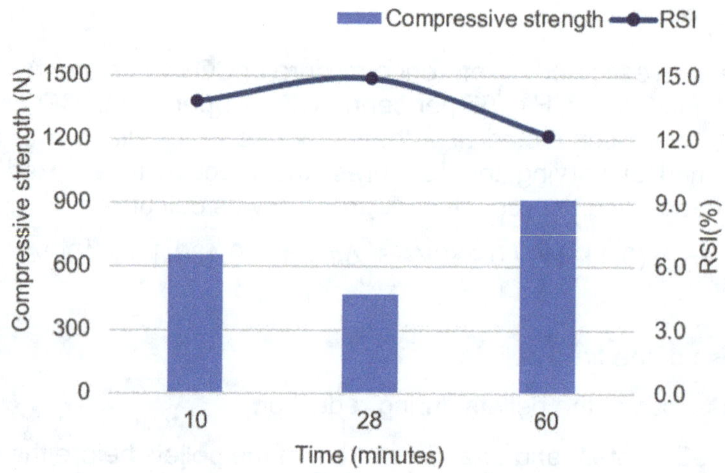

FIG 2 – Effect of time on RSI and compressive strength.

Figure 3 shows the evolution of pellet structure during reduction. Prior to reduction, the pellet is composed predominantly of hematite with a uniform microstructure, exhibiting minimal porosity and well-defined grain boundaries. Upon exposure to reducing gases, the outer shell of the pellet undergoes more rapid reduction than the core due to direct contact with the gas, which enhances gas–solid reaction rates at the periphery. After 10 mins of reduction, magnetite and wüstite phases are observed primarily in the outer region, indicating the initial transformation of hematite into intermediate oxides. By 28 mins, metallic iron formation becomes significant at the pellet periphery, coexisting with residual wüstite, reflecting the progressive reduction of intermediate phases toward the metallic state. At 60 mins, metallic iron is the dominant phase throughout both the outer shell and the central core, demonstrating near-complete reduction and the inward propagation of the reaction front. These observations confirm that the reduction process proceeds in a radial manner, with the periphery acting as the initial reaction zone and reduction gradually advancing toward the core. The microstructural evolution underscores the interplay between chemical reactions and gas diffusion, providing crucial insights into the temporal progression of hydrogen and CO-based reduction pathways in iron ore pellets. The mineral phase composition determined by XRD for the partially reduced pellets in Table 1 is consistent with the microstructural changes observed during pellet reduction.

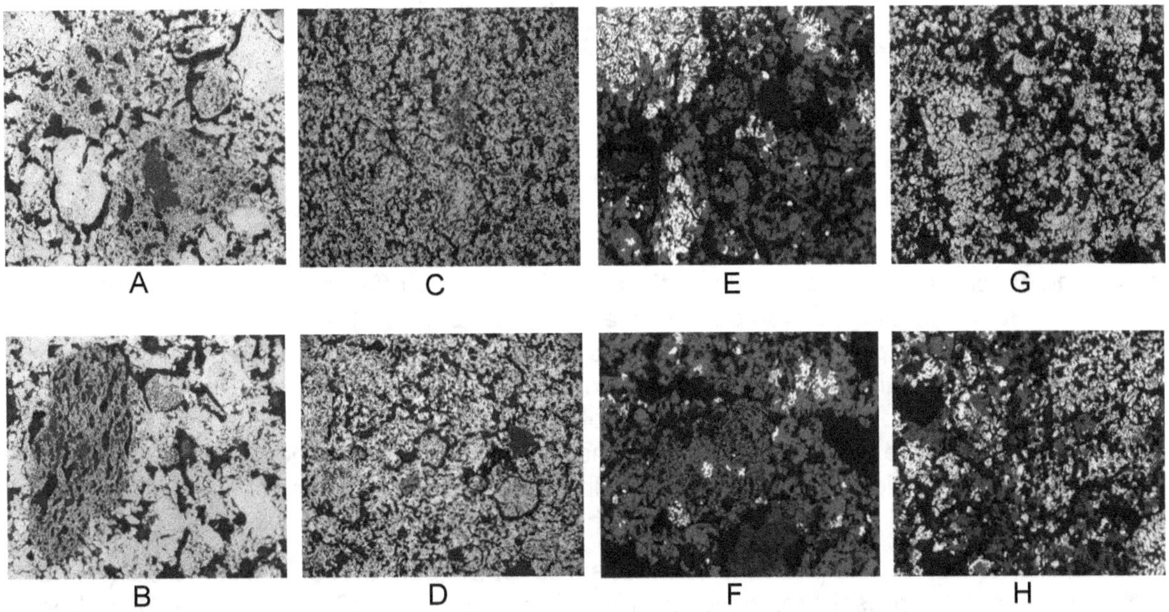

FIG 3 – Microstructure of pellets before and during the reduction. A: Fired pellet, Outer-Hematite/reacted silicate, 500×. B: Fired pellet, Core-Hematite/glass, 500×. C: Reduction 10 mins, Outer-Magnetite/FeO, 200×. D: Reduction 10 mins, Core-Magnetite, 500×. E: Reduction 28 mins, Outer-FeO/Fe, 500×. F: Reduction 28 mins, Core-FeO/Fe, 500×. G: Reduction 60 mins, Outer-Fe/FeO, 500×. H: Reduction 60 mins, Core- FeO/Fe, 500×.

TABLE 1

Mineral phase composition by XRD of partially reduced pellets.

Sample	Diopside	Fayalite	Hematite	Iron	Magnesioferrite	Magnetite	Quartz	Wüstite
Fired pellet	-	-	99	-	-	-	-	-
H_2 (55%)-CO (21%)-CO_2 (14%)-N_2 (10%) 10 mins	4	6	<1	7	-	20	1	62
H_2 (55%)-CO (21%)-CO_2 (14%)-N_2 (10%) 28 mins	6	9	<1	47	-	1	1	36
H_2 (55%)-CO (21%)-CO_2 (14%)-N_2 (10%) 60 mins	-	10	<1	82	5	<1	1	2

H_2 versus CO as a reducing agent

The reduction behaviour of pellets in H_2 and CO gas atmospheres was examined at 900°C for 60 mins with a gas flow rate of 15 L/min. Under a CO (30 per cent)–N_2 (70 per cent) atmosphere, the degree of reduction reached only 29.4 per cent. In contrast, a substantially higher reduction degree of 92.0 per cent was obtained under a mixed reducing gas atmosphere comprising H_2 (55 per cent), CO (21 per cent), CO_2 (14 per cent) and N_2 (10 per cent).

Table 2 illustrates the phase composition of pellets reduced under two different gas atmospheres. The H_2-rich mixture promotes the formation of metallic iron as the dominant phase, indicating a high degree of reduction. In contrast, the CO (30 per cent)–N_2 (70 per cent) atmosphere primarily produces wüstite and magnetite, consistent with a lower degree of reduction. Hydrogen penetrates the iron layer more readily than carbon monoxide because of its smaller molecular size and higher diffusion coefficient, thereby facilitating the reduction of pellets.

TABLE 2

XRD Comparison between two reducing gas atmospheres.

Gas composition	Hematite	Iron	Magnesioferrite	Magnetite	Quartz	Wüstite
Fired pellet	99	-	-	-	-	-
CO (30%)-N_2 (70%)	-	4	-	15	1	72
H_2 (55%)-CO (21%)-CO_2 (14%)-N_2 (10%)	<1	82	5	<1	1	2

The mechanical properties of reduced pellets were significantly affected by the gas atmosphere. For the pellets reduced in the CO (30 per cent) and N_2 (70 per cent) atmosphere, the compressive strength decreased sharply from 2373 N to 458 N likely due to their low degree of reduction. By contrast, the pellets reduced in the H_2 (55 per cent), CO (21 per cent), CO_2 (14 per cent) and N_2 (10 per cent) mixture showed a higher degree of metallisation resulting in a higher post-reduction strength of 915 N. These results demonstrate that a hydrogen-rich reducing atmosphere markedly enhances reduction efficiency and is likely to help pellets to restore their post-reduction mechanical integrity and dimensional stability more quickly than those reacted in a CO–N_2 system.

A relatively higher RSI of 17.2 per cent was measured for the pellets reduced in the CO (30 per cent) and N_2 (70 per cent) atmosphere, while the H_2 (55 per cent), CO (21 per cent), CO_2 (14 per cent) and N_2 (10 per cent) mixture led to a lower RSI of 12.2 per cent.

Effect of temperature on reduction degree

To evaluate the effect of temperature on the reduction degree of pellets tests were conducted at three temperatures 750, 850 and 950°C under a reducing gas mixture of H_2 (55 per cent), CO (21 per cent), CO_2 (14 per cent) and N_2 (10 per cent) at a flow rate of 15 L/min for a fixed reaction duration of 30 mins. As shown in Figure 4, the corresponding reduction degrees were 51.1 per cent, 64.2 per cent and 89.9 per cent, respectively. These results demonstrate a clear and progressive increase in reduction degree with increasing temperature, indicating that temperature plays a critical role in enhancing the kinetics and overall efficiency of the reduction process.

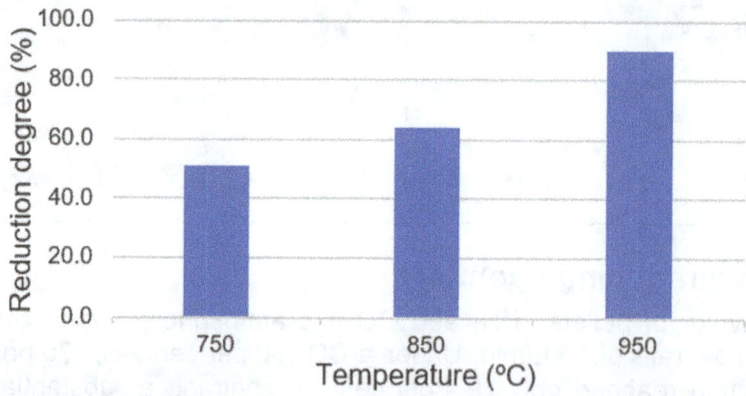

FIG 4 – Effect of temperature on reduction degree.

The relatively lower reduction degree of pellets reacted at 750°C suggests that the reaction is predominately limited by slower interfacial reaction kinetics. As the temperature increases, the reaction kinetics improve, resulting in a moderate increase in the reduction degree. When the temperature is further elevated to 950°C, the reduction degree increases sharply to 89.9 per cent, implying that the reduction reactions proceed more rapidly and extensively and that kinetic constraints are substantially reduced. These findings underscore the critical impact of temperature on reduction kinetics, with a marked increase in the reduction degree as the reaction temperature rises. These observations provide valuable insights into the design and optimisation of reduction processes, particularly in systems involving complex gas mixtures and metal oxide reduction.

Low temperature reduction disintegration index

RDI is an important quality parameter for iron ore pellets used in shaft furnaces, as it indicates the tendency of pellets to degrade into fines during low-temperature reduction. A low RDI is essential to preserve bed permeability and ensure stable furnace operation. The reduction disintegration behaviour of the pellets was investigated at 500°C using a reducing gas mixture consisting of H_2 (55 per cent), CO (21 per cent), CO_2 (14 per cent) and N_2 (10 per cent) at a total flow rate of 20 L/min for a duration of 2 hrs. The initial mass of the sample was 500 g, corresponding to approximately 156 pellets. After the reduction test, the partially reduced pellets were screened at 3.15 mm. The fine particles smaller than 3.15 mm were found to constitute 11.8 per cent of the reduced sample. The mechanical stability of the reduced pellets was further evaluated using a tumbling test. As shown in Figure 5, tumbling at a rotational speed of 18 rev/min for 50 mins increased the fraction of particles smaller than 3.15 mm to 17.0 per cent, while the number of unbroken pellets decreased to 124, indicating fragmentation caused by mechanical impact and abrasion during tumbling. To further assess pellet resistance under more severe mechanical stress, the tumbling speed was increased to 30 rev/min for an additional 30 mins. Under these conditions, the -3.15 mm fraction increased only marginally to 18.3 per cent and the number of intact pellets decreased slightly from 124 to 122. The limited generation of fines and minimal loss of intact pellets during the second tumbling stage demonstrate that the pellets exhibit good resistance to mechanical degradation, even under intensified tumbling conditions.

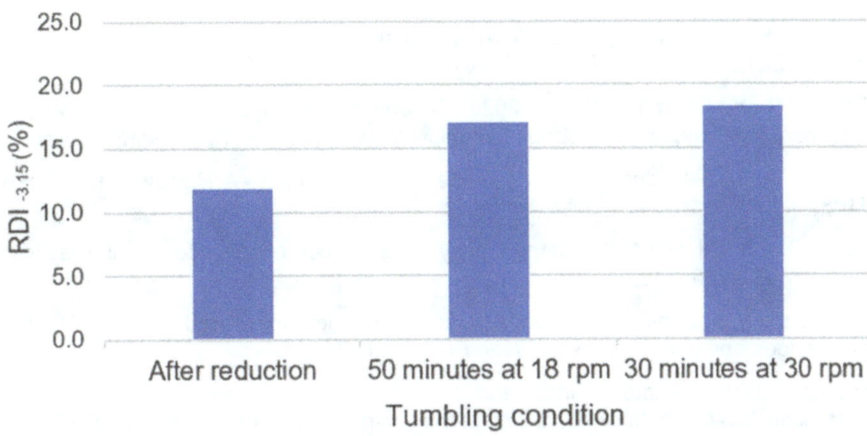

FIG 5 – The effect of tumble condition on $RDI_{-3.15}$.

CONCLUSIONS

The swelling characteristics, reduction behaviour and low temperature disintegration performance of pellets under various reducing conditions were systematically investigated.

The RSI initially increased during the early and intermediate stages of reduction, then decreased at longer reduction times, correlating with changes in compressive strength. Pellet swelling during the hematite → magnetite → wüstite transformations arose from phase changes and pore development, while slight shrinkage during wüstite → metallic iron stage resulted from iron grain coalescence and densification. The reduction degree increased steadily with reaction time, reflecting the progression from surface-controlled to nearly complete internal reduction.

The hydrogen rich mixed gas atmosphere exhibited a significantly higher reduction efficiency than the CO–N_2 system. Pellets reduced in the hydrogen mixed gas atmosphere retained relatively higher post-reduction compressive strength and exhibited a lower RSI, indicating superior mechanical integrity and dimensional stability.

Higher temperatures markedly accelerated reduction kinetics by enhancing gas diffusion and interfacial reactions.

Low-temperature reduction caused mild disintegration, with most pellets remaining above 3.15 mm demonstrating sufficient mechanical integrity without significant fragmentation.

The results demonstrate that hydrogen-rich reducing atmospheres enhance reduction efficiency while mitigating issues such as pellet swelling, strength degradation and fines generation. It provides guidance for optimising gas composition, temperature and reduction duration in shaft furnace ironmaking using hematite–goethite pellets.

ACKNOWLEDGEMENTS

The authors would like to gratefully acknowledge Hongliang Han and Tosin Aladejebi for their valuable contributions in preparing and providing the fired pellets used in this study. The authors also thank Rong Fan for his assistance and support with XRD analysis.

REFERENCES

International Energy Agency, 2021. Net Zero by 2050: A Roadmap for the Global Energy Sector [online]. Available from: <https://iea.blob.core.windows.net/assets/deebef5d-0c34-4539-9d0c-10b13d840027/NetZeroby2050-ARoadmapfortheGlobalEnergySector_CORR.pdf>

International Organization for Standardization (ISO), 2003. ISO 9516-1:2003 – Iron ores-Determination of various elements by X-ray fluorescence spectrometry.

International Organization for Standardization (ISO), 2015. ISO 4700:2015 – Iron ore pellets for blast furnace and direct reduction feedstocks — Determination of the crushing strength.

International Organization for Standardization (ISO), 2022. ISO 4698:2022 – Iron ore pellets for blast furnace feedstocks — Determination of the free-swelling index.

Purohit, S, Pownceby, M I and Guiraud, A, 2025. Sticking and swelling of iron ore pellets: mechanisms and controlling factors, *Journal of Sustainable Metallurgy,* 11:67–87.

Shahabuddin, M, Brooks, G and Rhamdhani, M A, 2023. Decarbonisation and hydrogen integration of steel industries: Recent development, challenges and technoeconomic analysis, *Journal of Cleaner Production,* 395:136391.

Spreitzer, D and Schenk, J, 2019. Reduction of Iron Oxides with Hydrogen—A Review, *Steel Research International*, 90(10):1900108.

Vogl, V, Åhman, M and Nilsson, L J, 2018. Assessment of hydrogen direct reduction for fossil-free steelmaking, *Journal of Cleaner Production,* 203:736–745.

Wang, R R, Zhao, Y Q, Babich, A, Senk, D and Fan, X Y, 2021. Hydrogen direct reduction (H-DR) in steel industry – An overview of challenges and opportunities, *J Clean Prod*, 329:129797.

World Steel Association, 2021. Climate change and the production of iron and steel [online]. Available from: <https://worldsteel.org/wp-content/uploads/Climate-change-production-of-iron-and-steel-2021.pdf>

SFCA-III and the effects of sintering variables on its formation

N A S Webster[1], M I Pownceby[2], S Huang[3], D Guja[4] and N Owen[5]

1. Research Group Leader, CSIRO Mineral Resources, Clayton Vic 3168. Email: nathan.webster@csiro.au
2. Senior Principal Research Scientist, CSIRO Mineral Resources, Clayton Vic 3168. Email: mark.pownceby@csiro.au
3. Research Projects Officer, CSIRO Mineral Resources, Clayton Vic 3168. Email: shu.huang@csiro.au
4. CERC Fellow, CSIRO Mineral Resources, Clayton Vic 3168. Email: dylan.guja@csiro.au
5. Senior Experimental Scientist, CSIRO Mineral Resources, Clayton Vic 3168. Email: nick.owen@csiro.au

ABSTRACT

SFCA (silico ferrite of calcium and aluminium), SFCA-I and now SFCA-III are well-known iron ore sinter bonding matrix phases. SFCA (n = 0), SFCA-I (n =1) and SFCA-III (n = 2) form a polysomatic series with general formula $M_{14+6n}O_{20+8n}$. Each structure consists of distinct ordering of spinel (S) and pyroxene (P) modules; SPSPSP for SFCA, SSPSSP for SFCA-I, and SSSPSSSP for SFCA-III. The SFCA-III phase, which has only been identified and described relatively recently, has general formula $M_{26}O_{36}$ – for example, $Ca_{2.99}Al_{4.56}Fe_{15.35}Si_{0.43}Mg_{2.67}O_{36}$ – and triclinic crystal structure (Kahlenberg, Krüger and Goettgens, 2019). Zöll *et al* (2018) reported investigations into FCAM-III which has similar structure to SFCA-III but in the SiO_2-free system. SFCA-III can be a significant component of industrial and pot-grate iron ore sinter (Webster and Pownceby, 2024; Clout *et al,* 2026), and more detailed and extensive fundamental knowledge about its compositional limits and crystal chemistry is important for the field of iron ore sintering research and characterisation.

In situ X-ray diffraction (XRD) based experimentation has revealed mechanistic information about SFCA-III formation and the effect of several variables on its formation. Experimentation conducted in the range 298–1623 K and at oxygen partial pressure (pO_2) of 5×10^{-3} atm has shown that SFCA-III forms during cooling as the first phase to crystallise from a $Fe_2O_3(FeO)$-CaO-SiO_2-Al_2O_3 melt, coexisting with Fe_3O_4 until SFCA-I and/or SFCA form at lower temperatures as cooling continues (Figure 1). Observation of SFCA-III in sinter has been suggested to be a possible indicator of localised high basicity and/or high MgO concentration within a sinter blend (Webster *et al,* 2019a). Increasing Al_2O_3 concentration also appears to favour SFCA-III formation (Webster *et al,* 2012), whereas increasing Ti results in the formation of SFCA being favoured over SFCA-III (Webster *et al,* 2019b).

SFCA and SFCA-I are both known to have Al_2O_3-free analogues (SFC and SFC-I, respectively), and in the case of SFC its compositional limits (Pownceby and Patrick, 2000; Cheng *et al,* 2020) and crystal structure (Liles, de Villiers and Kahlenberg, 2016) are reasonably well characterised. Both SFC and SFC-I are known to form in air (ie pO_2 = 0.21 atm). Figure 2 shows the *in situ* XRD results collected as part of the current study during heating and cooling of an Al_2O_3-free sample with bulk composition 79.7 wt per cent Fe_2O_3, 15.4 wt per cent CaO and 4.9 wt per cent SiO_2, in the range 298–1623 K and in air. Both SFC and SFC-I are observed to form during heating as expected, above ~1373 K, before they melt to form a Fe_2O_3 + melt phase assemblage above ~1525 K. During cooling, the β-CFF phase (nominal composition of $Fe_{14.85}Ca_{2.95}O_{25}$, trigonal crystal structure) formed as the first Ca-ferrite phase to crystallise from the melt, along with the α'_H polymorph of C_2S. β-CFF persisted to 298 K, and the C_2S phase had converted to the β polymorph below ~900 K. Fe_3O_4 was also observed in the range ~1623 to 1563 K during cooling, however Fe_2O_3 remained the dominant Fe-oxide phase throughout and below ~1563 K the Fe_3O_4 had converted back to Fe_2O_3. The Fe_3O_4 is likely stabilised by Ca^{2+} substitution for Fe^{2+}. In these new results, therefore, along with the recent results obtained for the same Al_2O_3-free sample heated and cooled at $pO_2 = 5 \times 10^{-3}$ atm (Webster *et al,* 2025), there was no evidence for the formation of SFC-III during cooling. And so, the fundamental questions of whether it is possible at all to form SFC-III and if is, under what conditions, remain to be answered. In addition, it is clear from Figure 1 that under the cooling rate used (2 K min^{-1}) the SFCA-III phase does do persist to 298 K whilst SFCA does. Therefore, the question of what cooling rate(s) is/are required for maintaining SFCA-III in the completely cooled sinter also remains to be answered.

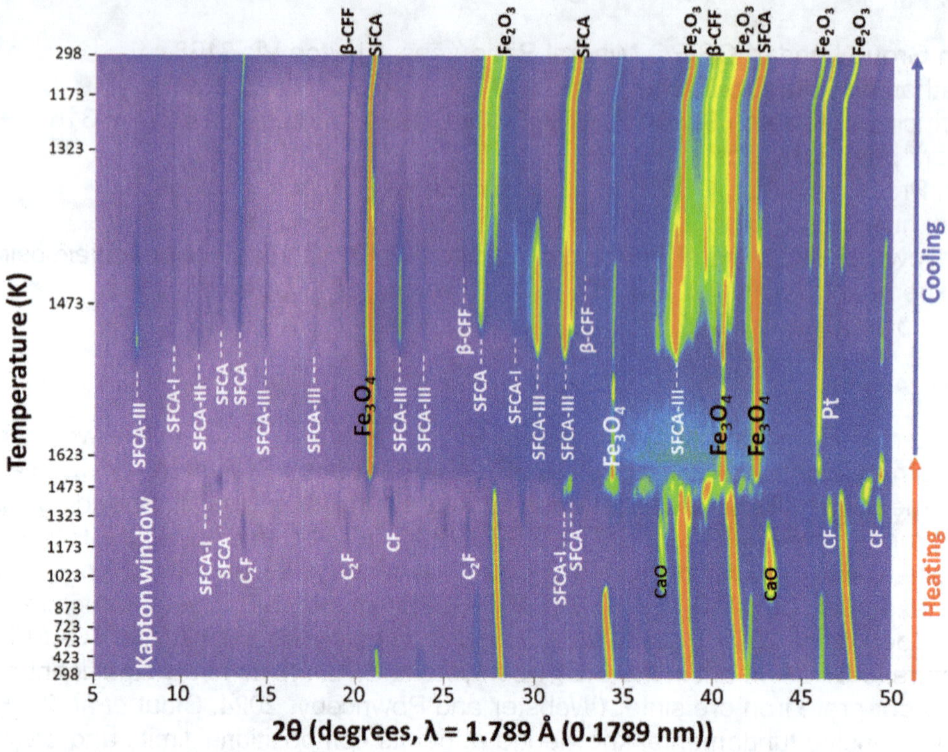

FIG 1 – *In situ* XRD data collected for a sample with bulk composition (in terms of wt per cent of oxides) 77.2 wt per cent Fe_2O_3, 14.2 wt per cent CaO, 3.6 wt per cent SiO_2 and 5.0 wt per cent Al_2O_3 in the range 298–1623 K at $pO_2 = 5 \times 10^{-3}$ atm. The colours are indicative of peak intensity, from purple (background intensity) to red (intense peaks).

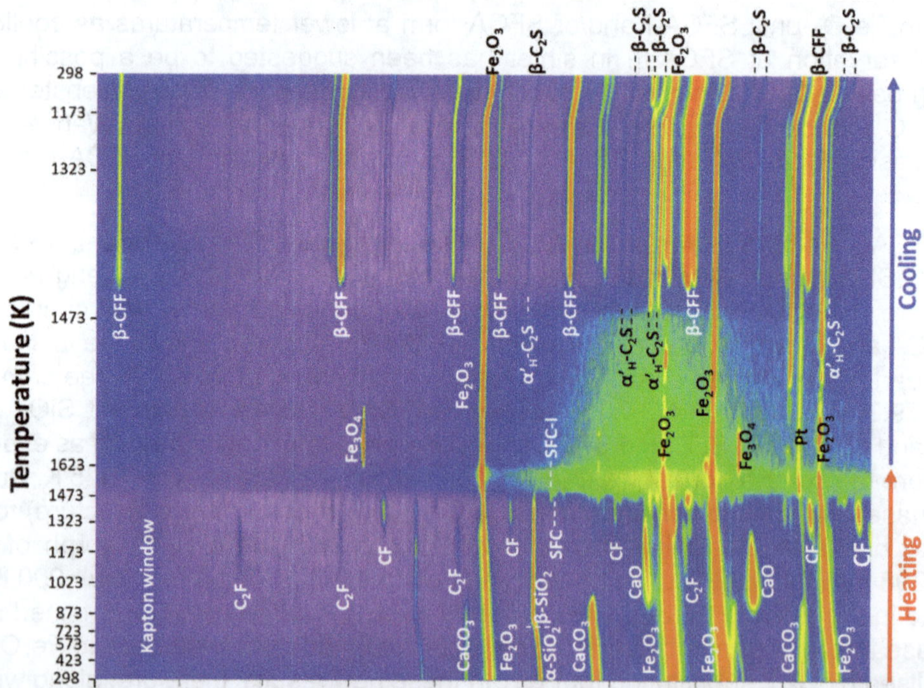

FIG 2 – *In situ* XRD data collected for a sample with bulk composition 79.7 wt per cent Fe_2O_3, 15.4 wt per cent CaO and 4.9 wt per cent SiO_2 in the range 298–1623 K at pO_2 = 0.21 atm.

REFERENCES

Cheng, S, Shevchencko, M, Hayes, P and Jak, E, 2020. Experimental Phase Equilibria Studies in the FeO-Fe_2O_3-CaO-SiO_2 System in Air: Results for the Iron-Rich Region, *Metallurgical and Materials Transactions B,* 51:1587–1602.

Clout, J M F, Ware, N A, Manuel, J R, Webster, N A S and Pownceby, M I, 2026. Fundamental Analysis of Sinter Solid Structure: Implications of Mineral Associations for Understanding Industrial Iron Ore Sinter Formation, *Minerals,* 16:129.

Kahlenberg, V, Krüger, H and Goettgens, V S, 2019. Structural elucidation of triclinic and monoclinic SFCA-III – killing two birds with one stone, *Acta Crystallographica,* B, 75:1126–1136.

Liles, D C, de Villiers, J P R and Kahlenberg, V, 2016. Refinement of iron ore sinter phases: a silico-ferrite of calcium and aluminium (SFCA) and an Al-free SFC and the effect on phase quantification by X-ray diffraction, *Mineralogy and Petrology,* 110:141–147.

Pownceby, M I and Patrick, T R C, 2000. Stability of SFC (silico-ferrite of calcium):solid solution limits, thermal stability and selected phase relationships within the Fe_2O_3-CaO-SiO_2 (FCS) system, *European Journal of Mineralogy,* 12:455–468.

Webster, N A S and Pownceby, M I, 2024. Exploring the Relationship Between Fe-rich SFCA and SFCA-III and Their Presence in Sinter Strand and Pot-grate Sinter, *ISIJ International,* 64:803–807.

Webster, N A S, Pownceby, M I, Huang, S, Owen, N D and Gu, Q, 2025. In-Situ X-ray Diffraction Analysis Reveals Complex Calcium Ferrite Phase Formation during Heating and Cooling of Silico-Ferrite of Calcium (SFC) Compositions, *ISIJ International,* 65:1588–1596.

Webster, N A S, Pownceby, M I, Madsen, I C and Kimpton, J A, 2012. Silico-ferrite of Calcium and Aluminium (SFCA) Iron Ore Sinter Bonding Phases: New Insights into Their Formation During Heating and Cooling, *Metallurgical and Materials Transactions B,* 43:1344–1357.

Webster, N A S, Pownceby, M I, Pattel, R, Manuel, J R and Kimpton, J A, 2019a. Fundamentals of Silico-Ferrite of Calcium and Aluminium (SFCA) Iron Ore Sinter Bonding Phase Formation: Effects of Basicity and Magnesium on Crystallisation during Cooling, *ISIJ International,* 59:263–267.

Webster, N A S, Pownceby, M I, Pattel, R, Manuel, J R and Kimpton, J A, 2019b. Fundamentals of Silico-Ferrite of Calcium and Aluminium (SFCA) Iron Ore Sinter Bonding Phase Formation: Effects of Titanium on Crystallisation during Cooling, *ISIJ International,* 59:1007–1010.

Zöll, K, Kahlenberg, V, Krüger, H and Tropper, P, 2018. Investigations on FCAM-III ($Ca_{2.38}Mg_{2.09}Fe^{3+}{}_{10.61}Fe^{2+}{}_{1.59}Al_{9.33}O_{36}$):A new homologue of the aenigmatite structure-type in the system CaO-MgO-Fe_2O_3-Al_2O_3, *Journal of Solid State Chemistry,* 258:307–319.

Metallurgical performance of oxidised pellets made from high proportion hematite-goethite ores in a hydrogen-based shaft furnace process

C C Yang[1], S J Qu[2], D Q Zhu[3], J Pan[4], Z Q Guo[5], S W Li[6] and L Lu[7]

1. Associate Professor, Central South University, Changsha Hunan 410083, China. Email: smartyoung@csu.edu.cn
2. PhD, Central South University, Changsha Hunan 410083, China. Email: qushijuan@csu.edu.cn
3. Professor, Central South University, Changsha Hunan 410083, China. Email: dqzhu@csu.edu.cn
4. Professor, Central South University, Changsha Hunan 410083, China. Email: pjcsu@csu.edu.cn
5. Professor, Central South University, Changsha Hunan 410083, China. Email: guozqcsu@csu.edu.cn
6. Associate Researcher, Central South University, Changsha Hunan 410083, China. Email: swli@csu.edu.cn
7. Senior Principal Research Scientist, CSIRO Mineral Resources, Kenmore Qld 4069, Australia. Email: liming.lu@csiro.au

ABSTRACT

The insufficient supply of high-grade pellet feeds and high production costs of oxidised pellets are important factors that significantly constraint the development and widespread application of the hydrogen-based shaft furnace direct reduction (H_2-DR) process. Utilisation of hematite-goethite iron ores for H_2-DR pellets production is one of effective measures to expand the source of pellet feeds and reduce production costs of both oxidised pellets and direct reduced iron (DRI). This work mainly investigated the balling and roasting performance of pellets made from high proportion of typical hematite-goethite ore fines (named as Ore G with 62 per cent Fe and a high LOI of 6.0 per cent) though a grate-kiln process, and further its metallurgical performance was evaluated under simulated HYL shaft furnace conditions. On this basis, the feasibility of using high-proportion Ore G in the preparation of H_2-DR pellets and some possible measures for improvement of its metallurgical performance were discussed. The results indicate that Ore G exhibits good grindability but inferior dewatering performance due to its mineralogical characteristics, which suggest that a dry grinding process may be more preferable. Qualified green pellets with sufficient mechanical strength and thermal shock temperature can be manufactured for pellet mixtures with 0~50 per cent Ore G. However, increasing the proportion of Ore G tends to reduce the thermal stability of green pellets. Through optimisation of pellet blends, acid pellets with binary basicity of lower than 0.3, cold compressive strength of more than 2500 N/P and tumble index of higher than 95 per cent can be manufactured for blends with 30~50 per cent Ore G under similar preheating and roasting conditions. Compared with the base case with no addition of Ore G, the roasted pellets made from 30~50 per cent Ore G demonstrate better reducibility and clustering performance. However, increasing proportion of Ore G tends to deteriorate the low-temperature reduction disintegration index (LTD) and causes a slight increase in the reduction swelling index (RSI) due to formation of a more porous pellet structure and weaker bonding of hematite recrystallisation. Despite the adverse impacts of using high proportion of Ore G, it is found that through appropriate blend structure optimisation and basicity adjustment, the overall pellet production energy consumption can be maintained at a comparable level at a lower raw material costs. The study demonstrates that the proper proportion of Ore G for H_2-DR pellet production is suggested to be no more than 30 per cent for a grate-kiln process.

INTRODUCTION

Under 'dual-carbon' strategy, iron ore oxidised pellets have become an important option for the low-carbon transformation of Chinese steel industry. Owing to their high iron grade, low levels of deleterious elements, and favourable metallurgical properties, the oxidised pellets are widely used in both blast furnace ironmaking and gas-based shaft furnace direct reduction processes. Compared with iron ore sintering, the pelletising process is more energy-efficient and environmentally friendly. The substitution of iron ore pellets for iron ore sinters in blast furnace can reduce energy consumption, coke rate, and CO_2 emissions, providing a practical pathway for sustainable ironmaking (Cai and Fan, 2023).

However, the limited availability of high-grade pellet feed in China restricts the development of pelletising and hydrogen-based shaft furnace processes (Zhu *et al,* 2024a; Liu and Lv, 2025). The demand for imported iron ore concentrates has increased significantly, with imports rising from 69.57 Mt in 2013 to 143 Mt in 2024. Meanwhile, the domestic magnetite resources are unevenly distributed and relatively expensive. Therefore, the use of low-cost hematite and limonite ores has attracted increasing attention as a means to reduce production costs (Hu, Qin and Yao, 2025; Liu, Zhang and Zhang, 2023).

Nevertheless, the limonite-type iron ores typically contain abundant goethite phases (FeOOH) with high content of combined water, resulting in poor pellet induration performance, higher energy consumption, lower productivity and inferior pellet quality. Thus, it cannot be usually used alone, and its proportion must be controlled. Achieving high-proportion utilisation of hematite–goethite ores while maintaining pellet quality is attractive when considering potential economical benefits but remains a key challenge. Previous studies show that optimising ore blending, partial pre-dehydration, and induration parameters can reduce dehydration load, suppress pellet cracking, and improve pellet strength. In addition, increasing magnetite content and adjusting pellet basicity are effective measures to promote hematite recrystallisation and improve the metallurgical performance in blast furnace (Yang *et al,* 2024; Zhu *et al,* 2018). Reportedly, the hematite–goethite pellets can be produced at industrial scale through both straight grate and grate–kiln processes. In straight grate systems, the proportion can reach 30–40 per cent or higher, while in grate–kiln systems it is usually below 20 per cent. However, these pellets are mainly used in blast furnace ironmaking, and the studies on their application in hydrogen-based shaft furnaces are still limited.

Therefore, this study mainly investigates the feasibility of using high proportion of one typical hematite–goethite ore fines (Ore G) for oxidised pellets applicable to hydrogen-based shaft furnace direct reduction (H_2-DR) process. The effects of blending ratio and ore-matching schemes on green pellet properties, induration behaviour, and metallurgical performance under HYL shaft furnace direct reduction conditions are examined. The results aim to support raw material diversification, process optimisation, and cost reduction in H_2-DR ironmaking.

MATERIALS AND METHOD

Raw materials characterisation

Chemical composition

The raw materials consist of eight types of iron ores, one type of bentonite, and one type of limestone for adjusting pellet basicity. Table 1 gives the chemical composition of the raw materials. Among iron ores, Ore A contains a highest iron grade of 70.41 wt per cent with an FeO content of 30.6 wt per cent, corresponding to a calculated magnetite content of 96.9 per cent. Ores B and C contain 27.12 wt per cent and 16.7 wt per cent FeO, respectively, and are typical magnetite and magnetite–hematite ores. In addition, Ores A, D, and E have similar iron grade of 66–67 per cent and relatively low contents of SiO_2, Al_2O_3, P, and S, whereas Ore C contains the highest SiO_2 content of 8.62 wt per cent.

In contrast, Ore G contains a lowest iron grade (~62 wt per cent) with highest loss on ignition (LOI) of 5.7 wt per cent. After deducting LOI, its effective iron grade increases to 65.7 wt per cent. This characteristic ensures that it can be potentially used as pellet feeds. However, Ore G contained relatively high Al_2O_3 (2.1 wt per cent) and P (0.12 wt per cent). Besides, addition of Ore G in pellet blend is likely to reduce the efficiency of wet grinding–sedimentation–filtration operations, deteriorate pellet induration behaviour due to high proportion of goethite ores, thereby adversely affecting pellet quality. Hence, the adverse impacts of using Ore-G should be taken into full consideration and its blending ratio must be strictly controlled (Yang, Hou and Pan, 2024; Yang, Hou, and Zhu, 2025; Yi, Jiang and Li, 2022; Pan, Ma and Zhu, 2021).

TABLE 1

Chemical compositions of raw materials/wt%.

Materials \ Compositions	TFe	FeO	SiO_2	Al_2O_3	CaO	MgO	P	S	K_2O	Na_2O	LOI
A	70.41	30.06	1.03	0.53	0.20	0.37	0.027	0.008	0.025	0.062	-2.98
B	66.00	27.12	4.99	0.65	0.10	0.33	0.020	0.010	0.140	0.020	-1.83
C	63.27	16.70	8.62	1.55	0.24	0.38	0.050	0.010	0.028	0.046	-1.23
D	66.61	5.59	4.21	0.24	0.34	0.37	0.017	0.017	0.009	0.043	0.22
E	66.80	0.23	2.89	0.62	0.10	0.02	0.040	0.003	0.030	0.030	0.86
F	64.62	0.80	2.88	1.04	0.09	0.05	0.074	0.019	0.019	0.009	2.88
G	62.01	0.17	3.49	2.10	0.07	0.09	0.120	0.017	0.010	0.018	5.70
Iron-bearing dust	55.12	4.90	6.74	1.68	5.94	1.54	0.047	0.061	0.061	0.091	4.15
Bentonite	7.21	-	54.11	15.05	1.89	2.48	0.86	0.015	1.018	2.112	8.08
Limestone	0.25	-	2.02	0.59	51.4	2.70	0.013	0.058	0.210	0.026	42.57

Particle size and specific surface area

The particle size and specific surface area (SSA) of the iron-bearing materials are listed in Figure 1. Ores C, G, F, and A were too coarse to be directly pelletised, and require pregrinding. In this study, they were pre-ground by wet ball milling to achieve 80–85 per cent passing 0.074 mm. In contrast, Ores E, D, and B are sufficiently fine, with over 80 per cent below 0.074 mm and specific surface areas of above 1500 cm^2/g.

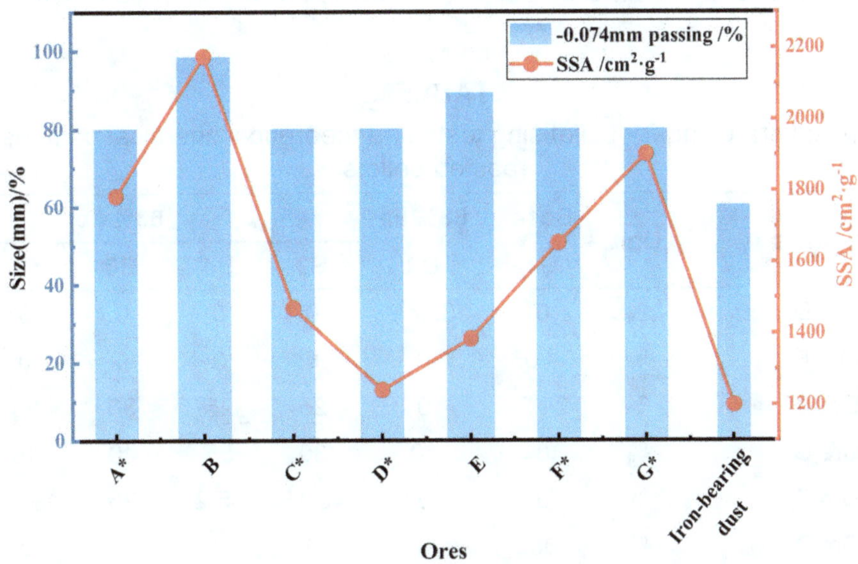

FIG 1 – Physical properties of iron-bearing raw materials (* Coarse iron ores were pre-ground to a proper fineness of approximately 80 per cent passing 0.074 mm fraction).

To improve balling and induration performance of pellet mixtures, the fine concentrates were first mixed with the pre-ground ores, and then pretreated by high-pressure grinding rolls (HPGR) to attain a proper SSA of 1500–1900 cm^2/g.

Bentonite and finely ground limestone possess suitable fineness with more than 90 per cent passing 0.074 mm, which can meet pelletising requirements.

Mineralogical characteristics

The XRD pattern, optical micrographs of mineral phases, and the distribution characteristics of Fe, Si, and Al elements of Ore G, are illustrated in Figure 2.

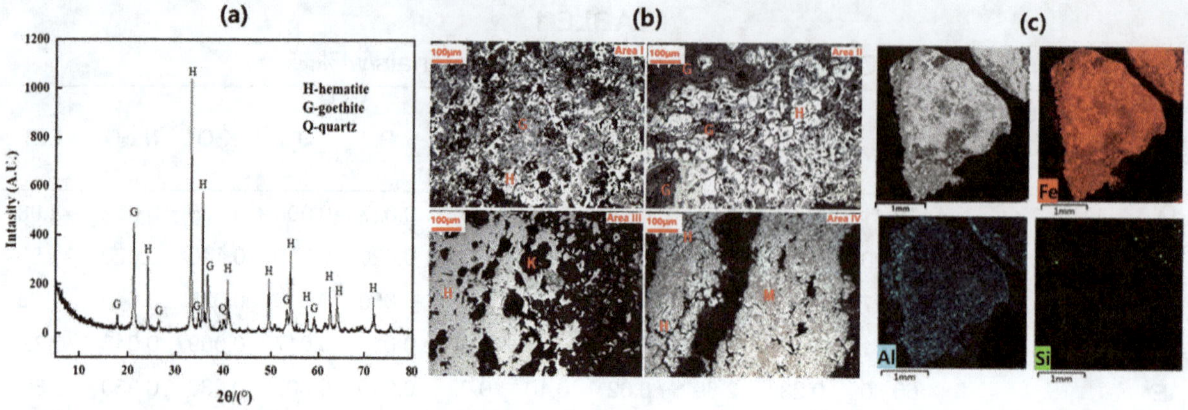

FIG 2 – (a) XRD phase analysis; (b) Optical microscopic photographs; (c) Elemental distribution characteristics of Ore G. H – hematite; G – goethite; M – magnetite; K – kaolinite.

The results indicate that Ore G is mainly composed of hematite, goethite, magnetite, kaolinite, and quartz, with approximate contents of 46 per cent, 46 per cent, 1 per cent, 3 per cent, and 3 per cent, respectively. Si is primarily present in the form of quartz and kaolinite. Quartz occurs in relatively low abundance and is loosely distributed. In contrast, Al is mainly associated with Al-substituted goethite, with a minor portion occurring as aluminosilicates. The Al-bearing phases are widely distributed and exhibit spatial distribution characteristics similar to those of iron.

Experimental

To investigate the influence of adding Ore G on green pellet properties, induration behaviour, and metallurgical performance, seven ore blends were designed, as shown in Table 2. B1 is the base case with 0 per cent Ore G. In B2 and B3, Ore G are fixed at 30 per cent, while B4–B7 contain 50 per cent Ore G.

TABLE 2

Experimental ore blend for pelletising tests and theoretical chemical compositions of roasted pellets.

Materials		Unit	Base	65% Fe	63% Fe				
			B1	B2	B3	B4	B5	B6	B7
Ore D		%	0	35.45	0	0	0	0	0
Ore F		%	39.05	0	0	0	0	0	0
Ore A		%	21.2	0	25	25	30	30	35
Ore G		%	0	30	30	35	40	45	50
Ore C		%	0	0	40.15	35.2	25	19.25	8.3
Ore D		%	35	0	0	0	0	0	0
Ore B		%	0	30	0	0	0	0	0
Bentonite		%	1	1	1	1	1	1	1
Limestone		%	1.25	1.05	1.35	1.3	1.5	2.25	3.2
Iron-bearing dust		%	2.5	2.5	2.5	2.5	2.5	2.5	2.5
Total		%	100	100	100	100	100	100	100
Theoretical chemical composition of pellets	Fe	%	65.70	64.77	63.32	63.48	63.87	63.76	63.88
	SiO_2	%	3.55	4.73	6.18	5.89	5.11	4.76	3.93
	Al_2O_3	%	0.85	1.22	1.24	1.32	1.39	1. 47	1.54
	Binary basicity	/	0.25	0.15	0.15	0.15	0.20	0. 30	0.50

All iron ore fines were pre-ground to obtain approximately 80 per cent passing −0.074 mm. The ores were blended according to Table 2 and subjected to high-pressure grinding rolls (HPGR) pretreatment to ensure the specific surface area of mixed concentrates reach 1500–1900 cm^2/g. For all cases, 1 wt per cent bentonite was added to improve the ballability while a specified amount of finely ground limestone to adjust pellet basicity. The mixtures were homogenised and pelletised using a disc pelletiser (diameter 1000 mm, rim height 300 mm). Four kg of dry material was used in each batch. The pelletising parameters were fixed at a rotation speed of 28 rev/min and a disc inclination angle of 47°, based on previous work (Yang *et al*, 2024; Zhu *et al*, 2024b). The balling duration was fixed at 10 min, with an extra 2 min for compaction.

The green pellets with diameters of 10–16 mm were selected for measurement of moisture content, compressive strength, drop strength, and thermal shock temperature. The quality criteria of green pellets is set as follows: compressive strength > 10 N/pellet, drop number > four times per 0.5 m, and thermal shock temperature > 300°C.

Green pellets with good performance were dried in an oven at 105°C for more than 4 hrs and then subjected to preheating and roasting tests simulating the grate–kiln process. The effects of preheating and roasting parameters on the cold compressive strength (CCS) of preheated and roasted pellets were examined to determine the optimal thermal regimes. Based on laboratory results, pilot-scale induration experiments were further conducted under simulated grate–kiln conditions.

In pilot-scale tests, the bed height of green pellets during drying and preheating was maintained at 220 mm, with 20–23 kg of pellets per batch. The pellets underwent updraft drying, downdraft drying, and downdraft preheating with controlled gas temperature, and gas flow rate. After preheating, the hot preheated pellets in different bed height were sampled for compressive strength and AC tumbler index measurements, and removed from the pot grate and directly charged into a rotary kiln (diameter 350 mm, length 1200 mm) for high-temperature induration. The gas temperature and atmosphere were controlled by adjusting the ratio of natural gas and air flow rates. After roasting, the roasted pellets were sampled, cooled to 500°C in the kiln, discharged, and air-cooled to room temperature.

The metallurgical properties of the roasted pellets under HYL conditions were evaluated, including reducibility (RI), reduction swelling index (RSI), low-temperature reduction degradation (LTD), and clustering index (CI). Industrial criteria were adopted as follows: RI (800°C) > 3, RI (950°C) > 4, RSI < 15 per cent, $LTD_{+6.3\ mm}$ > 80 per cent, $LTD_{-3.2\ mm}$ < 10 per cent, and CI < 20 per cent.

RESULTS AND DISCUSSION

Effect of Ore G proportion on green pellet properties

Figure 3 shows the green pellet properties obtained under corresponding optimal balling conditions for each blend. Under laboratory conditions, the proper moisture of green balls for blends B1–B7 ranges from 7.6 per cent to 8.1 per cent. Under these conditions, all green pellets exhibit good mechanical strength with high drop strength of exceeding five times/(0.5 m), compressive strength of above 15 N/pellet, and thermal shock temperature of higher than 350°C.

A clear trend is observed that the higher Ore G proportion, the better drop strength and compressive strength. This improvement is mainly attributed to the characteristics of Ore G. As a typical soft ore rich in goethite, Ore G exhibits strong hydrophilicity and water-retention capacity. Consequently, increasing its proportion raises the optimum moisture content.

During pelletisation, water also acts as a binder. It fills interparticle voids and forms liquid films and capillary bridges. As moisture increases, voids are progressively reduced and capillary forces become dominant. When the capillary water approaches saturation, capillary pressure rises. This promotes the migration of fine particles toward the pellet nucleus and enhances particle coalescence, resulting in higher pellet strength (Jiang, 2016).

However, excessive addition of Ore G, especially above 30 per cent, significantly reduces the thermal stability of green pellets, which however, is largely affected by the ore blend and characteristics. Under rapid heating, internal moisture cannot escape efficiently. This leads to a sharp increase in vapor pressure and internal stress. As a result, the thermal shock temperature decreases from 420–480°C to 350–360°C. Despite this decline, the values still meet industrial requirements.

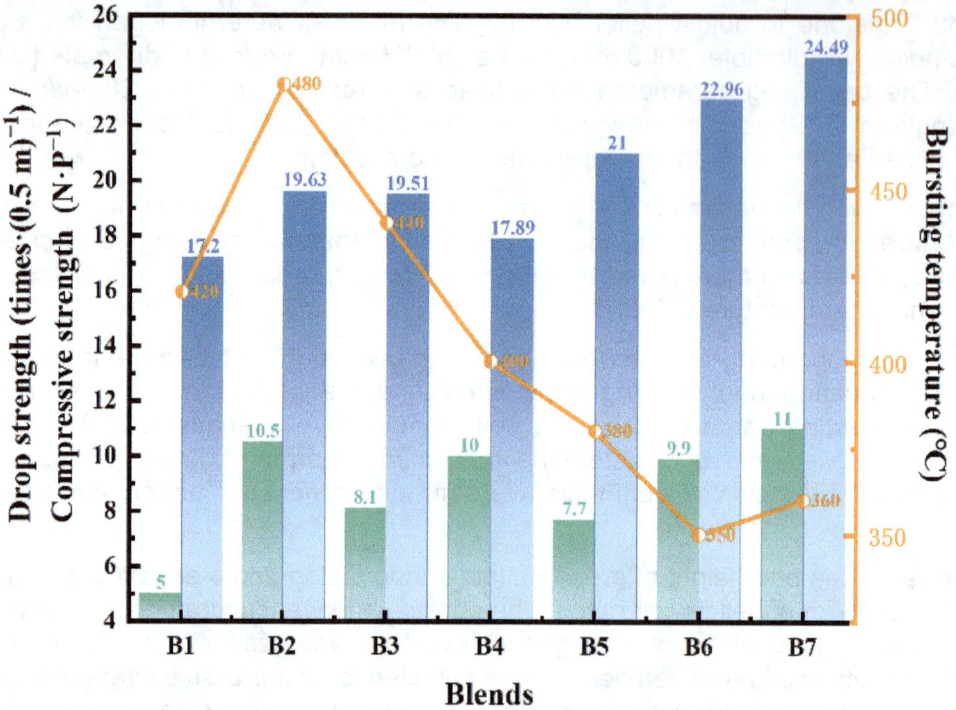

FIG 3 – Comparison of green pellets performance of different blends under corresponding proper balling conditions.

Effect of Ore G proportion on preheating and roasting behaviour of pellets

Effects of preheating temperature and time

Figure 4 shows the effect of preheating temperature and time on the CCS of preheated pellets for different blends. Increasing preheating temperature or extending preheating time are both beneficial to the improvement of the CCS of preheated pellets. This trend is attributed to an increase in the oxidation rate of magnetite to Fe_2O_3 and the initial bonding between newly formed and primary Fe_2O_3 particles, and thereby enhancing pellet strength (Li, Li and Zhang, 2009; Qiao *et al*, 2021; Lu *et al*, 2022).

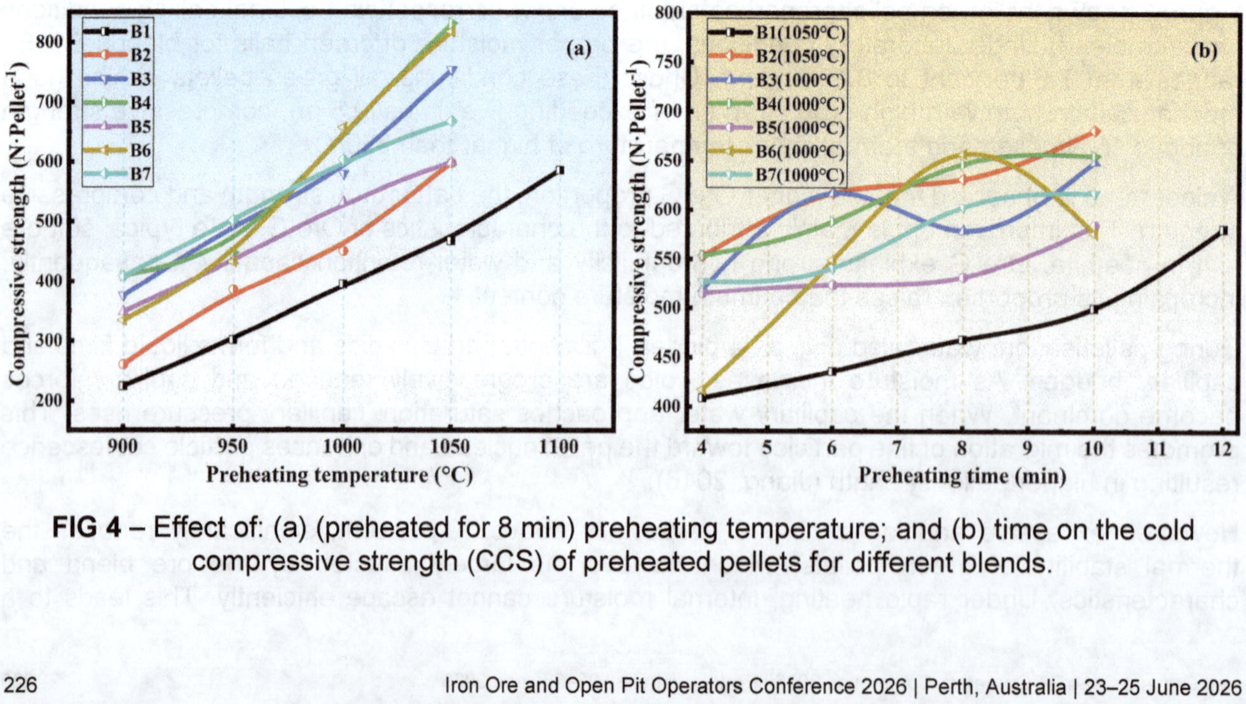

FIG 4 – Effect of: (a) (preheated for 8 min) preheating temperature; and (b) time on the cold compressive strength (CCS) of preheated pellets for different blends.

To ensure the CCS of preheated pellets attain above 500 N/pellet, the proper preheating temperature for B1 and B2 is 1050°C, with corresponding preheating time of 10–12 min and 6–8 min, respectively. In contrast, B3–B7 can reach the same strength at a lower temperature of 1000°C with a preheating time of 6–8 min.

Under the same preheating conditions, it is found that increasing the Ore G proportion leads to a rise followed by a decline in compressive strength. The maximum pellet strength is obtained at an Ore G proportion of 35 per cent–45 per cent. This indicates that an appropriate blending ratio can offset the negative effects associated with the high loss on ignition of Ore G.

Effects of roasting temperature and time

Figure 5 presents the effect of roasting temperature and time on the CCS of roasted pellets for different blends under proper preheating conditions. It can be observed that the compressive strength of roasted pellets increases markedly with the elevation of roasting temperature or prolongation of roasting time. However, the influence of the roasting temperature is more significant.

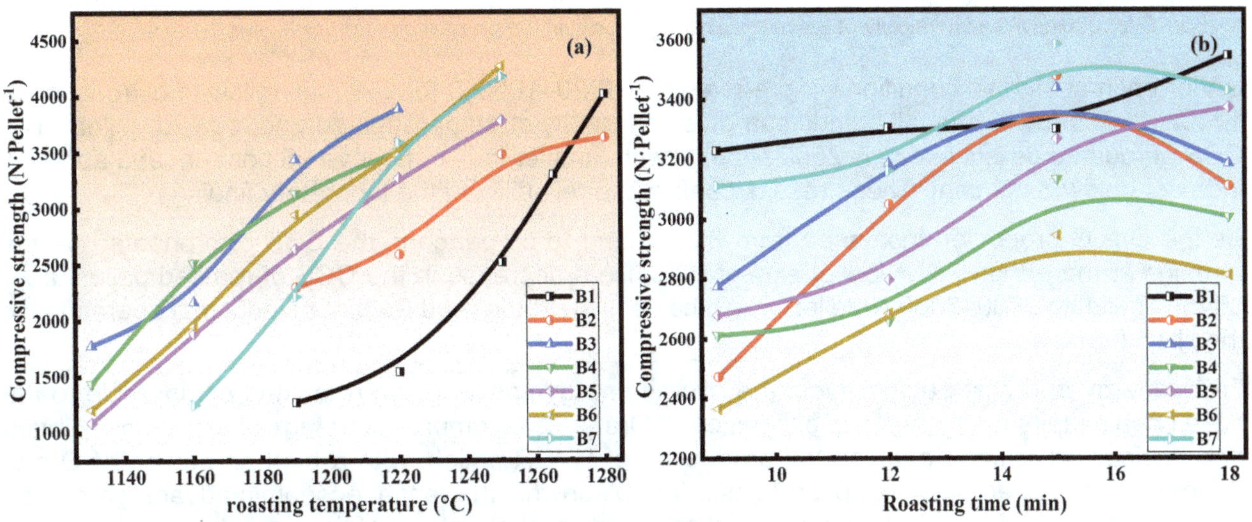

FIG 5 – Effect of: (a) roasting temperature; and (b) roasting time on compressive strength of roasted pellets.

The consolidation of roasted pellets during roasting is governed by solid-state sintering, characterised with substantial interparticle bonding forms through mass diffusion and neck growth at contact points. As temperature increases, mass diffusion and physicochemical reactions are accelerated. Meanwhile, hematite recrystallisation and grain growth promote pore spheroidisation and reduce total porosity, leading to enhanced pellet strength (Ran, Pan and Yang, 2025).

To attain a target CCS of roasted pellets, the preferable roasting temperature for B1 and B2 is 1250–1280°C for 9–15 min, while that for B3–B7 is 1190–1220°C for 12–15 min. The probable reason is that, although B3–B7 contain higher Ore G proportions, simultaneous increase in the proportion of magnetite concentrate and pellet basicity play an important role in improving roasting performance and thereby lowering or maintaining energy consumption.

Besides, for most blends, excessive roasting time leads to a decline in strength, except for B1 and B5. This deterioration is likely related to partial decomposition of Fe_2O_3 at prolonged high temperatures, which weakens intergranular bonding between hematite crystals (Forsmo, Samskog and Björkman, 2007).

Pilot-scale preheating and roasting tests

Based on the laboratory results, pilot-scale preheating and roasting tests were conducted under a simulated grate–rotary kiln process in terms of B1–B3, B5, and B7. The optimum parameters and corresponding results are summarised in Table 3.

TABLE 3

Optimal conditions and results of the pilot-scale tests for different blends.

Blend	Preheating		Roasting		Preheated pellets		Roasted pellets		
	Temp/°C	Time/ min	Temp/°C	Time/ min	CCS/ (N·P^{-1})	AC tumbler index/%, +3 mm	CCS/ (N·P^{-1})	TI/ %, +6.3 mm	AI/%, -0.5 mm
B1 (0% G)	1050	9	1265	15	829	96.84	3645	97.72	1.74
B2 (30% G)	1050	6	1250	15	453	97.00	2737	96.39	2.67
B3 (30% G)	1020	8	1250	15	817	97.04	2963	95.31	2.69
B5 (40% G)	1050	6	1250	15	752	96.92	2951	95.45	2.94
B7 (50% G)	1050	6	1250	15	875	97.79	2872	95.59	2.41

Note: The temperatures for updraft drying and downdraft drying were 250°C and 300°C, respectively, with corresponding drying time of 3 min and 4 min; the drying air velocity was 1.2 m s^{-1}. The transition preheating stage was conducted at 650°C for 3 min with a air velocity of 2.4 m s^{-1}. The air velocity in the preheating stage was also 2.4 m s^{-1}.

Under optimal thermal conditions of preheating at 1020–1050°C for 6–9 min, followed by roasting at 1250–1265°C for 15 min, all blends can give sufficiently strong preheated and roasted pellets. The CCS of roasted pellets exceeds 2500 N/pellet with tumbler index of above 95 per cent and abrasion index of below 3 per cent. These results confirm the reliability of the laboratory findings.

As the Ore G proportion increased from 30 per cent to 50 per cent, the CCS of preheated pellets remains at more than 700 N/pellet except B2 while a decrease in the CCS of roasted pellets from 3645 N/pellet to 2700–3000 N/pellet, which can still meet the requirement of industrial operations in the shaft furnace.

From energy and cost perspectives, the overall energy consumption remained comparable to the base case despite the higher loss on ignition of Ore G. The combined strategy of blend optimisation and basicity adjustment enables the production of H_2-DR pellets with a high proportion of Ore G under standard energy consumption constraints Meanwhile, potential economic advantage can be achieved due to a decrease in raw material cost and production cost of fired pellets.

Metallurgical properties of roasted pellets

Chemical composition

The total iron (TFe) content of the roasted pellets ranges from 64–66 wt per cent, while the FeO content varies between 0.1 and 0.6 wt per cent, indicating a high degree of oxidation. The residual FeO may exist in incompletely oxidised magnetite or be incorporated into the slag phase. For ideal oxide pellets, the FeO content should be minimised (typically ≤1 wt per cent) to promote extensive hematite recrystallisation and enhance reducibility of Fe_2O_3.

In terms of gangue components, the four blends containing 30–50 per cent Ore G exhibit higher SiO_2 and Al_2O_3 contents than the base case, reaching 3.7–5.5 wt per cent and 1.2–1.6 wt per cent, respectively. Meanwhile, the P content of the pellets increases with increasing Ore G proportion. The pellet basicity ranges from 0.1 to 0.5.

Reducibility (RI)

Figure 6a compares the reducibility of roasted pellets at 800°C and 950°C under simulated HYL conditions. At 800°C, the RI ranges from 2.0 to 3.5. When the temperature increases to 950°C, the RI exceeds 4.0 for all blends. Compared with the base case, the addition of Ore G improves RI of roasted pellets. This is mainly attributed to its high pellet porosity caused by higher proportion of goethite, which undergoes dehydration, promotes pore development during roasting and enhances gas diffusion. However, for B7, the higher basicity promotes liquid-phase formation, which partially blocks pore channels and restricts gas diffusion (Gao, Zhou and Shi, 2020; Li, Yang and Jiang, 2009).

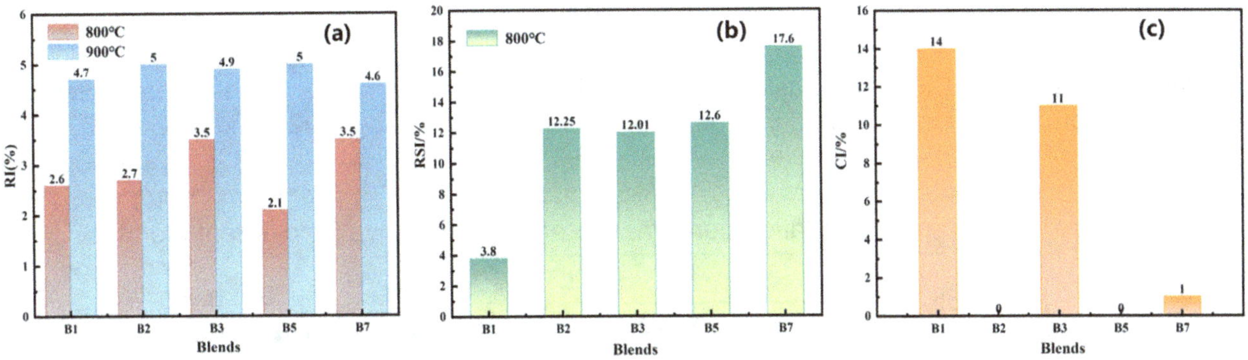

FIG 6 – Metallurgical properties of pellets: (a) RI; (b) RSI; (c) CI.

Reduction swelling index (RSI)

Figure 6b shows the RSI of roasted pellets at 800°C under simulated HYL conditions. It is found that all blends exhibit RSI values below 15 per cent except for B7,, satisfying operational requirements. Increasing Ore G proportion tends to increase the RSI of roasted pellets, which agrees well with the trend of pellet reducibility.

The reduction swelling of oxidised pellets is mainly related to several mechanisms, including phase transformation of hematite to magnetite (α-$Fe_2O_3 \rightarrow Fe_3O_4$), growth of iron whiskers and carbon deposit, all of which induce volume expansion (Jiang, 2016). Therefore, the reducibility of roasted pellets directly affects the rate of phase transformation, formation and growth of iron whiskers, which further influence the stress within the pellets. The relatively low RSI of B1 is attributed to the ore characteristics of raw materials because the combined use of hematite, specularite, and magnetite ores enhances inter-particle bonding and promotes hematite recrystallisation, resulting in a stable structure that can accommodate higher internal stress and hence suppresses expansion (Ran, Pan and Yang, 2025).

In contrast, blends containing Ore G show higher RSI due to improved reducibility and higher reduction rates, which generate greater internal stress. For B7, although higher basicity and magnetite content improve pellet strength, the densified structure reduces the ability to accommodate transformation stress. In addition, heterogeneous liquid-phase and pore distribution aggravates anisotropic swelling (Zhong, Guo and Ding, 2018; Li, Zong and Zhang, 2023).

Clustering index (CI)

Figure 6c compares the clustering performance of roasted pellets. All blends give low CI values of below 15 per cent. However, With increasing Ore G proportion, the CI decreases. This is attributed to the lower iron grade and enhanced liquid-phase formation, which promotes slag bonding and suppresses iron whisker growth. As a result, pellet sticking is reduced (Zhong, Guo and Ding, 2018; Yang *et al*, 2025).

Low-temperature reduction degradation index (LTD)

Table 4 shows the LTD results of roasted pellets. Through comparison, it is known that B1 exhibits the best performance, with $LTD_{+6.3}$ > 90 per cent, $LTD_{-0.5}$ < 5 per cent, and an unbroken pellet ratio of ~60 per cent.

The addition of 30–50 per cent Ore G leads to a general deterioration in LTD. Among these blends, B7 performs relatively better, with $LTD_{+6.3}$ > 80 per cent, $LTD_{-0.5}$ ~12 per cent, and an unbroken pellet ratio of ~67 per cent. B2 and B3 exhibit similar $LTD_{+6.3}$ (~77 per cent) but high $LTD_{-0.5}$ (16–18 per cent). Their unbroken pellet ratios differ significantly, ie ~55 per cent for B2 and ~68 per cent for B3. B5 shows the poorest LTD performance.

The low-temperature reduction degradation of oxidised pellets is primarily caused by the phase transformation of hematite to magnetite accompanied with crystal structural change from hexagonal to cubic and up to approximately 20 per cent volume expansion. As mentioned above, the superior LTD performance of the roasted pellets made from B1 is attributed to its reasonable ore structure, which accommodates stresses from phase transformation during reduction. In contrast, increasing

Ore G in pellet blends increases the pellet porosity and restricts the hematite recrystallisation, which weaken the resistance to phase transformation-induced stress, leading to uneven expansion and degradation. For B7, higher pellet basicity promotes liquid phase formation, which suppresses crack propagation and fines generation during reduction (Li, Zong and Zhang, 2023; Yang *et al*, 2025).

TABLE 4

Comparison of LTD degradation performance of roasted pellets for different blends.

Pellets	LTD/%			
	+6.3 mm	**0.5~6.3 mm**	**-0.5 mm**	**Intact pellet ratio**
B1 (0% G)	93.76	3.92	2.32	59.76
B2 (30% G)	77.63	5.72	16.65	54.55
B3 (30% G)	77.40	4.63	17.97	68.84
B5 (40% G)	60.81	2.60	36.59	53.52
B7 (50% G)	85.55	2.69	11.76	67.39

Mineralogical characteristics of roasted pellets

Figure 7 presents the microstructure of roasted pellets prepared from different blends. A comparative analysis shows that, with increasing Ore G proportion, both porosity and pore size of the roasted pellets exhibit a pronounced increasing trend. This microstructural evolution is beneficial to pellet reducibility by facilitating gas diffusion, however, it simultaneously weakens the resistance of the pellets to internal stresses associated with reduction swelling and disintegration.

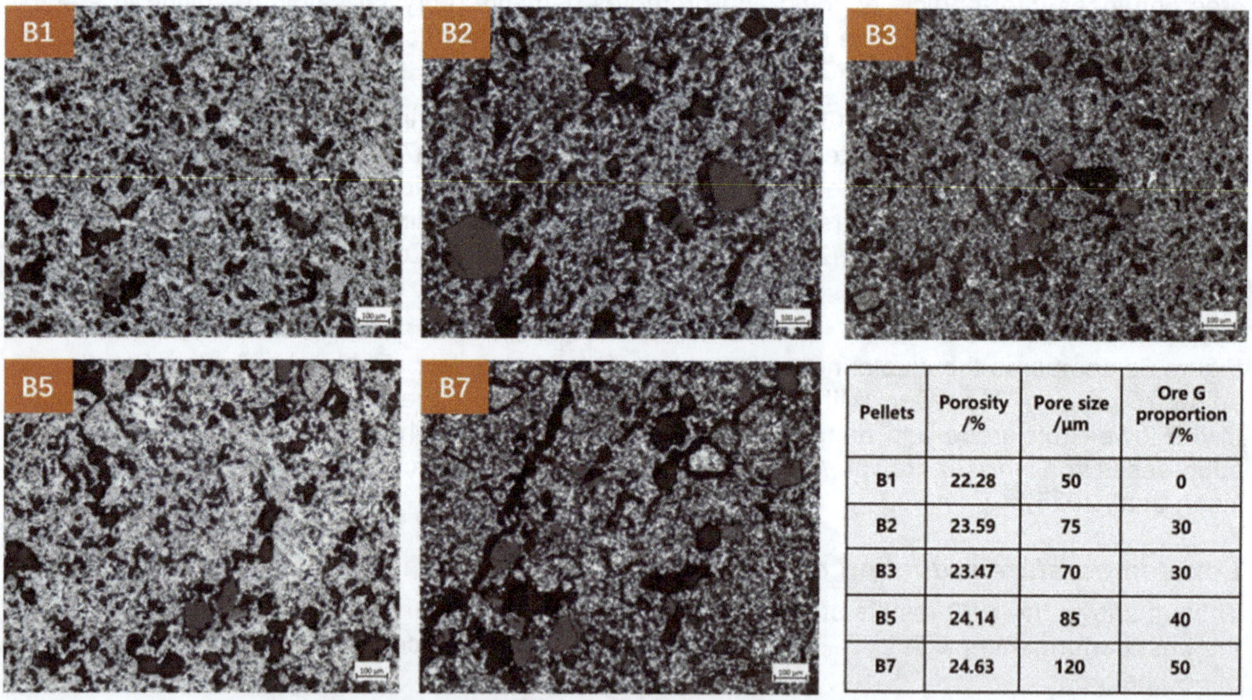

Pellets	Porosity /%	Pore size /μm	Ore G proportion /%
B1	22.28	50	0
B2	23.59	75	30
B3	23.47	70	30
B5	24.14	85	40
B7	24.63	120	50

FIG 7 – Microstructure of the roasted pellets for different blends.

The primary reason lies in the high goethite content of Ore G, which contains abundant structural water. During high-temperature roasting, the evaporation of combined water in goethite leaves behind additional pores, resulting in a more porous and loosely packed pellet structure. In contrast, pellets from B1, which are composed mainly of hematite, specularite, and magnetite with low loss on ignition, exhibit a significantly denser microstructure. For B5, the relatively high proportion of Ore G leads to the formation of microcracks, thereby deteriorating the low-temperature disintegration resistance. In the case of B7, the higher pellet basicity promotes liquid-phase formation during roasting, which enhances pellet consolidation and partially offsets the adverse effects of increased porosity.

CONCLUSIONS

- Increase the proportion of Ore G from 0 per cent to 30 per cent–50 per cent, qualified green pellets can be produced. The roasting performance can be improved through optimised blending, increased magnetite concentrate proportion, and appropriate basicity adjustment. Under grate–rotary kiln conditions, preheating at 1020–1050°C for 6–9 min followed by roasting at 1250–1265°C for 15 min yields pellets with compressive strength > 2500 N/pellet, tumbler index > 95 per cent, and abrasion index < 3 per cent.
- With increasing Ore G proportion, RI and CI improve, whereas RSI and LTD deteriorate under hydrogen-rich HYL conditions. This behaviour is attributed to increased porosity and pore size, which weaken the ability of pellets to accommodate stresses induced by hematite phase transformation during reduction.
- Despite the high LOI of Ore G, overall energy consumption remains comparable to the baseline due to optimised thermal regimes. For industrial application, the Ore G proportion should be controlled at ≤30 per cent to balance metallurgical performance and process stability. The combined strategy of blend optimisation and basicity adjustment provides a feasible route for utilising hematite–goethite ores in H_2-DR pellet production.

REFERENCES

Cai, D W and Fan, X H, 2023. Advances in low-carbon burden materials and green steelmaking technologies, *Sintering and Pelletizing*, 48(4):1–7.

Forsmo, S P E, Samskog, P O and Björkman, B, 2007. Mechanisms in oxidation and sintering of magnetite iron ore green pellets, *Powder Technology*, 183(2):247–259.

Gao, G F, Zhou, X L and Shi, Z, 2020. Study on induration process and influencing factors of pure limonite pellets, *Bulletin of the Chinese Ceramic Society*, 39(2):485–492.

Hu, S J, Qin, L H and Yao, Y K, 2025. Application of key low-cost technologies for binderless pellet production, *Sintering and Pelletizing*, 50(1):61–68.

Jiang, T, 2016. *Principle and technology of Agglomeration of iron ores* (Central South University Press: Changsha).

Li, G H, Li, X Q and Zhang, Y B, 2009. Induration mechanisms of oxidised pellets prepared from mixed magnetite-hematite concentrates, *Ironmaking and Steelmaking*, 36(5):393–395.

Li, Q, Yang, Y B and Jiang, T, 2009. Preparation technologies for siderite and limonite pellets, *Mining and Metallurgical Engineering*, 29(2):59–62.

Li, Z, Zong, Y B and Zhang, J L, 2023. Reduction process and metallurgical properties of high-silica magnesian fluxed pellets, *Iron and Steel*, 58(3):34–42+52.

Liu, H T and Lv, Z F, 2025. Research on domestic and international utilization of iron ore resources and development trends, *China Mining Magazine*, 34(7):315–323+2.

Liu, S G, Zhang, J J and Zhang, Q Y, 2023. Application and study of high proportion hematite in belt furnace pelletizing, *Metallurgical Information Review*, 60(4):32–35+13.

Lu, M, Chen, X, Y, Wang, X F, Zhang, J L, Liu, Z J, Wang, Y Z and Li, Z, 2022. Optimization of preheating and roasting process parameters for high-silicon magnetite concentrate pellets, *Sintering and Pelletizing*, 47(03):59–65.

Pan, J, Ma, W Z and Zhu, D Q, 2021. Effect of Al_2O_3 on iron ore sinter and blast furnace smelting, *Journal of Iron and Steel Research International*, 33(10):1095–1108.

Qiao, B, Luo, G P, Chai, Y F, Song, W and Wang, X T, 2021. Comparative analysis of preheating and induration performance of pellets prepared from different concentrates, *Sintering and Pelletizing*, 46(2):50–56.

Ran, W, Pan, J and Yang, C C, 2025. Effects of adding hematite–limonite on oxidation kinetics and induration performance of fluxed magnetite pellets, *The Chinese Journal of Process Engineering*, 25(5):510–521.

Yang, C C, Hou, D N and Pan, J, 2024. Wet grinding–settling–filtration behaviors of different types of iron ore fines and influencing factors, *Journal of Central South University (Natural Science Edition)*, 55(10):3627–3640.

Yang, C C, Hou, D N and Zhu, D Q, 2025. Mechanisms of iron ore fines' properties on wet grinding–settling–filtration performance, *Journal of Central South University (Natural Science Edition)*, 56(5):1711–1720.

Yang, C C, Liu, W, Zhu, D Q, Pan, J, Xia, G H and Qu, S J, 2025. Hydrogen-based direct reduction behavior of iron ore pellets with iron grades ranging from 59% to 68%. *International Journal of Hydrogen Energy*, 138:787–801.

Yang, C C, Qu, S J, Liang, H Y, Zhu, D Q, Pan, J and Guo, Z Q, 2024. Grinding characteristics of imported hematite–limonite mixed ores and their behavior in oxidised pellet preparation, *Journal of Iron and Steel Research*, 36(9):1152–1162.

Yi, L J, Jiang, L H and Li, J, 2022. Development and application of diversified, low-cost and clean production technologies for high-quality pellets, *Sintering and Pelletizing*, 47(1):95–103.

Zhong, S P, Guo, L and Ding, Z Y, 2018. Observation of iron whisker growth during gas-based direct reduction of iron ore fines, *Nonferrous Metals Science and Engineering*, 9(1):15–21.

Zhu, D Q, Guo, Z Q, Pan, J and Wang, Z Y, 2018. Insights on pretreatment of Indian hematite fines in grate-kiln pelletizing process: the choice of grinding processes, *Journal of Iron and Steel Research International*, 25(5):506.

Zhu, D Q, Li, X B, Pan, J, Guo, Z Q and Yang, C C, 2024a. Global iron ore companies' carbon reduction pathways and progress: review, *Iron and Steel*, 59(4):1–8.

Zhu, D Q, Yang, Y P and Song, Q S, Wang, W and Pan, J, 2024b. Preparation characteristics of PFC1 concentrate pellets under different pelletizing conditions, *China Metallurgy*, 34(6):44–54+73.

Effect of SiO_2 content on hydrogen-enriched reduction behaviour of high-grade hematite pellets

G Zhengqi[1], C Xing[2], Z Deqing[3], P Jian[4], Y Congcong[5] and L Siwei[6]

1. Professor, Central South University, Changsha Hunan 410083, China. Email: guozqcsu@csu.edu.cn
2. PhD, Central South University, Changsha Hunan 410083, China. Email: chxing@csu.edu.cn
3. Professor, Central South University, Changsha Hunan 410083, China. Email: dqzhu@csu.edu.cn
4. Professor, Central South University, Changsha Hunan 410083, China. Email: pjcsu@csu.edu.cn
5. Associate Professor, Central South University, Changsha Hunan 410083, China. Email: smartyoung@csu.edu.cn
6. Associate Researcher, Central South University, Changsha Hunan 410083, China. Email: swli@csu.edu.cn

INTRODUCTION

Hydrogen-rich gas-based direct reduction has become a promising low-carbon ironmaking route for decarbonising the steel industry. However, the shortage of high-grade hematite concentrates and the low SiO_2 content of such feeds make oxidised hematite pellets susceptible to swelling and fines generation during hydrogen-rich shaft-furnace reduction, thereby threatening stable DRI production. In this context, clarifying the role of SiO_2 in reduction behaviour and structural stability is essential for defining a suitable compositional window for shaft-furnace pellets. Therefore, this study systematically investigates the effects of SiO_2 content on the reduction kinetics, swelling behaviour, pellet strength, degradation performance, phase evolution, and microstructural development of high-grade hematite pellets under hydrogen-rich gas-based reduction conditions.

METHODOLOGY

High-grade hematite concentrate was used as the iron-bearing raw material, and flotation silica concentrate was added as a silica adjuster to prepare oxidised pellets containing 1 per cent, 2 per cent, 3 per cent, and 4 per cent SiO_2. A H_2–CO–CO_2–N_2 mixed gas was employed to simulate the atmosphere of a hydrogen-rich shaft furnace. Isothermal reduction experiments were conducted in the temperature range of 800–950°C, and additional tests with different H_2 fractions were performed at 950°C to evaluate the influence of gas composition. The reduction behaviour was assessed in terms of reducibility, reduction swelling index (RSI), compressive strength, and low-temperature degradation, while the underlying mechanisms were analysed through kinetic fitting and phase characterisation.

RESULTS AND DISCUSSION

Kinetic analysis showed that, within the investigated temperature range, pellets containing 2 per cent and 4 per cent SiO_2 exhibited only limited differences in reducibility. The reduction process was mainly controlled by the interfacial chemical reaction step (n=3), with apparent activation energies of 35.57 and 35.65 $kJ \cdot mol^{-1}$, indicating that SiO_2 had only a minor effect on the overall kinetic control regime under hydrogen-rich conditions. In contrast, temperature and atmosphere exerted much stronger effects on the macroscopic rate–swelling–strength relationship. As the temperature increased from 800 to 950°C, the time required to reach 96 per cent reduction decreased from 70 to 45 min, but the final compressive strength dropped sharply from 500–600 to 100–150 N per pellet, indicating severe high-temperature strength deterioration. Notably, the RSI reached its maximum at about 35 per cent reduction, while the compressive strength reached its minimum at nearly the same stage, identifying a critical phase-transformation stage that dominated structural damage.

At 950°C, increasing the H_2 fraction from 45 per cent to 100 per cent further shortened the reduction time from 45 to 15 min, confirming the strong promotional effect of hydrogen enrichment. However, swelling increased under all tested SiO_2 levels, and compressive strength first increased and then decreased, reaching its maximum at 82 per cent H_2, which suggests the existence of an optimum hydrogen-rich atmosphere window. Mechanistically, reduction proceeded inward according to the unreacted-core model, following the sequence $Fe_2O_3 \rightarrow Fe_3O_4 \rightarrow FeO \rightarrow Fe$. Around 35 per cent reduction, FeO was at its maximum while the metallic Fe network was still poorly developed, so the severe swelling and abrupt strength loss were mainly caused by volumetric mismatch and internal stress associated with iron-oxide phase transformations. Compared with its limited influence on kinetics, SiO_2 played a more important role in structural evolution: discrete quartz particles together with a minor liquid phase approaching Ca–Fe olivine composition formed a silica-bearing framework that buffered stress and regulated pore evolution. When this framework weakened, the pellets became more susceptible to structural loosening and residual strength loss, although the overall low-temperature degradation resistance remained relatively favourable.

TABLE 1

Summary of representative results in the results and discussion section.

Variable	Condition	Key result
SiO_2 content	2% and 4%	The difference in reducibility was limited; the apparent activation energies were 35.57 and 35.65 $kJ \cdot mol^{-1}$, respectively; the fitted kinetic parameter was n = 3.
Temperature	800–950°C	The time required to reach 96% reduction decreased from 70 min to 45 min.
H_2 fraction	45–100% H_2 at 950°C	The reduction time decreased from 45 min to 15 min; the compressive strength reached its maximum at 82% H_2.
Final strength	800 and 950°C	The final compressive strength was 500–600 N/pellet at 800°C, but only 100–150 N/pellet at 950°C.

CONCLUSION

This study demonstrates that SiO_2 content has only a limited effect on the kinetic control regime of hydrogen-rich reduction, but plays a decisive role in the structural stability of high-grade hematite pellets. A critical phase-transformation stage exists at approximately 35 per cent reduction, at which the swelling reaches its maximum and the compressive strength reaches its minimum because of the intense volumetric mismatch and internal stress associated with the sequential transformation of iron oxides. In addition, although increasing temperature and hydrogen fraction can markedly accelerate the reduction process, excessive thermal or hydrogen enrichment conditions aggravate swelling and may impair the residual strength of pellets. The silica-bearing framework, formed mainly by discrete quartz particles together with a minor Ca–Fe-olivine-like liquid phase, contributes to stress buffering and pore-structure regulation during reduction. These findings clarify the mechanism of irreversible structural damage during hydrogen-rich reduction and provide theoretical guidance for optimising pellet composition and atmosphere conditions in shaft-furnace DRI production.

Electrical conductivity measurement of slags for electric smelting ironmaking

X Zhou[1], Z Shi[2], R Brell[3], T Evans[4] and X Ma[5]

1. PhD Student, Julius Kruttschnitt Mineral Research Centre, Sustainable Minerals Institute, The University of Queensland, Indooroopilly Qld 4068. Email: xuefan.zhou@student.uq.edu.au
2. Research Fellow, Julius Kruttschnitt Mineral Research Centre, Sustainable Minerals Institute, The University of Queensland, Indooroopilly Qld 4068. Email: zmshi@uq.edu.au
3. Manager Process and Technology, Future Technologies, BlueScope, Melbourne Vic 3000. Email: ruwan.brell@bluescopesteel.com
4. Honorary Professor, Julius Kruttschnitt Mineral Research Centre, Sustainable Minerals Institute, The University of Queensland, Indooroopilly Qld 4068. Email: tim.evans@riotinto.com
5. Senior Research Fellow, Julius Kruttschnitt Mineral Research Centre, Sustainable Minerals Institute, The University of Queensland, Indooroopilly Qld 4068. Email: x.ma@uq.edu.au

INTRODUCTION

In the contemporary global context of carbon neutrality, it is imperative for the ironmaking industry to progressively transition away from the use of fossil fuels. The traditional blast furnace-basic oxygen furnace (BF-BOF) route, which remains the dominant pathway in steel production, is significantly reliant on coal. As a lower-carbon alternative, the direct reduced iron-electric arc furnace (DRI-EAF) route has been increasingly adopted. However, the EAF is highly sensitive to the level of impurities in DRI, which limits its applicability for processing low- to medium-grade iron ores.

In contrast, the electric smelting furnace (ESF) exhibits greater flexibility in raw material input and furnace composition adjustments. Owing to its capability for continuous and stable production, the ESF has been widely used in the production of ferroalloys and nonferrous metals and demonstrates great potential for future large-scale, low-carbon ironmaking. In the DRI-ESF route, BF grade ores can first be reduced using natural gas or hydrogen in shaft furnaces or fluidised beds to produce DRI, which is subsequently smelted in ESF using electricity. Currently, no commercial-scale ESF process operates using BF grade DRI. Therefore, a comprehensive understanding of the DRI behaviour in the ESF process, especially the properties of the ESF slag where DRI melting and smelting occur, is imperative for ESF design, process optimisation and control. Among these properties, the electrical conductivity of slag is of particular importance, as heat in ESF is generated through resistive heating of the slag, thereby directly influencing DRI melting and reaction behaviour.

In this study, the electrical conductivity of DRI smelting slags was systematically investigated as a function of slag composition and temperature using the four-electrode method. The experimental samples were synthesised from CaO, SiO_2, MgO, Al_2O_3, and 'FeO'. An orthogonal experimental design was employed to evaluate the effects of compositional variations and temperature on slag electrical conductivity. The underlying mechanisms governing slag electrical conductivity are discussed through correlations between the electrical conductivity, slag composition and temperature.

METHODOLOGY

The four-electrode method is extensively employed for measuring slag conductivity at high temperature due to its high accuracy and experimental simplicity. The fundamental calculation of resistivity and conductivity in an electrolyte cell is as follows:

$$\sigma = \frac{\kappa}{R_E}$$

Where:

σ	is the conductivity (S/m)
R_E	is the resistance (Ω)
κ	is the cell constant (m^{-1})

The cell constant was determined as a function of immersion depth and crucible geometries, and it can be measured using standard electrolyte with known conductivity. In this study, potassium chloride solutions with known conductivities at 25°C (1D KCl: 111.3 mS/cm; 0.1D KCl: 12.85 mS/cm) were used to calibrate the cell constant of the crucibles. The electrical resistance was measured by a Gamry 1010E potentiostat.

RESULTS AND DISCUSSION

Owing to the existence of gangue in silicon-containing phase, fluxes are introduced at the initial melting stage to improve slag properties, leading to the formation of 'CaO-SiO_2-MgO-Al_2O_3-FeO' slag in ESF. To systematically investigate the effects of different variables on slag conductivity, an orthogonal design was employed. The experimental design was based on an L_{16} (4^5) orthogonal array. The orthogonal array and corresponding results are detailed in Table 1.

TABLE 1

Orthogonal array.

No.	CaO (wt%)	SiO_2 (wt%)	Al_2O_3 (wt%)	MgO (wt%)	'FeO' (wt%)	Temperature (°C)	Electrical conductivity (mS/cm)
1	43.5	43.5	10	3	0	1450	27.56
2	44.33	38	13.33	4.33	0	1500	153.84
3	44.38	33.29	16.67	5.67	0	1550	101.43
4	43.8	29.2	20	7	0	1600	93.65
5	48	36	10	4.33	1.67	1600	242.77
6	49.2	32.8	13.33	3	1.67	1550	358.27
7	37.33	37.33	16.67	7	1.67	1500	106.40
8	39.13	33.54	20	5.67	1.67	1450	50.73
9	48.6	32.4	10	5.67	3.33	1500	77.49
10	43.62	32.71	13.33	7	3.33	1450	113.22
11	41.46	35.54	16.67	3	3.33	1600	160.55
12	36.17	36.17	20	4.33	3.33	1550	146.85
13	42	36	10	7	5	1550	379.09
14	38	38	13.33	5.67	5	1600	527.69
15	44.4	29.6	16.67	4.33	5	1450	188.97
16	41.14	30.86	20	3	5	1500	191.94

For the 16 samples, the electrical conductivity results of all groups sharing the same level of a given factor were averaged, and the four levels of that factor were subsequently plotted against their corresponding average conductivities (Figure 1). Range analysis was then applied to evaluate the influence of each factor, with the corresponding range values (D) also shown in Figure 1. The following conclusions can be drawn:

- MgO content and the CaO/SiO_2 ratio have minimal influence on the electrical conductivity, although slight fluctuations in the measurement values can still be observed.
- FeO content, Al_2O_3 content and temperature have a more pronounced effect on electrical conductivity. Specifically, when the FeO content increases from 0 to 5 wt%, the conductivity initially decreases and then increases. In contrast, Al_2O_3 content shows an opposite trend, with

conductivity first increasing then decreasing. In addition, the electrical conductivity consistently increases within rising temperature.

- Within the given conditions, FeO has the largest range value, which indicates the most significant influence on electrical conductivity. The effects of Al_2O_3 content and temperature are second only to that of FeO, whereas binary basicity and MgO content have a relatively minor impact within the studied range.

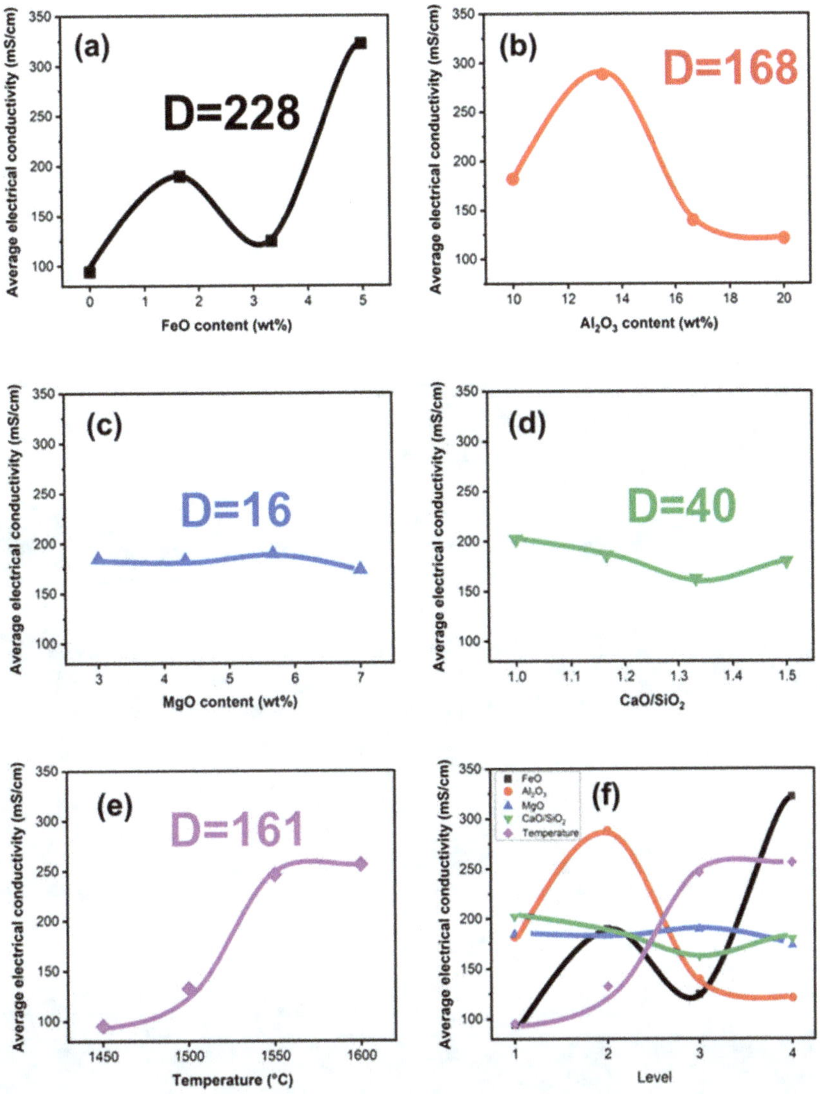

FIG 1 – Effect of various factors on conductivity: (a) FeO content; (b) Al_2O_3 content; (c) MgO content; (d) CaO/SiO_2; (e) Temperature; (f) Overlay of all factors.

CONCLUSION

- A reliable electrical conductivity measurement methodology was established to obtain fundamental conductivity data for ESF slags.
- The electrical conductivity of ESF slags increases with increasing FeO content and with increasing temperature over the range of 1673–1873 K.
- Within the ESF slags range, the influence of key factors on electrical conductivity follows the order: 'FeO'>$Al_2O_3 \approx$ Temperature > CaO/SiO_2 > MgO.

Iron Ore

Mining and processing

Influence of binder and sample pretreatment on the quality of green iron ore pellets

O A Aladejebi[1], H Han[2] and L Lu[3]

1. Research Officer, CSIRO Mineral Resources, Pullenvale Qld 4069.
   Email: tosin.aladejebi@csiro.au
2. Research Officer, CSIRO Mineral Resources, Pullenvale Qld 4069.
   Email: hongliang.han@csiro.au
3. Chief Research Consultant, CSIRO Mineral Resources, Pullenvale Qld 4069.
   Email: liming.lu@csiro.au

ABSTRACT

The characteristics of pellet feed and the type and dosage of binder used play an important role in the formation dynamics of green iron ore pellets and hence determine their quality. This study investigates the effects of binder and sample pretreatment on the quality of green pellets produced from a magnetite concentrate. The as-received pellet feed was initially tested using Binders 1 (bentonite) and 2. Increasing bentonite addition led to a noticeable improvement in drop strength number, whereas compressive strength only started to improve from 1.4 per cent bentonite. The Binder 2 dosage at 0.075 to 0.20 wt per cent improved both the drop strength number and compressive strength, demonstrating enhanced cohesiveness and structural integrity of the green pellets. A combined binder approach, incorporating 0.075 wt per cent Binder 2 with 1.0 to 1.4 wt per cent bentonite produced green pellets with higher drop strength number and compressive strength than those obtained with either binder alone. Overall, the as received concentrate requires a high binder dosage to achieve the green pellet quality.

The magnetite concentrate was then subjected to multiple high-pressure grinding rolls (HPGR) processing to enhance particle surface characteristics and increase specific surface area. The Blaine index of the pellet feed increased with each successive pass. Green pellets were made from the HPGR-processed pellet feed and compared with those from the as-received pellet feed. At a fixed bentonite dosage of 1.0 wt per cent, the drop strength number and compressive strength of the green pellets increase with increasing Blaine index. A substantial improvement in the drop strength number of the green pellets was achieved when HPGR pretreatment was combined with the addition of Binder 2 (0.075 wt per cent) together with bentonite (1.0 wt per cent). This enhancement is attributed to the increased specific surface area and irregular particle morphology induced by HPGR pretreatment, which facilitates stronger interparticle adhesion during seeding and growth. The findings demonstrate that integrating HPGR processing with optimised binder formulation provides a pathway to improve green pellet strength. This approach demonstrates the potential to reduce overall bentonite usage, minimise impurities associated with bentonite, and enhance the quality of green iron ore pellets.

INTRODUCTION

Iron ore deposits are commonly classified as high, medium, or low-grade based on their iron (Fe) content. With the progressive depletion of high-grade iron ore resources, the exploitation and processing of lower grade iron ores have become increasingly important to sustain global iron and steel production (World Steel Association, 2025). These lower-grade resources are characterised by lower total Fe contents and higher proportions of gangue minerals. To improve the iron grade and make these ores suitable for steelmaking applications, lower-grade iron ores are subjected to beneficiation processes. Beneficiation involves comminution to liberate iron-bearing minerals from gangue phases such as silica and alumina, followed by separation techniques (gravity concentration and magnetic separation) to produce iron-rich and gangue-rich streams. The iron-rich concentrate obtained from beneficiation is generally used as pellet feed in pelletising operations.

Pellet feeds that meet the required physical and chemical specifications are blended with additives such as inorganic binders (eg bentonite and hydrated lime), organic binders (eg carboxymethyl cellulose, sodium lignosulfonate and starch), and fluxing agents (Kawatra and Claremboux, 2021).

The homogenised mixture is then processed in a pelletising unit (rotating disc or drum) to produce green iron ore pellets of controlled size. Additives are used in iron ore pelletising to improve the balling behaviour of the pellet feed and enhance the quality (drop strength number and compressive strength) of the green pellets. Bentonite is used as an additive and binder in iron ore pelletising due to its excellent binding characteristics. The influence of bentonite on the quality of green pellets arises from its ability to form cohesive bridges between adjacent iron ore particles, thereby enhancing interparticle bonding within the pellet structure (Kawatra and Claremboux, 2022).

In iron ore pelletising operations, the particle characteristics of pellet feed can be modified through high-pressure grinding roll (HPGR) processing to promote enhanced interparticle bonding during pellet formation (Elahidoosta *et al,* 2022). HPGR treatment affects key pellet feed properties, including particle morphology, particle size distribution, and specific surface area (Blaine index) (van der Meer, 2015). These pellet feed modifications play a critical role in the pelletising operation. Previous studies have demonstrated that an increase in Blaine index positively influences the quality of green pellets (Pal, Ghorai and Venugopalan, 2020; Abazarpoor *et al,* 2020). Consequently, the enhanced green pellet strength observed in HPGR-processed feeds is attributed to improved particle-to-particle bonding arising from increased specific surface area.

The drop strength number and compressive strength of green pellets are critical parameters for evaluating pellet quality. Green pellets are expected to withstand a minimum of four drops per pellet in standardised drop strength tests and to exhibit a compressive strength greater than 9 N per pellet (Kawatra and Claremboux, 2021). Meeting or exceeding these thresholds indicates that the pellets possess adequate quality to endure handling and subsequent processing without significant degradation. This study investigates the effects of binders and HPGR pretreatment on the drop strength number and compressive strength of green pellets produced from a magnetite concentrate.

MATERIALS AND METHODS

Materials

The iron ore pellet feed used in this study was a magnetite concentrate characterised by its good Fe and high FeO contents. The chemical analysis of the pellet feed is given in Table 1. Binder 1 (bentonite) was used as a reference binder to assess the effect of Binder 2 on the quality of the green pellets.

TABLE 1

Chemical analysis of the pellet feed.

Composition	Fe	FeO	SiO_2	Al_2O_3	CaO	MgO	LOI_{1000}
wt (%)	65.49	24.40	4.90	0.16	1.13	1.04	-0.98

Experimental design and sample preparation

Table 2 shows the pelletising experimental design used in this work and outlines the key variables (binder and sample pretreatment) investigated to evaluate their influence on green pellet quality.

TABLE 2
Pelletising experimental design and the variables investigated.

Cases	Samples	Binding agent dosage (%) Binder 1	Binding agent dosage (%) Binder 2	Pellet feed pretreatment	Comment
I	A	1.0	0	No HPGR	Effect of Binder 1 addition
	B	1.2	0	No HPGR	
	C	1.4	0	No HPGR	
	D	1.6	0	No HPGR	
	E	1.8	0	No HPGR	
II	F	0	0.075	No HPGR	Effect of Binder 2 addition
	G	0	0.10	No HPGR	
	H	0	0.15	No HPGR	
	I	0	0.20	No HPGR	
III	J	1.0	0.075	No HPGR	Combined addition of binders
	K	1.2	0.075	No HPGR	
	L	1.4	0.075	No HPGR	
IV	M	1.0	0	No HPGR	Combination of HPGR-processing with bentonite
	N	1.0	0	HPGR (1 Pass)	
	O	1.0	0	HPGR (2 Pass)	
	P	1.0	0	HPGR (3 Pass)	
V	Q	1.0	0.075	No HPGR	Combination of HPGR-processing with combined addition of binders
	R	1.0	0.075	HPGR (1 Pass)	
	S	1.0	0.075	HPGR (2 Pass)	
	T	1.0	0.075	HPGR (3 Pass)	

The baseline condition for this work consisted of pellet feed mixed with Binder 1 (bentonite) as outlined in Case I. To evaluate the effect of Binder 2, it was used as the sole binding agent (Case II) and combined with bentonite (Case III). In addition to binder variation, pellet feed was processed using HPGR. The effect of pellet feed pretreatment with binding agents was evaluated using bentonite alone (Case IV) and a combined binder system (Case V). HPGR (1 pass), HPGR (2 pass), and HPGR (3 pass) denote the number of passes through the HPGR, respectively.

HPGR pretreatment of pellet feed

HPGR pretreatment modifies the pellet feed by refining the particle size distribution, altering particle morphology, and increasing the specific surface area. The physical characteristics and SEM micrographs of both as-received and HPGR-processed pellet feeds are illustrated in Figures 1 and 2, respectively. Compared to the as-received feed, HPGR processing improves particle surface characteristics by generating microcracks from attrition and shearing of particle surfaces and increasing the proportion of fine particles (van der Meer, 2015). As a result, the HPGR-processed feeds exhibit higher Blaine index values, with the value increasing with each successive pass through the rolls.

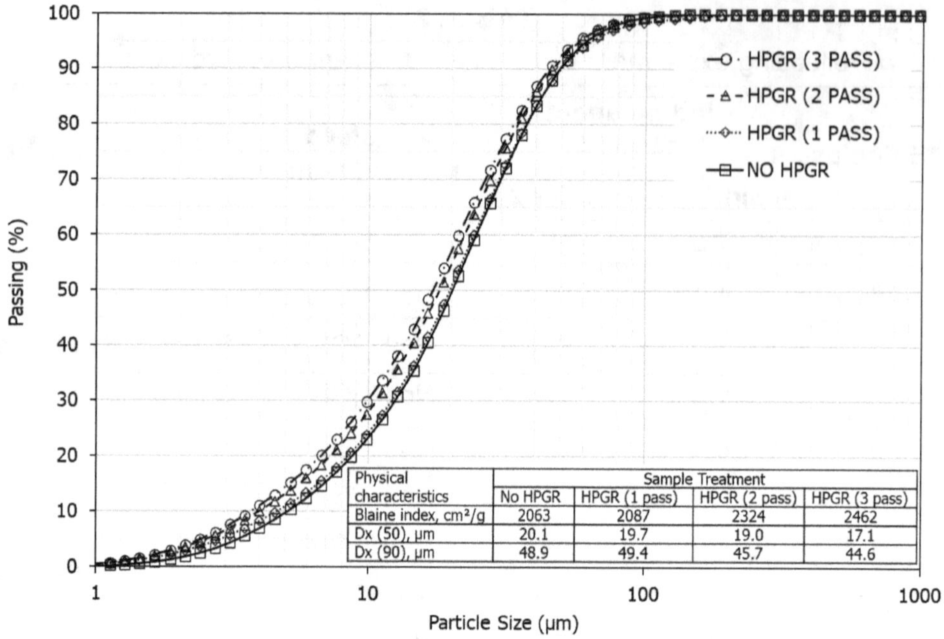

FIG 1 – The particle size distribution, size and Blaine index of the as-received and HPGR-processed pellet feeds.

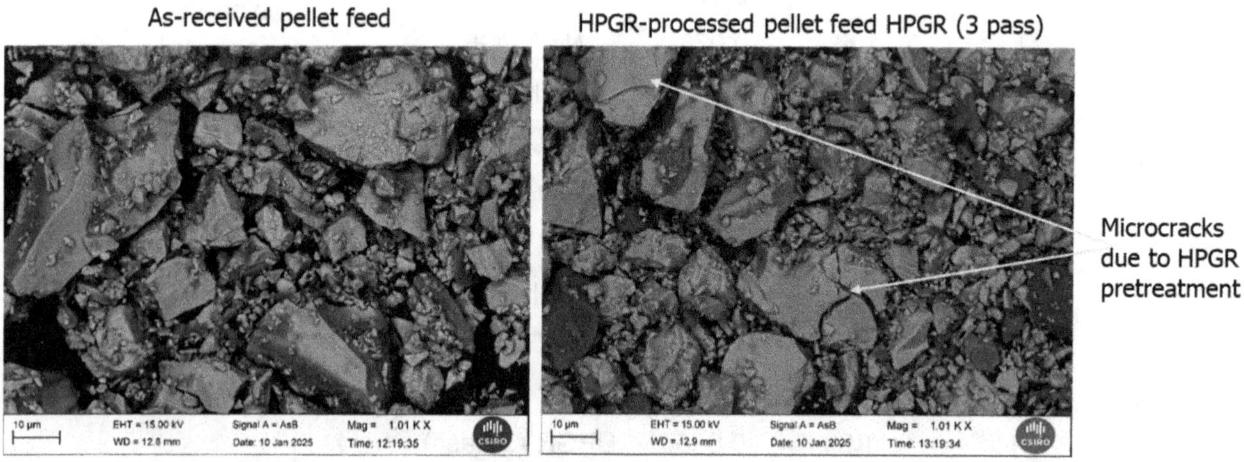

FIG 2 – SEM micrographs of as-received and HPGR-processed pellet feed.

Iron ore pelletising

Pellet feed and binding agents were thoroughly blended in an intensive mixer to achieve a homogeneous mixture and ensure uniform binder distribution throughout the material. The prepared mixtures were then fed into a laboratory-scale pelletising disc, with water added in a controlled manner to facilitate seeding and growth. The resulting green pellets were subsequently screened to a 10–12.5 mm size range and subjected to further characterisation.

Characterisation of pellet feed and green pellets

The particle size distribution of the as-received and pretreated pellet feeds was measured using Malvern Panalytical's Mastersizer 3000. The specific surface area of the pellet feeds, expressed as Blaine index, was determined using a Blaine apparatus. The quality of the green pellets was evaluated through drop strength and compressive strength. Drop strength was determined by repeatedly dropping the green pellets from a 46 cm height onto a steel plate until fracture or breakage occurred. Compressive strength measurements were performed using an Instron universal testing machine by compressing green pellets between steel plates at a deformation rate of 15 mm/min in compliance with iron ore pellets crushing method (ISO 4700:2015; International Organization for Standardization (ISO), 2015). The drop and compressive strengths were used to measure the resistance of the green pellets to impact and compressive loads, respectively, and provided an indication of their quality. The moisture content of the green pellets was determined by oven-drying the samples at 105°C for 24 hrs.

RESULTS AND DISCUSSION

Figure 3 provides an overview of the results obtained in this study. The results indicate that all green pellets produced exhibited compressive strengths exceeding 9 N per pellet, although most of the pellets recorded drop strength numbers below the standard minimum of four drops per pellet.

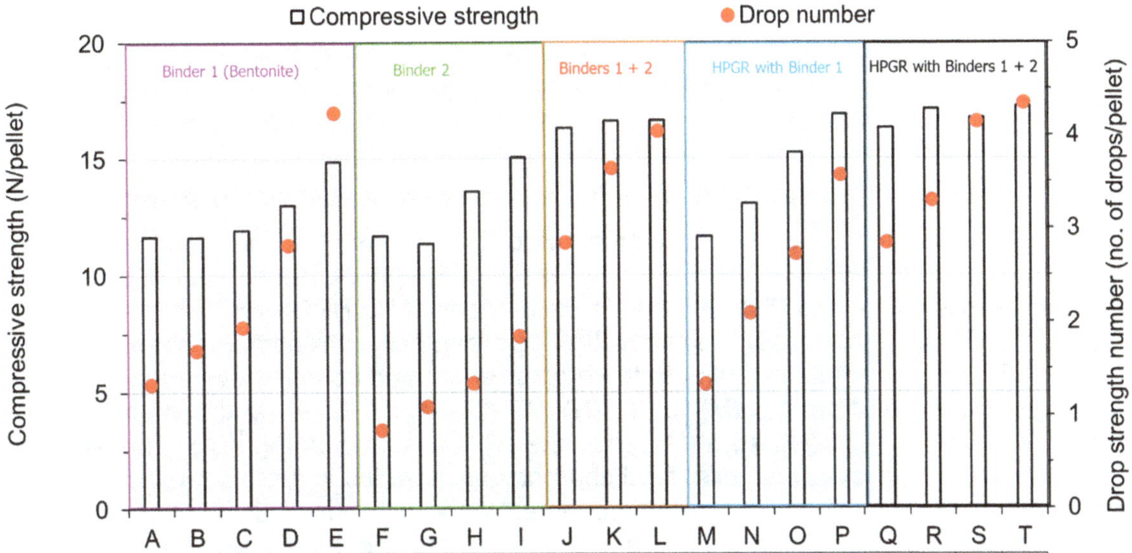

	A	B	C	D	E	F	G	H	I	J	K	L	M	N	O	P	Q	R	S	T
Bentonite	1	1.2	1.4	1.6	1.8	0	0	0	0	1	1.2	1.4	1	1	1	1	1	1	1	1
Commercial Binder	0	0	0	0	0	0.075	0.10	0.15	0.20	0.075	0.075	0.075	0	0	0	0	0.075	0.075	0.075	0.075

FIG 3 – Effect of binding agents and HPGR pretreatment on the quality of green pellets. A to T denotes sample conditions listed in Table 2.

Effect of bentonite addition on the quality of green pellets

The influence of bentonite addition on the quality of green pellets was evaluated using dosages ranging from 1.0 to 1.8 wt per cent (samples A to E), as shown in Figure 3. For bentonite dosages from 1.0 to 1.4 wt per cent (samples A to C), the compressive strength increased only marginally, with an obvious improvement observed at higher dosages of 1.6 wt per cent (sample D) and 1.8 wt per cent (sample E). The compressive strength values at all bentonite addition levels (1.0 to 1.8 wt per cent) exceeded the minimum requirement for industrial iron ore green pellets. The drop strength number of green pellets increased considerably with increasing bentonite content, from 1.3 drops per pellet at 1.0 wt per cent bentonite to 4.3 drops per pellet at 1.8 wt per cent. A bentonite dosage of 1.8 wt per cent was required to attain the drop strength number for industrial iron ore green pellets. Higher bentonite additions are known to reduce the total iron content of pellets through the introduction of gangue constituents, which may adversely affect their metallurgical performance (Zhu *et al*, 2015). Therefore, alternative solutions capable of reducing bentonite dosage while maintaining the acceptable green pellet quality were explored.

Effect of Binder 2 on the quality of green pellets

The effect of Binder 2 on the quality of green pellets was investigated in a dosage range from 0.075 to 0.20 wt per cent (samples F to I), as shown in Figure 3. The compressive strength values obtained at all dosages (0.075 to 0.20 wt per cent) exceeded the minimum standard for industrial iron ore green pellets. The drop strength number of green pellets increased with increasing binder dosage, from 0.9 drops per pellet at 0.075 wt per cent dosage to 1.9 drops per pellet at 0.20 wt per cent dosage. This improvement was not enough to meet the minimum industrial drop strength number requirement across the investigated dosage range. These observations suggest that while the binder contributes effectively to pellet cohesion under compressive loading, it is not effective in enhancing resistance to impact. Further investigation was undertaken to assess the combined effect of the binders on the quality of the green pellets.

Effect of combined addition of binding agents

The combined binder system, consisting of 0.075 wt per cent Binder 2 and 1.0 to 1.4 wt per cent bentonite, produced green pellets with higher drop strength number and compressive strength compared with those prepared using either binder individually at the same dosages, as shown in Figure 3. As observed for the individual binder additions (samples A to C for bentonite and samples F for Binder 2), the compressive strength of all pellets produced with the combined binder system (samples J, K and L) exceeded the minimum requirement for industrial iron ore green pellets. A drop strength number greater than four drops per pellet was achieved for sample L at Binders 1 and 2 dosages of 1.4 wt per cent and 0.075 wt per cent, respectively. Although the bentonite dosage for sample L (at 1.4 wt per cent) is less than that of sample F (at 1.8 wt per cent), a further reduction in bentonite dosage level was investigated by adopting a pellet feed pretreatment approach using the HPGR.

Effect of combined application of HPGR pretreatment with binding agents

For the HPGR-processed pellet feeds (samples M to P), with the bentonite dosage fixed at 1.0 wt per cent, both drop strength number and compressive strength increased with successive HPGR passes. Samples M, N, O, and P correspond to pellet feeds of No HPGR, HPGR (one pass), HPGR (two pass) and HPGR (three pass), respectively. The green pellet strength increased in the order M $< N < O <$ P. This trend is reflected in both the drop strength and compressive strength values, with drop strength increasing from 1.3 (M) to 2.1 (N), 2.7 (O), and 3.6 (P), and compressive strength rising from 11.7 (M) to 13.1 (N), 15.3 (O), and 16.9 (P), as shown in Figure 3. This suggests that HPGR pretreatment improves the pellet feed characteristics, which in turn enhances the qualities of the green pellets. Despite the observed improvements, the drop strength number obtained under these conditions remained below the minimum requirement for industrial green pellets. A substantial improvement in the drop strength number of the green pellets was achieved when HPGR pretreatment was combined with the addition of Binder 2 together with bentonite (samples Q to T). In this case, samples S and T attained a drop strength number greater than 4 drops per pellet. This improvement highlights the effect of HPGR pretreatment and optimised binder addition in achieving drop strength number, even at a reduced bentonite dosage of 1.0 wt per cent compared to 1.6 wt per cent in sample D and 1.8 wt per cent in sample E. The result of the work demonstrates the potential to reduce overall bentonite usage, minimising impurities associated with bentonite, while still meeting industry standards for green iron ore pellets.

Relationships between Blaine index and drop strength number

Figure 4 illustrates the relationship between the specific surface area of the pellet feed, expressed as the Blaine index, and the drop strength number of the green pellets.

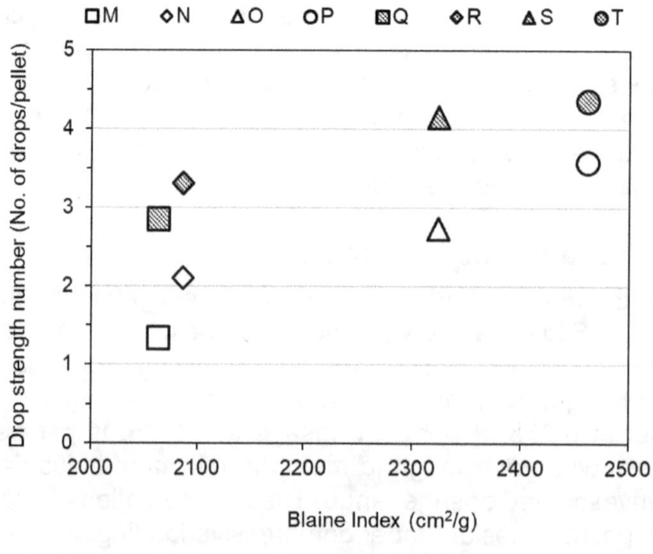

FIG 4 – Effect of Blaine Index on the drop strength number of green pellets.

An increase in the Blaine index, from 2063 cm^2/g for sample M and 2087 cm^2/g for sample N to 2324 cm^2/g for sample O and 2462 cm^2/g for sample P, led to an improvement in drop strength number of the green pellets from 1.3 (M) to 2.1 (N), 2.7 (O), and 3.6 (P). Furthermore, incorporating a small amount of Binder 2 (0.075 wt per cent) further increased the drop strength number of the green pellets in samples Q, R, S, and T compared to their corresponding samples M, N, O, and P, respectively. This trend is reflected in the values of 2.9 (Q), 3.3 (R), 4.2 (S), and 4.4 (T), which are higher than the corresponding values of 1.3 (M), 2.1 (N), 2.7 (O), and 3.6 (P), respectively. Samples S and T achieved a drop strength number greater than four drops per pellet. This enhancement can be attributed to the HPGR pretreatment, which promotes the generation of a finer pellet feed size distribution, as illustrated in Figure 1. At the microscopic scale, the HPGR-processed pellet feed exhibits microcracks that directly contribute to particle roughness. An improvement in the drop strength number of the resulting green pellets is in part attributable to enhanced particle–particle bonding within the green pellet structure. This observation is consistent with established findings in iron ore pelletising, where finer pellet feed provides increased specific surface area and more intimate interparticle contact, leading to stronger green pellets (Kawatra and Claremboux, 2021; Amiri and Abadi, 2022).

Implication of green pellet quality for pelletising operation

Iron ore green pellets are transferred from the pelletising disc to the induration furnace through a series of handling and transportation steps. During this process, the pellets are subjected to repeated mechanical stresses that include impact from multiple drops, pellet-to-pellet collisions, and compressive loading within the packed bed. These stresses can lead to deformation or breaking of weak pellets. Green pellets with adequate drop and compressive strengths are essential to withstand handling and transportation without excessive particle breakage. The quality of green pellets is a critical contributing factor to the quality of the fired pellet, such that an excessive pellet breakage negatively affects bed permeability and induration efficiency. Consequently, the quality of green pellets has a significant influence on induration performance, product yield, and the overall operational efficiency of the pelletising operation.

CONCLUSIONS

This study investigated the effects of bentonite addition, the type of binders, and HPGR pretreatment on the quality of green pellets produced from a magnetite concentrate. Compressive strength and drop strength number testing were used to assess the quality of the green pellets. The key findings are summarised as follows:

- All green pellets produced in this study showed compressive strengths greater than 9 N per pellet, indicating adequate resistance to static loading. However, many of the green pellets failed to achieve the minimum industrial requirement of four drops per pellet, highlighting limitations in impact resistance under certain conditions.
- The green pellets that met the minimum drop strength number of four drops per pellet were obtained under three specific conditions: (i) the use of bentonite at a dosage of 1.8 wt per cent, (ii) a combined binder system comprising 1.4 wt per cent bentonite and 0.075 wt per cent Binder 2, and (iii) the application of HPGR pretreatment in conjunction with a reduced binder system consisting of 1.0 wt per cent bentonite and 0.075 wt per cent Binder 2.
- The application of HPGR pretreatment and an optimised binder formulation demonstrates the potential to reduce total bentonite consumption, minimise impurities associated with bentonite, while still meeting industry standards for green iron ore pellets.

ACKNOWLEDGEMENTS

The authors would like to thank CSIRO Mineral Resources for permission to publish this paper and for the financial support of this work, and to acknowledge Dr Sarath Hapugoda for his valuable assistance with SEM imaging.

REFERENCES

Abazarpoor, A, Halali, M, Hejazi, R, Saghaeian, M and Zadeh, V S, 2020. Investigation of iron ore particle size and shape on green pellet quality, *Canadian Journal of Metallurgy and Materials Science,* 59:242–250.

Amiri, S H and Abadi, M I, 2022. Grinding iron ore concentrate by using HPGR and ball mills and their effects on pelletizing and reduction stages – a pilot-scale study, *Canadian Journal of Metallurgy and Materials Science,* 61:442–453.

Elahidoosta, H, Sheibania, S, Raygana, S, Hosseinib, L and Esmaeili, N, 2022. Mechanism of magnetite iron ore concentrate morphology affecting the pellet induration process, *Advanced Powder Technology,* 33:1–9.

International Organization for Standardization (ISO), 2015. ISO 4700:2015 Iron Ore Pellets for Blast Furnace and Direct Reduction Feedstocks – Determination of the Crushing Strength.

Kawatra, S K and Claremboux, V, 2021. Iron ore pelletization: Part I fundamentals, *Mineral Processing and Extractive Metallurgy Review,* 43:529–544.

Kawatra, S K and Claremboux, V, 2022. Iron Ore Pelletization: Part II, Inorganic Binders, *Mineral Processing and Extractive Metallurgy Review,* 43:813–832.

Pal, J, Ghorai, S and Venugopalan, T, 2020. Effect of High Blaine Iron Ore Fines in Hematite Ore Pelletization for Blast Furnace, *Mineral Processing and Extractive Metallurgy,* 129:299–307.

van der Meer, F P, 2015. Pellet feed grinding by HPGR, *Minerals Engineering,* 73:21–30.

World Steel Association, 2025. World Steel in Figures 2025 [online], World Steel Association. Available from: <https://worldsteel.org/data/world-steel-in-figures/world-steel-in-figures-2025> [Accessed: January 2026].

Zhu, D, Pan, J, Lu, L and Holmes, R J, 2015. Iron Ore Pelletization, in *Iron Ore: Mineralogy, Processing and Environmental Sustainability* (ed: L Lu), (Woodhead Publishing: Oxford).

Integration of the reflux classifier technology in Canadian iron ore operations

D Amariei[1] and S Gariépy[2]

1. Senior Metallurgist, Soutex, Québec QC G1N 4K8, Canada. Email: damariei@soutex.ca
2. Director of Growth, Soutex, Québec QC G1N 4K8, Canada. Email: sgariépy@soutex.ca

INTRODUCTION

Canadian iron ore is globally recognised for its high purity and its contribution to the production of green steel. The Labrador Trough represents the most important region for iron ore extraction in Canada. The Labrador Trough extends for about 1100 km south-east of Ungava Bay through both north-eastern Quebec and western Labrador. Further south, it turns south-west past the Wabush and Mont-Wright areas to within 300 km of the St. Lawrence River. The run-of-mine (ROM) ore in the Labrador Trough typically grades between 30 and 40 per cent Fe, requiring an extensive beneficiation process to produce high-grade concentrates. This process generally involves crushing, grinding, and physical separation, with spirals commonly employed to separate iron from its principal gangue mineral, quartz.

In 2010, the most common concentration flow sheet was composed of three stages of spirals arranged in a rougher–cleaner–recleaner configuration, in which the spiral concentrate is retreated twice, whereas the tailings of the cleaner and recleaner steps are recirculated to the rougher stage or grind feed. This configuration was present in the ArcelorMittal Mont-Wright concentrator, the Cliff–Bloom Lake concentrator (which in 2018 became Quebec Iron Ore) and partially in IOC's concentrator (Mills 12–14). The IOC concentrator also contains a circuit (Mill 11) where the milled feed is separated using hindered settlers into coarse and fines, followed by a two-stage spiral beneficiation in a rougher–cleaner configuration without tailings recirculation (Nevens, 2002).

The main challenge with these circuits is the loss of iron particles below 106 μm. The broad particle size distribution, the spiral separation mechanisms, as well as parameters such as wash water, are responsible for the loss of fine iron particles. This creates losses of valuable iron in spiral tailings and recirculating streams between roughly 150 and 45 μm. Against this backdrop, the Reflux Classifier (RC), originally developed for coal and industrial minerals, was identified as an intriguing but unproven option for Canadian iron ores.

The implementation did not follow a single deployment plan; instead, it progressed through a sequence of projects, each addressing specific questions that operators and engineers had about performance, scale, and robustness. The sections below describe this progression in the order it actually occurred.

CHAPTER ONE – PILOT RC™300 TRIALS AT COREM

The story begins in 2009 with COREM's pilot-scale evaluation of the RC300 unit (0.29 m × 0.29 m) on Canadian iron ore. Two streams from Labrador Trough operations were selected: a spiral tailings sample (A) and a screened spiral feed undersize at 106 μm (B), both containing hematite and magnetite as valuable minerals and quartz as gangue. These tests were deliberately targeted at the size range where spirals perform poorly namely 150–45 μm. The liberation of valuable and gangue minerals was pretty good at these sizes.

Under conditions of approximately 750 kg/h dry feed (1.5 t/(h*m^2)) at 37 per cent solids, the RC300 containing 6 mm inclined lamella channels produced, from spiral tailings grading 30 per cent Fe, concentrates with around 68–69 per cent Fe and 1–3 per cent SiO_2 at Fe recoveries exceeding 87 per cent, and tailings containing Fe grade below 7 per cent. For the screened spiral feed grading 42 per cent Fe, concentrate grades near 69–70 per cent Fe with about 1 per cent SiO_2 were achieved at Fe recoveries above 88 per cent, with tailings assaying below roughly 12 per cent Fe. Size-by-size analyses showed that fine fractions below 106 μm in the underflow approached theoretical Fe grades for pure hematite–magnetite mixtures and contained less than approximately 1 per cent SiO_2 (Amariei *et al*, 2013).

A comparison between the best RC tests and sink–float data showed that the RC matched the heavy-liquid 'ideal' separation for one sample and even out-performed it for the other. This early work did not yet implement the RC in a plant, but it provided the first quantitative evidence that Canadian ores could be treated successfully with the technology and justified taking the next steps.

CHAPTER TWO – MAKING THE RC USABLE FOR R&D (RC™100 SCALE-DOWN)

With pilot-scale feasibility established, attention shifted to how engineers could routinely test the RC on multiple ores and flow sheet options. The RC™300's capacity (approximately 1–3 t/h for iron ores) made it impractical as a standard laboratory/research tool, so a smaller RC™100 unit (0.10 m × 0.10 m) was acquired and evaluated using a screened spiral feed sample (B2) equivalent to the previous pilot sample (B1).

The scale-down was not straightforward. Differences in pressure-probe spacing, fluidisation system design and a relatively large underflow valve meant that simple geometric scaling did not preserve the same hydrodynamics. Initial tests using the smallest standard fluidisation perforations led an overall solids content during test work of roughly 18 per cent, shifting operation towards free settling, weakening the autogenous dense-medium effect and degrading density selectivity. The oversized ball-type underflow valve also caused recurrent collapse of the fluidised bed when opened beyond approximately 40 per cent.

Through several iterations, the team redesigned the fluidisation system with 16 perforations of 0.45 mm to restore water loading and superficial velocity consistent with RC™300 conditions, and adapted operating practices for the underflow valve. Once stabilised, the RC™100 delivered concentrates around 69–70 per cent Fe with about 3–4 per cent SiO_2, similar to the RC™300 concentrates.

Slightly lower weight yield and Fe recovery than the RC™300 (approximately 5–8 percentage points) were observed. As the sample evaluated with the RC™100 was from the same stream, but collected over different time intervals, sink–float tests were used to benchmark performance. The grade–recovery curves of the sink–float tests showed that samples B1 and B2 presented different metallurgical behaviour (sample B1 > sample B2). Comparison between the sink–float and RC results showed that the differences between the grade–recovery curves were very similar for both samples, therefore supporting the conclusion that the RC™100 gives the same results as the RC™300 under the conditions tested (Amariei, Lévesque and Hennessey, 2016).

Following these results, an extensive R&D program was initiated by Corem to understand the advantages and limitations of the Reflux Classifier technology and to guide Labrador Trough iron ore companies on its integration into their operations.

This chapter of the story made the RC practical as an R&D tool in Canada and gave metallurgists the confidence that lab-scale work could reliably support plant decisions.

CHAPTER THREE – PLANT-SCALE COMPARISON, PLANT IMPLEMENTATION AND INDUSTRIAL ADOPTION

The next turning point occurred when RC technology was taken into an operating plant and placed head-to-head against a more conventional option.

Iron Ore Company of Canada – Labrador City Operations

Iron Ore Company of Canada (IOC) is another iron ore producer that adopted RC technology. A pilot RC™2000 unit was installed on its Mill 11 circuit and tested against the existing spiral circuit. Tests conducted in 2018 have shown an increase in weight yield and iron recovery (RioTinto, 2018). The following phase consisted of demonstrating the full capabilities of the technology by installing two higher-capacity units (RC™3000). The success of this demonstration plant led to the transformation of the entire fine circuit with Reflux Classifiers.

The process of constructing, commissioning, and starting up the Reflux Classifier (RC) units was not without its difficulties. Achieving consistent metallurgical performance across the RC units proved to be a significant challenge. While testing one or two RC units provided valuable insights, scaling up to multiple units revealed a range of issues that had not been anticipated during the initial pilot phases. Maintaining uniform performance and ensuring reliable comparison between different units required careful attention and adaptation as new operational complexities emerged.

Despite these challenges, the installation of the Reflux Classifiers on Mill 11 led to a notable improvement in weight yield. As a result, a pilot plant was installed to assess the potential replacement of spiral concentrators with Reflux Classifiers in Mills 12–14 (Labrador Iron Ore Royalty Corporation, 2025).

This plant-scale comparison was the moment when RC technology transitioned from a promising candidate to a selected solution for fine-ore duties in a Canadian concentrator, closing the loop started by the early Corem trials.

Quebec Iron Ore – Bloom Lake Concentrator

At the Bloom Lake concentrator (property of Quebec Iron Ore), a pilot RC™300 and a pilot hydraulic classifier were installed at the scavenger-cleaner stage treating scavenger spiral concentrate, one of the finest and most challenging streams in the flow sheet. The plant team, supported by Soutex, modified a subset of spiral banks to feed the pilot units while keeping parallel banks as a control line.

Over several months, more than 300 sampling campaigns were executed, reconciled and normalised to allow fair comparison for given feed compositions and operating aims. The hydraulic classifier showed that it could produce low-silica concentrates in line with prior experience, but it required relatively high feed Fe grades (often >63 per cent Fe) and exhibited strong sensitivity to lithology, grind size and circuit disturbances, with Fe recoveries in the 45–80 per cent range when a ≤2.5 per cent SiO_2 product was targeted (Légaré *et al,* 2021).

The RC pilot, operating under similar feed conditions, delivered systematically higher Fe recoveries at any given product silica level and, crucially, maintained high recovery in the finest size classes. For low-silica concentrates, many RC campaigns achieved Fe recoveries between 80 and 99 per cent, and recovery remained strong even when the feed grade and lithology varied within normal plant ranges. Recovery–selectivity models developed from the pilot data and integrated into flow sheet simulations showed that including an RC at the scavenger-cleaner position gave the highest overall Fe recovery among the circuit options evaluated for both low- and standard-silica products.

This test work allowed to implement the Reflux Classifier technology at industrial scale. The phase 2 beneficiation circuit of Bloom Lake concentrator contains four Reflux Classifier units that process the concentrate produced by the spiral scavenger circuit (Allaire *et al,* 2023).

Kamistiatusset (Kami) Iron Ore Property

The Kamistiatusset Iron Ore project, property of Champion Iron Limited, is the most advanced development project in the Labrador trough and most probably the next new site to get in operation in the region. The Definitive Feasibility study is due for 2026.

The Kami project demonstrates that evaluating and integrating Reflux Classifier (RC) technology is now essential for new iron ore developments in Canada. Consequently, RCs were included in the beneficiation flow sheet for the project's prefeasibility study.

The optimised flow sheet for Kami replaces two conventional spiral stages—specifically, the cleaner and recleaner stages—with the RC for cleaning the rougher spiral concentrate. Test work conducted as part of the study demonstrated that the spiral rougher concentrate could be upgraded to a final concentrate containing less than 4.5 per cent SiO_2, while maintaining iron (Fe) recoveries exceeding 90 per cent (Allaire *et al,* 2024).

CONCLUSIONS

The implementation of Reflux Classifier technology in Canadian iron ore operations unfolded through successive, increasingly realistic steps: pilot trials demonstrating what was technically possible on Canadian fines, laboratory scale-down work that made the technology accessible for routine studies and clarified mineralogical behaviour, and finally plant-scale pilots that allowed direct comparison with incumbent equipment under real operating conditions.

Taken together, these steps form a coherent story of how RC technology moved from concept to practice in the Labrador Trough and explain why it is now considered a credible option for the iron ore recovery in Canadian concentrators.

These achievements have not only improved fine particle recovery and overall circuit performance but also paved the way for the global adoption of the technology in iron ore operations worldwide, including in Brazil, India, and Africa.

REFERENCES

Allaire, A, Dorval, A, Beaulieu, C, Girard, M, Paquette M-H, Millet, E, Farhanghi, S and Khoury, T, 2024. Pre-feasibility Study for the Kamistiatusset (Kami) Iron Ore Property, NI 43–101 Technical Report, Prepared for Champion Iron Limited. Available from: <https://www.championiron.com/wp-content/uploads/2024/03/2024-03-14-champion-iron-pre-feasibility-study-for-the-kamistiatusset-kami-iron-ore-property.pdf> [Accessed: 27 January 2026]

Allaire, A, Ouellet, B, Blanchet, V, Hamel, O, Ronald, E and Martin, J, 2023. Mineral Resources and Mineral Reserves for the Bloom Lake Mine, NI 43–101 Technical Report, Prepared for Quebec Iron Ore Inc. on behalf of Champion Iron Limited. Available from: <https://minedocs.com/25/Bloom-Lake-TR-01042023.pdf> [Accessed: 27 January 2026]

Amariei, D, Lévesque, S and Hennessey, C, 2016. Fine iron particle recovery using the Reflux Classifier: Assessment of the technology at pilot and lab-scale, Canadian Institute of Mining, Metallurgy and Petroleum, 13 p. Available from: <https://onemine.org/documents/fine-iron-particle-recovery-using-the-reflux-classifier-assessment-of-the-technology-at-pilot-and-lab-scale>

Amariei, D, Michaud, D, Paquet, G and Lindsay, M, 2013. The use of a Reflux Classifier for iron ores: Assessment of fine particles recovery at pilot scale, *Minerals Engineering*, (62):66–73.

Labrador Iron Ore Royalty Corporation, 2025. 2024 Results of Operations, Labrador Iron Ore Royalty Corporation,. Available from: <https://labradorironore.com/news/2024-results-of-operations-9982/> [Accessed: 27 January 2026]

Légaré, B, Lavoie, F, Bourassa, K and Buchard-Marchand, E, 2021. REFLUX vs Hydraulic classifier: A comparative analysis for hematite concentrate cleaning, in *Proceedings of the Iron Ore 2021 Conference,* 13 p (The Australasian Institute of Mining and Metallurgy: Melbourne).

Nevens, M, 2002. Concentrator upgrade program at the Iron Ore Company of Canada, Carol Lake Concentrator: concentrator testing program; concentrator flow sheet development, Canadian Institute of Mining, Metallurgy and Petroleum, 30 p. Available from: <https://onemine.org/documents/concentrator-upgrade-program-at-the-iron-ore-company-of-canada-carol-lake-concentrator-concentrator-upgrade-testing-program-concentrator-flowsheet-development>

RioTinto, 2018. *Mine to Port*, Fall 2018, IOC employee magazine. Available from: <https://www.ironore.ca/data/65-iocv5/ressources/documents/sys_docs/9502_ioc_mineauport_autumn_2018_v_final_eng.pdf> [Accessed: 27 January 2026]

Mineralogical drivers of the Blaine number at Grange Resources Savage River

J Apostol[1], D Olwagen[2], T Donald[3] and J Gerber[4]

1. Process Engineer, Hatch, Brisbane Qld 4000. Email: john.apostol@hatch.com
2. Quality and Process Improvement Manager, Grange Resources, Burnie Tas 7320. Email: dian.olwagen@grangeresources.com.au
3. Process Control Supervisor, Grange Resources, Burnie Tas 7320. Email: troy.donald@grangeresources.com.au
4. Mineral Process Engineer, Hatch, Brisbane Qld 4000. Email: johan.gerber@hatch.com

ABSTRACT

The Blaine number is a measure of specific surface area derived from a material's air permeability. It was originally developed for the cement and aggregates industry to quantify a powder's fineness which has implications for Portland cement's key characteristics. It has long been borrowed by the iron ore industry as this same measure of fineness has equally relevant implications for the filtering and sintering of iron concentrate during the pelletisation process. To this day, the industry utilises particle size (cumulative % passing 45 µm) as the prevailing factor that determines a concentrate's Blaine number and the effects of feed characteristics and mineralogy has been given little consideration.

This presents a particular challenge for mineral processing operations that commonly treat ore from several distinct sources but must deliver an iron concentrate of consistent and uniform Blaine. Grange Resources Savage River is one such operation that produces high-quality iron concentrate from two open pits (North and Centre Pit) of mineralogically distinct magnetite ores. This paper details the challenges of processing different feeds, particularly in achieving the target Blaine number, and the subsequent investigation that highlights the need for greater consideration for the effects of mineralogy on the final product's Blaine number.

INTRODUCTION

Grange Resources Savage River is an open pit magnetite mine situated in the north-west of Tasmania. It is among Australia's oldest operating iron ore mines, originally beginning production in 1967.

The Savage River Concentrator is comprised of two parallel Autogenous (AG) mill, ball mill, and pebble crushing (ABC) circuits. The primary circuit consist of the AG mill in closed circuit with pebble crushing followed by a single stage of rougher magnetic separation. Product from the primary circuit is then further treated in the secondary circuit comprising of a wet overflow ball mill in closed circuit with cyclones. Hydroseparators treat the cyclone overflow to reject low density silicious gangue prior to undergoing a final stage of upgrading in the finisher magnetic separators. Finisher screens at the back-end of the circuit classify the final concentrate to ensure no oversized material reaches the concentrate pipeline which delivers Savage River's final product 80 km north to the Port Latta pelletisation plant.

Besides maintenance works and modernisation of select equipment, the Savage River concentrator has remained largely unchanged from its original 1967 design. It continues to perform well, achieving excellent plant availability in the 92–96 per cent range while producing approximately 2.65 Mt/a of high-quality iron ore concentrate per annum. In recent years, Grange Resources have bolstered their efforts to drive process improvements in the concentrator and extend the Savage River mine life by up to an additional 10–15 years. Grange Resources first collaborated with Hatch through the 2023 Mine-to-Process (M2P) initiative to develop Savage River's first site-specific process models and debottleneck the processing plant. Collaboration continued through the 2024 Stage 2 project which aimed to identify further opportunities to increase the concentrator's production rates to 2.85 Mt/a of concentrate per annum. The Stage 2 project also supported process modelling to aid site in understanding the impact of transitioning to underground block caving on the concentrator's performance.

In parallel to all these initiatives, Hatch has been providing Savage River with on-site support, leveraging the process models and knowledge acquired from the M2P and Stage 2 projects to conduct further, more targeted investigations. One of the more recent challenges that Hatch has been assisting Savage River with is to do with the concentrator's product Blaine number. As of late, the concentrate Blaine has been highly variable, and operations has had limited success in controlling it. This paper details the investigations and findings into the Blaine number at Savage River.

HISTORY OF THE BLAINE NUMBER

The Blaine Number or Blaine Index is an indirect measure of a material's specific surface area or fineness, expressed in cm^2/g or m^2/kg, derived from its air-permeability as measured by the Blaine apparatus. It was first introduced in 1943 by the American National Standards Institute (ANSI) as a standardised quality-control metric for Portland cement as fineness has critical implications for the cements' strength development and setting behaviour (Ferraris, Peltz and Toman, 2018).

The Blaine method's ease and simplicity for measuring specific surface area contributed to its rapid adoption across several other industries, including the iron ore industry. The fineness of an iron ore concentrate has implications for the downstream pelletisation process ie filterability, pellet strength, sintering behaviour, furnace performance etc. For iron ores, and magnetite specifically, a final P_{80} grind size in the range of 45 µm is typically targeted as a proxy for the ideal Blaine. In general, P_{80}'s close to this range tends to result in Blaine numbers approximately 165–185 m^2/kg which is ideal for the pelletisation process.

BLAINE AT SAVAGE RIVER

Prior to 2023, when the plant was processing ore sourced exclusively from the mine's North Pit, concentrate Blaine remained relatively stable averaging approximately 169 m^2/kg. This was achieved by targeting a P_{80} grind size of 45 µm as a proxy for the Blaine number. However, as the mine transitioned towards processing Centre Pit ore from 2023 onwards, Blaine became increasingly variable and the average value decreased to 157 m^2/kg, despite the final grind size remaining relatively unchanged and no significant changes were made to the circuit or equipment. Consequently, the pelletisation process began encountering challenges with filter capacity and the resulting pellet strength (low Blaine values result in low pellet strength). Based on the conventional understanding of the index, operations began targeting a finer grind size with the Centre Pit ore in an attempt to increase the concentrate's Blaine number. This unexpectedly, however, did not result in an increase in Blaine. As the concentrator has transitioned to treating greater proportions of Centre Pit ore, the Blaine number has largely decreased and grown increasingly variable irrespective of the final grind size as shown on Figure 1. Furthermore, a t-test of the daily average Blaine number suggests that the change in Blaine across this period (pre- and post-2024) is significant as summarised by Table 1.

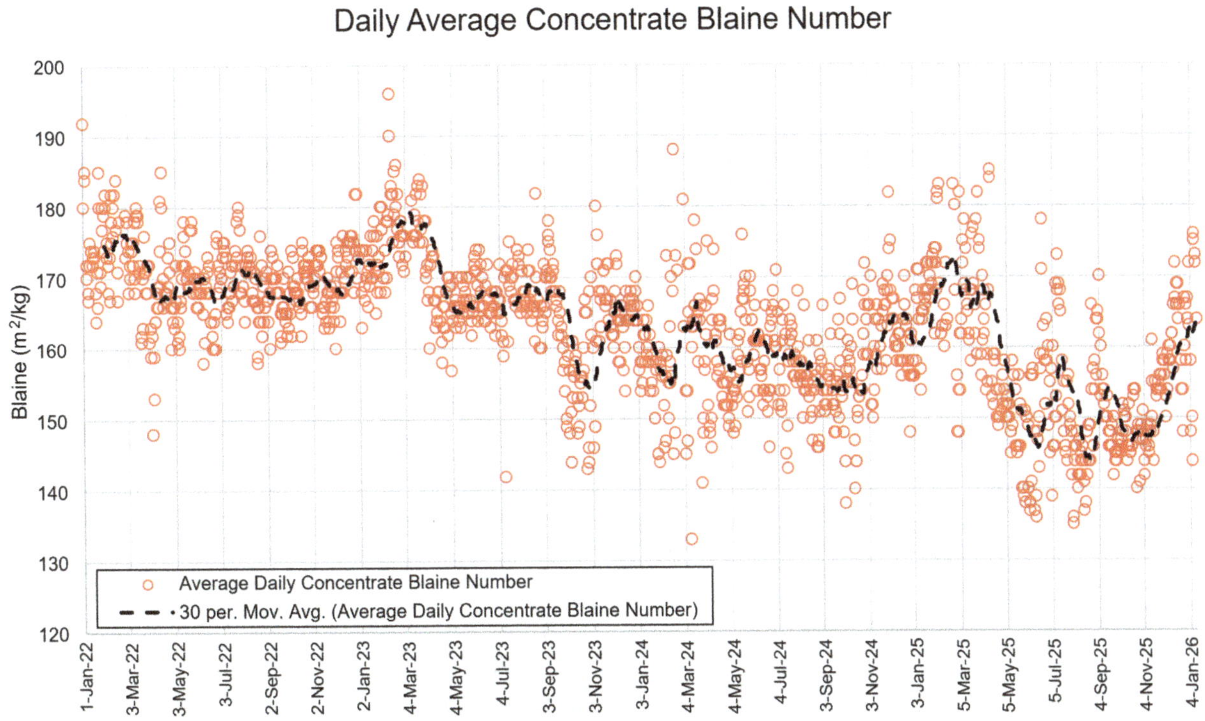

FIG 1 – Daily average concentrate Blaine number.

TABLE 1

t-test – Daily average concentrate Blaine number.

t-Test: Daily Average Concentrate Blaine

	Before Centre Pit Ore (2022 to 2023)	***Transition to Centre Pit Ore (2024 to Present)***
Mean	169	157
Variance	47	96
Number of Data Points	713	729
P(T<=t) two-tail	**8.3E-116**	**Difference in means is significant**

The historical understanding has always been that grind size is the critical parameter that drives the final Blaine number as this is the factor that has implications for the material's surface area. Little consideration has been given for the effects of mineralogy. These results however, prompted a deeper investigation into the factors affecting the Blaine number at Savage River. Production data from 2025 was analysed to evaluate the effects of the ore source (North versus Centre Pit ore) on the final concentrate Blaine number. While most of the feed during 2025 was sourced solely from the Centre Pit, material from the North Pit was still intermittently blended into the feed. The proportion of North Pit ore ranged mostly in the 20–30 w/w per cent range but reached >50 w/w per cent on a handful of days.

Grind size still has a notable impact on the resulting concentrate Blaine number. Although the correlation is weak ($R^2 < 0.3$), the overall trend is as expected, there is a positive correlation between Blaine and fineness ie as the per cent passing 45 µm increases, the Blaine number also increases. The presence of North Pit ore in the feed blend also seems to have a noticeable effect on the resulting concentrate Blaine number. Statistical analysis of the production data set suggests that the difference in average Blaine number between only Centre Pit ore and a feed blend of North and Centre Pit is significant as supported by Figure 2 and Table 2. This analysis seems to suggest that the final concentrate Blaine number is reliably higher when North Pit ore is present in the feed as opposed to treating solely Centre Pit material, regardless of final grind size (statistical analysis suggests that the difference in average grind size is not significant).

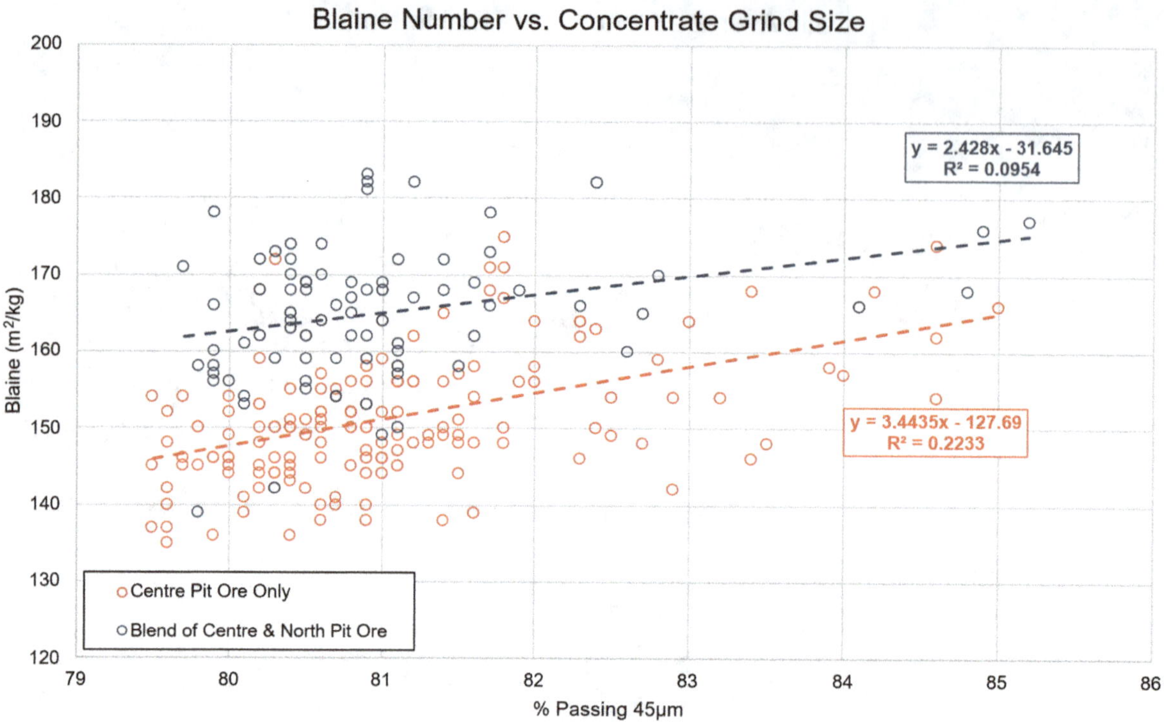

FIG 2 – Blaine number versus final concentrate grind size.

TABLE 2

t-test – Blaine and final grind size between centre pit only versus centre and north pit ore feed blend.

t-Test: Average Blaine and Final Grind Size (Only Centre Pite Ore vs. North & Centre Pit Ore Feed Blend)

	Only Centre Pit Ore	*Centre & North Pit Ore Blend*
Blaine		
Mean	154	165
Variance	134	74
Number of Data Points	178	82
P(T<=t) two-tail	**4.1E-16**	**Difference in means is significant**
Grind Size (% Passing 45 µm)		
Mean	81	81
Variance	1	1
Number of Data Points	166	82
P(T<=t) two-tail	**0.53**	**Difference in means is not significant**

As this is retrospective analysis of production data, the results were subject to a multitude of uncontrolled factors in the production environment. However, if operations were to continue with their current strategy of controlling Blaine with grind size, the data would indicate that the per cent passing 45 µm would have to be increased to the range of 85–86 per cent to meet site's ideal Blaine number while treating purely Centre Pit ore. With the current limitations around the comminution circuit, a final concentrate grind size this fine is simply not possible without adversely impacting throughputs. Therefore, the possible benefit of North Pit's mineralogy to Blaine was a worthwhile opportunity for operations to pursue.

Concentrate blending exercise

To better ascertain North Pit's mineralogical effects on the final concentrate Blaine in a better controlled environment, a simple exercise utilising two concentrate samples was completed (samples from 12/12/2019 and 17/10/2025 were chosen). These two samples represent the final concentrate from two days when the concentrator was treating purely North Pit material and purely Centre pit material respectively.

Five subsamples, comprising of different proportions (0, 25, 50, 75, and 100 w/w per cent) of North Pit concentrate was generated. The subsamples were then tested on the same Blaine apparatus available on-site. Additionally, the Blaine was also separately calculated by taking the weighted average of the two concentrate samples' Blaine number. Both these results are illustrated on Figure 3.

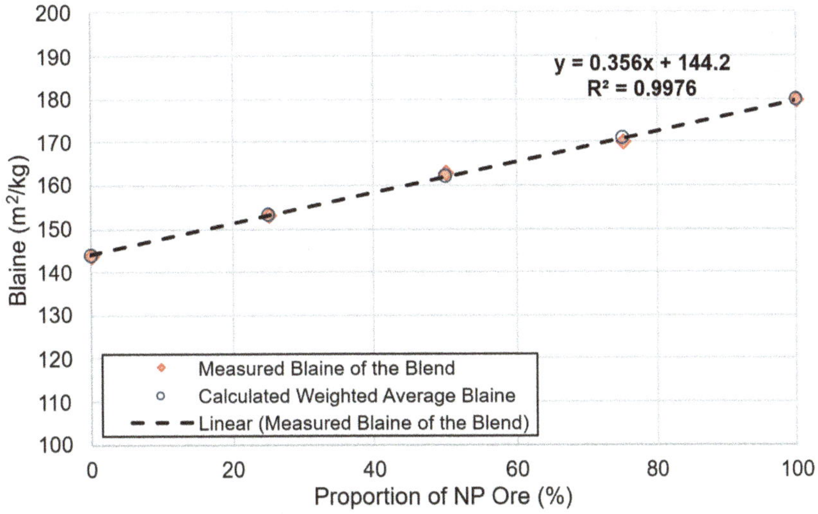

FIG 3 – Measured and calculated Blaine number for varying centre and north pit concentrate blends.

The Centre and North Pit concentrates had Blaine numbers of 144 and 180 m^2/kg respectively. It is important to note here that the North Pit sample from 2019 was finer than the Centre Pit sample from 2025 (83.0 versus 80.2 per cent cumulative per cent passing 45 µm). The Lab's long-term concentrate sample storage was searched for a North Pit sample of more similar grind size; however, no sample could be located that was from an appropriate period. Nevertheless, the analysis was continued as previous evaluations had already demonstrated that this difference in grind size is not enough to explain the significant difference in Blaine numbers.

As North Pit concentrate was blended in with the Centre Pit concentrate, the blended material's Blaine increased linearly in proportion to the amount of North Pit material in the blend. These results were somewhat expected, Blaine is a measure of a materials specific surface area so, blending in material with a higher specific surface area would be expected to also increase the combined blend's specific surface area. These results better suggest that North Pit concentrate has a unique mineralogical factor that yields better Blaine numbers compared to Centre Pit concentrate. Furthermore, the weighted average Blaine of the blended sample, calculated arithmetically, also closely aligns with the Blaine as measured by the Blaine apparatus. This seems to support the case that concentrate produced from the North Pit possesses something distinctly beneficial for Blaine that Centre Pit does not.

Published works exploring the relationship between mineralogy and Blaine is scarce, especially since mineralogy and feed characteristics are not as variable in the cement industry for which Blaine was originally developed for. Nasab and Sadeghi (2020) however, published some data that demonstrated that Blaine numbers for two different ore types could be drastically different despite having very similar particle size distributions. Those comparisons, however, were between hematite, magnetite, and clay materials. The results at Savage River further demonstrates that there may even be mineralogical distinctions within magnetite deposits that greatly impact the resulting Blaine. Based on the knowledge that the Blain apparatus is measuring air permeability, it is currently speculated that the likely factor contributing to the difference in Blain values is particle shape/grain shape. This directly impacts the materials surface area and packing characteristics which would have implications for the material's porosity and air permeability and thus the measured Blaine.

Blaine test work procedure

A final aspect worth noting is the Blaine measurement's sensitivity to sample handling. The sample preparation procedure explicitly states that the sample must be rolled twice prior to measurement in the Blaine apparatus. This step involves using a long cylindrical tool (similar to a kitchen rolling pin) to roll out the dry concentrate sample (like rolling out a bread dough). After being rolled once, the sample is to be mixed to form a pile and rolled once again for a second time. Failure to do so results in significantly different Blaine results.

To investigate the effects of rolling on the resulting Blaine in more detail, a single concentrate sample was used to generate 6 different subsamples. The subsamples differed only in the number of times they were rolled (0, 1, 2, 3, 4, and 5 times rolled) and all other factors were controlled (moisture, size distribution, Blaine apparatus etc). The result of this experiment is shown in Figure 4.

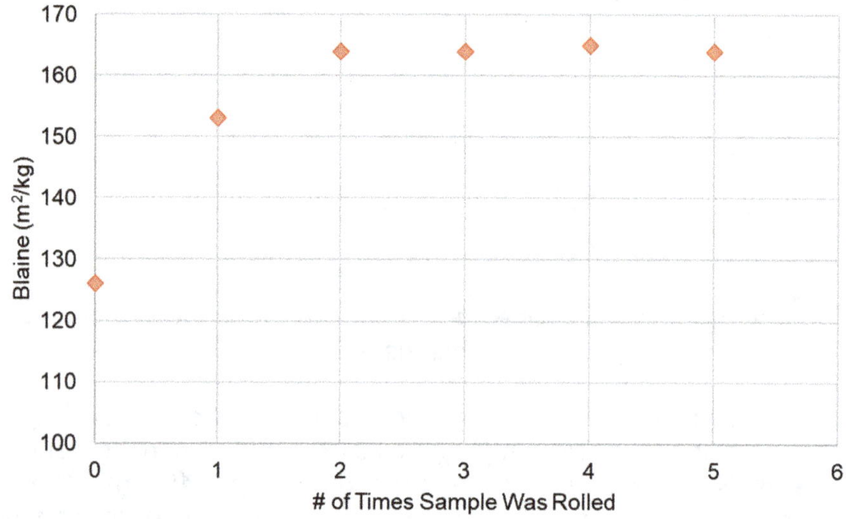

FIG 4 – Measured Blaine number versus the number of times the sample was rolled.

Not rolling the sample results in a very low Blaine number. The Blaine then progressively increases as the sample is rolled reaching its maximum value after being rolled twice. Additional rolling beyond this point seems to result in no further increase in Blaine.

It is not fully understood why rolling of the sample has such significant impacts on the resulting Blaine. It is very unlikely that the procedure is altering the sample's physicochemical characteristics to any capacity, so the large effect on Blaine is unexpected. Currently, it is speculated that the rolling process is aiding in deagglomeration and packing. It is possible that after magnetic separation and drying, the concentrate has agglomerated in such a way that is beneficial for the material's air permeability (hence the low Blaine values). Rolling of the concentrate may possibly be aiding in deagglomeration and aiding the material to become more tightly packed. This consequently results in a lower air permeability and therefore an increased Blaine number.

FUTURE WORKS

Further investigations, in the form of mineralogical test work is now being planned at Savage River. The knowledge and understanding from these investigations will aid operations in developing a more robust strategy for controlling Blaine and improving the quality and reliability of the concentrator's final product. A mineralogy-based blending strategy, for example, could assist operations in long-term planning and optimise the concentrator's throughput, concentrate quality, and Blaine number simultaneously. Furthermore, increased understanding of the relationship between Blaine and mineralogy may become increasingly critical as the Savage River mine plans to transition to an underground block cave operation in the near future. Accessing a new, deeper section of the deposit may yet introduce another factor that could potentially influence Blaine. As a starting point the following test works are planned:

- Mineralogical analysis of concentrate samples (XRD and QEMSCAN).

* Laboratory grinding and Blaine testing of concentrate samples.
* Mineralogical analysis, grinding, and Blaine testing of drill core samples for the underground ore.

CONCLUSION

Savage River has been producing high quality iron concentrate since 1967. In recent years, operations have increased their efforts to drive circuit-optimisation and process improvements to prepare the circuit for the mine's transition to underground block cave mining to further extend Savage River's life. As the open pits approach the end of their lives, Savage River has had to process a feed blend of increasingly variable ore characteristics which has adversely impacted the final product's Blaine number. Site has had limited success in implementing a control strategy for the Blaine while treating these new feed blends as the effects of mineralogical characteristics on the Blaine is not well understood. Initial investigations indicate that there may be mineralogical difference between the two ore sources (North and Centre Pit) that is driving the Blaine number. Savage River is now planning deeper investigations into these mineralogical differences to better understand the effects of mineralogy on the final Blaine of iron concentrate and to eventually implement an improved quality control strategy (mineralogy-based blending) for both the current and future underground operation.

ACKNOWLEDGEMENT

The authors from Hatch would like to acknowledge Grange Resources for the permission to publish this paper. Additionally, the authors would like to specifically thank the Savage River operational team whose support was critical for conducting these on-site investigations safely and effectively.

REFERENCES

Ferraris, C, Peltz, M and Toman, B, 2018. Certification of Standard Reference Material 46: Fineness of cement – Addition of Blaine and Particle Size Distribution, National Institute of Standards and Technology, US Department of Commerce.

Nasab, M and Sadeghi, M, 2020. Effect of particle size distribution and type of mineral on the Blaine number, *International Journal of Mining and Geo-Engineering*, 54:51–57.

Barite removal from goethitic ore – a sensor and magnetic approach

N Bottosso[1], H Turrer[2], L Senez[3], M Hoppe[4], J Greven[5] and K-H Becker[6]

1. Research Engineer, ArcelorMittal, Maizières-lès-Metz 57300, France. Email: nubia.bottosso@arcelormittal.com
2. Team Leader Mining Process, ArcelorMittal, Maizières-lès-Metz 57300, France. Email: henrique.turrer@arcelormittal.com
3. Research Engineer, ArcelorMittal, Maizières-lès-Metz 57300, France. Email: louis.senez@arcelormittal.com
4. Application Engineer, Steinert, Köln 50933, Germany. Email: meike.hoppe@steinert.de
5. Application Engineer, Steinert, Köln 50933, Germany. Email: jakob.greven@steinert.de
6. Application Engineer, Steinert, Köln 50933, Germany. Email: karl-heinz.becker@steinert.de

ABSTRACT

ArcelorMittal Maizières Research collaborated with Steinert to assess the feasibility of using sensor-based sorting and dry magnetic separation to beneficiate a Bosnian goethitic iron ore contaminated with barite. The goal was to reduce sulfur content – primarily from barite – to mitigate SOx emissions during sinter production for steelmaking.

An 1891 kg run-of-mine (ROM) sample was screened into size fractions: 40–100 mm, 20–40 mm, and 8–20 mm for sensor-based sorting, and <8 mm for dry magnetic separation, further subdivided into 3–8 mm, 1–3 mm, 0.5–1 mm, and <0.5 mm. No chemical characterisation was performed prior to testing; feed chemistry was back calculated from test results.

Sensor-based sorting employed four detection technologies — colour camera, 3D laser, induction, and XRT (X-ray transmission) — to identify optimal separation strategies. Results showed in Table 1 showed effective barite (and sulfur) removal from coarse fractions with minimal iron loss.

TABLE 1

Sensor-based sorting results summary.

Sample	Waste mass pull (%)	S grade (%)	S removal (%)	Fe recovery (%)
40–100 mm	21.9	2.76	88.5	80
20–40 mm	11.7	4.45	91.2	91.8
8–20 mm	9.1	2.85	82.9	92.7

Dry magnetic separation also proved effective for finer fractions, as depicted in Table 2.

TABLE 2

Dry magnetic separation results summary.

Sample	Waste mass pull (%)	S grade (%)	S removal (%)	Fe recovery (%)
3–8 mm	29.1	0.34	82.3	76.3
1–3 mm	26.7	0.19	58.1	78.3
0.5–1 mm	9.7	0.23	38.3	93.5
<0.5 mm	22.8	0.1	42.4	80.9

These findings suggest that sensor-based sorting at coarse sizes is an effective early-stage solution for barite removal, with dry magnetic separation enhancing sulfur rejection in finer fractions.

INTRODUCTION

Iron-based metallic materials have been used for millennia; first as natural iron metal and then extracted from iron ores (Muhamedagi and Oru, 2009). Ironmaking represents the first step in steelmaking. Steelmaking processes depend on different forms of iron as primary feed material. Traditionally, the main sources of iron for making steel were blast furnace hot metal and recycled steel in the form of scrap. Blast furnace ironmaking requires separate coke making and sintering plants (Ramakgala and Gwiranai, 2019).

Dust emissions are a problem in iron and steel production. Significant dust emissions come from sinter plants from processing raw materials all the way through to the finished product (Mohammad, Patra and Harichandan, 2023). In addition to the dust emission, iron ore sintering pollutants emissions (including SO_2, NO_x) are the primary source of polychlorinated *dibenzeno-p-dioxins* and *dibenzofurans* (PCDD/Fs) in iron and steel factories (Zhang, 2025).The SO_2 is one of the most important factors behind the formation of acid rain and haze, which seriously affects human health and ecological environment (Qie *et al,* 2017). Also, the presence of sulfur in iron ore has a detrimental effect on the smelting process reducing the quality of steel (Chen *et al*, 2025).

In this study, a greenfield project in the Ljubija region, in Bosnia, involving a goethitic iron ore contaminated by barite was targeted to be used in sintering process. Barite ($BaSO_4$), a sulfate of barium (Ba) is the major industrial source of barium, and the densest non-metallic mineral. $BaSO_4$ has the mass proportion of 66 per cent of BaO and 34 per cent of SO (Penaloza *et al*, 2023). The purpose was to reduce as much as possible the barite content, since the sulfur reaction with oxygen promotes the SO_2 formation, exceeding the established environmental thresholds as well as depleting de sinter quality. There was no specific target for the sulfur content: range reference of 0.02 per cent is acceptable while 0.2 per cent is considered high.

APPLICATION OF SENSOR-BASED SORTING

The most common beneficiation methods for barite comprise gravity separation, magnetic separation, and froth flotation after prior size reduction to increase mineral liberation (Ebunu *et al*, 2021). The goethite ores are generally low-grade types that require some beneficiation steps prior to be used in the iron and steel industry. Iron recovery technologies often employ flotation, physical separation, and magnetised roasting (Jaiswal, Soren and Jha, 2023). Considering the different physical properties of the minerals composing this ore, the use of sensor sorting based technology was investigated.

Optical Sorters using colour line-scan cameras have been around for over 20 years and used in various mineral applications as a pre-concentration method (Malhotra *et al,* 2009). Pre-concentration by sorting is seen as a method of improving the sustainability of mineral processing operations by reducing specific materials handling requirements, minimising energy consumption and water in grinding and concentration, and achieving more benign tailings disposal (Wills and Finch, 2016).

Wills and Finch (2016) also say that the key to successful operation lies in presenting a carefully selected and screened particle size range that ensures both adequate liberation and minimal fines, as excessive fine material can negatively impact availability. While the technical upper size limit is around 350 mm, liberation of barren waste often occurs below 100 mm. The lower size limit typically falls between 10 and 20 mm. Furthermore, the ratio of maximum to minimum particle size—known as the size range coefficient—should generally not exceed three.

There are two basic conditions for the efficient application of sensor-based sorting: first, the ore must be adapted to be sorted (correct size, for example) and, second, the particles must be singularly detected by the appropriate sensor (Li *et al*, 2026). To date, several different sensor technologies for sorting applications have been investigated: X-ray fluorescence, X-ray transmission, X-ray luminescence, radiometric, optical, near-infrared spectrometry, Laser (triangulation, induced breakdown spectroscopy, and induced fluorescence), visual spectrometry, and thermal infrared (Cardenas-Vera *et al*, 2022).

X-ray transmission (XRT) and near-infrared spectroscopy (NIR) are primarily applied for the pre-concentration of base metals and industrial minerals. XRT works by exploiting variations in atomic density among materials (Amar *et al*, 2023). These particles are distributed either on a chute or fast conveyor, and they are scanned and evaluated individually while passing. Hard X-ray radiation is emitted on one side of the material stream, and an image is recorded by a line scan detector on the other side (Robben, 2020). Because denser particles absorb more X-ray energy, they allow fewer rays to pass through, resulting in lower levels of transmitted X-rays detected. This contrast in absorption enables effective separation (Amar *et al*, 2023).

MATERIALS AND METHODS

The called 'ROM' sample of 1891.0 kg was screened in four size fractions (260.0 kg at -100 mm +40 mm; 206.0 kg at -40 mm +20 mm; 368.0 kg at -20 mm +8 mm; 1057.0 kg at -8 mm). Representative samples of each size fraction were taken to perform the SBS test and a sub-sample of <8 mm sample was screened in other four size fractions (-8 mm +3 mm; -3 mm +1 mm; -1 mm +0.5 mm and -0.5 mm). After blasting, This ROM sample only went through crushing in a double roll crusher before being shipped to testing.

This material was previously studied for other purposes and it was expected to have a head sample with around following grades: 40 per cent Fe, 1 per cent S, 10 per cent $BaSO_4$. Quartz, goethite, and barite are the dominant minerals with disparities in distribution. Goethite is concentrated in the coarse size fraction and quartz is present in the lower size fractions. It was also revealed that barite was uniformly distributed across all size ranges, posing a significant challenge for its separation. On the other hand, barite was liberated at least 80 per cent for all these size classes.

The target of the test work was to investigate the possibility of sorting barite from iron ore by sensor sorting using the coarser ranges (+40 mm -100 mm; +20 mm -40 mm; +8 mm -20 mm). And with the finer ranges (-8 mm +3 mm; -3 mm +1 mm; -1 mm +0.5 mm and -0.5 mm) to run exploratory dry magnetic separation tests to evaluate the separation from barite and goethite.

Sensor based sorting tests

The samples were scanned using a sensor sorter that collects information from four sensors: colour camera, 3D laser, induction and XRT. Based on the sensor data, the most promising technologies are selected and included into a test procedure.

To get the best understanding of the materials and sorter performance, multiple runs with different thresholds can be applied. The generated samples can then be dispatched for assaying. This procedure was used to find the best combination of settings. The sorting steps used during the test work does not represent the way the material would be sorted in an actual production. The findings of the trials will be combined in a single sorting program that could be used in a one-stage process.

As it is the best suited and most robust sensor for this test work the material was separated using the XRT sensor only. For this application, colour as well as laser sorting is a viable option. The advantages and disadvantages of laser and colour sorting in a mining application were discussed, where possible process options (washing, dust removal) were addressed. For a potential sorting using optical sensor, the material needs to be prepared upfront which includes more process steps and rising operating costs. In case if XRT alone is not sufficient to reach the process targets, additional steps with added optical sensors could be used to evaluate the potential. It is however required, to consider that optical systems required the right ambient conditions and are therefore more difficult to control than pure transmissive recognition.

The provided material was scanned separately on an industrial scale 1 m combination sensor sorter described as a KSS 100 XT CLI (Figure 1).

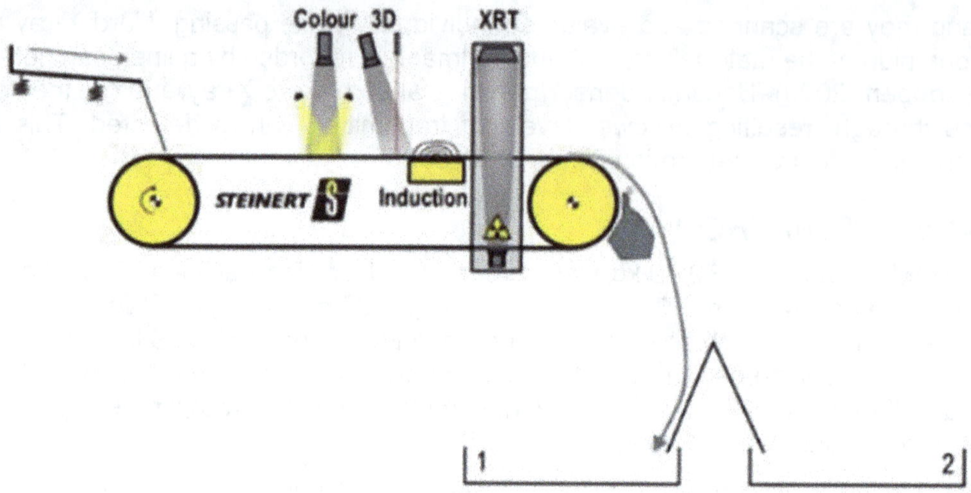

FIG 1 – Scanning of material through STEINERT's KSS XT CLI sorter.

The samples were scanned in the dynamic mode with a belt speed at 2.8 m/s, thus data acquisition is done at speeds equivalent to normal sorting operation. The data obtained during scanning was then uploaded onto STEINERT's proprietary software and analysed.

The selected sorting algorithm is based on the fraction 20–40 mm. This size fraction showed obvious trends in absorption tendencies which indicate a change in mineralogy. To get the best understanding of the material and sorter performance, three runs with different thresholds were applied. In this size fraction particularly a very high absorb fraction was outstanding compared to the rest of the material, this very absorbent material was likely to be the barite. Therefore, the first two sorting steps targeted the very high dense material. As a third step an additional proportion of the dense material was targeted. In one last step the lowest dense material was shot out.

The material was separated in a cascade test, where the drop after the first step is refed into the machine to separate another percentage of the potential waste material. This procedure is used to generate a grade-recovery curve and find the best combination of settings. The sorting steps used during the test work does not represent the way the material would be sorted in an actual production. Figure 2 depicts the test scheme.

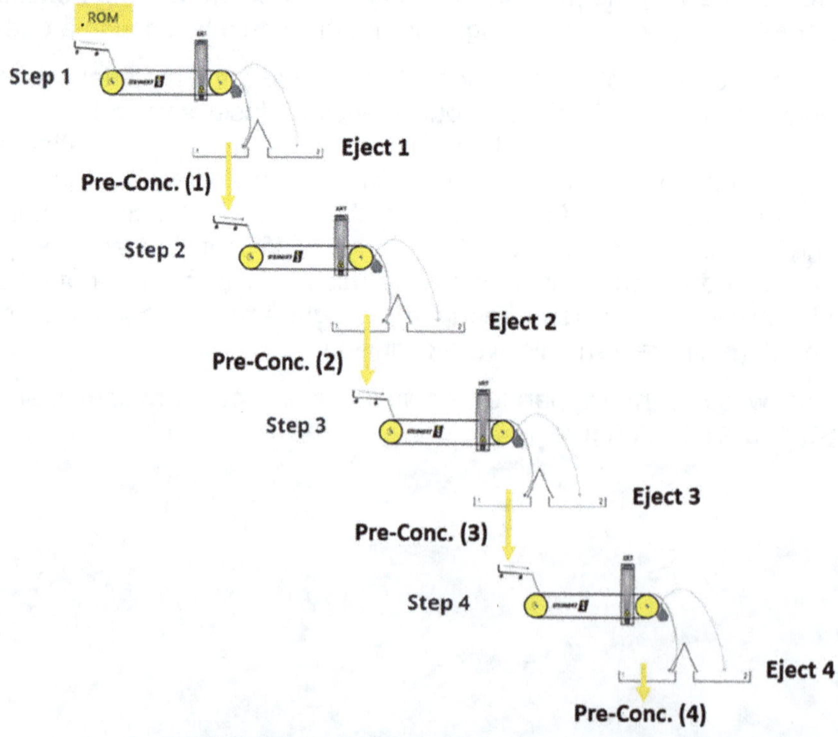

FIG 2 – Flow chart for particle sizes 20–40 mm.

For the larger size fraction +40 mm, the same test work approach was tested but with slightly different settings. Since the material had a poorer liberation – due to its size – more rocks consisted of both product and waste. Therefore, steps 1 and 2 for this grain size were combined into one, which means that only three steps were used for the +40 mm material, namely the dense material was ejected twice and – as for the 20–40 mm fraction – the lowest density was shot in the last step.

The situation was similar for the finest material 8–20 mm, but in comparison to the middle fraction, only the first two steps were carried out and dense material was ejected. The available data showed no trend for separating the material at low densities. Table 3 shows the sorting steps.

TABLE 3

Overview of the sorting steps.

Sample	Step	Target
+40 mm	1 and 2	Shoot high dense
+40 mm	3	Shoot low dense
20–40 mm	1, 2 and 3	Shoot high dense
20–40 mm	4	Shoot low dense
8–20 mm	1, 2 and 3	Shoot high dense

Table 4 refers to how the sensor-sorting results were interpreted. Various 'pre-concentrates' were back calculated to compare yield, metallurgical recovery, and sulfur grades if one or more eject steps were not included.

TABLE 4

Pre-concentrates scenarios calculation.

Sample	Ejects	Pre-concentrate name
+40 mm and 8–20 mm	1, 2, 3	Pre-conc (3)
+40 mm and 8–20 mm	1, 2	Pre-conc (2)
+40 mm and 8–20 mm	1	Pre-conc (1)
20–40 mm	1, 2, 3, 4	Pre-conc (4)
20–40 mm	1, 2, 3	Pre-conc (3)
20–40 mm	1, 2	Pre-conc (2)
20–40 mm	1	Pre-conc (1)

Due the cascade testing of the material, generated fines that would normally not be in the process were accumulated in the final drop of each trial. Therefore, those fractions were screened after the test at 10 mm to assay the fines separately. These fines were counted as a loss. Depending on the material an enrichment of product or waste material could occur in the fines.

Dry magnetic separation tests

Each grain size was sorted separately on Steinert's high-grade magnetic separators HGS and MSB-HG into a magnetic and a non-magnetic fraction. MSB-HG works on the same principle but is better suited for a coarser grain size.

RESULTS

Sensor based sorting tests

Since the sulfur bearing mineral was mainly barite, it was considered that all sulfur was attached to this mineral, as well as all iron linked to goethite. Figure 3 shows the screening results. The mass split that feeds the sensor sorter machine represents roughly 40 per cent of the head sample, on the other hand, it accounted for almost 80 per cent of the sulfur distribution, being the half (39 per cent of this distribution) on the coarser size range (+40 mm). The screening also segregated naturally the iron bearing particles, being the coarser ranges richer in iron contents. The 'Drop fines' in the plot is representing the sum of mass loss after each step in the sensor sorter.

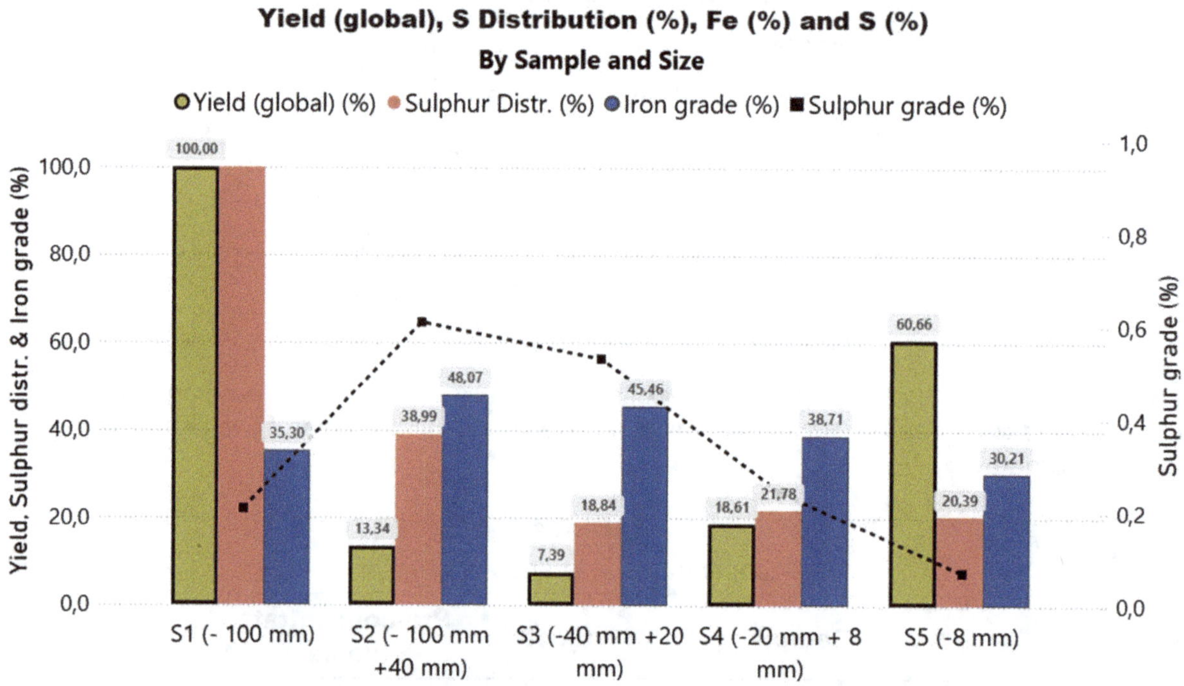

FIG 3 – Sensor based sorting screened sample.

Note that the plots (Figures 4, 7 and 8) consider yield, iron, and sulfur distributions globally (counting with the screening mass split), while the tables refer to the stage.

Figure 4 shows the mass, iron and sulfur splits and the sulfur grade by sample. These 'pre-concentrates' were estimations of quality if there were only one, two or more ejection steps. It is possible to see that only with one step (‹Eject 1›, Figure 5) it was possible to eliminate almost 88 per cent of the contained sulfur of the sample (31 per cent of the sulfur distribution), with a sulfur grade of 2.76 per cent. The final drop (Pre-conc (3), in Figure 6) presented an iron and sulfur contents of 51.4 per cent and 0.05 per cent, respectively, as indicated in Table 5.

If pre-concentrate 1 is considered, there was 21.9 per cent waste mass pull, at 2.76 per cent Sulfur grade, 88.5 per cent of Sulfur removed at 80.0 per cent iron recovery.

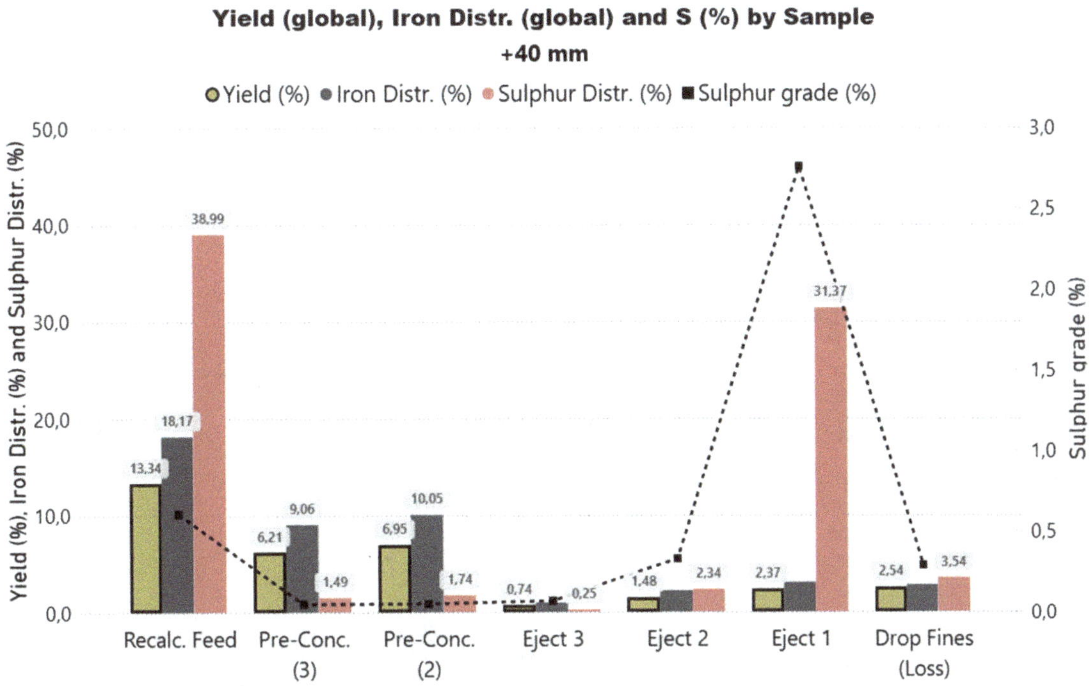

FIG 4 – +40 mm sensor based sorting results.

FIG 5 – Eject 1 from +40 mm sample.

FIG 6 – Final drop (pre-concentrate) from +40 mm sample.

TABLE 5

+ 40 mm fraction results.

Input size (mm)	Fraction	Net weight (kg)	Mass pull (%)	Fe grade (%)	S grade (%)	Fe Recovery (%)	S Recovery (%)
+40	Eject 1	40.0	21.9	45.8	2.76	20.0	88.5
+40	Eject 2	25.0	13.7	53.1	0.33	14.5	6.6
+40	Eject 3	12.5	6.8	47.3	0.07	6.5	0.7
+40	Drop	105.0	57.5	51.4	0.05	59.0	4.2
	Total	**182.5**	**100**	**50.2**	**0.68**	**100.0**	**100.0**
0–10	Screen	43.0		39.2	0.29		
	+ Fines	**225.5**		**48.1**	**0.61**		

The ‹+20 mm -40 mm› sample results are shown in Figure 7 and in Table 6. Differently from the +40 mm sample, this size range also significantly eliminated the barite in the second eject (5.42 per cent of the sulfur distribution, while that in the first eject around 10 per cent of the sulfur distribution was pulled out). The final drop presented an iron content of 48.8 per cent Fe and 0.03 per cent sulfur.

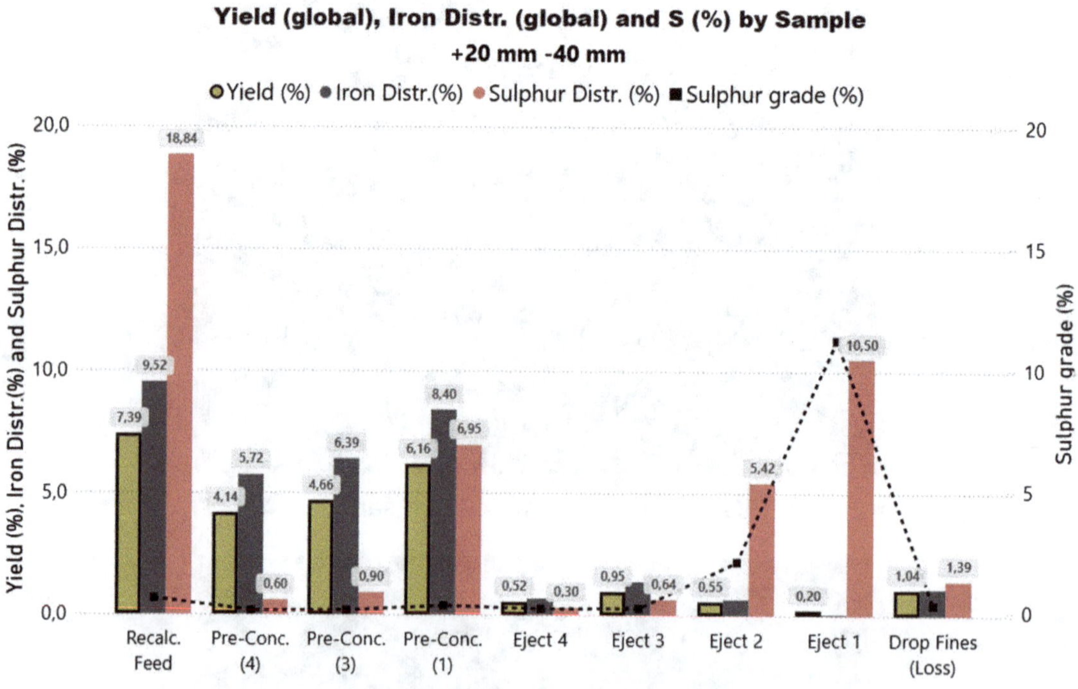

FIG 7 – +20 -40 mm sensor based sorting results.

TABLE 6
+20 mm -40 mm fraction results.

Input size (mm)	Fraction	Net weight (kg)	Mass pull (%)	Fe grade (%)	S grade (%)	Fe Recovery (%)	S Recovery (%)
20–40	Eject 1	3.30	3.1	8.6	11.20	0.6	60.2
20–40	Eject 2	9.30	8.7	41.4	2.05	7.6	31.0
20–40	Eject 3	16.00	14.9	50.6	0.14	16.1	3.6
20–40	Eject 4	8.80	8.2	45.5	0.12	7.9	1.7
20–40	Drop	70.00	65.2	48.8	0.03	67.8	3.4
	Total	**107.40**	**100.0**	**46.9**	**0.57**	**100.0**	**100.0**
0–10 mm	Screen	17.50		36.7	0.280		
	+ Fines	**124.90**		**45.46**	**0.557**		

If pre-concentrate 2 is considered, there was 11.8 per cent waste mass pull, at 4.45 per cent sulfur grade, 91.2 per cent of sulfur removed at 91.8 per cent iron recovery.

The ‹+8 mm -20 mm› sample results are shown in Figure 8 and in Table 7. Like the ‹+20 mm -40 mm› sample, most part of the sulfur was eliminated with two steps (sum of eject 1 and 2 represents around 15 per cent of sulfur distribution). However, the final drop (Pre-conc (3)) presented an iron ore content lower than the last eject (43.9 per cent and 45.6 per cent, respectively). Indicating that at this size range, the separation efficiency lowers.

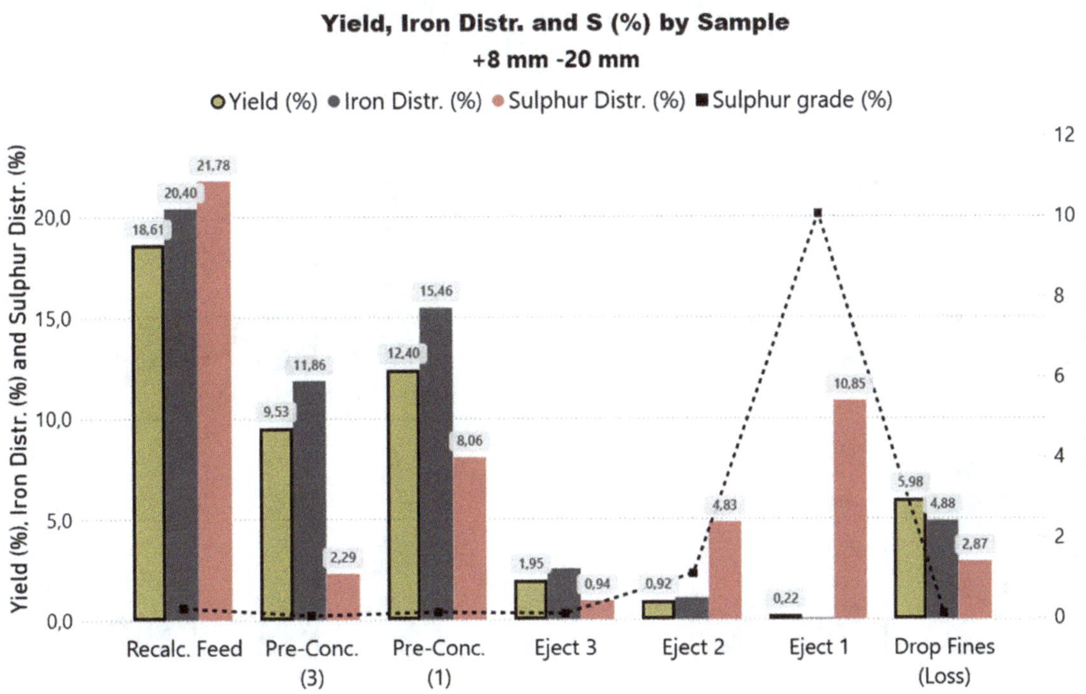

FIG 8 – +8 mm -20 mm sensor based sorting results.

TABLE 7

+8 mm -20 mm fraction results.

Input size (mm)	Fraction	Net weight (kg)	Mass pull (%)	Fe grade (%)	S grade (%)	Fe Recovery (%)	S Recovery (%)
8–20	Eject 1	3.80	1.8	10.1	10.05	0.4	57.4
8–20	Eject 2	15.60	7.3	41.1	1.09	6.9	25.6
8–20	Eject 3	33.00	15.5	45.6	0.10	16.3	5.0
8–20	Drop	161.00	75.4	43.9	0.05	76.4	12.1
	Total	**213.40**	**100.0**	**43.4**	**0.31**	**100.0**	**100.0**
0–10 mm	Screen	101.00	28.82	0.1			
	+ Fines	**314.40**	**38.71**	**0.2**			

If pre-concentrate 2 is considered, 9.1 per cent waste mass pull, at 2.85 per cent sulfur grade, 82.9 per cent of sulfur removed at 92.7 per cent iron recovery.

Dry magnetic separation tests

Coarser particles respond better to magnetic separation when analysing the sulfur removal (Figure 9), with maximum 82.3 per cent sulfur rejection in size range between 3 mm and 8 mm and minimum of 38.3 per cent in the size range +0.5 mm and -1 mm. The overall sulfur contents of the magnetic portions were maximum 0.05 per cent, as seen in Table 8.

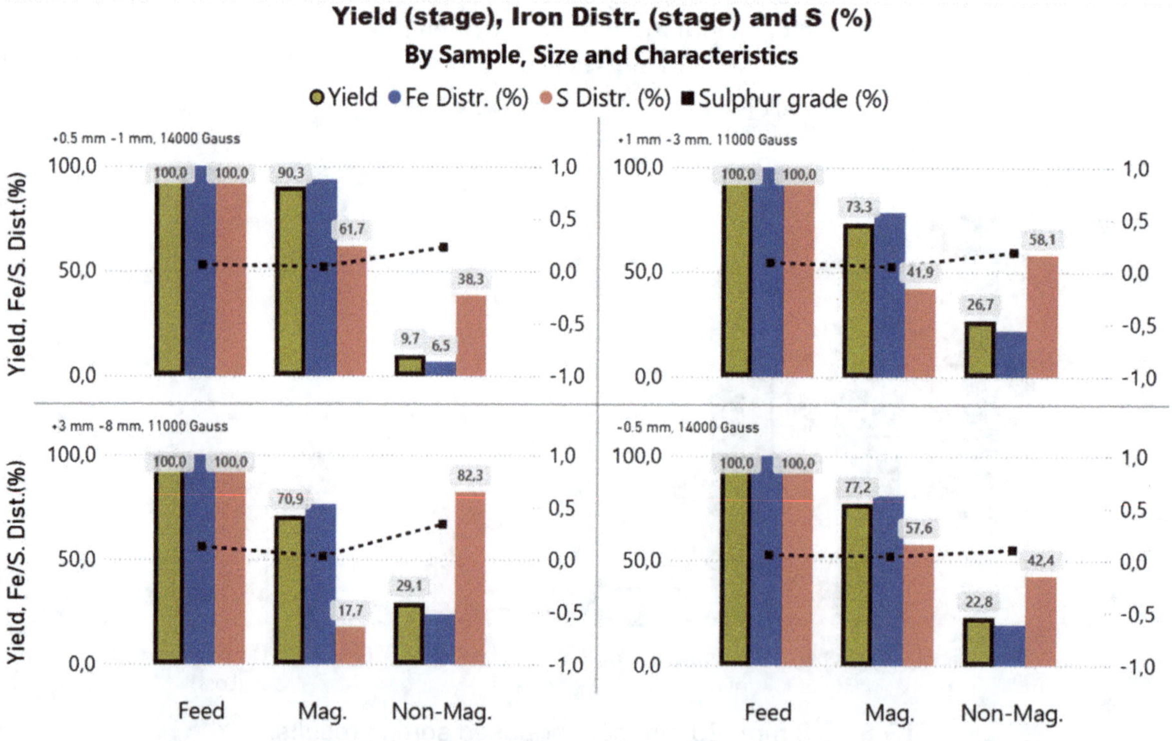

FIG 9 – Magnetic separation results.

TABLE 8

Dry Magnetic Separation results.

Input size (mm)	Fraction	Yield (%)	Fe grade (%)	S grade (%)	Fe recovery (%)	S recovery (%)
3–8	Non-mag	29.1	28.3	0.34	23.7	82.3
3–8	Mag	70.9	37.6	0.03	76.3	17.7
3–8	**Total**	**100.0**	**34.9**	**0.12**	**100.0**	**100.0**
1–3	Non-mag	26.7	25.5	0.19	21.7	58.1
1–3	Mag	73.3	33.5	0.05	78.3	41.9
1–3	**Total**	**100.0**	**31.4**	**0.09**	**100.0**	**100.0**
0.5–1	Non-mag	9.7	20.6	0.23	6.5	38.3
0.5–1	Mag	90.3	32.1	0.04	93.5	61.7
0.5–1	**Total**	**100.0**	**30.1**	**0.06**	**100.0**	**100.0**
<0.5	Non-mag	22.8	24.9	0.10	19.1	42.4
<0.5	Mag	77.2	31.1	0.04	80.9	57.6
<0.5	**Total**	**100.0**	**29.7**	**0.05**	**100.0**	**100.0**

CONCLUSIONS

The test-work conducted across all delivered grain-size fractions demonstrated that sulfur reduction in the run-of-mine ore can be achieved effectively through the applied separation technologies. X-ray transmission (XRT) sorting was identified as the most robust and technically appropriate sensor-based method for the material characteristics observed.

Alternative optical systems, such as colour-based and laser-based sorting, were examined conceptually; however, their successful application under mining conditions generally requires additional material-conditioning steps (eg washing or dust-removal), which increase process complexity and operational cost.

When considered together with the performance of magnetic separation, the results indicate that the selected combination of XRT sensor-based sorting and magnetic processing represents a technically viable and environmentally advantageous approach for sulfur removal under the tested conditions. Following results were delivered:

- Barite, and thus sulfur, can be effectively removed from coarser feed grain sizes at low yield ratios and minor iron losses by separation of high dense fractions using sensor-based sorting and XRT technology:
  - +40 mm: 21.9 per cent waste mass pull, at 2.76 per cent sulfur grade, 88.5 per cent of sulfur removed at 80.0 per cent iron recovery.
  - +20–40 mm: 11.7 per cent waste mass pull, at 4.45 per cent sulfur grade, 91.2 per cent of sulfur removed at 91.8 per cent iron recovery.
  - +8–20 mm: 9.1 per cent waste mass pull, at 2.85 per cent sulfur grade, 82.9 per cent of sulfur removed at 92.7 per cent iron recovery.
- At finer grain sizes, magnetic separation is also effective (at coarser grain sizes specifically) and allows to remove barite, and thus sulfur with non-magnetic fractions:
  - +3–8 mm: 29.1 per cent non-mag mass pull, at 0.34 per cent sulfur grade, 82.3 per cent of sulfur removed at 76.3 per cent iron recovery.

- +1–3 mm: 26.7 per cent non-mag mass pull, at 0.19 per cent sulfur grade, 58.1 per cent of sulfur removed at 78.3 per cent iron recovery.
- +0.5–1 mm: 9.7 per cent non-mag mass pull, at 0.23 per cent sulfur grade, 38.3 per cent of sulfur removed at 93.5 per cent iron recovery.
- -0.5 mm: 22.8 per cent non-mag mass pull, at 0.1 per cent sulfur grade, 42.4 per cent of sulfur removed at 80.9 per cent iron recovery.

ACKNOWLEDGEMENTS

ArcelorMittal R&D team appreciates the collaborative work with STEINERT and thanks each person involved in this project.

REFERENCES

Amar, H, Benzaazoua, M, Elghali, A, Taha, Y, El Ghorfi, M, Krause, A and Hakkou, R, 2023. Mine waste rock reprocessing using sensor-based sorting (SBS): novel approach toward circular economy in phosphate mining, *Minerals Engineering*, 204:108415.

Cardenas-Vera, A, Hesse, M, Möckel, R, Gerhard Merker, R, Heinig, T and Van Phan, Q, 2022. Investigation of sensor-based sorting and selective comminution for pre-concentration of an unusual parisite-rich REE ore, South Namxe, Vietnam, *Minerals Engineering*, 177:107371.

Chen, X, Ren, Y, Gao, W and Zhao, N, 2025. Study of the mechanism of high-efficient in-situ SO_2 fixation during oxidative roasting of high-sulphur iron ores and DFT calculation, *Fuel Processing Technology*, 267:108171.

Ebunu, A, Olanrewaju, Y, Ogolo, O, Adetunji, A and Onwualu, A, 2021. Barite as an industrial mineral in Nigeria: occurrence, utilization, challenges and future prospects, *Heliyon*, 7(6):e07365.

Jaiswal, R, Soren, S and Jha, G, 2023. A novel approach of utilising waste biomass in the magnetizing roasting for recovery of iron from goethitic iron ore, *Environmental Technology and Innovation*, 31:103184.

Li, S, Li, L, Zhang, H, Li, N, Wang, J, Mi, H, Liu, F, Liu, Z and Klein, B, 2026. Feasibility assessment of integrated optical and X-ray transmission (XRT) sorting for quartz-vein tungsten ores, *Minerals Engineering*, 205:109858.

Malhotra, D, Taylor, P, Spiller, E and LeVier, M (eds), 2009. 47.5 Sensors and measurement principles used in the mineral industry for sorting, in *Recent Advances in Mineral Processing Plant Design*, pp 476–489 (SME: Englewood).

Mohammad, S, Patra, S and Harichandan, B, 2023. Reductants in iron ore sintering: a critical review, *Fuel*, 332:126194.

Muhamedagi, S and Oru, M, 2009. Historical survey of iron and steel production in Bosnia and Herzegovina, *Materiali in Tehnologije*, 43(4):223–229.

Penaloza, I, Tita, A, McNew, E and Chu, P, 2023. Barite resources, production and recovery using froth flotation: a review, *Minerals Engineering*, 203:108327.

Qie, J, Zhang, C, Li, X, Guo, Y, Wang, H and Wu, S, 2017. Lower SO_2 emissions in the sintering process utilising the difference of sulphur contents of iron ores, *ISIJ International*, 57(12):2154–2161.

Ramakgala, C and Gwiranai, D, 2019. A review of ironmaking by direct reduction processes: quality requirements and sustainability, *Procedia Manufacturing*, 35:1083–1089.

Robben, C, 2020. X-ray transmission-based ore sorting at the San Rafael tin mine, *Minerals Engineering*, 146:106130.

Wills, B and Finch, J, 2016. Sensor-based ore sorting, in *Wills' Mineral Processing Technology*, 8th ed, pp 409–416 (Oxford: Butterworth-Heinemann).

Zhang, Y, 2025. Emission characteristics and removal efficiencies of SO2, NOx, particulate matter and dioxins in iron ore sintering after ultra-low emission transformation, *Atmospheric Pollution Research,* 16(2):101004.

From shell to schedule – evaluating the impact of direct block scheduling on iron ore mine planning

J Chung[1], B Maziarz[2] and D Rahal[3]

1. Global Technical Advisor – Optimization, Deswik, Perth WA 6000.
   Email: joyce.chung@deswik.com
2. Territory Manager – Western Australia, Deswik, Perth WA 6000.
   Email: ben.maziarz@deswik.com
3. Product Manager – Optimization and Caving, Deswik, Brisbane Qld 4000.
   Email: david.rahal@deswik.com

ABSTRACT

This case study evaluates how the choice of pit optimisation methodology—revenue factor shells versus direct block scheduling—influences strategic mine design and operational sequencing. This work will demonstrate the differences using a synthetic block model which is representative of the large, laterally extensive iron ore deposits which are typical of Western Australian, Pilbara-style deposits. The use of an artificial deposit will provide a controlled environment for comparing optimisation approaches, blending strategies, and capital allocation outcomes.

The base case establishes a conventional benchmark using a pit generated from revenue factor shells without blending constraints. The direct block scheduling approach, enhanced by the Bienenstock-Zuckerberg (BZ) algorithm, is then applied under two configurations—with and without blending constraints—to assess sensitivity to grade control, cut-off dynamics, and recovery thresholds. There are three primary configurations in the pit optimisation stage:

1. Polygonal group-release logic is used to emulate area clearance for overburden removal.
2. A dynamic cut-off grade mechanism allows additional material to be directed to the ore stream when processing capacity is available.
3. Capital expenditure limits are initially disabled to isolate the effects of algorithmic optimisation on pit geometry and value realisation.

Phase generation is then conducted based on mining-tonnage, with alternative configurations driven by consistent product tonnage and polygonal control boundaries. Each phase design is evaluated both with and without blending constraints to quantify its impact on feed consistency and schedule variability.

Once the phase designs are complete, two phase-bench, operational schedules, are developed to explore:

- Backfilling and flow modelling, where a ratio constraint and a parallel-flow model are applied to test in-pit waste placement efficiency.
- Constrained capital sequencing, where multiple deposits are analysed under maximum and limited CAPEX conditions to determine optimal mining order and investment prioritisation.

The results show clear differences between the strategic plans developed with revenue factor shells and direct block scheduling. There are clear differences in cut-off grade flexibility, lateral advance patterns, and inter-pit sequencing when capital constraints are included. The findings illustrate how integrating optimisation and scheduling workflows within a single environment can enhance realism and economic transparency in multi-deposit iron ore planning.

INTRODUCTION

Strategic mine planning for large iron ore deposits has long depended on revenue factor shells (RVS) to generate nested pit designs. This is done by applying economic multipliers to commodity prices (revenue factors) or costs. Although computationally efficient, this approach decouples pit shell generation from detailed scheduling. As a result, operational constraints such as mining capacity, processing limits, blending requirements, and capital availability are typically introduced only during subsequent scheduling stages. This sequential workflow can create inconsistencies between pit geometry and operational feasibility, often requiring iterative adjustments during later planning phases (Chung and Rahal, 2025). These limitations are especially pronounced in the laterally extensive, stratiform deposits characteristic of Western Australia's Pilbara region, where grade variability demands disciplined sequencing to maintain product specifications.

Recent advances in computational optimisation have enabled direct block scheduling (DBS) algorithms that integrate spatial precedence, equipment capacity, blending targets, and financial controls within a unified framework. The Bienstock–Zuckerberg (BZ) algorithm, as implemented in platforms such as Deswik.GO, provides an efficient way to model large numbers of extraction decisions/precedence decisions, allowing grade control and sequencing logic to shape pit geometry dynamically rather than as a post-optimisation step.

This study will compare plans derived from revenue-factor shells to those developed using DBS, with and without blending constraints. This will quantify differences in cut-off behaviour, lateral advance, and sequencing performance. The findings illustrate the broader implications of consolidating optimisation and scheduling within a single decision-making framework.

PROBLEM STATEMENT, APPLICATIONS AND DISCUSSIONS

A mid-tier iron ore producer operating in Western Australia's Pilbara region faces a strategic planning challenge common to multi-deposit operators: ensuring that project value is fully realised across three potential deposits. A critical question is whether their current static optimisation approach yields materially different strategic outcomes than temporally aware methods that incorporate dynamic cut-off grades, blending controls, and capital sequencing. This study addresses that question directly by comparing RVS and DBS planning frameworks. The producer has identified that cut-off behaviour, lateral advance, phase development, and the timing of capital expenditures will be critical to maximising project value.

The study begins by optimising the pit design. A RVS serves as the baseline, representing their current planning practice. Two DBS configurations are then examined – one without blending constraints and one incorporating blending controls. Both DBS configurations integrate polygon-based precedence and dynamic cut-off logic to route marginal material when processing capacity permits. Capital expenditure constraints are excluded at this stage to isolate the effects of algorithmic differences on pit geometry and extraction sequence.

Following pit optimisation, phase generation (PGN) is applied to convert the optimised block extraction sequence into practical mining phases. Three alternative approaches are tested: mining-tonnage-based phase delineation, product-tonnage-driven phase boundaries, and polygon-constrained phase definitions. Each configuration is evaluated both with and without inherited grade constraints to assess the impact on feed consistency and schedule variability.

Finally, phase-bench scheduling (PBS) is used to evaluate system-level operational strategies. Two strategic configurations are examined. The first incorporates in-pit waste management using ratio-based placement constraints and parallel material flow modelling to simulate backfilling behaviour. The second introduces capital expenditure constraints by imposing maximum investment limits for phase development across multiple deposits. These scenarios allow the study to quantify the sensitivity of extraction timing, material movement, and project value to operational and financial constraints within the mining system, illustrated in Figure 1.

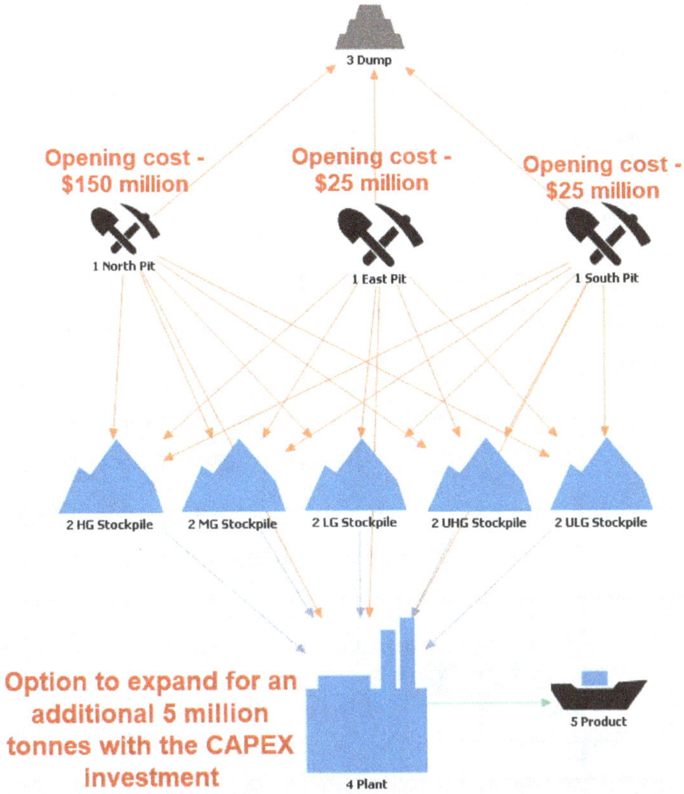

FIG 1 – Iron ore mining system flow network used in the optimisation model.

Stage 1 – pit optimisation

Stage 1 establishes the strategic extraction limits under production constraints. All DBS runs were subject to mining and processing capacity constraints, as well as vertical sink rate controls to regulate pit advance rates and reflect realistic operational advance. This ensures comparability between optimisation approaches while maintaining practical mining progression. In the DBS, it also includes the capital investment required to start mining each pit and the option to expand plant capacity, if required. These investment decisions cannot be considered within the conventional pit optimisation method.

Three optimisation configurations were evaluated:

- **Scenario 1a**: applied the traditional RVS methodology to generate nested pit limits without explicit integration of temporal capacity.
- **Scenario 1b**: implemented DBS without blending constraints, allowing extraction sequencing to respond to capacity and precedence while maximising discounted value.
- **Scenario 1c**: extended the DBS formulation by incorporating blending constraints to enforce grade consistency across processing streams, enabling assessment of the impact of quality control requirements on pit geometry and extraction sequence.

Figure 2 presents the outcome generated using the traditional RVS methodology. Such results are typical for iron ore deposits, where each incremental metal unit contributes economic value and the optimisation expands along favourable zones. However, while RVS defines an economic envelope, it does not provide meaningful guidance on practical extraction sequencing, as scheduling and blending constraints are not embedded within the optimisation process.

In contrast, the DBS results (Figure 3) provide clearer operational direction. The DBS formulation ensures that blocks selected for extraction are simultaneously feasible from a blending perspective, enabling mined material to be processed as compliant ore without requiring downstream reconciliation. When blending constraints are incorporated, the extraction strategy adjusts accordingly to satisfy grade targets, resulting in observable changes to sequencing patterns and pit extent relative to the unconstrained case. Blending, therefore, becomes an intrinsic driver of pit geometry rather than a post-processing adjustment.

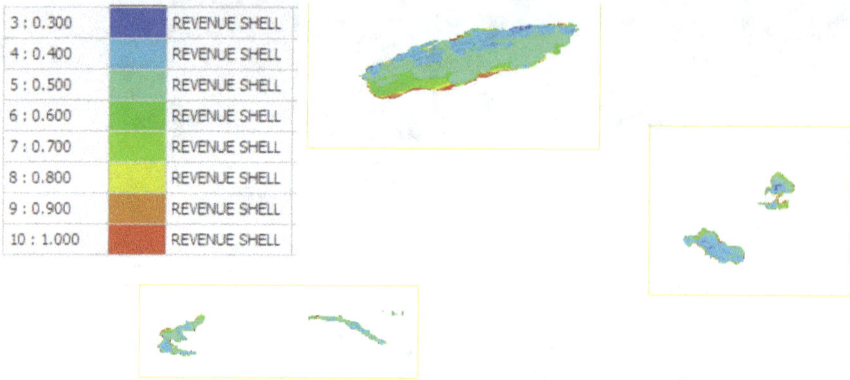

FIG 2 – Ultimate pit geometry generated using RVS.

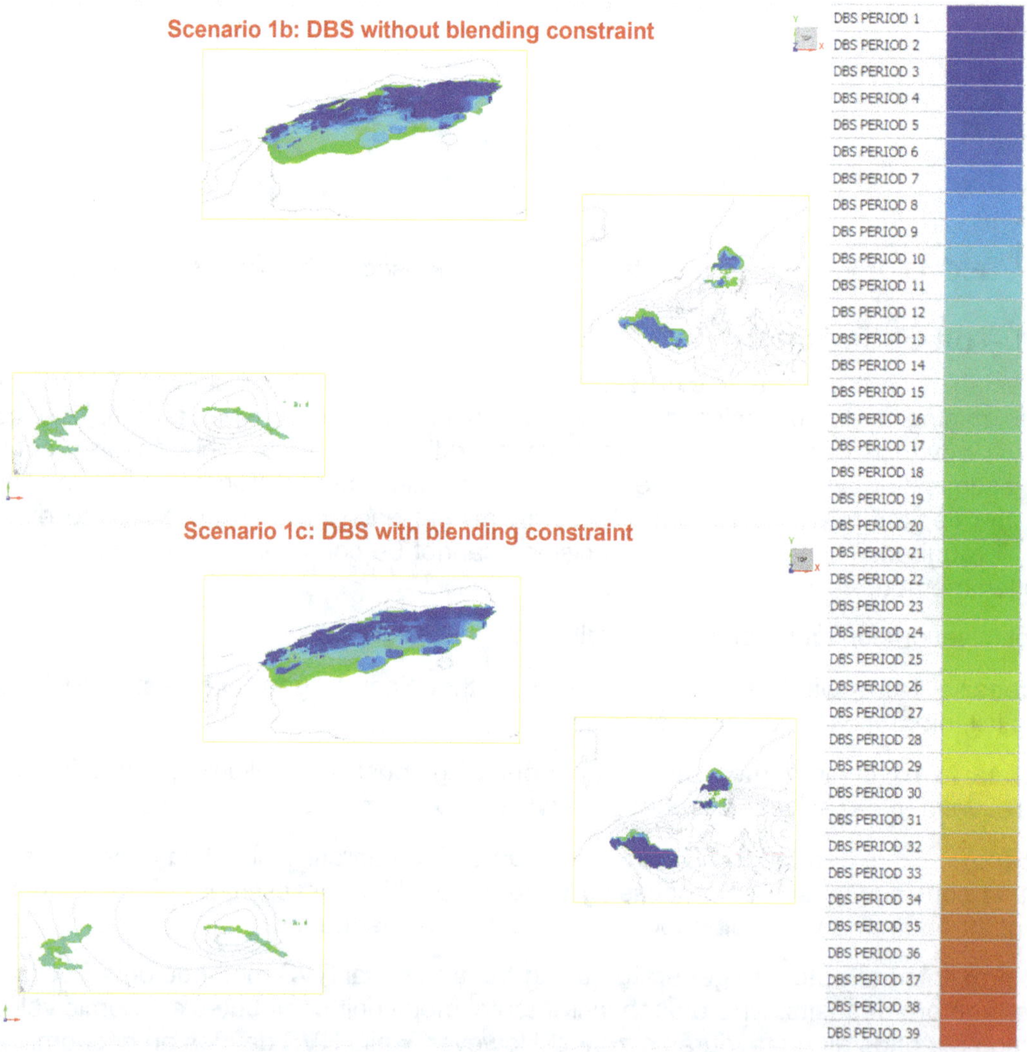

FIG 3 – DBS pit optimisation results with and without blending constraints.

Figure 4 demonstrates that cumulative discounted cash flow (DCF) remains comparable between the two DBS scenarios over the life-of-mine. Despite similar overall economic outcomes, the internal distribution of ore and waste differs across pits due to blending requirements. The blending-constrained scenario redistributes material movement to maintain grade compliance, leading to adjustments in ore extraction and waste stripping patterns across the East, North, and South-West pits. This indicates that blending constraints materially influence spatial and temporal material allocation without significantly compromising total value generation. Figure 5 also shows that grade distributions across periods are more consistent when blending constraints are applied in the DBS model.

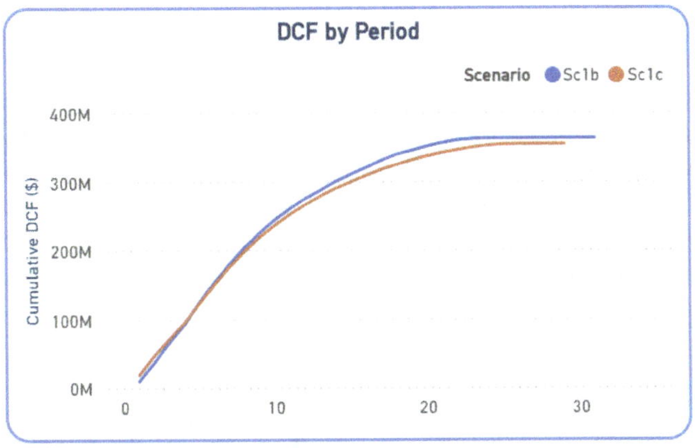

Scenario	Total Movement	Ore Tonnes Movement	Waste Tonnes Movement
Sc1c			
North Pit	512.4M	324.8M	187.6M
East Pit	24.5M	18.4M	6.2M
South Pit	17.8M	13.1M	4.7M
Sc1b			
North Pit	533.4M	298.3M	235.0M
East Pit	24.8M	17.9M	6.9M
South Pit	17.9M	12.6M	5.2M

FIG 4 – Comparison of discounted cash flow and material movement for DBS scenarios.

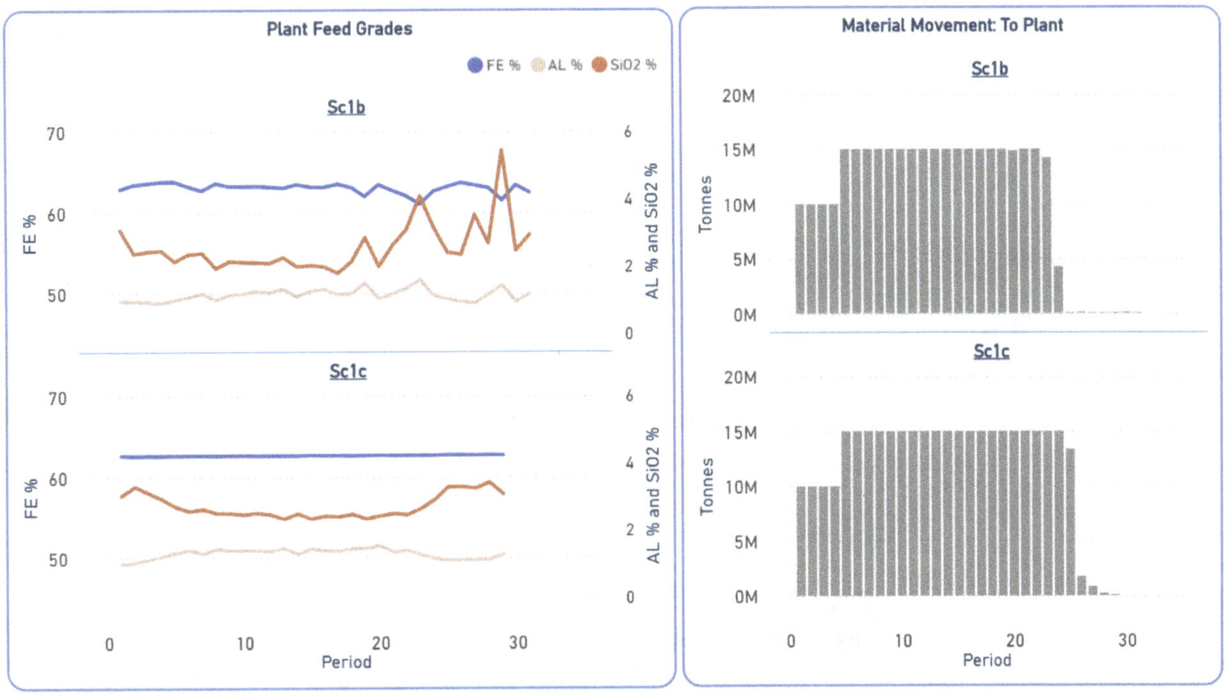

FIG 5 – Comparison of period grade distribution with and without blending constraints.

Stage 2 – mining phases

PGN applies a fuzzy clustering algorithm to group blocks into practical mining phases using the DBS extraction sequence as input. As a result, changes in the underlying DBS solution directly influence the resulting phase geometries, even when clustering parameters remain unchanged. In addition, the PGN output is sensitive to the selected target attribute (eg mining tonnes versus ore tonnes), as different targets drive alternative spatial aggregation behaviour.

To illustrate this dependency, four PGN configurations were generated based on alternative DBS inputs and clustering targets, with one additional case incorporating manual polygon adjustments. The manual polygon option is occasionally used to refine phase geometries to meet operational requirements, such as maintaining minimum mining widths or removing thin 'onion-skin' layers along pit boundaries.

The following scenarios were evaluated:

- **Scenario 2a**: Scenario 1b (DBS without blending) as input; PGN targeting mining tonnes.

- **Scenario 2b**: Scenario 1c (DBS with blending) as input; PGN targeting mining tonnes.
- **Scenario 2c**: Scenario 1b as input; PGN targeting ore tonnes.
- **Scenario 2d**: Scenario 1c as input; PGN targeting ore tonnes.
- **Scenario 2e**: Scenario 1c as input; PGN with manual polygon generation.

Figure 6 presents the PGN outcomes for Scenarios 2a–2d. The results demonstrate that phase geometries vary depending on both the DBS input (with or without blending constraints) and the selected clustering target attribute (mining tonnes versus ore tonnes). Although the clustering parameters remain consistent, differences in the underlying extraction sequence and optimisation drivers lead to distinct spatial aggregation patterns. Each PGN configuration can be considered internally optimal with respect to its DBS input and clustering objective, as the phases are generated directly from the value-consistent extraction sequence.

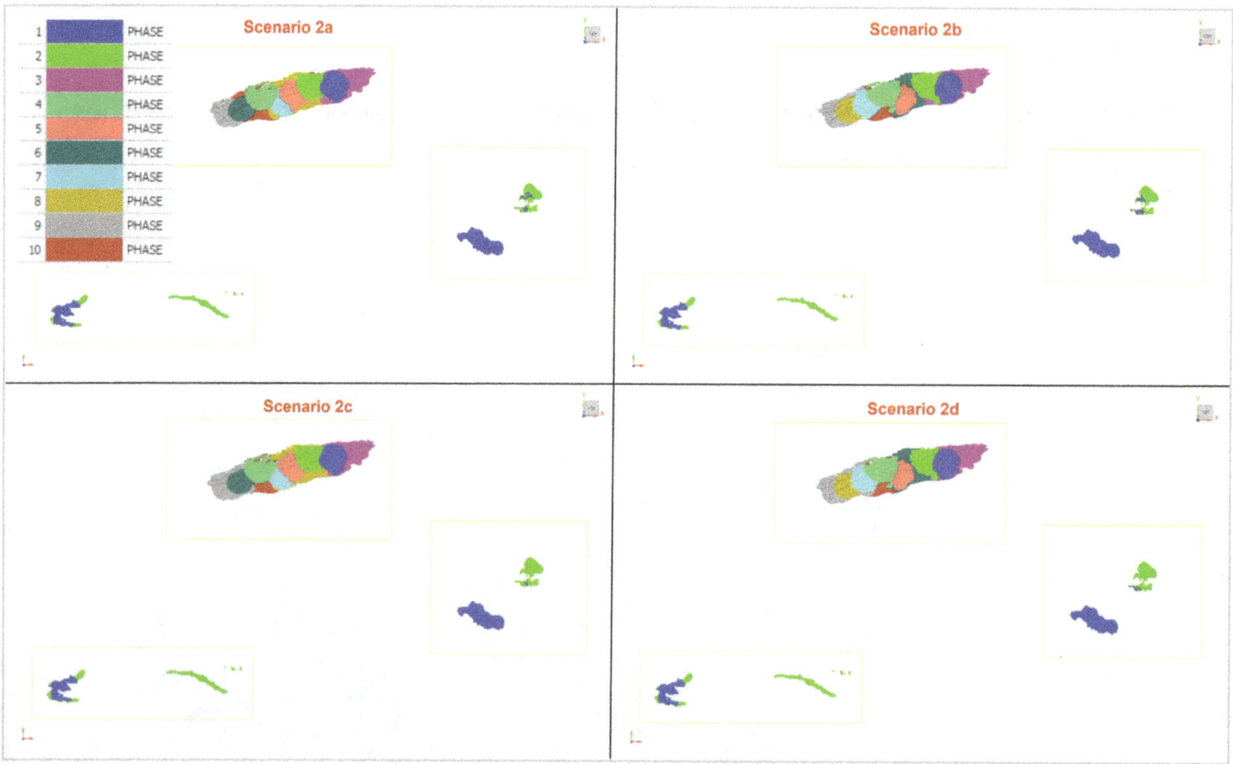

FIG 6 – PGN results for Scenario 2a – 2d.

Figure 7 illustrates Scenario 2e, where manual polygon generation is applied using the DBS solution as guidance. This approach demonstrates the flexibility to adjust phase boundaries to meet operational requirements—such as minimum mining widths or removal of thin edge strips—while still maintaining alignment with the optimised extraction strategy.

Figure 8 shows that the inventory within each phase will change when using different objectives in the clustering algorithm.

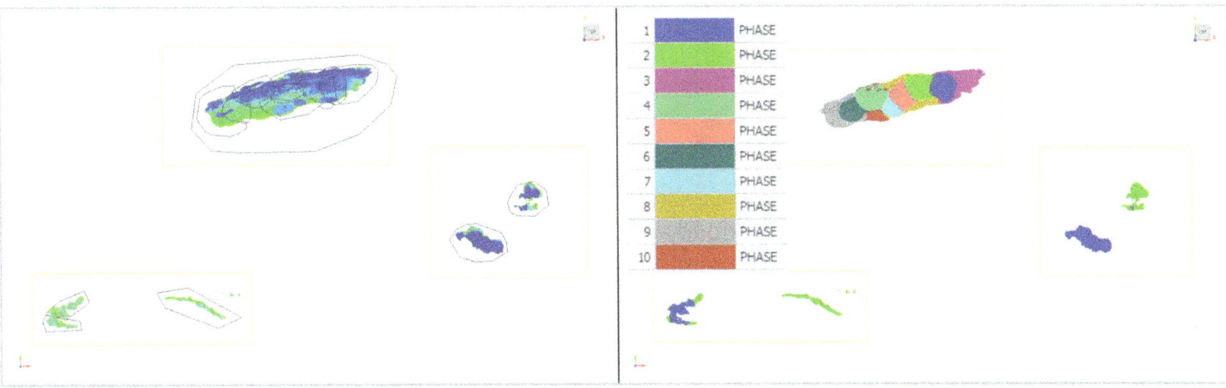

FIG 7 – PGN result using manual polygon control (Scenario 2e).

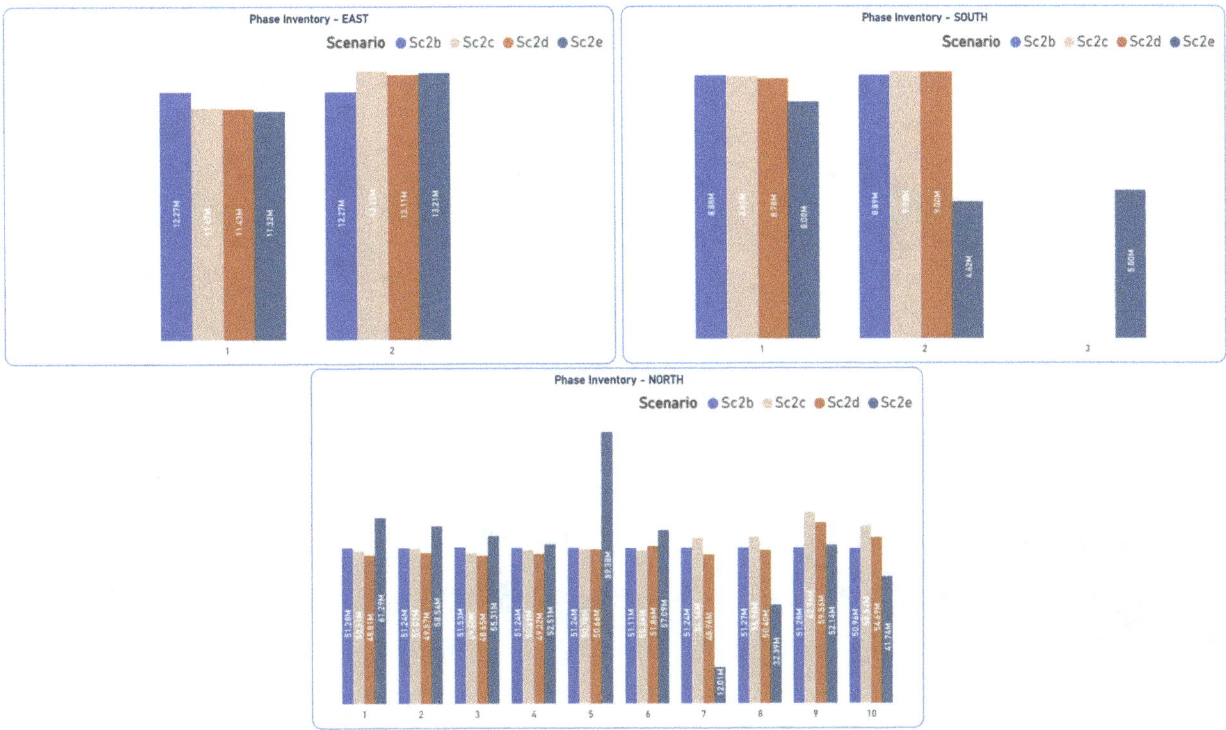

FIG 8 – Phase inventory comparison for alternative PGN clustering objectives.

Stage 3 – strategic optimisation

Stage 3 applies phase-bench scheduling (PBS) to evaluate strategic extraction sequencing under integrated operational and financial constraints. PBS operates on the phase geometries generated in Stage 2 and formulates a time-indexed optimisation problem that maximises discounted cash flow (DCF) while enforcing mining capacity, processing capacity, slope precedence, and inter-phase dependency constraints. By operating at the phase-bench level, PBS provides a more operationally realistic schedule and enables evaluation of system-wide trade-offs such as waste routing, backfilling strategies, and capital deployment across multiple deposits.

Three PBS configurations were analysed:

- Scenario 3a: PBS executed using Scenario 2e phase geometries with the same operational model applied in the DBS optimisation.
- Scenario 3b: PBS extended to include in-pit backfill modelling.
- Scenario 3c: Scenario 3b augmented with pit- and phase-level CAPEX variables and maximum expenditure constraints per period.

In Scenarios 3b and 3c, the East and South pits are designated as potential in-pit waste storage locations once mining activities in those pits have ceased. Introducing these pits as waste destinations modifies the material flow network by providing additional routing options for waste stripping. This configuration allows the optimisation to reduce haulage distances by redirecting waste material into available pit voids instead of transporting it to external dump locations.

Mining sequence

Because PBS aggregates blocks into phase-bench tasks, mining progression must follow predefined phase geometries. This reduces the flexibility of the optimiser relative to block-level DBS solutions and limits the ability to target isolated high-value blocks. As a result, changes between PBS scenarios are primarily expressed through the timing and activation order of phases, rather than through significant spatial differences in mining locations.

Figure 9 illustrates how the mining sequence changes when backfilling options are introduced. When the East and South pits become available as waste storage locations in Scenario 3b, the optimisation adjusts the extraction order to enable earlier completion of phases that create pit voids. This allows waste stripping to be redirected into these voids and reduces reliance on external dump locations. When CAPEX constraints are introduced in Scenario 3c, the mining sequence is further modified because capital availability limits the number of phases that can be activated simultaneously across the deposits.

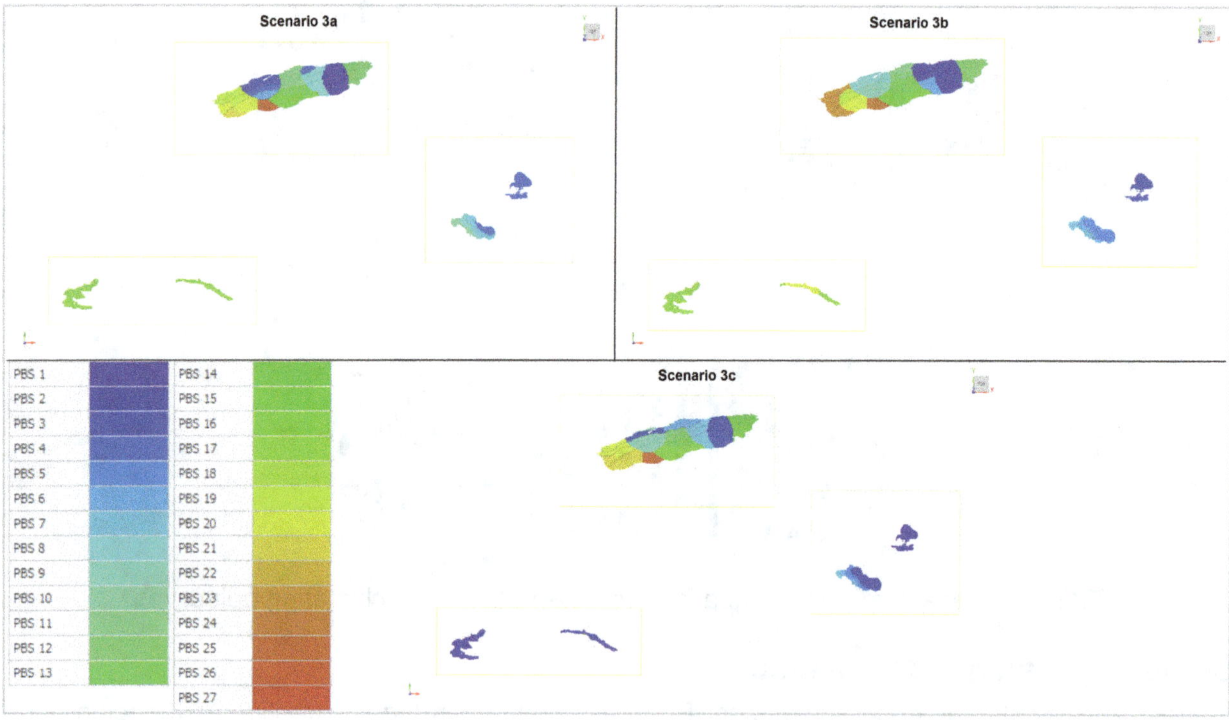

FIG 9 – Mining sequence comparison across PBS scenarios.

Material movement

Material movement statistics for the PBS scenarios are summarised in Figures 10 and 11. Across all scenarios, the North Pit dominates total material movement and accounts for the majority of both ore and waste extraction within the system. Although total material movement remains broadly consistent across scenarios, the distribution between ore and waste varies depending on the operational constraints applied.

In Scenario 3a, where no backfilling option is available, waste must be transported to external dump locations. When in-pit backfilling is introduced in Scenario 3b, the optimisation redistributes waste flows toward the East and South pits once these pits become available for storage. This modification reduces haulage requirements and improves waste management efficiency within the mining system. When CAPEX expenditure constraints are introduced in Scenario 3c, the timing of phase development changes, resulting in small adjustments to ore extraction and waste stripping across the pits.

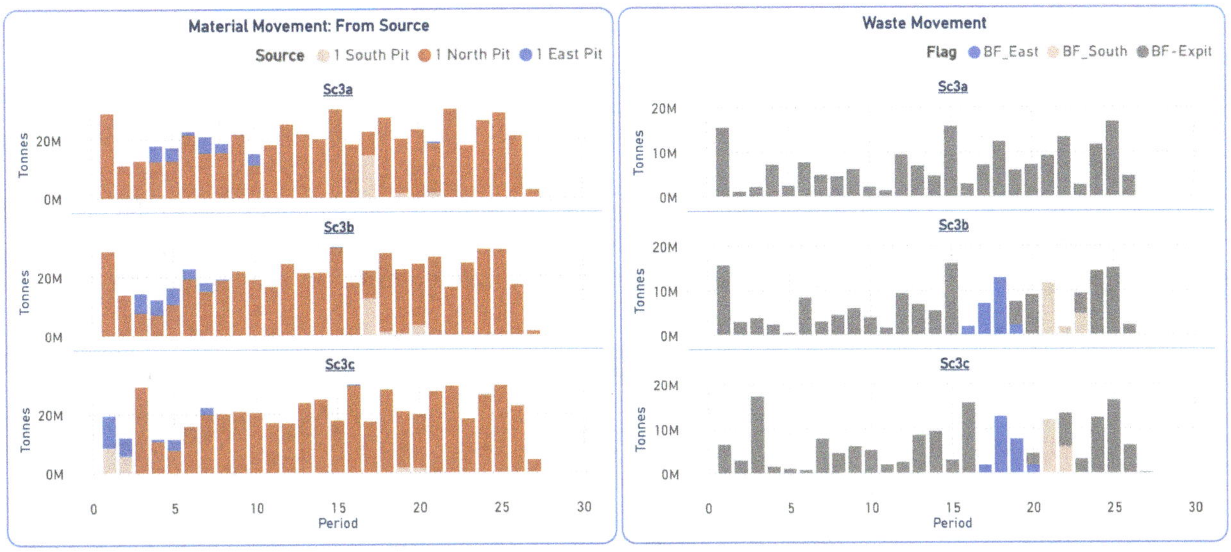

FIG 10 – Material movement and waste routing summary across PBS scenarios.

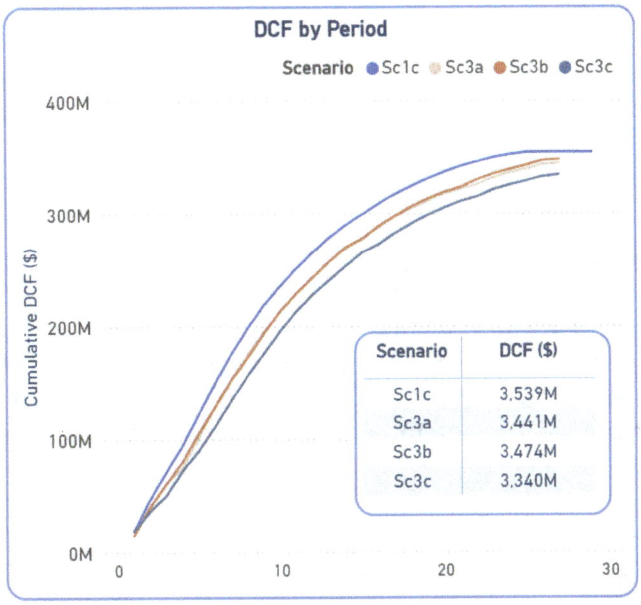

Scenario	Total Movement	Ore Tonnes Movement	Waste Tonnes Movement
Sc1c			
1 East Pit	25M	18M	6M
1 North Pit	512M	325M	188M
1 South Pit	18M	13M	5M
Sc3a			
1 East Pit	25M	18M	7M
1 North Pit	515M	341M	174M
1 South Pit	18M	13M	5M
Sc3b			
1 East Pit	25M	18M	6M
1 North Pit	515M	343M	171M
1 South Pit	18M	13M	5M
Sc3c			
1 East Pit	25M	18M	6M
1 North Pit	515M	342M	173M
1 South Pit	18M	13M	5M

FIG 11 – DCF comparison between DBS and PBS scenarios.

DCF

The economic performance of the optimisation scenarios is summarised in Figure 11. The direct block scheduling (DBS) solution from Stage 1 (Scenario 1c) provides the upper bound for project value, achieving a cumulative discounted cash flow (DCF) of $3539M. This result reflects the full block-level selectivity available in the DBS formulation, where individual blocks can be scheduled independently subject to precedence and capacity constraints.

When transitioning to phase-bench scheduling (PBS), a reduction in achievable value is observed due to the aggregation of blocks into phase-bench tasks. This aggregation introduces additional spatial constraints on extraction decisions and reduces the optimiser's ability to prioritise isolated high-value blocks. As a result, Scenario 3a produces a cumulative DCF of $3441M, representing the baseline PBS solution without backfill modelling.

The introduction of in-pit backfilling in Scenario 3b increases the cumulative DCF to $3474M. This improvement is primarily driven by reductions in waste haulage cost and improved material flow efficiency within the mining system. By enabling waste to be deposited in previously mined pit voids, the optimisation can reduce transport distances and improve overall stripping efficiency while maintaining ore supply to the processing plant.

In Scenario 3c, maximum CAPEX expenditure constraints are introduced to limit the rate of phase development. These financial constraints restrict the simultaneous activation of multiple pits and delay access to certain high-value mining areas. Consequently, the cumulative DCF decreases to $3340M. The reduction in project value reflects the delayed extraction of profitable material caused by capital availability constraints.

Figure 12 illustrates that the total capital expenditure remains below the specified constraint, confirming that the optimisation framework successfully enforces financial limits while adjusting the extraction schedule to maximise economic performance within the feasible solution space.

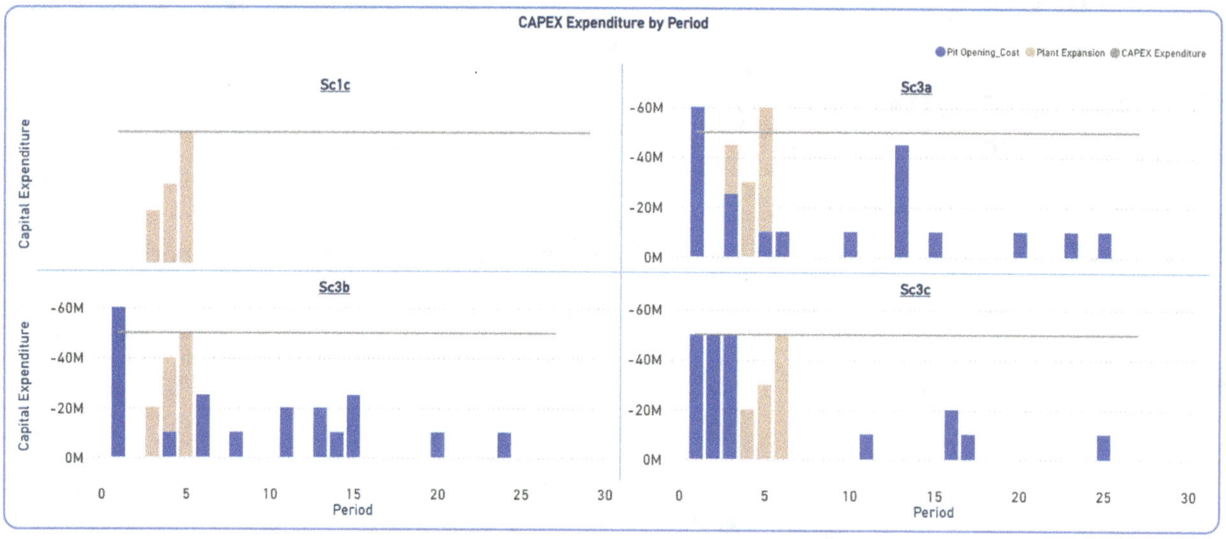

FIG 12 – CAPEX expenditure profile.

Overall, the results highlight the economic trade-off between optimisation flexibility and operational realism. While block-level DBS optimisation establishes the theoretical value envelope, the PBS scenarios demonstrate how operational factors—such as waste management strategies and capital expenditure limits—can influence both the timing of extraction and the achievable economic outcome.

CONCLUSIONS

This study evaluated the impact of optimisation methodology on strategic mine planning outcomes for a synthetic Pilbara-style iron ore deposit. By comparing conventional RVS optimisation with DBS and subsequent PBS, the analysis quantified how integrated optimisation frameworks influence pit geometry, extraction sequencing, material movement, and project economics within a multi-deposit mining system.

The results show that the traditional RVS approach provides an economic envelope for pit development but does not incorporate operational constraints, such as blending requirements, into the optimisation formulation. In contrast, the DBS framework embeds these constraints directly within the optimisation problem, allowing extraction decisions to be evaluated simultaneously against grade targets, capacity limits, and spatial precedence relationships. As a result, blending constraints influence both pit geometry and extraction sequencing rather than being addressed during downstream scheduling stages. Although the overall project value remains comparable between DBS configurations, the blending-constrained model produced more stable grade delivery across periods and redistributed ore and waste extraction across the pits.

At the strategic scheduling stage, the PBS results illustrate the operational and financial trade-offs associated with realistic mining system constraints. The block-level DBS optimisation (Scenario 1c) produced the highest theoretical project value with a cumulative DCF of approximately $3539 M. When the schedule was aggregated into phase-bench tasks in PBS, the achievable value decreased slightly due to reduced extraction selectivity, with the baseline PBS configuration (Scenario 3a) generating a DCF of $3441 M. Introducing in-pit backfilling (Scenario 3b) improved system efficiency and increased the DCF to $3474 M by reducing waste haulage distances and improving material flow within the mining system. When CAPEX expenditure limits were applied (Scenario 3c), the cumulative DCF decreased to $3340 M, reflecting delayed phase development and restricted access to high-value material due to capital constraints.

Overall, the results demonstrate that integrated optimisation workflows provide greater transparency when evaluating strategic mining decisions. Direct block scheduling establishes a value-consistent extraction framework, while subsequent phase and scheduling stages allow planners to evaluate how operational constraints—such as blending requirements, waste management strategies, and capital availability—affect extraction sequencing, material movement, and long-term project value.

ACKNOWLEDGEMENT

The authors would like to thank Deswik for supporting the development of this work and providing the time and resources to prepare and present this paper.

REFERENCES

Chung, J and Rahal, D, 2025. Enhancing pit optimisation with direct block scheduling (DBS) and the Bienenstock Zuckerberg (BZ) algorithm – maximising NPV and efficiency, in *Proceedings of APCOM 2025*, pp 47–62 (The Australasian Institute of Mining and Metallurgy: Melbourne).

Optimising sustainable iron ore beneficiation through multi-stage dry magnetic separation and advanced comminution technologies

O Cox[1], T Coventry[2], M Drechsler[3], J Robbeson[4], E Stocker[5] and G Abaka-Wood[6]

1. Mechanical Engineering Student, Adelaide University; Mechanical Engineering Intern, DryFlow Magnetics, Wingfield SA 5013. Email: engineering@dryflowmag.com
2. Mechanical Engineering Student, Adelaide University; Mechanical Engineering Intern, GRolls, Clovelly Park SA 5042. Email: tristan.coventry@grolls.com.au
3. Director, Gyratory Roller Solutions Pty Ltd, Adelaide SA 5051. Email: mark.drechsler@grolls.com.au
4. Chief Operating Officer, DryFlow Magnetics, Sydney NSW 2000. Email: jon@dryflowmag.com
5. Chief Technical Officer, DryFlow Magnetics, Sydney NSW 2000. Email: elena@dryflowmag.com
6. Research Fellow, ARC Centre of Excellence for Enabling Eco-Efficient Beneficiation of Minerals, Future Industries Institute, Adelaide University, Mawson Lakes SA 5095. Email: george.abaka-wood@adelaide.edu.au

ABSTRACT

This study investigates multi-stage dry magnetic separation beneficiation pathways centred on the DryFlow Magnetics system to improve the sustainability and efficiency of iron ore processing and beneficiation. The research focuses on reducing energy and water consumption, limiting over-grinding, and achieving high-grade iron recovery through progressive dry comminution and magnetic separation. To this end, this study aims to identify optimal combinations of comminution technologies and DryFlow magnetic separation configurations that minimise energy use while maximising recovery and product quality.

Ore samples from South Australian magnetite deposits are processed through sequential stages of size reduction using various dry comminution technologies including GRolls® crusher, vertical roller milling (VRM), and ball/rod milling, followed by successive stages of dry magnetic separation using the DryFlow dry magnetic separator. Each stage is analysed to determine how particle size distribution across coarse, intermediate, and fine fractions influences iron oxide recovery, grade, and overall process efficiency.

Results presented in this paper provide a comparative assessment between conventional dry and wet comminution technologies and magnetic separation processes and the emerging DryFlow enabled dry beneficiation route, respectively demonstrating how a fully dry flow sheet can materially reduce energy intensity, eliminate water dependency, and enhance the economic and environmental performance of future iron ore operations. The findings suggest that integration of DryFlow's proprietary magnetic separation technology enables accurate, water-free concentration of magnetite and hematite ores at each step, establishing a new benchmark for low-emission mineral processing.

INTRODUCTION

The iron ore industry is under increasing pressure to lower energy and water consumption while maintaining product quality as high-grade ores decline and mineralogical complexity increases. Magnetite-dominated iron ores are often found at lower grades than hematite deposits, meaning they often require fine grinding to achieve sufficient liberation for beneficiation, resulting in high energy intensity and significant water demand. Comminution alone accounts for around 1 per cent total global and up to 80 per cent of total mining operations energy consumption, making it a critical target for process optimisation (Engeco, 2021). In this context, dry beneficiation flow sheets that enable early gangue rejection, reduced grinding demand, and improved overall circuit efficiency are of growing interest.

This project presents an emerging multi-stage dry iron ore beneficiation concept that integrates novel dry magnetic separation with advanced dry comminution technologies. The approach focuses on processing multiple size fractions independently through a dry magnetic separator to maximise iron recovery and grade while reducing energy and water consumption across the beneficiation flow sheet.

BACKGROUND

Dry magnetic separation

Dry magnetic separation offers several advantages over conventional wet separation routes, including reduced water consumption, simplified tailings handling, and the ability to reject gangue early in the process flow sheet. Permanent dry magnetic separators have been widely applied for pre-concentration of iron ores, particularly in arid and water-constrained regions, and have demonstrated potential to increase feed grade to downstream processing while reducing grinding energy demand (Xie *et al,* 2022).

However, dry magnetic separation performance is strongly influenced by particle size, mineral composition, and particle morphology. Studies on wet low and high intensity magnetic separation have shown that recovery generally decreases with increasing particle size for most iron-bearing minerals other than magnetite, highlighting the importance of size-specific processing strategies (Bazin *et al,* 2024). Additionally, dry separators face operational challenges related to fine particles, electrostatic effects, and temperature limitations, particularly in rare-earth roll-type separators (Dobbins, Dunn and Sherrell, 2009). These limitations motivate the development and evaluation of alternative dry magnetic separation technologies and multi-stage processing strategies that can better accommodate a broad particle size distribution.

The DryFlow Magnetic Separator is an emerging magnetic separation technology being investigated for iron ore beneficiation applications. Rather than relying on a single-pass or single-size approach, this study conceptually explores the use of multiple size fractions processed independently through dry magnetic separation to improve overall circuit performance. By tailoring separation conditions to distinct particle size classes, the approach aims to mitigate known inefficiencies associated with broad feed size distributions in dry magnetic separators.

Comminution technologies

In conventional crushing and grinding circuits a combination of gyratory, cone, and ball milling stages are used to beneficiate magnetite. While these stages are effective for reaching liberation size, they are highly energy intensive and generate significant quantities of ultra-fine material, which adversely affect downstream separation (David, 2011). The need to reduce comminution energy has led to interest in alternative breakage mechanisms that promote preferential liberation at coarser sizes.

The Gyratory Roller Crusher (GRolls®) is an advanced dry comminution technology that applies gyrating compression and shear forces to particles. Studies comparing the GRolls® with conventional tumbling mills have demonstrated improved preferential breakage along mineral boundaries, producing better liberated particles at coarser sizes and reducing the generation of ultra-fines (Nyarko *et al,* 2026; Drechsler and Skinner, 2023). Importantly, GRolls® operates without grinding media and can be run in both dry and wet modes with minimal differences in size reduction performance (Drechsler, Skinner and Abaka-Wood, 2025a).

Energy comparisons indicate that GRolls® can achieve meaningful reductions in specific energy consumption relative to conventional milling for a range of ore types, positioning it as an intermediate comminution step between high pressure grinding rolls (HPGR) and fine/tertiary grinding mills (Drechsler, Skinner and Abaka-Wood, 2025b). This makes GRolls® particularly well suited to dry beneficiation flow sheets where minimising energy input and avoiding excessive fines generation are critical objectives.

Multi-stage beneficiation strategy

The central contribution of this study is a proposed multi-stage dry beneficiation strategy in which crushed ore is classified into distinct size fractions: coarse (+600 μm); grits (600–212 μm), and fines (-212 μm). Each fraction is processed separately through dry magnetic separation to generate a concentrate (Con), middlings (Mid) and tailings (Tails) for analysis. This approach builds on prior work demonstrating the benefits of dry cobbing and pre-concentration for process optimisation (Hejazi, Chehreh Chelgani and Hart, 2022).

Early rejection of barren or weakly magnetic gangue at coarse sizes can substantially reduce the mass flow to downstream comminution, thereby lowering energy consumption and operating costs. Intermediate size fractions can then be subjected to further size reduction using advanced comminution technologies such as GRolls®, producing particles with favourable liberation characteristics for subsequent dry magnetic separation. Fine fractions, which are traditionally challenging for dry separation, can be processed under tailored conditions to maximise recovery while avoiding the inefficiencies associated with treating a wide size distribution in a single separator.

Particle shape effects are also relevant in this context. Research has shown that comminution mechanisms influence particle morphology, which in turn affects magnetic separation performance (Inanloo, Moradi and Irannajad, 2024). Preferential breakage and elongated particle shapes produced by non-tumbling comminution technologies may enhance magnetic separation efficiency by improving mineral exposure and reducing gangue entrainment.

MATERIAL AND MINERALOGY

A South Australian Banded Iron Formation (BIF) magnetite was used for laboratory scale test work. The key characteristics of the ore are summarised in Table 1.

TABLE 1

Head assay and mineralogical summary of test feed material.

Name	BIF Magnetite
Location	Magnetite resource, Southern Middleback Ranges
Source geology	Banded Iron Formation (BIF) lower iron carbonate magnetite
Hardness	Moderate
Specific density	3.3–3.4 t/m^3
Maximum size	37.5 mm
Assay main oxides (%)	
SiO	19.22%
FeO	40.22%
Al_2O_3	1.72%
CaO	4.6%
MgO	7.03%
LOI	8.42%
Dominant mineralogy (approximate %)	
DTR Fe recovery 38 µm	Quartz 14%
DTR Mag Fe grade 38 µm	Magnetite 41%
	Goethite 5%
	Dolomite 15%
	Talc 7%
	Hematite 7%
	Siderite 6%
	Muscovite, amphibole, pyrophyllite and chlorite <5%
	71.84%
	70.2

GRolls® crusher

The GRolls® technology employed in this study is a laboratory-scale Alpha GRolls400 prototype located at Adelaide University. The GRolls400 consists of two intermeshing rollers, each measuring 400 mm in length and 250 mm in diameter. The primary roller is powered by a 7.5 kW direct-drive motor, while the opposing oscillating roller is driven by a 5 kW belt-driven motor. This oscillating roller incorporates an eccentric shaft with a 0.1 mm offset, operating at approximately 1000 rev/min. This configuration induces gyratory motion, resulting in cyclic compression, shear, and tensile forces within the multiple comminution chambers. Further details of the GRolls® technology and energy calculation methods are presented in Drechsler, Skinner and Abaka-Wood (2026).

DryFlow magnetic separator

The DryFlow magnetics TM-2020 is a lab scale magnetic separator which beneficiates magnetic iron ore through the production of three streams: Concentrate (high-grade product), Middlings (lower grade product), and Tailing (non-magnetic rejects). The TM-2020 uses steady state magnetics in a stacked conveyor design to achieve these streams. Due to IP reasons no further information can be publicly provided.

METHODOLOGY

The methodology adopted in this study comprises both an initial magnetic separation and crushing study and a subsequent controlled laboratory testing campaign (Campaign 1) to evaluate the performance of a staged dry beneficiation flow sheet for magnetite ore. The flow sheet incorporates conventional crushing, GRolls® crushing, and size-specific dry magnetic separation using the DryFlow Magnetic Separator.

The initial studies provided preliminary validation of the proposed processing concept under practical operating conditions, with portable XRF analysis utilised to assess indicative upgrading trends across key size fractions. Building on these outcomes, Campaign 1 was conducted under controlled laboratory conditions to generate detailed and repeatable data sets, including particle size distributions (PSD), mass balances, and separation performance metrics.

This staged approach enables both practical validation and rigorous quantitative assessment of circuit performance, with particular focus on size reduction behaviour, upgrading across discrete size fractions, and the integration of comminution and dry magnetic separation within a closed-circuit configuration.

Single-pass magnetic separation

As part of standard DryFlow testing procedures, the feed material was ground to a range of target particle sizes (P_{80} = 300 μm to 38 μm) and subjected to single-pass dry magnetic separation. The resulting mass distributions, iron grades, and recoveries are summarised in Table 2.

TABLE 2

Single pass magnetic separation results.

Grind size (P_{80})	Distribution			Fe recovery		Fe grade		
	Con	Mid	Tails	Con	Mid	Con	Mid	Tails
300 μm	57.0%	13.0%	30.0%	73.2%	9.5%	49.8%	28.5%	22.3%
150 μm	53.0%	13.6%	33.3%	71.5%	10.0%	52.0%	28.2%	21.5%
125 μm	53.0%	13.6%	33.3%	71.7%	10.1%	52.4%	28.7%	21.1%
50 μm	48.2%	13.6%	38.2%	69.9%	10.3%	56.4%	29.5%	20.3%
38 μm (VRM)	51.6%	10.1%	38.3%	74.4%	7.1%	59.2%	29.0%	21.8%

The results demonstrate a clear relationship between particle size and separation performance. As the grind size decreases, the iron grade of the concentrate increases from 49.8 per cent Fe at P_{80} =

300 µm to 56.4 per cent Fe at P_{80} = 50 µm, indicating improved mineral liberation at finer particle sizes. However, this increase in grade is accompanied by a reduction in concentrate mass yield and overall recovery, with Fe recovery in the concentrate decreasing from 73.2 per cent to 69.9 per cent across the same size range.

This trade-off reflects the fundamental limitation of single-pass magnetic separation: while finer grinding improves liberation and concentrate quality, it also leads to increased losses of magnetic material to middlings and tailings. Additionally, grinding to finer sizes is associated with significantly higher energy consumption, making such approaches less favourable from a sustainability perspective.

The relatively consistent middlings fraction across all grind sizes (~13–14 per cent) suggests that a proportion of the material remains only partially liberated, regardless of the degree of grinding. In a single-pass configuration, this material is not effectively recovered, representing a loss of potentially valuable iron units.

These results highlight the limitations of single-pass processing and reinforce the need for multi-stage beneficiation strategies. By enabling progressive liberation and targeted reprocessing of middlings, a multi-stage circuit can recover this partially liberated material more efficiently, reducing the reliance on excessive fine grinding while maintaining or improving overall recovery.

Conventional multi-stage circuit

The conventional circuit, designated Campaign 0, serves as the baseline for comparative performance assessment (Figure 1). It represents a simplified, linear implementation of multi-stage comminution and magnetic separation, consistent with established dry pre-concentration practices for magnetite ores.

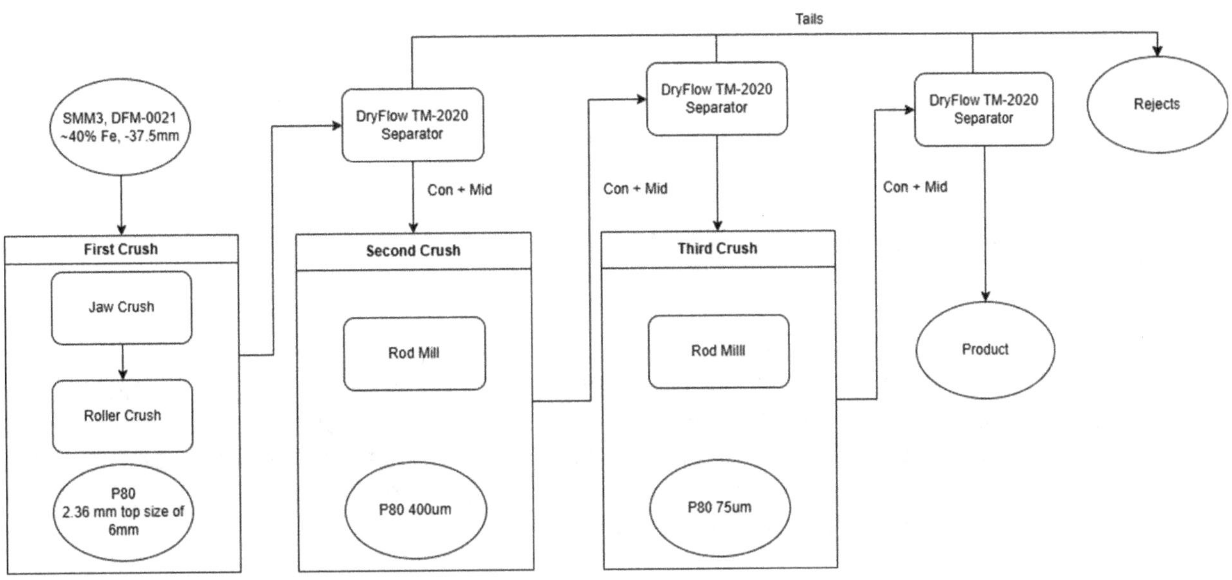

FIG 1 – Conventional dry beneficiation circuit flow sheet.

In this configuration, run-of-mine ore is subjected to sequential stages of size reduction followed by magnetic separation. The feed material (~40 per cent Fe, -37.5 mm) undergoes an initial laboratory jaw and roller crushing stage, after which the full composite stream is treated in a dry magnetic separator (DryFlow TM-2020). The combined magnetic product (concentrate and middlings) is then passed to subsequent rod milling stages for further size reduction, with magnetic separation applied after each stage.

Importantly, no size classification is employed at any point in the circuit. All material is processed as a composite stream, meaning that coarse, intermediate, and fine particles are subjected to identical separation conditions. While this approach simplifies circuit design and operation, it also introduces inefficiencies associated with size dependent magnetic response and liberation behaviour. In particular, the absence of classification may result in overgrinding of already liberated material and suboptimal recovery of coarser liberated particles.

All separation stages are conducted using dry magnetic separation, consistent with the use of DryFlow TM-2020 units, and no wet processing is involved. The final magnetic product from the last separation stage constitutes the circuit concentrate, while non-magnetic streams are rejected as tailings.

This linear, composite-feed approach provides a robust benchmark for evaluating the performance of the GRolls® circuit, particularly in terms of iron grade, recovery, and energy efficiency. Unlike more advanced flow sheets incorporating size classification and targeted regrinding, Campaign 0 reflects a traditional, low-complexity strategy in which liberation and separation are not independently optimised across particle size fractions.

The results of the conventional circuit presented in Table 3 demonstrate that a multistage approach increases iron grade while maintaining acceptable recovery. Compared to Davis Tube Recovery (DTR) benchmarks, the circuit achieves higher recovery due to the coarser sizes, as expected due to reduced liberation requirements. However, the absence of size classification leads to inefficiencies, including overgrinding and suboptimal separation performance across mixed particle sizes.

TABLE 3

Conventional dry beneficiation results.

Stream	Mass (%)	Fe Grade (%)	Fe Recovery (%)
Concentrate	45	62	68
Middlings	7	40	7
Tailings	48	22	25
Calculated Head	100	41	100
Product (cons + mids)	52	59	75

GRolls® – DryFlow multi-stage circuit

Size discovery

A size-by-size analysis was conducted (Table 4) to inform the selection of cut sizes for the three-stage beneficiation circuit in Figure 2. The mass distribution to concentrates, middlings, and tails was assessed across each size fraction to identify changes in separation behaviour. For size fractions coarser than 600 µm, a consistent distribution of mass to concentrates, middlings, and tails was observed, indicating similar selectivity across these size ranges. In contrast, the -600 µm fraction exhibited a distinct shift in mass partitioning between the three streams.

TABLE 4

Size-by-size analysis.

Stream	Mass (%)	Fe (%)
-6700 + 4750 µm		
Concentrate	37	48
Middlings	47	45
Tailings	16	26
Product	**84**	**46**

Stream	Mass (%)	Fe (%)
-2360 + 1180 µm		
Concentrate	29	47
Middlings	53	42
Tailings	18	25
Product	**82**	**44**

Stream	Mass (%)	Fe (%)
-600 + 212 µm		
Concentrate	51	45
Middlings	31	37
Tailings	19	24
Product	**81**	**42**

Cont...

Stream	Mass (%)	Fe (%)
-4750 + 2360 µm		
Concentrate	29	45
Middlings	54	37
Tailings	17	25
Product	**83**	**40**

Stream	Mass (%)	Fe (%)
-1180 + 600 µm		
Concentrate	30	51
Middlings	54	46
Tailings	16	25
Product	**84**	**48**

Stream	Mass (%)	Fe (%)
-212 µm		
Concentrate	58	66
Middlings	27	44
Tailings	14	25
Product	**86**	**58**

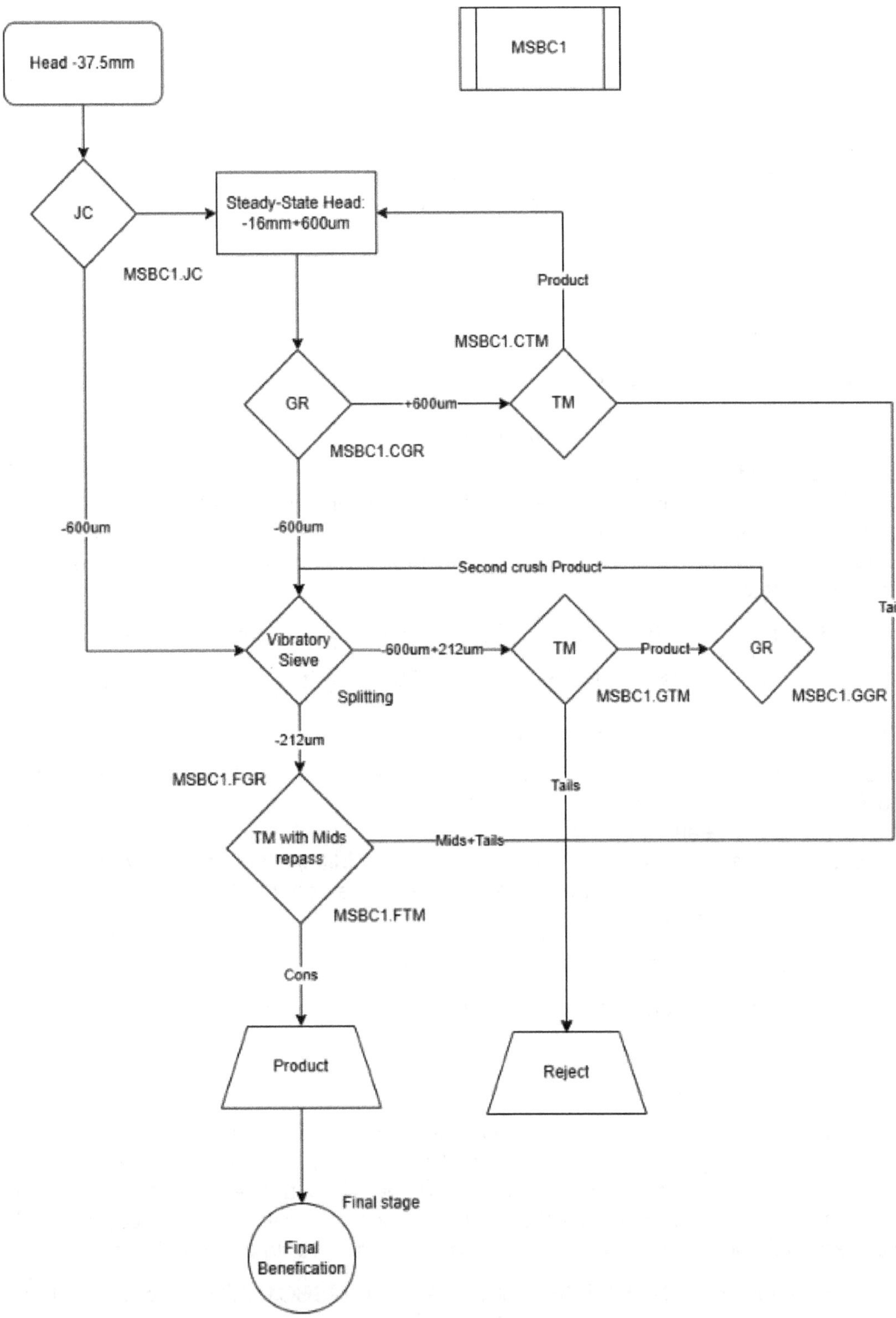

FIG 2 – MSBC1 multi-stage dry beneficiation circuit flow sheet (GRolls® circuit).

A further subdivision within the fines fraction identified 212 μm as an additional key cut size. While the -600 + 212 μm fraction showed only a modest increase in Fe grade in the concentrate stream, the -212 μm fraction demonstrated a pronounced increase in Fe grade relative to all coarser size fractions. Consequently, 600 μm and 212 μm were selected as the primary cut sizes for grits and fines.

Circuit overview

The GRolls® – DryFlow multi-stage dry beneficiation circuit (MSBC1) is designed to achieve size-specific comminution and magnetic separation across three distinct particle size fractions: coarse (+600 μm), grits (-600 μm +212 μm), and fines (-212 μm). The circuit integrates jaw crushing, GRolls® crushing, vibratory sieves, and multiple stages of DryFlow dry magnetic separation. A schematic of the full circuit is provided in Figure 2, with GRolls® identified as 'GR', jaw crusher as 'JC' and the DryFlow TM-2020 as 'TM', and each stage described in further detail following.

MSBC1.JC – jaw crushing stage

Head material (-37.5 mm) is jaw crushed to produce a primary crushed product. The crushed output is screened at 600 μm. The oversize fraction (+600 μm) recirculates to create the head material and is directed to the GRolls® comminution stage (MSBC1.CGR). The undersize fraction (-600 μm) contains a combined grits and fines fraction, which is forwarded to the vibratory sieving stage (Splitting):

- Inputs: head material (-37.5 mm).
- Outputs: +600 μm → MSBC1.CGR; -600 μm → Splitting (vibratory sieves).

MSBC1.CGR – GRolls® comminution stage

The +600 μm head material is fed to the GRolls® crusher for dry crushing, with the GRolls® product screened at 600 μm. The oversize (coarse fraction, +600 μm) is directed to the coarse magnetic separation stage (MSBC1.CTM). The undersize (-600 μm, combined grits and fines) is directed to the vibratory sieves (Splitting). Recycled magnetic product from MSBC1.CTM is recirculated into fresh feed which creates a steady state and higher-grade feed:

- Inputs: +600 μm from MSBC1.JC; recycled magnetic product from MSBC1.CTM.
- Outputs: +600 μm (coarse) → MSBC1.CTM; -600 μm (grits + fines) → Splitting.

MSBC1.CTM – coarse magnetic separation

The coarse fraction (+600 μm) from MSBC1.CGR is fed to the DryFlow dry magnetic separator configured for coarse duty. Magnetically susceptible particles (predominantly magnetite) report to the magnetic product stream and are recycled back to the GRolls® (MSBC1.CGR) for further crushing, as these coarse magnetic particles require additional breakage to reach the target liberation size. Non-magnetic material (gangue) is rejected as coarse tailings, representing the first gangue rejection step in the circuit:

- Inputs: +600 μm coarse material from MSBC1.CGR.
- Outputs: magnetic fraction (product) → recirculated to head; non-magnetic fraction → tailings rejection.

Splitting – vibratory sieve

Combined -600 μm material from both the jaw crusher (MSBC1.JC) and the GRolls® (MSBC1.CGR), together with recirculated grits from the grits GRolls® stage (MSBC1.GGR), are fed to a vibratory sieve. The vibratory sieve separates the combined feed into two size fractions: grits (-600 μm +212 μm) and fines (-212 μm). Each fraction is independently directed to its respective dedicated magnetic separation stage. In the ideal circuit an air classifier would be used instead:

- Inputs: -600 μm from MSBC1.JC; -600 μm from MSBC1.CGR; recirculated grits from MSBC1.GGR.

- Outputs: grits (-600 µm +212 µm) → MSBC1.GTM; fines (-212 µm) → MSBC1.FTM.

MSBC1.GTM and MSBC1.GGR – grits magnetic separation and regrind

The grits fraction (-600 µm +212 µm) from the vibratory sieves are fed to the DryFlow separator configured for grits duty. Magnetic product from this separation is fed into the GRolls® unit (MSBC1.GGR) for regrinding. The reground product is returned to the vibratory sieves (Splitting) for reclassification. Non-magnetic tailings are rejected. The grits regrind loop is a key design feature that progressively liberates partially locked magnetite particles within the grits size fraction without the energy penalty and ultra-fines generation associated with ball or rod milling:

- Inputs: grits from Splitting.
- Outputs: magnetic product → MSBC1.GGR (GRolls®) → Splitting; non-magnetic → tailings rejection.

MSBC1.FTM – fines magnetic separation

The fines fraction (-212 µm) from the vibratory sieves are fed to the DryFlow separator. Magnetic product from this stage constitutes the primary high-grade iron concentrate from the circuit and is accumulated as final product. Non-magnetic tailings material is rejected. Middlings (particles of intermediate magnetic susceptibility) are collected and assessed for further processing based on their mass proportion and iron grade. Depending on mine site requirements and final product grade targets, middlings may either be combined with the final product stream or forwarded to an optional final beneficiation stage:

- Inputs: fines (-212 µm) from Splitting.
- Outputs: magnetic product → high-grade iron final product; non-magnetic → tailings rejection; middlings → assessed on mass and Fe%.

Final stage (optional) – additional beneficiation

An optional final beneficiation stage may be employed depending on the iron grade and recovery achieved in the fine's separation stage, and on the specific commercial grade target for the operation. If additional upgrading is required, further milling to liberation size is performed (nominally using a rod or ball mill), followed by separation through one or more of: a dry magnetic separator (TM), a wet high-intensity magnetic separator (WHIMS), or flotation. This stage is ore and results dependent and is not expected to be required for all ore types:

- Inputs: concentrate from fines stage and/or middlings.
- Outputs: upgraded product → added to final product stream; tailings → rejected; optional recycle streams depending on performance.

RESULTS AND DISCUSSIONS

Particle size distributions

The particle size distributions (PSDs) for the MSBC1 circuit streams are summarised in Table 5 and presented in Figure 3. The results include the GRolls® product (MSBC1.CGR), the grits stream (MSBC1.GGR), and the final fines product (MSBC1.FGR). It is important to note that both the MSBC1.CGR and MSBC1.GGR data sets were generated under locked cycle conditions. As steady state was achieved and subsequent cycles produced effectively identical PSDs, only a single representative PSD is presented for each stream. This approach is considered appropriate as it reflects the stable operating condition of the circuit. Further published data of the GRolls® producing consistent PSDs for magnetite ore's is presented in Drechsler, Skinner and Abaka-Wood (2026).

TABLE 5

Particle size distribution summary for circuit products.

Passing	Head feed %	MSBC1.CGR discharge %	Grits head feed %	MSBC1.GGR discharge %	MSBC1.FGR discharge %
16 mm	100				
13.2 mm	94				
9.5 mm	76	100			
6.7 mm	64	99			
4.75 mm	58	99			
2.36 mm	47	95			
1.18 mm	22	75			
600 µm	0	50	100	100	
425 µm	0	39	52	72	
300 µm	0	33	22	40	
212 µm	0	28	0	20	100
150 µm	0	24	0	13	82
106 µm	0	21	0	11	69
75 µm	0	16	0	9	54
98%	15.08 mm	4.13 mm	592 µm	588 µm	205 µm
80%	10.36 mm	1.32 mm	525 µm	476 µm	144 µm
RR80		7.9		1.1	
50%	3.01 mm	600 µm	415 µm	340 µm	69 µm
RR50		5.0		1.2	
20%	1.13 mm	102 µm	292 µm	212 µm	28 µm
Slope (m)	0.63	0.54	2.36	1.71	1.41

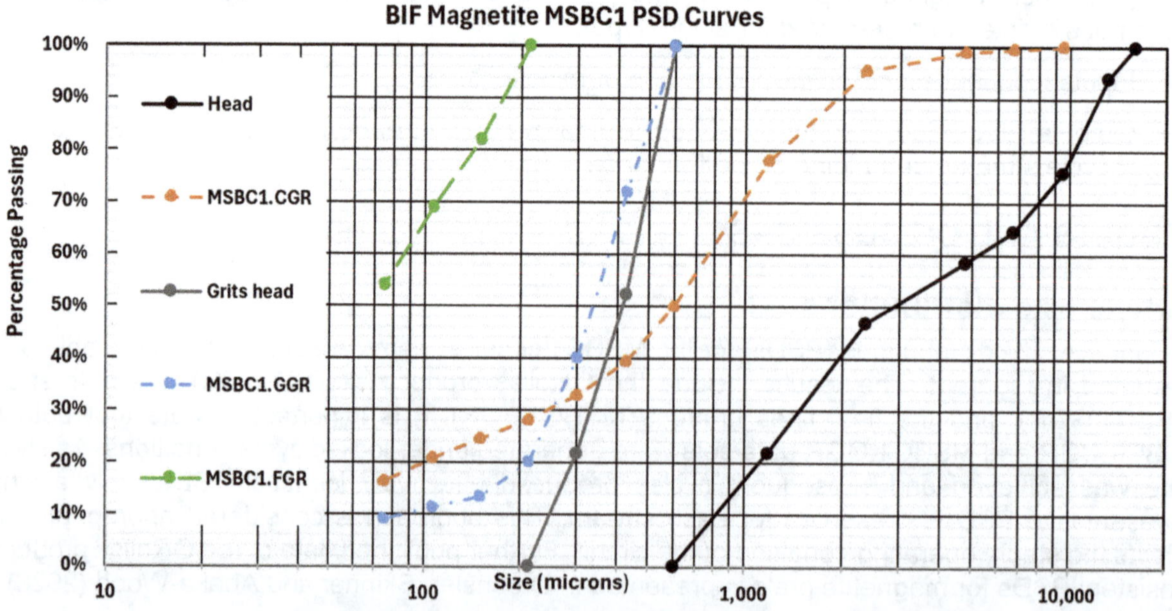

FIG 3 – Cumulative particle size distributions for key circuit products.

The head PSD for MSBC1.CGR corresponds to the steady-state circulating load rather than the initial feed. As such, it represents a realistic and internally consistent feed condition for assessing size reduction performance within the closed circuit. Similarly, the MSBC1.GGR head exhibits a tight and stable PSD across cycles, indicating minimal variation in classification performance and supporting its use as a representative steady-state stream.

The MSBC1.CGR discharge demonstrates significant size reduction relative to its steady-state head, with a P_{80} of approximately 1.32 mm compared to 10.36 mm for the head, corresponding to an RR80 of 7.9. The resulting PSD curve is relatively steep with a slope (*m*) of 0.54, indicating efficient breakage with limited production of excessive fines.

The MSBC1.GGR stream exhibits a much finer feed and discharge profile, with a P_{80} reduction from 525 µm to 476 µm (RR80 ≈ 1.1), indicating relatively limited additional breakage within this stage. However, the PSD remains narrow (slope (*m*) of 1.71), suggesting that this stage primarily acts to refine the particle size distribution rather than significant size reduction. The consistency of this stream across cycles further supports the stability of the recirculation processes.

The MSBC1.FGR product represents the combined fines fraction (-212 µm) extracted from both the MSBC1.CGR and MSBC1.GGR circuits. As this material has undergone multiple stages of processing and classification, it is inherently more refined than any single-pass product. This is reflected in the very fine PSD, with a P_{80} of approximately 144 µm.

Overall, the PSD results demonstrate that the MSBC1 circuit produces stable, repeatable, and relatively narrow particle size distributions under locked cycle conditions. The combination of efficient coarse breakage (MSBC1.CGR), and controlled refinement (MSBC1.GGR), highlights the circuit's ability to progressively reduce particle size while maintaining consistency and minimising excessive fines generation in individual stages.

Energy consumption

All energy calculations are normalised to the original 10 kg feed and reported as kWh per tonne (kWh/t) of final -212 µm (P_{80} 140 µm) product. Absolute energy values (kWh) are accumulated at each stage and normalised only at the conclusion of the analysis. Real time energy consumption at each of the stages is depicted in Figure 4.

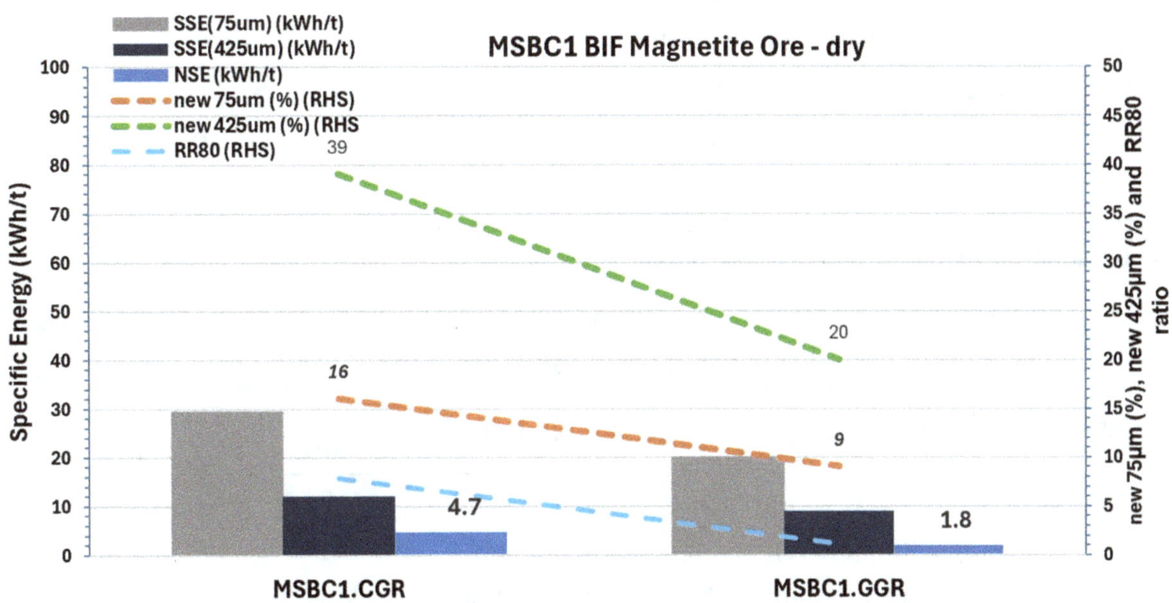

FIG 4 – Energy consumption at both stages.

General methodology

The energy consumed at each processing stage is calculated using the following general expression:

$$E_{Stage}(kWh) = \left(\frac{mass\ processed\ (kg)}{1000}\right) \times specific\ energy\ \left(\frac{kWh}{t}\right)$$

Coarse circuit (MSBC1.CGR) energy consumption

A specific energy of 4.7 kWh/t is applied to all coarse mill (MSBC1.CGR) stages, as locked-cycle testing is representative of steady-state operation.

Initial feed pass

For the initial 10 kg feed:

$$E_{CGR1} = \frac{10}{1000} \times 4.7 = 0.047\ kWh$$

Recirculation

Of the original feed, 4.2 kg is recirculated as coarse material. This represents 42 per cent of the original feed mass, and accordingly consumes 42 per cent of the initial pass energy:

$$E_{CGR2} = 0.42 \times 0.047 \ = \ 0.01974\ kWh$$

Total coarse circuit energy

$$\therefore E_{CGR} = 0.047 + 0.01974 = 0.06674\ kWh$$

Grits circuit energy consumption

Cycle mass tracking

The grits circuit begins with an initial grits mass of 3.3 kg and at every cycle mass is introduced to bring the mass backup to 3.3 kg (locked cycle). At each cycle:

- 80 per cent of the grits mass is retained for recirculation.
- 13 per cent of the retained mass is rejected.

The effective retention fraction per cycle is therefore:

- Retention factor = 0.80 × (1 - 0.13) = 0.80 × 0.87 = 0.696.

Resulting in an initial material remaining per cycle:

- Remaining factor =1 – (0.80 × 0.13) = 0.896.

This is multiplied by the previous cycles remaining material to determine how much total material is rejected.

At each cycle, 20 per cent of the current grits mass is converted to fines (-212 µm product). The mass balance for four cycles is summarised in Table 6.

TABLE 6

Grits circuit mass balance by cycle.

Cycle	Retention mass (kg)	Remaining (%)	Remaining (kg)	Fines generated (kg)
1	3.30	87.0	2.87	0.66
2	2.30 (= 3.30 × 0.696)	77.95 (87 × 0.896)	2.57	1.32
3	1.60 (= 2.30 × 0.696)	69.85 (77.95 × 0.896)	2.31	1.98
4	1.11 (= 1.60 × 0.696)	62.58 (69.85 × 0.896)	2.07	2.64

Number of cycles to equilibrium

to determine the fractional number of cycles required for the cumulative fines mass to equal the remaining grits mass, a linear interpolation is applied between cycles 3 and 4. Let *M* denote remaining grits mass and *F* denote cumulative fines mass at each cycle:

$$n = n1 + \frac{M1 - F1}{(F2 - F1) - (M2 - M1)} = 3 + \frac{2.31 - 1.98}{0.66 - (-0.23)} = 3.36$$

Accordingly, 3.36 cycles are required before the cumulative fines production equals the remaining grits mass. The fourth cycle is therefore applied as a partial cycle with a weighting factor of 0.36.

Energy per cycle

A specific energy of 1.8 kWh/t is applied to the grits mill at each cycle:

$$E_{Cycle} = \frac{m}{1000} \times 1.8$$

Cycle-by-cycle energy values and the accumulated total are presented in Table 7.

TABLE 7

Grits circuit energy consumption by cycle.

Cycle	Retention mass (kg)	Energy consumed (kWh)
1	3.30	0.00594
2	2.30	0.00414
3	1.60	0.00288
4	1.11	0.002

Total grits circuit energy

Summing across all cycles, with the fourth cycle weighted at 0.36:

$$E_{GGR} = 0.00594 + 0.00414 + 0.00288 + (0.36 \times 0.002) = 0.01368\ kWh)$$

Total circuit energy

The total energy consumed across both the coarse mill and grits circuits is:

$$E_{FGR} = E_{CGR} + E_{GGR} = 0.06674 + 0.01368 = 0.08042\ kWh$$

Size specific energy

Normalised to the original 10 kg feed are presented in Table 8:

$$SSE_{Fines} = \frac{0.08042\ (kWh)}{\left(\frac{10}{1000}\right)} = 8.04\ kWh/t$$

TABLE 8

Size specific energy (SSE) for product sizes.

Size (µm)	Mass (%)	SSE (kWh/t)
212	100	8.04
150	82	9.8
106	69	11.65
75	54	14.89

Iron grade and recovery (Fe%)

The results presented in Table 9 are based on preliminary elemental analysis obtained using portable X-ray fluorescence (pXRF), enabling rapid assessment of iron grade across multiple circuit streams during testing.

TABLE 9

Iron grade (Fe%) and recovery (%) by circuit stream – GRolls® circuit versus single pass circuit.

Stream	Mass (%)		Fe grade (%)		Fe recovery (%)	
	Multi (144 μm)	Single (38 μm)	Multi (144 μm)	Single (38 μm)	Multi (144 μm)	Single (38 μm)
Concentrate	41	51.6	59.3	59.2	59	74.4
Middlings	18	10.1	46.9	29	20	7.1
Tailings	41	38.3	21.1	21.8	21	18.5
Calculated head	100	100	41.4	41.8	100	100
Product	59	61.7	55.5	54.25	79	81.5

While pXRF provides reliable indication of relative trends, it is less accurate than a lab based XRF, with results influenced by factors including particle size, surface effects, and sample heterogeneity. As such, reported Fe grades should be considered indicative rather than absolute.

However, as all samples were analysed using a consistent methodology, the comparative assessment of grade, recovery, and mass distribution between single-pass, conventional, and GRolls® – DryFlow multi-stage circuits remains robust. Full laboratory assays are currently underway to validate these results, with no significant changes to overall trends expected.

The GRolls® – DryFlow multi-stage circuit demonstrates a clear improvement in overall beneficiation performance relative to both the conventional conventional-feed circuit and single-pass magnetic separation, despite operating at a significantly coarser final particle size. The GRolls® – DryFlow circuit produces a final fines product at -212 μm with a P_{80} of 144 μm, which is substantially coarser than the grind sizes typically required to achieve comparable results in a single-pass separation. At this coarser size, the circuit achieves a concentrate grade of 59.3 per cent Fe and an overall product grade of 55.5 per cent Fe at a recovery of 79 per cent.

By comparison, the conventional circuit achieves a slightly higher concentrate grade of 62 per cent Fe at a lower overall recovery of 75 per cent, while single-pass magnetic separation requires grinding to finer sizes (38 μm) to achieve similar concentrate grades (~59.2 per cent Fe). At coarser single-pass grind sizes (300–125 μm), concentrate grades remain significantly lower (approximately 49–52 per cent Fe), despite relatively high recoveries. This highlights a key limitation of single-pass processing: achieving high-grade requires extensive fine grinding, which is associated with increased energy consumption and material losses.

The ability of the GRolls® – DryFlow circuit to achieve comparable concentrate grades to single pass separation at 38 μm, while operating at a much coarser 144 μm, represents a significant performance advantage. This indicates that the staged approach enables more effective liberation and separation without the need for excessive size reduction, resulting in improved overall process efficiency.

This improvement is primarily attributed to the size-specific processing strategy. By separating coarse, grits, and fines fractions, the circuit avoids the inefficiencies associated with treating a broad particle size distribution under uniform separation conditions. Coarse gangue rejection in the MSBC1.CTM stage removes a significant proportion of non-magnetic material early in the process, reducing the mass flow to downstream comminution stages.

The grits regrind loop (MSBC1.GGR) plays a critical role in recovering partially liberated magnetite. Unlike conventional tumbling mills, the GRolls® mechanism promotes preferential breakage along mineral boundaries, enabling progressive liberation without generating excessive ultra-fines. This is reflected in the relatively stable middlings fraction and improved recovery of iron units that would otherwise be lost in both single-pass and conventional-feed configurations.

Energy efficiency is a key advantage of the GRolls® – DryFlow circuit. The total circuit energy consumption of 8.04 kWh/t (feed basis) is achieved while producing a fine product (P_{80} of 144 μm), demonstrating that staged comminution combined with early gangue rejection reduces the energy required for downstream size reduction. In contrast, single-pass separation requires grinding to significantly finer sizes (-38 μm) to achieve equivalent grades, implying substantially higher energy demand.

Importantly, the removal of quartz-rich gangue at coarse sizes reduces the overall hardness of the feed material reporting to the grits and fines stages, further lowering comminution energy requirements. Additionally, by rejecting material early in the circuit, the total mass subjected to repeated grinding cycles is significantly reduced, minimising unnecessary energy expenditure on already liberated or barren particles.

A key observation from the circuit performance is that the final fines product, while upgraded, is not yet at full liberation. The product stream is classified at -212 μm with a P_{80} of 144 μm, which remains significantly coarser than the magnetite ores liberation size (-38 μm). This indicates that a substantial proportion of magnetite remains partially locked within gangue minerals and confirms that the circuit is effectively operating as a pre-concentration stage rather than a full liberation circuit.

This represents a strategic advantage. The ability to reject gangue and recover iron units at relatively coarse particle sizes reduces both the mass and hardness of material reporting to downstream processing. As a result, subsequent fine grinding to achieve full liberation can be applied more efficiently, operating on a reduced and softer feed. This is expected to increase final iron grade beyond that achieved in the current study, while maintaining lower overall energy consumption compared to conventional or single-pass approaches.

Overall, the GRolls® – DryFlow circuit achieves a more balanced trade-off between grade, recovery, and energy consumption than both conventional and single-pass beneficiation strategies. The staged approach, combining coarse pre-concentration with targeted downstream grinding, provides a more energy-efficient pathway to achieving high-grade iron concentrates while maintaining high overall recovery.

IMPLICATIONS AND CONCLUSIONS

This study demonstrates that a multi-stage, size-specific dry beneficiation strategy can fundamentally improve the efficiency and sustainability of magnetite processing when compared to conventional multi-stage circuits. The results show that separating and treating discrete size fractions independently enables better alignment between comminution behaviour, liberation characteristics, and magnetic separation performance.

A key implication of this work is that effective early gangue rejection at coarse sizes (+600 μm) significantly reduces the mass reporting to downstream processing stages. This has a direct impact on circuit efficiency, as less material requires energy-intensive size reduction. The GRolls® crushing stages plays a critical role in this context, achieving substantial size reduction (RR80 = 7.9) while maintaining relatively narrow particle size distributions and limiting excessive fines generation. This supports the hypothesis that non-tumbling, compressive breakage mechanisms can promote controlled and consistent breakage, improving downstream separation conditions.

The identification of 600 μm and 212 μm as optimal cut sizes is a central outcome of the study. The results show that while coarser fractions exhibit similar separation selectivity, the -212 μm fraction delivers a marked increase in concentrate grade. This reinforces the importance of targeted processing of fines, rather than treating a broad size distribution under uniform conditions. By isolating and optimising each size fraction, the circuit can maximise both grade and recovery without relying on excessive fine grinding.

The particle size distribution results further highlight the stability and controllability of the multi-stage circuit. Across all stages, the circuit consistently produces narrow and repeatable PSDs under locked cycle conditions. This indicates that the system redistributes material into well defined size ranges rather than generating highly dispersed size distributions. Such behaviour is advantageous for magnetic separation, as it enables more predictable and efficient separation performance.

From an energy perspective, the staged circuit demonstrates a clear advantage over conventional approaches. By combining early rejection, selective regrinding of partially liberated material, and avoidance of unnecessary overgrinding, the circuit reduces the total energy required to produce a -212 µm product. The use of GRolls® in both coarse and grits stages enables progressive liberation with lower energy intensity compared to traditional milling, particularly by minimising the production of ultra-fine material that contributes little to recovery but significantly increases energy demand.

Importantly, the results also highlight the limitations of single pass and conventional multi stage circuits. Single pass magnetic separation was shown to exhibit a trade-off between grade and recovery, with consistent middlings generation indicating incomplete liberation. Similarly, the conventional multi stage circuit, while capable of achieving reasonable recovery, suffers from inefficiencies associated with overgrinding and sub-optimal separation across mixed particle sizes. These findings support the need for more advanced flow sheet designs that decouple size reduction and separation across discrete particle size classes.

Overall, this study establishes that the integration of GRolls® dry crushing with DryFlow dry magnetic separation in a multi-stage configuration offers a viable pathway toward fully dry, energy-efficient iron ore beneficiation. The proposed flow sheet reduces reliance on water, lowers energy consumption, and improves process selectivity, making it particularly attractive for arid regions and future low-emission mining operations.

Future work should focus on scaling the circuit to pilot and industrial levels, alongside further optimisation of operating parameters for each size fraction and evaluation across a broader range of ore types. Attention should also be given to the treatment and optimisation of middlings streams, as these present a clear opportunity to improve overall recovery and circuit performance. Following this dry pre-concentration stage, the upgraded material can be reduced to liberation size (38 µm) and subsequently beneficiated using either the DryFlow magnetic separator or a wet high-intensity magnetic separator, depending on the ore's separation response and the target product grade.

Ongoing research will further investigate energy consumption and evaluate the performance of varying flow sheet configurations to optimise the efficiency and sustainability of the proposed beneficiation approach.

ACKNOWLEDGEMENTS

The authors acknowledge the financial support of Gyratory Roller Solutions Pty Ltd and the loan of the laboratory GRolls400 unit by CBSM Mining Services Pty Ltd to the Future Industries Institute, Adelaide University, for research purposes and to support this honours and other PhD studies on comminution and downstream processing.

The authors acknowledge the financial support of DryFlow Magnetics and the use of their TM-2020 dry magnetic separator.

The authors acknowledge the PhD scholarship sponsored by the ARC Centre of Excellence for Enabling the Eco Beneficiation of Minerals, grant number CE200100009.

The authors thank the industry partners that have provided feed materials for the honours studies.

REFERENCES

Allen, M, 2021. Mining Energy Consumption 2021: A high-level study into mining energy use for the key mineral commodities of the future, Report commissioned by The Weir Group PLC, Coalition for Eco Efficient Comminution (CEEC) International Ltd, Coalition for Minerals Efficiency. Available from: <https://www.ceecthefuture.org/resources/mining-energy-consumption-2021>.

Bazin, C, Sista, R, Légaré, B, Caron, J and Lavoie, F, 2024. Investigation of the effect of mineral composition and size on particle separation in Wet low and high intensities magnetic separators: An industrial case, *Minerals Engineering*, 218:109003.

David, D J, 2011. Optimising crushing in an iron ore beneficiation process, PhD thesis (unpublished), University of Queensland, Brisbane.

Dobbins, M, Dunn, P and Sherrell, I, 2009. Recent advances in magnetic separator designs and applications, The 7th International Heavy Minerals Conference 'What next' (The Southern African Institute of Mining and Metallurgy).

Drechsler, M and Skinner, W, 2023. Commercialisation pathway for low energy wet/dry Gyratory Rolls Crusher comminution technology, *Minerals Engineering,* 204:108419.

Drechsler, M, Skinner, W and Abaka-Wood, G, 2025a. Comparing dry versus wet size reduction and energy consumption performance of innovative GRolls® crushing technology from a wide variety of industry ore feed materials, paper presented at Comminution25, Cape Town.

Drechsler, M, Skinner, W and Abaka-Wood, G, 2025b. Theoretically can the innovative GRolls® crushing technology replace SABC grinding circuits to improve economics, sustainability and mineral liberation of porphyry copper gold ore?', paper presented at Comminution25, Cape Town.

Drechsler, M, Skinner, W and Abaka-Wood, G, 2026. Dry comminution for future magnetite ores with the innovative GRolls® crushing technology, in *Proceedings of the Iron Ore and Open Pit Operators Conference 2026*, this volume, pp 303–314 (The Australasian Institute of Mining and Metallurgy: Melbourne).

Hejazi, R, Chehreh Chelgani, S and Hart, B, 2022. Application of dry cobbing pre-concentration and its impact on pit optimisation in iron ore mining, *Minerals*, 12(10):1253.

Inanloo, F, Moradi, I and Irannajad, M, 2024. Effect of particle shape produced from wet ball milling on magnetic separation concentrate by fractal dimension, *Particulate Science and Technology*, 42(6):962–973.

Nyarko, D, Amos-Judge, T, Drechsler, M, Abaka-Wood, G, Skinner, W, Zanin, M and Asamoah, R, 2026. Effect of Comminution Mechanism on Flotation of Coarse Composites, *Minerals Engineering*, 235:109851.

Xie, S, Hu, Z, Lu, D and Zhao, Y, 2022. Dry Permanent Magnetic Separator: Present Status and Future Prospects, MDPI AG, *Minerals*, 12(10):1251.

Dry comminution for future magnetite ores with the innovative GRolls® crushing technology

M Drechsler[1,2], W Skinner[3] and G Abaka-Wood[4]

1. PhD Student, ARC Centre of Excellence for Enabling Eco-Efficient Beneficiation of Minerals, Future Industries Institute, Adelaide University, Mawson Lakes SA 5095. Email: mark.drechsler@adelaide.edu.au
2. Director, Gyratory Roller Solutions Pty Ltd, Adelaide SA 5051. Email: mark.drechsler@grolls.com.au
3. FAusIMM, Research Professor, ARC Centre of Excellence for Enabling Eco-Efficient Beneficiation of Minerals, Future Industries Institute, Adelaide University, Mawson Lakes SA 5095. Email: b.skinner@adelaide.edu.au
4. Research Fellow, ARC Centre of Excellence for Enabling Eco-Efficient Beneficiation of Minerals, Future Industries Institute, Adelaide University, Mawson Lakes SA 5095. Email: george.abaka-wood@adelaide.edu.au

ABSTRACT

The mining industry is increasingly exploring comminution technologies for future mineral processing plants that can accommodate variable feed throughputs and fluctuating power sources, including renewable energy. These approaches aim to reduce operating costs and improve grade recoveries. In this context, dry comminution technologies such as the high pressure grinding rolls (HPGR) and vertical roller mills (VRM), offer significant opportunities for iron ore mineral processing flow sheets by reducing energy, water and grinding media consumption.

The innovative GRolls® crushing technology, currently being developed in South Australia, extends energy-efficient crushing mechanisms into dry comminution grinding applications. It could be integrated with downstream processes such as dry magnetic separation and coarse particle flotation to enhance the sustainability and efficiency of iron ore processing and beneficiation. GRolls® is characterised by its ability to generate steep, narrow, and consistent fine particle size distributions, even with variable feed sizes and throughput. Its short residence time within the crushing chamber minimise over-grinding, enabling dry operation without the need for grinding media.

This paper presents the results of an extensive laboratory testing program investigating the size reduction and energy performance of the laboratory scale Alpha GRolls400 unit for two magnetite ores, ranging from hard to soft, and one moderately hard hematite ore. The study provides critical performance data on the GRolls® crushing technology, supporting the identification of opportunities for future small to medium scale mobile and modular iron ore processing plants incorporating GRolls® units. Such flow sheets could offer the mining industry enhanced flexibility and energy efficiency for processing remote and complex magnetite and hematite orebodies, particularly in regions with limited energy and water resources.

INTRODUCTION

The International Energy Agency (2022) recommended the promotion of technical innovation throughout the mineral industry value chain and outlined future mining operations will need comminution technologies that can economically process orebodies that are becoming more complex, of lower grade, needing shorter development time and located in remote locations. Powell *et al* (2023) also proposes that different comminution technologies are required to grinding mills for future mineral processing plants with variable feed throughputs and variable power sources (renewable energy), technologies that can generate dry coarser products with consistent steep narrow size ranges for early gangue rejection flow sheets. Processing of future orebodies is also likely to be constrained by limited water resources, with the mining industry investigating dry comminution technologies into mining process flow sheets to reduce energy, water and media consumption.

High pressure grinding rolls (HPGR) technology has been studied and implemented into iron ore processing circuits replacing or augmenting tumbling mills with demonstrated reductions in energy, cost and media consumption (Dierx, van Rijswick and de Oliveira, 2025; Lovatt *et al,* 2023; Turrer, Kodukula and de Oliveira, 2025). Vertical roller mills (VRM's) are also being considered in future magnetite mineral processing flow sheets (David *et al,* 2024). Whilst HPGR's and VRM's are able to produce dry coarse products, their high capital and operating costs are providing opportunities for new novel comminution technologies to be developed and innovative circuits to be pursued by the mining industry to further optimise future beneficiation processes (Bearman, 2024).

The Gyratory Roller Crusher technology (GRolls®) is a novel comminution technology that proposes to extend crushing capabilities into primary and secondary grinding circuits utilising the best advantages of crushers and HPGR technologies to generate fine products from coarse feeds (Drechsler and Skinner, 2023). The GRolls® technology uses a combination of load mechanisms to simultaneously apply pulsed compression, tension and shear forces to a packed particle bed. The drive roller of the GRolls® 'pulls' material into the crushing chamber where the gyrating roller impacts the particles at high frequency, then particles are cleared of the crushing chamber by the continued rotary action of the drive roller. The laboratory scale Alpha GRolls400 unit built to prove the concept of the technology is shown in Figure 1.

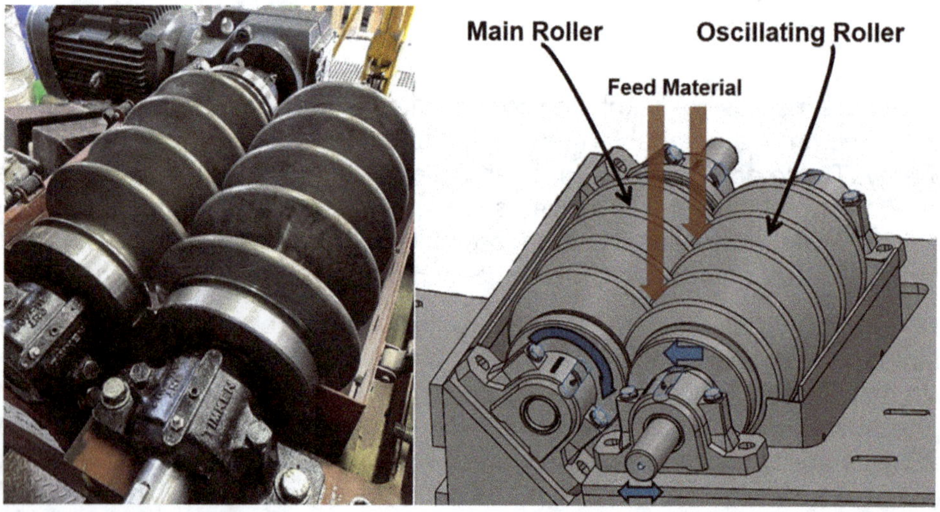

FIG 1 – GRolls400 photo and schematic drawing.

A PhD research study on the Gyratory Roller Crusher technology forms part of an ongoing project which seeks to identify novel and innovative technologies for downstream processing coarse particles and early gangue rejection strategies (Australian Government, 2025). Preliminary results on investigating the size reduction and energy performance of the GRolls400 unit on selected porphyry copper gold and magnetite ores (including wet testing) have been presented in the literature and the reader is referred to Drechsler, Skinner and Abaka-Wood (2025a, 2025b, 2026a, 2026b).

This paper presents the size reduction and energy performance results using a laboratory GRolls400 unit to test three iron ore feed materials, dry and without media. The study continues with further testing of the iron ore materials processed by the laboratory GRolls400 and attrition milling equipment to determine and compare particle breakage characteristics and mineral liberation performance. The potential benefits of the GRolls® crushing technology on downstream processes, mineral grade recoveries and early gangue rejection strategies will be published in the future.

MATERIALS AND METHODS

Feed materials

Three iron ore materials have been provided for comminution testing by the laboratory scale GRolls400 unit for the current study, comprising two magnetite ores and one hematite ore from South Australian sites, with detailed characteristics summarised in Table 1. Based on their Bond Ball Mill Work Index (BBMWI, kWh/t) results, the three iron ores represent a range from 'hard' to 'soft' ores which covers a large range of specific comminution energy recorded within mineral processing plants (Daniel, 2006).

TABLE 1
Detailed characteristics of the three iron ores in the study.

Name	BIF magnetite	BIF hematite	Braemar magnetite
Location	Magnetite resource, Southern Middleback Ranges, South Australia	Hematite resource, Northern Middleback Ranges, South Australia	Exploration resource, Braemar region, South Australia
Source geology	Banded Iron Formation (BIF) upper iron silicate magnetite	Banded Iron Formation (BIF) goethitic hematite	Magnetite-rich siltstone dominated tillite
Bond Ball Mill Work Index ($BBMW_I$)	15.7 kWh/t	12–13 kWh/t	8.6 kWh/t
Classification^	hard	moderate	soft
Specific Density	3.3–3.4 t/m^3		3.4 t/m^3
Maximum size	13 mm	37.5 mm	26.5 mm
Assay Main Elements (%)			
SiO	48.2%	45.4%	45.4%
FeO	31.9%	28.2%	21.2%
Al_2O_3	0.7%	8.7%	7.7%
CaO	0.4%	0.1%	3.6%
MgO	4.3%	0.3%	3.4%
S	0.1%	0.1%	0%
Dominant Mineralogy (approximate %)	Quartz 27% Magnetite 17% Goethite 14% Feldspar 12% Carbonate 8% Hematite 4% Muscovite, sericite, pyroxene and chlorite <5%	Iron Oxide 57% Quartz 22% Clay 6% Chlorite 6% Tourmaline Group 5%	Quartz 31% Plagioclase 16% Hematite group 11% Mica group 11% Dolomite group 11%

Note: ^ Classification as per Daniel (2006).

Crushing studies

The GRolls® technology used in the studies is a laboratory scale Alpha GRolls400 prototype located at Adelaide University. The GRolls400 unit is comprised of two interlocking rollers being 400 mm in length and 250 mm in diameter (Figure 1), with the main roller driven by a 7.5 kW direct drive motor and the oscillating roller driven by a 5 kW belt drive motor. The oscillating roller has a 0.1 mm offset shaft rotating at a frequency of around 1000 rev/min which generates the gyratory motion and pulsed compression, shear and tension loads within the multiple chambers. Further details of the GRolls® technology and applied crushing forces are presented in Drechsler and Skinner (2023).

The feed material is prepared, sieved to obtain varying size fractions, weighed and discharged onto a variable speed vibrating pan feeder which transfers the material into the GRolls400 crushing chamber. Each feed and discharge material are dry sieved from 37.5 mm to 75 μm using ISO 3310.1 (Standards Australia, 2016) nominal aperture test sieves to determine the particle size distribution (psd) and then compared to the feed material to determine size reduction performance (RR80, Equation 1).

$$RR80 = \frac{F_{80}}{P_{80}} \tag{1}$$

where F_{80} and P_{80} are the theoretical square mesh aperture sizes (mm) corresponding to the 80 per cent of cumulative undersize in the feed and product, respectively. A second metric for determining size reduction performance is the log/log slope of the psd curve of the product between 80 per cent and 20 per cent (P_{80} and P_{20} respectively), as determined by slope (m) (Equation 2), with higher m values reflecting steeper slopes.

$$\text{Slope (m)} = \frac{log20 - log80}{\log P20 - \log P80} \tag{2}$$

The laboratory GRolls400 unit is equipped with an electrical control panel which has variable speed drives for both the main and oscillating motors and a 3 phase power monitoring system for recording total power consumption (kW) as well as individual motor power consumption. The No Load Power Draw (kW) is determined by running the crusher empty with rollers disengaged for a minimum of 1 min (to facilitate bearings warming up) and then finding the average of the instantaneous power recorded in the 30 seconds before and after testing. The loaded power draw (Gross Power Draw, kW) with the rollers engaged was determined by averaging the power draws recorded during feed being discharged into the crushing chamber.

The Gross Power Draw (kW) minus the No Load Power Draw (kW) provides the Net Energy Consumption (kW) which is then converted to Net Specific Energy (NSE, kWh/t) based on the sample weight (kg) and duration (minutes) of each test (Equation 3).

$$NSE(kWh/t) = \frac{Gross\ power\ draw\ (kW) - No\ Load\ power\ draw\ (kW)}{Circuit\ tonnage\ (t/h)} \tag{3}$$

The Size Specific Energy, which refers to the energy required to produce target fractions, typically 75 µm to 600 µm (Ballantyne, Peukert and Powell, 2015) are determined for each test for comparison across the testing program and industry data using Equation 4.

$$SSE(xxx\mu m)(kWh/t) = \frac{NSE\ (kWh/t)}{(Product\ \%xxx\mu m\ - Feed\ \%xxx\mu m)/t} \tag{4}$$

Ballantyne, Peukert and Powell (2015) proposed the generation of -75 µm fraction as a simplified measure of specific energy, this size fraction in the middle of the typical range of 20 to 150 µm for fine flotation of base metals (Jameson, 2010). Furthermore, Ballantyne, Peukert and Powell (2015) also proposed that for circuits producing a coarser product, a coarser proxy size may also be more appropriate. Jameson (2010) proposed floating coarser particles up to 850 µm could be achieved depending on the mineral liberation characteristics of the ore, and early gangue rejection for coarse particles ranging from 100 to 600 µm have also been reported by Kohmuench *et al* (2018).

This study uses the -75 µm and -425 µm particle sizes for comparison with fine and coarse (early gangue rejection) gravity, magnetic and flotation separation downstream applications. The percentage of new fractions (discharge % passing minus feed % passing, represented as 'new 75 µm' and 'new 425 µm' criteria) generated in the GRolls400 crushing chamber is used to 'normalise' the particle size results between different feed sizes, ores and test conditions for statistical analysis and performance comparisons. As described in Equations 3 and 4, energy consumption data (NSE and SSE) is comprised of many data variables (and potential experimental variability and errors) derived during each test, therefore energy data are not statistically analysed but broadly discussed and compared in this paper.

The effect of different GRolls400 test configurations (feed size, feed rates, main drive speeds and moisture content) on size reduction and energy performance is presented in detail by Drechsler, Skinner and Abaka-Wood (2026b).

RESULTS AND DISCUSSIONS

All three iron ore feed materials (as provided) were crushed through the GRolls400 unit in a single pass at the same configuration (50 rev/min and 30 psi) at around 300–400 kg/hr, with the feed and single pass discharge psd curves presented in Figure 2. The size reduction of the BIF Hematite (coarsest feed) was the highest at RR80 of 11.9 and slope (*m*) of 0.68, with RR80 of 11.1 and *m* of 0.33 for the Braemar Magnetite and RR80 of 6.6 and *m* of 0.52 for the BIF Magnetite (finest feed). P_{80} ranged from 1.18 mm to 1.4 mm, generation of new 75 µm fraction from 10–18 per cent and new 425 µm fraction from 29–31 per cent, even though the maximum feed size ranged from 37.5 mm to 13 mm and ore hardness ranged from hard to soft.

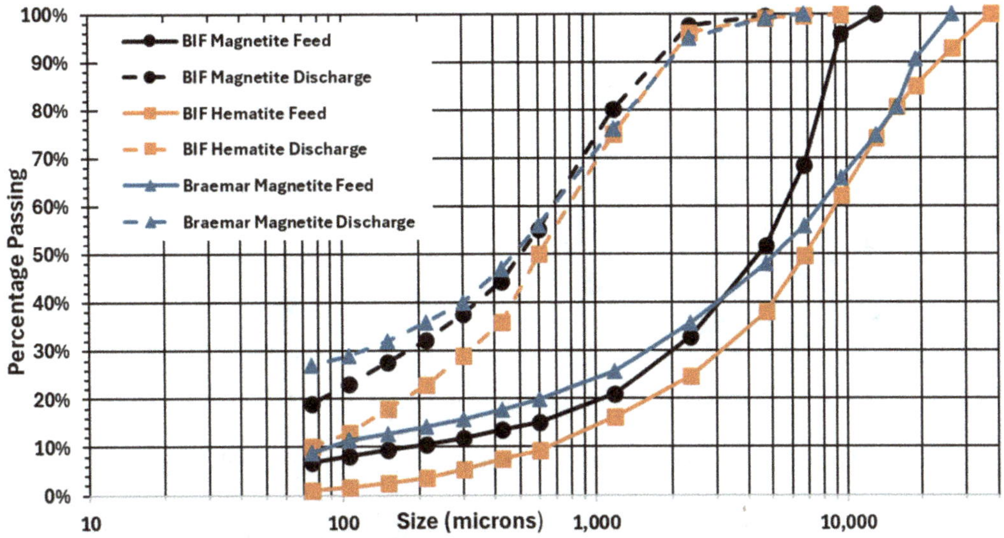

FIG 2 – Three iron ore feeds – single pass PSD curves.

The GRolls400 unit also generates consistent single pass discharge materials as demonstrated by repeat tests of the BIF Magnetite presented in Figure 3. Of the 13 tests conducted on the same feed material, the new 75 µm fraction ranged from 8 to 13 per cent, with 11 tests ranging from 11 to 13 per cent and two outlier tests of 8 per cent and 10 per cent. The new 425 µm fraction had a wider range from 22 to 33 per cent, with 11 tests ranging from 27 to 33 per cent and the two outliers of 22 and 24 per cent. The P_{80} ranged from 1.1 mm to 1.5 mm with RR80 ranging from 5.2 to 7.1 over those 13 tests.

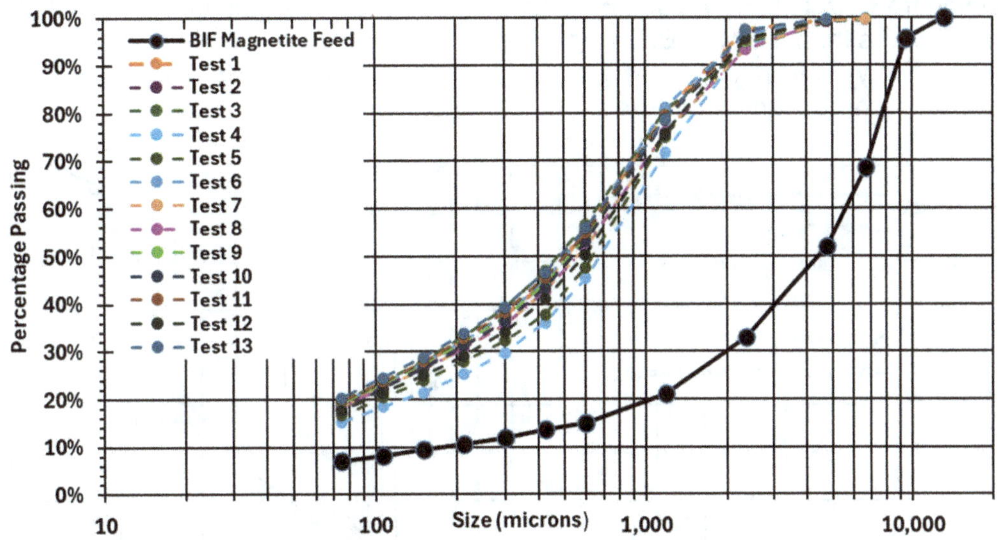

FIG 3 – BIF magnetite ore feed – single pass PSD curves for 13 tests.

As Figures 2 and 3 demonstrates, the GRolls400 generates a consistent narrow range of discharge materials even though the feed material may vary significantly. This ability to crush a wide size range of feed materials and produce consistent discharge products is further demonstrated in Figure 4 for Braemar Magnetite feed materials ranging from F_{80} of 22 mm down to 1.1 mm. The generation of new 75 µm particles was consistent from 17 to 20 per cent whilst the generation of new 425 µm particles increased from 27 to 45 per cent with reducing feed sizes. The size reduction performance of the GRolls400 ranged from a RR80 of 10.5 for the coarse feeds down to 1.4 for the very fine feeds.

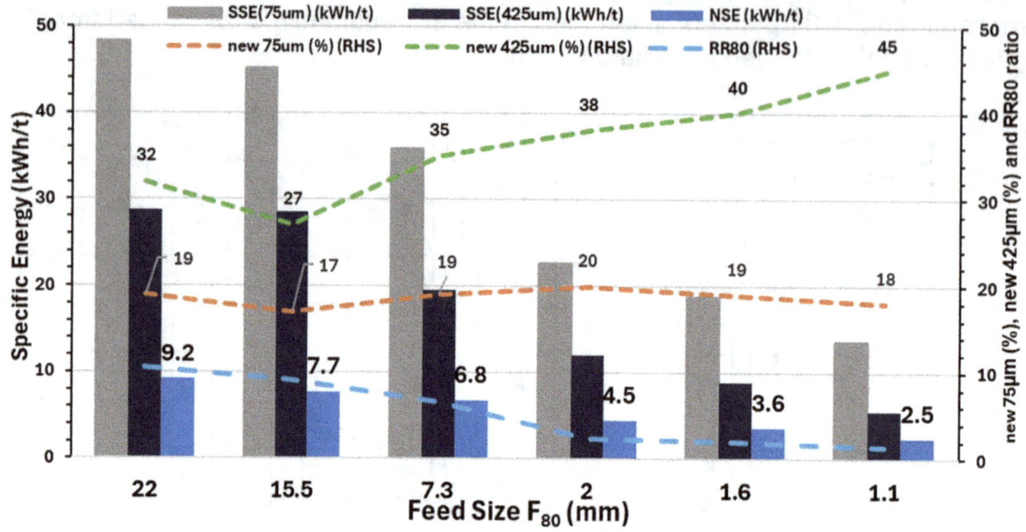

FIG 4 – Braemar magnetite – performance varying feed size (F_{80}).

As a result of these size reduction trends, the energy consumption of the GRolls400 decreased with decreasing feed sizes, ranging from NSE of 9.2 kWh/t for the coarse feed (F_{80} = 22 mm) down to 2.5 kWh/t for the very fine feed (F_{80} = 1.1 mm), with similar reductions in size specific energy (SSE) for the -75 µm and -425 µm fractions. These results confirm the high size reduction efficiency and the steep psd curves with a narrow size range of discharge materials by the GRolls400. Drechsler, Skinner and Abaka-Wood (2026b) describe how the unique configuration of the GRolls400 rollers with multiple crushing chambers facilitates a consistently low generation of new 75 µm particles and high generation of new 425 µm particles for all feed material sizes.

When comparing crushing a -9.5 mm feed size for all three iron ores (Figure 5) conducted at the same configuration around 310 kg/hr, all three ores had similar discharge curves +425 µm with RR80's ranging from 5.8 to 6.6. The soft Braemar magnetite ore resulted in finer (and flatter at slope *m* = 0.42) psd curves than the hard BIF magnetite and moderate BIF hematite ores (slope *m* = 0.52 and 0.63 respectively), reflecting generation of new 425 µm particles within the crushing chamber is higher for the softer ores than harder ores.

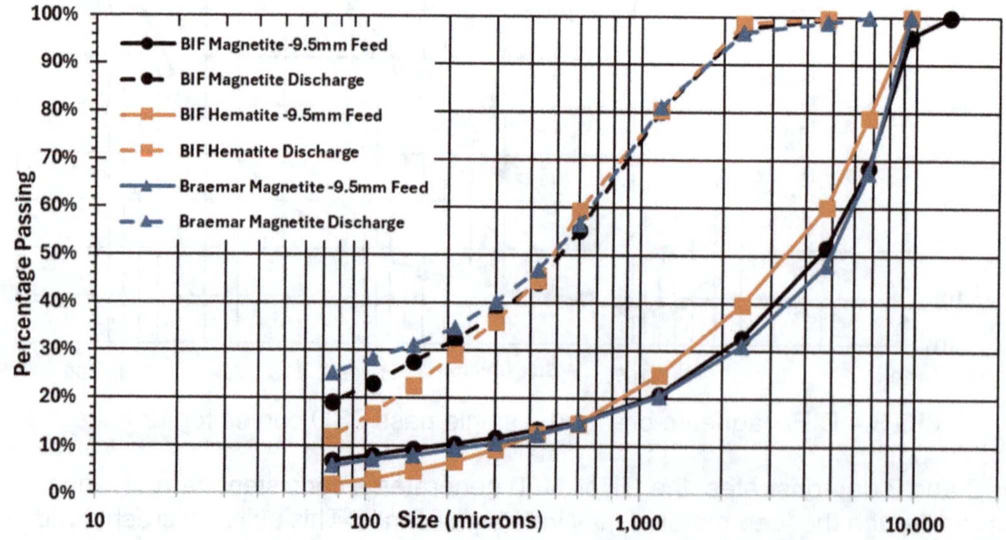

FIG 5 – Three iron ore -9.5 mm feeds – single pass PSD curves.

As presented in Figure 6, the GRolls400 generated 10–12 per cent new 75 µm fraction for both the 'hard' and 'moderate' ores and a slightly higher 19 per cent new 75 µm particles for the 'soft' Braemar magnetite ore. The higher fines generation (-75 µm and -425 µm fractions) for similar net energy consumption (NSE ranged from 6.3 to 6.8 kWh/t) results in lower SSE's for the soft Braemar magnetite compared to the BIF magnetite and BIF hematite ores. Figure 6 confirms the importance of normalising particle size data using new 75 µm and new 425 µm criteria when comparing different test conditions (ie feed and discharge psd curves such as Figure 5).

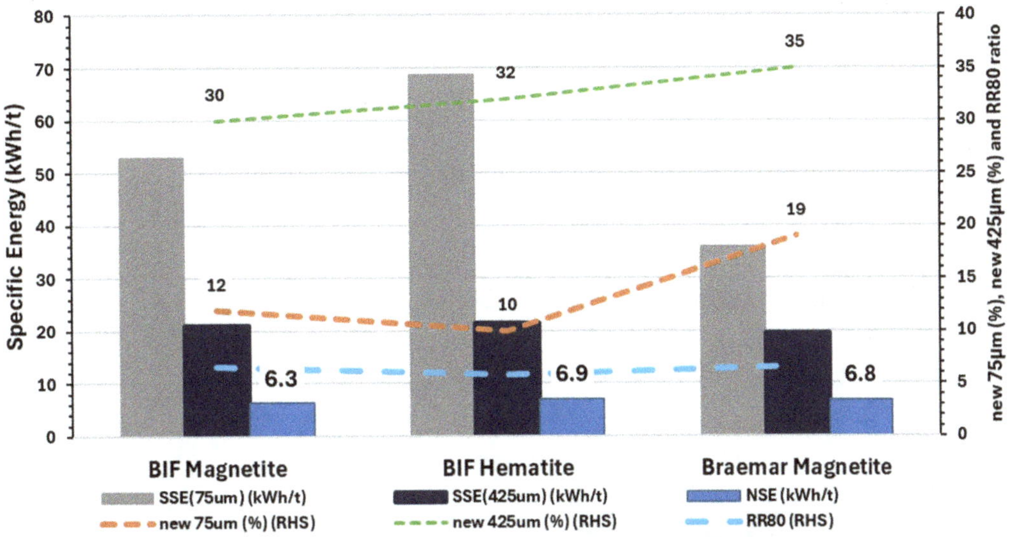

FIG 6 – Three iron ore -9.5 mm feeds – single pass comparative performance.

The new 75 µm results when crushing soft ores (as represented by the Braemar magnetite) compared to harder ores (BIF Magnetite and Hematite) suggests that high percentages of ultra fines (-75 µm) are unlikely in a single pass of the GRolls400 which may adversely impact downstream processes. The GRolls® technology has low residence time (seconds) of materials within the crushing camber, and when combined with the high frequency of the gyrating roller, is designed to limit generation of fines (-75 µm) and potential over grinding of soft particles during each pass.

A locked cycle test program was conducted on BIF Magnetite -6.7 mm to +425 µm (F_{80}=1.1 mm) feed to simulate a fine second stage (secondary grinding) closed circuit operation of the GRolls400, with the -425 µm fractions removed after each cycle and replaced with fresh -6.7 mm +425 µm feed in each subsequent cycle (Table 2). The -425 µm fraction was nominated for removal to simulate potential early gangue rejection strategies. As shown in Figure 7, the generation of both new 75 µm and new 425 µm material was found to be consistent across the last three cycles, resulting in relatively consistent NSE and SSE results and recirculating loads of around 300 per cent.

The locked cycle test demonstrated similar size reduction and energy consumption to single pass tests, steep curves with consistent new 75 µm generation (9–11 per cent) whilst generating high percentage of coarser fractions such as new 425 µm (20–27 per cent). These characteristics should result in high size reduction and lower recirculation cycles in closed circuits flow sheets.

TABLE 2

BIF magnetite ore -6.7 mm +425 µm locked cycle test.

Cycle #	1	2	3	4	5	6
Feed Mass (kg)	10.0	10.0	10.0	10.0	10.0	10.0
New Feed (kg)	10.00	2.83	3.09	3.20	3.24	3.32
Feed Rate (kg/hr)	545	474	474	439	419	429
+425 µm (kg)	7.17	6.91	6.98	6.77	6.68	6.74
-425 µm (kg)	2.83	3.09	3.20	3.24	3.32	3.26
new 75 µm generated (%)	9	10	9	11	11	11
new 425 µm generated (%)	20	21	21	26	27	26
NSE (kWh/t)	3.4	3.6	3.6	4.6	4.6	4.8
SSE(75 µm) (kWh/t)	38	36	40	41.7	42	44
SSE(425 µm) (kWh/t)	17.2	17.2	17.1	17.6	17.1	18.6

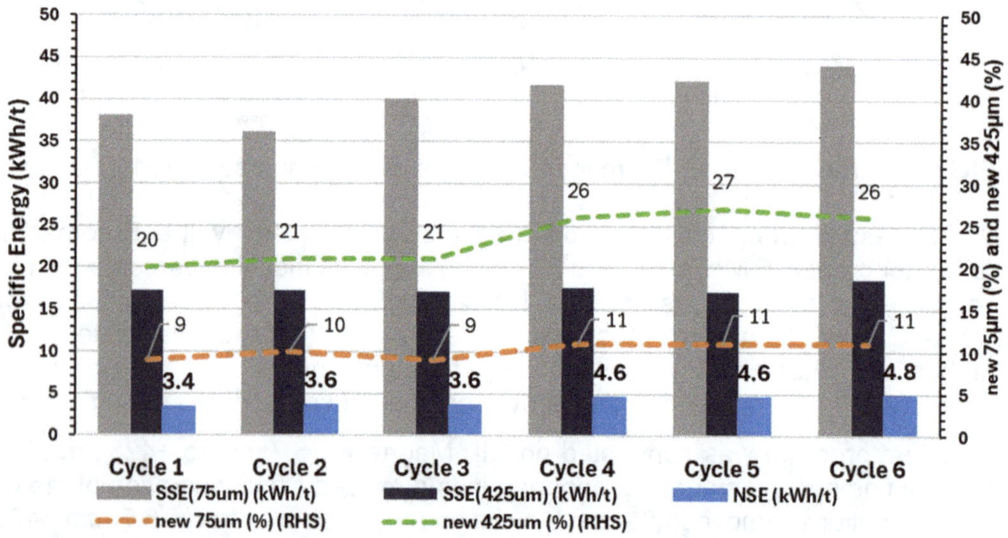

FIG 7 – BIF magnetite -6.7 mm +425 µm locked cycle test.

IMPLICATIONS OF STUDY

Dry testing by the GRolls400 has demonstrated for iron ore feeds with hardness ranging from 'hard' to 'soft' and for feed sizes up to 37 mm and down to 2.36 mm, a size reduction (RR80) ranged from 1.4 to 10.5 and NSE ranged from 2.5 kWh/t to 9.2 kWh/t. HPGR's in the same ore size range generally achieve size reduction (RR80) from 2 to 4 for NSE's ranging from 1 to 3 kWh/t (Daniel, Klein and Wang, 2019). These compare to dry crushers (+10 mm) with RR80 ranging from 3 to 6 and NSE's of 1 to 2 kWh/t, coarse grinding (10–1 mm) NSE's of 3 to 10 kWh/t and fine grinding (1 mm–100 µm) of 10–20 kWh/t (The AusIMM, 2012; Wills and Finch, 2016).

Baawuah *et al* (2020) conducted comminution studies on a BIF iron carbonate magnetite (BBMWi of 12–13 kWh/t) and their published laboratory HPGR size reduction data is presented in Figure 8 for comparison to the GRolls400 psd curve for the harder BIF iron silicate magnetite (BBMWi of 15.6 kWh/t) in this study. The GRolls400 achieved a much higher size reduction performance, with a RR80 of 6.6 compared to 1.8 for the HPGR, resulting in a steeper slope *m* of 0.52 compared to 0.39 and a higher percentage of new fines generated in a single pass (12 per cent new 75 µm compared to 6 per cent).

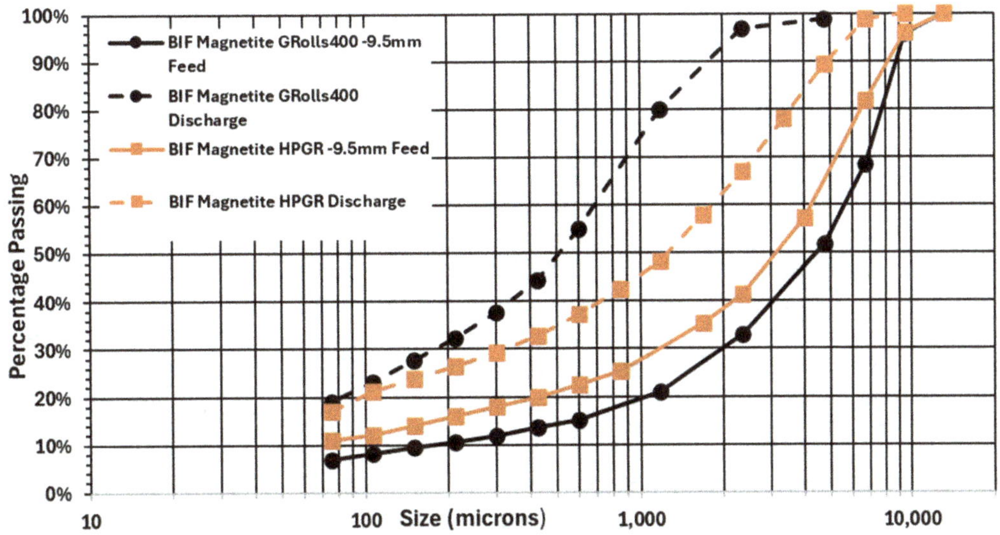

FIG 8 – BIF magnetite -9.5 mm feed – GRolls400 versus HPGR single pass PSD curves.

Putland, Lovatt and Byrnes (2025) analysed the PSD slopes (*m*) from their industry database of over 650 plant surveys from 170 comminution circuits across a range of commodities, deriving a mean slope (*m*) of 0.43 from 363 final milling circuit products and a steeper 0.52 for 16 HPGR-Ball circuit final products. Figure 9 presents a histogram comparing the industry wide slope (*m*) data (Putland, Lovatt and Byrnes, 2025) with GRolls400 data for 71 tests on BIF Magnetite and Braemar Magnetite in this study. The GRolls400 demonstrates steeper PSD slopes for both magnetite discharge products, with over 87 per cent of the GRolls400 discharge materials steeper than the industry average for final milling circuit products (*m* = 0.43) and 51 per cent of GRolls400 products steeper than the industry average for HPGR-Ball circuit final products (*m* = 0.52).

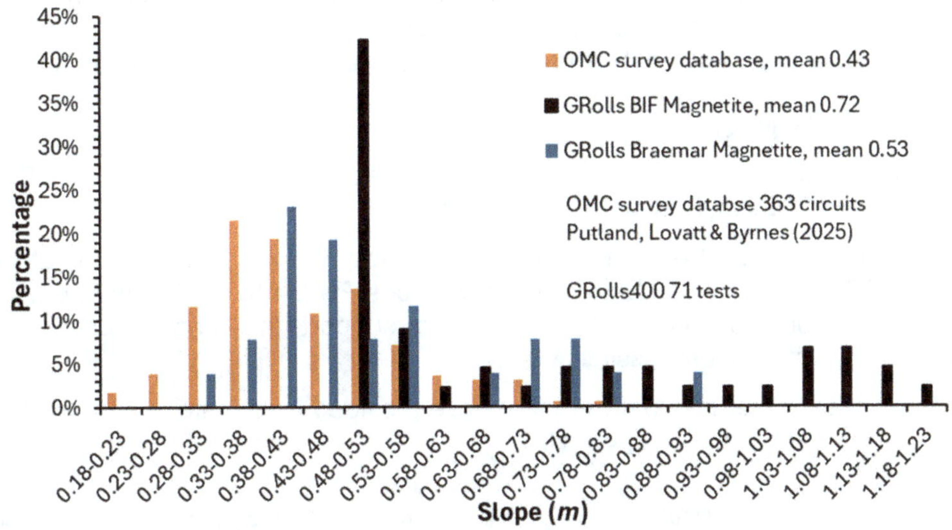

FIG 9 – Slope (m) – GRolls400 magnetite tests versus OMC survey database.

Testing of the laboratory GRolls400 unit with iron ore materials have demonstrated that the technology can generate a discharge product with a psd which is consistently steep and within a narrow size range in a single pass for a large range of dry feed sizes and ore hardness. The short residence time within the crushing chamber limits 'over-grinding' of 'soft' ores and particles. Additional studies are being undertaken to determine the impact of the dry coarse discharge from the GRolls400 with downstream processes, such as multi-stage dry magnetic separation (Cox *et al,* 2026), magnetic and electro-static separation and coarse particle flotation, to improve iron ore beneficiation.

The GRolls® technology is being progressed through its commercialisation pathway with the design, build and pilot plant operation of an upscaled 3–5 t/h GRolls1000 unit. The research studies are providing critical performance data on the GRolls® crushing technology and are identifying opportunities to match future GRolls® units with crushing and HPGR circuits within small to medium sized (<100 t/h) mobile to modular mineral processing plants that could provide the iron ore mining industry with flexibility and energy efficiency in achieving fine grinding performance for early gangue rejection flow sheets without media or water.

CONCLUSIONS

The current study was carried out to confirm the size reduction and energy performance of the laboratory GRolls400 unit using three iron ore feed materials ranging from 'hard' to 'soft'. The study has confirmed its high size reduction capacity, especially for coarse feed materials, and the generation of steep narrow psd curves featuring consistent generation of new 75 μm and new 425 μm in single pass and locked cycle (closed) tests. Size reduction performance (RR80) for the GRolls400 ranged from 1.4 to 10.5 and NSE results ranged from 2.5 kWh/t to 9.2 kWh/t.

The GRolls400 demonstrated the target for the novel comminution GRolls® crushing technology is dry secondary grinding applications ideally receiving 5 to 10 mm iron ore feed from crushers and HPGR's. Varied feed sizes and ore hardness have some impact, but not adversely, on the GRolls400 size reduction and energy consumption performance, and therefore is not expected to adversely affect downstream processes because of over-grinding and inconsistent discharge materials. These GRolls400 performance characteristics are considered ideal for early gangue rejection strategies such as multi-stage dry magnetic separation and coarse particle flotation technologies, able to generate consistent steep, narrow sized discharge products between 75 μm and 425 μm, even with variable feed, dry (or wet), without media, and with minimal opportunity for over-grinding due to short residence time of iron ore particles in the crushing chamber.

Research studies and technological development of the GRolls® is continuing to further explore the advantageous impacts of the dry, no media innovative comminution technology on particle shape, geochemistry, mineral liberation and downstream processes such as dry magnetic separation and coarse particle flotation technologies. When combined in small to medium sized mobile to modular plants, these innovative technologies could facilitate early gangue rejection strategies that improve grade recovery, reduce operating costs and unlock marginal iron orebodies of the future.

ACKNOWLEDGEMENTS

The authors acknowledge the PhD scholarship sponsored by the ARC Centre of Excellence for Enabling the Eco Beneficiation of Minerals, grant number CE200100009.

The authors also acknowledge the financial support of Gyratory Roller Solutions Pty Ltd and the loan of the laboratory GRolls400 unit by CBSM Mining Services Pty Ltd to the Future Industries Institute, Adelaide University, for research purposes and to support this and other PhD studies on comminution and downstream processing.

The authors thank the two industry partners that have provided feed materials for the PhD studies.

REFERENCES

AusIMM, The, 2012. *Cost Estimation Handbook*, second edition, Monograph 27, 528 p (The Australasian Institute of Mining and Metallurgy: Melbourne).

Australian Government, 2025. Annual Report 2024, Australian Research Council (ARC) Centre of Excellence for Enabling Eco-Efficient Beneficiation of Minerals (CoE Minerals). Available from: <https://www.coeminerals.org.au/about/annual-reports>

Baawuah, E, Kelsey, C, Addai-Mensah, J and Skinner, W, 2020. Comparison of the performance of different comminution technologies in terms of energy efficiency and mineral liberation, *Minerals Engineering*, 156:106454.

Ballantyne, G, Peukert, W and Powell, M, 2015. Size specific energy (SSE)—Energy required to generate minus 75 micron material, *International Journal of Mineral Processing*, 136:2–6.

Bearman, R, 2024. Changing the Face of Comminution – Insurmountable Hurdles or Just Speed-Humps on the Journey?, in *Proceedings of the XXXI International Mineral Processing Congress, IMPC2024*, pp 112–128 (Curran Associates, Inc: Red Hook).

Cox, O, Coventry, T, Drechsler, M, Robbeson, J and Abaka-Wood, G, 2026. Optimising sustainable iron ore beneficiation through multi-stage dry magnetic separation and advanced comminution technologies, in *Proceedings of the Iron*

Ore and Open Pit Operators Conference 2026, this volume, pp 285–302 (The Australasian Institute of Mining and Metallurgy: Melbourne).

Daniel, M, 2006. The Researcher Who Rummaged through Bond's Tool Box, in *Fourteenth Annual JKMRC Annual Conference and Second International JKMRC Conference*, pp 3–28 (JKMRC: Brisbane).

Daniel, M, Klein, B and Wang, C, 2019. High-pressure grinding roll technology, in *SME Mineral Processing and Extractive Metallurgy Handbook*, 443:2019.

David, D, Olwagen, D, Gerold, C, Schmitz, C, Baaken, S, Stanton, C and Everitt, M, 2024. Plant design comparison of dry VRM milling plus magnetic separation with AG and ball milling plus magnetic separation for Grange Resources' Southdown ore, in *Proceedings of Mill Operators Conference 2024*, pp 573–586.

Dierx, B, van Rijswick, R and de Oliveira, R N M, 2025. Next Leap in HPGR Flow sheet Design, paper presented at Comminution25, Cape Town.

Drechsler, M and Skinner, W, 2023. Commercialisation pathway for low energy wet/dry gyratory rolls crusher comminution technology, *Minerals Engineering*, 204:108419.

Drechsler, M, Skinner, W and Abaka-Wood, G, 2025a. Comparing dry versus wet size reduction and energy consumption performance of innovative GRolls® crushing technology from a wide variety of industry ore feed materials, paper presented at Comminution25, Cape Town.

Drechsler, M, Skinner, W and Abaka-Wood, G, 2025b. Theoretically can the innovative GRolls® crushing technology replace SABC grinding circuits to improve economics, sustainability and mineral liberation of a porphyry copper gold ore?, paper presented at Comminution25, Cape Town.

Drechsler, M, Skinner, W and Abaka-Wood, G, 2026a. Early gangue rejection using gyratory rolls crushing technology, paper presented at MetPlant 2026, Adelaide.

Drechsler, M, Skinner, W and Abaka-Wood, G, 2026b. Investigating the size reduction and energy performance of the innovative GRolls® crushing technology on porphyry copper gold and magnetite ore feed materials, *Powder Technology*, in review.

International Energy Agency (IEA), 2022. The Role of Critical Minerals in Clean Energy Transitions, World Energy Outlook Special Report, IEA. Available from: <https://www.iea.org/reports/the-role-of-critical-minerals-in-clean-energy-transitions>

Jameson, G, 2010. Advances in Fine and Coarse Particle Flotation, *Canadian Metallurgical Quarterly*, 49(4):325–330.

Kohmuench, J N, Mankosa, M J, Thanasekaran, H and Hobert, A, 2018. Improving coarse particle flotation using the HydroFloat™ (raising the trunk of the elephant curve), *Minerals Engineering*, 121:137–145.

Lovatt, I, Becker, M, Putland, B, Radford, R and Robinson, J, 2023. Trade-Off Realities in HPGR vs SAG Milling—A Practical Comparison of Tropicana and Gruyere Comminution Circuits, paper presented at SAG2023, Vancouver.

Powell, M, Reynolds, A, McCrae, S, Bracey, J, Way, D, Phasey, C, Littlechild, T and Manning, D, 2023. Can we shift to a new paradigm of flexibility in mining and processing to build mines powered exclusively by the variable input of renewables?, paper presented at World Mining Congress 2023, Brisbane.

Putland, B, Lovatt, I and Byrnes, J, 2025. Factors Influencing Fines Generation in Comminution Circuits, paper presented at Comminution25, Cape Town.

Standards Australia, 2016. ISO 3310.1-2016 — Test sieves — Technical requirements and testing — Part 1: Test sieves of metal wire cloth, August 2016.

Turrer, H, Kodukula, U and de Oliveira, R, 2025. Evaluation of HPGR Application for Size Reduction in Iron Ore Processing, paper presented at Comminution25, Cape Town.

Wills, B A and Finch, J A (eds), 2016. Comminution, in *Wills' Mineral Processing Technology*, eighth edition, pp 109–122 (Butterworth-Heinemann: Boston).

Effect of handling improver on sinter productivity under high-moisture conditions

S Fujiwara[1], S Kato[2], Y Kinugasa[3], S Awazu[4], Y Hirano[5], N Taniyama[6] and T Kikkawa[7]

1. Senior Researcher, JFE Steel Corp., Fukuyama Hiroshima 721-8510, Japan.
   Email: sh-fujiwara@jfe-steel.co.jp
2. Senior Management Staff, JFE Steel Corp., Chiyoda-ku Tokyo 100-0011, Japan.
   Email: shi-kato@jfe-steel.co.jp
3. Senior Management Staff, JFE Steel Corp., Fukuyama Hiroshima 721-8510, Japan.
   Email: y-kinugasa@jfe-steel.co.jp
4. Senior Management Researcher, JFE Steel Corp., Kawasaki Kanagawa 210-0855, Japan.
   Email: s-awazu@jfe-steel.co.jp
5. Chief Engineer, Kurita Water Industries Ltd, Akishima Tokyo 196-0005, Japan.
   Email: y.hirano59@kurita-water.com
6. Engineer, Kurita Water Industries Ltd, Akishima Tokyo 196-0005, Japan.
   Email: n.taniyama89@kurita-water.com
7. Senior Chief Engineer, Kurita Water Industries Ltd, Akishima Tokyo 196-0005, Japan.
   Email: t.kikkawa79@kurita-water.com

INTRODUCTION

Powdered raw materials such as iron ore frequently present operational challenges when moisture content increases or particle size becomes finer. Under high-moisture conditions, these powders can transition into slurry, resulting in severe handling difficulties during transport, storage, and processing (Shimizu and Sugihara, 2023). Such conditions often lead to operational disruptions, increased maintenance, and safety concerns in large-scale ironmaking operations. In addition to these operational issues, slurry-like raw materials are also known to deteriorate granulation behaviour, leading to reduced sinter bed permeability and lower sintering performance and productivity (Matsumura *et al,* 2009). Conventional strategies to reduce moisture or modify powder properties typically require significant capital investment, specialised equipment, and extended processing time, making them impractical for many plants. To address these issues in a simple and adaptable manner, a modification technique using handling improvers was developed. The principle of this approach lies in forming crosslinked structures between the additive and particle surfaces, which stabilises the bulk material and suppresses slurry formation under high-moisture conditions (Iveson *et al,* 2001; Qiu *et al,* 2003). This mechanism reduces the tendency for fluidisation without altering the fundamental chemistry of the raw material, ensuring compatibility with downstream processes and minimising environmental impact.

METHODOLOGY

A key technical property of the handling improver is its ability to increase the flow moisture point (FMP) of iron ore. FMP is a critical parameter that determines the threshold at which bulk materials begin to flow as a liquid under increasing moisture (Munro and Mohajerani, 2015). In this study, FMP is measured by flow table test. Figure 1 shows the result of laboratory measurements demonstrating that the addition of the handling improver increased the FMP of the tested iron ore by up to 5.5 percentage points, depending on the additive dosage. For example, the FMP increased from 9.9 per cent to 15.4 per cent as the dosage of the handling improver was raised. This improvement in FMP enables more stable handling and transport of high-moisture ore, providing a technical basis for further evaluation in the sintering process.

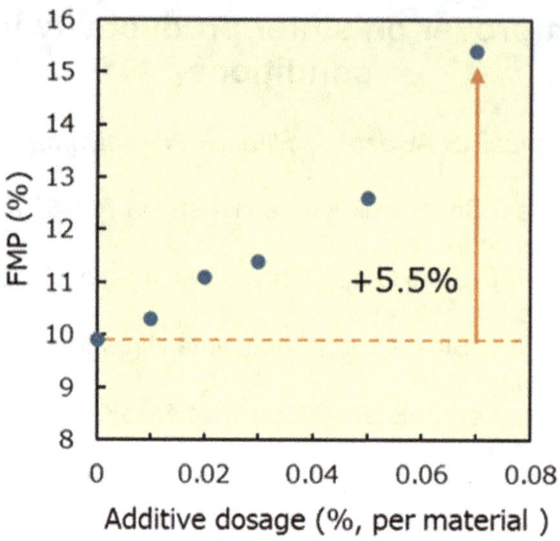

FIG 1 – Relationship between additive dosage and FMP.

The main focus of this study is the assessment of sinter productivity and process stability when using the handling improver under various moisture and dosage conditions. In the ironmaking process, the treated iron ore is granulated and charged into the sintering machine. Laboratory tests were conducted to examine the influence of additive application on granulation behaviour, bed permeability, and sinter quality indices. These tests provide insights into how the additive interacts with the sintering process, ensuring that improvements in handling do not compromise productivity or product quality.

Sinter pot tests were conducted to evaluate the effect of handling improver on sinter productivity under high-moisture conditions. For all test conditions, an ore mixture was prepared by blending two types of iron ore in equal proportions: one prone to handling problems ('problematic ore') and one less prone ('non-problematic ore'). The handling improver was added only to the problematic ore prior to mixing, while the non-problematic ore was used without any additive. The dosage of the handling improver was set at 0.01 per cent and 0.03 per cent by mass, calculated as a percentage of the total raw material weight.

The chemical composition of the sinter product was standardised to SiO_2 = 5.0 per cent and basicity (CaO/SiO_2) = 2.0 across all experiments, ensuring comparability. The ore mixture was prepared with moisture contents ranging from 6.0–12.0 per cent, simulating both optimal and challenging operational scenarios.

Granulation behaviour was assessed by measuring the mean granule size. The addition of the handling improver to the problematic ore within the mixture resulted in a decrease in mean granule size, attributed to enhanced water retention within particle clusters and reduced liquid bridging.

Subsequent sintering tests evaluated bed permeability, gas flow, and sinter yield. Figure 2 shows the relationship between sinter productivity and moisture content for different additive dosages. At lower moisture levels (6.0–7.5 mass per cent), the control condition (0 per cent additive) exhibited the highest productivity, and the addition of the handling improver resulted in a slight decrease in productivity. However, as the moisture content increased, the productivity of the control dropped sharply, while the mixtures containing the handling improver (0.01 per cent and 0.03 per cent) maintained higher productivity, especially at 9.0–10.5 mass per cent moisture. For example, at 10.5 mass per cent moisture, productivity with 0.03 per cent additive dosage was approximately double that of the control. These results indicate that while the handling improver may reduce productivity at lower moisture levels, it is highly effective in mitigating the negative impact of excess moisture and enables stable and efficient sintering operation under high-moisture conditions.

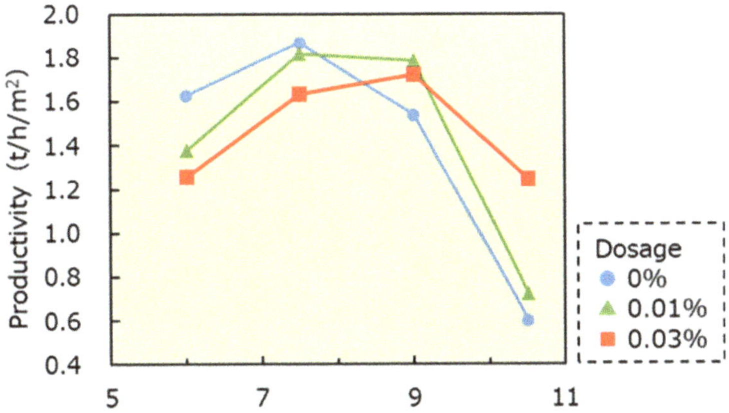

FIG 2 – Relationship between sinter productivity and moisture content in various additive dosages.

The mechanism underlying these improvements involves both the physical confinement of moisture within granules and the stabilisation of granule structure, leading to better airflow and higher yield. As a result, the granulated ore maintains its integrity during handling and sintering, even when moisture levels are elevated.

In regions with seasonal rainfall or where ore moisture exceeds optimal levels, the data suggest that handling improvers can enhance productivity and operational stability. The ability to maintain stable sintering operations under variable moisture conditions is particularly valuable for plants facing fluctuating raw material quality or environmental challenges. The approach described here offers a practical and scalable solution for ironmaking operations, enabling significant improvements in sinter productivity and process efficiency.

Importantly, the application of handling improvers does not introduce any adverse effects on the chemistry of the raw material or the quality of the sintered product. Laboratory analyses confirmed that the additive is compatible with existing process flows and does not interfere with downstream metallurgical reactions.

CONCLUSION

In summary, the use of handling improvers enables significant improvement in sinter productivity under high-moisture conditions, offering a practical and scalable solution for ironmaking operations facing moisture variability. This approach not only stabilises raw material handling and sintering but also enhances operational robustness and efficiency in challenging environments.

REFERENCES

Iveson, S M, Litster, J D, Hapgood, K and Ennis, B J, 2001. Nucleation, growth and breakage phenomena in agitated wet granulation processes: a review, *Powder Technology*, 117:3–39.

Matsumura, T, Maki, T, Amano, S, Sakamoto, M and Iwasaki, N, 2009. Effect of moisture absorption behaviour on optimal granulation moisture value of sinter raw material, *ISIJ International*, 49(5):618–624.

Munro, M C and Mohajerani, A, 2015. Determination of the transportable moisture limit of iron ore fines for the prevention of liquefaction in bulk carriers, *Marine Structures*, 40:193–224.

Qiu, G, Jiang, T, Li, H and Wang, D, 2003. Functions and molecular structure of organic binders for iron ore pelletisation, Colloids and Surfaces A: Physicochemical and Engineering Aspects, 224:11–22.

Shimizu, Y and Sugihara, M, 2023. A numerical approach on liquefaction of iron ore fines under roll motion of bulk carriers using microscopic fluid coupling scheme in discrete element method, *Ocean Engineering*, 283:114695.

Addressing 21st century challenges in iron ore beneficiation

K P Galvin[1]

1. Director, ARC Centre of Excellence for Enabling Eco-Efficient Beneficiation of Minerals, University of Newcastle, Callaghan NSW 2308. Email: kevin.galvin@newcastle.edu.au

ABSTRACT

With growing interest in iron ore beneficiation in Australia, it is important to consider the complex challenges that are likely to arise. What are the options in the Pilbara for addressing the state of the fines when mining below the water table? Here, some level of processing becomes essential, if only to prepare the material for utilisation in downstream smelters. What are the options for maximising grade and recovery of Pilbara ores in general, and perhaps securing limited sources of Direct Reduced Iron (DRI) grade? What are the options for establishing a DRI pathway for magnetite?

Key parts of the Australian industry will increasingly be impacted by the cost of producing concentrate. Here, Australian producers will need to compete on a more level playing field, contingent on the choice of beneficiation technology. The focus needs to shift to simple, highly efficient technology. One of the biggest costs is the loss of Fe units simply in delivering desliming. Then there is the considerable complexity in navigating the beneficiation of the ultrafines, maximising grade, while minimising the loss of the Fe units. Magnetite offers the prospect of DRI grades, but the circuits can require efficient milling to promote liberation at particle sizes as fine as 45 microns while ensuring there is no overgrinding.

This presentation reports on recent work concerning a suite of new capabilities aimed at addressing these challenges. These capabilities are central to developing highly functional circuits for processing sub 1 mm fines formed via classifying screens, or, through the preliminary breakage and comminution used to achieve mineral liberation. Gravity separation warrants a strategy that can target the recovery and concentration of hematite down to about 7 μm. Ideally the feed should be classified at a nominal size sufficiently fine to ensure success at the finer end of the distribution. This split creates two portions of material, the fine and the coarse particles. There will be many options available for addressing the coarse portion, dependent on the liberation, including additional comminution, gravity, reverse coarse particle flotation, and even magnetic separation. But the key is to ensure efficient classification, which means complete closure of the partition curve with no bypass (Awuye *et al,* 2025). Despite the effectiveness of high-frequency screens, hydrodynamic approaches will continue to play an important role, especially at classification sizes finer than 100 microns. The bypass produced by cyclones (Runge *et al,* 2024) needs to be addressed, especially in the context of comminution. This presentation examines recent advances in addressing these critical issues.

The finer portion presents by far the greatest challenge. Ideally the material is sufficiently liberated within this size range. Often the colloidal clays, the ultrafines well below 10 microns, generate significant additional viscosity. Beyond a critical feed concentration, the viscosity increases exponentially, meaning it may be necessary to lower the feed concentration. But if it is necessary to lower the feed concentration too far, the gravity separation may be deemed unproductive. It is then essential to extract the slimes responsible for the excessive viscosity. Hydrocyclones are largely ineffective in desliming at 10 μm due to the need to use a larger number of small cyclones, the considerable pressure drop that must be applied, and the strong tendency for significant bypass. A 30 per cent bypass still leaves excessive slimes that have exponential impact.

The Graviton™ Separator, which consists of modules of inclined channels in a spinning centrifuge, offers a natural point of desliming, retaining dense hematite while releasing the fine colloidal clays. The inclined channels create powerful segregation forces delivering a significant capacity advantage. The G forces of the centrifuge then multiply with the benefits of the inclined channels, leading to a capacity advantage greater than 1000-fold (Galvin and Dickinson, 2013). This presentation outlines the work to date in developing this technology, including direct upgrading to DRI grade by desliming (Carpenter, 2020). The Graviton™ Separator, which is currently under development, rejects the lower density slimes finer than about 20 microns, while retaining hematite down to about 7 microns. In some cases, it may be necessary to release 15 to 30 per cent of the solids to secure a slimes-free material.

The material that emerges from the underflow of the Graviton™ Separator is concentrated but importantly free of slimes-based viscosity (Carpenter, 2020). Thus, the feed volume is considerably smaller, with less low-density slimes, and less water. Ultrafine particles of hematite then segregate efficiently. The final phase of the circuit involves the upgrading of the Graviton™ Separator underflow. The volumetric feed rate can be reduced due to the high solids concentration available. This presentation presents recent advances in the single stage recovery of fine hematite utilising a device based on the Reflux Classifier known as the GradePro™ Classifier. This device consists of a lower fluidised bed of reduced cross-sectional area, incorporating a slow rotating rake. A system of inclined channels of larger cross-sectional area resides above. Single stage upgrade from 37 per cent Fe to 68.8 per cent Fe as hematite has been achieved consistently (Galvin, Zhou and Rodrigues, 2025), controlled directly to target specific grades via a set point setting. Thus, the gravity separation has been proven for hematite so will also work for magnetite.

In conclusion, this work describes a relatively simple but innovative flow sheet that is readily adapted for different scenarios. The above generic description involves coarse particle removal to maximise the performance of the fines upgrade. The fines then flow to the Graviton™ Separator. Here the Graviton™ Separator underflow is sent to a GradePro™ Classifier that covers the entire size range of the fine material to secure DRI grade. Similarly, the coarse portion could also undergo further classification, grinding, and upgrading. There is no need for additional stages of flotation and magnetic separation.

ACKNOWLEDGEMENTS

The authors acknowledge the funding support from the Australian Research Council for the ARC Centre of Excellence for Enabling Eco-Efficient Beneficiation of Minerals, grant number CE200100009.

REFERENCES

Awuye, D E, Starrett, J B, Amosah, M E, Rodrigues, A F D V and Galvin, K P, 2025. Hydrodynamic Classification of a Binary Density Feed with No By-Pass, *ChemRxiv*, preprint, https://doi.org/10.26434/chemrxiv-2025-kfhv9

Carpenter, J L, 2020. Gravity Separation and Desliming using Inclined Channels Subject to Different G-Forces, PhD Thesis, University of Newcastle.

Galvin, K P and Dickinson, J, 2013. Particle Transport and Separation in Inclined Channels Subject to Centrifugal Forces, *Chemical Engineering Science*, 87:294–305.

Galvin, K P, Zhou, J and Rodrigues, A F, 2025. Single stage production of ultra-high-grade iron ore using a novel fluidisation arrangement in a Reflux Classifier, *Minerals Engineering*, 233:109627. https://doi.org/10.1016/j.mineng.2025.109627

Runge, K, Frausto, J, Lisso, M and Jokovic, V, 2024. Importance of considering classification and liberation when optimising comminution and flotation, *Minerals Engineering*, 209:108612.

Grange Resources Savage River mine-to-process optimisation

J Gerber[1], D Olwagen[2], W Valery[3], A Jankovic[4], K Duffy[5], R Hayashida[6], E Tabosa[7], J Siong[8] and J Apostol[9]

1. Minerals Process Engineer, Hatch, Brisbane Qld 4000. Email: johan.gerber@hatch.com
2. Quality and Process Improvement Manager, Grange Resources, Burnie Tas 7320. Email: dian.olwagen@grangeresources.com.au
3. Global Director Consulting and Technology, Mining and Minerals Processing, Hatch, Brisbane Qld 4000. Email: walter.valery@hatch.com
4. Principal Consultant, Hatch, Brisbane Qld 4000. Email: alex.jankovic@hatch.com
5. Senior Process Consultant, Hatch, Brisbane Qld 4000. Email: kristy.duffy@hatch.com
6. Senior Consultant Drill and Blast, Hatch, Brisbane Qld 4000. Email: rodrigo.hayashida@hatch.com
7. Senior Process Engineer, Hatch, Brisbane Qld 4000. Email: erico.tabosa@hatch.com
8. Process Engineer, Hatch, Brisbane Qld 4000. Email: joyce.siong@hatch.com
9. Process Engineer, Hatch, Brisbane Qld 4000. Email: john.apostol@hatch.com

ABSTRACT

The Mine-to-Process integration and optimisation methodology utilises Hatch's expertise, industry validated modelling and simulation tools, and extensive database to maximise the profitability of the entire operation (from mine to plant) with minimal capital expenditure. This holistic approach has delivered throughput and efficiency benefits as well as cost reduction to many operations worldwide including Grange Resources' Tasmanian based Savage River operation.

This paper details the integrated Mine-to-Process optimisation completed at Grange Resources Savage River and how the mechanistic, mathematical models developed during the project have been continually leveraged to support site's successive optimisation and engineering design studies.

During the project, site-specific mechanistic models of the mine and plant (blast fragmentation, crushing, grinding, screening, cyclones, magnetic separation) were developed and calibrated based on historical data, audits and surveys conducted by Savage River and Hatch personnel. The models were used to simulate and determine the optimal strategy for achieving the objectives of maintaining concentrate production rates with the lower Fe feed grades expected towards the mine's end of life, improve final product quality and reduce costs.

Several opportunities to increase the comminution circuit throughput were identified and evaluated and as a result Savage River is now in the process of upgrading the pebble crushers in the primary grinding circuit and implementing several other recommendations. Additionally, Savage River is now also actively evaluating the upgrade of its secondary grinding circuit which was identified to be lacking capacity.

The site-specific models were again used in subsequent studies focused on the effects of transitioning from the current open pit to a block cave mining method. This study highlighted how ore characteristics and the Run-of-Mine (ROM) fragmentation from the block cave would differ from the current open pit operation and how it would likely evolve through the life-of-mine and impact the processing plant. The models were used to evaluate the effects of these changes on the performance of the comminution and concentration circuits and to identify opportunities to maintain concentrate production. Additionally, several other opportunities highlighted during the Mine-to-Process project were revisited and refreshed based on the updated information of Savage River's forward workplan.

INTRODUCTION

Grange Resources is an iron ore operation based in Tasmania, Australia. Its assets consist of the Savage River mine and Port Latta pellet plant, producing high-quality iron pellets as its final product to market. It is among Australia's longest operating iron ore operations having been first commissioned in 1967.

To date, Savage River has operated as a truck and shovel open pit mine that draws magnetite ore from two pits, North and Centre pit. The magnetite ore is fed to the Savage River concentrator (Figure 1) which currently consists of a single gyratory crusher, and two parallel grinding circuits Which are each comprised of a primary and secondary grinding stage. The primary stage consists of a 32-foot autogenous (AG) mill in closed circuit with pebble crushers. Product from the primary grinding stage is subjected to rougher magnetic separation with the concentrate stream reporting to the secondary grinding circuit which is comprised of a 13-foot ball mill in closed circuit with hydrocyclones. Product from the secondary grinding stage is then treated in hydroseparators to reject low density gangue (silica) before undergoing a final stage of magnetic separation in the finisher magnetic separators. Concentrate then undergoes two stages of screening before it reports to the concentrate thickener. The screens play a critical role of pipeline protection as concentrate must be pumped approximately 80 km north to the Port Latta pellet plant.

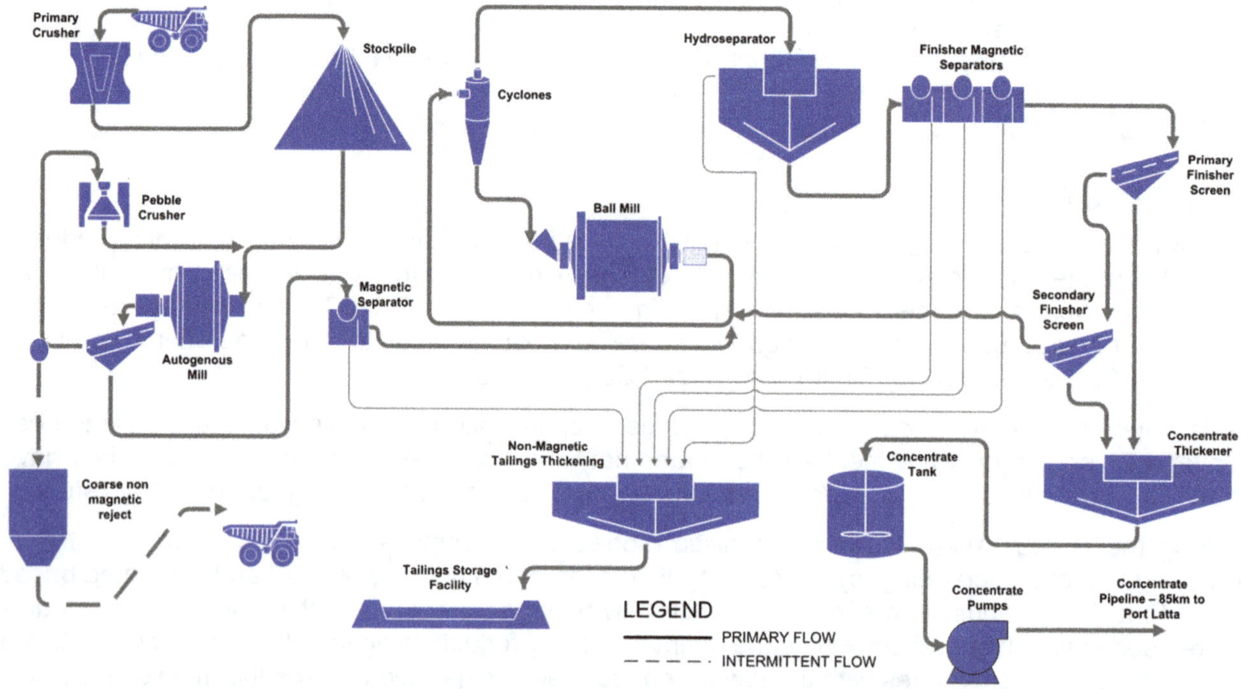

FIG 1 – Simplified flow sheet of the Savage River concentrator.

Hatch was engaged by Grange Resources in late 2022 to complete a Mine-to-Process (M2P) optimisation of its Savage River operations. The project developed site-specific, mechanistic models of the drill and blast, comminution, and magnetic separation processes. The models were integrated to evaluate the impacts and benefits of changes in the mine to the downstream plant to identify process improvements and optimised strategies to maintain concentrate production at lower feed grades with little to no capital expenditure. Several recommendations from the project are currently being implemented. In parallel, the models developed from the project along with the site-specific knowledge and understanding of the operation continue to be leveraged to support Grange Resources' initiatives to drive further process improvements at the Savage River concentrator and further extend the life-of-mine (LOM).

METHODOLOGY

Successful M2P optimisation considers the entire operation to form a strong fundamental understanding of the conditions, constraints, and limits of the operation. *In situ* ore characteristics significantly influence the performance of blasting and downstream processing (Kanchibotla, Valery and Morrell, 1999). Hence, a comprehensive understanding of the ore characteristics and distribution across the deposit is necessary, as well as an understanding of how these impact each stage of the mining and processing value chain.

Therefore, as part of the M2P project, the existing ore characterisation, mining, and processing practices were reviewed and a comprehensive drill and blast audit and plant survey with ore characterisation and comminution breakage testing was conducted. All collected data was used in conjunction with long-term historical site operating data along with Hatch's extensive industrial database to develop and integrate the site-specific models. Simulations were conducted to evaluate opportunities for improvement and optimisation in the mine and plant. For Savage River in particular, the primary focus of the M2P was threefold, maintain concentrate production at lower Fe feed grades, improve concentrate quality and recovery, and reduce operating costs. Several opportunities for each goal were recommended based on the findings of the simulations and wider M2P investigations:

- Maintain concentrate production at lower feed grades:
  - Optimised drill and blast design to produce finer ROM fragmentation.
  - Improved pebble crusher utilisation and operating strategy.
  - AG to semi-autogenous (SAG) conversion for finer and more friable feed materials.
  - Additional secondary grinding circuit power.
- Improve concentrate quality and recovery:
  - Improved hydroseparator instrumentation and operating/control strategy.
  - Improved finisher magnetic separation and finisher screen operation.
- Reduce operating costs:
  - Optimised drill and blast for waste mining and improved waste ore domain testing and characterisation.

These recommendations are detailed in the subsequent sections.

M2P PROJECT FINDINGS

Finer ROM fragmentation

Blasting is the first, and among the most cost-effective and energy efficient step of comminution. Blast optimisation can have significant benefits (increased throughput and reduced specific energy consumption) to downstream comminution circuits. However, the optimum ROM fragmentation is entirely dependent on the downstream comminution equipment and circuit configuration.

A review of the current drill and blast practice at Savage River along with simulations with the blasting model revealed that there may be potential scope to further optimise the blast design to improve the ROM fragmentation (increase the fines, -10 mm material content in the plant feed). The downstream AG mills however, require coarse material in the feed to act as grinding media. Consequently, new blast designs were not proposed for the current AG mill operation. However, preliminary blast designs were developed to aid during SAG operation which was also investigated and detailed in the subsequent section.

Waste blasting

The investigation of cost reduction was not in the original scope of the M2P project, however, it was later added, at site's request, as waste mining forms a large proportion of the mining operation at Savage River. The same blast models were leveraged to evaluate waste blast designs but due to its late addition to the scope, only a high-level assessment was completed. The rock strength for the waste ores was significantly variable, so the waste was further classified into distinct domains based on Point Load Test (PLT) results. The blast design for each domain was then optimised to reduce costs by relaxing the drill pattern and blast intensity respective to the domain's strength. It was estimated that the new designs could potentially lower waste blast costs by approximately 25 per cent or AUD300 000 per month.

AG to SAG conversion

The AG mills are the most frequent bottleneck at Savage River, reaching the installed power limit of the mills around 53–77 per cent of the time. Therefore, conversion of one or both AG mills to SAG milling is a potential opportunity to maintain or increase throughput. Additionally, the plant occasionally receives very fine, friable material from the Centre Pit which causes issues with maintaining load in the AG mills. More of this finer and friable ore is expected in the future, and conversion to SAG milling would make the comminution circuit more robust against these changes in ore characteristics.

The addition of steel grinding media to the AG mills (conversion to SAG) will increase breakage rates and power utilisation. However, steel media, which cannot be removed with a conventional belt magnet due to the magnetite ore, will require an alternative tramp metal removal system, else the pebble crusher will have to be retired completely.

Simulations completed in SAG operating modes indicated a throughput increase for all scenarios, but the product feeding the secondary circuit would coarsen and the resulting final grind size would be compromised unless additional power was installed.

Pebble crusher utilisation

The pebble crushers are currently operated at a wide range of power draw and closed-side setting (CSS) as they are used for 'fine tuning' of the AG mill circuit to balance the load between the primary and secondary grinding circuit. They also receive a high proportion of fine/product size material (due to the small 2.5 mm AG mill discharge screen aperture), which ideally should not be fed to the pebble crushers as it unnecessarily occupies volumetric capacity.

Due to the specific pebble crusher application described above only hypothetical simulations were carried out to assess potential pebble crusher impacts on the capacity of AG mill circuit: very fine pebble crushing, coarse pebble crushing and no pebble crushing

Simulations suggested that very fine pebble crushing (CSS of 10 mm) would increase capacity of the primary grinding circuit by just over 10 per cent. However, final product size would coarsen (P_{80} may increase by several microns) and the feed rate to the rougher magnetic separators (RMS) would exceed the current maximum capacity. Operation of pebble crushing with coarse CSS or operation without pebble crushing both decreased circuit throughput but resulted in a modestly finer final product size.

Rougher screen efficiency

Improved classification efficiency in the rougher screens would serve to alleviate both the AG mill and the RMS capacity by reducing the fines recycle. Currently, the screens operate at a very low classification efficiency of 58.8 per cent, leading to an AG mill total recirculating load of 155 per cent.

Simulations indicate that improving the classification efficiency of the rougher screens could potentially alleviate the recirculating load by approximately 30 per cent and decreases the feed to the RMS by roughly 80 t/h albeit at a slightly coarser product size.

Ball mill circuit optimisation

The ball mills were typically operating at about 1750 kW, well below the installed power of 2237 kW. Simulations investigated the potential benefits of operating at an increased ball charge (40 v/v per cent) to maximise power utilisation and smaller grinding media to improve grinding efficiency. However, both simulations indicated only minor improvements with respect to final product size.

Deeper investigations revealed that challenges with the ball mill circuit was largely due to high recirculating loads resulting in excessive volumetric flow rates. These flow rates were causing the mills to approach their volumetric capacity and resulted in low residence times and very high superficial slurry velocities (SSV). The residence time and SSV were estimated to be 0.03 h and 0.24 m/s respectively. When benchmarked against other industrial ball mills as shown in Figure 2, the residence time and SSV approaches the region in which operational problems are frequently encountered.

These high volumetric loads were contributing to the lower than expected power draws for the measured charge of the mills (Morrell, 2016). Follow-up efforts from the M2P therefore, focused on opportunities to reduce the recirculating loads in the secondary grinding circuit to alleviate these volumetric flow rates. A simple solution could not be easily implemented as the nature of the magnetite ore proved challenging for the hydrocyclones and finisher screens in terms of classification efficiency (Jankovic and Valery, 2017).

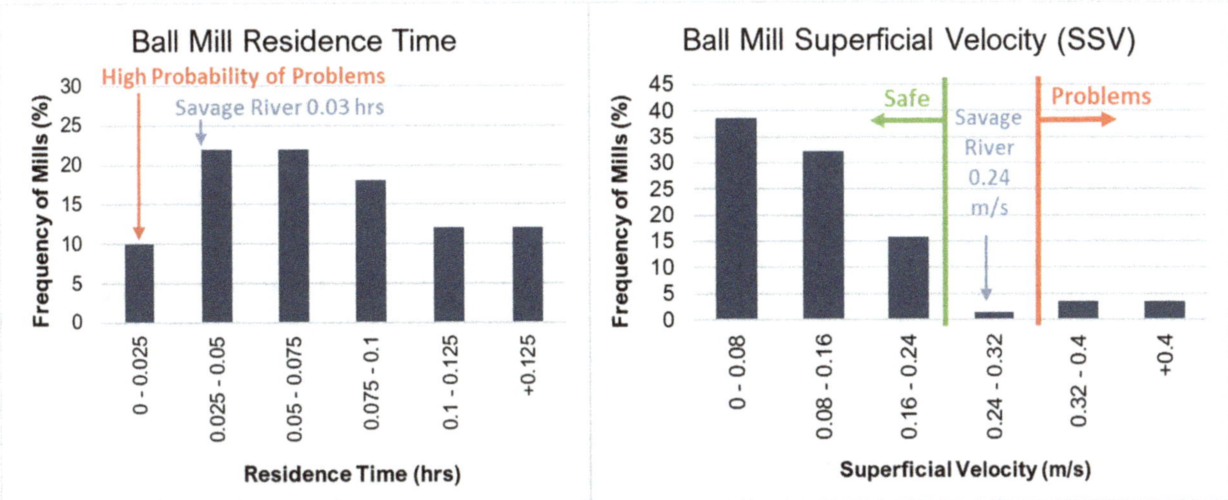

FIG 2 – Distribution of residence time and superficial slurry velocity in operating ball mills (Morrell, 2001).

Finisher circuit

Savage River operates with a high mass yield of 45 per cent and very good final concentrate grade (typically about 68 per cent Fe) compared to other magnetite operations in Hatch's database. However, due to downstream process constraints it is desirable to increase grade even further to target above 69 per cent Fe.

The mineralogical data of the rougher (RMS) and finisher (FMS) magnetic separation stages can identify the pathways for the loss of valuable minerals to the tailings stream as well as the recovery of gangue minerals to the final product stream. Mineralogy and liberation analysis was performed on these key streams from samples collected during the plant survey. The mass distribution of the major minerals in the key streams is shown in Figure 3.

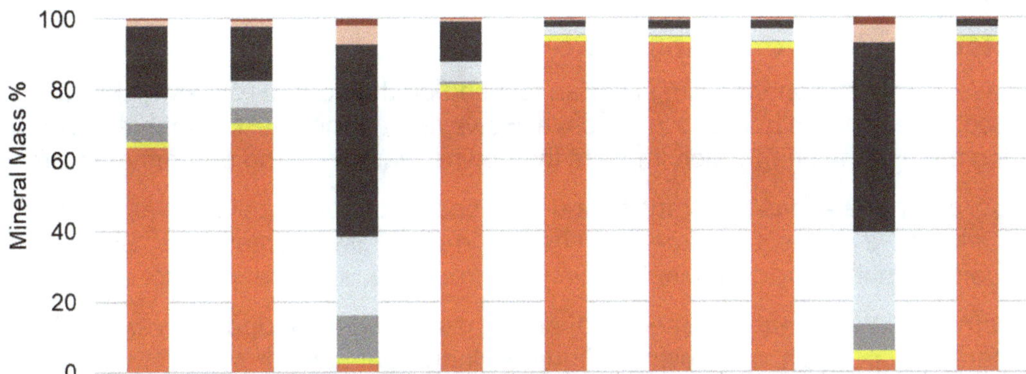

	AG Mill Feed	Rougher Magsep Feed	Rougher Magsep Tails	Rougher Screen U/S	Prim Finisher Screen U/S	Sec Finisher Screen U/S	Sec Finisher Screen O/S	Tail Tcknr Feed	Final Conc
Phosphates	0.50	0.54	1.89	0.21	0.04	0.05	0.04	2.15	0.05
Others	1.6	1.7	5.4	0.8	0.5	0.7	0.5	5.0	0.5
Other Silicates	20.1	15.4	54.2	11.1	1.9	2.3	2.6	53.5	2.1
Serpentine	7.2	7.5	22.4	5.6	2.3	2.0	3.5	26.0	2.4
Fe Sulphides	5.2	4.2	12.0	1.1	0.1	0.2	0.2	7.5	0.2
Ti Oxides	1.8	2.0	1.7	2.1	1.6	1.6	1.7	2.7	1.6
Fe Oxides	63.6	68.6	2.3	79.1	93.4	93.1	91.4	3.1	93.2

FIG 3 – Mass distribution of major mineral groups in key streams.

The AG Mill feed contains 63.6 per cent Fe Oxides, which includes undifferentiated magnetite, hematite and goethite. Fe Sulfides include pyrite and pyrrhotite and make up about 5.2 per cent of the feed. These are the major sources of Fe in the ore and account for about 97 per cent of the Fe deportment. The remainder of Fe is from Fe silicate minerals (Serpentine dominant in this deposit).

The major gangue minerals present in the feed are other silicates (including quartz, feldspar and others) and serpentine which make up 20.1 per cent and 7.2 per cent of the feed, respectively. The RMS removes a significant proportion of the silicates, which after the finishing stages is reduced to 2.1 per cent silicates in the final concentrate.

The nominal MgO limit in the final concentrate is 2.5 per cent, but levels above 3.0 per cent cause significant issues for the pellet plant due to the impact on energy balance and gas requirements. The minerals containing MgO are classified within the 'others' mineral class in Figure 3. A great proportion of MgO bearing minerals are removed in the RMS and following further upgrading through the finishing stages the 'others' mineral class only contributes about 0.5 per cent of the final concentrate mass.

The final concentrate also has a phosphorus limit of 0.08 per cent. Figure 3 illustrates that the AG mill feed contains about 0.50 per cent of phosphates. A significant amount of that is removed in the RMS stage with 1.89 per cent of phosphates in the RMS tails and this is further reduced through finishing stages to 0.05 per cent phosphate in final concentrate.

Liberation analysis on the final concentrate stream (Final Conc) indicated that a high percentage of the silicate minerals present are fully liberated, and this indicated improvements were needed in the rejection of this material. These improvements could increase the final product Fe.

TRANSITION TO BLOCK CAVE MINING

Following the M2P project, in early 2025 Grange Resources re-engaged Hatch to discuss the impacts of block cave mining on the concentrator. As part of those investigations the opportunity of increasing the capacity of the concentrator to 2.85 Mt/a from the current 2.4–2.65 Mt/a of concentrate production was also evaluated.

With the process models developed during the M2P study, Hatch worked with the Savage River team to recalibrate the models to reflect the most up-to-date plant performance and feed blends. The Hatch team then leveraged its industrial database to develop a ROM fragmentation for the underground block cave and subsequently developed simulations for the new fragmentation.

Expected block cave fragmentation

Block cave fragmentation modelling is often focused on the coarse rocks which is the primary factor that impacts ore flow and hold ups in the cave. However, it is the final fragmentation at the drawpoints (including the fine material) which impacts the downstream crushing and grinding circuits.

Data from other operations that transitioned from open pit to block caving was reviewed and used to benchmark and evaluate the expected change in ROM fragmentation at Savage River and the subsequent impact on the comminution circuits throughput and final product size.

Hatch compared the primary crusher product size distributions for open pit versus block cave mining for a reference operation, as shown in Figure 4. Individual belt cut size distributions are shown for the open pit and the overall range (also from belt cut samples) is shown for the block cave. The size distributions from the open pit operation falls within the block cave range but are less variable.

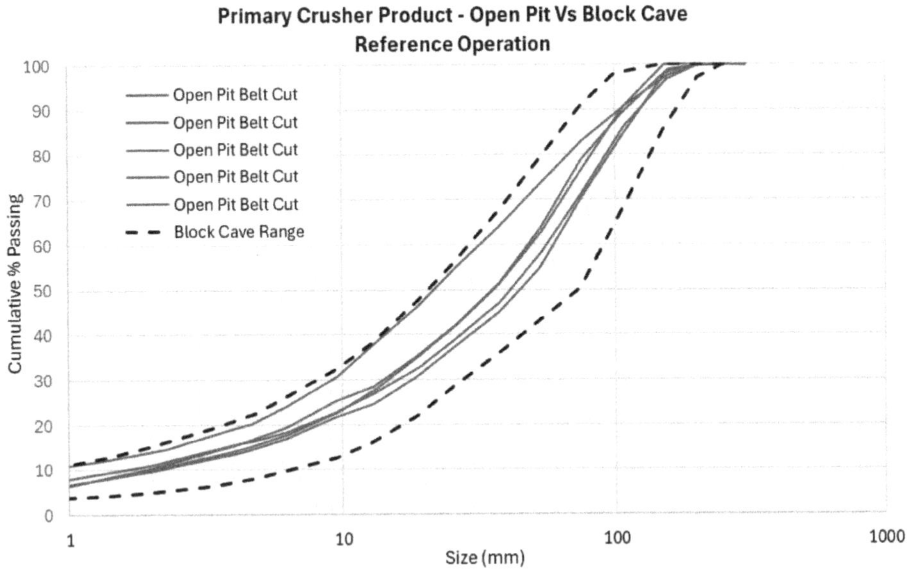

FIG 4 – Open pit versus block cave primary crusher product size example.

In an open pit operation, there are many drill and blast parameters that can be adjusted to control the resulting fragmentation (design in burden, spacing, explosive type and properties etc). However, in a block cave, the breakage occurs through natural processes and is affected by the geometry of the cave and other variables that cannot be freely adjusted to control the fragmentation. Breakage is driven by the inherent rock properties (structure and strength) and thus, the average fragmentation for the block cave is similar to the open pit, albeit with increased variability attributed to the decreased control. Additionally, fragmentation is typically coarsest at the start of a cave and progressively gets finer as the cave progresses and the height of draw (HOD) increases (Fuenzalida, Orrego and Ghazvinian, 2024). This needs to be considered in optimisation of downstream comminution circuits.

Revised base case for block cave

A new base case simulation was developed from the recalibrated site-specific model to achieve the target 2.85 Mt/a concentrate production based on the future ore characteristics and expected ROM fragmentation. An instantaneous throughput of 372 t/h per line is required (considering typical recovery and utilisation data for Savage River) to achieve the required annual concentrate production. This represents a 3 per cent throughput increase from the survey conducted in 2023 and suggested that the plant would be capable of 2.85 Mt/a if the block cave particle size distribution (PSD) was to remain similar to the current open pit ROM PSD. The base case simulation used the average size distribution for the block cave at a fresh feed rate of 372 t/h.

Additional simulations were conducted to evaluate the impact of the coarsest and finest ROM PSDs expected from the block cave on the performance of the concentrator. These indicated that the circuit would encounter several constraints while treating these extremes. These constraints would have to be addressed to ensure the plant could maintain 2.85 Mt/a production rates across the entire range of ROM fragmentation expected from the block cave. Several upgrades were recommended to prepare the circuit for the range of variability expected from the block cave.

PROPOSED UPGRADES

Upgrades were proposed to unlock throughput capacity and enable the concentrator to maintain 2.85 Mt/a production rates across the entire range of expected ROM PSDs:

- AG mill motor upgrade
- pebble crusher upgrade
- ball mill upgrade
- finisher screen upgrade.

Additionally, several other opportunities were identified (as discussed previously in the Finisher Circuit section) to potentially improve concentrate quality:

- hydroseparator efficiency and flow sheet position relocation
- instrumentation upgrades
- ore blending.

As the pre-eminent goal of the project was to increase production rates, these opportunities for improving concentrate quality detailed previously in the finisher circuit section were captured as potential projects that Grange could pursue in the future but will not be discussed any further in this paper.

AG mill motor upgrade

Simulations completed indicated that additional power for the AG mills would be required to maintain the transfer size to the secondary grinding circuit (ie, the primary grinding circuits final product grind size) when treating the coarsest material expected from the block cave.

The Savage River AG mills currently operate mostly below 30 per cent rock charge due to the 5 MW motor limit. A motor upgrade to 6 MW would allow the mills to operate closer to 35 per cent load, subject to mechanical integrity checks. A smaller motor upgrade to 5.5 MW was also considered and would allow mill loads up to about 34 per cent. However, this would have no contingency for operating variability and greatly limit the mill's flexibility with managing potential process upsets.

Additionally, the larger motor upgrade (to 6 MW) would be required for AG to SAG conversion as required to treat the finest feed expected from the block cave (expected later in the mine life with greater height of draw). Simulations demonstrated that the AG mill may not be able to operate at required production rates with this fine material due to the lack of coarse material to act as autogenous grinding media. The proposed solution would be the addition of steel grinding media and SAG operation, and this would require the 6 MW motor upgrade. If the mechanical/structural integrity of the mill does not allow the conversion to SAG, then alternative strategies would be necessary. These include looking at generating coarse material that can be blended with the feed or strategic blending if multiple caves of different ages (HOD) are available. The finer block cave feed is only expected at the latter stages of LOM, therefore there is time to investigate all possible options to process the finer block cave material.

Pebble crusher upgrade

Pebble crushing would not be applicable when operating in SAG mode but could offer benefits while operating in AG mode when treating the average and coarser block cave feed (earlier in the LOM).

The current Savage River operation has two Sandvik CH440 cone crushers in pebble crushing duty. These crushers process a proportion (approximately 40 per cent) of the AG mill screen's oversize stream before recycling back to the mills. A fraction of the pebbles bypasses the pebble crushers due to insufficient pebble crushing capacity.

Savage River currently plans to upgrade the current pebble crushers to Metso HP400 crushers, which have increased power capacity (315 kW versus 220 kW). The goal is to resolve various mechanical and operational challenges associated with the current pebble crusher's old age and to increase capacity. Simulations for both the average and coarser block cave feeds suggests that the upgraded crushers increased capacity and capability to operate at a finer CSS could potentially increase throughput with greater benefits expected for the coarse block cave feed which has a higher pebble rate. In both cases, the AG load and power draw is decreased. Simulations suggested that the target throughput could be achieved for the coarse ore without the AG mill motor upgrade. However, a mill motor upgrade would provide additional contingency against variability in feed characteristics (hardness, size distribution, throughput etc).

Increased pebble crushing is likely to result in a marginally coarser transfer size to the secondary grinding circuit. Therefore, without any modifications to the ball mill circuit, the final product size is likely to coarsen.

Ball mill upgrade

To maintain the final grind size and product quality at Savage River, increased grinding capacity in the secondary grinding circuit will be necessary.

The amount of additional grinding capacity required has been determined using both power-based methods and JKSimMet simulations. The basis for the analysis was the operating data and mass balance developed from the survey conducted and described earlier in the paper. It was concluded that an extra 900 kW of grinding power (per line) is required. The additional installed power could be fulfilled by either a ball or stirred mill.

Given the relatively small amount of extra grinding power required, the need for the extra energy efficiency factor of a stirred mill, in Hatch's opinion, is unwarranted. Additionally, due to the high flow rates from the circulating load, VertiMills or Tower mills would be the only stirred mills suitable for this application. HIG Mills would not be suitable due to the low reduction ratio and high volumetric flow rates.

The options evaluated by Hatch were discussed with Grange Resources and led to the decision of replicating the existing ball mills in both size and power. The rationale behind the decision centred primarily around operability in terms of commonality of spares, maintenance requirements, and operator experience. The addition of a third secondary grinding circuit would significantly reduce the volumetric flow rates through the existing mills improve overall efficiency by alleviating the residence time and SSV issues that were encountered during the original M2P project.

Finisher screen upgrade

The finisher screens at Savage River have a dual purpose. They provide pipeline protection from material larger than 106 μm and reject coarse silica bearing material to improve final concentrate quality.

The currently installed Johnson screens with a screen aperture of 125–150 μm is completing the duty as required, however, the screening efficiency is very poor and the amount of oversize material reporting back to the ball mill feed is excessive. During the original M2P survey, it was found that over 62 per cent of the material in the finisher screen oversize (O/S) is already final product sized. This adds to the already high flow rate through the ball mills and takes-up capacity in the finisher magnetic separators ultimately impacting concentrate quality.

Simulations completed with Derrick screens and higher screening efficiencies, indicates a reduced recycle load on the ball mills and improved capacity on the finisher magnetic separators.

ONGOING PARTNERSHIP

Hatch and Grange Resources have been working closely over the last four years to identify and improve the overall operation at Savage River. The Mine-to-Process study completed at the start has been instrumental in the decisions being made today to increase the capacity and efficiency of the concentrator. The models and knowledge from that study have been repeatedly utilised by Savage River in day-to-day operations and internal projects and has also been key for Hatch to provide effective on-site support for focused investigations.

Hatch is currently supporting not only the concentrator team at Savage River, but also the team at the pellet plant located at Port Latta in the value chain capacity increase to 2.85 Mt/a capacity that is associated with the planned transition of the mining from open pit to block cave. A study is currently underway to estimate the capital and operational costs associated with the required upgrades to unlock and maintain the value chain capacity at 2.85 Mt/a.

CONCLUSION

The Mine-to-Process optimisation at Savage River demonstrated that integrated, site-specific modelling can identify practical, largely capital-efficient strategies to sustain concentrate production, improve product quality, and control costs as feed grades decline.

The use of these models for the planned transition to block cave mining showed that the concentrator could achieve 2.85 Mt/a, provided targeted upgrades to the comminution and finisher circuits are implemented to manage increased variability in ROM fragmentation.

The ongoing partnership between Grange Resources and Hatch highlights the value of M2P frameworks not only as a one-off optimisation exercise, but as a long-term decision-support tool to de-risk strategic changes and extend the life-of-mine.

ACKNOWLEDGEMENTS

The authors from Hatch would like to thank Grange Resources for the permission to publish this paper and acknowledge the Savage River team's support which was instrumental for the success of the whole M2P project and their continued trust in the ongoing and subsequent projects.

REFERENCES

Fuenzalida, M, Orrego, C and Ghazvinian, E, 2024. Back analysis and forecast of cave fragmentation at Cadia East mine, in *Proceedings of the International Conference and Exhibition on Mass Mining, MassMin 2024* (eds: D Johansson and H Schunnesson), pp 679–691 (Luleå University of Technology, Luleå). https://doi.org/10.36487/ACG_repo/2435_D-19

Jankovic, A and Valery, W, 2017. Energy efficient ball mill circuit – equipment sizing considerations, *We are Metallurgists, Not Magicians*, Spectrum Series 23, pp 307–314 (The Australasian Institute of Mining and Metallurgy: Melbourne).

Kanchibotla, S, Valery, W and Morrell, S, 1999. Modelling fines in blast fragmentation and its impact on crushing and grinding, in *Proceedings of Explo'99 – A Conference on Rock Breaking,* pp 137–144 (The Australasian Institute of Mining and Metallurgy: Melbourne).

Morrell, S, 2001. Large Diameter SAG Mills Need Large Diameter Ball Mills – What Are The Issues?, in *Proceedings of the SAG 2001 Conference*, pp 179–193 (University of British Columbia – Mining and Mineral Processing Engineering: Vancouver).

Morrell, S, 2016. Modelling the influence on power draw of the slurry phase in Autogenous (AG), Semi-autogenous (SAG) and ball mills, *Minerals Engineering*, 89:148–156.

Preheating and roasting characteristics of pellets from a magnetite concentrate

H Han[1], *O Aladejebi*[2], *L Lu*[3] *and S Hapugoda*[4]

1. Research Officer, CSIRO Mineral Resources, Pullenvale Qld 4069.
   Email: hongliang.han@csiro.au
2. Research Officer, CSIRO Mineral Resources, Pullenvale Qld 4069.
   Email: tosin.aladejebi@csiro.au
3. Chief Research Scientist, CSIRO Mineral Resources, Pullenvale Qld 4069.
   Email: liming.lu@csiro.au
4. Senior Experiment Scientist, CSIRO Mineral Resources, Pullenvale Qld 4069.
   Email: sarath.hapugoda@csiro.au

ABSTRACT

The growing global demand for steel industry decarbonisation, alongside the rapid development of advanced and energy-efficient direct reduction iron (DRI) technologies, has significantly increased the demand for high-quality iron ore pellets. While concentrates of other ore types are being used to produce iron ore pellets, magnetite concentrates are widely employed as primary feedstock for iron ore pellet production. Magnetite is oxidised in the roasting process, which generates extra heat and reduces the roasting temperature, leading to reduced energy consumption.

The characteristics of iron ore concentrate plays an important role in the formation of green balls and the structure and final properties of fired pellets. In this study, the preheating and roasting behaviour of pellets from a magnetite concentrate are systematically investigated. Furthermore, the induration mechanisms governing the consolidation of the pellets are examined in detail. Based on experimental results and mechanistic insights, a series of technical strategies is proposed and tested to optimise the pellet quality.

INTRODUCTION

Iron ore pellets play a critical role in modern ironmaking processes, particularly in blast furnace and direct reduction iron (DRI) technologies. With the accelerating global drive toward steel industry decarbonisation and the rapid development of energy-efficient DRI processes, the demand for high-quality iron ore pellets with superior mechanical strength and consistent metallurgical performance has increased significantly (Parathodiel, Ahmed and Seetharaman, 2023; Vogl, Åhman and Nilsson, 2021). Among various iron ore types, magnetite concentrates are widely used as primary feedstock for pellet production due to their fine particle size, high iron content, and favourable oxidation behaviour during induration (Yin, Zhang and Li, 2022).

Compared with hematite ores, magnetite concentrates undergo an exothermic oxidation reaction during preheating and roasting, generating internal heat that can reduce external energy requirements and lower overall induration temperatures (Zhang *et al,* 2021). However, the successful utilisation of magnetite concentrates depends strongly on pellet induration behaviour, which is influenced by raw material characteristics, binder type, pretreatment methods, and thermal profiles (Fan *et al,* 2022). Inadequate control of these factors can lead to insufficient pellet strength, excessive cracking, or incomplete oxidation, thereby limiting industrial applicability.

Binders play a crucial role in pellet formation and induration. Inorganic binders such as bentonite provide robust high-temperature bonding but increase gangue content, while organic binders reduce impurities but tend to degrade at relatively low temperatures, potentially compromising pellet strength during preheating (Li *et al,* 2023). Composite binders combining inorganic and organic components offer a potential compromise, but their induration behaviour and strength development under different thermal conditions remain insufficiently understood. In addition, pretreatment technologies such as high pressure grinding rolls (HPGR) can modify concentrate particle morphology and reactivity, thereby influencing oxidation kinetics and pellet consolidation (Kawatra and Claremboux, 2021).

Despite extensive industrial experience, a comprehensive understanding of how binder type and pretreatment affect the preheating and roasting characteristics of magnetite pellets – particularly in relation to induration mechanisms and optimisation of thermal profiles – is still lacking. Therefore, this study systematically investigates the preheating and roasting behaviour of pellets produced from a magnetite concentrate using different binders and HPGR pretreatment. The evolution of pellet strength over a wide temperature range is examined, microstructural changes are analysed, and the governing induration mechanisms are elucidated. Based on these insights, optimised preheating and roasting strategies are proposed to achieve pellet strengths that meet industrial requirements while minimising energy consumption.

EXPERIMENTAL

Preparation of samples

The green pellets were made from as received (Pellets 1, 2, and 3) and HPGR pretreated magnetite concentrates (Pellets 4) under various binder additions. Pellets 1 and 4 were made using the same dosage of Binder 1, while Binder 2 was used in Pellet 2 and combined addition of Binder 1 and 2 in Pellet 3. The characteristics of these green and dry pellets are shown in Table 1.

TABLE 1

Characteristics of these green pellets and dry pellets used in this study.

	Conditions	Compression strength of green ball/N	Drop strength of green ball/times	Compression strength of dry ball/N
Pellet 1	Binder 1	12.51	1.45	58.77
Pellet 2	Binder 2	21.28	2.25	190.00
Pellet 3	Binders 1+2	24.28	1.80	91.08
Pellet 4	HPGR+Binder 1	13.26	2.60	65.87

Preheating and roasting tests

Preheating and roasting experiments were carried out in laboratory furnaces to investigate the effect of preheating temperature/duration and roasting temperature/duration. The compression strengths of the preheated and roasted pellets were measured using an Instron machine. Typically, ten pellets were measured for each testing condition and the average compressive values was reported and compared.

Microstructure analysis

Selected pellet samples roasted at different temperatures were further mounted in epoxy resin and polished to 1 μm to examine the microstructure of pellets using a reflected light optical microscope equipped with a digital camera.

PREHEATING AND ROASTING BEHAVIOUR OF PELLETS

Preheating behaviour

The effect of preheating temperature on pellet strength was consistent across all pellet samples (shown in Figure 1), strength increased with the preheating temperature from 850°C to 950°C. The most significant improvement occurred between 850°C and 900°C, suggesting that this temperature range is critical for strength development. The pellet strength continued to improve from 900°C to 950°C but at a slow pace, confirming that the pellet structure consolidated further at higher temperatures.

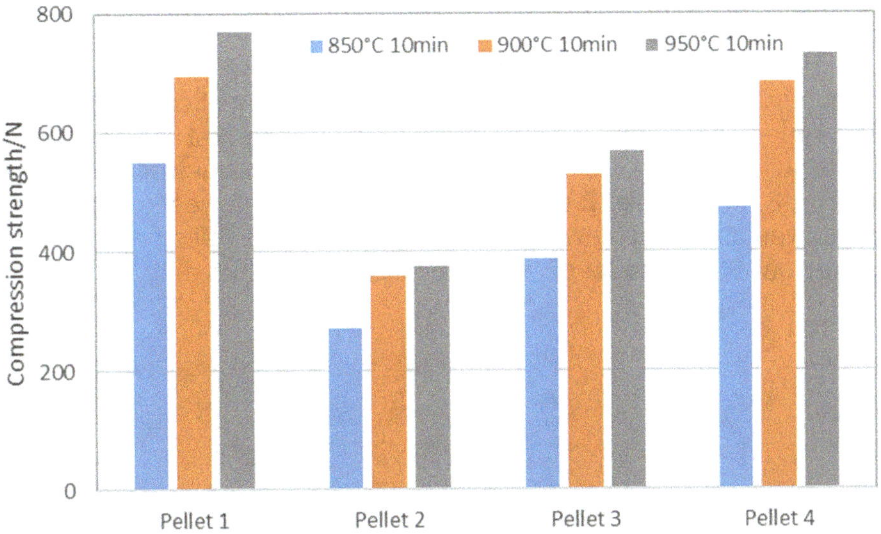

FIG 1 – Compression strength of pellets at different temperatures using different binders.

However, clear differences were observed among the pellets. Pellet 1 exhibited the highest strength at all temperatures, indicating superior consolidation. Pellet 4 showed comparable behaviour, particularly at higher temperatures, achieving strength values close to those of Pellet 1 at 900°C and 950°C. Pellet 3 displayed moderate strength, while Pellet 2 consistently recorded the lowest strength, indicating weaker particle bonding.

The effect of holding time at 900°C was displayed in Figure 2. All pellets demonstrated a substantial increase in strengths as the holding time increases from 3–10 mins. Again, Pellet 1 and Pellet 4 consistently showed the highest strength values, reaching 694N and 684N at 10 mins, respectively, while Pellet 2 displayed the lowest strengths throughout the holding times, with a maximum of 359N. Pellet 3 demonstrated intermediate strengths, rising from 188N at 3 mins to 528N at 10 mins. The biggest pickup in strength for all pellets occurred between 7.5 and 10 mins, suggesting that longer holding times at 900°C significantly enhance microstructural consolidation of the pellets.

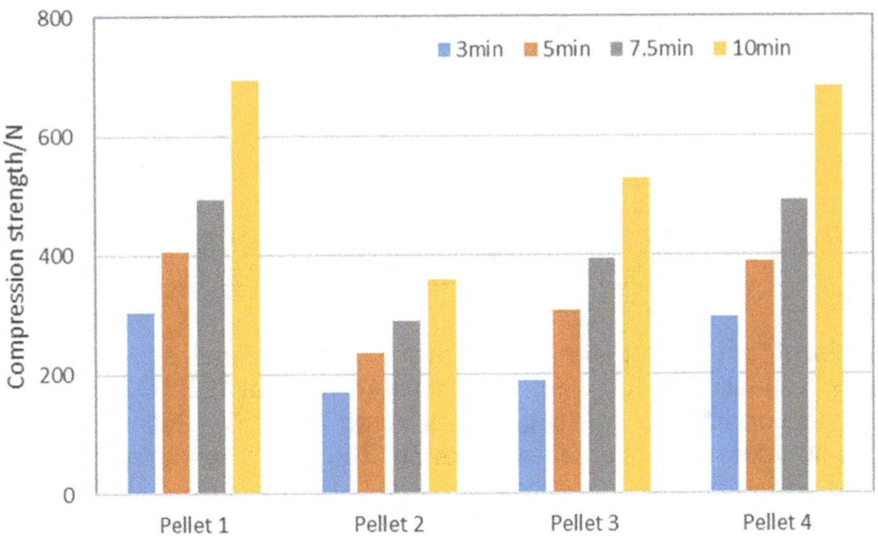

FIG 2 – Compression strength of pellets at 900°C with different holding time.

In summary, pellet strength increased with both preheating temperatures and holding times. The most significant improvements were observed at 900°C for 10 mins, where all pellets reached their highest strengths. Therefore, 900°C for 10 mins was selected in the subsequent roasting tests.

Roasting behaviour

High-temperature roasting results obtained at 1200°C, 1225°C, and 1250°C for a holding time of 15 mins were compared in Figure 3. An increase in strengths was observed for all pellets when the roasting temperature increased from 1200 to 1250°C, indicating that temperature is the dominant factor governing the pellet strength. Pellet 4 achieved the highest values at all temperatures, followed by Pellet 1 and Pellet 3, while Pellet 2 consistently recorded the lowest strength. Although a sharp increase in strength was observed between 1225°C and 1250°C, Pellets 1, and 4 already achieved sufficient strength below 1225°C to meet industrial strength requirements.

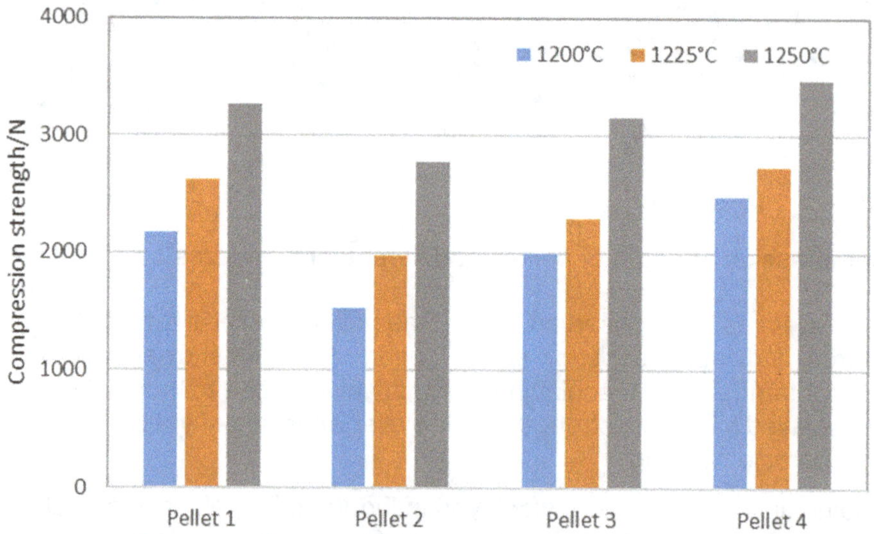

FIG 3 – Compression strength of pellets roasted at 1200°C, 1225°C and 1250°C for 15 min.

The effect of holding time on pellet strength at 1225°C was shown in Table 2. Strength increased substantially with increasing holding time from 10 to 20 mins. The strength increased from 2233N at 10 mins to 2615N at 15 mins, and further to 2878N at 20 mins, indicating that the extended holding time promoted stronger inter-particle bonding. However, the degree of improvement in strength between 10 and 15 mins was higher than that between 15 and 20 mins, suggesting that excessively long holding times were not favourable for energy efficiency.

TABLE 2

Compression strength of Pellet 1 at 1225°C with different holding time.

1225°C	10 min	15 min	20 min
Pellet 1	2233	2615	2878

In summary, Pellet 1 and Pellet 4 exhibited the highest overall strength and met industrial requirements after preheating at 900°C for 10 min and roasting at 1200°C for 15 mins or 1225°C for 10 mins. Pellet 2 required higher roasting temperatures (1250°C). Pellet 3 showed intermediate behaviour but did not meet strength requirements at 1200°C, indicating that roasting temperatures above 1200°C were necessary.

INDURATION MECHANISM OF PELLETS

Induration involves drying, preheating, and roasting green pellets to produce roasted pellets with desirable quality. The mechanical strength and metallurgical performance of roasted pellets depend on the mineralogy of the roasted pellets, which is formed through the reactions within and between ore particles as well as between ore particles and gangue or flux particles during induration.

The induration mechanism of magnetite type pellets is divided into two stages, the oxidation of magnetite and the formation of Fe_2O_3 crystallite bonds or recrystallisation of Fe_2O_3 newly formed in the oxidisation process (Prusti, Nayak and Biswal, 2017).

The oxidation of magnetite is the most important chemical reaction in pelletising of magnetite concentrates. The oxidation of magnetite begins at 200°C and ends at about 1100°C (Carter *et al,* 1983).

$$4Fe_3O_4 + O_2 \rightarrow 6\gamma\text{-}Fe_2O_3 \ (>200°C)$$

$$6\gamma\text{-}Fe_2O_3 \rightarrow 6\alpha\text{-}Fe_2O_3 \ (>400°C)$$

At first, the magnetite is oxidised to γ- Fe_2O_3 (magnetic hematite), a kind of hematite with the same cubic lattice structure as Fe_3O_4. However, γ- Fe_2O_3 is unstable, the crystal lattice is rearranged at higher temperatures, and oxygen ions are able to directly diffuse inside through the surface. Therefore, the transformation of lattice structures from the cubic to the hexagonal takes place. Moreover, the atoms on the surface of newborn Fe_2O_3 crystals have greater migrating capability, and consequently the Fe_2O_3 crystallites or recrystallisation of Fe_2O_3 are easily formed between particles (Kumar *et al,* 2018). In addition, the heat release by oxidation is generated within the pellets, and the oxidation results in low thermal requirements to complete the induration of the pellets.

In practical induration process, a shell of hematite is formed first around the pellet while the core is still magnetite. Oxidation of the outer shell of pellets is fast and controlled by the rate of the chemical reaction. After the fast superficial oxidation, the oxidation rate is claimed to be controlled by the diffusion rate of oxygen through the growing product layer (Forsmo *et al,* 2008). Recrystallisation of Fe_3O_4 (950°C) starts earlier compared to the hematite phase (1100°C). So the duplex structure forms when only the outer shell of pellets is oxidised before the roasting temperature is reached, and a distinct concentric oxidation front between the oxidised outer shell and the non-oxidised core is often found in pellets oxidised at temperatures above 1000°C. However, the magnetite core shrinks during induration, structural stress and concentric cracks form along the structural boundary, which affect the strength of pellets (Forsmo *et al,* 2008; Cho, Tang and Pistorius, 2014). An extreme solution would be complete oxidation of pellets before recrystallisation of Fe_3O_4 is started, so the preheating temperature is better below 950°C. In addition, the oxygen potential is also particularly important for the induration of magnetite pellets, high oxygen potential makes the oxidation of magnetite easily.

The microstructure of pellets preheating at 900°C under different holding times was shown in Figure 4. With increasing holding time, the oxidation of magnetite to hematite progressively advanced from the pellet surface toward the core, reflecting oxygen diffusion close to the core. At shorter holding times, incomplete oxidation resulted in magnetite-rich cores and a heterogeneous microstructure, which can limit inter-particle bonding. Prolonged holding times promoted a more uniform hematite distribution and improved contact between particles, contributing to enhanced pellet strength. At 900°C and a holding time of 10 mins, the microstructures at the pellet surface and core were very similar, indicating that preheating at 900°C for 10 mins effectively avoids the formation of a duplex structure.

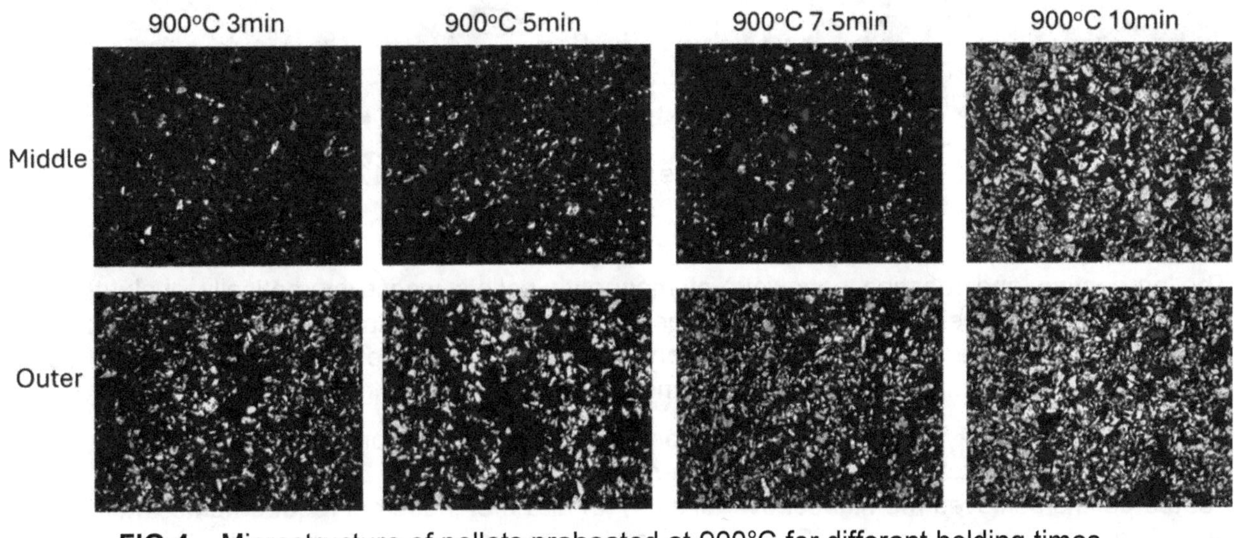

FIG 4 – Microstructure of pellets preheated at 900°C for different holding times.

With increasing temperature, pellet induration was governed by recrystallisation of Fe_2O_3 between particles to form crystallite bonds. Microstructure of pellets roasted at 1200°C and 1250°C with a holding time of 15 mins is shown in Figure 5. While the pellets roasted at 1200°C exhibited sufficient strength, their microstructure was not yet fully developed. At this temperature, induration was mainly controlled by initial Fe_2O_3 crystallite bonding and limited recrystallisation, resulting in incomplete neck growth, a less continuous hematite matrix, and relatively higher porosity.

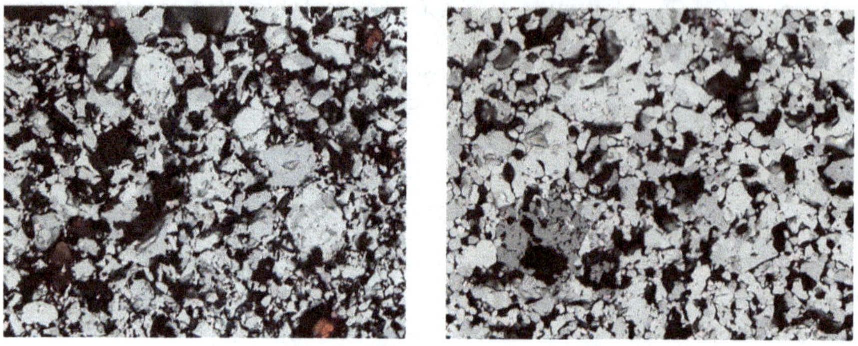

FIG 5 – Microstructure of pellets at 1200°C and 1250°C.

When the roasting temperature increased to 1250°C, atomic diffusion became much more intense. The Fe_2O_3 atoms migrated more rapidly, promoting recrystallisation and growth of Fe_2O_3 crystallites across particle boundaries. This resulted in stronger inter-particle bridges, grain coarsening, and the formation of a more continuous hematite matrix. At the same time, pores shrank, leading to a denser pellet structure.

However, the degree of consolidation in the magnetite pellets varies depending on the type of binder and the treatment process. As shown in Figure 6, in addition to the mechanisms discussed above, Pellet 1 formed low-melting silicate phases that promote slag bonding, resulting from the reaction between bentonite and the iron oxides and gangue minerals. In contrast, Pellet 2 exhibited high porosity due to the burnout of the binder at lower temperatures, which limited densification during roasting. Pellet 3 showed moderate densification with fewer pores compared to Pellet 2. Meanwhile, Pellet 4 exhibited a dense microstructure, where HPGR treatment enhanced Fe_2O_3 recrystallisation, resulting in strong inter-particle bridges and minimal porosity.

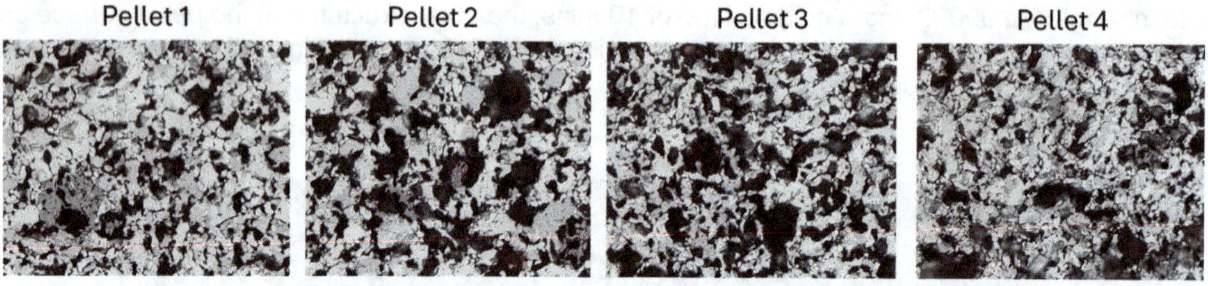

FIG 6 – Microstructure of different pellets at 1250°C.

CONCLUSION

The preheating and roasting behaviour of magnetite pellets was systematically evaluated as a function of binder type and HPGR pretreatment. Pellet consolidation was mainly governed by the oxidation of magnetite and the recrystallisation of hematite, but binder type and pretreatment of the material also significantly affected the pellet microstructure and strength.

Optimal preheating at 900°C for 10 mins promoted uniform oxidation of magnetite, prevented the formation of a duplex structure, and enhanced pellet consolidation. Bentonite-bonded and HPGR-treated pellets achieved the highest overall strength and met industrial requirements when roasted at 1200°C for 15 mins or 1225°C for 10 mins. In contrast, Pellet 2 required a higher roasting temperature (1250°C) due to strength loss from binder burnout during preheating, while Pellet 3 exhibited intermediate performance and failed to meet strength requirements at 1200°C, indicating the need for higher roasting temperatures.

These results highlight the importance of binder selection, pretreatment, and thermal profile optimisation for producing high-quality magnetite pellets under industrial and low-energy induration conditions.

ACKNOWLEDGEMENTS

The authors would like to thank CSIRO Mineral Resources for permission to publish the paper and for the financial support of this work.

REFERENCES

Carter, N B, Clark, K N, Thomas, C G and Bowling, K M, 1983. Major Variables Influencing Hematite Pellet Properties, *Ironmak Steelmak*, 10:243–252.

Cho, H J, Tang, M and Pistorius, P C, 2014. Magnetite Particle Size Distribution and Pellet Oxidation, *Metall Mater Trans B*, 45:1213–1220.

Fan, X, Zhou, Z, Wang, S, Gan, M, Ji, Z, Chen, X and Tian, Z, 2022. Mineralization characteristics of pellets prepared by ultrafine magnetite concentrate: influence on metallurgical property and improving method, *J Iron Steel Res Int*, 30:446–455.

Forsmo, S P E, Forsmo, S E, Samskog, P O and Bjorkman, B M T, 2008. Mechanisms in oxidation and sintering of magnetite iron ore green pellets, *Powder Technol*, 183:247–259.

Kawatra, S K and Claremboux, V, 2021. **HPGR** pretreatment effects on iron ore pellet feed quality and pelletisation performance, *Minerals Engineering*, 164:106822.

Kumar, T K S, Simonsson, M, Viswanathan, N N, Ahmed, H, Andersson, C, El-Geassy, A H A and Bjorkman, B, 2018. Establishing a Novel Methodology to Correlate the Macroscopic and Microscopic Degree of Sintering in Magnetite Pellets during Induration, *Steel Res Int*, 89.

Li, H, Zhou, M, Fan, X and Gan, M, 2023. Effect of organic and composite binders on induration and metallurgical properties of iron ore pellets, *Journal of Materials Research and Technology*, 24:512–523.

Parathodiel, A, Ahmed, H and Seetharaman, S, 2023. Developing iron ore pellets for hydrogen-based direct reduction: Binder selection and induration behaviour, *Sustainability*, 15:11415.

Prusti, P, Nayak, B K and Biswal, S K, 2017. Study of Temperature Profile in the Induration of Magnetite Iron Ore Pellets, *T Indian I Metals*, 70:453–462.

Vogl, V, Åhman, M and Nilsson, L J, 2021. Assessment of hydrogen direct reduction for fossil-free steelmaking, *Journal of Cleaner Production*, 280:124353.

Yin, G, Zhang, J and Li, Y, 2022. Oxidation and strength development of magnetite pellets during induration, *Powder Technology*, 401:117303.

Zhang, Y, Gan, M, Fan, X and Chen, X, 2021. Heat transfer and oxidation behaviour of magnetite pellets during preheating and roasting, *Ironmaking and Steelmaking*, 48(9):1012–1021.

Integrating model, plan and reality – a maturity framework for mine reconciliation in iron ore

R Hargreaves[1] and J Souza[2]

1. Principal Consultant, Datamine, Perth WA 6000. Email: rayleen.hargreaves@datamine.com
2. Production Consultant, Datamine, Santiago, Chile. Email: jady.souza@datamine.com

ABSTRACT

Mine reconciliation provides the analytical link between the geological model, the mine plan, and operational performance, enabling operations to understand and manage the causes of variance. For iron ore producers, where reconciliation practices are already well established, the next opportunity lies in strengthening integration and feedback across the value chain to further enhance predictability, accountability, and value delivery.

This paper builds on the *Universal Reconciliation* framework first presented by Hargreaves and Morley (2014), applying it as a maturity-based diagnostic framework to evaluate reconciliation performance in iron ore operations. It focuses on how mature sites can refine established systems by embedding reconciliation more deeply within planning, operational and governance processes. The approach emphasises multidisciplinary collaboration, process mapping, and the application of a reconciliation code of practice to ensure consistency, transparency, and continuous improvement.

A diagnostic Reconciliation Maturity Model is presented as a lens for assessing how effectively existing reconciliation systems close the loop between model, plan and reality. The framework highlights how well-defined reconciliation relationships, mapped data and material flows, and structured feedback loops can be used to benchmark current capability and identify areas for optimisation.

By integrating model, plan and reality within a maturity framework, mine reconciliation evolves from a compliance exercise into a strategic tool for operational assurance, one that enables iron ore producers to measure system effectiveness, anticipate risk, and drive continual value improvement across the mining value chain.

INTRODUCTION

Mine reconciliation has long been recognised as a fundamental control in mining operations, providing a means of comparing planned and actual outcomes and identifying sources of variance (Morley, 2003). Mine reconciliation is a fundamental part of how iron ore operations understand performance, manage risk and maintain confidence in reported outcomes. In large-scale iron ore operations, where production volumes are high, value margins are tight and product quality specifications must be met consistently at scale, the effectiveness of reconciliation has a direct influence on both operational performance and commercial outcomes. By comparing what was modelled and planned with what is actually mined, processed and delivered, reconciliation provides a critical link between technical assumptions and operational reality. For many iron ore producers, reconciliation practices are well established and embedded within routine reporting and review processes.

Despite this maturity, reconciliation does not always deliver the level of insight or assurance that operations expect. Variances are often well measured, but the feedback required to influence upstream assumptions, planning logic and operational controls is not always clearly defined or consistently applied. In complex iron ore value chains, factors such as blending, stockpiling and time delays can further obscure the source of variance, making it difficult to distinguish between normal operational variability and more systemic issues.

The Universal Reconciliation framework, first presented by Hargreaves and Morley (2014), addressed these challenges by promoting a multidisciplinary, value-chain-wide approach to reconciliation. The framework emphasised the importance of collaboration, process mapping, consistent reconciliation practices and ongoing monitoring as a means of improving transparency and reducing value loss. These principles remain highly relevant and are now reflected, to varying degrees, in many iron ore operations.

As reconciliation practices have matured, the focus is shifting from implementation toward effectiveness. The key question is no longer whether reconciliation is performed, but how well it is integrated into decision-making, governance and feedback processes across the mine value chain. This paper introduces a reconciliation maturity framework designed to support that shift. Rather than prescribing new reconciliation tasks, the framework provides a diagnostic lens for assessing how effectively existing reconciliation systems close the loop between model, plan and reality.

By applying this maturity-based approach, iron ore operations can better understand where reconciliation is supporting confident decision-making and where opportunities exist to strengthen integration, feedback and assurance. The framework is intended to support practical reflection and continuous improvement in mature operating environments, rather than to redefine reconciliation fundamentals.

The contribution of this paper is not a new reconciliation methodology, but a practical way of evaluating how well existing reconciliation systems function in mature iron ore operations. By reframing reconciliation maturity around integration, feedback and assurance, the framework addresses a gap between established reconciliation practice and its effective use in decision-making. This perspective is intended to support practitioners in diagnosing system effectiveness rather than implementing additional reconciliation activity.

BACKGROUND – UNIVERSAL RECONCILIATION FRAMEWORK

The Universal Reconciliation framework was developed in response to a recurring challenge observed across mining operations: despite well-defined technical processes, insufficient feedback between disciplines often resulted in models and plans that were poorly informed by operational reality. Geological modelling, mine planning, scheduling, mining and processing activities are inherently interdependent; however, in practice these functions are frequently managed within organisational silos, with reconciliation occurring in an ad hoc or reactive manner. Broader industry initiatives, such as AMIRA P754A (the audit/assessment framework that sits alongside the AMIRA P754 Code of Practice), also highlight the importance of integrated Mine to Mill reconciliation and consistent management of reconciliation processes across technical domains (AMIRA International, 2007).

Hargreaves and Morley (2014) introduced Universal Reconciliation as a multidisciplinary, value-chain-wide approach designed to address these shortcomings. Rather than treating reconciliation as a single comparison or reporting task, the framework positioned reconciliation as a system for understanding how information, assumptions and decisions propagate through the mining value chain. The objective was to identify where value is lost or opportunity created as material and data move from resource definition through to final product.

At the core of the Universal Reconciliation framework are four interrelated principles, shown in Figure 1. The first is the use of a *multidisciplinary team* to facilitate communication and shared understanding between geology, mine planning, operations and processing disciplines. By engaging stakeholders across these functions, reconciliation is grounded in a common view of how the operation actually works, rather than how it is assumed to work.

The second principle is *process mapping* across the mine value chain. This involves documenting physical material flows, information flows, work practices, roles and responsibilities, data sources and assumptions. Process mapping provides transparency around how decisions are made and highlights where inconsistencies, inefficiencies or data quality issues influence reconciliation outcomes.

The third principle is the documentation of a *reconciliation code of practice.* The code of practice establishes agreed definitions, methodologies, reconciliation relationships, data sources and reporting standards. Its purpose is to create consistency, embed good practice and ensure reconciliation processes are repeatable and sustainable over time, rather than dependent on individual knowledge or informal practices.

The fourth principle is the implementation of *monitoring systems* that provide visibility of reconciliation performance across key points in the value chain. Ongoing monitoring enables early identification of emerging issues and supports timely intervention before losses become systemic.

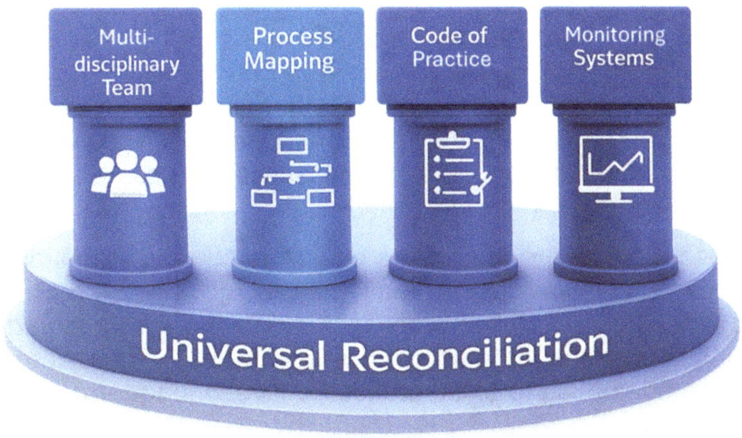

FIG 1 – Universal Reconciliation pillars and concepts.

Together, these principles reframed reconciliation as a proactive diagnostic and improvement framework rather than a retrospective reporting exercise. Universal Reconciliation demonstrated that value loss can occur at multiple points along the mining value chain and that meaningful improvement requires alignment between technical disciplines, processes and governance structures.

While the Universal Reconciliation framework remains highly relevant, the maturity of reconciliation practice within iron ore operations has continued to evolve. Many producers have adopted elements of the framework, including structured reconciliation relationships, standardised reporting and dedicated reconciliation roles. This evolution creates an opportunity to move beyond implementation of principles toward evaluating how effectively they are embedded. The maturity-based framework presented in this paper builds directly on these foundations, providing a structured lens for assessing integration, feedback and system effectiveness in modern iron ore operations. In this context, reconciliation provides the mechanism through which geological models, mine plans and operational reality are continuously tested and re-aligned.

RECONCILIATION CHALLENGES AND OPPORTUNITIES IN IRON ORE

The challenges associated with reconciling material and data across mining and processing stages, including time lags and inconsistent data definitions, have been well documented in Mine to Mill reconciliation studies (Lund, Lamberg and Lindström, 2013).

Iron ore operations are characterised by high production rates, relatively low unit margins, and a strong dependence on consistent product quality and supply chain reliability. As a result, reconciliation practices within the iron ore sector are generally more mature than in many other commodities, with established processes for tracking tonnes, grades and product movement from pit to port. Despite this maturity, iron ore producers continue to face challenges in ensuring that reconciliation outcomes are effectively integrated into planning, operational decision-making and governance processes.

One of the defining challenges in iron ore reconciliation is the complexity of material definition and product control. Variability in material types, moisture content, contaminants and lump–fines ratios can have a significant impact on both processing performance and product specification. While these factors are often well understood individually, reconciliation systems do not always provide a clear line of sight between how material is characterised in the geological model, how it is scheduled and mined, and how it ultimately performs through processing and shipping. This can result in variances being measured accurately but not fully understood or addressed upstream.

Blending and stockpile management present additional reconciliation challenges. Iron ore operations typically rely on multiple stockpiles to manage variability and meet product specifications, introducing time lags and material mixing that complicate reconciliation relationships. Without well-defined reconciliation boundaries and feedback loops, it becomes difficult to attribute the source of variance, and early indicators of product quality drift can be masked, particularly when reconciling short-term operational outcomes against longer-term planning assumptions.

Organisational and process-related factors also play a significant role. Although reconciliation outputs are commonly reported on a regular basis, the ownership of reconciliation outcomes and the mechanisms for acting on them are not always clearly defined. In some cases, reconciliation is treated as a reporting obligation rather than a decision-support tool, limiting its influence on planning adjustments, operational controls and governance processes. This can lead to a disconnect between measured variance and meaningful improvement.

From a systems perspective, iron ore operations often manage reconciliation data across multiple platforms and disciplines. While individual systems may be robust, integration between them is not always seamless. As a result, reconciliation processes can become fragmented, with different teams working from different assumptions, data sources or time frames. These fragmentation issues are particularly evident when attempting to close the loop between long-term models, short-term plans and daily operational execution.

Despite these challenges, iron ore operations are well positioned to advance reconciliation practice. The maturity of existing systems, availability of high-quality data and strong technical capability provide a foundation for moving beyond descriptive reconciliation toward a more integrated and evaluative approach. The opportunity lies in strengthening feedback mechanisms, clarifying reconciliation relationships and embedding reconciliation outcomes within multidisciplinary workflows and governance structures.

These challenges and opportunities highlight the need for a framework that allows mature iron ore operations to assess how effectively reconciliation is integrated across the value chain. A maturity-based approach provides a structured way to evaluate current practice, identify integration gaps and guide the evolution of reconciliation from measurement to proactive performance assurance.

A MATURITY FRAMEWORK FOR MINE RECONCILIATION

Rationale for a maturity-based approach

Maturity models are commonly used to assess how effectively processes and behaviours are embedded within organisations, providing a structured way to describe progressive capability without prescribing specific solutions (Fraser, Moultrie and Gregory, 2002). Maturity models are widely used in operational, safety and asset management contexts to describe progressive stages of capability and to guide continuous improvement.

A maturity-based framework also provides a structured lens through which reconciliation systems can be evaluated beyond simple compliance. Rather than asking whether reconciliation is performed, a maturity approach focuses on how effectively reconciliation outcomes influence upstream assumptions, planning logic and operational controls. This perspective is particularly relevant for iron ore operations, where reconciliation systems are often well established but not always fully integrated across disciplines or time horizons.

Structure of the reconciliation maturity framework

The Reconciliation Maturity Framework builds directly on the principles of Universal Reconciliation by describing how those principles are progressively embedded within an operation. The framework defines five levels of reconciliation maturity, representing a continuum from reactive, siloed practices to fully integrated, assurance-driven reconciliation systems.

The maturity levels are not intended to represent fixed states or to imply that all operations should aspire to the highest level in all circumstances. Instead, they provide a diagnostic structure that allows operations to understand their current capability and determine where targeted improvement will deliver the greatest value.

Reconciliation maturity levels

Level 1 – Ad hoc (Reactive)

At this level, reconciliation is performed inconsistently and is typically triggered by performance issues or reporting requirements. Processes are informal, data sources vary between disciplines and reconciliation outcomes are primarily backward-looking. Feedback into models and plans is limited, and reconciliation is often viewed as an explanatory exercise rather than a management tool.

Level 2 – Basic (Descriptive)

Reconciliation is performed on a regular basis, with defined comparisons between key points in the value chain. Reporting focuses on variance measurement, typically for tonnes and grade, but methodologies and assumptions are not consistently documented. While reconciliation outputs are visible, their influence on planning and operational decisions remains limited.

Level 3 – Structured (Standardised)

At the structured level, reconciliation practices are documented and standardised through a reconciliation code of practice. Reconciliation relationships, data sources, calculation methods and reporting frequencies are clearly defined. Accountability is established, and reconciliation results are trusted as a representation of operational performance. However, feedback mechanisms remain largely periodic rather than embedded within day-to-day workflows.

Level 4 – Integrated (Analytical)

Reconciliation is integrated across disciplines and time horizons. Process mapping is used to align data flows, assumptions and decision points across geology, planning, operations and processing. Reconciliation outcomes actively inform planning adjustments, operational controls and performance reviews. Variance trends are analysed to identify systemic issues, and feedback loops between model, plan and execution are well established.

Level 5 – Optimised (Assurance-Driven)

At the highest level of maturity, reconciliation functions as an operational assurance system. Reconciliation metrics are embedded within governance and risk management processes, providing early warning of performance drift and emerging risk. Feedback between model, plan and reality is continuous, enabling proactive intervention and ongoing optimisation. Reconciliation supports confidence in forecasts, reporting and strategic decision-making across the value chain.

Relationship to universal reconciliation principles

Each level of maturity, shown in Figure 2, reflects increasing depth in the application of the four Universal Reconciliation principles. Multidisciplinary collaboration evolves from informal interaction to structured, embedded workflows. Process mapping shifts from a diagnostic exercise to a mechanism for continuous alignment. The reconciliation code of practice matures from a static document to an active governance tool. Monitoring systems progress from periodic reporting to integrated performance assurance.

Viewed through this maturity lens, Universal Reconciliation provides the foundation, while the maturity framework offers a means to evaluate and evolve how effectively that foundation is applied within modern iron ore operations.

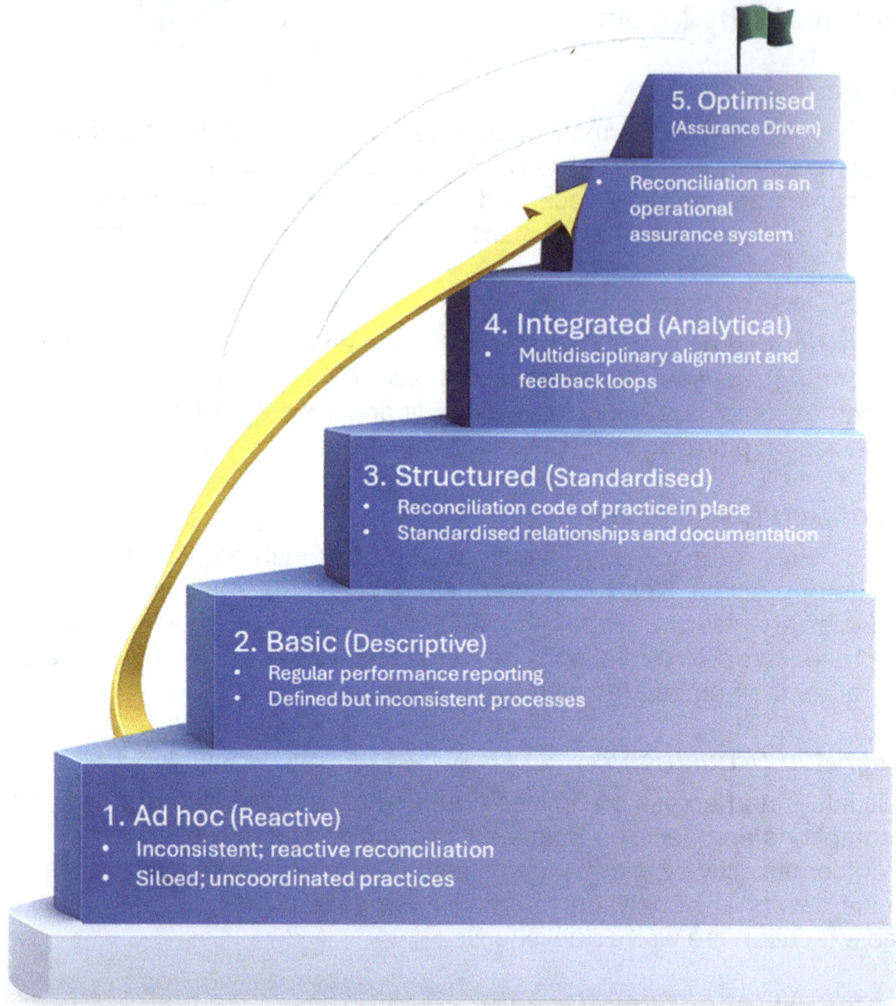

FIG 2 – Reconciliation maturity model for mining operations.

APPLYING THE FRAMEWORK AS A DIAGNOSTIC TOOL IN IRON ORE OPERATIONS

The framework does not prescribe reconciliation frequency, metrics or tools, but provides a way to assess whether existing reconciliation practices are effectively supporting integration, feedback and assurance.

The Reconciliation Maturity Framework is intended to be applied as a diagnostic lens rather than a prescriptive checklist. For iron ore operations with established reconciliation practices, its primary value lies in evaluating how effectively reconciliation outcomes are integrated across the mine value chain and embedded within decision-making processes.

Maturity is assessed based on how reconciliation behaviours, integration and feedback mechanisms are embedded across the value chain, rather than through quantitative scoring or compliance metrics.

Defining reconciliation relationships across the value chain

Effective diagnosis begins with clearly defining reconciliation relationships between key stages of the iron ore value chain. These relationships typically include reconciliation between the geological model and mine plan, the mine plan and mining execution, mining and processing, processing and product, and ultimately product against shipment or sale.

In many mature operations, these reconciliation points exist but are not always aligned in scope, timing or ownership. Applying the maturity framework encourages a review of whether reconciliation boundaries are clearly defined, whether assumptions are consistent between stages, and whether reconciliation outputs are used to inform subsequent planning cycles. This clarity is essential in iron ore operations, where time lags, blending and stockpiling can obscure the source of variance if relationships are poorly defined.

While reconciliation can be performed between any points in the value chain, this framework focuses on adjacent reconciliation relationships as the primary diagnostic mechanism, as these provide the clearest attribution of variance and most effective feedback into upstream assumptions. The primary, adjacent reconciliation relationships used as the core diagnostic focus of this framework are illustrated in Figure 3.

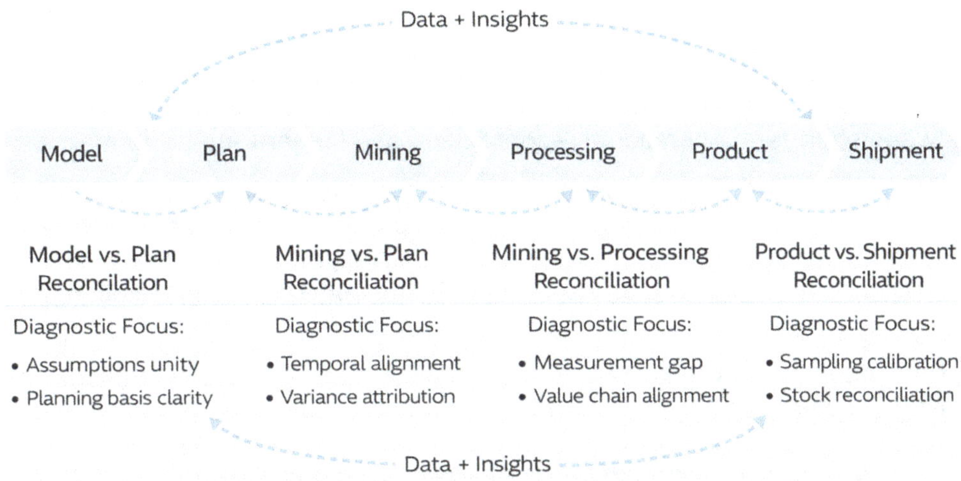

FIG 3 – Primary reconciliation relationships across the iron ore value chain, showing adjacent comparison points used as the core diagnostic mechanism for attributing variance and strengthening feedback between model, plan and reality.

As reconciliation maturity increases, operations are better positioned to interpret non-adjacent reconciliation relationships with confidence, using them as strategic assurance checks rather than primary diagnostic tools.

Mapping processes and feedback loops

In iron ore operations, reconciliation gaps often emerge at interfaces where high-volume material movement, blending and quality control intersect, making process mapping a foundational diagnostic activity within the framework. By mapping both physical material flows and associated data flows, operations can identify where reconciliation results are generated, where they are reviewed, and where feedback is expected to occur.

Using the maturity framework, process maps can be assessed to determine whether feedback loops are explicit, timely and actionable. At lower maturity levels, reconciliation outcomes may be reviewed periodically without triggering defined actions. At higher maturity levels, reconciliation insights are systematically fed back into geological interpretation, planning assumptions and operational controls. This distinction allows operations to diagnose whether reconciliation is functioning as a reporting mechanism or as an integrated management tool.

For example, consistent over-delivery to the crusher relative to plan may initially appear as a processing issue, when in reality it reflects misalignment between scheduling assumptions and mining execution. Similarly, underestimation of fines or moisture in the geological model can manifest downstream as blending challenges or product quality drift, while misalignment between grade control practices and mine planning assumptions can obscure the true source of variance.

Assessing integration across disciplines

A key diagnostic question posed by the maturity framework is the degree of multidisciplinary integration achieved through reconciliation. Iron ore operations often involve multiple teams working with different time horizons, priorities and performance measures. Applying the framework highlights whether reconciliation discussions are multidisciplinary in nature and whether accountability for variance resolution is shared or siloed.

At higher maturity levels, reconciliation forums involve geology, planning, operations and processing disciplines, with clear ownership of actions and outcomes. The framework enables operations to assess whether reconciliation results are actively influencing decisions beyond the reconciliation function itself.

Identifying opportunities for improvement

By applying the maturity framework across reconciliation relationships, process maps and governance structures, operations can identify specific areas where incremental improvement will yield the greatest benefit. These opportunities may include clarifying reconciliation definitions, strengthening feedback into planning cycles, formalising governance processes or improving alignment between short-term operational reconciliation and longer-term performance objectives.

Importantly, the framework does not assume uniform maturity across all aspects of an operation. Instead, it supports targeted improvement by recognising that different parts of the value chain may operate at different maturity levels. This targeted approach aligns well with the complexity and scale of iron ore operations and supports practical, staged enhancement of reconciliation practice.

Moving from a structured to an integrated maturity level may involve shifting reconciliation discussions from periodic reporting forums into regular planning and operational reviews, with defined actions and ownership linked directly to reconciliation outcomes.

Indicative diagnostic questions for mature iron ore operations

For operations with established reconciliation practices, the following questions are intended to prompt structured reflection on current reconciliation maturity:

- Are reconciliation outcomes routinely used to test and update upstream assumptions in geological models and mine plans, or are they primarily used to explain past performance?
- Are reconciliation discussions conducted on a multidisciplinary basis, with shared ownership of actions across geology, planning, operations, processing and shipping?
- Are feedback loops between reconciliation outcomes and planning or operational controls clearly defined, timely and consistently applied?
- Can sources of variance be confidently attributed despite the presence of blending, stockpiling and time lags typical of large-scale iron ore operations?
- Is reconciliation used as an input to governance and assurance processes, or treated largely as a reporting output?
- Are reconciliation outputs delivered in a consistent format (standard dashboards/templates) to avoid interpretation differences between teams and reporting cycles?
- Is there an auditable record linking reconciliation findings to corrective actions and verified outcomes over time, enabling evaluation of effectiveness?

These questions are not intended as a checklist, but as a means of prompting structured reflection on how effectively reconciliation is embedded within decision-making and assurance processes in mature operating environments.

BENEFITS OF A MATURITY-BASED APPROACH TO MINE RECONCILIATION

The Reconciliation Maturity Model presented in this paper is intended as a diagnostic framework rather than a scoring or compliance model. Maturity is assessed based on how effectively reconciliation behaviours, integration and feedback mechanisms are embedded across the mine value chain, rather than through quantitative metrics or prescribed implementation steps. Progression between maturity levels reflects observable changes in how reconciliation informs decisions, influences upstream assumptions and supports governance, rather than the adoption of specific tools or systems. This approach recognises that iron ore operations may demonstrate different levels of maturity across reconciliation relationships and disciplines, and that improvement is achieved through strengthening integration and feedback, not by repeating foundational reconciliation tasks. The use of reconciliation as an input to operational improvement, rather than solely as a reporting function, has been recognised as an important factor in improving performance outcomes (Coward, 2015).

In iron ore operations where reconciliation is already embedded, the most meaningful benefits of a maturity-based approach are realised through how reconciliation is used, rather than how it is performed. The framework helps distinguish between reconciliation that exists primarily as a reporting obligation and reconciliation that actively shapes decisions, priorities and risk management across the value chain.

One of the clearest benefits is improved confidence in the interpretation of variance. In more mature reconciliation environments, variance is not simply reported or explained after the fact but used to test assumptions and prompt earlier intervention. This is particularly important in iron ore operations, where high production volumes, blending and stockpiling can obscure the source of variance and delay visibility of emerging issues. When feedback is returned upstream in a timely and structured manner, reconciliation outcomes become more actionable and better aligned with operational reality.

Mature reconciliation also changes how disciplines interact. Rather than being owned by a single team, reconciliation becomes a shared reference point across geology, planning, operations and processing. Discussions shift away from reconciling numbers after the fact toward understanding system behaviour, trade-offs and the implications of decisions made at different points in the value chain. Over time, this improves alignment between short-term operational decisions and longer-term planning assumptions.

Another practical benefit is improved resilience of reconciliation practices. As maturity increases, reconciliation becomes less dependent on individual expertise and more embedded in documented processes, agreed definitions and governance routines. This consistency supports continuity through personnel changes and provides greater confidence in reported outcomes at both site and corporate levels, strengthening reconciliation as part of the overall reporting assurance framework.

A well-documented reconciliation basis provides a qualitative, evidence-backed record of identified issues, implemented actions and demonstrated outcomes over time. Rather than capturing variance only as a numeric result, it preserves the operational context behind reconciliation findings, including the rationale for decisions made, the interventions applied, and the lessons learned. This creates a site-specific library of learnings that supports continuity through personnel changes, strengthens decision traceability, and enables more consistent improvement in reconciliation effectiveness. In mature iron ore operations, where site conditions, constraints and operating practices can differ significantly, this accumulated record becomes a valuable source of practical knowledge that reflects the mine's particular characteristics and supports ongoing refinement of planning and operational controls.

From an operational risk perspective, mature reconciliation systems provide earlier visibility of performance drift. Patterns in reconciliation outcomes can highlight emerging issues before they translate into material value loss, product quality variability or shipment compliance risk. This proactive capability is particularly valuable in large-scale iron ore supply chains, where small deviations can compound quickly and have downstream commercial consequences.

For iron ore sites already operating at a high level of reconciliation maturity, the greatest gains are achieved through improved integration and feedback loops across the value chain, rather than through repeating basic reconciliation tasks. Taken together, these outcomes reflect a shift in the role of reconciliation, from a retrospective check to a practical system for operational assurance, supporting more confident decision-making, more reliable forecasts and a clearer understanding of how value is created and protected across the iron ore mine value chain.

FUTURE DIRECTIONS

In iron ore operations, reconciliation is no longer limited by a lack of data or measurement capability. Most sites already collect more information than they can practically absorb. The challenge now lies in how reconciliation information is used, how quickly it is fed back into decisions, and how confidently it is relied upon when performance begins to drift. In this context, maturity improvements will increasingly depend on reducing the cycle time between variance emergence, detection and corrective action, rather than increasing the volume of reporting.

One clear area of opportunity is the timing of reconciliation feedback. In many operations, reconciliation outcomes are still reviewed after the fact, once production periods have closed and explanations are required. As reconciliation maturity improves, there is scope to bring that feedback forward, using reconciliation insights more deliberately within short-term planning, scheduling and operational discussions. This does not require real-time systems, but it does require clarity around what signals matter and when they should prompt action. For iron ore operations, earlier signals related to product quality drift, such as fines generation, moisture behaviour or contaminant variability, may be as important as tonnes and grade variances in prompting timely intervention.

Reconciliation also has an increasing role to play in operational assurance. When viewed through a maturity lens, reconciliation provides context around whether observed performance reflects normal operational variability or indicates a breakdown in assumptions, processes or controls. Used in this way, reconciliation becomes a valuable complement to existing governance and performance review processes, rather than a parallel technical activity. A further opportunity lies in strengthening reconciliation closure across the full pit-to-port value chain, particularly where blending, stockpiling and logistics interfaces can delay visibility of issues.

There is also an opportunity to use reconciliation maturity to support more meaningful comparison across operations. As practices become more consistent, maturity provides a way to discuss performance beyond raw variance metrics. This is particularly relevant for organisations managing multiple sites, where understanding differences in process discipline and integration can be as important as understanding differences in orebody or operating conditions.

While advances in systems and analytics will continue to support reconciliation practice, experience suggests that sustained improvement depends far more on people and process than on technology alone. Clear ownership, shared understanding across disciplines and consistent governance remain the foundations of effective reconciliation. Without these, even the most sophisticated tools struggle to deliver lasting value. One practical future direction for mature operations is more structured management of planning assumptions, where reconciliation outcomes provide traceability of what changed, why it changed, and whether the change delivered the intended impact.

Looking ahead, reconciliation maturity provides a practical way for iron ore operations to reflect on how well their systems support decision-making and confidence. As reconciliation continues to move beyond reporting and into the management of risk and performance, its value will increasingly be measured not by the precision of the numbers produced, but by the quality of the decisions they inform.

The framework is intended to support reflective assessment in mature operating environments and does not replace the need for site-specific technical reconciliation methods.

CONCLUSION

Mine reconciliation has long been recognised as a fundamental control in iron ore operations, providing visibility of variance between what was modelled, planned and achieved. For many producers, reconciliation practices are already well established. The challenge is no longer whether reconciliation is performed, but how effectively it supports decision-making, governance and confidence across the iron ore mine value chain.

This paper has built on the Universal Reconciliation framework presented by Hargreaves and Morley (2014), applying a maturity-based lens to evaluate how reconciliation is embedded within modern iron ore operations. By focusing on integration, feedback and system effectiveness, the framework provides a practical way to distinguish between reconciliation that exists as a reporting requirement and reconciliation that actively informs performance management.

The Reconciliation Maturity Model presented offers a structured approach for assessing current practice and identifying opportunities to strengthen how reconciliation outcomes are used. When applied as a diagnostic tool, it helps clarify where variance is introduced, how effectively feedback is returned upstream, and whether reconciliation is influencing assumptions, plans and controls in a timely and meaningful way.

As iron ore operations continue to evolve, maintaining confidence in forecasts and performance will depend on how well reconciliation is integrated into everyday workflows, governance processes and multidisciplinary discussions. A maturity-based approach provides a practical reference point for sustaining that integration and ensuring reconciliation continues to support reliable decision-making and value delivery through improved integration between model, plan and reality across the iron ore mine value chain.

REFERENCES

AMIRA International, 2007. P754A: Mine to mill reconciliation and management, AMIRA International, Melbourne.

Coward, S, 2015. Reconciliation as a tool for operational improvement, *Journal of the Southern African Institute of Mining and Metallurgy*, 115(10):947–954.

Fraser, P, Moultrie, J and Gregory, M, 2002. The use of maturity models/grids as a tool in assessing product development capability, *Proceedings of the IEEE International Engineering Management Conference*, pp 244–249.

Hargreaves, R and Morley, C, 2014. Universal reconciliation - a multidisciplinary approach across the entire mining value chain to identify loss of value and maximise operational performance, in *Proceedings Ninth International Mining Geology Conference 2014*, pp 411–420 (The Australasian Institute of Mining and Metallurgy: Melbourne).

Lund, C, Lamberg, P and Lindström, B, 2013. Practices and challenges in mine-to-mill reconciliation, *Minerals Engineering*, 49:101–113.

Morley, C, 2003. Beyond Reconciliation — A Proactive Approach to Using Mining Data, in *Proceedings of the Fifth Large Open Pit Mining Conference*, pp 185–191 (The Australasian Institute of Mining and Metallurgy: Melbourne).

Optimising wet drum magnetic separator performance – the effects of slurry characteristics on magnetite recovery

C Lartey[1], N S Yap[2] and R Silva[3]

1. Technical Applications Specialist, Eriez, Melbourne Vic 3076. Email: clartey@eriez.com
2. Senior Global Product Manager, Eriez, Melbourne Vic 3076. Email: nyap@eriez.com
3. Global Process Development Manager, Eriez, Erie PA 16506-1402, USA. Email: rsilva@eriez.com

INTRODUCTION

Iron ore, a major raw material for steel production, plays a critical role in infrastructure development, with Australia contributing 38 per cent of global supply (Holmes, Lu and Lu, 2022; Lundaev *et al,* 2023). The beneficiation of magnetite, one of the most prevalent iron oxide minerals, relies heavily on wet drum separators (WDS) to separate magnetic from non-magnetic fractions. Despite decades of industrial application, WDS operations face persistent performance challenges. Recovery rates fluctuate between 75 per cent and 95 per cent, even within the same operation (Svoboda and Fujita, 2003) yet the mechanisms driving this variability remain poorly understood.

Efforts to improve WDS performance have followed two approaches. The first focuses on advancing separator designs through innovations in magnetic configurations. Whilst these developments show promise in controlled settings, new equipment cannot resolve process issues if the interactions between slurry properties and separation mechanisms remain poorly understood. The second approach emphasises process optimisation through adjustment of operational parameters (Chen *et al,* 2024; Dworzanowski, 2010, 2012). However, these studies examine individual parameters in isolation – particle size effects (Arol and Aydogan, 2004) or feed rate impacts (Wang *et al,* 2022) – without considering how multiple slurry properties collectively influence performance.

Recent works have applied modelling techniques to investigate the influence of slurry properties on separation efficiency (Tian *et al,* 2025; Zhao *et al,* 2026). While these studies represent progress in understanding WDS behaviour, further investigations remain necessary to elucidate how interactions between slurry properties influence WDS performance. Furthermore, models developed from industrial data sets face challenges in validation due to temporal inconsistencies, changing standards and instruments, changes in operators, raw material variations, and seasonal effects. These factors raise questions about whether models developed from historical plant data can reliably predict WDS performance under controlled or varying operational conditions, limiting confidence in their industrial application.

This study addresses these challenges by developing and validating a predictive model for WDS performance using complementary data sources. The study aims to: (1) develop a predictive model for the effects of four feed slurry variables (particle size (F_{80}), SiO_2, Al_2O_3, and total Fe grades) on WDS separation efficiency using historical plant data; (2) validate the developed model using independent survey data collected under controlled conditions. This approach ensures that the resulting models are both statistically rigorous, providing confidence that performance predictions derived from historical data can be reliably applied to guide operational decisions and process optimisation.

METHODOLOGY AND FINDINGS

The proposed solution involves utilising a tool that is sufficiently robust to mitigate many of the challenges of modelling historical data. This study employs Stat-Ease statistical software, which includes a dedicated module specifically designed for analysing and addressing common issues in historical data. Predictive models were developed to forecast yield, Y (%) and total Fe recovery, Fe_T Rec (%) based on slurry properties including particle size distribution (F_{80}), silica grade (SiO_2), alumina grade (Al_2O_3), and total Fe grades. Model validation was conducted using data from a plant survey of data from a three-stage industrial Steffensen self-leveling wet drum separator with 1200 mm diameter drums. All samples were analysed using standardised chemical analysis data for Fe, SiO_2, and Al_2O_3 content, and particle size analysis to determine F_{80} values. Mass flow rates were measured using calibrated electromagnetic flowmeters, and slurry densities were determined using Marcy scales to enable material balance calculations.

The data set used contains 116 data points from operation. Table 1 shows a basic description of the input variables. The data was analysed using a response surface methodology, which considers both the extreme levels of the variables as well as points in between. The evaluation assessed the variables on their own and examined the interactions among them. Table 2 shows a summary of the overall analysis and model information for the yield, Y (%) and total Fe recovery, Fe_T Rec (%) models.

TABLE 1

Basic description of input variables.

Factor	Name	Units	Type	SubType	Min	Max	Mean	Std Dev
A	Feed F_{80}	µm	Numeric	Continuous	12.55	98.96	39.90	10.44
B	Feed (Fe) total	%	Numeric	Continuous	17.02	61.26	54.01	11.61
C	Feed (SiO_2)	%	Numeric	Continuous	11.29	57.78	19.24	12.11
D	Feed (Al_2O_3)	%	Numeric	Continuous	0.54	3.01	0.9981	0.6558

TABLE 2

Stat-Ease modelling information.

File Version	25.0.1.0		
Study Type	Response Surface	**Subtype**	Randomised
Design Type	Import Data	**Runs**	116
Design Model	Quadratic	**Blocks**	No Blocks

The model effectively captured the interactive effects between slurry properties, particularly, the combined influence of particle size (lower F_{80}) and increased gangue (SiO_2) content on reducing separation efficiency. The results indicate that the performance of the model aligns with the results obtained from the plant survey data and accurately captures the slurry interactive effects. The model predicted a yield of 82.69 per cent and total Fe recovery of 95.69 per cent. These predictions closely matched the measured plant survey results of 82.73 per cent yield and 95.3 per cent total Fe recovery, representing a deviation of 0.05 per cent and 0.41 per cent, respectively, as presented in Table 3.

Table 3 also shows the results for assessing model robustness using coefficient of determination, R^2 with accuracies of 0.99 and 0.98 achieved for yield and total Fe recovery respectively. Also, a difference of 0.0016 between the adjusted and predicted R^2 values being less than 0.2 indicates that the model fits the data accurately and can reliably be used to interpolate.

TABLE 3

Modelling versus Plant survey results.

Description	Yield, Y	Fe_T Rec
Modelling results	82.69%	95.69%
Plant survey results	82.73%	95.30%
Performance Metrics		
R^2	0.9928	0.9870
Adjusted R^2	0.9925	0.9863
Predicted R^2	0.9909	0.9794
Adequate (Adeq) Precision	211.07	151.25

CONCLUSIONS

In conclusion, this study has investigated feed slurry properties including particle size distribution (F_{80}), silica grade (SiO_2), alumina grade (Al_2O_3), and total Fe grades for wet drum magnetic separator (WDS) performance. The validation results confirm the predictive capability of the developed models and demonstrate their practical applicability for industrial WDS operations. The close agreement between model predictions and plant survey measurements validates the methodology employed for modelling WDS performance from historical plant data. Critically, the validation establishes that interactive effects between slurry properties (particle size F_{80}, SiO_2, Al_2O_3, and total Fe grades) can reliably predict separation outcomes under operational conditions. Operations can utilise the model to predict performance based on feed characteristics and adjust operational parameters to optimise recovery or grade. A more advanced model could be developed by integrating a broader range of operational parameters, including % solids, feed throughput, and wash-water inputs.

REFERENCES

Arol, A I and Aydogan, A, 2004. Recovery enhancement of magnetite fines in magnetic separation, *Colloids and Surfaces A: Physicochemical and Engineering Aspects*, 232(2–3):151–154. https://doi.org/10.1016/j.colsurfa.2003.06.003

Chen, Y, Jiang, Y, Xian, Y and Chen, L, 2024. Enhancing Recovery of Ultra-Fine Magnetite from Low-Iron-Grade Cyanidation Tailings by Optimizing Flow Field Parameters of Low-Intensity Magnetic Separation (LIMS), *Separations*, 11(4). https://doi.org/10.3390/separations11040120

Dworzanowski, M, 2010. Optimizing the performance of wet drum magnetic separators, *Journal of the Southern African Institute of Mining and Metallurgy*, 110(11):643–653.

Dworzanowski, M, 2012. Maximizing the recovery of fine iron ore using magnetic separation, *Journal of the Southern African Institute of Mining and Metallurgy*, 112:197.

Holmes, R J, Lu, Y and Lu, L, 2022. Chapter 1 – Introduction: Overview of the global iron ore industry, in *Iron Ore* (ed: L Lu), second edition, pp 1–56 (Woodhead Publishing). https://doi.org/10.1016/B978-0-12-820226-5.00023-9

Lundaev, V, Solomon, A A, Le, T, Lohrmann, A and Breyer, C, 2023. Review of critical materials for the energy transition, an analysis of global resources and production databases and the state of material circularity, *Minerals Engineering*, 203. https://doi.org/10.1016/j.mineng.2023.108282

Svoboda, J and Fujita, T, 2003. Recent developments in magnetic methods of material separation, *Minerals Engineering*, 16(9):785–792. https://doi.org/10.1016/S0892-6875(03)00212-7

Tian, X, Wu, L, Zhao, Z, Dai, H, Qi, L and Wang, F, 2025. Modeling of separation performance prediction model for industrial low intensity magnetic separation equipment, *Minerals Engineering*, 228. https://doi.org/10.1016/j.mineng.2025.109311

Wang, F, Zhang, S, Zhao, Z, Gao, L, Tong, X and Dai, H, 2022. Investigation of the magnetic separation performance of a low-intensity magnetic separator embedded with auxiliary permanent magnets, *Minerals Engineering*, 178. https://doi.org/10.1016/j.mineng.2022.107399

Zhao, J, Wang, F, Wen, J, Dai, H, Wu, L, Guo, Y and Pu, J, 2026. Enhanced magnetite separation prediction by PLIMS multifactor coupling based on machine learning and SHAP interpretability analysis, *Powder Technology*, 467. https://doi.org/10.1016/j.powtec.2025.121546

Effect of roasting a South African low-grade iron ore on flotation performance

N Maistry[1] and A Singh[2]

1. Senior Engineer, Mintek, Randburg, Gauteng 2194, South Africa.
   Email: nicholem@mintek.co.za
2. Senior Technical Specialist, Mintek, Randburg, Gauteng 2194, South Africa.
   Email: ashmas@mintek.co.za

ABSTRACT

The future resilience of the iron ore industry is strongly dependent on developing efficient beneficiation circuits as iron ore is continuously in demand being primarily used within steelmaking industries worldwide, accounting for 98 per cent of iron ore consumption. Historically, high-grade iron ore was preferably mined as minimal beneficiation was required, however, due to depletion of high-grade Mineral Resources, iron ores with lower grade and complex mineralogy are being explored. A significant portion (~60 per cent) of remaining reserves are banded iron formations (BIF), which consist of iron-rich and silica-rich layers. Majority of BIF deposits are found in Australia, Brazill, Canada, India, Russia and South Africa. South Africa is the largest iron ore producer in Africa and is projected to have about 1.6 Bt (billion tons) of BIF material. Thus, to maintain profitability and sustainability, research into beneficiation of BIF material is needed to smoothly transition to low-grade materials. High-grade ores only require crushing, blending and screening prior to any steelmaking processes. Whereas, low-grade materials require more beneficiation steps, such as gravity separation, magnetic separation, and flotation, to maximise recovery while achieving a suitable grade. However, even when the above techniques are used, recoveries may still be compromised due to various physical properties, complex mineralogy and chemical composition. Thus, the objective was to perform reverse cationic flotation, using an ether amine, on physically separated pre-concentrates to generate a premium product suitable for industry (>63 per cent) at the maximum possible recovery.

The head grade of the ore was 32.9 per cent Fe, which increased to 39.3 per cent after roasting. Pre-concentrates were produced and comparatively assessed via shaking table, and roasting followed by low intensity magnetic separation (LIMS). The shaking table concentrate was graded at 55.1 per cent, with a 49.9 per cent recovery. Whereas the LIMS concentrate achieved a lower grade of 53.4 per cent, but at a significantly higher recovery of 85.9 per cent. Due to complex mineralogy, flotation has become the industry standard to effectively remove impurities from BIF material, however, it is an expensive process necessitating pre-concentration prior to flotation. The South African BIF used for the research is laminated and intergrown in nature, and contains hematite, hematite intergrown illite, illite, kaolinite and quartz. Unfortunately, the presence of certain impurities, such as illite and kaolinite, is detrimental to the flotation process and may result in poor recoveries. However, to ensure effective separation during flotation, in addition to pre-concentration, desliming was done before flotation. Reverse cationic flotation of pre-concentrates was performed using an ether alkyl amine, and a comparative analysis of the results was conducted to determine the feasibility of the processes and recommendations for further optimisation. The final flotation concentrates for natural and roasted hematite were graded at 63.2 per cent and 57.5 per cent, respectively. However, the Fe recovery of roasted hematite was 77.0 per cent, which is significantly higher in comparison to natural hematite, which achieved 32.0 per cent, justifying further optimisation of roasted hematite.

INTRODUCTION

Physical separation as a pre-concentration method, in conjunction with flotation, is becoming a more common method of beneficiation as high-grade iron ores become depleted (Ma, 2012). In the past, high-grade ores were mined and required simple processing to produce a premium grade product, typically ≥63 per cent Fe (Yongxiang, Kalevi and Lauri, 2014). However, as these resources are being consumed, remaining deposits are low-grade and have complex mineralogy, which require extensive processing to achieve a suitable product grade. Furthermore, ~60 per cent of these remaining reserves are banded iron formations (BIF), which has alternating layers of iron oxide and chert. Majority of BIF deposits are found in Australia, Brazil, Canada, India, Russia and South Africa. Specifically for South Africa, 66 per cent of the remaining Mineral Resources are BIF (Singh and Da Corte, 2021), thus, fine processing routes need to be explored and optimised to maximise resource utilisation and increase the life-of-mine. High-grade ores only require crushing, blending and screening prior to any steelmaking processes (Yongxiang, Kalevi and Lauri, 2014). However, due to complex mineralogy of lower more complex iron ores, flotation has become the industry standard to effectively remove impurities from BIF material. Flotation alone however, it is an expensive process necessitating prior pre-concentration. Physical separation methods, such as gravity and magnetic separation, are used to improve the grade of the ore, typically to a pre-concentrate grade (~55 per cent Fe) in an attempt to minimise reagent consumption during flotation. Flotation is then used to remove the remaining siliceous material to achieve a premium grade (~63 per cent Fe) product, which is required for downstream processing. If these grades are not achieved, downstream processing costs are significantly increased. Thus, beneficiation of low-grade iron ore is essential to the future resilience of the steel industry and this research aims to provide a sustainable, long-term solution to depleting high-grade resources.

This study was performed on a low-grade BIF hematite that previously underwent physical upgrade via gravity and magnetic methods to produce concentrates of ~55 per cent Fe. These concentrates required further processing to achieve a marketable concentrate grading at ~63 per cent Fe. Thus, reverse cationic flotation was performed on a pre-concentrate to perform a comparative analysis, in terms of mass yield, Fe grade and Fe recovery, between flotation of pre-concentrated natural and roasted hematite.

LITERATURE REVIEW

Physical properties of iron ores

There are different types of iron ores, such as hematite, magnetite, goethite, limonite and siderite, of which hematite and magnetite are the most abundant (Ohmoto *et al,* 2008). Each of these iron ores have a different composition, and consequently, different physical properties. Hematite (Figure 1) is typically reddish brown and is found in sedimentary, metamorphic and igneous rocks. It is usually formed in the presence of water, and it is associated with BIF (Dobrescu and Vasilescu, 2019). Magnetite (Figure 1) is grey to black in colour, it occurs in igneous and metamorphic rocks, and is sometimes associated with hydrothermal activity. Magnetite is also found in the form of sand on beaches near volcanic activity.

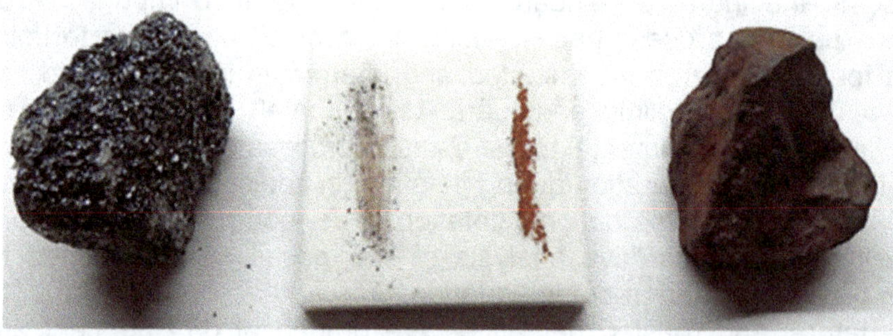

FIG 1 – Colour difference between magnetite (left) and hematite (right) (Waldron and Molinaro, 2024).

Hematite contains iron in the form of Fe_2O_3, whereas magnetite contain iron in the form of Fe_3O_4. The magnetic properties of a mineral depends on the presence and number of unpaired electrons within the mineral lattice, where magnetic performance increases with the number of unpaired electrons. Magnetite is ferromagnetic as it contains both ferrous (Fe^{2+}) and ferric (Fe^{3+}) ions with unpaired electrons, whereas hematite is paramagnetic as it only has ferric (Fe^{3+}) ions with unpaired electrons (Xiong, Lu and Holmes, 2022). Thus, the major difference between hematite and magnetite is their magnetic properties. Pre-concentration of magnetite is achieved via wet low intensity magnetic separation (WLIMS). Since hematite is paramagnetic, wet high intensity magnetic separation (WHIMS) would be required to achieve separation (Kallio, Tanskanen and Luukkanen, 2021). However, previous studies have shown that hematite is typically pre-concentrated via gravity methods, such as jigging or dense medium separation (DMS) and spirals (Makhija, Mukherjee and Ghosh, 2013), resulting in improved grades and recoveries when compared to (Medium Intensity Magnetic Separation (MIMS) and High Intensity Magnetic Separation (HIMS). In addition, in some cases, pyrometallurgical methods maybe be applied to ores to change their typical properties. When hematite is roasted, in the presence of Hydrogen (H_2) and Argon (Ar), hematite is reduced and it undergoes a phase transformation to magnetite at 565°C, which increases the magnetic susceptibility of the material. Roasted hematite can then undergo WLIMS, in an attempt to improve grades and recoveries during pre-concentration stages (Yu *et al,* 2017).

Physical separation of iron ores

Physical separation techniques are typically applied to lower grade hematite and magnetite ores to minimise gangue and produce a product suitable for ironmaking. Beneficiation is achieved by using physical properties of minerals to separate valuable mineral and gangue. Gravity separation depends on mineral density, whereas magnetic separation relies on magnetic susceptibility to achieve separation. BIF materials are more suitable for fine processing; gravity methods include laboratory shaking table, which is an indication of the ore amenability to processing via spirals, and magnetic separation methods include wet high-intensity and low-intensity magnetic separation, where higher intensities are needed for lower magnetic susceptibilities (Wills and Napier-Munn, 2006). Hematite ores require high-intensity magnetic separation, and for very fine materials a SLon® magnetic separator, which is a vertical, pulsating, high-gradient magnetic separator (Metso Corporation, 2024), is preferred. Alternatively, a more innovative approach involves magnetising roasting, which reduces hematite to magnetite and increases its magnetic susceptibility, followed by WLIMS. It is believed that this approach may improve Fe recovery. Furthermore, this process is more environmentally friendly as it uses hydrogen gas a s a reductant instead of carbon. Hydrogen reduction is also achieved at lower temperatures, minimising the energy and cost associated with magnetising roasting (Yu *et al,* 2017).

Flotation of iron ores

Flotation is usually used to further concentrate iron ore when the major gangue minerals are silicates, and it is typically processed at grinds below 149 μm. There are two types of flotation that may be applied, including direct flotation of iron oxides, and reverse flotation of silicates (which may either be cationic or anionic). Selection of the flotation type and the reagent are strongly dependent on the mineralogy of the ore. Low-grade ores are pre-concentrated via physical separation followed by reverse flotation of the pre-concentration, as the aim is to remove residual silicates present in the material (Zhang *et al,* 2021). Ether amines ($R\text{-}O\text{-}(CH_2)3\text{-}NH_2$) are widely used to remove silicates from iron ore as they act as both a collector and a frother, minimising reagent consumption. Other amines may also be used, however, literature indicates that high doses are required (Filippov, Filippova and Severov, 2010). Unfortunately, amines are not selective in floating silica from iron ores, thus, a depressant for iron oxide is required. Starch is the most common depressant for hematite, and it is also applicable to magnetite. The surface charge generation of various systems is thought to have several mechanisms. Oxide minerals include hematite, magnetite, quartz and alumina, and the origin of the electrical charge at the oxide surface/aqueous phase interface is known as protonation/deprotonation of the surface hydroxyls. This is given by the chemical reaction presented by (Rodrigues, Fernandes and Peres, 2024):

$$MOH_{(surf)} \leftrightarrow MO_{(surf)} + H_{(aq)}$$

$$MOH_{(surf)} + H_{(aq)} \leftrightarrow MOH_{2(surf)}$$

Furthermore, the mechanism of flotation of silicate and iron minerals is determined by the electrical properties and the solubility of the mineral, the charge and chain length of the collector, and stability of the salt metal-collector. The adsorptions of the depressant and dispersant are also related to mineral surface characteristics, such as the chemical composition, electrical charge distribution and solubility, the potential determining ions content in slurry, and the chemical and structural composition of the modifier agent (Rodrigues, Fernandes and Peres, 2024).

METHODOLOGY

Ore characterisation

A South African BIF, originating from the Northern Cape, was crushed to 100 per cent -1.18 mm. The ore is laminated and intergrown in nature. Representative sub-samples were removed for head assay and mineralogical analysis via Scanning Electron Microscopy (AutoSEM). The sample for mineralogy was milled to 80 per cent -106 µm to determine the liberation characteristics at a fine grind. Furthermore, the head assay of the roasted feed was performed to determine compositional changes that may have occurred during roasting.

Flotation

Previous test work on pre-concentration of hematite indicated that both gravity (shaking table) and magnetic (SLon®, which is a medium intensity magnetic separator) separation can achieve a pre-concentrate grade of ≥55 per cent Fe. However, the shaking table concentrate achieved a higher mass yield and recovery, thus, it was used for subsequent processing via flotation. Furthermore, the hematite sample underwent roasting and magnetic separation via WLIMS to generate an iron-rich concentrate.

Laboratory flotation test work was conducted on a non-roasted shaking table concentrate and a WLIMS roasted concentrate, where the scope of work is presented in Figure 2. The shaking table concentrate was milled to 80 per cent -75 µm, while the WLIMS concentrate was already milled to the same grind during pre-concentration, specifically the re-re-cleaner stage. Only the shaking table concentrate was deslimed prior to flotation due to limited sample mass of roasted product and to minimise losses of valuable material. Reverse cationic flotation tests were performed using a Denver D12 flotation machine in a 2 L flotation cell at an impeller speed of 800 rev/min. Each test was conducted with 600 g of material, which is equivalent to 30 per cent solids. Starch was added at a dose of 600 g/t to depress iron oxide, and conditioning took place for five mins. The pH was then increased by addition of sodium hydroxide (NaOH) to achieve a pH of 10.5. The total ether amine dosage was 300 g/t for each test. The ether amine was added at 100 g/t per stage and conditioned for 1 min.

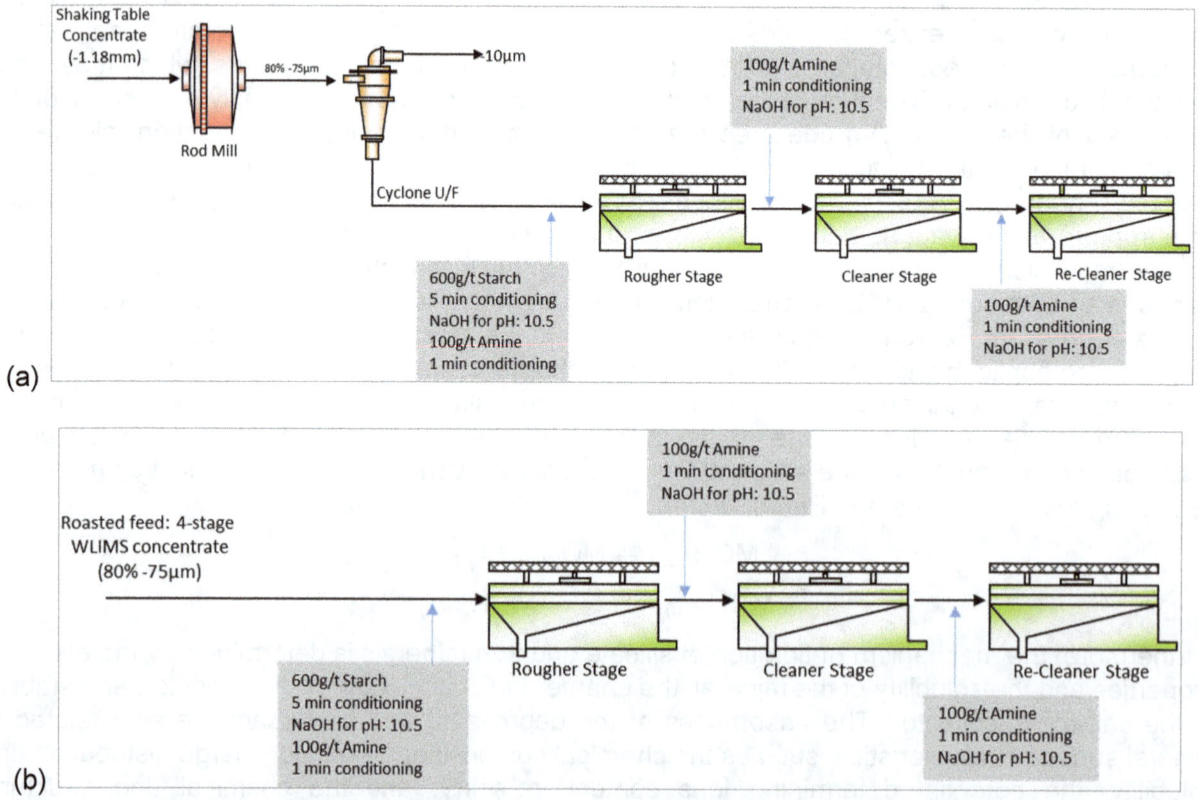

FIG 2 – Scope of work for: (a) shaking table concentrate; and (b) WLIMS roasted concentrate.

RESULTS AND DISCUSSION

Ore characterisation

Head grade analysis

Table 1 shows that the BIF hematite has a head grade of 32.9 per cent Fe and 17.2 per cent Si. After roasting, the head assay indicated a head grade of 39.3 per cent Fe, at 18.8 per cent Si. The head grades have been used to assess and reconcile the flotation data.

TABLE 1

Head grade of natural and roasted hematite.

Head assay	Grade (%)								
	Al	Ca	Fe	K	Mg	Mn	Na	P	Si
Natural hematite	4.81	0.38	32.9	2.13	0.21	0.11	0.21	0.10	17.2
Roasted hematite	5.10	0.32	39.3	1.17	0.18	0.12	0.04	0.05	18.8

Mineralogy – liberation characteristics

As shown in Table 2, mineralogical analysis indicated that five major mineral phases were present in the ore, namely hematite, hematite intergrown illite, illite, kaolinite and quartz. Unfortunately, the presence of certain impurities, such as illite and kaolinite, which are swelling and non-swelling clays are detrimental to the flotation process and may result in poor recoveries. Furthermore, Table 2 presents the liberation class ≥80 per cent per size fraction. Hematite clean, illite and quartz have similar liberation characteristics for all size fractions. Hematite grown illite makes up 26.9 per cent of the ore and is only well liberated in the -25 µm size range, indicating possibility of poor recovery as there may be poor selectivity during flotation as only moderate liberation is achieved in the +25 µm fractions. Kaolinite is not well liberated, even at -25 µm, also contributing to poor recovery as the presence of clays during flotation is associated with swelling behaviour, increased viscosity and slime coating (Taner and Önen, 2016).

TABLE 2

Mineral liberation for various size fractions.

Mineral phase	Mass %	≥ 80% Liberated class			
		-106+75 µm	-75+38 µm	-38+25 µm	-25 µm
Hematite clean	30.3	87.3	85.8	90.1	86.9
Hematite intergrown illite	26.9	68.2	42.7	48.7	79.0
Illite	14.4	84.6	83.0	83.2	86.8
Kaolinite	18.7	67.7	60.7	59.9	72.0
Quartz	7.74	80.1	82.8	82.5	85.9

Figure 3 shows AutoSEM images of the ore at different magnifications. Clean hematite has little to no impurities, which indicates good separation can be achieved. However, clean hematite only represent ~30 per cent of the material. Hematite intergrown illite, which is the second most dominant phase, is still present at very small sizes, ie 10–40 µm, indicating that separation of iron ore from gangue will be difficult. Furthermore, due to hematite and illite being intergrown, the surface chemistry is variable. Since flotation is dependent on surface chemistry, the behaviour of particles will depend on the exposed mineral. In some cases, there is displacement of valuable mineral into the tailings, and *vice versa*, decreasing the efficiency of the flotation process.

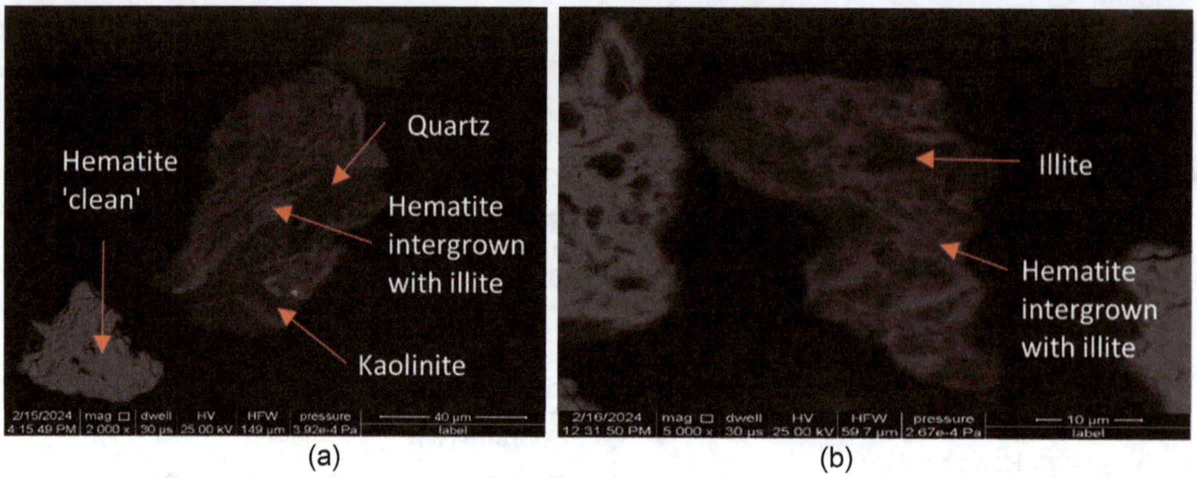

FIG 3 – AutoSEM images showing mineral phases present in the ore at: (a) 2000× magnification; and (b) 5000× magnification.

Pre-concentration of natural and roasted hematite pre-concentrates

The results for all pre-concentration methods are shown in Figure 4. A comparative analysis of pre-concentration of natural hematite indicates that the shaking table performed better than the SLon®, which correlates to literature where Makhija, Mukherjee and Ghosh (2013) found that gravity separation performs better than wet high intensity magnetic separation. When 55 per cent Fe is achieved, the mass yield of the SLon® product is less than 30 per cent (versus ~32 per cent for shaking table) and the Fe recovery is 46.0 per cent (versus 49.9 per cent or shaking table). Thus, only the concentrate produced by the shaking table and WLIMS underwent flotation.

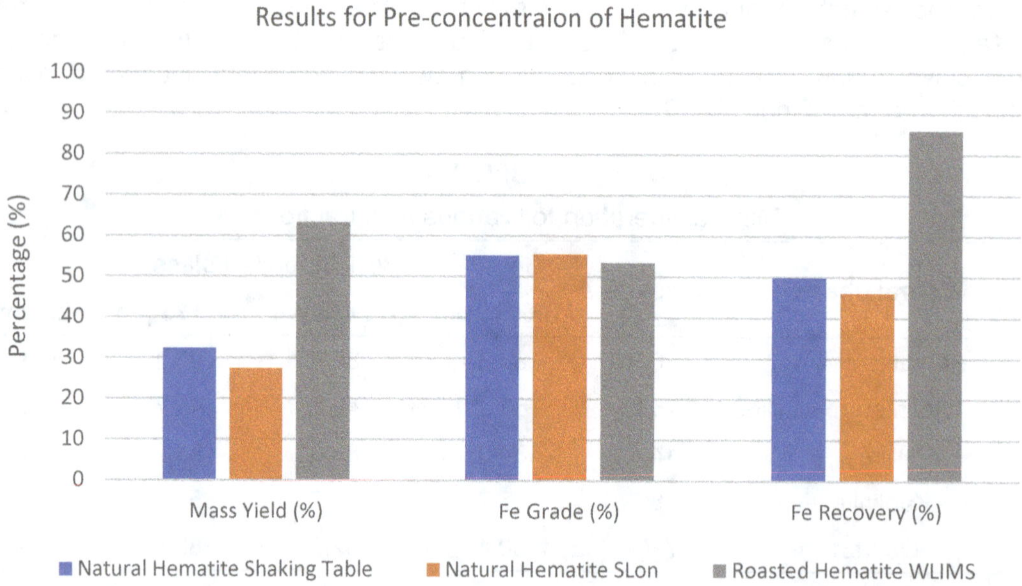

FIG 4 – Results for pre-concentration of hematite.

Flotation of natural and roasted hematite pre-concentrates

The flotation results for the shaking table concentrate is presented in Table 3.

TABLE 3

Flotation results for shaking table concentrate.
Test 1: Shaking table concentrate – 80% passing -75 µm.

Baseline flotation conditions	
Natural pH	8.22
Test pH	10.5

Reagent	Dosage	Conditioning time (min)	Flotation time (min)
Starch	600 g/t	5	
NaOH	100 g/t		
Amine	300 g/t		
Rougher concentrate	100 g/t	1	07:00
Cleaner concentrate	100 g/t	1	05:29
Re-cleaner concentrate	100 g/t	1	04:29

Product stream	Mass %	Discrete grade				Recovery (%)			
		Al	Fe	K	Si	Al	Fe	K	Si
Shaking table concentrate (feed)	32.4	2.67	55.1	1.62	7.40	13.9	49.9	10.7	10.0
Re-cleaner tailings	8.30	2.71	38.0	2.07	21.0	4.75	9.19	9.59	9.29
Re-cleaner concentrate	17.4	1.46	63.2	0.36	6.04	5.34	32.0	3.49	5.59
Cyclone overflow	6.70	2.67	55.1	1.62	7.40	3.80	10.8	6.11	2.66
Concentrate + cyclone overflow	24.1	1.80	60.9	0.71	6.42	9.14	42.8	9.60	8.25
Shaking table reject	67.6	6.04	24.4	2.14	22.9	86.1	48.0	80.8	82.5
Total	100					*100*	*100*	*100*	*100*
Head calculated		4.74	34.3	1.79	18.8				
Head measured		4.81	32.9	2.13	17.2				
Variance (%)		1.44	4.35	16.13	9.30				

The shaking table concentrate was milled and deslimed, with a representative sub-sample of the cyclone overflow being sent for assay and the remainder kept aside. The underflow was the flotation feed, which produced a re-cleaner tailings and re-cleaner concentrate as the final products. After the three stages of flotation, the re-cleaner concentrate was graded at 63.2 per cent, achieving an overall mass yield and a recovery of 17.4 per cent and 32.0 per cent, respectively. The tailings reported 38.0 per cent Fe, indicating misplacement of iron oxides into the tailings or hematite with intergrown illite reporting to the tailings. There is also a decrease in the content of Al, K and Si, indicating removal of unwanted minerals. Table 4 also shows that if the re-cleaner concentrate is combined with the cyclone overflow, the mass yield and recovery is increased to 24.1 per cent and 42.8 per cent, respectively. However, this is only achievable at a lower per cent Fe, specifically 60.9 per cent, which is only suitable for blending with high-grade ores.

Table 4 presents the flotation results for the roasted WLIMS concentrate.

TABLE 4

Flotation results for WLIMS concentrate.

Test 2: WLIMS concentrate – 80% passing -75 µm.

Baseline flotation conditions	
Natural pH	8.53
Test pH	10.5

Reagent	Dosage	Conditioning time (min)	Flotation time (min)
Starch	600 g/t	5	
NaOH	100 g/t		
Amine	300 g/t		
Rougher concentrate	100 g/t	1	02:45
Cleaner concentrate	100 g/t	1	02:00
Re-cleaner concentrate	100 g/t	1	01:42

Product stream	Mass %	Discrete grade				Recovery (%)			
		Al	Fe	K	Si	Al	Fe	K	Si
WLIMS concentrate (feed)	63.3	3.45	53.4	1.13	12.7	42.8	85.9	41.7	43.3
Re-cleaner tailings	10.6	6.00	32.3	2.10	19.5	12.5	8.73	13.1	11.1
Re-cleaner concentrate	52.7	3.04	57.5	0.92	11.5	31.4	77.0	28.4	32.5
WLIMS non-magnetics	36.7	7.80	15.3	2.73	28.5	56.2	14.2	58.6	56.4
Total	100					*100*	*100*	*100*	*100*
Head calculated		5.10	39.3	1.71	18.6				
Head measured		5.48	38.5	1.89	18.9				
Variance (%)		6.89	2.06	9.45	1.58				

Three stage flotation was performed on the WLIMS concentrate without any subsequent preparation, producing a re-cleaner tailings and re-cleaner concentrate. The re-cleaner concentrate achieved a mass yield of 52.7 per cent, at a grade of 57.5 per cent Fe and an overall Fe recovery of 77.0 per cent. The tailings contain 32.3 per cent Fe, indicating misplacement of iron oxides into the tailings or hematite with intergrown illite reporting to the tailings. The content of Al, K and Si in the re-cleaner concentrate did not decrease significantly from the feed, indicating poor removal of unwanted minerals.

Comparative analysis

Figure 5 presents a comparative overview of the flotation results, in terms of overall mass yield, Fe grade and overall Fe recovery. The results indicated that the shaking table concentrate performs better, only in terms of Fe grade and Si rejection. Approximately 63.2 per cent Fe was achieved, however, at a poor mass yield and recovery. Combining the cyclone overflow with the flotation concentrate results in an improved mass yield and recovery of 24.1 per cent and 42.8 per cent, respectively, but at a compromised grade (~60 per cent) only suitable for blending with high-grade ores. When compared to flotation results of the roasted WLIMS product, despite the desired grade not being achieved, the mass yield and Fe recovery is significantly higher. Although a premium grade (~63 per cent) was not achieved, the high mass yield and Fe recovery warrants further investigation to improve the grade. Thus, it is suggested that further flotation cleaning stages be done to achieve ~63 per cent Fe and, in turn, reduce the amount of Al, K and Si in the final concentrate whilst minimising iron losses.

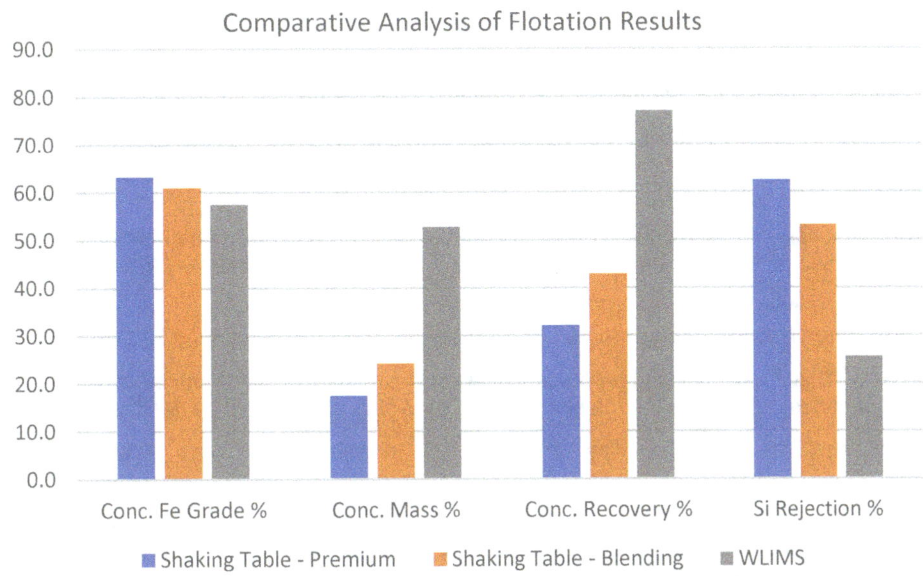

FIG 5 – Comparative analysis of flotation results.

CONCLUSIONS

A low-grade BIF originating in Northern Cape, South Africa reporting a head grade of 32.9 per cent Fe, with 17.2 per cent Si was utilised for research purposes. After roasting, the head grade increased to 39.3 per cent Fe, with 18.8 per cent Si. Mineralogical analysis via AutoSEM showed poor liberation of hematite intergrown illite and kaolinite, which is known to adversely affect flotation performance. Previous pre-concentration test work on 'as-is' hematite showed that the ore is more amenable to pre-concentration via gravity separation methods, and pre-concentration of roasted hematite (magnetite) is only achieved through magnetic separation. The shaking table concentrate achieved a grade of 55.1 per cent, with 49.9 per cent recovery. The LIMS concentrate was graded at 53.4 per cent, however, at a notably higher recovery of 85.9 per cent. The shaking table concentrate was able to achieve a grade of 63.2 per cent Fe after flotation, with a mass yield of 17.4 per cent and a Fe recovery of 32.0 per cent. The low mass yield and recovery may be improved to 24.1 per cent. and 42.8 per cent, respectively, if the flotation concentrate is combined with the cyclone overflow, to produce an iron ore grade (~60 per cent) that is suitable for blending with high-grade ores. The WLIMS concentrate was only able to achieve 57.5 per cent Fe after flotation. However, the mass yield and Fe recovery was significantly greater at 52.7 per cent and 77.0 per cent, respectively. While flotation of the WLIMS concentrate did not achieve the desired grade, the high mass yield and Fe recovery justifies further optimisation. Thus, it is recommended that additional flotation cleaning stages be done to achieve ~63 per cent Fe in an attempt to reduce the amount of gangue minerals in the final concentrate whilst minimising losses of iron.

ACKNOWLEDGEMENTS

The authors would like to acknowledge Mintek for funding the research project and providing the facilities and resources required to complete this study. The authors would also like to thank Mintek's pyrometallurgical, mineralogical and analytical divisions for their contribution to this research, as well as AusIMM for the opportunity to present these findings.

REFERENCES

Dobrescu, M and Vasilescu, M, 2019. Hematite Processing and Applications, *The Annals of 'Dunarea de Jos' University of Galati, Fascicle IX, Metallurgy and Materials Science*, 42(1):19–24.

Filippov, L, Filippova, I and Severov, V, 2010. The use of collectors mixture in the reverse cationic flotation of magnetite ore: The role of Fe-bearing silicates, *Minerals Engineering*, 23(2):91–98.

Kallio, R, Tanskanen, P and Luukkanen, S, 2021. Magnetic Preconcentration and Process Mineralogical Study of the Kiviniemi Sc-Enriched Ferrodiorite, *Minerals,* 11(9):966. https://doi.org/10.3390/min11090966

Ma, M, 2012. Froth Flotation of Iron Ores, *International Journal of Mining Engineering and Mineral Processing*, 1(2):56–61.

Makhija, D, Mukherjee, A K and Ghosh, T K, 2013. Preconcentration Feasibility of Gravity and Magnetic Techniques for Banded Hematite Jasper, *International Journal of Mining Engineering and Mineral Processing*, pp 8–15.

Metso Corporation, 2024. SLon® vertically pulsating high-gradient magnetic separator [online], Metso Plus, Metso Corporation. Available from: <https://www.metso.com/portfolio/vertically-pulsating-high-gradient-magnetic-separator/> [Accessed: 04/01/2026].

Ohmoto, H, Hoashi, M, Otake, T, Bevacqua, D and Watanabe, Y, 2008. Origins of hematite, magnetite and siderite in banded iron formations, *Geochimica et Cosmochimica Acta Supplement*, A699–A699.

Rodrigues, W, Fernandes, P and Peres, A, 2024. The effect of etheramine type on the hydrophobicity of quartz particles from iron ore, *Minerals Engineering*, 214:108769. https://doi.org/10.1016/j.mineng.2024.108769

Singh, A and Da Corte, C, 2021. Maximising resource utilisation in South Africa – the BIF story, in *Proceedings of the Iron Ore 2021 Conference*, 22 p (The Australasian Institute of Mining and Metallurgy: Melbourne).

Taner, H and Onen, V, 2016. Control of clay minerals effect in flotation, A review, Poland: Minerals Engineering Conference.

Waldron, J W F and Molinaro, D J, 2024. What the Geosphere is Made of: Minerals, in *Planet Earth* (ed: J W F Waldron), Open Education Alberta. Available from: <https://pressbooks.openeducationalberta.ca/planetearth/chapter/composition-of-the-earth-minerals/>

Wills, B and Napier-Munn, T, 2006. *Mineral Processing Technology,* 7th edn (Elsevier Science and Technology Books).

Xiong, D, Lu, L and Holmes, R, 2022. Chapter 9 – Physical Separation of Iron Ore: Magnetic Separation, in *Iron Ore* (ed: L Lu), 2nd edn, pp 309–322 (Woodhead Publishing).

Yongxiang, Y, Kalevi, R and Lauri, H, 2014. Chapter 1.1 – Ironmaking, in *Treatise on Process Metallurgy* (ed: S Seetharaman), pp 2–88 (Elsevier).

Yu, J, Han, Y, Li, Y and Gao, P, 2017. Beneficiation of an iron ore fines by magnetization roasting and magnetic separation, *International Journal of Minerals Processing*, 168:102–108.

Zhang, X, Gu, X, Han, Y, Parra-Alvarez, N and Clareboux, V K S, 2021. Flotation of Iron Ores: A Review, *Minerals Processing and Extractive Metallurgy Review*, 42(3):184–212.

Predictive modelling of a closed-circuit spiral flow sheet for low-grade iron ore using rougher spiral data and shaking table benchmark test data

T Mokgomola[1], A Singh[2] and V Sibanda[3]

1. Engineer, Mintek, Randburg 2125, South Africa. Email: tebogomo@mintek.co.za
2. Snr Technical Specialist, Mintek, Randburg 2125, South Africa. Email: ashmas@mintek.co.za
3. Professor, University of the Witwatersrand, Gauteng 2050, South Africa.
   Email: vusumuzi.sibanda1@wits.ac.za

ABSTRACT

The mineral processing industry has been using shaking tables as a precursor to spiral concentration for decades. While a single spiral cannot replicate the performance of a single shaking table, an optimised closed circuit spiral flow sheet can achieve comparable results. This demonstrates that a single shaking table can mirror the separation performance of a well-designed spiral flow sheet. This study investigates shaking table performance under varying operating parameters and compares it with multiple spirals flow sheets producing similar grades, with the aim of developing a predictive model that can determine achievable grades and recoveries in closed-circuit spirals. The proposed model uses data from a single batch (rougher) spiral stage and shaking table tests to predict the performance of a full spiral flow sheet, thereby reducing the experimental and pilot scale workload of testing full spiral flow sheets. The robustness of the model will be tested by varying feed rates and particle size distributions. D-grade banded iron ore formation (BIFs) with a head grade of 33 per cent Fe was used for the test work. With B-grade and C-grade iron ores projected to be depleted in the next decade, the development of a viable beneficiation flow sheet for D-grade resource is important to extend the iron ore resource within South Africa.

The BIF ore was upgraded to grades of 55 per cent and 60 per cent by both shaking table and spirals for a coarse feed with a P_{80} = 533 μm. Shaking table outperformed spirals with high yields and Fe recovery suggesting the need for spirals optimisation. Rao's stochastic model was adopted as the partition model to describe the probability of particles reporting to either concentrate, middlings, or tailings. Given that fine and narrow particle size distributions were tested, the effects of size were neglected, and density was used as the dominant attribute. Model parameters for each spiral stage that will be used to perform stage wise mass balancing and predict closed circuit spiral performance were determined.

INTRODUCTION

The mining and mineral industry has advanced South Africa (SA)'s socio-economic development for over a century and remains a key contributor to the national economy. In the first three quarters of 2025, SA's mining sector grew by 3.7 per cent, contributing 0.2 per cent to SA's overall GDP growth of 0.8 per cent (Motsa, 2025). Despite this contribution, the iron ore sector has faced significant pressure. Iron ore prices declined by 30 per cent in 2025, resulting in job losses and restructuring within the iron ore and steel industries (Sepheku and Mabunda, 2025). Rising domestic energy costs increased production expenses, while weakened steel demand in China further reduced global spot prices.

South Africa is currently the seventh largest iron ore producer globally. However, the reserves of high-grade (A (>60 per cent Fe) and B (50–59 per cent Fe)-grade) and medium grade (C (40–49 per cent Fe)) iron ores are steadily declining. Within the next decade, it is projected that primarily low-grade (D (<40 per cent Fe)) resource will remain which makes up 65 per cent of the iron ore reserve (Singh and da Corte, 2021). D-grade ores are made up of banded iron ore formations (BIFs), which are sedimentary rock formations made up of alternating iron-rich layers and silica-rich layers. These ores have complex mineralogy and require fine grinding to achieve liberation which lends current beneficiation methods less effective. This poses a challenge in the iron ore and the steelmaking industries as this might necessitate importing raw iron ore for steelmaking purposes if viable beneficiation methods are not developed. The demand for iron ore is projected to rise to three billion by 2035 (Research and Markets, 2023), further indicating the need to extend the life-of-mine through improved processing of low-grade resources (Singh and da Corte, 2021).

Mineral processing plays a critical role in the development, expansion, and sustainability of the mining sector. Increasing commodity price volatility, declining grades, and rise in production costs have showed the need for strategic minerals processing techniques that are cost-effective, energy-efficient, and generate less environmental impact. The most established and robust technique known to humankind is gravity separation using spirals concentrators and shaking tables.

Gravity separation exploits the differences in densities to bring about separation between high density particles and gangue particles in fluids such as water under the influence of gravitational and centrifugal forces (Myburgh and Nortje, 2014). These techniques are widely applied to commodities such as iron ore, coal, chrome, beach sands, platinum group metals, gold, tin, and manganese. Shaking table and spirals are two pieces of equipment widely used in mineral processing because they are simple, reliable, and have high selectivity (Wills and Napier-Munn, 2006).

The shaking table utilises a flowing film of slurry to recover denser mineral particles into high-grade products (Wills and Napier-Munn, 2006). According to Manser, Barley and Wills (1991), factors that drive separation on a Wilfley shaking table are wash water to effect separation, feed pulp density, optimum deck inclination per given mineral, amplitude, and feed rate. The principle of separation on the shaking table is the motion of the particles according to their specific gravity and size moving in the pulp across the inclined shaking table oscillating back and forth at right angles to the slope (Bahrami, Hosseini and Shafaee, 2008). Vertical stratification on the shaking table occurs behind rifles and runs parallel to the longitudinal part of the shaking table until it dies down at the end of the rifles when the table smoothens out and the particles are separated into concentrate, middlings, and tailings. Shaking tables are low-capacity units but they are important in mineral processing as they offer high selectivity and are reliable in producing high-grade products as operators can visually identify separation and adjust parameters accordingly (Wills and Napier-Munn, 2006).

Shaking tables have been used to test if certain materials will be amenable to gravity separation before processing on large scale gravity concentrators like spirals. As a result, shaking table tests can be performed on feed mass from 2 kg upwards, making them perfect for project testing or production of a small amount of finished product for offtake acceptance tests. They are ideal for lab-scale gravity testing because the parameters can easily be tuned to test response of different ore types, shapes, sizes, at different solids feed rate, slurry density, and water. They provide theoretical grades, yields and recoveries that can be expected when processing on a large-scale pilot unit like spirals (Bahrami, Hosseini and Shafaee, 2008).

Spirals are inexpensive devices with no mechanical moving parts used to concentrate on various ore types in industry owing to their capacity (Wills and Napier-Munn, 2006). According to Gawad *et al* (2015), spirals were introduced by Humphreys in 1941 and have successfully been used as a cost-effective means of concentrating a variety of ores because of attributes like high throughput, low energy requirement, high wear resistance, less water consumption, low space requirements, easy operability, and are environmentally friendly using only water as a separation medium. Furthermore, spirals separation efficiency can be monitored visually and thus operation can be altered online. According to Wills (1992) and Matthew, Fletcher and Patridge (1999), spirals consist of an open down trough that twists downward in a helical configuration about a central axis. The feed material mixed with water acting as the medium of separation is introduced at the top with a solids weight percentage ranging between 15–45 per cent and a solids feed size ranging between a top size of 2 mm and bottom size of 38 μm (Wills and Napier-Munn, 2006). Recent research studies by Bornman, Foisy and Naude (2020) and Ceko, Ramotsabi and Molehe (2022) presented that traditional spirals can treat down to 45 μm and the new ultrafine processing spirals can treat down to 10 μm. Since their introduction there have been attempts to optimise and to improve separation efficiency of spirals for different feed particle sizes and minerals by looking at spiral design and tuning of parameters using different types of modelling tools (Mokgomola, Singh and Sibanda, 2025).

Holland-Batt (1995) discussed spiral separation and concluded that fluid dynamics is the main drive of separation on the spirals. It consists of a primary (down trough from the top) flow component followed by a complicated secondary flow circulation, which operates radially across the spiral trough (Matthew, Fletcher and Patridge, 1999). Two motions occur on the spiral, one is towards, and the other is away from the spiral centre column. The spiral trough is divided into two sections with a transition region in the centre of the trough. All three regions exhibit different aspects of fluid flow. Particle behaviour on the inward part of the spiral is directed by the creation of a slow-moving bed of particles overlaid by a free motion region dictated by the secondary circulation. As velocity increases in the inward zone of the spiral, there is a particle size limit imposed in the inward zone forcing finer particles to be lifted into the upper levels of the flow and are transported by water to the tails stream resulting in radial stratification along the spiral radius. Fuerstenau and Han (2003), presented that the following mechanisms apply for particles separation in a fluid; hindered settling, asymmetrical acceleration, and flowing film concentration resulting in particles experiencing the following forces: gravity, centrifugal force (spirals), vibration force (shaking tables), drag, buoyancy, and friction forces.

In the later 1990s, spiral optimisation looked at the treatment of fine particles conducted by Holland-Batt (1995) and Richards *et al* (2000). To modify the conventional spirals to process fine material, there is a need to control the Reynolds number (turbulence) across the trough to ensure optimised flowing film and particle settling. Fine particles have low setting velocity making separation longer than coarse particles (Holland-Batt, 1995). These studies were carried out by experimentation requiring a lot of empirical input and using trial and error which is expensive. The solution was using computational fluid dynamics (CFD) which was presented by Matthews *et al* (1998). The CFD model was successful in simulating fluid flow characteristics and performing particulate analysis at both low and high concentrations to examine the influence of particle-particle interactions and the hydrodynamic processes.

In this current study, modelling spirals was revisited by using shaking table to investigate to what extent laboratory shaking table data and batch(rougher) spiral data can be applied to develop a predictive model for a pilot scale continuous closed spiral circuit. To investigate the performance of gravity separators in terms of separation efficiency, the probability of a particle with a certain size and density reporting to one of the product streams (concentrate or tails) is used. Separation efficiency can be expressed as a curve which can produce a product particle distribution if feed distribution is used. Although it is advisable to use only particle density for efficiency curves, the effect of particle size on efficiency curves shows that as the particle size decreases, efficiency also decrease gradually showing a requirement for consideration of the effects of both particle size and density. This makes separation efficiency a bivariate parametric representation of the partition curve called partition surface (Rao, 2004).

For spirals and shaking table or other industrial applications the feed is divided into multiple products ranking from high-grade to low-grade making density the attribute with the highest effect on separation compared to size due to the narrow size distribution employed across these units in general. The partition surface can either use a constant particle size or not consider particle size (Rao, 2004).

The use of partition curves to examine and model the performance of gravity separators was first introduced by Napier-Munn (1991) with the aim of extending two existing classification models by Lynch and Rao (1975) and Plitt (1971) from one attribute to two attributes (size and density). Lynch and Rao (1975) used Whiten's regression model to describe the classification curve of a hydrocyclone by size as:

$$Y = \frac{\exp(\alpha x) - 1}{\exp(\alpha x) + \exp(\alpha) - 2} \tag{1}$$

Plitt (1971) assumed that the turbulence in a classifier resulting from mixing is responsible for imperfect separation. The Plitt model is expressed as follows:

$$Y = 1 - \exp[-x^m \ln(2)] \tag{2}$$

For both models, Y shows corrected efficiency of separation which defines the corrected fraction of feed reporting to the underflow, $x = d/d_{50}$ is the ration of particle size d to particle size corresponding to 50 per cent separation d_{50}. Sharpness of separation is defined by parameters m and α. On both models x can also be substituted for particle density.

Both models are adequate to represent the partition for only one attribute of the particle which is not convenient for gravity separation since both particle size and particle density have an effect on separation efficiency. However, direct extension of Equations 1 and 2 to include two attributes would require the knowledge of d_{50} for each density class. To solve this matter Scott and Napier-Munn (1992) used the pivot point phenomenon which extended the Lynch and Rao (1975) Model to a size-density partition curve. Pivot phenomenon imposes conditions such that all size based partition curves pass through a pivot point called Y_p which has a particle density of ρ_p. All partition curves of different sizes are related to the partition surface using an empirical law of Ecart probable which is expressed in terms of size as:

$$Ep = kd^n \tag{3}$$

Where *k* is the proportionality constant representing the effects of viscosity and strength of forces in the separator and *n* represents the flow conditions in the separator. The pivot phenomenon extends Equation 1 into a size-density partition surface given by:

$$Y = \frac{100}{1 + \exp[\ln(Y_p^{-1} - 1) + 1.099(\rho_p - \rho)/(kd^n)]} \tag{4}$$

The parameters Y_p, ρ_p, *k* and *n* are estimated from experimental data. Parameters such as d_{50} and p_{50} derived from Equation 4 when Y = 0.5.

The Scott and Napier-Munn (1992) model was developed and validated on dense medium cyclones and not on other gravity separators such as spirals and shaking tables. To model size-density curves without invoking the pivot phenomenon and to include other gravity separators, Rao (2004) proposed a stochastic model for size density partition surfaces which was versatile compared to other existing model and can be used to derive the Scott and Napier-Munn (1992) model shown in Equation 4. The foundation of the model simplified particle flows in gravity separators as flowing from an arbitrary height h=0 and the particles moving down the separator as a result of gravity, hydrodynamic forces from fluid and random forces coming from particle-particle interactions. The influence of these combinations of forces moves particles either to the underflow (concentrate) or the overflow (tailings). The flowing fluid plays a role in preventing particles from settling. The velocity of a particle can be given as:

$$\frac{dh}{dt} = V_s - V_1 + V_r \tag{5}$$

Where V_s, V_1 and V_r represent setting velocity, drift velocity due to fluid flow and random velocity in the separator. Equation 5 can be used to calculate the potion of a particle h(t) on a spiral at time t. Velocity V_r which denotes a random variable that determines the probability of a particle reporting to either concentrates or tailings can be written as:

$$V_r = \frac{h(t)}{t} + V_1 - V_s \tag{6}$$

Rao (2004) used the central limit theorem and justified that the random velocity is a normally distributed with a mean of zero and variance σ^2. The cumulative distribution functions in the form of random velocity after substituting Equation 6 can be given as:

$$Y(h) = \frac{1}{2}\left[1 + \operatorname{erf}\left(\frac{V_r}{2^{0.5}\sigma}\right)\right] \quad -\infty \leq h \leq \infty \tag{7}$$

Equation 7 shows the transient partition surface at position h and time t. It can be observed that for $Y(-\infty) = 0$ and $Y(\infty) = 1$ fulfilling the conditions for a distribution function. Under steady state conditions, Johnson, Li and Logan (1996) settling velocity equation can be substituted into Equation 7 and after lumping variance and other constants into terms A and B Rao's stochastic model is defined as:

$$Y = 50\left[1 - \operatorname{erf}\left(Ad^c(\rho - \rho_p) - B\right)\right] \tag{8}$$

A and B can be expressed as:

$$A = \frac{g}{18\sqrt{2}\mu\sigma} \tag{9}$$

$$B = \frac{v_1}{\sigma\sqrt{2}} \tag{10}$$

Parameters A, B, c, and p_p describe the size-density partition surface. The combination of A and c in Equation 8 account for fluid viscosity, hindered settling, particle shape, and turbulence in the system.

Rao (2004) stochastic model was validated using 16 sets data from literature including the data from Ferrara and Schena (1987) and Scott and Napier-Munn (1992). The best fit parameters for each set were determined from the partition curves and sum of squares errors (SSE) using Levenberg-Marquardt least squares minimisation technique. It was found that flow conditions parameter c lies within a range between 0.46 and 1.63 and the closer it is to 2 the flow conditions are laminar. When parameter c decreases the flow transitions to turbulent.

Rao's (2004) model relates to spirals and shaking tables because the probability of a particle of a certain size or density reporting to either concentrate, middlings or tails is driven by a normally distributed random process associated with the velocity of a given particle. This agrees with the work by Holland-Batt (1995) stating that fluid flow on the spirals drives the separation mechanism. In spirals and shaking table, bulk performance of the particles defined by grade and density will be looked at thus the density attribute has more influence than size. In this case a modified version of Equation 8 will be used using only the density attributes shown below:

$$Y = 0.5\left[1 - \mathrm{erf}\left(A(\rho - \rho_p) - B\right)\right] \tag{11}$$

In this study, a multi-stage spiral and shaking table experimental data using low-grade BIF iron ore as feed will be used to find the best fit model parameters. The target of the spiral flow sheet is to achieve grade and SG similar to that of the BIF iron ore shaking table results. Data from each spiral stage will be modelled to determine model parameters for future predictions. It is envisaged that the model parameters determined from the multistage-stage spiral on iron ore can be validated using different commodities at varying grades.

MATERIALS AND METHODS

Materials

Low-grade BIF iron ore with a head grade of 32.9 per cent Fe at a crushed top size of -1.18 mm was used for research in this study.

Methodology

Figure 1 presents the schematic diagram of a workflow sheet showing experimental procedure followed for the BIF iron ore feed at both lab and pilot scale.

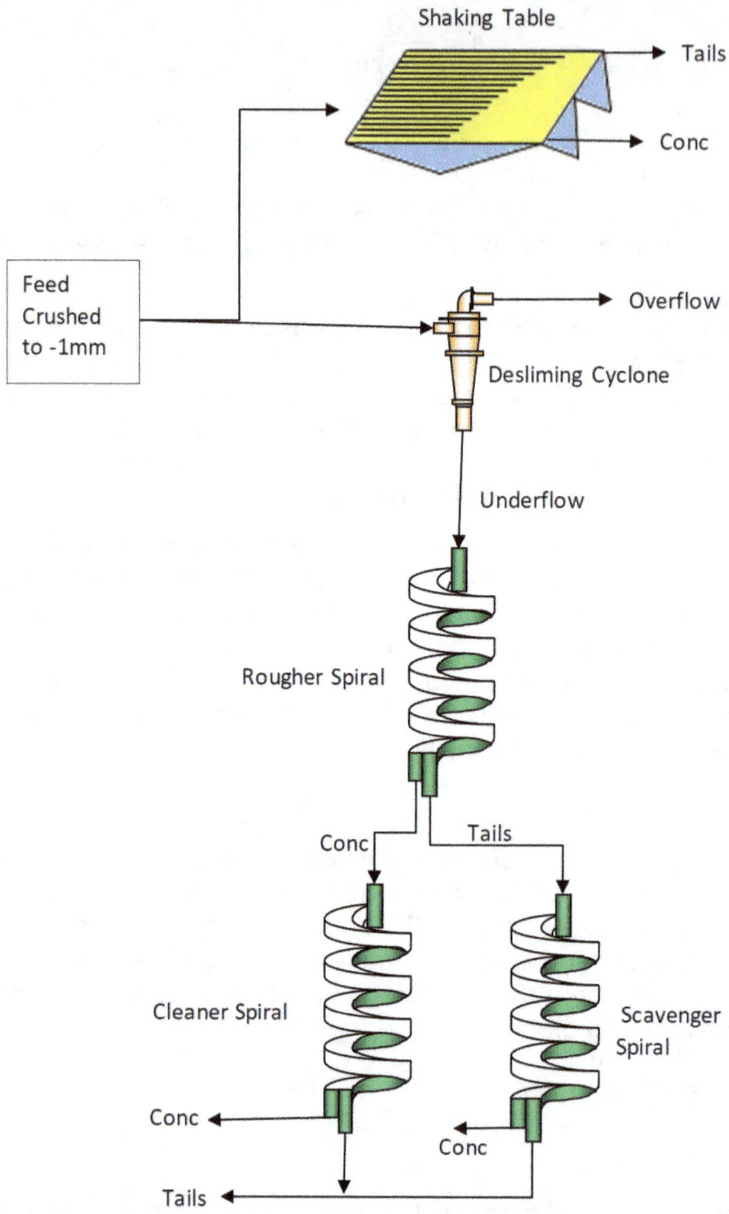

FIG 1 – Schematic test workflow sheet.

Sample preparation

The as-is sample was crushed to three different grinds, 100 per cent passing 1.18 mm with P_{80} of 533 µm. The samples were sub-sampled using cone and quartering followed by rifle splitting to remove samples representative of the original feed. The sub-samples were analysed for the deportment of iron (Fe) and major gangue elements to determine the head analysis. Gravity separation using shaking tables, desliming cyclone and spirals was conducted as shown in Figure 1.

Chemical analysis

The XRF11 method was used for iron (Fe) and other elements such as silica (SiO_2) chemical analysis on the feed and test work products. The process requires fluxing a sample with a mass of 0.5 g followed by fusing at high temperatures >1000°C and analysing using XRF11 method. The lower detection limit of the method for Fe is 0.04 per cent.

Shaking table, desliming, and spiral processing

Iron ore feed with approximately 10 kg was tested using the Wilfley shaking table to determine the expected grade profile, recovery and yield before pilot processing using spirals. The test work produced nine products (four concentrates, two middlings, two tails and slimes) which were prepared and submitted for chemical analysis. The shaking table has its own desliming mechanism in which all slimes are washed to the discard launder while spirals processing relies on deliming cyclones.

A multistage spirals test work was conducted with the aim of scavenging and recycling to meet the target grade, mass yield and recovery closer to that of the shaking table for each size fraction tested. According to Wills and Napier-Munn (2006), gravity separators such as spirals are highly sensitive to ultrafine (<25 μm) material which can increase viscosity, reduce sharpness of separation and obscure visual cut points. It is important to remove particles less than 25 μm.

The feed before spirals was processed through a desliming cyclone targeting a cut point of 25 μm. The cyclone overflow was analysed, and the underflow was processed through spirals (rougher, cleaner, scavenger, recleaner, and re-recleaner) stages.

Modelling

Rao's Stochastic model was used to define parameters for each stage and the output partition number (yield from one stage was used as input to the next stage). Rao's model will be defined by Equation 12 for each stage as:

$$Y_i = 0.5[1 - \mathrm{erf}(A_i(\rho - \rho_{p,i}) - B_i)] \tag{12}$$

Where i = the number of stages from Rougher to re-recleaner spiral.

RESULTS AND DISCUSSION

Feed characterisation

Table 1 and Figure 2 show feed characterisation by chemical analysis and particle size distribution after crushing to 100 per cent passing 1.18 mm.

TABLE 1

Feed head grade (%).

Fe	SiO_2	Al_2O_3	MgO	MnO	CaO
32.89	36.77	9.09	0.34	0.14	0.53

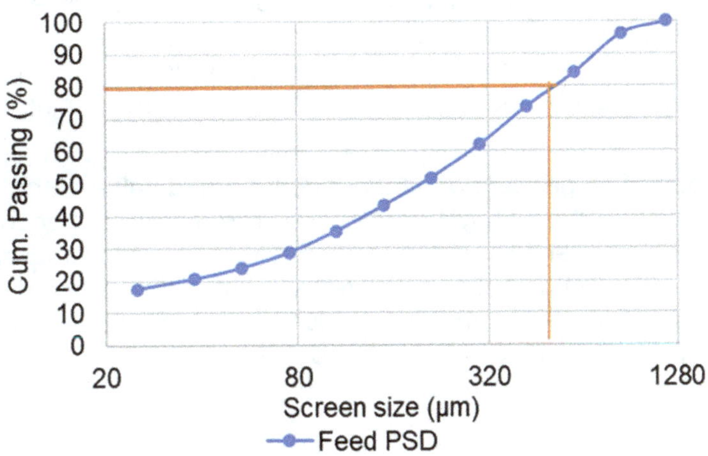

FIG 2 – Feed particle size distribution. P_{80} = 533 μm.

The head grade of the feed was 32.89 per cent Fe and 36.77 per cent SiO_2 with a specific gravity of 3.73. The mineralogy of the feed sample showed that the sample consists mainly of hematite (49.6 mass per cent) with illite and quartz as the major gangue minerals present. Only 67 mass per cent of feed of hematite is liberated. Hematite contributes 97 mass per cent of Fe to the total iron of the sample.

Particle size distribution of the feed showed that 80 per cent of the feed passes 533 μm (P_{80} = 533 μm) suggesting a course grind. According to Singh and da Corte (2021), the alternating iron-rich and silica rich layers means BIFs ores require finer grinds to achieve Fe liberation. As a result, finer grinds of (P80<300 μm) will be tested in an attempt to liberate more iron units.

Gravity separation (shaking table versus spirals)

Tables 2 and 3 show shaking table and spiral separation results summaries to compare the performance of both separators.

TABLE 2

Shaking table (P_{80} = 533 μm).

Gravity summary	Target Fe grade (%)	Density (g/cm^3)	Cum. yield (%)	Cum. recovery (%)
Concentrate 1	60	4.7	11.4	19.6
Concentrate 2	55	4.4	24.5	38.7

TABLE 3

Spirals (P_{80} = 533 μm).

Gravity summary	Target Fe grade (%)	Density (g/cm^3)	Cum. yield (%)	Cum. recovery (%)
Concentrate 1	60.74	4.78	2.37	4.57
Concentrate 2	55.43	4.55	9.65	10.28

The results showed that both spiral and shaking table achieved grades of 55 per cent and 60 per cent Fe with shaking table having a higher yield and recovery. This is because Shaking tables have the ability to operate under controlled operating conditions while Spirals on the other hand are less selective on a single stage and require multiple stages with recycling to achieve a performance similar to that of shaking tables (Fuerstenau and Han, 2003). To reduce the recovery and grade gap between the two devices, there is a need to add more spiral stages with recycling of tails, increasing the experimental load. Developing a model addresses the gap by simulating a multi-stage closed circuit spiral which allows the modelling of the partition behaviour of each spiral using Rao's stochastic model described by Equation 12. The model provides final yields, grade and recovery progressively as multiple spiral stages are added showing how low efficient rougher and scavenger separation will be compensated by the downstream cleaner and recleaning stages thus improving overall yield and recovery.

Currently, products grade of 55 per cent and 60 per cent Fe add value to the available resource as they can be used to blend with the high-grade resource. The results further confirm that finer grinds (P_{80} <300 μm) be tested in an attempt to optimise recovery of Fe as this improves valuable minerals liberation. Chaurasia and Nikkam (2015) beneficiated low-grade hematite ore from 50 per cent Fe to 65 per cent Fe using fine grinds (-75 μm) using a Multi Gravity Separator (MGS) unit showing the potential to upgrade low-grade iron ores at finer grinds.

Model parameters

The stochastic model in Equation 12 was used to determine the relationship between the partition number and SG for spiral test work via ten stages (rougher, rougher cleaner, rougher middlings cleaner, rougher middlings recleaner, scavenger, scavenger cleaner, scavenger recleaner, rougher cleaner scavenger, and final recleaner). Figure 3 shows the comparison between model generated partition curves to experimental data showing partition number versus density for each spiral stage. The least-square minimisation was used to determine the best fit model parameters for each spiral stage. In general, the variation between experimental and model data is small with most of the curves superimposed on each other. Table 4 shows the model parameters after fitting stochastic model in Equation 12 to the experimental data.

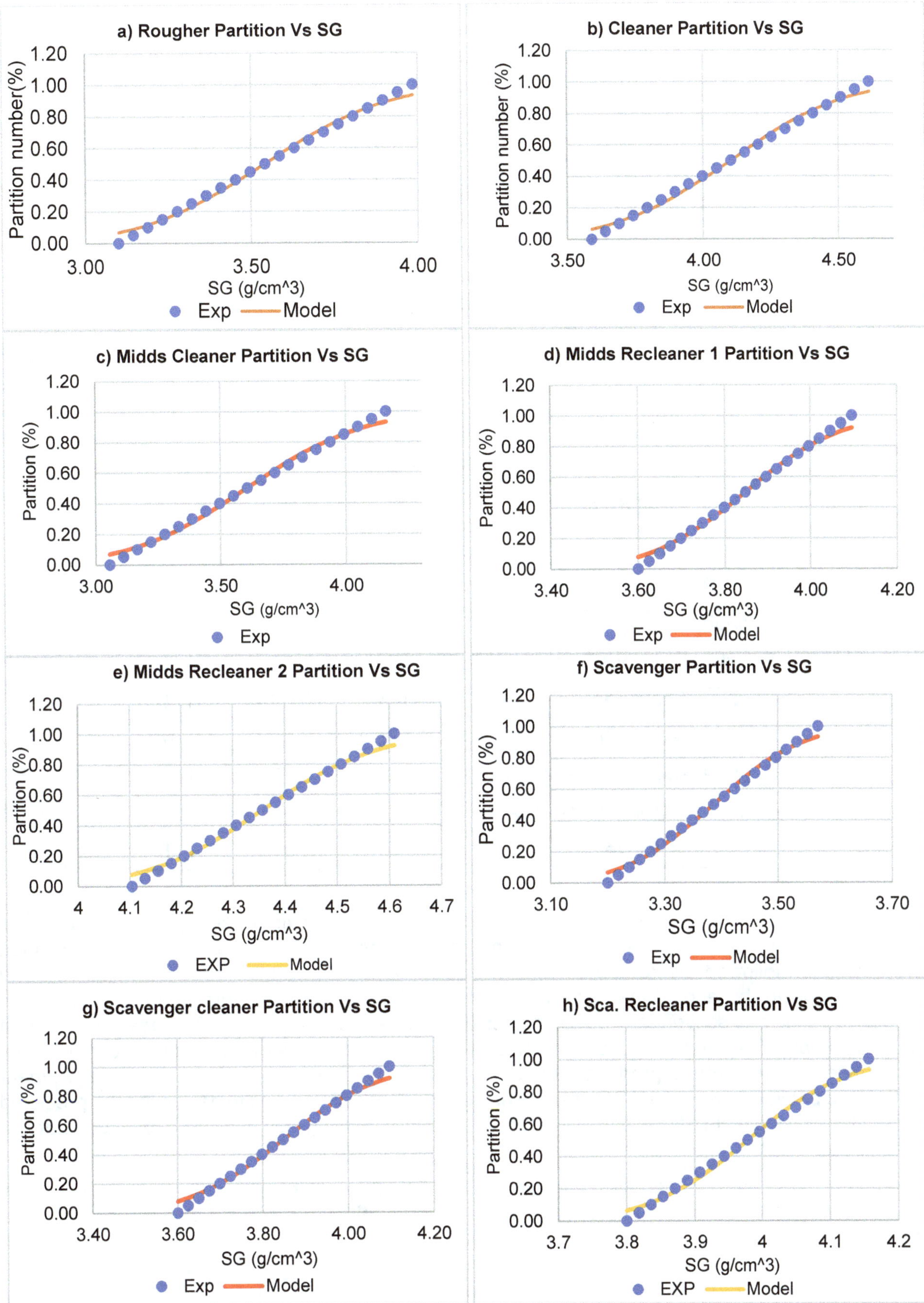
a) Rougher Partition Vs SG
Partition number(%)
SG (g/cm^3)
Exp
Model
b) Cleaner Partition Vs SG
Partition number (%)
SG (g/cm^3)
Exp
Model
c) Midds Cleaner Partition Vs SG
Partition (%)
SG (g/cm^3)
Exp
d) Midds Recleaner 1 Partition Vs SG
Partition (%)
SG (g/cm^3)
Exp
Model
e) Midds Recleaner 2 Partition Vs SG
Partition (%)
SG (g/cm^3)
EXP
Model
f) Scavenger Partition Vs SG
Partition (%)
SG (g/cm^3)
Exp
Model
g) Scavenger cleaner Partition Vs SG
Partition (%)
SG (g/cm^3)
Exp
Model
h) Sca. Recleaner Partition Vs SG
Partition (%)
SG (g/cm^3)
EXP
Model

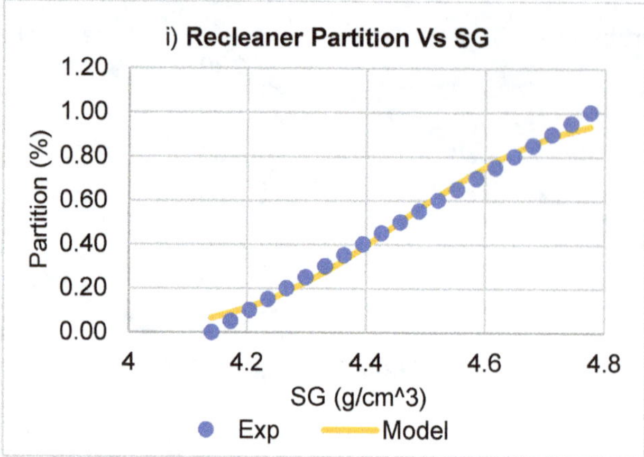

FIG 3 – Comparison of experimental data to model fit of different spirals – each is defined by letters (a) to (i).

TABLE 4
Model Parameters.

Spiral Stage	**A**	**B**	**ρ_p**
Rougher	-2.38	0.00	3.54
Rougher Cleaner	-2.09	0.08	4.14
Rougher Middlings Cleaner	-1.91	0.00	3.61
Rougher Middlings Recleaner	-4.00	1.20	4.15
Scavenger	-5.69	0.00	3.38
Scavenger Cleaner	-4.00	0.09	3.87
Scavenger Recleaner	-6.00	0.07	3.99
Rougher Cleaner Scavenger	-4.00	0.07	4.37
Final Recleaner	-3.37	0.07	4.48

The curves in Figure 3 show probability of each density on the x-axis reporting to the concentrate increasing with cumulative mass to the concentrate. The concentrate of the rougher spiral was fed to the cleaner followed by recleaner. Similar steps were followed for rougher middlings and scavenger and in the end a final recleaner spiral combining products from rougher concentrate cleaner, middling and scavenger were combined to produce the final product. The goal with BIF iron ore is to combine Rao models generated for each spiral, input rougher data and predict how many spiral stages are required to produce a density distribution similar to that of shaking table. A linear relationship between grades and SGs can be established and will be used to determine the predicted grade. The shaking table data is used to determine the theoretical grade, yield and recovery that can be achieved.

The model parameters in Table 4 determined after fitting model to experimental data shown in Figure 3 were used to predict the product SGs for each spiral stages and the percentage error was compared in Table 5.

TABLE 5
SG model prediction Vs Experimental SG prediction.

Spiral stage	SG exp	SG model	% Error
Rougher	3.98	3.70	7.10
Rougher cleaner	4.61	4.32	6.33
Rougher middlings cleaner	4.16	3.81	8.48
Rougher middlings recleaner	4.10	3.93	3.94
Scavenger	3.57	3.45	3.32
Scavenger cleaner	4.10	3.95	3.50
Scavenger recleaner	4.16	4.04	2.70
Rougher cleaner scavenger	4.61	4.45	3.52
Final recleaner	4.78	4.58	4.10

Sensitivity analysis showed that the model can predict concentrate SGs for each spiral stage with percent error varying between 2.7 per cent and 8.5 per cent, with the highest deviations observed in the rougher stages. Error percentages above 5 per cent result in a horizontal shift of the partition curve leading to misplacement of particles near the separation boundary. Since the spiral circuit is a closed and continuous system, the errors are not confined to the rougher stages but can be transferred to the downstream spiral stages resulting in cumulative deviations in the predicted final grade. This shows that although the model can provide a robust framework to predict continuous spirals behaviour, the accuracy is sensitive to accurate SG estimation especially in rougher stages where feed variability and liberation have a huge effect on separation. To improve performance in the prediction it is important to improve SG estimation in the rougher stages. To address this, a prediction based on full mass balance connecting the spirals stages based on yields is currently being undertaken to reduce the overall percentage error. Despite these sensitivities, the model for each stage however is robust and shows alignment with experimental data.

CONCLUSIONS

- The head grade of the feed was reported at 32.89 per cent Fe and 36.77 per cent SiO_2.
- Mineralogy revealed that the feed was mainly hematite (49.6 mass per cent) with illite and quartz as the major gangue minerals.
- Shaking table and spirals test work were conducted producing Fe grades of 55 per cent and 60 per cent Fe. The shaking table outperformed spirals by higher yield and recovery of Fe of 20 per cent and 39 per cent for both target grades.
- The upgraded Fe grade although not premium quality, can be used as a blend to the current high-grade resource.
- A stochastic partition model using density as the main parameter was derived to represent the relationship between density and partition number using a ten-stage spiral flow sheet.
- The prediction for partition number versus density produced from the model was compared to those of experimental data and model parameters were consistent across different spiral stages.
- A model equation that can predict the grade of downstream stages using rougher data to obtain grades like shaking table can be developed by combining the partition curves of one spiral to another using mass fractions obtained experimentally.

ACKNOWLEDGEMENTS

Mintek for offering funding for studies and facilities to perform experiments. Professor Vusumuzi Sibanda and Dr Ashma Singh for supervising the work.

REFERENCES

Bahrami, A, Hosseini, M R and Shafaee, S Z, 2008. The Shaking Table Concentrator: Influence of Table Parameter on Manganese Separation, Beijing.

Bornman, F, Foisy, R and Naude, N, 2020. Spiral conditions for the recovery of ultra-fine particles, Cape town, International Mineral Processing Congress.

Ceko, R, Ramotsabi, C and Molehe, R, 2022. UX7 spiral: Ultrafine processing made possible, Multotec Process Equipment (Pty) Ltd, Spartan, South Africa.

Chaurasia, R C and Nikkam, S, 2017. Beneficiation of low-grade iron ore fines by multi-gravity separator (MGS) using optimization studies, *Particulate Science and Technology*, 35(1):45–53. https://doi.org/10.1080/02726351.2015.1124161

Ferrara, G and Schena, G, 1987. Cycloning in dense media separation, in *Proceedings of the 3rd International Conference on Hydrocyclones* (ed: P Wood), pp 101–110 (Oxford).

Fuerstenau, M and Han, K, 2003. *Principles of Mineral Processing* (Society for Mining, Metallurgy and Exploration: CO).

Gawad, A, Gamal, M A M, Doheim, M A and Rizk, M, 2015. Numerical simulation of particulate-flow in spiral separators (15% solids), Barcelona: Afinidad.

Holland-Batt, A, 1995. Some design considerations for spiral separators, *Minerals Engineering*, 8(11):1381–1395.

Johnson, C P, Li, X and Logan, B, 1996. Settling velocities of fractal aggregates, *Environ Sci Technol*, 30(6):1911–1918.

Lynch, A J and Rao, T C, 1975. Modelling and scale-up of hydrocyclone classifiers, in *Proceedings of the 11th International Mineral Processing Congress*, pp 245–269 (Università, Istituto di arte mineraria e preparazione dei minerali: Cagliari, Sardinia).

Manser, R, Barley, R and Wills, B, 1991. The shaking table concentrator - the influence of operating, *Minerals Engineering*, 4(3/4):369–381.

Matthew, B, Fletcher, C and Patridge, T, 1999. Particle flow modelling on spiral concentrators: benefits of dense media for coal processing, *Engineering, Environmental Science*, pp 1–6.

Matthews, B W, Holtham, P, Fletcher, C A, Golab, K and Partridge, A C, 1998. Computational and Experimental Investigation of Spiral Concentrator Flows, in *Proceedings of the First Australasian Coal Operators' Conference* (eds: N Aziz and B Kininmonth), pp 415–421 (The Australasian Institute of Mining and Metallurgy: Melbourne).

Mokgomola, T, Singh, A and Sibanda, V, 2025. Predictive Modelling of Pilot Spiral Continuous Circuit Using Rougher Batch Spiral Data and Laboratory Shaking Table Data as Input, Proceedings of the 11th World Congress on Mechanical, Chemical and Material Engineering (MCM'25).

Motsa, 2025. Mining led South Africa's modest Q2 2025 GDP growth, Minerals Council, South Africa.

Myburgh, H and Nortje, A, 2014. Operation and performance of the Sishen jig plant, *The Journal of The Southern African Institute of Mining and Metallurgy*, pp 569–574.

Napier-Munn, T, 1991. Modelling and simulating dense-medium separation processes – a progress report, *Min Eng*.

Plitt, L R, 1971. The analysis of solid – solid separations in classifiers, *CIM Bulletin*, 64(708):42–47.

Rao, B, 2004. General gamma representation for product particle split in gravity, *The European Journal of Mineral Processing and Environmental Protection*, pp 84–93.

Research and Markets, 2023. Global iron ore mining to 2035 [online]. Available from: <https://www.researchandmarkets.com/reports/4726740/global-iron-ore-mining-to-2035?utm_source=GNE&utm_medium=PressRelease&utm_code=87r8xv&utm_campaign=1931860+-+Iron+Ore+Mining+Industry+Analysis%3a+2030+Global+Forecast+Reflects+Growth+and+Emerging+Trends&u> [Accessed: 24 January 2026].

Richards, R G, MacHunter, D M, Gates, P J and Palmer, M K, 2000. Gravity separation of ultra-fine (-0.1mm) minerals using spiral separators, *Minerals Engineering*, pp 65–77.

Scott, I A and Napier-Munn, T J, 1992. A dense-medium cyclone model based on the pivot phenomenon, *Transactions of the Institute of Mining and Metallurgy: Section C Mineral Processing and Extractive Metallurgy*, 101:C61–C76.

Sepheku, A and Mabunda, X, 2025. South Africa's Ailing Mining Industry, SITEF and CO [online]. Available from: <https://sitef.co.za/south-africas-ailing-mining-industry/#:~:text=In%20addition%2C%20mineral%20sales%20at,and%20chromium%20ore%20(23.2%25)> [Accessed: 04 November 2025].

Singh, A and da Corte, C, 2021. Maximising resource utilisation in South Africa – the BIF story, in *Proceedings of Iron Ore Conference 2021*, 21 p (The Australasian Institute of Mining and Metallurgy: Melbourne).

Wills, B and Napier-Munn, T, 2006. *Mineral Processing Technology* (Elsevier Science and Technology Books).

Wills, B, 1992. *Mineral Processing Technology: An Introduction to the Practical Aspects of Ore Treatment and Mineral Recovery*, 5th edn (Pergamon Press Ltd: Oxford).

Reducing oversize and improving operational safety through modular secondary crushing stations in a large-scale iron ore plant

E Nunes[1], D Silva[2], I Sales[3], L Leme[4] and R Machado[5]

1. Sales Support Manager, Metso Corporation, Sorocaba SP 18087-125, Brazil. Email: edis.nunes@metso.com
2. Complex Project Delivery Director, Metso Corporation, Sorocaba SP 18087-125, Brazil. Email: daniel.nagano@metso.com
3. Sales Engineer, Metso Corporation, Belo Horizonte MG 30350-250, Brazil. Email: igor.sales@metso.com
4. Project Engineer, Metso Corporation, Sorocaba SP 18087-125, Brazil. Email: luiz.leme@metso.com
5. Engineering Manager, Metso Corporation, Sorocaba SP 18087-125, Brazil. Email: renato.machado@metso.com

ABSTRACT

A common and major issue encountered in large mining operations is the presence of excessive slabs in the primary crushing station product. This problem is especially prevalent in the iron ore mining utilising large jaw crushers as the first stage of crushing. Due to the nature of jaw crusher operation, combined with large feed opening dimensions and large closed-side-settings, particles with one dimension up to 600 mm can frequently be observed on the primary crusher product conveyor. Mining sites experience equipment damage, belt rips, transfer chute blockages and/or silo/bin obstructions. These events significantly impact operations, as they often result in prolonged stoppages for repairs and material clearing. Additionally, some of these interventions may expose personnel to unsafe working conditions, such as working at height or in confined space.

A relatively straightforward solution is to implement a secondary crushing stage directly after the primary jaw crusher station. In recent case studies, a cone crusher has been used to reduce the oversize and slabby particles down to a size and shape suitable for transport. However, in most cases the available footprint to install a secondary building is limited, and the expected plant shutdown make project feasibility difficult to justify. This paper describes a case study at a large iron ore plant in Brazil where three secondary crushing circuits were successfully implemented using a modular design with agile assembly and reduced foundation infrastructure costs. The total installation time frame, considering all mechanical and electrical activities, was 56 days – effectively 28 per cent shorter downtime than the original project schedule – and immediately mitigated the unplanned downtime and maintenance work that the oversize slabs caused on the downstream equipment.

INTRODUCTION

A large-scale iron ore mining plant located in northern Brazil has been operating three mobile primary crusher (herein referred to by the plant naming of 'MCR') stations. Each of these three lines consists of a loading hopper, an inclined apron feeder, vibrating grizzly scalper, and a jaw crusher. The three MCR units discharge onto the same conveyor belt, and the material is then transported via a 9.5 km long-distance conveyor belt (LDCB) for downstream processing. Figure 1 illustrates the three lines of the MCR stations.

FIG 1 – Three mobile crushing units (MCR).

The minimum closed-side setting (CSS) recommended for this jaw crusher size-class stated by the manufacturer is 200 mm. In practice, the CSS often opens to >250 mm due to the jaw dies wearing into a profile often referred to as the 'bathtub shape', which creates uneven gaps along the width of the jaw. The combination of a large jaw crusher CSS and feed material that is naturally prone to slabby breakage frequently generates very coarse MCR discharge, with measured blocks having up to 500–600 mm dimensions in one or two dimensions.

The operation reported a high incidence of unplanned downtime due to frequent conveyor belt rips caused by the lamellar shape of the slabs. Blockages and lockups in the transfer chutes were also reported, resulting in significant maintenance costs and production losses. These repairs or replacement of long-distance conveyors belts require extended downtime at an excessively high expense.

The number of shutdowns at this operation had increased significantly in the last few years since the iron ore is getting coarser, harder and more abrasive. The iron ore being mined previously had solid density above 4 t/m^3, high-grade, low compressive strength (UCS) values (100–150 MPa), and very fine feed. However, some mine areas had now reached low-grade iron ores, with solid density reduced to 3.0 t/m^3, UCS up to 400 MPa hardness, and blocky feed characteristics. The shift in feed material was quickly identified to be primarily responsible for the reduction in plant availability and drove the decision to plan the implementation of a dedicated recrushing operation. This observation emphasises the importance of considering the mineralogical composition of rocks when evaluating their hardness and potential comminution behaviour, as it can provide valuable insights into the efficiency of the extraction process and the overall performance of the mining operation (Wills and Napier-Munn, 2016).

SECONDARY CRUSHER STATION

The proposed solution to eliminate oversized material was to install three additional recrushing lines, each positioned between a MCR line and the conveyor that discharge the material onto the LDCB. Each recrushing plant consists of a single deck banana vibrating screen (Metso MF-2467-1), a secondary cone crusher (Metso GP500S), feed and product conveyor belts, continuous metal extraction, along with the steel structures to support the new equipment, as shown in the Figure 2.

FIG 2 – Three recrushing plants.

Equipment selection was defined based on the process data provided by the customer, along with material property tests executed at the Metso-Brazil laboratory in Sorocaba. Using these as the basis for process design, crushing simulations performed using the Metso proprietary Bruno™ software completed the equipment selection and mass balance validation.

In order to economically justify the solution, not only did the solution need to be show strong return on investment by saving on lost production and downtime costs, it had to also show justifiable capital costs and minimal shutdown of the plant operation with lost production. Three key customer requirements governed the project:

1. A single Original Equipment Manufacturer (OEM) to supply the complete system as an integrated package, reducing interface risks and facilitating performance guarantees.
2. Fast engineering and manufacturing, favouring pre-designed modular concepts.
3. Rapid assembly and commissioning, requiring minimal intervention in existing MCR equipment to avoid production losses.

MODULAR CONCEPT – FIT™ STATIONS

Each recrushing plant was designed following a modular concept, trade-named FIT™ station supported by a metal chassis, referred to as skids, which eliminate the civil foundation. Each FIT™ station was composed of four modules as shown in Figure 3.

FIG 3 – FIT™ station on skid base – the screening module (green), the crushing module (blue), access and maintenance platforms module (yellow) and skid base module (red).

The adoption of a modular product architecture lies in the core of any effort to utilise modularisation in the production system. A modular product architecture differs from integral architecture as the former includes a one-to-one mapping from the functional elements to the physical components of the product and specifies de-coupled interfaces between components (Ulrich, as cited in Peltokorpi *et al*, 2018).

A detailed engineering process subdivided the standard modules into smaller sub-modules for manufacturing and transportation. The purpose of the sub-modules was to allow for pre-assembly of those components to reduce the number of components to be assembled in the field. To speed any field assembly and installation that was necessary, subcategories of drawings were created for each assembly stage, along with an animated assembly methodology – complete with lifting plans – was developed to reduce time and errors in the various module positioning. Significant effort was placed in the design and planning stage to minimise time needed in the field. For example, all the sub-modules were fully welded at manufacturing shop to be bolted and connected at site.

FINAL ASSEMBLY AND INSTALLATION

The final assembly and installation were based on pre-modularisation (partial pre-assembly) in an area separate-but-near-to the final operating position, avoiding major stoppages in the existing operation. Small sub-modules were then combined into larger modules, ultimately forming full-scale modules (Figure 3).

The planned schedule to complete the three plants pre-modularisation was 145 days. Following this, two stoppages of the existing MCR plants were scheduled: 48 days to execute the final assembly and installation of the first recrushing line (January 2025) and 61 days for the installation of the remaining two lines (March 2025), totalling 109 days of total work.

Pre-modularisation was completed in 115 days, 30 days ahead of the original plan. A major factor contributing to the speed of the assembly works and reduction in this project schedule was determined to be the absence of dimensional deviations of the structures and components – of which these deviations and non-conformities frequently cause delays in site works – since the sub-modules were pre-assembled and fit-checked in the fabrication facility prior to shipping to site.

Total field assembly for the three lines, including all mechanical and electrical activities was completed in 56 days; this was 53 days less than the project plan of 109 days. The main enablers for this massive reduction in working time was the efficiency and reduction of unplanned delays due to proper planning, fully assembling modules in the pre-modularisation area, and transporting them and placing them as complete units. Figure 4 shows a fully assembled module being transported to its final position.

FIG 4 – Recrushing module in transit to final position.

COMMISSIONING AND OPERATION

Once in place, the stations needed to be commissioning and validated for operation. Each recrushing plant was designed for a nominal capacity of 2200 t/h with the requirement to reduce top size from 500 to 150 mm. The MF 2467-1 banana screen media aperture was defined as 150 mm and the GP500S cone crusher was initially adjusted to 90 mm CSS. Bruno™ simulations predicted a P99 of 125 mm, which was within the target and warranted parameters.

Start-up and commissioning were executed in just 15 days, again ahead of schedule. During initial operation, it was observed that the GP500S was operating with partially empty chamber, even with the plant operating at the nominal throughput; this resulted in reduced crusher efficiency and concentrated liner wear. This initial low loading is commonly observed during the start-up phase, given that the design specifications can often use conservative/coarse feed assumptions, while actual feed conditions during start-up might have less oversize material to feed the crusher that the conservative design requirements predicted.

The crusher CSS was reduced from 90 to 75 mm to achieve a finer product and minimise slabby particles. Unlike the rectangular shape of a jaw crusher discharge, the GP500S has a conical/curved discharge opening that hinders the slabby material flow and effectively controls the passage of slabs. Most importantly for the success of the project, the transfer size of the ore onto the long-distance conveyor belt has now been controlled with effectively no slabs/oversize that caused the unplanned downtime and costs previously encountered at this site.

Due to the reduction in particle size and the mitigation of the lamellar particle characteristics, there was no further reported cases of unplanned stoppages due to conveyor belt rips. In addition, chute and hopper blockages caused by material interlocking ceased, and no further chute clogging events occurred in this circuit. This had drastically reduced maintenance activities in confined spaces, significantly improving operational safety for maintenance teams, which later occurred exclusively during scheduled shutdowns, without isolated emergency interventions that could compromise the safety of the personnel involved.

A few minor aspects had to be changed in the start-up phase and are considered as lessons learned for future cases. The conveyor belts start-stop sequence had to be adjusted in commissioning stages to avoid belt spillages or overloads. The metal extractor was installed only a short distance (450 mm height) from the conveyor belt bottom trough (belly); thus, on initial operation, a few large blocks were able to contact the metal extractor, and the engineering team had to modify the height of this magnet to eliminate potential equipment damage.

The Figure 5 shows the final plant layout. The front-end-loader and MCR are shown on the left side while the complete FIT™ station – with the feed conveyor belt, the Banana vibrating screen and the GP500S are shown on the right size.

FIG 5 – The FIT™ during the commissioning phase.

CONCLUSIONS

A modular recrushing plant, using FIT™ stations, was successfully implemented at a large-scale iron ore operation. The modular design and assembly-process was considered to be the key driver to satisfy the customer requirements of fast engineering, manufacturing, delivery, assembly and installation phases.

Engineering, manufacturing and delivery were completed in 9 months, which was 36 per cent faster than traditional or conventional methods, which typically require near to 14 months for a similar project. This reduction was achieved due to the module's standardisation, reduced amount of basic and detailed engineering and parallel fabrication with multiple shops and suppliers.

Assembly and installation in total were completed in 6 months, while the original plan was estimated for 8 months (28 per cent reduction), enabled by skid-mounted structures requiring no conventional foundations and minimal rework due to pre-modularisation.

Overall plant availability was projected to increase by 10 to 15 per cent with the improved particle transfer size and control of slabby particles reaching the long-distance conveyor belt. While a conclusive estimate of availability improvement will only be measured after years of operation, initial operation validated the recrushing plants' ability to control the major cause of downtime at the plant in recent years.

ACKNOWLEDGEMENTS

The authors would like to thank the Metso Corporation for supporting this research project.

REFERENCES

Peltokorpi, A, Olivieri, H, Granja, A D and Seppänen, O, 2018. Categorizing modularization strategies to achieve various objectives of building investments, *Construction Management and Economics*, 36(1):32–48.

Wills, B A and Napier-Munn, T J, 2006. *Mineral processing technology: an introduction to the practical aspects of ore treatment and mineral recovery*, 7th edn (Elsevier: Oxford).

Characterisation of low-grade and complex iron ore stockpile (Thabazimbi, South Africa) and identification of possible beneficiation strategies

G Omaga[1], L Filippov[2], I Filippova[3] and H Turrer[4]

1. PhD candidate, Université de Lorraine, CNRS, GeoRessources, 54000 Nancy, France. Email: gillian-nikka.omaga@univ-lorraine.fr
2. Professor, Université de Lorraine, CNRS, GeoRessources, 54000 Nancy, France. Email: lev.filippov@univ-lorraine.fr
3. Professor-Researcher, Université de Lorraine, CNRS, GeoRessources, 54000 Nancy, France. Email: inna.filippova@univ-lorraine.fr
4. Team Leader, ArcelorMittal R&D Maizieres Process, Maizières-lès-Metz Cedex, Voie Romaine-BP 30320-57280, France. Email: henrique.turrer@arcelormittal.com

ABSTRACT

As high-grade iron ore reserves continue to decline, attention shifts towards the beneficiation of low-grade, complex iron ores to supply the growing demand for steel. This study presents a comprehensive characterisation of low-grade iron ore deposits using optical microscopy, X-ray diffraction (XRD), geo-elemental analysis, and liberation size determination through SEM-based automated mineralogy (AMICS). The results reveal hematite and goethite as the major iron-bearing minerals, with quartz and kaolinite as the principal gangue phases, and notable occurrences of Fe-bearing silicates and carbonates such as aegirine, biotite, and siderite. Bulk chemical analysis indicates an average composition of approximately 40 wt per cent Fe, 32 wt per cent SiO_2, and 3 wt per cent LOI. Liberation analysis shows that effective mineral liberation occurs below 250 μm and 80 μm for the two ore types examined. Based on these findings, a beneficiation flow sheet was proposed, incorporating spiral concentration as a pre-concentration stage followed by reverse cationic flotation as the primary concentration step. The developed process produced concentrates containing over 60 wt per cent Fe and less than 10 wt per cent combined SiO_2 + Al_2O_3, demonstrating its potential for generating high-quality iron ore suitable for direct reduction (DRI) feed.

INTRODUCTION

Global demand for iron ore has surged significantly over the past 20 years (Hellmer and Ekstrand, 2013; Song, Wang and Yu, 2019; Yellishetty, Ranjith and Tharumarajah, 2010), following the rise of steel consumption, due to the continuous development of the global economy. Iron ore, the raw material for the steel industry, is critical to a wide range of applications including infrastructure, transportation, green technology etc (Yellishetty, Ranjith and Tharumarajah, 2010). High-grade and easily processed iron ore resources are now being depleted hence, there is a growing need to find a suitable processing options for low-grade, complex and refractory iron ore resources (Nunna *et al,* 2022). However, these resources exhibit complex mineralogy and contain higher levels of silica and alumina impurities that can be detrimental to the quality of iron in steel production (Clout and Manuel, 2015).

Precambrian banded iron formations (BIF) provide more than 90 per cent of the world's iron ore production. These formations can be classified into three classes: (1) unenriched primary iron ore formations, which contain 30–45 wt per cent Fe; (2) martite-goethite ores formed by supergene processes characterised by abundant hydrous iron oxides and contain 56–63 wt per cent Fe; and (3) high-grade hematite-rich ores of hypogene or metamorphic origin, with 60–68 wt per cent Fe.

The major mineral compositions of BIFs are chert or quartz, magnetite, hematite, carbonates (siderite and Fe-dolomite to ankerite series), and a few silicates. While BIFs hold massive amount of iron, the ore is generally too low-grade to be economically viable. Iron formations are only mined when coarse-grained metamorphogenic magnetite is present which can be easily concentrated by magnetic separation (eg 'taconites' mined in the Superior region of the United States), or when metamorphosed and subsequently strongly weathered into saprolite, composed of coarse-grained iron oxides and quartz, which can be separated by gravity separation (eg 'itabirites' in Brazil).

The continuous development of technology in the processing plants has enabled the processing of deposits previously considered impractical, including the reprocessing of stockpiles and tailings dams.

Originally considered mining waste, stockpiles or waste rock are now viewed as potential resources due to developments in recovery methods and the need for low-grade ores to meet the growing demands for raw materials (Kasmaee, Tinti and Bruno, 2018).

The selection of beneficiation strategy depends on the nature of the gangue present and its association with the valuable mineral. A comprehensive characterisation of these complex ore is thereby a very important first step to obtain information on the mineralogy and textual attributes of the ore and their bearing on mineral processing.

With this outlook, Thabazimbi mine located in Limpopo province in South Africa has remaining stockpiles that has potential for beneficiation. The deposit consists of hard hematite, carbonate hematite and supergene altered ores. In this context, a detailed characterisation of the Thabazimbi sample has been undertaken and possible routes of beneficiation are being identified.

MATERIALS AND METHODS

Material

Four bulk samples used in this study were collected from two areas: the Kumba Pit and the Van der Bijl Pit. Samples from the Kumba Pit were taken from three different locations and are referred to as BIF 1, BIF 2, and BIF 3. These materials correspond to remaining stockpiles from the former mining operations and were not previously processed because their grades were below the operational cut-off grade. In contrast, the Van der Bijl deposit represents a newly explored area.

Analytical methods

Detailed chemical compositions of both bulk and size-by-size samples were determined by Inductively Coupled Plasma–Atomic Emission Spectroscopy (ICP-AES) using an iCAP 7000 Series instrument (Thermo Scientific, Waltham, MA, USA). Prior to analysis, the samples were subjected to alkali fusion with lithium metaborate, followed by acid dissolution, before being introduced into the instrument's nebuliser. The chemical analyses were performed at the PGM laboratory of GeoRessources (Nancy, France).

The particle size distribution analysis of the crushed samples was carried out using a 20 cm diameter sieve mounted on a Rotap sieve shaker, under wet conditions. The sieve apertures were selected considering the top size of the sample and adopting the Tyler mesh series.

The mineralogical composition of the major phases was evaluated by powder X-ray diffraction (XRD) using a Bruker D2 Phaser diffractometer equipped with a goniometer in Bragg-Brentano geometry, a Co Kα source (1.789 Å), and a Lynxeye 1D silicon-drift detector. X-ray patterns were acquired as coupled 2θ/θ scans in the 7–80°2θ range, with a step size of 0.02° 2θ, and a dwell time of 1.0 to 1.5 seconds per step. Phase identification was performed using Bruker's EVA software with the ICDD PDF-2 database. Prior to the acquisition, the samples were crushed using a ring mill.

In addition, the same samples were examined by Scanning Electron Microscopy (SEM) using TESCAN VEGA 3 LM coupled to a 30 mm^2 Bruker XFlash 6 Energy Dispersive Spectrometers (EDS) located at. The equipment produced backscattered electron (BSE) images and energy dispersive X-ray (EDS) analyses to make textural observations and extract various qualitative and quantitative mineralogical information. The SEM is equipped with a tungsten filament gun and the imaging analysis was maintained at 15kV.

Mineral liberation analysis was performed using a software package for automated mineral identification and quantification called AMICS (Advanced Mineral Analysis and Characterization System). It enables Bruker's energy dispersive X-ray analysis systems on SEM's BSE images to become a fully automated mineral liberation analyser. Mineral quantification was estimated using mineral surface area densities. Both SEM and AMICS were performed at the SCMEM laboratory of GeoRessources (Nancy, France).

RESULTS AND DISCUSSION

Elemental composition of the bulk

As shown in Table 1, The Fe and SiO_2 contents range from 37–45 wt per cent and 29–36 wt per cent, respectively, confirming that the material represents a low-grade iron ore with significant amount of impurities (Holmes and Lu, 2015; Mesquita *et al,* 2024; Ozcan *et al,* 2020).

TABLE 1

Bulk elemental analysis of Thabazimbi samples measured using ICP-MS.

	Fe_{tot}	SiO_2	Na_2O	Al_2O_3	K_2O	MgO	MnO	CaO	P_2O_5	S	C	LOI
BIF 1	39.6	31.8	0.92	1.94	0.35	1.07	1.05	0.77	0.17	0.01	1.4	3.7
BIF 2	44.7	29.3	0.66	1.01	0.33	0.23	0.91	0.21	0.09	0.01	0.6	1.8
BIF 3	37.1	35.5	2.07	2.21	0.86	0.91	0.84	0.46	0.16	0.03	1.4	2.9
VDB	41.4	35.4	<0.05	2.01	0.05	0.25	0.27	0.25	<0.05	0.01	0.4	2.0

The BIF samples contain minor amounts of Na, Mg and K, while the VDB sample is free of these elements, suggesting a more complex mineralogy in the BIF samples. This suggests that the BIF samples are more strongly contaminated by silicate minerals, likely aluminosilicates (eg clays or feldspars). Both BIF and VDB samples exhibit evidence of carbonate phases, as supported by the presence of C, CaO, MgO, and elevated LOI values.

These differences between the BIF and VDB samples likely reflect varying degrees of hydrothermal alteration and supergene modification across the deposit (Netshiozwi, 2002).

Mineralogical composition thru X-ray diffraction

Powder XRD and AMICS (SEM-based automated image analysis) were used to identify and quantify the mineralogy of the Thabazimbi samples. Although both techniques provide reliable quantitative results, uncertainties may arise from improper sample preparation, mineral misidentification, and methodological limitations (Rahfeld *et al,* 2018). AMICS data are particularly influenced by sample preparation, such as the preferential settling of heavy iron oxides, and analytical constraints, including mixed spectra from fine grains, the two-dimensional nature of the measurement, and the limited lateral resolution of EDX, which can cause spectral overlap between adjacent minerals. In contrast, XRD quantification becomes challenging when mineral assemblages are complex or contain amorphous phases, and when peak overlap occurs among minerals with similar crystal structures (Lund, Lamberg and Lindberg, 2013; Rahfeld *et al,* 2018).

For consistent geometallurgical interpretation, mineral phases were grouped by lithological type, and the detailed modal mineralogy obtained from both techniques is provided in the supplementary document. Overall, XRD and AMICS can determine the mineralogical composition of the Thabazimbi samples within analytical errors of approximately ±10 per cent. However, for simplicity and representativeness, XRD results were used for compositional interpretation, whereas AMICS data were used in the liberation study to evaluate free surfaces and mineral associations.

Table 2 summarises the mineralogical composition of the samples determined by XRD and quantified using Rietveld refinement. In contrast to the BIF samples, the VDB sample displays a simpler mineralogy characteristic of hard, massive hematite ore with limited supergene alteration. The BIF samples, by comparison, contain a more complex assemblage of gangue minerals derived from both hydrothermal alteration and supergene modification (Netshiozwi, 2002). Hematite is the dominant iron oxide, ranging from 38.1 per cent to 55.6 per cent in the BIF samples and 49.5 per cent in VDB. Goethite occurs in minor amounts (4.2–10.5 per cent), and magnetite appears only in trace to low proportions (0.7–2.2 per cent). Quartz is the major gangue phase, varying from 26.4 per cent to 30.8 per cent in the BIF samples and increasing to 42.2 per cent in the VDB sample.

TABLE 2

Quantitative phase analysis of head samples determined by XRD Rietveld.

Group	Mineral	Formula	BIF 1	BIF 2	BIF 3	VDB
Iron oxides	Hematite	Fe_2O_3	39.9	55.6	38.1	49.5
	Goethite	$FeO(OH)$	10.5	5.5	6.2	4.2
	Magnetite	Fe_3O_4	0.7	1.2	2.2	1.3
Tectosilicate	Quartz	SiO_2	30.8	28.2	26.4	42.2
	Orthoclase	$KAlSi_3O_8$	0.5	0.4	1.7	-
Phyllosilicates	Kaolinite	$Al_2Si_2O_5(OH)_4$	1.9	0.5	2.9	2.0
	Chlorite	$(Mg,Fe,Al)_6(Si,Al)_4O_{10}(OH)_8$	3.3	1.1	3.0	-
Pyroxene	Aegirine	$NaFeSi_2O_6$	8.0	5.9	15.3	-
Carbonates	Siderite	$FeCO_3$	1.3	0.8	3.2	-
	Dolomite	$CaMg(CO_3)_2$	2.9	0.4	0.9	0.8
Mn-Oxide	Pyrolusite	MnO_2	0.4	0.2	0.1	-

Phyllosilicates such as kaolinite, chlorite, and biotite occur in low quantities (<4 per cent), suggesting limited supergene weathering. Aegirine, a Fe-bearing silicate, shows notable variation among samples, from 5.9 per cent in BIF 2 to 15.3 per cent in BIF 3. Carbonates such as siderite, dolomite, and calcite occur in small amounts (≤3 per cent). Mn-oxide (pyrolusite) occurs only as a trace component in the BIF samples, and its negligible abundance is unlikely to influence processing performance.

Optical and electron microscopic studies

The SEM–EDS and optical microscope were used to characterise the grain shape, texture, and chemistry of the minerals present in the samples. Representative photomicrographs from the optical microscope, SEM analysis and chemical mapping are all provided in the supplementary document.

The BIF samples consist primarily of iron oxides, mostly hematite and goethite, with minor traces of magnetite. The remaining minerals include silicates (most likely quartz and aegirine) and carbonates (ankerite or siderite, and dolomite). In contrast, the VDB sample exhibits a much simpler mineralogy, composed mainly of iron oxides and quartz.

Hematite occurs predominantly as pseudomorphic hematite after magnetite (martite) and was observed in several textural forms, including porous, patchy, specular, and platy hematite that commonly cross-cuts quartz. Goethite is most likely derived from the alteration of high-grade hematite ores through supergene weathering processes and appears brown to yellowish in reflected light. Quartz grains frequently enclose dusty and small subhedral to euhedral hematite inclusions, producing a distinct reddish colouration. Aegirine, an important Fe-bearing gangue mineral, is identified by its shard-like, elongated morphology with visible fibrous structures. It also appears to be highly brittle and tends to accumulate in the fine fraction during comminution.

Liberation studies

Liberation sizes were determined using AMICS by area-grade and free surface. In this study, liberation size by area-grade is defined as the 80 per cent of iron oxides liberated at an area-grade of more than 90 per cent while liberation size by free surface is defined as 80 per cent of the iron oxides are liberated at a free surface coverage of more than 50 per cent. BIF samples are liberated in the 50–80 μm size range, whereas in VDB samples they are liberated in the 160–250 μm size range. Furthermore, as shown in Figure 1, more than 30 per cent of quartz and aegirine are liberated in BIF samples while over 50 per cent of quartz is liberated in VDB samples at coarse size fractions.

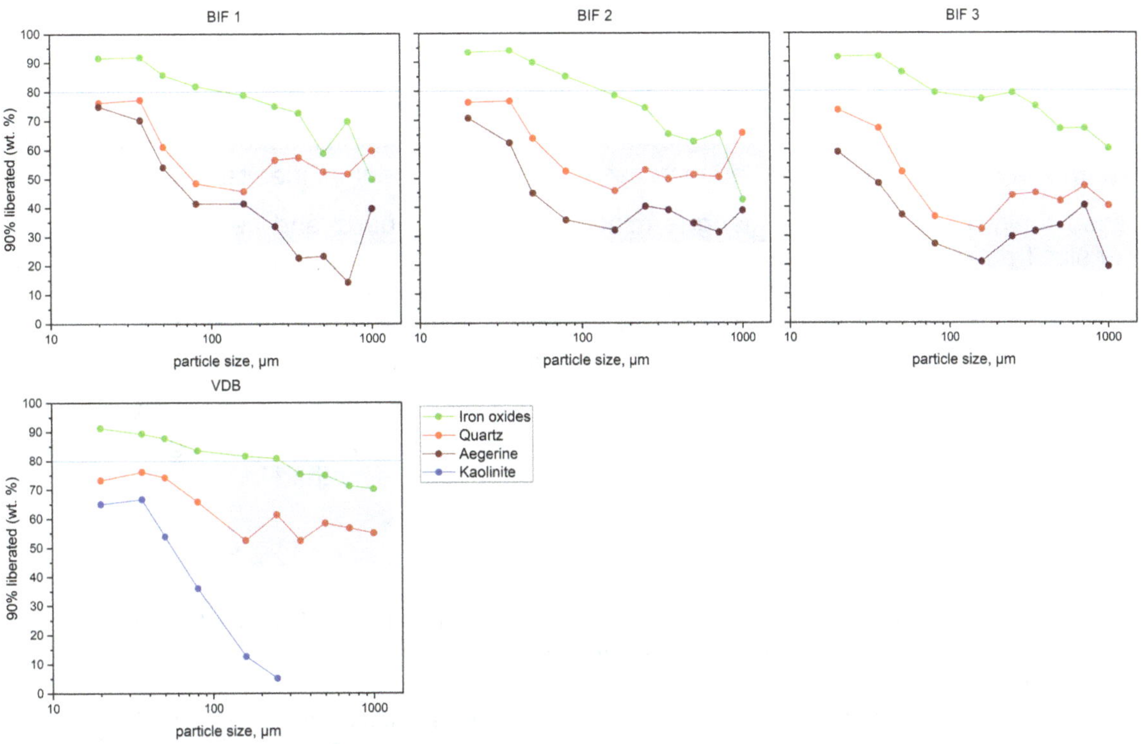

FIG 1 – wt per cent of 90 per cent liberated minerals present in Thabazimbi ore.

Beneficiation test

Previous test assessments demonstrated that gravity concentration performs better as a pre-concentration step, while flotation is more effective as the main concentration method thus, the beneficiation flow sheet proposed as seen in Figure 2. The beneficiation process was conducted in two distinct size fractions: 75–1000 µm, and 25–75 µm for Blend BIF samples and 212–1000 µm, and 25–212 µm for the VDB sample, to assess the potential liberation of iron oxides and gangue minerals within these size ranges. The coarser size fractions were used for pre-concentration to remove a large volume of gangue minerals prior to final concentration. Since the main concentration method is reverse cationic flotation, numerous studies have highlighted the importance of desliming (Filippov, Severov and Filippova, 2014; Lima *et al,* 2020). Therefore, preliminary desliming tests were conducted, and the results showed that desliming at a cut-off of 25 µm produced acceptable Fe grade and recovery.

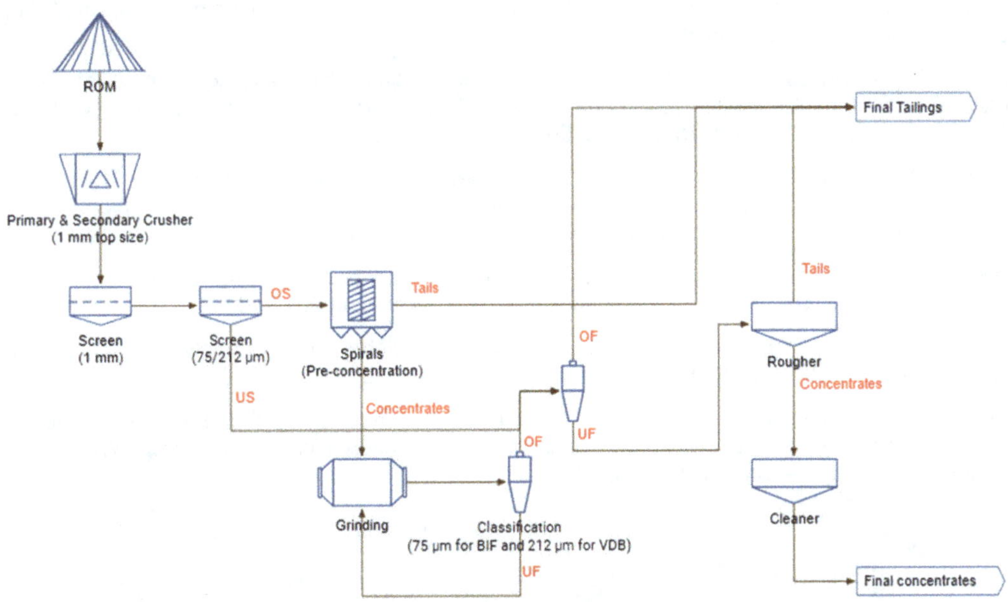

FIG 2 – Suggested flow sheet to concentrate the Thabazimbi iron ore.

CONCLUSION

This study characterised the mineralogy and liberation behaviour of low-grade iron and complex stockpiles from the Thabazimbi mine. The BIF samples displayed more complex mineralogy with the presence of Fe-bearing silicates and carbonates, while VDB sample exhibited a simpler mineralogy. Despite this, both samples showed an increase in kaolinite content in finer particle sizes.

Liberation analysis indicated that gangue minerals, such as quartz and aegirine, are liberated in coarser size fractions. This suggests the need for a pre-concentration stage to remove these gangue minerals early in the process and to reduce the volume of sample that needs processing. Since hematite and goethite are denser and feebly magnetic in nature, gravity separation or magnetic separation are recommended as potential pre-concentration strategies.

Moreover, desliming is a critical step due to the occurrence of kaolinite in finer size fractions. Its presence could affect downstream processes, making fine particle management essential. Additionally, previous studies have identified flotation as a suitable method for recovering iron oxides, and this will be the focus of future research.

REFERENCES

Clout, J M F and Manuel, J R, 2015. Mineralogical, chemical, and physical characteristics of iron ore, in *Iron Ore – Mineralogy, Processing and Environmental Sustainability*, pp 45–84 (Elsevier). https://doi.org/10.1016/B978-1-78242-156-6.00002-2

Filippov, L O, Severov, V V and Filippova, I V, 2014. An overview of the beneficiation of iron ores via reverse cationic flotation, *International Journal of Mineral Processing*, 127:62–69. https://doi.org/10.1016/j.minpro.2014.01.002

Hellmer, S and Ekstrand, J, 2013. The iron ore world market in the early twenty-first century—the impact of the increasing Chinese dominance, *Miner Econ*, 25:89–95. https://doi.org/10.1007/s13563-012-0021-1

Holmes, R J and Lu, L, 2015. Introduction: Overview of the global iron ore industry, in *Iron ore: Mineralogy, Processing and Environmental Sustainability,* pp 1–42 (Woodhead Publishing). https://doi.org/10.1016/B978-1-78242-156-6.00001-0

Kasmaee, S, Tinti, F and Bruno, R, 2018. Characterization of metal grades in a stockpile of an Iron Mine (Case study – Choghart iron mine, Iran), *Mining-Geology-Petroleum Bulletin*, 33:51–59. https://doi.org/10.17794/rgn.2018.2.5

Lima, N P, Silva, K, Souza, T and Filippov, L, 2020. The characteristics of iron ore slimes and their influence on the flotation process, *Minerals*, 10(8):675. https://doi.org/10.3390/min10080675

Lund, C, Lamberg, P and Lindberg, T, 2013. Practical way to quantify minerals from chemical assays at Malmberget iron ore operations: An important tool for the geometallurgical program, *Minerals Engineering*, 49:7–16. https://doi.org/10.1016/j.mineng.2013.04.005

Mesquita, J, Belissont, R, Badawi, M, Turrer, H and Foucaud, Y, 2024. Integrated approach for Mont Reed iron ore: From characterization to ore processing (preprint), *SSRN*. https://doi.org/10.2139/ssrn.4933160

Netshiozwi, S, 2002. Origin of high-grade hematite ores at the Thabazimbi Mine, Limpopo province, South Africa, Doctoral dissertation, University of Johannesburg.

Nunna, V, Suthers, S P, Pownceby, M I and Sparrow, G J, 2022. Beneficiation Strategies for Removal of Silica and Alumina from Low-Grade Hematite-Goethite Iron Ores, *Mineral Processing and Extractive Metallurgy Review*, 43:1049–1067. https://doi.org/10.1080/08827508.2021.2003353

Ozcan, O, Aghlmandi, A, Orhan, E and Ergun, L, 2020. Beneficiation and flowsheet development of a low-grade iron ore: a case study, *Bulletin of the Mineral Research and Exploration,* 165:1–24. https://doi.org/10.19111/bulletinofmre.834182

Rahfeld, A, Kleeberg, R, Möckel, R and Gutzmer, J, 2018. Quantitative mineralogical analysis of European Kupferschiefer ore, *Minerals Engineering,* 115:21–32. https://doi.org/10.1016/j.mineng.2017.10.007

Song, Y, Wang, N and Yu, A, 2019. Temporal and spatial evolution of global iron ore supply-demand and trade structure, *Resources Policy*, 64:101506. https://doi.org/10.1016/j.resourpol.2019.101506

Yellishetty, M, Ranjith, P G and Tharumarajah, A, 2010. Iron ore and steel production trends and material flows in the world: Is this really sustainable? Resources, *Conservation and Recycling*, 54:1084–1094. https://doi.org/10.1016/j.resconrec.2010.03.003

Unlocking nameplate performance in large iron ore operations with radar-controlled crushing

A Phillips[1] and D Uhlhorn[2]

1. Business Development Manager, Hexagon, Perth WA 6000. Email: alex.phillips@hexagon.com
2. Radar Solutions Product Manager, Hexagon, Bochum, North Rhine-Westphalia, Germany. Email: daniel.uhlhorn@hexagon.com

ABSTRACT

A large iron ore operation was unable to achieve nameplate capacity across its secondary and tertiary crushing circuit despite having sufficient installed equipment. The primary instrumentation constraint was a set of legacy single-point (one-dimensional, or 1D) radar and ultrasonic level devices on critical surge bins, which produced erratic or missing readings at low fill levels and under high throughput in dusty operating conditions. One-dimensional sensors return a single distance measurement along the beam axis; two-dimensional (2D) radar scanners sweep a beam across the full bin cross-section to generate a spatial surface profile and calculated volume.

This lack of reliable visibility into bin behaviour forced conservative operating set points, increased the frequency of crusher trips and belt protection events, and reduced operator confidence in the control system. The net effect was a highly variable feed to the secondary gyratory and tertiary cone crushers, lower average throughput than design, and elevated unplanned downtime in the crushing area.

The operation engaged a technology partner to implement a radar-based instrumentation and control upgrade across the secondary and tertiary crushing circuit. The solution comprised two complementary elements: high-speed 2D radar scanners installed on each critical surge bin to provide continuous fill profiles and calculated volumes, and upgraded high-frequency 1D radar sensors fitted inside the secondary crusher cavity to enable cavity-level-driven speed control. Both measurement streams were integrated into the plant distributed control system (DCS) to drive automated feeder and conveyor set points and sustain choke-fed operation across the circuit.

Each scanner provides a true 2D fill profile and calculated volume for bins feeding the secondary gyratory and tertiary cone crushers, as well as for key recirculation and classification bins. This data is used to drive automatic feeder and conveyor set points, stabilising crusher feed conditions and enabling choke-fed operation across the circuit. The approach builds on proven use of radar in harsh bulk materials handling applications, extending that reliability into the comminution control layer.

Within the air classification and crushing areas, integration of 2D radar measurements into the supervisory control and data acquisition (SCADA) and control system gave operators a real-time, intuitive view of bin behaviour and material movement inside previously opaque vessels. With trustworthy, high-resolution level and volume data, control set points were progressively lifted from highly conservative values toward the true design envelope of the equipment.

This shift from reactive, operator-driven control to stable, instrument-driven control enabled the operation to achieve consistent nameplate throughput across multiple crushing stages while materially reducing downtime events associated with empty bins, overfills, and crusher protection trips. In practical terms, the circuit moved from operating below nameplate capacity under conservative conditions to sustained operation at nameplate with controlled and predictable risk.

INTRODUCTION

Hematite iron ore operations, particularly direct shipping ore (DSO) plants in Western Australia, are typically configured as multi-stage crushing and screening circuits with several secondary and tertiary crushers interlinked by surge bins, screens and stockpiles. The relatively simple 'crush-screen-ship' flow sheet belies the capital intensity of these installations and the high installed power concentrated in the secondary and tertiary crushing area, where sustained operation at or near nameplate capacity is essential to maintain low unit costs and competitive shipped tonnes (Napier-Munn *et al,* 1996; Wills and Finch, 2015). In practice, many large hematite plants exhibit a persistent gap between the theoretical capacity of their crushing trains and realised throughput, with unstable behaviour in intermediate bins, highly variable recirculating loads, and unreliable level and flow instrumentation across multiple crushers and bins frequently identified as primary constraints.

Secondary and tertiary crushing stages in hard rock operations are designed to operate in choke-fed conditions, where a stable column of material within the crusher cavity promotes inter-particle crushing, consistent product sizing and optimal energy efficiency (Evertsson, 2000; Napier-Munn *et al,* 1996; Wills and Finch, 2015). Achieving and maintaining choke-fed operation requires accurate, real-time measurement of surge bin levels and automated control of feed rates to each crusher. Legacy single-point level sensors, including ultrasonic and mechanical devices, are prone to interference from dust, material build-up and signal dropout under high throughput conditions, forcing operators to adopt conservative set points and manual intervention strategies.

This paper presents results from an iron ore operation where two complementary radar technologies were implemented to improve crushing circuit control. Two-dimensional (2D) radar scanners (instruments that sweep a microwave beam across the full bin cross-section to generate a surface profile) were installed on critical surge bins feeding the secondary gyratory and tertiary cone crushers. In parallel, high-frequency one-dimensional (1D) radar sensors (single-beam, non-contact distance measurement devices) were installed inside the secondary crusher cavity to enable continuous cavity-level-driven speed control. It is important to distinguish between legacy 1D point devices, which are basic single-point switches or low-rate ultrasonic sensors, and the upgraded high-speed 1D radar sensors used in this application: both are 1D in geometry but differ substantially in measurement rate, noise filtering capability, and suitability for closed-loop control. Performance data demonstrate measurable improvements in throughput, stability and availability, validating radar-enabled control as a pragmatic approach to unlocking capacity from existing crushing infrastructure.

SITE BACKGROUND AND CIRCUIT DESCRIPTION

The operation is a large-scale iron ore producer targeting multi-million tonnes per annum of lump and fines output. The comminution circuit comprises:

- The primary crusher reduces run-of-mine ore to nominal 200 mm.
- Secondary gyratory crusher in open circuit, fed via apron feeder from primary surge bin.
- Tertiary cone crushers in closed circuits with vibrating screens, producing nominal 12 mm product for downstream grinding.
- Screenhouse bins provide surge capacity between crushing stages and recirculation paths for oversize material.

During commissioning and ramp-up, the crushing circuit consistently operated below design throughput. Investigation identified several contributing factors including infrastructure bottlenecks in material handling, progressive wear of chutes and transfer points, and critically, unreliable level sensing on key surge bins. Existing instrumentation comprised single-point ultrasonic and basic 1D Radar level devices, which exhibited frequent signal dropout, erratic readings at low fill levels, and poor performance in dusty conditions.

The lack of reliable bin level data forced operations personnel to maintain wide safety margins on bin operating levels, run feeders in manual or semi-automatic mode, and accept frequent crusher trips and belt protection events. The net effect was highly variable feed to both secondary and tertiary stages, constraining average throughput and increasing specific energy consumption.

CHALLENGES IN SURGE BIN LEVEL MEASUREMENT

Accurate level measurement in iron ore crushing surge bins presents several practical challenges:

- **Dust and fines:** Hematite and goethite fines generated during crushing, screening and transfer create persistent dust clouds that attenuate or scatter ultrasonic signals and obscure optical devices.

- **Material characteristics:** Crushed hematite ore is dense and abrasive, imposing high mechanical loads, impact and vibration on bin structures and mounted instruments during filling and discharge.
- **Surface irregularities:** Bulk ore forms highly uneven surface profiles within bins due to segregation, funnel flow and mass flow patterns, and build-up or hang-ups on bin walls, making single point measurements unrepresentative of true inventory.
- **Operating environment:** Crushing and screening plants are inherently loud, dusty and subject to temperature variation and vibration from adjacent equipment, with fine material progressively coating internal surfaces and sensor faces.
- **Maintenance access:** Surge bins located within multi-level crusher and screen buildings often have constrained access and limited shutdown windows for safe inspection, cleaning or replacement of level instruments.

Legacy instrumentation on many hematite plants has consisted of ultrasonic level transmitters and mechanical point devices such as diaphragm or paddle switches. In practice, ultrasonic units have suffered from false echoes and signal loss in dusty conditions, while mechanical switches have provided only intermittent indication and demanded frequent maintenance. Together, these technologies have proven inadequate as primary process variables for reliable closed loop control of crusher feed rates, forcing operators to revert to conservative set points and manual intervention.

RADAR LEVEL MEASUREMENT TECHNOLOGY

Non-contact radar level measurement employs frequency modulated continuous wave (FMCW) technology to determine distance to material surfaces. A radar antenna emits microwave signals that reflect from the material surface, with distance calculated from the time delay and frequency shift of the returned signal (Brasseur, 2001). Two-dimensional radar scanners extend this principle by mechanically or electronically sweeping the radar beam across the bin cross-section, generating a spatial map of the material surface.

Key advantages of 2D radar technology for crushing circuit applications include:

- Non-contact operation eliminates mechanical wear and material build-up on sensor surfaces.
- Microwave signals are minimally affected by dust, temperature variation or air currents.
- Surface profile data enables calculation of material volume, compensating for uneven surfaces and segregation effects.
- High measurement update rates (1 to 5 seconds) support real-time process control.
- Robust industrial hardware designed for vibration, temperature extremes and harsh environments.

The 1D radar sensors selected for this implementation operate at a frequency of 76 GHz, with a beam width of approximately 3° and a measurement rate of 8000 Hz, providing spatial resolution of 0.1–0.2 cm within the crushers, compared to the old radar measurement rate of 1 Hz. Each sensor outputs material level, via Modbus TCP protocol for integration with the site distributed control system (DCS).

The selected 2D 360° scanning radar operates at 76 GHz with measurement rates exceeding 8000 Hz, enabling target separation down to 2.0° and single-target localisation of 1 mm, well suited to high-dust environments such as surge bins. Each sensor delivers up to ten level outputs across the profile via Modbus TCP to the DCS.

IMPLEMENTATION APPROACH

Site survey and instrumentation design

A typical site survey identifies seven critical surge bins for radar instrumentation (see Table 1).

TABLE 1
Radar instrumented surge bins in crushing circuit.

Bin location	Diameter (m)	Height (m)	Service
Primary dump pocket	6.0	10.0	Jaw crusher feed (truck dump)
Primary product bin	5.0	8.0	Secondary cone crusher feed
Secondary scalping bin 1	4.5	7.0	Tertiary cone feed (train 1)
Secondary scalping bin 2	4.5	7.0	Tertiary cone feed (train 2)
Screen oversize bin	4.0	6.5	Recirc to tertiary crushers
Product screen bin (lump)	5.0	8.0	Lump stockpile conveyor
Product screen bin (fines)	5.0	8.0	Fines stockpile/train loadout

Each bin required structural modifications to accommodate radar mounting flanges, with installation occurring during scheduled maintenance shutdowns. Radar scanners were positioned centrally on bin roofs with line-of-sight to the full bin volume, avoiding interference from internal structures or discharge chutes.

Control system integration

Radar outputs were integrated into the existing DCS using the following architecture:

- Ethernet IP network connecting radar scanners to Radar Processing units via industrial switches.
- Radar processing units connecting to DCS via Modbus TCP, sending all data via a mapped telegram.
- Custom function blocks in DCS logic are mapped to the telegram input provided.
- PID (proportional–integral–derivative) control loops using radar-derived signals to modulate apron feeder and belt conveyor speeds.
- Alarm and interlock logic based on bin high-high, high, low and low-low thresholds.
- Operator interface screens displaying real-time bin fill profiles and volume trends.

Control loop tuning followed a staged approach, initially running radar signals in parallel with existing instrumentation to validate measurement accuracy. Once confidence was established, control loops were progressively transferred to radar-based set points, with legacy sensors retained as backup.

Operational ramp-up

Following commissioning, a three-month ramp-up period focused on optimising control parameters and building operator confidence in the new instrumentation. Key activities included:

- Validation of radar measurement accuracy using proven Site Acceptance Testing criteria and comparisons against existing sensors.
- Tuning of PID control loop parameters (proportional gain, integral time, derivative time) to achieve stable feeder response.
- Progressive increase in bin operating set points from conservative levels toward design envelope.
- Training of operations and maintenance personnel on radar system operation and diagnostics.
- Establishment of performance measurement framework to quantify throughput and stability improvements.

RESULTS AND PERFORMANCE ANALYSIS

Throughput and stability

Comparison of crushing circuit performance before and after radar implementation demonstrates measurable improvements across multiple metrics. The data presented in this section cover a weeklong parallel-running trial in which measurements from the Hexagon iLDR-M-1D high-frequency radar sensor were compared against those from the OEM-supplied 1D sensor it replaced in the secondary crusher cavity. The iLDR-M-1D operates at 76 GHz with a measurement rate of 8000 Hz and onboard noise filtering; the legacy OEM sensor operated at approximately 1 Hz with no equivalent filtering.

As shown in Figure 1, the principal measurable improvement was a reduction in the variability of crusher motor current draw during operation. The higher measurement rate and onboard noise filtering of the upgraded sensor produced a smoother level signal, which in turn enabled tighter PID control of the crusher feed and reduced the amplitude of current draw excursions.

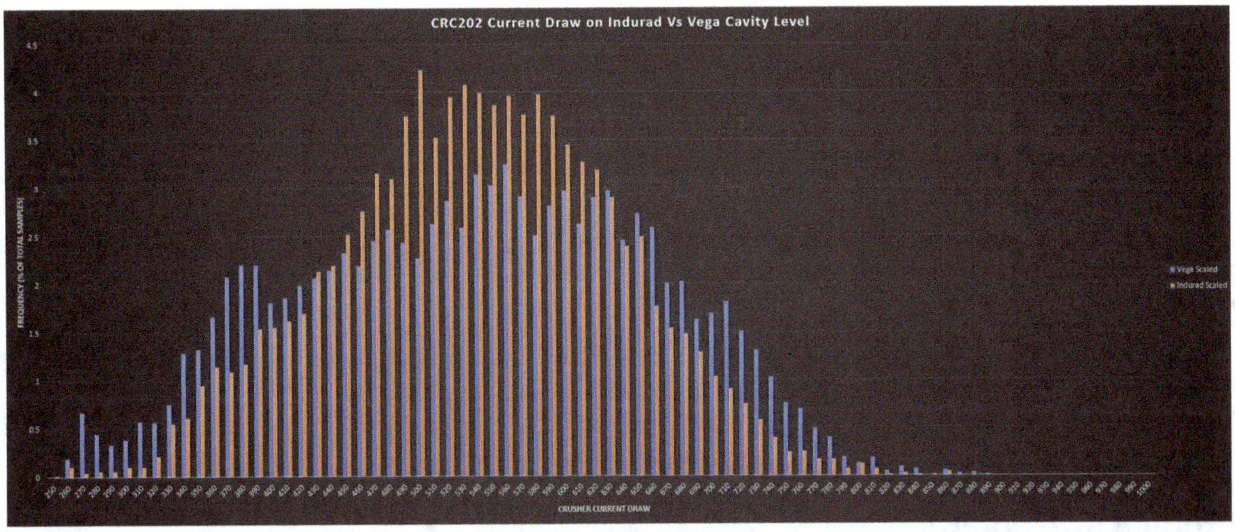

FIG 1 – Secondary crusher motor current draw: comparison of OEM 1D sensor versus upgraded high-frequency 1D radar.

This outcome is consistent with findings reported elsewhere for high-speed sensors used in place of low-rate OEM-supplied devices: when the raw measurement signal requires extensive back-end filtering logic to be usable, control loop performance is limited by processing latency rather than the underlying process dynamics.

One instance, as part of a larger improvement project, the radar was used to help re-write the hopper-to-tertiary crusher feeder control logic, by removing any restrictions or contingencies that limited crusher output. This resulted in a >40 per cent uplift of nine times Tertiary crushers, going from 540 tph to >800 tph, whilst providing increases with liner life and lower power consumption over time.

Figure 2 illustrates how the high-measurement-rate 1D radar enables level measurement to track feeder speed changes with minimal lag, supporting a tighter closed control loop. The smoother signal transmission of the upgraded sensor, compared with the OEM-supplied device, reduces noise from dust, fly rock and water spray without requiring additional back-end filtering in the DCS.

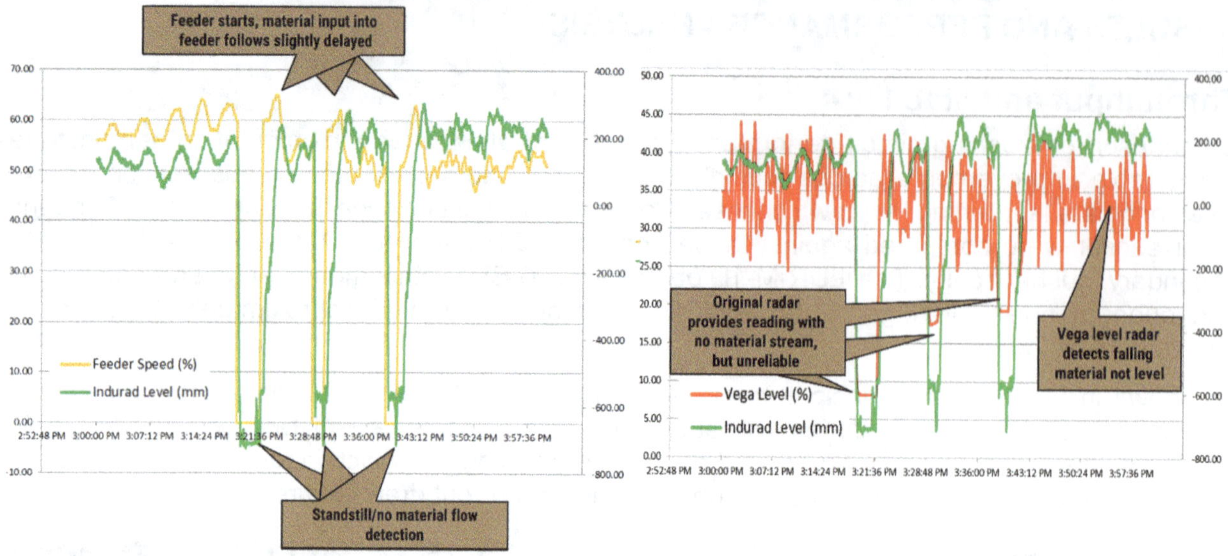

FIG 2 – Secondary crusher level signal and feeder speed synchronisation: OEM 1D radar versus high-frequency 1D radar.

Availability and reliability

Unplanned downtime events in the crushing circuit decreased significantly following radar implementation. The frequency of crusher trips due to empty bin conditions decreased from an average of 3.2 events per week to 0.6 events per week. Belt protection trips associated with bin overfill or material spillage decreased from 2.1 events per week to 0.4 events per week.

Mean time between failures (MTBF) for the crushing circuit increased from 68 hrs to 142 hrs when considering all unplanned stoppages. This improvement reflects both the direct impact of reliable bin level measurement and the indirect benefit of more stable operating conditions reducing mechanical stress on equipment.

Product size control

Product size distribution from the tertiary crushing stage exhibited improved consistency following radar implementation. The P_{80} product size (80th percentile) varied within ±1.5 mm of target specification for 89 per cent of operating time, compared with 64 per cent of time in the pre-implementation period. Tighter product size control supports downstream grinding efficiency and overall plant recovery.

Energy efficiency

Energy consumption in comminution circuits is strongly influenced by feed stability and degree of choke feeding, as described in classical comminution theory (Napier-Munn *et al,* 1996; Wills and Finch, 2015). Specific energy consumption in the crushing circuit decreased by approximately 8 per cent following radar implementation, attributed to more consistent choke-fed operation and reduced idle running time. Historically, secondary gyratory specific energy usage has decreased from 0.95 kWh/t to 0.87 kWh/t. Tertiary cone specific energy decreased from 1.12 kWh/t to 1.04 kWh/t.

Economic analysis

The radar instrumentation project required capital expenditure of approximately A$2 million, comprising hardware procurement (five radar scanners and associated mounting hardware), structural modifications to bin roofs, electrical and communications infrastructure, and engineering services for design and commissioning.

Annual operating cost impact was minimal, with radar systems requiring negligible routine maintenance beyond occasional cleaning of antenna faces during scheduled plant shutdowns. No additional operating labour was required, with routine monitoring incorporated into existing control room operations.

The simple payback period was calculated to be less than one quarter (three months), based on increased production revenue alone. When additional benefits such as reduced power consumption and maintenance costs are considered, the payback period may be shorter.

Case study evidence – surge bin level control for secondary and tertiary crushers

The aggregate performance improvements described above were achieved through two distinct but complementary instrumentation upgrades at this site. The following subsections document the specific context, implementation approach and observed outcomes for each, beginning with the surge bin level control system that forms the foundation of the circuit's automated feed management.

Importance of optimised bin level control

Optimised bin level control ahead of secondary crushers is critical because it is the primary lever for maintaining continuous, choke-fed operation. The transition from basic 1D point devices to 2D profile scanners materially improves the accuracy, robustness and controllability of that feed. By delivering a full surface profile and multiple independent outputs per bin, 2D radar enables more sophisticated control strategies that reduce feed variability, unlock higher throughput and improve equipment utilisation compared with single-point switches or simple continuous probes.

Role of surge bin level in secondary crusher performance

Surge bins between crushing stages are designed to buffer upstream variability and provide stable feed conditions to downstream equipment, a principle fundamental to comminution circuit design (Wills and Finch, 2015).

Surge bins between the primary and secondary stages are intended to decouple upstream blasting and primary crushing variability from the steady, controlled feed that secondary cones require to operate at or near choke conditions. In practice, the effectiveness of this buffer is governed almost entirely by the quality of level measurement and the control philosophy used to manipulate feeders and conveyors based on that signal. Where bin levels are accurately controlled, the crusher sees a narrow band of cavity levels and a relatively constant mass flow, which supports higher average throughput, smoother power draw and reduced risk of trips from empty-bin or overfill conditions.

Conversely, poor level visibility forces operators to run with wide safety margins, either holding bins artificially low to avoid overflows or cycling between low and high extremes, both of which propagate as large, slow disturbances into the secondary crusher. This results in periods of under-feeding interspersed with short over-loads, undermining choke-fed operation, increasing specific energy consumption and contributing to unplanned downtime events in the crushing area.

Limitations of 1D point level devices

Traditional 1D point level devices such as mechanical switches, basic capacitance probes and single ultrasonic point sensors are widely used in iron ore surge bins due to their low capital cost, but they exhibit significant limitations under real plant conditions. In high-dust, abrasive environments typical of magnetite and hematite operations, these instruments are prone to false trips, signal dropout, or complete loss of indication due to dust coating, material build-up, ratholing and bridging. Point switches also provide, by design, only discrete 'on/off' information at one or two elevations, which is insufficient for closed-loop control of feeder speed or for detecting asymmetric filling patterns in large diameter bins.

As a result, many operations fall back to manual or semi-automatic feeder control, using conservative set points and operator experience rather than tight, instrument-driven automation. This control mode limits the effective utilisation of installed surge capacity and makes it difficult to maintain a stable bin operating window, directly impacting crusher feed stability and plant throughput.

Advantages of 2D profile level measurement

Two-dimensional radar scanners overcome these limitations by sweeping a narrow microwave beam across the full cross-section of the bin, generating a high-resolution surface profile rather than a single distance measurement. From this profile, the system calculates not only minimum and maximum levels, but also average level, material volume and, where bulk density is known, an estimate of contained tonnage. Because radar energy is largely unaffected by dust, temperature variation and air currents, these scanners maintain reliable measurements even during heavy filling, water spray operation and high throughput conditions, where ultrasonic and optical devices often fail.

Vendors of 3D and multi-point scanning solutions for mining applications report that this approach is particularly valuable for solids that pile irregularly, segregate, or form cone-up/cone-down shapes, as it captures the true material topography rather than inferring level from a single point. In large iron ore surge bins feeding secondary crushers, this capability allows operators and control systems to detect preferential filling near the discharge, side-wall build-up, or emerging ratholes early, and respond before these translate into erratic discharge and crusher feed excursions.

Figure 3 shows an example of the 2D radar measuring the profile of the material in a hopper, then via the processing unit, splitting the profile into six segments (maximum of ten), providing significantly improved visualisation of the material state within the area.

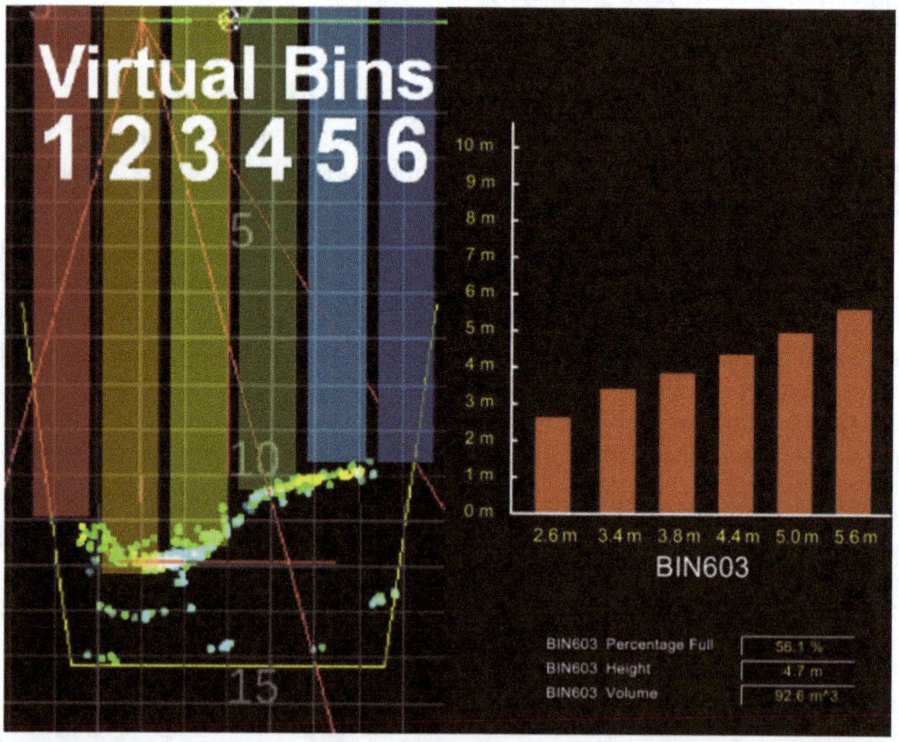

FIG 3 – Surge bin 2D radar profile measurement and SCADA segment visualisation.

Each segment has its own high-high and low-low limits that are triggered via the telegram Booleans allowing for a less conservative DCS operating window while enabling replication of the profile view on operator SCADA screens.

The same 2D scanning radar platform used on the surge bins in this case study is also deployed at truck- and rail-based ore receiving stations, where volumetric measurement of a car dumper or dump pocket is required. Figure 4 shows the software visualisation interface for a car dumper application, included here to illustrate the generality of the measurement and profile-segmentation approach described above; the core sensor technology, DCS integration architecture and multi-output control strategy are directly transferable from car dumper pocket monitoring to the surge bins described in this paper.

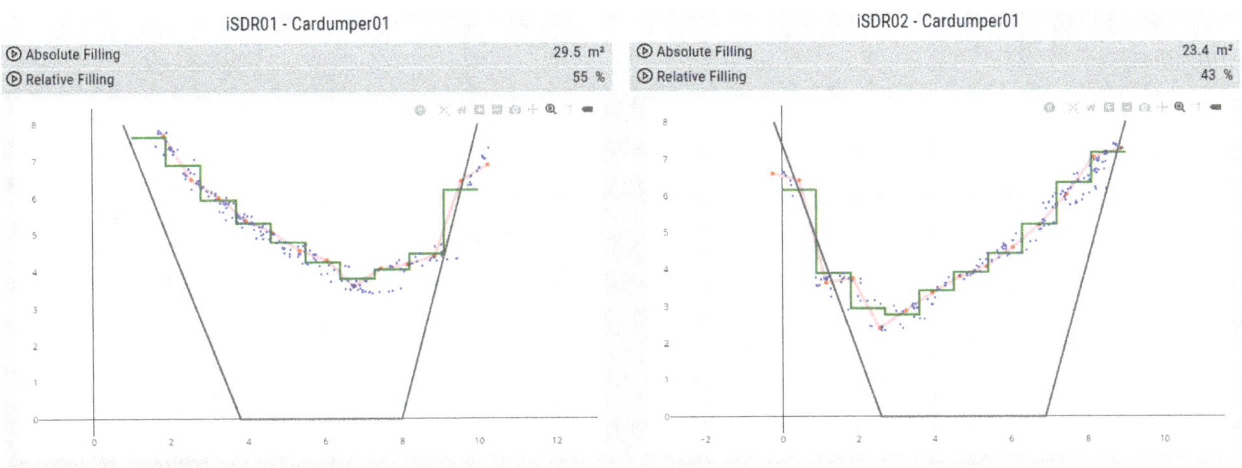

FIG 4 – 2D radar software visualisation for a car dumper application, illustrating the multi-segment profile and DCS integration approach used in the surge bin implementation.

Multi-output sensors and advanced bin control

A key enabler for process optimisation is the availability of multiple independently configurable outputs from a single 2D scanner, each representing a different aggregate of the underlying profile (for example, peak level, average level, and level in specific regions of interest). In a surge bin feeding a secondary cone, these outputs can be mapped to distinct control and protection functions: one output can drive the primary feeder speed to maintain a target average inventory; another can provide high-high and low-low alarms based on peak level for overfill and empty-bin protection; and additional outputs can monitor regions near the outlet or walls for early indication of asymmetric flow or build-up.

By exposing up to ten logical level signals per bin from a single instrument, the control system can implement a layered strategy that simultaneously maximises utilisation of bin capacity, protects against spills and starvation, and stabilises the effective head of material presented to the secondary crusher. In documented magnetite operations, integrating 2D radar profiles into feeder PID loops and interlocks has delivered significant increases in secondary and tertiary crusher feed rates, large reductions in feed-rate variability, and large reductions in crusher trips related to empty or overfull surge bins.

Beyond throughput gains, this more granular visibility into bin behaviour supports improved long-term asset health by reducing the frequency and amplitude of surge loading events that drive fatigue in chutes, feeders and crushers, as well as enabling more proactive maintenance planning based on stable, repeatable operating conditions. For iron ore producers seeking to close the gap between installed and realised crushing capacity without major capital expansions, upgrading from 1D point level devices to multi-output 2D profile measurement on critical surge bins represents a high-leverage, relatively low-risk intervention.

Case study evidence – crusher cavity radar and variable speed control

Having established stable, automated bin level control across the surge bins, the second element of the upgrade addressed variability within the secondary crusher itself. A high-frequency 1D radar sensor was installed inside the crusher cavity to provide continuous measurement of the rock column height, enabling a closed-loop control scheme that modulates eccentric speed to maintain a target cavity level. The following subsections document the implementation approach, observed throughput and stability outcomes, early liner wear findings, and product size distribution improvements at this site.

Throughput and operating stability

The introduction of radar-based cavity level control with variable crusher speed has enabled sustained operation at, and in periods above, the secondary crusher's nameplate throughput while materially improving stability of power draw and cavity level. Early operational evidence and industry experience suggest this more controlled operating regime is also favourable for liner life, long-term mechanical integrity and tighter product size distribution, although these benefits are still being quantified.

Prior to the upgrade, the cone operated with highly variable cavity levels driven by inconsistent bin discharge and manual speed adjustments, resulting in frequent excursions below choke feed and under-utilisation of installed power. The radar sensor now provides a continuous, high-confidence measurement of the rock column inside the crushing chamber, which is used as the primary process variable in a closed-loop control scheme that modulates eccentric speed to maintain a target cavity level. This shift to cavity-level-driven speed control has allowed the crusher to run closer to its design power limit for extended periods, smoothing short-term feed disturbances and delivering a step change in average hourly throughput compared with the pre-implementation baseline.

Consistent choke-fed operation is recognised as a key determinant of cone crusher performance, promoting inter-particle crushing, improved product shape and higher volumetric capacity. By actively regulating speed to hold the chamber just above the spider, rather than relying solely on upstream bin levels or fixed speed settings, the upgraded control delivers a more stable operating point across a range of ore hardness and feed conditions. Operators report a marked reduction in short stops related to low cavity level, overload events and protection trips, translating into higher utilisation of the crusher and more predictable feed to downstream screening and grinding circuits.

Liner wear and long-term operating conditions

Although a full liner life cycle has not yet been completed under the new control regime, early inspections indicate a more even wear pattern between the feed and discharge zones and between opposing sides of the chamber. This is consistent with literature showing that true choke feeding reduces localised impact loading and promotes rock-on-rock attrition over the full liner surface, which in turn tends to extend liner campaigns and delay profile-related loss of capacity. The ability to hold cavity level within a narrow band also reduces the frequency and magnitude of power and mechanical load swings, a factor known to influence fatigue life of crusher components such as main shafts, bearings and frames.

From a maintenance and reliability perspective, the radar hardware itself is non-contact and designed for dusty, high-vibration environments, requiring only periodic visual inspection and cleaning during scheduled shutdowns. As a result, the overall intervention burden is shifted away from reactive responses to crusher trips and liner damage toward scheduled condition monitoring and planned liner change-outs, supporting higher long-term availability. Formal measurement of liner mass loss per operating hour, in combination with power and cavity-level histories, is planned to quantify the wear benefits over multiple campaigns.

Product size and downstream performance

Stabilising the relationship between cavity level, eccentric speed and closed side setting has already produced visibly tighter and more repeatable crusher product size distributions, with fewer periods of excessively coarse or fine material reported by downstream operators. Research on cone crusher control shows that coordinated adjustment of speed and CSS (Closed-Side-Setting) around a stable cavity level can be used to trade-off throughput against fines generation in a controlled manner, improving yield of the target size fraction and reducing recirculating load. In the present application, the radar-driven speed control has reduced the incidence of oversize reporting to the screens while maintaining or improving the proportion of material near the target P_{80}, which is expected to translate into more efficient grinding once longer-term data sets are available.

At this stage, product size improvements are being tracked primarily through shift-based PSD sampling and operator feedback, with plans to integrate online size measurement to more rigorously link cavity-level stability, speed set points and product curve behaviour. Nonetheless, the combination of higher and more stable throughput, emerging evidence of improved liner utilisation, and observed tightening of the size distribution strongly indicates that reliable cavity level radar, coupled with speed-based control, provides a robust pathway to improving both the short-term productivity and long-term operating economics of secondary cone crushers in iron ore service.

Applicability to other hard rock operations

While this case study focuses on a large-scale iron ore operation, the technical challenges and solution approach are applicable to hard rock mines globally. Common characteristics across porphyry copper, iron ore, gold and base metal operations include:

- Requirement for stable, choke-fed operation of secondary and tertiary crushers.
- Surge bins operating in dusty, abrasive environments with limited maintenance access.
- Control systems dependent on accurate, real-time level measurement for closed-loop automation.
- Economic pressure to maximise throughput from existing crushing infrastructure before committing to major capacity expansions.

The radar instrumentation and control architecture presented in this paper can be replicated across different ore types, crushing equipment configurations and site conditions. Key design principles include:

- Focus radar deployment on critical surge bins that directly control crusher feed rates.
- Integrate radar outputs into existing control systems using standard industrial protocols.
- Implement staged commissioning to validate measurement accuracy and tune control loops before full deployment.
- Establish quantitative performance metrics to measure and sustain improvements.
- Design for scalability, enabling replication across multiple bins and circuits.

Several hard rock operations have subsequently adopted similar radar-based control strategies, reporting comparable improvements in crushing circuit performance (Powell, Govender and McBride, 2008; Hulthén and Evertsson, 2011).

CONCLUSIONS

Implementation of 2D radar level scanners on critical surge bins and smarter 1D radar sensors inside conical crushers of a large iron ore crushing circuit delivered substantial improvements in throughput, stability and availability. Radar-based measurement overcame limitations of legacy instrumentation in dusty, high-throughput conditions, enabling automated control of crusher feed rates and sustained choke-fed operation.

Performance data demonstrate up to 40 per cent improvement in secondary crusher throughput, approximately 50 per cent reduction in feed rate variability, and greater than 40 per cent decrease in reliability events. Economic analysis indicates simple payback of less than one quarter, with net present value exceeding 58 million Australian dollars over ten years.

The solution architecture is applicable to hard rock operations globally, providing a systematic approach to unlocking latent capacity from existing crushing infrastructure. Radar-enabled control represents a pragmatic alternative to major capital expansions, delivering high returns with minimal operational disruption.

The detailed analysis of both bin-level 2D radar control and crusher cavity radar control demonstrates that optimised level measurement at multiple points in the crushing circuit—from surge bins through to the crusher chamber itself—provides complementary benefits that compound to deliver step-change improvements in circuit performance. The transition from basic 1D point devices to high-resolution 2D profile scanners in surge bins, combined with continuous cavity level monitoring inside crushers, creates a robust measurement and control foundation that enables operations to realise the full potential of their installed crushing capacity.

Future work should examine integration of radar-based bin monitoring with advanced process control systems and machine learning algorithms to further optimise comminution circuit performance.

ACKNOWLEDGEMENTS

The authors acknowledge the site operations, engineering and maintenance teams for their support during implementation and commissioning of the radar instrumentation system.

REFERENCES

Brasseur, G, 2001. Radar level measurement: an established technique makes progress, *Sensors*, 18(11):36–42.

Evertsson, C M, 2000. Cone crusher performance, Doctoral dissertation, Chalmers University of Technology, Göteborg, Sweden.

Hulthén, E and Evertsson, C M, 2011. Algorithm for dynamic cone crusher control, *Minerals Engineering*, 24(8):778–784.

Napier-Munn, T J, Morrell, S, Morrison, R D and Kojovic, T, 1996. Mineral Comminution Circuits: Their Operation and Optimisation, 413 p, Julius Kruttschnitt Mineral Research Centre.

Powell, M S, Govender, I and McBride, A T, 2008. Applying DEM outputs to the unified comminution model, *Minerals Engineering*, 21(11):744–750.

Wills, B A and Finch, J A, 2015. *Wills' Mineral Processing Technology: An Introduction to the Practical Aspects of Ore Treatment and Mineral Recovery,* 8th ed, 512 p (Butterworth-Heinemann).

Integrating flotation hydrodynamics into process control strategies for iron ore production

M Safari[1,2], *F S Hoseinian*[2], *D Deglon*[3], *L Leal Filho*[4] *and T C Souza Pinto*[5]

1. Minerals Processing Division, Mintek, Johannesburg 2000, South Africa.
   Email: mehdis@mintek.co.za
2. Faculty of Engineering and the Built Environment, University of the Witwatersrand, Johannesburg 2000, South Africa.
3. Centre for Minerals Research, Department of Chemical Engineering, University of Cape Town, Cape Town 7700, South Africa.
4. Mining and Petroleum Department, Polytechnic Engineering School, University of São Paulo, São Paulo SP, Brazil.
5. Mineral Development Center – CDM/Vale, Santa Luzia MG, Brazil.

ABSTRACT

The efficiency of reverse flotation in iron ore beneficiation is highly dependent on hydrodynamic parameters that govern stability, selectivity, and metallurgical performance. Building on insights from comparative evaluations of self-aerated and forced-air mechanical flotation cells, this study explores how hydrodynamic indicators can be integrated into process control strategies to enhance iron ore production. Real-time monitoring of parameters such as air flow rate, bubble size, and velocity distribution can provides valuable feedback for optimising reagent dosage, froth depth, and air control systems. These parameters directly affect concentrate grade, tailings recovery, and energy consumption—key variables in automated process control. By linking hydrodynamic behaviour to operational data, predictive control algorithms can be developed to stabilise recovery and improve process efficiency. The integration of such feedback-driven control strategies represents a step toward smarter, data-informed iron ore flotation circuits with higher selectivity and reduced operational variability.

INTRODUCTION

Flotation is a selective separation technique used to separate mineral species based on differences in their surface properties. The efficiency of flotation separation is governed by the recovery of valuable and gangue minerals through both true flotation and entrainment mechanisms (Hoseinian *et al,* 2021; Neethling and Cilliers, 2009; Wang *et al,* 2015). Both chemical variables and hydrodynamic conditions within flotation cells significantly influence mineral recovery and separation selectivity (Hoseinian *et al,* 2018; Hoseinian *et al,* 2019a; Nazari *et al,* 2023). In iron ore beneficiation, cationic reverse flotation of quartz is widely employed to selectively remove silica and produce high-grade pellet feed (Safari *et al,* 2018). However, reverse flotation performance is often limited by two key challenges: the reporting of coarse quartz particles to the tailings and the unintended entrainment of fine iron oxide particles into the concentrate (Safari *et al,* 2021; Vieira and Peres, 2007).

Flotation is inherently a size-dependent process, in which fine, intermediate, and coarse particles exhibit distinct flotation behaviours due to differences in collision, attachment, and detachment mechanisms (Farrokhpay and Fornasiero, 2017; Safari *et al,* 2022). Recovery–particle size curves are commonly used to evaluate the flotation response and to identify the maximum particle size that can be efficiently recovered under specific chemical and hydrodynamic conditions. The flotation efficiency of coarse particles is typically limited by reduced attachment efficiency and increased detachment in highly turbulent regions of flotation cells, despite relatively high collision probabilities (Farrokhpay and Fornasiero, 2017; Safari and Deglon, 2020). Conversely, adhesion probability generally increases with decreasing particle size. However, in reverse cationic flotation of iron ores, a high proportion of fine particles leads to increased collector consumption and reduced selectivity (Filippov, Severov and Filippova, 2014). Vieira and Peres (2007) investigated the effects of collector type and concentration, pH, and particle size on quartz flotation in mechanical cells, demonstrating that quartz particles across a wide size range can be effectively recovered through appropriate reagent selection. Specifically, ether diamines were found to be more effective for the flotation of medium to coarse quartz particles, whereas ether monoamines were better suited for fine quartz flotation (Vieira and Peres, 2007). Comparative batch flotation studies by Ma, Marques and Gontijo (2011) further showed that ultrafine metal losses are higher in reverse cationic flotation than in reverse anionic flotation, primarily due to the entrainment of ultrafine hematite particles, while reverse anionic flotation exhibited slightly inferior performance in the coarse particle size range.

The recovery of fine particles due to the entrainment can be related to the water recovery (Cilek, 2009). It can be explained by the mechanisms of the hydrodynamic boundary layer, the wake and swarm effect of rising bubbles (Hoseinian *et al,* 2020). The size and rising velocity of bubbles are depended to operational factors including the type and concentration of frother, the froth removal rate, froth residence time, airflow rate and impeller speed (Hoseinian, Irannajad and Safari, 2017; Hoseinian *et al,* 2019b). In addition, the flotation cell hydrodynamics determined the probability of bubble particle collision, as well as the probability of bubble particle detachment, which is strongly depend on the particle size, bubble size and the flotation cell turbulence (Cilek and Yılmazer, 2003; Safari *et al,* 2016). Therefore, it is apparent that the entrainment and true flotation strongly depends on the flotation cell hydrodynamic conditions. By far the majority of iron flotation research, particularly with respect to chemical condition such as reagent selection for investigate the flotation performance, is carried out using mechanical cell (Araujo, Viana and Peres, 2005; Mowla, Karimi and Ostadnezhad, 2008; Pavlovic and Brandão, 2003; Yuhua and Jianwei, 2005).

Previous test work done by other researchers on a number of Brazilian iron ores (Lima *et al,* 2016; Lima, Valadão and Peres, 2013; Ma, Marques and Gontijo, 2011) indicates that while the mechanical flotation cell was efficient at cleaning quartz particles between 10 and 150 µm, the collection efficiency decreased significantly for larger particles (Hassanzadeh *et al,* 2022). In this study, the effects of various hydrodynamic operating parameters including solid percentage, particle size, impeller speed, froth height and superficial gas velocity on the hematite recovery with a view to evaluate the entrainment during the process were investigated. Recent advances in flotation research increasingly emphasise the integration of hydrodynamic characterisation with operational data to enable predictive and adaptive process control. Hydrodynamic variables such as gas dispersion, bubble size distribution, and turbulence intensity have been shown to exert a first-order influence on particle–bubble collision efficiency, froth stability, and overall metallurgical performance (Gorain, Franzidis and Manlapig, 1997; Mesa and Brito-Parada, 2019). When these parameters are monitored in real time and linked to operating variables such as air rate, froth depth, and reagent dosage, they provide a robust foundation for the development of predictive control algorithms capable of stabilising recovery and improving selectivity (Shean, Hadler and Cilliers, 2018; Maldonado, Bergh and Yianatos, 2021). Such feedback-driven control strategies represent a shift from empirical, operator-dependent control toward smarter, data-informed flotation circuits, reducing operational variability while enhancing energy efficiency and concentrate quality in iron ore reverse flotation systems (Wills and Finch, 2016; Forbes, 2020).

MATERIAL AND METHODS

The iron sample was obtained from the rougher feed from Vale, Brazil. To ensure that all samples were representative, feed sample were blended and divided into individual feed samples for the flotation tests using two different rotary sample dividers in two different steps. The elemental composition of the iron sample was determined by X-ray diffraction (XRD) and inductively coupled plasma mass spectrometry (ICP) analysis (64.4 per cent Fe_2O_3 and 34.9 per cent SiO_2). Ether dodecyl amine (EDA) was received in liquid form and used as collector. MIBC (methyl isobutyl carbinol) was used as a frother. Starch was received in powder form and used as a depressant. Reagents were prepared on a daily basis using deionised water and adjusted to the appropriate pH by NaOH and HCl, if required. All reagents were of the analytical grade quality. The conditions for the flotation experiments are presented in Table 1.

Flotation tests were carried out using a Leeds laboratory flotation cell. Several concentrates were collected at the specified times during the flotation experiments. At the end of the flotation experiment, the concentrate samples and tail were separately filtrated, dried, and weighed for mineral recovery calculation, screening in seven different size fractions and sizing using a Malvern Mastersizer™ 2000. The flotation products were subjected to a chemical analysis for the determination of the content of total Fe and SiO_2.

TABLE 1

Experimental conditions for flotation experiments.

Condition/Parameter	Value
Feed mass (Hematite/Quartz Ore)	1–3 kg
Froth height (Relative to cell height)	5.60 to 33.30%
Impeller speed	600 to 1500 rev/min
pH	10.5

RESULT AND DISCUSSION

Particle size and density play a critical role in determining the flotation behaviour of hematite and quartz particles. To assess the influence of particle size on hematite flotation efficiency, flotation experiments were conducted under various operating conditions, during which concentrate samples were collected at different flotation times for chemical analysis and particle size distribution measurements. A representative result from one flotation test is presented in Figure 1, illustrating that both mass recovery and water recovery increased progressively with flotation time. Particle size was identified as a dominant factor influencing flotation efficiency. Accordingly, the impact of hydrodynamic conditions on flotation performance was evaluated with particular emphasis on particle size through the analysis of recovery–particle size relationships.

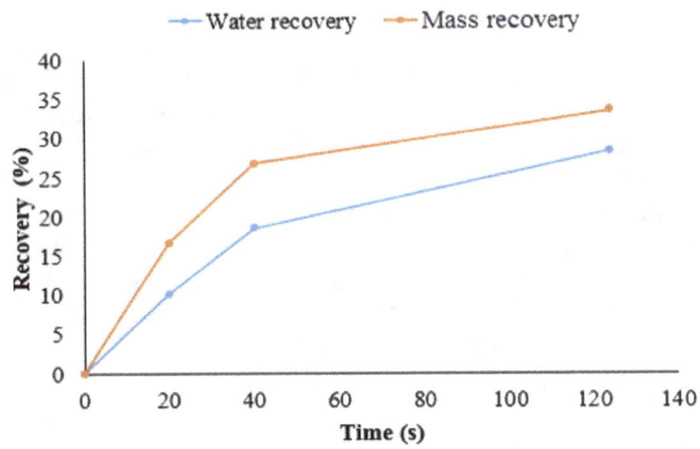

FIG 1 – Effect of particle size on the entrainment of hematite.

Froth height

To investigate the effect of froth height on quartz recovery, flotation experiments were conducted at varying froth heights while maintaining constant impeller speed, flotation time, and reagent dosages (collector, depressant, and frother). The results, presented in Figure 2, show that both mass recovery and water recovery decreased almost linearly with increasing froth height. This trend is attributed to the longer froth residence time at higher froth depths, which enhances liquid drainage and promotes the return of both particles and water from the froth phase to the pulp phase (Wang *et al,* 2015). Analysis of size-by-size quartz recovery revealed that the highest recovery for fine particles (31 µm) is primarily associated with entrainment effects, whereas the recovery peak observed for coarser particles (126 µm) is attributed to a combination of increased attachment efficiency and reduced detachment probability within this size fraction. Overall, coarse quartz recovery decreased nearly linearly with increasing froth height, with the reduction being more pronounced for coarse particles than for fine particles. Enhanced bubble drainage at higher froth heights promotes the detachment of coarse, high-density quartz particles, reducing their likelihood of reporting to the concentrate and increasing their return to the pulp phase.

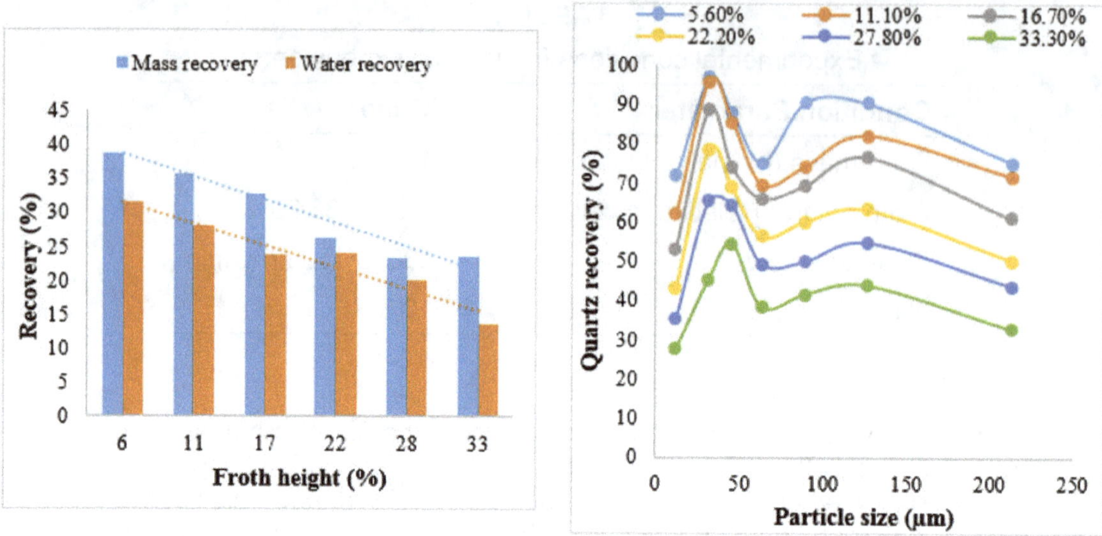

FIG 2 – Effect of froth height on the recovery of quartz particles (impeller speed = 1200 rev/min, Jg = 0.6 cm/s, solid percentage = 25 per cent, depressant dosage = 1000 g/t, collector dosage = 100 g/t).

Impeller speed

The impeller speed relate directly to sub-processes of the air dispersion into bubbles, particle suspensions in the pulp, the probabilities of collision, attachment and detachment of particle–bubble. To determine the effect of impeller speed on the process, flotation experiment were performed as a function of impeller speed at a fixed froth height, flotation time, collector dosage, depressant dosage and frother dosage. The results are shown in Figure 3. The results show that the impeller speed has a strong effect on the recovery of quartz particles. The increase of impeller speed significantly increased particle–bubble detachment of coarser particles but has a lesser effect on fine particles. The pulp phase detachment of solid particles from rising particle-bubble aggregates plays an important role in the reduction in recovery of coarse quartz particles. Mass recovery and water recovery increased with the increase in impeller speed from 600 rev/min to 1200 rev/min, while it decreased at impeller speed of 1350 rev/min. The recovery of coarse quartz particles increased with increasing impeller speed to an optimum, after which they decrease. It is clear from this figure that increasing impeller speed beyond the optimum leads to very large decreases in coarse quartz recovery (Wang *et al*, 2015). It is clear from this figure that there is an optimum impeller speed range for coarse quartz particle recovery, beyond which recovery is very poor (<10–20 per cent).

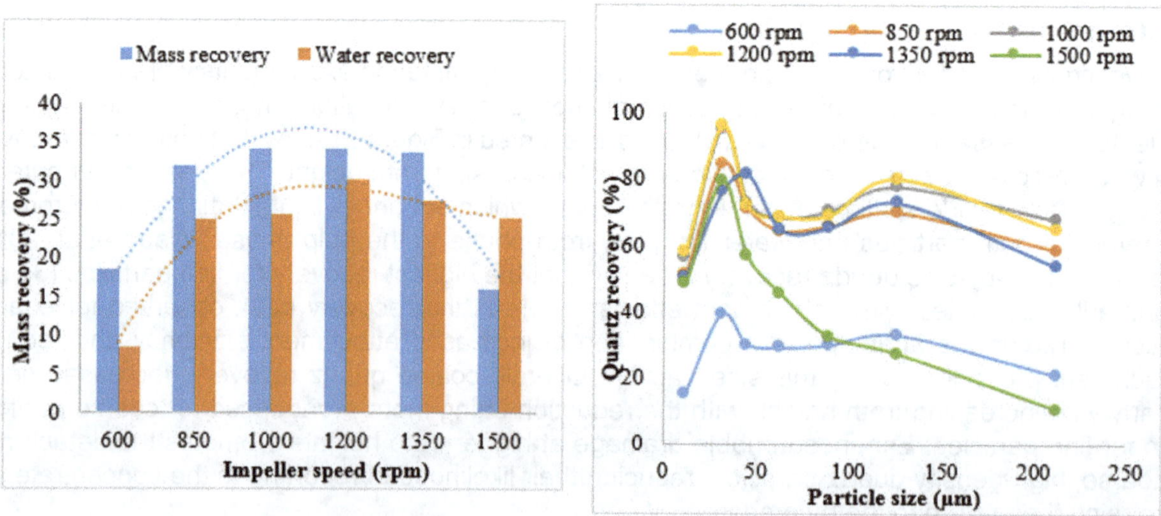

FIG 3 – Effect of impeller speed on the recovery of coarse quartz and fine ore particles (Froth height = 16.7 per cent, Jg = 0.6 cm/s, solid percentage = 25 per cent, depressant dosage = 1000 g/t, collector dosage = 100 g/t).

The higher impeller speed enhances the attachment efficiency by increasing the overcoming of the energy barrier to particle-bubble attachment that led to increase of recovery. However, the high impeller speed causes an increase in detachment and disruption of bubble-particle aggregates, especially for coarser particles and also may create more turbulence to the froth and affect the froth stability. It leads to decrease of recovery of coarse quartz particles at impeller speed higher than 1200 rev/min. The results demonstrated that increasing impeller speed generally increases the recovery of coarse quartz particle to an optimum.

Superficial gas velocity

The effect of superficial gas velocity on coarse quartz recovery is illustrated in Figure 4. At a superficial gas velocity of 0.9 cm/s, the maximum recovery of coarse quartz particles (126 µm fraction) reached 83.76 per cent. As shown in Figure 4, increasing the superficial gas velocity from 0.4 cm/s to 0.9 cm/s resulted in higher mass recovery, accompanied by a corresponding increase in water recovery. Elevated superficial gas velocities enhanced the recovery of coarse quartz particles, which are relatively dense and therefore require stronger hydrodynamic conditions to achieve effective transport to the froth phase. An increase in superficial gas velocity leads to the generation of a larger number of bubbles, promoting the entrainment of additional water from the pulp into the froth. In addition, superficial gas velocity influences the extent of water drainage within the froth phase, thereby affecting both particle recovery and overall froth stability.

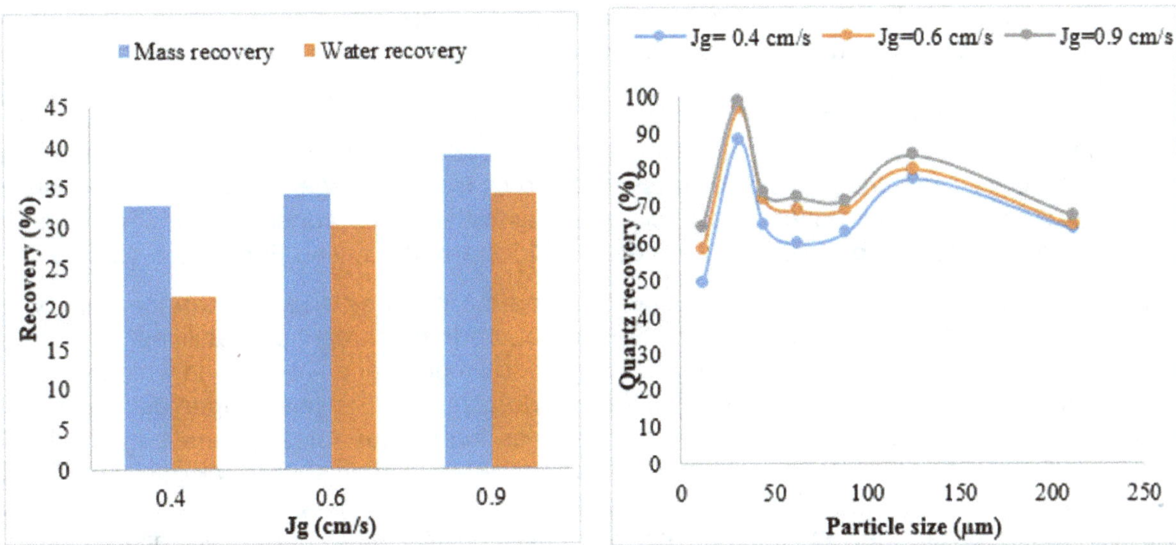

FIG 4 – Effect of superficial gas velocity on the recovery of coarse quartz and fine ore particles (impeller speed = 1200 rev/min, Froth height = 16.7 per cent, solid percentage = 25 per cent, depressant dosage = 1000 g/t, collector dosage = 100 g/t).

Solid percentage

Pulp solids concentration has a pronounced influence on both mass recovery and water recovery, as illustrated in Figure 5, which presents the combined effects of solids percentage, froth height, and particle size on recovery by true flotation. Solid dispersion in the pulp below the pulp–froth interface is governed by parameters such as particle size and impeller speed, which together control the extent of solid entrainment into the concentrate. A decrease in the recovery of quartz particles larger than 126 µm was observed, indicating that coarse quartz particles tend to settle within the flotation cell rather than report to the froth phase. Similarly, coarse hematite particles exhibit reduced entrainment into the concentrate, further highlighting the size-dependent limitations of flotation under varying pulp solids conditions.

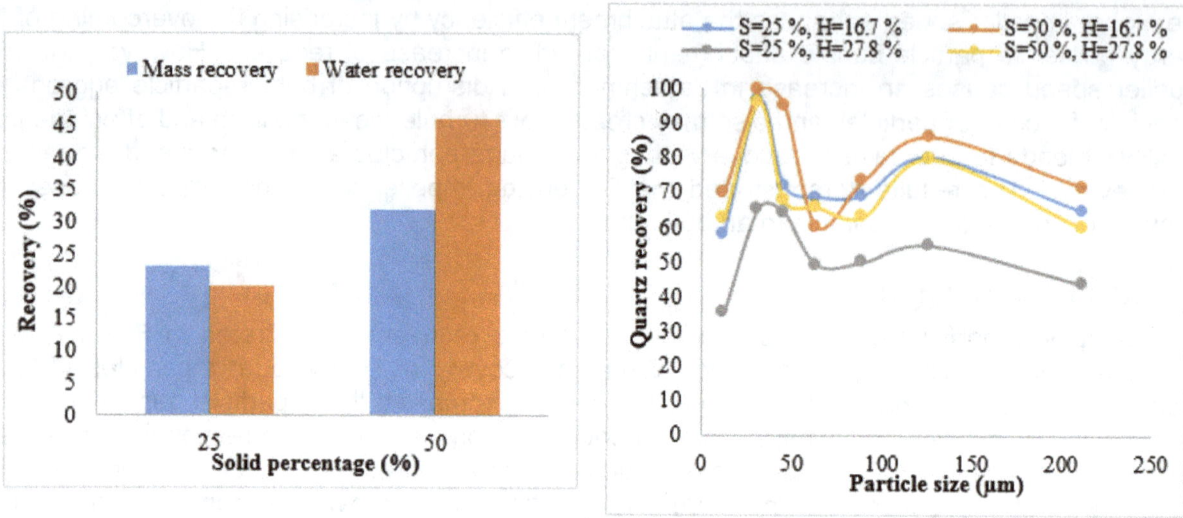

FIG 5 – Effect of solids percentage on the recovery of coarse quartz and fine ore particles (impeller speed = 1200 rev/min, Jg = 0.6 cm/s, depressant dosage = 1000 g/t, collector dosage = 100 g/t).

DISCUSSION

The results demonstrate that hydrodynamic operating parameters play a decisive role in determining flotation performance, particularly for coarse quartz and hematite particles. Froth height was observed to have a negative impact on both mass and water recovery, with increasing froth depth leading to longer froth residence times and enhanced liquid drainage. These effects contributed to reduced coarse particle recovery due to detachment and backflow into the pulp phase. The particle size dependency is clear: fine particles are less affected by froth drainage, whereas coarse particles exhibit more pronounced decreases in recovery at higher froth heights, consistent with earlier studies on size-dependent flotation behaviour (Wang *et al,* 2015; Farrokhpay and Fornasiero, 2017).

Impeller speed was shown to control the dispersion and mixing of air and solids in the pulp phase, directly influencing particle–bubble collision, attachment, and detachment probabilities. An increase in impeller speed enhanced coarse particle recovery up to an optimal point, beyond which excessive turbulence may lead to bubble coalescence or particle detachment. Similarly, superficial gas velocity exhibited a positive effect on coarse quartz recovery, as higher gas rates generate more bubbles and enhance particle transport to the froth phase. However, higher superficial gas velocity also increases water entrainment, highlighting the trade-off between recovery and water handling. Pulp solids concentration further influenced recovery through its effect on solid dispersion and entrainment. Higher solids percentages improved fine particle recovery but exacerbated settling of coarse particles, which have lower entrainment potential and are more sensitive to hydrodynamic conditions. Collectively, these observations demonstrate that coarse quartz and hematite flotation is highly sensitive to the combined effects of froth height, impeller speed, superficial gas velocity, and solids concentration, with particle size mediating the response. Importantly, these findings provide a foundation for linking hydrodynamic behaviour to operational data. By integrating measurements of froth height, impeller speed, gas velocity, and pulp solids into predictive models, it becomes possible to develop feedback-driven control strategies that stabilise recovery, optimise selectivity, and reduce variability. Such an approach represents a step toward data-informed, automated iron ore flotation circuits, capable of maintaining consistent performance despite variations in particle size distribution or feed composition, while also minimising energy consumption and reagent use. The representative trends of coarse quartz recovery as a function of these key hydrodynamic parameters are illustrated in Figure 6, summarising the combined effects discussed above.

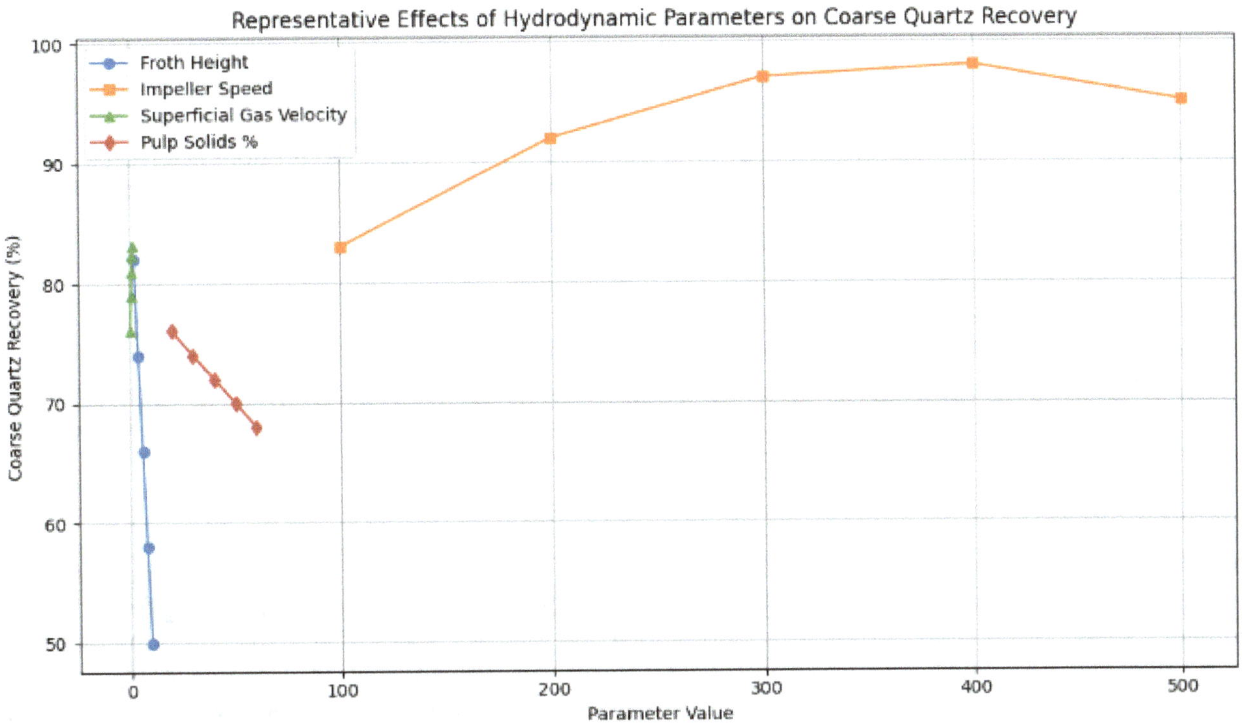

FIG 6 – Demonstrative graph showing the representative trends of coarse quartz recovery with respect to key hydrodynamic parameters: froth height, impeller speed, superficial gas velocity, and pulp solids percentage. (Multi-scale graph).

Key trends illustrated:

- Froth height: Recovery decreases linearly with increasing froth height.
- Impeller speed: Recovery increases to an optimum, then plateaus or slightly decreases.
- Superficial gas velocity: Recovery increases asymptotically with higher gas flow.
- Solids percentage: Recovery of coarse particles decreases at higher solids.

Here are some possible interactions between the parameters, which can provide deeper insight beyond individual trends:

- Froth Height × Superficial Gas Velocity: At higher gas velocities, the detrimental effect of increased froth height may be partially mitigated due to enhanced bubble flux and froth mobility. However, excessive froth height combined with high gas rates may lead to instability and increased entrainment.
- Impeller Speed × Superficial Gas Velocity: Increasing impeller speed improves bubble dispersion and particle–bubble collision, but at high gas velocities, excessive turbulence may promote bubble coalescence or detachment, reducing recovery. An optimal balance between these two parameters is therefore critical.
- Impeller Speed × Solids Percentage: At higher solids concentrations, increased impeller speed may be required to maintain adequate suspension and mixing. However, excessive agitation in high solids systems can increase particle detachment and reduce coarse particle recovery.
- Froth Height × Solids Percentage: Higher solids content, particularly with fine or clay-rich material, can lead to more stable froth. When combined with increased froth height, this may reduce drainage efficiency and negatively impact selectivity and recovery.
- Superficial Gas Velocity × Solids Percentage: While increasing gas velocity generally enhances recovery, at high solids loading the system may experience reduced bubble–particle attachment efficiency due to hindered settling and increased slurry viscosity.

The observed trends and interactions between these parameters highlight how hydrodynamic parameters—froth height, impeller speed, superficial gas velocity, and pulp solids—directly influence coarse quartz recovery and overall flotation performance. By systematically monitoring and controlling these variables, it is possible to maintain optimal recovery and selectivity across varying feed conditions. These relationships provide the foundation for predictive, feedback-driven process control strategies, where real-time data on hydrodynamic conditions can be used to adjust operating parameters dynamically. Implementing such strategies in iron ore flotation circuits can enhance concentrate quality, reduce operational variability, and improve energy and reagent efficiency, moving toward smarter, data-informed production systems. The interactions between these parameters highlight the complexity of flotation systems and the need for an integrated modelling approach. While the current study outlines key individual trends and potential interactions, a comprehensive predictive model incorporating these variables and their coupled effects is under development. This model, including its validation and application to process optimisation, will be presented in detail in a subsequent publication.

CONCLUSION

The influence of particle size on true flotation during the reverse flotation of hematite was investigated to evaluate the role of key hydrodynamic operating parameters, including froth height, impeller speed, superficial gas velocity, and pulp solids concentration. The results demonstrate that hydrodynamic conditions exert a strong influence on the efficiency of hematite reverse flotation. Increasing froth height led to reductions in both mass and water recovery as a result of extended froth residence time and enhanced liquid drainage, while the recovery of coarse quartz particles decreased with increasing froth height due to intensified bubble drainage and particle detachment. In mechanical flotation cells, impeller speed governs air and solids dispersion in the pulp phase and directly affects particle–bubble collision, attachment, and detachment probabilities. An increase in impeller speed improved coarse quartz recovery up to an optimum value. Similarly, higher superficial gas velocities and increased pulp solids concentrations enhanced quartz recovery. Collectively, these findings highlight the strong sensitivity of reverse flotation performance—particularly for coarse quartz—to hydrodynamic conditions. By quantitatively linking hydrodynamic behaviour to operational data, predictive control algorithms can be developed to stabilise recovery and improve process efficiency. The integration of such feedback-driven control strategies represents a critical step toward smarter, data-informed iron ore flotation circuits with improved selectivity and reduced operational variability.

REFERENCES

Araujo, A, Viana, P and Peres, A, 2005. Reagents in iron ores flotation, *Minerals Engineering*, 18(2):219–224.

Cilek, E and Yılmazer, B, 2003. Effects of hydrodynamic parameters on entrainment and flotation performance, *Minerals Engineering*, 16(8):745–756.

Cilek, E C, 2009. The effect of hydrodynamic conditions on true flotation and entrainment in flotation of a complex sulphide ore, *International Journal of Mineral Processing*, 90(1–4):35–44.

Farrokhpay, S and Fornasiero, D, 2017. Flotation of coarse composite particles: Effect of mineral liberation and phase distribution, *Advanced Powder Technology*, 28(8):1849–1854.

Filippov, L, Severov, V and Filippova, I, 2014. An overview of the beneficiation of iron ores via reverse cationic flotation, *International Journal of Mineral Processing*, 127:62–69.

Forbes, E, 2020. Froth flotation control: A review of conventional and emerging technologies, *Minerals Engineering*, 157:106534.

Gorain, B K, Franzidis, J-P and Manlapig, E V, 1997. Studies on impeller type, impeller speed and air flow rate in an industrial scale flotation cell—Part 1: Effect on bubble size distribution, *Minerals Engineering*, 10(6):615–635.

Hassanzadeh, A, Safari, M, Khoshdast, H, Güner, M K, Hoang, D H, Sambrook, T and Kowalczuk, P B, 2022. Introducing key advantages of intensified flotation cells over conventionally used mechanical and column cells, *Physicochemical Problems of Mineral Processing*, 58.

Hoseinian, F S, Irannajad, M and Safari, M, 2017. Effective factors and kinetics study of zinc ion removal from synthetic wastewater by ion flotation, *Separation Science and Technology*, 52(5):892–902.

Hoseinian, F S, Rezai, B, Kowsari, E and Safari, M, 2018. Kinetic study of Ni (II) removal using ion flotation: Effect of chemical interactions, *Minerals Engineering*, 119:212–221.

Hoseinian, F S, Rezai, B, Kowsari, E and Safari, M, 2019a. Effect of impeller speed on the Ni (II) ion flotation, *Geosystem Engineering*, 22(3):161–168.

Hoseinian, F S, Rezai, B, Kowsari, E and Safari, M, 2020. The effect of water recovery on the ion flotation process efficiency, *Physicochemical Problems of Mineral Processing*, 56.

Hoseinian, F S, Rezai, B, Safari, M, Deglon, D and Kowsari, E, 2019b. Effect of hydrodynamic parameters on nickel removal rate from wastewater by ion flotation, *Journal of Environmental Management*, 244:408–414.

Hoseinian, F S, Rezai, B, Safari, M, Deglon, D and Kowsari, E, 2021. Separation of nickel and zinc from aqueous solution using triethylenetetramine, *Hydrometallurgy*, 202:105609.

Lima, N P, de Souza Pinto, T C, Tavares, A C and Sweet, J, 2016. The entrainment effect on the performance of iron ore reverse flotation, *Minerals Engineering*, 96:53–58.

Lima, N P, Valadão, G E S and Peres, A E C, 2013. Effect of particles size range on iron ore flotation, *Rem: Revista Escola de Minas*, 66(2):251–256.

Ma, X, Marques, M and Gontijo, C, 2011. Comparative studies of reverse cationic/anionic flotation of Vale iron ore, *International Journal of Mineral Processing*, 100(3–4):179–183.

Maldonado, M, Bergh, L and Yianatos, J, 2021. Advanced process control in flotation: A review of control strategies and future trends, *Minerals Engineering*, 164:106810.

Mesa, D and Brito-Parada, P R, 2019. Scale-up in froth flotation: A state-of-the-art review, *Minerals Engineering*, 140:105887.

Mowla, D, Karimi, G and Ostadnezhad, K, 2008. Removal of hematite from silica sand ore by reverse flotation technique, *Separation and Purification Technology*, 58(3):419–423.

Nazari, S, Hoseinian, F S, Li, J, Safari, M, Khoshdast, H, Li, J and He, Y, 2023. Synergistic effect of grinding time and submicron (nano) bubbles on the zeta potential state of spent lithium-ion batteries: A gene expression programming approach, *Journal of Energy Storage*, 70:107942.

Neethling, S and Cilliers, J, 2009. The entrainment factor in froth flotation: Model for particle size and other operating parameter effects, *International Journal of Mineral Processing*, 93(2):141–148.

Pavlovic, S and Brandão, P R G, 2003. Adsorption of starch, amylose, amylopectin and glucose monomer and their effect on the flotation of hematite and quartz, *Minerals Engineering*, 16(11):1117–1122.

Safari, M and Deglon, D, 2020. Evaluation of an attachment–detachment kinetic model for flotation, *Minerals*, 10(11):978.

Safari, M, Harris, M, Deglon, D, Leal Filho, L and Testa, F, 2016. The effect of energy input on flotation kinetics, *International Journal of Mineral Processing*, 156:108–115.

Safari, M, Hoseinian, F S, Deglon, D A, leal Filho, L d S and Pinto, T C d S, 2021. Investigating the effect of operational parameters on the flotation of itabirite iron ore in a pneumatic flotation cell, Proceedings of the XXX International Mineral Processing Congress, IMPC 2020 (International Mineral Processing Council (IMPC): Carlton).

Safari, M, Hoseinian, F, Deglon, D A, leal Filho, L d S and Pinto, T C d S, 2018. Investigation of the reverse flotation of hematite in three different types of laboratory flotation cells, Proceedings of the International Mineral Processing Congress, IMPC 2018 (International Mineral Processing Council (IMPC): Carlton).

Safari, M, Hoseinian, F, Deglon, D, Leal Filho, L and Pinto, T S, 2022. Impact of flotation operational parameters on the optimization of fine and coarse Itabirite iron ore beneficiation, *Powder Technology*, 408:117772.

Shean, B J, Hadler, K and Cilliers, J J, 2018. The effect of gas rate on bubble size and flotation performance, *Minerals Engineering*, 121:29–38.

Vieira, A M and Peres, A E C, 2007. The effect of amine type, pH and size range in the flotation of quartz, *Minerals Engineering*, 20(10):1008–1013.

Wang, L, Peng, Y, Runge, K and Bradshaw, D, 2015. A review of entrainment: Mechanisms, contributing factors and modelling in flotation, *Minerals Engineering*, 70:77–91.

Wills, B A and Finch, J A, 2016. *Wills' Mineral Processing Technology: An Introduction to the Practical Aspects of Ore Treatment and Mineral Recovery,* 8th edn (Butterworth-Heinemann).

Yuhua, W and Jianwei, R, 2005. The flotation of quartz from iron minerals with a combined quaternary ammonium salt, *International Journal of Mineral Processing*, 77(2):116–122.

Air-decks as a cost lever – numerical modelling to production rollout in iron ore

G Sampaio Lopes[1], T McRae[2], J Fern[3], P Klaric[4], J Hawinkel[5] and N Bodley[6]

1. Engineering Solutions Manager, MTi Group, Perth WA 6019.
   Email: gustavo.sampaio@mtigroup.com.au
2. Senior Drill and Blast Engineer – Solomon, Fortescue, Perth WA 6019.
   Email: tom.mcrae@fortescue.com,
3. Specialist Drill and Blast Operations, Fortescue, Perth WA 6019.
   Email: josh.fern@fortescue.com
4. Principal Drill and Blast Operations, Fortescue, Perth WA 6019.
   Email: paul.klaric@fortescue.com
5. Global Business Manager – MTi Group, Perth WA 6019.
   Email: johan.hawinkel@mtigroup.com.au
6. Chief Executive Officer – MTi Group, Perth WA 6019. Email: nick.bodley@mtigroup.com.au

ABSTRACT

Balancing drill-and-blast cost is often constrained by the risk that lower explosive energy will degrade fragmentation, reduce dig rates and/or processing productivity, or introduce adverse blast behaviours that propagate into broader mining KPIs. At Solomon, a Western Australia Iron Ore mine, a structured cost-reduction initiative targeted improvements in cost-effectiveness in drill and blast operations with the explicit requirement that downstream performance remains within planning targets. Air-decking was adopted as the key enabler to reduce overall powder factor in two geological domains lowering both cost per tonne and the associated carbon footprint while keeping every blast compliant with operational requirements.

Different alternatives for blast design optimisation were explored, evaluating its potential results but also its implementation complexity. The chosen pathway was to improve the stemming efficiency, through airdecks, enabling a reduction in the charge per blasthole. The new designs were model led rather than just empirical as they were developed using insights from a numerical model previously developed.

The numerical model provided key insights into the blast designs demonstrating that the airdeck alters how detonation energy is consumed and implications for the stemming efficiency. These insights were the foundation of the initial blast design tested during the field trials and, once initial results were confirmed, that airdecks could deliver the improvement program's objectives.

The program progressed from a POC to expanded trials to production rollout. Performance was quantified through a rigorous data workflow: D&B design parameters, explosive consumption and digging performance data and blast reports were consolidated; a criterion and scoring system was developed for assessing digging performance to identify key contributors (geology, explosive type/powder factor, blast size, diameter, airdeck length).

Across the 22 production blasts the new designs achieved a 15.9 per cent powder-factor reduction, saving 1039 t of explosive and 1053 t of stemming material, corresponding to 196.4 t less CO_2-equivalent emissions. Furthermore, this powder factor reduction did not compromise the balance of the mining value chain: no top size over specifications and digging rates were ~5 per cent above the targets on average.

INTRODUCTION

Iron ore producers are operating in a rapidly changing business environment. EY highlights in its annual report of risks and opportunities for mining and metals a climate of instability that is driving a stronger short-term emphasis on cost reduction and capital preservation, alongside persistent cost pressures. Furthermore, the miners surveyed also identified operational complexity as a leading risk for the sector (EY, 2025).

Confirming the EY findings, iron ore prices have shown strong swings over recent years: peak monthly averages moving from US$217/t in July 2021 to US$83/t in September 2024 and recently ranging around US$110/t (Trading Economics, 2026).

This volatility reinforces the need for cost-reduction initiatives that can be deployed with low operational disruption, enabling more resilient operations. However, changes that improve one step of the mining value chain can also create instability or performance loss elsewhere. In drill and blast, fragmentation quality, and particularly the control of coarse material, is an important performance consideration because it affects downstream loading performance and operational stability. Previous studies have linked suboptimal fragmentation and oversize with degraded loading performance, including reduced bucket fill and longer loading cycles (Doktan, 2001). Similarly, blast energy characteristics and charge design have been shown to influence shovel digging time and power consumption, highlighting the trade-off between blasting outcomes and downstream loading performance (Brunton *et al,* 2003).

Therefore, streamlining blasting operations requires an in-depth understanding of downstream impacts. Improving cost efficiency is not simply a matter of reducing drilling or explosive consumption; it also requires deploying resources more effectively. In order to achieve this effectively, blast design optimisation must consider: (i) variable ground conditions and geologic domains, and their breakage requirements; (ii) operational interdependencies between planning, drilling, and blasting; (iii) explosive energy confinement and utilisation; and (iv) the change-management burden associated with modifying well-tuned production systems.

For Solomon, an iron ore operation in the Pilbara, Australia, the challenge was the same: navigate through price volatility finding opportunities to lift its cost efficiency, specifically in drill and blast without upsetting the existing balance of the mining value chain. The site was already delivering blasts that met planning expectations, so new initiatives needed to be low-disruption and repeatable, while continuing to meet planning and production targets. Success was therefore defined as reducing powder factor, explosive consumption, cost and associated CO_2-e emissions while maintaining digging performance and achieving zero out of specification blasting fragmentation.

This study describes the approach used to streamline drill and blast activities at Solomon by improving stemming efficiency through airdecking. A high-fidelity numerical model was used to better understand how airdecks influence confinement and energy use, and to translate that understanding into practical charge designs that reduce explosives per hole. The designs were then implemented in stages: model led design, proof of concept (POC) blast, extended trials and production rollout to confirm that the cost-efficiency gains were not compromising safety, compliance or productivity.

SITE AND OPERATIONAL CONTEXT

Solomon is a large-scale iron ore mine in the Pilbara, Australia. It is part of Fortescue's integrated mining system, with a production capacity between 68–70 Mt/a of Hematite. Solomon is a major Fortescue iron ore operation with a workforce of over 2200 and two processing facilities (Fortescue, 2026).

Solomon sits within the Hamersley Province and comprises a sequence of banded iron formation (BIF) units and associated iron ore horizons, interbedded with shale and other sedimentary lithologies. The geology is classified into six domains, varying from softer rocks with average UCS values of around 20 MPa to up to 120 MPa. Two of these domains, channel iron deposit (CID) and detrital iron deposit (DID), are the focus of this study (Fortescue, 2022).

Blasting is carried out using standard patterns within each geology domain, typically using 229 mm blasthole for production and 165 mm and 203 mm blastholes for contour and special applications. Explosive selection is tailored to the requirements of the rock mass and adjusted accordingly, using products such as ANFO, emulsions, and heavy ANFO. All blasts are initiated using electronic detonators. Besides maintaining satisfactory digging rates, a key blast compliance requirement at Solomon was that no out of specification fragmentation be generated, defined as blocks larger than 1 m × 1 m × 1 m, together with no safety incidents such as excessive ejection or flyrock.

Loading is performed by the primary hydraulic fleet comprising of Liebherr 996Bs, R 9250s and R9400 and CAT 6040. The loading performance is assessed using instantaneous digging rates (t/h) as the principal compliance metric and planning targets are equipment-class specific, considering a 90-day moving averages across all geological domains, which provides a stable reference and helps normalise performance by smoothing short-term noise. Hauling is carried by 98 trucks where more than 88 per cent of the fleet is fully autonomous. Product is railed 300 km to the Herb Elliott Port in Port Hedland for export (Fortescue, 2026).

Furthermore, only a small proportion of material is processed as direct feed; most orepasses through blended stockpiles and feed streams. In a typical 12-hour plan, multiple benches are mined by 12 shovels, and seven ROM areas contribute to the feed. This creates numerous possible ore-destination and blending combinations, meaning crusher performance reflects the response of an integrated blending system rather than the outcome of a single blast.

Given the strong existing operational performance in the CID and DID domains, their relatively soft rock characteristics and lower blasting risk, and their coverage of approximately 30 to 40 per cent of the Solomon complex, these domains were selected for a drill and blast optimisation program aimed at improving cost efficiency. The Table 1 describes the CID and DID geological parameters of the two domains chosen to conduct this work and the baseline blast designs.

TABLE 1

DID and CID properties and the current blast designs – dry conditions.

Rock type	DID	CID
UCS (MPa) – average	40.8	26.6
Young's modulus (GPa) – average	38.9	27.4
Hole diameter (mm)	229	229
Burden (m)	6.5	6.5
Spacing (m)	7.5	7.5
Subdrill (m)	0.0	0.0
Charge length (m)	8.0	8.0
Stemming length (m)	4.0	4.0
Dry Explosive	ANFO	Heavy ANFO
Density (g/cc)	0.82	0.97
Powder factor (kg/m^3)	0.46	0.55

Given this operating context, digging performance was selected as the most reliable indicator of blasting outcomes at Solomon, as direct fragmentation measurements are not in place. Prior to this study, blast performance in the CID and DID domains consistently supported digging rates at or above planning targets, indicating that blasting was not a limiting constraint in the mining value chain. Table 2 summarises the 90-day moving average instantaneous digging rates for the entire mine recorded before program implementation across the Solomon complex.

TABLE 2

90-day moving average instantaneous digging rates by equipment type for the whole mine prior to the program implementation.

Equipment type	Instantaneous digging rate (t/h)
996B	8132
R9400	4481
R9250	3676
CAT 6040	3527

Overall, baseline performance indicated clear headroom to improve drill-and-blast cost efficiency in the CID and DID domains without destabilising downstream operations, as these domains were consistently compliant and supported digging rates at or above planning targets.

SOLUTION DEVELOPMENT APPROACH

The program was intentionally designed for a multi-stage rollout to manage implementation risk and build confidence through progressive validation before broader deployment. Its objective was to improve drill-and-blast cost efficiency at Solomon while preserving downstream loading performance and keeping blasts within specification. In practical terms, this meant pursuing measurable savings in explosives cost without impacting digging performance and generating fragment sizes over specification limits. It began by identifying and screening feasible cost levers, then translating the preferred concept into an initial blast design. This model-informed design was progressed to a proof of concept (POC) so the operational team could validate the expected field response under controlled, production-relevant conditions. The POC results were used to refine the first designs, ensuring that any required adjustments were before scaling deployment.

Once the concept performed as predicted, the program transitioned to extended trials to test robustness across normal -changing operational conditions (eg geology, charging conditions, and execution noise) and to quantify performance drivers with higher confidence. These trials ran for six months under standard operating conditions, providing a sufficiently large data set to separate signal from noise and confirm that the cost-efficiency gains could be achieved consistently prior to full production rollout.

The data set was consolidated and analysed through a consistent workflow that linked design inputs, blast outcomes, explosive consumption and digging performance. Different equipment types were considered to verify that the new designs maintained downstream compliance while delivering measurable savings. This data also served as input for further analysis to identify key performance indicators and provide insights for the implementation of the new designs.

EVALUATING DRILL AND BLAST COST LEVERS AND POTENTIAL SOLUTIONS

Different solutions were identified as potential pathways to lift drill and blast's cost efficiency, differing in investment requirements, operational disruption, and how directly they affect blasting outcomes and downstream processes. The main options were screened against Solomon's constraints before selecting the most practical pathway for detailed development.

For example, switching explosive types can change delivered energy, explosive mass per hole and VOD, influencing fragmentation and sometimes allowing pattern expansion or productivity gains (Wilkinson *et al,* 2015). Key operational requirements include new product availability, changes on MMUs and logistics, changes on quality control processes, and updated risk controls.

Another option is to change hole diameter and/or optimise the drill pattern. Adjusting burden, spacing and diameter can influence oversize and overall fragmentation and shift the cost balance between drilling and explosives (Leng *et al,* 2020). However, such changes also need to be considered in the context of wider technical and economic uncertainties affecting drilling and blasting performance, including operating cost and asset management factors (Abbaspour *et al,* 2018).

A third lever assessed was improving stemming efficiency and reducing charge per hole. Stemming governs confinement and gas retention at the blasthole collar, directly affecting energy retention and utilisation. It also plays a major role in top-size generation in the collar zone, which is one of the strongest disruptors of digging and crusher productivity. According to Monjezi *et al* (2023), airdecking is defined as the intentional creation of an air gap within the explosive column of a blasthole, located either at the bottom, middle, or top of the charge. Historically, the approach was used in surface blasting to achieve a more uniform distribution of explosive energy throughout the rock mass. More recently, based on theoretical and modelling work, showed by introducing an airdecks between the explosive charge and the stemming can improve explosive energy use effectiveness by changing the breakage mechanisms in the upper column and reducing upward loading on the stemming column (Lopes, Hawinkel and Furtney, 2025).

In the end, improving stemming efficiency through the introduction of an airdeck in the upper explosive column was selected as the preferred pathway for further development at Solomon. Compared with changes to explosive products or drilling geometry, airdecking could be deployed with minimal modification to established drill patterns, fleet utilisation, and short-term planning processes. Although it introduces an additional step during charging, this operational overhead was judged manageable relative to the scale of potential explosive reduction and the low disruption to production scheduling.

BLAST DESIGN DEVELOPMENT

To ensure that the new solution is feasible and efficient, advanced modelling tools were deployed to test several blast design alternatives considering airdecks. They were built on the insights established in Lopes, Hawinkel and Furtney (2025). In particular, the analysis leveraged three key mechanisms associated with the airdeck:

1. Reduction in inefficient work in the air column, which eliminates the shear stresses in the blasthole's surrounds, increasing the fragmented volume when compared to traditional stemming.
2. The replacement of stemming by air yields higher near field vibration levels, impacting fragmentation and fracturing.
3. Reduced upward pressure on the stemming column, improving confinement and lowering the likelihood of premature ejection and energy loss while still supporting rock breakage.

For each design analysed, model outputs were post-processed to quantify resultant vibration (reported as peak particle velocity or PPV) together with the corresponding change in upward pressure at the stemming column. Predicted PPVs were then benchmarked against the critical peak particle velocity (PPVc), the vibration threshold associated with the onset of fresh fracturing in intact rock. According to PPVc theory, PPV levels approaching the PPVc value are expected to initiate new fractures, while PPVs reaching approximately four times the PPVc threshold are typically sufficient to fragment the rock, forming the fractured and fragmented volumes respectively (McKenzie, 2016).

With drilling geometry and diameter unchanged several three designs were developed for each geology, based on the potential reduction of explosives and the SDoB guidelines (ISEE, 2011) and dry bench conditions. Each design was simulated and the results were assed accordingly to the resultant PPVs and the upward pressure, compared to the current designs (_cur). Table 3 details the main blast design parameters chosen and Figure 1 details the modelling outputs.

TABLE 3

Simulated blast designs for DID and CID.

	DID_cur	DID_1	DID_2	CID_cur	CID_1	CID_2
Rock type	DID	DID	DID	CID	CID	CID
Diameter (mm)	229	229	229	229	229	229
Burden (m)	6.5	6.5	6.5	6.5	6.5	6.5
Spacing (m)	7.5	7.5	7.5	7.5	7.5	7.5
Subdrill (m)	0.0	0.0	0.0	0.0	0.0	0.0
Charge (m)	8.0	6.5	7.5	8.0	6.5	7.5
Stemming length (m)	4.0	3.5	3.0	4.0	3.5	3.0
Airdeck (m)	0.0	2.0	1.5	0.0	2.0	1.5
Dry Explosive	ANFO	ANFO	ANFO	Heavy ANFO	Heavy ANFO	Heavy ANFO
Density (g/cc)	0.82	0.82	0.82	0.97	0.97	0.97
Powder factor (kg/m^3)	0.46	0.38	0.43	0.55	0.44	0.51

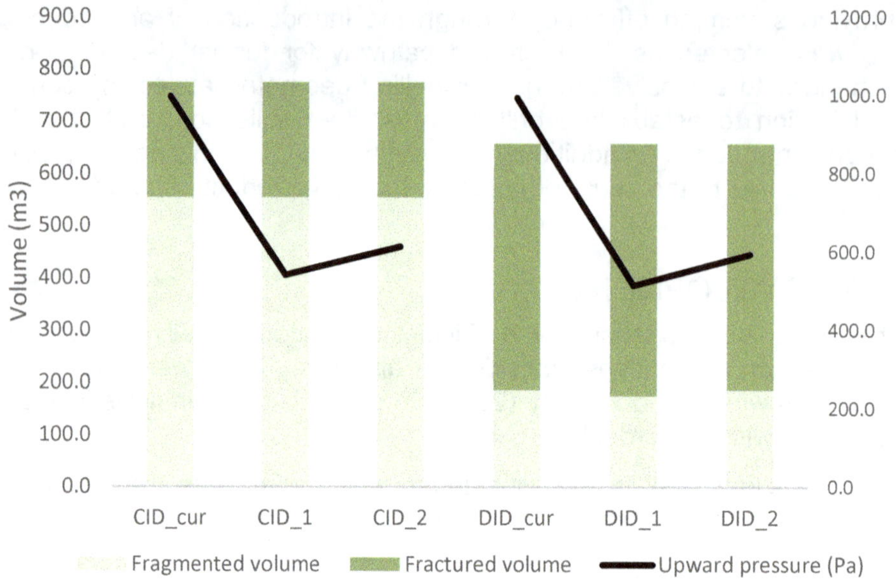

FIG 1 – Simulation outputs of the tested scenarios.

For the CID geology, the numerical model predicts essentially invariant breakage volumes across the designs: the fragmented volume is almost flat in 553.5 m^3 in all cases and the fractured volume remains unchanged. The similarity is CID's rock properties presenting a low UCS and Young's modulus and consequently, low PPVc. For the DID, the new designs will potentially reduce the fragmented volume by 6–7 per cent and the fractured volume by 2–3 per cent.

Another important analysis is the upward pressure at the stemming. The simulations indicate a sharp reduction ranging from 52 per cent (DID_1) to up to 61 per cent (CID_2) on the upward pressure on the stemming in the new blast designs. this reduction is expected to improve confinement, reduce the likelihood of stemming ejection and flyrock, ensuring the safety of the trials.

Although there is a reduction on the fragmented and fractured volumes in the DID, the simulations' outputs were considered within the acceptable range and provided enough evidence to move forward to a proof of concept to evaluate the impacts on fragmentation and the downstream processes.

FIELD VALIDATION PROGRAM

Proof of concept (POC) – testing modelling outcomes

Following the blast design development, the POC was planned as a controlled production trial to test whether introducing an airdeck could reduce explosive consumption while maintaining acceptable blast outcomes and digging performance. Rather than relying on isolated trial shots, the POC was structured as side-by-side polygons executed under comparable operating conditions to support a fair, production-relevant comparison.

POC 1 – first trial and operational limitations

The first trial (POC trial 1) was conducted in the DID and the intent was to test the DID_1 and DID_2 designs against the site current design (DID_cur) splitting the blast polygon in three regions. However, operational constraints imposed a mix of 203 mm and 229 mm hole diameters used across the polygons and some charge variations affecting the geometry and powder factor across the bench, which was recorded as a limitation for interpreting results.

No major issues in deploying the airdecks were reported and the blast was delivered safely. Although parts of the muck pile were reported as tighter, no issues with hard digging or excessive boulders were reported during excavation. POC 1 therefore confirmed safe and practical deployment but was not treated as the primary controlled comparison because of hole-diameter variability.

POC 2 – building confidence to scale trials

Given the diameter variability and associated comparability challenges at the first POC, a second trial (POC 2) was conducted to provide a more systematic assessment, with improved control of conditions and a single hole diameter aligned with the simulated designs. This POC 2 compared the current charge (DID_cur) against DID_2 charge profile, taking into account a smaller drill pattern using 5.6 m burden and 6.5 m spacing for both sides due to operational requirements for that polygon. In total, 2480 holes were blasted where 1594 considering DID_cur charge profile and 886 DID_2.

The blast designs were implemented successfully, and the powder factor was reduced by 6.5 per cent at the DID_2 polygon (from 0.62 kg/m^3 to 0.58 kg/m^3), Figure 2 illustrate the full blast area and Table 4 details the results in terms of the dig rates for both polygons.

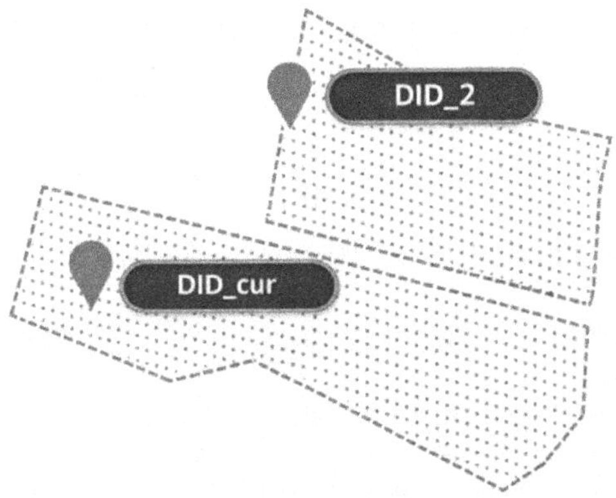

FIG 2 – POC 2 polygons illustrating the blasting area considered for each design.

TABLE 4

Instantaneous digging rates for the POC.

Flitch	DID_cur		DID_2	
	Dig rate (t/h)	Sum of payload (t)	Dig rate (t/h)	Sum of payload
Bottom	8165	550 000	7939	200 000
Middle	9030	580 000	9360	700 000
Top	8985	660 000	6714	70 000
Average*	8748	1790 000	8876	970 000

*Average values are payload-weighted. The lower DID_2 top-flitch rate should be interpreted with caution as the top-flitch payload was significantly smaller, 70 000 t, compared with 700 000 t in the middle flitch.

The average digging rates in the second trial were essentially equivalent, with a small uplift on the DID_2 side (~1.5 per cent). By flitch, DID_2 performed higher in the middle flitch and also carried a higher share of the total payload in that flitch, while DID_cur was higher in the bottom and top flitches. Overall, the differences observed across flitches are more impacted by operational and volume distribution between flitches and their effects than by the blast design changes, indicating broadly comparable blasting outcomes between the two designs.

The POC was successful for the purposes of scaling up the program. It demonstrated that the new designs could be deployed safely and practically while meeting expected fragmentation targets. Furthermore, the results observed between current practice and DID_2 at the second trial and no major issues at trial 1 provided confidence to proceed to extended trials aimed at confirming whether explosive reductions could be achieved without compromising downstream productivity. POC 2 therefore served as the primary controlled comparison, confirming a powder factor reduction with broadly comparable digging performance.

Extended trials

After the results of the POC, the program progressed to an extended trial designed to test the selected blast designs under a broad range of operating conditions and to quantify impacts on downstream loading performance. The objective was to confirm at production scale that the proposed designs (including the ones not trialled during the POC phase) and hence explosive and powder-factor reductions, could be achieved while maintaining acceptable blasting outcomes and digging performance. In total, the extended trials covered 22 production blasts: 23 381 blastholes and an estimated 11 043 414 m^3 of blasted rock over five months.

The trials intentionally spanned a wide variation of blast parameters considering airdecks (Table 5), including different geology compositions and operational conditions, such as water presence and explosive changes, contour holes with smaller diameters and charge adjustments due to implementation deviations, affecting both blasting execution and outcomes. This variability was crucial to build a relevant and robust data set to evaluate of whether the design behaved consistently beyond the initial controlled environments and under the usual operational conditions and challenges faced by Solomon's teams.

TABLE 5

Blast parameters applied during the extended trials.

Blast number	Geology	Diameter (mm)	#holes	Explosive	Stemming (m)	Airdeck (m)	Powder factor (kg/m^3)
1	DID	229	842	ANFO	4	2	0.45
2	CID	229	465	Emulsion	3.5	2.5	0.42
3	CID	229	354	Emulsion	3.5	2	0.41
4	CID	229	1165	Heavy ANFO	3.5	2	0.54
5	HC/CID	229	2029	Emulsion	3.5	1.5	0.72
6	DID	165/229	1412	ANFO	3.5	2	0.4
7	DID	165/229	1707	ANFO+Emulsion	3.5	2	0.44
8	DID/BID	165/229	911	ANFO+Emulsion	3.5	2	0.48
9	CID	229	332	Emulsion	3.5	2	0.39
10	CID	229	840	Heavy ANFO	3.5	2	0.44
11	DID	165/229	1146	ANFO	3.5	2	0.62
12	DID/HC	229	430	Emulsion	3	1.5	0.6
13	DID/HC	229/203	944	ANFO/Emulsion	3	1.5	0.66
14	DID	229	1135	ANFO	3.5	2	0.4
15	DID	229	942	ANFO	3.5	2	0.4
16	DID	229	1163	ANFO+Emulsion	3.5	2	0.46
17	DID	229	841	ANFO+Emulsion	3.5	2	0.35
18	DID/HC	229	1830	Heavy ANFO	3.5	2.5	0.59
19	DID	229	1670	ANFO	3.5	2	0.4
20	DID	229	424	ANFO	3.5	2	0.4
21	DID	229	905	ANFO	3.5	2	0.4
22	DID/BID	140/229	1894	Heavy ANFO +Emulsion	3	2	0.99

Performance data were sourced from instantaneous digging rates recorded in the site dispatch system and loading sheets, complemented by blast crew reports based on drone footage and visual inspections. These reports captured key post-blast observations, including oversize boulders and muck pile heave. Together, these inputs were the basis for the key performance indicators assessed in this study: digging-rate performance relative to planning targets and baseline conditions, variability by equipment type and flitch, explosive consumption and cost relative to the baseline blast design, and out of specification fragmentation.

These data were analysed at blast level through a structured workflow to generate a consistent KPI set for each blast, linking blast design parameters, post-blast observations and downstream loading results, as follows:

- Instantaneous digging rates by equipment: for each loading unit calculated as: instant digging rate (t/h) = total mass dug/cycle time and then aggregated to an average rate per equipment class/unit over the trial data set.
- Instantaneous digging rates by flitch and equipment: the same metric was calculated within individual flitches for each equipment unit according to the excavation strategy used.
- Comparison against planning targets: results were benchmarked against the planning targets defined for each equipment class to quantify relative performance to the entire mine and the baseline target to allow direct cross-unit comparison.
- Distribution and deviation analysis: rate distributions were evaluated to understand variability and consistency, including the proportion of observations above/below expected performance and the tendency for positive or negative deviations.
- Explosive consumption: explosive consumption was obtained from consolidated charging sheets extracted from the mine's system. These recorded quantities were then compared against an estimated baseline consumption calculated for the current designs.
- Out of specification fragmentation: any block over 1 m × 1 m × 1 m cannot be fed into the crusher and the appearance of these were recorded and represent poor fragmentation/performance of the blast design.

Crushing productivity was not assessed in this study because although it would be a relevant downstream comparison in principle, ROM's flow at Solomon does not provide a direct one-to-one link between an individual blast and crusher feed performance.

Review of the post-blast reports indicated safe deployment of the airdecking technique and no safety incidents during the trial period as no flyrock events were recorded, and no instances of excessive ejections or heave were reported. All blasts remained within site specification limits, with no out of specification fragmentation across the data set. Table 6 summarises the high-level results of the extended trials.

TABLE 6

High level results of the extended trials.

Number of blasts	22	**Total explosive saved (t)**	1039
Total blastholes	23 381	**Overall powder factor (kg/m^3)**	0.49
Out of specification fragmentation (#boulders)	0	**Powder factor/explosive reduction (%)**	15.9%
Expected explosive consumption (t)	6543	**CO_2-e reduction (t)**	196.4
Expected overall powder factor (kg/m^3)	0.59	**Total stemming saved (t)**	1053
Total explosive consumed (t)	5504	**Overall cost reduction (%)**	12%

Overall, the average digging rate during the trials was approximately 5 per cent higher than the post-trial planning targets, although the uplift varied by equipment class. The strongest result was recorded for the CAT 6040 fleet, which performed about 10 per cent above target. In most cases, both the trial-average instantaneous digging rates and the planning targets were broadly consistent with, or slightly above, the pre-trial target rates. The only exception was the CAT 6040, for which the pre-trial target was comparatively low and likely influenced by site-specific operating constraints and conditions. Table 7 summarises the consolidated results by equipment class and compares them with the post-implementation planning targets.

TABLE 7

Instantaneous digging rate by equipment class against planning target – extended trials.

	Average (t/h)	Target post (t/h)	Variation	Target pre (t/h)	Variation (targets)
996B	8256	7904	+4%	8132	-3%
CAT 6040	4889	4431	+10%	3527	+25%
R9250	3924	3748	+5%	3676	+1.9%
R9400	4808	4523	+6%	4481	+0.9%

At the flitch scale, performance remained above target in nearly all cases: all equipment and flitch combinations exceeded target slightly except two: bottom flitch for R9250 and top flitch for R9400. Figure 3 illustrates the instantaneous digging rates per equipment class per flitch.

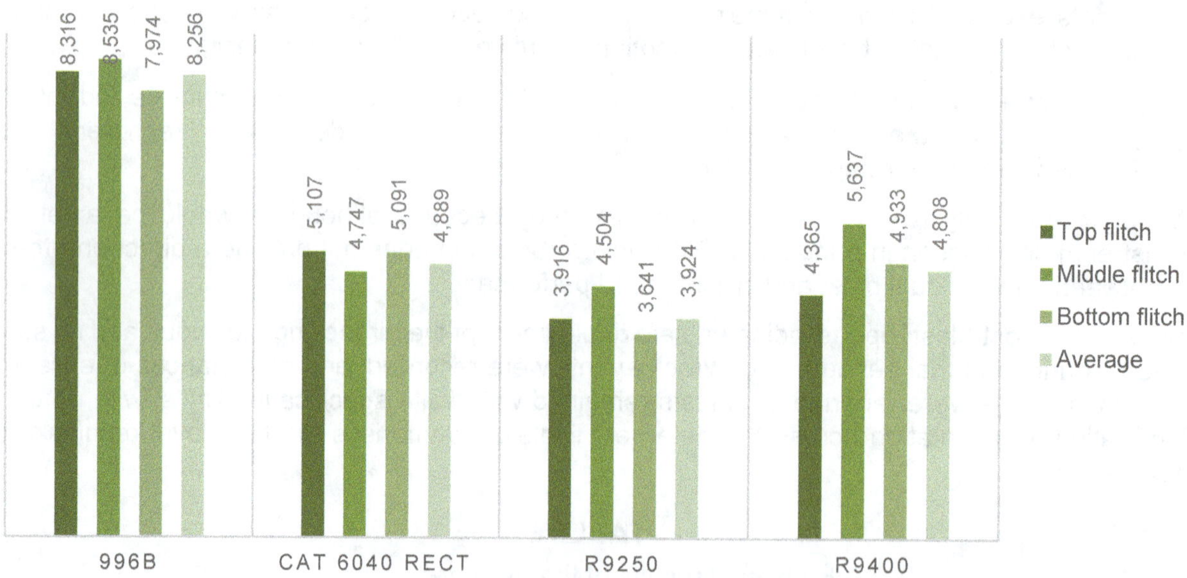

FIG 3 – Instantaneous digging rate (t/h) by equipment class and flitch – extended trials.

KEY PERFORMANCE FACTORS

To identify the primary drivers of downstream loading performance and provide insights after the full roll out of the new practices, overall instantaneous digging rates per blast were calculated within each equipment class and converted to a unitless 1 to 10 score, where 1 corresponds to the lowest digging rate recorded for that equipment across all analysed blasts and 10 to the highest. This scoring system was used to normalise the digging rates allowing comparisons between the blasts analysed.

The digging-rate score was then compared against blast design inputs (powder factor, explosive type, powder factor, blast size/number of blastholes and airdeck length), predominant geology domain (DID versus CID), and blast outcomes (vertical heave and ejection). The comparisons were also repeated while splitting the data by geology and muck pile heave/ejection profiles, to check whether outcome effects changed by domain. Tables 8 and 9 detail the results obtained.

TABLE 8

Digging scores (1= lowest digging rate, 10 = highest) per equipment class against parameters and outcomes.

Parameter	996B	CAT 6040	R9250	R9400
DID	7.2	6.8	5.3	4.4
CID	3.9	4.1	3.9	2.2
ANFO	7.3	6.0	5.5	5.2
Emulsion	6.4	6.0	5.0	2.8
PF <0.5	6.8	5.1	6.1	4.1
PF >0.5	6.7	7.3	3.9	3.4
<1000 holes	7.0	6.1	6.9	3.5
>1000 holes	6.8	5.9	3.8	4.2
1.5 m airdeck	6.8	7.6	3	2.5
2 m airdeck	7	5.3	5.4	4.2
2.5 m airdeck	9	3.6	7.7	3.5
Low ejection	7	5.8	6	4.1
Aggressive ejection	8.3	6.7	2.1	1
Great heave	7.6	6.5	4.2	3.4
Low heave	6	5.1	7	7

TABLE 9

Digging scores per equipment class, geology, heave and ejection.

	Parameter	996B	CAT 6040	R9250	R9400
Overall	Great heave	7	6.8	2.3	3.3
	Low heave	6	5.1	7	4.8
DID	Great heave	8.3	6.8	1.7	3.8
	Low heave	6.3	6.6	7	5.4
CID	Great heave	3.8	6.8	3.9	2.5
	Low heave	4	2.9	-	1.9
Overall	Low ejection	7	5.8	6	4.1
	Aggressive ejection	8.3	6.7	2.1	1
DID	Low ejection	7	6.7	6.6	5.1
	Aggressive ejection	8.3	6.7	1.7	1
CID	Low ejection	3.9	4.1	3.9	2.2
	Aggressive ejection	n/a	n/a	n/a	n/a

The results indicate that the relationship between muck pile movement (heave and ejections) and loading performance is equipment dependent. In DID blasts, strong heave and more aggressive ejections corresponded with better digging outcomes for the 996B and CAT 6040, yet this same behaviour did not translate to the R9250 and R9400, which performed better under more moderate heave conditions. This divergence is practically significant: a single 'optimal' muck pile movement is unlikely to suit all loading units, and the best performing blast may depend on the equipment working the material.

Explosive type also shows a repeatable association with digging performance. ANFO is linked to higher digging rates than emulsion for every equipment class assessed, with the strongest uplift observed for the 996B and R9250. While the trials do not isolate causality, the consistency across equipment class indicates that explosive selection remains a meaningful lever in digging performance, usually associated with muck pile swell and heave.

For air-deck length, the current data set is not yet sufficient to draw firm conclusions at the longer lengths tested: only three blasts at 1.5 m and two at 2.5 m are available. Given the small sample size, any apparent trends would be highly sensitive to geology, blast size, and operational variability. Additionally, it wasn't possible to evaluate how aggressive ejection plays in the CID as no blasts were recorded with aggressive ejection in CID.

DISCUSSION

The staged, model led approach adopted in this program proved effective for translating a theoretical optimisation concept into a practical production outcome. By progressing from numerical modelling to proof of concept, extended trials and finally production rollout, the program reduced implementation risk while generating a sufficiently robust data set to assess both technical performance and operational feasibility. This validation logic is important to the interpretation of the results as the POC established feasibility and comparable controlled performance, and the extended trials demonstrated that those outcomes could be sustained under normal production variability.

A key indicator of the scale of improvement achieved is the 15.9 per cent reduction in powder factor recorded during the extended trials. At production scale, this represents a material reduction in explosive input per unit volume, equivalent to 1039 t of explosive saved, an estimated 12 per cent reduction in blasting cost, and 196.4 t less explosives-related CO_2-e emissions. Importantly, these savings were achieved without generating out of specification fragmentation and safety issues.

The close alignment between trial-average instantaneous digging rates and planning targets, together with overall digging performance remaining approximately 5 per cent above the post-trial targets, shows that the reduced-charge designs preserved loading productivity at production scale. Importantly, the post-trial planning targets were, for most equipment classes, broadly consistent with or slightly above the pre-trial targets, which means the trial results cannot be explained by a relaxation of benchmark expectations. In practical terms, this shows that the reduced-charge designs preserved loading performance relative to both historical operating conditions and current planning requirements.

At flitch level, performance was also generally maintained above target, with only two exceptions across the data set: the bottom flitch for the R9250 and the top flitch for the R9400. This is particularly relevant for interpretation of the top-flitch results, because the upper portion of the bench was the zone most directly modified by the introduction of the airdeck. Although one top-flitch exception was observed for the R9400, the absence of a broader top-flitch penalty across the data set provides an important cross-check on the overall digging-rate results, indicating that the changes made to the upper charge column did not produce a systematic loss of loading performance. Taken together, these outcomes provide strong evidence that the air-decked designs improved blast cost efficiency without materially compromising the established balance of the mining value chain, and they are supported by a large, operationally representative data set that provides a robust basis for these findings.

Furthermore, the observed field response was consistent with the program intent and underlying design rationale. Placement of an airdeck within the upper charge column was intended to improve stemming efficiency and reduce inefficient energy losses near the collar, thereby enabling a lower charge per hole. In this study, support for that mechanism is indirect, reflected in maintained overall and flitch-level digging performance together with the absence of out of specification fragmentation, rather than in direct fragmentation measurements. Even so, the field response is consistent with more efficient use of explosive energy, in line with the modelling basis and previous work by Lopes, Hawinkel and Furtney (2025).

A key outcome of the extended trials is the confirmation that geological domain is a significant input while determining blast design parameters and the consequent loading productivity. Blasts in the DID domain generally returned higher digging-rate performance across the fleet, whereas CID blasts were consistently lower for all machines, with the largest excavators most affected. This domain dependence justified segregating the analysis by geology: pooling CID and DID would mask systematic domain-driven differences, inflate apparent scatter, and increase the likelihood of misattributing natural geological variability to blast design changes.

Two operational implications follow. First, performance interpretation and benchmark target rates should be defined within geological domains prior to evaluation of secondary factors such as explosive type, powder factor, air-deck length, blast size, or observed muck pile. Second, tuning should explicitly consider equipment-dependent sensitivity to muck pile movement. The relationship between heave/ejection and digging performance is not uniform across a mixed fleet; consequently, a single 'optimal' muck pile is unlikely to maximise performance for all loading units. In summary, this supports a domain-based, fleet-aware tuning approach.

A further outcome of the program was confirmation of implementation feasibility at production scale. Incorporating an airdeck within the stemming column was executed through standard charging practice, with low additional operational effort and limited change-management requirements. The approach did not require changes to drilling geometry or downstream operating practices, and no disruptions to the mining value-chain were apparent, as the monitored downstream loading metrics remained within planning targets throughout the production rollout.

Key limitations should be noted. The conclusions are grounded in the domains and operating envelope represented in the data set. Downstream assessment also relies on instantaneous digging rate as the primary compliance metric, it does not directly quantify fragmentation distributions, crusher metrics, or full-cycle productivity effects. Future work should extend controlled rollout into additional domains and incorporate complementary fragmentation or downstream process indicators to refine tuning and broaden transferability.

CONCLUSIONS

This work documents a staged, model led improvement program at Solomon that translated air-deck mechanics into production-ready charge designs and validated them through progressive field testing, from early trials to a production rollout supported by a consolidated data workflow. In practical terms, the POC confirmed deployability and comparable controlled performance, while the extended trials confirmed that these outcomes were repeatable under production-scale operating variability.

The program aimed to make its operations more cost efficient and resilient given the changing state of the iron ore industry. This study established a replicable and robust workflow for developing and implementing air-decked blast designs aimed at improving cost efficiency. The work was intentionally structured as a staged program so that the selected solution could be tested in varied operational conditions and its implications could be assessed with confidence.

The extended trials delivered 1039 t of explosive savings, corresponding to a 15.9 per cent reduction in powder factor, an estimated 12 per cent reduction in blasting cost, and 196.4 t less explosives-related CO_2-e emissions. Downstream loading compliance was maintained throughout the production data set, with fleet-average digging rates remaining at or above planning targets and no out of specification fragmentation and safety incidents recorded.

Beyond demonstrating that downstream loading performance was maintained, the study enabled separation of systematic effects from short-term operational variability. By structuring the analysis across equipment, flitch, geology, and blast design attributes, the data set supported identification of the dominant contributors to digging-rate response and reduced the risk of attributing normal operating scatter to design changes.

The reliability of the conclusions is underpinned by the trial's scale, the robustness of the data set and by a consistent analytical approach that integrated design parameters, blast outcome observations, and downstream digging performance against planning targets.

Overall, the program was comprehensive enough to justify full implementation in the geological domains studied, as it closed the loop from design intent to production verification, while also building an evidence base on why performance behaves the way it does, supporting previous studies on airdecking applications. For Solomon, this means lower explosive consumption, lower explosives-related CO_2-e emissions and a material reduction in blasting costs without disturbing the existing mining value chain balance.

REFERENCES

Abbaspour, H, Drebenstedt, C, Badroddin, M and Maghaminik, A, 2018. Optimized design of drilling and blasting operations in open pit mines under technical and economic uncertainties by system dynamic modelling, *International Journal of Mining Science and Technology*, 28(6):839–848.

Brunton, I D, Thornton, D M, Hodson, R and Sprott, D, 2003. Impact of blast fragmentation on hydraulic excavator dig time, in *Proceedings Fifth Large Open Pit Mining Conference*, pp 39–48 (The Australasian Institute of Mining and Metallurgy: Melbourne).

Doktan, M, 2001. Impact of blast fragmentation on truck shovel fleet performance, in *Proceedings 17th International Mining Congress and Exhibition of Turkey – IMCET 2001*, pp 375–379.

Ernst & Young (EY), 2025. Top 10 business risks and opportunities for mining and metals in 2026, EY, 16 October. [Accessed: 29 May 2026].

Fortescue, 2022. FY22 Mineral Resources and Ore Reserves Update [online], ASX Market Update. Available from: <https://www.marketindex.com.au/asx/fmg/announcements/fy22-mineral-resources-and-ore-reserves-update-6A1106307>

Fortescue, 2026. Our operations [online], Fortescue Ltd. Available from: <https://www.fortescue.com/en/what-we-do/our-operations/Solomon>

International Society of Explosives Engineers (ISEE), 2011. *Blasters' Handbook*, 18th edn, International Society of Explosives Engineers.

Leng, Z, Fan, Y, Gao, Q and Hu, Y, 2020. Evaluation and optimization of blasting approaches to reducing oversize boulders and toes in open pit mine, *International Journal of Mining Science and Technology*, 30(3):373–380.

Lopes, G, Hawinkel, J and Furtney, J, 2025. Using Air Decks To Achieve Superior Blast Outcomes – The Mechanics Behind The Air Decking, in *Fragblast – Proceedings of the 14th International Symposium on Rock Fragmentation by Blasting*, pp 489–495.

McKenzie, C, 2016. Blasting near open pit walls, in *APSSIM 2016: Proceedings of the First Asia Pacific Slope Stability in Mining Conference* (ed: P M Dight), pp 83–94 (Australian Centre for Geomechanics: Perth).

Monjezi, M, Amiri, H, Seyed Mousavi, M N, Khademi Hamidi, J and Khandelwal, M, 2023. Comparison and application of top and bottom air decks to improve blasting operations, *AIMS Geosciences*, 9(1):16–33.

Trading Economics, 2026. Iron Ore [online], Trading Economics. Available from: <https://tradingeconomics.com/commodity/iron-ore>

Wilkinson, D, Rigby, G, Norgard, J and Thomson, S, 2015. Higher Energy Bulk Explosives – Matching Products to Rock Types Using an Energy Map Concept, in *Proceedings of the 11th International Symposium on Rock Fragmentation by Blasting*, pp 579–584 (The Australasian Institute of Mining and Metallurgy: Melbourne).

Breaking boundaries – novel approach to pre-concentration of low-grade BIF from South Africa

A Singh[1] and N Maistry[2]

1. Senior Technical Specialist, Mintek, Johannesburg, South Africa. Email: ashmas@mintek.co.za
2. Senior Engineer, Mintek, Johannesburg, South Africa. Email: nicholem@mintek.co.za

ABSTRACT

Iron (Fe) ore is recognised as a critical strategic commodity for South Africa (SA) with SA ranking as the seventh largest producer and third largest exporter. The life-of-mine (LOM) forecast for primary Fe ore production in SA suggests that high-grade material will be depleted within the next decade in the absence of further exploration and discoveries. Thus, with the depletion of high-grade iron ore resources, there has been a drive to explore low-grade hematite resources in a bid to extend life-of-mine (LOM) and maximise resource utilisation. Banded Iron Formations (BIF) makes up approximately two thirds of future reserves in SA. At low Fe head grades (<40 per cent Fe) overall, Fe product grades and recoveries are compromised. In most instances due to liberation of valuable mineral from gangue, fine grinding (<1 mm) is required. The technologies applicable for pre-concentration are also limited at finer liberation grind sizes. Innovative research leading to a practical demonstration of concept is vital to the future of mining within the iron ore industry, especially due to imminent changes in orebodies requiring beneficiation and overall production cost concerns. Thus, the aim of the research study involved production of a pre-concentrate grading at approximately 55 per cent Fe with potential for alternative methods such as cationic reverse flotation to be tested on the pre-concentrate for further upgrading to >63 per cent Fe.

The industrial benefit of pre-concentration is evident in upfront gangue reduction and in its ability to improve production capacity. In the case of the current study, >60 per cent mass can be rejected upfront ensuring more valuable material and hence an upgraded product is processed through the downstream plant resulting in smaller re-circulating loads and buildup. This results in a reduction in process costs ensuring that the upgrading process is economically feasible. Research included a comparative analysis of gravity separation versus magnetic separation, the latter involving: (i) medium intensity magnetic separation to recover paramagnetic hematite; versus (ii) conversion of hematite to magnetite via roasting and magnetic separation on a low intensity magnet to recover converted paramagnetic minerals.

INTRODUCTION

The iron ore industry in South Africa is facing a confluence of challenges with declining grades, high operational cost margins, skills shortage and volatility in pricing acerbating the crisis in the industry. Banded iron formation (BIF) represents the future life-of-mine (LOM), however declining grades pose processing challenges with efficiency of upgrade and recovery being compromised (Eriksson, Hattingh and Altermann, 1995; Eriksson *et al,* 2004; Creamer Media, 2020).

The current research aims to beneficiate low-grade BIF in an attempt to extend LOM and maximise resource utilisation (Basson *et al,* 2018). Modifying magnetic susceptibility of hematite via reduction roasting to magnetite is potentially a novel approach to pre-concentrate low-grade material to a feasible grade for further downstream processing. Magnetic separation is thus explored assessing the upgrade potential prior to and after roasting (Wills and Napier-Munn, 2005).

MATERIALS AND METHODS

Sample characterisation – head analysis and mineralogical evaluation

Approximately 200 kg of low-grade feed material at a top size of 100 per cent passing 30 mm was utilised for research purposes. The sample was stage crushed to 100 per cent passing 1.7 mm and thereafter screened at 1.18 mm with the oversize further crushed to -1.18 mm, this was done to avoid excessive generation of ultrafines/slimes.

Representative sub-samples were removed for head grade analysis, particle size distribution analysis, mineralogical modal characterisation and magnetic separation test work with and without roasting.

Roasting

Test work was conducted to determine the feasibility of roasting low-grade hematite using H_2 gas as the reductant, towards decarbonisation of the downstream smelting process. The objective of the magnetising roasting test work was to selectively and partially reduce hematite to magnetite to increase susceptibility to magnetic separation. The roasting tests were completed using a fluidised bed furnace (Spreitzer and Schenk, 2020). The scope of work included: (i) thermochemical simulations for the determination of the magnetising roasting temperature; (ii) cold tests to determine the minimum fluidisation gas velocity of a 200 g batch of low-grade hematite feed sample using argon (Ar) gas; and (iii) roasting of hematite using H_2 gas (Tawane and Goso, 2024).

Initially cold tests were completed for the determination of the fluid dynamics of the concentrates using Ar gas. The tests assisted with determining the minimum flow rate at which the solid particles would start fluidising. The cold test involved weighing a sample, carefully placing it in a dedicated quartz reactor, and placing the charged reactor inside a cold furnace. A flow of Ar gas was slowly and carefully opened allowing it to flow-through the cold sample until the sample started fluidising. The minimum flow rate for fluidising the concentrate was determined. This is the total flow rate that was used in the test, which was also achieved by mixing Ar and H_2. The gas distribution used in the test was 50/50.

A laboratory-scale fluidised bed furnace set-up was used to complete the roasting test work as represented in Figure 1. The set-up is made up of a box furnace with SiC heating elements and equipped with a vertically configured 80 mm quartz reactor. The reactor has a porous frit that is made from SiO_2. During test work, the porous frit holds the solid sample and acts as a distributor of the reaction gas. The reactor is equipped with a K-type thermocouple that is positioned such that it sits in the sample bed for continuous monitoring of the sample temperature throughout the test duration. The outlet of the reactor is connected to a scrubbing system through which the gas emissions pass before they are released into the atmosphere.

FIG 1 – Laboratory scale fluidised bed furnace set-up at Mintek, South Africa.

Three roasting tests were completed. Each test involved the weighing of 200 g mass of hematite. The weighed sample was transferred into the quartz reactor, which was subsequently placed in the box furnace. The sample inside the quartz reactor was fluidised throughout the heating cycle using Ar. The furnace was heated to 500°C at 10°C/min. Once the furnace and the sample reached the operation temperature, the Ar flow was reduced to the calculated ratio, then H_2 was introduced into the system. The introduction of the H_2 reflects the start of the experiment. The gases were allowed to flow in the system at the test temperature until the test duration of 2 hrs elapsed. At the end of the test, only H_2 gas was closed. The sample was allowed to cool down under fluidisation using the Ar atmosphere inside the furnace. The cooled roast products were utilised for subsequent magnetic separation tests.

Magnetic separation test work

Davis tube recovery (DTR) tests

Davis Tube Recovery (DTR) tests were attempted on the -1.18 mm feed sample 'as is' at 1, 2 and maximum 2.3 A. Magnetics were not obtained even at the highest intensity of 2.3 A. The feed was then run through a matrix of SLon tests at higher intensities.

The roasted feed underwent DTR tests at three intensities namely 0.6, 0.7 and 1 A. At 1 A, a significant proportion of the mass reported as magnetics (>85 per cent mass) and hence lower current settings of 0.6 and 0.7 A were tested. The DTR magnetic field strength (G) to current (A) conversion is represented in Table 1.

TABLE 1

DTR current (A) to magnetic field strength (G) conversion.

Intensity (Amps)	Gauss strength (Gauss)
0.6	1380
0.7	1630
1.0	1990
2.0	3870
2.3	4100

A Davis Tube is classified as a laboratory magnetic separation unit, used to characterise the magnetic response of materials to various, magnetic field strengths. The Davis Tube is a laboratory magnetic separator that is always operated wet. The apparatus consists of an electromagnet with pointed poles and a glass tube with the feed inlet at the top and the outlet for the non-magnetic fraction at its end.

The tube is placed between the poles of the magnet, and it agitates the water flowing through with short, vertical strokes. The apparatus is illustrated in Figure 2.

FIG 2 – Davis Tube.

The electromagnet attracts strongly magnetic material. The sample passes down the water-filled inclined tube and then encounters the magnetic field. This is the point between the two poles of the electromagnet whereby the magnetic material is held at this point and the gangue (non-magnetic material) is washed out. The magnetic field that the slurry will encounter on a Davis Tube can be varied by varying the current (A). The tube oscillates and the non-magnetic material is washed down the tube with a gentle flow of water. Once the non-magnetic fraction has been collected, the current is switched off and the magnetic fraction is collected.

The resulting magnetics and non-magnetic product samples are dried, weighed and prepared for chemical analysis.

SLon wet high intensity magnetic separation (WHIMS) on feed 'as is' sample

The feed sample being paramagnetic was processed initially through a Davis Tube however, the maximum achievable magnetic field strength was 4100 G which did not yield any magnetics. The sample was then processed through a matrix of SLon magnetic separation tests varying the pulse amplitude, magnetic field intensity and pulsation frequency.

The laboratory-scale SLon is a magnetic separator that separates paramagnetic minerals from gangue or other non-magnetic minerals. In this case, the SLon was used to recover hematite to the magnetic stream. The laboratory scale SLon operates in a similar manner to the Wet High Intensity Magnetic Separator (WHIMS) with one distinguishing factor, namely, the incorporation of a pulsating mechanism. The pulsating mechanism (pulsating frequency and amplitude) drives the slurry up and down within the separating zone. As such, the particles are maintained in a loose state in the separation zone. This facilitates the retention of magnetic particles by the matrix. Conversely, non-magnetic particles are more easily washed through the matrix. Figure 3 represents the SLon-100 magnetic separator utilised for test work.

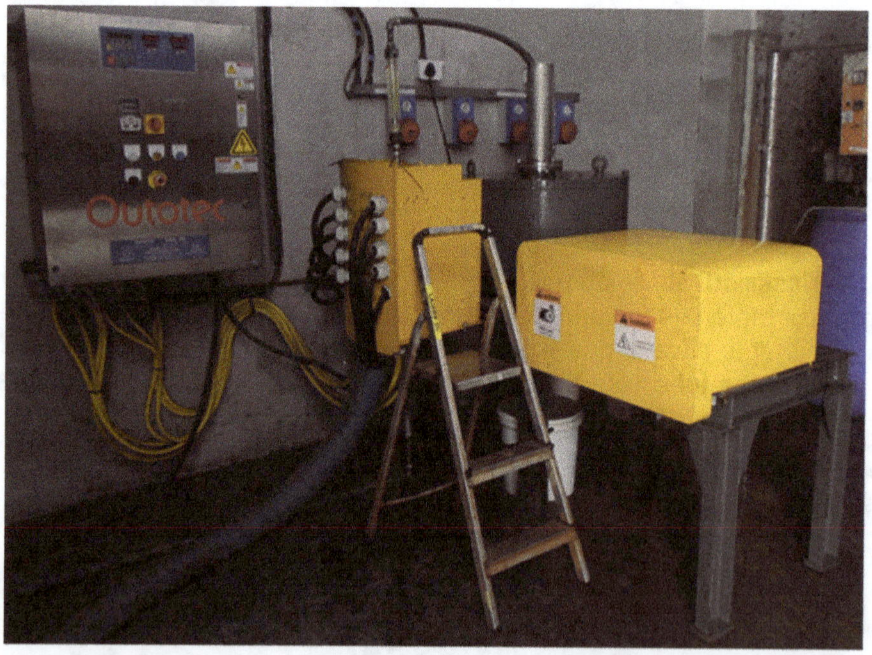

FIG 3 – SLon-100 magnetic separator at Mintek, South Africa.

Eight 100 g sub-samples of the -1.18 mm feed sample were prepared for laboratory scale SLon rougher optimisation test work. Table 2 indicates the parameters considered during the optimisation testing as per historical testing at Mintek (Singh and da Corte, 2019).

TABLE 2

Optimisation parameters for SLon tests.

Test	Pulse amplitude (mm)	Intensity (Gauss)	Pulse frequency (Hz)
1	5	6000	28
	10	6000	28
3	5	8000	28
8	10	8000	28
5	5	6000	35
6	10	6000	35
7	5	8000	35
8	10	8000	35

Fixed operating parameters included the matrix type, cooling water pressure and feed water flow rate. The pulse amplitude and matrix rod size were selected based on the top size of feed, whilst the remaining variables tested remained the same. The fixed operating conditions and their ranges are summarised in Table 3.

TABLE 3

SLon fixed operating conditions.

Matrix type	Rods
Matrix rod diameter (mm)	2
Cooling water pressure (psi)	0.1
Slurry flow rate (L/hr)	600

Wet low intensity magnetic separation (WLIMS) on roasted feed

Wet Low Intensity Magnetic Separators (WLIMS) are used primarily to remove ferromagnetic material such as magnetite (Fe_3O_4) from material with low magnetic susceptibility. WLIMS typically utilises a drum with permanent magnets which generate a magnetic field of approximately 1000 Gs at the surface of the drum. The drum rotates and the magnetic fraction that adheres to the drum will move either co-currently, or counter-currently, with reference to the feed, and are removed with a scraper from the surface of the drum opposite from the feed in an area where the magnetic attraction ends as in Figure 4. Co-current operation is most effective for producing an extremely clean magnetic concentrate and is widely used in both dense media and ore concentration systems.

Counter-current operation which was applied as in Figure 5 is desirable when magnetic losses are to be minimised, and a less clean concentrate is satisfactory to maximise recovery and produce a pre-concentrate for further upgrade. Feed is introduced at the opposite end to the magnetic discharge, and flows in the direction of drum rotation. Magnetic material is collected by the magnetic element and is agitated to free entrapped non-magnetic particles as the material is transported by the revolving drum surface. Non-magnetic material is discharged at the bottom through the tailings discharge port. The magnetic particles are held up to the drum surface and passed through an opening between the drum shell and the squeeze pan where excess water is forced out. The magnetic material is discharged at the concentrate discharge lip.

FIG 4 – Laboratory WLIMS unit at Mintek, South Africa.

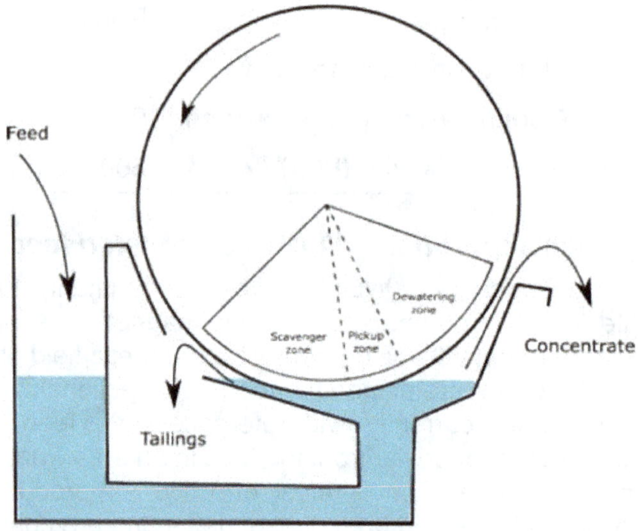

FIG 5 – Schematic view of counter-current WLIMS.

Mineralogy methods

Petrography of the crushed feed sample was conducted and comprised of scanning electron microscopy (SEM) and Electron microprobe analysis (EMPA) techniques to determine the bulk modal mineralogy, hematite chemistry, as well as hematite grain size, liberation and mineral association.

The data obtained was validated against the bulk chemical composition of the sample for quality control and assurance.

Representative polished sections were prepared of the feed for electron microprobe analysis. The feed was screened into different size fractions, and the fractions were mounted into polished sections for analysis by QEMSCAN.

Chemical analysis

The head sample and test work products were prepared and submitted for chemical analysis via Inductively Coupled Plasma – Optical Emission Spectrometry (ICP-OES) analysis as shown in Table 4.

TABLE 4

Description of analytical methods for solid samples.

Analysis requirements	Analysis method	Description	Elements	Elemental detection limits
Fe_Ore ICP	ICP	Fusino followed by acid dissolution in HCL/HNO_3	Al, Ca, Cr, Cu, Fe, Mg, Mn, Si, Ti, V, Zn	0.0005
LOI	Wet chem-loss on ignition @ 900°C	LOI – determined gravimetrically by weighing a sample portion before and after heating at 900°C (or agreed temp) to constant mass	LOI	N/A

The head samples were also submitted to an alternative laboratory for XRF analysis as XRF analysis was the preferred analytical method. In addition, this was used as a quality check for the ICP method.

RESULTS AND DISCUSSION

Head grade analysis of 'as is' feed

The head grade of the feed reported at 32.9 per cent Fe, 17.2 per cent Si, 4.8 per cent Al, 0.20 per cent Mg, 0.11 per cent Mn and 0.38 per cent Ca. Figure 6 represents the crushed feed particle size distribution (PSD) prepared in duplicate for test work purposes as well as mineralogical evaluation. The PSD analysis conducted in duplicate correlate well with the curves superimposed on each other.

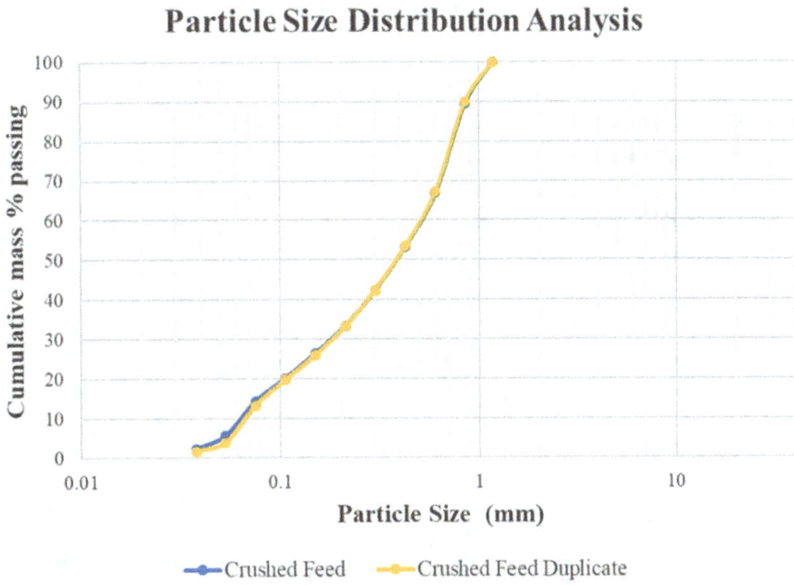

FIG 6 – Crushed feed particle size distribution.

Bulk model mineralogy

The low-grade feed sample consists predominantly of hematite (49.6 mass per cent) with illite and quartz as the major gangue minerals present (Table 5). Illite is a clay mineral contained within the mica group, with a lattice structure that does not expand on absorption of water. Other minerals are present in trace amounts (<5 mass per cent) and include mostly kaolinite.

TABLE 5

Bulk modal mineralogy of the crushed feed sample.

Mineral	Chemical formula	Mass %
Hematite	Fe_2O_3	46.9
Illite	(K,H_3O) $(Al,Mg,Fe)_2(Si,Al)_4O_{10}[(OH)_2,(H_2O)]$	23.0
Quartz	SiO_2	19.6
Kaolinite	$Al_2Si_2O_5(OH)_4$	5.5
Calcite	$CaCO_3$	0.5
Dolomite	$CaMg(CO_3)_2$	0.1
Barite	$BaSO_4$	0.2
Other oxides	-	1.0
Other Silicates	-	0.6
Total		100.0

Other oxides include rutile and ilmenite. Other silicates include pyroxene, amphibole, and chlorite however both are in trace amounts.

Magnetic separation test work on 'as is' feed

Davis Tube tests performed on the 'as is' -1.18 mm feed sample at 1, 2 and maximum 2.3 A did not yield any magnetics as hematite is paramagnetic requiring higher intensities for separation, thus the feed 'as is' was only processed through the SLon wet high intensity magnetic separator.

SLon wet high intensity magnetic separation optimisation

Eight SLon optimisation tests, using the equipment settings in Table 2, were conducted and the results are shown in Figure 7. A calibration curve was developed from assays and specific gravity values of product and feed streams and used to verify the grades with the optimum test result (Test 6) further verified by chemical analysis (Table 6).

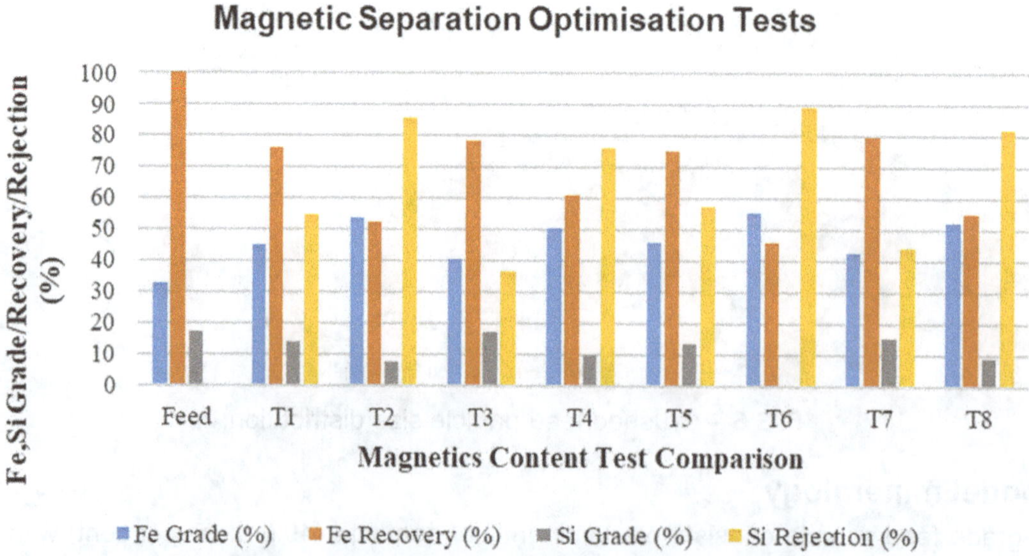

FIG 7 – SLon magnetic separation test results.

TABLE 6

Results of optimum SLon magnetic separation test.

Optimum test	Stream	Mass yield (%)	Grade (%)		Recovery (%)	
			Fe	Si	Fe	Si
Test 6	Mags	27.35	55.35	6.75	46.03	10.74
	Non-mags	72.65	24.43	21.12	53.97	89.26
Total measured (%)		100.00	32.89	17.19	100.00	100.00

Low pulse amplitudes of 5 mm does not yield Fe upgrades above 46 per cent. A higher pulse amplitude of 10 mm along with 6000 G (393 A) field intensity seems to yield highest Fe upgrade, as reflected by Test 2 and Test 6 with only Test 6 meeting the >55 per cent Fe target specification for a pre-concentrate. Test 6 is also considered optimum with approximately 90 per cent of Si being rejected. This route may be used as a waste rejection stage prior to upgrading to higher marketable Fe grades via flotation. Of concern however, is the low overall Fe recovery of 46 per cent.

Head grade analysis of roasted feed

The head grade of the roasted feed reported at 38.5 per cent Fe, 18.9 per cent Si, 5.5 per cent Al, 0.03 per cent Mg, 0.12 per cent Mn and 0.46 per cent Ca.

Magnetic separation test work on roasted feed

Davis tube recovery tests

The roasted feed having increased its magnetic susceptibility was initially subjected to 1 A resulting in a majority of the mass reporting to the magnetics (87 per cent). The intensity was thereby reduced to 0.7 and 0.6 A for comparative purposes and to optimise grade and recovery. The results obtained are represented in Figure 8 and Table 7.

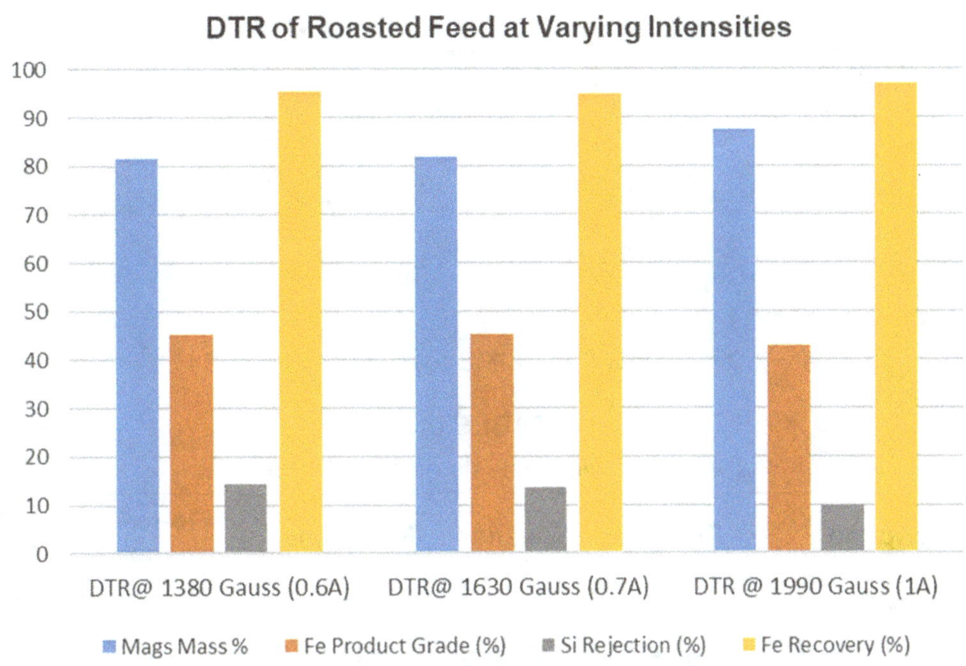

FIG 8 – Davis Tube test work on roasted feed.

TABLE 7

Results of DTR tests on roasted feed.

Description	DTR@ 1380 Gauss (0.6A)	DTR@ 1630 Gauss (0.7A)	DTR @ 1990 Gauss (1A)
Mags mass %	81.49	81.89	87.35
Fe product grade (%)	45.18	45.08	42.82
Si rejection (%)	14.38	13.69	9.83
Fe recovery (%)	95.45	94.70	96.95

The lowest current setting of 0.6 A produced the highest Fe upgrade at 45.2 per cent at a Fe recovery of 95.5 per cent. The roasted feed was further subjected to wet low intensity magnetic separation (WLIMS) tests introducing multiple cleaning stages to ascertain if the pre-concentrate grade could be improved to >55 per cent Fe whilst minimising mass losses.

WLIMS tests on roasted feed

Roasting resulted in the reduction of paramagnetic hematite to ferromagnetic magnetite. The results for WLIMS rougher, cleaner re-cleaner and re-re-cleaner tests are represented in Figure 9 and Table 8. The re-cleaner magnetics was first milled to 80 per cent passing 75 µm in an attempt to liberate magnetite from gangue (primarily Si) and thereafter subjected to the re-re cleaner tests.

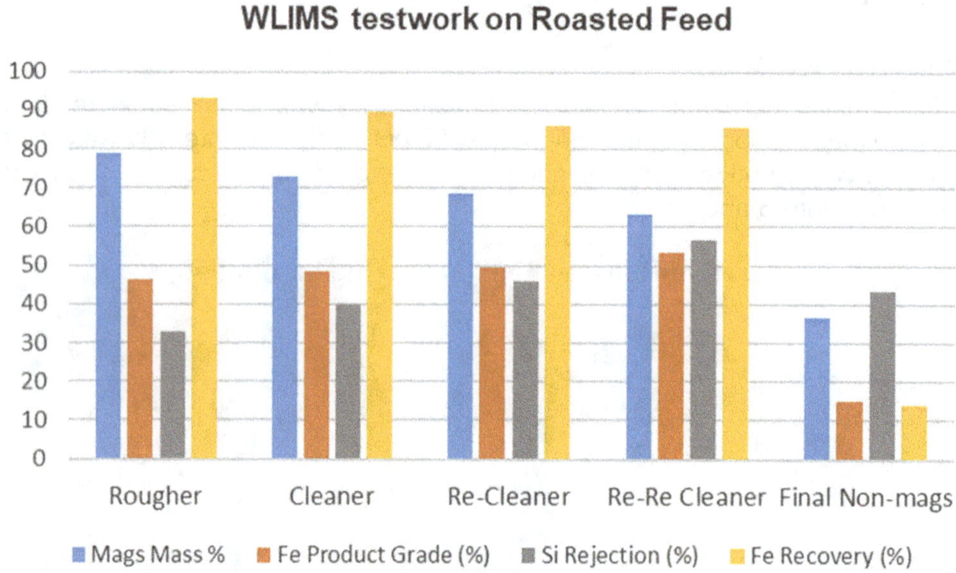

FIG 9 – WLIMS test work.

TABLE 8

Results of WLIMS tests on roasted feed.

Description	Rougher	Cleaner	Re-cleaner	Re-re-cleaner	Final non-mags
Mags mass %	79.05	72.97	68.53	63.27	36.73
Fe product grade (%)	46.40	48.30	49.40	53.40	15.25
Si rejection (%)	32.90	40.10	46.08	56.53	43.47
Fe recovery (%)	93.13	89.48	85.95	85.78	14.22

The highest obtainable Fe grade obtained by magnetic separation on a roasted feed was 53.4 per cent Fe at an overall recovery of 85.6 per cent, this is in line with a higher mass yield to product of 63.3 per cent. An overall Si rejection of 56.5 per cent was obtained.

It is therefore possible with further cleaning stages to attain a 55 per cent Fe pre-concentrate whilst improving Si rejection and sacrificing Fe recovery marginally.

CONCLUSION AND RECOMMENDATIONS

With the decline in iron ore grade in SA there is a need to unlock sterile low-grade resources in an attempt to extend LOM and maximise resource utilisation. BIF material makes up approximately 63 per cent of the future life-of-mine. Due to the banded/complex nature of low-grade BIF, fine processing is required. The ore researched reported a head grade of approximately 33 per cent Fe with quartz and illite as the major gangue minerals present. SA is high on the production cost curve and as such processes proposed need to be cost sensitive. Pre-concentration of low-grade material prior to downstream refinement has numerous benefits in terms of upfront gangue reduction and ensuring a more uniform higher grade feed to the final beneficiation stage. The study focused on pre-concentration via magnetic separation on a feed crushed to -1.18 mm prior to and after reduction roasting.

The highest obtainable Fe grade obtained by reduction roasting was 53.4 per cent Fe for the re-re-cleaner stage, slightly lower than that obtained without roasting which was 55.4 per cent Fe. Roasting resulted in an overall silica rejection of 56.5 per cent much lower than 89.3 per cent obtained without roasting. A huge benefit of roasting prior to magnetic separation however is the overall attainable recovery of Fe which was 85.8 per cent as opposed to only 46.0 per cent without roasting.

It is therefore possible with further cleaning stages to attain a 55.0 per cent Fe pre-concentrate with roasting whilst improving Si rejection and sacrificing Fe recovery marginally. The economics of the process should be evaluated to determine the feasibility of reduction roasting prior to magnetic separation as a pre-concentration stage taking into account that the pre-concentrate will further require flotation upgrade to produce a premium saleable concentrate.

ACKNOWLEDGEMENTS

The author would like to acknowledge Mintek for providing the facilities, resources and funding required for the research. Also, South Africa's primary iron ore producers for part funding and collaboration. In addition, the invaluable contribution of Mintek's pyrometallurgical, analytical science and mineralogy division for their support.

FUNDING SOURCES

The work was supported by the Department of Minerals and Petroleum Resources (DMPR) in South Africa. The current research did not receive any specific funding from commercial sectors.

REFERENCES

Basson, I J, Thomas, S A J, Stoch, B, Anthonissen, C J, McCall, M J and Britz, J L, 2018. The structural setting of mineralisation at Kolomela Mine, Northern Cape, South Africa, based on fully-constrained, implicit 3D modelling, *Ore Geology Reviews*, 95:306–324. https://doi.org/10.1016/j.oregeorev.2018.02.032

Creamer Media, 2020. Iron-Ore 2020: A review of the iron-ore sector, Creamer Media, Engineering News. Available from: <https://www.creamermedia.co.za/article/iron-ore-2020-a-review-of-the-iron-ore-sector-pdf-report-2020-06-26>

Eriksson, P G, Altermann, W, Nelson, D R, Meuller, W U and Catuneanu, O, 2004. The Precambrian Earth: Tempos and Events, in *Precambrian Geology*, pp 378–379 (Elsevier Science and Technology).

Eriksson, P G, Hattingh, P J and Altermann, W, 1995. An overview of the geology of the Transvaal Sequence and Bushveld Complex, South Africa, *Mineralium Deposita*, 30(2):98–111. https://doi.org/10.1007/BF00189339

Singh, A and da Corte, C, 2019. Beneficiation of Ultra-fine low grade Fe Ore as feedstock for the Fe and Steel Value Chain, internal report, Mintek.

Spreitzer, D and Schenk, J, 2020. Fluidization behavior and reducibility of iron ore fines during hydrogen-induced fluidized bed reduction, *Particuology*, 52:36–46. https://doi.org/10.1016/j.partic.2019.11.006

Tawane, G and Goso, X, 2024. Magnetizing roasting of MPD's hematite, Mintek, internal report.

Wills, B A and Napier-Munn, T, 2005. *Wills' Mineral Processing Technology*, An Introduction to the Practical Aspects of Ore Treatment and Mineral Recovery, seventh edition. https://doi.org/10.1016/B978-0-7506-4450-1.X5000-0

Cold briquetting of iron ores and the performances assessment

Z Tong[1], *B Gu*[2] *and T Chun*[3]

1. Master student, Anhui University of Technology, Maanshan Anhui Province 243002, China. Email: 1071160754@qq.com
2. Phd candidate, Anhui University of Technology, Maanshan Anhui Province 243002, China. Email: gubshu@126.com
3. Professor, Anhui University of Technology, Maanshan Anhui Province 243002, China. Email: tjchun@ahut.edu.cn

ABSTRACT

Cold briquetting of iron ores represents a significant pathway toward low-carbon metallurgy, so cold briquettes were manufactured and their performances evaluated. The results show that a compressive strength of 2048 N was obtained under the conditions of mixing the iron ores with 2.55 per cent binder, briquetting at 3 per cent moisture with a pressure of 20 MPa, and drying at 105°C for 4 hr. As the burden for a blast furnace, the dried briquettes presented excellent indexes with $RDI_{+3.15\ mm}$ of 87.68 per cent, TI of 81.83 per cent and RSI lower than 5 per cent. For direct reduction process, such as the burden for Midrex, the metallisation rate was 88.12 per cent, the $RDI_{DR-3.15\ mm}$ was 28.98 per cent and the compressive strength after reduction was 970 N. As the burden for HYL, the $LTD_{+6.3\ mm}$ was 75.89 per cent, the RSI was lower than 5 per cent, the reducibility was 3.67 per cent when reduced at 950°C, and the compressive strength of the dried briquettes was 1212 N after reduction at 950°C.

INTRODUCTION

Driven by the global carbon neutrality goal, the metallurgical industry, as a core sector with high energy consumption and high emissions, has witnessed its low-carbon transformation becoming an inevitable trend for industry sustainability (Guo *et al*, 2024). The development of low-energy-consumption and low-emission iron ore pre-treatment and smelting technologies has become a research hotspot. Traditional iron smelting processes rely on high-temperature sintering, coking, and oxide pelletisation for burden preparation, which not only consumes a large amount of energy but also emits significant pollutants such as CO_2 and SO_2, contradicting the concept of low-carbon development and failing to meet the demands of the green industrial transformation (Liu *et al*, 2022).

Against this backdrop, iron ore cold-pressing pelletisation technology, which requires no high-temperature roasting, featuring a short process flow sheet, low energy consumption, and low carbon emission intensity, has emerged as an important technical pathway for the efficient utilisation of iron ore fines in low-carbon metallurgical systems. It provides high-quality raw material options for mainstream ironmaking processes such as blast furnace smelting and direct reduction (Halt, Roache and Kawatra, 2015; Shen and Sha, 2003). This technology involves mixing iron ore fines with a binder at room temperature, followed by press moulding and low-temperature drying. Its core advantage lies in achieving carbon reduction at the raw material pre-treatment stage while effectively utilising fine iron ore powder and improving resource utilisation efficiency (Zheng *et al*, 2025). It can also adapt to the raw material requirements of different ironmaking processes, supplying high-quality feedstocks for mainstream metallurgical routes such as blast furnace smelting and direct reduction (Yan, 2020).

In recent years, scholars at home and abroad have conducted extensive research on iron ore pelletisation preparation. Sha Steel adopted microwave curing, achieving an energy consumption of 120 kWh/ton and a strength retention rate of 92 per cent, thereby shortening the curing period (Gan *et al*, 2014). Wang *et al* (2017) developed composite additives to improve thermal performance based on key indicators such as the reduction disintegration index. Michigan Technological University (MTU) in USA developed the MTU method for cold-press moulding of iron-containing dust using cement-based binders, achieving stable compressive strength after steam curing and demonstrating suitability for non-sintering processes (Halt, Roache and Kawatra, 2015). Russian researchers employed ultra-high pressure moulding combined with low-grade ores, achieving low costs and meeting blast furnace strength requirements (Mao and Sloof, 2017). These studies have confirmed that binder dosage and moisture content are key factors affecting the physical properties of pellets (Li, 1996). Furthermore, the development and practice of low-temperature metallurgical theory has provided theoretical support for the application of cold-pressed pellets in low-temperature reduction smelting processes (Liu *et al,* 2019).

However, existing research still has obvious deficiencies. First, it mainly focuses on optimising burden performance for a single ironmaking process, while there is a lack of systematic research on the adaptation of cold-pressed briquettes to different processes such as blast furnace, Midrex, and HYL technologies. Second, the optimisation of process parameters primarily targets basic physical properties, and research on the coordinated regulation of briquette or pellet metallurgical performance remains insufficient. At the same time, existing cold-pressed briquette preparation processes still face challenges such as high binder addition levels and some metallurgical performance indicators failing to meet industrial application standards, which restricts the large-scale promotion of this technology.

Based on this background, this paper aims to study the influence of process parameters on the physical properties of iron ore cold-pressed briquettes for low-carbon metallurgical applications, and to systematically determine the optimal process conditions. It also evaluates the key metallurgical properties of cold-pressed briquettes as raw materials for blast furnace, Midrex and HYL processes, and explores their adaptability in different low-carbon ironmaking processes. The purpose of this paper is to clarify the coupling relationship between process parameters and the comprehensive performance of iron ore cold-pressed briquettes, to obtain a preparation technology with excellent performance, to provide theoretical support and technical reference for the industrial promotion of this technology, and to help the metallurgical industry achieve carbon reduction at the raw material pre-treatment stage while promoting the advancement of low-carbon ironmaking technology systems.

EXPERIMENTAL MATERIALS AND METHODS

Iron ore concentrate

Chemical characteristics

The iron ores used include Iron ore fines and iron ore concentrate. The chemical compositions of the iron ores are listed in Table 1. The iron grade of the iron ore fines was over 62 per cent, whilst the concentrate was over 65 per cent Fe.

TABLE 1

Chemical compositions of iron ores/%.

Item	Iron ore fines	Iron ore concentrate
Fe	62.73	65.07
FeO	0.50	0.61
SiO_2	6.28	4.28
Al_2O_3	1.69	1.56
CaO	0.27	0.17
MgO	0.14	0.08
MnO	0.11	0.09
S	0.01	0.01
P	0.06	0.05
LOI	2.09	1.00

Physical characteristics

The size distributions of the iron ores are tabulated in Table 2. The particle size distribution of the iron ore fines is primarily distributed between 1–3 mm. The particle size of the iron ore concentrate is 84.88 per cent passing 0.075 mm.

TABLE 2

Particle size distributions and moisture content of the iron ores/%.

Size/mm	Iron ore fines	Iron ore concentrate
>8.0	0.00	0.00
5.0–8.0	0.05	3.02
3.0–5.0	30.27	1.64
1.0–3.0	51.88	2.00
0.5–1.0	6.51	0.00
0.15–0.5	5.19	0.56
0.075–0.15	0.28	7.90
< 0.075	5.82	84.88
Moisture/%	1.43	1.02

Binder

Figure 1 shows the TGA-DSC analysis of the composite adhesive. The thermal stability of the composite adhesive is good. The main weight loss is the evaporation of water, the decomposition reaction of organic substances, the removal of hydroxyl and carboxyl groups, and the combustion of organic carbon. The final weight loss reaches 26.17 per cent. From Figure 1, there is an endothermic peak at 91.3°C, which is mainly caused by the evaporation of water in the adhesive. The endothermic peak at the peak temperature of 197.2°C is mainly caused by the removal of hydroxyl and carboxyl groups, and the peak at 313.2°C is mainly a heat release reaction caused by the cross-linking and curing of the adhesive. The glass transition temperature TG of the cross-linking agent is 446.2°C, and is mainly the process of the adhesive softening from the solid state. The endothermic peak at the peak temperature of 851°C is mainly caused by the melting of some impurities.

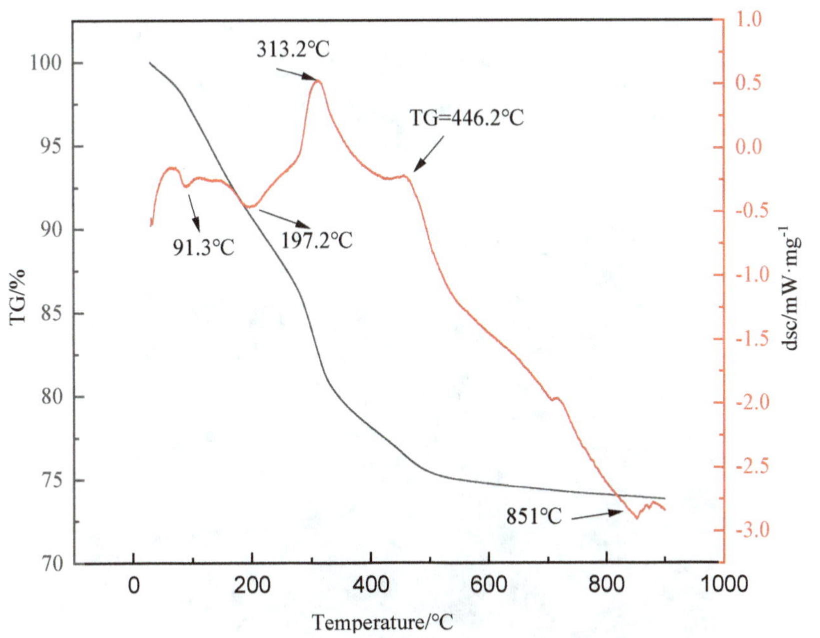

FIG 1 – Analysis of composite binder TGA-DSC.

Experimental methods

The experimental flow sheet in Figure 2. The iron ores and binder were mixed at a given ratio and prepared for briquetting. Green briquettes were made in a briquetting machine and followed by a drying process.

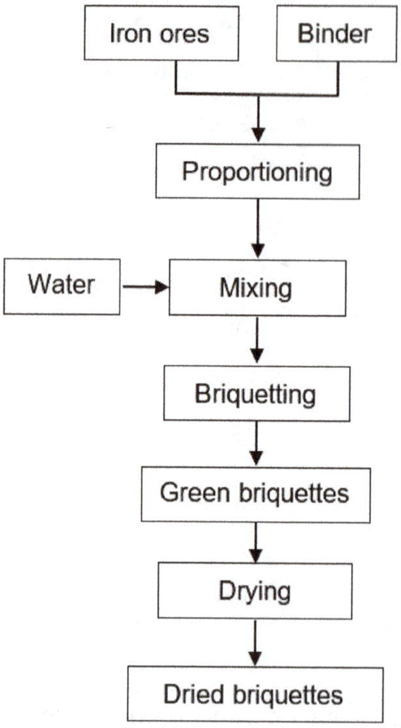

FIG 2 – The experimental flow sheet.

The experiment used a twin-roll briquette machine (Figure 3) to prepare cold-bonded briquettes. Three kg of iron ore concentrate was weighed and thoroughly mixed with 2.55 per cent composite binder and a small amount of water (the moisture content of the briquettes was controlled at 3 per cent). After secondary mixing, a Zhongding 360-type twin-roll briquette machine was used to produce elliptical cold-bonded briquettes with a length of 40 mm, a width of 28 mm, and a thickness of 20 mm.

FIG 3 – Briquetting machine.

RESULTS AND DISCUSSION

To study the influence of different amounts of binder on the strength of the cold-cured briquettes and the compressive strength of the dry briquettes under the conditions of a fixed 3 per cent moisture and particle size composition, cold-cured briquettes were prepared using the roller compaction process. If the binder ratio was too low, the compressive strength would be insufficient; if it was too high, it would easily cause sticking and significantly increase the production cost (De Moraes *et al*, 2020). Therefore, in this study, the binder addition amount was controlled within the range of 2–3 per cent, and the influence on briquette performance was mainly investigated. The experimental results are shown in Figure 4.

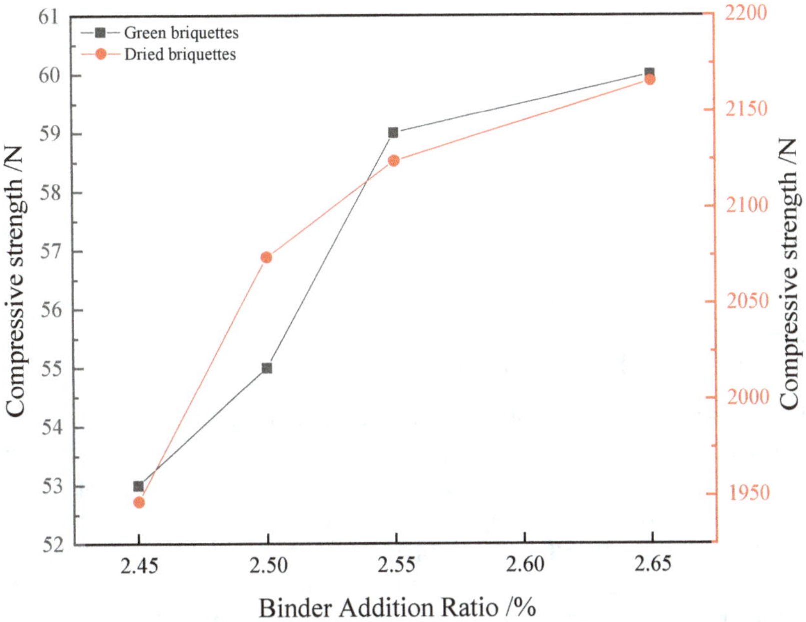

FIG 4 – Influence of different binder dosages on the cold strength of the briquettes.

Influence of binder addition amount on the cold-state strength of the briquettes

As the binder addition amount increased from 2.45 per cent to 2.65 per cent, the drop strength of the wet briquettes increased from 4.2 times/0.5 m to 5.0 times/0.5 m, the compressive strength of the wet briquettes increased from 53 N to 60 N, and the compressive strength of the dry briquettes also increased from 1947 N to 2166 N. The interpretation is that the binder effectively increases the friction resistance between particles by constructing a multi-component bonding network through physical bonding, chemical adsorption, and crystal bridge connections, promoting the densification and arrangement of the materials (Bizhanov and Zagainov, 2021). Based on this synergistic mechanism, the strength performance of cold-cured briquettes show a significant positive correlation with the binder addition amount. When the binder addition amount was 2.55 per cent, the drop strength of the wet briquettes reached 5.1 times/0.5 m, the compressive strength of the wet briquettes was 59 N, and the compressive strength of the dry briquettes was 2124 N (see Table 3). All indicators met the requirements of actual production (Li, 1996). Considering both strength performance and economic cost, the appropriate binder addition ratio was determined to be 2.55 per cent.

TABLE 3

Effect of binder dosage on the performances of the briquettes.

Binder	Green briquettes		Dried briquettes
Dosage/%	Drop numbers times/0.5 m drop	Compressive strength/N	Compressive strength/N
2.65	5.0	60	2166
2.55	5.1	59	2124
2.50	4.8	55	2074
2.45	4.2	53	1947

Evaluation as the burden for the blast furnace

In evaluating the suitability of cold-pressed briquettes for different low-carbon ironmaking processes, three key indicators were selected: low-temperature reduction disintegration index (RDI, specifically the fraction below 3.15 mm after reduction), reduction swelling index (RSI), and low-temperature reduction decrepitation (LTD). These indices are widely recognised in the industry as critical parameters for assessing the performance of ferrous burden materials in shaft furnaces, including blast furnace (BF), Midrex, and HYL processes.

The $RDI_{-3.15\ mm}$ is particularly important for BF and Midrex processes, where the material is subjected to moderate reducing conditions at temperatures between 500°C and 900°C. High disintegration generates fines, which reduce bed permeability and can lead to operational upsets. The internationally accepted threshold for BF applications, as defined by ISO 4696-2 (2022) and widely adopted in industry practice, is an $RDI_{-3.15\ mm}$ of less than 20 per cent for good performance. For Midrex and HYL direct reduction processes, the requirement is even stricter, typically below 15 per cent due to the greater sensitivity of gas-based shaft furnaces to dust generation. Similarly, LTD measures decrepitation under rapid heating, which is critical for processes with abrupt temperature changes. A common industrial standard for $LTD_{-3.15\ mm}$ is below 5 per cent for both BF and direct reduction processes. RSI reflects the volume change during reduction; excessive swelling causes sticking and disrupts burden descent. The acceptable RSI range is generally 15-20 per cent for BF and below 15 per cent for Midrex and HYL processes, as swelling beyond these values leads to operational difficulties.

The performance of the dried briquettes as the burden for blast furnaces is listed in Table 4. The excellent indexes with compressive strength of 2048 N, shatter index (SI) of 97.75 per cent, reduction degradation index ($RDI_{+3.15}$) of 87.68 per cent, tumble index (TI) of 81.83 per cent, and Reduction Swelling Index (RSI) lower than 5 per cent were obtained. The Reducibility index (RI) is not very good, being lower than 50 per cent. The briquettes have good thermal structural stability in the blast furnace, but their reduction performance is known to be poor (Huang *et al,* 2013).

TABLE 4

Performance of the dried briquettes as the burden for a blast furnace.

Compressive strength/N	SI +10 mm/%	RI/%	RSI/%	TI +6.3 mm/%	RDI/%		
					+6.3 mm	+3.15 mm	-0.5 mm
2048	97.75	42.38 43.05	4.24 4.53	81.83	86.53	87.68	4.98

The softening and melting performances of the dried briquettes are shown in Table 5. The softening temperature is 1196°C and the dripping temperature is 1328°C with a melting zone of 132°C.

TABLE 5

Softening and melting performances of the dried briquettes.

T_{10}/°C	T_{40}/°C	Softening zone/°C	T_s/°C	T_d/°C	Melting zone/°C	Max ΔP/kPa
1029	1158	129	1196	1328	132	6.3

The compressive strength after the waterproof test is shown in Table 6. The compressive strength of the briquettes was 1069 N after the dried briquettes were soaked in water for 8 hrs, while the compressive strength increased to 1904 N when the soaked briquettes were dried again. After the briquettes were dried again, the compressive strength was 1904 N, which was slightly lower than that of the briquettes before soaking, but still they had a relatively high compressive strength. This indicates that the briquettes had good water resistance and good structural stability.

TABLE 6

Compressive strength after the dried briquettes waterproof test.

Item	Compressive strength/N
In the water for 8 hrs	1069
In the water for 8 hrs and then drying	1904

Evaluation as the burden for direct reduction

The performance of the dried briquettes as the burden for a Midrex process is listed in Table 7. The metallisation rate was 88.12 per cent, the degree of reduction was 62.33 per cent, the $RDI_{DR\text{-}3.15\ mm}$ was 28.98 per cent, and the compressive strength after reduction was 970 N.

TABLE 7

Performance of the dried briquettes as the burden for a Midrex process.

Metallisation rate/%	Breaking ratio/%	Degree of reduction R_{90}/%	$RDI_{DR\text{-}3.15}$/%	Compressive strength after reduction (800°C)/N	TI/%	
760°C for 300 min		800°C for 90 min	760°C for 300 min		+6.73 mm	-0.595 mm
88.12	25	62.33	28.98 30.56	970	79.49	11.84

The performance of the dried briquettes as the burden for a HYL process is listed in Table 8. The low temperature reduction degradation ($LTD_{+6.3\ mm}$) was 75.98 per cent, and the $LTD_{-3.2\ mm}$ was 22.44 per cent, the Reduction Swelling Index (RSI) was lower than 5 per cent, the reducibility was 3.67 when reduced at 950°C, and the compressive strength of the dried briquettes after reduction at 950°C was 1212 N, while it was 507 N when the dried briquettes were reduced at 800°C.

TABLE 8

Performances of the dried briquettes as the burden for HYL process.

LTD/% 500°C for 2hrs		RSI/%		Reducibility/%		Compressive strength after reduction/N		TI/%	
+6.3 mm	-3.2 mm	950°C	800°C	950°C	800°C	950°C	800°C	6.3 mm	-3.2 mm
75.98	22.44	4.03	4.91	3.67	2.28	1212	507	81.83	23.67

Compared with the Midrex method, the compressive strength of the briquettes after reduction at 800°C using the HYL method significantly decreased (from 970 N to 507 N). This is because the reduction rate of the briquettes is faster in a hydrogen-rich atmosphere, and the lattice expansion is therefore thought to be more intense. Additionally, the reduction degree in the HYL method is higher. When the reduction degree is greater than 60 per cent, the strength shows a strong negative correlation with the reduction degree (Yi, 2013). For every 10 per cent increase in reduction degree, the compressive strength after reduction decreases by approximately 15 per cent to 25 per cent (Mao and Sloof, 2017).

Due to the ample reduction time at 950°C, FeO rapidly transforms into metallic Fe. The metallic iron grains grow and form a crystal network, and the pores are bridged by iron crystals, while cracks are 'sewn up'. At the same time, 950°C is close to the low-melting-point eutectic temperature of FeO and gangue (approximately 1000°C) (Wang, 2009). A small amount of liquid phase is generated and bonds the particles, which is the reason for the recovery of the compressive strength of the reduced pellets.

The chemical composition of the dried briquettes and those reduced at 950°C are presented in Table 9. The iron grade of dried briquettes was 61.43 per cent and Na_2O was 0.47 per cent. The iron grade of the reduced briquettes was 81.91 per cent and metallic iron was 77.25 per cent with a metallisation rate of 94.31 per cent.

TABLE 9

Chemical compositions of the dried and reduced briquettes.

Item	TFe	MFe	FeO	SiO_2	Al_2O_3	CaO	MgO	MnO	S	P	Na_2O	LOI
Dried briquettes	61.43	-	0.49	7.33	1.65	0.26	0.13	0.11	0.01	0.06	0.47	2.45
Reduced briquettes	81.91	77.25	5.99	9.64	2.22	0.36	0.17	0.15	0.01	0.08	0.52	-

From an industrial application and scale-up perspective, the results demonstrate that cold briquetting technology, using a relatively low binder addition (2.55 per cent) and moderate drying temperature (105°C), can produce briquettes with satisfactory metallurgical properties for both blast furnace and direct reduction processes. The simplicity of the process—room-temperature mixing, press moulding, and low-temperature drying—offers significant advantages in terms of energy savings and reduced CO_2 emissions compared to traditional sintering or oxide pelletisation (Song *et al*, 2008). For scale-up, key considerations should include maintaining consistent moisture control and briquetting pressure in large-scale roller presses, optimising binder distribution in industrial mixers, and ensuring uniform drying conditions to avoid uneven strength development. Additionally, the slightly high $RDI_{DR\ -3.15\ mm}$ value for the Midrex application suggests that further pilot-scale trials may be needed to fine-tune the formulation or curing conditions for specific direct reduction routes. Overall, the cold briquetting technology developed in this work presents a viable and industrially promising low-carbon feedstock option for modern ironmaking.

CONCLUSIONS

- Under the conditions of 2.55 per cent binder addition, 3 per cent moisture, 20 MPa forming pressure, and 105°C drying for 4 hrs, the dried briquettes have a compressive strength of up to 2048 N, meeting the basic strength requirements for transportation and furnace loading.
- As blast furnace feedstock, the dried briquettes exhibit excellent metallurgical performance. The $RDI_{+3.15\ mm}$ was 87.68 per cent and the TI was 81.83 per cent, indicating that the briquettes display strong anti-friability during blast furnace smelting. The low reduction expansion rate (RSI <5 per cent) indicated good thermal stability of the briquettes, which can effectively improve the permeability of the furnace charge and promote efficient operation of the blast furnace.

- In the direct reduction process, the briquettes also show good adaptability. When used as Midrex raw material, the metallisation rate of the briquettes reached 88.12 per cent, and the direct reduction degradation rate ($RDI_{-3.15\ mm}$) was 28.98 per cent, remaining at a good level; the compressive strength after reduction was 970 N. This demonstrates the excellent reduction performance and structural stability of the briquettes in the Midrex process. When used as HYL furnace charge, the low-temperature reduction degradation rate ($LTD_{+6.3\ mm}$) was 75.89 per cent, the reduction expansion rate was low (RSI <5 per cent), and the thermal strength was good; after reduction at 950°C, the compressive strength was 1212 N, demonstrating excellent structural stability and a high reduction rate (RI of 3.67 per cent/min).

ACKNOWLEDGEMENTS

The authors are thankful to the Projects of Natural Science Foundation China (52374315) and Natural Science Foundation of Anhui Province (2408085Y030) for sponsoring the study.

REFERENCES

Bizhanov, A M and Zagainov, S A, 2021. Tests of briquettes for mechanical strength, *Metallurgist*, 65(3):247–256.

De Moraes, S L, Lima, J R B D, Neto, J B F, Fredericci, C and Saccoccio, E M, 2020. Binding mechanism in green iron ore pellets with an organic binder, *Mineral Processing and Extractive Metallurgy Review*, 41(4):247–254.

Gan, Y, Wang, J K, Bao, C J, Luo, W B, Lu, G H and Zhang, Z Y, 2014. Research progress of cold-pressing pelletizing technology in metallurgy, *Multipurpose Utilization of Mineral Resources*, 35(1).

Guo, P M, Wang, L, Kong, L B and Zhao, P, 2024. Low-temperature metallurgy theory, technology and future development, *Iron and Steel*, 59(9):71–83.

Halt, J A, Roache, S C and Kawatra, S K, 2015. Cold bonding of iron ore concentrate pellets, *Mineral Processing and Extractive Metallurgy Review*, 36(3):192–197.

Huang, Z C, Zhao, L Z, Yi, L Y and Jiang, T, 2013. Research on powdering behavior of iron ore oxidized pellets during direct reduction with gas at low temperature, *Metal Mine*, 42(3):69–72+76.

International Organization for Standardization (ISO), 2022. ISO 4696-2:2022 – Iron ore and direct reduced iron – Second part: Determination of low-temperature reduction-disintegration by static method[S].

Li, X W, 1996. A brief discussion on the cold solidification granulation production process, *Baosteel Technology*, (2):60–64.

Liu, H X, Wu, L M, Zhao, L, Cui, Y, Wang, S Q, Si, H L, Zhou, Y N, Zhou, H and Xue, J C, 2022. Analysis of CO_2 emission characteristics and treatment technologies in the steel industry, *Sintering and Pelletizing*, 47(1):38–47.

Liu, Z J, Wang, R R, Liu, Y R, Li, Z G, Zheng, A Y and Zhang, J L, 2019. Optimization of preparation parameters for cold solidification pellets using polymeric adhesive, *Iron and Steel*, 54(2):15–19.

Mao, W C and Sloof, W G, 2017. Reduction kinetics of wüstite scale on pure iron and steel sheets in Ar and H_2 gas mixture, *Metallurgical and Materials Transactions B*, 48(3):1427–1438.

Shen, Z B and Sha, Y Z, 2003. Research on the cold solidification pelletization process of steel dust, *Iron and Steel*, 38(12):1–5.

Song, X, Yang, J P, Zhao, Z C and Chen, Z J, 2008. Cold-pressed briquettes: an important means of energy saving and emission reduction, *Ferroalloys*, (5):32–34.

Wang, R R, Zhang, J L, Liu, Z J, Liu, Y R, Liu, X L and Li, N Y, 2017. Mechanism of the effect of new bonding agents on cold solidification pelletization, *Iron and Steel*, 52(11):9–14.

Wang, Z C, 2009. *Basic research on direct reduction of oxidized pellets in gas-based shaft furnace*, PhD thesis (unpublished), Northeastern University.

Yan, L X, 2020. Feasibility analysis of cold pressing and pelletizing of returned materials from sintering for furnace admission, *Metallurgical Management*, (21):9–10.

Yi, L Y, 2013. *Basic research on direct reduction of iron ore pellets with CO-H_2 mixed gas in gas-based shaft furnace*, PhD thesis (unpublished), Central South University.

Zheng, X H, Chen, T J, Wan, J Y, Zhou, X L, Liu, J W and Deng, Y, 2025. Research progress on low-carbon multi-form waste-iron ore powder cold solidification pelletization, *China Metallurgy*, 35(7):132–142+179.

Engineering an integrated process for efficient Fe–V–Ti separation and utilisation from vanadium–titanium magnetite

Q Zhang[1], J Pan[2], D Q Zhu[3] and S W Li[4]

1. PhD student, School of Minerals Processing and Bioengineering, Central South University, Changsha 410083, China. Email: vqianzhang@csu.edu.cn
2. Professor, School of Minerals Processing and Bioengineering, Central South University; Low Carbon and Hydrogen Metallurgy Research Center, Central South University, Changsha 410083, China. Email: pjcsu@csu.edu.cn
3. Professor, School of Minerals Processing and Bioengineering, Central South University; Low Carbon and Hydrogen Metallurgy Research Center, Central South University, Changsha 410083, China. Email: dqzhu@csu.edu.cn
4. Associate Researcher Fellow, School of Minerals Processing and Bioengineering, Central South University; Low Carbon and Hydrogen Metallurgy Research Center, Central South University, Changsha 410083, China. Email: swli@csu.edu.cn

ABSTRACT

Vanadium–titanium magnetite (VTM) is a typical complex multi-metal iron ore that contains significant amounts of iron, vanadium, and titanium. Owing to the strong mineralogical coupling among these elements, efficient separation and comprehensive utilisation of VTM remain major technical challenges in ironmaking and extractive metallurgy. Conventional blast furnace and direct reduction routes primarily focus on iron production, often leading to inefficient vanadium recovery and limited utilisation of titanium, which is usually concentrated in slag with low added value. With the increasing demand for vanadium in high-strength steels and energy storage applications, as well as the growing importance of titanium resources, the development of an integrated and industrially feasible process for the coordinated recovery of Fe, V, and Ti from VTM has become increasingly important. In this study, an integrated metallurgical route combining sodium salt roasting, hydrometallurgical vanadium recovery, pellet conditioning, coal-based pre-reduction, and electric-furnace smelting separation was developed and systematically investigated. The proposed process aims to achieve high-efficiency vanadium extraction while maintaining favourable physical and metallurgical properties of the solid residue for downstream high-temperature treatment. Particular emphasis was placed on addressing key engineering challenges associated with residual alkali, Ti-rich phases, and pellet integrity, which often limit the applicability of sodium salt roasting–leaching routes in industrial practice.

During sodium salt oxidative roasting, vanadium present mainly in the V(III) state is selectively oxidised and converted into water-soluble vanadate species. The effects of sodium carbonate dosage and roasting temperature on pellet strength and vanadium leachability were systematically evaluated. Under optimised conditions, with 4–6 wt per cent Na_2CO_3 addition, preheating at 900°C followed by roasting at 1260°C, the roasted pellets exhibited a compressive strength exceeding 1700 N per pellet, indicating sufficient mechanical stability for subsequent handling. Water leaching achieved a vanadium extraction efficiency over 90 per cent, demonstrating the effectiveness of the roasting–leaching step for vanadium recovery. The vanadium-rich leachate was subsequently treated by chemical precipitation and calcination to produce high-purity V_2O_5 with a purity of approximately 99 per cent, meeting the requirements for high-value vanadium applications. Meanwhile, the water-leached residue retained most of the iron and titanium but also contained residual alkali and Ti-rich phases, which significantly affected pellet behaviour during further thermal treatment. Experimental observations showed that residual Na_2O readily reacted with moisture, leading to pellet pulverisation and deterioration of mechanical strength. In addition, the formation of pseudobrookite (Fe_2TiO_5) solid solution during roasting inhibited crystal growth and diffusion bonding between iron oxide grains, further weakening pellet consolidation.

To overcome these limitations, a mixed-acid dealkalisation strategy was introduced to modify the water-leached residue. A combination of H_2SO_4 and HF with an optimised molar ratio of approximately 1.42 was employed to effectively remove residual alkali and partially dissolve aluminosilicate gangue minerals. As a result, the Na_2O content in the treated pellets was reduced to below 0.3 wt per cent, while the combined content of SiO_2 and Al_2O_3 decreased to around 3 wt per cent. This treatment significantly improved pellet stability and created favourable conditions for subsequent reduction and smelting. Furthermore, the addition of 20–30 wt per cent ultrapure magnetite was adopted as a pellet strengthening measure. The introduced magnetite acted as a reactive bonding phase during reheating, promoting solid-state sintering and enhancing diffusion bonding among iron oxide particles. Consequently, the compressive strength of the treated roasted pellets increased to above 2000 N per pellet, satisfying the mechanical requirements for reduction and electric-furnace smelting.

Coal-based pre-reduction of the strengthened pellets was carried out to convert iron oxides into metallic iron prior to smelting. Pre-reduction experiments conducted at 1150°C with a C/Fe molar ratio of approximately 0.3 and a holding time of 60–90 min yielded reduced pellets with a metallisation degree of 85 per cent or higher. The high metallisation level ensured efficient melting and separation during the subsequent smelting step while minimising iron losses to slag. Electric-furnace smelting was performed at 1600°C for 45–60 min to achieve effective separation of iron and titanium. The smelting results demonstrated that molten iron with a total iron content (Fe_{Total}) of at least 96 per cent and an iron recovery exceeding 94 per cent could be obtained. Simultaneously, a high-titanium slag containing 60–66 wt per cent TiO_2 was produced. This slag composition is suitable for use as a high-quality feedstock in the production of ferrotitanium alloys or other titanium-based materials, thereby enhancing the overall economic value of the process.

In summary, the proposed integrated route enables efficient separation and comprehensive recovery of Fe, V, and Ti from vanadium–titanium magnetite. High-purity V_2O_5 is obtained as a valuable by-product through hydrometallurgical treatment, while iron and titanium are effectively recovered through pre-reduction and smelting. The systematic mitigation of residual alkali effects, control of Ti-rich phase behaviour, and enhancement of pellet strength highlight the engineering feasibility of the process. The results demonstrate significant process innovation and strong industrial scalability, providing a promising technical pathway for the high-value utilisation of complex vanadium–titanium magnetite resources.

Iron Ore

Ore characterisation and geometallurgy

A comparative study of blastcone sampling methods and flitch distributions at the Roy Hill Mine

D J Clauson[1], J Batchelor[2], A Huremovic[3] and S Bladon[4]

1. Specialist Digital Orebody, Hancock Iron Ore, Perth Airport WA 6105. Email: derek.clauson@hio.com.au
2. Manager Digital Orebody, Hancock Iron Ore, Perth Airport WA 6105. Email: james.batchelor@hio.com.au
3. Specialist Digital Orebody, Hancock Iron Ore, Perth Airport WA 6105. Email: adin.huremovic@hio.com.au
4. Manager Geology Operations, Hancock Iron Ore, Perth Airport WA 6105. Email: stuart.bladon@hio.com.au

ABSTRACT

Blasthole cone sampling is widely used in iron ore mining to increase grade control resolution, supplementing the coarser reverse circulation (RC) grid. While the debate between the effectiveness of RC and cone sampling continues in the literature, the benefits of greater lateral sample density and lower cost with cone sampling are recognised. At Hancock Iron Ore (HIO), a comprehensive three-part study was conducted at the Roy Hill (RH) mine to evaluate different blastcone sampling methods, with the potential to apply findings across all HIO operations.

Roy Hill is mined in 4 m flitches within a 12 m bench, with material stockpiled by flitch. In contrast, McPhee, a newer HIO mine, is mined as two 4 m flitches for an 8 m bench but is stockpiled at the bench scale. This operational difference means RH requires flitch-level sampling, while bench-level sampling suffices at McPhee. The study aimed to characterise the performance and uncertainty of both bench and flitch sampling methods.

Three sampling techniques were trialled: flitch-scale shovelling, bench-scale shovelling, and bench-scale auger sampling. The auger method was hypothesised to offer more repeatable, less biased, and ergonomically favourable sampling, though it is limited to bench-scale. Shovelling, while more physically demanding, is simpler and allows for flitch-scale sampling, provided the flitch distribution within the cone is understood.

To appropriately evaluate the methods, a large study was conducted in three phases:

- **Phase 1** provided an opportunity to demarcate and measure blastcone flitch heights as they were drilled. This allowed for characterisation of flitch distribution within 79 blastcones to determine representative flitch ratios and optimal sampling locations.
- **Phase 2** compared shovel and auger sampling methods across 70 cones using nearest (≤5 m) pre-existing RC as a reference.
- **Phase 3** comprised a highly controlled trial on 30 cones incorporating shovel and auger samples, whole-cone reference samples (~1 t) and RC twin holes drilled within 1 m. This phase allowed for a comprehensive analysis to be completed.

Results demonstrate pronounced lateral heterogeneity within individual blastcones, reinforcing the value of multi-point sampling. Multi-sample (four samples) auger averaging achieved the strongest agreement with whole-cone assays (Fe: R = 0.87, MAE = 1.02), approaching the performance of near-offset RC. Single-point shovel bench sampling performed poorest (Fe: R = 0.52, MAE = 1.49), while shovel flitch sampling provided intermediate performance where flitch-scale resolution is required. The relative performance rank of the methods was consistent comparing both raw assays and results from interpolation. Monte Carlo analysis on auger sample count indicates that two to three auger samples capture most performance gains, with continued improvement but diminishing returns thereafter.

Uncertainty envelopes were quantified for each method using two standard deviations of paired differences to whole-cone assays, demonstrating that auger averaging approximately halves Fe uncertainty relative to shovel techniques. The study provides a practical framework for selecting blastcone sampling strategies based on operational context, supporting auger multi-sampling for bench-scale applications and empirically constrained shovel flitch sampling where flitch control is mandatory.

INTRODUCTION

A comparative study of blasthole cone sampling techniques ('blastcone sampling') was undertaken at the Roy Hill mine to understand relative sampling performance of the techniques and associated uncertainty margins. Given Hancock Iron Ore sites are mined on both the flitch- and bench-scale, both bench- and flitch-scale sampling techniques were assessed. The assessed techniques were:

- shovel-bench (single-point) sampling
- shovel-flitch (single-point per flitch) sampling
- auger-bench (multi-point) sampling.

This was a multi-component study conducted in three phases:

- Phase 1: A blastcone flitch demarcation study was completed on over 70 blastholes to better understand average flitch distribution (and variation) within blastcones, and optimal sampling points to minimise flitch delimitation bias.
- Phase 2: A preliminary comparative sample technique study with data collected from 108 blastholes (culled to 70 for analysis) comparing auger versus shovel was conducted, using pre-existing ex-pit reverse circulation drill holes (RC) as reference data set. This phase collected data across holes drilled by both Pit Viper 271 (Epiroc) and D65 (Epiroc) rig types. Additional data like subjective cone quality and cone wetness were collected.
- Phase 3: Another comparative sampling technique phase involved a detailed study of 30 blastholes, collecting auger and shovel samples from each hole, the whole-cone sample (as a robust authoritative reference), and drilling new in-pit RC drill hole twins within 1 m (secondary reference). This robust data set allowed for comprehensive comparison of the techniques against a solid reference set.

The study aimed to characterise sampling uncertainty, quantify technique performance, and provide practical operational guidance for the selection and deployment of sampling methods across differing mine configurations, geological conditions, and business constraints. Figure 1 provides a schematic overview of the bench, flitch, blasthole, blastcone, sampling techniques and sampling points relevant to the study.

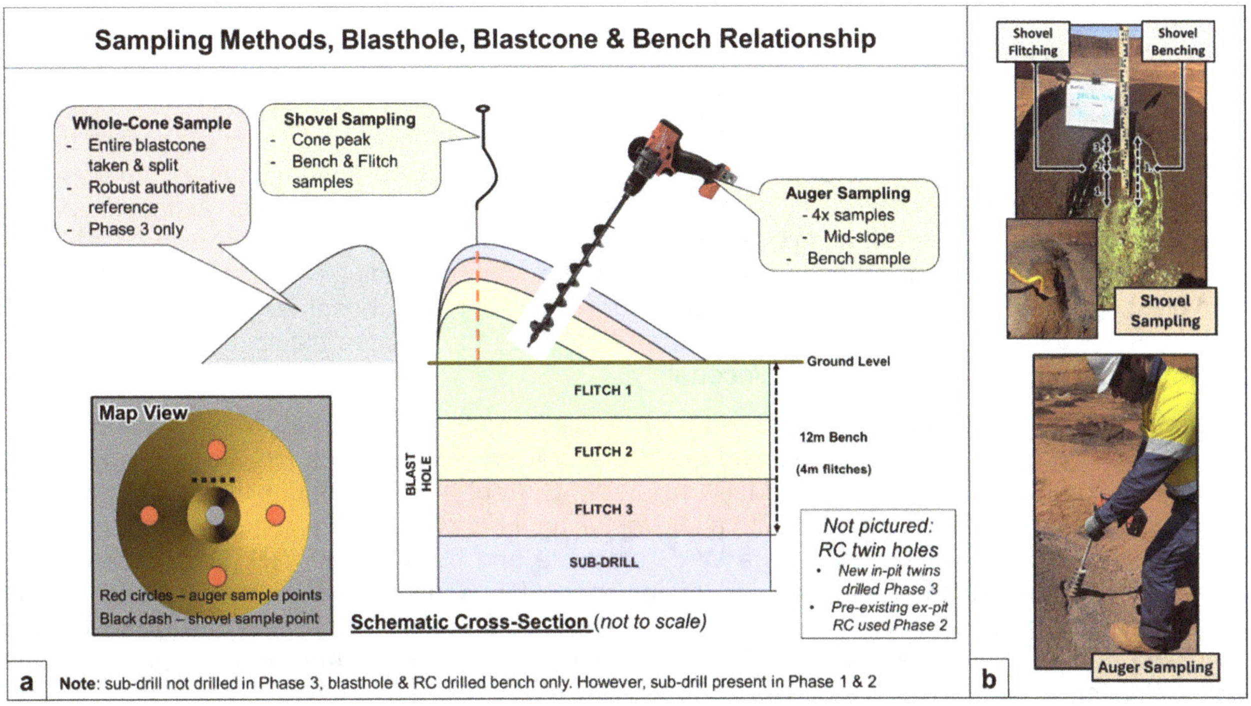

FIG 1 – Sampling methods, and schematic relationship between bench, blasthole, blastcone and sample points. Only sample type not shown are the RC twin holes used in the study. (a) Provides schematics of the above. Map view sub-inset indicates shovel sampling point near peak of cone. Red dots on map view cone indicate roughly equidistant multi-point auger sampling, taken at the mid-slope cone position. (b) Demonstrates example photos of sampling in action (of Pit Viper 271 blastcones). It also makes the distinctions between shovel flitch samples (single point sample per flitch, per cone) and shovel bench (single point bench sample). All points from the figure are relevant to Phase 2, except for the taking of whole-cone samples and true shovel bench samples. The sub-drill demonstrated in the bench and the blastcone was not present in the Phase 3 study component, as all holes were control drilled to bench depth (hence no sub-drill needed removal from blastcone). Sub-drill was a consideration in the Phase 1 and 2 though, as discussed in next sections.

BACKGROUND AND OPERATIONAL CONTEXT

The Roy Hill iron ore mine is located approximately 115 km north of Newman in Western Australia's Pilbara region. The project targets a world-class, low-phosphorous Neoarchean Marra Mamba Iron Formation orebody with a defined resource exceeding 2 Bt above 50 per cent Fe. Though the orebody is predominantly a Marra Mamba-hosted bedded deposit, there are minor, overlying Cenozoic-aged channel iron and other detrital deposits. The operation currently produces over 60 Mt/a, with ore transported via a purpose-built 344 km heavy haul railway to a two-berth port facility at Port Hedland.

The McPhee Creek iron ore mine is located approximately 100 km north of the Roy Hill mine, and 30 km north of Nullagine. While geologically separate from the primarily Marra Mamba–hosted orebody mined at Roy Hill, the Cleaverville Formation hosted McPhee Creek orebody is strategically positioned to complement Roy Hill's existing integrated mine–rail–port supply chain and long-term production outlook. It is Hancock Iron Ore's newest mine, with first dirt broken in 2024 and projected annual production of 8 Mt/a.

At the Roy Hill mine, blasthole cone sampling is used to provide additional granularity to the regular 25 m × 25 m RC drill hole grid. Flitch mining is undertaken on 12 m benches, divided into 3 × 4 m flitches. Blastcone sampling therefore aims to flitch sample, with these samples both sent for assay, and the flitch material logged by the mine geologist to aid prediction in processing behaviour for the grade control block (GCB). These samples are collected using a shovel, with a sampling face created near the peak of the cone (removing estimated sub-drill and cone material distal to the targeted sampling face). Assumptions around flitch ratios within the cone are used to aid delimiting the individual flitches, with these ratios being based off legacy conceptual blastcone modelling rather than empirical measurements.

The newer Hancock Iron site – McPhee Creek, commenced mining overburden in 2024. This site has a much coarser RC grid than Roy Hill (~50 m × 50 m RC spacing – increasing at depth due to hole angle), with a more structurally complex orebody. Being a more marginal mine, further spend to increase RC density is challenging. Given the RC density at McPhee was seen as insufficient for adequate grade control detail, additional sampling was required to fill the gap. After a cost-benefit study that weighed different options to increase sample density (including RC, downhole assay tools and blast rig sampling kits), traditional field blastcone sampling was chosen as the preferred option. This was the catalyst that drove a Hancock Iron Ore study to compare blastcone sample method performance, and associated method uncertainty.

Operationally, McPhee Creek has an 8 m bench divided into 2 × 4 m flitches. Though the bench is mined in two passes (flitch), it is blocked out as, and stockpiled as, a bench. So unlike Roy Hill, bench-scale sampling is adequate at McPhee Creek. Current blastcone sampling at McPhee (and other bench-scale HIO sites) is done with a shovel taking a bench sample of the cone. McPhee feed is to be delivered to Roy Hill mine as a WAG stockpile and blended with RH material at the RH crusher. As a result, ROM WAG grade and tonnes cannot be used to determine the feed and product impacts related to McPhee feed. This further underscores the importance of achieving greater sampling density at McPhee, and maximising sampling method performance/accuracy within the constraints of budget.

Blastcone sampling has traditionally had a reputation as being inferior to RC sampling, especially when critiqued against the principles of Peirre Guy's Theory of Sampling (Gy, 1992; Pitard, 1993, 2005). Workers such as Pitard (2008) stated 'blasthole sampling cannot provide representative samples' and compiled a comprehensive list of the disadvantages of blasthole sampling versus the advantages of RC drilling. Pitard (2008) also highlighted both the lateral and vertical heterogeneity of blastcones, increasing sampling difficulty. Though he concluded blasthole (BH) sampling could not be representative, he offered methods to maximise blasthole sampling performance, such as radial bucket sampling and electric augering that mimicked radial bucket sampling. Meanwhile, authors specifically working on sampling in iron ore, such as Engström and Esbensen (2017b), McArthur, Jones and Murphy (2010), and Abzalov *et al* (2007), were more positive on blastcone sampling versus RC. Engström and Esbensen (2017b) found blasthole sampling to be useful and fit-for-purpose for grade control as a complement to RC. In fact, Engström and Esbensen (2017a) found that careful manual shovel sampling of a blastcone achieved similar performance to more intensive optimised blastcone sampling (sectorial cutter sampling like the radial bucket sampling suggested in Pitard (2008)). Abzalov *et al* (2007), when comparing full blastcone assay results to RC, found that 'RC drilling does not guarantee an improvement in the quality of the sample' versus blastcone. In modelling their study area, they also found that not using the increased density offered by blastholes (5 × 5 m BH versus 25 × 25 m RC), could result in a 2× increase in the number of misidentified (Dual Class System – Ore versus Waste) Selective Mining Unit (SMU) sized blocks (in the worst case). McArthur, Jones and Murphy (2010) described that despite the theoretical sampling flaws of blastcone sampling, it had historically proven robust at Rio Tinto Iron Ore in estimating key analytes required for product builds. The ultimate message that one may draw from existing literature is that RC sampling is ideal for several reasons, but in iron ore mining thoughtful and well-executed blasthole sampling can be a value-adding complement to RC. Ultimately budget may determine the RC:BH ratio. Another re-occurring theme is that different site/region studies display differences in results/recommendations, and this highlights the utility of site-specific studies.

Given the increased reliance on blasthole sampling at McPhee Creek, the lost genealogy from GCB to processing plant (co-fed to crusher with RH material as highlighted above), and the learnings from existing literature – maximising blasthole sampling performance justified specific study. As bench-scale sampling was adequate for this site, the auger sampling technique was flagged as an alternative that may provide greater accuracy and less operator-induced bias than currently used shovel sampling. Additional potential auger benefits included less ergonomically-taxing sampling and faster sample collection. Whilst putting in the effort to study blastcone sampling, it was also seen as prudent to further explore and assess flitch-scale sampling used at Roy Hill.

METHODOLOGY

This study was conducted in three key work efforts/study phases:

1. **Flitch Ratio/Distribution** Characterisation (Phase 1):
   - designed to understand the distribution of flitches and sub-drill within the blastcone to best inform:
     - Average ratio of each flitch and sub-drill to better guide flitch sampling.
     - Best sampling point for each sample technique.
2. **Preliminary Testing** of Sampling Techniques against pre-existing RC (Phase 2):
   - This phase aimed to collect a large bulk of data comparing shovel versus auger technique against pre-existing RC.
   - This sample count would not be possible in next study phase (Detailed Study), where extra time and money was committed collecting whole-cone samples and drilling RC twin holes.
   - This phase also collected data on different blast rig types, auger sample count per blastcone, subjective cone condition and cone wetness, to examine impacts of these properties to sampling performance.
3. **Detailed Study** of Sample Techniques against Whole-cone Samples and RC Twin Holes (Phase 3):
   - All three sample techniques (shovel bench, shovel flitch, auger bench) collected from each blastcone.
   - Whole-cone sample taken.
   - RC twin hole drilled within 1 m of each blasthole.

Phase 1 – Flitch distribution characterisation

The aim of this study phase was to:

- Establish average flitch ratios within Roy Hill blastcones drilled by the Pit Viper 271 fleet, to aid flitch delimitation in flitch sampling of blastcones.
- Determine best sampling point for auger: the cone peak as used for shovelling – or the mid-point of the cone slope?

To do this, technicians demarcated the bench floor beneath the blastcone, and each flitch as it was drilled (see Figure 2).

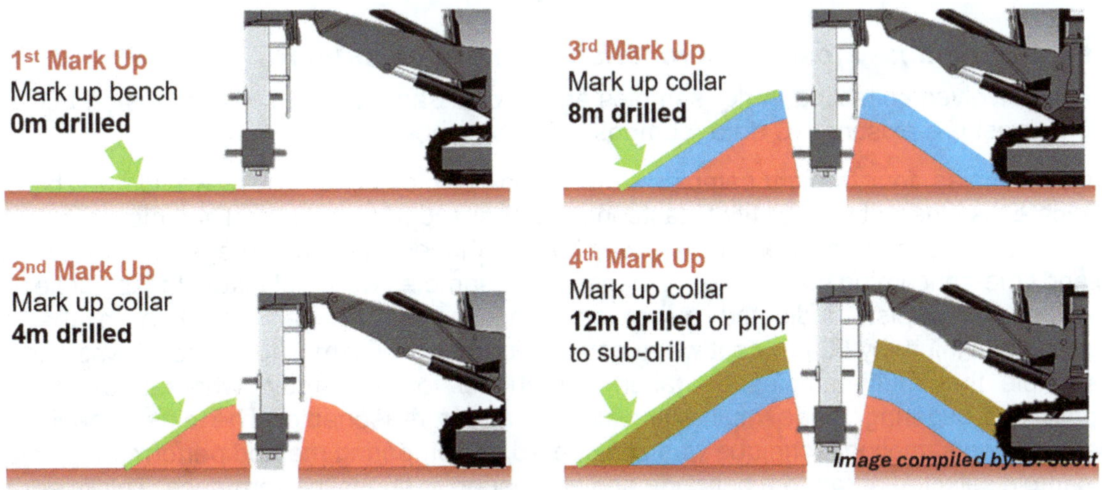

FIG 2 – Demonstration of bench and flitch mark-up process for each blastcone as part of flitch distribution characterisation study phase. NB – D65 depicted in figure for ease of demonstration, but Pit Viper 271 drilled all holes in study.

Demarcation step-by-step:

- Technician demarcated the bench floor before the PV moved onto the hole location.
- PV then drilled first flitch, shutdown. Technician then demarcated top of first flitch.
- PV then drilled second flitch, shutdown. Technician then demarcated top of second flitch.
- PV then drilled third flitch, shutdown. Technician then demarcated top of third flitch.
- This became the final demarcation between the uppermost flitch and the sub-drill (which the PV then drilled to finish the hole).

Both the cone peak, and the midpoint of the slope of the cone were faced off as flitch height measurement locations. Each flitch and sub-drill was then measured and recorded for both locations (cone peak and mid-point) as per Figure 3. Photographs were taken of the cones, and additional details of the cone were recorded – cone quality, wetness, etc.

FIG 3 – Example blastcone measurement. Picture a demonstrates mid-slope and cone peak measurement locations for each cone. Picture b demonstrates measurement of flitch and sub-drill height in the cone peak position, with flitch and sub-drill sections labelled. Both images show green ceremonial blast powder used in demarcation.

This process was repeated for over 79 holes across six pit areas (to capture potential variability in cone morphology due to geologic variability across pit areas). This data could then be analysed to determine average flitch ratios that should be assumed in flitch sampling and compare differences in flitch ratios between cone peak and mid-slope sampling points.

Phases 2 and 3 – Comparative sampling trials

Phase 2 – Preliminary testing: comparisons to pre-existing RC

This phase involved collecting auger samples and shovel samples from blastcones and comparing them to nearest pre-existing RC data (composited to blasthole).

For auger samples, four to eight samples were taken per blastcone. These would later be used to help indicate the inherent variability of blastcone material and assess optimal sample count per cone. For shovel samples, one sample per cone was taken. Shovelling is on average more ergonomically taxing and time-consuming than augering, hence more samples were not taken. In fact, assessments by site physiotherapists indicated auger sampling led to a total reduction in the risk of musculoskeletal injury by 52 per cent when compared to shovel sampling. Hence it was considered unreasonable that multi-samples be taken per shovel for the study, whereas the favourable ergonomics of augering made multi-sampling via auger reasonable. The difference in sample number between sampling technique was considered by comparing method performance of just one auger sample selected at random per cone and comparing to shovel performance. This is addressed in the results section of study.

For both sample methods, estimated sub-drill was removed before sampling. The multiple auger samples were taken at mid-slope locations on the cone, taken roughly equidistant from each other around the cone where possible (cone-condition dependent). The auger insertion angle was ~perpendicular to the cone slope. Each auger sample point was put into separate sample bags for separate assay results (to aid constraining blastcone variability). Where sample mass was insufficient from one auger insertion, an additional auger insertion was taken as close to the original sample point as possible until adequate sample weight was achieved.

The shovel sample was taken from the cone peak. The sample face was exposed by removing material from the outside of the cone, and then the sampler estimated flitch proportions to take a sample that attempted to estimate the bench. Unlike the next phase, Phase 3, the shovel samples could be considered a hybrid between shovel flitch and single-point shovel bench samples. Hybrid because the sampler attempted to take extra time to identify each flitch and put it in the one sample bag for the bench, whereas true single-point shovel bench sampling would not attempt to identify and correct for flitches. In true flitch sampling, each flitch would be put in a separate sample bag (as per next phase). This ensures that differences in sample weight between flitch samples doesn't bias the bench to over- or under-represent the flitches. Ultimately, each shovel sample was put in one sample bag per cone for this phase. Excess manual handling was aimed to be avoided in this sample phase, hence less sample bags collected, given Phase 3 would test both shovel techniques robustly.

Qualitative field data was also collected for each blastcone on cone condition (Good, Moderate, Poor condition) and cone wetness (Dry, Moist, Wet). Blast rig type was also recorded.

Data from 108 holes were originally collected in this phase, but the data set was culled to 70 holes to remove blasthole to RC twin hole distances >5 m and removing holes that were missing at least four auger samples and one shovel sample. This ensured paired/matched analysis across the data set without excessive RC to blasthole spacing.

Phase 3 – Detailed study: comparison to whole-cone samples

In this phase, an RC rig initially drilled 30 holes twinned to planned blasthole locations (within 1 m), on an ore shot in the Sierra 4 region of Roy Hill mine. The RC holes were drilled before the blastholes, as due to the compressed air and potential communication between twin holes, safety procedure for RC drilling requires a minimum offset distance from existing open holes (far beyond 1 m). The RC holes were sampled at 1 m intervals.

A Pit Viper 271 then entered the pattern and drilled the twinned blastholes to match the RC depth (12 m).

Once the Pit Viper had moved off, auger bench (four per blastcone), shovel bench/single point bench (one per blastcone) and shovel flitch (one per flitch for a total of three samples per blastcone) samples were taken from each blastcone. Augering workflow was identical to that discussed in the previous comparative phase.

The whole-cone sample was then collected and drummed, with each blastcone typically weighing around 1 t. This material was then sent off-site to International Mineral Standards (IMS), where it was crushed down to a top-size of 10 mm, homogenised and split into a primary and duplicate sample (performed separately for each blastcone). These sample bags were then sent back to the Roy Hill Mine Lab, where the samples were prepared and analysed as per all other samples collected (auger, shovel, RC samples). Figure 1 presents an overview of the sampling. Figure 4 demonstrates RC and Pit Viper twin hole drilling, along with whole-cone sample collection.

FIG 4 – Phase 3 RC and Pit Viper drilling, and whole-cone sample collection.

DATA ANALYSIS and RESULTS

Phase 1 – Flitch ratio/distribution

This phase of the study was inspired by McArthur, Jones and Murphy (2010), who also undertook flitch measurements in blastcones to determine average flitch ratios (at RTIO's Mesa K deposit). A key point of differentiation between the two studies is: their study involved 8 m benches with 2 × 4 m flitches covering a channel iron deposit (CID), whereas this study involves a 12 m bench for 3 × 4 m flitches covering primarily bedded material. Both this study and their study took flitch height measurements at the cone peak, but this study also took measurements at the mid-slope of the cone (typical auger sample-point at other operations).

Figure 5 demonstrates the ratio of the flitches and sub-drill as drilled (in-ground) for the blastholes used in this study. The average depth of the blastholes was 13.5 m, with each of the three flitches representing 29.4 per cent of the hole, with sub-drill on average representing 11.7 per cent.

For the flitch thickness measurements taken at the peak of the blastcones (Figure 6), notable variability was observed between blastcones. But on average Flitches 1 and 2 made up a similar proportion of the blastcone height (34 per cent and 32 per cent respectively), whilst Flitch 3 represented a reduced proportion (22 per cent). The sub-drill on average represented 12 per cent of the cone height (which was close to the *in situ* average).

In contrast, measurements taken at the mid-slope location of the blastcone (Figure 7), found Flitches 1, 2 and 3 had average contributions of 26 per cent, 33 per cent and 26 per cent respectively. Though Flitch 2 is the most dominant on average, overall this location displays a more even contribution of each flitch once sub-drill is removed compared the cone peak location.

As the mid-slope location displays a more equitable distribution of flitches in the blastcone, this is considered favourable for a blind bench-sampling technique like augering. This ensures the bias of certain flitches are minimised (on average) in a sample of the bench. However, for a technique like shovel flitch sampling, the mid-slope provides little sample face to aid splitting of flitch-samples and obtaining adequate flitch sample weight (average face thickness of mid-slope = 25.8 cm versus 34.65 cm for cone peak sans sub-drill). This is where the cone peak is considered a better sampling location for shovelling.

With further regards to shovel-flitch sampling, the ratios for the cone peak location were carried forwarded into the Detailed Study phase of this study for assessing this sampling technique.

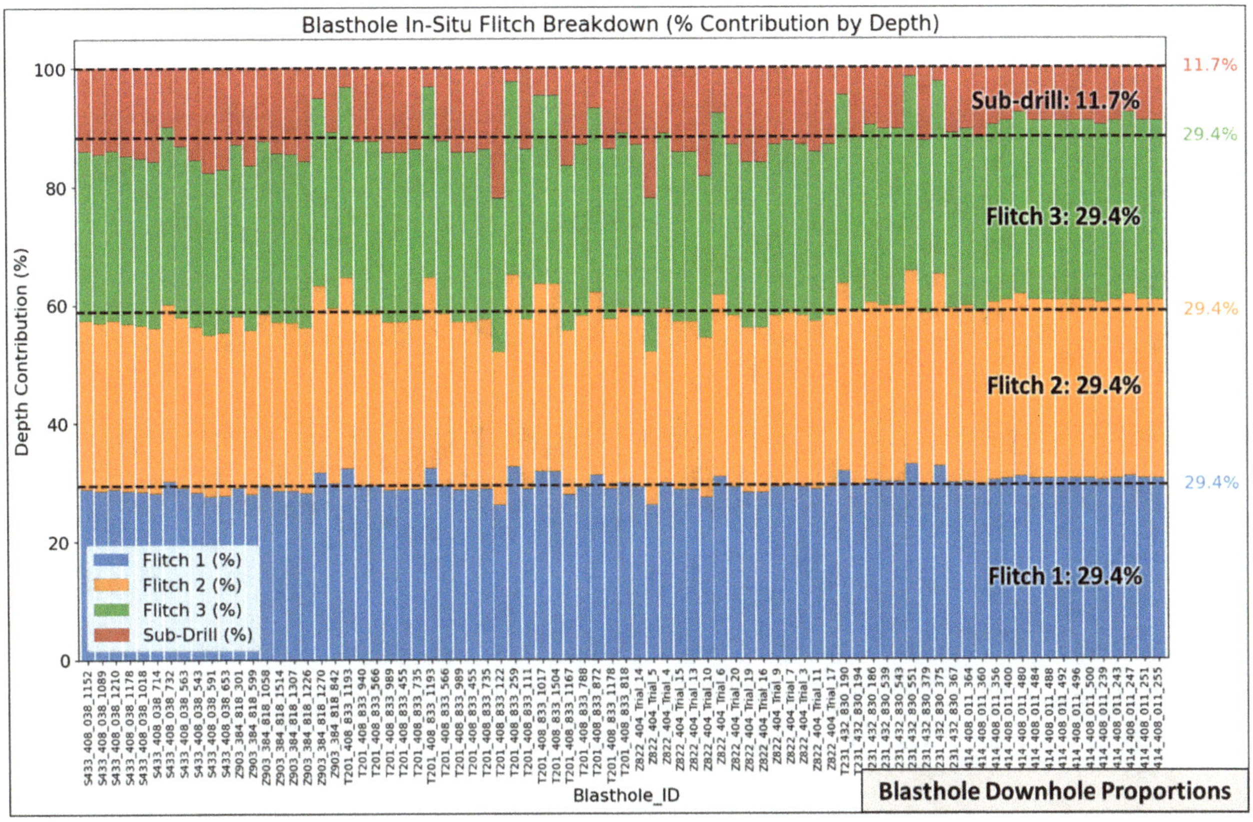

FIG 5 – Downhole/*in situ* proportions of flitches and sub-drill for the blastholes used in this study phase. These average in-ground ratios can be directly compared to the average ratios found in the blastcone measurements.

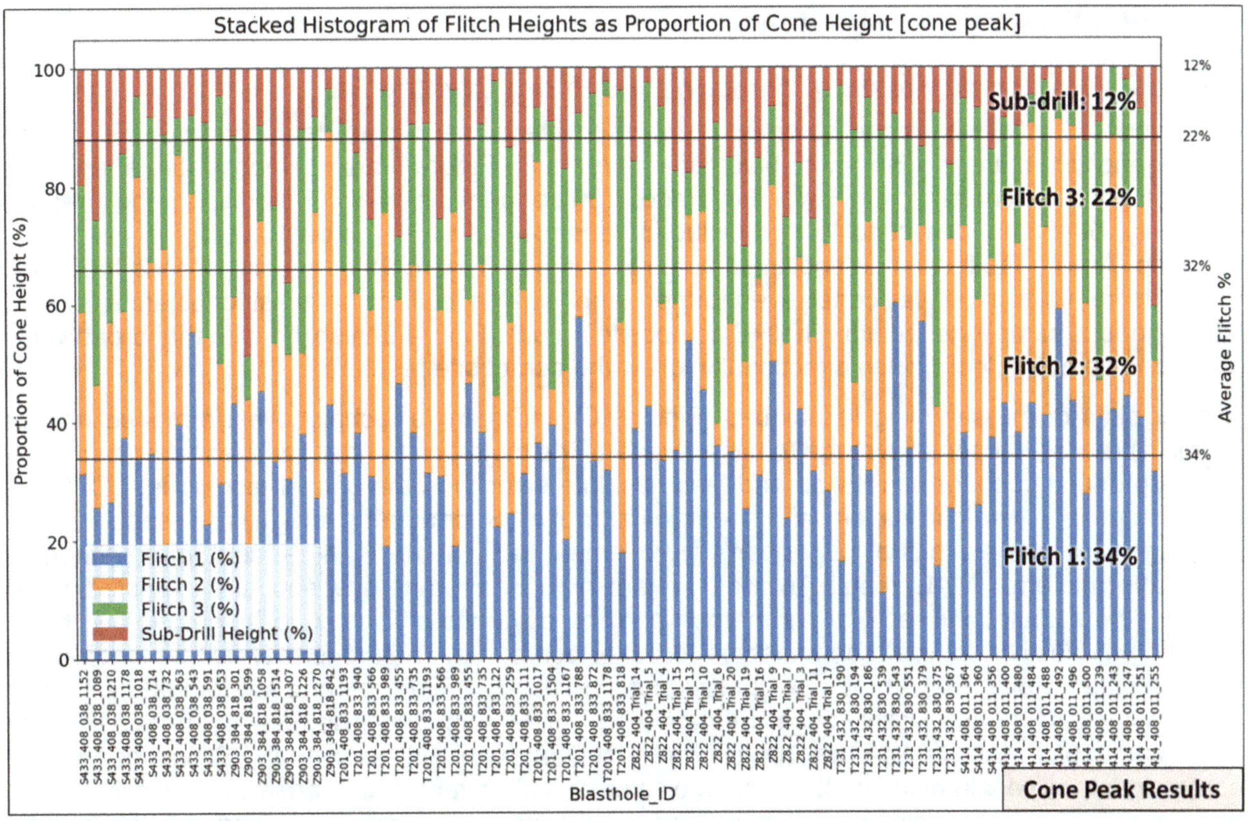

FIG 6 – Blastcone flitch and sub-drill measurements taken at the peak of the blastcone. Notable variability is found, with the average ratios of the blastcone components indicated on right of figure.

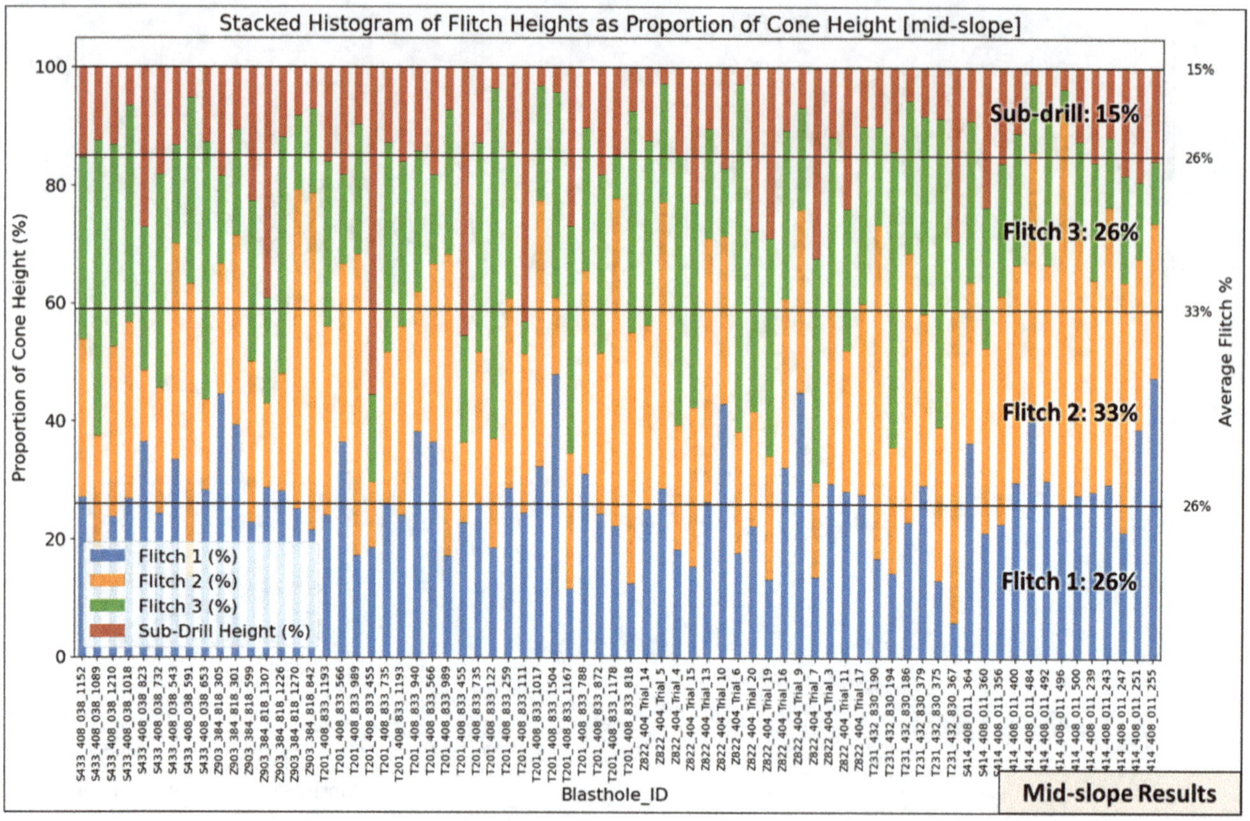

FIG 7 – Blastcone flitch and sub-drill measurements taken at the mid-slope location of the blastcone. Again, a high degree of variability is observed, but once sub-drill is removed a slightly more equitable ratio of flitches is observed versus cone peak ratios.

Sampling method performance

Phase 2 – Preliminary testing: comparisons to pre-existing RC

Analysing the multiple auger samples taken per blastcone alone demonstrated the spatial heterogeneity in material inherent to blastcones (Figure 8). Figure 8 also demonstrates a tendency towards greater variability with lower Fe grade. <50 per cent Fe blastcones display the greatest variability, 50–55 per cent tend toward reduced variance, with >55 per cent averaging the lowest variance. Adding the blastcone shovel samples to this plot (Figure 9), generally extends the range in results from individual blastcones even further, emphasising the fact that blastcones are a laterally heterogenous sampling medium. The RC composite results (Figure 9) also indicate that broadly, the blastcone sampling techniques directionally follow the composite results.

Scatter plots of the sampling techniques (Figure 10) and sample technique performance metrics (Table 1) reveal that auger average (cone average of the multi-point auger samples taken per cone) results produce superior correlation and less absolute error than the shovel technique. LOI is the only exception, where correlation values are similar (but error for auger average is still superior). For Fe, auger average exhibits a 12 per cent improvement in R, and a 10.4 per cent improvement in mean absolute error (MAE) result. An improvement in auger average over shovel is expected, given the multiple samples utilised via this method and the evidence of blastcone heterogeneity. Multi-sampling of blastcones would be expected to better combat this heterogeneity versus single-point cone samples. However, using the auger random derivative, which uses one random auger sample from each blastcone (the terms 'auger random' and 'single-point auger sample' will herein be used interchangeably), results are still better than shovel samples (especially R values of key contaminants silica, alumina and Mn). Mean bias error (MBE) values are consistently positive for Fe and negative for contaminants. This is across all sample techniques. Therefore all sample techniques are conservative in their sampling (samples on average underestimate Fe, and overestimate contaminants versus authoritative reference). This bias is further observed in the next phase of the study, despite the authoritative reference being different (whole-cone) – and is discussed in the next sections.

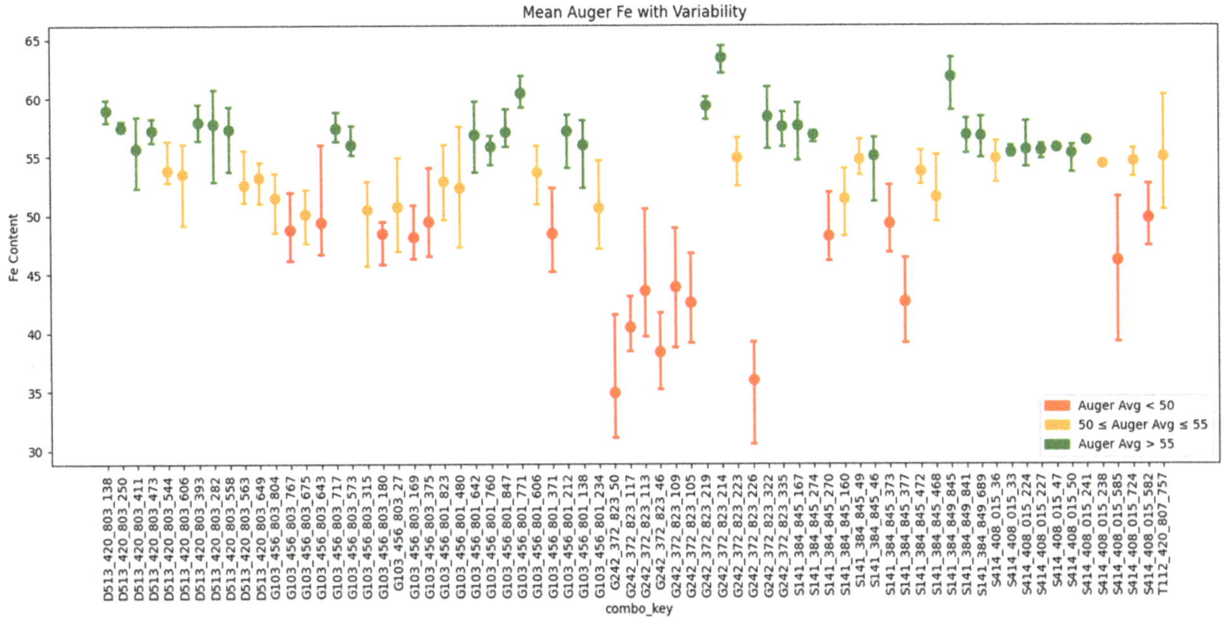

FIG 8 – Blastcone Fe variability, as demonstrated by the four to eight samples taken per blastcone. Mean auger Fe value is represented by the circle, with the range in values found shown by line. Results are coloured by mean auger Fe content (combo_key = blasthole).

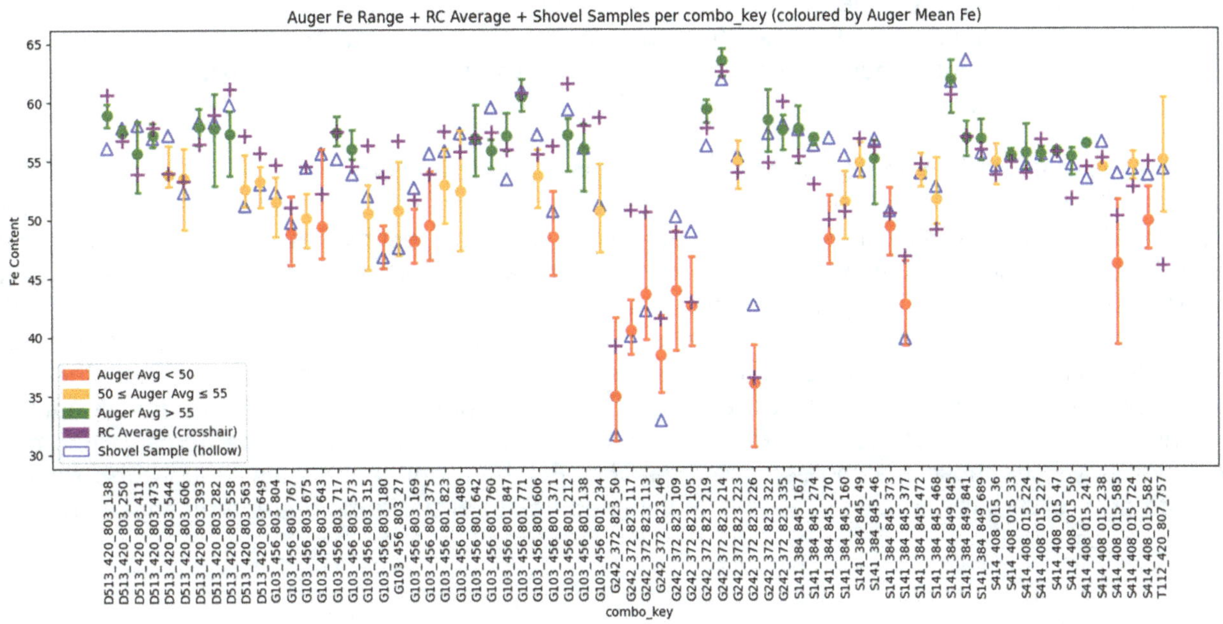

FIG 9 – Auger Fe range as per previous figure, with shovel sample results (hollow blue triangles) and pre-existing RC composite results (purple crosshairs – authoritative reference for this study phase) added. The addition of the shovel samples further emphasises the heterogeneity evident in the study blastcones. Generally, the blastcone sampling techniques directionally follow the RC composites.

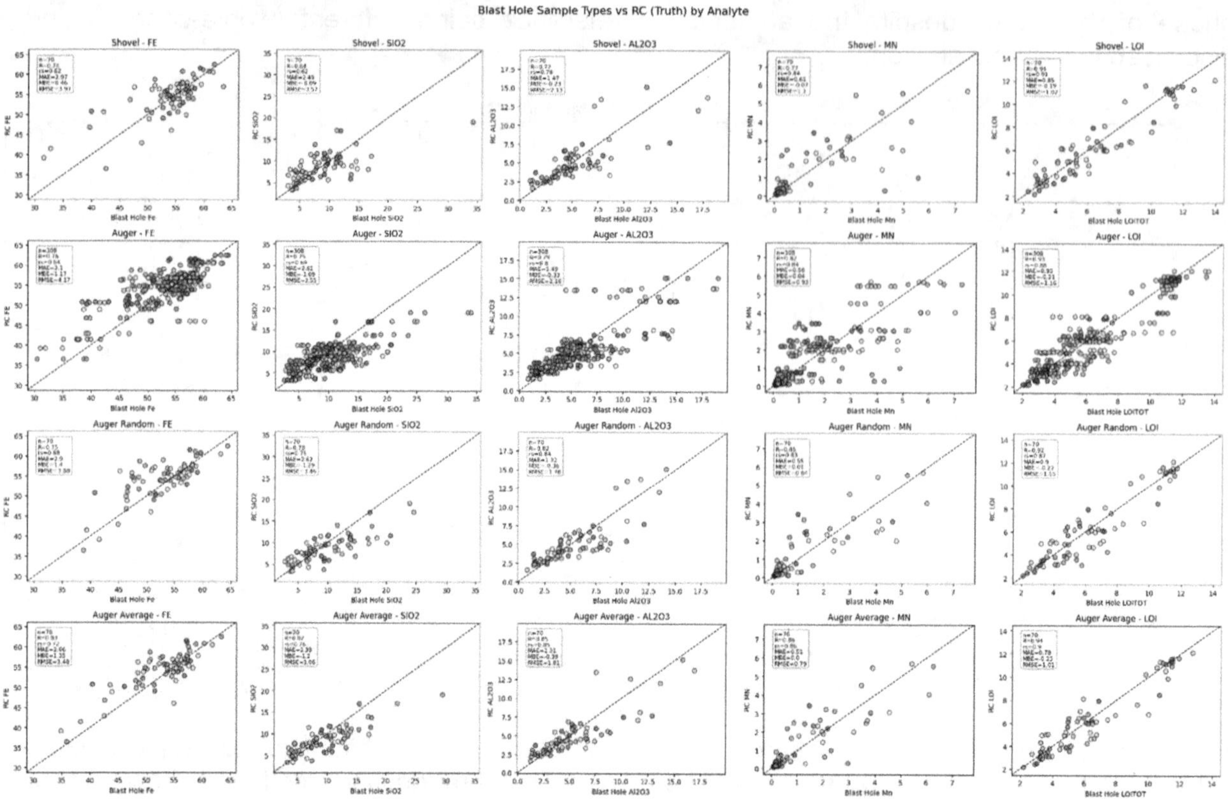

FIG 10 – Performance of shovel versus auger sampling techniques per analyte. 'Auger Random' represents one auger sample selected at random per blastcone (single point auger sample per cone), whilst 'Auger Average' represents the average result of all auger samples taken from a blastcone. Data point colours indicate blasthole ID – this further highlights blastcone variability under the 'auger' result). All blasthole to RC spacing ≤5 m.

Further observations from this study phase of note were regarding cone wetness and drill type. Cone wetness (classes: dry, moist, wet) generally demonstrated that the moist/damp cones (n = 32) performed better than dry cones (n = 29), across all sampling techniques. The performance was even better for wet cones, however the sample count was so low that this observation should be taken with caution (n = 3). The count may have also been low because wet cones were not amenable to sampling (shape/low aspect ratio) and were frequently skipped. If sampling was forced on those wet cones, it potentially could have brought performance down. Nonetheless, at a minimum the data indicates better performance if the cones are at least damp/moist rather than dry for all sample techniques. As for drill rig used, the Pit Viper demonstrated better performance across all sample techniques versus those blastholes drilled via D65. For both drill techniques though, the relative performance of all sampling techniques is maintained (ie auger performs better than shovel for both drill rig types).

TABLE 1

Summary of performance metrics per sample technique, per analyte. R = Pearson correlation coefficient, rs = Spearman's rank correlation coefficient, MAE = mean absolute error, MBE = mean bias error, RMSE = root mean squared error. Correlation coefficient's are coloured by green to red scale; 1 = dark green, 0 = dark red (negative values not present in this data set/no negative correlations). Error metrics are coloured by a relative scale of purple – white that is based off the limits of the data per analyte, per error metric. Dark purple = greatest error, white = lowest error. Pre-existing RC composites taken as reference with BH to RC distance ≤5 m.

Analyte	Sample Method	R	rs	MAE	MBE	RMSE
FE	Shovel	0.74	0.62	2.97	0.46	3.97
FE	Auger	0.76	0.64	3.10	1.17	4.17
FE	Auger Random	0.75	0.68	2.90	1.40	3.88
FE	Auger Average	0.83	0.72	2.66	1.35	3.48
SIO2	Shovel	0.64	0.62	2.49	-0.09	3.57
SIO2	Auger	0.75	0.69	2.61	-1.09	3.55
SIO2	Auger Random	0.78	0.75	2.62	-1.29	3.46
SIO2	Auger Average	0.82	0.76	2.39	-1.20	3.06
AL2O3	Shovel	0.77	0.78	1.47	-0.23	2.13
AL2O3	Auger	0.79	0.80	1.49	-0.33	2.16
AL2O3	Auger Random	0.82	0.84	1.32	-0.36	1.78
AL2O3	Auger Average	0.85	0.85	1.31	-0.39	1.81
MN	Shovel	0.77	0.84	0.61	-0.07	1.10
MN	Auger	0.82	0.84	0.58	0.04	0.93
MN	Auger Random	0.85	0.83	0.55	0.01	0.84
MN	Auger Average	0.86	0.86	0.51	0.00	0.79
LOI	Shovel	0.94	0.91	0.85	-0.19	1.07
LOI	Auger	0.93	0.88	0.85	-0.21	1.16
LOI	Auger Random	0.92	0.87	0.90	-0.22	1.15
LOI	Auger Average	0.94	0.90	0.78	-0.23	1.01

A small sub-set of the sampled blastcones were taken as eight auger samples instead of four (only n = 7 blastcones after cull). This was to aid testing of optimal sample count for augering. To gain insight into the impact of varying sample count on auger sampling performance, Monte Carlo simulation was run on the auger blastcone data set. The process iterated through different sample counts per blastcone, with randomly selected combinations of samples (1000 iterations). Average impact to MAE and R were reported as performance metrics per sample count. Figure 11 demonstrates that the greatest performance improvements are found between two to three samples per blast cone. However, performance improvement for both R and MAE continues to improve through to eight samples, the maximum sample count, though at a reduced rate. Given this was a limited data set, the results were cross-referenced against the four auger samples per blastcone data set (n = 63). The results of the simulation of this data set are found in Figure 12. This corroborates the results found in Figure 11, performance metrics continue to improve through to the maximum sample count of these blastcones (four), with the greatest performance improvement seen at two samples. The results of both simulations would argue that a minimum of two to three samples should be taken per blastcone, but that performance continues to improve the more samples taken (with sample count above eight not tested).

A word of caution on the results from this phase – use of the pre-existing RC drill holes is a sub-optimal authoritative reference. Firstly, the lateral offset of the holes (up to 5 m), introduces the impact of geological variability to the analysis (with greater impact the greater the offset). Secondly, the composites are not perfectly aligned with the blastholes in the vertical/Z direction. As the pre-existing RC was drilled from original topography, not the bench, the 2 m RC sample intervals will not perfectly align with the sampled blasthole intervals. This is why the results from the next phase, which obtains whole cone samples, and newly drilled RC from the bench, should be taken as a superior analysis. Nonetheless, due to the increased cost and time spent on the next phase, the data set size is smaller, and it could not capture as many variables as this phase (eg different mine sub-regions, blast rig types etc).

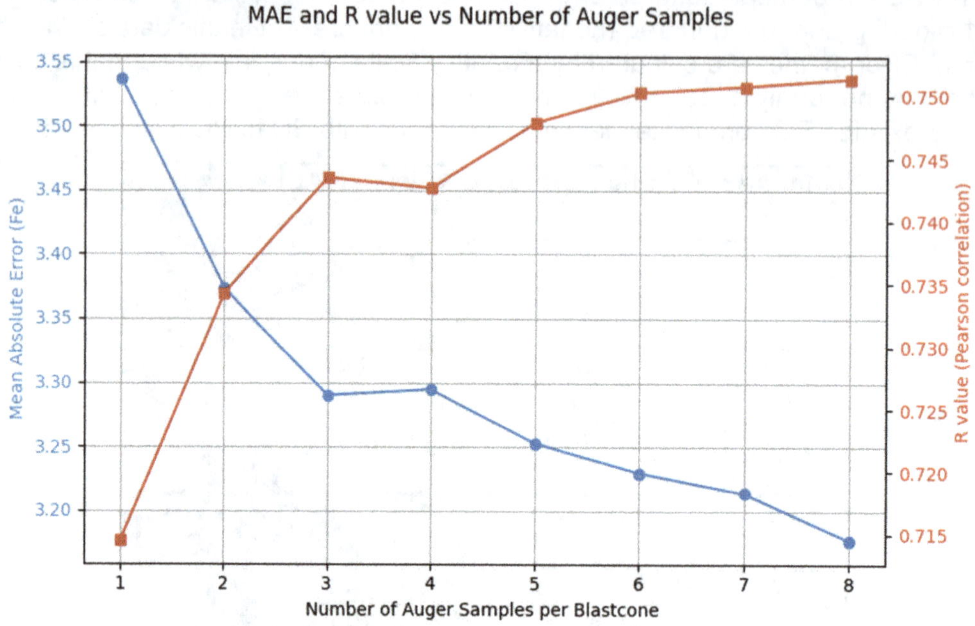

FIG 11 – Summary of Monte Carlo simulation run on eight auger sample count blastcones (n = 7). MAE and R score used to evaluate performance. Performances continues to improve for both metrics right up to the maximum sample count. Greatest performance improvement between two to three samples, with reduced rate of improvement thereafter. Simulation iterations = 1000.

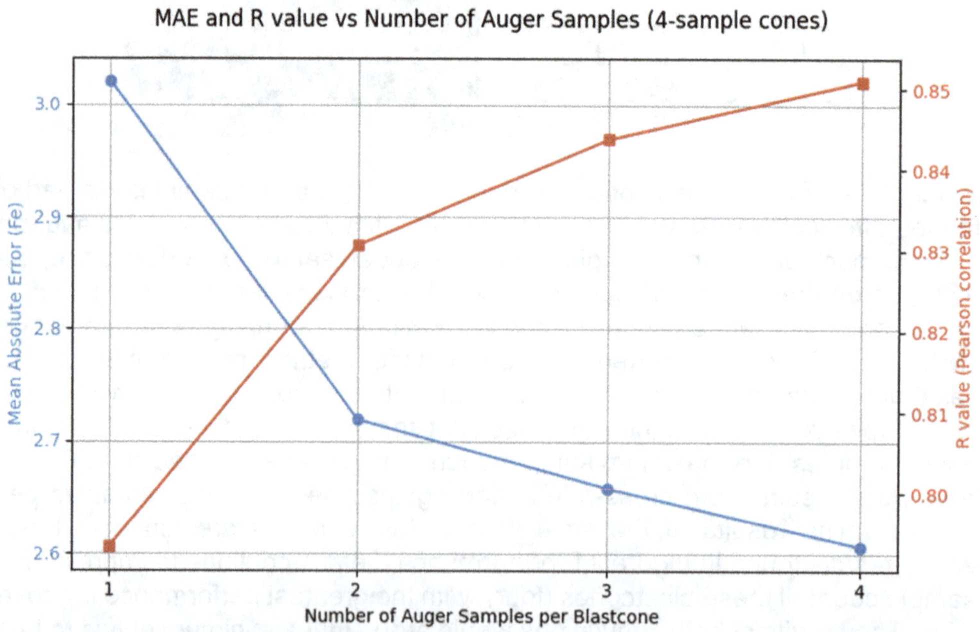

FIG 12 – Summary of Monte Carlo simulation on four auger sample count blastcones (n = 63). Improving performance right through to maximum sample count (four). Simulation iterations = 1000.

Phase 3 – Detailed study: comparisons to whole-cone sample and RC twin holes

The results of this phase of the study were considered the most reliable given the taking of whole-cone samples, and the drilling of dedicated, in-pit, minimally offset RC twin holes to the blastholes as authoritative references. Whole-cone samples were seen as the most robust authoritative reference to assess performance of the sampling methods, given the sampling methods are direct sub-samples of the blastcone. The RC twin holes were an additional (secondary) validation source. Given the whole-cone samples have such significant sample mass (~1 t), it was important to validate the splits were representative. Figure 13 demonstrates the results of primary and duplicate sample comparisons for each blasthole taken from splits of the whole-cone sample (after reducing to a top-size of 10 mm and homogenisation process). Results demonstrate good agreement between primary and duplicate results for key analytes, with the R and R^2 value of all analytes being >0.98, with low MAE.

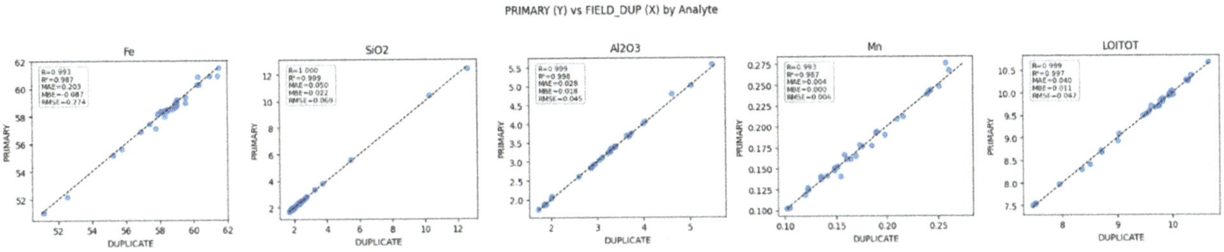

FIG 13 – Comparison of whole-cone primary and duplicate samples per blastcone for key analytes. Results demonstrate good agreement between both primary and duplicate samples.

Summary statistics of all sample types are displayed in Table 2. All sample types (including RC twin holes) have mean Fe values that are lower than whole-cone sample values. For most analytes, including Fe, the 95 per cent confidence interval of whole-cone sample and RC twin hole data has the closest overlap/agreement of all the sample types. This is expected given they are primary and secondary reference data sets in this experiment but is nonetheless a positive validation. Fe values for whole-cone sample demonstrate a range of values from low-grade to high-grade content. Silica and alumina present a range from low to high values. LOI shows a tighter range, generally towards the higher end of Roy Hill grade bins. Manganese demonstrates the tightest range, and very low values. The values in this sample set are not ideal for evaluating Mn sampling performance due to the very low values, with a mean value of 0.18 and a low dispersion with a standard deviation of 0.05. All other key analytes are adequate.

Table 2 also highlights the differences in dispersion between sample types. Again, whole-cone samples and RC twin holes show the closest agreement, with similar standard deviation. Shovelling techniques (bench and flitch) display the greatest variability of all techniques, with shovel flitch having a 30 per cent larger standard deviation than whole-cone samples. Auger techniques demonstrate the least dispersion. This difference in dispersion may at least be partially attributable to the volume-variance effect, with the whole-cone and RC samples having the largest sample support/volume. Meanwhile, the auger and shovel samples have similar sample support, so the direct comparison of these samples should be minimally affected by bias due to differing sample support.

TABLE 2

Summary statistics of all sample types. CI = 95 per cent confidence interval, lower and upper bounds.

Analyte	Sample Type	count	mean	median	var	std	min	max	CI_lower	CI_upper
Fe	Whole_Blastcone	30	58.22	58.58	5.29	2.3	51.01	61.47	57.37	59.08
Fe	RC_Twin	30	57.64	58.19	4.46	2.11	51.72	60.23	56.85	58.42
Fe	Shovel Bench	30	57.16	57.97	7.65	2.77	49.26	61.44	56.13	58.19
Fe	Shovel Flitch	30	57.77	58.46	8.8	2.97	43.88	61.07	56.66	58.87
Fe	Auger	120	57.53	57.88	3.37	1.84	49.37	60.83	57.19	57.86
Fe	Auger Random	30	57.63	57.88	2.64	1.62	53.44	60.83	57.02	58.23
Fe	Auger_Average	30	57.53	57.84	2.8	1.67	53.18	60.39	56.9	58.15
SiO2	Whole_Blastcone	30	2.96	2.2	5.98	2.45	1.67	12.47	2.05	3.87
SiO2	RC_Twin	30	3.18	2.16	6.7	2.59	1.69	13.13	2.21	4.14
SiO2	Shovel Bench	30	3.27	2.34	5.68	2.38	1.58	12.19	2.38	4.16
SiO2	Shovel Flitch	30	3.29	2.06	17.13	4.14	1.61	24.24	1.74	4.83
SiO2	Auger	120	3.08	2.42	3.03	1.74	1.65	10.75	2.76	3.39
SiO2	Auger Random	30	3.06	2.45	3.28	1.81	1.7	10.61	2.39	3.74
SiO2	Auger_Average	30	3.08	2.46	2.66	1.63	1.83	8.53	2.47	3.69
Al2O3	Whole_Blastcone	30	3.22	3.26	0.86	0.93	1.75	5.57	2.87	3.57
Al2O3	RC_Twin	30	3.52	3.47	0.9	0.95	2.0	5.72	3.17	3.88
Al2O3	Shovel Bench	30	3.9	3.64	2.25	1.5	1.27	9.05	3.35	4.46
Al2O3	Shovel Flitch	30	3.43	3.51	0.71	0.84	1.49	5.49	3.11	3.74
Al2O3	Auger	120	3.72	3.66	0.93	0.97	1.66	6.86	3.55	3.9
Al2O3	Auger Random	30	3.64	3.58	0.79	0.89	2.08	5.54	3.31	3.98
Al2O3	Auger_Average	30	3.72	3.64	0.78	0.88	2.07	5.42	3.39	4.05
Mn	Whole_Blastcone	30	0.18	0.17	0.0	0.05	0.1	0.28	0.16	0.19
Mn	RC_Twin	30	0.17	0.18	0.0	0.05	0.09	0.27	0.15	0.19
Mn	Shovel Bench	30	0.17	0.16	0.0	0.06	0.1	0.41	0.15	0.2
Mn	Shovel Flitch	30	0.18	0.16	0.0	0.07	0.08	0.37	0.15	0.2
Mn	Auger	120	0.17	0.16	0.0	0.05	0.1	0.3	0.16	0.18
Mn	Auger Random	30	0.17	0.16	0.0	0.05	0.11	0.27	0.15	0.19
Mn	Auger_Average	30	0.17	0.16	0.0	0.04	0.11	0.28	0.15	0.18
LOITOTAL	Whole_Blastcone	30	9.38	9.71	0.81	0.9	7.5	10.68	9.04	9.72
LOITOTAL	RC_Twin	30	9.68	9.85	0.68	0.83	7.84	10.74	9.37	9.99
LOITOTAL	Shovel Bench	30	9.89	10.06	0.61	0.78	7.7	11.29	9.6	10.18
LOITOTAL	Shovel Flitch	30	9.49	9.67	0.69	0.83	7.15	10.65	9.18	9.8
LOITOTAL	Auger	120	9.76	9.97	0.67	0.82	6.33	11.0	9.61	9.91
LOITOTAL	Auger Random	30	9.75	9.92	0.53	0.73	7.9	11.0	9.48	10.03
LOITOTAL	Auger_Average	30	9.76	9.91	0.58	0.76	8.07	10.73	9.47	10.04

Scatter plots of all samples versus whole-cone samples (Figure 14) highlight aspects already mentioned with regards to summary statistics. Single-point shovel bench samples demonstrate significant dispersion, and distance from the 1:1 line when plotted against the reference (whole-cone samples). Shovel flitch samples, though variable, display a better affinity for the 1:1 line than single-point shovel bench samples. Auger samples again highlight the heterogeneity of each blastcone, with an observable spread between the four samples per blastcone (some holes more variable than others). When looking at auger results, versus single-point auger sample (auger random) or auger average results, one may notice the data appears to have greater sampling variance ('noisier') when all auger points are plotted. This is because each data point contains a component of *truth + error (noise/variance contribution)*. By plotting all auger results as individual data points, the weight of this variance contribution is increased. By averaging the results, the variance contribution is reduced. Also, even single-point auger samples (auger random) can present less noise, by reducing the error contribution from the other three auger points per blastcone. This is also dependent on random selection (ie which combination of sample points are selected).

The scatter plots of Figure 14 graphically demonstrate the performance metrics found in Table 3. The shovel sampling methods show the poorest affinity to the 1:1, whilst the auger techniques, especially with averaging, displays the best performance. In terms relative rank, single-point shovel bench (Fe R = 0.52, MAE = 1.49) is the poorest performer and should be avoided as a sampling technique where possible. Shovel Flitch is the second worst performing sampling technique (Fe R = 0.71, MAE = 1.14). Auger results plotted in raw form (no averaging of results per blastcone) is the third ranked performer (Fe R = 0.78, MAE = 1.11). Again, the unnecessary impact from the additive variance contribution of the individual samples is working against the technique. The auger's superior performance versus shovel techniques is highlighted by reducing to one auger sample per blastcone selected at random. This also allows for a direct one sample to one sample comparison with the single-point shovel bench sample method (same sample count per blastcone). Single-point auger sampling delivers a result of Fe R = 0.83, MAE = 0.99, comfortably ahead of the shovel flitch sampling. Auger Average yields the best blastcone sample method performance (Fe R = 0.87, MAE = 1.02). This supports the intended use of the tool: to quickly and ergonomically collect multiple samples around a blastcone to capture heterogeneity with less sampling bias.

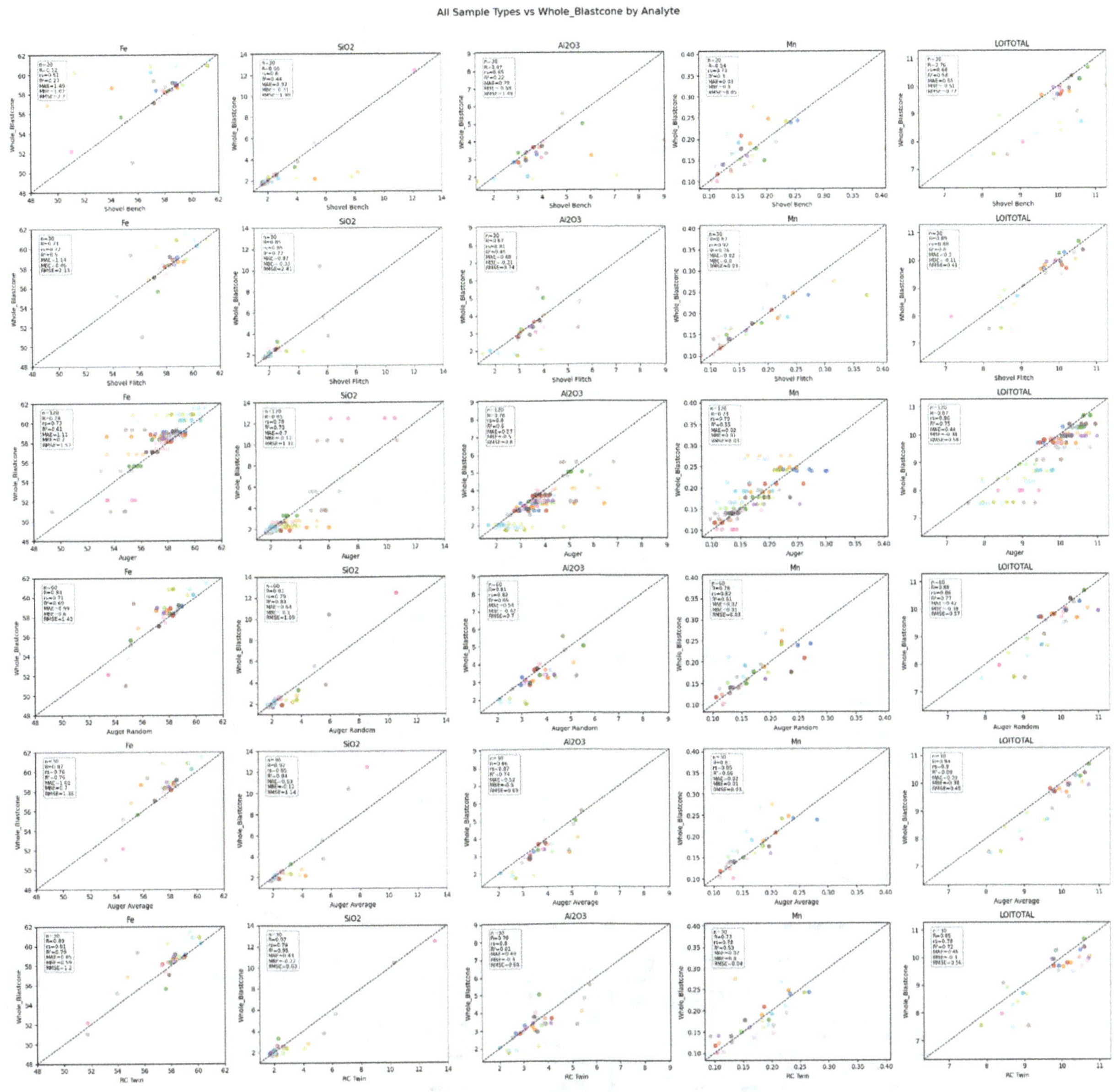

FIG 14 – Scatter plots of whole-cone samples versus single-point shovel bench sample, shovel flitch, auger and RC twin hole samples. Auger scatter plots highlight variability in spot samples across a blastcone. Auger Random (single-point auger sample) sample type represents one random auger sample selected at random per blastcone (versus total of four per blastcone). Auger Average represents the average of all four auger samples per blastcone. Scatter plots are coloured by blasthole ID. Equivalence/1:1 line is plotted in dash through middle of graph. Data pairs appearing above the line indicate sample under-estimating whole-cone assay, whilst those underneath the line indicates sample type overestimates analyte. In general, the sample methods tend to underestimate Fe, and overestimate contaminants.

TABLE 3

Detailed Study/Phase 3 performance metrics per sample type. Whole-cone sample is the authoritative reference used to evaluate performance. R = Pearson correlation coefficient, rs = Spearman's rank correlation coefficient, MAE = mean absolute error, MBE = mean bias error, RMSE = root mean squared error. Correlation coefficients are coloured by green to red scale; 1 = dark green, 0 = dark red (negative values not present in this data set). Error metrics follow a red to green colour scale, but scales are determined by the value range by metric, by analyte (re-scales for each analyte). MBE is a symmetrical scale around 0, to account for negative and positive values, but is again guided by the limits of the data. Mn is uncoloured, due to issues with the Mn data as detailed in the main text.

Sample Type	Analyte	R	rs	MAE	MBE	RMSE
Shovel Bench	Fe	0.52	0.51	1.49	1.07	2.70
Shovel Flitch	Fe	0.71	0.72	1.14	0.46	2.13
Auger	Fe	0.78	0.73	1.11	0.70	1.57
Auger Random	Fe	0.83	0.71	0.99	0.60	1.43
Auger Average	Fe	0.87	0.76	1.02	0.70	1.35
RC Twin	Fe	0.89	0.81	0.85	0.59	1.20
Shovel Bench	SiO2	0.66	0.80	0.92	-0.31	1.98
Shovel Flitch	SiO2	0.85	0.85	0.87	-0.33	2.41
Auger	SiO2	0.85	0.78	0.70	-0.12	1.31
Auger Random	SiO2	0.91	0.79	0.64	-0.10	1.09
Auger Average	SiO2	0.92	0.85	0.63	-0.12	1.14
RC Twin	SiO2	0.97	0.74	0.43	-0.22	0.63
Shovel Bench	Al2O3	0.47	0.65	0.79	-0.68	1.49
Shovel Flitch	Al2O3	0.67	0.81	0.48	-0.21	0.74
Auger	Al2O3	0.78	0.80	0.57	-0.50	0.80
Auger Random	Al2O3	0.81	0.82	0.54	-0.42	0.70
Auger Average	Al2O3	0.86	0.87	0.52	-0.50	0.69
RC Twin	Al2O3	0.78	0.80	0.49	-0.30	0.68
Shovel Bench	Mn	0.54	0.73	0.03	0.00	0.05
Shovel Flitch	Mn	0.87	0.92	0.02	0.00	0.03
Auger	Mn	0.74	0.79	0.02	0.01	0.03
Auger Random	Mn	0.78	0.82	0.02	0.01	0.03
Auger Average	Mn	0.81	0.85	0.02	0.01	0.03
RC Twin	Mn	0.73	0.78	0.02	0.00	0.04
Shovel Bench	LOITOTAL	0.76	0.68	0.55	-0.51	0.77
Shovel Flitch	LOITOTAL	0.89	0.88	0.30	-0.11	0.41
Auger	LOITOTAL	0.87	0.86	0.44	-0.38	0.58
Auger Random	LOITOTAL	0.88	0.86	0.42	-0.38	0.57
Auger Average	LOITOTAL	0.94	0.90	0.39	-0.38	0.49
RC Twin	LOITOTAL	0.85	0.78	0.46	-0.30	0.56

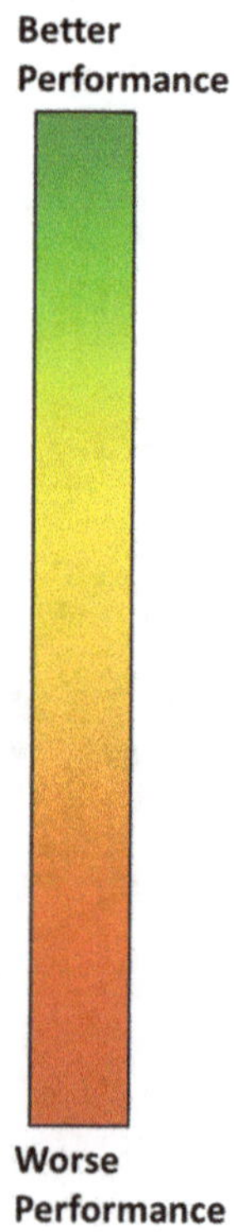

By inspection of Figure 14, one may argue that some of the Fe shovel data points are outliers and require removal. But these samples make up a significant portion of the data (~20 to 30 per cent), and the authors would argue that these are irreducible error associated with this sampling technique. That is, if this sampling technique is used, similar dispersion and error will continue to be observed. Spearman's rank correlation coefficient (rs) was also used as a cross-check to Pearson correlation coefficient. Spearman's coefficient attenuates the negative impact of outliers on correlation by sorting data pairs into relative ranks and then calculating correlation. If a data set contained significant outliers, the rs value should be higher than the R value, as the impact of the outliers is reduced. Table 3, shows for Fe shovel samples, the R and rs values are similar, and rs does not outperform R. This does not support significant outlier presence for Fe. For nearly all other analytes, despite differences in R and rs, the relative performance rank is generally maintained between sample techniques (single-point shovel bench sample worst performer, auger average best performer).

Figure 15 presents two panels. The upper panel a pseudo-whisker plot (Fe) similar to that presented in the previous Phase 2 analysis section (Figures 8 and 9). The lower panel a difference plot between authoritative reference (whole-cone) and single point shovel bench sample. The upper panel again highlights the heterogeneity inherent to blastcones, with a tendency towards greater heterogeneity the lower the grade. Both panels visually highlight the repeatability of the auger versus the shovel techniques. The auger average results more closely and consistently track the whole-cone value. Meanwhile the shovel techniques, especially the poorest performing single-point shovel bench sample technique in the lower panel, demonstrate large deviations from the authoritative reference value. When the shovel samples are off, they have a greater tendency to present much larger magnitude errors than the auger. The root mean squared error (RMSE) of the shovelling techniques highlighted in Table 3 also corroborates this (RMSE is useful as it emphasises samples in a data set with larger error deviations, whereas MAE just reflects the straight-line average). A spatial element can be noticed in the Figure 15 plots – that is both auger and shovel tended to encounter larger errors in spatial groups of holes. Figure 16 highlights these spatial groupings and presents cross-sections covering them – with RC twin hole Fe assay results plotted. Using the RC twin hole data allows for the smallest sampling interval (1 m) to be presented to observe downhole assay variability. Comparing the Fe difference plot to the cross-sections, the areas of higher sampling error do (in general), coincide with holes with greater downhole variability. Holes with less downhole variability, and greater consistency, tend to produce less error. Blastcones already present variability across individual blastcone samples, this can be observed by looking at the pseudo whisker plot and comparing to RC twin holes that have relatively consistent downhole material. But a tendency can be expected that the greater the downhole variability, the variability of the blastcone will likely increase in line. This flags the increased importance of capturing heterogeneity (ie multi-samples) in blast patterns that are expecting greater variability (ie holes crossing strata/domains/material types versus holes following strata in steeply dipping orebodies or otherwise expecting homogenous material).

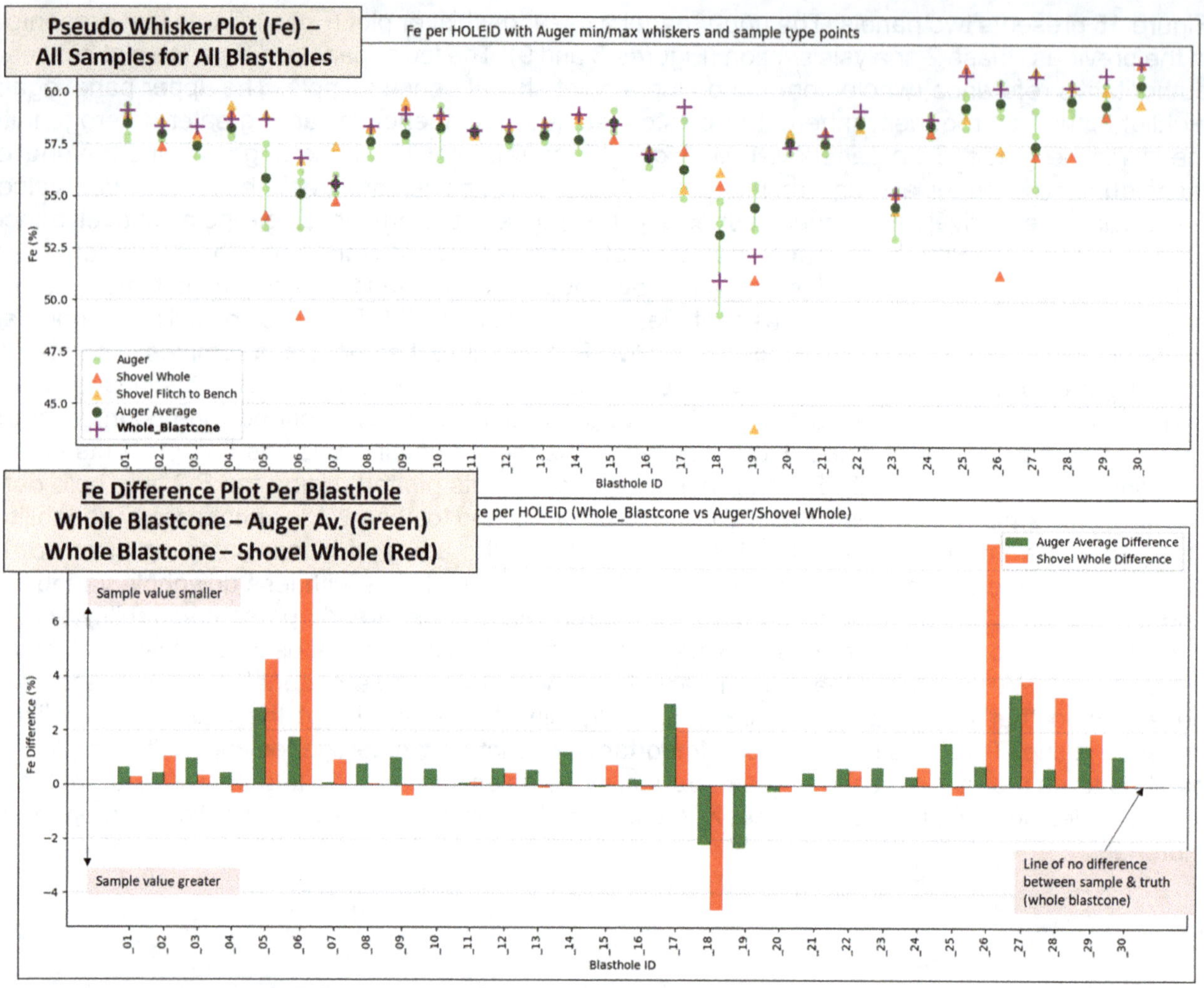

FIG 15 – Dual panel figure demonstrating blastcone variability and auger versus shovel sampling differences (Fe). Upper panel = pseudo whisker plot that demonstrates whole-cone assays (authoritative reference/truth) versus blastcone sample types. Individual auger values per blastcone are shown in light green. The average of these four auger values per blastcone are represented by dark green circles. Lower panel = Fe difference plot comparing Auger Average to Shovel Bench ('shovel whole') results per blasthole. Line of no difference displayed that represents no difference between whole-cone and sample method. Together, both panels demonstrate that augering tends to greater repeatability and proximity to truth values. Shovelling techniques show a tendency that when deviations occur, they tend towards large magnitude deviations. Data also shows spatial clustering indicating areas associated with greater error and variability, which is further explored in Figure 16. Shovel Whole = Shovel Bench, Shovel Flitch to Bench = Shovel Flitch.

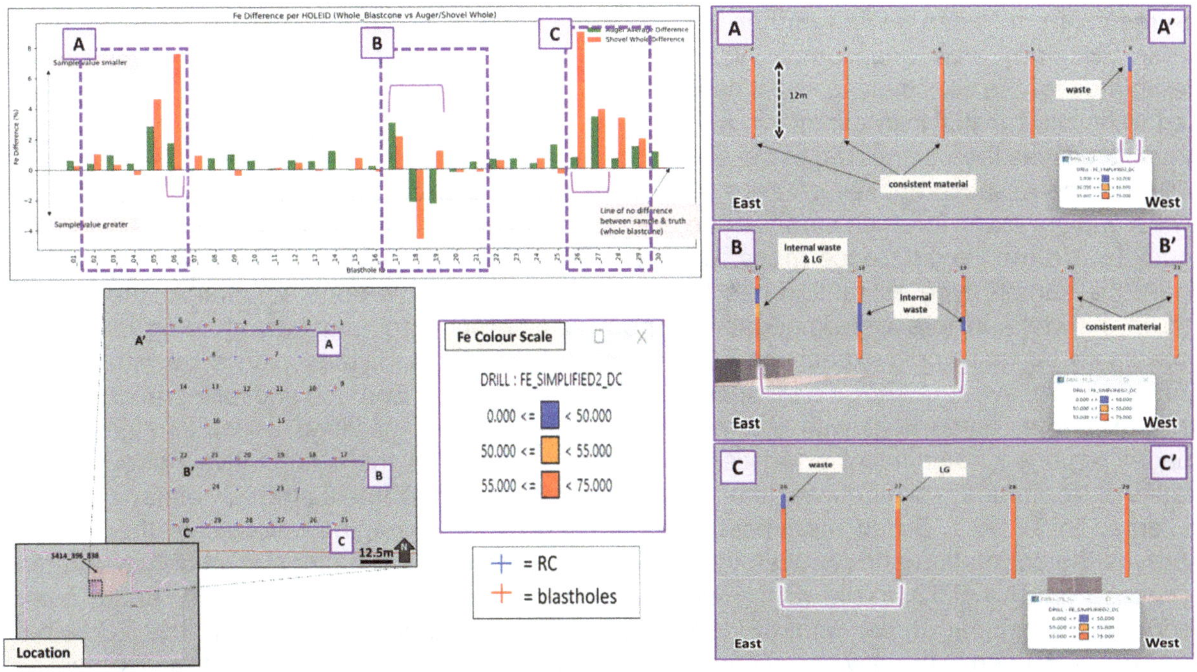

FIG 16 – Cross-sections of RC twin holes examining spatial grouping of three areas with greater error and variability (A, B and C) indicated in last figure. The cross-sections indicate a tendency that blastholes with greater downhole variability tend to show larger error and variability in blastcone sampling.

Table 3, Figures 14 and 15 all demonstrate that all sample techniques (including RC twin holes) underestimate Fe and overestimate contaminants when compared to the whole-cone sample. For this data set the samples produce results that are conservative in general (understate the analyte of economic interest and over-state the impurities). The mean bias error (MBE) values displayed in Table 3 support this conclusion. For MBE – positive values indicate the sample is underestimating the authoritative reference source (whole-cone), with negative values indicating overestimation. Fe MBE are positive values for all sample types, ranging from 0.46 to 1.07 Fe wt per cent. For all the contaminants (except Mn), the MBE values are consistently negative. Mn is close to zero, but as previously mentioned, Mn values in this data set are simply too low to draw reliable conclusions. As the Phase 3 study area covered a limited area of the mine (one sub-region of an orebody that is over 20 km in strike length), reference to Table 1 from Phase 2 (previous section) will allow for cross-reference to results over the large mine area. Again, although this earlier study phase uses a sub-optimal authoritative reference, and not the whole-cone samples, it provides a larger data set across multiple sub-regions. The results in Table 1 indicate that the same observations of MBE are present. That is, sampling techniques underestimate Fe, and overestimate contaminants versus reference (though different references are used between phases). For Mn, the results are mixed with shovel showing negative values and all auger modalities showing positive. However, all are near zero regardless. An interesting observation is that for the Phase 3 data set in Table 3, RC values all demonstrate the same bias relative to whole-cone sample results as the blastcone sampling techniques. Considering the Phase 2 data presented in Table 1 is based on comparisons to RC data (as no whole-cone samples were taken in this phase), it may be expected that the biases would be minimal given the reference already showed the same directional bias in Phase 3 when compared to whole-cone samples. Perhaps if Phase 2 were compared to whole-cone samples instead of RC results, the MBE may be even more biased across analytes. Nonetheless, Phase 2 lends further strength that the Phase 3 analyte biases may be persistent across site and extend beyond the detailed study area. But it requires further study to validate (whole-cone samples taken across more sub-regions). No sizing data was taken as part of this study. A good understanding of size fraction tendency per sample technique versus reference (and assays of size fractions), would help to understand the source of the bias. This is a potential optimisation of future study design that could help to isolate the size fraction delta between sample technique and reference responsible for the bias.

The results from Phase 3 were further scrutinised by interpolating the results for each sample type across the study area and comparing the resulting analyte heat maps. Each sample type was interpolated for the bench using radial basis function as the estimation algorithm. A cell grid was then used to divide the study area into 1 m × 1 m cells. The values within each of these cells for a given sample type was then compared to the authoritative reference map (whole-cone) cells. Performance could then be assessed based on the same correlation and error metrics. Figure 17 represents the raw Fe analyte maps for each sample type, including the authoritative reference map – whole-cone. Figure 18 are the Fe difference maps, with the convention being sample type minus whole-cone. Where the sample type estimation has zero difference with whole-cone estimation, the difference map will be white. Where there is negative or positive difference, red and blue shading will be shown respectively. As one would expect, the RC twin hole map displays the least difference with the whole-cone, but auger average shows a surprisingly similar agreement. Both the shovel techniques visually show the worst agreement with the whole-cone, introducing localised estimation artifacts and broader regions of significant disagreement. Table 4 illustrates the quantitative performance. All relative performance ranks between the sample methods are the same as the raw assay comparison presented in Table 3. But in many instances, the power of averaging involved in the estimation improves the correlation and reduces the error. For instance, single-point shovel bench sample Fe improves from R = 0.52 and MAE = 1.49 to values of R = 0.65 and MAE = 0.99. Auger average produces results near on par with, or in the case of LOI and alumina better than, the RC twin hole results. In fact, even in the raw assay comparison, auger average performs better than RC twin holes for these analytes. This may suggest that for this pattern at least, geologic variability is greater for these analytes than the other analytes. That is, the 1 m offset of the RC is enough to vary the result for alumina, that the R score is notably diminished versus directly sub-sampling the blastcone. On the other side of this, silica potentially displays the least geologic variability, with the RC twin holes scoring R = 0.97 in raw assay comparison and R = 0.98 in the estimation comparison, indicating minimal geologic spatial variability for this analyte.

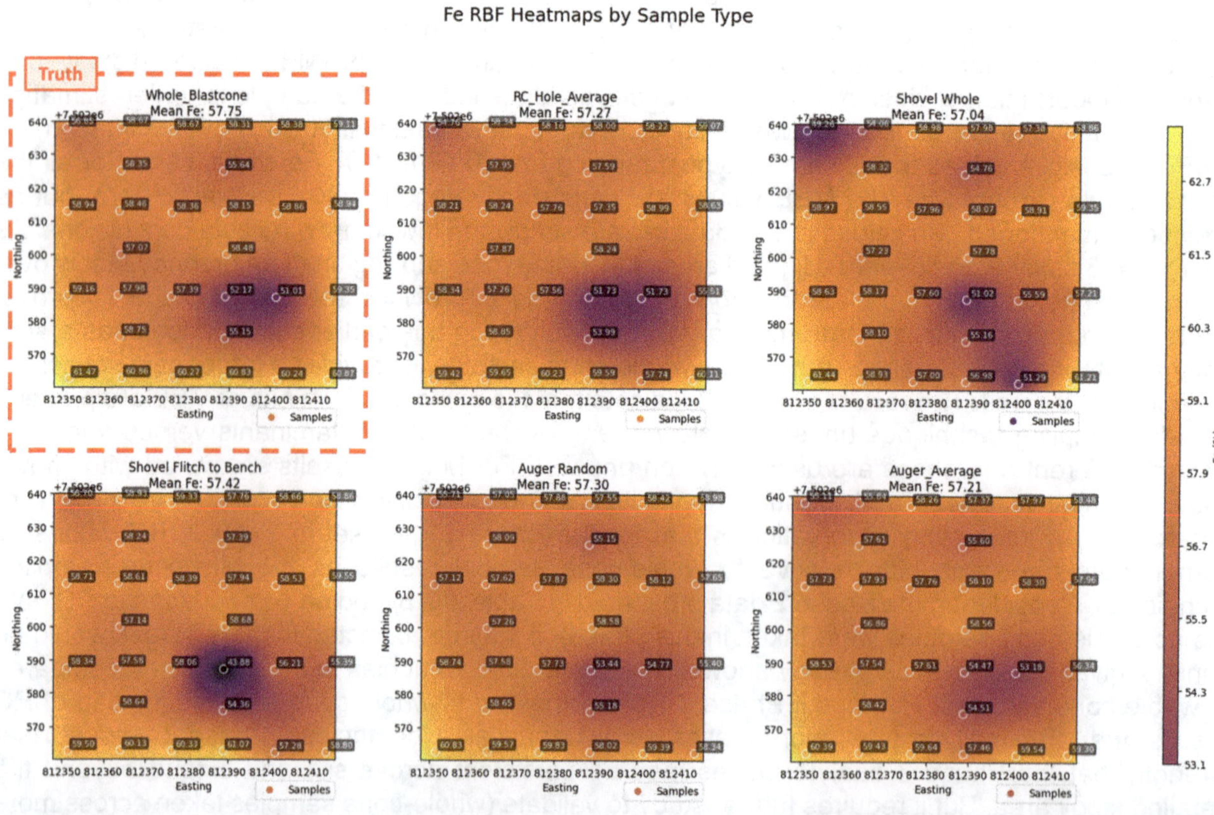

FIG 17 – Fe estimation heat maps produced for each sample type. Maps can be directly compared to the primary authoritative reference (whole-cone samples/whole blastcone – top left) to visually assess difference. RC twin hole = 'RC_Hole_Average', Shovel Bench = 'Shovel Whole', Shovel Flitch = 'Shovel Flitch to Bench'.

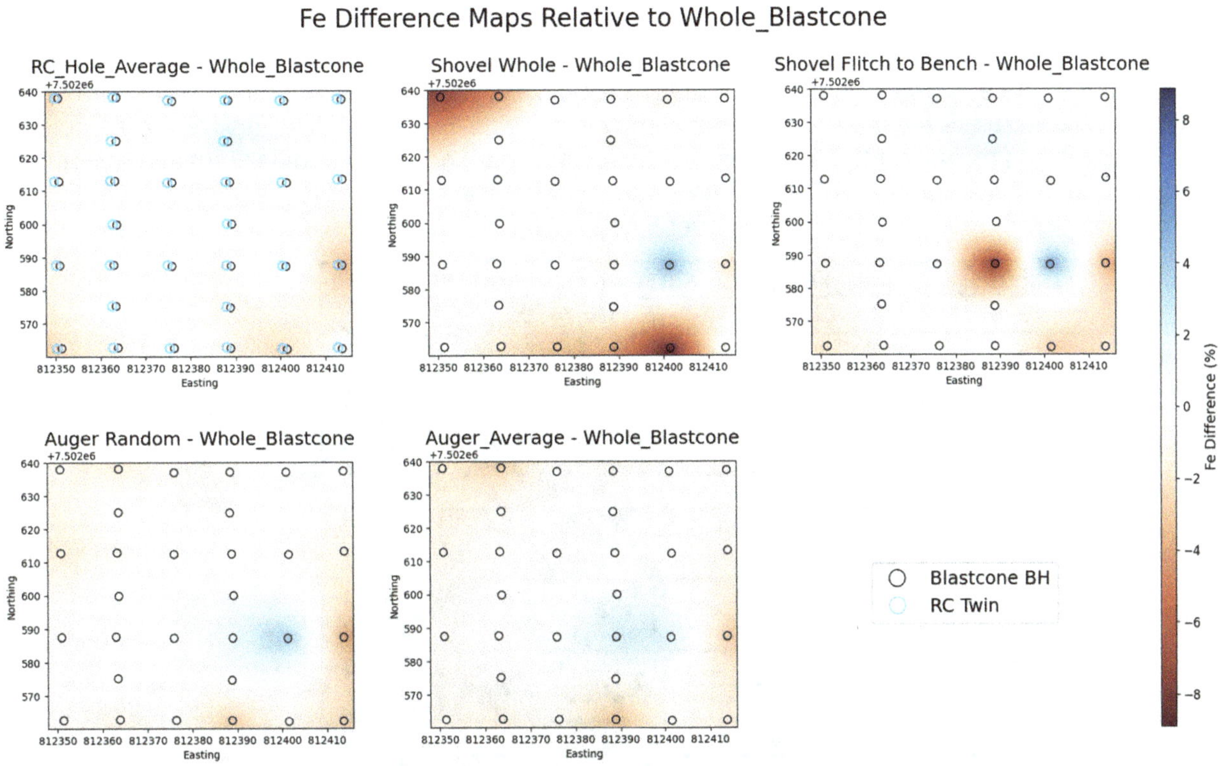

FIG 18 – Fe difference maps relative to whole-cone (truth). These maps are based on the estimated Fe maps displayed in Figure 17. Convention is sample type – whole-cone. RC twin and auger results show the least difference to the whole-cone results, with the shovel sampling techniques introducing noticeable error. RC twin hole = 'RC_Hole_Average', Shovel Bench = 'Shovel Whole', Shovel Flitch = 'Shovel Flitch to Bench', Whole-cone = 'Whole_Blastcone'.

TABLE 4

Summary of performance of producing estimations for each sample type by analyte and comparing to the whole-cone estimates as the authoritative reference. Most correlation and error values have improved versus raw assay comparisons shown in Table 3. The relative rank of the sample techniques has remained the same as raw assay comparison, despite differences in absolute performance. RC twin hole = 'RC_Hole_Average', Shovel Bench = 'Shovel Whole', Shovel Flitch = 'Shovel Flitch to Bench'.

Analyte	Sample Type	R	rs	MAE	MBE	RMSE
Fe	Shovel Whole	0.65	0.70	0.99	-0.73	1.75
Fe	Shovel Flitch To Bench	0.85	0.86	0.70	-0.34	1.18
Fe	Auger Random	0.90	0.83	0.75	-0.46	1.00
Fe	Auger_Average	0.92	0.87	0.79	-0.55	0.98
Fe	Rc_Hole_Average	0.92	0.85	0.66	-0.49	0.88
SiO2	Shovel Whole	0.75	0.83	0.69	0.06	1.31
SiO2	Shovel Flitch To Bench	0.92	0.96	0.54	0.26	1.30
SiO2	Auger Random	0.95	0.90	0.48	0.01	0.77
SiO2	Auger_Average	0.95	0.92	0.49	0.01	0.81
SiO2	Rc_Hole_Average	0.98	0.86	0.35	0.21	0.48
Al2O3	Shovel Whole	0.51	0.70	0.56	0.54	0.98
Al2O3	Shovel Flitch To Bench	0.70	0.73	0.33	0.17	0.50
Al2O3	Auger Random	0.81	0.75	0.42	0.38	0.54
Al2O3	Auger_Average	0.86	0.82	0.44	0.44	0.55
Al2O3	Rc_Hole_Average	0.74	0.67	0.43	0.27	0.55
Mn	Shovel Whole	0.65	0.73	0.02	-0.00	0.04
Mn	Shovel Flitch To Bench	0.90	0.94	0.01	-0.00	0.02
Mn	Auger Random	0.82	0.80	0.02	-0.00	0.02
Mn	Auger_Average	0.85	0.86	0.01	-0.01	0.02
Mn	Rc_Hole_Average	0.77	0.79	0.02	-0.00	0.02
LOITOTAL	Shovel Whole	0.77	0.78	0.47	0.46	0.65
LOITOTAL	Shovel Flitch To Bench	0.93	0.91	0.20	0.09	0.29
LOITOTAL	Auger Random	0.89	0.90	0.36	0.34	0.48
LOITOTAL	Auger_Average	0.95	0.94	0.35	0.35	0.42
LOITOTAL	Rc_Hole_Average	0.92	0.88	0.29	0.24	0.37

Better Performance

Worse Performance

As per the last phase, a Monte Carlo simulation was run on the number of auger samples, to test whether there was a point where additional samples returned no uplift. The observations from this phase are arguably more reliable for this than the Phase 2 results, as more reliable authoritative reference are utilised. Hence more trust can be put in the error and correlation metrics ouptut from this simulation. The last phase (Phase 2) took majority four samples per blastcone, with a smaller subset of eight samples. As a reminder, simulations ran on this data indicated increasing improvements in R and MAE right up to the sample limit (and potentially beyond), but the greatest uplift in results came between two to three samples. The results from the simulation of the Phase 3 data are shown in Figure 19. The results generally agree with the Phase 2 sample number simulations, with two samples displaying the greatest rate of performance improvement, but continued improvement through to four samples (just at a reduced rate).

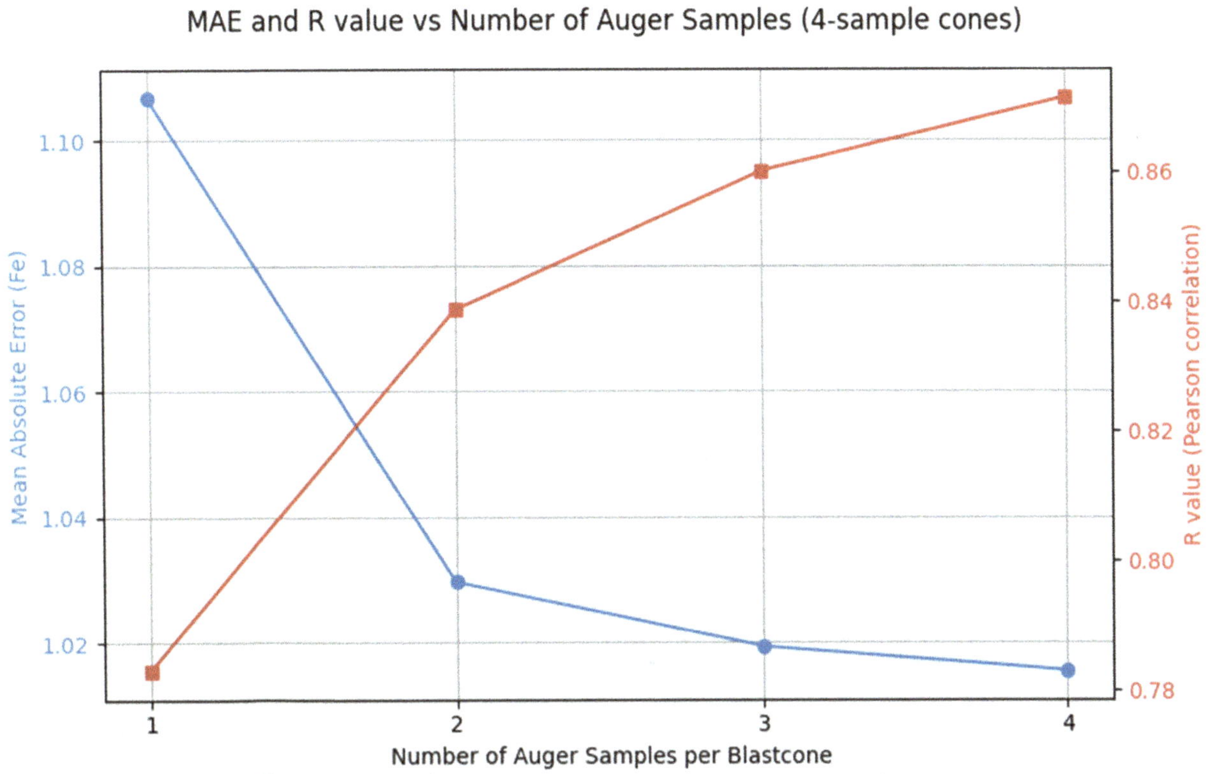

FIG 19 – Phase 3 Monte Carlo simulation on four auger sample count blastcones (n = 30) in the Detailed Study. Improving performance right through to maximum sample count (4). Simulation iterations = 1000.

Uncertainty margins

To support both comparison and communication of the performance of each sampling technique, uncertainty margins were established to describe the range of outcomes that an assay from each method could represent. To do this the difference between each sampling technique and the reference assays (whole-cone) was calculated for all data pairs. ±2× the standard deviation of the difference was then used to establish uncertainty margin for each technique (the commonly used 2σ convention [note σ formally refers to the population standard deviation, but in this context it is referring to the sample standard deviation *s*]). This was used for several reasons. The underlying assay populations are often approximately Gaussian, but the paired difference distributions show a broader range of behaviours, from near-normal to non-normal. Despite this, for this data set the ±2σ approach was considered the more conservative operational representation of uncertainty than a typical low-side/high-side percentile approach such as P_{10}/P_{90}. In addition, the use of ±2σ is easy to communicate and understand for field staff turning around grade control processes under time duress.

One could argue that a percentile-based approach would be appropriate to characterise uncertainty for this difference data. However, a percentile-based approach would directly inherit the previously discussed bias present in the data set. In this study, that bias is conservative: Fe is generally under-predicted, while contaminants are generally over-predicted relative to whole-cone assays. As a result, percentile-based bounds would produce an asymmetrical range that implies a relatively optimistic upside correction (eg add much more Fe in the high-side than it will take away in the low-side). This is acceptable if the conservative bias in the sampling techniques is present across all sub-regions of the Roy Hill mine and not localised to the Phase 3 study area. Although Phase 2 data supports the conservative bias, until it is further proven with more whole-cone analysis across more sub-regions, the conservative approach provided by ±2σ of the paired difference is preferred.

The uncertainty margins for each sample technique are presented in Table 5, with an example of a difference distribution and associated uncertainty in Figure 20. For Fe and silica, shovel techniques exhibit around twice the uncertainty of augering. In fact, for Fe, the uncertainty margin for Auger

Average approaches that observed for closely-spaced RC twin holes, highlighting its superior performance to shovelling techniques.

TABLE 5

Summary of the difference between whole-cone (authoritative reference) and other sampling modalities. ±2 standard deviations of the paired difference were used to characterise uncertainty. This highlights that the uncertainty margin associated with both shovel sampling techniques is materially higher than other sampling methods. Auger Average shows an Fe uncertainty margin around half that of the shovel techniques. In fact, for Fe, it achieves a similar uncertainty margin to near-offset RC twin hole data. Uncertainty colour scale: red to green scale; red = largest uncertainty, green = smallest uncertainty margin. Colour scale limits reset for each analyte. 'Std Difference' = standard deviation of the difference between authoritative reference and sample.

Analyte	Sample Type	n	Mean Difference	Std Difference (±)	2x Std Difference (±)
Fe	Shovel Bench	30	-1.07	2.52	5.05
Fe	Shovel Flitch	30	-0.46	2.11	4.22
Fe	Auger	120	-0.7	1.41	2.83
Fe	Auger Random	30	-0.6	1.32	2.64
Fe	Auger_Average	30	-0.7	1.18	2.35
Fe	RC Twin	30	-0.59	1.06	2.13
SiO2	Shovel Bench	30	0.31	1.99	3.97
SiO2	Shovel Flitch	30	0.33	2.42	4.85
SiO2	Auger	120	0.12	1.31	2.62
SiO2	Auger Random	30	0.1	1.1	2.2
SiO2	Auger_Average	30	0.12	1.15	2.3
SiO2	RC Twin	30	0.22	0.6	1.21
Al2O3	Shovel Bench	30	0.68	1.35	2.69
Al2O3	Shovel Flitch	30	0.21	0.72	1.45
Al2O3	Auger	120	0.5	0.63	1.26
Al2O3	Auger Random	30	0.42	0.57	1.14
Al2O3	Auger_Average	30	0.5	0.48	0.96
Al2O3	RC Twin	30	0.3	0.62	1.24
LOITOTAL	Shovel Bench	30	0.51	0.59	1.18
LOITOTAL	Shovel Flitch	30	0.11	0.41	0.81
LOITOTAL	Auger	120	0.38	0.44	0.89
LOITOTAL	Auger Random	30	0.38	0.44	0.88
LOITOTAL	Auger_Average	30	0.38	0.31	0.62
LOITOTAL	RC Twin	30	0.3	0.48	0.97

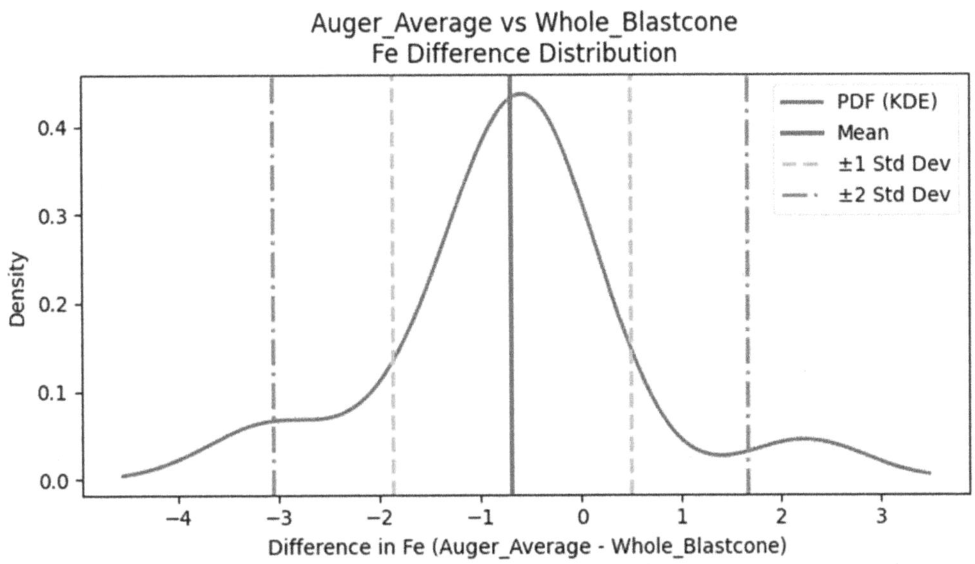

FIG 20 – Example difference distribution (sample minus authoritative reference) with 2× standard deviation plotted (green dash-dot lines). This uncertainty range is robust to the majority of outcomes and covers 90 per cent of the data (a perfect Gaussian distribution would cover ~95 per cent with two standard deviation). Covering 90 per cent of the outcomes is more conservative than a conventional percentile approach, where the commonly used low-side to high-side P_{10}–P_{90} spread covers 80 per cent of outcomes. Using a percentile approach would also provide asymmetrical low-side to high-side uncertainty range, due to the bias present in the paired difference data. Until this bias is proven valid across site, it is considered prudent to use the conservative ±2σ approach (detailed in main text).

DISCUSSION

This study demonstrates that significant lateral spatial heterogeneity exits within blastcones. That is, the material is not distributed evenly throughout the blastcone as each flitch is drilled, and different lateral sampling points throughout the blastcone can return notably different assay results. There is a tendency towards greater heterogeneity the more variable the drilled downhole geology, as one would expect.

Across all trials, multi-point auger sampling with averaging ('auger average') delivered the highest correlation to whole-cone assays and the lowest uncertainty, outperforming shovel-based techniques even when restricted to a single sample per cone. This superior performance was also observed in running estimations based on the sampling techniques and comparing to the whole cone estimation. Monte Carlo simulations indicate that two to three auger samples capture the majority of achievable improvement, with continued but diminishing benefit at higher counts.

Given the demonstrated lateral heterogeneity of blastcones, and the improvements in multi-point versus single point sampling for auger, it is reasonable to assume multi-point sampling would improve shovel sampling performance also. This was not directly tested in this study. One could argue that shovel flitch sampling is a form of multi-point testing, given multiple samples taken per blastcone (one per flitch). However, this is targeted at vertical divisions to minimise delimitation bias (of the flitch samples), not lateral heterogeneity around the cone previously discussed. Nonetheless, the more accurate capture of vertical heterogeneity (via reducing delimitation bias) is likely why shovel flitch sampling performs better than single-point shovel bench sampling. Again, it is likely that also capturing lateral heterogeneity through multiple samples across the cone laterally would further improve shovel performance. In terms practicality, shovelling can be more manually intensive than the auger technique – so capturing additional samples presents extra ergonomic burden for a site team. Nonetheless, if greater accuracy is required, and shovel sampling is the chosen approach, it is likely that at least two laterally offset shovel samples per blastcone will improve performance as demonstrated by Monte Carlo simulation of auger sampling data.

Single-point shovel bench sampling consistently produced the weakest results and should be avoided where alternatives exist. Shovel flitch sampling provides a viable option when flitch-scale resolution is operationally required, provided that site-specific flitch ratios are first established to minimise delimitation bias. As a further pragmatic example, if auger is selected as primary sampling technique at a bench-mined operation and a backup technique is required (ie edge cases: blastcone is too indurated for auger, auger temporarily not available – battery etc), shovel flitch would be preferred over single-point shovel bench sampling method given the much-improved performance (assuming average flitch ratios for the sampling face have been established).

The flitch ratio study phase established empirical-based average flitch ratios to use in flitch sampling at Roy Hill; and highlighted preferred sampling points on the blastcone per method. It found the cone peak location was most suitable for shovel sampling, as it presented a realistic sample face height to obtain adequate flitch samples, compared to a mid-slope cone sampling location. In contrast, the mid-slope location was the best sampling point for auger, as it presented the most even ratio of flitches for a representative sample of the bench.

All auger samples were taken from the mid-slope of the cone throughout this study as suggested by the Phase 1 results and achieved good performance – comparable to near offset RC. In effect, both auger and shovel samples were taken from fixed-points on the cone: mid-slope and cone peak respectively. Previous workers have suggested that auger sampling should be designed to simulate radial cuts throughout the blastcone (cuts covering the whole cone cross-section rather than fixed point). This would involve around 12 auger sample points taken in triangle-like patterns spread in four channels on the cone (Pitard, 2008). This is suggested as the best way to combat lateral pile segregation/lateral heterogeneity. In this study, four samples in the mid-slope location were enough to achieve good performance. Also, as mentioned in an earlier section, Engström and Esbensen (2017a) found manual, fixed-point sampling of iron ore blastcones to be equally as reproducible as radial cuts of the blastcone. Amongst other potential reasons, this minimal difference between radial cut and fixed-point sampling could be because in iron ore there are less pronounced density differences in material versus other commodities. In other commodities, where larger density differences are seen – lateral pile segregation driven by density may be more impactful. For example – the denser materials would be expected to concentrate closer to the blasthole in the cone, with less dense materials found in greater amounts more distal in the cone (as they can travel further). In this instance, methods like radial cuts may become more important to ensure lateral density zonation/bias in the cone is accounted for. Nonetheless, whether this is unique to iron ore (achieving similar reproducibility and reasonable performance at set points on the cone versus radial-cut/cross-sectional sampling), and is not as applicable to other commodities, could be the subject of further study.

Phase 2 of the study indicated that some moisture presence in blastcones improved both auger and shovel sampling performance (in a qualitative three class system – dry, moist, wet). Sampling performance was better in moist blastcones versus dry. Sampling performance was even better in wet cones, but sample size was so low that those results should be considered unreliable. At minimum, the conclusion that can be drawn is if blastcones are moist/damp better performance can be expected versus dry. The mechanism as to why this occurs could be better sample capture as the sample is more coherent, and by extension, perhaps the blastcone holds better size and shape with some additional moisture. Relative blast rig performance was also tested in Phase 2 with the Pit Viper performing better for all samples techniques versus D65. This could be because in general the Pit Viper produces a better quality, larger cone more amenable to sampling. Even though the D65 produced poorer performance, the relative rank of the sampling methods remained the same between the two rig types (auger average best performer, shovel sampling worst for both).

The uncertainty margins established in the study provide guidance to aid sample method selection, and to flag the range of outcomes an assay result could be. If multiple sample techniques are used on a site, it allows the technical team some further insight in to how to treat the data from each technique, and manage expectations downstream (ie scheduling, planning, processing). If the blasthole sampling data is being combined with RC data in short-term models, it puts further pressure on selecting the method with the tightest uncertainty margin. It could also aid the creation of low-side and high-side short-term models (whether fed into an over-arching probabilistic assessment or not), to further constrain and characterise product risk.

Limitations of the study, and scope for future work if the study were expanded, include the following. The comparison of whole-cone reference assays to the different sampling technique assays is limited to one area of the Roy Hill site. Expanding this effort to more pit areas of the site would provide a stronger basis for conclusions – given a greater variety of material types and cone conditions could be tested. Unfortunately, this effort is the most time-consuming and costly (especially if trying to obtain an adequate data set size per sub-region), and hence why its scope was limited. Further, including a program of particle size data collection across sampling techniques and reference samples, with assays of the size fractions, would help to identify the root cause of bias mentioned in earlier sections. For example, the sizing assay data could help identify if the sampling techniques under- or over-represent a particular size fraction, relative to the reference. Sampling design could then potentially be optimised to try to correct any issues with particle size capture, or an informed correction factor developed. Again, this adds more cost, complexity and time to the study, but helps to identify sources of bias and correct for them.

While the observations from this study can help guide sample methods at other sites, some form of validation of the results of this study is suggested (ie a condensed validation study at minimum). The individual characteristics of a particular mine – local geology, bench and flitch configuration, blast rig type, orebody position relative water table, blast design, etc are all elements that could influence the results of blastcone sampling and may lead to different outcomes. Nonetheless, this study stands as a comprehensive, three-phased exploration of blasthole sampling at the Roy Hill mine. This work demonstrates that well-designed blastcone sampling, particularly multi-sample augering, can provide fit-for-purpose grade control information approaching closely-spaced RC at lower cost.

CONCLUSIONS AND RECOMMENDATIONS

- Blastcones exhibit significant lateral heterogeneity, and multi-point sampling is preferred to mitigate the impacts of this on sampling performance.
- At least two to three samples per blastcone is recommended (results supported by Monte Carlo analysis of sample number).
- In bench-mined operations, it is recommended to adopt multi-point auger sampling to achieve the most accurate results.
- For flitch-mined operations, retain the shovel flitching method, but establish empirical-based flitch ratios to guide flitch sampling. The addition of multi-point lateral sampling will further increase accuracy by mitigating blastcone lateral heterogeneity.
- Local site testing of selected sampling techniques is recommended to verify performance and constrain uncertainty margins against local conditions.
- Blastcone sampling can be a cost-effective grade control complement to RC, especially for sites with low RC density.

ACKNOWLEDGEMENTS

The site teams are thanked for their immense contributions to the study. Mine Geology and Ore Definition for their help in organisation, coordination and intensive sample collection campaigns and dedication to the three study phases. Special mention to Edward Sykes, Samy Nimalanathan, Martin Heydari, Darshan Panchal, Kate Schubach, Zak Clancy, Reyce Douglas and Sivarak Phachansiri. Drill and Blast are thanked for their help in organising the flitch ratio study and twinning phase, and the patience involved in giving up some productive drilling time to allow the business to better understand sampling of the orebody. Thank you to Hancock Iron Ore for the opportunity to publish this work.

REFERENCES

Abzalov, M, Menzel, B, Wlasenko, M and Phillips, J, 2007. Grade control at the Yandi iron ore mine, Pilbara Region, Western Australia – comparative study of the blastholes and reverse circulation holes sampling, in *Proceedings of Iron Ore 2007*, pp 37–44 (The Australasian Institute of Mining and Metallurgy: Melbourne).

Engström, K and Esbensen, K H, 2017a. Blasthole sampling (replicate and variographic experiments) in LKAB open pit iron ore mines – fit-for-purpose representativity?, in *Proceedings Eighth World Conference on Sampling and Blending 2017*, pp 85–96 (The Australasian Institute of Mining and Metallurgy: Melbourne).

Engström, K and Esbensen, K H, 2017b. Optimal grade control sampling practice in open pit mining – a full-scale blast hole versus reverse circulation variographic experiment, *Transactions of the Institutions of Mining and Metallurgy: Applied Earth Science,* 126(4):176–187.

Gy, P, 1992. *Sampling of Heterogeneous and Dynamic Material Systems: Theories of Heterogeneity, Sampling and Homogenizing*, eBook (Elsevier Science).

McArthur, G, Jones, C and Murphy, M, 2010. Blasthole cone sampling experiments for iron ore flitch mining, in *Proceedings of Sampling 2010,* pp 131–136 (The Australasian Institute of Mining and Metallurgy: Melbourne).

Pitard, F F, 1993. *Pierre Gy's Sampling Theory and Sampling Practice*, second edition (CRC Press).

Pitard, F F, 2005. Sampling correctness—A comprehensive guideline, *Proceedings of Second World Conference on Sampling and Blending 2005*, pp 55–66 (The Australasian Institute of Mining and Metallurgy: Melbourne).

Pitard, F F, 2008. Blasthole sampling for grade control—the many problems and solutions, in *Proceedings Sampling 2008*. pp 15–22 (The Australasian Institute of Mining and Metallurgy: Perth).

Towards quantitative nuclear magnetic resonance of magnetite at laboratory and bulk scale

M Crowther[1], R Yong[2], A Curtain[3] and T C L Ly[4]

1. Research Technician, CSIRO, Lucas Heights NSW 2234. Email: matthew.crowther@csiro.au
2. Team Leader, CSIRO, Lucas Heights NSW 2234. Email: richard.yong@csiro.au
3. Research Technician, CSIRO, Lucas Heights NSW 2234. Email: andrew.curtain@csiro.au
4. Research Technician, CSIRO, Lucas Heights NSW 2234. Email: thai.ly@csiro.au

ABSTRACT

A custom-built laboratory nuclear magnetic resonance (NMR) spectrometer was used to quantify the magnetite grade of 35 blended magnetite samples. The samples were approximately 2 g each and were mixed with quartz to a range of magnetite grades from approximately 10–80 wt per cent. Quantitative X-ray diffraction was also performed on these samples to independently determine the grade. An excellent linear correlation between magnetite grade as determined by quantitative X-ray diffraction, and NMR spin echo signal was achieved with $R^2 = 0.98$ for the blended samples.

An NMR spectrometer intended for use on a primary iron ore belt was built and two bulk samples from Braemar iron formation were measured. The spectrometer aperture was built to 1.2 m as a manageable scale for our laboratory and could measure over 70 kg of magnetite ore. The NMR spin echo signal as a function of the mass of magnetite present in the sensor was determined. An excellent correlation between magnetite mass and NMR spin echo signal was achieved for both bulk samples. A lower detection limit was inferred based on the results and was calculated to be 2.07 kg of pure magnetite in 50 sec of measurement time. As such, magnetic resonance of magnetite presents an excellent opportunity for bulk quantification or beneficiation processes.

INTRODUCTION

Demand for iron ore in the steel industry continues to rise due to rapid global industrialisation. Typically, this demand has been met by hematite (Fe_2O_3, 69.9 wt per cent iron) as it occurs naturally in high concentrations of around 60 wt per cent iron. These ores are known as direct shipping ores (DSO) and are preferred as they require little to no beneficiation (Davies and Twining, 2018). However, these high-grade sources of iron ore are being steadily depleted and lower grade iron ores are required to meet the demand of the steel industry.

Another iron-bearing mineral used in the industry is magnetite (Fe_3O_4), which occurs naturally as an ore ranging from 15–40 wt per cent magnetite. However, pure magnetite contains 72.4 wt per cent iron meaning that after beneficiation, it can produce higher iron concentrations than hematite (Bulayani, Raghupatruni, Mamvura, and Danha, 2024). The higher-grade resultant product can be sold at a premium and typically contains less deleterious impurities such as phosphorous and sulfur (Summerfield, 2020). The lower impurity content and higher iron content also provides another benefit as the world progresses towards decarbonisation: it is suitable for use in direct reduced iron (DRI) technologies and electric arc furnaces (EAF). The carbon emissions from these processes are significantly lower than that of traditional blast furnace methods (Horngren *et al*, 2023). As magnetite is magnetic, it is much easier to beneficiate to the required higher purities than hematite, which presents a growing demand for magnetite to help reach carbon emission targets (Eames, 2021).

The beneficiation process can still be expensive and energy intensive. As such, methods to reduce processing of gangue material are desirable. This can be achieved through bulk ore sorting, if the desirable mineral content of bulk ore can be quantified. In the ideal case, uneconomic minerals can be removed from the processing line before crushing and/or milling reducing processing costs, energy and water use, and increasing metal recovery (Cetin, Klein and Futcher, 2023).

Many quantification methods exist in the minerals industry, such as X-ray diffraction and fluorescence (XRD and XRF), and laser-induced breakdown spectroscopy (LIBS) (Kurth, 2013; Lessard, Bakker and McHugh, 2014). However, these methods are not suitable for bulk analysis as they are either for small samples, limited to surface measurements, or too slow for a sorting application (Duffy *et al,* 2015).

One widely used method of non-destructive bulk measurement is prompt gamma neutron activation analysis (PGNAA). This technique uses neutrons to irradiate a bulk sample and characterise it based on the energies of the photons consequently emitted (Peukert, Xu and Dowd, 2024). The energy spectrum is characteristic of the elements activated in the process. PGNAA measures the total iron content and is unable to specifically discriminate magnetite, which limits applicability in cases where the beneficiation process is mineral specific. PGNAA also requires extensive radiation shielding to limit exposure of personnel to both gamma rays and neutrons (Abdelnour *et al,* 2025).

Another bulk measurement technique is direct magnetic sensing, an example of which is magnetic susceptibility sensing. There exist commercialised examples of this technique which have been installed on iron ore conveyor belts for real time measurement (Ultra-Dynamics Pty Ltd, 2019). However, a limitation of the technique is that it measures the total magnetic content of the ore, from which the magnetite content is inferred. Thus, over-estimation of magnetite content is possible when other sources of magnetic materials are present.

In this paper we present the application of NMR for the quantification of magnetite at the laboratory scale and the bulk scale for conveyor belts. NMR is well suited to this task as it employs radiofrequencies (RF) for measurement which is non-ionising and capable of penetrating bulk samples. For the case of magnetite, NMR is specific only to the ^{57}Fe nuclei in magnetite providing further benefit over non-specific measurement techniques. Similar technology has been demonstrated by the Commonwealth Scientific and Industrial Research Organisation (CSIRO) for measurement of copper in chalcopyrite (Bennett, Miljak and Khachan, 2009) including scaling to conveyor belt analysis and sorting applications (Coghill, Miljak and Williams, 2015; Coghill and Simpemba, 2024). Applying this technology to the measurement of magnetite represents further advancement of this technique and opens opportunities for improving beneficiation in the iron ore mining industry.

NUCLEAR MAGNETIC RESONANCE IN MAGNETITE

The basics of NMR

NMR is a form of spectroscopy employing radiofrequencies to probe for a specific nucleus within a sample. To facilitate this measurement the nucleus of interest must have a non-zero nuclear spin ($I > 0$) and the sample must be placed in a magnetic field B_0. In these conditions, the nuclear spin will precess about the direction of the magnetic field with resonant angular frequency $\omega = \gamma B_0$, where γ is the gyromagnetic ratio – specific to the nuclear isotope, and $\omega = 2\pi f$ where f is the resonant frequency in Hz. By applying an RF pulse of strength B_1 at the same frequency, the sample will emit an RF signal proportional to the number of nuclei in the sample.

An NMR measurement is typically performed by placing the sample inside a solenoidal coil capable of transmitting and receiving RF signals. The B_0 field is usually generated by either permanent magnets or an electromagnet surrounding the coil. The RF signal is supplied to the coil at the resonant frequency as a 'spin echo' pulse sequence (Hahn, 1950), as shown in Figure 1. If NMR sensitive nuclei are present in the sample, a signal will be produced in the form of a spin echo, which is detected as a voltage V_S in the coil. The inter-pulse spacing τ dictates the characteristics of the spin echo signal, the centre of which occurs at time 2τ. The duration of the RF pulses, p_1 and $p_2 = 2p_1$, and the RF field B_1 are carefully chosen such that the maximum NMR signal is achieved. The spin echo pulse sequence parameters are specified in the form $[p_1 - \tau - p_2 -$ acquire$]_n$, where n is the number of measurements for the purpose of this paper.

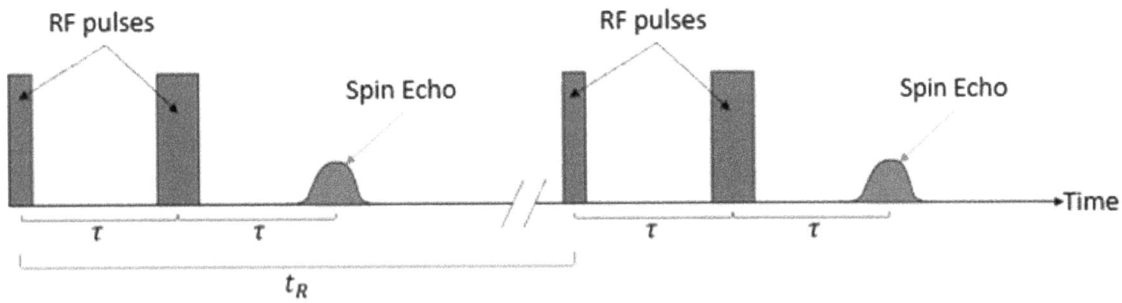

FIG 1 – A 'spin echo' pulse sequence, characterised by two pulses with the second being twice the duration of the first. The inter-pulse spacing τ dictates the characteristics of the spin echo. The pulse sequence is repeated after the repetition time t_R to improve the signal to noise ratio.

However, there exists an inherent thermal noise across the coil terminals and there may also be external sources of electromagnetic interference which compete with the NMR signal voltage. It is important to maximise the signal-to-noise ratio $SNR = \frac{V_S}{V_N}$, where V_N is the total noise voltage, to facilitate quantitative NMR with minimal statistical error. As a single measurement of a spin echo typically has a low SNR, signal averaging is used by repeating the pulse sequence. Assuming a constant signal between measurements, and that the thermal noise is random with a mean of zero, the SNR will increase proportionally with $\sqrt{n}$. Therefore, the pulse sequence is repeated after the repetition time t_R to facilitate signal averaging of the spin echo. If one considers a larger NMR coil, say in comparing NMR performed in the laboratory compared to on a primary conveyor belt, the SNR will increases in proportion to the coil volume as long as the volume fraction of the sample remains constant (Hoult and Richards, 1976).

As the signal received is directly proportional to the amount of the target nuclei in the sample, the spin echo signal can be calibrated to deliver a quantitative measurement of the number of target nuclei in the sample. This is achieved by comparing the signal from the unknown sample to that of a sample of known grade.

NMR of magnetite

For the NMR analysis of magnetite, the nucleus of interest is ^{57}Fe. Due to the ferrimagnetic structure of magnetite, the iron atoms experience an intrinsic magnetic field at the iron nucleus which negates the need for an applied B_0 to perform NMR, simplifying the measurement significantly. This results in a resonant frequency f = 67.6 MHz at room temperature as measured by (Bastow and Trinchi, 2009), which is specific to ^{57}Fe in magnetite only. Also, other materials in a sample are invisible to NMR at this frequency, which is ideal for beneficiation processes as they are usually specific to the ore species.

The magnetic domain structure that exists in a ferrimagnet further benefits NMR of magnetite by providing a phenomenon known as the ferromagnetic enhancement factor, η. It was first observed in NMR measurements performed on elemental iron (Stearns, 1967). The enhancement factor improves the measurement sensitivity in two ways:

1. The magnitude of the applied RF magnetic field, experienced at the iron nuclei, is multiplied such that they experience a magnitude $B_{obs} = \eta B_1$.
2. The NMR signal voltage induced in the NMR coil, V_0, is also enhanced, such that the measured signal is $V_S = \eta V_0$.

The enhancement factor can be on the order of 10^3 making the RF field strength (and power) required to optimally excite the iron nuclei extremely low and signal averaging times very quick. Without the ferromagnetic enhancement factor, measurement of ^{57}Fe in magnetite by NMR would be very poor, due to its low natural isotopic abundance of 2.12 per cent (Meija *et al,* 2016) and otherwise weak sensitivity to NMR.

Towards a quantitative measurement

Unlike XRD or PGNAA, ferromagnetic NMR is not an industrially established quantitative measurement method. As such, the NMR signals gathered in this work cannot be compared to a database of spectra from which the unknown grade of the magnetite can then be extracted. To achieve a quantitative measurement, a calibration is performed by first measuring a suite of samples with known grade from the same locality as the unknown samples. A linear correlation analysis between the mass normalised magnetite NMR spin echo signal magnitude and an externally measured magnetite assay, such as that from quantitative XRD, is used as the calibration. A spin-echo is then obtained from each of the samples with unknown grade which is compared to the calibration line to determine the magnetite grade. Certain NMR characteristics of the calibration samples are also determined as part of the calibration, such as the inhomogeneous linewidth of the NMR signal and nuclear relaxation times. These characteristics are not necessarily the same between samples of magnetite from different localities, hence the requirement to perform a calibration for samples from different mines.

A simplified example of the workflow to calibrate for a quantitative measurement for a mine installation would be as follows:

- Install NMR sensor at the desired operating point on the conveyor belt.
- Measure a suite of magnetite samples using NMR, then have these sub-sampled and assayed by an on-site lab/external laboratory.
- Determine the linear correlation from the linear regression between the NMR data and external assay.

At this point, a sample of unknown magnetite grade can be quantitatively assayed using NMR on the moving conveyor belt. A weight measurement, taken just before the ore passes through the sensor, is used to mass-normalise the NMR signal.

In this work, the samples under investigation were either synthesised or from a new locality not previously measured before. As such, a correlation analysis was performed against the external assay as a full quantitative analysis, as just described, would have required a calibration against samples previously obtained from said locality. Consequently, the correlation analysis can then be used as a quantitative calibration for future samples of unknown grade that used the same synthesis method or are from the same locality.

LABORATORY NMR OF MAGNETITE

Methods

There were 35 × 2 g samples prepared by blending magnetite ore concentrate with quartz to form mixtures with different magnetite concentrations. The blended magnetite samples were prepared to arbitrary concentrations between 10 wt per cent and 80 wt per cent magnetite. As the exact grade of the magnetite concentrate was not known, the samples were subjected to quantitative XRD to accurately determine the final magnetite content.

A 170 mL NMR sample coil was constructed by wrapping a sheet of copper around a PVC former. This design is called a loop-gap coil and is ideal for high frequency NMR due to its low inductance and high quality factor (Hornak, Ceckler and Bryant, 1986). The coil was tuned to the desired resonant frequency, and impedance matched using a combination of ceramic capacitors and Comet high-voltage variable capacitors in a symmetric L-network configuration. The pulse sequence was synthesised by a custom made NMR pulse programmer board which was amplified by a TOMCO Technologies BT8K-GammaS power amplifier. The amplified RF was passed through a passive, quarter-wave transmit/receive (T/R) switch connected to the sample coil. Preamplification of the received NMR signal was performed with a Miteq AU1583 low-noise amplifier (LNA) and a homemade variable gain amplifier (VGA). The amplified signal was detected in quadrature using a digital down converter, which was digitised and stored on the NMR pulse programmer board. A $[20\ \mu s - 275\ \mu s - 40\ \mu s - \text{acquire}]_n$ spin echo sequence was used with $t_R = 20$ ms. The experimental set-up is shown in Figure 2. The correlation between the resulting spin echoes and the quantitative XRD determination of the magnetite grade was then sought.

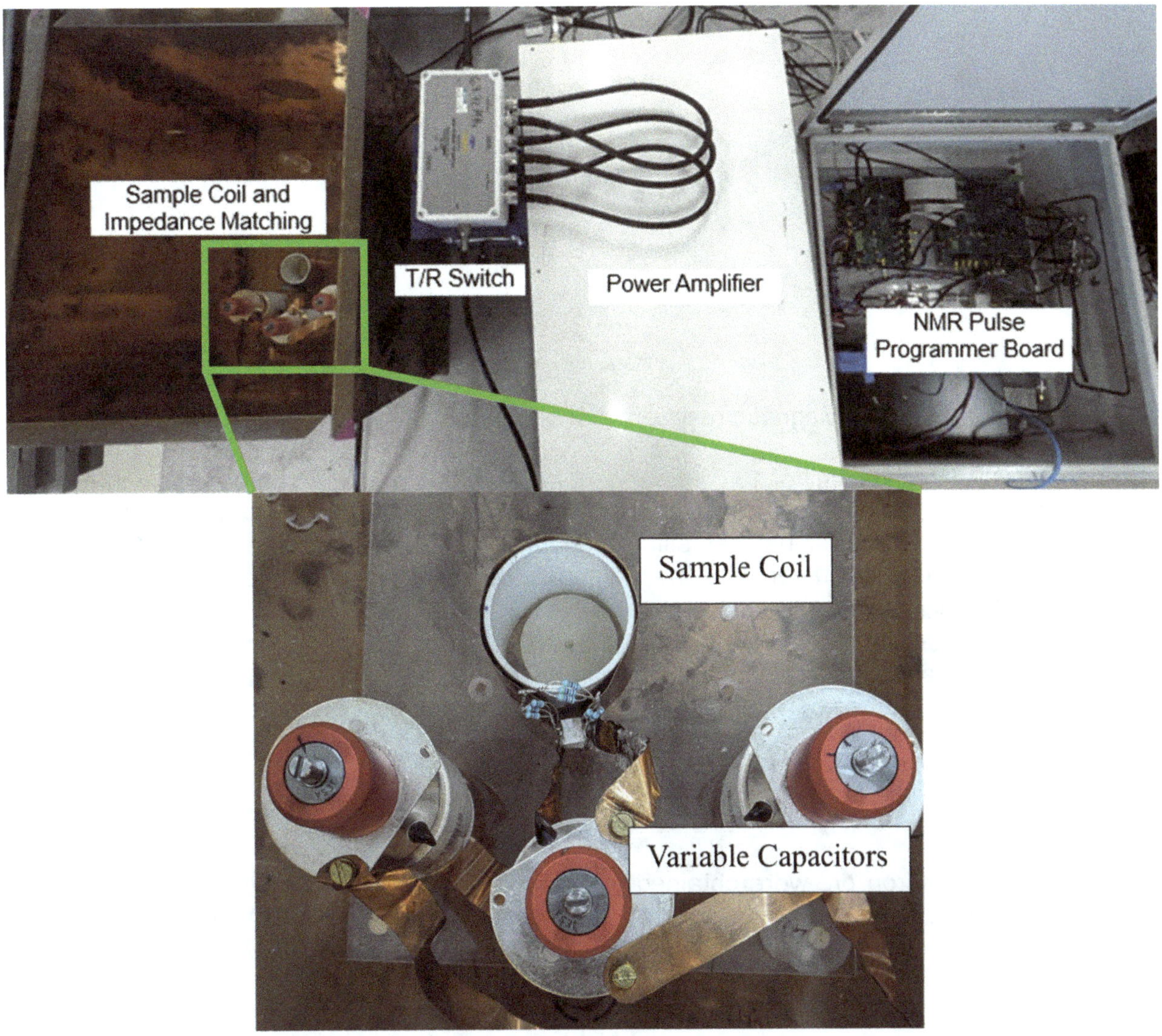

FIG 2 – Top: Laboratory set-up for ^{57}Fe NMR of magnetite. The box containing the NMR board also contains the preamplification chain. Bottom: Zoomed-in image of the loop gap sample coil and impedance matching circuit.

Results

Preliminary measurements on a high-grade sample of magnetite determined the resonant frequency to be (67.72 ± 0.02) MHz. The discrepancy from the resonant frequency as determined by (Bastow and Trinchi, 2009) arose from the strong temperature dependence of the magnetite resonance and the natural drift in temperature of the samples in the laboratory in which the measurements were taken. The maximum spin echo signal was recorded for an input RF magnetic field strength of $B_1 =$ 0.7 μT. The power required to produce this magnetic field in the laboratory coil was of the order of tens of milliwatts.

The blended magnetite samples were subjected to $n =$ 1000 acquisitions (20 sec measurement time) to achieve a satisfactory SNR. The magnetic resonance signals were normalised to the mass of the samples and then directly compared to the XRD assay. A line-of-best-fit was calculated via a linear regression to the data resulting in a coefficient of determination $R^2 =$ 0.98 (2 dp). The data are shown in Figure 3, where the NMR data has been normalised to the largest observed spin echo signal.

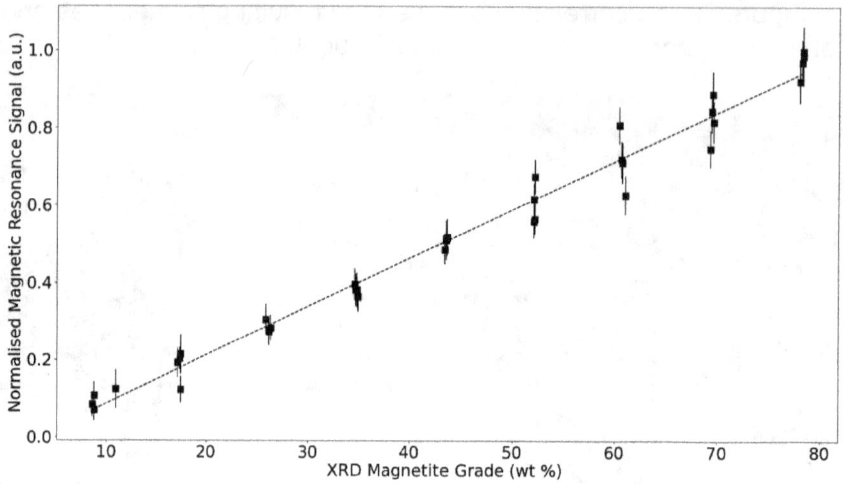

FIG 3 – Magnetic resonance of blended magnetite samples.

A 3σ-confidence detection limit of 13.2 wt per cent magnetite in a 2.02 g sample was achieved in a 20 sec measurement time. A few of the measured samples reported grades below this threshold. For these samples an NMR signal was still discernible but was below the required SNR for a 3σ-confidence detection. This is reflected in the magnitude of the individual error bars in Figure 3. Since the counting statistics of the white noise scale as $\sqrt{n}$, the SNR and thus the detection limit can be easily improved by using longer measurement times. Conversely, a quicker measurement time can be achieved at the sacrifice of a higher detection limit. The results achieved in the laboratory suggest good applicability for the use of NMR to measure magnetite at a bulk scale.

BULK NMR OF MAGNETITE

Methods

Two bulk samples of iron ore were obtained from the Braemar Iron Formation, South Australia – denoted 'Braemar 1' and 'Braemar 2'. Both samples were crushed to approximately 20 mm. A 'grab sample' of magnetite was obtained for each and independently examined by XRD to determine the approximate grade. The results are collated in Table 1.

TABLE 1

Magnetite grade of Braemar iron ore samples.

Sample	XRD Magnetite Grade (%)
Braemar 1	19
Braemar 2	12

To facilitate the bulk NMR measurement of magnetite, a sensor was designed and built intended to represent use on a conveyor belt and is shown in Figure 4. The NMR coil resides within the central copper box (not visible) and surrounds the 1.2 m wide aperture. RF power is supplied to the coil using an inductively coupled, series-resonant drive coil. The resonant frequency of the NMR coil was adjusted using Comet variable capacitors such that an impedance of $Z \approx (50 + j0)\Omega$ was presented at the drive coil input at the desired frequency. The apertures of the sensor were connected to steel shields that acted as waveguide attenuators to mitigate external electromagnetic interference. The NMR coil was enclosed in a copper structure to limit exposure to external sources of RF noise. This embodiment of the NMR sensor is very similar to that used in real world implementations of copper sensing using NMR (Coghill, Miljak and Williams, 2015; Coghill and Simpemba, 2024).

FIG 4 – Conveyor sensor for bulk NMR measurement of magnetite.

The RF generation and amplification, experimental parameters, and data analysis were performed using the same components as the laboratory measurements. With only two samples available, it was decided to measure the linearity of the magnetite NMR signal as a function of the mass of sample within the sensor. This was done by adding discrete masses of magnetite into the sensor, up to approximately 70 kg, and recording a spin-echo each time mass was added. This was performed instead of a direct measurement of the grade, since a two-point comparison would not have been sufficient to determine a correlation. A [20 μs – 475 μs – 40 μs – acquire]$_n$ spin echo sequence was used with $t_R = 50$ ms with $n = 1000$ acquisitions.

Results

Preliminary measurements determined that approximately 50 W of RF power was required to produce the necessary B_1 field strength for a maximum spin echo signal. The samples were loaded one bag at a time into the sensor, with each bag weighing between 5 kg and 10 kg. The sensor had to be tuned back to the appropriate resonant frequency with the addition of more magnetite due to the conductivity and permeability of the magnetite samples affecting the electrical loading of the NMR coil. In real-world applications, automatic tuning and matching circuits are implemented to do this process autonomously. Results of these measurements are shown in Figure 5.

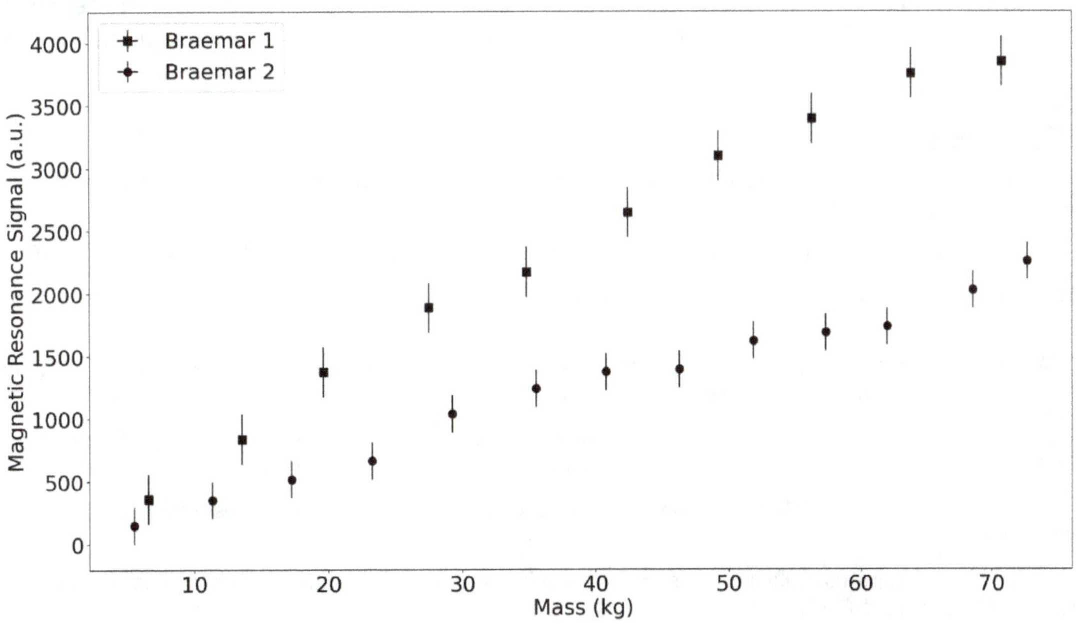

FIG 5 – Magnetic resonance results for bulk magnetite samples.

As the mass of magnetite ore in the sensor increased, the magnetic resonance signal also increased linearly for both samples which demonstrated a good, linear correlation between the magnetite spin-echo magnitude and the magnetite mass. For the Braemar 1 sample, a 3-σ detection limit of 2.07 kg of magnetite was achieved in a 50 sec measurement time. The same method applied to the Braemar 2 samples gives a detection limit of 1.95 kg of magnetite in 50 sec. The difference between the two values likely arises from a slight difference in external electromagnetic interference experienced by the sensor across the days that the measurements took place.

The measurements were conducted in a warehouse where the ambient temperature was not controlled. As such, the temperature of the magnetite sample, and thus the resonant frequency of the magnetite, were constantly fluctuating. This was observed qualitatively during measurements and was partially mitigated by performing the entire set of mass measurements as quickly as possible. However, it was not possible to completely mitigate the shifting resonance frequency which led to uncertainty in the magnitude of the measured magnetic resonance signal. This can be compensated for by automatically detecting the peak resonance frequency and adjusting the coil tuning frequency as it changes. However, it was not performed in these measurements due to a limitation of available resources. Such a strategy has been successfully implemented by (Coghill, Miljak and Williams, 2015; Coghill and Simpemba, 2024) for implementations at copper mines.

The amount of sample used in this work was much lower than the masses typically seen on a magnetite production belt. If one assumes a tonnage of 4000 t/hr, and a potential NMR sensor with a length of 1 m along the belt direction, that equates to 1100 kg of magnetite ore within the sensor volume at any given time. Extrapolating a detection limit of ~2 kg of pure magnetite represents a grade of ~0.2 wt per cent magnetite for a 50 sec measurement of 1100 kg of ore in the sensor.

CONCLUSIONS

Measurements of 35 × 2 g samples of magnetite using NMR in the laboratory showed good agreement to the grade as measured by XRD for the blended magnetite samples, with a correlation coefficient $R^2 = 0.98$. A bulk implementation using an NMR sensor at a suitable size scale for a primary conveyor belt showed good correlation of the NMR spin echo signal to magnetite mass. The 3-σ detection limit was determined to be 2.07 kg and 1.95 kg of pure magnetite – based on results from two different samples from the Braemar region in a 50 sec measurement. Thus, bulk ore analysis via NMR is an attractive proposition for beneficiation of magnetite due to its specificity, scalability, speed, and lack of sample preparation.

ACKNOWLEDGEMENTS

The authors would like to thank Heta Lampinen for both supplying and performing the XRD on the blended samples. The XRD work was funded by the CSIRO Mineral Resources Discovery Research Program.

REFERENCES

Abdelnour, M R, Liu, J, Hossny, K, Wajid, A M, Li, W and Liu, Z, 2025. Prompt gamma neutron activation analysis: A review of applications, design, analytics, challenges and prospects, *Radiat Phys Chem*, 234:112693.

Bastow, T J and Trinchi, A, 2009. NMR analysis of ferromagnets: Fe oxides, *Solid State Nucl Magn Reson*, 35:25–31.

Bennett, D, Miljak, D and Khachan, J, 2009. The measurement of chalcopyrite content in rocks and slurries using magnetic resonance, *Miner Eng,* 22:821–825.

Bulayani, M M, Raghupatruni, P, Mamvura, T and Danha, G, 2024. Exploring Low-Grade Iron Ore Beneficiation Techniques: A Comprehensive Review, *Minerals*, 14(8):796. https://doi.org/10.3390/min14080796

Cetin, M C, Klein, B and Futcher, W, 2023. A conceptual strategy for effective bulk ore sorting of copper porphyries: Exploiting the synergy between two sensor technologies, *Miner Eng,* 201:108182.

Coghill, P J and Simpemba, E, 2024. A bulk sorting trial of copper ore using a magnetic resonance analyser, *Miner Eng,* 210:108664.

Coghill, P, Miljak, D and Williams, E, 2015. Consequences of fractal grade distribution for bulk sorting of a copper porphyry deposit, *Geosci Front,* 6:477–480.

Davies, M and Twining, M, 2018. Magnetite: South Australia's resource potential, *MESA J*, 86:30–44.

Duffy, K-A, Valery, W, Jankovic, A, Holtham, P and Valle, R, 2015. Bulk ore sorting for pre-concentration: What, how, why?, in Proceedings of the 11th International Mineral Processing Conference, Procemin 2015, (Gecamin).

Eames, M, 2021. Where will future iron ore supply come from?, Magnetite Mines. Available from: <https://www.magnetitemines.com/_files/ugd/8fe33d_905c6ad509394b8c90ecc391364c5ccb.pdf>

Hahn, E L, 1950. Spin Echoes, *Phys Rev*, 80:580–594.

Hornak, J P, Ceckler, T L and Bryant, R G, 1986. Phosphorus-31 NMR spectroscopy using a loop-gap resonator, *J Magn Reson*, 68:319–322.

Horngren, T, Leach, T, Maxwell, R, Graham, P, Kelly, R, Turner, K and Malos, A, 2023. Pathways to industrial decarbonisation: positioning Australian industry to prosper in a net zero global economy, Phase 3, Climateworks Centre. Available from: <https://climateworkscentre.org/resource/pathways-to-industrial-decarbonisation-positioning-australian-industry-to-prosper-in-a-net-zero-global-economy/>

Hoult, D I and Richards, R E, 1976. The signal-to-noise ratio of the nuclear magnetic resonance experiment, *J Magn Reson*, 24:71–85.

Kurth, H, 2013. Real Time On-Belt Elemental Analysis For Advanced Process Control In Iron Ore, in *Proceedings of the 43rd Ironmaking and Raw Materials Seminar and 14th Brazilian Symposium on Iron Ore,* pp 364–373 (ABM – Brazilian Association of Metallurgy, Materials and Mining).

Lessard, J, Bakker, J D and McHugh, L, 2014. Development of ore sorting and its impact on mineral processing economics, *Miner Eng*, 65:88–97.

Meija, J, Coplen, T B, Berglund, M, Brand, W A, Bièvre, P D, Gröning, M, Holden, N E, Irrgeher, J, Loss, R D, Walczyk, T and Prohaska, T, 2016. Isotopic compositions of the elements 2013 (IUPAC Technical Report), *Pure Appl Chem,* 88:293–306.

Peukert, D, Xu, C and Dowd, P, 2024. Validation of a PGNAA sensor model for bulk material sorting, *Miner Eng,* 217:108950.

Stearns, M B, 1967. Spin-Echo and Free-Induction-Decay Measurements in Pure Fe and Fe-Rich Ferromagnetic Alloys: Domain-Wall Dynamics, *Phys Rev,* 162:496–509.

Summerfield, D, 2020. Australian Resource Reviews: Iron Ore 2019, Geoscience Australia, Canberra. http://dx.doi.org/10.11636/9781925848670

Ultra-Dynamics Pty Ltd, 2019. ironSCAN On Conveyor Magnetite Monitor, Ultra-Dynamics Pty Ltd. Available from: <https://ultradynamics.com.au/process-measurement-products-guide/magnasat-magnetic-susceptibility-meter/ironscan-on-conveyor/> [Accessed: April 2026].

OreAI – bringing next generation AI to iron ore production and planning

L Deakin[1], M Knackstedt[2], R Sok[3], U Troitzsch[4], T Evans[5] and L O'Connor[6]

1. Software Lead, OreAI program, Department of Materials Physics, The Australian National University, Canberra ACT 2601. Email: lachlan.deakin@anu.edu.au
2. Technical Lead, OreAI program, Department of Materials Physics, The Australian National University, Canberra ACT 2601. Email: mark.knackstedt@anu.edu.au
3. OreAI program, Department of Materials Physics, The Australian National University, Canberra ACT 2601. Email: rob.sok@anu.edu.au
4. Lead Mineralogist, OreAI program, Department of Materials Physics, The Australian National University, Canberra ACT 2601. Email: ulrike.troitzsch@anu.edu.au
5. Chief Engineer R&D, Iron Ore Technical Marketing, Rio Tinto, Perth WA 6000. Email: tim.evans@riotinto.com
6. Principal Advisor R&D, Iron Ore Technical Marketing, Rio Tinto, Perth WA 6000. Email: louisa.o'connor@riotinto.com

ABSTRACT

Future steelmaking technologies and the gradual degradation of ore quality are forcing producers to change current mined iron ore workflows and product offerings from the Australian Pilbara region. Globally iron and steelmaking industries are working towards addressing the challenge of reducing greenhouse gas (GHG) emissions. There are many potential pathways to achieve lower emission steel production including hydrogen direct reduced iron with electric smelting furnace (DR-ESF), electrolysis, and other innovative technologies. Advancing the industry along any of the possible processing pathways via these process routes will require the assessment of the suitability of current ores, the design of new ore product streams and the extensive development of downstream steelmaking technology.

OreAI aims to deliver a disruptive 4D product analysis workflow for iron ore to produce bespoke Pilbara iron ore products most suited to emerging supply chain requirements. OreAI builds proprietary 3D X-ray CT imaging techniques integrating them with bulk and 2D microanalysis data, and combining these with machine learning and artificial intelligence to create a novel workflow applicable across the mining value chain—from pit to furnace.

An example of OreAI's application for supporting orebody characterisation is described in the present study where 1200 lump samples of iron ore from a Pilbara asset are utilised. The results show that OreAI can be used to extend microanalytical information from 1D assay data and 2D microanalytical image data to 3D X-ray images of the ore material on a large batch of ore material across a range of material types. OreAI uses image analysis tools and AI to map and quantify porosity (connected and disconnected), textural information, mineral phase distributions and associations, density measures (apparent dry, wet, moist and skeletal) in 3D volumes. Tests of the model illustrate OreAI's ability to accurately propagate high resolution 2D data to 3D and offers high quality mapping and an ability to define uncertainties in predictions. Detailed 3D distributions of mineralogy, textures and physical properties across material types can be derived for the orebody.

INTRODUCTION

The global iron and steelmaking industries are under growing pressure to dramatically reduce greenhouse gas (GHG) emissions, given that the sector currently accounts for around 7–9 per cent of global CO_2 emissions. Several potential technological pathways are being pursued to address this. These include hydrogen-based direct reduced (DR) with electric arc furnace (EAF) route for converting high-grade iron ores directly to steel or hydrogen-based DR with electric smelting furnace (ESF) for converting mid and low-grade iron ores to iron to remove their significant gangue contents before steelmaking. Other low-carbon pathways being developed include various forms of electrolysis, and flash ironmaking processes. Each pathway requires not only downstream innovation in iron and steelmaking technology but *also the upstream development of new iron ore product streams tailored to these processes*.

The transition to low-emissions steelmaking presents a structural challenge to the Australian iron ore sector. Currently ores that are exported from the Pilbara are best suited to the blast furnace (BF) which includes the sintering process, where mid to low-grade fines ores are agglomerated into lump sinter for use in the BF. Processes such as hydrogen-based DR-EAF require higher-grade iron ores and although the ESF allows use of mid and low-grade iron ores, the DR process requires different ore metallurgical properties than those currently exported from the Pilbara. Without significant innovation in iron ore-product design and beneficiation, Australia risks losing its competitive advantage to higher-grade producers in Brazil and Guinea, who are already positioning to supply 'lower-carbon-ready' feedstocks.

Pilbara ores typically consist of finely intergrown goethite–hematite systems with variable impurity levels, making them less suitable for hydrogen-electric processing routes without beneficiation, blending or pretreatment. Iron ore product development incurs large costs and significant development time is required before confidence in product performance would entertain conducting any large-scale industrial test which costs tens of millions of dollars. Also, this approach often yields limited insight into ore behaviour across different processing conditions. Primary outcomes are more based on 'suck it and see' – the test either worked or it didn't. As a result, there is no current systematic capability to predict how Australian ores will perform in low-emissions steelmaking routes, nor any integrated framework linking ore characteristics in the ground to downstream crushing, beneficiation processing and smelting performance.

To remain globally competitive, Australia must build a predictive capability that enables its iron ore producers to develop new product streams suitable for emerging low-emissions processes. The OreAI program is attempting to deliver such a product analysis workflow. The workflow utilises advanced 3D characterisation of ore materials and process experiments mimicking downstream processes to offer an understanding of the downstream performance potential of the ore as it progresses along the value chain from the mine through to the smelter. Generating data on large scales combined with artificial intelligence (AI) and machine learning (ML) enables one to understand and predict ore behaviour across the full mining-to-steelmaking value chain.

The current team at the Australian National University (ANU) had previously pioneered this capability through its Digital Rock Technology (DRT) platform and patented data-fusion methods in the energy industry (Arns *et al,* 2005; Sok *et al,* 2009). The DRT platform integrates 1D, 2D, 3D, and process data to capture rock texture, porosity, and composition and undertakes *in situ* experiments of multiphase fluid displacement at cm and metre scales. This led to improved predictions of production in hydrocarbon recovery processes and storage in carbon capture projects. This DRT platform illustrated the importance of incorporating realistic geological/sedimentological structure on the prediction of multiphase flow properties and the technology has delivered an improved confidence in predictions of flow and recovery at field scales (Bagheri *et al,* 2021).

Building on this foundation, the new OreAI™ platform is applying and extending the DRT platform and, combined with machine learning, offering advanced 3D ore characterisation and downstream performance potential. In this paper we detail an OreAI characterisation program for over 1200 lump particulate ore from the Pilbara. Bulk data measurement, data acquisition in 3D and microanalytical 2D imaging of a larger volume of material is described. The workflow enables the mapping of 2D image data across microanalytical platforms to the equivalent area of the 3D image volume. This data enables training of the OreAI model. The AI model is then deployed to directly use 3D image volumes to accurately map porosity, mineral phase, density measures (apparent dry, wet, moist and skeletal) on each particle. Tests of the model illustrate the ability to propagate the 3D information offers high quality mapping and an ability to define uncertainties in predictions. The paper concludes with a vision how this can further enable product analysis across the supply chain including productivity (eg optimal processing and beneficiation conditions) and develop low carbon ironmaking ore streams (eg identify ores most suited to DRI processing).

EXPERIMENTAL TECHNIQUES AND TRAINING METHODOLOGY

Pilbara Ore

A suite of 1078 +6.3 mm-10 mm sized particles were selected from a beneficiation/density separation trial on Pilbara lump ore. The sized particles were typed by apparent density values derived from laboratory water pycnometry measures. Over 20 apparent separation density bins were acquired with particle densities reporting from 1.8–1.9 g/cc through to 4.0–4.1 g/cc. All particles were also classed via a standard geometallurgical protocol based on likely/majority mineral content and apparent hardness. This paper will describe the full analysis suite and comparisons acquired on a subset of the full data trial, reporting predictions on over 100 particles.

Sample preparation and bulk analysis

Sample intake

Particles were individually barcoded and eachparticle was weighed using a Sartorius Entris 2 analytical balance (accurate to 0.1 mg) and the value was entered into the database. The longest axis of each particle was measured in millimetres using callipers. All samples were photographed from multiple angles using a custom 360° imaging set-up.

Helium pycnometry

To prepare samples for skeletal density measurements, particles were arranged on trays and dried in an oven overnight at 80°C. The dried particles were then allowed to cool to laboratory temperature in a desiccator for 2 hrs. Skeletal density was determined by gas pycnometry using Micromeritics AccuPyc III (later also AccuPyc 1330) with helium as the analysis gas.

3D image acquisition

Basic scanning of tube packs

Particles were packed into custom peek sample scanning tubes (17 mm dia, 125 mm deep), with foam used as a spacer between each particle keeping the particles separate and in packing order. Between 15 and 30 particles were packed into each tube depending on particle size (average of 20 particles packed per tube). The packed tubes were scanned on a Thermofisher Heliscan MK2 micro-CT system with an additional source aperture (RT1 system) for 3 hrs per scan. Scan geometry, source settings, acquisition time and filtering were kept the same for all low-resolution tube scans, resulting in consistent image quality and the same voxel size of 15.4 micron. The identity, particle properties and packing order of each particle packed into a tube were tracked using a custom software user interface during packing and unpacking, which allowed the 3D tomographic data from the CT scan, to be digitally separated into the correct individual particles automatically for further analysis.

Fixed low resolution scan settings used were:

- X-ray energy and current: 150kV, 50uA.
- Sample and detector distances: 19.3 mm and 350 mm.
- Exposure time: 0.25 sec per frame with 1800 projection per pitch.
- Detector: 1.5K pixels (binned by 2).
- Total scan time: ~8 mins per low resolution particle.
- Scan voxel size: 15.4 micron.

High-resolution scanning (dry and attenuated)

A subset of particles was embedded in a foamed porous mount and placed in a custom cell for high resolution scanning under dry conditions and after the addition of an X-ray attenuating fluid.

Between 10–12 particles were mounted in each cell and they were assembled in three 40 mm stacks with plastic separators to allow for split reconstruction to be used. Again, the identity, properties and packing order of each particle packed in the cell were tracked using a custom software user interface during packing and unpacking, allowing 3D tomographic data to be automatically separated and processed. The packs were scanned on the CTlab universal high resolution micro-CT (ANU3) using a *region of interest (ROI)* scanning technique which uses an overview scan to improve the quality of a region of interest scan. Each packed cell was scanned as a pair of ROI scans for difference imaging, first in a 'Dry' state, then in the attenuated state.

The scan geometry, source settings, acquisition time and filtering were optimised for the second Xenon scan and kept the same for all High-Resolution scans. Each ROI scan consisted of a 40 hr scan and a 5 hr overview scan (ie 46 hrs for the Dry scan and then another 46 hrs for the attenuated scan). The long scan time reduces the noise in the image resulting in high image quality and a voxel size of 5.7 micron. Fixed High resolution scan settings used were:

- X-ray energy and current: 140kV, 50uA.
- Sample and detector distances: 20 mm and 486 mm.
- Exposure time: 0.4 sec per frame with ten accumulations and 3600 projection per pitch.
- Detector: 3K pixels.
- Total scan time: ~4 hrs per high-resolution particle scan.
- Scan voxel size: 5.7 micron.

Totals

- **961 particles** went through the *basic analysis* and *low-resolution scanning*:
  - Average scan time per particle for low resolution scanning was **7–9 mins**.
- **117 particles** went through the *high-resolution* scanning and *SEM/mineral analysis*:
  - Average scan time per particle, scanned in a pair (dry and Xenon) of high-resolution scans for micro porosity analysis was **8 hrs** (4 hrs 'dry' + 4 hrs 'attenuated').

Polished section preparation

To prepare the SEM samples, porous embedded samples were cut into two subsamples using a diamond blade saw. The smaller subsamples were then impregnated with epoxy resin Struers EpoFix under vacuum and polished to 1 micron finish using a Struers Tegramin-30 polisher.

2D image acquisition and mineral analysis

Automated Mineral Mapping was carried out using a FEI Quanta QEMSCAN®, a Field Emission Scanning Electron Microscope (FESEM) equipped with twin Bruker 30 mm^2 Energy Dispersive Spectrometers. The analytical working distance was 13 mm, acceleration voltage 15 kV and beam current 10 nA. Tile size (field of view) was 1024 × 884 μm, image resolution 500 nm (for BSE) and 5 μm (EDS), dwell time 10 μs (BSE) and 6 ms (EDS), aiming for an average of 4500 X-ray counts per analysis.

To verify the accuracy of mineral identifications in the QEMSCAN maps in the early stages of the project, quantitative mineral compositions were analysed for 35 of the polished samples using a FESEM Hitachi 4300 SE/N Schottky FE. Acceleration voltage was 15 kV, beam current 600 pA, working distance 18 mm, and analysis time 60 sec. Analyses were calculated using ZAF corrections. Standard-based analysis including oxygen with tight beam current control provided sufficient accuracy to allow discrimination of magnetite, hematite, goethite and hydro-hematite through calculation of Fe^{2+}/Fe^{3+} and water/hydroxyl content.

2D–3D data fusion and model training

The tomograms for the dry and attenuated runs were first registered to one another using a 3D-to-3D registration technique summarised in Latham, Varslot and Sheppard (2008). Image registration brings into geometric alignment two or more images, taken at different times, from different orientations and/or by different instruments. In the case of 3D-to-3D registration, the entire overlapping blocks of particle voxels are aligned to one another in all three dimensions, so that all voxels within an image coincide in both tomograms. Regions in the dry image can be directly compared to their counterparts in the fluid attenuated image. Regions with large differences in attenuation between the image pairs contain the attenuating fluid and represent regions with porosity or microporosity. A 3D porosity map is therefore generated in 3D to locate and quantify the open porosity and effective microporosity of each voxel in the particle.

Details of the microporosity in the 3D image can be directly correlated with high-resolution microscopy data using 2D-to-3D image registration. The polished face in 2D is registered to the corresponding slice of the difference tomograms. In the case of 2D-to-3D registration, the software searches through the volumetric image for the slice that matches the SEM. The SEM provides a higher-resolution image of the particle (down to 500 nm scales in this case) and allows the evaluation of the pore-structure and associated pore filling material in the microporous regions.

The same method is used to populate solid voxels in the tomogram. The 2D polished QEMScan image offers high resolution 2D mineralogy which can be mapped onto the solid voxels in the image enabling a mapping between 3D X-ray attenuation and 2D mineralogy for the same slice.

AI model training is undertaken by training the registered 3D mineral and porosity maps on the matching slice data to the BSEM/SEM-EDS data acquired on slices. A suite of samples is used for model training and deployment is then undertaken using the training data to infer the mineral and porosity maps directly on the coarse resolution 3D tomograms (see Figure 1)

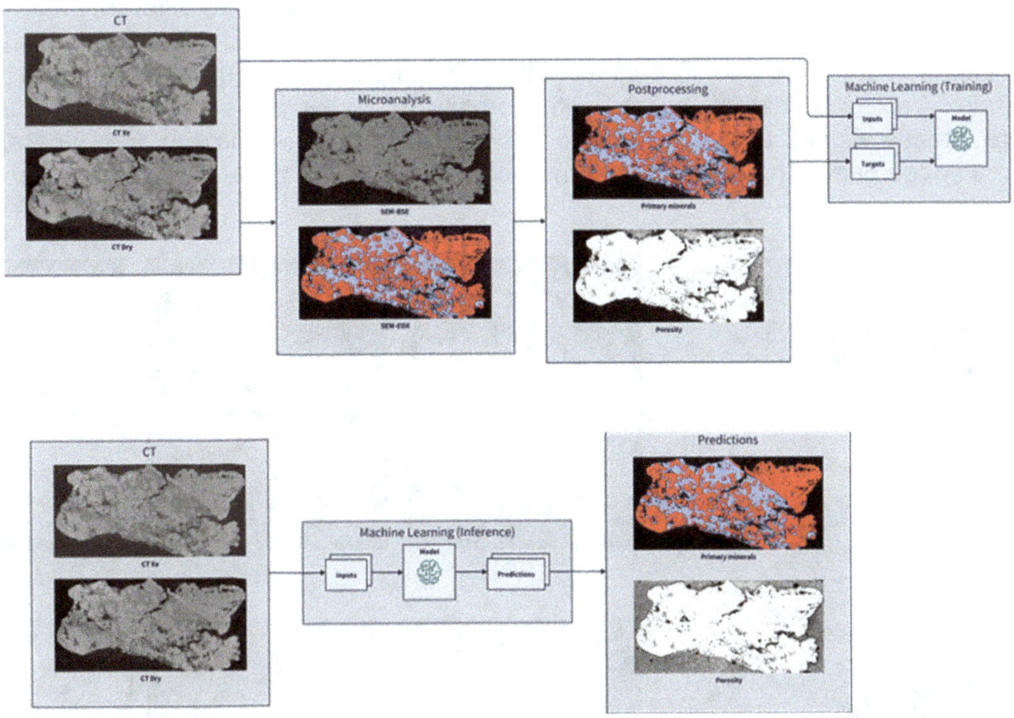

FIG 1 – Technical outline of the data fusion/image registration pipeline for building training data on particulate samples of iron ore. Upper row: CT Images are acquired on 3D image volumes and polished sections acquired for the sample. The training data is acquired by comparing the data acquired on the equivalent slice of the CT data to the Postprocessed 2D microanalytical data. Porosity, mineralogy is mapped directly. Lower row: Once sufficient training data is acquired one can use the OreAI training data to predict mineralogy, porosity, density properties on the 3D particle. In the specific example described in this paper training data is acquired for 28 samples were used for training and over 100 particles acquired at high resolution are analysed.

Bulk mineral analysis for calibration

Bulk mineral identification and quantification was undertaken on a subset of 62 samples via Powder X-ray Diffraction (XRD) complemented by X-ray Fluorescence Spectrometry (XRF). The plan was to utilise the powder samples analysed by XRD to be subsequently reused for XRF majors analysis. Note that the limited sample size (0.35–2.8 g) posed a challenge with respect to subsample representativity for XRF analysis. In particular not enough material was available to obtain reliable and stepped loss on ignition (LOI) data so XRF data is not presented here.

The 62 particles set in resin and polished for SEM analysis used a small part of the particle. The remaining, larger volume of the particle for XRD was milled by hand with acetone in an agate mortar.

Quantitative powder XRD analyses were carried out on the samples with a Malvern Pananalytical Empyrean Series 3 which is equipped with Bragg-Brentano HD divergent beam optic and a PIXcel3D detector (1D scanning mode, 3.347° active length), using CoKα radiation. The samples were analysed over a range of 4–90°2θ, with step width of 0.0262606°2θ and a total dwell time of 97 s/step. Due to instrument outage and time-pressure, 21 samples were analysed on a second diffractometer, a Malvern Pananalytical Aeris with CuKa radiation, 5–90°2θ, with step width of 0.022°2θ and a total dwell time of 198 s/step. All samples were suspended on low-background holders (single crystal Si or quartz). XRD data were interpreted with the software HighScore Plus 5.3a 2025 (Degen *et al,* 2014) and International Centre for Diffraction Data (ICDD) PDF-2 database (2004).

RESULTS

Training data was acquired on a set of 30 particles which included over 20 different material classes varying from Goethite-rich, Hematite-rich and many particles identified by a geometallurgist as a mixture of Hematite and Goethite. Examples of the 2D and 3D data acquisition for image training on three complex (mixed Hematite/Goethite with porosity) particles are shown in Figure 2. The 2D images are directly aligned to the 3D data to enable training of the porosity and mineral mapping between the image pairs. The predictions from the trained data for three particles are also shown in Figure 2.

(a)

(b)

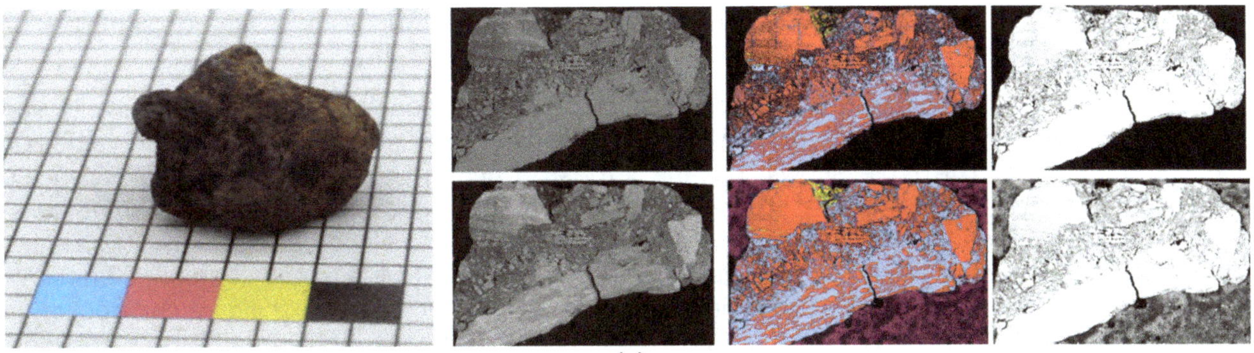

(c)

FIG 2 – Particle photographs of three medium hardness to friable particles (Labelled A, B, C) characterised from geometallurgist examination as Hematite/Goethite mixtures. The upper row for each particle shows (from left to right) the BSE image acquired at 250 nm resolution, the QEMScan map of the mineralogy of the slice (with hematite red, goethite blue, kaolinite yellow and resolved porosity black) and the porosity map from the SEM image (white solid, black pores)—identifying visible pores at 250 nm pixel sizes. The lower row shows (from left to right) the slice of the 3D volume image registered to the 2D microanalytical image, the predicted mineral map on that image from training and the porosity map trained on the SEM image porosity.

With the training data acquired across a range of particles one can then directly infer the porosity and mineral map on all other 3D image acquired samples and test predictions by comparing to 2D data not used in the training. Samples B and C from Figure 3 were used for building the training data, but Sample A was **not** included in the training data so the lower row in Figure 2a show mineral and porosity information inferred from the ML data. The visual match for mineralogy and porosity is good.

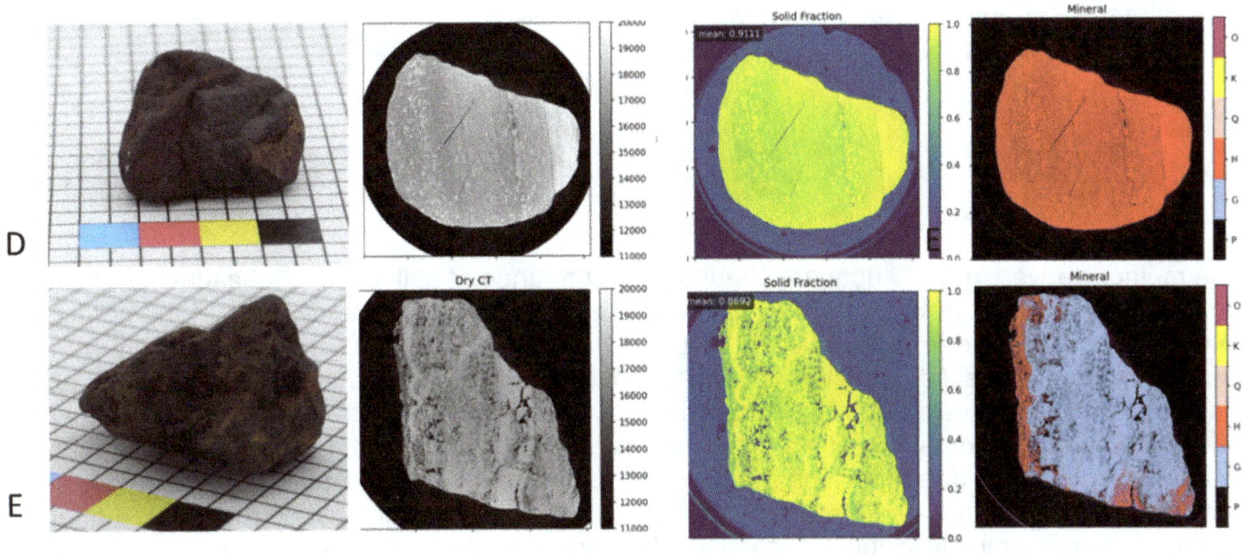

FIG 3 – From left to right: Particle photographs of Samples D and E. (d) a hematite rich particle and (e) a goethite rich particle. The images then show a slice from the CT (input data), and the inferred porosity and mineral map from that same slice using Ore AI analysis.

The prediction of the porosity and mineralogy are not limited to the 2D slice but naturally extend to the full 3D image so one can utilise the image data to predict to generate 3D maps and define a range of bulk ore characterisation metrics—eg mineral content, porosity, apparent dry density, apparent wet density. Data acquired in the basic bulk analysis are used to test the predictions from image data in 3D. In Figure 3 slices from a 3D CT image for which no training data is used—the result shows the inference of mineralogy and porosity using solely the 3D image data is shown. A photo of Particles D and E is shown along with a slice from the 3D image and the inferred porosity/solid phase map and mineral map for the same slice. Particle D was identified by the geometallurgist as a hard hematite particle and E as a medium hardness goethite rich sample. The inference from the 3D CT images aligns is consistent with this interpretation.

Table 1 details the prediction of the 3D porosity, apparent dry density, apparent wet density and skeletal density for the Particles A–E shown above and the comparison to bulk measured values on the same samples. The table shows that the largest error from the image prediction is in the prediction of particle porosity—measured > predicted. The difference is due to a mismatch between the bulk porosity measured by He pycnometry which has access to all pores regardless of size and the training data which is trained on the porosity identified in the SEM at a resolution of 500 nm. We have used higher resolution BSEM data and mercury pycnometry to note significant contribution of pores below 500 nm are present in the particles– most notable in many of the goethite rich particles. In future developments an augmented training data set will be generated to better estimate the true particle porosity.

TABLE 1

Measured data on particles A–E is given—the original particle set was binned based on Archimedes based apparent density. Measured data indicates: Mass, skeletal density from Helium pycnometry, %H:%Gt by weight from XRD, Porosity from apparent volume versus skeletal volume. Prediction based on OreAI inference from 3D CT image data. The sample A highlights a trend that smaller particles offer larger measured errors. Results tend to show a low prediction of porosity due to the training model being based on porosity only resolved in BSEM above 500 nm voxels. Subsequent work has shown that significant pore volumes can be associated with smaller pores particularly in some Goethite rich types.

	Measured					Prediction				
Sample	Pycnometry Density Bin	Mass(g)	Skeletal Density (g/cc)	%H:Gt	Porosity(%)	Skeletal Density (g/cc)	Dry Density (g/cc)	Wet Density (g/cc)	$H:Gt	Porosity (%)
A	2.5-2.6	0.46	4.69	65:35	18.9	4.86	3.84	4.07	63:37	22.1
B	3.7-3.8	1.77	4.70	56:44	14.0	4.73	4.21	4.42	62:38	11.1
C	3.2-3.3	1.42	4.48	50:48	15.1	4.57	4.21	4.35	53:45	13.5
D	3.7-3.8	1.99	5.23	96:4	13.4	5.24	4.77	4.86	98:2	9.7
E	3.3-3.4	2.23	4.09	11:89	15.6	4.04	3.58	3.73	13:86	11.5

The metrics derived from 3D data are used to predict a range of bulk particle measures for a subset of 60–110 particles from the high-resolution data set in Figure 4. Figure 4a Mass of the particle on the image is estimated from the skeletal volume/porosity maps and mineral identification from the image—mass of the solid goethite phase is estimated as 3.9 g/cc and solid hematite as 5.1 g/cc. Particulate mass shows a good match between 3D image and measurement. Figure 4b as noted above the porosity comparison curve shows a consistent mismatch with He porosity larger than 3D image-based porosity due to training the CT data down to only pore sizes of 500 nm. We do note a small number of particles do show higher porosity than He pycnometry based measures—analysis of these specific (mostly Goethite rich) particles indicates the presence of disconnected pores. Figure 4c shows apparent dry density versus porosity illustrates the broad distribution of porosity across the ore types and the variability of the apparent density—this illustrates the well-known difficulty in density-based separation processes for beneficiation of the Pilbara ore types. Particles rich in hematite and low in gangue content span a broad range of apparent densities. A mineralogy comparison from XRD on part of the particles and 3D CT images is shown in Figure 4d. An apparent overestimation of hematite content is noted in the particles made up of fine grained mixed mineral system. Careful studies note that finely interspersed goethite:hematite textures tend to be predominant in these particles and indicate a lack of sufficient data training on these complex fine grained textural materials. Given the training data presented here is limited to only 28 particles the results indicate a need to train data on more samples to improve the mineral identification of fine complex textured ores. Note also that the training data was limited to the two primary iron ore types (hematite and goethite). Insufficient sample numbers were available to train for predictive mapping of hydro-hematite, and different grades of goethite (clean, medium gangue content, dirty). Work is ongoing on larger sample volumes to improve the model.

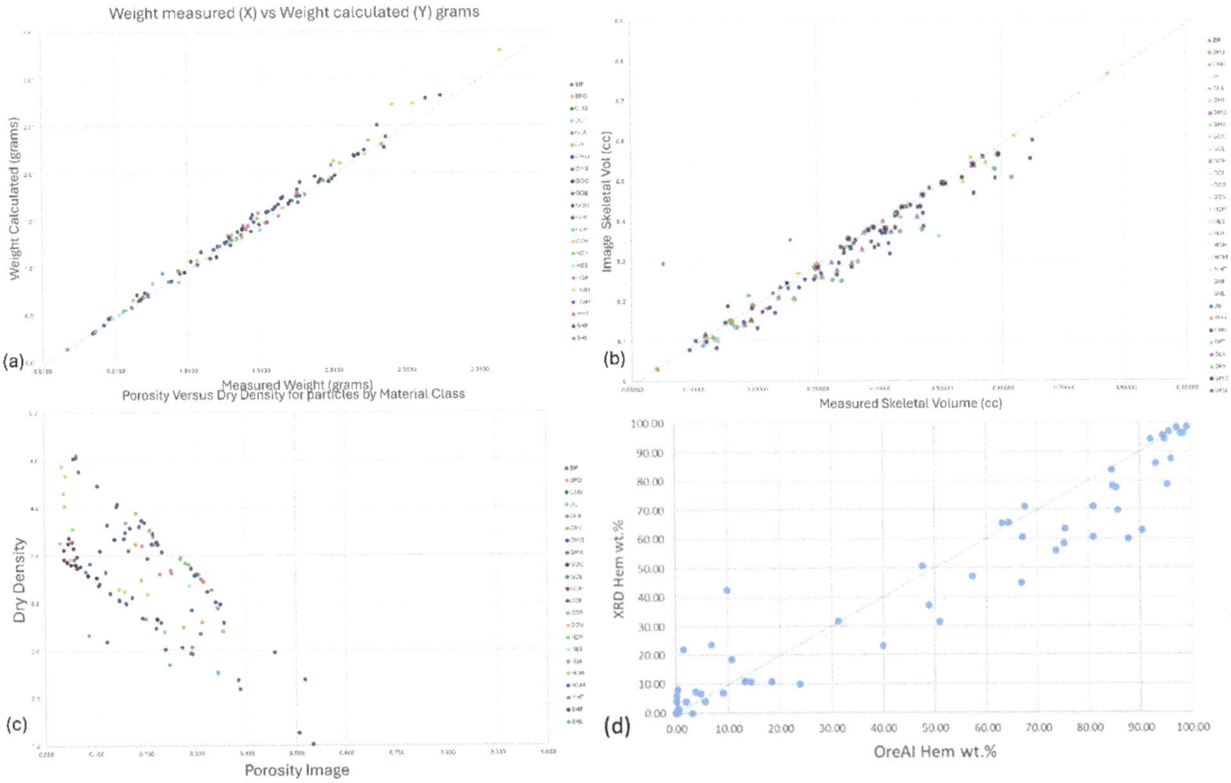

FIG 4 – Summarises data bulk data inferred from 3D images (OreAI data) versus lab measurement. (a) Particle mass; (b) He pycnometry porosity versus inferred from SEM resolution images; (c) Apparent particle density versus porosity; and (d) XRD-based estimates of %H:%Gt versus inference from 3D grey scale data.

CONCLUSION

An example of OreAI's application for supporting orebody characterisation is described for a set of lump samples of iron ore from a Pilbara iron ore asset. The results show that OreAI can be used to extend microanalytical information from 1D assay data and 2D microanalytical image data to 3D X-ray images of the ore material on a large batch of ore material across a range of material type classes. OreAI uses image analysis tools and AI to map and quantify porosity, textural information, mineral phase distributions and associations, density measures (apparent dry, wet, moist and skeletal) in 3D volumes. Mineralogy is mapped in 3D for every particle and comparisons of the overall content with bulk mineralogy data is encouraging. Pore mapping is based on training on BSEM data at resolutions of 500 microns—this leads to an underestimation of porosity in the model. This can be corrected via the use of higher resolution pore mapping data for building training data. Tests of the model illustrate OreAI's ability to accurately propagate high resolution 2D data to 3D and offers high quality mapping and an ability to define uncertainties in predictions.

Additionally detailed information of downstream performance for each particle can be derived and used to enhance orebody knowledge. As an example, one can use the particles characterised in 3D via this method to undertake breakage or reduction analysis of the particles and start to correlate material type/texture/porosity/mineralogy/size to product outcomes. In other published work we have shown the ability to undertake breakage (Francois *et al*, 2022), separation (Zhang *et al*, 2024) and reduction (Kiy *et al*, 2026) using the OreAI platform. Ongoing studies are supporting work on safety (eg slope stability), productivity (eg optimal processing and beneficiation conditions) and low carbon steelmaking (eg identify ores most suited to DRI processing).

REFERENCES

Arns, C H, Bauget, F, Sakellariou, A, Senden, T J, Sheppard, A P, Sok, R M, Ghous, A, Pinczewski, W V, Knackstedt, M and Kelly, J C, 2005. Digital Core Laboratory: Petrophysical Analysis From 3d Imaging Of Reservoir Core Fragments, *Petrophysics*, 46(04):SPWLA-2005-V46n4a1.

Bagheri, M, Knackstedt, M, Dance, T and Raab, M, 2021. Reducing the risk of CO2 storage projects through cost effective reservoir characterisation: application of digital core analysis to a CO2 injection reservoir, *The APPEA Journal*, 61(1):146–162.

Degen, T, Sadki, M, Bron, E, Koenig, U and Nenert, G, 2014. The HighScore Suite, *Journal of Powder Diffraction*, 29(S2):13–18. http://doi.org/10.1017/S0885715614000840

Francois, N, Cruikshank, R, Herring, A, Kingston, A, Webster, S, Knackstedt, M and Saadatfar, M, 2022. A versatile microtomography system to study in situ the failure and fragmentation in geomaterials, *Reviews of Scientific Instruments*, 93:083704.

International Centre for Diffraction Data (ICDD), 2004. PDF-2 database, International Centre for Diffraction Data, Newtown Square, USA.

Kiy, A, Shadwell, M, Evans, T and Knackstedt, M, 2026. Hydrogen reduction of iron ore – insights from 3D analysis, in *Proceedings of the Iron Ore and Open Pit Operators Conference 2026*, this volume, pp 151–154 (The Australasian Institute of Mining and Metallurgy: Melbourne).

Latham, S, Varslot, T and Sheppard, A P, 2008. Image Registration: Enhancing and Calibrating X-ray Micro-CT Imaging, *International Symposium of the Society of Core Analysis*, paper 35, 12 p.

Sok, R M, Varslot, T, Ghous, A, Latham, S, Sheppard, A P and Knackstedt, M A, 2009. Pore Scale Characterization of Carbonates at Multiple Scales: Integration of Micro-CT, BSEM and FIBSEM, *International Symposium of the Society of Core Analysis*, paper 18-SCA-2009, 12 p.

Zhang, Y, Francois, N, Deakin, L, Barron, A, Evans, C, Bensley, S, Loesel, P, Kandula, N, Henley, R and Knackstedt, M, 2024. Improved particle separation, characterisation and analysis for ore beneficiation studies using 3D X-ray micro-CT, *Minerals Engineering*, 216:108835.

Magnetoimpedance for iron ore sensing and sorting applications

L Q Huston[1], S Nguyen[2], R Yong[3], P Coghill[4] and D Miljak[5]

1. Research Scientist, CSIRO Minerals Resources, Lucas Heights NSW 2234. Email: larissa.huston@csiro.au
2. Engineer, CSIRO Minerals Resources, Lucas Heights NSW 2234. Email: sy.nguyen@csiro.au
3. Project Scientist, CSIRO Minerals Resources, Lucas Heights NSW 2234. Email: richard.yong@csiro.au
4. Principal Research Scientist, CSIRO Minerals Resources, Lucas Heights NSW 2234. Email: peter.coghill@csiro.au
5. Research Program Director, CSIRO Minerals Resources, Lucas Heights NSW 2234. Email: david.miljak@csiro.au

INTRODUCTION

The ability to differentiate between iron ore types is highly desirable in the iron ore industry. The ore type may affect bulk handling, dust generation, impurity levels, pelletising, blast furnace performance and ultimately ore marketability. Measuring of iron ore types, whether real-time or 'offline', may provide opportunities to lift the productivity of iron ore mining. For instance, on-belt ore sorting measures the ore directly on a conveyor belt, using a method that discriminates between high-grade and low-grade material. This enables the overall grade to be upgraded by discarding the low-grade fraction. For iron ore applications, sorting may also be based on problematic ore types containing, for example, shales, high manganese, or other deleterious elements at high levels. Sorting problematic ore types prior to stockpile formation can mitigate the loss of ore value that would otherwise result during the stockpiling process. For ore sorting applications, measurements must be non-contact and be representative of the ore being measured. The ore measurement method must also be scalable—both in size, from laboratory testing set-ups to large on-belt sorting equipment, and in time to ultimately enable real-time operation.

Magnetoimpedance is a magnetic property of iron ore that shows promise for ore sorting techniques. Magnetoimpedance refers to two effects; the change in the sample electrical resistivity due to an applied magnetic field which is known as magnetoresistance, and the change in magnetic susceptibility which is typically influenced by magnetic hysteresis. CSIRO has developed a new non-contact laboratory-based measurement system that measures magnetoimpedance. This new method allows for the bulk volume of the sample to be measured, providing a representative characterisation of the material and eliminating the need for surface preparation while accommodating uneven or rough surfaces. This method is suitable for measurement of granular materials for particle sizes between 0.1–200 mm, making it potentially useful for hard rock mining and mineral processing applications.

METHODOLOGY

Figure 1a shows a photograph of the CSIRO-developed laboratory sensor used to measure the magnetoimpedance of mineral samples. The sensor contains a measurement circuit that applies a weak 10 MHz radio-frequency magnetic field to the sample positioned within the sensing region, enabling probing of the sample's electromagnetic response. An external electromagnet simultaneously applies a 47 mT sinusoidally oscillating magnetic field. Together, these fields allow the system to measure both the resistive and magnetic-susceptibility components of the sample's magnetoimpedance. Further details of the measurement system are provided in Huston *et al* (2025). Using this method, measurements were performed on iron ore minerals such as hematite, as well as iron ore samples with known textural classifications including martite, itabirite, goethite, phyllite, and clay.

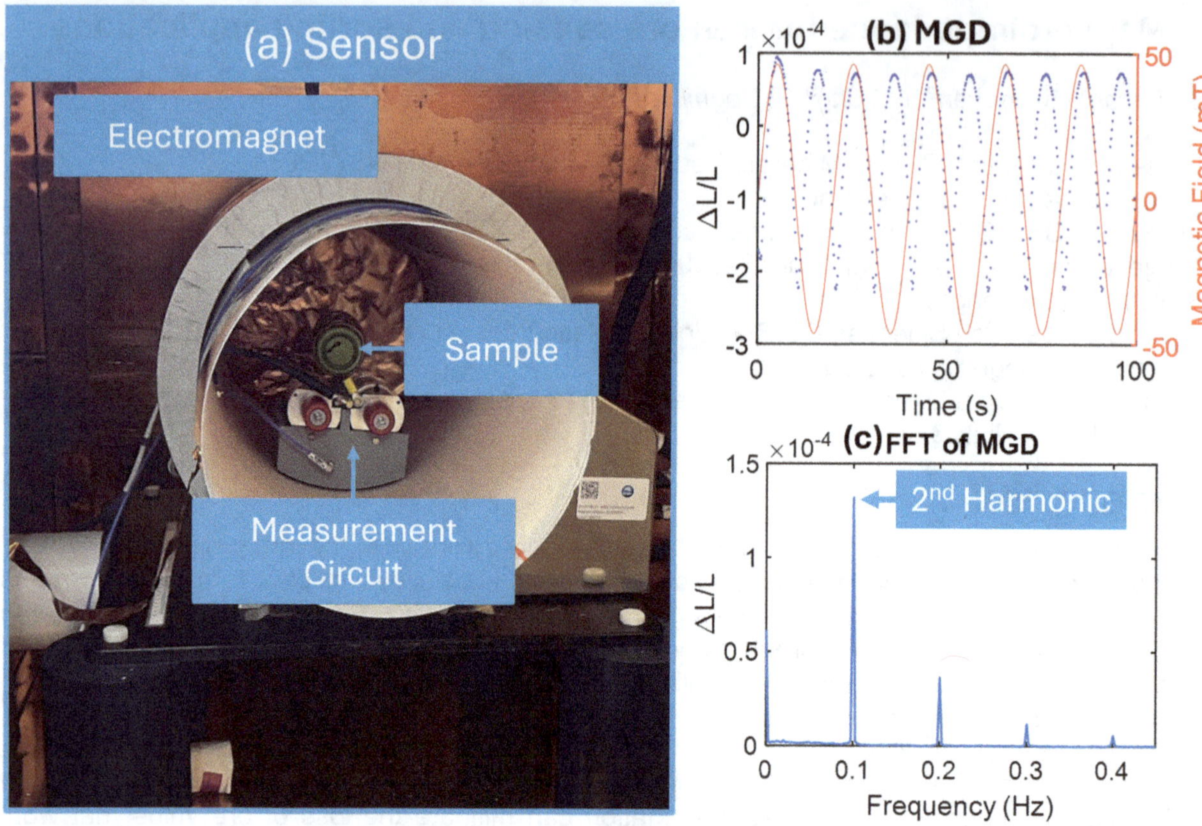

FIG 1 – (a) Photograph of the CSIRO-developed laboratory sensor used to measure the magnetoimpedance of minerals consisting of a measurement circuit that contains the sample and an electromagnet. (b) The change in magnetic susceptibility signal ($\Delta L/L_0$) from a martite sample. (c) Fourier Transform of the change in magnetic susceptibility signal ($\Delta L/L_0$) from a martite sample.

RESULTS

Figure 1b shows an example of magnetic susceptibility related signal that appears as a change in inductance ($\Delta L/L_0$) for a martite sample (containing 78 per cent hematite) after digital processing of the raw sensor output. Each material tested exhibits a distinct $\Delta L/L_0$ response, as well as a unique magnetoresistance signal ($\Delta r/r_0$). The Fourier transform of the $\Delta L/L_0$ signal, which reveals the frequency components of the signal, shows several peaks at multiples of twice the modulation frequency of the magnetic field. The peak at twice the modulation frequency of the magnetic field, also known as the second harmonic, is labelled. It was found that the peak intensities and their relative amplitudes differ for each material, offering a method for distinguishing between materials, with each material having a characteristic spectrum.

Figure 2 compares the magnitude of the second harmonic of both $\Delta r/r_0$ and $\Delta L/L_0$ for a range of iron ore samples with established textural classifications. Using the magnetoimpedance method, it was found that the resistive and magnetic-susceptibility components of the response can be used to identify the corresponding ore textural group. Martite (M2H) and the martite–goethite samples (MGD, MGF, MGM, and MGH) occupy a distinct region near the lower centre of the plot, characterised by lower $\Delta L/L_0$ values and moderate $\Delta r/r_0$ responses. Gangue materials, such as CLA (clay) and PHY (phyllite), show relatively weak signals in both components, making them easily distinguishable from ore-bearing minerals. For the goethite samples, GOL (goethite–limonite) exhibits a comparatively weak response in both $\Delta r/r_0$ and $\Delta L/L_0$, whereas GOV (vitreous goethite) shows a stronger signal; however, this enhanced response is likely influenced by the presence of approximately 5 per cent hematite in the sample. The itabirite samples (IGF, IHC, and IHF) exhibit strong signals in $\Delta L/L_0$ and moderate values of $\Delta r/r_0$, forming a clearly identifiable cluster. These itabirite samples contain 7–11 per cent magnetite and 41–60 per cent hematite, minerals that are relevant to Australian iron ore.

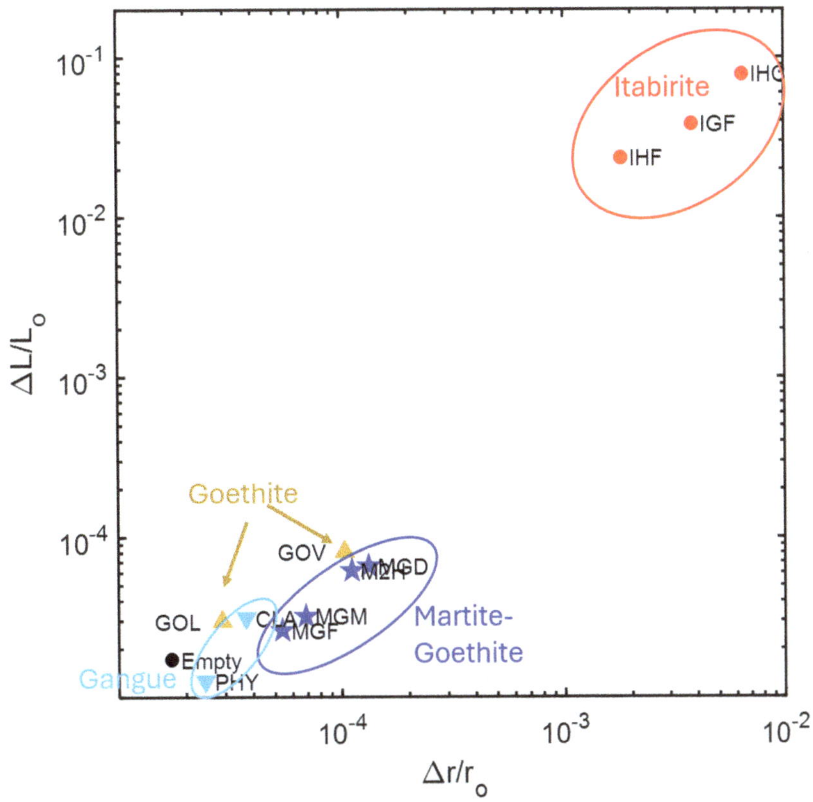

FIG 2 – Comparison of the second-harmonic components of $\Delta r/r_0$ and $\Delta L/L_0$ for iron ore samples of varying textural classifications.

CONCLUSION

We have developed a sensing method that shows strong potential for magnetoimpedance-based measurement in analytical applications and ore-sorting processes for iron ore. We anticipate that measurement times can be reduced to below 1 sec and the measurement scale increased, enabling the technique to be adapted for detection of iron bearing minerals in drill cores and drill chips, as well as for real-time on-belt ore-sorting. In the near future measurement of many well characterised samples (such as drill core) needs to be undertaken to better determine the discrimination of the new method.

REFERENCE

Huston, L Q, Yong, R, Nguyen, S, Coghill, P and Miljak, D, 2025. Contactless Magnetoresistance and Magnetic Hysteresis Measurement Method for Industrial Applications, *IEEE Open Journal of Instrumentation and Measurement*, 4:1–11.

Sustainable dephosphorisation of goethitic iron ores via Fe(II)-catalysed recrystallisation – the role of crystallinity, substitution and amorphous phases

N Karimian[1], M I Pownceby[2], A J Frierdich[3] and N A S Webster[4]

1. Research Scientist, CSIRO Mineral Resources, Clayton Vic 3168. Email: niloofar.karimian@csiro.au
2. Senior Principal Research Scientist/Team Leader – Iron Ore Geometallurgy, CSIRO Mineral Resources, Clayton Vic 3168. Email: mark.pownceby@csiro.au
3. Senior Lecturer, School of Earth, Atmosphere and Environment, Monash University, Clayton Vic 3800. Email: andrew.frierdich@monash.edu
4. Principal Research Scientist and Group Leader, Microbeam and Diffraction, CSIRO Mineral Resources, Clayton Vic 3168. Email: nathan.webster@csiro.au

ABSTRACT

Phosphorus (P) is a major impurity in goethite-rich iron ores and significantly reduces steel quality while increasing the energy and carbon costs of downstream processing. As global iron ore production increasingly relies on goethitic resources, there is growing interest in low-temperature, low-carbon approaches capable of selectively removing structurally bound P. Fe(II)-catalysed recrystallisation has recently emerged as a promising mechanism for mobilising impurities from Fe(III) oxides through coupled electron transfer, atom exchange, and mineral restructuring. This study investigates the effectiveness of Fe(II)-catalysed recrystallisation for dephosphorisation of two contrasting Pilbara goethitic ore microtypes: a relatively crystalline, Fe-rich goethite (GOL) and a structurally disordered, Al-Si-rich vitreous goethite containing significant amorphous material (GOV). Ore mineralogy and impurity speciation were characterised using X-ray diffraction (XRD), X-ray fluorescence (XRF), Mössbauer spectroscopy, Raman spectroscopy, synchrotron X-ray absorption spectroscopy (XAS), and aqueous geochemistry before and after reaction with aqueous Fe(II) under anoxic circumneutral conditions. Both ores exhibited rapid Fe(II) uptake and transient P mobilisation; however, the extent of P redistribution was strongly controlled by mineralogical architecture. The more crystalline GOL ore underwent greater Fe atom exchange and sustained recrystallisation, resulting in substantial transfer of P into alkali-extractable pools. Following 30 days of reaction, approximately 80 per cent of total P became extractable by 1 M NaOH, compared with ~25 per cent in untreated controls. In contrast, the more Al-Si-substituted and partially amorphous GOV ore exhibited rapid P re-sequestration and achieved only ~40 per cent extractable P. Aluminium remained largely immobile (<10 per cent extractable) in both systems, demonstrating selective mobilisation of P relative to Al. These results demonstrate that Fe(II)-catalysed recrystallisation functions as a mineralogical conditioning step that redistributes P from relatively inaccessible hosts into alkali-extractable forms. Ore crystallinity, substitutional chemistry, and amorphous content exert primary controls on dephosphorisation efficiency, providing a mineralogy-informed framework for sustainable upgrading of goethitic iron ores.

INTRODUCTION

Phosphorus (P) is widely recognised as one of the most problematic impurities in iron ores. Even at trace concentrations, P degrades steel quality, increases slag generation, and significantly elevates the carbon footprint of ironmaking, contributing up to ~80 kg of additional CO_2 per tonne of hot metal during refining (Karimian *et al,* 2024; Pownceby *et al,* 2019). As global iron ore supply increasingly shifts toward goethite-rich resources, particularly in the Pilbara region of Western Australia, this challenge has intensified because goethite commonly hosts elevated P concentrations alongside Al and Si. In such ores, P is not merely surface-associated but is structurally incorporated within the goethite lattice, distributed within nanocrystalline domains, or sequestered in amorphous Al-Si-rich materials (Dukino, England and Kneeshaw, 2000; Fisher-White, Lovel and Sparrow, 2012). These modes of occurrence render P highly resistant to conventional beneficiation methods, creating a pressing need for low-temperature, low-carbon, Fe-preserving dephosphorisation strategies (Kaczorek, Brümmer and Sommer, 2009). Fe(II)-catalysed recrystallisation has emerged as a mechanistically robust pathway for mobilising impurities from Fe(III) oxides. In this process, aqueous Fe(II) adsorbs to Fe(III) mineral surfaces, transfers electrons to structural Fe(III), and promotes coupled dissolution–reprecipitation reactions that enable atom exchange between the solid and aqueous phases (Boland *et al,* 2014; Hansel *et al,* 2003). Our previous work on synthetic P- and P+Al-substituted goethite demonstrated that this mechanism can render up to ~68–83 per cent of structurally incorporated P extractable following Fe(II) treatment and alkaline extraction, while total Al removal remains comparatively limited (~17–27 per cent), highlighting strong selectivity for P under controlled mineralogical conditions (Karimian *et al,* 2024). These synthetic systems establish an upper benchmark for Fe(II)-driven dephosphorisation efficiency under idealised crystallinity and compositional homogeneity. However, the applicability of these findings to natural iron ores remains poorly constrained due to their inherent mineralogical heterogeneity, variable crystallinity, substitutional complexity, and amorphous components. Here, we extend these synthetic-phase insights to natural goethitic iron ores, explicitly testing how mineralogical attributes modulate the effectiveness of Fe(II)-catalysed recrystallisation and selective P mobilisation.

EXPERIMENTAL APPROACH

Two contrasting Pilbara goethitic ore microtypes were investigated: GOL, an Fe-rich, moderately P-enriched limonitic goethite characterised by high crystallinity and low Al-Si substitution; and GOV, a vitreous, Al-Si-rich goethite exhibiting pronounced structural disorder, significant amorphous content, and tightly incorporated P (with bulk P concentrations of ~1100 ppm in GOL and ~350 ppm in GOV). These differences are consistent with their contrasting mineralogical character, with GOL representing a more crystalline, low-substitution goethite and GOV reflecting a more disordered, Al-Si-rich and partially amorphous system. Mineralogical and chemical characterisation using X-ray diffraction (XRD), X-ray fluorescence (XRF), Fe and P K-edge X-ray absorption spectroscopy (XAS), Mössbauer spectroscopy, Raman microspectroscopy, and solution-phase inductively coupled plasma–mass spectroscopy (ICP-MS) provided a multiscale assessment of crystalline, nanocrystalline, and amorphous components, as well as the coordination environments of Fe, P, and Al prior to Fe(II) exposure. Batch reactions were conducted under anoxic, circumneutral conditions (pH ~7.6) at ~0.5 mM Fe(II) and a solids loading of ~2 g L^{-1}, without Fe(II) replenishment (Figure 1).

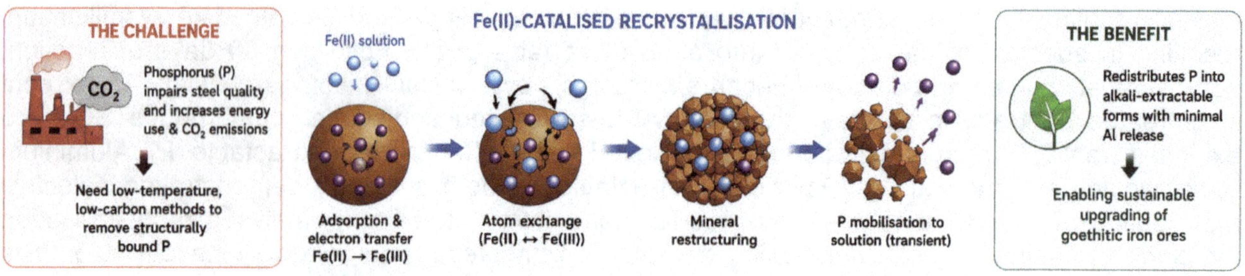

FIG 1 – Fe(II)-catalysed recrystallisation pathway and P mobilisation.

RESULTS AND DISCUSSION

Despite both samples being dominated by goethite, their mineralogical architectures differ fundamentally. GOL comprises ~89 wt per cent predominantly well-crystalline goethite with minor hematite and low substitution levels, as confirmed by XRD, Raman, and Mössbauer analyses. In contrast, GOV contains two distinct goethite populations, with refined lattice parameters indicating ~5.3 mol per cent Al substitution and ~10–11 per cent amorphous material dominated by aluminosilicate-rich and poorly crystalline Fe phases. Mössbauer spectra reveal a pronounced paramagnetic component in GOV, consistent with nanophase Fe(III) domains and extensive structural disorder, while Fe K-edge XAS corroborates these observations. P K-edge X-ray absorption near edge structure (XANES) shows that P is overwhelmingly associated with Fe phases in both ores; however, P speciation differs markedly. GOL hosts crystalline Fe(III) phosphate phases (eg strengite and rockbridgeite) alongside P-bearing goethite and ferrihydrite-like components, whereas GOV exhibits a more diverse P assemblage including strengite, vivianite, P-incorporated ferrihydrite, and minor apatite.

Thirty-day batch experiments conducted under strictly anoxic, ambient-temperature conditions demonstrate that Fe(II) reactivity and impurity mobility are governed primarily by mineralogical controls, with microstructural features exerting a secondary influence on reaction kinetics and persistence. Both ores exhibit rapid initial Fe(II) consumption, reflecting strong sorption and oxidation at reactive Fe(III) sites; however, their subsequent behaviour diverges markedly. Fe(II) was rapidly removed from solution and incorporated into the solid phase, with negligible net Fe loss, indicating that recrystallisation proceeds via atom exchange and short-range structural reorganisation rather than bulk dissolution. In GOL, aqueous P increases progressively during early reaction stages, peaks at ~24 hrs, and remains detectable over extended periods, indicating sustained Fe(II)-catalysed recrystallisation and effective P liberation. In contrast, GOV exhibits a short-lived P pulse followed by rapid re-sequestration into Fe(II)/Fe(III) phosphate phases, reflecting restricted Fe(II) access to reactive Fe sites and strong structural retention of P.

Aluminium behaviour further differentiates the two ore microtypes and constrains the mechanisms governing phosphorus extractability. In GOL, Al is released slowly and continuously, consistent with limited disruption of minor Al-bearing sites during Fe(II)-catalysed restructuring of well-crystalline goethite. In contrast, GOV exhibits earlier but consistently lower Al release, reflecting stronger structural incorporation of Al within Al-Si-rich, poorly ordered Fe phases that are resistant to Fe(II)–Fe(III) atom exchange. The absence of coupled P-Al release in both systems indicates that phosphorus mobilisation is controlled primarily by transformations within Fe-bearing phases rather than by aluminosilicate dissolution.

These trends are reinforced by the results of 1 M NaOH extraction experiments. In GOL, Fe(II) exposure increases alkali-extractable P approximately fourfold, rendering ~80 per cent of total P extractable compared with ~25 per cent in non-Fe(II) controls. In contrast, Fe(II)-treated GOV reaches only ~40 per cent extractable P despite substantial relative increases. This lower extractability reflects the glassy, Al-Si-rich and partially amorphous nature of GOV, which restricts sustained Fe(II)-driven recrystallisation and promotes rapid re-stabilisation of released P into newly formed Fe(II)/Fe(III) phosphate phases rather than persistent labile pools. Importantly, Al extractability remains below 10 per cent across all treatments, confirming preferential mobilisation of P and demonstrating that Fe(II) treatment fundamentally redistributes structurally bound P rather than merely removing a small, pre-existing adsorbed fraction (Figure 2).

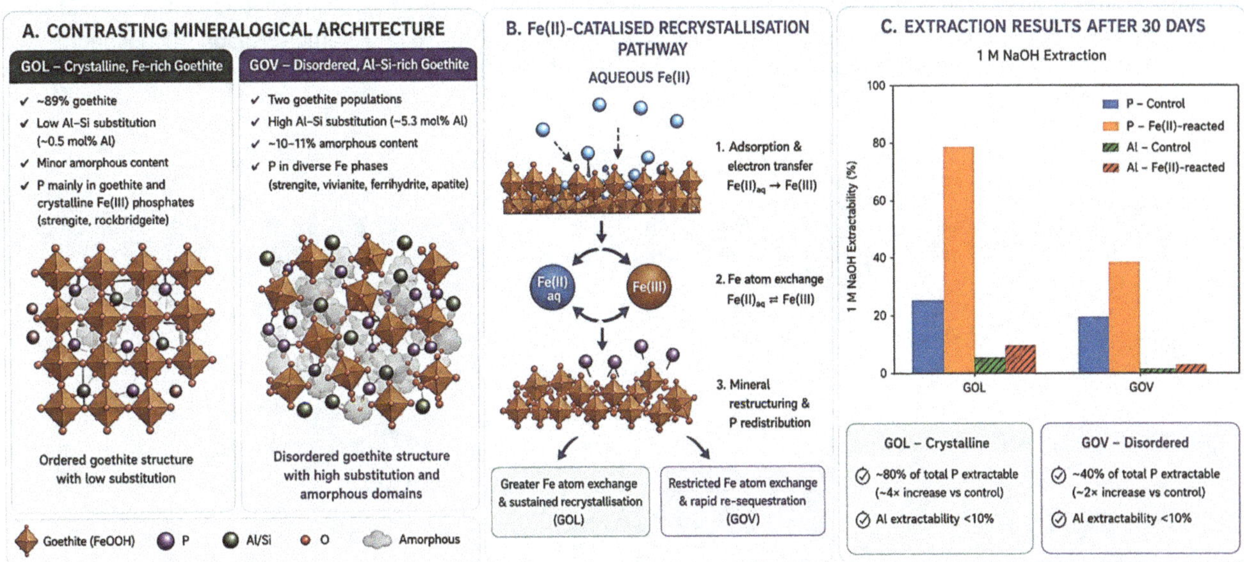

FIG 2 – Mineralogical controls on Fe(II)-catalysed phosphorus redistribution in natural goethitic iron ores. (a) Conceptual representation of the contrasting mineralogical architectures of the two Pilbara ore microtypes. (b) Schematic illustration of Fe(II)-catalysed recrystallisation. (c) Comparison of P and Al extractability after 30 days of reaction with aqueous Fe(II).

CONCLUSIONS

Collectively, these results provide direct mechanistic insight into Fe(II)-catalysed dephosphorisation of natural goethitic iron ores. Goethite with high crystallinity and low Al-Si substitution permits effective Fe(II) access to reactive Fe sites, enabling recrystallisation-driven P mobilisation and sustained removal. Conversely, extensive substitution, structural disorder, and amorphous components restrict Fe(II) interaction and limit dephosphorisation efficiency. These findings establish a mineralogy-informed framework for low-temperature, low-carbon beneficiation strategies with direct implications for geometallurgy, ore classification, and decarbonised iron ore processing.

ACKNOWLEDGEMENTS

The authors gratefully acknowledge financial support from CSIRO. We thank the staff of the Australian Synchrotron (ANSTO), particularly Dr Negin Foroughimehr (MEX2) for assistance with P K-edge XANES data collection and Dr Priya Ramkissoon (MEX1) for assistance with Fe XAS measurements. Steve Peacock (CSIRO Mineral Resources, Clayton) is gratefully acknowledged for XRF analyses.

REFERENCES

Boland, D D, Collins, R N, Miller, C J, Glover, C J and Waite, T D, 2014. Effect of Solution and Solid-Phase Conditions on the Fe(II)-Accelerated Transformation of Ferrihydrite to Lepidocrocite and Goethite, *Environmental Science and Technology*, 48(10):5477–5485.

Dukino, R, England, B and Kneeshaw, M, 2000. Phosphorus distribution in BIF-derived iron ores of Hamersley Province, Western Australia, *Applied Earth Science*, 109(3):168–176.

Fisher-White, M J, Lovel, R R and Sparrow, G J, 2012. Phosphorus removal from goethitic iron ore with a low temperature heat treatment and a caustic leach, *ISIJ International*, 52(5):797–803.

Hansel, C M, Benner, S G, Neiss, J, Dohnalkova, A, Kukkadapu, R K and Fendorf, S, 2003. Secondary mineralization pathways induced by dissimilatory iron reduction of ferrihydrite under advective flow, *Geochimica et Cosmochimica Acta*, 67(16):2977–2992.

Kaczorek, D, Brümmer, G W and Sommer, M, 2009. Content and Binding Forms of Heavy Metals, Aluminium and Phosphorus in Bog Iron Ores from Poland, *Journal of Environmental Quality*, 38(3):1109–1119.

Karimian, N, Pownceby, M I, Burton, E D, Wells, M and Frierdich, A J, 2024. Fe(II)-Catalyzed Recrystallization Drives Phosphorus and Aluminum Release from Goethite, *Environmental Science and Technology*, 58(42):19016–19026.

Pownceby, M I, Hapugoda, S, Manuel, J, Webster, N A S and MacRae, C M, 2019. Characterisation of phosphorus and other impurities in goethite-rich iron ores – Possible P incorporation mechanisms, *Minerals Engineering*, 143:106022.

CSIRO iron ore proficiency program

R Maree[1] and S Birch[2]

1. Experimental Scientist, CSIRO, Adelaide SA 5065. Email: rachel.maree@csiro.au
2. Principal Experimental Scientist, CSIRO, Adelaide SA 5065. Email: sally.birch@csiro.au

ABSTRACT

The Proficiency Test for Chemical Analysis of Iron Ore was established in 1992 at the 17th meeting of the ISO/TC 102/SC2 (ISO, 2008) committee. This program which is coordinated by CSIRO has now been active for 33 years, providing participants with valuable assessment of the analytical performance of their laboratory. This proficiency test is unique in that most samples distributed are not certified reference materials but are real production materials, prepared in a production laboratory and donated to the program by iron ore producers. Variation in the physical properties of real-world samples, such as particle size distribution, can influence digestion efficiency and homogeneity. Compared to certified reference materials, these samples present an analytical challenge to participating laboratories and are a true test of the robustness of their analytical methods with regards to sample preparation.

Throughout the course of the program, the geographical distribution of participants has closely matched the geographical distribution of industry stakeholders (both iron ore producers and customers). Also, the country of origin for the samples provided for each round has closely matched the export market share distribution worldwide. With data provided from all major stakeholders and all major sample sources accounted for, this means that this data set can be considered strongly indicative of industry analytical performance and iron ore quality over time. As part of the program, most participants specify the analytical methods that they use. Comparison of the performance of each of these methods separately can further be made to determine the most precise method for each analyte and the difference between methods.

The data set accumulated over the course of the program contains over 850 000 data points (ie number of contributing labs × number of samples analysed × number of analytes × number of replicates). Data used for assessment was limited to the concentration ranges for major and minor elements outlined in the scope of ISO 9516–1 (International Organisation for Standardization (ISO), 2003) (see Table 1). Samples with concentrations for analytes outside of these ranges are not considered representative iron ores for the purposes of comparison with ISO 9516–1 (ISO, 2003). Of the 335 testing rounds over the course of the program, 103 were removed. For this paper, therefore, analysis was only performed on this representative data set, which was reduced to 500 000 data points, 41 000 of which are for iron, spread across the remaining 232 rounds.

There are two main analytical methods reported by participants for the analysis of iron in the program; titration and X-ray fluorescence (XRF). ISO 2597-1 (ISO, 2006) is the most common standard method used by participants for titration results and is an established referee method suitable for use in commercial settlement. ISO 9516 (ISO, 2003) is the most common standard method used by participants for XRF results. In addition to iron by titration and XRF, iron is calculated by the convenor by subtracting the sum of all elements other than iron as well as the loss on ignition (LOI) from 100 per cent. There is currently no standard method for calculated iron. To be considered as a referee method, there are strict criteria for trueness (accuracy) and precision tests as outlined by the ISO TC102/SC2 Iron ore and Direct Reduced iron – Chemical Analysis Technical Committee (ISO, 2008). The purpose of these mathematical tests is to ensure the results provided by the proposed referee method are not significantly different or significantly less precise than an already established referee method.

TABLE 1
Concentration ranges used to determine representative sample data for this assessment (ISO, 2003).

Component element	Concentration range for analysis (%)
Fe	38 to 72
Si	0.2 to 6.5
Ca	0.019 to 12.7
Mn	0.02 to 0.82
Al	0.1 to 3.5
Ti	0.016 to 4.7
Mg	0.2 to 2.0
P	0.006 to 0.6
S	0.007 to 0.6
K	0.012 to 0.45

Application of the accuracy and precision tests to interlaboratory test data gathered during development of ISO 9516 (1990–1992; ISO, 2003) determined that the XRF method did not meet the criteria required to be considered suitable for use as a referee method for iron. Unfortunately, these tests cannot be directly applied to this data set as not all samples are certified reference materials. However, useful information can still be obtained from the data set. An accuracy assessment can be made by performing a statistical comparison of the mean results for titration and XRF. As result set is not normally distributed, a nonparametric statistical test is required. A Mann-Whitney Rank sum test was performed, and the result of this test shows there is no significant difference in the accuracy of XRF compared to titration (P=0.783). There is also no significant difference in accuracy of the calculated iron result compared to titration (P=0.9718).

For precision, an assessment can be made by comparing the reproducibility standard deviation (SD) obtained for each method. The reproducibility SD of a method is a measure of the variability observed when different laboratories apply the method to identical samples. If the maximum reproducibility SD of a proposed test method does not exceed the maximum repeatability SD of a published referee method, then it may be considered suitable for classification as a referee method itself. Analysis of representative samples from the set of data for the proficiency test that has been collected in the years since the initial assessment shows that the imprecision observed during initial development of ISO 9516 (ISO, 2003) remains, despite improvements in XRF technology and software. Statistical comparison of the reproducibility SD data using the Mann-Whitney Rank sum test showed that both XRF and calculated iron compared to titration have a significant difference (P=<0.0001 for both). As shown in Figure 1, the average reproducibility SD is higher for XRF (0.21 per cent) than for titration (0.14 per cent). The average reproducibility SD is lower for calculated iron (0.11 per cent), and no value exceeds that of titration.

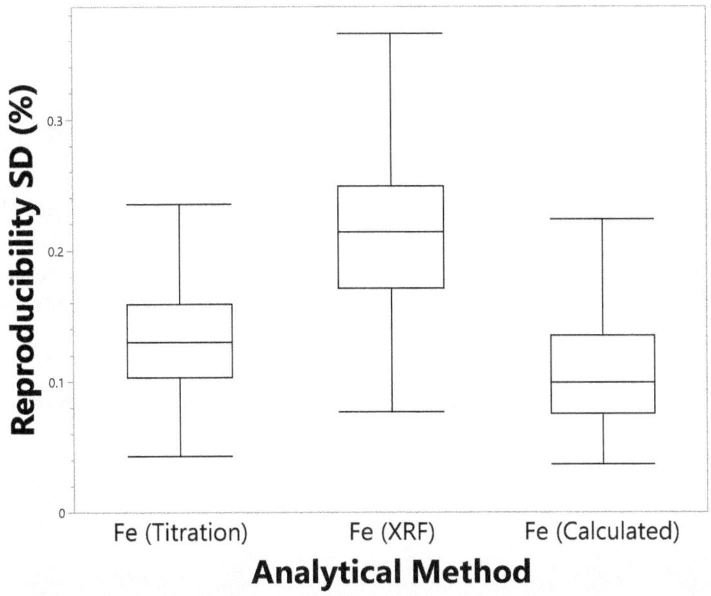

FIG 1 – Comparison of Reproducibility SD for iron obtained in representative rounds using each main analytical method.

As an indirect measurement, the accuracy and precision of the calculated iron result rely heavily on the quality of the analysis provided for the 10+ analytes that form part of the calculation. Despite this apparent flaw, calculated iron is very accurate and precise, providing results that are not significantly different to that of titration in terms of accuracy or precision. Therefore, calculated iron has the potential of meeting the requirements of a referee analysis.

Particle size distribution is particularly important for analysis of iron ore as the different grinding characteristics of the variety of mineralogical components can lead to imprecision for some elements due to segregation and inhomogeneity. Analysis of samples for which sizing data is available (n=54) shows a correlation between the particle size distribution and the repeatability standard deviation of results for SiO_2. This in turn has a direct effect on the precision of calculated iron as SiO_2 contributes a large amount of bias and uncertainty to the final calculated iron (Figure 2).

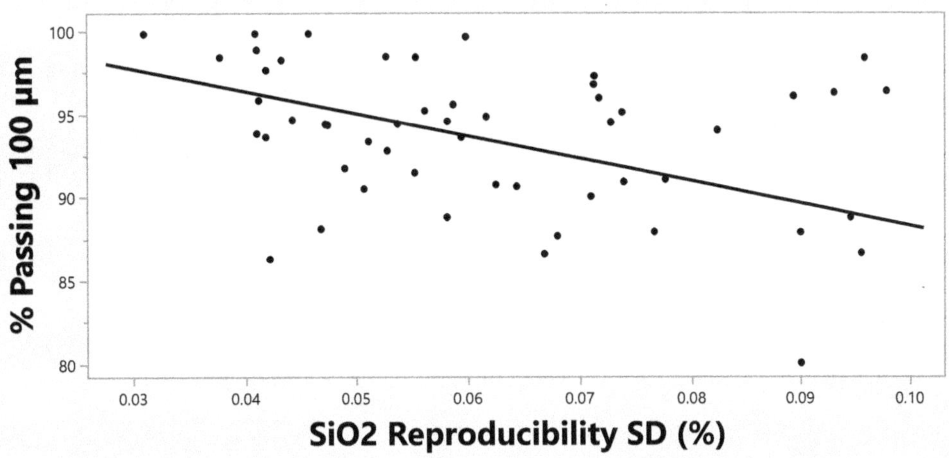

FIG 2 – Reproducibility SD of SiO_2 (All Methods) compared to % passing 98 µm (n=54).

From this single observation, the value of the data set obtained by this proficiency test is clear. Interlaboratory testing as part of standard method development uses certified reference materials that are prepared to a very high standard. Consistency of sample preparation to control analytical parameters can obscure important observations. The robustness of the calculated iron method in real world applications with industry typical samples is proven with the data obtained during the proficiency test. Inclusion of certified reference materials in the proficiency test would provide the data required for the justification of calculated iron as an accepted referee method.

REFERENCES

International Organisation for Standardization (ISO), 2003. ISO 9516-1:2003 – Iron ores – Determination of various elements by X-ray fluorescence spectrometry – Part 1: Comprehensive procedure, April 2003.

International Organisation for Standardization (ISO), 2006. ISO 2597-1:2006 – Iron ores — Determination of total iron content, Part 1: Titrimetric method after tin(II) chloride reduction, May 2006.

International Organisation for Standardization (ISO), 2008. ISO/TC 102 N 610 E – ISO/TC 102 – Iron ore and direct reduced iron – TC 102 Technical Committee Report No.5 – Procedure for statistical evaluation of analytical data resulting from international tests for chemical analysis, March 2008.

Fundamental characterisation studies examining the relationship between goethite textural types, crystallinity and contained impurities

E Morton[1], M Chen[2], A J Whitworth[3], N A S Webster[4] and M I Pownceby[5]

1. Postgraduate Student, CSIRO Mineral Resources and Royal Melbourne Institute of Technology, Melbourne Vic 3169. Email: eleanor.morton@csiro.au
2. Team Leader, Leaching Fundamentals, CSIRO Mineral Resources, Clayton Vic 3168. Email: miao.chen@csiro.au
3. CERC Postdoctoral Fellow, CSIRO Mineral Resources, Clayton Vic 3168. Email: anne.whitworth@csiro.au
4. Characterisation Group Leader, CSIRO Mineral Resources, Clayton Vic 3168. Email: nathan.webster@csiro.au
5. Senior Principal Research Scientist, CSIRO Mineral Resources, Clayton Vic 3168. Email: mark.pownceby@csiro.au

ABSTRACT

Globally, high-grade, low impurity iron ores are being rapidly depleted. The decline in ore quality is particularly problematic for Australian producers as a large proportion of Australia's future Pilbara iron ore resources are composed of lower grade, impurity-rich hematite-goethite and goethite-rich ores. Aluminium (Al), silicon (Si), and phosphorus (P) are the most commonly occurring and troublesome impurities found in these ores with previous work indicating that these impurities are commonly associated with discrete aluminosilicate gangue mineral phases while P, Al and Si are associated with goethite. The distribution and incorporation mechanism of the impurities, particularly those directly associated with goethite, differ greatly across different goethitic-containing ore types. A current lack of knowledge regarding the mineralogical and textural controls on the distribution and concentration of impurities in goethite has resulted in immense pressure to develop a fundamental understanding of impurity deportment in goethite-rich ores.

In this study, the Pilbara ore group classification scheme, which identifies and defines several texturally different goethite ore groups, is used to investigate the possible links between goethite textural type and impurity distribution. The impurity deportment was examined using a combination of standard geochemical and mineralogical analysis methods such as X-Ray Fluorescence (XRF) spectroscopy and X-Ray Diffraction (XRD), as well as more advanced analysis and imaging techniques such as X-Ray Absorption Near Edge Structure (XANES) spectroscopy and Electron Probe Microanalysis (EPMA). The characterisation techniques were applied to both individual ore particles and whole-of-ore samples with results used to identify any relationships between the predefined goethitic textural ore groups and possible impurity incorporation mechanisms and controls.

Two samples of goethite reference textural types, VG (Vitreous Goethite) and OG (Ochreous Goethite), were examined. The bulk chemistry of the two samples was determined by XRF. The VG material contained low iron (77.2 per cent Fe_2O_3) and high levels of the impurities SiO_2 (6.58 per cent) and Al_2O_3 (3.48 per cent). Phosphorus was also a significant impurity in VG at 0.106 per cent P_2O_5. In comparison, the OG sample was higher in iron (83.8 per cent Fe_2O_3) and contained significantly less SiO_2 (2.61 per cent) and Al_2O_3 (2.37 per cent) impurities. The phosphorus level in the OG sample was very high at 0.294 per cent P_2O_5.

XRD phase identification was performed using PANalytical Highscore Plus© software (V4.1) and quantitative phase analysis (QPA) was carried out via the Rietveld method using the software, TOPAS V5.The goethite peaks in the XRD data sets collected for the OG (Figure 1a) and VG (Figure 1b) samples were modelled using two separate goethite phases, goethite(1) and goethite(2) to improve the fits, with the final goethite concentration in Table 1 being the combined contribution of these.

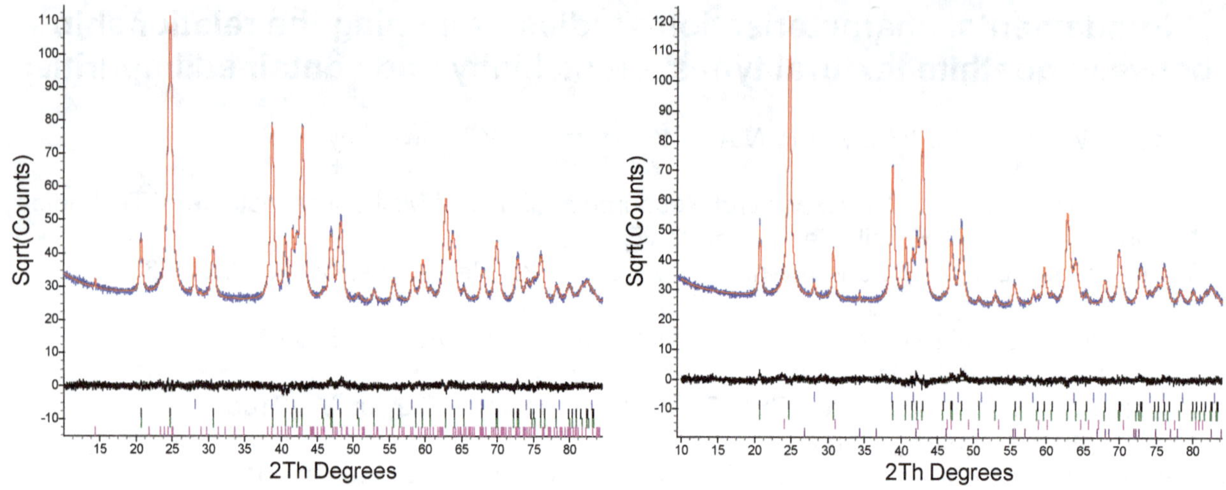

(a) (b)

FIG 1 – Output from Rietveld analysis of OG (A) and VG (B) where the blue line is the observed data, the red is the calculated pattern, and the black line is the difference. The vertical lines are the Bragg peak markers for (top to bottom) hematite, goethite(1), goethite(2) and kaolinite.

TABLE 1

Results of QPA (relative wt% of crystalline phases).

Sample	Hematite	Goethite	Kaolinite	Quartz	Calcite
OG	10.9	88.6	0.5		
VG	6.9	92.6		0.2	0.3

To calculate amorphous/unaccounted-for phase content both the internal and external standard methods were used. A highly crystalline (99 ± 1 wt per cent) corundum sample was used as the standard. Table 2 shows the results of the amorphous/unaccounted-for phase content calculations. There is reasonable agreement between the values calculated using the internal versus external standard methods. VG appears to contain more amorphous/unaccounted-for material than OG.

TABLE 2

Results of the amorphous/unaccounted-for phase content calculations.

Sample	Internal standard	External standard
OG	7.7	6.0
VG	11.3	10.6

A comparison of the data sets collected for the OG and VG samples showed that the peak shapes at low-2θ are markedly different. The refined c lattice parameters for the goethite(2) phase in each sample suggested the presence of Al-goethite, although in the OG sample the Al substitution appears to be a at very low levels (~0.5 mol per cent), with the Al-goethite in the VG sample containing a greater degree of Al substitution (~5.3 mol per cent). The latter is consistent with the higher levels of Al_2O_3 recorded in the bulk assay for the sample. This will be confirmed through follow-up SEM and/or quantitative EPMA test work.

Further work using DTA/TGA is also in progress to examine the effects of crystallinity and impurity incorporation on dehydroxylation together with EPMA mapping of the two samples to measure the distribution of P, Al and Si within the two goethite types. In addition, we have recently conducted synchrotron-based soft X-ray spectroscopy on the samples to determine possible P incorporation mechanisms in goethite.

Improved understanding and identification of the occurrence and distribution of deleterious impurities like Al, Si, and P will be essential for the successful transition from high-grade hematite-rich ores to lower-grade, higher-impurity goethite and hematite-goethite ores. This information can be used to provide a more informed basis for development of specific beneficiation strategies that factor in the relationship between goethite texture and composition.

Purpose-built downhole assay solutions for iron ore – design and application

T J Neville[1], B Birt[2], R Griffiths[3] and L Folan[4]

1. Principal Research Fellow, Orica Digital Solutions, Welshpool WA 6106. Email: tom.neville@orica.com
2. Senior Petrophysicist, Orica Digital Solutions, Welshpool WA 6106. Email: benjamin.birt@orica.com
3. Principal Research Fellow, Orica Digital Solutions, East Melbourne Vic 3002. Email: roger.griffiths@orica.com
4. Senior Technology Manager, Orica Digital Solutions, Welshpool WA 6106. Email: lee.folan@orica.com

ABSTRACT

This paper presents the development of a downhole assay tool specifically engineered for mining environments, with a strong focus on iron ore resource evaluation. Moving beyond the constraints of repurposed oilfield technologies and design philosophies, this tool leverages the relatively benign conditions encountered in mining boreholes to optimise measurement accuracy, precision, and operational efficiency. A key design innovation is the use of non-metallic housings to eliminate interference from iron-bearing tool components, ensuring reliable detection and quantification of iron and associated gangue elements critical to ore characterisation.

The tool employs pulsed fast and thermal neutron activation–gamma ray spectrometry to temporally separate gamma rays generated by distinct neutron interactions—fast neutron interactions and thermal neutron interactions—and from radioactive decay of unstable nuclides created from such interactions. This approach enhances element identification and quantification through cross-validation of elemental responses across interaction types. Monte Carlo nuclear modelling guided key design decisions, including detector configuration, shielding geometry, and material selection to minimise tool background. The resulting tool is compact and lightweight, while maintaining robustness and reliability for high-volume mining operations.

The tool design enables efficient measurement acquisition across all stages of mining. It is engineered to handle the deeper boreholes encountered during early exploration and resource development, while also optimised for the rapid cycle time between holes critical in production environments. To meet the demands of blasthole logging, the tool incorporates a standby capability that facilitates streamlined movement between holes, supporting high productivity while maintaining safe operating practices.

By integrating advanced nuclear measurement principles with mining-specific design priorities, this technology delivers accurate, reproducible, and timely compositional data suitable for modern resource evaluation, with potential applications in near-real-time decision support during drilling, grade-informed blast design, load and haul optimisation, grade control, and geometallurgical modelling.

INTRODUCTION

This paper outlines the design choices made to optimise a downhole assay tool based on pulsed fast and thermal neutron activation–gamma ray spectrometry (PFTNA–GRS) for use in mining applications, particularly in iron ore. To highlight the importance of these choices, an introduction to the physics of measurement is provided along with a brief historical review of the development of wireline-conveyed logging tools using similar principles. The paper then discusses the constraints imposed by design for mining in terms of measurement, environmental, and operational requirements and the design choices that these have both imposed and enabled. Finally, acquisition and preprocessing of spectral data is described.

PHYSICS OF MEASUREMENT

Isotopic and accelerator-based neutron sources generate fast neutrons (neutrons with kinetic energies above 0.1 MeV). These neutrons interact with atomic nuclei through a variety of scattering and absorption reactions (Lamarsh and Baratta, 2013), as illustrated in Figure 1. During these interactions, the neutrons lose energy and are finally absorbed into a nucleus with absorption typically occurring once thermal energy levels (~ 0.025 eV) are reached.

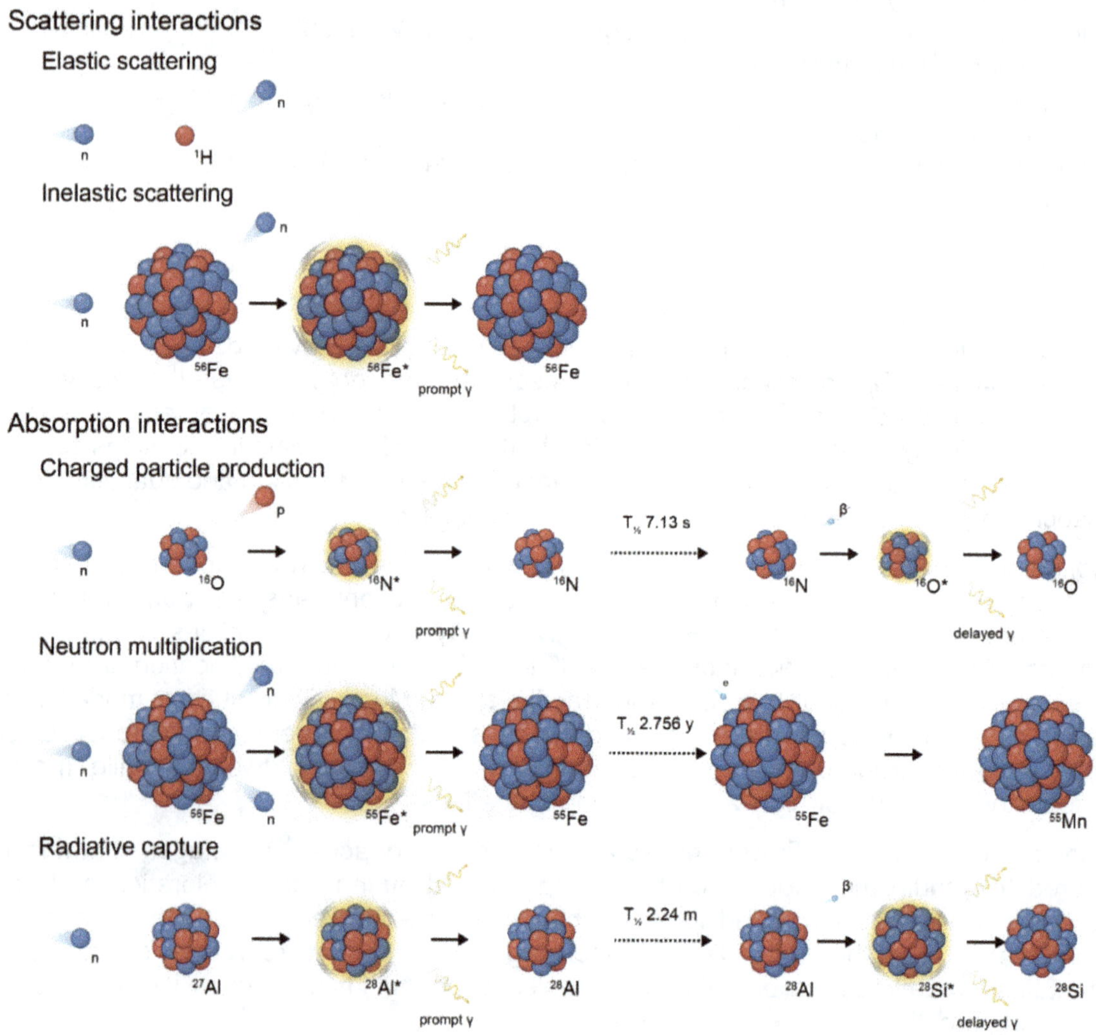

FIG 1 – Neutron interactions.

Neutron energy loss occurs through scattering interactions. Elastic scattering involves a neutron colliding with an atomic nucleus, transferring energy to the nucleus while the total kinetic energy of the system is conserved. This mechanism is most effective when the atomic nucleus involved has a similar mass to the incident neutron, such as hydrogen. Elastic scattering with hydrogen nuclei is the main mechanism for moderation (slowing) of neutrons—reducing their kinetic energy to thermal levels. Inelastic scattering is a similar process; however, the total kinetic energy of the system is not preserved and some of the energy of the incident neutron is converted into excitation of the nucleus. The resulting nucleus then undergoes prompt de-excitation, with a half-life typically in the picosecond range ($1e^{-12}$ s). De-excitation is the process by which a nucleus disposes of its excitation energy; typically, but not always, through gamma-ray emission.

Free neutrons are incorporated into atomic nuclei through absorption interactions. The most common form of absorption is radiative capture, in which a neutron is absorbed into a nucleus, causing an excited compound nucleus, which then undergoes prompt de-excitation. Neutron absorption can also be accompanied by the emission of charged particles from the nucleus, such as protons or alpha particles, or by the emission of multiple neutrons through neutron multiplication. These interactions also result in excited residual nuclei that undergo prompt de-excitation.

In radiative capture and neutron multiplication, the residual nucleus is a new isotope, while in charged particle emission the residual nucleus is of a new element. The new nuclide may be, and in the case of neutron multiplication and charged particle emission typically is, unstable and undergoes radioactive decay. The decay commonly involves emission of gamma rays, in this case delayed gamma rays, over a period dictated by the half-life of the new nuclide.

The final type of absorption reaction is fission, which is not relevant to this measurement, and is not illustrated above, but is important in other downhole assay technology such as prompt fission neutron (PFN) logging for uranium quantification.

The rate at which these different interactions take place is given by:

$$R = \Phi \cdot \Sigma = \Phi \cdot \mathrm{n} \cdot \sigma = \Phi \cdot \frac{\rho \cdot N}{M} \cdot \sigma$$

where:

R	is the interaction reaction rate ($cm^{-3} \cdot s^{-1}$)
Φ	is the neutron flux ($cm^{-2} \cdot s^{-1}$)
Σ	is the macroscopic cross-section of the material for the interaction (cm^{-1})
n	is the number density of target nuclei (cm^{-3})
σ	is the microscopic cross-section of the target nuclei for the interaction (cm^2)
ρ	is the mass density of the material ($g \cdot cm^{-3}$)
N	is Avogadro's Number (6.02214076e23 mol^{-1})
M	is the molar or atomic mass of the material ($g \cdot mol^{-1}$)

The cross-section is therefore the fundamental nuclear property that dictates reaction rate, with the microscopic cross-section σ representing the number of reactions that take place per target nuclide per incident neutron in a unit cross-sectional area, and the macroscopic cross-section Σ representing the total number of reactions that take place per incident neutron travelling through a unit length of a material. Cross-sections vary for each nuclide, and hence material, and for each interaction. Cross-sections are also energy-dependent, and this dependency controls under what conditions the different interactions become important. Figure 2 shows Σ as a function of incident neutron energy for a hematite-water mixture for the two scattering and three absorption reactions discussed above.

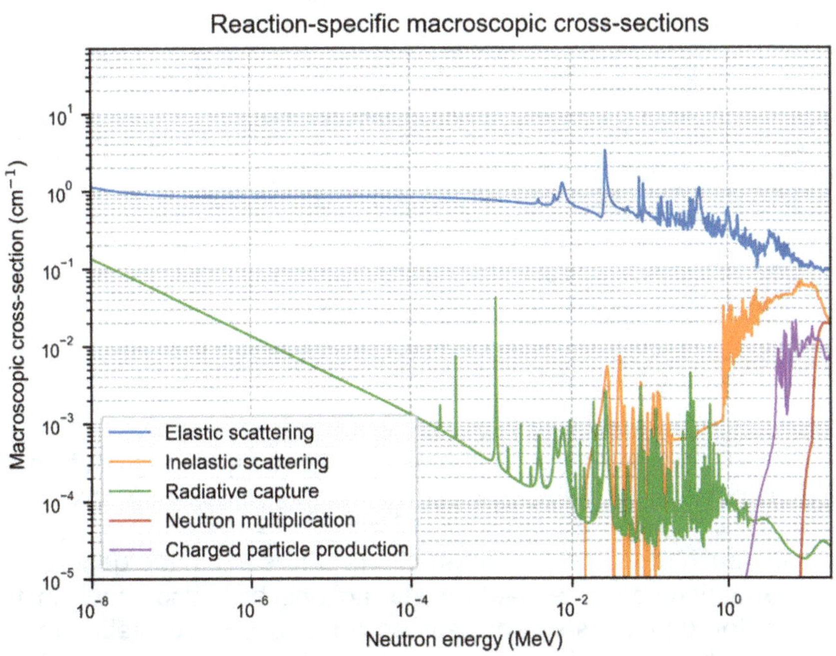

FIG 2 – Neutron interaction macroscopic cross-sections Σ for a mixture of 80 per cent by volume hematite and 20 per cent by volume water.

Elastic scattering takes place with almost equal likelihood across the full energy range from fast to thermal energy levels. The likelihoods of inelastic scattering, charged particle emission, and neutron multiplication are high for fast neutrons, but then fall abruptly. These are all threshold reactions in that the incident neutron needs to have kinetic energy exceeding a certain value for the interaction to take place and are therefore generally limited to fast incident neutrons. The likelihood of radiative capture progressively increases as neutron energy decreases and becomes a dominant interaction for thermal incident neutrons.

As discussed above, neutron sources generate fast neutrons, typically with energies above 1 MeV. These neutrons lose energy through scattering interactions, both elastic and inelastic scattering for fast neutrons and exclusively elastic scattering for lower energy neutrons. Eventually the neutrons are absorbed. Fast neutrons may be absorbed resulting in charged particle emission or neutron multiplication, but it is most likely that neutrons survive until they reach thermal energy levels at which time they undergo radiative capture. Figure 3 displays the time-dependent neutron energy behaviour expected in the hematite-water mixture introduced above, as well as the time-distribution of the different interaction types. These results are obtained using the Monte-Carlo N-Particle transport code MCNP®, version 6.3 (by Los Alamos National Laboratory, Los Alamos), which models neutron transport and interaction processes through stochastic (Monte Carlo) methods and evaluated nuclear data.

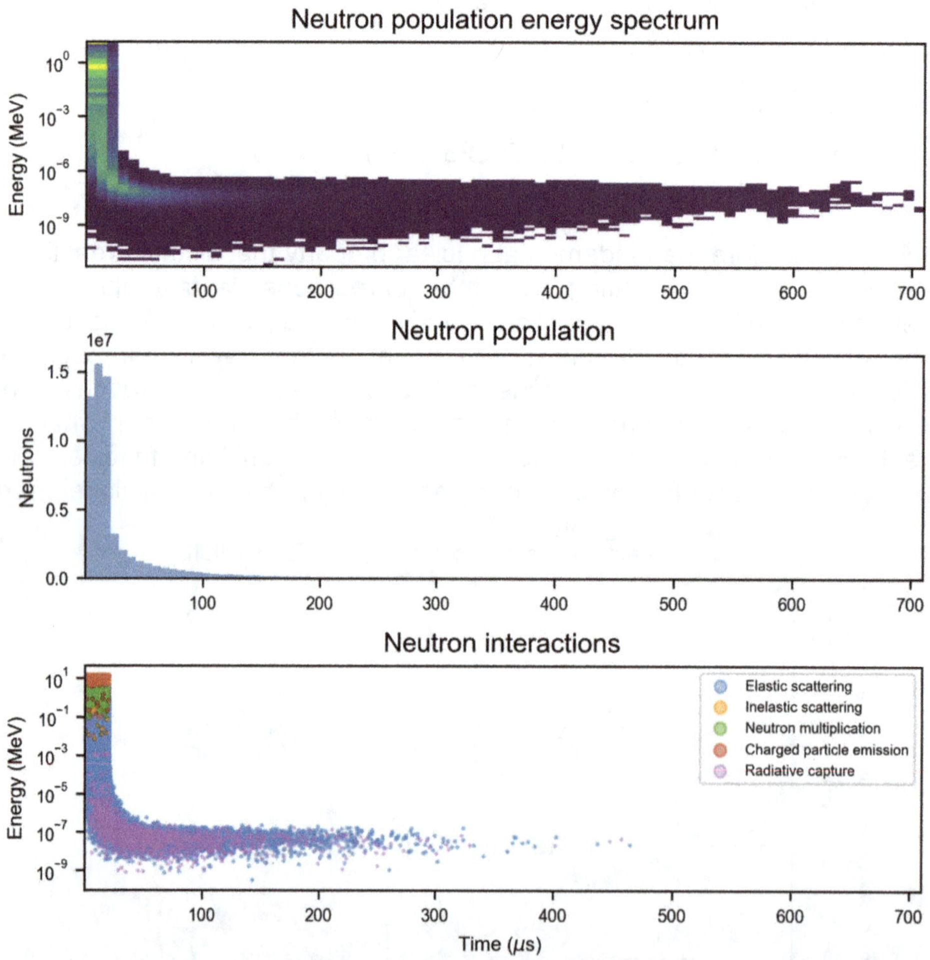

FIG 3 – Time-dependent neutron energy distribution and corresponding neutron interactions for a 20 μs pulse of 14 MeV neutrons within a hematite-water mixture.

Inelastic scattering and each of the capture interactions can result in the generation of gamma rays through prompt de-excitation, and the capture interactions can also result in the generation of delayed gamma rays through subsequent radioactive decay of unstable residual nuclei. The energies of these gamma rays are unique for each element and interaction combination. Measuring and analysing the energy spectrum of gamma rays produced from these interactions allows the identification and quantification of the elemental composition of the material with which the neutrons have been interacting.

A BRIEF HISTORY OF NEUTRON ACTIVATION LOGGING

Wireline-conveyed logging tools designed to measure gamma ray energy spectra resulting from neutron interactions have been available for around 75 years. The first commercial tools, developed in the late 1950s, targeted the measurement of energy spectra resulting from delayed gamma rays, using a pulsed neutron generator (PNG) accelerator-based neutron source. This allowed for a burst of neutrons to be created and then, after waiting sufficient time for all prompt interactions to have ceased, the delayed gamma rays to be measured. This measurement was primarily used to distinguish oil from water by looking for a reduction in the presence of oxygen, which is not found in hydrocarbons but is found in water. The oxygen content was inferred from the gamma rays released during radioactive decay of ^{16}N produced from charged particle production interactions with ^{16}O.

By the late 1970s commercial tools had been developed targeting the gamma ray energy spectra coming from fast neutron interactions. Due to the high neutron flux generated by a PNG when it is operating, and subsequent high interaction rates, measuring usable energy spectra at the resulting high gamma ray counts required technical developments that were not available earlier. The primary focus of such tools continued to be determination of oil saturation, in this case by using the gamma ray signatures resulting from inelastic scattering by carbon and oxygen to produce a carbon/oxygen ratio that relates to the volume of oil versus water in the formation. Saturation evaluation is still the primary use of this type of technology in oilfield applications today.

In the 1990s the first commercial tools were developed with the primary focus of measuring the chemistry of the rock matrix, both for oilfield and mining use. These tools measured gamma ray energy spectra resulting from radiative capture and primarily used isotopic neutron sources. These sources produce continuous streams of slightly lower energy neutrons than PNGs, which reduces the magnitude of the response coming from fast neutron interactions, and so these employ prompt gamma neutron activation analysis (PGNAA) principles.

In the 2010s, a new generation of inorganic scintillators for gamma ray detection became available, led by cerium-doped lanthanum bromide ($LaBr_3$:Ce). These scintillators have properties suitable for handling both the high count-rate requirements of measuring energy spectra resulting from fast neutron reactions and the high-resolution required to maximise the information that can be extracted from energy spectra resulting from thermal neutron reactions. This led to the development of a new generation of tools, both for oilfield and mining use, using PNG neutron sources and based on pulsed fast and thermal neutron activation (PFTNA) principles.

DESIGN FOR MINING

As the previous discussion suggests, a large body of knowledge has developed over the past 75 years on the construction of wireline-conveyed logging tools for the measurement of chemical composition of subsurface formations, particularly through the development of tools for oilfield applications. In building a new tool, it is important to build on this knowledge. However, developing a tool designed specifically for mining allows typical oilfield tool design paradigms to be discarded and every aspect of the tool to be rethought to explicitly deliver the best possible tool to address mining requirements.

Designing a new logging tool specifically optimised for mining imposes three sets of requirements and specifications:

1. Measurement requirements.
2. Environmental specifications.
3. Operational requirements.

Measurement requirements and implications

The most obvious measurement requirement for a downhole assay tool is to provide fit-for-purpose elemental composition of subsurface formations. This immediately differentiates such a tool from the common oilfield application of carbon/oxygen logging and imposes a stricter requirement for accuracy and precision than is commonly required in oilfield applications.

From a physics perspective, two factors have a significant impact on the accuracy and precision with which a particular element can be quantified. All nuclear measurements are statistical in nature, and so the precision with which a particular element can be characterised is related to the number of counts, in this case gamma rays, which are generated from neutron interactions with that element and are subsequently recorded by the tool. Both precision and accuracy are influenced by how unique an element's spectral response is, compared to other elements, which dictates how effectively the response from different elements can be distinguished.

The number of gamma rays generated by a particular element from a particular interaction is a function of the mass concentration of the element, and the mass-normalised cross-section $\left(\frac{N}{M} \cdot \sigma\right)$ (Ellis and Singer, 2008). Table 1 details mass-normalised cross-sections for non-elastic interactions (inelastic scattering, charged particle production, and neutron multiplication) involving 14 MeV neutrons and radiative capture of thermal neutrons for key elements of interest in iron ore, as well as the cross-sections normalised to that of iron for each of the interaction types.

TABLE 1

Mass-normalised cross-sections for gamma ray producing interactions involving 14 Mev and thermal neutrons for key iron ore elements.

Element	Mass-normalised 14 MeV non-elastic cross-section ($cm^2 \cdot g^{-1}$)	Normalised to iron	Mass-normalised thermal neutron radiative capture cross-section ($cm^2 \cdot g^{-1}$)	Normalised to iron
Hydrogen	0.0000321	0.00211	0.199	7.15
Carbon	0.024	1.58	0.000192	0.00692
Oxygen	0.0239	1.58	0.00000645	0.000232
Sodium	0.0254	1.67	0.0138	0.498
Magnesium	0.0299	1.97	0.00156	0.0562
Aluminium	0.0215	1.42	0.00521	0.187
Silicon	0.0234	1.54	0.00354	0.127
Phosphorus	0.0179	1.18	0.00329	0.118
Sulfur	0.0183	1.21	0.00964	0.347
Potassium	0.0163	1.08	0.0321	1.16
Calcium	0.0187	1.23	0.00658	0.237
Titanium	0.0162	1.07	0.0806	2.9
Manganese	0.0146	0.964	0.146	5.24
Iron	0.0152	1	0.0278	1

A review of this table shows the wide variation in thermal neutron radiative capture cross-sections, with some elements, such as iron, engaging in radiative capture and subsequent gamma ray generation much more readily than others. Coupled with the high concentration of iron in an iron ore, this means that iron is the dominant contributor of gamma rays generated by radiative capture, with other elements contributing significantly less, both due to their lower concentration and their lower probability of being involved in a radiative capture interaction.

On the other hand, the cross-sections for non-elastic interactions of 14 MeV neutrons are much more consistent, and iron has one of the lower values. This means that the fraction of gamma rays coming from each element will be more proportional to the mass concentration of that element.

Figure 4 displays the energies and likelihoods of gamma rays generated from inelastic scattering of 14 MeV neutrons and radiative capture of thermal neutrons for the same elements included in Table 1, as well as delayed gamma rays from unstable residual nuclei formed through neutron interactions with these elements. The diversity of gamma rays generated by different elements undergoing radiative capture demonstrates why this has historically been the preferred interaction to analyse for downhole assay. However, there are some critical uniqueness issues, most notably between iron and aluminium, which produce radiative-capture-induced gamma rays of similar energy. The responses of these elements are more readily distinguished when viewing the gamma rays generated by inelastic scattering of 14 MeV neutrons, and aluminium is also notable as being a source of delayed gamma rays. Other elements show similar characteristic behaviours that differ between the interaction types, allowing the elements to be readily distinguished by cross-validation of the gamma ray spectra resulting from each interaction type.

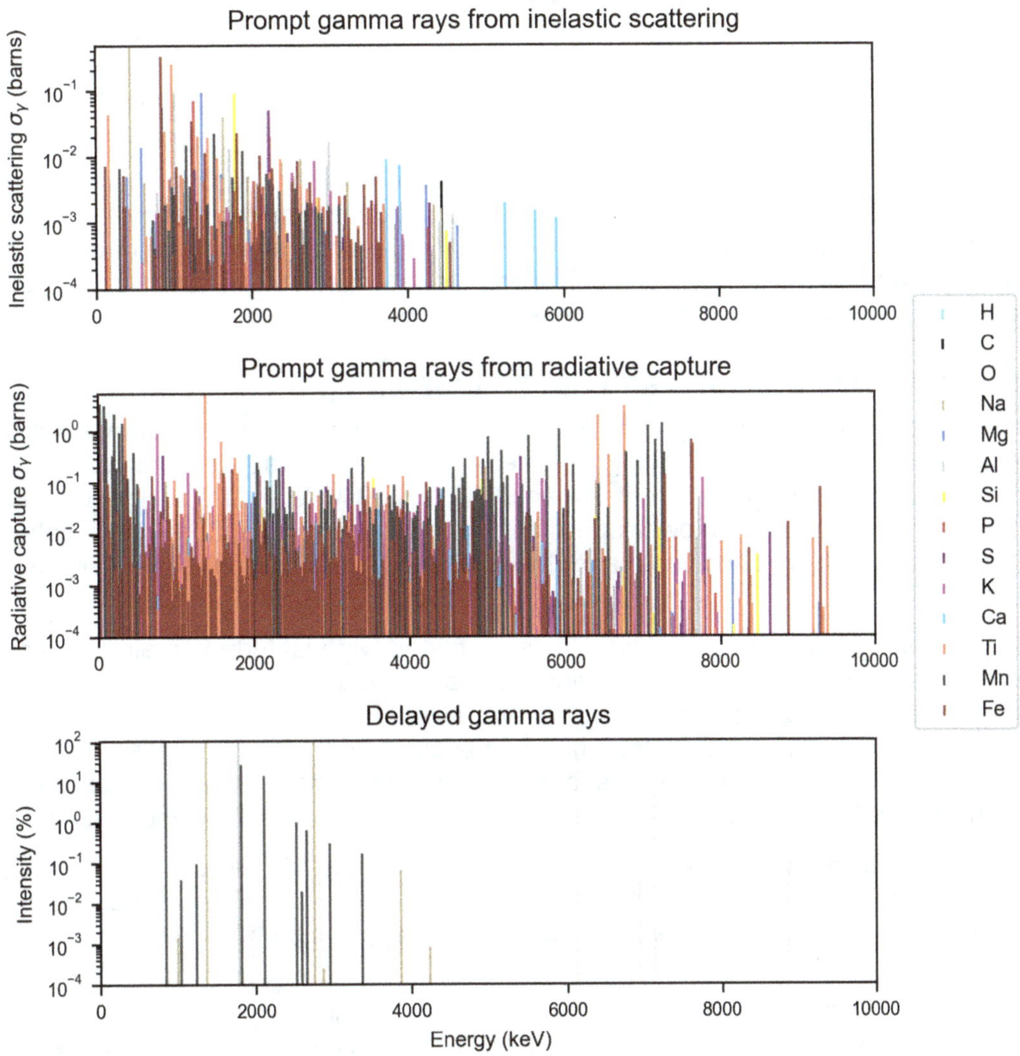

FIG 4 – Gamma rays resulting from inelastic scattering of 14 MeV neutrons (top panel), radiative capture of thermal neutrons (middle panel), and radioactive decay of unstable nuclides (lower panel) for key iron ore elements.

Therefore, it is beneficial to leverage the different sensitivity of elements to the various neutron interactions described above in improving the accuracy with which they are identified and quantified. This imposes one of the most fundamental design choices, use of a PFTNA-style measurement as opposed to PGNAA-style measurement. This in turn imposes the requirement for an accelerator-based neutron source rather than an isotopic source.

Figure 3 suggests how these responses can be separated using temporal sampling of the generated gamma rays. When the PNG is generating neutrons, fast neutron interactions dominate initially, although neutrons quickly lose energy and start to undergo radiative capture. When the PNG is turned off, the remaining fast neutrons are quickly moderated, and the resulting thermal neutrons undergo radiative capture, with a progressive reduction in the neutron population. Once all thermal neutrons have been absorbed the only source of gamma rays is decay of unstable nuclides.

To maximise gamma ray counts from fast neutron interactions a longer PNG 'on' period is preferred, however as the PNG 'on' duration increases the proportion of thermal neutron interactions also increases, so the PNG 'on' timing needs to be optimised to balance the two constraints. Once the PNG stops producing fast neutrons, the fast neutron population decreases almost instantaneously, and the thermal neutron population also begins to diminish. To maximise gamma ray counts coming from thermal neutron interactions, measurements should be made very soon after the PNG stops producing neutrons, but definitively after the PNG stops producing neutrons. The PNG therefore has an important role to play in maximising the measurement quality as the time it takes at the end of each PNG 'on' period for the neutron output to diminish to zero influences how long gamma rays from fast neutron interactions continue to appear in the measurement at the same time as the number of gamma rays from thermal neutron interactions is diminishing. A PNG with a fast neutron production ramp-up and ramp-down is therefore desirable both to improve measurement purity and increase measurement precision. The PNG used in this tool has a ramp-down duration of under 500 ns.

An emphasis on measuring prompt gamma rays from fast neutron interactions and delayed gamma rays has other tool design implications. A review of Figure 4 demonstrates that most of these gamma rays are of low energy. To minimise shielding effects that result in attenuation of these low energy gamma rays, tool and detector housings consisting of low density and low atomic number materials are critical in maximising the usefulness of these responses.

Environmental specifications and implications

In contrast to oilfield wireline logging tools, which are built to typical environmental specifications of 175°C and 138 MPa, most wireline logging tools designed for mining applications have more modest environmental specifications, generally less than 80°C and 20 MPa. A tool designed specifically for blasthole logging may have even lower specifications, with the temperature specification dictated largely by surface working conditions and the pressure specification essentially atmospheric pressure.

These relatively benign environmental conditions permitted wide latitude in material selection for a mining-focused tool, with the primary selection criteria based on impact on measurement quality. Neutron interactions take place in all atoms in and around the tool, including the formation, the borehole, and the tool itself. As the elemental composition of the formation is the measurement objective, one way to increase the 'usable' counts and hence precision of the formation measurement is to reduce the response coming from the tool, known as the tool background.

To reduce the tool background, the neutron flux in the vicinity of the tool can be reduced and materials can be selected that will contribute fewer gamma rays to the measured energy spectra through neutron interactions. These design considerations apply in both the fast and thermal neutron energy ranges.

The properties and distribution of fast neutrons and thermal neutrons are quite different, and hence different strategies are required to reduce the flux of each in the vicinity of the tool. For fast neutrons, the PNG acts as a point source and the neutrons largely travel directly away from the source. Fast neutrons travelling directly from the source towards the vicinity of the detector are the major contributors to tool background. For thermal neutrons, the main source is neutrons being scattered back from the formation and borehole into the tool in the vicinity of the detector. Therefore, fast neutron shielding is required along the tool axis to minimise direct neutron paths, while thermal neutron shielding is required around the tool detector section. Extensive modelling using the MCNP® code has been used to optimise the position, geometry, and choice of materials for both fast and thermal neutron shielding, as well as the overall tool design (Figure 5). Decisions have been validated by measurements such as those depicted in Figure 6, which compares the radiative capture tool background response for the tool with no thermal neutron shielding, to the response with the final shielding design. These measurements demonstrate that tool background has been reduced by approximately 80 per cent through optimised shielding design.

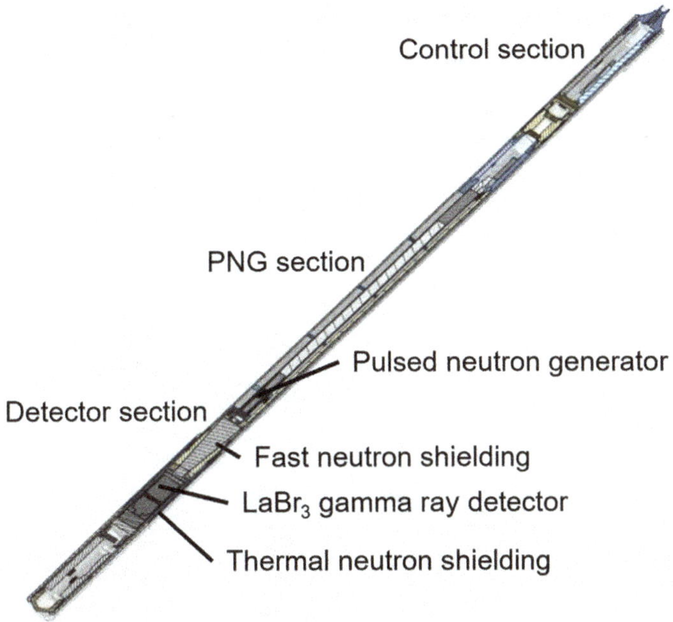

FIG 5 – New tool design showing major components.

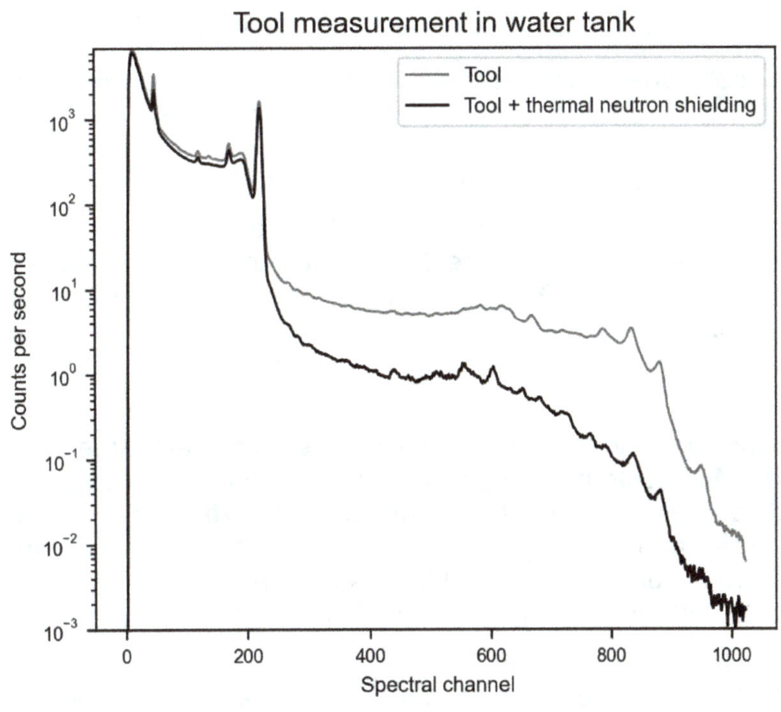

FIG 6 – Comparison of radiative capture tool background response with (black) and without (grey) thermal neutron shielding based on a tool measurement in a water tank. The prominent peak near channel 200 and the majority of the response to the left of the peak is from hydrogen in water. The response to the right of the peak is from the tool.

Figure 7 displays a plot of 14 MeV neutron non-elastic cross-section versus thermal neutron radiative capture cross-section for the naturally occurring elements. Materials dominated by elements plotting towards the bottom-left of this diagram will show lower levels of neutron interactions and resulting gamma ray production compared to those containing elements plotting towards the top-right. Using these considerations as a guide, again validated with extensive modelling, a specially engineered composite has been used for the tool housing, while the internal chassis is made of a high purity alloy of low cross-section elements. This, in conjunction with the shielding discussed above, has resulted in the exceptionally low tool background observed in measurements made with the new tool (Figure 6), while still achieving environmental specifications of 80°C and 20 MPa. A useful byproduct of these material choices is a total tool mass of 35 kg. The housing material also supports the requirement for minimal gamma ray shielding to maximise the number of low energy gamma rays that make it to the detector, enhancing the low energy response of the measurements.

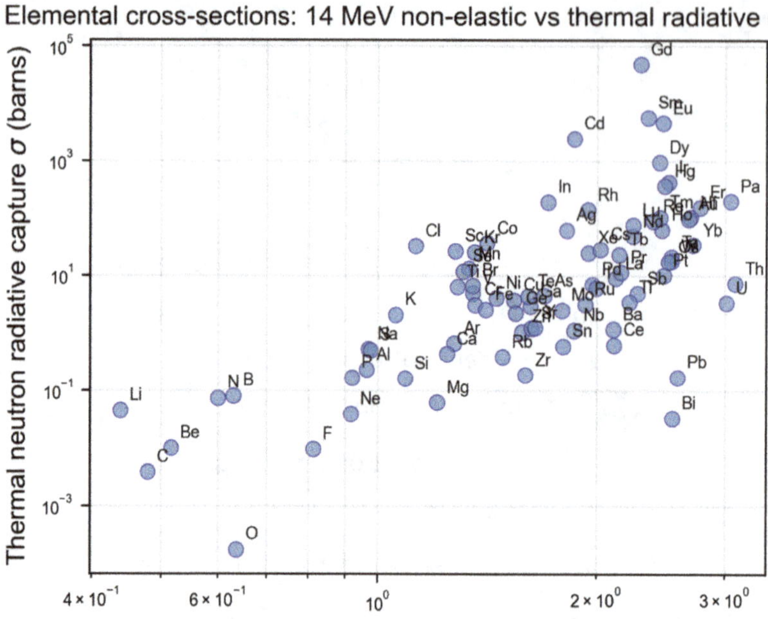

FIG 7 – 14 MeV neutron non-elastic cross-section versus thermal neutron radiative capture cross-section for naturally occurring elements.

Operational requirements and implications

The stable operation of two tool components—the PNG and the detector—is key in obtaining high quality data from a PFTNA-GRS tool. In oilfield operations, the common operational sequence is to power on a tool and allow the PNG and detector to stabilise before logging for multiple hours continuously in one borehole. In a mining environment, and particularly in logging of production blastholes, a single borehole may take only minutes to log and there is a need to cycle tools between boreholes again on a time frame of minutes.

The PNG neutron source used in most PFTNA-GRS tools uses beam-target fusion of deuterium and tritium to generate 14 MeV neutrons. This process involves three steps. Firstly, deuterium is ionised to produce charged particles. These deuterium ions are then extracted to the low-voltage end of the accelerator region. Then a high voltage difference accelerates the deuterium ions into a tritium-containing target where the deuterium and tritium fuse to form helium with the resulting emission of high energy neutrons. Pulsing of the source can be achieved by manipulating any of these three operations.

Preparation of a PNG to generate neutrons can take several minutes from a cold start. Both the formation of deuterium ions and the process of the high voltage power supply coming up to operating voltage are time consuming. Safely switching off a PNG can also take several minutes. Such time frames are not operationally acceptable in a rapid cycle time environment such as blasthole logging. Therefore, a standby mode has been implemented for the PNG used in the new tool. This freezes the operating condition of the PNG but removes its ability to generate neutrons, allowing the tool to be safely transported between boreholes, while at the same time allowing a fast transition to full neutron output again.

When the PNG is operating, a stable neutron output is desirable as this improves the interpretability of the measured gamma ray energy spectra. PNG output can vary due to short-term effects such as temperature variations and long-term effects such as PNG aging. To ensure constant source strength, a diamond fast-neutron detector is used to continuously monitor PNG neutron output and adjust operating parameters accordingly.

The detector in most modern spectrometry tools consists of an inorganic scintillator crystal, $LaBr_3$:Ce in this tool, and a photomultiplier tube (PMT). The scintillator converts incoming gamma rays to optical photons. The PMT in turn converts these photons into an amplified electrical signal. The detector system has an intrinsic proportionality constant relating the energy of an incoming gamma ray to the voltage of the output electrical signal. This proportionality constant can drift with changes in count rate and temperature, which causes a shift in peak positions in the measured gamma ray energy spectra. A gain control feedback loop is typically employed to adjust the amplification applied by the PMT to keep the spectrum stable.

As the PNG neutron output increases when it is powered on or taken out of standby mode, the count rate measured by the detector increases rapidly, causing significant changes in detector response and hence peak positions on a timescale of seconds. A very fast acting gain control feedback loop is therefore required to stabilise the detector response prior to acquisition.

Figure 8 shows tool operation indicators for a 45 min period of blasthole logging with the new tool. This was part of a trial in which 202 passes were acquired across 191 blastholes during three shifts, including non-operational time due to blasting, with an average cycle time of 6.5 mins per blasthole including vehicle movement. The uppermost track shows tool depth, with blue shading for the time intervals over which depth passes were acquired in the seven blastholes logged during this period.

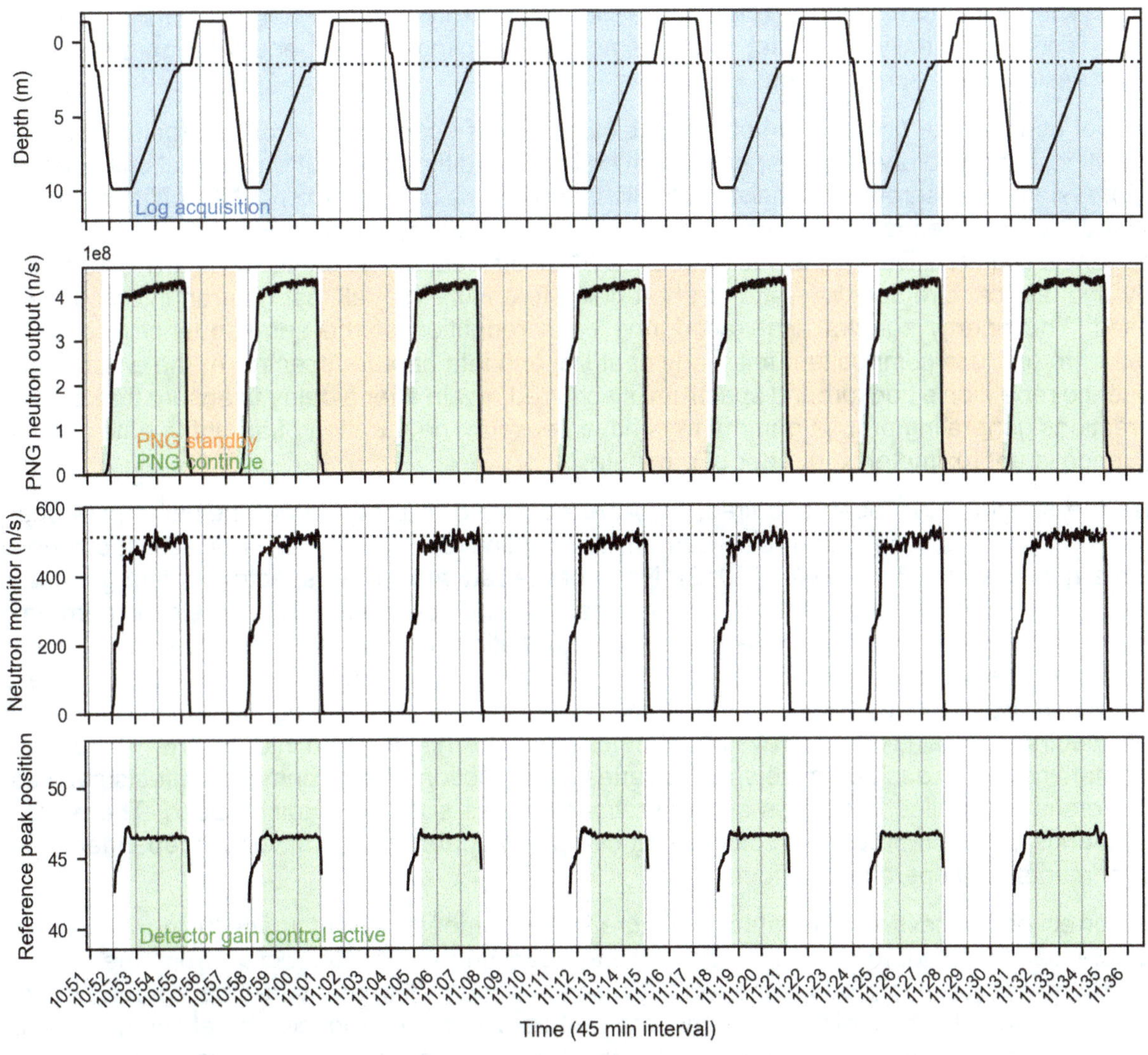

FIG 8 – Tool operation indicators from 45 mins of blasthole logging with the new tool.

PNG standby and continue are indicated by orange and green shading in the second track, with full standby and continue modes achieved when the shading spans the entire track. As the PNG is taken out of standby mode, the ramp up in PNG neutron output can be seen in the second track and the increase in counts recorded by the fast-neutron detector can be seen in the third track, with full neutron output obtained in under 45 sec.

The impact of this increase in count rate on detector response can be seen in the variation in reference peak position (lowermost track) over the same intervals. When at full neutron output, the detector gain control loop is enabled (green shading) and detector stabilisation is achieved in under 15 sec. These processes can be started as the tool is running in hole, so that the tool is ready to start acquiring data once on bottom.

When the tool reaches its minimum safe working depth near surface the PNG is put into standby mode, which also takes approximately 15 sec, after which the tool can be recovered to the surface and the logging vehicle moved to the next borehole.

SPECTRAL DATA ACQUISITION AND PREPROCESSING

As outlined previously, leveraging the differing sensitivity of elements to the various neutron interactions described above helps improve the accuracy with which the elements can be identified and quantified. Clean separation of prompt gamma ray response from fast neutron interactions (inelastic scattering, charged particle emission, and neutron multiplication), prompt gamma ray response from thermal neutron interactions (radiative capture), and delayed gamma ray response is crucial. The role of the neutron source in delivering accurate and fast neutron pulses has already been described. Equally important is the temporal sampling of the generated gamma rays and preprocessing of the spectral data.

This tool uses a data acquisition sequence similar to that used in some oilfield tools optimised for acquisition of gamma ray energy spectra from both fast and thermal neutron interactions (Rose *et al,* 2015). A single measurement consists of multiple repetitions of a PNG 'on', PNG 'off' sub-sequence, followed by an extended PNG 'off' period. During the sub-sequence PNG 'on' period, a burst spectrum is acquired. During the sub-sequence PNG 'off' period, two capture spectra are acquired—early and late capture spectra—after a short delay to ensure that all fast neutron interactions have ceased. The energy spectra acquired during each repetition of the sub-sequence are summed together to produce composite burst, early capture, and late capture spectra. At the end of the sub-sequence repetition a background spectrum is acquired, again after a delay to ensure that all neutron interactions generating prompt gamma rays have ceased. The duration of an entire data acquisition sequence is approximately a quarter of a minute.

Once the spectra have been acquired, they are corrected to produce net inelastic (plus charged particle emission and neutron multiplication), net (radiative) capture, and net activation (delayed gamma ray) spectra (Radtke *et al,* 2012). None of the raw measured spectra contain gamma rays resulting from only one type of interaction. The background spectrum includes delayed gamma rays from the various interactions, as well as intrinsic radiation from the detector itself which includes lanthanum, one isotope of which is naturally radioactive. The capture spectra include delayed gamma rays from already decaying radionuclides created by previous acquisition sequences. The burst spectrum contains gamma rays from both fast and thermal neutron interactions. The proportion of 'contamination' of each of these spectra varies as a function of environmental effects such as air-filled versus liquid-filled boreholes and operational effects such as logging speed. Therefore, this correction is critical in producing spectra that can be evaluated quantitatively, independent of such environmental and operational effects.

The correction process is illustrated in Figure 9, for a measurement made using an aluminium characterisation standard. The intrinsic detector background is subtracted from the measured background spectrum to produce the net activation spectrum. The measured background spectrum is subtracted from the sum of the early and late capture spectra to produce the net capture spectrum. The early capture spectrum is subtracted from the burst spectrum to produce the net inelastic spectrum. The subtraction factors α, β, and γ are a function of count rate, pile-up, and acquisition time for each spectrum.

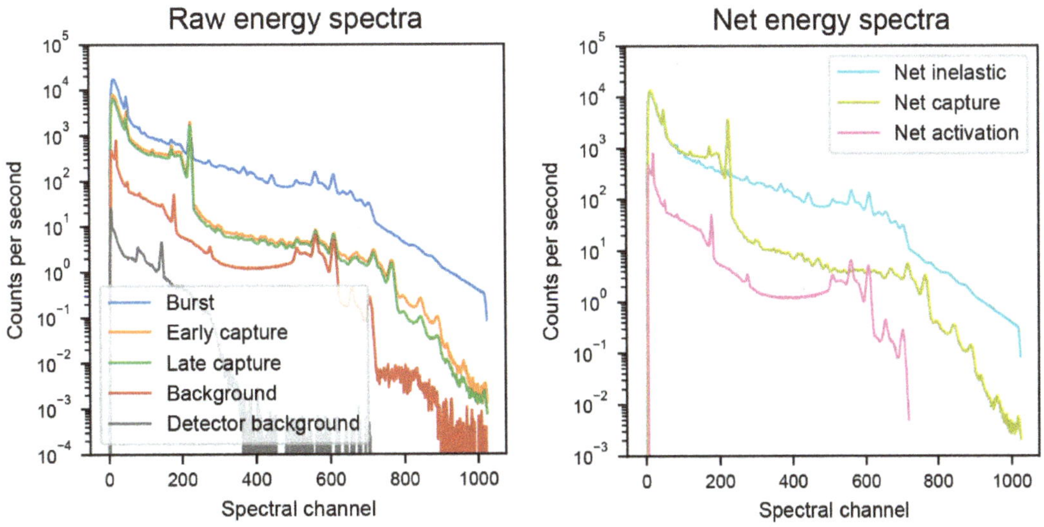

burst - α • early capture = net inelastic

(early capture + late capture) - β • background = net capture

background - γ • detector background = net activation

FIG 9 – Raw spectrum correction process to produce net energy spectra.

CONCLUSIONS

A new wireline-conveyed downhole assay tool has been developed customised to mining requirements. Building on electronically controlled neutron generator and high efficiency detector technologies developed over decades for oilfield applications, the new tool leverages the lower pressure and temperature environment found in mining applications to deliver a light, fast tool with measurement capabilities optimised for the elements of interest to mining, particularly in iron ore.

Optimised for measurement quality, the tool has also been designed for operational efficiency with the implementation of a neutron generator standby mode that allows the tool to be safely moved from hole to hole without requiring a lengthy shutdown and restart cycle. This permits rapid deployment facilitating the logging of many holes per shift as demonstrated during recent field testing.

ACKNOWLEDGEMENTS

The authors would like to take this opportunity to acknowledge all the past and present members of the ASSAYTrack™ project team who have contributed to the development of this new tool. The authors would also like to thank the management of Orica Limited for approval to publish this work, and to all the companies that have assisted in supporting field testing of this tool.

REFERENCES

Ellis, D V and Singer, J M, 2008. *Well Logging for Earth Scientists,* 2nd edn (Springer Science + Media, B V: Dordrecht).

Lamarsh, J and Bararra, A, 2013. *Introduction to Nuclear Engineering,* 3rd edn (Pearson Education: London).

Radtke, R J, Lorente, M, Adolph, B, Berheide, M, Fricke, S, Grau, J, Herron, S, Horkowitz, J, Jorion, B, Madio, D, May, D, Miles, J, Perkins, L, Philip, O, Roscoe, B, Rose, D and Stoller, C, 2012. A New Capture and Inelastic Spectroscopy Tool Takes Geochemical Logging to the Next Level, *SPWLA 53rd Annual Logging Symposium,* paper SPWLA-2012–103 (Society of Petrophysicists and Well Log Analysts: Houston).

Rose, D, Zhou, T, Beekman, S, Quinlan, T, Delgadillo, M, Gonzalez, G, Fricke, S, Thornton, J, Clinton, D, Gicquel, F, Shestakova, I, Stephenson, K, Stoller, C, Philip, O, La Rotta Marin, J M, Mainier, S, Perchonok, B and Bailly, J-P, 2015. An Innovative Slim Pulsed Neutron Tool, *SPWLA 56th Annual Logging Symposium*, paper SPWLA-2015-XXXX (Society of Petrophysicists and Well Log Analysts: Houston).

Managing ore variance through continuous ore characterisation – real-time PGNAA for process control in iron ore

R Sheehy[1], J Jones[2] and N Kennedy[3]

1. Chief Technology Officer, Real Time Instruments, Mackay Qld 4740.
   Email: robin.sheehy@realtimegrp.com
2. Performance Assurance Manager, Real Time Instruments, Mackay Qld 4740.
   Email: jake.jones@realtimegrp.com
3. Marketing Specialist, Real Time Instruments, Brisbane Qld 4006.
   Email: nathan.kennedy@realtimegrp.com

ABSTRACT

Ore variability remains one of the most persistent challenges in iron ore operations, influencing blending efficiency, plant stability, recovery performance, and final product specification. Variations in iron grade and impurity levels arise from geological heterogeneity, mining process, and stockpile handling practices. Traditional ore characterisation relies heavily on in process sampling and laboratory analysis, which, while typically unbiased, provides delayed feedback, limited temporal resolution and may have worse precision than required. As a result, short-term variability within ore streams is often averaged out, detected only after it has already affected downstream processing or missed entirely.

This paper presents a methodology for managing ore variance through continuous ore characterisation using on-belt Prompt Gamma Neutron Activation Analysis (PGNAA). The approach enables high-frequency measurement of key elemental parameters, including Fe, SiO_2, and Al_2O_3, directly on the conveyor belt. Real-time visibility of ore composition allows operations to detect and respond to grade variability as it occurs, supporting proactive process control and stockpile management decisions.

INTRODUCTION

Ore variability and its impact on downstream processing performance remains a central challenge in iron ore operations. The ability to characterise and respond to this variability in real time is critical for maintaining product quality, improving recovery, and ensuring operational stability.

ORE VARIABILITY AND OPERATIONAL IMPACT

Analysis of calibration and operational data sets from multiple iron ore installations highlights the extent of short-term variability that can exist within a single feed. Observed Fe values ranged from approximately 59.9–63.8 per cent, while SiO_2 concentrations varied between approximately 2.0 per cent and 5.1 per cent over relatively short time frames. Such fluctuations are sufficient to affect blending strategies, recovery efficiency, and product compliance if unmanaged.

In conventional sampling regimes, this variability may be masked by composite samples or delayed laboratory results, limiting the ability to distinguish between transient and sustained grade changes. This often leads to reactive adjustments to blending or process parameters after deviations have already propagated through the plant. Continuous characterisation addresses this gap by providing a persistent, time-aligned data stream that reflects the true variability of material moving through the system.

METHODOLOGY – CONTINUOUS ORE CHARACTERISATION USING PGNAA

PGNAA technology provides non-contact, cross-belt measurement of bulk material composition by detecting characteristic gamma emissions from elements within the ore. When deployed on conveyors, the sensor enables continuous monitoring without interrupting production.

Calibration against laboratory assay data is central to the methodology and accounts for site-specific ore characteristics, belt loading conditions, and material presentation. Accuracy is assessed using Root Mean Square Deviation (RMSD), reflecting combined analyser, sampling, and laboratory uncertainty. Typical calibration results demonstrated RMSD values of approximately 0.47 per cent for Fe and 0.53 per cent for SiO_2, confirming strong alignment with laboratory reference data.

Sampling error remains an inherent consideration in any on-belt analysis system, with uncertainty of approximately 0.3–0.4 per cent expected even under ideal conditions. The methodology therefore emphasises trend analysis and data reconciliation over time rather than reliance on individual point measurements.

Operational experience from multiple deployments indicates that temporal resolution is often the dominant value driver for continuous ore characterisation. High-frequency measurement enables identification of short-term variability patterns that are not visible through discrete sampling. Rather than relying on averaged values, operators can observe the rate, persistence, and magnitude of grade fluctuations, providing a clearer understanding of ore behaviour as it moves through the system.

In several deployments, multiple analysers were positioned at strategic locations such as plant feed conveyors and stockpile reclaim points. This configuration supports cross-validation of data and improves traceability of ore characteristics across mining, stockpiling, and processing stages.

INTEGRATION WITH PROCESS CONTROL AND STOCKPILE MANAGEMENT

Continuous ore characterisation becomes most effective when integrated into plant control and monitoring systems. Real-time elemental data visualised alongside operational parameters allows operators and metallurgists to interpret grade variability in context and adjust blending, reclaim strategies, and process set points accordingly.

The availability of continuous elemental data supports a shift from conventional grade-based control toward variance-aware control strategies. Process decisions are informed not only by mean Fe or SiO_2 values, but also by feed stability and consistency. This enables differentiation between transient fluctuations and sustained changes in ore quality, reducing unnecessary corrective actions and supporting more stable operation.

Continuous characterisation also alters the functional role of stockpiles. Rather than acting solely as passive averaging buffers, stockpiles become actively characterised inventories when supported by real-time elemental data. Reclaim and blending strategies can therefore be adjusted based on observed material quality rather than historical assumptions, improving feed predictability and strengthening reconciliation between planned and actual material movement across mining, processing and loadout stages.

QUALITATIVE OUTCOMES AND OPERATIONAL LEARNINGS

While site-specific quantitative outcomes vary, implementation of continuous ore characterisation has demonstrated measurable improvements in feed consistency and operational awareness. Across monitored conveyors, reductions in short-term variability of Fe and SiO_2 were observed following integration of real-time data into blending and process control decisions, typically in the range of 0.5 per cent to 5 per cent depending on ore type, operating conditions, and the maturity of control integration. Continuous visibility supported tighter control of feed composition compared to reliance on delayed laboratory sampling alone. Operators also reported measurably improved confidence in feed chemistry, greater awareness of short-term variability, and more consistent decision-making across shifts, reducing the need for reactive blending interventions.

Effective implementation depended on calibration being treated as an ongoing operational process rather than a one-off activity. Continuous data highlighted when ore characteristics drifted beyond the original calibration domain, providing early indication of recalibration requirements and supporting sustained confidence in sensor outputs.

Human factors were also critical. Clear data visualisation, alignment with existing workflows, and collaboration between technology providers and site personnel ensured that continuous characterisation data informed routine operational decisions rather than existing in parallel to established practices.

CONCLUSIONS

The methodology presented provides a transferable framework for operations seeking to integrate continuous ore characterisation into geometallurgically informed process control strategies. Continuous ore characterisation using on-belt PGNAA provides a practical framework for managing ore variance in iron ore operations. High-frequency elemental data for Fe, SiO_2, and Al_2O_3 delivers real-time visibility of variability often obscured by conventional sampling approaches. When integrated with process control and stockpile management systems, this methodology strengthens the application of geometallurgical principles and supports more stable, predictable, and informed iron ore processing operations.

New insights into the formation of (secondary) skeletal hematite and its association with gangue rich nuclei

N A Ware[1], J M F Clout[2], J Manuel[3] and S Hapugoda[4]

1. Senior Experimental Scientist, CSIRO Mineral Resources, Pullenvale Qld 4069
   Email: natalie.ware@csiro.au
2. WA School of Mines, Curtin University, Bentley WA 6102. Email: john.clout@curtin.edu.au
3. Senior Experimental Scientist, CSIRO Mineral Resources, Pullenvale Qld 4069.
   Email: james.manuel@csiro.au
4. Senior Experimental Scientist, CSIRO Mineral Resources, Pullenvale Qld 4069.
   Email: sarath.hapugoda@csiro.au

ABSTRACT

The partial degradation of iron ore sinter that occurs during reduction in the shaft section of the ironmaking blast furnace detrimentally affects the permeability of the ferrous burden, leading to increased coke consumption, lower productivity and ultimately higher greenhouse gas emissions. The propensity of sinter to breakdown during reduction is measured by the reduction degradation index (RDI). Previous studies have observed a correlation between the presence of so called 'secondary' skeletal hematite (SSH) and SFCA containing high levels of Al_2O_3 located near micropores, high Al_2O_3 levels in sinter blends associated with use of Australian iron ores and the presence of fine pores due to high levels of -1.0 mm limestone and -0.5 mm coke. Secondary skeletal hematite is widely considered to have formed by oxidation of high temperature skeletal 'multi-component' (high Al, Si, Ca) magnetite during cooling, hence the use of the 'secondary' preface term.

During a recent study identifying and characterising the properties of sinter structure, it was discovered that skeletal hematite, in association with glass, forms immediately adjacent to remnant quartz-rich nuclei in the absence of SFCA/SFCA III. Experiments were carried out by embedding coarse quartz particles in a -1 mm matrix blend A of SFCA composition and firing the samples using the CSIRO developed laboratory scale sintering method to replicate this condition. Preliminary SEM/EDS analysis, quantitative X-ray diffraction analysis and optical microscopic observation was carried out on samples fired under typical low (1290°C) and high (1330°C) temperature and pO_2 5 × 10^{-3} atm sintering conditions. The results demonstrate the formation of skeletal hematite in association with localised high-silica (low basicity) melts under both conditions. Comparable morphology and mineral associations were observed in plant and pot grate sinter.

Skeletal hematite was found to be associated with glass in areas of low (<0.8) basicity (matrix melt) as well as from low basicity melt attack of hematite nuclei for the two temperatures. Examination of previous phase diagram studies at typical sintering pO_2 5 × 10^{-3} atm conditions indicates that hematite (ie skeletal hematite) with liquid (glass) is stable in the presence of low (<0.8) basicity melts whereas magnetite with liquid is stable above >0.8 basicity, without the need to invoke oxidation of skeletal magnetite to skeletal hematite during cooling, or crystallisation at different times, pO_2 or temperatures. Magnetite, both skeletal and non-skeletal, was found to be much higher in gangue elements (Si, Ca, Al, Mg) than skeletal hematite, which discounts the possibility that skeletal hematite is 'secondary' formed from oxidation of pre-existing skeletal magnetite. Micro-spatial variations in melt basicity are therefore able to explain the apparently anomalous close spatial association of skeletal hematite-glass and magnetite-glass. Whilst potentially oxidation of skeletal magnetite during cooling could form SSH, it was not observed in the current study, with the element data demonstrating that skeletal magnetite has consistently much higher levels of gangue element (Al, Si, Ca and Mg) levels than skeletal hematite. In the absence of compelling evidence, it is therefore recommended that the interpretive term 'secondary' be dropped in favour of the simple descriptive term 'skeletal hematite'.

The formation of skeletal hematite directly from a low basicity melt as well as from low basicity melt attack and assimilation of primary/secondary hematite nucleus, simultaneous with magnetite (glass) precipitation is a simple explanation but fundamental change in thinking on skeletal hematite formation and implies a more complex relationship with RDI than previously understood. This suggests the need for further in-depth investigation to better understand the drivers for RDI.

INTRODUCTION

Significant volumes of Australian fine iron ore are exported to Asia but first require agglomeration by sintering before they can be fed to an ironmaking blast furnace, with sinter accounting for the majority (>60 per cent) of the blast furnace ferrous feed (Lu *et al,* 2022). Sinter consists of a fine matrix of silicoferrite of calcium and aluminium (SFCA, SFCA-III and SFCA-I), magnetite, glass and minor larnite which encloses unreacted hematite ore nuclei and macro- micro-pores (Clout *et al,* 2026). The sintering process is a complex process, occurring at high (1270–1330°C) temperature, using coke as a fuel source and various fluxes (limestone, dolomite) which also balance the desired chemistry (CaO/SiO_2 ratio and MgO) for the blast furnace.

One of the key quality issues for the typical +5 -50 mm iron ore sinter product is the partial degradation during reduction in the shaft section of the blast furnace, producing potentially as much as 35 wt per cent -2.8 mm dust, based on laboratory tests (Loo and Bristow, 1998a, 1998b, 1998c; Lu, Holmes and Manuel, 2007; Lu and Lu, 2022). Dust from reduction breakdown, largely from sinter, results in partial blocking of pores and air channel ways in the ferrous burden, leading to reduced gas permeability, less solid/gas contact, which in turn results in less indirect reduction in the furnace shaft, and a lower blast rate, leading to increased coke consumption and lower productivity (Loo and Bristow, 1998a; Peacey and Davenport, 1979). In addition, increased reduction degradation of sinter and thermal shock decrepitation of lump leads to more dust, especially -0.5 mm dust, reporting to the top gas which then requires subsequent capturing and recycling back to the sinter plant (Loo and Bristow, 1998a, 1998c; Lu and Lu, 2022). The recycling of more dust back to the sinter plant and the impact of reduction degradation dust increases coke consumption in the blast furnace, leading to higher greenhouse gas emissions for the blast furnace and the sinter plant.

The propensity of sinter to breakdown during reduction is measured in the laboratory by various reduction degradation index (RDI) tests (eg ISO-4696-1 and ISO-4696-2; International Organization for Standardization (ISO), 2015a, 2015b), although the laboratory tests at best simulate but do not replicate the known differences in temperature-time-gas composition profiles between the centre and walls of a blast furnace (Loo and Bristow, 1998a, 1998b). Sinter breakdown increases significantly in the temperature range of 500 to 750°C, especially in the presence of hydrogen, and is also the temperature at which hematite undergoes reduction to magnetite (Loo and Bristow, 1998b; Mizutani *et al,* 2020).

Previous studies have drawn correlations between higher sinter RDI and the presence of secondary skeletal hematite and SFCA containing high levels of Al_2O_3 located near micropores (Inazumi, Shinada and Kawabe, 1982; Shigaki, Sawada and Norio Gennai, 1986), high Al_2O_3 levels in a sinter blend associated with use of Australian iron ores (Yamaguchi *et al,* 1994) and the presence of fine pores due to high levels of -1.0 mm limestone and -0.5 mm coke (Hosotani *et al,* 1996). Researchers have discussed the low fracture toughness of skeletal hematite as a driver in the breakdown of the sinter (Peterson and Manuel, 2023). A recent synchrotron X-ray study of reduced sinter has concluded that magnetite formed around reduction-induced microcracks in SFCA has a fine-grained structure, with the microcracks being the trigger sites for further reduction and mechanical degradation, and that reduction breakdown favours hematite and SFCA exposed to open pores (Takeichi, Murao and Kimura, 2023). Secondary skeletal hematite is thought to have formed in sinter during cooling and oxidation of earlier formed Al-Si-Mg-rich 'multi-component' magnetite, with 'secondary' skeletal hematite (SSH) being structurally unstable and prone to microcracking during low temperature reduction in the shaft section of the blast furnace (Inazumi, Shinada and Kawabe, 1982). However, further work is required to better understand the primary drivers for RDI, in particular the origin of skeletal hematite.

Clout *et al* (2026) has recently identified that sinter solid structure is composed of seven mineral associations (MA's) each characterised by relative differences between key minerals, textures, and relationships between porosity and hematite nuclei. The typical abundances of each MA were determined using reflected light microscopy (RLM) optical image analysis, as described by Donskoi, Poliakov and Manuel (2022), with the total modal mineralogy cross-checked by quantitative X-Ray diffraction (QXRD) using the Rietveld refinement method. MA-A comprises abundant SFCA-I microplates with hematite; MA-B consists of disseminated fine-grained magnetite in a network of SFCA-III microplates; MA-C is similar to MA-B but contains patches of dendritic SFCA-III with larnite and minor glass; MA-D comprises magnetite surrounded by coarse prisms of SFCA within glass; MA-Ds, an important sub-type of MA-D, includes SFCA with skeletal hematite; MA-E consists of anhedral to skeletal magnetite or hematite in a matrix of glass; and MA-N is comprised of unmelted primary hematite nuclei from iron ore feedstock. The work included the impact of cooling rate, bulk chemical composition, coke, quartz, different MgO sources and gas atmosphere on the formation of the mineral associations. An important finding of this work was the discovery that skeletal hematite with glass and larnite forms immediately adjacent to remnant quartz-rich nuclei in the absence of SFCA/SFCA-III. Similar examples were evident in pot grate and industrial plant sinter samples suggesting partial to complete assimilation of silica-rich hematite nuclei forms skeletal hematite, either in MA-Ds (partially assimilated MA-N) or MA-E (fully assimilated MA-N).

Further work is required to better understand skeletal hematite formation mechanisms and what may assist with minimising it's development and hence lower RDI. The aims of the current investigation were to provide preliminary evidence for the spatial location of skeletal hematite in plant and pot grate sinter and to understand the deportment of Al, Si, Mg and Ca in hematite, skeletal hematite, magnetite and glass with distance from quartz nuclei.

EXPERIMENTAL

Quartz particles in the form of +1 mm silica sand, were embedded into 4 g compacted cylindrical tablets (compacts) enclosed by a -1 mm matrix comprising Blend A with a SFCA bulk chemical composition (Table 1), prepared using -1 mm ores and pulverised fluxes. Samples were fired using the CSIRO developed laboratory scale sintering method in a horizontal tube furnace with a low oxygen partial pressure of 5×10^{-3} atm using a heating profile designed to simulate the sintering process (Clout and Manuel, 2003). The firing process included a preheat holding step to 400°C, which minimises potential explosive degradation of the compacted tablet due to steam release from dehydroxylation of goethite below 400°C and steam from free moisture. Following the end of preheating, after about 13 min, the heating and cooling sintering temperature profile (Figure 1) was within the range of typical variations in plant and laboratory sintering profiles (eg Dawson, Ostwald and Hayes, 1984, 1985; Loo and Dukino, 2014). Samples were set in epoxy resin, sectioned, polished and imaged. High magnification optical reflected light microscopy (RLM) plane-polarised images were taken of the remnant quartz material and adjacent phases using a Zeiss Axio Imager Z2M petrographic microscope with an Axiocam HRc digital camera and AxioVision v2.6 software (Carl Zeiss, Oberkochen, Germany). The phase identification and typical abundances of each MA were determined using RLM optical image analysis, as described by Donskoi, Poliakov and Manuel (2022) and relies on differences in reflectivity, colour and grain shape. QXRD results for compact sinter Blend A matrix and the plant and pot grate samples were previously reported by Clout *et al* (2026). Scanning electron microscope (SEM)/energy dispersive spectrometry (EDS) analysis was carried out with a Zeiss Gemini SEM (Carl Zeiss Microscopy GmbH, Jena, Germany) with Bruker XFlash 7 energy dispersive X-ray detector (Bruker Nano GmbH, Berlin, Germany) to compare the key element compositions of key phases from the centre to outer region of the relevant sample cross-sections and to re-confirm the RLM and QXRD identification of phases.

TABLE 1

Blend A bulk matrix SFCA composition of compacts.

Fe_2O_3	SiO_2	Al_2O_3	CaO	MgO	TiO_2	CaO/SiO_2
74.2	6.1	3.7	15.2	0.37	0.10	2.49

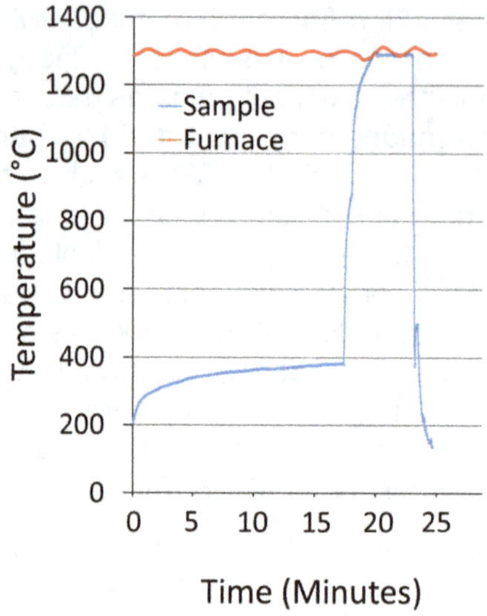

FIG 1 – Furnace and sample time-temperature profiles.

RESULTS

Plant and pot grate sinter observations

Skeletal hematite is commonly observed in both pot grate and plant sinter (Figure 2), with high magnification images showing its presence adjacent to what is interpreted as matrix melt that now comprises various mineral associations (Clout *et al,* 2026), particularly MA-B, MA-Ds, as well as both remnant and assimilated ex-hematite/goethite ore nuclei (MA-N).

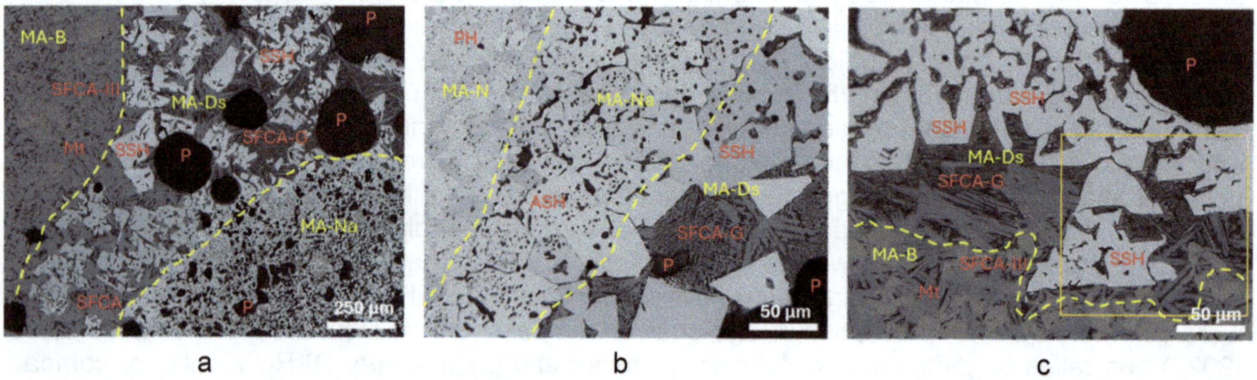

FIG 2 – RLM optical photomicrographs showing examples of skeletal hematite ('SSH') in plant (ab) and pot (b and c) sinter) showing the presence of skeletal hematite ('SSH', anhedral-skeletal hematite (ASH), magnetite (Mt) glass (G), pores (P) silicon-ferrite-calcium-aluminium phase (SFCA, SFCA III) and various adjacent mineral associations- (MA-Ds, MA-B, MA-N) and assimilated hematite nucleus (MA-Na). Distinct regions of various compositions are segregated by yellow dash lines. The orange rectangle denotes an area of enlargement shown in Figure 3.

Figure 2a is an example of plant sinter and shows a zone of skeletal hematite-SFCA-glass (MA-Ds) is bounded by assimilated hematite nucleus (MA-Na) on the right and magnetite-SFCA-III (MA-B) on the left, with the distinct boundaries of MA-Ds accentuated by yellow dashed lines.

Figure 2b and 2c are examples of pot grate sinter; Figure 2b shows a dense hematite nucleus (MA-N, at left) showing a rim of recrystallised hematite (MA-Na) with secondary micropores and minor interstitial SFCA. At right, there is a transition to a matrix melt assemblage (MA-Ds) dominated by coarse grains of skeletal hematite surrounded by dendritic SFCA, with interstitial glass and larnite. There is a micropore (P, black) at far right. Figure 2c shows MA-Ds, composed of skeletal hematite with SFCA-III/SFCA and glass, adjacent to a pore at upper right. The lower portion of this image is composed of MA-B with disseminated magnetite, SFCA-III, SFCA and glass. This type of skeletal hematite occurrence is often seen adjacent to larger pores and voids in pot grate/plant sinter where there is no direct association with remnant hematite nuclei (MA-N or MA-Na).

The area bordered by the orange rectangle in Figure 2c shows an area which has been extracted and enlarged in Figure 3, to emphasise the composition of the matrix surrounding skeletal hematite, including micro-skeletal SFCA and micro-skeletal hematite/magnetite (MSH), intergrown with glass, while the lowermost region is a zone of MA-B with disseminated magnetite intergrown with SFCA-III, SFCA and glass. The precipitation of MSH around the rim of the skeletal hematite grains implies a lack of available Ca in the adjacent melt to further form SFCA and a locally Si-rich, Ca-deficient low basicity melt composition.

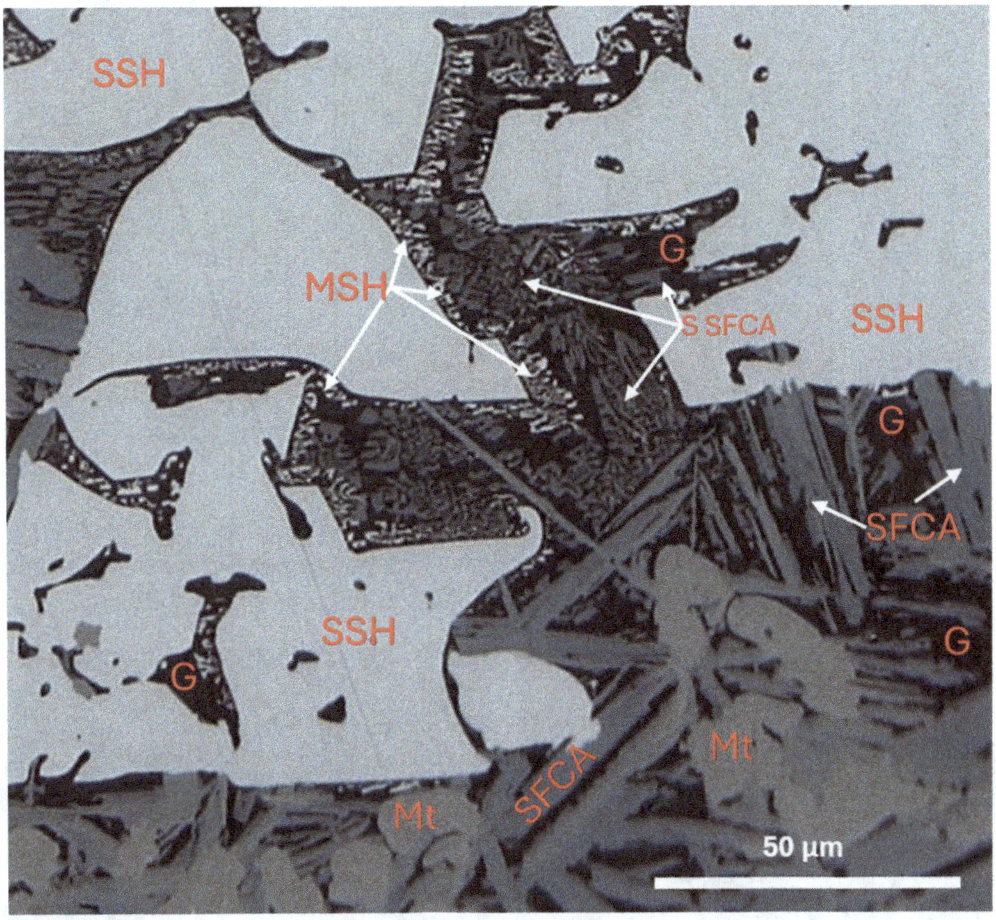

FIG 3 – Enlargement of Figure 2c RLM image of plant sinter showing details of matrix surrounding skeletal hematite ('SSH', including skeletal SFCA (S SFCA) and micro-skeletal hematite/magnetite (MSH) intergrown with glass (G).

In both Figure 2a and Figure 2b, it is likely that skeletal hematite is formed directly from hematite derived from nucleus assimilation and partial reaction with the adjacent melt.

These examples illustrate skeletal hematite forms in conjunction with several other minerals, in pilot and plant scale sinter, and may be associated with micro-skeletal forms of SFCA, magnetite and hematite in glass.

Laboratory scale sinter observations

The mineralogy of the bulk matrix blend A compacts within the temperature range of 1270°C to 1330°C was determined previously using QXRD by Clout *et al* (2026), with the predominant phase being SFCA-I with minor SFCA, SFCA-III, glass, magnetite and hematite (Figure 4).

Images of fired laboratory samples of SFCA bulk sample composition with embedded ex-quartz particles are shown in Figures 5a and 5b and Table 2 summarises the major phases present, relative to distance from the quartz particles. Optical RLM combined with detailed EDS spectra was used to confirm phase identification as well as to detect the presence of minor to trace levels phases below the detection limit for QXRD. A combination of optical microscopy and analysis of EDS chemical composition were used as cross-checks for identification of phases present, particularly in the outer region and provided in previous QXRD results by Clout *et al* (2026). Figures 5a and 5b indicate evidence of skeletal hematite similar to that seen in Figures 2 and 3. The original quartz is now most likely, but not confirmed as, in the form of β_2-tridymite.

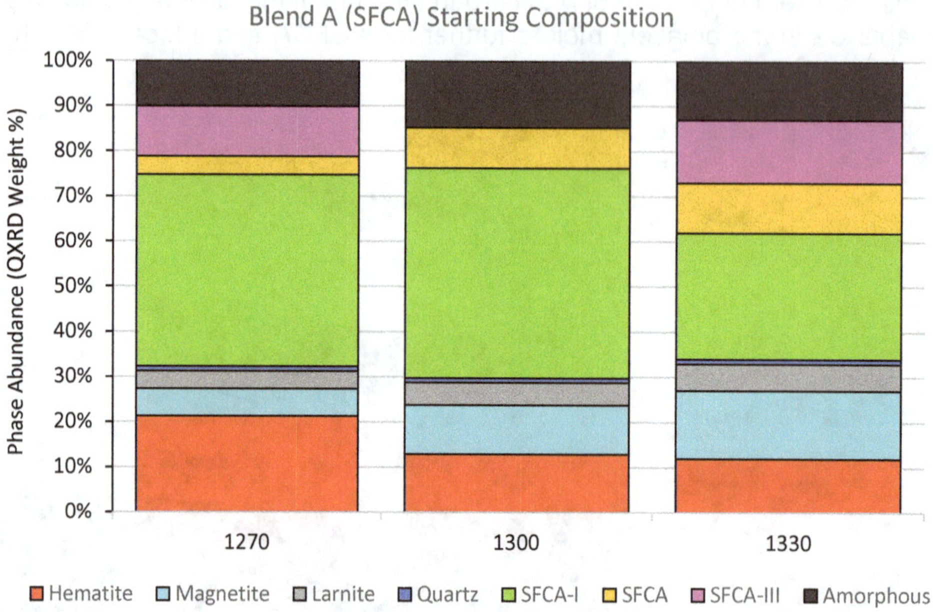

FIG 4 – Quantitative XRD of the compact matrix with an SFCA bulk composition fired to various temperatures with 3 mins holding time at peak temperature (as per Figure 1). The amorphous component represents glass. Modified from Clout *et al* (2026).

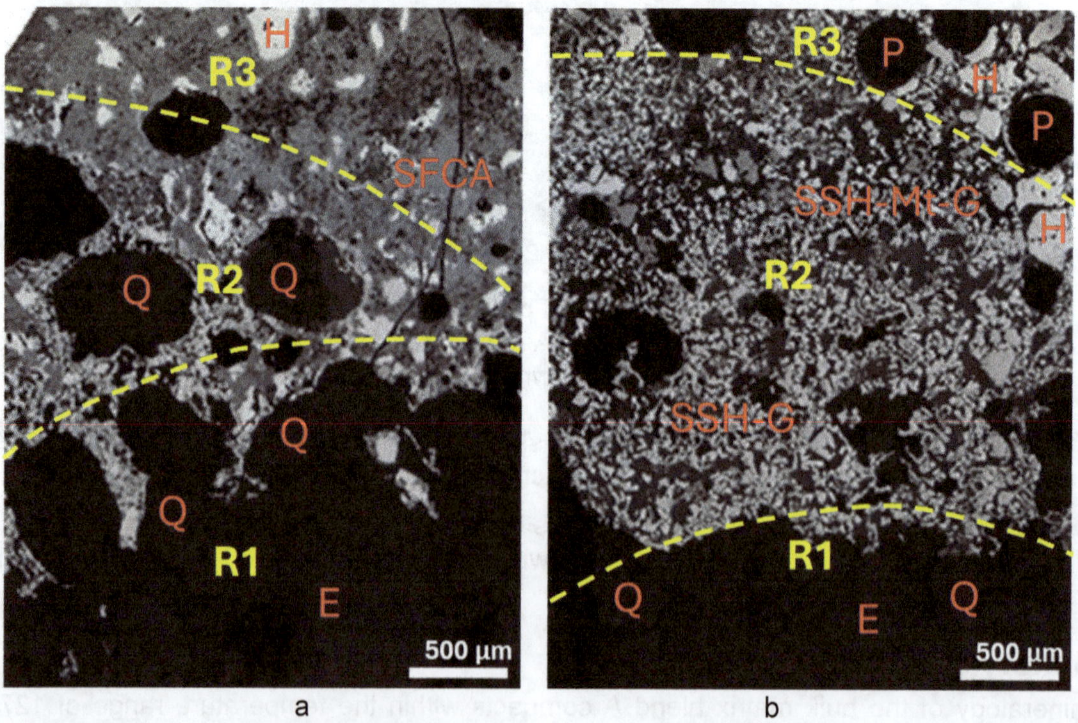

FIG 5 – Optical RLM plane-polarised images of compacts with quartz (Q) grains at their cores and fired to (a) 1290°C and (b) 1330°C. Each image has been designated into regions R1, R2 and R3, based on homogeneity of sinter microstructure. E = epoxy resin.

TABLE 2

Phases identified in each region by optical RLM, corresponding mineral association and number of points analysed by SEM EDS. Note region 3 has QXRD confirmation in Figure 4 over the temperature range.

	1290°C (Figure 5a)	No. points	MA	1330°C (Figure 5b)	No. points	MA
Region 1	quartz, glass ± SH	12	-	hematite, SH, glass	35	MA-E
Region 2	hematite, SH, glass., SFCA-I, SFCA-III	50	MA-Ds	glass, glass + ferrite, hematite/magnetite, SFCA-III, magnetite	30	MA-Ds
Region 3	SFCA-I, SFCA III, SFCA, trace calcium ferrite, glass, larnite, magnetite, hematite as nuclei	52	MA-B	SFCA-I, SFCA-III, SFCA, trace calcium ferrite, glass, larnite, magnetite, hematite as nuclei	35	MA-E

Compact samples fired at both 1290°C (low temperature sintering conditions) and 1330°C (high temperature/magnetite sintering conditions) are shown in Figure 5 and three distinct regions of different microstructures have been visually identified, segregated with dashed yellow lines. In each case, the embedded quartz is at the bottom of the image, moving perpendicularly away from the centre of the section towards the top of the image. Visually, the mineralogical composition and structure of the sinter are seen to change relative to distance from the remnant quartz particles. Generally, region 1 (R1), nearest the compact core, shows unreacted ex-quartz grains, surrounded by porosity (filled with epoxy resin, E), region 2 (R2) contains a mixture of skeletal hematite and glass with minor SFCA, whilst region 3 (R3) is comprised of a matrix of SFCA-I/SFCA-III/SFCA-magnetite – hematite and appears unaffected by the more distant proximity to quartz (Table 2). There are some slight differences evident between the two temperatures: at 1290°C (Figure 5a), skeletal hematite really only occurs immediately adjacent to quartz, while at 1330°C (Figure 5b), R1 shows evidence of skeletal hematite – glass and in R2, a more extensive layer of skeletal hematite – magnetite – glass formed, where considerably more reaction with SiO_2-rich low basicity melt derived from quartz has occurred.

EDS point analysis (n = 214) of key phases was conducted across regions 1, 2 and 3 and representative SEM images from each region are shown in Figure 6 (1290°C) and Figure 7 (1330°C). Point EDS analyses were carried out on key sinter phases within each region and the data has been collated and summarised in the plots in Figures 8 and 9. The aim was to establish any change in phase composition and mobility of major and minor elements with increasing distance from the unreacted ex-quartz at the centre of the sample. Individual SEM analyses were normalised to 100 per cent on an elemental basis and arithmetic mean values were taken from multiple analysis points. These values are plotted in Figures 8 and 9.

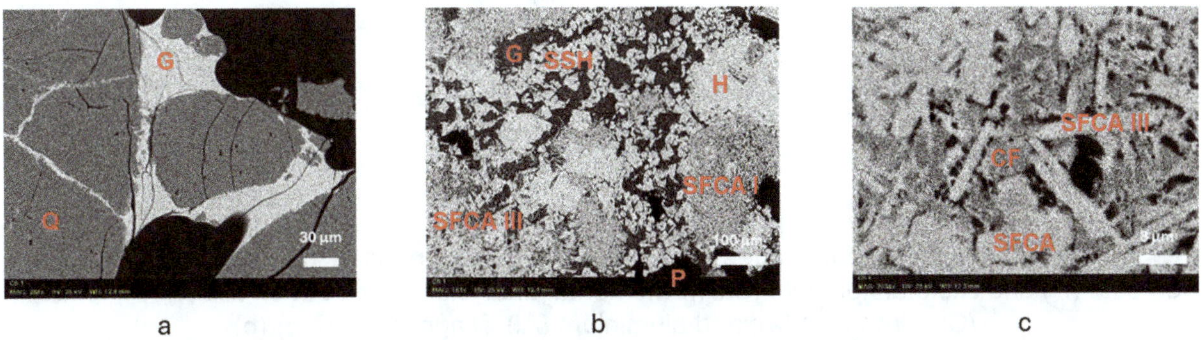

FIG 6 – Representative SEM images of sample fired at 1290°C, representing 'typical' structure of region 1 (a), region 2 (b) and region 3 (c).

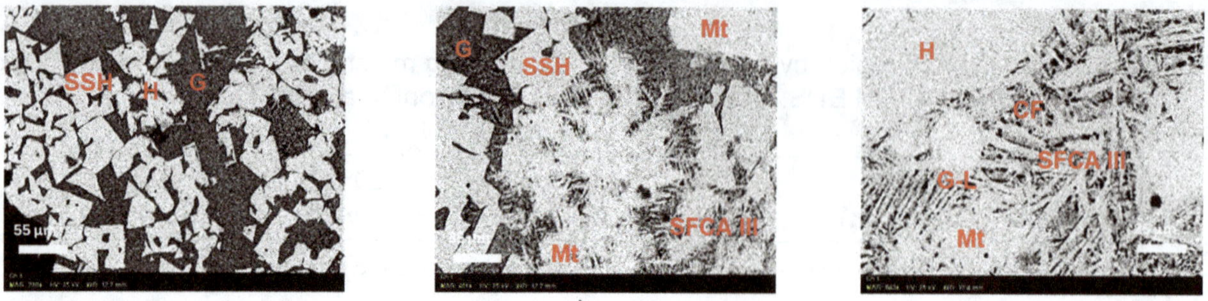

a b c

FIG 7 – SEM images of sample fired at 1330°C, representing 'typical' structure of region 1 (a), region 2 (b) and region 3 (c). N.B. SEM analysis conducted on the zone 3 representative image was reported as oxides and elemental analysis was determined via calculation.

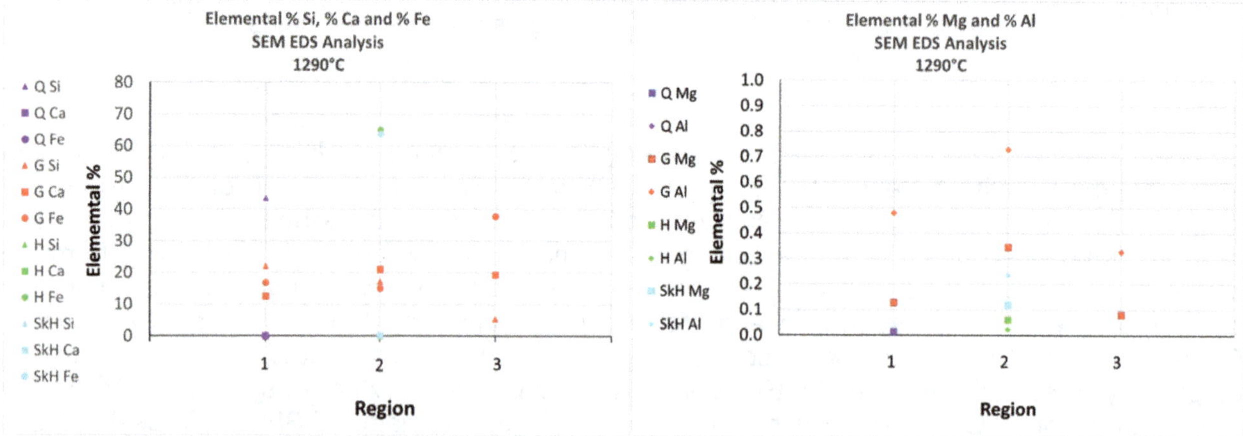

a b

FIG 8 – SEM EDS analysis for images shown in Figure 5a for phases; quartz (Q), glass (G), hematite (H) and skeletal hematite (SkH). Elemental analysis carried out on silicon (Si), calcium (Ca), iron (Fe) (a); and aluminium (Al), magnesium (Mg) (b).

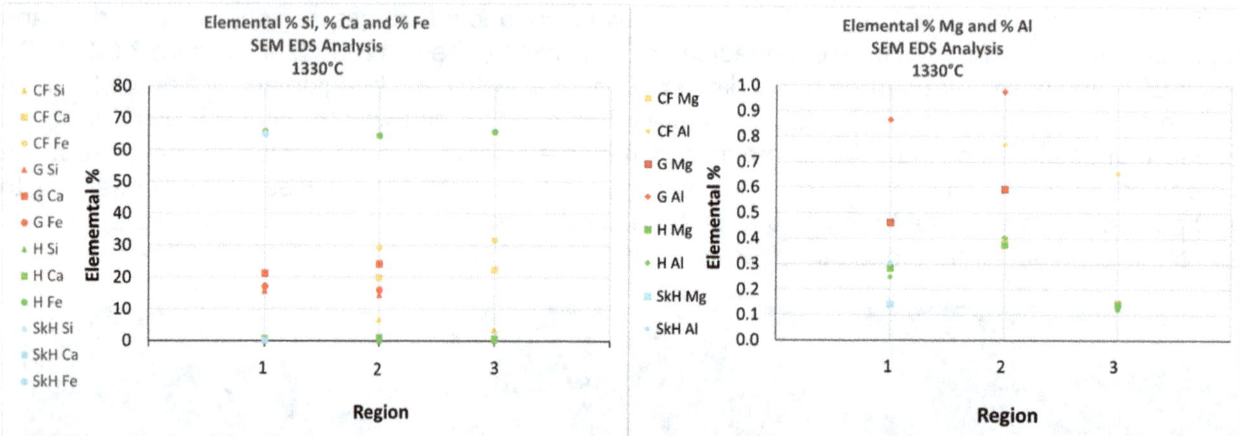

a b

FIG 9 – SEM EDS analysis for images shown in Figure 5b for phases; quartz (Q), glass (Glass), hematite (H) and skeletal hematite (SkH). Elemental analysis carried out on silicon (Si), calcium (Ca), iron (Fe) (a); and aluminium (Al), Magnesium (Mg) (b).

The key results in Figures 6 to 9 are:

- No significant change in the Si content of hematite or skeletal hematite was observed across the three zones (ie no significant level of Si present at 1330°C, with a similar result at 1290°C (R2).
- A decrease in Si content of glass from R1 to R3 at 1290°C and a corresponding increase in Ca content and a significant increase in basicity (0.3 to 1.7) was seen from R1 to R2, with Si slightly lower and significantly higher (2.3–4.0, mean 3.1) basicity in R3. Mg and Al (minor content) show similar trends. At 1330°C, similar changes were observed in glass from R1 to R2, with a corresponding decrease in Si and Fe content and a corresponding increase in basicity, Mg and Al.
- Glass composition did not change significantly between R1 (basicity of 1.1–1.7, mean 1.4) and R2 (basicity of 1.6–3.3, mean 2.4) at 1330°C, but melting was more extensive in this case than at 1290°C. The Al and Mg in glass is more elevated in R2 at 1330°C.
- At 1290°C, Al and Mg increased slightly from R1 to R2, then decreased to their lowest level in R3. This may be attributed to partitioning of Al and Mg into SFCA, rather than glass in R3, away from immediate proximity to quartz.
- Mg and Al were slightly elevated in skeletal hematite with respect to 'regular' hematite at 1290°C. This was not observed at 1330°C.

In addition to the EDS analysis data summarised above, it was noted when carrying out the analysis that whereas skeletal hematite typically contained from ~0.2 to 0.4 per cent Al, magnetite analyses often showed significant levels of Ca and/or Mg (up to ~2.5 per cent) and up to ~0.5 per cent Al. Further work will be carried out to differentiate the compositions of these phases and to better understand the range of impurity levels found within skeletal hematite and constraining its formation from primary/secondary hematite or magnetite.

DISCUSSION

Comparison of equilibrium assemblages versus those formed during compact sintering conditions

The most obvious apparent anomaly in the laboratory experimental results is the presence of skeletal hematite-glass and magnetite-glass/SFCA/SFCA-III/SFCA-I as discrete assemblages in the one compact sample, often in close (5–100 μm) proximity to each other, even though fired under the same temperature-pO_2 conditions, including for firings at 1290°C and 1330°C. In order to better understand this apparent anomaly, it is essential to investigate further through an examination of the thermodynamic stability of the phases from previous studies which are relevant to the experimental conditions (composition, pO_2 and temperature) used in this study.

The previous theory is that SSH forms by oxidation during cooling of earlier formed skeletal impurity-rich formed magnetite ('multi-component' magnetite, ie impure/high Al, Si) (Inazumi, Shinada and Kawabe, 1982). This theory is not supported in the current study since all magnetite was found to have much higher impurity levels (Al, Si, Ca and Mg), whilst the skeletal hematite has lower impurity levels (see Figures 8 and 9) and so indicates an origin for skeletal hematite unrelated to simple oxidation of earlier formed magnetite.

Early work by Hsieh and Whiteman (1989) established that the majority of the sintering reaction process occurred under a near-neutral pO_2 of 5×10^{-3} atm that maximised the formation of Ca-rich ferrites (which include the SFCA family of phases) whilst still producing mineral assemblages similar to those found in industrial sinter. Investigations of equilibrium phase relations in air for the Fe_2O_3-rich portion of the system FeO-Fe_2O_3-CaO-SiO_2 (FCS) (eg Phillips and Muan, 1959; Nicol, Jak and Hayes, 2019a, 2019b; Chen *et al,* 2019; Cheng *et al*, 2020, 2021) and the pseudo-ternary 'Fe_2O_3'-CaO-SiO_2 system (Pownceby, Clout and Fisher-White, 1998) in the temperature range of 1240–1300°C have been enlightening by demonstrating the large extent of the hematite + liquid stability field at and above 1270°C. Experimental phase equilibria studies by Cheng *et al* (2021) in the '$FeO_{1.5}$'-CaO-$AlO_{1.5}$ system, also in air, between 1210°C and 1260°C, used a novel equilibration/quenching/EPMA measurement approach, enabling quenching from any temperature to determine the direct quantification of liquid and solid phases compositions in equilibrium at high temperature. Of more direct relevance are the Nicol, Jak and Hayes (2019a, 2019b) studies which used a similar equilibration/quenching approach, but more importantly replicated formation of the skeletal to dendritic hematite texture with intergrown larnite-glass-calcium ferrite (Nicol, Jak and Hayes, 2019a; Figures 4 and 6), on cooling from 1350°C to 800°C, at 275°C/s in air using a high Fe-high basicity (3.46) starting composition (72.74 wt per cent Fe_2O_3, 6.11 wt per cent SiO_2 and 21.14 wt per cent CaO). The 3.46 basicity used is quite dissimilar to the current study with a very low 0.1–1.2 basicity range from EDS analysis of glass that encloses skeletal hematite in regions 1 and 2 (MA-Ds). Skeletal hematite was not observed in the current study in region 3, which had a starting basicity of 2.49 and where glass surrounding SFCA-(I, III)-larnite-(magnetite) has a basicity range of 1.7–3.9 with a mean of 3.05 (n = 24). However, the above research and a significant number of other phase relations studies (see review by Heikkinen, Iljana and Fabritius, 2020) were conducted in air and so are not directly applicable to the near-neutral pO_2 of 5×10^{-3} atm used in the current study.

In contrast, work by Pownceby and Clout (2000), under pO_2 of 5×10^{-3} atm, is believed to more able to explain the phases formed during the industrial sintering process and is therefore more relevant to the current investigation and the compositional ranges tested. In addition, hot stage *in situ* XRD work by Webster *et al* (2013) demonstrated the significant impact that oxygen partial pressure has on the formation mechanisms of Ca-rich ferrites, with magnetite + liquid field found to be stable above 1259°C at $pO_2 = 5 \times 10^{-3}$ atm whereas hematite + liquid is stable above 1321°C at $pO_2 = 0.21$ atm. Clout *et al* (2026) used the same blend A (SFCA) bulk composition and pO_2 used in the current study with a compact preparation sequence of 2 hr firings at temperature, then water quenched, pulverising, re-compaction, firing for 2 hrs followed by water quenching, for a total of 4 hrs at peak temperature. The aim was to approximate quasi-equilibrium under Scheil–Gulliver cooling conditions, that is not allowing for any re-equilibration within solids after solidification. QXRD of the fired blend A samples revealed largely magnetite with larnite at 1330°C and 1300°C but> 70 wt per cent SFCA with minor magnetite-SFCA-III and larnite at 1270°C (Figure 10). Hematite was absent in the XRD results and optical RLM examination of the polished sections. This work is also consistent with other *in situ* XRD studies indicating magnetite + liquid (glass), rather than hematite + liquid (glass), is most commonly associated with the typical SFCA-family sinter bonding phases (Webster *et al*, 2012, 2014; Webster, Pownceby and Madsen, 2013; Scarlett *et al,* 2004a, 2004b). Nicol, Jak and Hayes (2019a) noted that the main limitation of the *in situ* XRD method is the inability to quench the samples during the solidification sequence to determine the processes taking place during solidification rather than just during heating or slow cooling. The data in Figure 10 therefore provides further insight than the published *in situ* XRD results.

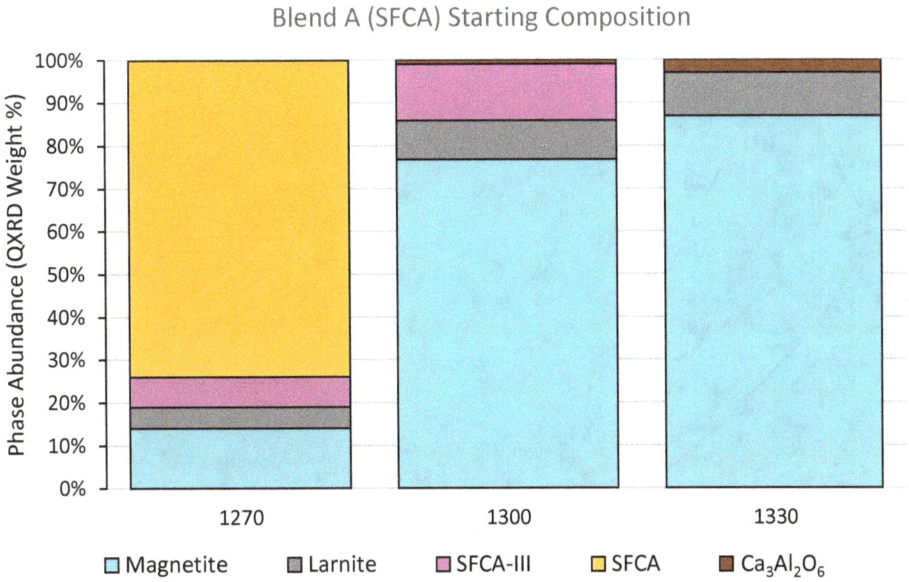

FIG 10 – Quantitative XRD of the compact matrix blend A with an SFCA bulk composition fired to various temperatures but each having a total of 4 hr holding time at peak temperature, water quenched. Modified from Clout *et al* (2026).

The glass compositions for regions 1 (adjacent to quartz sand), region 2 (intermediate distance) and region 3 (furthest distance from quartz sand) (Figure 5) were determined by a total of 82 spot EDS analyses and projected onto the Fe-rich corner of the Fe_2O_3-CaO-SiO_2 system for 1270°C (1290°C EDS data) and 1300°C (1330°C EDS data) (Figure 11). The glass compositions for region 1 (R1-G) for both temperatures lie within the hematite + liquid (cristobalite) stability field, consistent with the optical RLM observations in Figure 5 which indicate skeletal hematite in contact with glass. In contrast, region 2 (R2-G) glass compositions have a dominate trend in the hematite + liquid field (basicity <0.8) along the red line, but also a smaller number of glass results with a basicity range of 0.8–1.2 which plot in the magnetite + liquid field, straddling the approximate magnetite-hematite phase boundary (Figure 11). This explains the mixed skeletal magnetite-glass/skeletal hematite-glass assemblages observed in region 2 as a simple localised variation in basicity rather than oxidation of earlier formed magnetite or change in pO_2. The lower trend of region 2 glass compositions along the 1.0 basicity line are interpreted as variable Fe depletion by magnetite precipitation from a melt kept at constant basicity due to lack of element fractionation into other phases (eg larnite). No relevant phase relation data for the same pO_2 was found for 1330°C but it would be expected that the magnetite + liquid and liquid-only fields (Figure 11, grey shading) would both increase in size at higher temperature, the hematite + liquid field would persist, but at higher SiO_2 values. Glass compositions for region 3 (R3-G) plot predominantly in the expected dicalcium silicate (larnite, Ca_2SiO_4 or C_2S) + liquid, liquid and dicalcium ferrite (C2F)/SFCA fields (Figure 11). Glass for region 3 is typical of high basicity glass compositions in industrial iron ore sinter. The presence of skeletal hematite with glass is therefore explained by the phase diagram by a local liquid basicity, which then plays a key role in determining if skeletal hematite (basicity <~0.8) or magnetite precipitates(>~0.8 basicity) from melt, hence allowing magnetite to be locally present near skeletal hematite without the need for micro-temperature or pO_2 variations, precipitation occurring at different temperatures or times, or formed by oxidation of earlier formed magnetite. It is therefore recommended that the 'secondary' interpretive term be dropped and simply refer to as skeletal hematite.

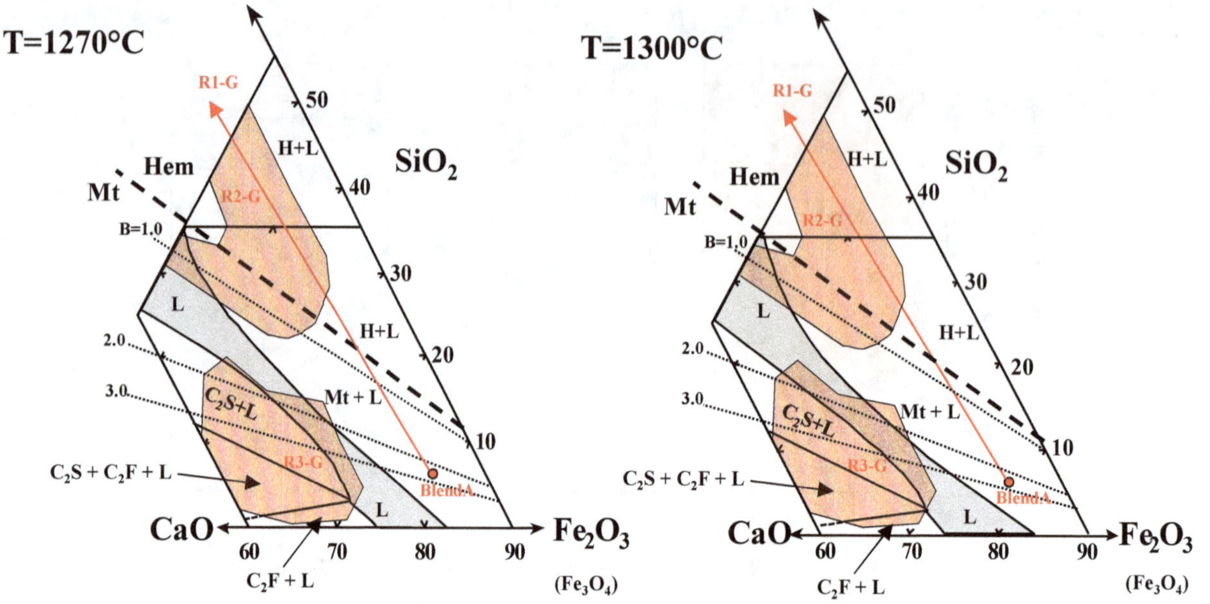

FIG 11 – Diagrams illustrating change in phase relations (at $pO_2 = 5 \times 10^{-3}$ atm) at 1300°C and 1270°C, red shaded regions showing locations of normalised EDS probe glass compositions for regions 1 (R1-G, extending from the red arrow tip to the 100 per cent SiO_2 apex), 2 (R2-G) and 3 (R3-G) as well as the bulk composition for blend A. Note that R-1 glass compositions plot outside of the phase region shown and range from 100 per cent to 53 per cent SiO_2. R2 has two data trends, one around the red line from blend A towards R1-G and the second around basicity = 1.0 line. Phase boundaries represented by solid lines and lightly stippled areas indicate regions where melting occurs. Approximate position of magnetite–hematite (Mt/Hem) boundary indicated by dashed line. Lines of constant basicity (also dashed lines) plotted. Modified from Pownceby, Clout and Fisher-White (1998).

Comparison with pot grate and plant sinter mineralogy

The laboratory compact sintering results outlined above demonstrate the formation of skeletal hematite in association with localised high-silica (low basicity, 0.1–0.8) melts under both low (1290°C) and high (1330°C) temperature conditions, with comparable morphology to that observed in plant and pot grate sinter. In both laboratory compact sinter (Figure 5) and plant sinter (Figures 2 and 3), skeletal hematite-glass have a close spatial association with magnetite-glass-larnite/SFCA that can be explained simply by micro-spatial differences in basicity/melt composition rather than local micro-temperature or pO_2 variations or formation at different times during the cooling or heating process. There is a clear transition in the compact samples from an skeletal hematite -glass association (region 1), in close proximity to partially reacted quartz, through a transition zone with localised/patchy formation of fine SFCA III and/or SFCA I (region 2), to a magnetite/SFCA III matrix (with remnant unreacted hematite) and only minor interstitial high basicity glass/larnite (region 3), as expected for a blend with SFCA composition (see blend A in Clout *et al,* 2026). At higher temperature, the extent of glass formation is increased due to the greater reaction of quartz and effective 'dilution' by the primary Ca-rich melt, reflected in lower SiO_2/higher basicity of EDS analyses of glass composition at increasing distance from the remnant quartz.

The laboratory sinter results also clearly show that the skeletal hematite -glass (MA-E) association, which is typically considered characteristic of high temperature sintering, may also be formed locally at a relatively low sintering temperatures (1290°C) under localised low (<0.8) basicity compositions.

Examples from pot grate and plant sinter show that skeletal hematite is also formed around the rim of partially assimilated hematite ore nuclei (Figure 2b), again in association with low-Ca/high-Si low basicity glass and strongly implies that skeletal hematite may be formed by near-equilibrium reaction/recrystallisation from low basicity melt attack/assimilation of the outer rims of primary hematite particles. This low basicity formation process interpretation is believed to be a more plausible alternative route to the more widely accepted re-oxidation of (skeletal) 'multi-component' magnetite to SSH during cooling from peak sintering temperature, as proposed by other studies (eg Inazumi,

Shinada and Kawabe, 1982). This has implications for impurity profiles in skeletal hematite formed in different locations, given previously proposed mechanisms of enhanced reduction breakdown due to elevated Al levels in skeletal hematite formed by re-oxidation of magnetite adjacent to pores and voids.

'Secondary' skeletal hematite has been associated with low temperature reduction degradation of sinter, with evidence of cracking due to volume change during reduction of SSH to magnetite, but there is some uncertainty as to whether breakdown is primarily influenced by skeletal hematite morphology, or due to hematite lattice defects resulting in strain generation during recrystallisation (Inazumi, Shinada and Kawabe, 1982; Mizutani *et al,* 2020).

There was a low Al level in skeletal hematite (Figures 8b and 9b) formed by melt-quartz reaction, and this is also likely to be the case for skeletal hematite formed around assimilated nuclei with Si-rich/low basicity infill. At least for the current study, there is as yet no evidence that skeletal hematite has formed by oxidation of magnetite in matrix areas, evidenced by the high impurity levels of magnetite versus the lower impurity levels of skeletal hematite. In the hematite nuclei- skeletal hematite case, skeletal hematite is formed directly by reaction/recrystallisation of primary/secondary hematite (low impurity levels), hence would explain the low skeletal hematite impurity levels. In the case of skeletal hematite formed from matrix melt, it is simpler to explain the low levels of impurity elements in skeletal hematite being due to the preference for hematite to not take up these elements from the melt during crystallisation whereas the magnetite crystal structure and substitution allows for greater uptake of impurity levels (including Al, Si, Ca and Mg) to form 'multi-component' magnetite. The phase diagram, local micro-spatial variations in basicity and substitution preferences adequately explain the closer than expected association of skeletal hematite and magnetite. In the case of nucleus assimilation, SFCA formation is not possible due to local low melt basicity. In the case of where skeletal hematite is present, SFCA formation is limited due to dependence of higher basicity required to form any of the SFCA family members.

The laboratory sintering results show evidence of limited element spatial mobility within melts and variable local melt composition (eg basicity), interpreted to result in different mineral associations. The relative proportions of skeletal hematite associated with low basicity melt surrounding hematite nuclei or skeletal hematite direct precipitation from high SiO_2/low basicity-rich gangue melt, respectively, may go some way to explaining anomalous results with respect to partial correlations of total skeletal hematite level in sinter with RDI values. Note that this will also be influenced by ore type, for example, the proportions of reactive versus dense/unreactive nuclei versus SiO_2-rich porous reactive goethite/hematite nuclei) and local melt basicity.

These observations raise the potential importance of further work on skeletal hematite formed by low basicity melts and the relative influence of lattice impurities (particularly Al and Ti) in skeletal hematite, versus morphology, in crack generation promoting breakdown during low temperature reduction.

CONCLUSIONS

Skeletal hematite can be produced in the laboratory from low (<0.8) basicity melts by embedding coarse quartz-silica sand particles in a -1 mm matrix of SFCA bulk composition. Skeletal hematite is also observed in association with partially assimilated SiO_2-rich hematite nuclei in pot grate and plant sinter.

Evidence shows that skeletal hematite in contact with glass may be formed from localised low (<0.8) basicity melts whilst an adjacent localised intermediate (>0.8–1.3) basicity melt forms magnetite-glass or high (1.7–3.9) a basicity melt results in magnetite at peak temperature followed during cooling by SFCA-larnite-glass. The skeletal hematite phase has been observed in pot and plant sinter samples adjacent to various mineral associations, including MA-B, MA-D/Ds, MA-E and MA-N. Skeletal hematite is also associated with partially assimilated hematite nucleus particles, formed directly from low basicity melt attack of hematite nuclei over a wide temperature range, as well as in areas of low basicity matrix melt. Whilst potentially oxidation of skeletal magnetite during sinter cooling could form SSH, it was not observed in the current study, with the element data demonstrating that skeletal magnetite has much higher gangue element levels than skeletal hematite. The common factor in both assimilated nuclei-skeletal hematite and matrix melt formed skeletal hematite is a SiO_2-rich/low basicity, Ca-depleted ± variably Fe-depleted melt. EDS analysis results of glass in laboratory sinter compacts show evidence of limited element spatial mobility within melts and micro-local high variability in basicity and overall melt composition.

In the absence of compelling evidence, it is therefore recommended that the interpretive term 'secondary' be dropped in favour of a non-genetic but descriptive 'skeletal hematite'.

The role of Al and Ti lattice distortion effects versus morphology in determining the influence of skeletal hematite on RDI warrant further investigation, based on the results presented.

These results suggest that the accepted correlation between skeletal hematite occurrence and low temperature reduction breakdown may be more complex than previously understood. Further work is required to understand the drivers for RDI, and a fundamental study needs to be undertaken with a systematic approach to the investigation of variables known to affect blast furnace RDI.

REFERENCES

Chen, J, Schevchenko, M, Hayes, P and Jak, E, 2019. A Phase Equilibrium of the Iron-rich Corner of the CaO–FeO–Fe_2O_3–SiO_2 System in Air and the Determination of the SFC Primary Phase Field, *ISIJ International*, 59(5):795–804.

Cheng, S, Shevchenko, M, Hayes, P C and Jak, E, 2020. Experimental phase equilibria studies in the FeO-Fe_2O_3-CaO-SiO_2 system in air: Results for the iron-rich region, *Metall Mater Trans B,* 51:1587–1602.

Cheng, S, Shevchenko, M, Hayes, P C and Jak, E, 2021. Experimental phase equilibria studies in the FeO-Fe_2O_3-CaO-Al_2O_3 system in air, *Metall Mater Trans B,* 52:2416–2429.

Clout, J M F and Manuel, J R, 2003. Fundamental investigations of differences in bonding mechanisms in iron ore sinter formed from magnetite concentrates and hematite ores, *Powder Technology*, 130(1–3):393–399.

Clout, J M F, Ware, N A, Manuel, J R, Webster, N A S and Pownceby, M I, 2026. Fundamental analysis of sinter solid structure: Implications of mineral associations for understanding industrial iron ore sinter formation, *Minerals*, 16:129:1–42.

Dawson, P R, Ostwald, J and Hayes, K M,1984. The influence of the sintering temperature profile on the mineralogy and properties of iron ore sinters, in *The AusIMM Proceedings 1984*, pp 163–169.

Dawson, P R, Ostwald, J and Hayes, K M,1985. Influence of alumina on development of complex calcium ferrites in iron ore sinters, *Trans IMM Sect C, Min Process Extr Met,* 94:71–78.

Donskoi, E, Poliakov, A and Manuel, J R, 2022. Automated optical image analysis of natural and sintered iron ore, in *Iron Ore: Mineralogy, Processing and Environmental Sustainability* (ed: L Lu), 2nd edn, ch 4, pp 127–178 (Woodhead Publishing: Cambridge).

Heikkinen, E P, Iljana, M and Fabritius, T, 2020. Review on the Phase Equilibria in Iron Ore Sinters, *ISIJ International*, 60(12):2633–2648.

Hosotani, Y, Konno, N, Yamaguchi, K, Orimoto, T and Inazumi, T, 1996. Reduction properties of sinter with fine dispersed pores at high temperatures of 1273K and above, *ISIJ International*, 36:1439–1447.

Hsieh, L H and Whiteman, J A, 1989. Effect of oxygen potential on mineral formation in lime-fluxed iron ore sinter, *ISIJ International,* 29:625–634.

Inazumi, T, Shinada, K and Kawabe, M, 1982. Influence of mineralogical characteristic of hematite in self-fluxed sinter on its degradation during the reduction, *Tetsu-To-Hagane*, 68:143–150.

International Organization for Standardization (ISO), 2015a. ISO 4696-1:2015, Iron Ores for Blast Furnace Feedstocks—Determination of Low-Temperature Reduction-Disintegration Indices by Static Method—Part 1: Reduction with CO, CO_2, H_2 and N_2. International Organization for Standardization (ISO), Geneva.

International Organization for Standardization (ISO), 2015b. ISO-4696-2:2015, Iron Ores for Blast Furnace Feedstocks—Determination of Low-Temperature Reduction-Disintegration Indices by Static Method—Part 2: Reduction with CO and N_2. International Organization for Standardization (ISO), Geneva, Switzerland.

Loo, C E and Bristow N J, 1998b. Properties of iron bearing materials under simulated blast furnace indirect reduction conditions: Part 2 - Reduction degradation, *Ironmaking and Steelmaking,* 25(4):287.

Loo, C E and Bristow, N J, 1998a. Properties of iron bearing materials under simulated blast furnace indirect reduction conditions: Part 1 - Review and experimental procedure, *Ironmaking and Steelmaking,* 25(3):222.

Loo, C E and Bristow, N J, 1998c. Properties of iron bearing materials under simulated blast furnace indirect reduction conditions: Part 3 - Reducibility, *Ironmaking and Steelmaking,* 25(5):366.

Loo, C E and Dukino, R D, 2014. Laboratory iron ore sintering studies, 2, Quantifying flame front properties, *Trans IMM C Miner Process Extr Metall,* 123:197–203.

Lu, L and Lu, Y, 2022. Blast furnace ironmaking and its ferrous burden quality requirements, in *Iron Ore: Mineralogy, Processing and Environmental Sustainability* (ed: L Lu), 2nd edn, ch 17, pp 579–603 (Woodhead Publishing: Cambridge).

Lu, L, Holmes, R J and Manuel, J R, 2007. Effects of alumina on sintering performance of hematite iron ores, *ISIJ International,* 47:349–358.

Lu, L, Ishiyama, O, Higuchi, T, Matsumura, M and Higuchi, K, 2022. Iron ore sintering, in *Iron Ore: Mineralogy, Processing and Environmental Sustainability* (ed: L Lu), 2nd edn, ch 15, pp 489–537 (Woodhead Publishing: Cambridge).

Mizutani, M, Nishimura, T, Orimoto, T and Higuchi, K, 2020. In-situ evaluation for crack generation behavior of iron ore agglomeration during low temperature reduction by applying acoustic emission method and analysis of reduction disintegration behavior, Nippon Steel Technical Report No. 123, March 2020, pp 48–54.

Nicol, S, Jak, E and Hayes, P C, 2019a. Microstructure evolution during controlled solidification of 'Fe_2O_3'-CaO-SiO_2 liquids in air, *Metall Mater Trans B*, 50:2706–2722.

Nicol, S, Jak, E and Hayes, P C, 2019b. Controlled solidification of liquids within the SFC primary phase field of the 'Fe_2O_3'-CaO-SiO_2 system in air, *Metall Mater Trans B*, 50:3027–3038.

Peacey, J G and Davenport, W G, 1979. *The Iron Blast Furnace Theory and Practice* (Oxford: Pergamon Press).

Peterson, M J and Manuel, J R, 2023. Using a particle-based approach (sinter microtypes) for the geometallurgical characterisation of iron ore sinter, *Mineral Processing and Extractive Metallurgy*, 132:1–12.

Phillips, B and Muan, A, 1959. Phase equilibria in the system CaO-Iron Oxide-SiO_2 in air, *J Am Ceram Soc*, 42:413–23.

Pownceby, M I and Clout, J M F, 2000. Phase relations in the Fe-rich part of the system Fe_2O_3(-Fe_3O_4)-CaO-SiO_2 at 1240–1300°C and $pO_2=5\times10^{-3}$ atm: Implications for iron ore sinter, *Miner Process Extract Metall (Trans Inst Min Metall C)*, 109:36–48.

Pownceby, M I, Clout, J M F and Fisher-White, M J, 1998. Phase equilibria for the Fe_2O_3-rich part of the system Fe_2O_3-CaO-SiO_2 in air at 1240°–1300°C, *Miner Process Extract Metall (Trans Inst Min Metall C)*, 107:1–10.

Scarlett, N V Y, Madsen, I C, Pownceby, M I and Christensen, A, 2004b. In situ X-ray diffraction analysis of iron ore sinter phases, *J Appl Cryst*, 37:362–68. https://doi.org/10.1107/S002188980400353X

Scarlett, N V Y, Pownceby, M I, Madsen, I C and Christensen, A, 2004a. Reaction sequences in the formation of silico-ferrites of calcium and aluminium in iron ore sinter, *Metall Mater Trans B*, 35B:929–936.

Shigaki, I, Sawada, M and Norio Gennai, N, 1986. Increase in low-temperature reduction of iron ore sinter due to hematite-alumina solid solution and columnar calcium ferrite, *Transactions of the Iron and Steel Institute of Japan,* 26(6):503–511.

Takeichi, Y, Murao, R and Kimura, M, 2023. Micromechanism of heterogeneous reduction of iron ore sinters investigated by synchrotron X-ray multimodal analysis, *ISIJ International,* 63(12):2017–2022.

Webster, N A S, Pownceby, M I and Madsen, I C, 2013. In situ X-ray diffraction investigation of the formation mechanisms of Silico-Ferrite of Calcium and Aluminium-I-type (SFCA-I-type) complex calcium ferrites, *ISIJ International*, 53:1334–1340.

Webster, N A S, Pownceby, M I, Madsen, I C and Kimpton, J A, 2012. Silico-ferrite of calcium and aluminium (SFCA) iron ore sinter bonding phases: New insights into their formation during heating and cooling, *Metall Mater Trans B*, 43B:1344–1357.

Webster, N A S, Pownceby, M I, Madsen, I C and Kimpton, J A, 2013. Effect of Oxygen Partial Pressure on the Formation Mechanisms of Complex Ca-rich Ferrites, *ISIJ International*, 53:774–781.

Webster, N A S, Pownceby, M I, Madsen, I C, Studer, A J, Manuel, J R and Kimpton, J A, 2014. Fundamentals of silico-ferrite of calcium and aluminium (SFCA) and SFCA-I iron ore sinter bonding phase formation: Effects of CaO:SiO_2 ratio, *Metall Mater Trans B*, 45:2097–2105.

Yamaguchi, K, Ueno, H, Kawaguchi, T, Matsunaga, S, Oda, H and Amano, S, 1994. Influence of Al_2O_3 on reduction-meltdown behavior of sinter in blast furnace, *ISIJ International*, 34(12):964–972.

Mastering hide and seek – the power of combining multiple characterisation techniques to understand how phosphorus is hiding in goethite-rich iron ores

A J Whitworth[1], M Chen[2], M I Pownceby[3], M J Peterson[4], S Peacock[5], N C Wilson[6] and R Fan[7]

1. CERC Postdoctoral Fellow, CSIRO Mineral Resources, Clayton South Vic 3169.
   Email: anne.whitworth@csiro.au
2. Senior Principal Research Scientist, CSIRO Mineral Resources, Clayton South Vic 3169.
   Email: miao.chen@csiro.au
3. Senior Principal Research Scientist, CSIRO Mineral Resources, Clayton South Vic 3169.
   Email: mark.pownceby@csiro.au
4. Senior Experimental Scientist, CSIRO Mineral Resources, Pullenvale Qld 4069.
   Email: michael.peterson@csiro.au
5. Senior Experimental Scientist, CSIRO Mineral Resources, Clayton South Vic 3169.
   Email: steve.peacock@csiro.au
6. Principal Research Scientist, CSIRO Mineral Resources, Clayton South Vic 3169.
   Email: nick.wilson@csiro.au
7. Senior Experimental Scientist, CSIRO Mineral Resources, Urrbrae SA 5064.
   Email: rong.fan@csiro.au

ABSTRACT

Australian iron ore producers are turning to lower grade, goethite-rich iron ores containing phosphorus (P) to meet steel demand as high-grade deposits are depleted. These iron ores are considered less desirable when P exceeds certain thresholds due to the negative impact P has on steel quality and production. Iron ore producers currently blend high-P and low-P ores to dilute P concentrations and limit penalties. This strategy will be challenged as new supply enters the market, and as steel makers increasingly turn to higher iron (Fe), lower gangue products, in response to the drive to decarbonise the steel industry (MCA, 2021). Iron ore dephosphorisation prior to steelmaking offers an alternative pathway to develop a high Fe, low P product that is competitive in the global market. Successful development of these processes relies on having a comprehensive understanding of the deportment of P within iron ores, and there are several physical, chemical, mineralogical, and textural analytical techniques that are suitable for characterising iron ores.

This study investigates the deportment of P (and associated impurities) in one medium-P (head concentration of 0.08 wt per cent P) and three high-P (head concentrations ≥ 0.10 wt per cent P) iron ore samples from the Pilbara region of Western Australia. One iron ore sample, designated AUS-2-HP, consists of crushed drill core chips from an iron ore deposit. The other three samples, designated AUS-3-MP, AUS-4-HP, and AUS-5-HP(W), consist of iron ore blends. Representative sub-samples were generated using a combination of rotary and riffle splitters. These sub-samples were then subjected to size fraction analysis (involving dry ± wet screening), X-ray Fluorescence spectroscopy (XRF), X-ray Diffraction (XRD), optical microscopy, Electron Probe Microanalysis (EPMA), and X-ray Absorption Near Edge Structure (XANES) spectroscopy.

Size fraction analysis shows that the mass distributions of AUS-2-HP, AUS-3-MP, and AUS-4-HP are skewed towards coarser particle sizes, while for AUS-5-HP(W) it is skewed towards the -38 µm size fraction. XRF analysis shows that the main impurities present in the samples are P (0.081–0.188 wt per cent in head fractions), aluminium (Al: 0.74–0.147 wt per cent in head fractions), and silicon (Si: 1.18–2.13 wt per cent in head fractions), although significant concentrations of manganese (Mn) and titanium (Ti) are present in AUS-2-HP (head fraction 1.56 wt per cent Mn) and AUS-4-HP (head fraction 0.106 wt per cent Ti). XRF analysis also shows that there are strong to very strong, negative correlations between Fe and Al, and Fe and Si, for all samples. There is also a strong, negative correlation between Fe and P for AUS-2-HP, but this correlation is significantly weaker for the other three samples. XRF and size fraction analysis shows that the impurity distribution typically reflects mass distribution for each size fraction, except for AUS-5-HP(W), where P distribution is significantly higher than mass distribution in the -38 µm size fraction.

XRD analysis and optical microscopy shows that the main iron (oxyhydr)oxide phases present in the samples are goethite, hydrohematite, hematite, and martite, and the main accessory phases present are kaolinite, quartz, and anatase. XRD analysis also shows that the samples contain a significant (13–21 wt per cent) concentration of amorphous (poorly crystalline) material, and XRF and XRD analysis suggests that this material is comprised of Fe-bearing and non-Fe-bearing phases. The presence of kaolinite and quartz likely explains the XRF correlations observed between Fe and Al, and Fe and Si.

Optical microscope point counting of particles using the CSIRO iron ore classification scheme developed by Clout (2003) shows that AUS-2-HP and AUS-3-MP are predominantly comprised of moderately hard to hard hydrohematite-bearing and goethitic ore types. In contrast, AUS-4-HP and AUS-5-HP contain a wider range of ore types containing goethite, hydrohematite, hematite, and martite. However, AUS-4-HP is also comprised of predominantly moderately hard to hard particles, while AUS-5-HP(W) is comprised of moderately hard to friable particles. This likely explains why AUS-5-HP(W) has a high proportion of fines. Examination of goethite particles shows that the samples contain ochreous and brown/vitreous goethite, which are frequently co-mingled with hydrohematite, hematite, martite and/or kaolinite. Goethite textures are diverse, with colloform and botryoidal textures observed in all samples, cellular goethite observed in AUS-3-MP, and goethite pseudomorphs after silicate, carbonate, and quartz present in some samples. Hydrohematite-goethite pisolites are also present in AUS-3-MP which, together with the cellular goethite textures, suggests that AUS-3-MP is comprised of channel iron deposit (CID) ± detrital iron deposit (DID) material. Optical microscopy combined with XRF analysis suggests that there is no clear correlation between goethite texture type and P concentration.

EPMA shows that the samples are predominantly comprised of iron (oxyhydr)oxides, and that P, Al and Si are all hosted within goethite (although Al and Si are primarily associated with kaolinite and quartz gangue phases). Phosphorus distribution maps of high-P grains show that P is finely disseminated within both ochreous and brown/vitreous goethite types, and that the level of P enrichment can vary on a micron scale. The finely disseminated distribution of P within goethite suggests that P is incorporated in the structure of goethite, and this is supported by P XANES analysis. Phosphorus is also present as rare phosphate minerals (likely wavellite and hydroxyapatite) in AUS-4-HP and AUS-5-HP(W).

The main findings from this study are that:

1. The main impurities present in the samples are P, Al, and Si, and goethite can host all three impurities.
2. Both ochreous and brown/vitreous goethite are present, and are frequently co-mingled with other phases.
3. P incorporation does not appear to favour one goethite type.
4. P appears to be structurally incorporated within goethite.

Dephosphorisation processes will need to account for the incorporation of P in the structure of goethite, as well as the textural and mineralogical diversity of goethite-bearing particles.

This study demonstrates the power of using a range of characterisation techniques when examining the deportment of P (and other impurities) in iron ores. It also illustrates the physical, chemical, and textural variability of different Pilbara iron ores. This variability means that new dephosphorisation processes may need to be tailored to individual iron ore types and blends.

REFERENCES

Clout, J M F, 2003. Upgrading processes in BIF-derived iron ore deposits: implications for ore genesis and downstream mineral processing, *Applied Earth Science*, 112:89–95.

Minerals Council of Australia (MCA), 2021. *Iron Ore: When Quality Meets Opportunity, Best in Class: Australia's Bulk Commodity Giants*, Minerals Council of Australia.

Blueprint for representative drill hole and sample selection workflows for orebody characterisation – an industry-focused example from an apatite-iron deposit

B B G Zanderink[1], P Krolop[2], A-J Miles[3] and T Seifert[4]

1. PhD Candidate, TU Bergakademie Freiberg, Freiberg SN 09599, Germany. Email: bas.zanderink@mineral.tu-freiberg.de
2. Specialist and Technical Development, LKAB, Kiruna Norrbotten 98186, Sweden. Email: patrick1.krolop@lkab.com
3. Research Scientist, Mineral Resources, CSIRO, Perth WA 6000. Email: jo.miles@csiro.au
4. Professor, TU Bergakademie Freiberg, Freiberg SN 09599, Germany. Email: thomas.seifert@mineral.tu-freiberg.de

ABSTRACT

The Per Geijer (PG) Iron Oxide-Apatite (IOA)-REE deposit near Kiruna, northern Sweden, comprises a major exploration target for the Swedish mining company LKAB. Elevated *in situ* apatite ($Ca_5(PO_4)_3(OH,F,Cl)$) content (up to 27 wt per cent modally) and associated rare earth elements (REEs) distinguish PG from other IOA systems, positioning it as an alternative source of critical elements to the global economy. This variability also has important implications for iron ore processing, and potentially recovery of REEs and P as by-products. For apatite-hosted phosphorus (P) and REEs to be considered by-products of iron ore mining, it is essential to characterise the variable textures, chemistry and affiliation with different iron oxides, ore domains and gangue minerals. These attributes directly influence beneficiation, concentrate quality and geometallurgical domain definition. In this contribution, an industry-ready drill hole and sample selection workflow is presented, designed to reduce large assay data sets into a representative, processing-relevant sampling suite. Drill hole selection integrates assay, lithological and mineralogical data sets from over 50 drill holes, and sampling targeted apatite-bearing intervals for thin-section preparation. The workflow ensures that sampling captures ore variability relevant to processing, particularly apatite distribution across magnetite-dominated, mixed and hematite-dominated domains. This supports downstream mineralogical analysis, liberation assessment and metallurgical test work planning. Although every ore deposit is unique, this workflow provides a transferable approach for geometallurgical characterisation and early beneficiation studies in other IOA and Fe-P systems.

INTRODUCTION

A comprehensive deposit-scale mineral characterisation study is essential for the development of a mineral resource. For industry, such characterisation underpins geometallurgical modelling, process design and optimisation and risk reduction in mine planning. At the typical stage when an advanced mineral characterisation study takes place, projects usually have extensive drill core data sets. However, time and resources in an operational environment are limited. Therefore, efficient and representative drill core selection is critical – not only for geological understanding but for predicting ore behaviour during applied beneficiation steps and potential by-product recovery. In this contribution, an integrated workflow is presented showcased for apatite in an Iron-Phosphorus (Fe-P) deposit. The Per Geijer (PG) iron oxide-apatite (IOA)-REE deposit near Kiruna, in northern Sweden functions as a case study that can be transferred to other Fe-P resources.

The rising demand for mineral fertiliser, together with the global shift towards renewable energy, places the PG deposit in a favourable position to support Europe's future supply of these critical elements. Since 2018, PG is a major exploration target for the Swedish mining company LKAB. In 2023, a PERC-CRIRSCO compliant mid-year Mineral Resource Estimation (MRE – a combined indicated- and inferred-resource) of 734 Mt grading 47.3 per cent Fe, and 2.3 per cent and 0.15 per cent apatite-hosted P and Total Rare Earth Oxides (TREO) has been reported and increased to over 1 Bt in 2025. This positions the PG deposit among the largest known resources of REEs in Europe. It's elevated apatite and related REEs content differentiates it from typical IOA ores in northern Sweden.

Understanding how apatite varies across domains is essential for understanding and predicting processing performance, designing beneficiation flow sheets, and constructing robust geometallurgical domains. Consequently, it is essential to characterise the variable apatite textures, chemistry and affiliation with different iron oxides, ore domains and gangue minerals throughout the deposit. This resulted in a detailed apatite mineral characterisation study and forms the key motivation for the drill core selection and sampling workflow presented here. This workflow is transferable to all Fe-P deposits, including IOA-types, and Banded Iron Formation (BIFs).

PART 1 – EVALUATION OF THE MINERAL RESOURCE AND AVAILABLE DATA

A mineral characterisation study builds upon a thorough evaluation of the existing mineral resource and all available supporting data. In case (internally) available, this includes assessing the distribution and grades of the ore minerals within the resource model, as well as practical considerations such as drill core availability and physical access to drill cores for sampling. These factors are project specific and vary depending on the project's state of development.

Available drill core (data) for the Per Geijer case study

For the PG apatite characterisation study, work commenced with the most recent mineral resource estimate (2023 mid-year MRE) available (LKAB public summary technical MRE report; Lauri, Miles and Baker, 2023). This estimation encompassed a total of 72 assayed drill cores. The envisaged sampling campaign could utilise 52 of the 72 drill cores due to ongoing exploration efforts. At the time of sampling, the lateral extent of the PG deposit was well constrained and therefore, the 52 drill cores were considered representative of the entire PG deposit. Altogether, they encompass 6288 bulk assay intervals with an average interval length of 2.46 m (SD=0.78).

The Per Geijer resource model

As of the 2023 mid-year MRE, the PG resource model subdivides mineralisation using Fe grade shells and Fe-oxide mineralogy. Rock containing Fe>15 per cent, is considered ore, whilst everything below, even though minorly mineralised, is described as waste rock. For mineralised areas in the deposit, further iron ore grade shells were constructed and subdivided into low-grade (Fe=15–30 per cent), medium-grade (Fe=30–50 per cent) and high-grade (Fe>50 per cent) shells. Besides the Fe-ore grade shells, the resource and geological model separate the Fe-ore into 3 distinctive domains based on Fe-oxide mineralogy, ie magnetite (Fe_3O_4)-dominated, hematite (Fe_2O_3)-dominated and mixed magnetite-hematite. This is based on the ratio between magnetite and hematite, also named MagDivHem (Table 1). This ratio and the mineral domaining is based on Krolop (2022), and specific ratios and relative abundances of each domain within the 2023 MRE are summarised in Table 1. Whilst the 2023 mid-year MRE did not discriminate the mixed magnetite-hematite domain into a magnetite-rich magnetite-hematite mixed ore and a hematite-rich magnetite-hematite mixed ore domain, this subdivision was considered for this apatite characterisation study as is assumed to have significant influence on downstream processes.

TABLE 1

Iron ore mineral domains of the Per Geijer deposit and their relative abundance based on the MagDivHem (magnetite divided by hematite) ratio determined by Krolop (2022). Mt=Magnetite and Hm=Hematite.

Ore domain	Magnetite-Hematite ratio (MagDivHem)	Relative abundance of Fe in 2023 mid-year PERC MRE in %
Magnetite dominated	Mt/Hm>4 (ie >80% Mt)	76.6
Mixed magnetite-hematite	Mt/Hm=0.25–4	16.8
Magnetite rich	*Mt/Hm=1–4*	*Unspecified*
Hematite rich	*Mt/Hm=0.25–1*	*Unspecified*
Hematite dominated	Mt/Hm<0.25 (ie >80% Hm)	6.6

General apatite content based on assay interval grade data

In the beginning of the apatite characterisation study, the 2023 PG MRE did not include any grade shells or ore domaining for apatite-associated phosphorus and REEs. However, by integrating apatite major element (Ca and P) data from assays with observations from drill core logs, it was possible to make informed interpretations of the general apatite content and distribution. When correlating Ca and P with apatite content, it is important to consider other minerals that may incorporate these elements. Several Ca-bearing minerals (mostly calcite and dolomite) occur frequently and can be present in significant quantities. As a result, Ca is not a reliable indicator of apatite abundance.

Other P-bearing minerals are rare and, when present, occur only in minor abundance as monazite $(REE)PO_4$ and xenotime $Y(PO_4)$. Then, P contributes to just 10–20 per cent of their mass. Therefore, P grades provide a well-supported proxy for apatite content at the assay-interval scale, and any overestimation of apatite caused by the presence of monazite and xenotime is expected to be minimal. Apatite content was thus calculated based on the total P concentrations from assay interval data and assuming ideal (fluor-)apatite stoichiometry. Figure 1 illustrates that the relative amount of low-grade P (P=0–0.2 per cent) assay intervals is most frequent. This means the majority of Fe-oxide assay intervals analysed contain a low apatite content (~<0.2 per cent P). However, the broad spread in P values also demonstrates that intervals with significantly higher apatite content can occur frequently.

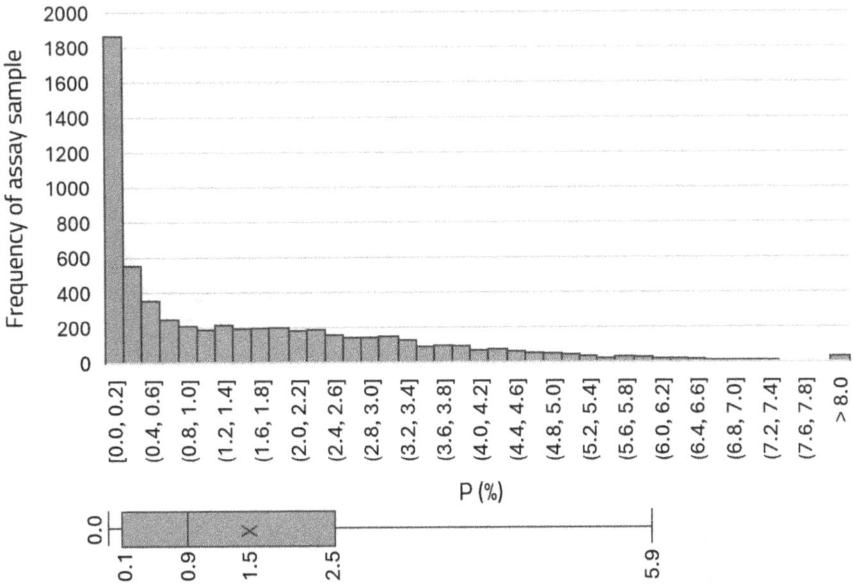

FIG 1 – P-grade distribution of all the available drill core assay data from the 52 drill cores available for the PG deposit apatite characterisation study. The bottom box-and-whisker plot indicates from left to right the lower extreme (0.0), quartile 1 (0.1), median line (0.9), average cross (1.5), quartile 3 (2.5) and the upper extreme (5.9). The average assay interval length is 2.46 m, with SD=0.78. Please note that the average P concentration from the 2023 MRE is higher due to the incorporation of a broader data set, the varying assay interval lengths and modelling effects.

Further analysis of assay interval grade data

Combining P with other element grades from the same assay intervals allows further interpretation of how apatite content relates to Fe-grade shells, ore domains and gangue minerals. A clear understanding and statistical description of these relationships are essential for selecting representative drill core and for sampling apatite-bearing intervals. Specifically, the relationship between Fe-oxides and apatite is critical, as Fe-oxides form the primary product and are the main driver for resource modelling and mine development at PG.

Relation between P grades, Fe grades, ore domains and grade shells

The dominance of Fe-oxides over other Fe-bearing gangue minerals in the PG deposit also allows for more meaningful interpretations of these elemental relationships. Figure 2 presents an overview of the relationship between P and Fe grades (column A) and the distribution of P (column B), for each ore domain (rows 1–4). The histograms and box-and-whisker plots in Figure 2 build-upon Figure 1 and indicate that P grades are lowest in the magnetite- and hematite-dominated ore domains and higher in the mixed ore domains. The magnetite- and hematite-dominated ore domains have a large proportion of low-P (P=0–0.2 per cent) intervals (which also translates to Figure 1), whereas the mixed domains exhibit a broader spread.

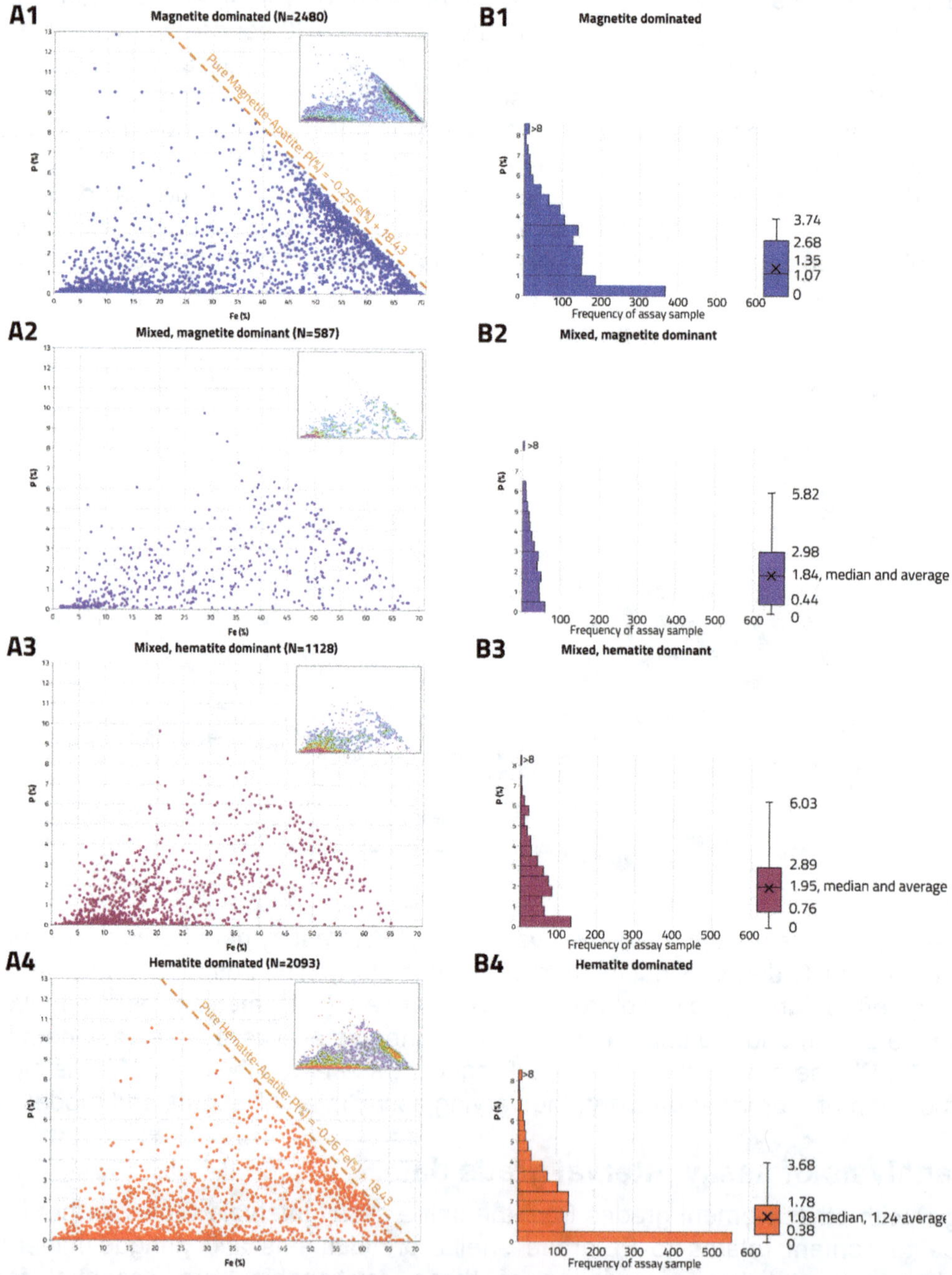

FIG 2 – Breakdown of assay interval P grades, in conjunction with Fe grades (column A) and statistically represented as histograms and box-and-whisker plots (column B), per ore domain (rows 1–4). The insert in column A figures shows a point density plot of the corresponding Fe-P grade plot. The dashed orange lines in A1 and A4 calculated for the theoretical limit of varying Fe-P grades when the assay interval would be fully composed out of magnetite-(fluor)apatite and hematite-(fluor)apatite, respectively. Fluorapatite was specifically used as approximation for apatite based on earlier preliminary studies on PG apatite (eg Krolop, 2022).

Figure 3a summarises column A of Figure 2 by plotting all Fe-P assay interval grades from all ore domains together. From the corresponding point-density overlay plot (Figure 3b) three main clusters can be observed:

1. A cluster of near-zero P grades for Fe grades between ≈0–30 per cent.
2. A cluster of intermediate P grades that forms a decreasing linear trend with increasing Fe grades for Fe grades between ≈50–63 per cent.
3. A near-zero P grade cluster for Fe grades between ≈65–70 per cent.

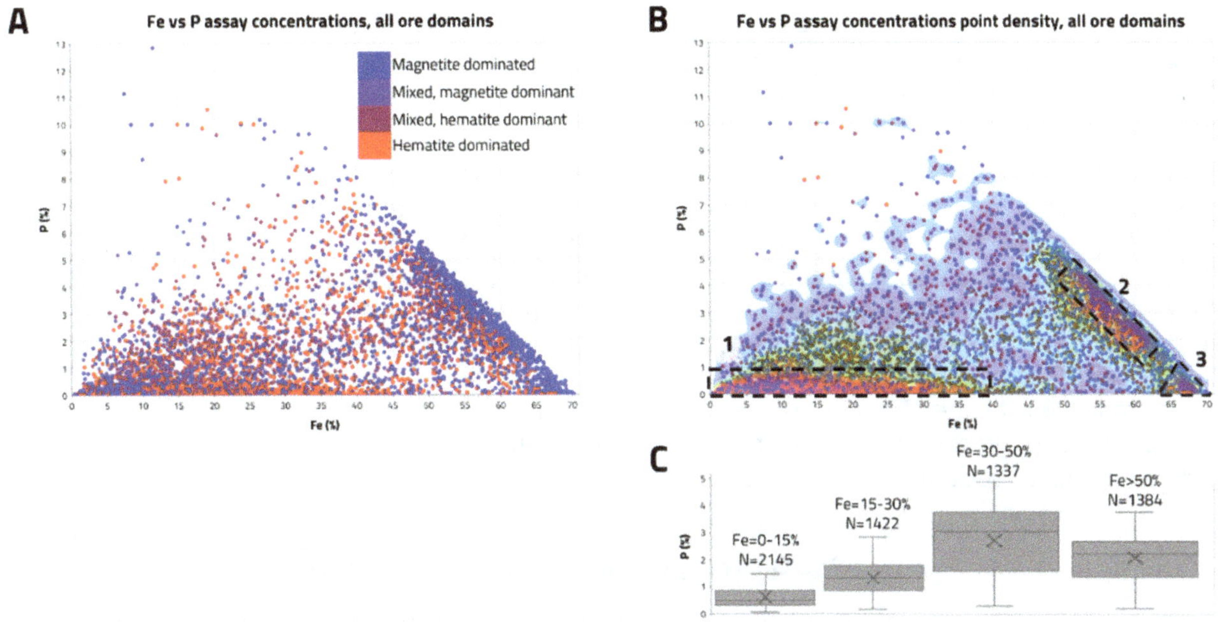

FIG 3 – Summary of Figure 2, with Fe-P concentrations of all assay intervals from all ore domains plotted together in A, the corresponding point density plot in B, and box-and-whisker plot of P concentration per defined Fe grade shell in C.

Cluster one represents the majority of low-P (P=0–0.2 per cent) assay intervals observed in the data set (Figure 1), to which the magnetite- and hematite-dominated ore domains contribute most significantly (Figure 2). These intervals likely contain high proportions of side rock, or a mix of ore and side rock material.

The data forming cluster two, approach a linear trend consistent with an assay interval composed almost entirely of Fe-oxide-apatite, assuming Fe occurs as magnetite and/or hematite and P as fluorapatite. The individual linear functions for magnetite-(fluor)apatite and hematite-(fluor)apatite are indicated by dashed orange lines in Figure 2 A1 and Figure 2 A4, respectively. Data points above these lines are physically impossible, as they would imply a total sum of minerals exceeding 100 per cent. Therefore, all data points plotting close to these trends represent material that is almost fully composed of magnetite-apatite, hematite-apatite or mixtures of these oxides with apatite.

Cluster three represents the Fe-oxide-dominated endmember of cluster two, with near-zero P concentrations and Fe concentrations approaching the theoretical maximum in magnetite (Fe=72.36 per cent) or hematite (Fe=69.94 per cent). These intervals are therefore interpreted as almost pure Fe-oxides. Near-zero P intervals plotting between cluster one and three (with higher Fe concentrations) are interpreted as Fe-oxide-dominated intervals without apatite but containing significant gangue minerals.

Intervals with higher P concentrations that extend from cluster two along the Fe=oxide-apatite line represent similar material to cluster two, but with proportionally more apatite content and lesser Fe-oxides. All remaining assay-interval data reflect variable mixtures of Fe-oxides, apatite and gangue minerals/side rock. Assay intervals with low Fe and high P are effectively absent.

Figure 3c shows that the most sampled intervals fall within the lower Fe grade domains (Fe=0–15 per cent and Fe=15–30 per cent) and that P concentrations increase with increasing Fe concentrations, up to Fe≈50 per cent. The Fe=30–50 per cent range with relatively high P mostly reflects intervals containing significant gangue mineral/side rock component, whereas the Fe>50 per cent range largely comprises intervals of almost pure Fe-oxide-apatite.

Relation between P grade and other gangue minerals

In mineral systems with a homogeneous and limited gangue mineralogy, the major elements can be used to estimate mineral proportions through mass-balancing. It was previously shown that this works well for P for apatite and Fe for magnetite and hematite in the PG deposit as the relative content of P almost exclusively relates to apatite and Fe to the Fe-oxides. However, the gangue mineralogy is complex in the PG deposit and variable in terms of multiple Ca- or Mg-bearing minerals that may occur together within a single assay interval. This makes it difficult to draw detailed conclusions about apatite-gangue mineral relationships using simple element-element analysis.

However, a known P:Ca ratio in ideal (stoichiometric) apatite does, however, allow an estimate of how much of the total Ca in an assay interval relates to apatite versus other Ca-bearing minerals. Figure 4 illustrates this for PG, showing in 25 per cent increment intervals (dashed lines and colour groups) the proportion of Ca attributable to apatite versus other Ca-bearing minerals. It is shown that most apatite and other Ca-bearing minerals occur separately (green and yellow groups), with mix cases occurring less frequently, indicated by the blue and purple groups, respectively. Drill core logging confirms the main Ca-bearing minerals besides apatite are calcite and dolomite. Using their approximate stoichiometric mass proportions of Ca (1:1:2 for apatite:calcite:dolomite), Figure 4 can aid in estimating the content of these minerals in the plotted assay intervals, especially when calcite or dolomite occur separate from each other.

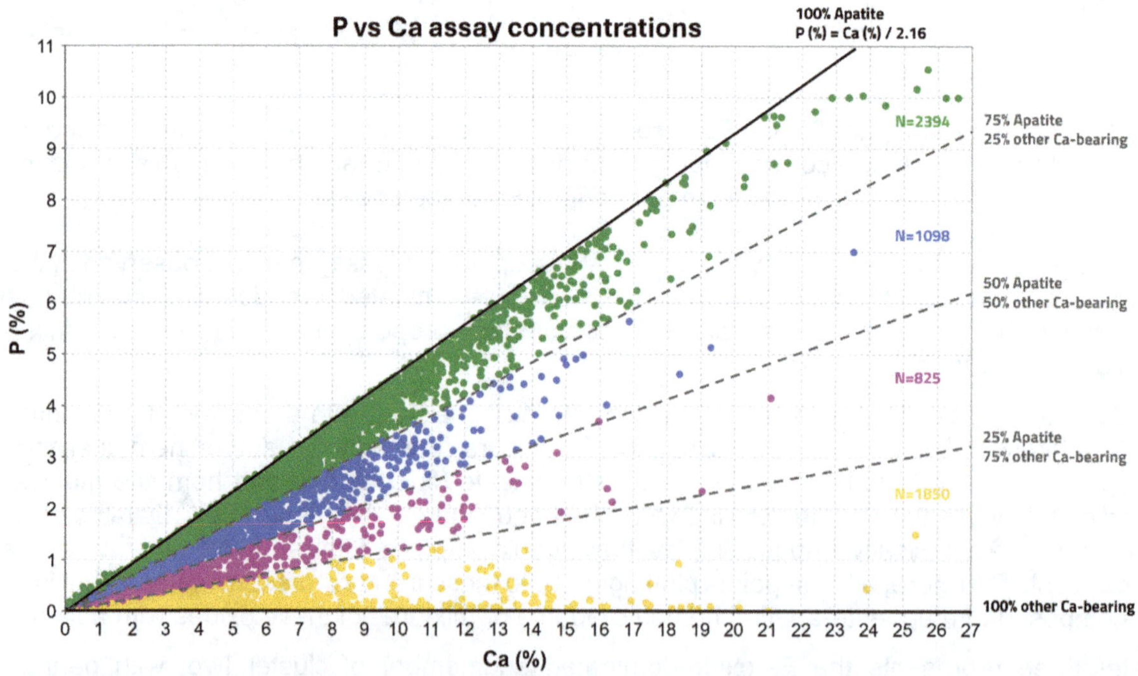

FIG 4 – P versus Ca assay concentrations plot of all 6288 assay intervals. The black line indicates the ideal proportion of P to Ca referring to ideal (stoichiometric) apatite (indicated as 100 per cent apatite). The grey dashed lines and colour groups indicate, in 25 per cent increment intervals, what proportion of the reported Ca assay concentrations belongs to apatite and what proportion to all other Ca-bearing minerals.

These aspects apply well to PG because almost all P in the assay is sourced from apatite, as evidenced by the near absence of points above the 100 per cent apatite line (Figure 4). If additional P-bearing and Ca-bearing minerals are present together, the Ca:P ratio could shift below the

100 per cent apatite line. Given the size of the data set (~6288 assay intervals), this scenario is considered unlikely.

PART 2 – THE DRILL CORE AND SAMPLE SELECTION WORKFLOW

Establishment of drill core and sample selection criteria

After evaluating the mineral resource and available data for drill hole selection, the next step in the proposed drill core and sample selection workflow is to establish clear drill core and sample selection criteria. Although these criteria depend mainly on project-specific objectives and practical constraints, it is important for any deposit-scale mineral characterisation study to define them upfront as a framework for justified drill core selection (where required) and for consistent sampling.

For the PG apatite characterisation study, the project-specific criteria focused on characterising and quantifying the different apatite types (textural and chemical), mineral associations, and textures from micro- to deposit-scale. This information is intended to support future mineral processing and mine planning (eg what can be expected in the processing feed and when) as well as geological investigations into ore-forming processes of apatite in PG. The sampling campaign targeted apatite-bearing drill core intervals for subsequent preparation of thin-, thick and fluid inclusion sections used for (micro-)analytical work. Based on the evaluated PG mineral resource and available data, the sampling procedure focused on the following aspects:

- **The samples should comprise the various apatite textures** observed in the available PG drill core. The main focus is on apatite textures within the Fe>15 per cent ore shell, but intervals within the <15 per cent Fe shell were also considered because they can provide additional geological context, amenability for potential future mining (dilution), and may contain economically meaningful quantities of P (see Figure 3b) and REEs.
- **Each sample should be representative of the texture in the corresponding drill core assay interval.** The aim is to sample apatite textures across the three iron ore mineral domains roughly proportional to the reported PERC-compliant 2023 mid-year MRE tonnages of the different domains (Table 1). A small overrepresentation of samples from the (shallow) hematite dominated assay intervals is preferred, as these zones are likely to be mined first and are currently of particular interest.
- **Most samples should represent apatite textures that occur laterally or spatially across the deposit.** This requires an appropriate drill core logging inspection. Unique apatite textures that may provide additional insights into the genesis of the PG deposit should also be included. Representativeness in terms of apatite content is important for future mineral processing studies, with potential to upscale observations from micro (thin-section) to macro (drill core) scales.
- **As a practical project consideration (time frame, resources, purpose), the target is to select approximately 100–130 samples**, distributed over a total of 10–13 drill cores. Considering point 2 above, this means approximately 77–92, 7–8 and 17–20 magnetite dominated-, hematite dominated- and mixed magnetite-hematite ore samples, respectively (Table 2).

TABLE 2

Breakdown of P grades within and outside the mineralised ore shells (Fe>15 per cent and Fe<15 per cent, respectively), for the 52 initial drill cores.

Concentration (%)	Fe>15	Fe<15	Total
P≤2	37.64	29.56	67.20
P>2	30.81	1.99	32.80
Total	68.44	31.56	100

Drill core selection procedure

The established drill core and sample selection criteria form the framework for guiding the drill core selection procedure. Before starting the drill core selection, it is useful to consider a limitation in the number of drill cores and samples beforehand. This number should offer a reasonable trade-off between practical feasibility and conforming the project specific and subsequent drill core and sample selection criteria. This can be done empirically based on drill core images, logs and the resource model, but can also include a more statistical approach based on the drill core assay data (in case available).

The drill core selection approach for PG includes a statistical and empirical component and criteria to narrow down to 10–13 drill cores in total. A first screening step was a statistical approach based on the available assay interval data. Drill cores were evaluated to determine how many assay intervals conform to the following two filters:

1. >15 per cent Fe grade.
2. P>2 per cent.

The first filter complies with the first drill core and sample selection aim and evaluates the number of assay intervals per drill core that are mineralised. As previously described the mineralised assay intervals contain on average the highest P concentrations, meaning these intervals provided the highest chance to encounter apatite to sample (Table 2). Filter 2 was applied to specifically screen for assay intervals that have the highest chance of containing macroscopically visible apatite. Based on initial comparison between P concentrations with visible apatite, this often turned out to be challenging for P concentrations <2 per cent, hence the set filter at P>2 per cent.

Generally, the more assay intervals in a drill core meeting the selection filters, the more extensive its intersection with mineralisation and the greater the likelihood that it contains abundant and diverse apatite types for sampling. Therefore, the drill cores with the highest number of assay intervals were selected and specifically considered for further evaluation, whilst most drill cores with the lowest number of assay intervals (at least up until the median) were excluded (Figure 5).

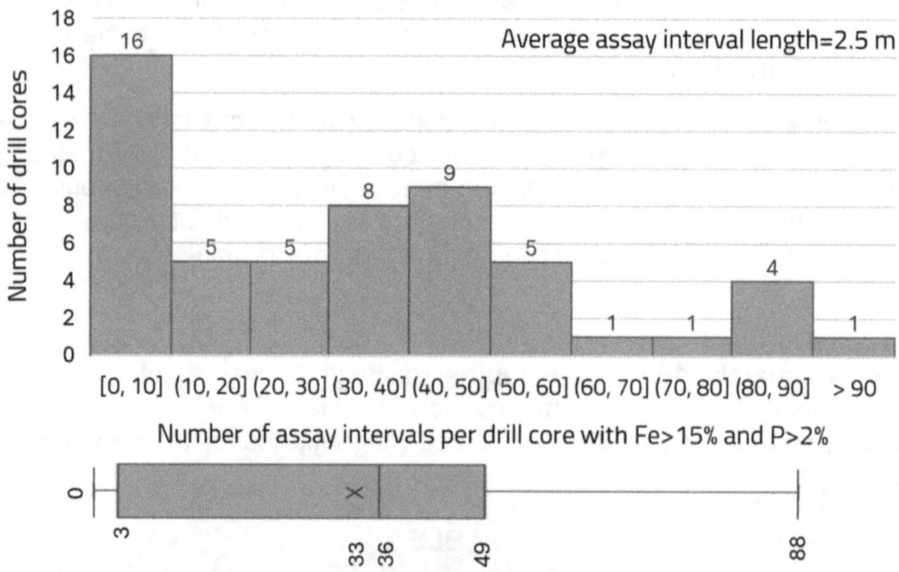

FIG 5 – Distribution of assay intervals per drill core that comply with the drill core selection procedure filters.

The evaluation of the 13 drill cores with the highest number of assay intervals that complied with the set filters, showed that all are located in the centre of the southern part of the deposit. As this region and the drill cores therein comprise a significant portion (roughly 50 per cent) of the PG deposit and because all drill cores intersect with the three different ore domains that include assay intervals with P>2 per cent, five out of the aimed total of 13 drill cores were chosen for further sampling (drill core 1–5, Figure 6). These five drill cores all have >80 assay intervals complying with the two filters (Figure 4). The remaining eight drill cores were reselected for representativeness and to ensure better spatial coverage (aim three). Therefore, drill core 6 (dark blue) and 7 (dark blue) were chosen to specifically target apatite bearing samples associated with the deepest magnetite and shallow hematite in the south of the deposit. That allows for sampling over the full range of depth (Figure 5). Drill cores 8–10 were chosen to represent the middle part of the deposit, drill core 12 and 13 for the northern part and drill core 11 for connecting both parts. In summary, all ore domains are represented by at least 11 drill cores (excluding drill cores 6 and 7), ensuring that apatite-bearing intervals and the full range of representative apatite textures are sampled within each domain. Drill cores further towards the south and north were unavailable for sampling, due to ongoing drilling campaigns at the time.

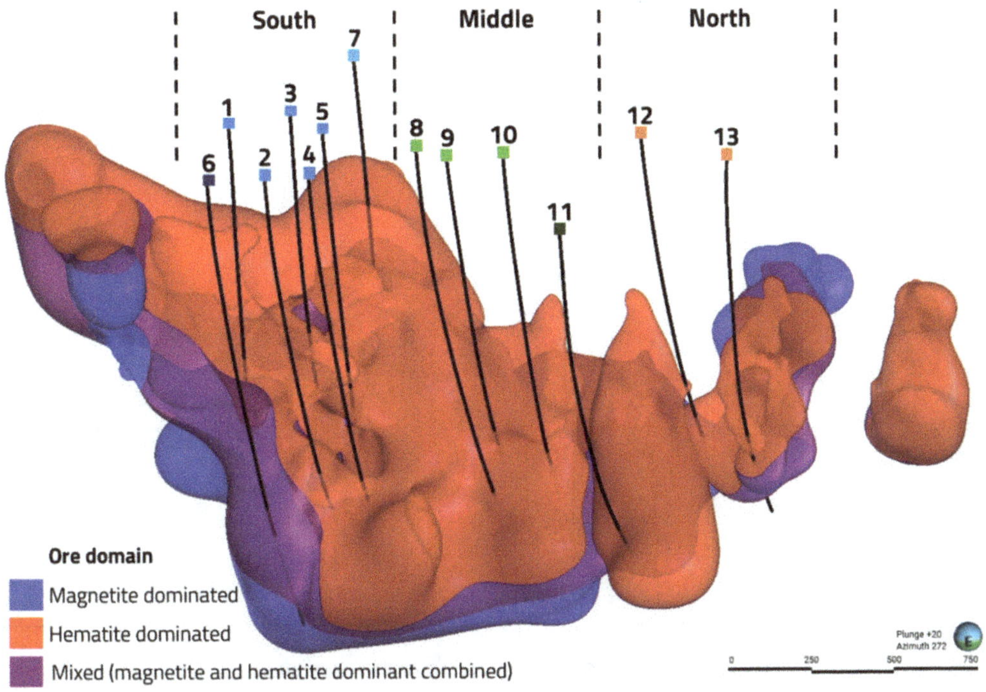

FIG 6 – The 2023 mid-year Per Geijer resource model and final 13 drill cores selected in the drill core and sample selection workflow for the apatite characterisation study. For convenience, the resource model is divided into a southern, middle and northern part, annotated in shades of blue, green and orange colours, respectively (modified after Lauri, Miles and Baker, 2023).

Sample selection procedure

The final step after the drill core selection procedure is the actual sampling of drill core specimens for the mineral characterisation study itself. As for the drill core selection procedure, the sampling procedure depends on the project and corresponding drill core and sample selection criteria. Before sampling begins, it is useful to develop a general understanding of the geological features in drill core. This can (depending on project criteria) include descriptions of stratigraphy, stratigraphic contacts, and knowing where the different ore domains are located. This forms the framework for detailed sampling and can be used as guidance for representative coverage of different lithologies, mineralogy and ore domains. Sampling itself is done hands-on and empirically.

In the case of the PG apatite characterisation study, the targeted drill core intervals were pre-selected after the drill core selection and before the actual sampling. This minimised practical efforts by preventing the movement of unnecessary drill core boxes from the core repository onto the drill core tables. The selected drill core intervals typically started with the first and last assay sample interval that fulfilled the Fe>15 per cent filter. The start and end points are independent of the ore domain and include all drill core intervals between them, including those that did not meet the two drill core selection filters. This approach provides sufficient opportunity to sample intervals with P concentrations ≤2 per cent and ensures that within mineralised (Fe>15 per cent) intervals, the full range of relevant lithological and mineralogical relationships can be continuously observed and documented. In practice, one or two side-rock boxes (approximately 4–10 m) with Fe<15 per cent and/or P<2 per cent were also placed on the logging tables to capture and document important side-rock relationships (Figure 7).

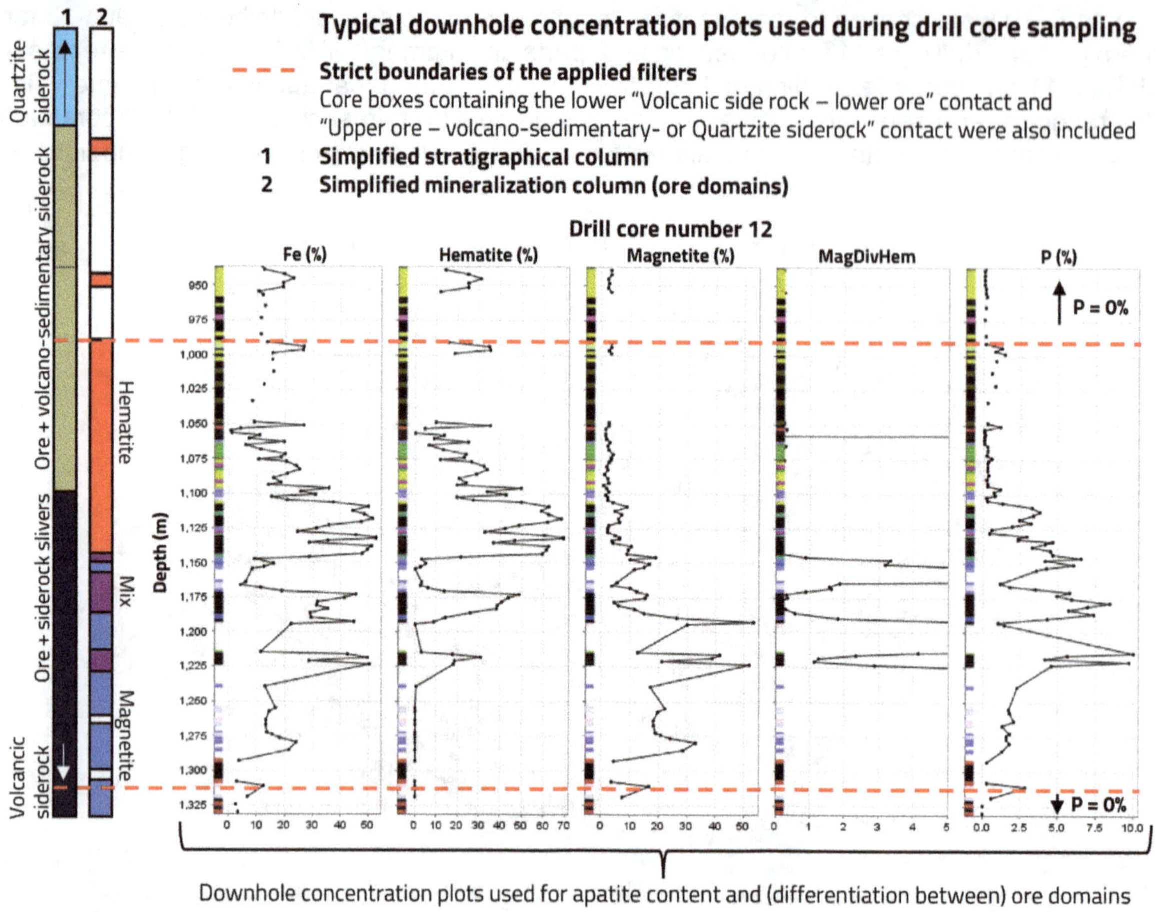

FIG 7 – Representative example (drill core number 12) of a selected drill core interval for sampling (dashed red lines plus side-rock), plotted on the simplified stratigraphic- and mineralisation column and downhole concentration plots. The x-axes of the downhole concentration plots correspond to documented assay concentration tables and MagDivHem calculation, available for the drill core and sample selection procedure.

Again, the established drill core and sample selection criteria (2.1) served as a framework for the actual sampling itself. The (number of) samples taken per drill core depended on the total length of the drill core intervals of interest, the observable different macroscopic apatite types and textures, and the representativeness of the respective types and textures. Downhole concentration plots were used as guidance and to target specific intervals of interest by providing a quantitative overview of the different ore domains and iron oxide and apatite content (Figure 7). Additional concentration and ratio plots were applied to further steer sampling and to indicate side-rock content and type, stratigraphic contacts and alteration. Eventually, 128 apatite-bearing samples were taken from the drill cores to meet the sample selection criteria and prepared as thin sections for further analysis. Sampling was performed on NQ2 (=50.6 mm) half-core, with sample lengths being around 10–20 cm. From the samples standard size thin sections of 26 mm × 46 mm × 30 μm were produced for further analysis.

DISCUSSION

This contribution illustrates an effective selection workflow from drill hole to drill core and specimen sample for mineralogical characterisation of Fe-P deposits with the PG deposit as a case study. Figure 8 summarises and illustrates the workflow in four steps as described before.

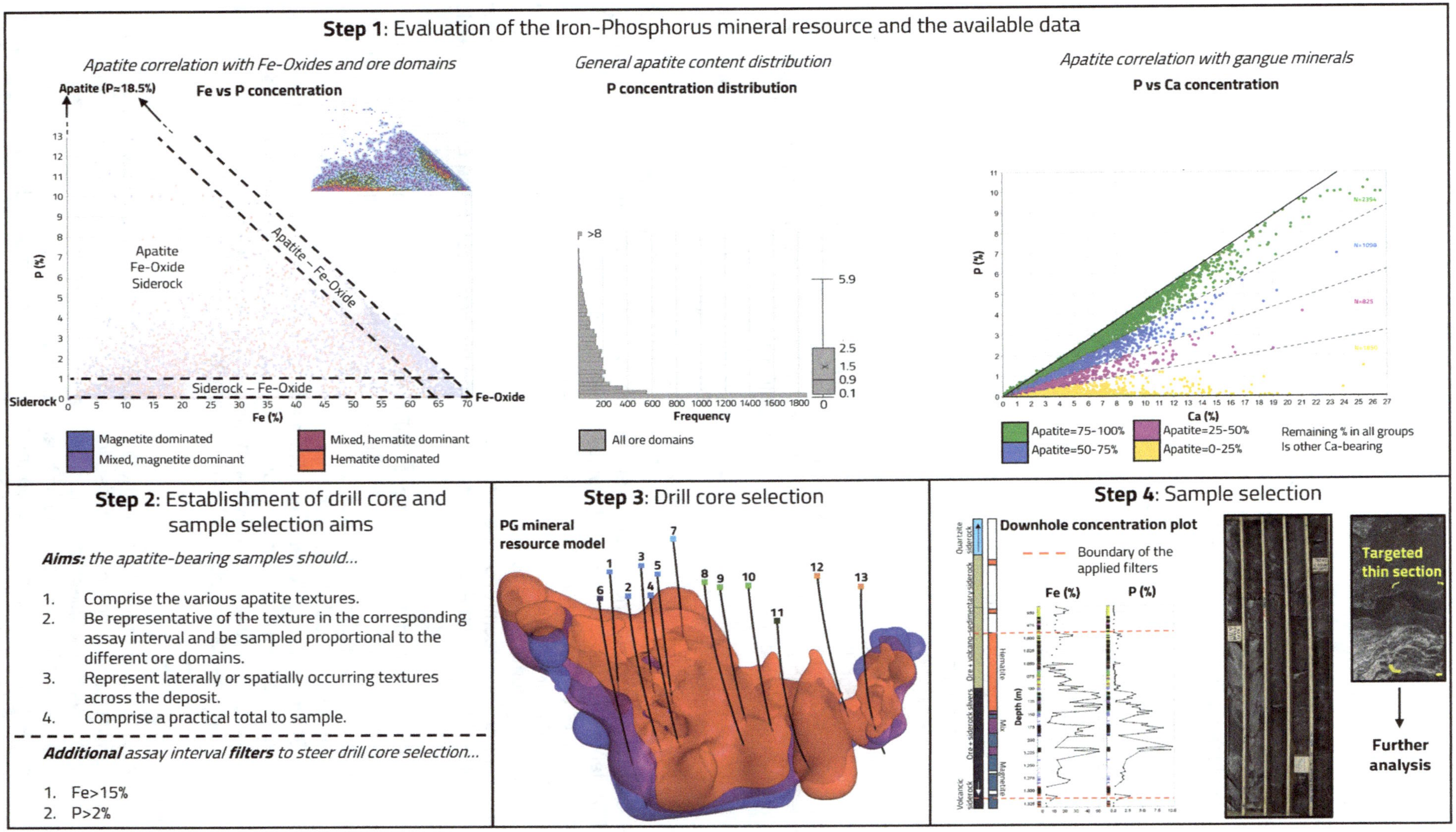

FIG 8 – Summary overview of the applied drill hole, drill core and sample selection procedure, exemplified for the apatite characterisation study in the PG deposit.

With a sustained vital demand for iron ore and an increasing demand for phosphorus and REEs to the global economy, this workflow offers a tangible framework to characterisation studies applicable to other Fe-P deposits and beyond. Based on the commonality of the ore minerals in Fe-P commodities, this workflow is transferable to other iron ore deposits that are considering by-product recovery such as Banded Iron Formations (eg Wells and Ramanaidou, 2011).

The outlined drill core and sample selection workflow in this contribution specifically focuses on apatite and is based on a comprehensive evaluation of the PG mineral resource and available drill core data. Noteworthy, some key conditions (data availability, geological and process-mineralogical knowledge) and assumptions need to be highlighted as they are crucial for a successful application of the workflow and may differ in each project.

- **Conditions**: Before starting the apatite characterisation study, the PG exploration project was already in an advanced state with a substantial amount of validated drill core data available. This included assay data, detailed mineralogical- and lithological drill core logs, an existing and validated mineral resource model with different ore domains and a general idea on the existing macroscopic apatite textures. These formed key conditions for performing an efficient evaluation of the general apatite content, its distribution throughout the deposit and its association with the different ore domains and gangue minerals. This knowledge was key for representative drill core selection and sampling.
- **Assumption**: The key assumption made throughout the evaluation of the PG mineral resource was that all (or at least the majority of) P was apatite hosted. This enabled a direct link between P concentration data from assay intervals to apatite content, and thus to correlate apatite content further eg with Fe-oxide content, the different ore domains and other Ca-bearing (gangue) minerals. For the PG deposit this assumption was carefully validated through analysing the drill core mineralogical and lithological logs, photographs and internal LKAB knowledge. Besides, Figure 4 illustrates that, based on P-Ca concentrations, the presence of significant contents of other P-bearing minerals is unlikely.

In most Iron Oxide-Apatite (IOA) deposits, apatite is the primary and most dominant P (and REE-bearing) phosphate mineral present (Yan and Liu, 2022). This accounts for most major IOA deposits globally, in for example deposits in the Kiruna district in Sweden, Bafq district in Iran, and the Andean mountains in Chile and Peru (eg Yan and Liu, 2022; Reich *et al*, 2022; Meigoon *et al*, 2021; Mirzababaei *et al,* 2021). Monazite and xenotime are the most important other phosphates that might be present in IOA deposits besides apatite, but typically in lower contents and predominantly formed as an alteration product of the primary apatite (Harlov, 2015; Harlov *et al*, 2002; 2016). Thus, the methods applied here, eg linking P assay interval concentrations to apatite and correlating this with for example Fe and Ca domains, is considered highly transferable to other IOA deposits. A straightforward P versus Fe concentration plot could even directly be used to discriminate between apatite, Fe-oxide, and side-rock content, as illustrated by the dashed areas in the P versus Fe plot in Figure 7.

Like IOAs, the bulk of P in Banded Iron Formations (BIFs) is mainly occurring in apatite (Wells and Ramanaidou, 2011; Trendall and Blockley, 1970). Phosphorus in BIFs might also occur within other P-rich phases (eg monazite, xenotime, vivianite, strengite and wavellite) or as iron-oxide bound P (mostly P-rich goethite), but in much lower contents compared to apatite (eg Bekker *et al*, 2010; Smith, Beukes and Gutzmer, 2013; Wells and Ramanaidou, 2011). Despite P in BIFs being considered as deleterious element and overall P concentrations being much lower compared to IOAs (typically <0.5 wt per cent), the illustrated workflow in this contribution might, for the same reasons as described for IOAs, also be transferable for apatite characterisation in BIFs.

CONCLUSIONS

This contribution presents a comprehensive drill hole and sample selection workflow that is used as a fundamental framework for orebody characterisation, with the PG Fe-P deposit used as a case study. The workflow follows four efficient and practical steps that incorporates orebody evaluation, establishment of project-specific criteria, and the actual drill core and sample selection itself (Figure 7). Through Fe versus P concentration plots, histograms, box-and-whisker plots and sophisticated attributing of assay interval data, general relations between Fe-ore, ore domains and apatite can be efficiently uncovered (Figures 2 and 3). A Ca versus P concentration plot can be utilised to estimate proportions between apatite and other Ca-bearing minerals, which in case of the PG deposit forms a group of important gangue minerals (Figure 4). A thorough understanding of the above-described relations can, in turn, aid in developing specific criteria of and for sampling drill core specimens for subsequent analyses (steps two to four). The fundamental evaluation of the mineral resource (step 1) is considered highly transferable to any other Fe-P deposit, when a comparable amount of assay interval data is available, and that P concentrations can be directly linked to apatite content.

ACKNOWLEDGEMENTS

We thank LKAB's field exploration geology team for their help in the drill core repository.

REFERENCES

Bekker, A, Slack, J F, Planavsky, N, Krapez, B, Hofmann, A, Konhauser, K O and Rouxel, O J, 2010. Iron formation: the sedimentary product of a complex interplay among mantle, tectonic, oceanic and biospheric processes, *Economic Geology*, 105(3):467–508.

Harlov, D E, 2015. Apatite: A fingerprint for metasomatic processes, *Elements*, 11(3):171–176.

Harlov, D E, Andersson, U B, Förster, H J, Nyström, J O, Dulski, P and Broman, C, 2002. Apatite–monazite relations in the Kiirunavaara magnetite–apatite ore, northern Sweden, *Chemical Geology*, 191(1–3):47–72.

Harlov, D E, Meighan, C J, Kerr, I D and Samson, I M, 2016. Mineralogy, chemistry and fluid-aided evolution of the Pea Ridge Fe oxide-(Y+ REE) deposit, southeast Missouri, USA, *Economic Geology*, 111(8):1963–1984.

Krolop, P, 2022. The Per Geijer iron ore deposits: Characterization based on mineralogical, geochemical and process mineralogical methods, Doctoral thesis, TU Bergakademie Freiberg.

Lauri, L, Miles, A-J and Baker, H, 2023. A summary technical report on the mineral resources of LKAB, Sweden – Per Geijer Iron Ore Apatite deposit, Sweden, LKAB, June 2023.

Meigoon, M S, Lotfi, M, Afzal, P, Nezafati, N and Razi, M K, 2021. Detection of rare earth element anomalies in Esfordi phosphate deposit of Central Iran, using geostatistical-fractal simulation, *Geopersia*, 11:115–130.

Mirzababaei, G, Yazdi, M, Behzadi, M and Reza Rezvanianzadeh, M, 2021. REE-Th mineralization in the Se-Chahun magnetite-apatite ore deposit, central Iran: interplay of magmatic and metasomatic processes, *Ore Geology Reviews*, 139:104426. https://doi.org/10.1016/j.oregeorev.2021.104426.

Reich, M, Simon, A C, Barra, F, Palma, G, Hou, T and Bilenker, L D, 2022. Formation of iron oxide–apatite deposits, *Nature Reviews Earth and Environment*, 3(11):758–775.

Smith, A J, Beukes, N J and Gutzmer, J, 2013. The composition and depositional environments of Mesoarchean iron formations of the West Rand Group of the Witwatersrand Supergroup, South Africa, *Economic Geology*, 108(1):111–134.

Trendall, A F and Blockley, J G, 1970. The Iron-Formations of the Precambrian Hamersley Group, Western Australia, with Special Reference to the Associated Crocidolite (Bulletin 119), Geological Survey of Western Australia.

Wells, M A and Ramanaidou, E R, 2011. Occurrence and mineralogical association of phosphorus in Australian bedded iron ore deposits, in *Proceedings of the Iron Ore Conference*, pp 11–13 (The Australasian Institute of Mining and Metallurgy: Melbourne).

Yan, S and Liu, W, 2022. Rare earth elements in the iron-oxide apatite (IOA) deposit: Insights from apatite, *International Geology Review*, 64(22):3230–3247.

Open Pit

AI, automation and autonomous systems

AI-powered mine reconciliation – enhancing operational efficiency

S Bream[1]

1. Principal Geologist, AMC Consultants, Perth WA 6000. Email: sbream@amcconsultants.com

ABSTRACT

Mine reconciliation is a critical process for aligning predicted Mineral Resource Estimates or Ore Reserve estimates, as defined under JORC (The Joint Ore Reserves Committee of The Australasian Institute of Mining and Metallurgy, Australian Institute of Geoscientists and Minerals Council of Australia), with actual production outcomes across the mining value chain. It provides quantitative measures of operational performance, identifies sources of ore loss and dilution, and underpins confidence in mine planning, reporting, and decision-making. However, reconciliation workflows are traditionally spreadsheet-based and delivered on month-end cycles, constraining operational teams' ability to respond to emerging issues within relevant operational time frames. This paper compares how an AI assistant processes and reports F2 reconciliation data relative to conventional, human-driven methodologies. It presents an AI-assisted reconciliation workflow focused on F-style factors at the mine interface, including F2, and demonstrates how increasing reporting cadence from monthly to weekly and daily intervals improves operational agility without compromising governance, accountability, or auditability.

A de-identified open pit copper data set was used to demonstrate the workflow, comprising daily records of mined tonnes and grade compared with modelled values for sulfide and marginal material. Results show that short-cadence reporting reveals trends and deviations that may be masked by month-end averages, enabling earlier investigation and intervention. The workflow provides a practical blueprint for implementing AI-assisted mine reconciliation within existing assurance frameworks while supporting future scalability.

AI-assisted F2 reconciliation represents a step change in how mining operations validate data, maintain resource accountability, and translate reconciliation outcomes into operational action. When reconciliation calculations remain deterministic, governed, and auditable, AI can be safely applied to high-effort tasks such as data handling, validation support, variance triage, and report generation within a structured human-in-the-loop framework. In the case study, the AI assistant completed the F2 reconciliation in approximately ten minutes, compared with two hours using a traditional manual workflow, producing clear, decision-oriented outputs that highlighted key variances. Although human oversight continued to play a crucial role, this method significantly boosted productivity by lessening manual work and shortening reporting delays. As a result, professionals could dedicate their attention to identifying root causes and implementing corrective measures, which improved operational efficiency. With appropriate data pipelines, uncertainty frameworks, and governance controls, AI-assisted F2 reconciliation offers a scalable bridge between spreadsheet-based methods and near-real-time operational control while preserving technical rigour.

INTRODUCTION

Reconciliation systematically compares ore estimates and measurements at various stages of the mining value chain to assess how accurately planned tonnes and grades are converted into actual product. Mine reconciliation aligns Mineral Resource or Ore Reserve estimates (as per JORC) with real production results, offering quantitative performance insights, identifying ore loss and dilution, and supporting mine planning, reporting, and decisions. Traditional reconciliation relies on monthly spreadsheet workflows, limiting timely operational responses. It delivers both operational feedback on variance sources and corporate assurance regarding forecast reliability. The widely used F-series factors, as shown in Figure 1 was formalised by Parker (2011), provide ratio-based comparisons for quick variance detection and communication.

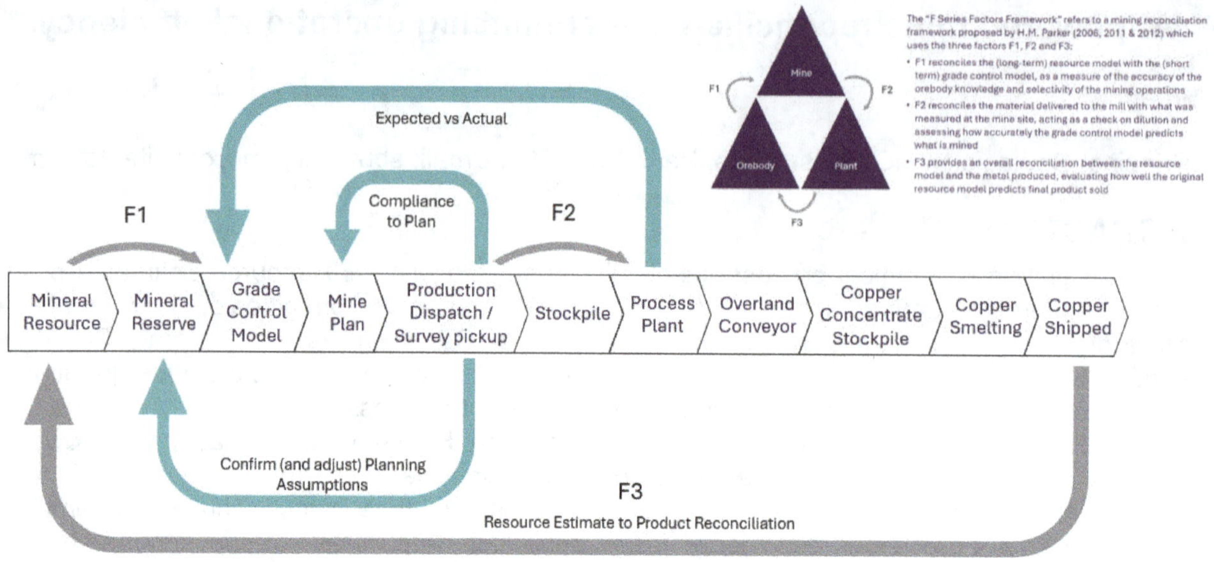

FIG 1 – AMC Reconciliation value stream map.

Despite its importance, many operations still perform reconciliation using manual data collation and spreadsheet workflows, typically reported at month end. Month-end reporting aligns to financial close but often mismatches operational decision cycles (daily/weekly), making it harder to respond to emerging variance drivers (misclassification, dilution, ore loss, sampling bias, grade control issues) before they accumulate. In addition, manual workflows tend to embed fragile logic, inconsistent definitions and single-person dependency issues repeatedly identified in reconciliation codes of practice and guidance papers.

Recent advances in data engineering, machine learning and generative AI create an opportunity to move reconciliation reporting closer to short-term operational time frames, provided governance and auditability are preserved.

This paper provides an example of an AI-assisted reconciliation workflow designed to:

- Increase reporting cadence (daily, weekly, monthly).
- Reduce manual effort through automation and repeatable analytics.
- Preserve accountability via deterministic calculations and human-in-the-loop controls for AI-generated content.

RECONCILIATION FRAMEWORK AND SCOPE (F2 FOCUS)

Reconciliation as a value-chain control system

Reconciliation is most effective when implemented as a value-chain control system, where each handover point is clearly defined (what is measured, by whom, in what units, and on what time basis). Macfarlane (2015) describes reconciliation and metal accounting as a governance requirement that depends on clearly understood control points, variability and loss mechanisms across the value chain.

A common failure mode is unclear ownership and inconsistent definitions across geology, mining, survey and processing, which results in repeated debate about 'whose number is right' rather than rapid resolution of 'why numbers differ' and 'what to do next'. Establishing a reconciliation code of practice helps standardise definitions, responsibilities and calculations so reconciliation becomes business-as-usual rather than a bespoke spreadsheet exercise.

Factor-based reconciliation (F-series)

Factor-based reconciliation expresses comparisons between estimate/measurement pairs as ratios. Parker (2011) framework describes how factors can be used to isolate where variance is introduced and to quantify the impact of orebody knowledge, selectivity, dilution and ore-loss.

In the example presented here, the data set supports an F2 comparison at the mine interface using paired modelled and mined values (tonnes, grade and contained metal) for defined material classes. While the classic Mine to Mill F3 often incorporates stockpile movement and plant-measured feed, the approach remains consistent in principle: a controlled comparison of two independent (or semi-independent) sources with clear boundaries and validation rules.

DATA AND DETERMINISTIC METHODOLOGY

Data set (example)

The example data set, shown in Table 1 comprises daily records from an open pit copper operation, including a bench identifier and the following fields for two material classes (sulfide ore and marginal ore):

- mined volume, tonnes and Cu (%)
- corresponding 'block model' volume, tonnes and Cu (%)
- pre-calculated reconciliation differences (block model−mined) for tonnes and grade.

TABLE 1

An example F2 data set used in the case study.

		Mined Sulfide Ore			Block Model Sulfide Ore			Mined Marginal Ore			Block Model Marginal Ore		
Date	Bench	Volume (bcm)	Tonnes (Mt)	Cu (%)	Volume (bcm)	Tonnes (Mt)	Cu (%)	Volume (bcm)	Tonnes (Mt)	Cu (%)	Volume (bcm)	Tonnes (Mt)	Cu (%)
1/09/24	1173	4037	11 183	0.83	1000	2763	1.67	70 282	194 682	0.3	1125	3108	0.34
2/09/24	1163	67 242	186 260	1.01	68 000	187 878	1.19	165 223	457 668	0.39	37 750	104 272	0.37
3/09/24	1153	30 768	85 227	0.94	76 875	212 391	1.21	114 421	316 946	0.31	49 375	136 385	0.38
4/09/24	1143	31 473	87 181	0.77	89 625	247 609	1.18	169 011	468 160	0.35	72 375	199 925	0.38

Deterministic calculations (audit-friendly)

Deterministic reconciliation replaces manual spreadsheet interpretation with fixed, repeatable rules for calculating metal, aggregating periods, and computing F2 ratios—reducing errors, improving auditability, and enabling daily/weekly reporting at operational speed.

All reconciled values are calculated deterministically from source fields. For each day and reporting period:

- **Contained copper (metal)**

  $$M_{\mathrm{Cu}} = T \times \frac{G}{100}$$ where (T) is tonnes and (G) is grade in per cent.

- **Period aggregation**

  Tonnes and metal are summed across days; period grade is computed as tonnage-weighted grade: $$G_{\mathrm{period}} = \frac{\sum M}{\sum T}\times 100$$.

- **F2-style factors**

  $$F2_{\mathrm{tonnes}}=\frac{T_{\mathrm{model}}}{T_{\mathrm{mined}}},\quad F2_{\mathrm{grade}}=\frac{G_{\mathrm{model}}}{G_{\mathrm{mined}}},\quad F2_{\mathrm{metal}}=\frac{M_{\mathrm{model}}}{M_{\mathrm{mined}}}$$
  calculated separately for sulfide, marginal and total ore (sulfide + marginal).

Deterministic validation checks

To support operational trust and auditability, the workflow applies explicit validation rules, including:

- missing-day detection (gaps in time series)
- non-physical values (negative tonnes, zero tonnes with non-zero metal)
- grade bounds checks (eg 0–100 per cent)
- reconciliation closure checks where pre-calculated recon columns exist (confirming they equal model−mined).

These checks are executed before any AI assistance is applied, ensuring AI does not 'explain' unreliable data.

AI-ASSISTED ANALYTICS AND REPORTING WORKFLOW

Design principle – AI supports insight, not numbers

- To preserve governance and auditability, the workflow confines AI to the 'insight layer' only.
- All reconciled numbers (tonnes, grade, metal, factors) are computed deterministically.
- AI is used for:
  - variance triage (what to look at first)
  - anomaly detection (what is out-of-pattern)
  - narrative drafting (consistent operational commentary)
  - summarisation for different cadences (daily/weekly/monthly).

Anomaly detection (example approach)

The daily F2 time series exhibits natural variability due to operational sequencing, selective mining, sampling variance, and data capture error. The objective is not to eliminate variability but to rapidly identify days that are out of pattern and warrant investigation.

A practical and robust approach uses control-limit logic based on robust statistics (median ± 3 × MAD-derived sigma). Days outside control limits are flagged for review and linked to the underlying components (tonnes and grade by material class, bench, or another operational dimension). This provides a 'short list' for daily/weekly review rather than requiring manual scanning of the full data set.

Narrative generation with human-in-the-loop controls

Generative AI is used to draft report commentary based solely on computed metrics (eg 'F2 tonnes below target driven by sulfide tonnes under-call'). Controls include:

- fixed reporting template (headings, tone, scope)
- no external data ingestion (only computed metrics and approved metadata)
- mandatory human review and approval before release
- retained traceability (commentary references the tables/figures used).

These controls reduce the risk of hallucination or over-interpretation and ensure accountability remains with the reconciliation owner and functional leads.

Reporting products and cadence

The workflow produces three report types:

1. Daily report:
   - prior-day F2 tonnes/grade/metal, by class (sulfide/marginal) and total

- anomaly flags and validation issues
- 'top drivers' list (largest variance contributors).

2. Weekly operational summary report:
   - trend charts and rolling means
   - list of out-of-control days and associated drivers
   - actions taken and status (decision log).
3. Monthly reconciliation report:
   - period factor summary table (tonnes, grade, metal)
   - consolidated commentary and recommendations
   - governance signoffs and version control.

CASE STUDY

This case study applies an F2 reconciliation of a de-identified open pit copper data set that had been formatted to allow AI analysis, spanning September 2024 to June 2025. The data set contains daily paired records of mined versus block model values for two ore categories (sulfide and marginal), with fields for volume, tonnage, and Cu grade (%), enabling direct variance assessment by material type.

Two reconciliation workflows were executed using the same input data set and calculation logic. First, F2 reconciliation was completed using a traditional human methodology (spreadsheet-based collation, checking, calculation, and preparation of deliverables). Second, the identical data set was processed through an AI-assisted workflow. The resulting reconciliation outputs were compared against the human-prepared results to assess consistency of calculated outcomes and differences in reporting efficiency and presentation.

Throughout this section, variance = mined–model estimated. Negative variance therefore indicates a shortfall/under-mining relative to the block model expectation (for volume and tonnage), or a lower mined grade than predicted (for Cu grade). This convention is applied consistently across sulfide and marginal material streams.

AI F2 reconciliation results

AI sulfide ore reconciliation

Volume variance (sulfide)

Sulfide volume reconciliation indicates a persistent negative bias. The average variance is -92 035 blocks, with a range from -339 602 to 167 913 blocks and a standard deviation of 98 740 blocks.

Interpretation: sulfide ore volumes reported as mined are consistently below the block model expectation, indicating systematic under-recovery, boundary definition effects, or operational loss/dilution mechanisms acting at the mine interface.

Tonnage variance (sulfide)

Sulfide tonnage shows the same directional bias, with an average variance of -106 262 t, ranging from -543 480 to 134 300 t, and a standard deviation of 108 431 t.

Interpretation: the sulfide stream exhibits a sustained tonnage shortfall relative to modelled expectations, consistent with the negative volume variance and suggestive of systematic drivers rather than random noise.

Copper grade variance (sulfide)

Sulfide grade variance is predominantly negative, with an average variance of -0.546588 per cent Cu, a range of -1.426830 per cent to 0.505417 per cent Cu, and a standard deviation of 0.246899 per cent Cu.

Interpretation: mined sulfide grades are generally lower than predicted by the block model, reinforcing the conclusion that sulfide reconciliation issues are not limited to mass recovery alone but extend to grade performance.

AI results – marginal ore reconciliation

Volume variance (marginal)

Marginal ore shows a smaller average bias but increased dispersion: average variance -14 188 blocks, with a range of -357 743 to 254 194 blocks and standard deviation of 111 230 blocks.

Interpretation: marginal ore exhibits lower systematic bias than sulfide, but much higher variability, indicating less consistent performance and greater sensitivity to local conditions (eg boundary placement, classification, or mining selectivity).

Tonnage variance (marginal)

Marginal tonnage variance is comparatively balanced on average (-16 773 t) but highly variable, ranging from -503 832 to 327 520 t, with a standard deviation of 130 662 t.

Interpretation: while the mean bias is modest, the variability suggests intermittent over- and under-recovery that could be driven by localised ore/waste contacts, short-range geological variability, or operational decisions affecting material routing and classification.

Copper grade variance (marginal)

In contrast to sulfide, marginal grade variance is positive on average: +0.255764 per cent Cu, ranging from -0.802606 per cent to 1.097647 per cent Cu, with standard deviation 0.245258 per cent Cu.

Interpretation: marginal material tends to reconcile at higher grades than predicted, indicating a directional grade bias opposite to sulfide and strongly suggesting a classification boundary or modelling bias between material types.

Comparative interpretation (sulfide versus marginal)

Figures 2, 3 and 4 display monthly AI-generated F2 variance charts, along with monthly trends in sulfide and marginal overtime. Over the course of the study, four primary reconciliation behaviours are observed:

1. **Systematic sulfide under-performance:** Sulfide exhibits consistent negative variance in volume, tonnage, and grade, indicating a persistent shortfall relative to the block model.
2. **Marginal material is less biased but more volatile:** Marginal ore shows smaller mean bias, but substantially greater scatter in both mass and grade outcomes.
3. **Opposing grade behaviour:** Sulfide grades reconcile below the model on average, whereas marginal grades reconcile above the model on average, implying a likely classification boundary effect (eg sulfide ↔ marginal partitioning) and/or estimation bias at the domain interface.
4. **Variability contrast:** Marginal ore demonstrates greater overall variability but reduced systematic bias compared to sulfide. Reported coefficients of variation highlight this contrast, with sulfide volume variance materially steadier than marginal volume variance (107.3 per cent CoV versus 784.0 per cent CoV, respectively).

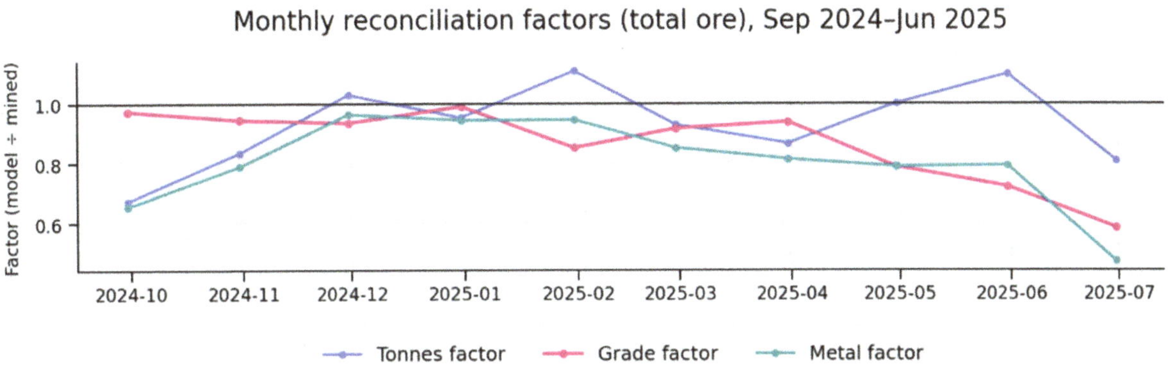

FIG 2 – Monthly grade factors by total ore (model ÷ mined), Sep 2024 – Jun 2025.

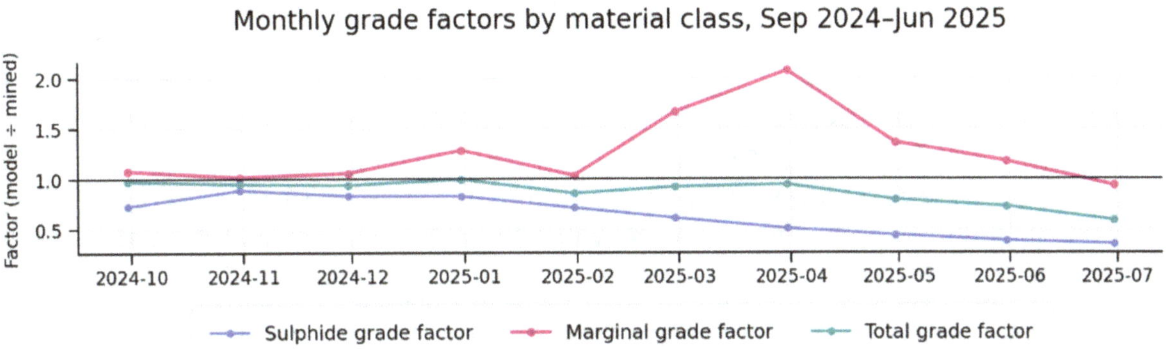

FIG 3 – Monthly grade factors by material class (model ÷ mined), Sep 2024 – Jun 2025.

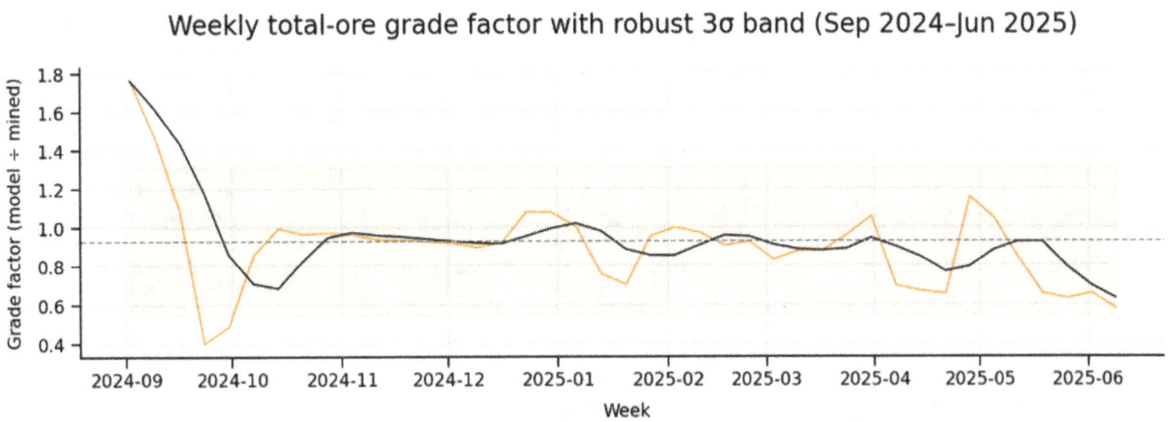

FIG 4 – Weekly total-ore grade factor with robust 3σ band for anomaly flagging, Sep 2024 – Jun 2025.

Bench-level findings (spatial/vertical performance)

Bench group performance

When grouped by bench ranges, reconciliation performance varies systematically with depth:

- **Mid-High benches (783–973): Best performer** – sulfide volume variance improves to approximately **-38 400 blocks** (vs **-92 000** overall), indicating the most balanced reconciliation response.
- **Low-Mid benches (393–583): Worst performer** – strongest sulfide under-mining (approximately **-122 011 blocks**) and concurrent marginal over-mining, consistent with boundary definition challenges.
- **Mid benches (593–773): Persistent under-mining** – sulfide under-mining remains elevated (approximately **-114 213 blocks**) with high variability.

- **Lowest benches (193–383): Near-average behaviour** – consistent sulfide under-mining broadly aligned with the overall data set mean.
- **Highest benches (983–1173): Moderate volume performance but large tonnage deficit** – tonnage variance reaches approximately **-193 828 t**, indicating additional deep-zone complexity or selectivity constraints.

Critical patterns and operational implications

As illustrated in Figure 5, several distinct patterns can be observed from the bench analysis:

- **Bench-height effect:** Performance improves at depth, plausibly reflecting stronger geological definition and/or more stable operating conditions, while shallow benches show increased susceptibility to ore loss/dilution and deeper benches show selectivity constraints.
- **Sulfide under-mining persists at all benches:** Although the magnitude varies by bench band, the bias remains negative throughout.
- **Marginal ore instability:** Marginal performance is comparatively unpredictable across benches; the occurrence of over-mining in low-mid zones supports the hypothesis of boundary placement or classification effects.
- **Temporal stability with minor seasonal modulation:** Variance signatures remain broadly stable through time, with some seasonal variation but no evidence that the bias reverses or resolves naturally.

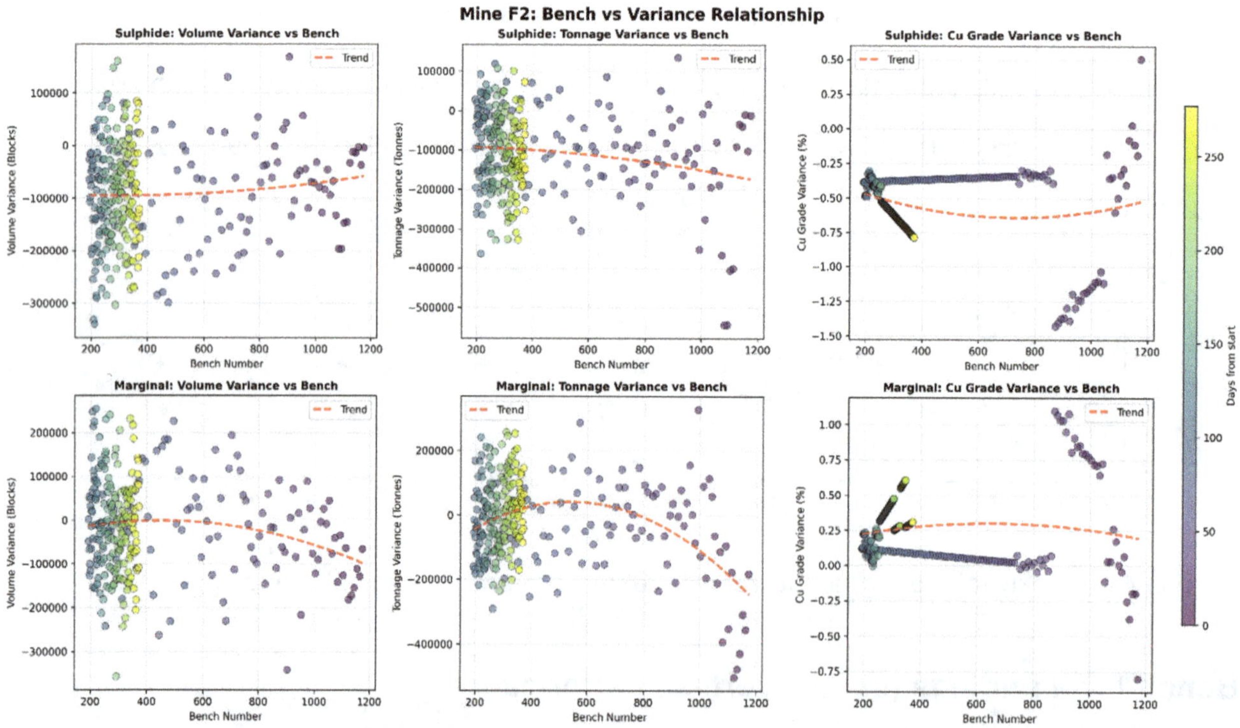

FIG 5 – Bench verses F2 reconciliation variance scatter plot analysis.

AI-assisted recommendations (actionable follow-up)

Based on the reconciliation data observed, the AI-assisted workflow prioritised the following actions:

- **Review sulfide domain assumptions in the block model:** Persistent negative sulfide reconciliation across mass and grade suggests a systematic modelling and/or domain definition issue.
- **Investigate grade divergence across ore types:** The opposing sulfide (negative) versus marginal (positive) grade variances warrant targeted review of the sulfide–marginal classification boundary and estimation controls at domain contacts.

- **Interrogate spatial/temporal drivers:** Assess whether specific panels, pits, or time windows disproportionately contribute to the variance (eg changes in dig limits, blast movement, sampling representativity, or haulage/routing rules).
- **Improve marginal tonnage predictability:** High variability in marginal tonnage indicates uncertainty that may be reduced through refined domain modelling, tighter classification controls, or improved operational reconciliation at the ore–waste boundary.
- **Implement focused grade QA/QC:** Because grade behaviour diverges by ore type, increased emphasis on grade control and domain validation (particularly around the marginal–sulfide boundary) is recommended.

DISCUSSIONS

What AI changed (and what it did not)

The primary benefit of AI assistance in reconciliation is speed-to-insight and consistency of communication, rather than changing the mathematics of reconciliation. Deterministic calculations and validation rules preserve auditability; AI reduces the time spent scanning data and drafting commentary, enabling analysts and operational teams to spend more time investigating causes and implementing actions. This supports the continuous improvement of reconciliation systems.

Example daily and weekly reporting use

Daily flash reporting, as shown in Figure 6 highlights out-of-pattern days for immediate triage, while weekly summaries smooth operational noise and isolate sustained variance. By linking flagged periods to benches and material classes, teams can prioritise targeted investigation of dilution, ore loss and grade-control predictability before month-end close.

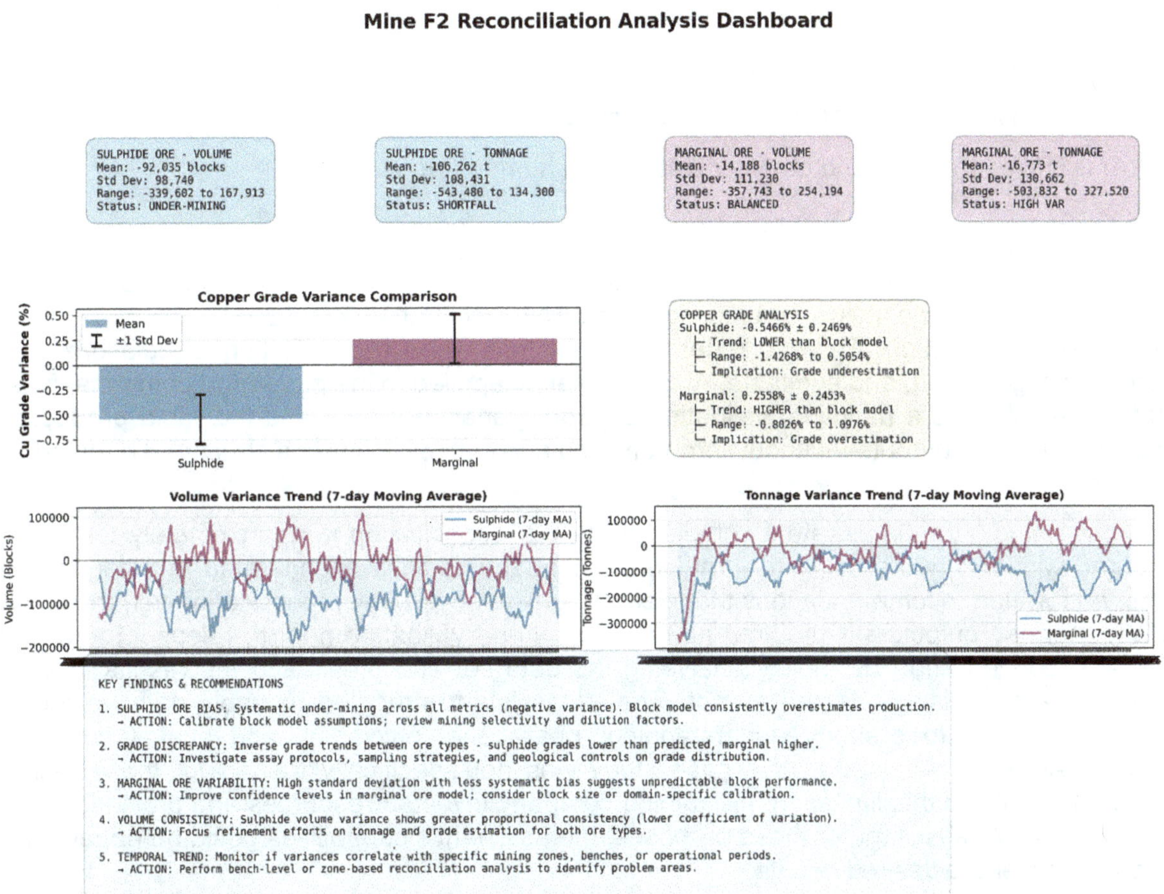

FIG 6 – AI reconciliation report dashboard, showing daily to weekly results.

Governance and risk controls

Data security is a material consideration when using an AI assistant for reconciliation reporting because reconciliation data sets can contain commercially sensitive information (eg production rates, grade distributions, ore-loss/dilution drivers, reserve performance indicators, contractor productivity, and forward-looking planning assumptions). If prompts, attachments, or outputs are processed in unmanaged or consumer-grade AI services, there is a risk of unauthorized disclosure, data residency non-compliance, over-retention, or secondary use (eg logs retained beyond required periods), as well as operational risks such as prompt injection (where malicious or contaminated inputs cause the assistant to reveal restricted information) and inadvertent leakage through copied narrative text. To mitigate these risks, AI use should be confined to an approved enterprise environment with tenant isolation, role-based access control, encryption in transit and at rest, controlled connectors to source systems, explicit data classification rules, and audit logging of user queries and outputs; additionally, outputs should be treated as controlled technical records with defined storage locations and retention policies.

Reconciliation is often undermined by weak governance—unclear ownership, inconsistent definitions, and uncontrolled spreadsheets. Establishing a reconciliation code of practice and a governed data model mitigates these risks by defining calculation methods, validation checks, responsibilities, and sign-off points. For generative AI, additional controls are required to address hallucination and overreach. The workflow described here restricts AI to computed inputs, requires human review, and maintains traceability from narrative statements to the underlying metrics.

Practical transferability to other open pit operations

While the example data set pertains to copper and features two material classes, this methodology is applicable to a range of open pit operations where reconciliation is influenced by challenges in data integration and reporting delays.

The enabling factors are:

- agreed definitions and boundaries
- deterministic calculations with validation checks
- AI is limited to triage and communication with human approvals.

CONCLUSIONS

AI-assisted F2 reconciliation represents a fundamental shift in how mining operations can validate material movement, maintain resource accountability, and translate reconciliation outcomes into timely operational action. This work demonstrates that when reconciliation calculations remain deterministic, governed, and auditable, AI can be safely applied to the high-effort components of the workflow, such as data handling, validation support, variance triage, and the rapid production of consistent reporting outputs—without compromising accountability under a structured human-in-the-loop model.

In the case study, the AI assistant completed the F2 reconciliation in approximately ten minutes, compared with two hours using the traditional manual workflow. It generated a suite of graphs, tables, and written recommendations that clearly highlighted the key reconciliation variances. While the AI-generated outputs still required human review and validation prior to release, the assistant materially reduced effort by rapidly analysing the data set and presenting the results in a clear, structured, and decision-oriented format. The AI assistant struggled to apply detailed technical knowledge related to orebody understanding, operational efficiency, and spatial compliance. However, since the AI used in this case study was not specifically trained for these tasks, its performance reflects this limitation. If a mining company invested resources and provided targeted data to train an AI assistant for these purposes, the assistant's capabilities could be enhanced and refined to deliver the desired results.

This represents a material productivity gain, saving hours that would otherwise be spent on manual data collation, analysis, formatting, validation, and compilation of deliverables. For mining companies, these efficiencies translate into direct cost savings by reducing the labour and time required for routine reporting. Critically, by freeing mining professionals to focus on addressing and investigating root causes, AI-assisted workflows support improvements in operational efficiency by helping identify and mitigate issues that drive cost and value leakage, such as operational delays, stockpile grade variability, and material movement or measurement errors that adversely impact reconciliation performance.

Looking forward, the approach provides a scalable bridge between traditional spreadsheet-based methods and more integrated, near-real-time operational controls. To fully realise this potential, three foundations remain essential: (i) robust, low-latency data pipelines that unify fleet, survey, and plant data into a consistent reconciliation model; (ii) uncertainty and tolerance frameworks that quantify confidence and distinguish meaningful deviations from noise; and (iii) strong governance controls—auditable templates, source-linked outputs, and escalation rules—to ensure reconciliation remains an assurance process rather than a black box. With these elements in place, AI-assisted F2 reconciliation can improve both the timeliness and accessibility of reconciliation insights while preserving the technical rigour required for effective mine control and future autonomous operating environments.

FUTURE APPLICATION

The strategic opportunity, however, extends far beyond incremental reporting improvements. As mines transition toward autonomous operations, AI-driven reconciliation can evolve into an integrated management system that serves as the operational backbone linking all mine technologies. In this future state, real-time data streams from automated haulage fleets, load-haul-dump equipment, survey drones, grade sensors, and geotechnical monitoring systems feed a centralised reconciliation engine that operates within a people-light environment. The system continuously monitors critical metrics material movement against geological blocks, grade factor variability, resource accounting, operational spatial compliance, and reconciliation performance while simultaneously directing autonomous equipment dispatch and routing based on reconciliation feedback. Survey drones autonomously capture updated pit geometry and material placement data, creating a closed-loop system where physical reality is continuously validated and integrated. Rather than generating fragmented reports, all reconciliation outputs, equipment status, drone surveys, and anomaly alerts converge into a centralised data repository accessible from remote operations centres, enabling field and office personnel to review curated insights and drill into source data sets from a single, governed digital space. In this architecture, reconciliation transcends compliance documentation, becoming the connective tissue linking equipment decisions, geological understanding, and strategic planning while maintaining full audit traceability.

ACKNOWLEDGEMENT

The author acknowledges the anonymous operation that provided a de-identified data set for demonstrating the reconciliation workflow.

REFERENCES

Macfarlane, A S, 2015. Reconciliation along the mining value chain, *Journal of the Southern African Institute of Mining and Metallurgy*, 115(8):679–690.

Parker, H M, 2011. Reconciliation principles for the mining industry, in *Proceedings of the 35th APCOM Symposium 2011* (The Australasian Institute of Mining and Metallurgy: Melbourne).

Autonomous haulage meets open pit gold mine – lessons from engineering and operation at Newmont Boddington

S Ling[1], S Westwood[2] and W Nalder[3]

1. Short-term Production Engineer, Newmont Corporation, Melbourne Vic 3004. Email: samuel.ling@newmont.com
2. Short-term Production Engineer, Newmont Corporation, Perth WA 6065. Email: scott.westwood@newmont.com
3. Production Scheduler/Mine Coordinator (M3), Newmont Corporation, Perth WA 6232. Email: warrick.nalder@newmont.com

ABSTRACT

This paper presents the key engineering and operational lessons learned from the implementation of autonomous haulage systems (AHS) at Newmont Boddington, with a focus on two practical case studies. The intent of the paper is to share real, site-based learnings that highlight how short-term planning, design adaptation, and operational discipline are critical to achieving safe and productive autonomous operations in a deep open pit mine.

The first case study focuses on the use of widened drop cuts and modified spur designs to support continuous mining during bench turnover into the next reduced level. At Boddington, drop cut development is a major bottleneck due to limited space at the bottom of deep, spatially constrained pits. Conventional drop cut designs proved insufficient under AHS conditions, leading to congestion, controller interventions, and reduced productivity. By increasing drop cut width and incorporating engineered spur access, the operation was able to maintain shovel productivity, improve truck flow, and safely manage bench turnover activities without stopping production. These short-term design changes proved effective in de-bottlenecking critical areas and improving overall system performance.

The second case study examines the application of spine roads on an oxide waste dump operating under AHS. It demonstrates how 50 m wide spine roads, 110-degree intersections, dump optionality, and disciplined planning practices allowed autonomous haulage to operate reliably through the wettest part of the year. The spine road set-up and winter operating strategy implemented on the 11WD (11 waste dump) in 2025 significantly reduced the need to divert trucks to alternate dumps as well as controller intervention to recover if stuck. As a result, cycle time variability was reduced and the gap between planned and actual performance was minimised.

Together, these case studies show that AHS success relies on thoughtful design, operational flexibility, and proactive planning rather than technology alone.

INTRODUCTION

The rise of artificial intelligence (AI) is pushing many industries around the world toward more autonomous systems, and mining is now moving in the same direction. In open pit mining, autonomous haulage systems (AHS) promise better safety by removing people from high-risk areas and improving consistency compared to a normal conventional fleet. But AHS cannot reach its full potential just by replacing an operator with a machine. To make autonomy work properly, planning, design standards, and day-to-day practices must improve and adapt.

In 2021, Boddington became the first gold mine worldwide to deploy a fully autonomous haulage system across its open pit truck fleet (Mining Technology, 2022; Society for Mining, Metallurgy & Exploration (SME), 2021; Newmont Corporation, 2021). This marked a global milestone: while autonomy had previously been trialled and implemented in iron ore operations in Western Australia by Rio Tinto, BHP and FMG, Boddington was the first to extend the system to a large-scale gold operation. The transition required integrating Caterpillar's Command for Hauling system, modifying haul roads, and adapting mine planning processes to meet autonomous operating requirements.

This paper presents the key engineering and operational lessons learned from AHS implementation at Boddington, with a focus on ramp and haul road design. Several design and operational elements were reviewed and updated for autonomy, including ramp width, shovel mining behaviour, and short-term planning approaches. Changes were also made to pit development planning to improve interactions between autonomous trucks, drill and blast activities, and field personnel.

Case Study 1 highlights the trade-off between increased road maintenance and improved productivity. By understanding AHS limitations during shovel loading, design changes were introduced that allowed autonomous trucks to achieve higher productivity compared to conventional approaches. Case Study 2 shows that AHS requires proactive planning, including increased dump optionality and disciplined planning practices, to operate reliably through the wettest part of the year.

BODDINGTON PIT LAYOUT

Boddington operates two main pits: the North Pit and the South Pit (Figure 1). At the current stage of mining, the South Pit is operating in hard rock, while the North Pit is predominantly stripping oxide. The 11WD waste dump is currently mostly a oxide dump and dedicated to the North Pit, whereas the 10WD waste dump is allocated to the South Pit.

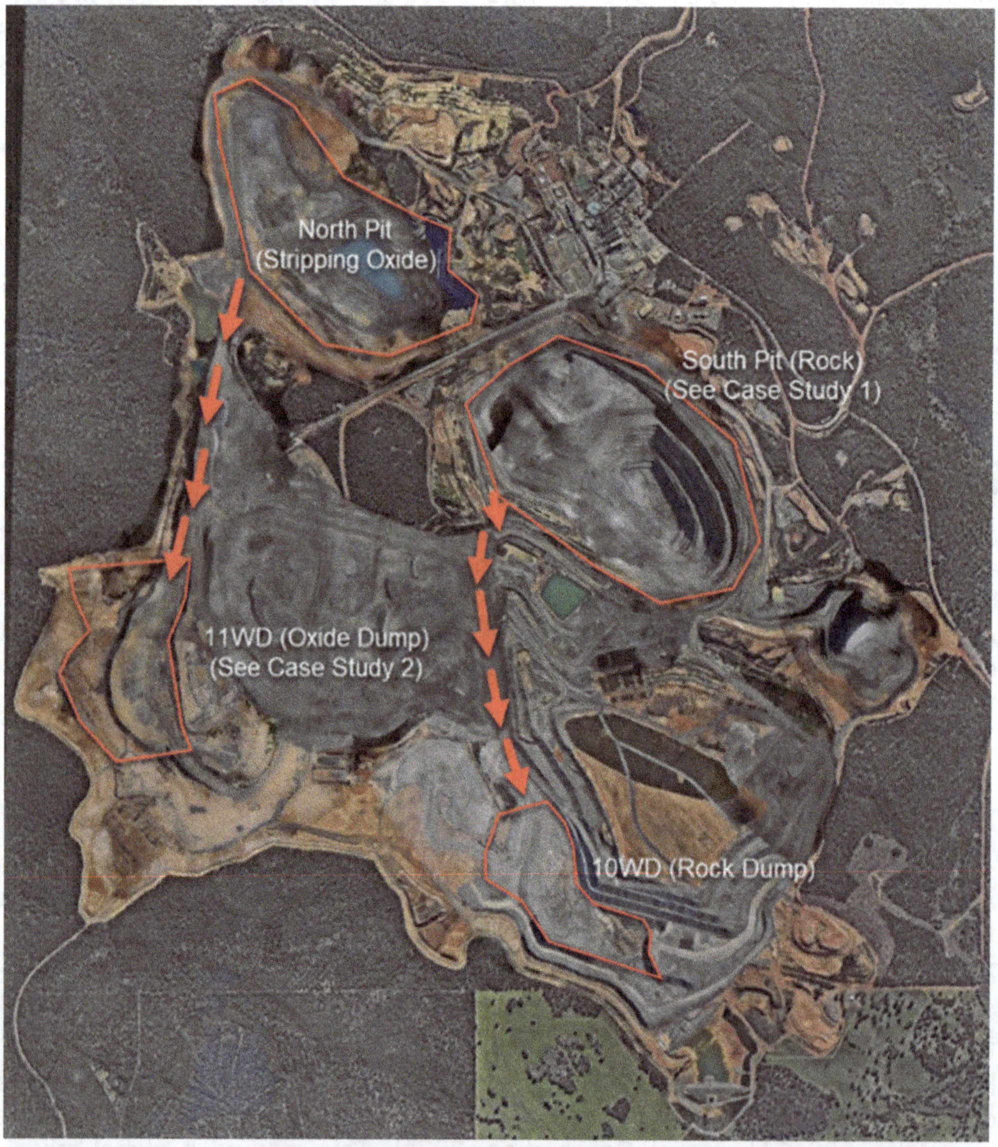

FIG 1 – Boddington gold mine layout.

OPERATIONAL CONTEXT AND PIT SEQUENCING CONSTRAINTS AT BODDINGTON

At Boddington, bottlenecks directly control how the pit sequencing and how productive the mine can be. Across the wider mining industry, many operations have the option to move to another pit while a drop cut is being established. This option is not available at Boddington. As a deep, constrained open pit, production must continue while new access is developed. As a result, drop cuts often need to be completed while access to the bench above is maintained, allowing all loading units and truck fleets to continue operating at high utilisation (refer to Figure 2).

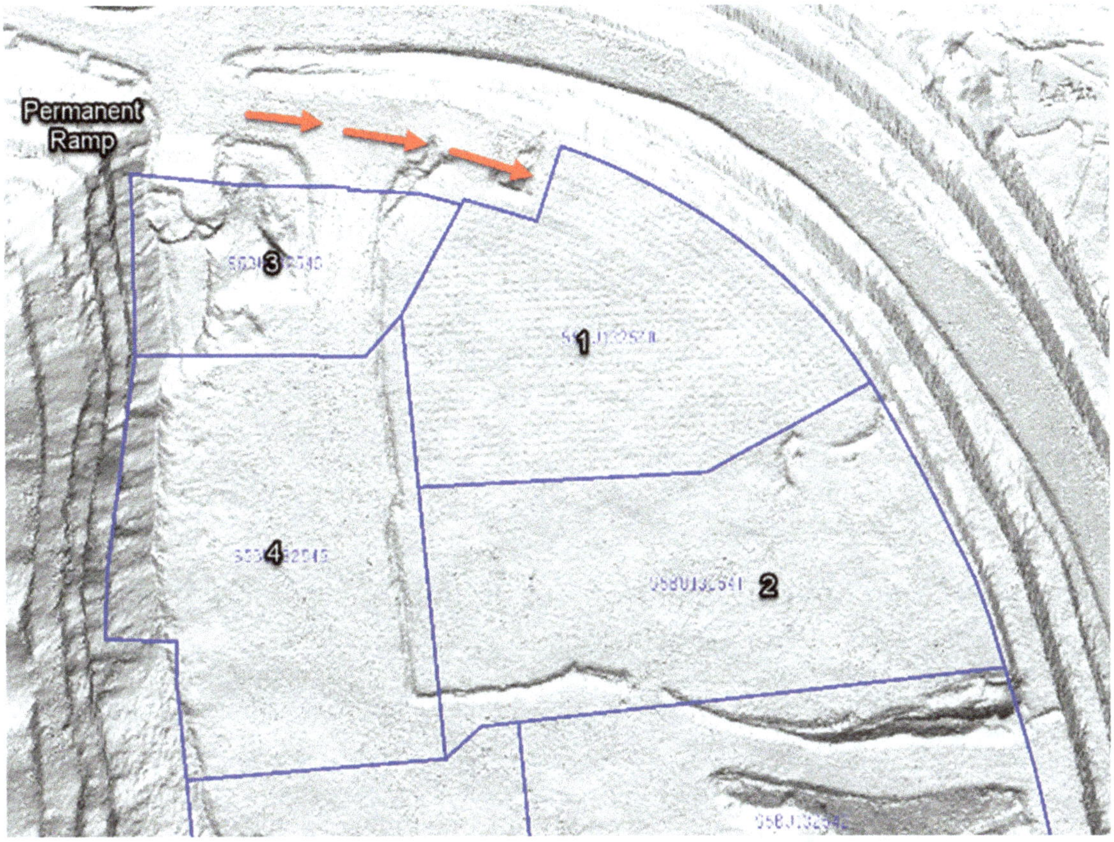

FIG 2 – Typical pit sequence for the S05B cutback.

In practice, pit sequencing at Boddington often requires a drop cut into the next bench while approximately 40 per cent of material remains on the upper bench (refer to Figure 2). With only two main deep pits and limited working space at the pit bottom, maintaining access to the upper bench during this transition is critical. To support this requirement, a spur ramp is typically constructed to provide temporary access to the upper bench while the drop cut into the next level is underway. This approach allows mining activities to continue in parallel without interrupting production.

Another key bottleneck at Boddington is highwall bench turnover (BTO). To manage this constraint, the preferred sequencing strategy is to expose the highwall face as early as possible. This enables development equipment to access the bottleneck area sooner and allows critical activities such as scaling and rock breaking to progress without delay.

Due to the depth of the pit, available space at the pit bottom is extremely limited. Under these conditions, narrow drop cuts quickly become a constraint, leading to AHS stoppages, manual overrides, and congestion around the dig face. To maintain autonomous truck movement and sustain dig unit productivity while following the required mining sequence, wider drop cuts are often necessary. The additional space allows bench turnover activities and continuous haulage to occur in parallel without disrupting production.

DIG UNITS AND TRUCK UNIT SPECIFICATION

At Boddington, the two main types of loading units are currently used: electric shovels and hydraulic excavators.

The two electric shovels in operation are Cat 7495 Series. These are high-production electric mining shovels designed for large-scale operations. They typically operate with an 80 t dipper and are capable of loading 200–250 t haul trucks in three to four passes. With a working weight of over 1200 t, these shovels are well suited for heavy-duty mining conditions.

In addition, two Caterpillar 6060 hydraulic excavators are used on-site. The Cat 6060 is an industry-leading hydraulic excavators, weighing approximately 600 t and optimised for loading 240 t haul trucks, in four to five passes.

Haulage is primarily carried out using Caterpillar 793F haul trucks. These trucks have a nominal payload capacity of approximately 220–231 t and The 793F is designed for high-volume mining operations and operates at speeds of up to approximately 60 km/h.

Based on the specifications of the loading and haulage equipment outlined above, permanent ramp designs at Boddington are developed to safely and efficiently accommodate these machines.

CASE STUDY 1 – 100 m DROP CUT PLUS LARGE SPUR WITH WING

The shift from conventional haulage to autonomous haulage at Boddington brought one of the most significant changes to short-term mine design and sequencing: autonomous trucks required much more controlled space and geometry than human-operated fleets. Unlike a human operator, who can visually assess hazards, adjust speed, or steer around uneven spots, autonomous trucks rely on Caterpillar MineStar's hard safety envelope, obstacle detection, berm pressure detection. While this improves safety for both personnel and equipment, it also reduces flexibility in confined areas such as drop cuts and lower benches.

Problem – why the conventional drop cut geometry created AHS bottlenecks

Temporary wide conventional drop cuts at Boddington were typically designed at around 60 m width, which was sufficient to maintain a reasonable dig rate for conventional operations. However, when the same 60 m drop cut was used for autonomous haulage, a clear reduction in productivity was observed.

This reduction was mainly related to how the shovels mine at the corners of the dig face and how the AHS system manages truck positioning during loading. In a conventional operation, operators can visually assess hazards, adjust speed, and steer around uneven berms. Autonomous trucks are capable of this, but there are limitations.

One of the key differences between autonomous and conventional mining is how sensitive autonomous trucks are to positioning during loading. In particular, autonomous trucks are highly sensitive to differences in left and right strut pressures when spotting at the shovel. If ramp or loading area geometry does not follow the recommended 10 per cent elevation or flatter, manual recoveries are often required. From site experience, gradients of 12 per cent or steeper consistently begin to cause operational issues.

When an autonomous truck sits across a gradient, with one side higher than the other, the resulting strut pressure difference can trigger the truck's berm detection system. This inbuilt safety system uses changes in strut pressure to determine whether the truck has backed onto material such as a windrow, tip head, or the toe of a dig face. When berm detection is triggered, the truck may stop before reaching its designated loading position, leading to delays and controller intervention.

If the truck does not reach the correct loading position, it will remain in spotting mode instead of switching to loading mode. In this state, the truck will not record dipper data, grade, or payload information. The shovel operator is also unable to send the truck once loading is complete, and a second truck cannot be called into the cusp.

As a result, mine controller intervention is required to recover the truck and manually assign the correct mining block and loading information. This increases controller workload, creates delays at the dig face, and introduces a higher risk of data errors during material tracking.

This manual intervention increases controller workload and creates risk. Incorrect block assignment or missing material modifiers can occur, which increases the chance of oversize or ground engaging tools (GET) being sent to the crusher.

As the shovel progresses down the drop cut, a 'tail' of material is often left along the edge of the dig face (refer to Figure 3). This further reduces the effective running width. The narrower running surface increases the risk of obstacle detection as the spacing decrease, results in more controller interventions, slows truck movement, and ultimately reduces overall productivity.

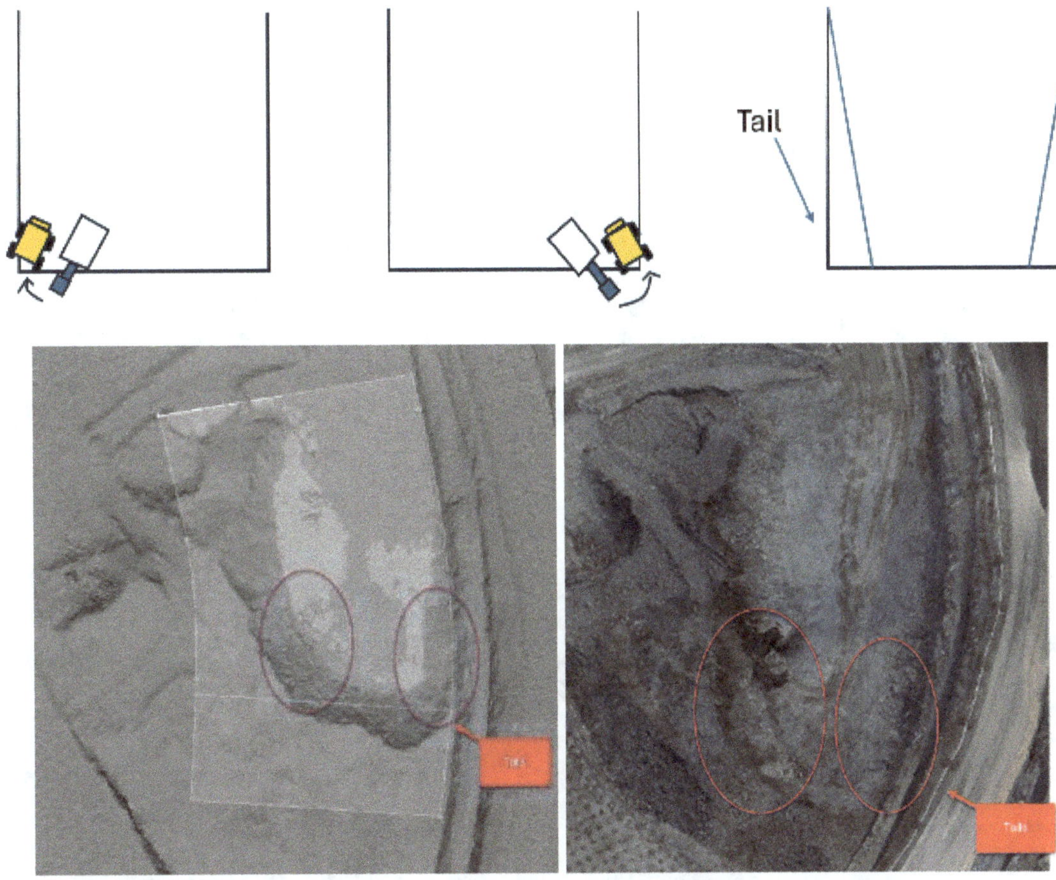

FIG 3 – Example of tail left behind by shovel while mining drop cut.

Solution – designing a 100~120 m drop cut plus spur with wing

A faster and more practical approach was required to maintain production targets. At Boddington, this led to a solution that was deliberately designed around the known limitations of AHS, instead of attempting to force autonomous trucks to operate within conventional designs.

The solution was to increase the temporary drop cut width to 100–120 m, depending on the situation. It was already understood that AHS trucks struggle near the edges of a drop cut and that some loss of effective running width is unavoidable. By accepting this behaviour and allowing a 'tail' of material to remain along the edge, productivity could still be maintained. With a 100 m wide drop cut, even if the tail occupied 20–30 per cent of the width by the toe of the drop cut, there was still approximately 70 m of effective running width available for autonomous trucks. Although this approach required additional dozer and grader effort to maintain the wider temporary drop cut, experience at Boddington showed a clear improvement in overall productivity.

The increased drop cut width also has sequencing benefits. As discussed in the earlier sequencing context, Boddington often requires a spur to be built on the bench above to maintain temporary access while mining and BTO activities are completed. The additional width of the drop cut naturally provided sufficient material to construct this spur. In addition, the widened area before the spur is used for drill and blast access, supporting ongoing development activities. In this way, material left behind due to AHS limitations was repurposed to support multiple functions. The widened drop cut also allowed better segregation between BTO work along the highwall and AHS truck movements. Once BTO down the drop cut was completed, a service windrow is installed and part of the temporary access converted into the final ramp. The remaining width then continued to serve as access for mining the previous spur, drill and blast activities, or future spur development.

Moreover, the increased width improves truck cusping flexibility, which further increases truck productivity and reduces congestion, ultimately improving both safety and productivity.

This approach demonstrates how Boddington's short-term planning team turned an AHS limitation into a practical, multi-use solution that improved productivity, maintained access, and supported safe sequencing in a deep, space-constrained pit.

Figure 4 shows the step-by-step process used to apply this approach at Boddington.

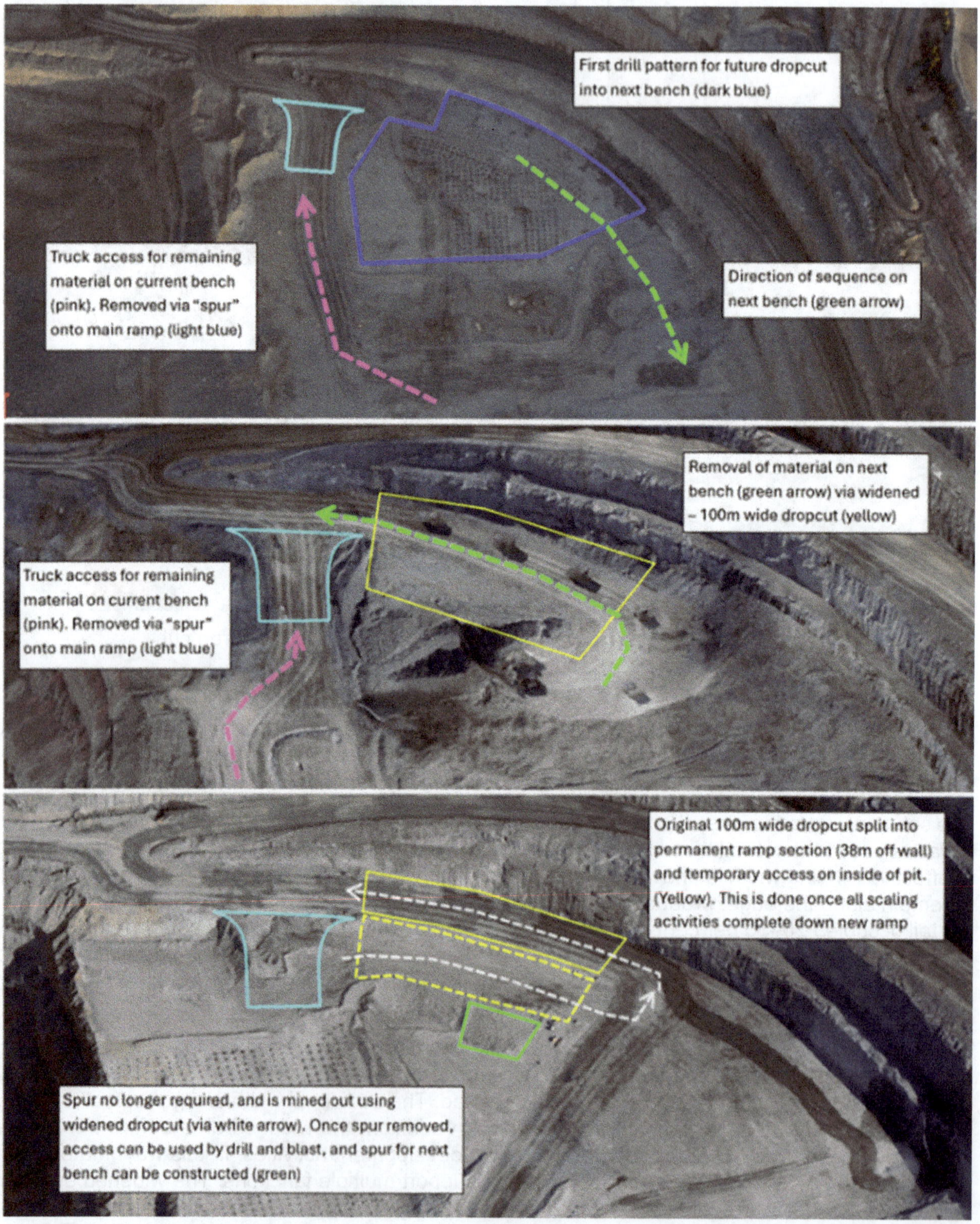

FIG 4 – Flow of the 100 m wide drop cut and spur with massive wing.

Lesson learned

The transition from conventional haulage to autonomous haulage at Boddington clearly showed that traditional drop cut designs are no longer fit for purpose when applied to AHS operations. While a 60 m wide drop cut was sufficient for conventional fleets, it created productivity losses, frequent controller interventions, and congestion when used by autonomous trucks. These issues were mainly driven by AHS sensitivity to edge conditions, shovel interaction at the dig face, and reduced effective running width as mining progressed.

By increasing temporary drop cut widths to 100–120 m, Boddington was able to provide the controlled space and geometry required for AHS to operate reliably. Even with the accepted formation of a material 'tail' along the dig face, sufficient running width was maintained to support consistent truck flow. Importantly, this additional width was not wasted. It was repurposed to support spur development, drill-and-blast access, BTO segregation, and future ramp construction.

This case study demonstrates that wider drop cuts are not inefficiencies but critical enablers of autonomous productivity, safety, and sequencing flexibility in deep, single-pit operations like Boddington.

CASE STUDY 2 – SPINE ROAD ENGINEERING FOR AHS RELIABILITY ON OXIDE WASTE DUMPS

Boddington faces a specific operational challenge during the winter months. As indicated by the red square in Figure 5, the site typically receives around 271–312 mm of rainfall over a three-month winter period (Bureau of Meteorology, 2026), which leads to soft and unstable ground conditions on oxide waste dumps and the rapid formation of ruts. On a tiphead in conventional operations, haul truck operators can manually adjust their routes, avoid soft areas, or change driving behaviour to suit conditions. Autonomous trucks do not have this flexibility.

Year	Jan	Feb	Mar	Apr	May	Jun	Jul	Aug	Sep	Oct	Nov	Dec	Annual
Graph													
2011												23.6	
2012	10.8	4.2	0.2	31.2	60.8	110.2	26.6	83.6	97.4	11.0	43.2	31.4	510.6
2013	7.2	8.6	44.2	6.8	110.4	21.6	93.8	79.6	118.8	18.4	6.0	1.8	517.2
2014	0.6	0.0	7.0	9.2	121.2	53.0	120.6		69.0	21.4	22.6	0.0	
2015	4.0	2.4	13.6	53.8	38.8	45.6	52.8	74.0	29.4	14.4	17.0	24.2	370.0
2016	33.6	0.8	35.2	31.4	82.8	53.6	47.4	96.2	49.8	34.8	4.6	27.2	497.4
2017	1.2	117.8	21.2	0.0	78.0	90.4	119.8	117.8	59.2	26.8	7.6	31.4	671.2
2018	120.2	14.4	7.6	1.8	49.0	91.8	128.8	101.2	14.4	32.8	7.2	3.6	572.8
2019	1.4	0.0	15.4	15.2	20.4	122.6	68.8	76.6	25.6	23.8	17.2	0.6	387.6
2020	0.2	29.0	11.0	6.6	101.0	81.4	60.2	64.6	73.6	16.6	59.6	1.4	505.2
2021	9.0	42.0	118.0	48.8	53.2	50.8	193.2	68.2	65.8	69.0	5.2	3.6	726.8
2022	0.0	1.8	5.8	25.2	70.2	66.4	87.0	137.0	65.0	36.4	16.2	18.0	529.0
2023	1.8	0.0	87.8	79.4	13.4	96.6	84.6	48.6	62.2	6.2	0.0	5.2	485.6
2024	4.2	1.6	0.2	0.0	107.2	82.4	119.8	102.0	29.8	25.0	12.4	0.4	485.0
2025	3.0	9.6	49.0	21.6	28.0	47.2	118.0	133.6	52.6	24.8	25.0	5.4	517.8
2026	3.4	1.4	75.2										

FIG 5 – Boddington monthly rainfall (Bureau of Meteorology, 2026).

Problem – wet-weather constraints on AHS haul roads

AHS trucks operate on predefined haul routes and approved running surfaces. Outside of winter, rainfall at Boddington is generally low and road conditions remain stable. However, during winter, short but intense rainfall events can occur. Empirically, rainfall of around 15 mm over a short period is enough to cause road performance issues.

Under these conditions, haul roads can degrade rapidly, increasing the likelihood of AHS stop-faults, bogging (Figure 6), and controller interventions. Unlike conventional trucks, autonomous trucks cannot adjust their driving behaviour to suit changing ground conditions, making them more sensitive to wet and unstable surfaces.

To manage this risk and maintain productivity during wet weather, Boddington developed a structured engineering solution known as the AHS Spine Road System.

FIG 6 – Bogged truck due to no rock spine road on oxide dump.

Solution – purpose built AHS spine roads

To reduce winter-related delays, the engineering and operations teams worked together to build strong, consistent, and load-bearing access routes across the waste dumps. These routes are referred to as Spine Roads.

Spine Road Specification (Boddington Standard: refer to Figure 7). Along the waste dump, a spine road is constructed every 100~120 m with the following features:

- 50 m road width.
- A crowned base layer constructed from waste material (approximately 150 mm particle size), built to a maximum height of 0.75 m, with a 100 mm roadbase wearing surface on top 3 per cent crown.
- Designed to suit the weight and operating behaviour of Cat 793F AHS trucks.

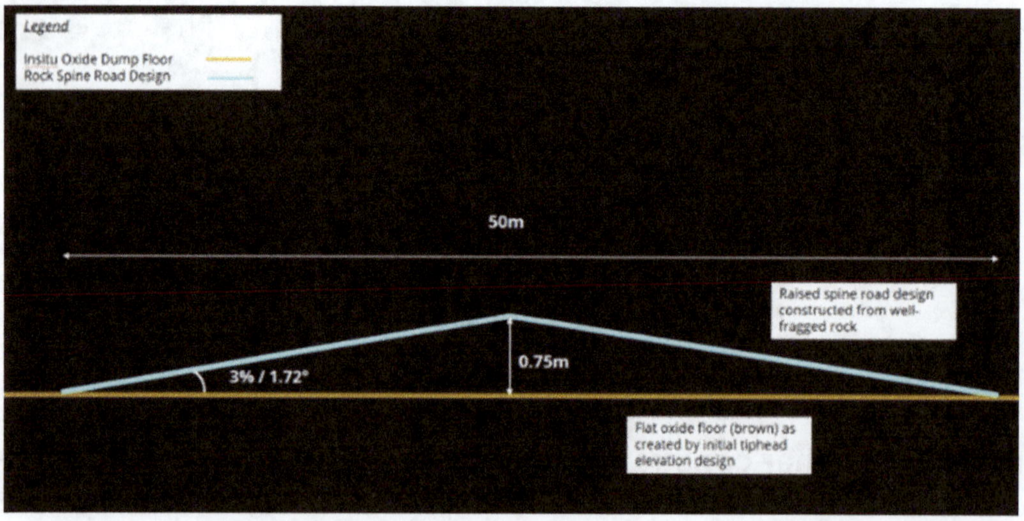

FIG 7 – Cross-section (typical spine road design).

These spine roads function as primary all-weather haul routes, allowing autonomous trucks to continue operating reliably through wet conditions.

The spine roads provide AHS trucks with a consistent and predictable running surface. This helps improve drainage, reduce the risk of bogging, and lower the number of automatic stop events caused by unstable ground. The improved road condition also allows safer interaction between autonomous trucks, dozers, and graders. With stable geometry and sufficient material strength, AHS trucks can commit to haul cycles with fewer interruptions. This also gives the operation more flexibility, with multiple haul routes available, which reduces congestion and limits interaction (Figure 8). One main road was established with spine roads branching off approximately every 120 m to service each dump cell. Six dump cells were set-up in total, each with its own spine road positioned at the midpoint. This provided enough flexibility to keep trucks dumping to other cells while road maintenance was carried out. Spine road intersections were built at ~110 degrees to minimise rutting, which also matches the site SOP (Standard Operating Procedure) limit.

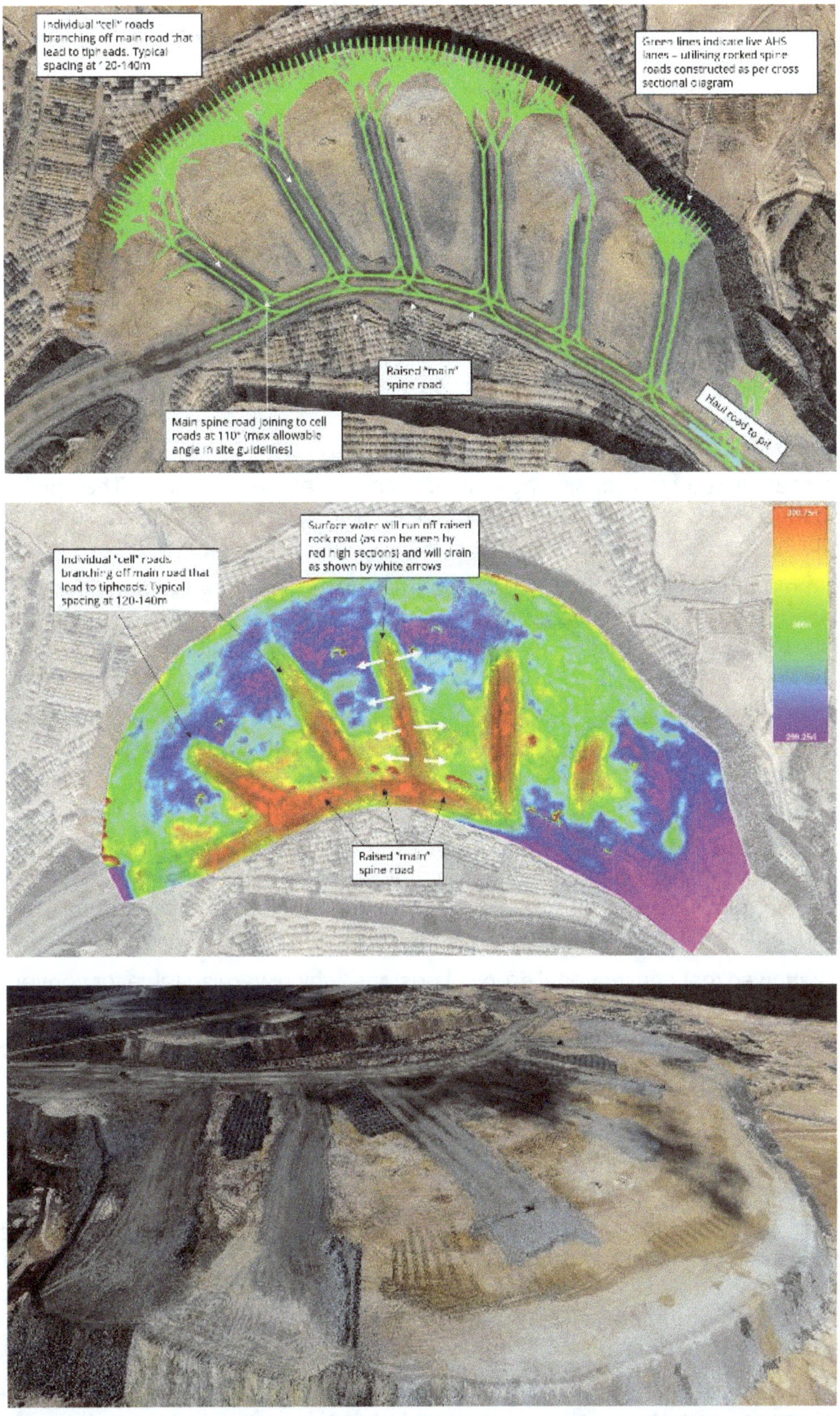

FIG 8 – Spine road design, elevation heat map and actual spine road.

At Boddington, rainfall outside of June, July, and August is significantly lower than during the winter months (Figure 8). Outside the main wet season is the most suitable time for construction activities. By completing major road construction during this period, the impact of wet road conditions during winter can be significantly mitigated, or in some cases eliminated, which ultimately improves overall mining productivity.

From a scale point of view, as shown in Figure 8, the oxide dump supported by the spine road system has a total capacity of approximately 40.9 Mt. The total amount of rock material used to construct one main arterial road and six spine roads was approximately 164 kt. In simple terms, every one tonne of road construction material enabled around 250 t of oxide to be hauled on a stable and efficient running surface.

In worst-case scenarios, and as shown by the pit layout, trucks hauling from the oxide pit were forced to travel to the rock waste dump instead of the planned oxide dump. This resulted in cycle times increasing to nearly 1.5 times the planned haul distance. With proper planning and management of the oxide dump using spine roads, this additional travel and associated productivity loss can be avoided.

After spine roads were implemented, Boddington saw a significant reduction in AHS bogging events, more predictable cycle times, fewer manual recovery interventions, and improved effectiveness of dozers and graders. Traffic flow across the waste dump became safer and more controlled. Most importantly, the operation was able to maintain consistent production through winter, instead of experiencing the usual seasonal productivity drop.

Lesson learned

Autonomous haulage requires consistency rather than operator adaptability. The spine road system clearly shows that good waste management combined with well-engineered road structures is critical for AHS reliability.

By investing most of the year in building sufficient spine roads for winter conditions, the operation avoids thousands of hours of recovery and delay during wet weather. The experience also shows that consistent road geometry is just as important as width and gradient for autonomous haulage performance.

As a result, Boddington's spine road approach has become a strong example of how waste dumps can be designed to support autonomous haulage in regions with seasonal rainfall.

CONCLUSION

The implementation of Autonomous Haulage Systems at Newmont Boddington has demonstrated that autonomy is not simply a technology upgrade, but a fundamental change to how open pit mines must be designed, planned, and operated. The two case studies presented in this paper highlight that AHS performance is strongly influenced by space, geometry, and planning discipline.

Case Study 1 showed that conventional drop cut designs, which were suitable for conventional fleets, became a productivity bottleneck under autonomous operation. The introduction of wider temporary drop cuts and modified spur designs provided the additional space required for AHS trucks to operate consistently during shovel loading and bench turnover activities. Although these designs required more upfront dozer and grader effort, they significantly reduced controller interventions, congestion, and unplanned stoppages. Most importantly, they allowed mining to continue efficiently in a deep, narrow-pit environment where alternative access options are limited.

Case Study 2 demonstrated that AHS reliability on oxide waste dumps is highly sensitive to ground conditions, particularly during wet weather. The development of engineered spine roads, combined with increased dump optionality and disciplined planning practices, allowed autonomous trucks to operate consistently through the wettest part of the year. Rather than reacting to seasonal disruptions, Boddington adopted a proactive approach that focused on consistency, predictable geometry, and long-term dump development.

Together, these case studies show that successful autonomous haulage depends on designing for system limitations rather than expecting the system to adapt to conventional practices. Wider access, consistent geometry, and flexible planning options are not inefficiencies; they are enablers of safety, productivity, and operational resilience. The lessons from Boddington provide practical guidance for other operations considering or expanding AHS, particularly in complex, deep open pit environments.

ACKNOWLEDGEMENT

This paper summarises the work of a dedicated group at Newmont Boddington Operation, and the authors acknowledge their contributions to its success. The authors also extend their appreciation to Newmont for its support and permission to publish this work.

REFERENCES

Bureau of Meteorology, 2026. Monthly rainfall – Boddington North, Bureau of Meteorology, Commonwealth of Australia. Available from: <https://www.bom.gov.au/jsp/ncc/cdio/weatherData/av?p_nccObsCode=139&p_display_type=dataFile&p_stn_num=109516>

Mining Technology, 2022, August 5. Boddington Gold Mine (BGM), Australia, Mining Technology, GlobalData Plc. Available from: <https://www.mining-technology.com/projects/boddington/>

Newmont Corporation, 2021, October 5. Newmont Provides Update on Boddington Mine, Newmont Corporation. Available from: <https://www.newmont.com/investors/news-release/news-details/2021/Newmont-Provides-Update-on-Boddington-Mine>

Society for Mining, Metallurgy & Exploration (SME), 2021, October 6. Newmont launches first autonomous haulage system at Boddington Mine, Society for Mining, Metallurgy & Exploration (SME). Available from: <https://me.smenet.org/newmont-launches-first-autonomous-haulage-system-at-boddington-mine>

GLOSSARY

AHS (Autonomous Haulage System):

- A system that enables haul trucks to operate without onboard operators, using sensors, software, and control systems to autonomously navigate, load, and haul material.

AMT Berm Detection:

- A safety monitoring system that ensures Autonomous Mining Trucks (AMTs) do not drive over open areas or pit edges by detecting berm presence and integrity.

AMT Obstacle Detection:

- A critical, multi-layered safety system designed to identify, classify, and avoid obstacles in harsh and often unstructured mining environments. It relies on a combination of sensors and artificial intelligence to prevent collisions without human intervention.

AMT Truck Strut Pressure:

- An onboard measurement system in AMT trucks that monitors suspension performance by comparing internal nitrogen gas and hydraulic fluid pressures across all wheels.

Bench Turnover (BTO):

- The set of preparatory and development activities required to safely transition mining operations from one bench or reduced level (RL) to the next, in accordance with regulatory and site safety standards.

Bottleneck:

- A constraint or condition within the mining sequence or operation that restricts material flow and causes delays to production or equipment utilisation.

Controller (Dispatch):

- Personnel responsible for monitoring, managing, and intervening in Autonomous Haulage System operations from a control room environment.

Cusp:

- A truck position state where the vehicle is in reverse, preparing to back into a designated area such as a loading or dumping location.

Drop Cut:

- A temporary in-pit ramp constructed to provide access between benches or reduced levels during pit deepening and sequencing activities.

Ground Engaging Tools (GET):

- Wear components attached to excavation equipment—such as bucket teeth, cutting edges, and adapters—that directly interact with the ground during digging or loading.

Loading Spot Point:

- A truck operational mode indicating it is in a loading position, receiving feedback from the shovel or digger, including tonnes loaded, material grade, and any hazardous material information.

MineStar:

- A Caterpillar automation and fleet management system used to control, monitor, and manage autonomous and semi-autonomous mining equipment.

Oxide:

- Material that has undergone oxidation; in mining, this typically refers to very soft material that can easily become muddy during rain events, degrading road conditions and reducing productivity.

Ruts:

- Deep grooves or depressions in the ground formed by repeated passes of heavy equipment wheels.

Service Windrow:

- A flat-topped windrow designed to support services such as electrical cables, water pipelines, or communications infrastructure. These windrows are typically wider than standard safety windrows.

SOP (Standard Operating Procedure):

- An approved document that outlines the standard method for carrying out a task or process.

Spine Road:

- A haul road constructed higher than the surrounding contour to improve running conditions, particularly during wet season operations.

Spur:

- A short, temporary haul road or ramp branching from a main ramp or haul road, used to provide interim access to a working face, bench, or temporary mining area. Spurs are commonly used during bench turnover or drop cut development.

Staging Point:

- A truck operational mode indicating travel toward a designated area. This mode occurs prior to entering the loading spot point.

Toe of Drop Cut:

- The lowest reduced level (RL) at the base of a drop cut where the ramp transitions into a flat working area.

Combining artificial intelligence and simulation to optimise haulage

A J Price[1], A O Warren[2] and A Dahmen[3]

1. Product Manager, RPMGlobal, Brisbane Qld 4000. Email: aprice@rpmglobal.com
2. Senior Consultant, World Wide Technology Inc, Denver CO USA.
   Email: sasha.warren@wwt.com
3. Business Solution Architect, Freeport Minerals Corporation, Phoenix AZ USA.
   Email: adahmen@fmi.com

ABSTRACT

Haulage is one of the most significant cost drivers and productivity levers in open-pit mining. Even modest improvements in truck–shovel interactions can generate substantial operational gains. Yet most planning processes still rely on static assumptions that fail to reflect the variability and constraints of real-world haulage systems. As a result, mine plans are often difficult for operations teams to execute, and the disconnect between planning and day-to-day fleet control can lead to inefficiencies, bottlenecks, and divergence from intended outcomes.

Traditional schedules typically assume a fixed dig rate for each shovel and derive truck requirements only as a post-processing 'check'. These simplified methods overlook the effect of queuing, delays, routing choices, and the dynamic interactions that shape actual haulage performance. Compounding this challenge, the fleet assignment or autonomous haulage systems used in operations are rarely 'plan-aware'. Plans are often communicated through complex documents that do not align with the input formats controllers require, whose decisions and experience significantly influence productivity.

This paper introduces a practical, integrated approach to bridging this long-standing gap between planning and operations. By combining Discrete Event Simulation (DES) with Artificial Intelligence (AI), planners and can evaluate haulage strategies in a realistic virtual environment and identify optimal configurations before passing them to operations. A DES model (Figure 1) replicates the natural variability of truck–shovel operations, incorporating shift constraints, equipment availability, road networks, and expected delays. An AI optimisation engine then explores hundreds of potential haulage scenarios, systematically adjusting parameters to find solutions that best satisfy production, efficiency, and cost objectives.

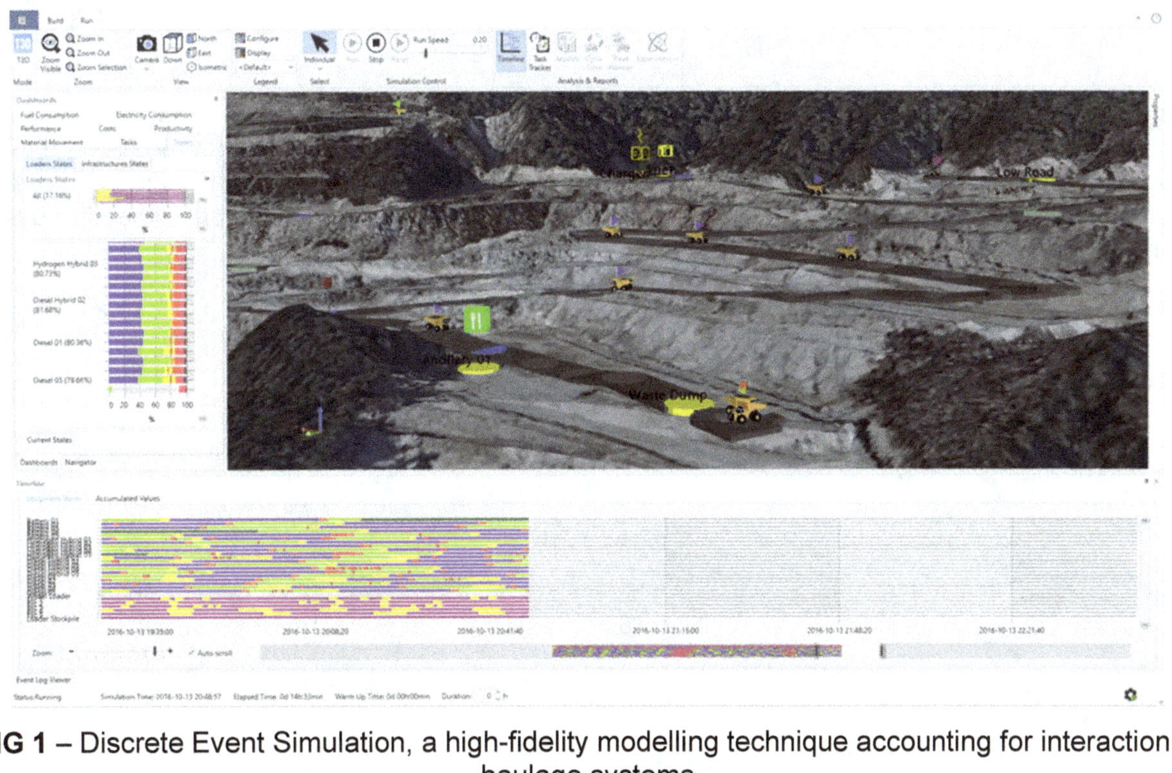

FIG 1 – Discrete Event Simulation, a high-fidelity modelling technique accounting for interaction in haulage systems.

The integration of DES and AI enables planners to test alternative approaches in a risk-free environment, quantify the impact of operational changes, and uncover performance opportunities that may not be obvious through traditional analysis. Additionally, this integration provides operations with a direct expression of the plan in settings relevant to fleet assignment software. A case study demonstrates how this method increased throughput and reduced queuing using an existing fleet—achieving measurable productivity gains without additional capital investment.

Cloud-based simulation technology strengthens this workflow by making high-fidelity DES models accessible on demand. Planners can publish a single, authoritative plan (Figure 2) that incorporates material movements, destinations, downtime, road closures, and routing details. Once published, this plan is available through the web to all personnel on-site who require access, ensuring a single source of truth. This multi-dimensional plan ensures alignment across teams. AI-driven scenario evaluation and model runs can be executed automatically in the cloud, enabling rapid iteration during planning cycles.

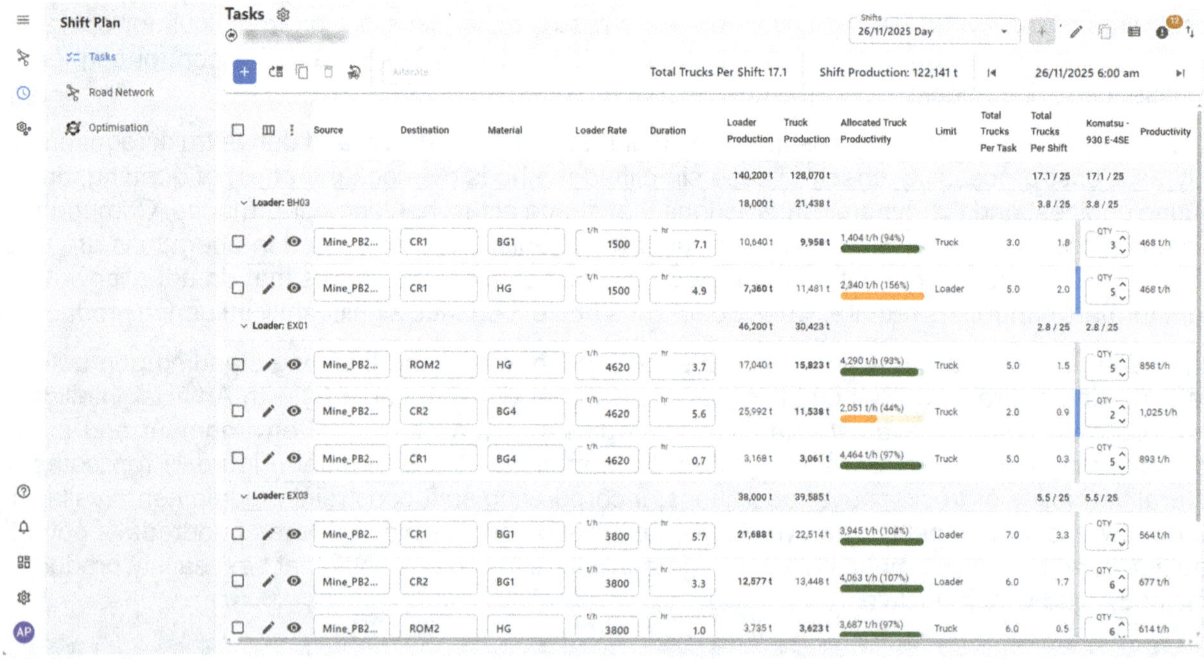

FIG 2 – Published haulage plan.

Many of the calculations underpinning the haulage plan and the DES rely on a road network that is up to date and configured to accurately reflect current site conditions. This is another strength of the adopted cloud-based approach. The road network, along with other variables used to calibrate the calculations, is stored in a single location for the site. The master road network can be edited by any user with appropriate permissions directly in the web browser (Figure 3).

FIG 3 – A single source of up-to-date data is essential. The road network is a key input.

For each shift, the inputs—plan details, road conditions, equipment status, and more—are used to generate the simulation model. The AI engine begins with the planner's baseline and explores modifications through parallel simulation runs with iterative improvement. Success is measured using an operation-specific objective function; for example, maximising high-value crusher feed and total material movement while penalising deviations from planned mining sequences.

The diagram below (Figure 4) shows how the AI system can identify improvements to the plan that have a material impact. Note that there are many simulation results that were worse than the initial 'Baseline' simulation. These simulations, along with all of the other simulations represented in the diagram, were used to 'find' the best achievable plan for the day.

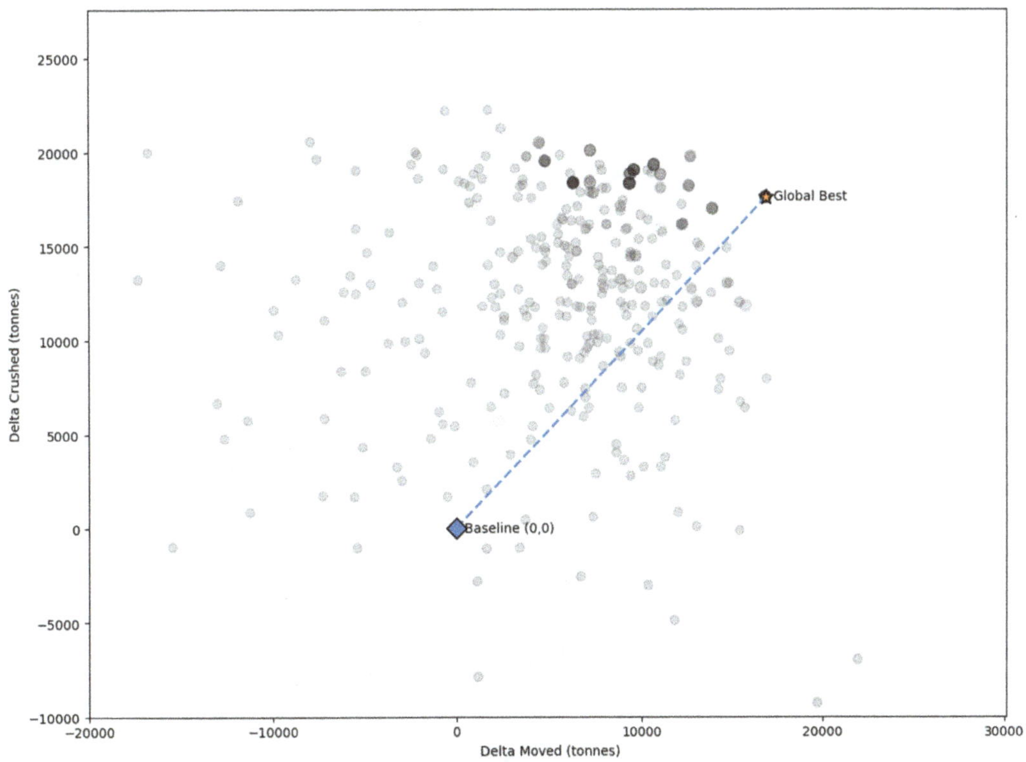

FIG 4 – Simulation results showing improvement from baseline.

By identifying both improvement opportunities and infeasible plans before execution, this approach promotes consistency, reduces operational risk, and increases trust in the planning process. The combined use of simulation and AI delivers a more agile, data-driven method for planning and executing haulage, supporting faster decision-making, more reliable plans, and tangible gains in fleet productivity. It ensures that the improvement gains seen when a 'good' operator is controlling the fleet are available every shift.

Open Pit

Capability development

The last line of defence – integrating psychological capital into the hierarchy of controls for psychosocial risk

R N James[1], G Farr-Wharton[2], T Bentley[3] and Y Brunetto[4]

1. PhD Candidate, Edith Cowan University, Perth WA 6000. Email: rnjames@our.ecu.edu.au
2. Senior Lecturer and Director of Engagement, Edith Cowan University, Perth WA 6000. Email: g.farrwharton@ecu.edu.au
3. Professorial Chair Mining, Health and Safety; Edith Cowan University, Perth WA 6000. Email: t.bentley@ecu.edu.au
4. Professor, Southern Cross University, Gold Coast Qld 4225. Email: yvonne.brunetto@scu.edu.au

ABSTRACT

The Western Australian mining sector stands at a critical regulatory juncture. While the industry has continued progress on engineering controls for physical hazards, the rising prevalence of psychosocial risks coupled with the new Code of Practice obligations (AICD and KWM, 2025) exposes a significant gap in our safety architecture. Traditional 'Swiss Cheese' models, designed for tangible hazards, often fail to contain invisible stressors such as production pressure, fatigue, and role ambiguity (Stuckey-Clarke *et al*, 2026). When organisational systems falter, the frontline worker's psychological state and personal capacities become the 'last line of defence' against an incident.

This research proposes a reconceptualisation of **Psychological Capital (PsyCap),** comprising Hope, Self-Efficacy, Resilience, and Optimism (HERO), not merely as a well-being metric, but as a quantifiable safety control. Furthermore it challenges the industry's reliance on generic resilience training. Instead, it posits that without a supportive Psychosocial Safety Climate (PSC) and high-quality supervisor relationships, individual resilience can become a liability leading worker to 'tough out' unsafe conditions (the 'Dark Side' of PsyCap).

We hope to offer Mining leaders a new perspective in that PsyCap is a malleable, measurable state that can be developed and audited. By treating human psychological factors with the same rigour as engineering controls, the mining sector can move from reactive mental health support to proactive psychosocial risk management, ensuring the 'last line of defence' is robust, supported, and sustainable.

REFERENCES

Australian Institute of Company Directors (AICD) and King and Wood Mallesons (KWM), 2025. Governing WHS psychosocial risks: A primer for boards, AICD. Available from: <https://www.aicd.com.au/regulatory-compliance/work-health-and-safety-act/reporting/governing-whs-psychosocial-risks-a-primer-for-boards.html>

Stuckey-Clarke, P, Le Blond, J, Moran, J and Sutton-Kunc, A, 2026, January 15. Psychosocial Safety at Work: 2026 will require a different approach, Dentons. Available from: <https://www.dentons.com/en/insights/alerts/2026/january/15/psychosocial-safety-2026>

Open Pit

Critical minerals

Geotechnical criteria for safe saprolite extraction in valley-confined ion-adsorption rare earth deposits

D Eloi[1], G Junqueira[2], K Rocha[3], F Costa[4] and M Pedrosa[5]

1. Senior Geotechnical Consultant, WSP Canada, Vancouver BC V6Z2M1, Canada. Email: danilo.eloi@wsp.com
2. Researcher, PUC Minas, Minaçu GO 76450-000, Brazil. Email: gabriel.junqueira@live.com
3. Researcher, UNOPAR, Palmas TO 77015-012, Brazil. Email: kassionegrao47@gmail.com
4. Geotechnical Manager, Vale S.A., Belo Horizonte MG 30575-000, Brazil. Email: filipe.costa@vale.com
5. Project Tailings Engineer, WSP Australia, Adelaide SA 5000. Email: matt.pedrosa@wsp.com

ABSTRACT

Ion-adsorption rare earth deposits developed on deeply weathered granitic massifs represent a distinctive mineralisation style associated with tropical saprolite profiles. In regions such as the Serra Dourada Granite in central Brazil, prolonged chemical weathering produces a thick regolith composed of laterite, saprolite, and transitional W2–W3 horizons that host clay-bound rare earth elements (REEs). These materials originate from *in situ* granite decomposition, preserving relict mineralogy while exhibiting high degrees of chemical alteration, with chemical index of alteration (CIA) values commonly exceeding 70–80 per cent. Their geotechnical behaviour is controlled by intense fracturing, high porosity, variable cementation, and pronounced vertical stratification inherited from the weathering profile.

Because ion-adsorption clays are extracted at shallow depths, mining operations typically rely on the formation of temporary working platforms within narrow valleys or amphitheatre-shaped depressions. The geometry and geomechanics of the saprolite impose strict constraints on the stability of these operating surfaces. The material displays a heterogeneous fabric, ranging from loose, dispersive zones in upper horizons to denser, partially cemented saprolite near the bedrock interface. Variations in structure and suction conditions lead to distinct responses in shear strength, deformation, and moisture-induced softening. Under wet seasonal conditions, reductions in matric suction may cause significant strength losses, particularly in clay-rich horizons.

To ensure safe excavation and trafficability, valley-bottom platforms must often be founded on more competent saprolite or weathered granite, with removal of weak colluvial or lateritic layers. Numerical assessments and field experience indicate that shallow platforms are sensitive to pore-pressure generation and deformation under operational loading and seasonal infiltration. Effective stabilisation measures include excavation to competent saprolite, drainage control, localised compacted pads, and continuous monitoring of pore pressure and displacement.

Overall, the geotechnical behaviour of granitic saprolites reflects the interaction between tropical weathering, clay-ion adsorption processes, and shallow mining geometry. Understanding this coupled system is essential for designing stable extraction platforms and ensuring safe exploitation of rare-earth-bearing saprolitic horizons.

INTRODUCTION

Ion-adsorption rare earth element (REE) deposits developed in deeply weathered granitic terrains represent one of the most important global sources of heavy rare earth elements. These deposits form through prolonged chemical weathering processes, during which rare earth elements released from primary minerals are progressively adsorbed onto the surfaces of clay minerals, particularly kaolinite and illite, within saprolitic horizons. This process results in thick weathering profiles composed of laterite, saprolite, and partially weathered granite, typically preserved in tropical and subtropical climates (Bao and Zhao, 2008; Chi and Tian, 2015).

From a geotechnical perspective, these deposits are characterised by materials that occupy a transitional domain between soil and rock. Granitic saprolites often preserve relic rock structures, including joints, mineralogical banding, and textural heterogeneities, while simultaneously exhibiting typical soil-like behaviour such as high porosity, low stiffness, and strong sensitivity to moisture variations (Lambe and Whitman, 1969; Butt, Lintern and Anand, 2000). The result is a highly heterogeneous and state-dependent material, whose mechanical response is governed not only by shear strength parameters but also by hydraulic conditions, structural anisotropy, and degree of weathering.

Field observations and geological mapping of the study area indicate that the saprolitic profile is developed over granitic bedrock within a geomorphological setting composed of narrow, confined valleys with intermittent drainage channels and steep valley sides. These valleys are characterised by rocky or gravelly beds with large granitic blocks, typical of headwater basins, and exhibit seasonal groundwater conditions, with water present during the wet season and disappearing during prolonged dry periods. In addition, talus deposits containing granite boulders with diameters on the order of several metres are commonly observed along the slopes, forming highly heterogeneous ground conditions.

Such geomorphological confinement directly influences the hydraulic regime of the saprolite, promoting high moisture contents, shallow groundwater conditions, and localised zones of saturation. The presence of less permeable horizons, partially weathered rock cores, and embedded boulders creates hydraulic barriers and preferential flow paths, resulting in complex pore pressure distributions and locally elevated hydraulic gradients. These conditions are particularly critical during mining operations, where excavation and traffic loads are imposed on materials whose strength is strongly dependent on matric suction and drainage conditions.

Extraction of ion-adsorption deposits is typically carried out through shallow, top–down strip mining methods, involving successive excavation stages and the construction of temporary working platforms. In valley-confined geometries, the available working width is often limited, and the mining sequence must adapt to narrow, elongated strips bounded by natural slopes and drainage features. Under these conditions, stability is rarely governed by classical deep-seated failure mechanisms. Instead, operational issues are dominated by localised shear deformation, progressive bearing capacity failures, excessive rutting, and loss of trafficability.

These problems are further aggravated by the intrinsic characteristics of granitic saprolites. The materials often exhibit sandy to silty textures, high porosity, and weak or variable cementation, combined with the presence of large granite boulders embedded within the weathered matrix. The interaction between these rigid inclusions and the surrounding saprolite produces strong contrasts in stiffness and permeability, leading to stress concentrations, differential settlements, and localised zones of instability, particularly under wet conditions.

Numerous studies have shown that granitic saprolites are highly sensitive to changes in moisture conditions, with substantial reductions in shear strength and stiffness associated with the loss of matric suction during rainfall events or infiltration (Lambe and Whitman, 1969; Butt, Lintern and Anand, 2000). In strip mining operations, the progressive exposure of fresh saprolite surfaces to precipitation, combined with limited drainage capacity in confined valleys, creates conditions conducive to rapid strength degradation and localised failures of temporary platforms.

In this context, the safe and efficient extraction of ion-adsorption saprolites requires geotechnical criteria that explicitly consider the interaction between material behaviour, groundwater conditions, structural heterogeneity, and mining geometry. Conventional approaches based solely on peak shear strength parameters or idealised drained conditions are often insufficient to capture the dominant mechanisms observed in these deposits.

This paper presents a geotechnical framework and engineering criteria for the safe extraction of saprolite-hosted ion-adsorption rare earth deposits in valley-confined environments. The focus is placed on the mechanisms of instability associated with elevated groundwater conditions, sandy saprolitic textures, embedded granite boulders, and the sequential geometry of strip mining operations. The proposed approach integrates geological, geomorphological, hydraulic, and mechanical aspects to support safer and more predictable mining practices in these complex tropical weathering environments.

GEOLOGICAL, GEOMORPHOLOGICAL, AND HYDROGEOLOGICAL SETTING

Geological context of the ion-adsorption deposit

The rare earth deposit analysed in this study is developed over deeply weathered granitic rocks, forming a thick lateritic–saprolitic profile typical of ion-adsorption systems. In such deposits, rare earth elements are released during prolonged chemical weathering and subsequently adsorbed onto the surfaces of clay minerals, particularly kaolinite and illite, within the saprolitic horizons. This process produces mineralisation that is distributed throughout the weathering profile rather than concentrated in discrete veins or orebodies.

The resulting profile typically comprises a lateritic cap, transitional horizons, and a structured saprolite overlying partially weathered granite. These materials preserve relic structures from the parent rock, including joints, veins, and textural variations, resulting in a heterogeneous mass with pronounced mechanical and hydraulic anisotropy. Such features strongly influence both the strength and permeability of the saprolite, particularly at interfaces between horizons of different weathering grades.

Within the deposit, the saprolite is predominantly sandy to silty in texture, with variable clay content depending on the degree of weathering and mineral alteration. The material exhibits relatively high porosity and significant spatial variability in density and stiffness. Localised zones of partially weathered granite occur as embedded boulders within the saprolitic matrix, creating sharp contrasts in stiffness and permeability. These heterogeneities play a critical role in stress redistribution, water flow paths, and the development of localised deformation during mining operations.

Typical weathering profiles in granitic terrains evolve from lateritic caps at plateau positions to saprolitic and colluvial materials along hillsides and valley bottoms, reflecting long-term geomorphological processes and erosion. A conceptual representation of this evolution is shown in Figure 1.

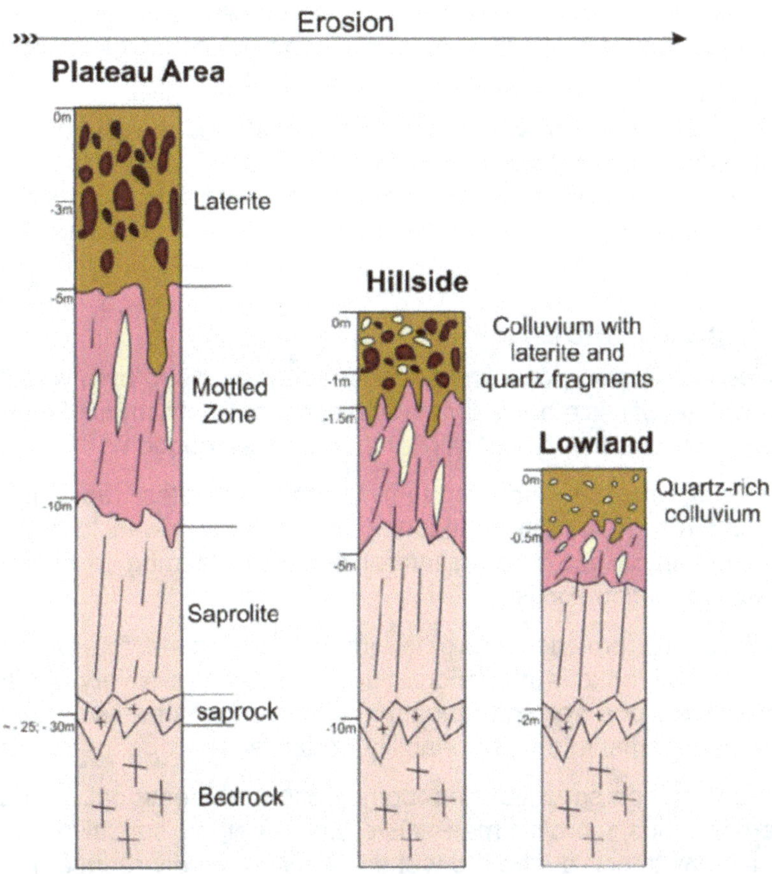

FIG 1 – Conceptual lateritic–saprolitic weathering profile across plateau, hillside, and valley settings, illustrating the geomorphological control on saprolite thickness, colluvial deposits, and groundwater conditions.

Geomorphological configuration and valley confinement

The deposit is situated within a valley-confined geomorphological setting, where the saprolitic profile has been preserved in topographic lows and along gentle slopes. This configuration imposes important constraints on both groundwater flow and mining geometry. A general view of the valley setting and the distribution of saprolitic materials is presented in Figure 2.

FIG 2 – General view of the valley-confined geomorphological setting, showing the saprolitic terrain and the approximate limits of the mining area.

Valley confinement limits lateral drainage and promotes the concentration of surface and subsurface water within the saprolitic horizons. During rainfall events, water tends to accumulate along the valley axis and infiltrate into the exposed mining surfaces. The presence of less permeable horizons and embedded granitic cores further restricts vertical drainage, leading to localised zones of elevated saturation.

From an operational perspective, the valley geometry constrains the layout of mining strips, often resulting in narrow, elongated working platforms with limited options for lateral drainage. This configuration increases the sensitivity of the operation to rainfall and groundwater conditions, as even modest increases in moisture can significantly reduce shear strength and bearing capacity of the saprolite.

Local hydrogeological conditions

The project area is located in a tropical climatic region characterised by a well-defined wet and dry seasonal cycle. The wet season is marked by several consecutive months of high precipitation, while the dry season is associated with very low rainfall and reduced humidity.

During the wet season, intense and frequent rainfall promotes rapid infiltration and rising groundwater levels within the saprolitic horizons. The valley-confined topography enhances this effect by concentrating run-off and limiting lateral drainage, leading to sustained high moisture contents and localised saturation zones.

In contrast, the dry season provides more favourable conditions for excavation and platform preparation, as lower rainfall reduces infiltration and allows partial recovery of matric suction in near-surface materials. However, localised saturation may still occur in areas where drainage is restricted by geomorphological confinement, low-permeability horizons, or embedded granitic cores.

From a hydrogeotechnical perspective, the combination of seasonal rainfall, confined valley geometry, and heterogeneous saprolitic materials creates a highly sensitive environment. Elevated groundwater levels and infiltration during the wet season can rapidly reduce matric suction, leading to significant decreases in shear strength and stiffness of the saprolite. These conditions directly influence the stability of temporary mining platforms, shallow cuts, and strip-mining geometries.

Geotechnical implications of the geological and hydrogeological setting

The integrated geological, geomorphological, and hydrogeological characteristics of the deposit result in a foundation material that is structurally heterogeneous, with relic rock fabric and embedded boulders, predominantly sandy saprolite with variable fines content, highly sensitive to moisture variations and loss of matric suction, subjected to seasonal groundwater fluctuations; and located within a valley-confined system with restricted drainage.

These conditions lead to a geotechnical environment in which stability is strongly governed by transient hydraulic conditions rather than by static strength parameters alone. The interaction between strip-mining geometry, rainfall infiltration, groundwater levels, and material heterogeneity forms the basis for the instability mechanisms discussed in the subsequent sections and motivates the development of specific geotechnical criteria for safe saprolite extraction.

GEOTECHNICAL CHARACTERISTICS OF SAPROLITIC MATERIALS RELEVANT TO MINING

Grain size distribution and index properties

Granitic saprolites associated with ion-adsorption rare earth deposits typically exhibit sandy to silty textures, with variable proportions of gravel and clay depending on the degree of weathering and the proximity to residual rock cores. The materials generally occupy an intermediate domain between transported soils and residual weathered rock, preserving relic structures while exhibiting soil-like mechanical behaviour.

Typical grain size distributions observed in the study area indicate predominantly sandy matrices with varying amounts of fines. Coarse fractions are often represented by gravel and cobble-sized particles, associated with partially weathered granite fragments or embedded boulders. In representative profiles, sandy fractions commonly range between approximately 40 per cent and 60 per cent, with silts and clays contributing between 25 per cent and 45 per cent, depending on the horizon. A representative grain size distribution for the saprolitic materials is shown in Figure 3.

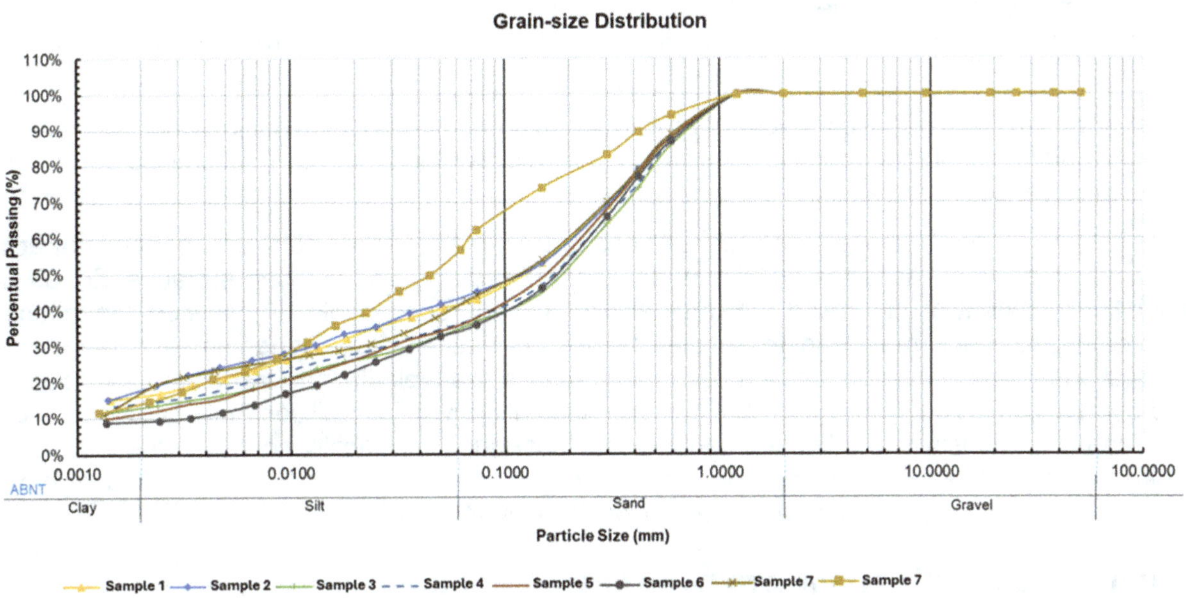

FIG 3 – Representative grain size distribution of the saprolitic materials, showing predominantly sandy matrices with variable fines content.

Natural unit weights for saprolitic and colluvial materials are typically on the order of 16 to 18 kN/m^3, with dry unit weights ranging between 14 and 16 kN/m^3. Measured void ratios generally fall between 0.6 and 0.9, indicating loose to medium-dense conditions consistent with highly weathered, porous materials. Natural moisture contents vary significantly, typically ranging from about 4 per cent in sandy alluvial materials to more than 12–15 per cent in finer colluvial or saprolitic horizons, reflecting the influence of seasonal groundwater conditions and local drainage patterns.

These index properties are consistent with a loose, heterogeneous and state-dependent material, whose mechanical response is strongly influenced by density, moisture content and structural fabric.

From an operational perspective, grain size distribution and index properties are fundamental for anticipating the behaviour of saprolitic materials under environmental exposure, particularly during rainfall events. Sandy-dominated matrices with variable fines content tend to exhibit high erodibility, increasing the potential for sediment mobilisation and downstream transport during wet periods. This behaviour is especially critical in valley-confined environments, where run-off concentration and limited drainage can amplify erosion processes. As a result, these materials require careful management through the implementation of surface drainage systems, sediment control structures, and localised protection measures to reduce environmental impacts and maintain operational continuity.

Strength and deformability parameters

In situ penetration tests and laboratory triaxial tests indicate relatively low penetration resistance and moderate shear strength, typical of loose to medium-dense sandy saprolites.

Standard penetration test results commonly fall within the range of approximately 1 to 15 blows, with average values near 7 to 8, indicating materials of low to moderate density and significant variability along the profile. This variability reflects the heterogeneous nature of the saprolite, including transitions between colluvial soils, residual horizons and partially weathered rock.

Shear strength parameters derived from correlations and triaxial tests indicate effective friction angles typically ranging between approximately 27° and 34°, with representative values around 30°. In some triaxial tests on sandy alluvial horizons, lower friction angles on the order of 22° have been obtained, reflecting loose conditions and contractive behaviour. Apparent cohesion values are generally low, commonly around 0 to 20 kPa, and are primarily associated with weak cementation or partial saturation effects.

Deformability parameters obtained from laboratory and *in situ* correlations indicate relatively low stiffness. Young's modulus values derived from triaxial tests typically range from approximately 5 MPa at low confinement levels to about 20 MPa at higher confining stresses. Correlations with penetration resistance generally yield average modulus values between 10 and 15 MPa for near-surface horizons.

Overall, the materials exhibit contractive behaviour under shear, particularly at low confining stresses and high moisture contents, which is consistent with the observed tendency for localised deformation and bearing capacity problems in field operations.

The relatively low shear strength and stiffness of saprolitic materials, combined with their contractive behaviour, directly influence the stability of shallow excavations and working surfaces. Zones with higher fines content or increased moisture are particularly susceptible to localised reductions in shear strength, which may lead to small-scale instabilities, surface sloughing, and progressive deformation along excavation faces. These mechanisms are typically not associated with classical deep-seated failures but rather with shallow, operationally driven instabilities that can evolve rapidly under changing moisture and loading conditions. Consequently, stability control in these materials requires close attention to excavation geometry, exposure time, and moisture management, particularly during periods of increased rainfall or groundwater rise.

Hydraulic conductivity and drainage characteristics

Hydraulic conductivity values measured in the near-surface sandy horizons generally fall within the range of approximately 7×10^{-8} to 1×10^{-6} m/s, with typical median values around 4×10^{-7} m/s. These values indicate moderately permeable materials, consistent with sandy or silty-sandy textures.

Despite these moderate permeability values, the overall drainage behaviour of the saprolitic profile is strongly influenced by stratification, the presence of less permeable horizons, and the occurrence of partially weathered granite cores. These features act as local hydraulic barriers, promoting the development of perched water tables and zones of localised saturation.

In valley-confined environments, lateral drainage is often restricted, and groundwater levels tend to fluctuate seasonally. During the wet season, the combination of shallow groundwater, limited drainage capacity and continuous exposure of fresh saprolite surfaces results in elevated moisture contents and reduced effective stresses in the working platforms.

For operational controls, the hydraulic conductivity and drainage characteristics of saprolitic materials are key factors controlling trafficability and the performance of working platforms. Despite moderate permeability values, the presence of stratification, less permeable horizons, and valley confinement often leads to restricted drainage and localised saturation. Under these conditions, the combination of high porosity and moisture sensitivity results in a rapid reduction in effective stress and bearing capacity when the material becomes wetted. This directly affects equipment mobility, increasing the risk of rutting, bogging, and operational delays. Therefore, effective drainage control and minimisation of water infiltration are essential to preserve platform integrity and ensure consistent operational performance, particularly during the wet season.

Implications for mining behaviour

The combined geotechnical characteristics of the saprolitic materials, loose structure, sandy to silty textures, moderate permeability, low stiffness and contractive response under shear, create conditions highly sensitive to changes in moisture and loading.

Under dry or partially saturated conditions, matric suction contributes to the apparent strength and stiffness of the material. However, this contribution can be rapidly lost during rainfall events or groundwater rise, leading to abrupt reductions in shear strength and bearing capacity.

In strip mining operations, where successive excavation stages expose new saprolite surfaces, the interaction between traffic loads, moisture variations and structural heterogeneity becomes the dominant control on stability. As a result, localised shear deformations, excessive rutting and progressive platform failures are more common than classical deep-seated slope failures.

These characteristics highlight the importance of incorporating hydraulic conditions, material state and operational sequencing into the geotechnical design criteria for safe saprolite extraction in valley-confined ion-adsorption deposits.

STRIP MINING GEOMETRY, OPERATIONAL STABILITY AND GEOTECHNICAL CONTROL CRITERIA

Extraction of ion-adsorption rare earth deposits in granitic saprolites is typically performed through shallow, top–down strip mining methods. In this approach, successive excavation stages are carried out to form temporary working platforms for equipment traffic, material handling, and leaching operations. Unlike mining in competent rock masses, where slope stability is governed primarily by well-defined shear strength parameters and structural discontinuities, the stability of saprolitic materials is strongly influenced by moisture conditions, material state, and operational sequencing.

The saprolitic materials described in the previous section exhibit sandy to silty textures, relatively low stiffness, moderate permeability, and contractive behaviour under shear. These characteristics produce a material that is workable under dry or partially saturated conditions, but whose strength and trafficability degrade rapidly when saturation increases or matric suction is lost. As a result, excavation geometry and drainage conditions become primary controls on operational stability.

In valley-confined environments, strip geometry is strongly controlled by the natural topography. Working platforms are typically narrow and elongated, laterally bounded by natural slopes and drainage channels. This configuration illustrates the operational context of the excavation areas, where the limited working width and shallow cuts are imposed by the valley geomorphology, as shown in Figure 4, which presents the slope mining conditions, and in Figure 5, which illustrates the strip-mining operations on the plateaus.

FIG 4 – Strip mining in valley-confined slopes, illustrating narrow working platforms, restricted drainage, and increased sensitivity to saturation and localised instability due to rock boulders.

FIG 5 – Strip mining operations on plateau areas, showing wider working platforms and more favourable drainage and stability conditions.

Under such conditions, lateral drainage is restricted and groundwater tends to concentrate along the excavation floor, producing high moisture contents and locally saturated zones within the saprolite.

Because of the relatively loose structure and moderate permeability of the saprolitic materials, the stability of working platforms is rarely governed by classical deep-seated failure mechanisms. Instead, the dominant operational issues include localised shear deformation, progressive bearing capacity failures, excessive rutting, and loss of trafficability. These mechanisms are consistent with the contractive behaviour observed in laboratory testing and with the low stiffness values measured *in situ*.

Strip mining geometry plays a dual role in this context. On one hand, the sequential excavation in narrow strips helps to control groundwater conditions by limiting the exposed area at any given time. This reduces the potential for widespread infiltration and allows localised drainage measures to be implemented more effectively. On the other hand, the narrow geometry also restricts surface water management options and increases the sensitivity of operations to localised zones of saturation.

The presence of partially weathered granite cores and large embedded boulders further complicates excavation stability. These rigid inclusions introduce abrupt contrasts in stiffness and permeability, creating stress concentrations and preferential flow paths. During excavation, the need to cut around or expose these features often produces irregular platform geometries and variable saprolite thicknesses. Under wet conditions, these contrasts lead to differential settlements and localised shear zones, particularly adjacent to the rock–saprolite interfaces.

Field experience indicates that many operational instabilities develop at shallow depths, often within the upper saprolitic horizons. These failures are typically progressive and associated with loss of matric suction, localised saturation, or inadequate drainage. In such cases, the material does not fail through well-defined rotational mechanisms, but rather through gradual deformation and reduction of bearing capacity under traffic loads.

From a geotechnical standpoint, the geometry of the strip mining sequence becomes a primary control variable for stability. The depth of each cut, the width of the strip, and the rate of lateral advance must be defined in accordance with the hydromechanical response of the saprolite. Excavation stages that are too deep or too wide may expose loose, moisture-sensitive horizons for prolonged periods, increasing the risk of strength loss and deformation.

Practical geotechnical criteria for safe saprolite extraction can therefore be defined based on the interaction between excavation geometry, drainage conditions, and material properties. Maximum excavation depths should be selected to limit the exposure of loose, moisture-sensitive saprolite and, where possible, to reach denser or partially cemented horizons that provide improved support for working platforms. In zones with highly sensitive materials, strip widths should be reduced and the duration of exposure before loading should be minimised, particularly during the wet season.

Control of surface and subsurface water constitutes the most critical element of operational stability. Surface drainage systems, such as longitudinal channels along the strips and perimeter ditches in valley bottoms, are essential to reduce direct infiltration into excavation surfaces. In areas with persistent saturation or low-permeability horizons, localised subsurface drainage may be required to relieve pore pressures and prevent prolonged saturation of the saprolite. Figure 6 shows the mining operations during the rainy season.

FIG 6 – Mining platforms during the rainy season, highlighting the effects of high moisture content and restricted drainage on operational stability.

Preparation of working platforms is equally important. Removal of loose colluvial or highly weathered surface materials, followed by localised compaction or improvement layers, helps reduce differential settlements and improves load distribution. In areas adjacent to granite boulders, special treatments may be required to mitigate stiffness contrasts and prevent concentration of shear strains.

Finally, construction sequencing should be treated as a primary geotechnical control rather than merely an operational decision. Rapid or simultaneous advances in multiple strips may significantly increase the risk of instability, particularly under wet conditions. Conversely, a controlled advance, combined with continuous observation of deformations, groundwater conditions, and traffic performance, allows for adaptive adjustments in excavation depth, strip width, and drainage measures.

In this framework, strip mining geometry becomes an integral component of geotechnical design. Stability is achieved not through reliance on peak strength parameters alone, but through the combined control of excavation geometry, groundwater conditions, platform preparation, and adaptive operational sequencing. This integrated approach provides a practical basis for safe and efficient extraction of saprolitic ion-adsorption rare earth deposits in valley-confined environments.

CONCLUSIONS

This study presented an integrated geotechnical framework for the safe extraction of ion-adsorption rare earth deposits developed in granitic saprolites within valley-confined environments. The analysis combined geological, geomorphological, hydrogeological, and geotechnical aspects with the operational characteristics of top–down strip mining, emphasizing the interaction between material behaviour, groundwater conditions, and excavation geometry.

The results demonstrate that granitic saprolites associated with ion-adsorption deposits constitute structurally complex and highly state-dependent materials. These deposits are characterised by heterogeneous weathering profiles, relic rock structures, sandy to silty textures, moderate permeability, and strength strongly influenced by matric suction and moisture conditions. The presence of partially weathered granite cores and large embedded boulders introduces additional heterogeneity, producing abrupt contrasts in stiffness and permeability that control both deformation patterns and local groundwater flow.

In valley-confined geomorphological settings, the hydraulic regime becomes a dominant factor governing material behaviour. Shallow groundwater levels, restricted lateral drainage, and the presence of less permeable horizons promote localised saturation and elevated pore pressures, particularly during the wet season. Under these conditions, saprolitic materials may experience rapid reductions in shear strength and stiffness, leading to progressive deformation, loss of trafficability, and localised bearing capacity failures rather than classical deep-seated slope instabilities.

The top–down strip mining method inherently creates temporary geometries that strongly influence the hydromechanical response of the material. Narrow, elongated platforms and sequential excavation stages modify local drainage conditions, stress distributions, and exposure times of moisture-sensitive horizons. As a result, operational stability becomes more dependent on excavation sequencing, drainage control, and platform preparation than on conventional peak strength parameters alone.

The study highlights that the dominant instability mechanisms in these deposits are associated with reduction of shear strength due to loss of matric suction under wet conditions, localised shear deformation and progressive bearing capacity failures in working platforms, differential settlements induced by stiffness contrasts between saprolite and embedded rock cores, concentration of pore pressures along stratigraphic interfaces and less permeable horizons.

These mechanisms are strongly controlled by the interaction between excavation geometry, groundwater conditions, and the intrinsic heterogeneity of the saprolite.

From a design and operational perspective, the findings demonstrate that stability in saprolitic strip mining is achieved primarily through hydromechanical control rather than through reliance on global slope stability criteria. The most effective mitigation measures are those that directly influence moisture conditions, stress distribution, and deformation control within the working platforms. These include limiting excavation depths to avoid prolonged exposure of loose, moisture-sensitive horizons, adjusting strip widths and sequencing to maintain partially drained conditions, implementing efficient surface and, where necessary, subsurface drainage systems, preparing working platforms through removal of weak materials and localised compaction, adopting adaptive operational strategies based on continuous monitoring of groundwater levels, deformations, and traffic performance.

The results also demonstrate that strip mining geometry can be used as an active geotechnical control mechanism. Sequential excavation in narrow strips reduces the exposed area, facilitates localised drainage, and limits the development of widespread saturation. When properly sequenced, the mining method itself contributes to the control of pore pressures and the preservation of adequate strength conditions within the saprolite.

From a broader perspective, the mechanisms and criteria identified in this study are applicable to a wide range of ion-adsorption deposits developed in deeply weathered granitic terrains across tropical and subtropical regions. Although specific material properties may vary from site to site, the fundamental controls associated with residual structure, moisture sensitivity, groundwater confinement, and excavation geometry remain consistent.

Overall, the findings reinforce that the safe extraction of saprolite-hosted ion-adsorption rare earth deposits requires a geotechnical approach explicitly tailored to the state-dependent and hydraulically controlled nature of these materials. The integration of geological characterisation, hydrogeological understanding, material behaviour, and operational sequencing provides a more realistic basis for stability assessment than approaches relying solely on conventional strength parameters.

The framework presented in this work offers practical engineering criteria for improving operational safety, reducing the occurrence of localised instabilities, and increasing the predictability of performance in strip mining operations within valley-confined saprolitic deposits. This integrated approach contributes to the development of safer and more sustainable rare earth mining practices in deeply weathered tropical environments.

REFERENCES

Bao, Z and Zhao, Z, 2008. Geochemistry of ion-adsorption type rare earth element deposits, *Ore Geology Reviews*, 33(3–4):519–535.

Butt, C R M, Lintern, M J and Anand, R R, 2000. Evolution of regoliths and landscapes in deeply weathered terrain: implications for geochemical exploration, *Ore Geology Reviews*, 16(3–4):167–183.

Chi, R and Tian, J, 2015. *Weathered Crust Elution-Deposited Rare Earth Ores* (Nova Science Publishers: New York).

Lambe, T W and Whitman, R V, 1969. *Soil Mechanics* (John Wiley and Sons: New York).

Value of mineralogical analysis using XRD for strategic mine planning and sustainable energy management in critical minerals mining and processing

U König[1], M Pernechele[2], L Negrão[3], I Duran[4] and Y Gomes[5]

1. Business Development Mining/Metals, Malvern Panalytical, Almelo, The Netherlands. Email: uwe.koenig@malvernpanalytical.com
2. Application Specialist, Malvern Panalytical, Almelo, The Netherlands.
3. Application Specialist, Spectris do Brasil, São Paulo, Brazil.
4. Application Specialist, Spectris Canada Inc., Quebec, Canada.
5. Hydrometallurgy Analyst, Borborema Recursos Estratégicos, Belo Horizonte, Brazil.

ABSTRACT

The accelerating global demand for critical minerals such as lithium, rare earth elements (REE), and nickel is transforming the mining industry and placing increasing pressure on operations to improve efficiency, reduce energy consumption, and ensure long-term sustainability. In this context, advanced mineralogical tools are essential for reducing geological, metallurgical, and economic risk across the entire mining value chain. X-ray diffraction (XRD), when integrated with advanced statistical and multivariate methods such as Rietveld refinement, cluster analysis, and Partial Least Squares Regression (PLSR), provides a robust and predictive framework for mineralogical domain definition, geometallurgical modelling, process optimisation, and sustainable energy management.

This paper presents four integrated case studies demonstrating the strategic value of XRD in critical minerals mining: (i) nickel laterite deposits, (ii) hard rock lithium ores, (iii) REE exploration samples, and (iv) prediction of mill energy consumption through Bond Work Index (BWi) modelling. In nickel laterites, XRD enables detailed characterisation of mineralogical zonation within saprolite and laterite horizons, supporting grade control, ore blending, and optimised processing routes. In lithium systems, XRD provides accurate quantification of lithium-bearing phases and supports prediction of processing performance and mineral transformation efficiency. In REE exploration, XRD combined with clustering techniques allows mineralogical domain classification, improving geological modelling and exploration targeting. Beyond resource characterisation, XRD-derived mineralogical data are directly linked to energy-intensive processes such as comminution, where mineralogy controls grindability and energy demand. By integrating XRD with PLSR and BWi modelling, predictive energy models can be developed to support Mine to Mill optimisation and energy-efficient circuit design. Collectively, these case studies demonstrate that XRD-based mineralogical analysis is not merely descriptive, but forms a strategic foundation for geometallurgy, mine planning, process optimisation, and sustainable energy management in critical minerals mining and processing.

INTRODUCTION

Critical minerals such as nickel, lithium, and rare earth elements (REE) are fundamental to the global energy transition, electrification, and digital infrastructure (Crundwell *et al,* 2011; Bradley *et al*, 2013). Their rapidly growing demand places increasing pressure on mining operations to improve resource efficiency, reduce energy consumption, and minimise environmental impact, while maintaining economic viability and long-term supply security. As the mining sector moves toward decarbonisation and sustainability targets, energy-intensive unit operations such as comminution, thermal processing, and hydrometallurgical extraction have become key focus areas for optimisation.

Traditional mine planning and process design have relied primarily on bulk chemical assays and empirical metallurgical testing. While these approaches provide valuable information on grade and metallurgical performance, they do not adequately capture mineralogical variability, phase associations, crystallinity, and structural complexity, which fundamentally control processing behaviour, metallurgical recovery, equipment wear, and energy consumption. As a result, variability in mineralogy is often only detected during processing, leading to suboptimal performance, increased operational risk, and inefficient energy use.

X-ray diffraction (XRD) provides direct access to crystalline phase information and quantitative mineralogical composition. When integrated with advanced statistical and multivariate techniques, XRD evolves from a descriptive analytical method into a predictive geometallurgical tool. This paper demonstrates how XRD-based mineralogical analysis forms a foundational pillar for critical mineral exploration, mining, processing, and energy optimisation, enabling strategic mine planning and sustainable energy management across the entire Mine to Mill value chain, Figure 1.

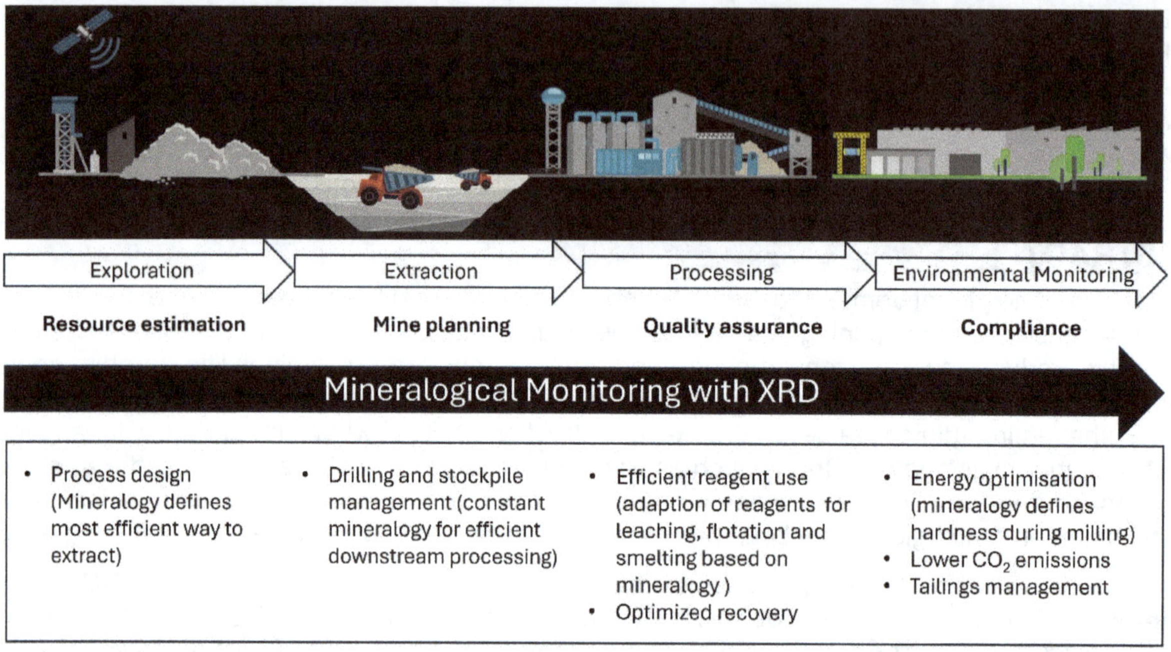

FIG 1 – Conceptional integration of XRD in mining with exploration, mine planning processing and environmental management.

METHODS

X-ray diffraction (XRD)

XRD measurements were performed on powdered samples prepared using standardised milling and pressing procedures to minimise particle size effects, preferred orientation, and preparation bias. Diffraction patterns were collected using a Malvern Panalytical Co-anode Aeris Minerals compact diffractometer, featuring measurement times of about 5 min per sample. Phase identification and quantitative analysis were carried out using HighScore software (Degen *et al,* 2014).

Quantitative analysis (Rietveld refinement)

Full-pattern Rietveld refinement was applied to obtain quantitative mineral phase compositions (Rietveld, 1969). This approach enables accurate phase quantification, deconvolution of overlapping reflections, structural discrimination (eg α- versus β-spodumene), and consistent comparison across ore types and processing streams. Rietveld-based quantification provides a reproducible and statistically robust mineralogical data set suitable for multivariate modelling and predictive analytics.

Statistical and multivariate tools

Cluster analysis was applied to identify mineralogical domains and sample groupings prior to detailed quantification. This approach enables objective domain definition based on mineralogical similarity rather than lithological assumptions. Partial Least Squares Regression (PLSR) models were developed to correlate mineralogical variables with chemical grades, metallurgical performance, process parameters, and energy consumption indicators, enabling predictive modelling across the Mine to Mill value chain.

CASE STUDIES

Nickel laterites

Nickel laterite deposits exhibit strong mineralogical zonation between lateritic and saprolitic horizons, which fundamentally controls grade distribution, processing behaviour, and metallurgical performance (König, 2021). XRD enables robust discrimination between ore types, quantitative determination of key mineral phases, and correlation of mineralogy with nickel grade, reducibility, and smelting or leaching performance.

Cluster analysis reveals distinct mineralogical domains corresponding to laterite and saprolite horizons, Figure 2. Lateritic domains are characterised by high goethite contents, while saprolitic zones are dominated by lizardite, olivine, and pyroxene assemblages. Transitional zones are resolved using fuzzy clustering, capturing gradual mineralogical transitions rather than imposing artificial lithological boundaries. This enables more realistic geological modelling and improved ore blending strategies.

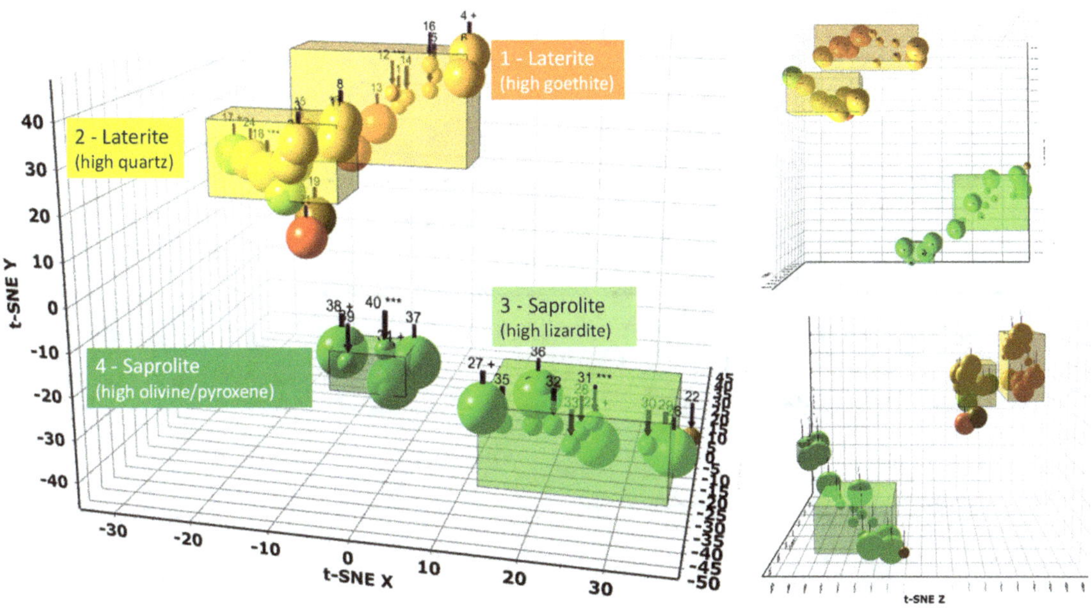

FIG 2 – Cluster separation of lateritic versus saprolitic ores based on XRD data (König, 2021).

XRD-based mineralogical monitoring was extended across multiple processing stages, including roasting, pre-reduction, matte formation, and slag generation (Crundwell *et al,* 2011). This enables improved control of phase stability, slag chemistry, reducibility, refractory interaction, and metal recovery, Table 1. Mineralogical information supports optimisation of ore blending, energy input management, furnace performance, and recovery of nickel and cobalt, demonstrating the value of XRD throughout the entire nickel laterite value chain.

TABLE 1

Value and parameter to increase the profitability of nickel laterite processing (König, 2021).

Value	XRD – Tool
Optimisation of ore blends from various nickel laterite deposits	Cluster analysis
Adjustment of superheat in the feed and optimisation of energy costs	Mineralogy of ore blend
Control of mineralisation acidity	Silicate composition
Prevention of highly corrosive slag causing erosion of the refractories	Silicate composition
Better reducibility in the furnace	Olivine content
Boost nickel recovery rates and reduction of metal loss in slag	Slag composition
Increase of cobalt recoveries	Co-bearing minerals

Lithium

XRD enables accurate identification and quantification of lithium-bearing minerals, including spodumene, lepidolite, and petalite, as well as structural discrimination between α- and β-spodumene polymorphs (London, 2008; Pöllmann and König, 2021). This capability is essential for monitoring mineral transformations during mining, beneficiation, calcination, and downstream processing.

Ore mineralogy directly controls flotation performance, reagent selection, and spodumene recovery. XRD therefore provides critical feedback for process optimisation and metallurgical control, Figures 3 and 4. Downstream, XRD monitoring of calcination processes allows direct tracking of the α- to β-spodumene transformation, enabling optimisation of thermal treatment efficiency, energy input, and lithium recovery. Residue and by-product mineralogy further supports process optimisation and recovery improvement.

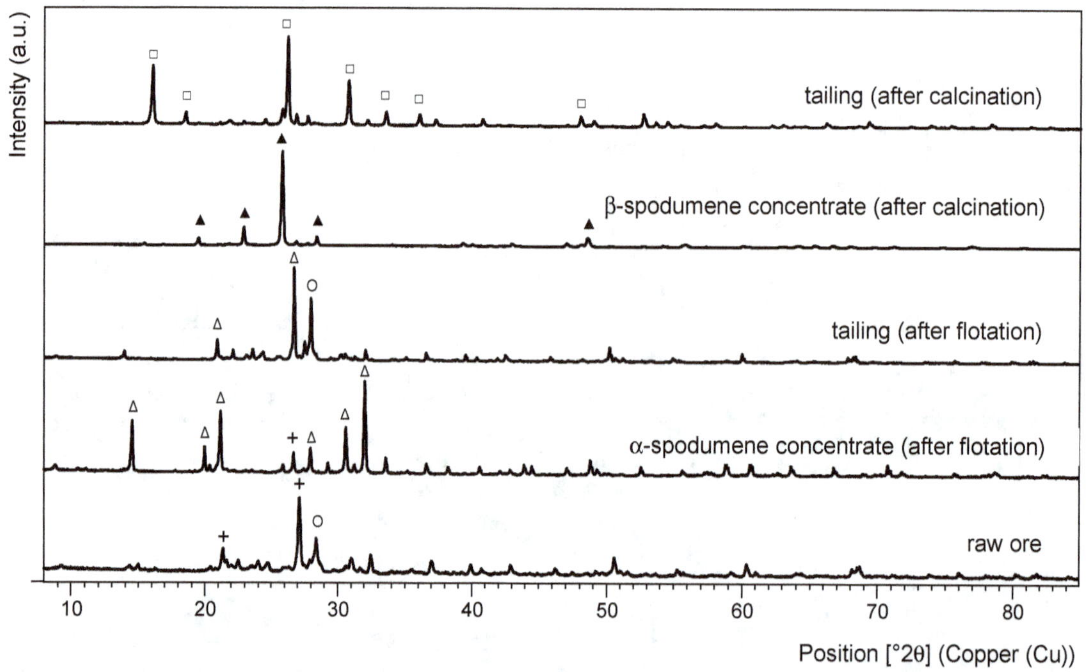

FIG 3 – XRD patterns from lithium ore, concentrates and tailing after flotation and calcination; Δ – α-spodumene, ▲ – β-spodumene, O – albite, + – quartz, □ – analcime, modified after Norberg *et al* (2020).

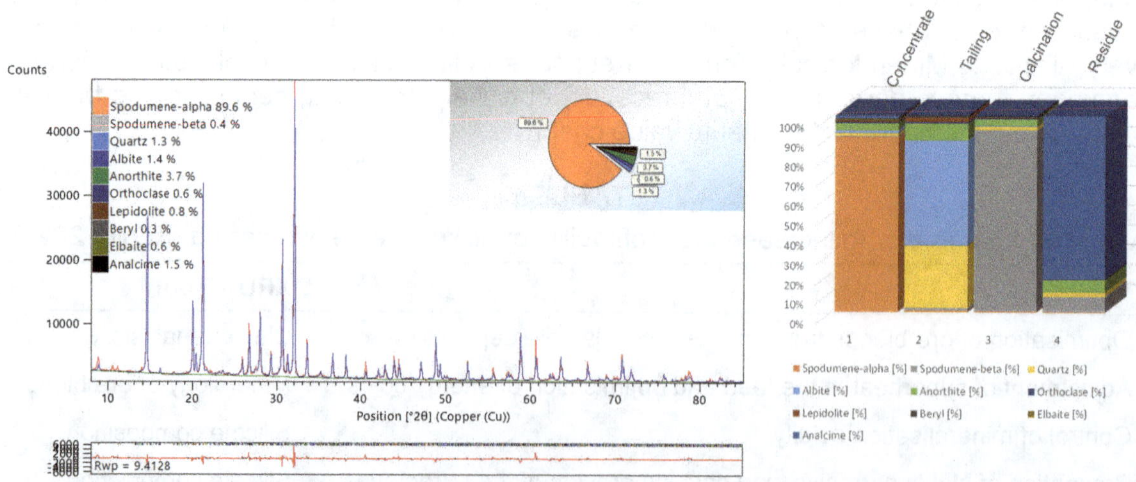

FIG 4 – Example quantification of a spodumene concentrate (left) and composition of concentrate, tailing, calcination product and residue (right) (Pöllmann and König, 2021).

Quantitative mineralogical analysis of lithium pegmatite ores reveals complex assemblages dominated by spodumene, quartz, feldspars, and accessory lithium minerals. Rietveld refinement provides precise mineral proportions, enabling definition of mineralogical domains within the orebody and supporting ore sorting and blending strategies across different geological units.

Rare earth elements (REE)

REE projects increasingly require detailed mineralogical understanding to reduce geological, metallurgical, and economic risk from exploration through processing. XRD-based mineralogical analysis was applied to REE ores from a high-concentration deposit in Brazil, representing one of the most prospective emerging REE provinces globally, Figure 5.

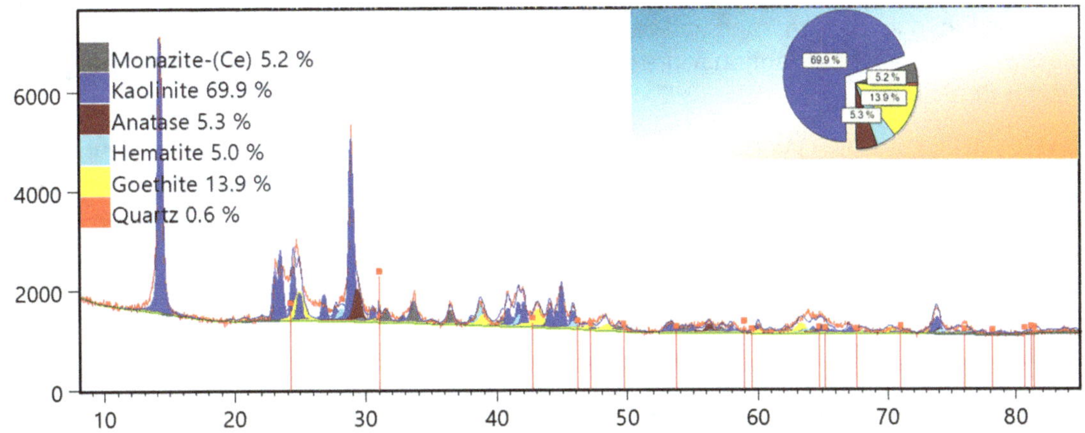

FIG 5 – Quantitative REE mineralogy based on a Rietveld refinement.

Samples covering a wide range of REE grades, from ultra-high to low total rare earth oxides (TREO), were analysed to capture mineralogical variability. Quantitative analysis identified monazite as the dominant REE-bearing phase in one portion of the deposit, with additional phosphate and silicate minerals contributing to REE deportment. Florencite was identified as a secondary REE carrier, highlighting a previously under-recognised contributor to overall REE distribution.

Cluster analysis defined distinct mineralogical domains, including an ultra-high-grade monazite-rich domain, Figure 6. The mineralogical patterns align with the regional geological framework and provide a robust basis for exploration targeting, resource modelling, and process design. Understanding REE mineral hosts, crystallinity, and phase variability supports optimisation of beneficiation strategies, reagent selection, and recovery forecasting, enabling efficient and sustainable REE resource development.

Kaolinite Al2Si2O5(OH)4	Quartz SiO2	Orthoclase $KAlSi_3O_8$	Phlogopite KMg3(Si3Al)O10(F,OH), Phengite KAl(Fe, Mg)	Monazite Ce(PO4)	Florencite CeAl3(PO4)2(OH)6	Hematite Fe2O3	Goethite FeOOH	Ilmenite FeTiO3	Anatase TiO2	
49	21	17	13							Low REE
18	33	28	21							
37	5				4	15	30	1	8	
57	12				3	9	16		3	
85	12		1	<1	<1	2				High REE (goethite, high florencite)
51	44			2	2	1				
67	27			3	2	1				
65				7		5	18		5	
70	1			5		5	14		5	Very high REE (monazite)
21				43		22			14	

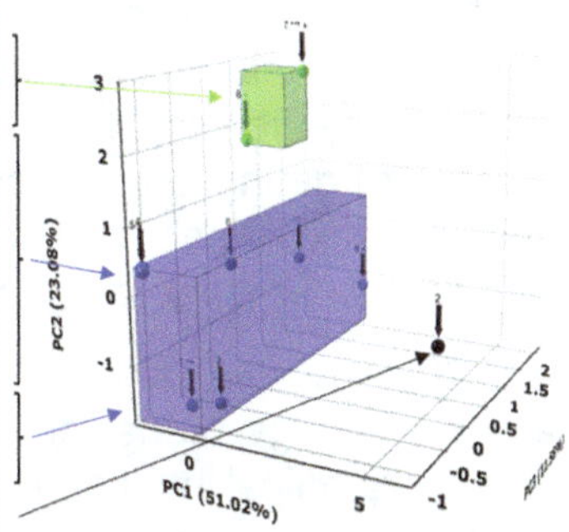

FIG 6 – REE mineralogical domain classification using cluster analysis, numbers represent % of the minerals quantified.

Bond work index (BWi) and energy modelling

The BWi is a fundamental parameter for predicting grinding energy consumption and designing comminution circuits (Bond, 1961). Traditional BWi testing is time-consuming, costly, and operationally intensive, making large-scale geometallurgical energy mapping impractical.

XRD provides a rapid and low-cost alternative by supplying detailed mineralogical data that can be statistically correlated with grindability behaviour. When integrated with PLSR, XRD data enable reliable prediction of BWi values across large data sets (Duran *et al*, 2024).

In a case study involving 130 representative samples, XRD – PLSR modelling achieved a prediction error of only 0.6 kWh/t, demonstrating strong predictive performance, Figure 7. This approach enables deposit-scale geometallurgical energy mapping and supports Mine to Mill optimisation, energy-efficient circuit design, and sustainable grinding strategies. The direct linkage between mineralogy and energy consumption represents a major advancement in predictive energy modelling for mineral processing.

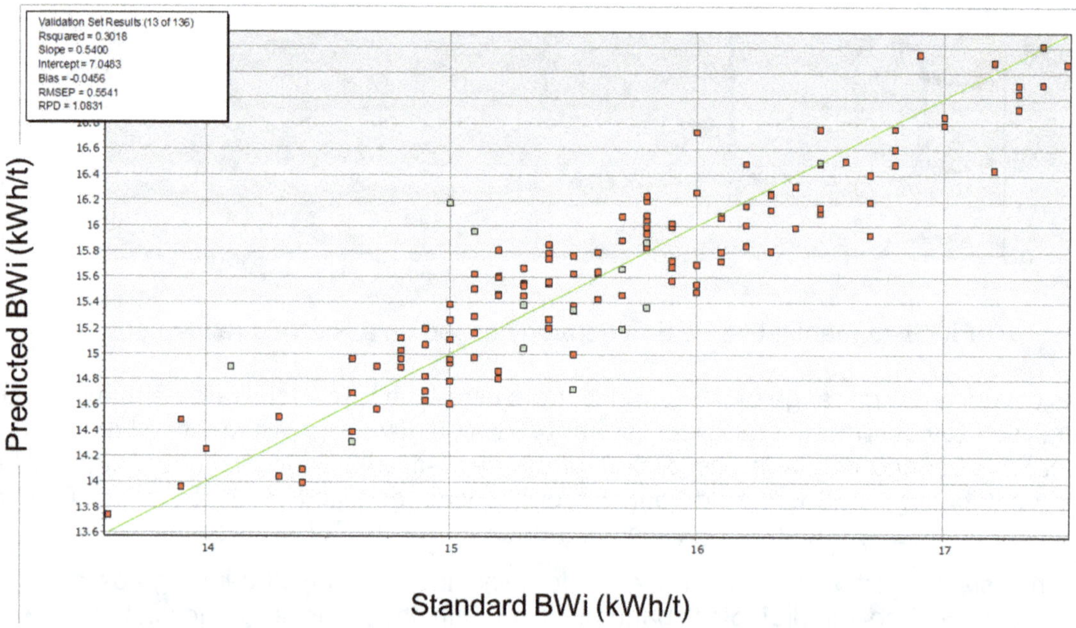

FIG 7 – Predicted (XRD) versus measured BWi values (Duran *et al*, 2024).

CONCLUSIONS

XRD is not an auxiliary analytical technique, but a strategic foundation for critical minerals exploration, mining, and processing. Through four integrated case studies – nickel laterites, lithium hard rock ores, REE systems, and BWi-based energy modelling – this paper demonstrates the central role of XRD in linking mineralogical knowledge to geometallurgical modelling, process optimisation, and sustainable energy management.

In nickel laterites, XRD enables definition of mineralogical domains that control smelting and leaching behaviour. In lithium systems, it supports prediction of grindability, liberation, and mineral transformation efficiency. In REE projects, it provides mineralogical domain classification that guides exploration targeting and process design. When integrated with multivariate modelling and BWi correlations, XRD enables reliable prediction of grinding energy requirements, supporting energy-efficient circuit design and Mine to Mill optimisation.

While XRD is not a real-time process control tool, it is indispensable for strategic mine planning, sustainable energy management, and responsible development of critical mineral resources. By linking mineralogical intelligence to predictive models, XRD empowers mining operations to optimise energy use, reduce risk, and make informed decisions that support long-term sustainability and operational efficiency in critical minerals mining and processing.

REFERENCES

Bond, F C, 1961. *Crushing and Grinding Calculations*, British Chemical Engineering.

Bradley, D, Munk, L A, Jochens, H, Hynek, S and Labay, K A, 2013. A Preliminary Deposit Model for Lithium Brines, Open-File Report 2013–1006, US Geological Survey. https://doi.org/10.3133/ofr20131006

Crundwell, F K, Moats, M S, Ramachandran, V, Robinson, T G and Davenport, W G, 2011. *Extractive Metallurgy of Nickel*, Cobalt and Platinum Group Metals, Elsevier.

Degen, T, Sadki, M, Bron, E, König, U and Nénert, G, 2014. The HighScore suite, *Powder Diffr*, 29:S13–S18.

Duran, I R, Provencher, M E, Pernechele, M, König, U, Vosoughi, E and Bergeron, Y, 2024. New tools to increase mill efficiency during gold ore processing – X-ray diffraction (XRD) and chemometrics, 56th Annual Canadian Mineral Processors Conference (CMP 2024), Ottawa, Canada.

König, U, 2021. Nickel laterites—Mineralogical monitoring for grade definition and process optimization, *Minerals*, 11:1178.

London, D, 2008. *Pegmatites*, Canadian Mineralogist Special Publication.

Norberg, N, König, U, Bhaskar, H and Narygina, O, 2020. Accurate mineralogical analysis for efficient lithium ore processing, in *Proceedings of ALTA 2020, 25th Annual Lithium & Battery Technology Conference, including Trends in Battery Technology Forum*, pp 88–93 (ALTA Metallurgical Services: Melbourne).

Pöllmann, H and König, U, 2021. Monitoring of lithium contents in lithium ores and concentrate—Assessment using X-ray diffraction (XRD), *Minerals*, 11:1058.

Rietveld, H M, 1969. A profile refinement method for nuclear and magnetic structures, *Journal of Applied Crystallography*, 2:65–71.

Open Pit

Drill and blast – design, execution and scale

Biodegradable air deck application in mine surface blasting

T Azarel[1], *R Heryadi*[2] *and H Utama*[3]

1. Student, SMA Pradita Dirgantara, Boyolali, Central Java 57375, Indonesia. Email: teresa.azarel@praditadirgantara.sch.id
2. AAusIMM, Lecturer, Science and Technology Faculty, Universitas PGRI Kalimantan, Banjarmasin, South Kalimantan 70122, Indonesia; COO, PT Suma Yogara Sejahtera, Pancoran, South Jakarta 12810, Indonesia. Email: rudy.heryadi@breny.my.id
3. MAusIMM, Mining Manager, PT Indo Muro Kencana, Murung Raya, Central Kalimantan 73961, Indonesia. Email: heru.utama@imkgold.co.id

ABSTRACT

The term 'air deck' can be defined as a tool that creates an air gap in a blasthole to reduce bulk explosive consumption and improve fragmentation in surface mine blasting operations. In addition, the primary reason for air deck application in mine blasting is to reduce bulk explosive consumption, decrease blasting costs, and also improve blasting fragmentation. In general, there are three types of air decks used in surface mine blasting: top, middle, and bottom. The most common air deck types used in blasting practices in surface mining operations in Indonesia are bottom air decks, stemming plugs, and gas bags. In addition to commercial and technical benefits, the air deck can also help reduce GHG emissions from blasting activities. However, the conventional plastic material used for the air deck still contributes to life cycle GHG emissions from its production and transportation. Plastic debris has, over time, harmed the environment; therefore, research on air deck materials with lower life cycle GHG emissions and greater biodegradability is urgent. It is proposed that biodegradable material can be used as an air-deck alternative material either in the top column of the blasthole or the bottom column of the blasthole. Further, this study evaluates a sustainable alternative: a bio-composite air deck device fabricated from Polylactic Acid (PLA) reinforced with 40 per cent soybean husk agricultural waste. By conducting a desk study of an Indonesian limestone quarry, the study compares the proposed bio-composite with industry-standard plastic devices. The results indicate that the transition to Soy-PLA achieves a 78.6 per cent reduction in device-related GHG emissions, lowering the carbon footprint from 98.8 kg to 21.1 kg CO_2-eq per blast. Mechanically, despite the inclusion of waste filler, the device retains a compressive Safety Factor exceeding 1400 under static borehole loads, confirming its structural integrity during charging. Crucially, the bio-composite utilises a programmed obsolescence mechanism driven by hydrolytic degradation. Unlike conventional plastics, which persist for centuries, the proposed device degrades into harmless biomass within 12 to 24 months of burial. This transition eliminates approximately 972 kg of permanent plastic waste per site annually. This research demonstrates that valorising agricultural by-products into blasting consumables offers a viable engineering pathway toward zero-waste mining without compromising operational safety or fragmentation efficiency.

INTRODUCTION

The mining industry stands at a critical time, navigating the tension between the escalating demand for mineral resources driven by urbanisation and the green energy transition, and the urgent mandate to decarbonise industrial operations. Mining is an energy-intensive process, with rock comminution (drilling, blasting, and crushing) accounting for a significant percentage of the world's industrial energy consumption (Mudd, 2010; Norgate and Haque, 2010; Azadi *et al,* 2020). As the sector moves toward zero-waste frameworks and Environmental, Social, and Governance (ESG) compliance, the focus has historically been placed on macro-scale interventions, such as the electrification of haul trucks, renewable energy grids for processing plants, and tailing management. However, the environmental footprint of auxiliary consumables, particularly single-use materials used in the drilling and blasting cycle, remains largely unquantified and overlooked. Traditionally, the primary environmental concern in blasting has been the chemical release of Greenhouse Gases (GHG) and toxic fumes (NOx, CO) resulting from the detonation of Ammonium Nitrate Fuel Oil (ANFO) and bulk emulsion explosives (BEE). The stoichiometric reaction of ANFO generates approximately 0.189 kg of CO_2 per kilogram of explosive detonated, contributing significantly to a mine's Scope 1 emissions (Stantec Consulting, 2015; Heryadi, Utama and Prasmoro, 2024). This paper addresses that gap by interrogating the material life cycle of air decking devices.

Drilling and blasting constitute the primary phase of rock fragmentation in hard rock mining and quarrying. The efficiency of this stage dictates the downstream economics of loading, hauling, and crushing. To mitigate both explosive consumption and ground vibration, the industry has adopted the Air Decking technique. Based on the principle of enhanced shock wave reflection, air decking introduces a specific void space within the explosive column (either at the bottom, middle, or top of the blasthole) (Melnikov *et al,* 1978; Jhanwar, 2011). The mechanics are well documented; the air gap allows the detonation gases to expand into the void, prolonging the gas pressure phase and generating secondary stress waves that enhance microfracturing in the rock mass. By decoupling the charge, mines can reduce the specific powder factor by 10 per cent to 45 per cent (Hayat, Alagha and Ali, 2019; Monjezi *et al,* 2022), without compromising the fragmentation size distribution.

In contemporary quarrying operations, including the reference site for this study (Heryadi, Utama and Aprisko, 2025), these barriers are mechanical devices consisting of a 'cup' and a 'stand' (or spacer) located at the bottom of the blasthole. While air decking successfully reduces the blast's chemical footprint, it introduces a material paradox: the proliferation of non-biodegradable solid waste. The available air decking devices at the reference site are manufactured from Polypropylene/PP (cup) and include Polyvinyl Chloride/PVC (stand) components to enhance rigidity (Fröhlich, Liebich and Volz, 2016; Fröhlich, Ludmann and Liebich, 2023; Heryadi, Utama and Aprisko, 2025). The environmental persistence of PP and PVC is well established, with degradation timelines to achieve their half-life spanning 780 years for PP to more than 2500 years for PVC in terrestrial environments (Chamas *et al,* 2020).

In a typical large-scale quarry firing 100 to 200 holes per blast multiple times a week, the cumulative introduction of plastic into the lithosphere is substantial. Upon detonation, the plastic device is not vaporised; rather, it is shattered into thousands of irregular fragments ranging from macroplastics to microplastics, with sizes less than 5 mm (Rochman, 2018). These fragments are inextricably mixed into the muck pile. While larger pieces may be screened out during processing, the microplastic fraction remains in the crushed product or is washed into the mine's sediment ponds and surrounding soil matrix. Furthermore, the production of these petrochemical polymers has a high embodied carbon footprint, with approximately 3.5 kg CO_2-eq/kg for PP and nearly 8.0 kg CO_2-eq/kg for PVC (Climatiq, 2021; Liang and Yu, 2023).

To resolve this paradox, material science must intersect with mining engineering. This study proposes the theoretical substitution of petrochemical plastics with a Biodegradable Composite specifically made for the high-stress environment of a blasthole. The proposed material matrix utilises Polylactic Acid (PLA) reinforced with Soybean Husk waste.

Soybean husks, by-products of soybean processing activities such as tempe making, are produced in large quantities. Its seed coat, or epidermis, is composed mainly of 20 per cent cellulose, 50 per cent hemicellulose, lignin, protein, ash, and extractives, making it a potential bioplastic material suitable for stem plugs (Riaz, 2015). Using the Tukey HSD analysis, sorbitol is shown to maintain bioplastic structure-retaining qualities. However, a stronger, biodegradable thermoplastic is needed for stem plug use.

PLA is a thermoplastic polyester derived from renewable biomass, typically fermented plant starch (corn or sugarcane) (Štimac *et al,* 2021; Kuru and Kaya, 2023). It is fully biodegradable under industrial composting conditions and significantly degradable in soil environments over time. However, pure PLA can be brittle and cost-prohibitive for single-use industrial applications. The cost-prohibitive and brittle nature of PLA necessitates the inclusion of a filler. Soybean husk, the outer covering of the soybean seed, is an abundant agricultural byproduct (Mohanty, Misra and Drzal, 2002; Sakti *et al,* 2025). In many agro-industrial regions, it represents a waste stream that requires disposal. By valorising this biomass as a reinforcement filler in the PLA matrix, two synergistic effects are achieved:

1. Cost and Carbon Reduction, where the husk acts as a cost-reducer, displacing the volume of the more expensive PLA resin, and has almost neutral carbon footprint.

2. Accelerated Degradation, where the hydrophilic nature of natural fibres (cellulose and hemicellulence in the husk) promotes moisture absorption from the soil.

The critical constraint for any air deck device is mechanical integrity. The device must possess sufficient compressive strength to support the stemming and explosives load, the weight of the crushed stone aggregate or drill cuttings used to seal the top of the hole, and, in bottom-deck applications, to resist the hydrostatic pressure of the explosive column above it. If the device collapses before detonation, the air gap is lost, and blast efficiency is compromised.

Theoretical analysis suggests that a PLA itself can achieve compressive strengths of 40–65 MPa (Garlotta, 2002), which is well within the safety factor required for an 89 mm diameter blasthole with a typical stemming height of 2 to 3 m (Heryadi, Utama and Aprisko, 2025). While a PLA-soybean husk composite study is still lacking, some similar studies suggest moderate compressive strength for PLA-soybean husk composite (Nyambo, Mohanty and Misra, 2010; Dananjaya *et al,* 2025; Krawczyk *et al,* 2026). Therefore, the substitution is not merely environmentally desirable but operationally feasible.

This paper presents a theoretical comparison of the Life Cycle Assessment (LCA) of standard PP/PVC air deck devices with that of the proposed Soybean husk-PLA bio-composite. Utilising blasting parameters and geometric data from a reference limestone quarry operations as detailed in Heryadi, Utama and Aprisko (2025), this study aims to:

- Quantify the reduction in 'Scope 3' Greenhouse Gas emissions achieved by switching device materials.
- Calculate the mass of permanent plastic waste eliminated from the mine site annually.
- Establish a theoretical degradation timeline for the bio-composite in a tropical soil environment.
- Demonstrate that the transition to biodegradable accessories supports the broader industry goal of sustainable, low-carbon blasting without necessitating changes to drill patterns or explosive charging protocols.

By integrating agricultural waste streams into the mining supply chain, this research advocates for a circular economic model in which the waste of one industry (agriculture) becomes a sustainable resource for another (mining), ultimately reducing the anthropogenic scar left by mineral extraction.

MATERIAL AND METHODS

This study employs a theoretical comparative modelling approach to evaluate the environmental efficacy of substituting conventional polymer blasting devices with a bio-composite alternative. The methodology is divided into four distinct phases:

1. Reference Site Characterisation and Blasting Design.
2. Theoretical Formulation of the Bio-Composite Material.
3. GHG Emission Calculation Model.
4. Waste Accumulation Modelling.

To ensure the theoretical simulation reflects real-world operational conditions, baseline data were adopted from an active limestone quarry operation in Central Sulawesi, Indonesia, as detailed in recent industry studies (Heryadi, Utama and Aprisko, 2025). The reference site utilises a conventional drilling-and-blasting method for overburden removal and limestone production.

The blasting geometry selected for this simulation utilises a predetermined pattern designed for 89 mm diameter blastholes. The specific design parameters are tabulated in Table 1.

TABLE 1
Specific design parameters of quarry.

Parameter	Value/description
Burden (B)	3.0 m
Spacing (S)	4.0 m
Average hole depth	3.6 m
Stemming height	2.0–2.5 m
Explosive type	Ammonium Nitrate Fuel Oil (ANFO) (Density (approximately 0.85 g/cm^3)
Initiation system	Electric detonators

In the baseline scenario (Scenario A, Control Group), the site employs a bottom-hole air decking technique utilising a commercial cup-stand type assembly (Figure 1). This assembly consists of two distinct polymer components: a cup (gas retention barrier) and a stand (spacer). The conventional material composition is identified as Polypropylene (PP) for the cup and Polyvinyl Chloride (PVC) for the rigid stand. The total combined weight of one conventional unit is standardised at 0.12 kg. Scenario B is the scenario in which a bio-composite material is used, with a mixture of PLA and soybean husk forming the air deck. This approach aims to compare the qualities of conventional and biodegradable stemming plugs.

FIG 1 – Bottom air deck (Sysdeck™ Blastfren).

Theoretical formulation of the bio-composite material

The experimental variable in this study is the material composition of the air deck device. The bio-composite is engineered to match the mechanical specifications of the conventional device while introducing biodegradability.

Matrix and reinforcement

The proposed material is a particle-reinforced polymer composite consisting of:

- Matrix: Polylactic Acid (PLA) resin. PLA is a thermoplastic aliphatic polyester derived from renewable resources, specifically fermented plant starch. It serves as the binder, providing compressive strength and hydro-resistance.
- Reinforcement: Pulverised Soybean Husk. The husk is selected because it is an abundant agricultural waste product. It acts as a low-density filler that reduces the device's overall cost and accelerates degradation by introducing hydrophilic cellulose fibres (Bledzki and Gassan, 1999) into the hydrophobic polymer matrix.

Theoretical mix ratio and density

For this simulation, a mass fraction ratio of 60 per cent PLA/40 per cent Soybean Husk is utilised. Density Implications: Standard PP has a density of 0.89–0.92 g/cm^3; for the calculation, a value of 0.90 g/cm^3 was used, and standard PVC density is 1.38 g/cm^3 (Fröhlich, Liebich and Volz, 2016; Fröhlich, Ludmann and Liebich, 2023), whereas pure PLA is denser at 1.25 g/cm^3 (Garlotta, 2002). However, soybean husk has a low bulk density (0.4 g/cm^3) (Kg-m3.com, 2020). The composite is theoretically modelled to achieve a bulk density of approximately 0.95–1.0 g/cm^3, ensuring the device remains lightweight enough for field handling while maintaining sufficient structural wall thickness.

Mechanical load and safety factor modelling

To ensure the proposed material functions without failure, the static load exerted by the explosive column (bottom-deck configuration) or the stemming material must be calculated. The static pressure P_{static} acting on the device is derived using Equation 1:

$$P_{Static} = \rho \; x \; g \; x \; h \qquad (1)$$

Where:

ρ	= Density of the material above the device (ANFO; approximately 850 kg/m^3)
g	= Gravitational acceleration (9.81 m/s^2)
h	= Height of the column above the device (2.5 m)

The material competency is then evaluated by calculating the Safety Factor (SF) using Equation 2. A Safety Factor > 1.0 indicates the material will not yield; however, industrial applications typically require SF > 2.0 to account for dynamic stemming impact (Budynas and Nisbett, 2015).

$$SF = \frac{\sigma_{yield}}{P_{Static}} \qquad (2)$$

Where:

σ_{yield}	= Compressive yield strength of the material (MPa)
P_{Static}	= Calculated static pressure (MPa)

GHG emission calculation model

The environmental impact is quantified using a 'Cradle-to-Gate' emission factor approach (World Resources Institute (WRI) and World Business Council for Sustainable Development (WBCSD), 2005; Pryshlakivsky and Searcy, 2013), focusing specifically on the global warming potential (GWP) of the material manufacturing process. The total GHG reduction is calculated by comparing the embodied carbon of the conventional fleet with that of the bio-composite fleet for a single typical blast of 135 holes. The calculation utilises Equation 3.

$$E_{Total} = N_{holes} \; x \; m_{device} \; x \; EF_{material} \qquad (3)$$

Where:

E_{Total}	= Total GHG emissions (kg CO_2-eq)
N_{holes}	= Number of blastholes per blast (135)
m_{device}	= mass of the air deck device (0.12 kg)

$EF_{material}$ = Emission Factor of the specific material (see Table 2)

The emission factor (EF) utilised is listed in Table 2.

TABLE 2

Greenhouse Gas (GHG) emission factors utilised for the theoretical Cradle-to-Gate Life Cycle Assessment.

Material component	Emission factor (EF)	Basis of derivation	Reference
Polypropylene (PP)	3.576 kg CO_2-eq/kg	Standard polymerisation energy inputs for virgin resin	(Fröhlich, Liebich and Volz, 2016)
Polyvinyl Chloride (PVC)	7.83 kg CO_2-eq/kg	High factor due to energy-intensive chlorine electrolysis and rigidising additives	(Fröhlich, Ludmann and Liebich, 2023)
Soybean-PLA Composite	1.30 kg CO_2-eq/kg	Weighted average calculation: • PLA Resin (60%): ~1.8 kg CO_2-eq/kg • Soybean Husk (40%): ~0.1 kg CO_2-eq/kg • *Calculation:* (0.6 × 1.8) + (0.4 × 0.1) + 0.18 (Safety Margin)	(Vink and Davies, 2015) (Pishvaee and Razmi, 2012)

Waste accumulation modelling

Solid waste accumulation is calculated as a linear function of blast frequency. The annual accumulation W_{annual} is determined by Equation 4:

$$W_{annual} = N_{holes}\ x\ m_{device}\ x\ f_{blast} \tag{4}$$

Where:

f_{blast} = Frequency of blasts per annum (assumed 60 blasts/annum)

The degradation timeline is qualitatively modelled based on the literature on hydrolytic degradation rates of PLA-natural fibre composites, in which soil moisture absorption by the husk causes swelling and matrix fracturing, allowing microbial enzymatic breakdown (Tokiwa *et al,* 2009; Karamanlioglu, Preziosi and Robson, 2017).

RESULTS AND DISCUSSION

The theoretical simulation yields significant variations in environmental performance between the baseline (Conventional PP/PVC) and the proposed (Soybean Husk-PLA) scenarios. The results are categorised into material mechanical competency, greenhouse gas (GHG) emission abatement, and solid waste dynamics.

Theoretical mechanical competency and integrity

The primary operational prerequisite for the air deck device is structural integrity under borehole loading. Using the parameters defined in Section 2.2, the static pressure exerted by the explosive column was calculated. Subsequently, the Safety Factors (SF) for both the conventional PVC/PP assembly and the proposed Soybean husk-PLA Composite were derived using Equation 2.

The results, presented in Table 3, demonstrate that although the inclusion of 40 per cent agricultural waste filler nominally reduces the compressive yield strength of the material compared to virgin PVC, the structural capacity remains excessive for the application. With a Safety Factor exceeding 1400, the bio-composite is theoretically fully capable of supporting the stemming load without deformation or collapse. This result validates the study's assumption of a 1:1 functional performance ratio.

TABLE 3

Comparative mechanical properties and theoretical safety factors calculated under static borehole loading conditions.

Property	Unit	Conventional (PP/PVC)	Proposed (Soybean-PLA)
Material density	g/cm^3	0.90–1.38 (Fröhlich, Liebich and Volz, 2016; Fröhlich, Ludmann and Liebich, 2023)	0.95–1.05 (Kg-m3.com, 2020; Garlotta, 2002)
Compressive yield strength	MPa	40–65 (Fröhlich, Liebich and Volz, 2016; Fröhlich, Ludmann and Liebich, 2023)	30–45 (Nyambo, Mohanty and Misra, 2010; Farah, Anderson and Langer, 2016)
Simulated static borehole load	MPa	0.021	0.021
Safety factor (static)	Ratio	> 2 500	> 1400
Hydro-resistance	-	Impermeable	Impermeable (Short-term)
Deformation risk (2 hours)	-	Negligible	Negligible (Tokiwa *et al*, 2009)

The data indicate a significant divergence in material strength, yet a convergence in operational suitability. The conventional PP/PVC stand possesses a high compressive yield strength, resulting in a Safety Factor (SF) exceeding 2500 against the static load of the explosive column. This SF calculation suggests that current industry-standard devices are vastly 'over-engineered' for shallow-hole quarrying applications. They possess a mechanical capacity far exceeding the requirements of the specific task.

While pure PLA exhibits a yield strength of approximately 40–65 MPa (Garlotta, 2002), the addition of 40 per cent untreated biomass filler creates stress concentration points that reduce the material's load-bearing capacity. Experimental data from Nyambo, Mohanty and Misra (2010) demonstrate that high-loading agro-composites experience a 30–50 per cent reduction in mechanical strength due to interfacial debonding. Consequently, the compressive yield strength of the proposed Soy-PLA composite is conservatively estimated to be 30–45 MPa, which remains sufficient for the calculated borehole static loads. However, despite this reduction, the calculated Safety Factor for the bio-composite remains above 1400. In material engineering contexts, a Safety Factor of 1.2 to 4.0 is typically considered sufficient for static loads (Callister Jr and Rethwisch, 2003). Consequently, an SF of 1400 confirms that the bio-composite is effectively rigid under the specific loading conditions of the reference quarry.

A critical consideration in blasting is the environment within the borehole. In tropical surface mines, such as the reference site in Central Sulawesi, borehole temperatures can exceed 40°C, and water may be present. Pure PLA has a glass transition temperature of approximately 55°C (Garlotta, 2002), which is dangerously close to extreme field temperatures. However, the inclusion of 40 per cent solid agrofiller acts as a thermal stabiliser, can increase the matrix's crystallinity and resistance to thermal creep (Nyambo, Mohanty and Misra, 2010). The simulation suggests that for the short duration between loading and firing (typically less than 2 hrs), the device will maintain its dimensional integrity. While the long-term goal is degradation, the short-term requirement is resistance. The theoretical density of the composite (0.95–1.05 g/cm^3) ensures the device remains denser than loose ANFO prills but light enough to be handled easily. The stand component, which bears the weight, is predicted not to soften significantly within the operational window, ensuring the air gap remains at the designed 0.5 m length. The length of the void that remains the same validates the premise that the *physics* of the blast, which is shock wave reflection and gas expansion, will remain identical to the baseline scenario/scenario A.

Comparative greenhouse gas emissions

The most significant deviation between the two scenarios lies in 'Scope 3' emissions, which are embodied in the consumables. While air decking reduces Scope 1 emissions (by reducing the need for explosives), the device's manufacturing process incurs a carbon footprint.

The application of the Cradle-to-Gate Life Cycle Assessment model (Equation 3) reveals a substantial reduction in the embodied carbon profile of the blasting operation. In the baseline scenario (scenario A), the use of high-emission PVC results in a significant carbon footprint. Conversely, the proposed bio-composite leverages the low carbon intensity of the soybean husk filler to dilute the overall footprint.

Table 4 details the total emissions per blast event. The transition to the bio-composite reduces the GHG emissions associated with device manufacturing by approximately 78.6 per cent. This reduction is achieved solely through material substitution, independent of the explosive savings typically associated with air decking.

TABLE 4

Calculated Greenhouse Gas (GHG) emissions per blast event and annualised projection.

Parameter	Unit	Scenario A: conventional (control)	Scenario B: bio-composite (test)	Variance (reduction)
Air deck device mass per unit	kg	0.12	0.12	0%
Weighted emission factor	kg CO_2-eq/kg	6.10	1.3	-78.60%
Total emissions per blast	kg CO_2-eq	98.8	21.1	-77.7 kg
Annual emissions (60 blasts)	tonnes CO_2-eq	5.93	1.27	-4.66 t

The analysis reveals a dramatic 78.6 per cent reduction in GHG emissions. The fundamental chemical origins of the materials contributed to this reduction. Scenario A, which uses a PP/PVC air deck, relies heavily on PVC for the rigid stand. PVC production is energy-intensive, requiring the cracking of naphtha or natural gas to produce ethylene, and the electrolysis of brine to produce chlorine (Fröhlich, Ludmann and Liebich, 2023). The resulting emission factor (7.83 kg CO_2-eq/kg) is among the highest of common thermoplastics. For every blast of 135 holes, the mine effectively emits nearly 100 kg of CO_2 to manufacture the plastic spacers. While in scenario B, the proposed composite fundamentally shifts the carbon equation. The PLA matrix is derived from fermented plant sugars, which sequester atmospheric CO_2 during the biomass growth phase (Vink and Davies, 2015). More significantly, the reinforcement phase involves adding Soybean Husk, an agricultural byproduct. By utilising a waste stream (with a near-zero processing footprint of ~0.1 kg CO_2-eq/kg) to displace 40 per cent of the polymer mass, the composite achieves a dilution effect. The resulting weighted emission factor of 1.3 kg CO_2-eq/kg represents a radical improvement in material efficiency.

While a reduction of 4.66 t of CO_2 annually per site may appear minor in comparison to diesel combustion from haulage fleets, it represents low-hanging fruit in the hierarchy of decarbonisation. Unlike the electrification of heavy machinery, which requires massive capital expenditure (CAPEX) and infrastructure upgrades, switching to bio-composite air decks requires only a change in procurement policy. Furthermore, if extrapolated across a mining holding company operating 20 similar quarries, the cumulative reduction approaches 100 t of CO_2-eq annually. This contribution helps offset hard-to-abate emissions elsewhere in the operation, thereby improving the mine's carbon-intensity metrics for ESG reporting.

Solid waste dynamics and degradation analysis

The most distinct operational advantage identified is the alteration of the solid waste stream. Using the accumulation model, the annual plastic burden was quantified. Table 5 presents the comparison between the permanent pollution generated by the conventional system and the transient waste of the bio-composite.

TABLE 5

Comparative analysis of solid waste generation and environmental persistence.

Environmental indicator	Conventional system (PP/PVC)	Bio-composite system (Soybean-PLA)
Plastic mass per blast	16.2 kg	16.2 kg
Waste classification	Permanent inorganic waste	Transient organic waste
Annual accumulation	972 kg/annum (cumulative)	0 kg/annum (post-cycle)
Persistence timeline	> 780 years (Chamas *et al,* 2020)	12–24 months (Kale *et al,* 2007)
Degradation mechanism	Photo-oxidation/fragmentation (Lucas *et al,* 2008)	Hydrolytic/enzymatic (Tokiwa *et al,* 2009)
End-of-life product	Microplastics (<5 mm) (Rochman, 2018)	CO_2, water, humus (Karamanlioglu, Preziosi and Robson, 2017)

While the conventional method adds nearly one tonne of permanent plastic to the mine's lithosphere annually, the bio-composite system effectively operates as a zero-waste consumable. The degradation mechanism, driven by the hydrophilic expansion of the soybean fibre, ensures that the material loses structural integrity rapidly after use, eventually mineralising into biomass within the modelled 12–24 month time frame (Kale *et al,* 2007).

The analysis of Scenario A (baseline) reveals a disturbing long-term trend. The detonation pressure of ANFO, which can reach 5.41 GPa (Biessikirski *et al,* 2024), does not vaporise a large portion of the plastic device; instead, it shatters the brittle PVC stand and shreds the ductile PP cup. Though plastic can be vaporised by explosives, in practice, some plastics are difficult to process with explosives, especially those with chlorine content, such as PVC (Siberian Federal University, 2023). Since a single device is reduced to hundreds of irregular fragments, the accumulation happens, with an annual input of nearly 1 metric tonne (972 kg) of plastic, a mine with a 20-year life-of-mine (LOM) will permanently deposit approximately 20 t of plastic into the waste dump or the final void. These plastic polymers are hydrophobic and bio-inert; they do not degrade but only weather (Lucas *et al,* 2008). Over centuries, it breaks down into microplastics measuring less than 5 mm and nanoplastics (Rochman, 2018). These particles can leach additives (plasticisers from PVC) into groundwater or be incorporated into the final aggregate product. This shows the importance of scenario B substituting for scenario A to reduce environmental disturbances.

The analysis of scenario B, where the Soybean Husk-PLA composite operates on a mechanism of programmed obsolescence. The discussion of this mechanism is critical to validating the Zero Waste claim. Unlike glass fibre or talc fillers, soybean husk is hygroscopic. When the blasted rock containing the device fragments is moved to a waste dump or used in reclamation, the husk fibres absorb soil moisture. This absorption causes the fibres to swell (Dhakal *et al,* 2013). Further, the hygroscopic swelling of natural fibres creates internal stress, leading to matrix microcracking (Dhakal *et al,* 2013). These microcracks increase the material's specific surface area, allowing soil microbes to penetrate its deep structure (Karamanlioglu, Preziosi and Robson, 2017). The microbes secrete enzymes that hydrolyse the ester bonds of the PLA, breaking the long polymer chains into shorter water-soluble oligomers and monomers (Tokiwa *et al,* 2009; Karamanlioglu, Preziosi and Robson, 2017). Finally, these products are metabolised by the microbes into carbon dioxide, water, and biomass (Karamanlioglu, Preziosi and Robson, 2017). The theoretical model predicts that within 12 to 24 months in a soil environment (Kale *et al,* 2007), the device will have lost structural integrity, transforming from a pollutant into a soil conditioner.

The theoretical benefits of the bio-composite must be weighed against operational realities. A frequent barrier to the adoption of green technologies in mining is the disruption of established workflows or prohibitive costs. The simulation confirms that the physical dimensions and functional density of the bio-composite device mirror those of the commercial PP/PVC air deck. Consequently, no retraining of the blast crew is required. The charging procedure, by lowering the device to the bottom, checking the depth, and loading explosives, all remain identical. This drop-in replacement characteristic is vital for adoption, as it ensures that the blasting cycle time is not negatively impacted.

Moreover, Soybean husk is a beneficial, economical material. Soybean husk is an agricultural residue often treated as waste. Its cost is primarily associated with transport and grinding, which is significantly lower than the cost of virgin polymer resin. The composite can be processed using standard injection moulding machinery (with adjusted temperature profiles). This processing similarity implies that existing manufacturers of air deck devices can transition to this material without scrapping their capital equipment.

A distinct operational challenge identified in the analysis is storage stability. Because the material is designed to degrade in the presence of moisture, long-term storage in the humid, open-air warehouses typical of Indonesian mine sites poses a risk of premature degradation. If devices absorb moisture before use, their compressive strength may degrade, potentially leading to downhole failure. To ensure the safety factor greater than 1400 is maintained, the bio-composite devices must be packaged in vacuum-sealed, moisture-barrier bags (similar to the packaging of welding electrodes or food products). Once opened on the bench, they should be used within 24 hrs. This procedural adjustment is a minor trade-off for the significant environmental benefits realised.

Scenario A, which uses the PP/PVC air deck device, achieves rock fragmentation efficiency but externalises environmental costs to the lithosphere (plastic waste) and the atmosphere (manufacturing emissions). Scenario B, where Bio-Composite is used, internalises the sustainability goals. It achieves the *same* rock fragmentation efficiency (theoretical validation by mechanical competency analysis) while simultaneously decoupling device manufacturing from the petrochemical industry, reducing the carbon intensity of the blasting consumable supply, and eliminating the generation of persistent microplastics.

Theoretical analysis indicates that commercial amorphous PLA exhibits a baseline yield strength of approximately 59–70 MPa (Farah, Anderson and Langer, 2016). The substitution of 40 per cent of the polymer mass with soybean husk introduces stress concentration points within the matrix, which theoretically reduces the ultimate compressive strength to an estimated 30–45 MPa. However, literature on natural fibre-reinforced PLA composites demonstrates that while excessive filler loading can decrease tensile ductility, the rigid lignocellulosic structure of agricultural waste ensures the composite retains substantial compressive rigidity (Nyambo, Mohanty and Misra, 2010). Furthermore, this cellulosic filler acts as an effective nucleating agent, increasing the crystallinity of the PLA matrix and thereby improving its resistance to thermal creep in high-temperature boreholes.

Regarding water-resistance, while the composite is designed to eventually degrade via long-term moisture absorption and micro-cracking (Dhakal *et al*, 2013), the inherent hydrophobicity of the PLA matrix ensures short-term impermeability sufficient to survive the standard 2 hr operational window between charging and detonation. Consequently, even at the lower strength threshold of 30 MPa, the bio-composite maintains a compressive Safety Factor exceeding 1400 under static borehole loads, confirming its structural viability. However, as this study serves as a theoretical proof-of-concept, transitioning to industrial application requires empirical validation. Future field studies must physically test optimal filler ratios, precise compressive degradation curves in saturated boreholes, and heat-resistance limits to meet rigorous mining standards.

Soybean Husk-PLA air deck represents a theoretically superior technological standard. It bridges the gap between the extractive nature of mining and the regenerative principles of the circular economy. By leveraging agricultural waste to extract mineral wealth, the industry can take a definitive step toward zero-waste mining without compromising blast efficiency.

CONCLUSION

This study articulates a potential critical pathway for the extractive industries to align operational efficiency with environmental stewardship. By theoretically evaluating the substitution of conventional Polypropylene/PVC air deck devices with a Soybean Husk-reinforced Polylactic Acid (PLA) composite free from conventional plastic materials, this research demonstrates that sustainable blasting is achievable without compromising rock fragmentation or blasting safety.

The comparative Life Cycle Assessment (LCA) yields two definitive environmental advantages. First, adopting the bio-composite results in substantial decarbonisation of the supply chain, achieving a 78.6 per cent reduction in device-related GHG emissions (from 98.8 kg to 21.1 kg CO_2-eq per blast). Second, and perhaps more visibly, the material switch eradicates the generation of persistent solid waste. Where the conventional baseline introduces approximately 972 kg of permanent plastic into the lithosphere annually per site, the proposed solution utilises a programmed obsolescence mechanism, ensuring the device degrades into harmless biomass and water within 12 to 24 months of burial.

Mechanically, the theoretical simulation confirms that the composite, despite utilising 40 per cent agricultural waste filler, still retains a compressive Safety Factor exceeding 1400 under standard borehole conditions. This Safety Factor value indicates that the transition to biodegradable consumables requires no modification to existing drill patterns, explosive charging protocols, or initiation timing.

Ultimately, this research serves as a proof-of-concept for the circular economy within mining. By valorising agricultural by-products (soybean husk) to solve an engineering challenge, mining operations can effectively mitigate their Scope 3 emissions and eliminate microplastic pollution. Future research should prioritise field trials to empirically validate the shelf-life stability of these bio-composites under high-humidity tropical storage conditions, thereby bridging the gap between theoretical modelling and industrial scalability.

REFERENCES

Azadi, M, Northey, S A, Ali, S H and Edraki, M, 2020. Mining to enable climate change mitigation, *Nature Geoscience*, 13(3):180–184. https://doi.org/10.1038/s41561-020-0531-3

Biessikirski, A, Pytlik, M, Nachlik, S and Stopkowicz, A, 2024. Impact of nitromethane additive on ANFO performance, *Scientific Reports*, 14(1):1–8. https://doi.org/10.1038/s41598-024-79524-7

Bledzki, A K and Gassan, J, 1999. Composites reinforced with cellulose based fibres, *Progress in Polymer Science*, 24(2):221–274. https://doi.org/10.1016/S0079-6700(98)00018-5

Budynas, R G and Nisbett, J K, 2015. *Mechanical Engineering Design,* 10th edn (McGraw-Hill Education: New York).

Callister Jr, W D and Rethwisch, D G, 2003. *Characteristics, Application and Processing of Polymers, Materials Science and Engineering – An Introduction* (John Wiley and Sons).

Chamas, A, Moon, H, Zheng, J, Qiu, Y, Tabassum, T, Jang, J H, Abu-Omar, M M, Scott, S L and Suh, S, 2020. Degradation rates of plastics in the environment, *ACS Sustainable Chemistry & Engineering*, 8(9):3494–3511. https://doi.org/10.1021/acssuschemeng.9b06635

Climatiq, 2021. Emission Factor: PP (polypropylene), Climatiq. Available from: <https://www.climatiq.io/data/emission-factor/8f6dd6e1-31a3-43a3-b58e-5b7a21e10685> [Accessed: 9 March 2026).

Dananjaya, V, Yang, M, Zheng, Y and Abeykoon, C, 2025. Effect of filler characteristics and processing route on bamboo powder-reinforced PLA composites, *Journal of Cleaner Production*, 537:147193. https://doi.org/10.1016/j.jclepro.2025.147193

Dhakal, H N, Zhang, Z Y, Bennett, N M and Lopez-Anido, R, 2013. Development of flax/carbon fibre hybrid composites for enhanced properties, *Carbohydrate Polymers*, 96(1):1–8. https://doi.org/10.1016/j.carbpol.2013.03.074

Farah, S, Anderson, D G and Langer, R, 2016. Physical and mechanical properties of PLA and their functions in widespread applications — A comprehensive review, *Advanced Drug Delivery Reviews*, 107:367–392. https://doi.org/10.1016/j.addr.2016.06.012

Fröhlich, T, Liebich, A and Volz, S, 2016. Eco-profiles and environmental product declarations of the European plastics manufacturers – Polypropylene (PP), updated December 2016, PlasticsEurope. Available from: <https://www.researchgate.net/publication/318864901_Eco-profiles_and_Environmental_Product_Declarations_of_the_European_Plastics_Manufacturers_-_Polypropylene_PP> [Accessed: February 2026].

Fröhlich, T, Ludmann, S and Liebich, A, 2023. Eco-profiles and environmental product declarations of the European plastics manufacturers – Vinyl chloride (VCM) and polyvinyl chloride (PVC), revised June 2023, PlasticsEurope / The European Council of Vinyl Manufacturers (ECVM). Available from: <https://pvc.org/wp-content/uploads/2023/06/230628_Eco-profile-PVC_june23.pdf> [Accessed: February 2026].

Garlotta, D, 2002. A literature review of poly(lactic acid), *Journal of Polymers and the Environment*, 9(2):63–84. https://doi.org/10.1023/A:1020200822435

Hayat, M B, Alagha, L and Ali, D, 2019. Air Decks in Surface Blasting Operations, *Journal of Mining Science*, 55(6):922–929. https://doi.org/10.1134/S1062739119066307

Heryadi, R, Utama, H and Aprisko, L, 2025. Reducing a quarry's greenhouse gas emission through air deck implementation, *Application of Computers and Operations Research in the Minerals Industry*, pp 1–9.

Heryadi, R, Utama, J D H and Prasmoro, A V, 2024. Cost and fragmentation prediction of ammonium nitrate fuel oil versus bulk emulsion explosives, *IOP Conference Series: Earth and Environmental Science*, 1422(1). https://doi.org/10.1088/1755-1315/1422/1/012001

Jhanwar, J C, 2011. Theory and Practice of Air-Deck Blasting in Mines and Surface Excavations: A Review, *Geotechnical and Geological Engineering*, pp 651–663. https://doi.org/10.1007/s10706-011-9425-x

Kale, G, Karkhanis, S, Barkach, S and Almenar, E, 2007. Compostability of bioplastic packaging materials: An overview, *Macromolecular Bioscience*, 7(3):255–277. https://doi.org/10.1002/mabi.200600168

Karamanlioglu, M, Preziosi, R and Robson, G D, 2017. Abiotic and biotic environmental degradation of the bioplastic polymer poly(lactic acid): A review, *Polymer Degradation and Stability*, 137:122–130. https://doi.org/10.1016/j.polymdegradstab.2017.01.009

kg-m3.com, 2020. Soybean, hulls density [online], All about density – kg-m3.com. Available from: <https://kg-m3.com/material/soybean-hulls> [Accessed: 5 February 2026).

Krawczyk, M B, Przybylski, R, Figiel, P and Kwiatkowski, K, 2026. Investigation of the influence of natural fillers on the properties of the samples obtained using additive manufacturing technology, *Advances in Science and Technology*, 20(2):494–503. https://doi.org/10.12913/22998624/213203

Kuru, Z and Kaya, M A, 2023. Poly(Lactic Acid) / Polyester Blends: Review of Current and Future Applications, *The European Journal of Research and Development*, 3(1):175–199. https://doi.org/10.56038/ejrnd.v3i1.259

Liang, Q and Yu, L, 2023. Assessment of carbon emission potential of polyvinyl chloride plastics, *E3S Web of Conferences*, 393:1–3. https://doi.org/10.1051/e3sconf/202339301031

Lucas, N, Bienaime, C, Belloy, C, Queneudec, M, Silvestre, F and Nava-Saucedo, J-E, 2008. Polymer biodegradation: Mechanisms and estimation techniques – A review, *Chemosphere*, 73(4):429–442. https://doi.org/10.1016/j.chemosphere.2008.06.064

Melnikov, N V, Marchenko, L N, Zharikov, I F and Seinov, N P, 1978. Blasting methods to improve rock fragmentation. *Acta Astronautica*, 5(11–12):1113–1127. https://doi.org/10.1016/0094-5765(78)90014-0

Mohanty, A K, Misra, M and Drzal, L T, 2002. Sustainable Bio-Composites from Renewable Resources: Opportunities and Challenges in the Green Materials World, *Journal of Polymers and the Environment*, 10:19–26. https://doi.org/10.1023/A:1021013921916

Monjezi, M, Amiri, H, Seyed Mousavi, M N, Hamidi, J K and Khandelwal, M, 2022. Comparison and application of top and bottom air decks to improve blasting operations, *AIMS Geosciences*, 9(1):16–33. https://doi.org/10.3934/geosci.2023002

Mudd, G M, 2010. The Environmental sustainability of mining in Australia : key mega-trends and looming constraints, *Resources Policy*, 35(2):98–115. https://doi.org/10.1016/j.resourpol.2009.12.001

Norgate, T and Haque, N, 2010. Energy and greenhouse gas impacts of mining and mineral processing operations, *Journal of Cleaner Production*, 18(3):266–274. https://doi.org/10.1016/j.jclepro.2009.09.020

Nyambo, C, Mohanty, A K and Misra, M, 2010. Polylactide-based renewable green composites from agricultural residues and their hybrids, *Biomacromolecules*, 11(6):1654–1660. https://doi.org/10.1021/bm1003114

Pishvaee, M S and Razmi, J, 2012. Environmental supply chain network design using multi-objective fuzzy mathematical programming, *Applied Mathematical Modelling*, 36(8):3433–3446. https://doi.org/10.1016/j.apm.2011.10.007.

Pryshlakivsky, J and Searcy, C, 2013. Fifteen years of ISO 14040: A review, *Journal of Cleaner Production*, 57:115–123. https://doi.org/10.1016/j.jclepro.2013.05.038

Riaz, M N, 2015. *Soy Beans: Processing, Encyclopedia of Food and Health*, 1st edn (Elsevier Ltd). https://doi.org/10.1016/B978-0-12-384947-2.00641-3

Rochman, C M, 2018. Microplastics research – from sink to source. *Science*, 360:28–29. https://doi.org/10.1126/science.aar7734

Sakti, A M, Darmadi, D B, Gapsari, F, Widodo, T D and Ganda, A N F, 2025. Enhancing the structural and antibacterial properties of soybean hull fibers via alkali treatment and zinc oxide nanocoating, *Cleaner Waste Systems*, 11:100289. https://doi.org/10.1016/j.clwas.2025.100289

Siberian Federal University, 2023. Scientists Figure Out How to Destroy Plastic Waste with Explosion, Siberian Federal University. Available from: <https://www.sfu-kras.ru/en/news/27708> [Accessed: 5 February 2026).

Stantec Consulting, 2015. Greenhouse Gas Emission Inventory Summary for KGHM Ajax Mine, Burnaby BC.

Štimac, B, Škrlec, V, Dobrilović, M and Sućeska, M, 2021. Numerical modelling of non-ideal detonation in ANFO explosives applying Wood-Kirkwood theory coupled with EXPLO5 thermochemical code, *Defence Technology*, 17(5):1740–1752. https://doi.org/10.1016/j.dt.2020.09.014

Tokiwa, Y, Calabia, B P, Ugwu, C U and Aiba, S, 2009. Biodegradability of plastics, *International Journal of Molecular Sciences*, 10(9):3722–3742. https://doi.org/10.3390/ijms10093722

Vink, E T H and Davies, S, 2015. Life Cycle Inventory and Impact Assessment Data for 2014, *Ingeo*, 11(3):167–180. https://doi.org/10.1089/ind.2015.0003

World Resources Institute (WRI) and World Business Council for Sustainable Development (WBCSD), 2005. The GHG Protocol for project accounting, Greenhouse Gas Protocol Initiative report. Available from: <https://pdf.wri.org/ghg_project_accounting.pdf>

Blast movement simulation of stratiform orebodies, a case study from a layered mafic intrusion

C J Gniel[1], B C Gower[2] and G Theron[3]

1. MAusIMM, Product Owner, Augment Technologies, Subiaco WA 6008.
   Email: cael@augmenttechnologies.ai
2. Chief Revenue Operations Officer, Augment Technologies, Subiaco WA 6008.
   Email: brice@augmenttechnologies.ai
3. Resource Geologist, Tharisa, Bryanston 2191, South Africa. Email: gtheron@tharisa.com

ABSTRACT

Accurate characterisation of material movement during blasting is critical for maintaining grade control and ensuring that extracted material remains within specification. Conventional blast-movement approaches typically provide limited representation of displacement in the vertical (Z) direction, instead relying on simplified horizontal vectors or sparse point-based measurements. This limitation is particularly consequential in stratiform mineral systems, including layered mafic intrusions, where mineralisation occurs as laterally extensive, sub-parallel layers with sharp ore–waste contacts that are highly sensitive to vertical displacement.

In layered mafic intrusions, even modest vertical movement across these contacts can result in significant dilution, ore loss, and misclassification of material during post-blast mark-out and excavation. Misunderstanding the true three-dimensional nature of blast-induced movement therefore represents a material risk to both grade control accuracy and downstream reconciliation.

Recent advances in computing power, remote sensing, and data integration have enabled the development of simulation-based tools capable of modelling blast-induced material behaviour in three dimensions. Augment Technologies' Ore Movement Policy™ (OMP) is an AI-driven 3D blast movement solution that integrates pre- and post-blast surface surveys, drill and blast design data, and the grade control block model to generate a high-resolution Muckpile Block Model™ (MBM) at 1 m^3 scale.

This paper presents the results of a campaign of high-density ore-waste contact field measurements across five blasts in an operating mine. The results reveal the true nature of material movement due to blasting. Each blast is also simulated and quantitative comparisons are made between field measurements, simulation results and mark-outs. Implications for adopting simulations are discussed including the effect on ore loss and dilution and additional operational benefits enabled by blast movement simulation including personnel safety. Practical strategies to mitigate adverse blast movement impacts are discussed, including a novel change in mark-out practices and the application of dynamic flitch heights.

INTRODUCTION

Layered mafic intrusions represent a globally significant class of mineral deposits, hosting stratiform PGE, Chromium, Nickel and copper mineralisation. In these settings, ore–waste boundaries are typically sub-horizontal to moderately dipping, laterally extensive, and sharply defined (Latypov, Chistyakova and Mukherjee, 2017). As a result, grade control outcomes are particularly sensitive to vertical material movement during blasting.

Blasting induces complex three-dimensional displacement within the rock mass, driven by explosive energy distribution, confinement, geological structure, and rock mass properties. The resultant movement has negative impacts on ore recovery and dilution (La Rosa and Thornton, 2011). Despite this, blast movement correction practices in many operations remain heavily simplified, commonly relying on horizontal displacement vectors or sparse point measurements that are not well suited to capturing vertical movement. While such approaches may be sufficient in massive or vertically continuous orebodies, it is unclear if they suited to stratiform systems where small vertical offsets can have material impacts on dilution and ore recovery.

This paper focuses on the application of 3D blast movement simulation within a layered mafic intrusion, with the objective of improving understanding of vertical material displacement across stratiform ore–waste contacts and its implications for grade control. The work builds on recent advancements in data acquisition and simulation, integrating UAV-derived surface models with blast design and geological information to deliver a spatially continuous representation of post-blast material distribution.

GEOLOGICAL SETTING

The case study is located at the Tharisa Mine, situated on the south-western limb of the Bushveld Complex, South Africa. The Bushveld Complex is a world-class layered mafic intrusion that hosts laterally extensive, stratiform chromitite and platinum-group element (PGE) mineralisation developed through prolonged magmatic differentiation and cumulate layering (Eales and Cawthorn, 1996; Latypov, Chistyakova and Mukherjee, 2017).

At Tharisa, mining targets the Middle Group (MG) chromitite reefs, specifically the MG0 to MG4 horizons. These reefs occur as laterally continuous, tabular chromitite layers hosted within a sequence of mafic and ultramafic cumulates. Stratiform chromitite layers within the Bushveld Complex are well documented as regionally extensive units with sharp and laterally persistent ore–waste contacts (Naldrett *et al*, 2011; Cawthorn, 2010; Emsbo *et al,* 2016). The MG chromitite reefs are typically sub-horizontal to gently dipping and are bounded by anorthositic, noritic, and pyroxenitic host rocks, resulting in a vertically constrained but laterally extensive orebody geometry.

BLAST MOVEMENT MODELLING

The sharpness and continuity of the MG chromitite reefs make grade control outcomes particularly sensitive to vertical displacement during blasting. Even minor blast-induced movement across reef boundaries can therefore lead to ore loss, dilution, or misclassification of material over large plan areas. The relatively shallow dip of the MG reefs, combined with bench-scale open pit mining methods employed at Tharisa, further underscores the importance of accurately accounting for blast movement. These geological and operational characteristics make the Tharisa Mine an appropriate and representative setting for evaluating three-dimensional blast movement simulation and material-centric mark-out methodologies in stratiform deposits.

Conventional blast movement practices

Conventional blast movement correction practices at many operations are typically based on simplified representations of material displacement. These commonly include horizontal movement vectors derived from sparse measurements or uniform translations applied to geological contacts. While such approaches may provide a first-order correction in some settings, they are inherently limited in their ability to represent three-dimensional material behaviour.

At the Tharisa mine blast movement was historically addressed by translating the top-of-reef contact upward by a specified, uniform amount. The magnitude of this translation varied by reef and was derived from accumulated field observations, supplemented by real-time input from ore spotters during mining. This approach implicitly assumed that vertical displacement was uniform across the blast and that lateral variability in movement was negligible.

In practice, however, blast-induced movement is highly heterogeneous, influenced by blast design, confinement, local geology, and rock mass properties. Uniform vertical translations therefore fail to capture localised zones of exaggerated movement. Furthermore, the reliance on ore spotters to refine mark-outs introduces additional sources of uncertainty. Spotter-based adjustments are inherently subjective, can vary between operators, and are particularly vulnerable to reduced accuracy during night shifts or under poor visibility conditions. Recognising these limitations, the operation sought to move away from spotter-dependent workflows towards a more objective, data-driven approach to blast movement correction.

Ore movement policy

Ore Movement Policy (OMP) is a three-dimensional, simulation-based blast movement framework designed to explicitly model material displacement throughout the blasted volume. Rather than applying movement corrections to *in situ* geological contacts, OMP focuses on tracking material itself as it moves during blasting and subsequent muck pile formation.

The system integrates multiple data sources, including pre- and post-blast surface surveys, drill and blast design parameters, and the grade control block model. High-resolution pre-blast topography establishes the initial spatial configuration of material, while post-blast surveys—typically acquired using unmanned aerial vehicle (UAV) technology—capture the geometry of the resulting muck pile. Drill and blast data are used to inform the spatial distribution of explosive energy and confinement conditions, which influence the magnitude and direction of material movement.

These inputs are combined within the OMP framework to generate a Muckpile Block Model™ (MBM) at 1 m^3 resolution. Each block within this model represents a discrete volume of material whose post-blast location is explicitly simulated in three dimensions. This enables continuous tracking of material across ore–waste contacts, including vertical displacement that is not readily observable using conventional methods.

A key distinction between OMP and traditional approaches is the way mark-outs are generated. Rather than translating an *in situ* mark-out based on assumed movement vectors, mark-outs are created directly on the simulated muck pile. Geological domains and grade attributes are carried forward into the post-blast configuration, allowing ore–waste boundaries to be redefined based on the actual spatial distribution of material after blasting.

By generating mark-outs on the moved material, OMP provides a more precise basis to inform mark-out decisions. The resulting mark-outs can therefore better reflect local variability in blast movement, improving confidence in material classification, reducing dilution, and supporting more consistent grade control outcomes. For stratiform deposits such as layered mafic intrusions, where vertical displacement is a primary source of risk, this material-centric approach represents a potential step-change in approach.

Novel markout methodology

A further outcome of generating a high-resolution MBM is the ability to adopt alternative, more efficient approaches to post-blast mark-out creation. Traditional mark-out workflows typically rely on manually digitised wireframes constructed in serial sections. While effective, this approach is time-consuming, operator-dependent, and poorly suited to rapid iteration.

In this study, a novel mark-out methodology was implemented using Datamine Studio RM's implicit modelling functionality. Each block belonging to the target strata within the simulation was treated as a discrete point in three-dimensional space. These points collectively define the spatial extent of the mineralised material after blasting and the target for the mark-out. An enclosing isoshell surface was generated around the point cloud representing the target strata. This process enables the rapid creation of a continuous wireframe that conforms to the simulated post-blast distribution of material, without the need for manual section-by-section digitisation.

Once generated, the implicit isoshell wireframes provides an immediate basis for rapid tonnage and grade calculations directly from the MBM. This allows grade control personnel to quickly evaluate the impact of alternative mark-out scenarios, based on varied isoshell parameters to generate the desired grade and tonnage. Importantly, the isoshells may be edited in section after creation to fine-tune the outcomes and adjust for diggability.

Compared to conventional serial section workflows, this approach significantly reduces mark-out turnaround time and supports a more iterative and optimised grade control process.

Case study methodology

This case study includes five production blasts, resulting in a validation data set comprising 578 independently collected ore–waste contact observations. These observations provide a sufficiently large and diverse sample to support quantitative assessment of model performance across varying blast conditions and local geological contexts.

Field observations were collected through ore spotting, where experienced ore spotters used differential GPS (dGPS) units to record the position of the top-of-reef contact, and where operationally feasible, the bottom-of-reef contact following blasting. Observations were made by progressively removing overburden material with an excavator until the blasted reef was exposed, at which point contact positions were logged.

Quantitative comparison between field observations and the blast movement simulation outputs was conducted using a multi-step workflow. From the MBM, hanging wall (HW) and footwall (FW) reference points were derived for all relevant stratigraphic horizons by selecting the uppermost (HW) and lowermost (FW) blocks within each unique XF–YF (post-blast coordinates) column and extracting their cell centres as vertices. This provided a consistent block-scale representation of the post-blast position of ore–waste contacts.

The observation points were then compared directly to the derived MBM HW and FW reference points. Distances between each observation point and the nearest HW and FW MBM cell centres were calculated and statistically summarised to assess block-scale agreement between simulated material movement and field observations.

In parallel, the implicit isoshell mark-outs generated from the MBM were used to assess performance at the operational mark-out scale. Each observation point was projected onto the corresponding isoshell surface, and the vertical projection distances were recorded. This enabled direct comparison between observed contact positions and the smoothed, continuous surfaces used for mark-out generation.

Field observations were subject to exposure conditions and operational and safety constraints resulting in spatial clustering of the data. In addition, individual observation points are subject to uncertainty related to operator judgement. Despite these data limitations and uncertainties, the aggregated data set is considered representative for ensemble-based analysis.

RESULTS

Ore spotting results

Ore spotting observations were visualised in Datamine Studio RM by converting them into a digital terrain model (DTM) representing the post-blast ore–waste topography. This surface provides a continuous approximation of the reef geometry after blasting, constrained by field observations.

The post-blast DTM was visualised alongside the *in situ* geological model of the ore horizon (Figures 1 to 3).

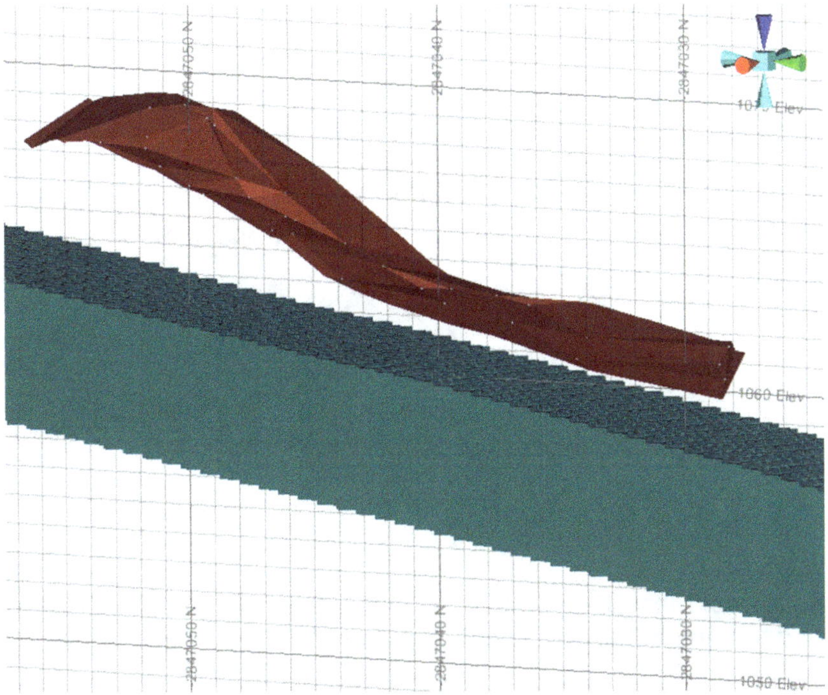

FIG 1 – Oblique section of Blast B0518 modelled *in situ* MG2CR (green – block model) and top-of-reef ore spotting points showing complex post-blast ore topography.

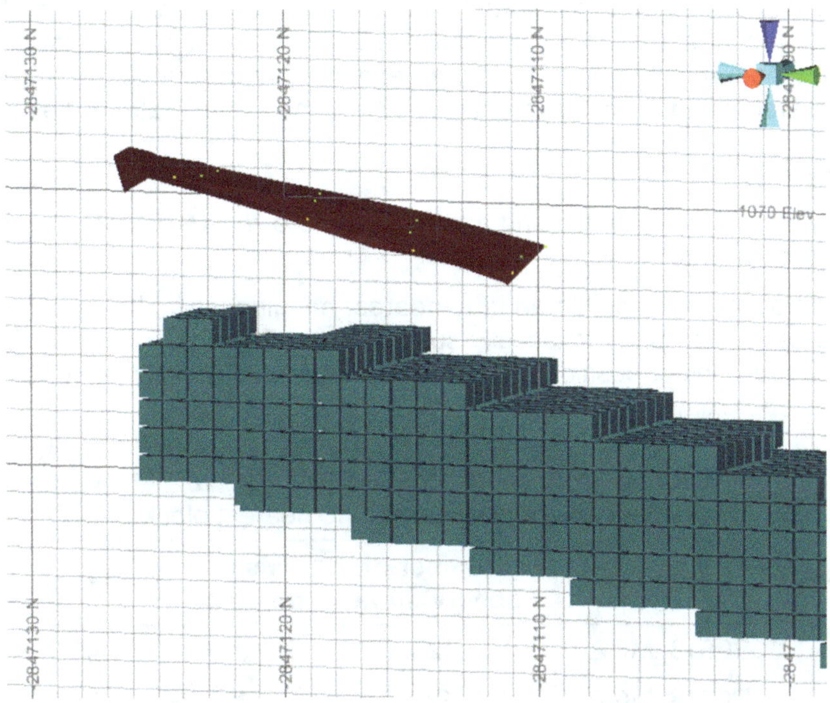

FIG 2 – Oblique section of Blast B0818 modelled *in situ* MG2CR reef (green – regularised block model) and top-of-reef ore spotting points showing post-blast ore topography.

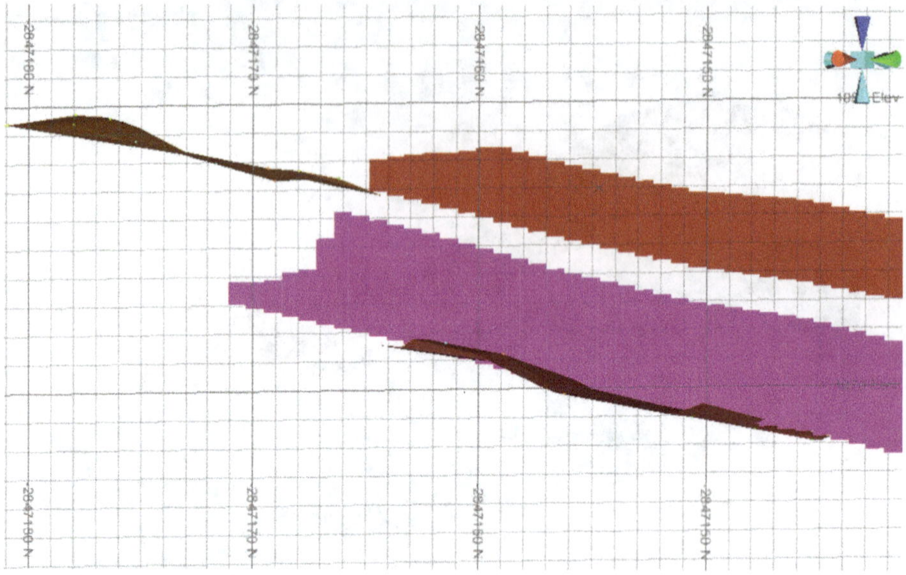

FIG 3 – Oblique section of Blast B1418 modelled *in situ* MG4CR and MG4ACR reefs (pink and red respectively) and top and bottom ore spotting points for MG4CR reef showing post-blast ore topography.

Figures 1 and 2 highlight the two end members of post-blast topography observed across all five blasts. Blast B0518 has considerably more complex post-blast ore topography with distinct peaks and troughs. This this case the ore topography dipped up to 30° towards the north compared to a uniform 15° *in situ*. In contrast, the results from Blast B0818 show a more subtle undulous nature of the post-blast ore topography. Figure 3 shows an example where due to the movement of ore, the top of reef contact first mined is not directly above where the *in situ* material was blasted.

Simulation results

Quantitative analysis was performed using a total of 578 ore–waste contact observations collected across five production blasts.

Distances between observation points and the nearest hanging wall (HW) and footwall (FW) MBM cell centres were calculated and statistically summarised (Table 1). The absolute mean distance between the observed contact positions and the estimated position of the corresponding HW and FW MBM blocks is approximately 1.15 m.

TABLE 1

Distance from observations to nearest block.

Blast	Number of observations	Mean distance	Std.Dev
B0518_MG2	123	1.02	0.49
B0818_MG2	94	1.18	0.57
B1417_MG3_HW	135	0.98	0.53
B1417_MG3_FW	75	1.56	0.89
B1418_MG4	48	1.41	0.69
B1418_MG4_FW	48	0.85	0.46
B1517_MG3	55	1.27	0.51

Statistical summary of the vertical projection distances indicates an average absolute mean distance of approximately 1.05 m between observed contact positions and the isoshell-derived mark-outs (Figure 4 and Table 2).

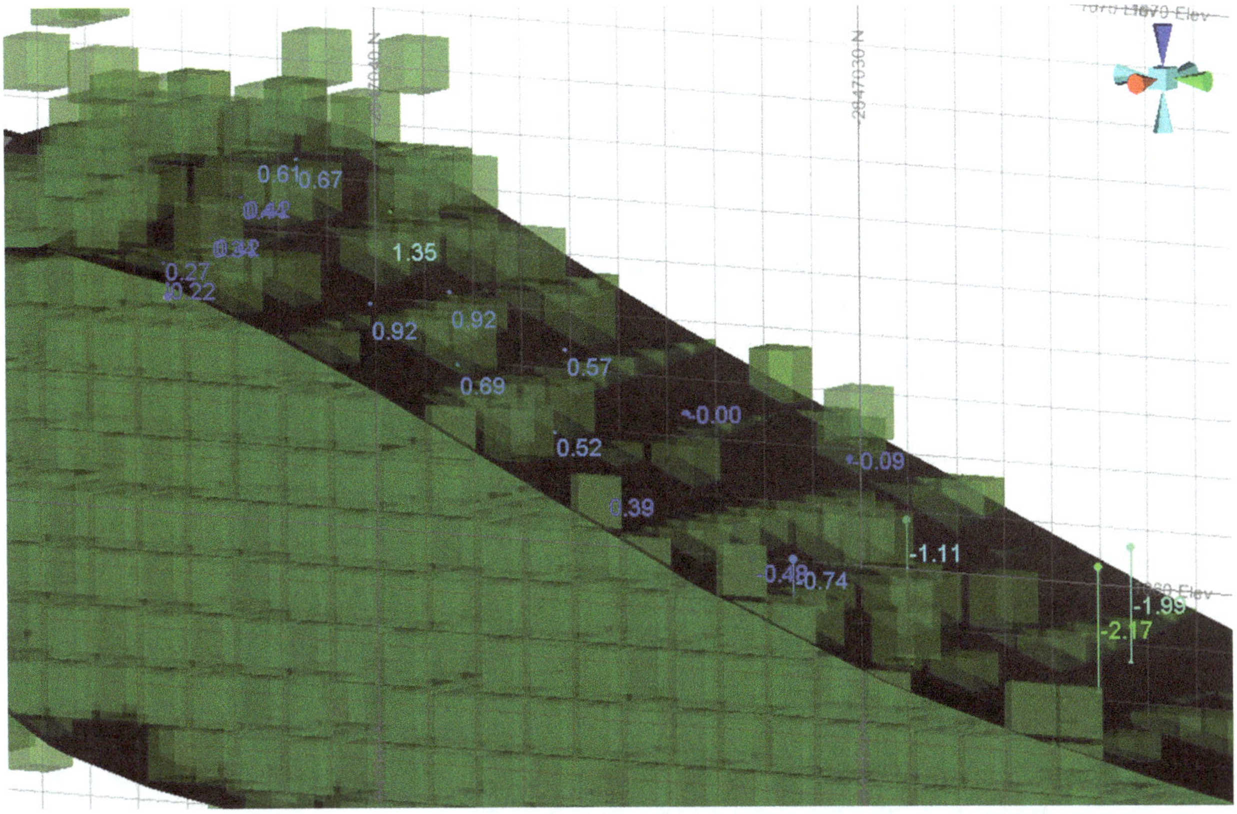

FIG 4 – B0518 ore spotting data visualised against the MBM and mark-out showing distance to top of mark-out.

TABLE 2

Distance from observations to mark-out.

Blast	n. observations	Mean distance	Std.Dev
B0518_MG2	94	0.89	0.82
B0818_MG2	61	0.9	0.6
B1417_MG3_HW	120	0.83	0.63
B1418_MG4	94	2.65	1.72
B1418_MG4_FW	61	0.58	0.52
B1517_MG3	120	1.57	0.85

This reduction in mean distance reflects the effect of three-dimensional smoothing inherent in the implicit modelling process, which mitigates local block-scale variability while preserving the overall geometry of the moved material.

DISCUSSION

Post blast ore topography

The results of this study demonstrate that blasting produces complex, three-dimensional movement patterns that can result in highly variable post-blast ore–waste topography in shallow-dipping, stratiform orebodies such as layered mafic intrusions. The ore spotting-derived surfaces show that post-blast ore geometry commonly departs from the relatively smooth and laterally continuous form implied by *in situ* geological models. This complexity reflects the inherently heterogeneous response of the rock mass to blasting and is driven by a combination of blast design parameters, including powder factor, initiation timing, blast morphology, confinement conditions, and local geological variability.

Traditional versus simulated blast movement

Traditional blast movement correction approaches that rely on simple translation of *in situ* material cannot reproduce this observed complexity. Even in cases where post-blast ore topography appears relatively simple at a broad scale, the assumption of a uniform vertical translation is unlikely to match reality at the local scale. The results indicate that vertical displacement varies spatially across the blast, meaning that any single translation distance represents, at best, an average approximation. Consequently, such methods are unlikely of capture the short wavelength variability evident in the observed post-blast contacts.

While the use of blast vector indicators (BVIs) can improve understanding of material movement, the spatial resolution provided by these devices is typically insufficient to capture the level of variability observed in this study. Point-based measurements, even when accurate, require interpolation and assumption to be applied across the blasted volume, limiting their ability to resolve localised vertical displacement in stratiform systems.

With limited exception, the quantitative results show good agreement between the simulated blast movement outputs and the observed ore–waste contact positions derived from ore spotting. This level of agreement supports the use of blast movement simulation as a realistic representation of post-blast material distribution.

Operational considerations

Generating mark-outs based on the simulated post-blast position of ore, rather than translating *in situ* geometry, provides several operational benefits. These include a reduced reliance on ore spotters, enabling faster and more consistent digging rates, whilst removing people from the muck pile and therefore increasing safety. In addition, the precision of these methods enables a reduction in ore loss and dilution through improved mark-out accuracy.

Furthermore, the simulation-based approach also enables greater flexibility in mining method selection. By explicitly modelling vertical and lateral displacement, operations can tailor flitch heights or excavation strategies to the observed movement for individual blasts. For example, where a standard 3 m flitch height would intersect mixed ore and waste due to vertical smearing, simulation results may indicate that a 4 m flitch would reduce dilution for a specific blast. This allows informed, blast-by-blast operational decisions rather than reliance on fixed standards.

Limitations

The accuracy of the blast movement simulation is inherently dependent on the quality and completeness of input data and user-defined parameters. Key inputs such as pre- and post-blast surfaces, drill and blast design, and the definition of the *in situ* blast volume directly influence the resulting MBM. Errors or inconsistencies in these inputs can propagate through the simulation and impact the reliability of predicted material movement. In particular, the creation of strings defining the blast extent remains a user-defined step that can introduce potential variability. As with any geological or mining tool, data inputs and simulation outputs should therefore be subject to appropriate review and validation prior to informing operational or economic decisions.

The ore spotting data set is spatially clustered due to operational and safety constraints, with observations preferentially collected in accessible areas of the muck pile. This clustering introduces potential sampling bias, as certain zones of the blast, particularly those that are less accessible may be underrepresented in the data set. As a result, the validation may disproportionately reflect movement characteristics from specific portions of the blast rather than providing fully uniform spatial coverage. While the data set is considered suitable for ensemble-based analysis, future trenching programs should be designed to minimise clustering through more systematic spatial sampling, such as targeting predefined locations across the blast extent or incorporating stratified sampling approaches to ensure representative coverage of different movement domains.

CONCLUSIONS

This study demonstrates that blast-induced material movement in stratiform chromitite deposits hosted within layered mafic intrusions is inherently three-dimensional and spatially variable. High-density ore–waste contact observations collected across five production blasts at the Tharisa Mine show that post-blast ore topography commonly departs from the smooth, laterally continuous geometry implied by *in situ* geological models and previously established mark-out methodologies. Quantitative comparison between field observations and simulated blast movement outputs shows generally good agreement, with average absolute distances of approximately 1.15 m at the block scale and 1.05 m when using implicit isoshell mark-outs derived from the MBM. The improved agreement achieved through implicit surface generation highlights the benefit of smoothing block-scale variability while preserving the overall geometry of the moved material.

Importantly, generating mark-outs directly on simulated post-blast material, rather than translating *in situ* geology, provides a more physically representative basis for grade control decision-making in stratiform orebodies. This approach reduces reliance on subjective ore spotting, improves safety by limiting personnel exposure on the muck pile, and supports more consistent digging rates while reducing ore loss and dilution. In addition, explicit modelling of vertical displacement enables blast-by-blast optimisation of flitch height and excavation strategy. Collectively, these findings demonstrate that three-dimensional blast movement simulation represents a substantive advancement over conventional blast correction practices for layered mafic intrusions and similar stratiform mineral systems.

ACKNOWLEDGEMENTS

Augment would like to thank Tharisa PLC for their involvement in the study.

REFERENCES

Cawthorn, R G, 2010. The Bushveld Complex, *Minerals*, 2:3–16.

Eales, H V and Cawthorn, R G, 1996. The Bushveld Complex, in *Layered Intrusions – Developments in Petrology* (ed: R G Cawthorn), 15:181–229 (Elsevier: Amsterdam).

Emsbo, P, Seal, R R, Foose, M P and Gottlieb, E S, 2016. Stratiform chromite deposit model, US Geological Survey Scientific Investigations Report 2010-5070-E, 90 p.

La Rosa, D and Thornton, D, 2011. Blast movement modelling and measurement, *Mining Technology*, 120(2):79–88.

Latypov, R, Chistyakova, S and Mukherjee, R, 2017. A novel hypothesis for the origin of massive chromitites in the Bushveld Igneous Complex, *Journal of Petrology*, 58(10):1899–1940.

Naldrett, A J, Wilson, A, Kinnaird, J and Chunnett, G, 2011. PGE tenor and metal ratios within stratiform PGE–chromitite deposits of the Bushveld Complex, *Mineralium Deposita*, 46:403–423.

The largest single mining blast – optimising blasting techniques to ensure long-term stability of adjacent pit walls

J M Kabuya[1], A Kamalibandpey[2], A Caron[3] and J Atalla[4]

1. Mining Engineering Services Manager, ArcelorMittal Mining Canada, Fermont QC, Canada. Email: josephmukendi.kabuya@arcelormittal.com
2. Senior Rock Mechanics Engineer, ArcelorMittal Mining Canada, Fermont QC, Canada. Email: abbas.kamalibandpey@arcelormittal.com
3. Rock Mechanics Engineer, ArcelorMittal Mining Canada, Fermont QC, Canada. Email: antoine.caron@arcelormittal.com
4. Head of Section – Drill and Blast, ArcelorMittal Mining Canada, Fermont QC, Canada. Email: joe.atalla@arcelormittal.com

ABSTRACT

In November 2025, ArcelorMittal's Mining Exploitation in Canada carried out a major single mining blast at the Mont-Wright mine in northern Quebec. A total of 8.2 million tonnes (Mt) of hard rock was successfully fragmented in a single drilling pattern, representing an unprecedented scale in surface mining. The blast required 2855 drill holes and over 3 million kilograms of emulsion, covering an area of 1000 m × 335 m and breaking more than 3.1 million m^3 of rock. Proximity to Fermont City and vulnerable quartz zones made vibration control and pit wall stability critical concerns. To address these challenges, advanced 4D blast design technology was implemented to optimise emulsion loading, positioning, and timing, ensuring efficient rock fragmentation while minimising vibration amplitudes. The blast timing was distributed in an inverse V configuration, leveraging two free faces to enhance rock breakage and mitigate the concentration of vibration on the final pit wall. A comprehensive instrumentation and monitoring program was deployed to assess the geotechnical response of the pit walls. Time Domain Reflectometry (TDR) and piezometer (PZ) measurements were recorded both before and after the blast to evaluate stress redistribution, pore pressure changes, and micro-deformation within the rock mass. Additionally, post-blast slope stability monitoring with slope stability radar (SSR) provided continuous real-time measurements of slope displacement, confirming the structural integrity of the walls and enabling immediate validation of the blast performance. In this regard, numerical analyses were performed using UDEC and FLAC3D to evaluate the stability of the rock slope and to compare the predicted displacements with radar monitoring data, while accounting for the effects of the excavation damage zone (EDZ). The event demonstrates that modern mining methods can safely manage large-scale hard rock blasting while maintaining pit stability and minimising vibration impacts.

INTRODUCTION

Drilling and blasting are the primary excavation methods in hard rock open pit mining, particularly in strong and brittle lithologies such as quartz-rich formations, where blast performance controls fragmentation efficiency, productivity, and downstream processing (Persson, Holmberg and Lee, 2018). The detonation of explosives generates stress waves that propagate through the rock mass, producing ground vibrations that can influence pit wall stability and surrounding infrastructure. These vibrations are commonly evaluated using peak particle velocity (PPV) and dominant frequency, which have been widely linked to excavation damage zone (EDZ) in rock mass and slope response (Dowding, 1985). Quartz-rich formations pose blasting challenges due to high stiffness and low damping, while quartz–gneiss contacts act as weak planes that concentrate stress. These conditions increase the likelihood of vibration amplification and localised damage near pit walls, making vibration control a critical design consideration in such geological environments (Hoek and Brown, 2012; Ullah *et al,* 2025). In addition, repeated exposure to blast vibrations may lead to progressive degradation of rock mass properties along the contact zone and within the surrounding rock mass, even in the absence of immediate instability, highlighting the importance of conservative design and continuous monitoring. Advances in blast design technology have significantly improved the ability to balance fragmentation efficiency with vibration mitigation. The widespread adoption of electronic detonators, precise delay sequencing, and optimised explosive loading has enabled more controlled energy release and reduced destructive wave superposition. Initiation strategies that utilise multiple free faces and spatially distributed timing—such as inverse V firing patterns—have been shown to improve rock breakage while limiting vibration transmission toward sensitive pit walls (Khandelwal and Singh, 2009; Monjezi, Ghafurikalajahi and Bahrami, 2011). More recently, three-dimensional and four-dimensional (4D) blast design approaches have been introduced to explicitly account for both spatial and temporal aspects of detonation, allowing for refined control of energy distribution and reducing adverse effects on slope stability.

4D™ is an advanced bulk blasting system that supports both surface and underground blasting operations and adapts to varying drilling conditions to achieve the desired outcomes. It leverages proprietary emulsion chemistry and intelligent technologies. This enables precise, automated, and efficient energy control. This results in improved ore recovery, reduced overall explosive consumption, and enhanced control of fumes, vibrations, and nitrate leaching. The system can adapt in real time to varying drilling conditions to achieve the desired blasting outcomes. Using this system, a consistent emulsion density can be maintained along the borehole, improving charge uniformity and performance (Mont-Wright, 2025). Numerical modelling such as finite difference method (FDM) and distinct-element method (DEM) are commonly employed to evaluate factors of safety, potential failure surfaces, and the influence of weak planes such as lithological contacts between quartz and gneiss (Duncan, 1996; Hoek and Bray, 1981; Cundall and Strack, 1979). Complementing numerical analyses, field instrumentation and monitoring are critical for validating blast performance and ensuring geotechnical safety. Integrated monitoring systems, including installation of geophones, Time Domain Reflectometry (TDR), piezometers, and slope stability radar (SSR), allow comprehensive evaluation of both transient vibration effects and longer-term rock mass response (Read and Stacey, 2009; Dick *et al,* 2015, 2020). These combined approaches enable the assessment of stress redistribution, pore pressure variation, and slope deformation, providing a robust basis for post-blast evaluation and calibration of numerical models.

This paper examines the technical aspects and outcomes of this landmark blasting operation in detail. A brief overview of the mine site context is presented, followed by blasting optimisation techniques as well as instrumentation program and geotechnical numerical simulations. Results are then discussed as to why this blasting event posed an important challenge but ended up an engineering success for ArcelorMittal's mining team.

SITE DESCRIPTION

The Mont-Wright mining operations in Canada is ArcelorMittal's largest mining operation as well as its biggest iron ore concentrate supplier. The site is therefore of importance for the company to ensure a complete and effective value chain, from Iron Ore to steel finished products.

Location and mining context

The Mont-Wright mining complex is located near Fermont, Québec, Canada. Located north of the 53rd parallel, the town is at approximately 8.75 km from the nearest part of the mining operation, in straight-line distance. The studied blast is located at 11.06 km from the city limits. Figure 1 presents the localisation of the Mont-Wright operations in relation to the city of Fermont as well as the blast localisation. ArcelorMittal operates six open pits within the Mont-Wright property and two within the Fire Lake property, pit A being the largest at the moment. The pit A operation consist of a pushback on both south and north walls, as well as opening on the east side for a ramp exit. The goal is to open the pre-existing pit shell in order to be able to mine deeper into the iron ore formation.

In this case, multiple benches of Quartz formation must be excavated to reach the iron ore bearing formation underneath. The reaching of this ore is of paramount importance because it is the main source of iron ore for the next years to come. The excavation of the waste rock above the ore is therefore a sensitive operation and must be done quickly in order to minimise the value chain disruption that an iron ore shortage would cause. It is why ArcelorMittal's mining team is trying to minimise its operation delays associated with blast evacuation. The aim is therefore to minimise blasting events.

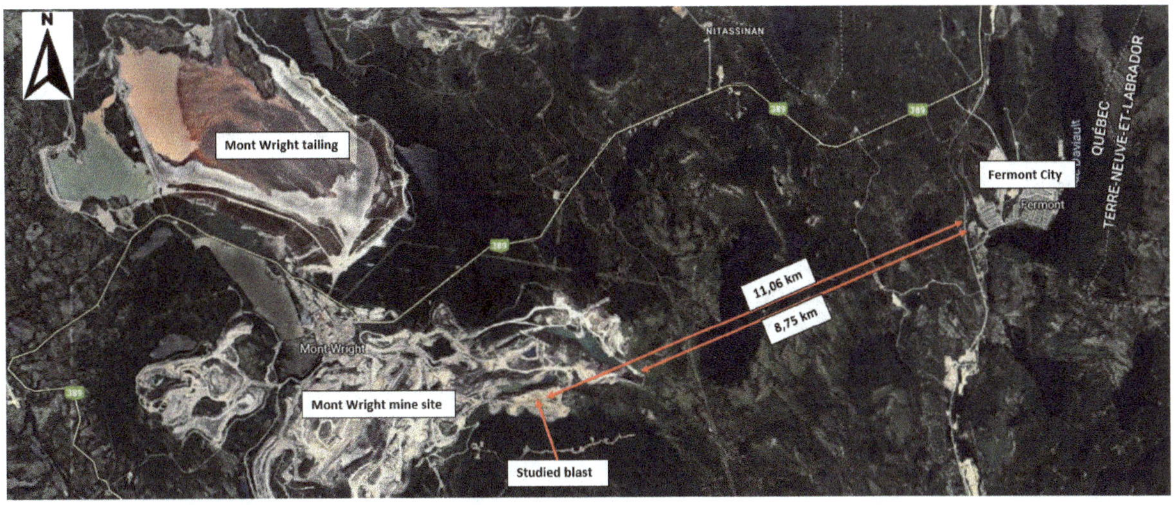

FIG 1 – Mont-Wright operations and studied blast in relation to the city of Fermont.

In fact, the majority of AMEM's shovelling and drilling equipment are powered by electricity. Therefore, cable management is a tedious and time-consuming task. To minimise the frequency of blasting events, blast pattern is increased in size. By having less but bigger blasts, operations delay are reduced by having less evacuation time, production equipment downtime, less clean up on the adjacent bench floors and less rearrangement of electric cables and set-up. It is the main reason that pushed the engineering and operations teams to reach the creation of the A-798-86, the biggest single mining blast ever recorded.

Geological and geotechnical settings

Mont-Wright geological and geotechnical settings are complex by the intense structural deformation that has occurred over time, as well as by the important lithologies variation within a single pit. Mont-Wright's complex and heterogenous geology represent the primary challenge faced by rock mechanics engineers on-site. Five main lithologies are found in Mont-Wright: Gneiss (GN), Amphibolite (AMP), Quartzite (QR), Quartz Rock Mica Schist (QRMS) and Iron Ore Formation (IF). Except for the Gneiss, IF, QR and QRMS formed through sedimentation and subsequently underwent metamorphism. These lithologies are presented in Figure 2.

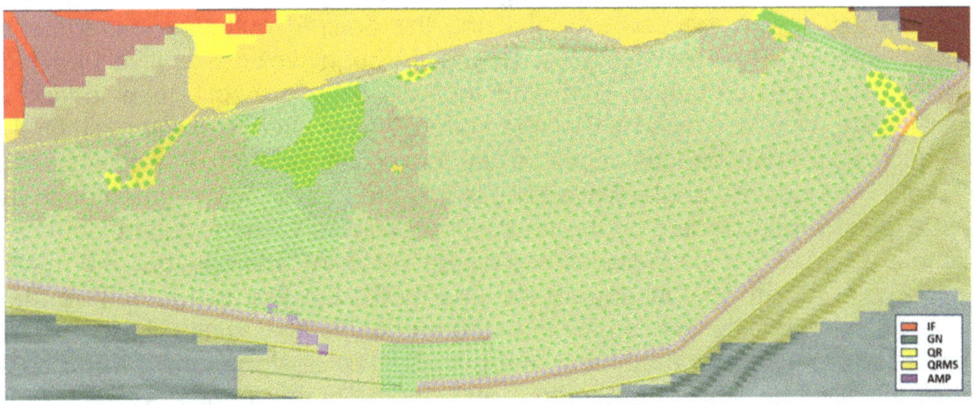

FIG 2 – Lithological plan view of the studied area.

The designed pit walls drive mainly through the QR unit. However, some areas expose QRMS lenses on a few benches scale wise. GN is the footwall and is not expected to be exposed, according to the geological model and the designed pit shell. Quartzite is occurring above the Gneiss unit. It is mechanically strong and blocky. In fact, the QR unit is characterised by a low Rock Quality Designation (RQD) in this area where crushed zone can occur, issued from shearing corridors. The QR unit then gradually transition into the QRMS unit as the content of mica (Phlogopite and Muscovite) increase. This high content in micas gives the QRMS a lower internal cohesion than the QR, as well as a weaker compressive strength. QRMS can be punctually interlocked between the

GN and the QR units creating slip pathways, or it can also be more in-pit towards the Iron Ore Formation. The later is the predominant QRMS location, relatively speaking. Both units are favourable for high hydraulic conductivity because of their high joint frequency, respectively 9.93e^{-6} m/s and 1.47e^{-7} m/s for QR and QRMS. Laterally (west–east) the superposition of the GN-QR contact is wavy but of good quality. QR-GN and QR-QRMS lithological contacts create a high heterogeneity complexity within the rock mass. All the lithological units have their bedding dipped towards the pit. The bedding acts as the principal anisotropy vector within the rock mass and it is the main contributor to potential slope instability. These aspects constitute engineering challenges to be considered in the design of blasting operations in order to ensure the long-term stability of the pit walls.

BLASTING OPTIMISATION

The process of large-scale blasts

Through pioneering the efficiency of iron ore extraction at ArcelorMittal's operations, a monumental blast emerges through engineering stewardship and execution. The process undergoes different technical phases, reshaping the entire process for the blast execution. Through rigorous blast planning with commitment to advance geotechnical analysis, in order to produce throw fragmentation through integration of blast sequencing and predictable energy distribution. Hence, various technical conditions were considered to produce the blast design, such as mine planning, geological conditions, geotechnical aspects, rock type, design parameters, timing, and operational execution.

Geology and geotechnical aspects

The mine orebody is composed of a variety of rock mass domains with mechanical behaviour. From hard competent rock to interbedded sediments with weak and jointed layers, the blast falls in the weak rock – called QRMS in-pit A, whose structural domain impacts the blast design planning. Thus, the rock structural complexity encompasses a variety of geological features, such as joint sets and domains, which impact fragmentation through *in situ* block size and rock dip direction. Acknowledging the rock characterisation and geological conditions, tailored the blast design parameters.

Blast design parameters

Blast design relies on key components, such as geometry design, loading process, explosive density and timing. The blast design focused on achieving the desired outcome in fragmentation, blast heave, and blast vibration. The area of interest is mined through two consecutive 14 m high benches, followed by a 14 m wide catch bench. Some 3.5 inch diameter pre-split holes loaded with 40 mm diameter cartridge explosives, spaced 1.8 m apart laterally, are used to shear each final wall. These cartridges produce a 1.2 kg/m load factor, with a decoupling ratio of 0.45. Pre-split is used to achieve a linear wall profile at a desired angle, while diminishing the *in situ* rock fragmentation by creating a shield against explosive gas created by the production explosion. The regular blastholes, 34.93 cm of diameter, are positioned in a staggered shape dispersed in V-Pattern layout (9.5 m × 11 m) – with crest design of 6.5 m × 8 m, to concentrate the energy for the first two row. This maximises the fragmentation by eliminating the risk coming from the preconditioned *in situ* block. Two free surfaces were also available and used for the timing of the blast. The designed blast includes 2855 holes, triggered electronically with Ikon III detonators and 450 g Pentex boosters. Each hole is double primed, where the first detonator is located at the bottom of the hole (above the 1.8 m subdrill) and the second is at mid-hole height. Regular production holes were loaded with Fortis Clear 100 Emulsion at a 1200 kg/m^3 density. Buffer holes were loaded at a density of 1000 kg/m^3 (Relative Bulk Strength (RBS) of 130) using 4D technology. The RBS 130 product played a major role in vibration control and wall control. Two buffer holes rows were designed. The first one (T) is 4 m away from the pre-split, while the second one (N) is at 9 m. The 10 inch diameter T holes are charged with 650 kg of emulsion while the 13¾ inch diameter N holes are charged with 1230 kg. With RBS 130 optimisation, the explosive quantities were adjusted to 500 kg and 1080 kg, respectively. A uniform density was maintained along the entire holes. The charge per delay was adjusted to a time envelope of 8 ms, with a maximum charge per delay of 6100 kg. A geophone was installed 650 m in front of the first initiated hole. The firing sequence led to a peak particle velocity (PPV) of 31.6 mm/s. However, the results show that the primary frequency of the blast was 31 Hz. This aligns with the Multiple Seed Wave study, which suggests that Mont-Wright rock responds well to high frequency to produce enhanced fragmentation and to minimises vibration (Kabuya *et al,* 2026). Hence, at 7 HZ of frequency, the blast-induced vibrations decreased significantly following the blast. The initiation timing plan started at 25 ms/m and then increased to 28 ms/m for the last five rows. This helped relieving the burden and minimising the vibrations in the first part of the blast.

Rock fragmentation

A total of 8.2 Mt of hard rock – primarily quartz formation- was successfully fragmented in a single drilling pattern. FRAGTrack™ shovel-based configuration system is used at Mont-Wright mine. It delivers online fragmentation data, benchmarks and insights needed to ensure sustainable improvements in blast performance. Blast fragmentation is measured using image analysis with FRAGTrack™ of photos collected at post blast muck piles. Based on data collected, rock fragmentation was improved, ensuring operational efficiency. The P_{80} is equal to 320 mm, and the shovel achieved an average performance of 6580 t/h. The Figure 3 shows the blasted rocks.

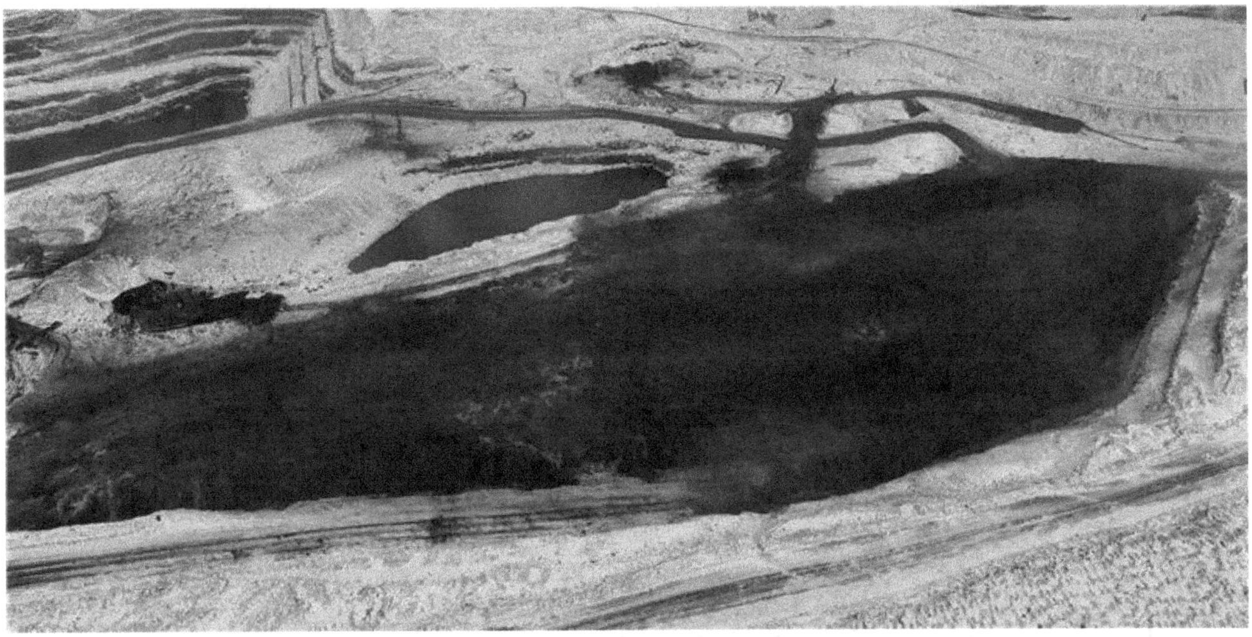

FIG 3 – Blast result in the study area.

INSTRUMENTATION AND MONITORING PROGRAM

Multiple geotechnical monitoring devices are used to monitor Pit A south-west wall. Pre-existing geotechnical concerns in the area, formed by the lithological units' contacts position explained in the geology section, are the main reason driving the implementation of such exhaustive instrumentation program in the area These geotechnical concerns, coupled with the important blasting event of A-798-86, creates a challenge that rock mechanics engineer need to prepare in order to assure safe reopening of the area for operation after the blast. Boreholes were drilled from the pit surface in order to install both piezometer and TDR equipment. The area is also monitored using Slope Stability Radar (SSR) and geophones to record blast vibrations.

Monitoring program

Because of the scale of the blasting event, TDR and Piezometer data were acquired beforehand in the case where instruments could be damaged because for the induced vibrations or slip within the rock mass. The same data was then acquired afterwards in order to conduct relative comparison of the effect of the blasting event on the instrumentation readings, and therefore the slope stability. Readings were also conducted in the subsequent days to analyse the mid-term behaviour of the rock mass. The TDR was installed in an area where the GN–QR contact lies within 30 m of the pit design and may act as a slip pathway; the installation crosses this contact to monitor deformation at the interface (Figure 4). The TDR is installed in a 247 m deep borehole that dips towards the open pit. Nodes spaced at 1 m intervals along the TDR cable record real-time extension or contraction, indicating pit wall displacement relative to the slip surface. The instrument can help monitor slow deformation in the rock mass that could be associated with multiple benches scale failure and at the same time, pinpoint where the slip surface is located in the pit wall.

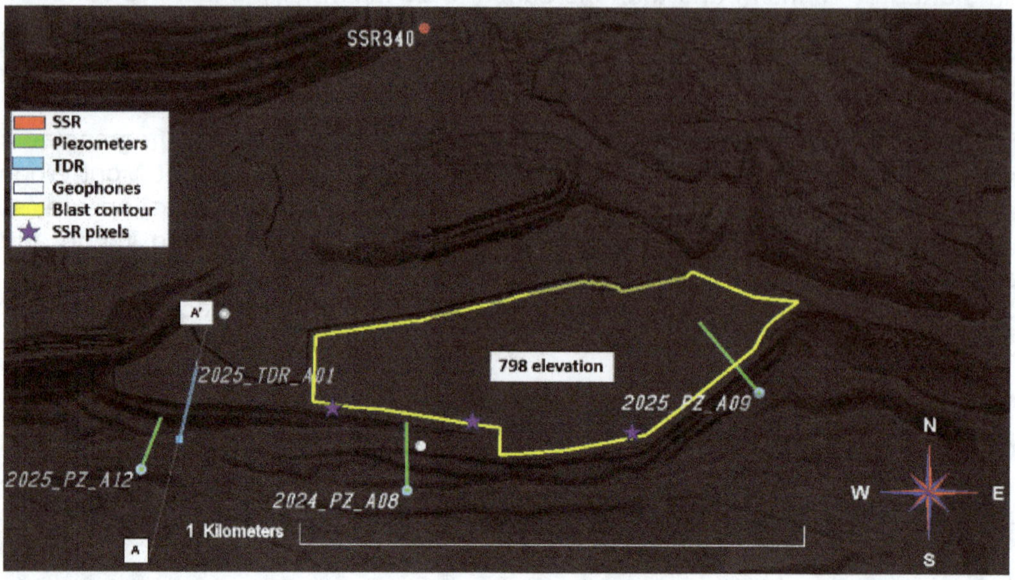

FIG 4 – Monitoring instruments locations.

Three piezometers are also installed in boreholes dipping towards the pit. The piezometer can measure the position of the phreatic surface within the rock mass, which is helpful to conduct any type of stability analysis in the area. The position of the phreatic surface is also a good indicator whether a slip surface has open in the pit wall or not, where water pressure could decrease and the phreatic surface could be lower because of water dissipation in newly created pathways in the pit wall. The sensors of piezometers 2024_PZ_A08, 2025_PZ_A09 and 2025_PZ_A12 are respectively installed at elevations 749 m, 632 m and 712 m (Figure 4).

The SSR is located on the north wall, on the other side of the pit, where it has a clear view of the area of interest which is Pit A's south wall (Figure 4). The SSR monitors any deformation on the wall surface. These deformations could indicate a progressive failure which could be linked to a failure on multiple benches or smaller failures like wedges, toppling and planar failure on a range of scales. In this case, the SSR is specifically placed to monitor a potential semi-circular failure sliding along the GN–QR contact and exiting through QR anisotropy dipping toward the pit centre. The monitoring of the deformation is in real time 24/7. Whenever deformation thresholds are crossed, site engineer is advised of the deformation. Proper decision-making is therefore possible with this information, regarding operations in the area.

Two geophones were installed at strategic positions relatively to the blast and used to record vibrations during the blasting event. One was placed on the 784 m bench, north-west of the blast and one was placed in the ramp just south of the blast at the 818 m elevation (Figure 4). They are used to check the predictive model accuracy as well as to establish and improve a vibrations database relative to lithology and blasting parameters such as borehole diameter, hole charges and drilling pattern.

Slope stability monitoring results

Data analysis after the blasting event generally indicates that there was no major deformation. Figure 5a presents the piezometric elevation derived from the three piezometers that are near the blast. Except for 2024_PZ_A08 and 2024_PZ_A09 which indicates respectively water level drops of 4.3 m and 2 m, the water level remained constant in the days prior and after the blasting event. This drop could be caused by fracture opening within the rock mass, which could increase hydraulic conductivity by increasing the fracture network connectivity. The sensor could be located in the EDZ, beyond the blast damage zone (BDZ) and the de-stressed rock mass, which increased in-depth after the large-scale blast. In fact, the piezometric sensor was close to the blast, with the distance to the pre-split wall being 78 m. Regarding vibration monitoring, the geophone on the 784 m bench recorded a PPV of 31.60 mm/s, while the geophone in the ramp above the blast recorded 18.98 mm/s PPV. Both values sit within the accepted range for building damages.

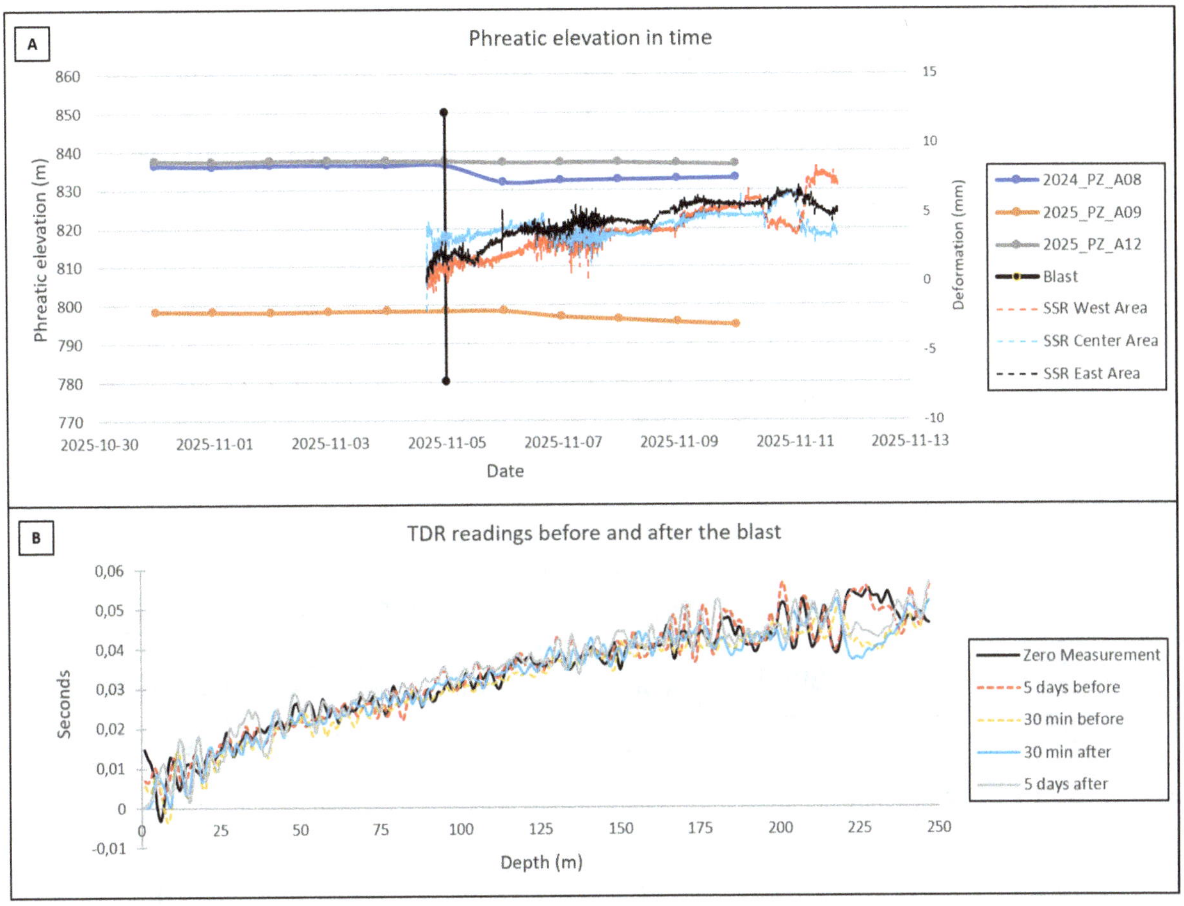

FIG 5 – (a) Piezometric levels around the blast area and SSR derived wall deformation before and after blast A-798–86; (b) TDR data over different time intervals relative to the blast, with zero measurement as reference.

Both TDR and SSR reported little to no deformation in the days after the blasting events. Figure 5b shows that no changes in deformation along the TDR cable were recorded before and after the blasting event. Therefore, it is possible to conclude that the GN-QR lithological contact was not mobilised because of the blast. It is also possible to observe a small accumulation of deformation on the south wall surface, which is presented by Figure 5a. In fact, the three zones monitored by the radar presented no more than 8 mm of cumulative deformation. This is well within the range of normal deformation observed in the area during regular operations and throughout the mine site over the course of the year.

NUMERICAL SIMULATION AND PRE-BLAST ANALYSIS

This section presents the rock mass parameters, the effect of blasting on the rock slope, and the numerical analysis methods performed in this study. Rock mechanical parameters were obtained from laboratory tests on rock core and then subsequently used as input data for numerical simulation. The discontinuum and continuum behaviours of the rock mass were simulated using UDEC (Itasca, V.7) and FLAC (Itasca, V.7) software, respectively.

Rock mass parameters

Rock mass parameters were determined based on investigation conducted between 2020 and 2025, during which many boreholes were drilled in Pit A and comprehensive rock mechanics laboratory tests were performed. The rock slope in this zone consists of GN, QR, and weakened QRMS units. The geotechnical parameters of intact rock, rock mass, joint sets, and geological contacts are summarised in Tables 1 and 2. *In situ* stress equilibrium was achieved by modelling the initial topography before excavation using the rock mass parameters of the AMP, IF, and QRMS units.

TABLE 1

Intact and rock mass parameters of different lithologies.

Lithology	Density (kg/m^3)	Poisson Ratio	UCS (MPa)	GSI	mi	ERMS D=0 (GPa)	ERMS D=1 (GPa)
AMP	3037	0.28	96	80	13	47.27	17.5
QR	2665	0.18	124	80	38	62.5	23.1
QRMS	2691	0.23	78	71	13	34.2	10.24
IF	3486	0.31	187	76	39	63.7	21.6
GN	2721	0.35	82	79	13	42.15	15.27
WZqrms*	2691	0.23	19	45	13	1.48	0.34

*WZqrms: QRMS weak zone.

TABLE 2

Rock mechanics parameters of joint and geological contacts.

Geological contact/joint	Cohesion (Kpa)	Friction angle (°)	JKn (MPa/mm)	JKs (MPa/mm)
GN-QR	100	35	57	19.46
QR-QRMS	30	28	48	17.95
GN-WZqrms/QR-WZqrms	30	28	14	7.78
QR-J1: 51/351	75	30	57	19.46
QR-J2: 69/217, J3:89/314	75	35	57	19.46
GN- J1: 56/015	150	30	33.33	13.43
GN- J2: 29/243	200	35	33.33	13.43

Influence of blasting on rock slope

Blasting excavation and stress relief create an excavation damage or disturbance zone (EDZ) in rock slopes. In other word, EDZ refers to the rock mass beyond the BDZ and the de-stressed rock mass domain, which experiences post-blast dilation due to unloading, new crack formation, and dilation of pre-existing cracks as the pit deepens. Ignoring the EDZ leads to significant overestimation of slope stability, in some cases exceeding 50 per cent (Yang *et al,* 2020). Previous studies have shown that the deformation modulus within the EDZ is reduced to approximately 52–90 per cent of that of undisturbed rock (Malmgren *et al,* 2007). Therefore, accurate prediction and understanding of blast-induced vibrations in rock masses are essential. Research indicates that radial PPV is the most suitable parameter for vibration control (Hu *et al,* 2018). Based on PPV monitoring data from Pit A, the recorded radial PPV values are below the recommended safety vibration limit (<20 cm/s). However, despite meeting vibration criteria, the presence of an EDZ must still be accounted for in numerical analyses. The depth of the EDZ is influenced by blasting parameters and the availability of free surfaces, while rock mass properties also play a critical role.

The BDZ ranges from about 0.5H for controlled pre-shear blasting to 2.5H or more for uncontrolled production blasting, where H is the blasthole depth. In fair ground with moderate *in situ* stress, the EDZ width is estimated at 10–30 per cent of pit depth (Stacey *et al,* 2003). The presence of discontinuities significantly increases EDZ depth, with more fractured rock masses experiencing greater damage intensity. Consequently, EDZ depth and damage intensity are highly site dependent. In the QR unit, three joint sets with high persistence are observed, and the rock bridge percentage is less than 13 per cent. Based on the recommendations of Evert Hoek (2012), in the case of large and confined production blasts with little or no control, the thickness of the excavation damage zone (EDZ) can be estimated as 2 to 2.5H. Although large-scale production blasting with controlled techniques and consideration of various blasting aspects has been implemented, the site-specific conditions—particularly the existence of persistent JS1 joints dipping out of the slope—justify adopting this category of Hoek's recommendations. Therefore, the worst-case scenario from Hoek (2012) has been considered, and the EDZ thickness is assumed to be 28 m, H being the height of each blasted bench (14 m). It is also important to note that, due to the large scale of blasting and uncertainties in ground conditions, such as the strength parameters of QR-GN and GN-WZqrms/QR-WZqrms contacts, the disturbance factor (D) has been conservatively taken as 1.

Numerical simulation

For the numerical analysis, UDEC and FLAC3D software packages were used. The 86 m high slope contains QR and GN. QR and GN exhibit foliation and therefore behave as anisotropic materials. According to Table 1, the joint set and foliation in QR and GN are parallel to the slope face, dipping at about 50–55°. To account for this behaviour, the Improved Unified Constitutive Model (IUCM) was adopted. Strength anisotropy in the IUCM is explicitly represented through a ubiquitous joint model, which incorporates both the strength of the intact rock matrix and the reduced strength associated with the presence of an anisotropy plane (Vakili, 2016). This model is implemented in the explicit finite-difference code FLAC3D. In UDEC, discontinuities within the QR and GN, as well as the geological contact, influence the response of the intact rock by reducing its strength and inducing non-linear and anisotropic stress–strain behaviour. The material model most commonly used in UDEC to explicitly represent joints is a linear elastic–plastic model. The stress field is predominantly gravity-controlled, and the *in situ* stress condition was defined using a horizontal-to-vertical stress ratio of 1.5:1. Figure 6a illustrates the groundwater level and the extent of the EDZ. Figure 6b presents the velocity distribution within the model. Lorig and Varona (2000) reported that velocities below 1×10^{-6} indicate stability in FLAC and FLAC3D, whereas velocities exceeding 1×10^{-5} indicate instability. Based on the authors' modelling experience, velocities greater than 1×10^{-5} represent a progressive failure phase, while velocities between 1×10^{-7} and 1×10^{-5} correspond to a regressive but slow-moving phase. Except for zones near the slope surface, velocities in the remainder of the model are less than 1×10^{-7}. Similar behaviour was also observed in the UDEC simulations. Model displacements are shown in Figures 6c and 7a. The displacement predicted by UDEC (12.4 mm) is lower than that predicted by the FLAC3D model (50 mm) and shows better agreement with radar monitoring records (Figure 5a). Due to the two-dimensional nature of the slope modelling, the dip direction of foliated domains (eg GN and QR) was assumed to be aligned with the pit slope. This assumption is conservative and results in a more cautious assessment of slope stability.

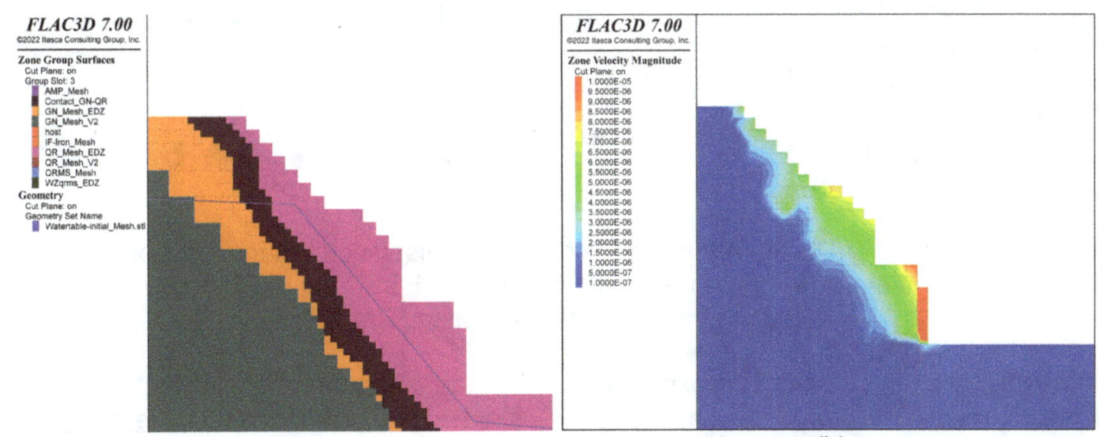

(a) (b)

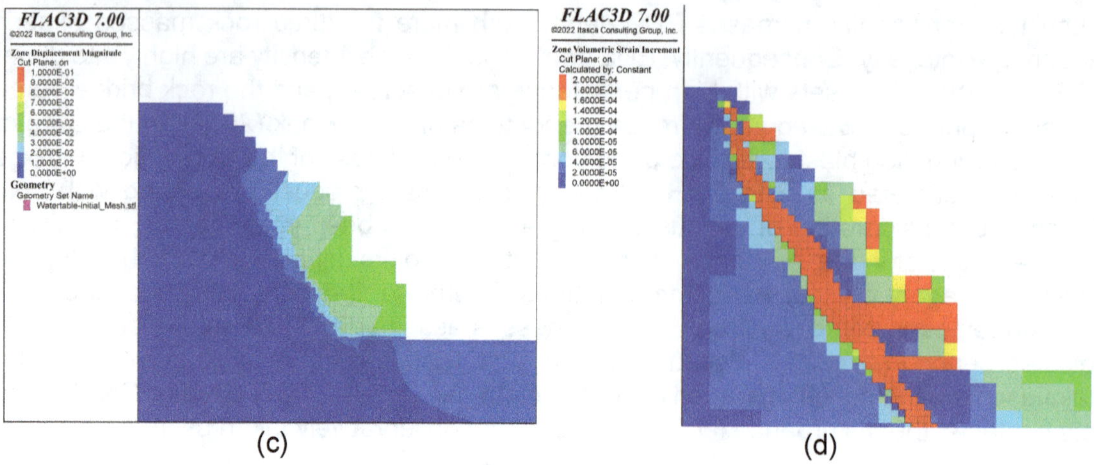

FIG 6 – The results of FLAC3D model: a) The geometry of model with the extension of EDZ and water level, b) Distribution of velocity magnitude, c) Distribution of displacement, and d) Volumetric strain increment in the model.

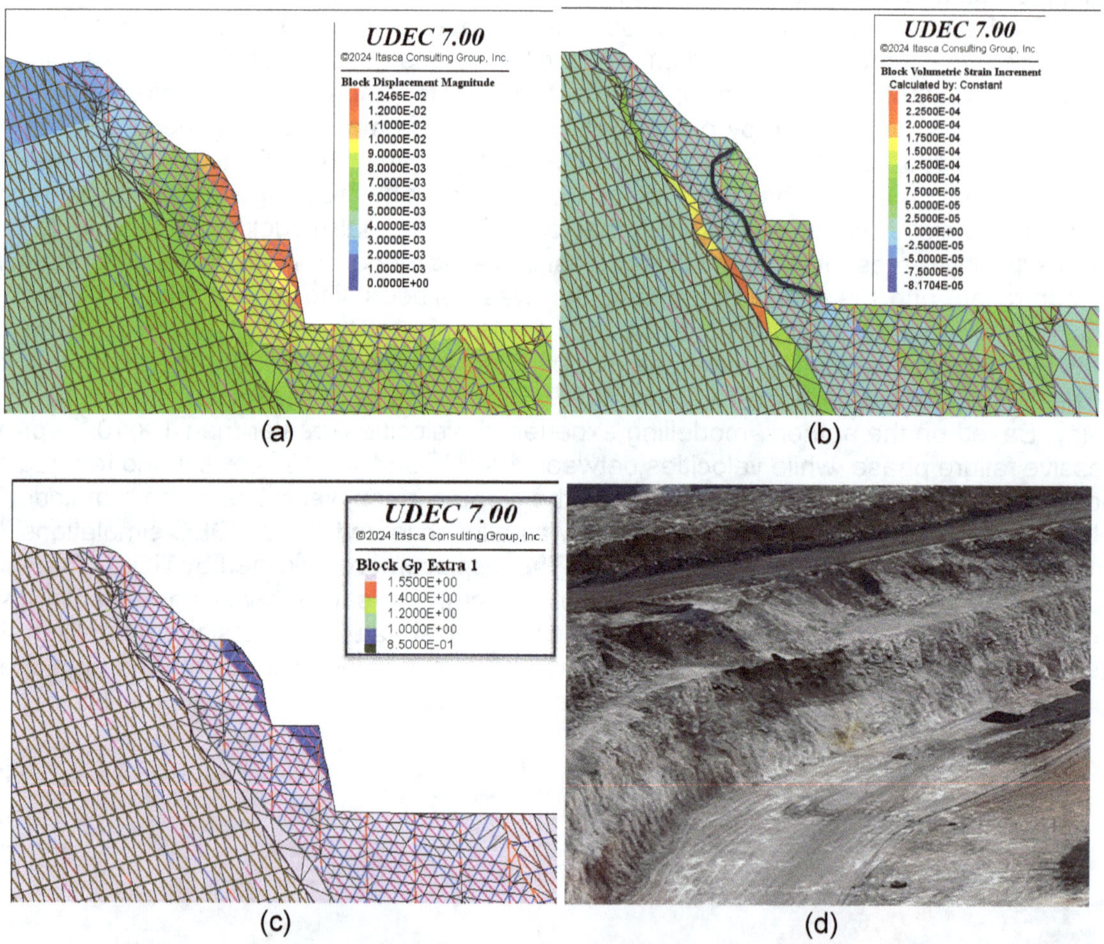

FIG 7 – The results of UDEC model: a) Distribution of displacement in rock slope, b) Block volumetric strain increment in the model, c) Distribution of FOS, and d) Plane failure along Joint Set J1 in the quartzite, validating Figure 7c.

Volumetric (dilatational) strain (VS) is expected to better identify areas prone to dilation and reduced confinement. The VS distributions obtained from the two numerical methods are presented in Figures 6d and 7b. According to Vakili, Albrecht and Sandy (2014), zero visible damage corresponds to volumetric strains in the range of 0.0–0.5 per cent. In both modelling approaches, the calculated VS remains below 0.5 per cent, indicating no visible damage. However, as shown in Figures 7d and 6b, most of the VS is observed along the QR–GN contact, highlighting it as a weak plane. Slope stability was evaluated using industry-accepted Factor of Safety (FoS) criteria, consistent with the guidelines presented by Read and Stacey (2009). The FoS calculated using FLAC3D is 3.5, indicating a high level of overall stability. In UDEC, the FoS was calculated using a FISH function, and the overall slope FoS exceeds 1.5 (Figure 7c). However, due to the presence of joint sets—particularly Joint Set J1 in the quartzite, which dips towards the pit centre—the FoS of some localised blocks near the slope surface is less than 1. This behaviour is consistent with observed ground conditions where crest lost are observed due to the brittle and fractured rock mass (Figure 7d). Overall, the results from both numerical models were cross-checked against TDR, radar monitoring data and visual inspection of rock slope, and the predicted displacements show good agreement with the recorded field measurements.

Discussion

The potential instability of Pit A's south wall slope was confirmed to be structurally controlled. The quartz–gneiss (QR–GN) contact forms a mechanically weak plane, promoting stress concentration and differential deformation under loading. Quartz-dominated formations with three joint sets present challenges for large-scale blasting due to their high stiffness, low damping capacity, and tendency to transmit stress waves over long distances. Excessive vibrations can activate discontinuities, causing cumulative degradation in brittle rock masses and highlighting the importance of vibration control and geotechnical risk management.

Numerical modelling has become an essential tool for assessing the stability of slopes under blast-induced disturbances prior to field implementation. Thus, numerical modelling using UDEC and FLAC3D provided detailed insights into slope behaviour. UDEC explicitly represented discontinuities and block interactions, while FLAC3D used the Improved Unified Constitutive Model (IUCM) to capture anisotropic GN and QR behaviour along foliation planes dipping 50–55° parallel to the slope face. Global slope stability was confirmed, with FLAC3D predicting a factor of safety (FoS) of 3.5 and UDEC showing FoS > 1.5 overall. Localised blocks near the slope surface, associated with joint set J1 in quartzite, exhibited FoS < 1, consistent with displacements observed in the field. The slope instability in this small scale at pit A was confirmed through analysis as being structurally controlled. Predicted displacements from UDEC (12.4 mm) closely matched radar monitoring data, whereas FLAC3D predicted 50 mm, reflecting the small-scale instability. Volumetric strain (VS) remained below 0.5 per cent across the slope, with concentrations along the QR–GN contact, identifying it as a weak plane.

The record-breaking blast at Mont-Wright required careful integration of advanced 4D™ blast design, numerical modelling, and monitoring to manage large-scale blasting in a quartz-rich environment near sensitive pit walls and urban infrastructure. Pre-blast simulations refined blast sequences, assessed wave interactions, and verified compliance with vibration thresholds. Post-blast monitoring – including TDR, SSR, piezometers, and geophones – showed minimal deformation. Cumulative surface displacements did not exceed 8 mm, which sits well with the UDEC pre-blast simulation. Although some piezometers indicated a decrease in water pressure, overall water levels remained largely stable. Peak particle velocities (PPVs) recorded near the blast (31.6 mm/s and 18.98 mm/s) were within safe limits and aligned well with model predictions. These results highlight the effectiveness of combining calibrated numerical models with extensive instrumentation and 4D™ blast optimisation. The Mont-Wright case demonstrates that modern mining techniques can safely handle unprecedented rock volumes while preserving slope integrity, mitigating vibration risks, and generating valuable data for future large-scale blasting projects.

CONCLUSIONS

Given the proximity of Fermont City and the known geotechnical vulnerabilities of adjacent pit walls, vibration control and a good geotechnical instrumentation program were critical. In facts, those elements, coupled with good pre-blast numerical analysis, were key factors that allowed for the safe resumption of mining activities in the area after the blasting event. The study showed that by having a good and comprehensive knowledge of the geological conditions and a good understanding of the geotechnical risk, a good monitoring program can be implemented on the field. This program is one of the main elements of the risk assessment and mitigation plan for the realisation of the blast. Advanced blasting techniques such as 4D loading and pre-blast vibrations modelling proved to be a key element that gave the engineering department the confidence to produce such an important blast, while maintaining ArcelorMittal's primary value: Safe mining operation. Overall, the historical blasting event was a complete success. Advanced engineering solutions proved to be essential for the realisation of the historic event, which delivered important gains for the mining operation.

ACKNOWLEDGEMENTS

The authors would like to thank ArcelorMittal Mining Canada and Orica for their support of this project. Many thanks to Mining One and Cavroc for their support in preparing the FLAC3D models and for their valuable input in discussing the results.

REFERENCES

Cundall, P A and Strack, O D L, 1979. A discrete numerical model for granular assemblies, *Geotechnique*, 29(1):47–65.

Dick, G, Eberhardt, E, Cabrejo Liévano, A G, Stead, D and Rose, N D, 2015. Development of an early warning time of failure analysis methodology for open pit mine slopes utilizing ground based slope stability radar monitoring data, *Canadian Geotechnical Journal*, 52(4):515–529.

Dick, G, Nunoo, S, Smith, S, Newcomen, W, Kinakin, D, Stilwell, I and Danielson, J 2020. Monitoring and managing large deformation pit slope instabilities at a British Columbia copper mine, in *Slope Stability 2020: Proceedings of the 2020 International Symposium on Slope Stability in Open Pit Mining and Civil Engineering* (ed: P M Dight), pp 439–452 (Australian Centre for Geomechanics: Perth), https://doi.org/10.36487/ACG_repo/2025_25

Dowding, C H, 1985. *Blast vibration monitoring and control* (Prentice Hall).

Duncan, J M, 1996. State of the art: Limit equilibrium and finite-element analysis of slopes, *Journal of Geotechnical Engineering*, 122(7):577–596.

Hoek, E and Bray, J W, 1981. *Rock Slope Engineering*, 3rd edn (Institution of Mining and Metallurgy: London).

Hoek, E and Brown, E T, 2012. The Hoek–Brown failure criterion, *Journal of Rock Mechanics and Geotechnical Engineering*, 45:981–988.

Hoek, E, 2012. Blast damage factor D, Technical note for RocNews, Rocscience. Available from: <https://www.academia.edu/38668181/Blast_Damage_Factor_D>

Hu, Y G, Liu, M S, Wu, X X, Zhao, G and Li, P, 2018. Damage–vibration couple control of rock mass blasting for high rock slopes, *International Journal of Rock Mechanics and Mining Sciences*, 103:137–144.

Kabuya, J M, Atalla, J, Proulx, R, Yang, R, Etulle, L and Desrochers, J, 2026. Multiple Seed Waveform Vibration Model of blasts Established at Mont-Wright Mine, International Society of Explosives Engineers Conference, Reno, Nevada.

Khandelwal, M and Singh, T N, 2009. Prediction of blast-induced ground vibration using artificial neural network, *International Journal of Rock Mechanics and Mining Sciences*, 46:1214–1222.

Lorig, L and Varona, P, 2000. Practical slope stability using finite difference codes, in Proceedings of Slope Stability in Surface Mining (eds: W A Hustrulid, M K McCarter and D J A Van Zyl), (Society for Mining, Metallurgy and Exploration: Englewood).

Malmgren, L, Saiang, D, Töyrä, J and Bodare, A, 2007. The excavation disturbed zone at Kiirunavaara mine Sweden by seismic measurements, *Journal of Applied Geophysics*, 61:1–15.

Monjezi, M, Ghafurikalajahi, M and Bahrami, A, 2011. Prediction of blast-induced ground vibration using artificial neural networks, *Tunnelling and Underground Space Technology*, 26:46–50.

Mont-Wright, 2025. ORICA 4D™ Technology at Mont-Wright (phases 1–2 overview), Internal technical report.

Persson, P A, Holmberg, R and Lee, J, 2018. *Rock Blasting and Explosives Engineering* (CRC Press).

Read, J and Stacey, P, 2009. *Guidelines for Open Pit Slope Design* (CSIRO Publishing: Melbourne).

Stacey, T R, Xianbin, Y, Armstrong, R and Keyter, G, 2003. New slope stability considerations for deep open pit mines, *Journal of the Southern African Institute of Mining and Metallurgy*, 103(6):373–389.

Ullah, S, Ren, G, Ge, Y, Memon, M B, Kinyua, E M and Ndayiragije, T, 2025. Dynamic slope stability assessment under blast induced ground vibrations in open pit mines: A pseudo static limit equilibrium approach, *Sustainability*, 17(14):6642.

Vakili, A, 2016. An improved unified constitutive model for rock material and guidelines for its application in numerical modelling, *Computers and Geotechnics*, 80:261–282.

Vakili, A, Albrecht, J and Sandy, M, 2014. Rock strength anisotropy and its importance in underground geotechnical design, in *Proceedings AusRock 2014: Third Australasian Ground Control in Mining Conference*, pp 167–180 (The Australasian Institute of Mining and Metallurgy: Melbourne).

Yang, J, Dai, J, Yao, C, Jiang, S, Zhou, C and Jiang, Q, 2020. Estimation of rock mass properties in excavation damage zones of rock slopes based on the Hoek–Brown criterion and acoustic testing, *International Journal of Rock Mechanics and Mining Sciences*, 126:104192.

Open Pit

Mine planning

Key performance indicators to drive productivity of truck-shovel mining systems in surface mines

D Rose[1], E Topal[2], G Barakos[3] and S Akdag[4]

1. Director at KPMG Australia; PhD candidate, Department of Mining Engineering, WA School of Mines, Curtin University, Bentley WA 6102.
2. Professor, Department of Mining Engineering, WA School of Mines, Curtin University, Bentley WA 6102. Email: e.topal@curtin.edu.au
3. Professor, Department of Mining Engineering, WA School of Mines, Curtin University, Bentley WA 6102. Email: george.barakos@curtin.edu.au
4. Lecturer, Department of Mining Engineering, WA School of Mines, Curtin University, Bentley WA 6102. Email: sela.akdag@curtin.edu.au

ABSTRACT

Truck and shovel mining remains the dominant earthmoving method in open pit mining, typically accounting for 60 per cent of site costs. Despite extensive performance reporting, many mines still struggle to convert operational data into sustained productivity improvement.

In pursuit of productivity improvements, the industry had deployed 1068 autonomous trucks globally as of 2022 and as of July 2025 this increased to 3832 units (GlobalData, 2022, 2025). Comparison of performance indicators of productivity and operational efficiency before and after autonomous haulage system implementation broadly show labour cost savings and improvements in utilisation of truck fleets in line with expectations, but unfulfilled potential for productivity improvement measured in tonnes per hour (in like-for-like pit conditions).

Insights from manufacturing and processing industries emphasise the importance of carefully selected process-oriented KPIs to drive operational productivity, which begs the question: is there a comprehensive set of KPIs for a generic truck-shovel system which can be used by mine management to drive productivity performance using conventional management practice?

A key tenet of this research is that inappropriate and/or incomplete KPIs are often used by mines in part due to an inadequate theoretical understanding of the complex productivity drivers of truck-shovel systems, in particular the difficulty of defining KPIs which fairly reflect the system-wide interactions and variability of truck-shovel operations. The very feature of truck-shovel systems that makes it the preferred mining system in open pit mines – its flexibility – is also what makes it difficult to predict and measure.

The study will explore the opportunity to develop a mathematically rigorous KPI set which can be integrated into an overall productivity and efficiency metric for the truck-shovel system. This would enable mine management to drive significant productivity improvements in open pit mines simply by aligning KPIs with critical success factors and organisational roles. This will enhance operational decision-making and structured business improvement, thereby addressing the unfulfilled potential of truck-shovel systems in open pit mining.

REFERENCES

GlobalData, 2022. Global autonomous mining truck population tops 1,000 [online], GlobalData. Available from: <https://www.globaldata.com/data-insights/automotive/global-autonomous-mining-truck-population-tops-1000/> [Accessed: 10 July 2025].

GlobalData, 2025. Development of Autonomous Trucks in the Global Mining Sector, report: GDMINEQ021-TH. Available from: <https://www.globaldata.com/store/report/development-of-autonomous-trucks-in-mining-market-analysis/>

Open Pit

Mine planning and strategic optimisation

Technical evaluation of open pit to underground transition using integrated optimisation methods

M Flanagan[1], T Pelech[2], F Grobler[3] and N Barros[4]

1. Director/Global Lead Open Pit, AMC Consultants Pty Ltd, Melbourne Vic 3000.
   Email: mflanagan@amcconsultants.com
2. Technical Lead Strategy Optimization, AMC Consultants Pty Ltd, Melbourne Vic 3000.
   Email: tpelech@amcconsultants.com
3. Principal Mining Consultant, AMC Consultants Pty Ltd, Melbourne Vic 3000.
   Email: fgrobler@amcconsultants.com
4. Mine Sustainability Lead, AMC Consultants Pty Ltd, Melbourne Vic 3000.
   Email: nbarros@amcconsultants.com

ABSTRACT

As open pit mines approach their economic limits, there is increasing impetus to consider transitioning to underground mining, either in parallel or as a subsequent phase to open pit operations.

Various authors have developed approaches to optimise the transition from open pit to underground mining (Whittle *et al,* 2016; MacNeil and Dimitrakopoulos, 2017) with acknowledged limitations to define the preferred transition point.

This paper explores an integrated optimisation approach that enables mining decision-makers to assess the full spectrum of technical (all mining methods including underground caving), economic (revenue, capital and operating costs), and operational factors (mine sequencing, production rates, ESG targets and constraints) influencing open pit to underground transition decisions. Further expanding on conventions through extensive practical experience, case studies and advanced modelling techniques, the authors demonstrate the substantial impact that an integrated approach can deliver.

INTRODUCTION

The decision to transition from open pit to underground operations is a multifaceted challenge, devoid of a one-size-fits all solution.

For greenfield projects, the choice between open pit and underground mining typically hinges on orebody depth, geometry, and the trade-off between capital investment and early cash flow. In contrast, for existing mining operations, deciding if-and-when to transition to open pit or underground mining is more complex. Traditionally, such decisions focus on economic and technical risks and can be biased toward continuing the open pit extending beyond its optimal point.

Deciding where and when to transition from an open pit to an underground operation (or *vice versa*) is a widely debated subject. Whittle *et al* (2016) focused on the cost differential of extracting a block of material from either open pit or underground mining methods, referred to as an opportunity cost. The underlying assumption is that if a block can be mined by either method, if it is not mined by open pit, then it will be mined by underground. The restricted (opportunity cost) pit is typically smaller than the lower limit pit. It should be noted that the restricted pit does not necessarily align with a particular phase in the ultimate pit, as the basis for calculating the block values is quite different. This approach provides a guide on potential transition zones, but has limitations associated with simultaneous mining using open pit and underground methods, the influence of discounted cash flows on value, and the consideration of execution factors.

Some of these limitations have been considered by MacNeil and Dimitrakopoulos (2017) in techniques where a set of open pit and underground extraction schedules are defined for different transition depths. These techniques predict discounted operating cash flows for each open pit and underground schedule, and then sum together, with the combined discounted operating cash flow that represent the most favourable transition decision. Geological uncertainty is considered in this approach using stochastic methods to understand the potential cash flow risk profile of different mining sequences and transition points, and is also discussed by Grobler (2016). This technique provides an improved view of discounted operating cash flow but does not consider the full spectrum of factors.

A wide range of technical and non-technical factors that might influence the decision include the level of knowledge about the deeper parts of the orebody, the availability of underground mining skills, the availability and feasibility of obtaining ore feed from other sources, sustaining capital considerations of an open pit operation, the size and timing of capital expenditure to develop an underground operation, the suitability of the current plant to process potentially different ore types at a lower feed rate, and the time frame for developing an underground operation. Furthermore, sustainability (ESG) considerations now play a more prominent role, adding new dimensions to the decision-making process (Hall *et al,* 2025).

These factors demand a higher level of sophistication in determining the preferred development and operating strategy for an open pit and underground operation. Strategy optimisation offers a solution and differs from previous optimisation techniques in that it models all components of the mining system to achieve overall long-term goals. Whereas typical mine optimisation focuses on optimising a component of the mining system such as the mine schedule, to achieve a specific objective such as to maximise NPV, strategy optimisation models all the practical options (strategic choices over which we have control), under all the possible scenarios (things we can't control such as metal prices), and reports as many different metrics simultaneously as desired. By modelling all the relationships within the mining system, the trade-offs between achieving different goals ie financial, ESG, production etc can be fully understood.

Case studies demonstrate the benefits of this integrated approach. For instance, a conventional pit-only optimisation might select a different transition depth than a fully integrated open pit – underground analysis, highlighting the value of simultaneous optimisation. Multi-objective analysis shows that optimising for multiple goals (eg maximising NPV while meeting operational constraints) can alter the preferred transition strategy. Factors like underground cut-off grade and inclusion of secondary products also strongly influence outcomes. By capturing these effects, an integrated approach yields more robust and value-maximising mine plans than traditional methods, enabling an informed selection of the strategy that best meets the corporate objectives.

METHODOLOGY

The integrated methodology for evaluating the open pit to underground transition comprises several steps, which together form a comprehensive optimisation workflow, shown in Figure 1.

FIG 1 – Strategy optimisation framework.

Strategy objectives and input parameters

A key component in determining an optimal open pit to underground transition is aligning the strategic objectives of the asset (or portfolio) with the asset owner. The agreed strategic context is used to identify a diverse range of opportunities, then converged to define and frame priority opportunities to be further evaluated in consultation with stakeholders, depicted in Figure 2.

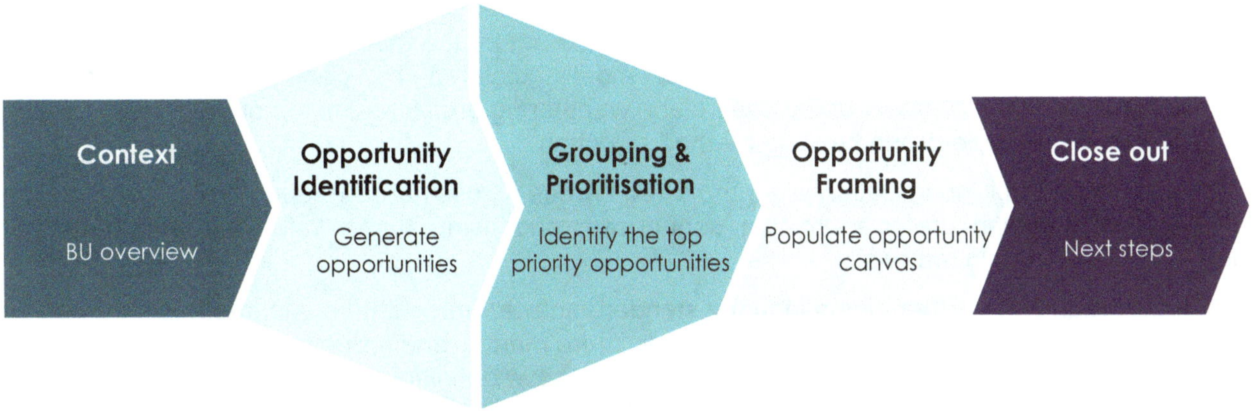

FIG 2 – Strategy objective alignment.

Examples of strategic objectives and options that can be considered include:

- Inclusion or exclusion of ore sources and timing of mining.
- Impact of including other internal (eg low-grade or Inferred Mineral Resources) or external ore sources.
- Alternative mining methods.
- Production quality and quantity targets.
- Range of mining or processing throughput rate.
- Materials handling or processing method(s).
- Optimal combination of mining zone-specific cut-off grade or cut-over value.
- Availability and timing of capital (ie applied capital constraints).
- Sustainability targets such as decarbonisation, water abstraction, resources stewardship etc.
- Best strategic outcomes when considering alternative or multiple objectives eg NPV, free cash flow, minimised capex, mine life, footprint etc.

After establishing the strategy objectives, input parameters are collated for the range of options being considered. Many mineral assets will have data available for the current operational conditions, for example the metallurgical recoveries derived from most common combination of minerals produced at the mine, costs reported to stock exchanges, and data demonstrating the overall productivity of the mine. The information that is probably not available, is the data representing conditions that are different from the current or historical settings. This missing data represents the region of the optimisation problem space that must be modelled based on either benchmarked data, known physical or observed relationships (such as the Six-Tenths rule for scaling costs according to equipment capacity, (Lanz and Seabrook, 2012)), and first-principles derivations.

It is important that, rather than having a single data point available that represents only one operating point, numerical relationships (eg derived regression functions) are built into the optimisation model that consider all the operating points that should be tested to generate candidate solutions. The exact nature of important relationships will be unique to every asset, and a period of consultation is required to determine their appropriate makeup.

Mining inventory definition and transition zone identification

The conversion of raw geological and engineering data into a set of mining inventories is a foundational step in optimising open pit and underground mining.

The process begins by defining mining shapes for the deepest viable open pit and the shallowest feasible underground level, based on geometry and geotechnical constraints. The lower open pit limit is set by the deepest pit that can be economically mined without considering an underground operation. While the upper underground mining limit is set by the minimum depth the orebody can be extracted using underground methods, mostly dictated by geotechnical and stability considerations. This is typically at the base of the weathering profile (ie the top of fresh rock) but can differ according to underground mining method selection.

Well known open pit optimisation algorithms (eg Lerchs–Grossmann, Pseudo-flow) delineate the ultimate pit limit and preferred mining sequences, prioritising higher-value mineralisation and deferring lower-margin material.

For underground mines there is no industry standard software/algorithm comparable to the open pit optimisation process and consequently is typically more manual and arduous. Potential mining limits are usually established using mine shape optimiser or cave definition software to generate practical stope or caving shapes according to applicable underground mining methods based on a range of factors, schematically represented in Figure 3.

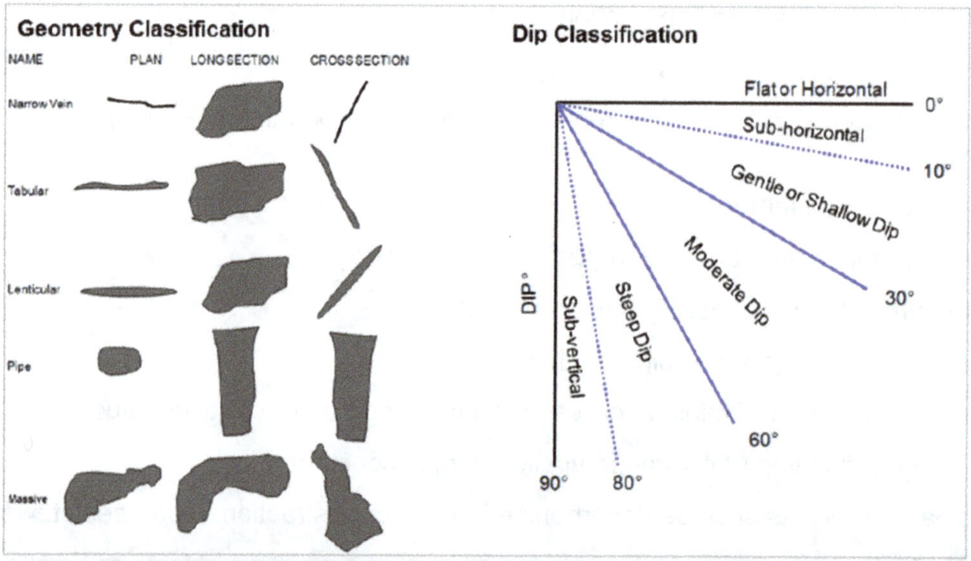

MINING METHODS		DEPOSIT CLASSIFICATION	
GENERIC GROUPING	ATTRIBUTE	DIP	GEOMETRY
Caving Methods	**Low Cost High Production Low Selectivity & Control**		
Block Caving		Vertical to Sub-Vertical	Massive to Pipe
Sub-level Caving		Vertical to Sub-Vertical	Massive to Lenticular
Core and Shell		Vertical to Sub-Vertical	Massive to Lenticular
Open Stoping Methods			
Sub-level Open Stoping **		Vertical to Sub-Vertical	Massive to Lenticular
Longhole Open Stoping **		Sub-Vertical to Moderate	Lenticular to Tabular
Benching **/ Avoca / Modified Avoca		Sub-Vertical to Moderate	Tabular to Narrow Vein
Shrinkage / Vertical Retreat		Sub-Vertical to Moderate	Tabular to Narrow Vein
Drifting Methods			
Room and Pillar		Flat to Sub-Horizontal	Lenticular to Tabular
Post Pillar Cut and Fill *		Sub-Horizontal to Moderate	Lenticular to Tabular
Mechanised Cut and Fill *	**High Cost Low Production High Selectivity & Control**	Flat to Vertical	Lenticular to Narrow Vein
Drift and Fill *		Flat to Vertical	Lenticular to Narrow Vein

* Backfilling is incorporated into the method

** Backfilling can be used to increase mining recovery

FIG 3 – Underground mining method selection.

Hall (2014) provides a comprehensive examination of strategy optimisation with a focus on cut-off grades, underscoring that decisions should not be made in isolation. Rather, integrated with other key variables such as mining sequences, production rates, stockpiling strategies, and capacity expansions to maximise overall project value. This holistic perspective recognises cut-off grades as dynamic outcomes of the optimisation process rather than fixed inputs, ensuring alignment with corporate strategy objectives. Accordingly, a set of underground inventories are developed for a range of mining methods and cut-off grades, using relevant input parameter ranges (Table 1).

TABLE 1

Underground mining method parameter example ranges.

Parameter	Unit	Caving (BC, SLC, VCR)	Open stope (SLOS, LHOS)	Drift (CAF, DAF, R&P)
Mining cost	US$/t	10–30	30–100	100–200
Dip range	Degrees	60–90	40–90	20–90
Shape strike length	m	16–22	10–40	10–20
Minimum transverse width	m	15–30	5–10	3–4
Shape height	m	50–500	15–60	4–5
Minimum transverse pillar width	m	NA	5–10	4–5
Total ELOS overbreak	m	60°–70° cone	1–3	0.5–1
Vertical advance	vmpa	30–60	40–80	30–50
Production rate	Mt/a	2–20	1–6	0.25–1.50
Hydraulic radius	m	10–50	5–10	1–5
Pre-production capex	US$ M	750–3000	300–1000	30–400
Ore tonnes per development metre	t/dev m	500–2000	150–500	50–150
Mining recovery	%	70–85	85–95	40–95

These inventory sets are not mutually exclusive, with the preferred underground mining approach, or approaches, being evaluated in subsequent schedule optimisation. Each inventory set is indexed by a range of cut-off grade values, creating a multi-dimensional array of pre-computed inventory options.

Conceptual mine designs can be developed specific to the asset, outlining layouts for access, infrastructure, and extraction sequences. To manage complexity in large-scale operations, parametric relationships can also be established, described by Pelech and Grobler (2025), with mineable shapes and designs aggregated into strategically relevant mining zones, which represent logical groupings based on spatial proximity, geological similarity, operational dependencies, or phased development stages. For instance, in an underground inventory set, zones might correspond to distinct open pit phases, underground mining levels, panels, or zones that can be developed and mined semi-independently.

To maintain computational efficiency while preserving essential strategic insights, exclusive open pit and underground mining zones can be developed according to the potential transition zone, adopting a modified approach as defined by Whittle *et al* (2016). Mineable shapes of the underground inventory sets can be incorporated into the geological block model and evaluated in an open pit optimisation, an example application using Geovia Whittle software is shown in Figure 4.

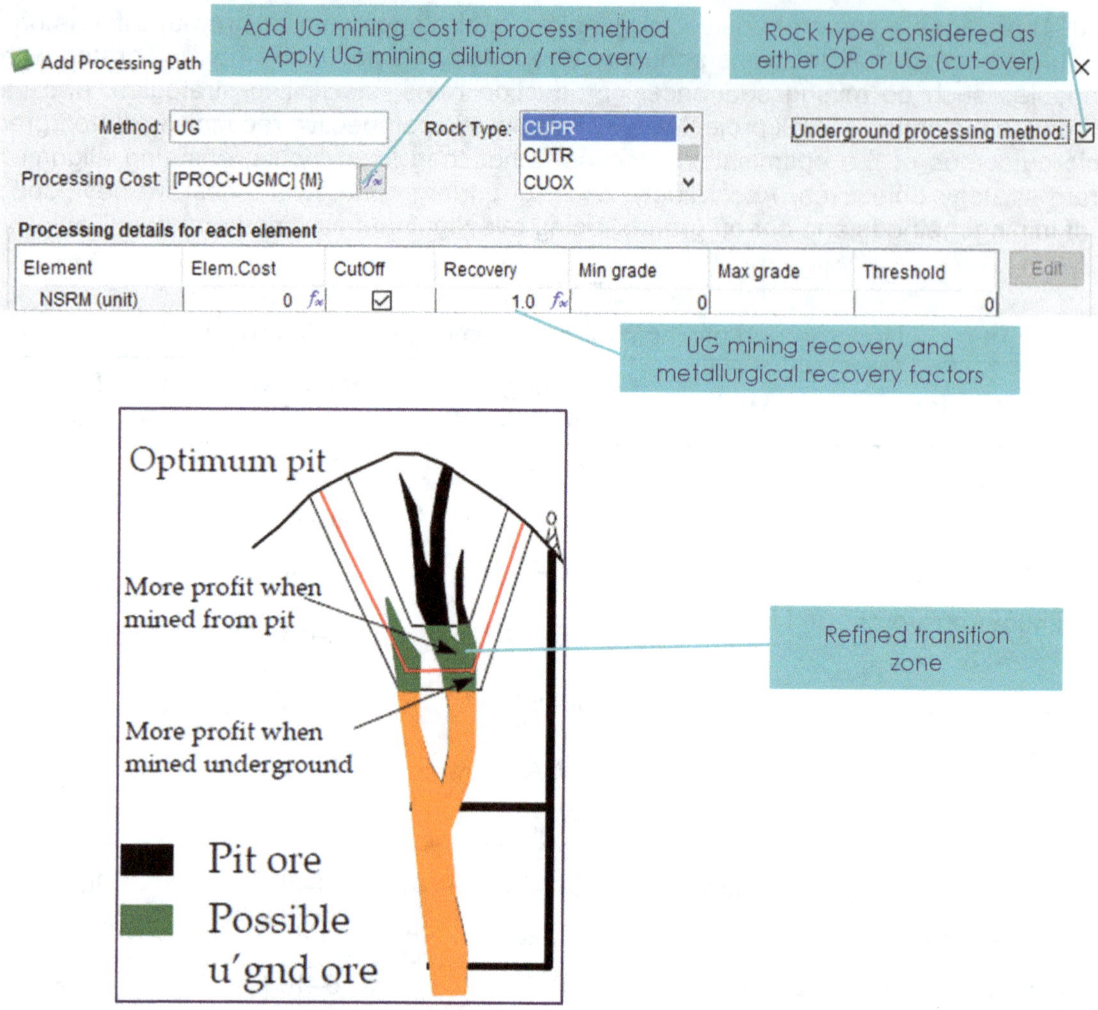

Source: Geovia Whittle

FIG 4 – Transition zone identification application.

Each strategic zone is characterised by a set of attributes that enable the creation of representative production schedules. These attributes include, but are not limited to, vertical rate of advance, development metres (eg lengths of drifts, raises, and declines required for access), ore tonnages (total extractable material), average grades (concentrations of valuable minerals), backfill quantities (for stability in stoping methods), production rate potential, associated costs (capital and operating expenditures), and any other customised metrics relevant to the asset, such as environmental impacts or water management requirements. This attribute-based representation forms the backbone for downstream scheduling and strategy selection, allowing the schedule optimisation to simulate realistic cash flows and operational timelines without delving into tactical-level details like individual equipment assignments.

Schedule optimisation

The schedule optimisation step presented in the integrated strategy optimisation approach is deeply rooted in real options analysis (ROA) thinking (Dimitrakopoulos and Abdel Sabour, 2007; Damodaran, 2009; Trigeorgis and Reuer, 2016; Chandra and Hartley, 2024; Pelech, Du Preez and Hall, 2026); acknowledging that the highest-NPV static plan is frequently not the most valuable when managers can adapt to new information.

This is achieved by implementing a rapid scheduling approach designed to create efficient, high-level representations of mine assets that balance accuracy with computational speed. Scheduling models must be capable of replicating the current operational plan (if applicable), as well as alternative cases that arise from varying strategic choices. The mining inventories are integrated with cost models (both capital and operating) to generate flexible periodic cash flows, which serve as the primary inputs for traditional financial valuation metrics like NPV and IRR.

Various scheduling approaches are employed, all sharing the common goal of transforming geological and inventory data into strategic plans. For example, constraint-based optimisation might prioritise production from high-grade zones while respecting capacity limits on processing or haulage. Simplification is important to enable computation of the industrial-scale of strategic open pit and underground optimisation problems. Rather than modelling every operational detail, the process focuses on core drivers while abstracting secondary elements. Non-primary production tasks, such as mining area preparation, grade control, drilling and blasting, ventilation upgrades, ground support installation, or backfilling in stopes, are accounted for in the cost model ensuring their financial impacts are captured but are not explicitly scheduled. Similarly, primary mining activities like ramp and level construction are assumed to be completed in a timely manner ahead of production, eliminating the need for granular sequencing of these tasks. This approach maintains strategic fidelity while allowing the model to rapidly iterate through thousands of options.

Multi-objective optimisation is a powerful technique employed in the strategy optimisation process, allowing balanced decision-making for various strategy goals. Beyond the conventional focus on financial metrics like NPV, this approach can incorporate broader policy objectives such as fulfilling social responsibilities (eg community engagement or environmental stewardship), extending mine life to maximise resource utilisation, and facilitating resource discovery through sustained exploration investments. By considering these goals and more, multi-objective optimisation enables robust strategic plans that perform well under various uncertain conditions, including market volatility, geological uncertainty, or regulatory shifts while observing overarching policy decisions. This resilience is achieved by evaluating strategies not just for peak performance in ideal scenarios but for their ability to adapt and maintain value across probabilistic outcomes.

The process employs the Non-dominated Sorting Genetic Algorithm III (NSGA-III), as described by Deb and Jain (2014), which is tailored for high-dimensional, many-objective problems. As an evolutionary algorithm, NSGA-III operates by initialising a population of potential solutions each representing a unique combination of decision variables such as cut-off grades, production rates, or investment timings and iteratively evolving them through a process that mimics natural selection. The algorithm adjusts input parameters, evaluates the resulting modelled outcomes against the defined objectives, retains and develops the fittest solutions. Over successive generations, it converges to a set of near-optimal trade-offs, ensuring coverage across the objective space rather than converging to a single point.

Strategy selection

Post-processing techniques, or data analysis is then undertaken on the NSGA-III output and can be represented using techniques such as the Hill of Value® process (Pelech and Grobler, 2025). This analysis allows stakeholders to gain valuable insights from the schedule optimisation process. The AMC Hill of Value® process is being unified into a dedicated software application leveraging decades of strategy optimisation development, strengthened by expert guidance and judgement.

An example output is the definition of a Pareto Front (or Optimum Frontier), the line of optimal trade-off combinations between two or more objectives. Along this front, all the model input parameters and variable data can be extracted from the results and analysed to see how those model parameters are changing along the Front. This kind of insight is very valuable to understand the consequences of changes in operational settings, such as mining methods, transition zones, mining rates, mine sequencing, cut-off grade, processing rates, and ore selection (amongst others).

This kind of analysis allows stakeholders to understand the importance of each of these potential parameters in the context of achieving their strategy, allowing greater tactical or managerial flexibility to achieve those outcomes in practice (in the case where change in settings is required). The delivered integrated strategy optimisation with these enhancements included is now far more nuanced and expansive than the conventional output based on a simple 3D surface chart comprising three variables.

CASE STUDIES

Case study 1 – base metals operation

Case Study 1 considers a large, relatively shallow underground base metals operation. The mine is nearing the end of its planned life and is no longer able to fill its current processing capacity. This leads the operation to be positioned higher on the cost curve due to higher fixed costs per unit of product. There remains a large amount of low-grade mineralisation (below underground cut-off) still available, which may have economic value with the right strategy.

A conversion to an open pit operation has been investigated in the past via traditional deterministic methods. The open pit option was dismissed at the time due to the relative value of the high-grade mineralisation underground, perceived execution risks, and closure costs introduced by an open pit.

The operation has a large closure liability component upon cessation of mining, which is driven by waste dump rehabilitation costs. An open pit will significantly increase those costs, although also delay their realisation with an extension of the mine life. These factors and others turn a simple scenario analysis into a complex trade-off problem space which can only be solved algorithmically.

An AMC Strategy Optimisation study was commissioned to investigate this complex options space. Being an established underground operation, the primary value driver was expected to be the ore-waste cut-off grade used as an input into mine plans. The underground inventory was parametrised according to that assumption. For this operation, modifying the cut-off grade also impacts operational costs in the form of modified development, backfill and ground support requirements. The dynamic nature of these relationships is commonly overlooked in traditional cut-off calculations. In this case, the closure costs and fixed costs of extending the mine life are important factors to balance.

Open pit mining, inclusive of associated equipment and infrastructure and environmental and social impacts, must be considered against the option of mining the mineralisation above the underground cut-off via the established underground infrastructure, effectively competing for the same material. This is not a classic open pit to underground transition problem, but the inverse, although the method applied to this problem shares many similarities.

This options analysis space was parametrised at a conceptual level as follows:

1. Depletion of the geological model to the start of the underground schedule.
2. Creation of a diluted mining block model for open pit mining.
3. Generation of potential open pit mining limits using pit optimisation algorithms.
4. Selecting and labelling a series of open pit phases that represent appropriate size mining areas, collated into vertical bench intervals.
5. Mapping out and labelling the underground inventory (for all cut-offs) that overlaps with each open pit phase and bench interval (this can include provision for crown pillars but didn't in this case study).
6. Overlaying mining parameters (mining method, phase, bench, zone etc) to the relevant blocks in the block model.
7. Binning the material blocks by a range of cut-off grades and aggregating to a strategic resolution.
8. Integrating the mining inventory block model into the Hill of Value® strategy optimisation model.

A visual representation of the Hill of Value® inventory for this case study is shown in Figure 5.

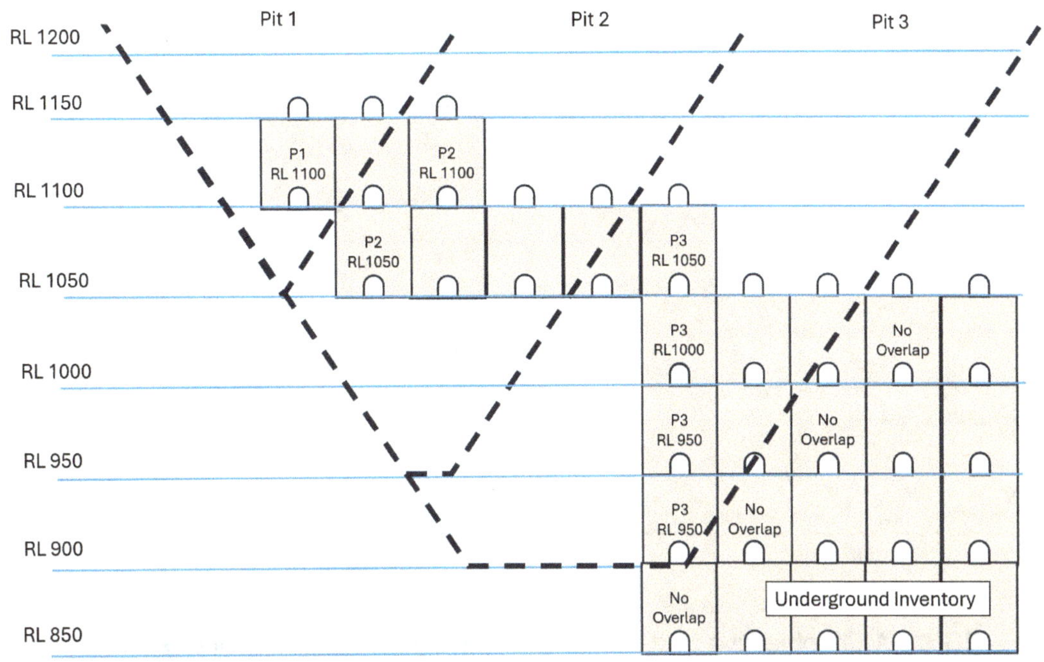

FIG 5 – Underground and open pit inventory parametrised into one problem space.

This method of inventory representation allows two (or more) different mining methods to be considered simultaneously within one problem space for various cut-off grades. Each unique section of the inventory can be included in the schedules, but not both overlapping sections. The open pit inventory is coded into a hierarchy to ensure that all overlaps in Pit 1 are respected when also mining Pit 2 and 3. When mining of a pit phase is commenced, it is assumed that underground mining for that zone is stopped permanently.

With this inventory parametrisation, it is possible to solve for the desired objectives using an evolutionary algorithm. For this case, the algorithm iteratively tested a wide range of strategic input combinations such as the ultimate pit limit, the cut-off grades for each block and bench, and sequence priorities until the output schedule results converge to an optimum which can then be used for more detailed mine planning.

For each period in the strategy optimisation output schedule, mineralisation depletion data is tracked for every underground block. This is necessary to ensure that depleted mineralisation is not counted twice when considering a later pass of open pit mining. This also allowed evaluation of the impact of choosing to preserve open pit inventory rather than mining underground as previously planned and then initiating an open pit.

By exploring and testing the vast solution space, the evolutionary algorithm can find the optimal transition period between underground and open pit mining considering the cost differences between mining methods (including capex deployment), cut-off grades and their impact on the available inventory, dynamic resource depletion, the step-change in closure costs and the value of delaying those costs by extending the mine life, if any. Beyond that, there are many other input variable relationships relating to the strategic decisions, simultaneously assessed in the strategy optimisation model.

The process included a secondary objective of reducing the operational intensity (mining rate) to increase production reliability. A multiple objective optimisation was conducted, allowing the open pit plant feed to reduce the demand on the underground mine. The trade-off solution space is shown in Figure 6. The operational intensity for the underground is represented by the number of annual stopes mined per annum (x-axis), which correlates to complexity in coordinating works, a proxy for execution risk.

The results demonstrated that a significant reduction in operational intensity for the underground mine is possible, with minimal NPV erosion, when considering the open pit in conjunction with cut-off grade and sequence optimisation.

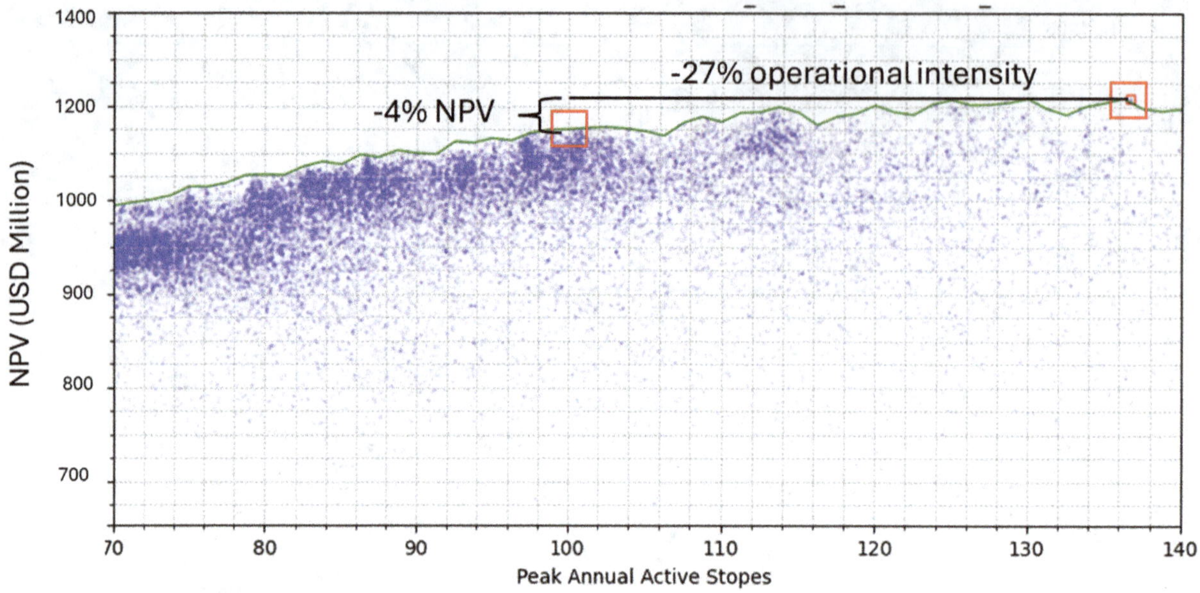

FIG 6 – Multi-objective trade-off between NPV and intensity of operations.

Figure 7 shows the monthly stope turnover rate for selected strategic cases. The existing mine plan is shown in a dotted line, a conceptual representation of that schedule in the Hill of Value® strategy optimisation model is shown in yellow. The optimised underground-only schedule is shown in blue, and the green shows the reduction in the number of underground stopes completed per month when an open pit option is considered.

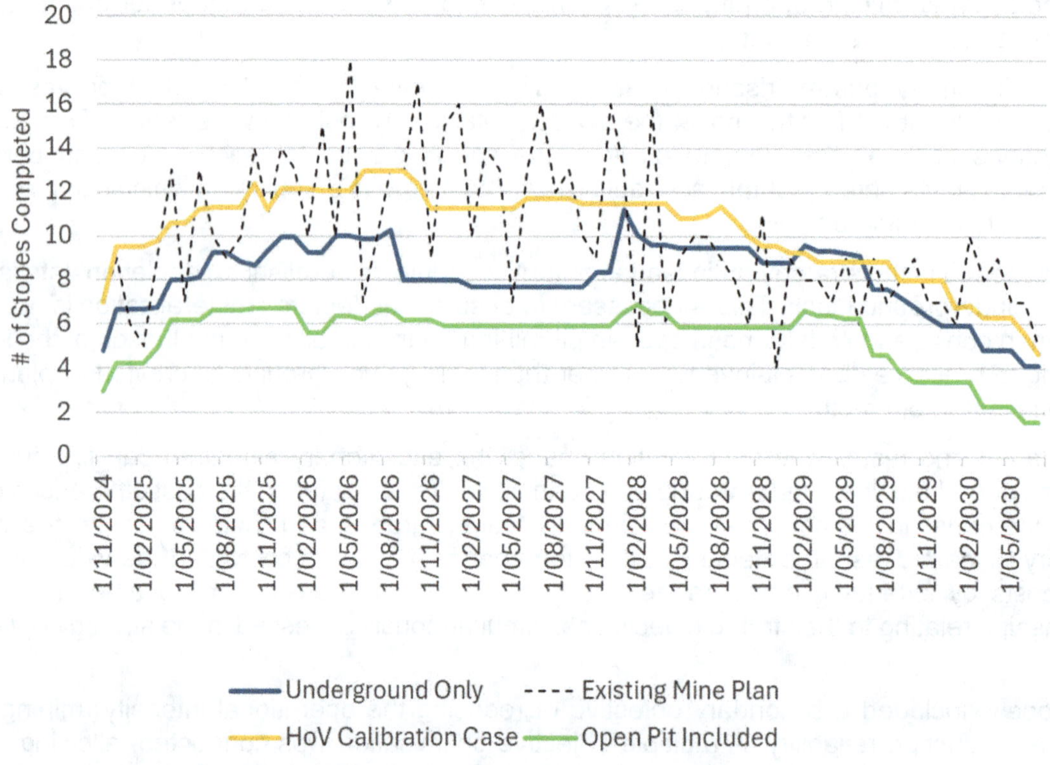

FIG 7 – Intensity of underground operations for selected strategic cases.

The NPV outcome when evaluating the underground to open pit transition options was 19 per cent higher than the optimised underground-only plan, and 43 per cent higher than the original plan. The strategy to transition to open pit mining also enabled increased extraction of the large, low-grade mineralisation, extending the mine life by approximately ten years, whilst maximising the use of existing process plant infrastructure to again reach nameplate capacity and improve the operations' cost curve position. Finally, the secondary objective of increasing operational reliability was achieved by providing feed from an open pit source and reducing execution risk for the complex underground operation. This insight assisted immediately with the strategic direction of the near-term underground mine plans, in preserving the open pit option for further study.

Case study 2 – large Cu-Au operation

Case Study 2 considers a large operating open pit copper-gold mine. The operation has a large mineral endowment with deep extensions.

Mining has progressed through five open pit stages, with strategic options for mining the remaining three open pit stages (PH06, PH07 and PH08) or transitioning to underground mining methods being considered.

Additionally, the strategy is considering other parameters (more than 200) including mine expansion, potential additional revenue from a new mineral processing plant and secondary product, limited capacity for tailings storage, and several underground mining methods and cut-off grades (sub-level stoping, sub-level caving, and block caving).

Due to the complexity of the case study, a deterministic scenario or conventional single-objective optimisation approach is not feasible. A multiple objective optimisation was completed targeting maximising NPV and minimising Scope 1 CO_{2-e} emissions.

A strategy optimisation model was developed with both economic and ESG parameters modelled. The strategy optimisation results in Figure 8 illustrate the impact on NPV and Scope 1 CO_{2-e} emissions of transitioning from open pit to underground mining methods under three scenarios of mining an open pit progressively deeper to the end of PH06, PH07 and PH08 respectively, with corresponding underground mining below the open pit phases.

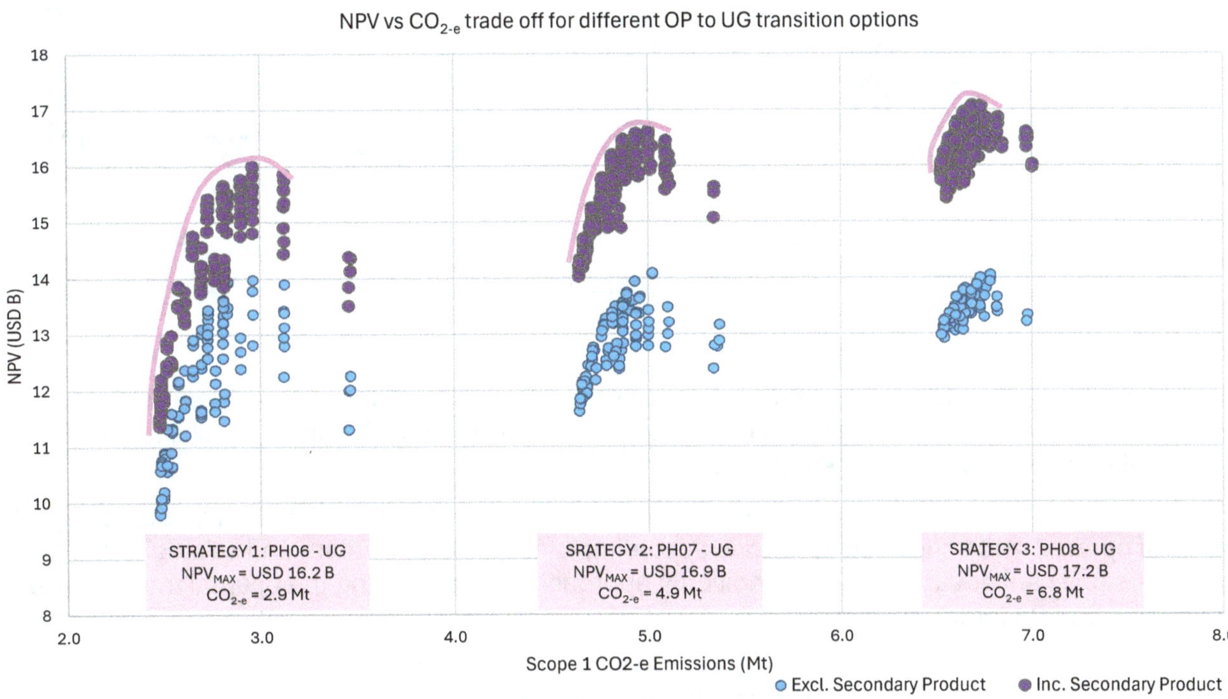

FIG 8 – Trade-off between different OP to UG transition points.

An optimisation 'front' is produced shown by the pink line. Two distinct populations are observed in the results, representing scenarios with and without the secondary product.

In this case, the maximum NPV was achieved by extending the open pit to PH08, with significantly higher total Scope 1 CO_{2-e} emissions. However, a life-of-mine Scope 1 CO_{2-e} emissions reduction of 57 per cent can be achieved by mining to a smaller open pit (PH06) and transitioning earlier to underground mining at the cost of a 16 per cent reduction in NPV.

The multiple objective optimisation demonstrates that interrogation beyond NPV outcomes, may lead to different strategic decisions when considering other important optimisation objectives.

Using the output data from the multiple objective strategy optimisation, where each point on the chart represents a strategic plan with associated parameter settings, it is possible to analyse the importance of each parameter in the optimisation against each objective. Selected parameters are shown in Figure 9, of most importance that effect NPV were underground cut-off grade value and the sale of secondary product, whereas the parameters that most effected Scope 1 CO_{2-e} emissions were the underground cut-off and the open pit maximum phase. Identifying parameters with co-existing relationships that impact both NPV and CO_{2-e} emissions (eg underground cut-off value) allows targeted strategies to be deployed focusing resources and corporate objectives.

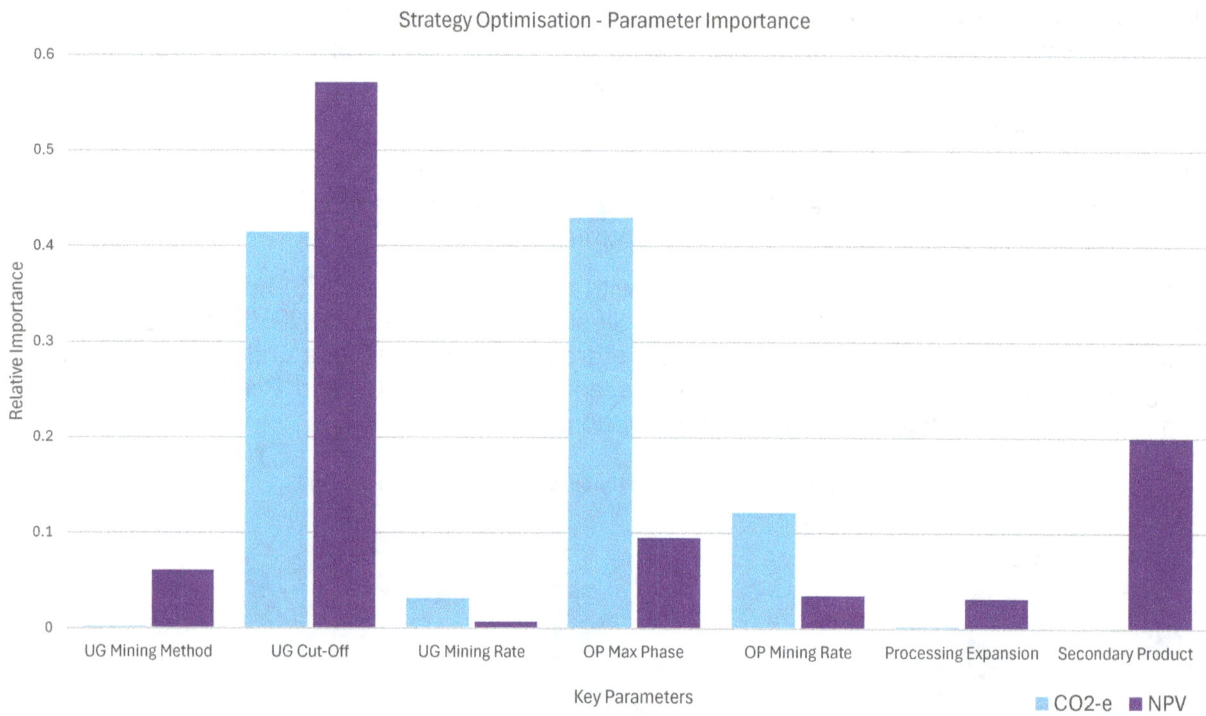

FIG 9 – Parameter importance.

CONCLUSIONS

Determining the optimal transition from open pit to underground mining (and *vice versa*) is a complex, multi-faceted decision that must balance economic returns with technical feasibility, execution practicality, and increasingly mine sustainability.

This paper has demonstrated that an integrated strategy optimisation process that evaluates a wide range of open pit and underground options according to all input parameters and strategic objectives provides significant advantages over traditional approaches of evaluating the optimal transition point of open pit and underground mining.

Case Study 1 has shown the benefits of parametrising an underground and open pit inventory within the same problem space, allowing the right balance between cash flow, closure costs and execution risks to be found algorithmically using the Hill of Value® process.

Case Study 2 has demonstrated the capability of multi-objective optimisations in uncovering the impacts of a maximal NPV strategy on other objectives such as environmental sustainability. This approach demonstrates enables increased visibility for decision-makers on trade-off decisions. Importantly, that a small sacrifice in NPV may lead to a large benefit in another objectives.

The implementation of this approach provides mine operators or project owners the opportunity to maximise the success of their assets through continued optimisation.

Development is ongoing with the AMC Hill of Value® process being unified into a dedicated software application providing a platform to further enhance the open pit to underground optimisation.

ACKNOWLEDGEMENTS

The authors acknowledge AMC Consultants Pty Ltd for supporting this publication.

REFERENCES

Chandra, A and Hartley, P R, 2024. Sequential investment decisions for mining projects using compound multiple volatility real options approach, *Resources Policy*, 97:105241.

Damodaran, A, 2009. *The dark side of valuation: Valuing young, distressed and complex businesses* (Ft Press).

Deb, K and Jain, H, 2014. An Evolutionary Many-Objective Optimization Algorithm Using Reference-Point-Based Nondominated Sorting Approach, Part I: Solving Problems with Box Constraints, *IEEE Transactions on Evolutionary Computation*, 18(4):577–601.

Dimitrakopoulos, R G and Abdel Sabour, S A, 2007. Evaluating mine plans under uncertainty: Can the real options make a difference?, *Resources Policy*, 32(3):116–125.

Grobler, F, 2016. Optimised Decision-Making under Grade Uncertainty in Surface Mining, PhD thesis, Curtin University, Perth.

Hall, A, Lee, D, Flanagan, M, Sharman, T and Grobler, F, 2025. The impact of enhanced ESG outcomes on the optimal open pit to underground transition, in *Proceedings of the Underground Operators Conference 2025*, pp 371–386 (The Australasian Institute of Mining and Metallurgy: Melbourne).

Hall, B (ed), 2014. Cut-Off Grades and Optimising the Strategic Mine Plan, in *Cut-off Grades and Optimising the Strategic Mine Plan*, Spectrum Series 20 (The Australasian Institute of Mining and Metallurgy: Melbourne).

Lanz, T and Seabrook, W, 2012. Operating Cost Estimation, in *Cost Estimation Handbook*, second edition, Monograph 27, ch 5, pp 57–82 (The Australasian Institute of Mining and Metallurgy: Melbourne).

MacNeil, J and Dimitrakopoulos, R G, 2017. A stochastic optimization formulation for the transition from open pit to underground mining, *Optim Eng*, pp 793–813. https://doi.org/10.1007/s11081-017-9361-6

Pelech, T and Grobler, F, 2025. AMC's Hill of Value®: Modelling sequential decisions to transfer strategy optimization into an operational mine plan, in *Proceedings of the Applications of Computers and Operations Research in the Minerals Industry (APCOM)*, pp 833–845 (The Australasian Institute of Mining and Metallurgy: Melbourne).

Pelech, T, du Preez, D and Hall, A, 2026. AMC Hill of Value®: Strategic Mine Optimization and Real Options Analysis at Industrial Scale, in *Proceedings of MINEXCHANGE 2026, the SME Annual Conference and Expo*, pp 64–74 (Society for Mining, Metallurgy and Exploration: Englewood).

Trigeorgis, L and Reuer, J J, 2016. Real options theory in strategic management, *Strategic Management Journal*, 38:42–63.

Whittle, D, Brazil, M, Grossman, P A, Rubinstein, J H and Thomas, D A, 2016. Determining the Open Pit to Underground Transition – a New Method, in *Proceedings of MassMin 2016* (eds: C Carr and G Chitombo), pp 731–741 (The Australasian Institute of Mining and Metallurgy: Melbourne).

ACKNOWLEDGEMENTS

[illegible]

REFERENCES

[illegible]

Parametric pit design – turning optimised shells into practical mine plans

A Phillips[1] and C Fong[2]

1. Product Manager, RPM Global, Brisbane Qld 4000. Email: aphillips@rpmglobal.com
2. Senior Solutions Executive, RPM Global, Perth WA 6000. Email: cfong@rpmglobal.com

INTRODUCTION

By applying a parametric modelling approach to pit design, experienced mining engineers can spend less time on routine drafting tasks required to produce safe, practical, and geotechnically compliant designs. This improved efficiency enables engineers to evaluate more options, generate greater value, and respond more quickly when circumstances change.

This paper presents the parametric techniques for open pit mine design developed by RPM Global for their MINEPLANNER software. It summarises the key challenges encountered when creating a parametric model tailored to open pit geometry and how they were addressed. It also outlines the benefits reported by early adopters of the new approach.

BACKGROUND

The application of Parametric Modelling to design workflows has become standard practice across many engineering disciplines over the past three decades. In architecture, for example, tools such as Revit have fundamentally transformed not only the design process itself but also downstream activities such as quantity surveying, cost estimation and change management. In contrast, open pit mine design remains largely dependent on methodologies introduced more than 40 years ago.

PROBLEM STATEMENT

The optimised pit shells used to guide the design process are financial models that define the economically viable portion of the deposit. They do not consider access requirements and only loosely honour geotechnical constraints. Open Pit Design focuses on bridging the gap between the 'optimised pit shell' and a practical pit design.

Traditional Open Pit Design is typically performed using complex 3D CAD systems that require engineers to manually construct the mines geometry – they must essentially '*draw the pit*'. Automation has been introduced to assist with specific tasks like wall projection and ramp insertion and over the last five years, these automated 'wizards' have become more sophisticated. But despite the increasing use of automation that reduces the time design engineers spend moving and re-projecting lines, the outcome is nevertheless a series of triangulated strings.

Such representations encode shape and geometry but not meaning or intent. While an engineer may understand that specific strings represent crests, toes or ramps, this information is not embedded within the design itself. The design contains no semantic information describing the role of each component, the relationships between them or the constraints that govern them.

This reliance on passive geometric artefacts lies at the core of traditional pit design's most significant limitation. Even small or localised changes to key parameters can invalidate core design requirements and render otherwise complete designs obsolete, triggering extensive rework.

Producing or modifying a single pit design can require weeks of iterative effort, making design a bottleneck that limits the alternatives that can be evaluated and reduces responsiveness to changing circumstances. When design modifications are inevitably required, planners must pivot quickly and the resultant design may not achieve the deposits maximum potential.

PARAMETRIC MODELLING OF OPEN PIT DESIGNS

Parametric modelling takes a fundamentally different approach to open pit design. Instead of representing the pit as a passive, syntactically ignorant triangulation derived from strings, it creates a detailed mathematical representation of the design and all its components.

This approach does not automate the design process or remove the need for competent engineers; rather, it reduces the routine draughting and rework typical with manual design, making the entire process more interactive. This shifts the focus to engineering judgement and the evaluation of alternative options, while reducing the risk of error and helping to ensure the design remains compliant with the parameters and constraints defined by the engineer.

The mathematical representation consists of a large, tightly integrated network of smart objects that represent discrete functional components of the design. Each object encapsulates both geometry and behaviour, including the design rules, constraints, and dependencies that govern how it can be generated and modified. Together, these objects form an explicit model of the pit that can be interrogated, validated, and interactively modified. Rather than a static drawing that must be manually reworked, the parametric model regenerates itself whenever changes are requested by the engineer.

Each smart object has a set of parameters to describe its geometry and behaviour. For example, a ramp segment is characterised by parameters such as width, grade and minimum turning radius. When these parameters are modified, the ramp geometry is automatically regenerated in a manner consistent with the defined design rules, while maintaining continuity to adjacent segments and intersections where applicable.

Importantly, these smart objects are not independent: they also encode relationships with other components within the pit. Ramps and pit walls, for example, are aware of their spatial and logical dependencies, so movement of a wall is propagated to affected ramps (and vice-versa), with the system re-evaluating geometry against the relevant constraints. Mathematical topologies are used to describe key relationships between smart objects and model the complexity of a typical open pit.

MODELLING THE BEHAVIOUR OF PIT WALLS

Pits are typically divided into pushbacks (stages), and each stage may include multiple nested sub-pits. Parametric modelling captures these hierarchical relationships explicitly, so modifications to walls and ramps within a stage or sub-pit are propagated consistently across all dependent elements. This reduces the common challenge of maintaining geometric consistency between pushbacks as designs evolve.

Parametric modelling also removes the need for users to specify projection directions explicitly. When an intermediate crest or toe is adjusted, the model determines whether wall geometry must be projected upward, downward, or in both directions, to satisfy the design rules. This lets the engineer focus on the intended design change, rather than the mechanics of how to implement it. It also allows the projection logic to be optimised, ensuring only impacted walls are pre-projected.

Changes to pushback and sub-pit geometry can invalidate other aspects of the design. Dragging a wall inwards, for example, can reduce the width of lower benches to a point where they must be removed. Conversely, dragging a sub-pit wall outwards can cause upper benches to overlap adjacent sub-pits following projection, requiring portions of those sub-pits to be merged.

Wall projections are governed by geotechnical constraints that can vary across different areas of the pit. When a wall is moved, its design parameters (like berm width and batter angle) may need to be updated to reflect the requirements at the wall's new location. This applies both to the wall segment the engineer explicitly moves and to any additional wall segments adjusted during re-projection.

Managing wall behaviour as it transitions between geotechnical domains is another critical aspect of a pits design. Frequent changes to slope profiles should ideally be avoided and in most situations, it's preferable for the wall to transition smoothly between the two profiles. All transitions must be constructed in a manner that ensures the slope criteria are never exceeded in either domain.

Geotechnical requirements are often refined over time as additional information becomes available and with traditional techniques, this typically requires a complete re-design. But when the design is modelled parametrically, users can simply change the geotechnical model and the entire design will re-generate itself.

MODELLING OF RAMPS AND RAMP NETWORKS

Ramps are not modelled as isolated features applied to crest and toe strings; instead, they are represented as part of an interrelated ramp network that includes switchbacks, intersections, flat sections, and parameter transitions. As a result, local changes are consistently resolved across the network, including downstream impacts to intersections and connected ramps.

In traditional design approaches, ramps are one of the most common drivers of rework when modifications are required. By modelling the ramp system as an integrated component of the design, changes to any portion of the ramp layout—whether made directly by the user or indirectly via changes to other design components—are propagated through all connected sections, reducing the need to manually re-create individual ramps.

The behaviour of parametric design objects must explicitly account for the implications of change and the possibility that user intent is geometrically invalid or conflicted. For example, when a new ramp segment is inserted, its projection may intersect an existing ramp, making it impossible to realise the engineer's original intent while maintaining the applicable geometric and operational constraints.

A parametric system must resolve these conflicts in a robust, predictable manner while preserving underlying intent where possible. Where intent cannot be satisfied immediately, it should be retained as a latent design condition so that, if the conflicting ramp is later removed or relocated, the original ramp can be regenerated in a form consistent with the engineer's initial objective.

CASE STUDY – USING PARAMETRIC PIT DESIGN AT A LARGE IRON ORE MINE

In 2025, A large iron ore operator evaluated multiple pit design tools for its Pilbara iron ore mines, comparing a parametric design approach with several conventional alternatives. We thank this operator for sharing the findings of this evaluation at the RPM Global *Mine Planning Forum* in August 2025.

They observed that conventional pit design tools have seen limited advancements over the past two decades, despite the introduction of some AutoPit functionality. In the previous year, the large iron ore operator completed more than 500 designs, with significant time devoted to repetitive refinement and iterative stakeholder review.

With a parametric design approach, approximately 80 per cent of the time typically spent on iterative design refinement was returned to mine designers, enabling deeper exploration and improvement of design options. Stakeholders were more engaged and there were substantially fewer multi-step design review cycles.

During the evaluation, several limitations were identified, generally relating to the scale of the pits being designed. Working in partnership with RPM, the large iron ore operator was able to reduce the time required to convert an optimal pit shell into a geotechnically compliant initial pit from many weeks to less than an hour. Significant performance improvements were also realised in other aspects of the design process.

The large iron ore operator's objective is to become the 'Best Operator for Mine Design'. They identified the parametric approach to design as an opportunity to maintain a 'living' pit design that evolves through progressively more detailed planning horizons, rather than being redesigned at each stage. They also see scope to extend parametric methods to multi-stage designs with integrated backfill, as well as to shorter-term, operational planning time frames.

CONCLUSIONS

While parametric design has been used in other engineering disciplines for the past 30 years, open pit design has been slower to adopt this approach, relying instead on the traditional CAD-based workflow of drawing pits using line strings and projections. Although more sophisticated tools can automate some tasks, they still typically produce a static geometric representation of the pit – a drawing. As a result, creating and updating pit designs remains labour-intensive and time-consuming, often taking weeks for a single design iteration. Modifying these traditional pit designs is similarly time-consuming and often requires the entire design to be re-created.

Parametric design takes a fundamentally different approach that is based on a detailed mathematical representation of the design, rather than a drawing that captures shape and geometry, but not meaning.

The network of inter-related smart objects that form this mathematical model represent discrete functional components of the mine. A set of parameters describe the geometry and behaviour of the objects, along with their relationship with other objects in the design. This ensures that changes to one aspect of the design can be correctly propagated to other components impacted by the change.

Parametric design facilitates a fundamental shift in how open pit design is conducted in the mining industry. Designs can be completed in a fraction of the time required by traditional methods, and they can be edited while remaining compliant with geotechnical and other design constraints. This enables engineers to respond more quickly to change and creates more opportunity to evaluate alternative design options.

DBS-BZ, phase generation and phase bench scheduling for an operational growth program

D Rahal[1], J Chung[2] and B Maziarz[3]

1. Product Manager, Deswik, Brisbane Qld 4000. Email: david.rahal@deswik.com
2. Global Technical Advisor – Optimization, Deswik Consulting, Perth WA 6000. Email: joyce.chung@deswik.com
3. Territory Manager – Western Australia, Deswik Consulting, Perth WA 6000. Email: ben.maziarz@deswik.com

ABSTRACT

Miners seeking sustainable growth face a dual challenge: proving near-term operability while credibly extending value through additional capital investments. This case study presents a structured pathway to develop a strategic mining schedule for a three-pit copper–nickel system. The goal is to align funding decisions with realistic timelines and constrained operations.

This work presents an alternative to conventional pit-shell workflow using an integrated direct block scheduling (DBS) approach leveraging the Bienenstock–Zuckerberg (BZ) algorithm. DBS incorporates time-dependent constraints such as limited sink rates, mining boundaries, earliest start dates, material blending, and staged capital expenditures. The advantage of DBS lies in its ability to embed these constraints early in the optimisation process and preserve the time value of money. The result is a schedule that is based on dynamic rather than static cut-off grades.

The DBS optimised extraction sequence is then translated into operational phases using clustering-based phase generation (PGN), with outputs that vary depending on phase targets—for example, smoothing mine production or stabilising ore feed to the plant.

These operational phases (pushbacks) are converted into an executable strategic schedule using phase bench scheduling (PBS). The PBS model converts the optimised extraction sequence (block based) into operational mining units, benches within each phase. In addition to the DBS constraints, limitations on annual CAPEX expenditure, precedence between purchases, and the replenishment of capital items are included in the optimised operational schedule.

The contribution is a practical, capital planning ready workflow: start with DBS (BZ accelerated), derive clustered phases, enforce PBS under realistic sink rates and blend constraints, and evaluate incremental plant investments and logistics options. For mine operators seeking funding for expansion, this workflow connects optimisation to execution and demonstrates credible value uplift with defensible assumptions.

INTRODUCTION

Mid-tier mining operations seeking sustainable growth are often challenged to provide near-term operability while credibly extending value through capital investment (eg additional pits and plant investments.) Traditional mine planning workflows struggle to bridge this gap, typically relying on nested pit shell optimisation followed by manual scheduling adjustments. This sequential approach does not capture the complex interplay between time-dependent constraints, capital phasing, and material blending requirements. These interactions are particularly important for in multi-pit mining systems. The result is often overoptimistic strategic plans that prove fragile when tested against realistic operational constraints. This overvaluation often undermines stakeholder confidence and jeopardises funding decisions.

The fundamental limitation lies in separating value optimisation from enforcing operational constraints. By defining extraction limits first and imposing constraints later, conventional workflows fail to capture the time value of money. Dynamic cut-off grades cannot be effectively integrated into static pit shell definitions. This limitation occurs because manual interpretation is required to translate the revenue shells to executable mining plans. This introduces inefficiency and risks discarding the value gained through block optimisation (Chung and Rahal, 2025).

This paper presents an integrated workflow addressing these limitations. It outlines the use of:

- Block sequencing through direct block scheduling (DBS) enhanced by the Bienstock-Zuckerberg (BZ) algorithm.
- Clustering-based phase generation (PGN).
- Phase bench scheduling (PBS).

This approach embeds operational constraints directly into the optimisation process rather than treating pit definition, phase design and production scheduling as separate, sequential problems. This maximises both economic value and operational feasibility throughout the planning process. The case study examines a three-pit copper-nickel system where a mid-tier operator must secure additional funding by providing a credible expansion plan. The workflow connects optimised strategic extraction sequences to value generating operational plans. It does this using defensible assumptions that allow technical reviewers and financial decision-makers to make more informed decisions.

PROBLEM STATEMENT, METHODOLOGY, APPLICATIONS AND DISCUSSIONS

A mid-tier mining company has approval to mine from two pits, Corvid and MV. They are seeking to determine the viability of a third pit, Wolfpass. It is expected that this pit will take at least five years to fund and develop and a study is required to provide a go-no-go decision. As part of this process, the company wants to identify the most effective approach to processing their ore. Their options are: expand the tailings capacity of their existing plant and/or constructing a new plant to reduce haulage costs. The integrated, three stage methodology will allow them to translate their current resources (block models) into a service of optimised schedules (scenarios). Figure 1 illustrates the flow model for the mining system.

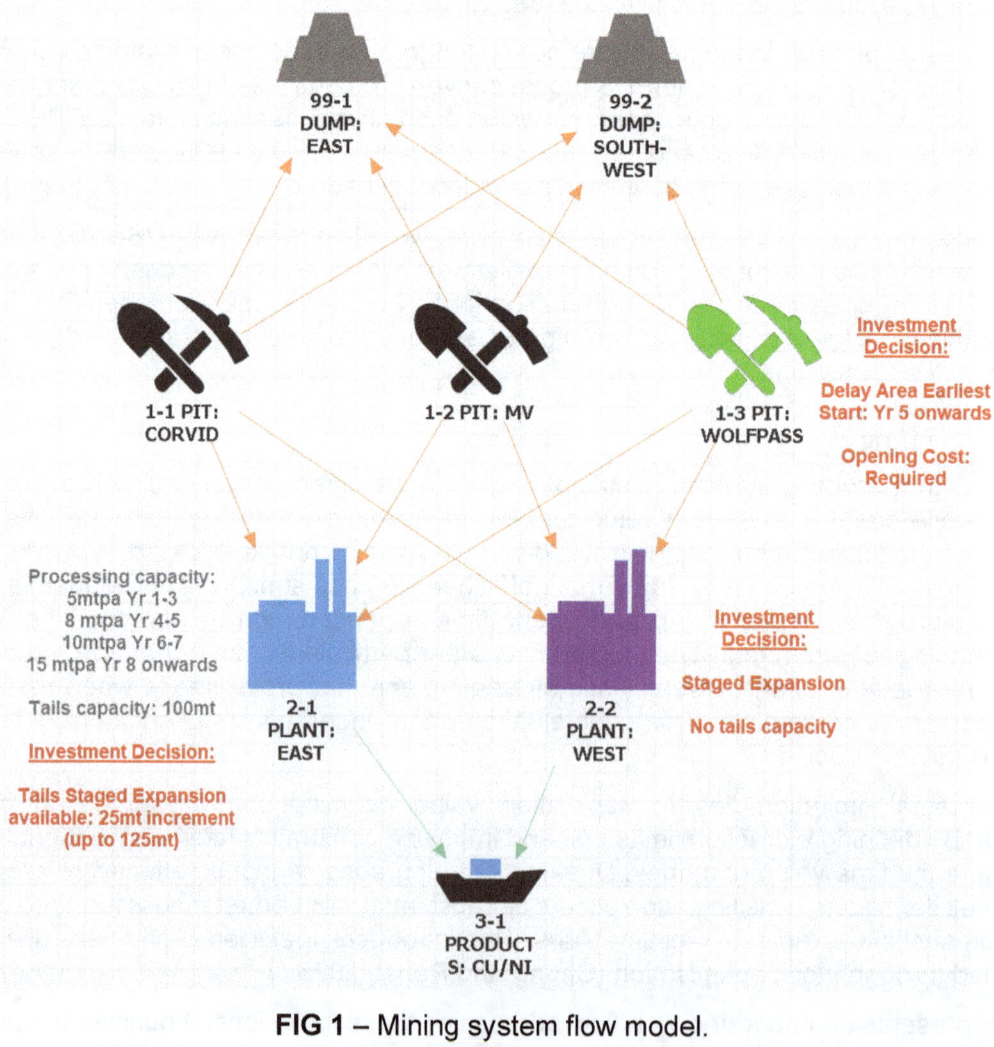

FIG 1 – Mining system flow model.

The evaluation process begins with determining the optimised block extraction sequences using DBS.

Stage 1 – Direct block scheduling

DBS with BZ is used to develop a practical extraction strategy that reflects how the separate pits would realistically be developed over time within the constraints on the mining system. This offers advantages compared to the traditional method (shells then schedule), since DBS evaluates the mining sequence, processing capacity, and capital decisions simultaneously.

The key operational constraints that drive mining strategy in this stage are:

- Separate mining rate limits for each pit.
- Restrictions on when each pit can start mining.
- Processing capacity constraints at the existing plant (Plant East).
- Material blending requirements linked to metallurgical performance.
- Capital investment decisions for:
  - Tailings expansion.
  - Additional processing capacity.

By modelling these elements directly within the optimisation, the resulting mining strategy naturally adapts to temporal changes in system capacity and investment timing. Cut-off grades are not fixed at the outset. They vary with time in response to available processing capacity and capital commitments. This allows the model to prioritise high value material when capacity is constrained and defer marginal material until capital expansions become economically justified.

Three DBS scenarios were created to understand how different investment options impact mining strategy and project value:

- **Scenario 1a (Base Case)**: Mining from two pits (Corvid and MV) supplying Plant East, with no additional capital investments.
- **Scenario 1b**: Include tailings capacity expansion at Plant East and the development of the Wolfpass pit.
- **Scenario 1c**: Provide additional processing capacity (and reduce processing costs) by commissioning Plant W.

The following figures illustrate the impact of alternative investment options on mining sequence, value realisation, and material movement. Figure 2 shows clear differences between mining sequences as additional investment is introduced. The optimised extraction sequence changes both spatially and temporally in response to capital expenditure. In a similar way, Figure 3 shows that staged capital expansion increases project value. The cumulative discounted cash flow comparison, by period, demonstrates that higher value is achieved by incorporating staged capital investment.

Figures 4 and 5 illustrate how ore and waste movement changes with time, subject to blending and processing constraints. The first shows this for mine production while the second shows plant feed. Note that only 'ore' is fed to the plant while waste is directed to the dump. Collectively, these results show that capital investment alters the timing and location of mining. It also has an impact on where and when ore is processed (eg the use of Plant West in Scenario 1c). These differences result in significantly different pit shape progression, block extraction sequences, and production profiles. This reinforces that mining strategy and capital planning must be treated as a single, coupled problem rather than addressed through sequential optimisation steps.

The Stage 1 output is a constrained, value-maximising extraction sequence that serves as a direct input to downstream phase design and scheduling. By establishing credible pit limits and extraction priorities, it provides a robust foundation for defining pit pushbacks and evaluating additional mining constraints and capital funding decisions.

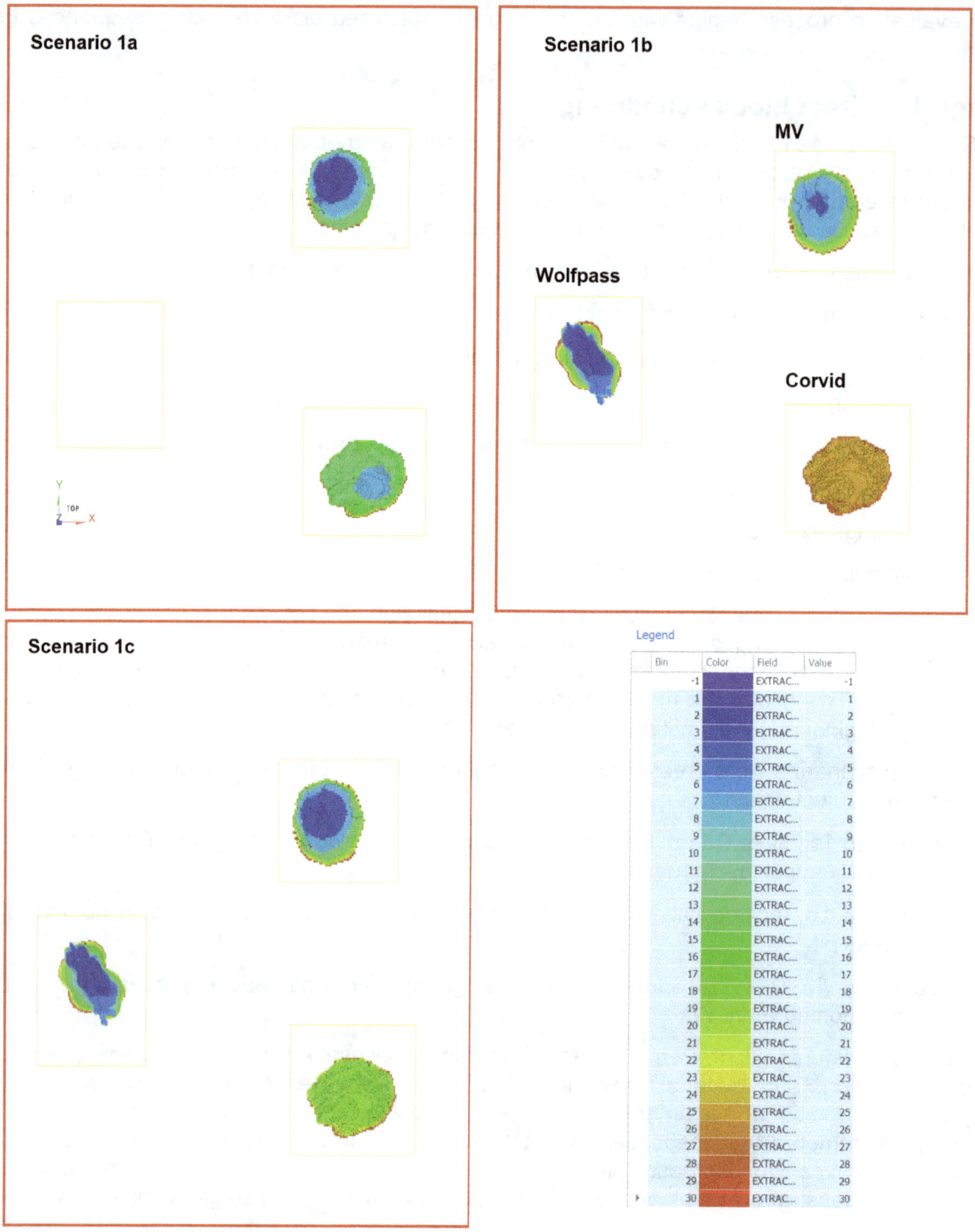

FIG 2 – Extraction sequences (by period) for the three DBS scenarios.

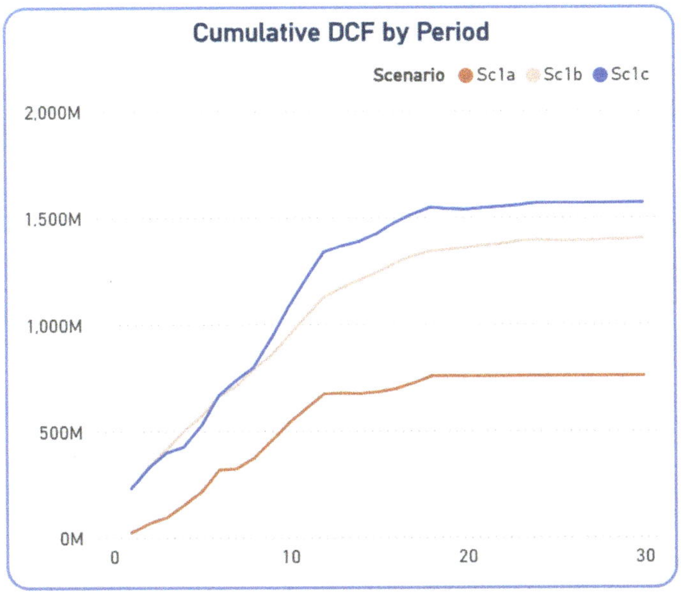

Scenario	Total Movement	Ore Tonnes Movement	Waste Tonnes Movement
Sc1a			
Corvid	128M	7M	121M
MV	454M	169M	285M
Sc1b			
Corvid	128M	8M	120M
MV	487M	163M	324M
WP	366M	149M	217M
Sc1c			
Corvid	128M	8M	120M
MV	486M	185M	302M
WP	377M	158M	218M

FIG 3 – DBS discounted cash flow and production comparisons.

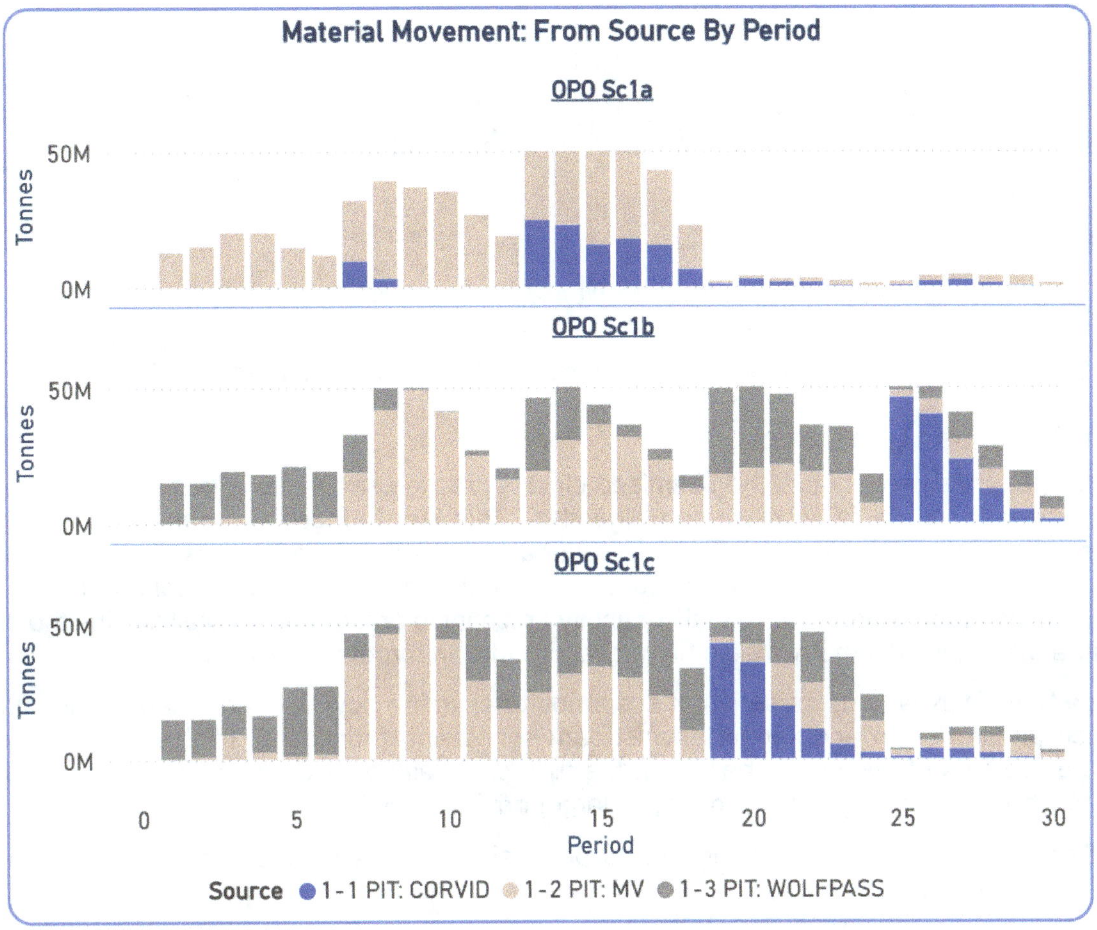

FIG 4 – Material movement into the mining network by source.

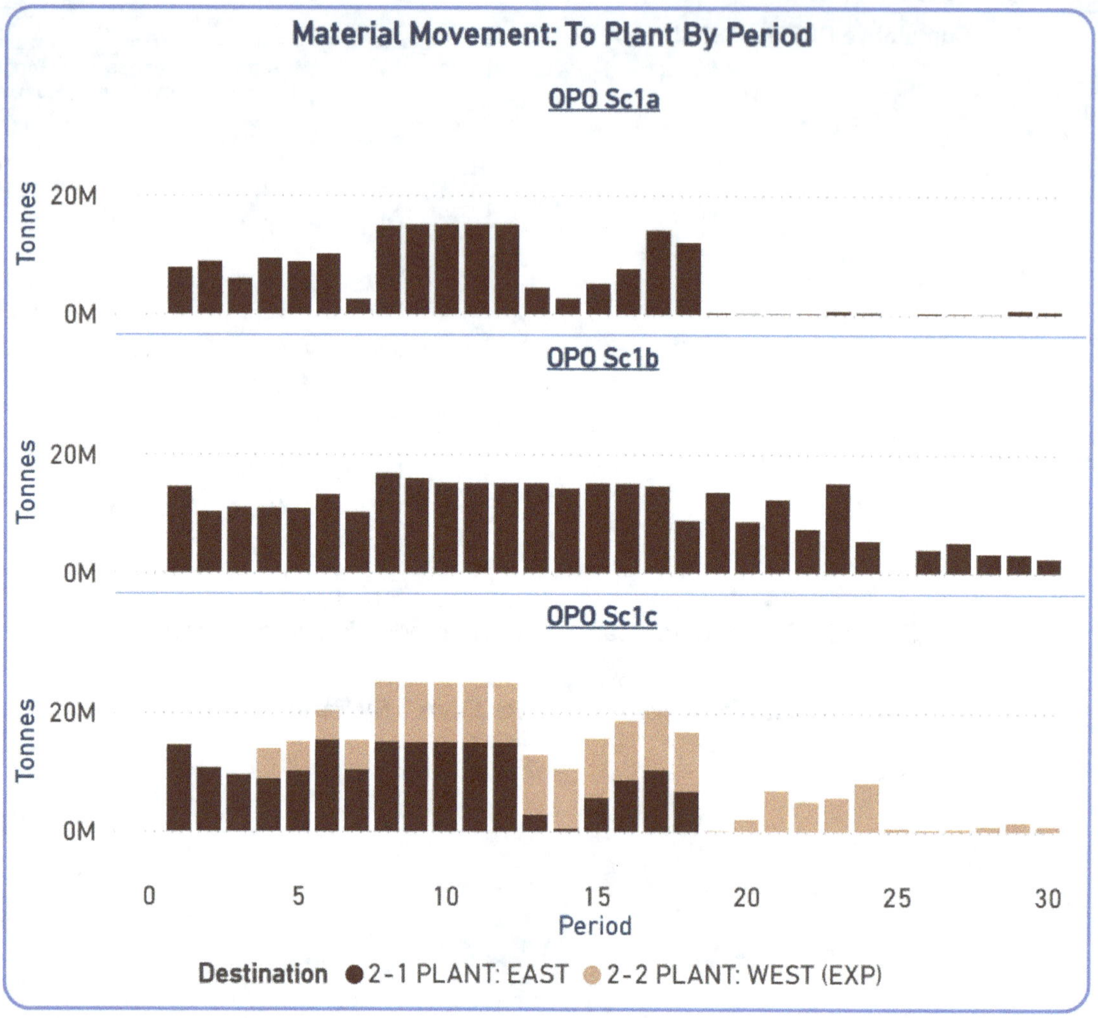

FIG 5 – Plant feed material by period (Plant East and Plant West).

Stage 2 – phase generation

In this stage, phase generation, the optimised block extraction sequence is translated into practical mining phases using a fuzzy clustering algorithm. This step bridges the gap between a value-optimised block sequence and a series of operationally workable mining phases. To do this, the mine blocks are aggregated into three-dimensional clusters that can be mined as coherent pushbacks. The clustering process enforces spatial continuity and mining precedence while allowing the planner to ensure that phase geometry meets high-level production targets.

The goal with PGN is to use a series of scenarios to examine how different operational objectives influence phase design and impact downstream scheduling processes. In this example, we will generate two sets of pushbacks based on the fully constrained DBS scenario (Scenario 1c). The difference between the two is the production target driving PGN:

- **Scenario 2a**: Target total mining tonnes, prioritising steady equipment utilisation and smoother total material movement.
- **Scenario 2b**: Target ore tonnes, with the objective of stabilising mill feed and supporting consistent plant operation.

The resulting phase geometries, illustrated in Figure 6, show that although both scenarios follow the same underlying DBS extraction sequence, the phase layouts differ substantially within the pits. Targeting mining tonnes produces phases that are more evenly distributed and volumetrically balanced, supporting consistent mining effort across periods. In contrast, targeting ore tonnes results in phases that preferentially expose ore-bearing regions earlier, particularly in pits supplying the plant, leading to more selective phase advancement. These differences in phase geometry and sequencing propagate directly into bench-level schedules and capital deployment in the subsequent

PBS stage, demonstrating that PGN is an active planning decision rather than a neutral translation step.

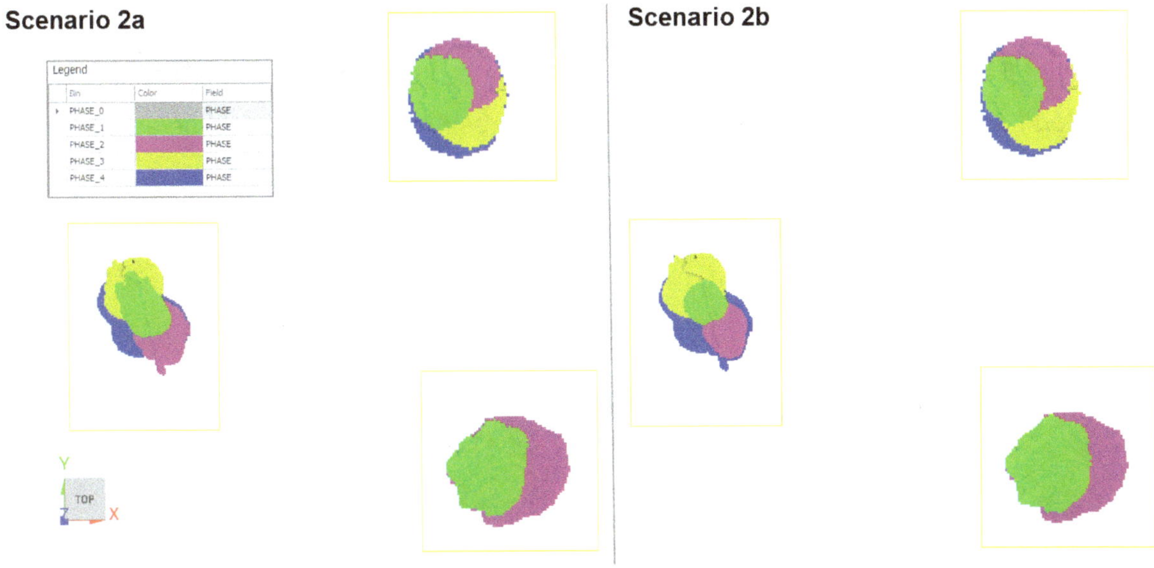

FIG 6 – Phase designs which maintain steady mine production (2a) and ore production (2b).

Stage 3 – Phase bench scheduling

PBS converts the mining phases into an operational schedule by aggregating the mined blocks into larger mining units based on their extraction phase and bench elevation. The aggregated tasks are designated as phase-benches, and the extraction sequence follows the phase order and bench extraction sequence from top to bottom. The PBS optimisation model is an expanded version of DBS Scenario 1c, in which all original constraints are maintained. Additional, realistic capital-planning constraints are introduced to define capital expenditure limits per period, provide alternate capital payment schedules, and replenish consumed capital items such as service haulage equipment. The PBS model also incorporates pit opening costs and incremental expansion opportunities for the processing facility, ensuring that the optimised schedule respects both physical mining constraints and available capital. The output is a period-by-period schedule of tonnage, grade, and capital expenditure timing, providing the basis for informed funding evaluation and operational planning.

Four PBS scenarios were evaluated to assess the impact of progressively more realistic capital constraints on project value and mining sequence. All scenarios inherit the DBS Scenario 1c extraction sequence and the Scenario 2a (mined tonnes) phase design as inputs.

A consistent set of economic and operational parameters is applied across all PBS scenarios. Mining cost increases with depth but does not vary between scenarios. Overhead costs and a 5 per cent royalty rate are held constant, as are copper and nickel prices. Nickel recovery is fixed, while copper recovery varies by block as a block model attribute. Processing costs differ between the two plants, with Plant East operating at A$60/t and Plant West at A$50/t.

The four PBS scenarios are constructed incrementally, each adding a layer of capital constraint to the previous:

- **Scenario 3a (Base Case)**: Applies the same constraint set as DBS Scenario 1c, with no maximum CAPEX limit and no additional tailings capacity. This scenario establishes the baseline PBS value and quantifies the loss of selectivity from aggregating blocks into phase-bench tasks.
- **Scenario 3b**: Introduces a maximum CAPEX expenditure limit of $500M per period and allows tailings capacity at Plant East to be expanded in 25 Mt increments, with up to five increments available.
- **Scenario 3c**: Builds on Scenario 3b by adding the option to commission Plant West, which offers a lower processing cost than Plant East. Two funding contracts are modelled: Option 1

($500M spread over five payments) and Option 2 ($550M spread over ten payments), allowing the optimiser to select the most economical payment schedule.

- **Scenario 3d**: Applies the same CAPEX limit and tailings increments as Scenario 3b, but it adds a capital precedence constraint requiring that at least two tailings expansion increments at Plant East are purchased before Plant West can be commissioned.

The updated flow model and an overview of the PBS investment decisions can be seen in Figure 7.

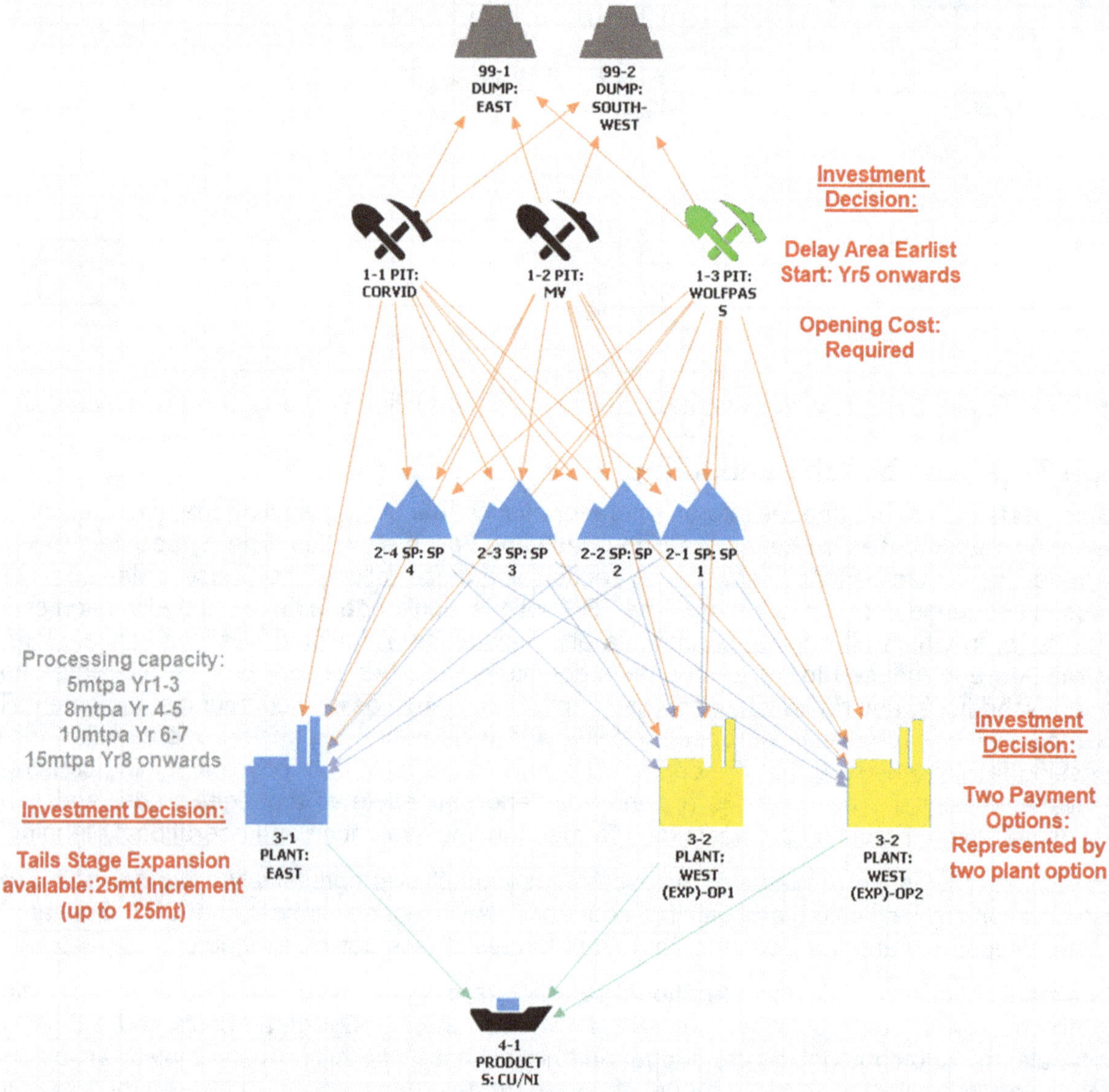

FIG 7 – PBS flow model and constraint overview.

Mining sequence

Aggregating blocks into phase-bench tasks reduces selectivity compared to DBS. Because mining is constrained to follow the predefined phase geometry, there is less variation in where mining occurs across scenarios when constraints are changed. The primary driver of change between scenarios is therefore capital expenditure timing rather than differences in the underlying mining sequence. Notable exceptions (shown in Figure 8) include more restricted early production at the Wolfpass pit in Scenario 3c (smaller dark blue area), and differences in overburden removal at Corvid early in Scenarios 3a and 3d.

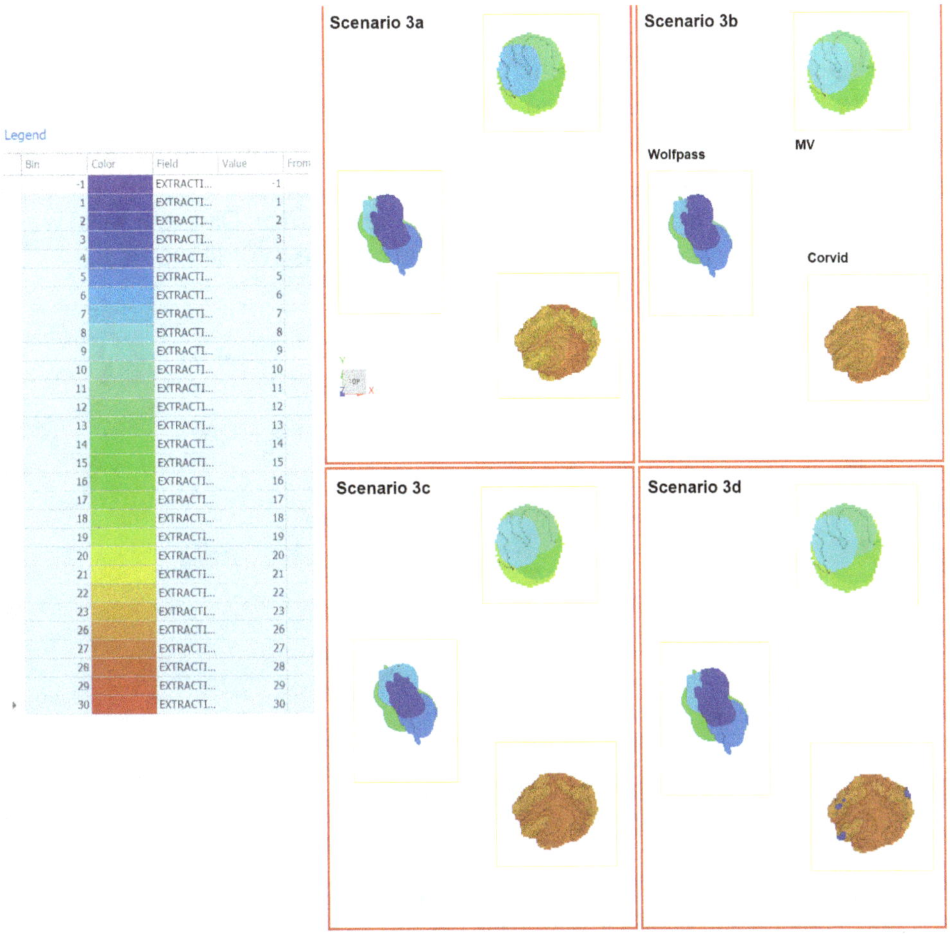

FIG 8 – Phase-bench mining sequences (by period).

Discounted cash flow

The transition from block-level DBS (Scenario 1c, \$1.57B) to phase-bench scheduling (Scenario 3a, \$1.15B) results in an approximate 25 per cent reduction in NPV, as illustrated in Figure 9 (left). The material movement through the mining system is shown in Figure 9 (right). It can be seen in the four scenarios that the amount of material reporting to the dump (waste) increases in PBS compared to the DBS baseline, Scenario 1c (Figure 10). The drop reflects the loss of selectivity and flexibility inherent in aggregating blocks into larger mining units. Additional refinement of the phase design and material classification could reduce this gap. For example, it is unlikely that an operation would use only four phases to mine more than one Bts of material. Increasing the number of phases will increase selectivity by increasing the number of mining tasks.

Introducing a \$450M per-period CAPEX limit alongside staged tailings expansion (Scenario 3b, \$1.17B) has a modest positive effect relative to Scenario 3a. The effect of adding the CAPEX limit to Scenarios 3b is shown in Figure 11 (top right). The ability to increase tailings capacity unlocks additional ore processing. The graph also confirms that total capital expenditure is held below the \$450M threshold. In the next scenario, Scenario 3c, taking the option to process ore at a lower cost in Plant West improves value further with a total NPV of \$1.20B. In doing this, Option 1 was selected by the optimiser in preference to the Option 2 payment schedule (\$500M and \$550M respectively). The difference between the two payment schedules is shown in Figure 12.

The last scenario, Scenario 3d, adds a precedence constraint to the use of Plant W. Plant W cannot be commissioned until the tailings capacity of Plant East has been increased by 50 Mt. Adding this constraint reduces the total value back to \$1.16B (Scenario 3d). The definition of the precedence relationship between the two plants and the impact of this relationship is shown in Figure 13 and Figure 14. The latter shows how the cumulative East tailings capacity limit is progressively filled over time (left) and the total capacity of Plant West is constrained with time. It also shows that Plant West only comes online once the second tailings increment is purchased in Period 4.

Material movement

The reduction in selectivity at the phase-bench level has a direct impact on the amount of material directed to waste (eg see Figure 10). Corvid shows the most consistent ore-to-waste distribution across scenarios.

In contrast, both MV and Wolfpass, show an increase waste sent to the dump with a corresponding reduction in ore reporting to the plant (compare to DBS Scenario 1c). Differences in waste movement between PBS scenarios are small, at less than 1 per cent of the overall mean.

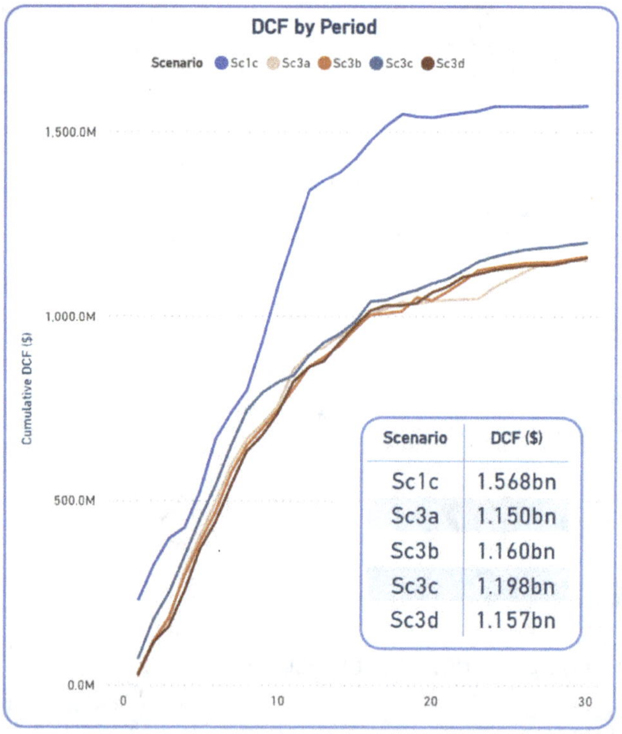

Scenario	Total Movement	Ore Tonnes Movement	Waste Tonnes Movement
⊟ **Sc1c**			
Corvid	128M	8M	120M
MV	486M	185M	302M
WP	377M	158M	218M
⊟ **Sc3a**			
Corvid	128M	7M	121M
MV	486M	214M	273M
WP	377M	187M	189M
⊟ **Sc3b**			
Corvid	128M	7M	121M
MV	486M	215M	272M
WP	377M	190M	187M
⊟ **Sc3c**			
Corvid	128M	9M	119M
MV	486M	221M	266M
WP	377M	188M	189M
⊟ **Sc3d**			
Corvid	128M	6M	122M
MV	486M	220M	267M
WP	377M	190M	187M

FIG 9 – PBS discounted cash flow and production comparisons.

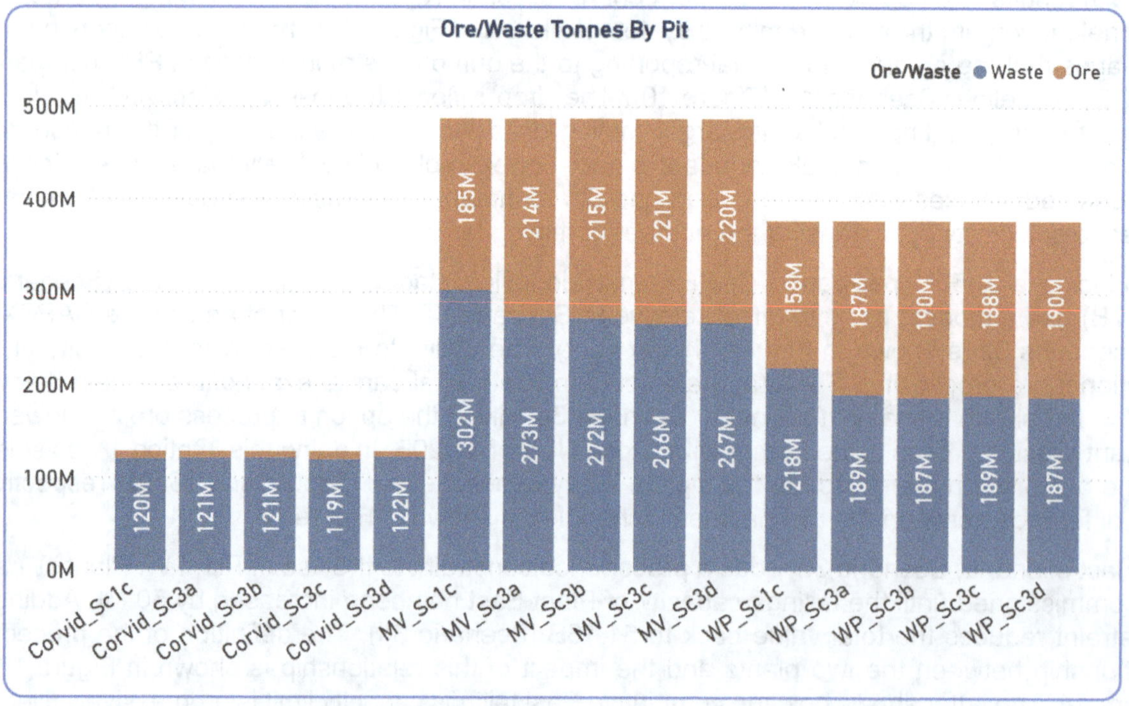

FIG 10 – Material movement (ore/waste) in the PBS scenarios.

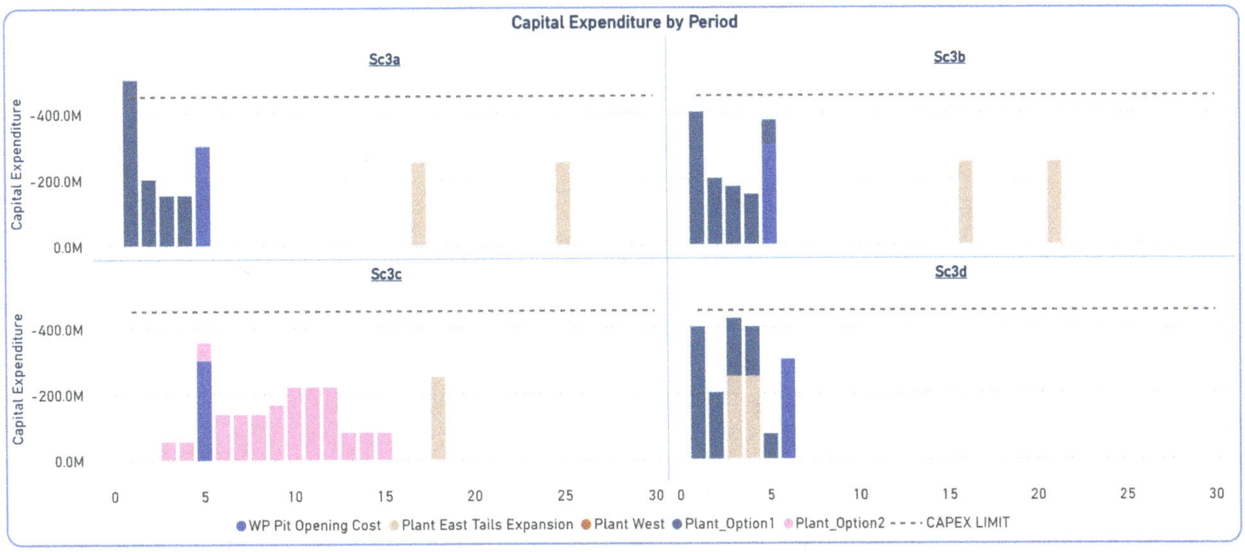

FIG 11 – Maximum capex per period limits capital expenditure, Scenarios 3b, 3c, and 3d.

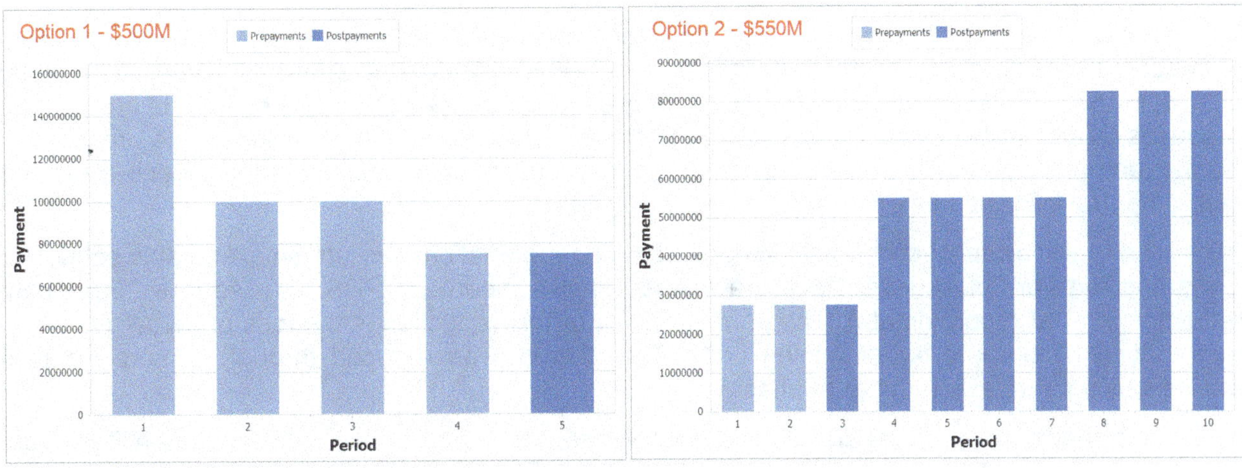

FIG 12 – Plant West expansion payment schedule options.

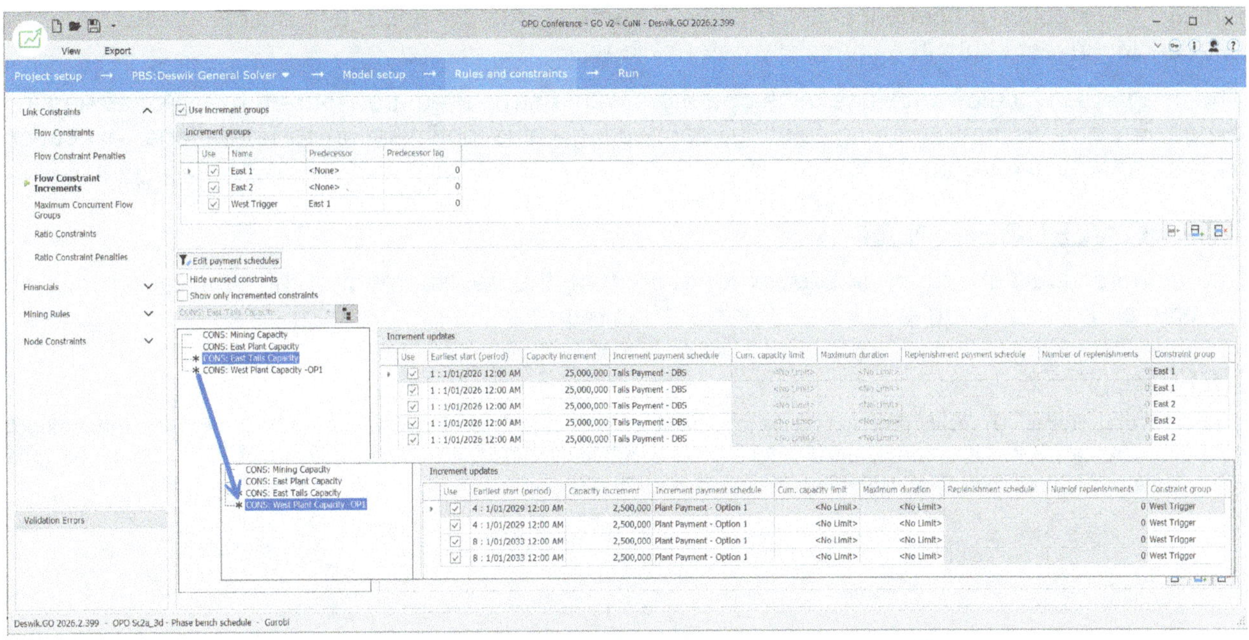

FIG 13 – Capital increment and precedence set-up.

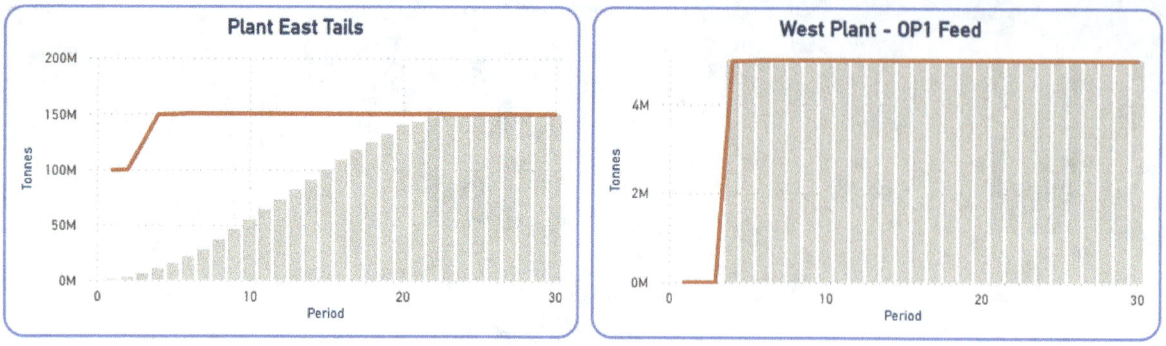

FIG 14 – Capital expansions to increase Plant East and Plant West capacity.

CONCLUSIONS

This paper has presented an integrated three-stage workflow: DBS with BZ acceleration, clustering-based PGN, and PBS. This sequence is well suited to developing a credible, capital-ready strategic mining schedule. By embedding operational and capital constraints at each stage, the workflow avoids the limitations of traditional, sequential planning approaches.

DBS demonstrated that capital investment decisions materially alter pit shape progression, extraction sequences, and production profile. The fully constrained scenario (1c) achieved a total DCF of $1.57B. PGN showed that the choice of production target, mined tonnes versus ore tonnes, produces substantially different phase geometries. This will have a measurable impact on downstream scheduling. It also highlights that phase design is an active planning decision rather than a neutral translation step.

PBS used realistic capital constraints to convert the operational phases into an executable schedule. The transition from block-level DBS to a phase-bench schedule in PBS resulted in a 25 per cent reduction in NPV (approximate). The primary driver of this reduction is the loss of selectivity when aggregating blocks into larger mining units. This gap can be reduced through a more detailed analysis using a greater number of phases.

It was also shown that project value could be increased with capital investment in both the Plant East tailings capacity and commissioning with Plant West. The discounted cash flow with these two expansions (Scenario 3c) was $1.22B.

Finally, the capital precedence constraints (Scenario 3d) provided a practical mechanism for sequencing investments in a financially defensible way.

The integrated workflow provides mine operators with a structured, transparent basis for evaluating expansion options and securing funding, connecting value optimisation to operational execution through defensible assumptions.

ACKNOWLEDGEMENTS

The authors would like to thank Deswik for supporting the development of this work and providing the time and resources to prepare and present this paper.

REFERENCES

Chung, J and Rahal, D, 2025. Enhancing pit optimisation with direct block scheduling (DBS) and the Bienenstock Zuckerberg (BZ) algorithm – maximising NPV and efficiency, in *Proceedings of APCOM 2025*, pp 47–62 (The Australasian Institute of Mining and Metallurgy: Melbourne).

Enhancing mining value chain optimisation – a linear programming approach with practical stockpiling constraints

T Whitfield[1] and N Jarrott[2]

1. Optimisation Engineering and Delivery Manager, Deswik, Brisbane Qld 4000.
   Email: tony.whitfield@deswik.com
2. Senior Optimisation Software Engineer, Deswik, Brisbane Qld 4000.
   Email: nancy.jarrott@deswik.com

BACKGROUND AND MOTIVATION

Material destination scheduling for mining value chains is commonly optimised using linear programming (LP) formulations derived from the multi-commodity network flow problem. In practice, these models take a mining schedule as input and optimise an economic objective (often net present value) through decisions such as stockpiling, processing, blending, and transportation across time and infrastructure constraints. A key modelling challenge is representing stockpiles in a way that is both computationally tractable and operationally realistic.

A common approach is to model stockpiles by adding time-based storage arcs, allowing material to remain at a node across time steps. While this is convenient, it effectively turns a stockpile into a warehouse: the optimiser can 'cherry-pick' parcels of material regardless of the order in which they were stacked. This behaviour can violate real operational constraints, where reclaim is constrained by how material is physically placed (eg stacking and reclaiming as a queue or a stack). The gap between the modelled and physically achievable reclaim behaviour can reduce plan fidelity and lead to schedules that are difficult to execute.

Beyond feasibility, reclaim-order realism can also improve downstream quality forecasting. When a model can cherry-pick parcels, predicted blends (eg forecast Fe on a vessel) may rely on selective reclaim that is not physically achievable. Enforcing FIFO/LIFO ties the planned reclaim sequence to executable stockpile behaviour, making grade predictions more credible.

A very simple network flow formulation is provided below for context, omitting the objective function which would generally involve costs/revenues associated with flows on arcs.

Multi-commodity network flow

K	set of commodities
L	set of locations
$T := \{1, \dots, \lvert T \rvert\}$	set of time steps
$N := L \times T$	space-time node set
A^{tr}	transport arcs
A^{st}	storage arcs
$A := A^{\text{tr}} \cup A^{\text{st}},\ A^{\text{tr}} \cap A^{\text{st}} = \emptyset$	all arcs partitioned into transport/storage.

For each $n \in N$, let $A^I(n)$ and $A^O(n)$ denote the incoming and outgoing arcs. The decision variables are nonnegative flows by commodity on each arc:

$$f_{a,k} \in \mathbb{R}^+ \quad \forall a \in A,\ \forall k \in K.$$

Flow conservation enforces mass balance at each space-time node (with net supply/demand $b_{n,k}$):

$$\sum_{a \in A^I(n)} f_{a,k} - \sum_{a \in A^O(n)} f_{a,k} = b_{n,k} \quad \forall n \in N,\ \forall k \in K.$$

Figure 1 illustrates the storage arcs A^{st} carrying inventory forward in time, while transport arcs A^{tr} move material between locations across the network.

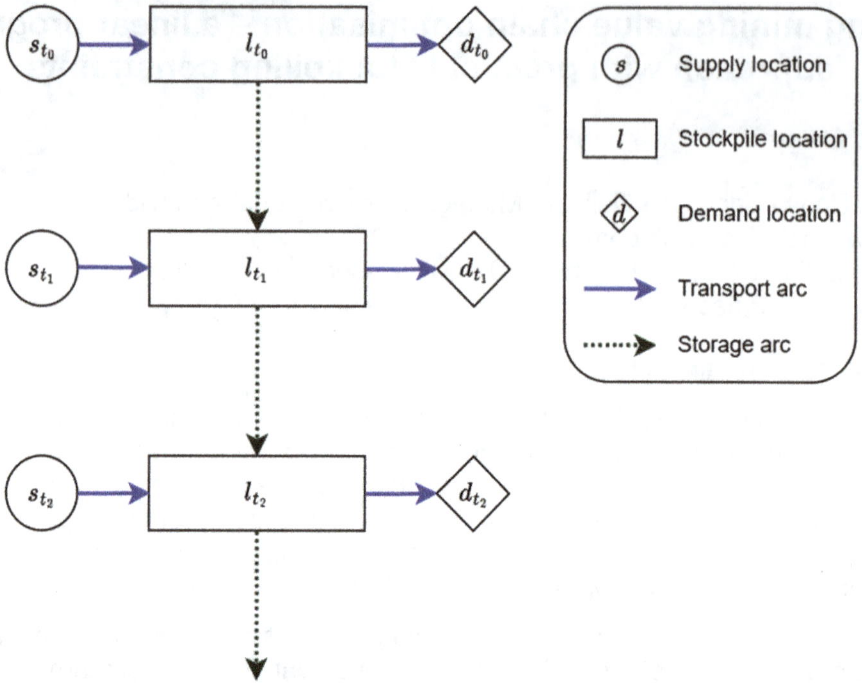

FIG 1 – Basic stockpile space time network representation.

PROBLEM STATEMENT AND OBJECTIVES

Problem statement

Standard network flow-based LP models represent stockpiles with storage arcs that permit arbitrary offtake from inventory, enabling unrealistic parcel selection and undermining operational feasibility.

Objectives

- Extend the LP (Linear Programming) formulation to incorporate operationally realistic stockpile reclaim ordering, specifically first-in, first-out (FIFO) and last-in, first-out (LIFO).
- Preserve tractability and compatibility with existing destination scheduling workflows (stockpiling, processing, blending, transportation) while improving schedule fidelity.
- Produce schedules that better reflect physical constraints and therefore support more resilient decision-making under changing prices, logistics, and strategic priorities.

CONTRIBUTIONS

- Introduce an extension compatible with MILP (Mixed-Integer Linear Programming) to network flow destination scheduling that enforces operationally realistic stockpile reclaim behaviour under FIFO and LIFO policies.
- Show how replacing unconstrained storage arcs with reclaim-order constraints prevents 'cherry-picking' from inventory while preserving integration with stockpiling, blending, processing, and transportation decisions.
- Provide an implementable formulation suitable for standard solvers (eg Gurobi), with a discussion of model size and tractability implications relative to the baseline storage-arc approach.
- Demonstrate (on representative instances) the feasibility and planning-fidelity gains of FIFO/LIFO reclaim, and quantify the associated objective and runtime trade-offs.

APPROACH

Our approach subdivides a stockpile into multiple internal slots indexed by stacking time. The 'warehouse-like' reclaiming behaviour from a stockpile is replaced with an ordering mechanism that can *disable reclaim* from a given internal storage slot whenever reclaim from other slots must occur first. At each time period, if inventory is detected in a set of 'blocking' slots, then transport (reclaim) out of the current slot is not permitted. By choosing the blocking set as predecessor slots we enforce FIFO; choosing successor slots enforces LIFO.

For each (time, slot) pair, a binary switch activates when there is any storage flow in the blocking set at that time, and when active it forces the relevant reclaim/transport flows to zero.

Incompatible storage-transport constraint

For a given time period $t \in T$ and a storage slot $s \in L$.

Let $B(s) \subseteq L$ be the set of blocking slots for s (FIFO: predecessors; LIFO: successors).

Define the local arc subsets:

$$\begin{aligned} A^{\text{st}}(l,t) \quad &:= \{a \in A^{\text{st}} \mid a \text{ stores at } l \text{ over } t \to t+1\} \qquad &&\text{(storage arcs)} \\ A^{\text{tr,O}}(s,t) \quad &:= \{a \in A^{\text{tr}} \mid a \text{ reclaims from } s \text{ at time } t\} \qquad &&\text{(reclaim arcs).} \end{aligned}$$

Introduce a binary switch $\varphi_{s,t} \in \{0,1\}$ indicating whether *any* blocking inventory exists at time t.

(1) Storage-detection (switch) constraint:

$$\sum_{l \in B(s)} \sum_{k \in K} \sum_{a \in A^{\text{st}}(l,t)} f_{a,k} \leq M_{s,t}^{\text{st}} \varphi_{s,t}$$

(2) Transport-banning constraint (reclaim is only allowed when switch is off):

$$\sum_{k \in K} \sum_{a \in A^{\text{tr,O}}(s,t)} f_{a,k} \leq M_{s,t}^{\text{tr}} \left(1 - \varphi_{s,t}\right).$$

Here $M_{s,t}^{\text{st}}$ can be taken as a capacity bound on blocking storage at time t, and $M_{s,t}^{\text{tr}}$ as an upper bound on outbound flow (eg total supply).

FIFO example

Figure 2 illustrates the incompatible storage-transport constraint at time t_1 and slot s_{t_1}.

Storage flow along $f_{a_{st}}$ activates the binary switch and transport flow along $f_{a_{tr}}$ is banned.

Slot s_{t_0} in the stockpile stores material stacked at time t_0. In FIFO, this older material must be reclaimed before newer material in slot s_{t_1} (stacked at time t_1) can be reclaimed.

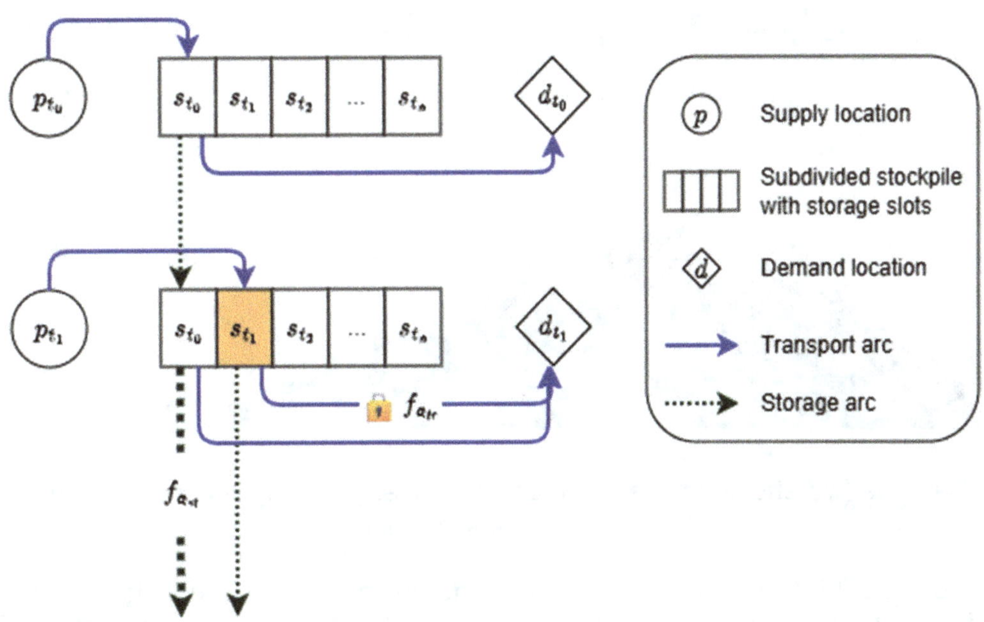

FIG 2 – FIFO stockpile example.

IMPLEMENTATION AND EXPERIMENTAL SET-UP

We implemented FIFO/LIFO reclaim behaviour in Deswik BOLT, a cloud-based destination scheduling platform that uses Gurobi 12 as the underlying solver. Experiments were performed on an iron ore scenario, modelling material movements from pit ROMs through the value chain to shiploaders.

A key focus is the crusher ROM stockpile, where multiple grades of ore accumulate up to a possible capacity of 60 000 t. In the baseline scenario, this inventory is represented using standard storage arcs, allowing warehouse-style reclaim (ie arbitrary parcel selection). In experimental scenario 1, we add a FIFO reclaim constraint on this stockpile, and in scenario 2, we change this to be LIFO. The three scenarios are otherwise identical.

All three scenarios were compared using two evaluation metrics: (i) reclaim ordering of stacked material under FIFO, LIFO, or warehouse behaviour; and (ii) solve time.

RESULTS

Table 1 summarises the experimental results. Across all scenarios, the model remained feasible. Introducing reclaim ordering constraints increased solve time relative to the baseline, with FIFO incurring the largest solve time increase in this instance.

TABLE 1
Evaluation summary.

Scenario	Feasible	Solve time	Observed behaviour
Baseline	True	17 s	Cherry-pick/warehouse
FIFO	True	2 min 11 s	FIFO/queue
LIFO	True	44 s	LIFO/stack

Observed reclaim behaviour

Figure 3 illustrates the baseline (warehouse) behaviour. Reclaim is not constrained by stacking order, and the optimiser cherry-picks parcels from inventory. Material E32 (blue) is selectively reclaimed from the stockpile on 28 Mar ahead of other available parcels. Material D4 (turquoise) is also reclaimed on 30 Mar ahead of other available parcels.

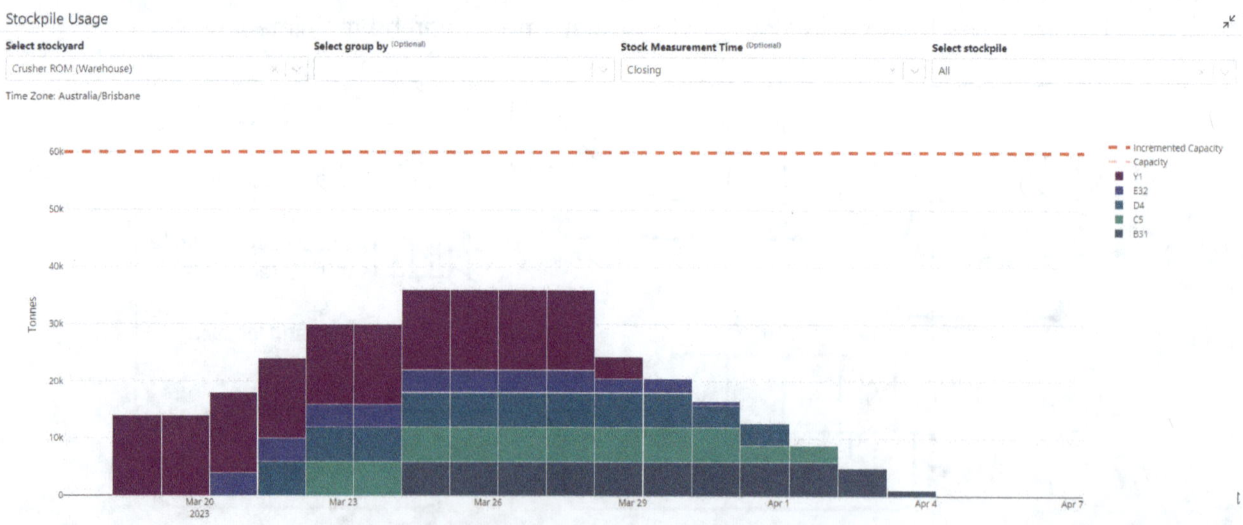

FIG 3 – Baseline (warehouse) reclaim: parcel-level cherry-picking is possible, deviating from physical reclaim order.

Figure 4 shows the FIFO scenario. Instead of cherry-picking, reclaim follows first-in, first-out dynamics. Material Y1 (purple) is stacked in the stockpile first, followed by material E32 (blue).

Material Y1 (purple) is reclaimed first on 28–29 Mar. Material E32 (blue) is reclaimed next on 30 Mar. All materials are reclaimed in their stack order.

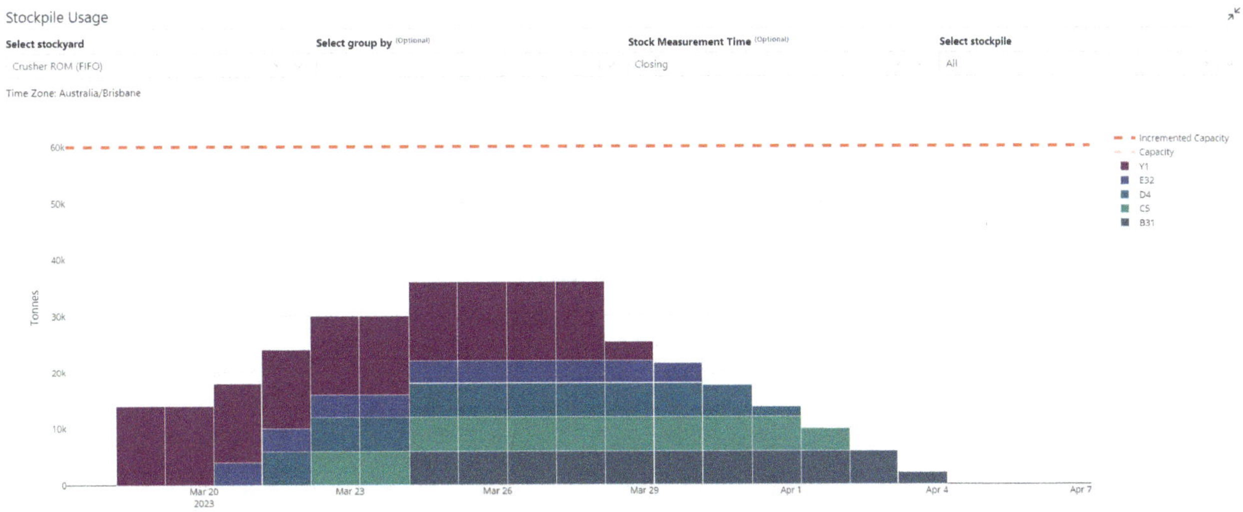

FIG 4 – FIFO reclaim: early-arriving parcels (eg Y1 and E32) are reclaimed first, consistent with queue-like behaviour.

Figure 5 shows the LIFO scenario, exhibiting stack-like behaviour. In particular, material B31 (grey) is stacked last (24 Mar) and is reclaimed first on 28 Mar, consistent with last-in, first-out reclaim.

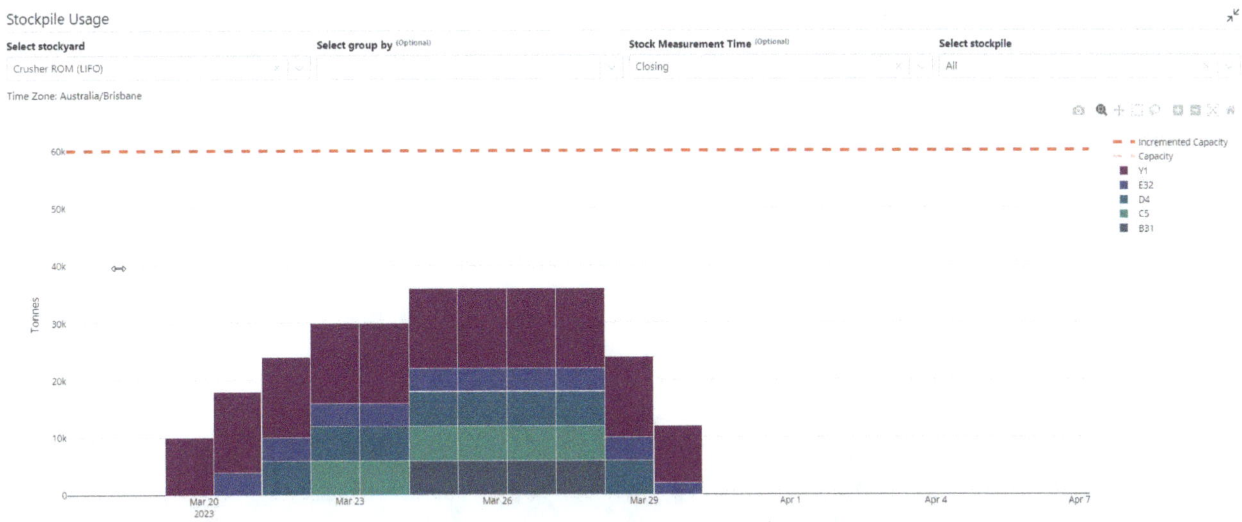

FIG 5 – LIFO reclaim: late-arriving parcels (eg B31) are reclaimed first, consistent with stack-like behaviour.

DISCUSSION

The results show that reclaim-order enforcement can be integrated into a standard destination-scheduling network flow model in a way that reliably prevents warehouse-style cherry-picking and produces clearly FIFO/LIFO-consistent schedules. This closes an important feasibility gap between optimiser behaviour and physical stockpile operation, improving plan realism and downstream quality forecasting without requiring detailed geometric stockpile simulation.

The main limitation is computational. The binary switch variables $\varphi_{s,t}$ convert the base LP network flow model into a mixed-integer linear program (MILP), introducing combinatorial complexity and solver branching. As the time horizon grows, the number of ordering interactions (and associated binary decisions) can expand rapidly (often on the order of $O(|T|^2)$ per stockpile), so solve time increases can be substantial and nonlinear. In practice, this motivates careful horizon selection (shorter time horizons or longer time periods will run much faster), selective application to high-impact stockpiles, and tighter bounds/valid inequalities to improve tractability.

CONCLUSION

We present a practical MILP extension to network flow destination scheduling that enforces FIFO/LIFO reclaim on stockpiles, eliminating unrealistic warehouse-style cherry-picking. On an iron ore case study in Deswik BOLT, the formulation produces the intended reclaim behaviour for both FIFO and LIFO, at the cost of increased (and horizon-sensitive) solve times. Future work includes quantifying the trade-offs between schedule fidelity and computational tractability across a wider range of instances and assessing the value of improved reclaim-order realism in terms of schedule robustness and downstream quality forecasting accuracy.

Open Pit

Operational data, analytics and optimisation

Using AI machine vision and load sensors to determine the optimal P_{80} for a real-life Australian mine

W Alker[1], A Kuyuk[2] and R Wilkinson[3]

1. Area Manager, Weir Motion Metrics, Lytton Qld 4178. Email: will.alker@mail.weir
2. Former Commercial Manager, Weir Motion Metrics, Vancouver BC V5X 3E3, Canada. Email: camilo.sanchez@mail.weir
3. Customer Success Manager, Weir Motion Metrics, Vancouver BC V5X 3E3, Canada. Email: reid.wilkinson@mail.weir

INTRODUCTION

Optimising rock fragmentation is critical for efficient load and haul operations in hard rock mining. Poor fragmentation leads to reduced shovel productivity, increased energy consumption, and higher maintenance costs. This study investigates the relationship between particle size distribution and shovel bucket payload using AI-enabled machine vision and load sensors installed on a HITACHI EX3600B backhoe shovel. The goal is to identify the optimal P_{80} (80 per cent passing size) that maximises shovel performance.

To meet this goal, an important parameter monitored after each blast was the uniformity index. The uniformity index is commonly used as an indicator of fragmentation distribution, describing how uniformly particle sizes are distributed within the blasted rock pile and the extent of size variability.

Fragmentation monitoring and particle segmentation are illustrated in Figure 1.

(a) (b)

FIG 1 – Fragmentation monitoring with image processing and segmentation of particles: (a) not segmented; (b) segmented.

METHODOLOGY

The study utilised the ShovelMetrics™ Gen 3 system from Weir's MOTION METRICS™, which integrates image-based fragmentation analysis and sensor-based payload monitoring. Data was collected over six weeks and included shovel payload, dig duration, and cycle time. A Python-based data pipeline merged sensor and fleet management system data using timestamp proximity and equipment IDs to ensure accurate event correlation. The analysis focused on identifying the particle size range that consistently yielded high payloads and efficient loading cycles.

The data processing and integration workflow is shown in Figure 2.

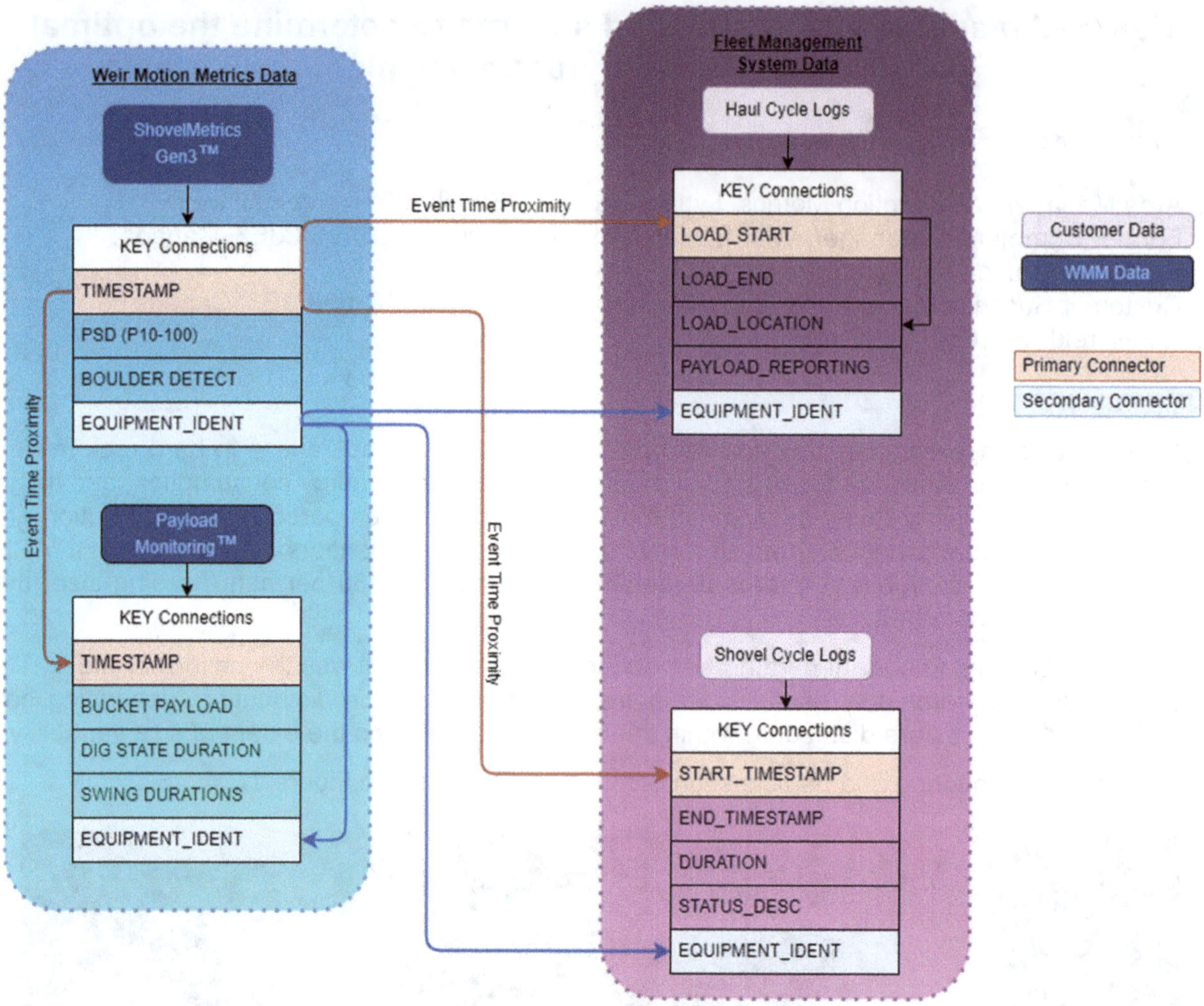

FIG 2 – Data processing and merger pipeline.

RESULTS

The analysis revealed that the shovel achieved optimal performance when particle size distribution P_{80s} were within the 20–25 cm range. Specifically, a 35 t dipper reached over 90 per cent capacity in this range, while payload dropped to 60 per cent for P_{80s} larger than 32 cm.

When data was binned into increments of 2 t for payload correlation, the results showed a corresponding relationship between decreasing median P_{80} (cm) and increasing payload values.

The evaluation criteria used in this analysis are summarised in Table 1.

TABLE 1

Evaluation criteria based on particle size.

High payload criteria	20–35 tonnes
Cycle time criteria	≤30 seconds
Dig duration criteria	≤2.5 seconds

This relationship shows that bucket fill correlates directly to fragmentation conditions, with optimal bucket fill consistency observed as the particle size approaches the ~22 cm range.

Productivity gain potential was estimated to be in the realm of an additional ~8 per cent material movement per month based on the established relationship, coupled with the payload and fragmentation baselines established across the six week data collection window. See the following figures for examples of high and low payload cases and their corresponding P_{80} values.

CONCLUSION

This study confirms that maintaining fragmentation within optimal P_{80} ranges can significantly enhance shovel productivity. The integration of AI-enabled vision systems and load sensor data, coupled with Python-based data merging, provides a reliable method for both establishing and continuously monitoring these targets.

Both short-term productivity gains and long-term operational sustainability rely on optimising blast outcomes and improving downstream loading efficiency.

CONCLUSION

[illegible]

[illegible]

Bit by bit – unravelling the MWD mystery

L Karlson[1]*, V Jonsson*[2]*, M Jessell*[3]*, M Lindsay*[4]*, G Pirot*[5]*, I Minniakhmetov*[6] *and M Keep*[7]

1. Superintendent Systems and Data Management, BHP, Perth WA 6000.
   Email: lance.karlson@bhp.com
2. Senior Geotechnical Engineer, BHP, Perth WA 6000. Email: valdimar.jonsson@bhp.com
3. Professor, Centre for Exploration Targeting, University of Western Australia, Crawley WA 6009.
   Email: mark.jessell@uwa.edu.au
4. Science Leader, CSIRO Mineral Resources, Kensington WA 6151.
   Email: mark.lindsay@csiro.au
5. Senior Research Fellow, Centre for Exploration Targeting, University of Western Australia, Crawley WA 6009. Email: guillaume.pirot@uwa.edu.au
6. Principal Global Modelling and Data, BHP, Perth WA 6000.
   Email: ilnur.minniakhmetov@bhp.com
7. Professor, University of Western Australia, Crawley WA 6009. Email: myra.keep@uwa.edu.au

ABSTRACT

Measure while drill (MWD) data from blastholes offer enormous potential for improving the resolution of geological models in open cut mines. However, the practical application of MWD data is limited by high variability, which persists even in well-calibrated drilling systems. Without identifying and correcting these sources of variance, analytical methods such as machine learning cannot deliver reliable lithological interpretation.

This study addresses a critical but previously overlooked factor influencing MWD parameters: drill bit wear. Using data from more than 4000 bits across ten Pilbara mine sites, we estimate the within-bit wear response by applying a fixed-effects design with a penalised spline over cumulative metres, while controlling for lithology and site differences. Contrary to conventional assumptions, the fleet-level curve indicates that penetration rates increase as bits wear, a trend consistently observed across mine sites.

These results challenge established expectations about drilling performance and highlight drill bit wear as a systematic source of variance in MWD data. Recognising, recording and correcting for this effect is important for any geological interpretation using MWD data. While this preliminary study focuses on isolating the within-bit wear trend rather than providing a complete correction formula, the findings offer a practical basis for more reliable use of MWD in stratigraphic interpretation, blast planning, and geotechnical risk assessment.

INTRODUCTION

Measure while drill (MWD) data from blasthole rigs include rate of penetration (ROP), weight on bit, torque, rotation speed and air pressure variables. These variables are widely collected in the mining industry to monitor drilling performance and can also be used to model subsurface geology at production scale, complementing widely spaced exploration drill holes (Rai *et al,* 2014; Navarro *et al,* 2021). However, routine use of MWD by geologists and mining engineers remains challenging because signals reflect overlapping mechanical, geological and operating influences that cause uncertainty unless properly normalised (Rai *et al,* 2014; Goldstein, Aldrich and O'Connor, 2024). Recent studies demonstrate both the promise and pitfalls of MWD-based stratigraphic interpretation and blast design, underscoring the need for careful correction of influences before using the data for geological interpretation (Karlson *et al,* 2023; Arnold, 2023; Goldstein *et al,* 2025).

A long-standing assumption from both operators and mining engineers is that penetration rates decline as a drill bit wears. This is an expectation derived from classic drilling mechanics and ROP models that relate cutting efficiency to drill bit condition, energy delivery and rock strength (Teale, 1965; Bourgoyne and Young, 1974). Teale's foundational work on specific energy suggests that more energy per unit volume is required as cutting becomes less efficient, while regression-based ROP models incorporate bit wear as a factor that should progressively reduce ROP (Bourgoyne and Young, 1974; Sobhi, Dobbi and Hachana, 2022).

In practice, the drill bit wear-ROP relationship is made more complex by fragmented bit histories, imperfect record-keeping at bit change-outs, and variability between mine sites. Tricone wear studies show that using bit life or ROP alone to trigger change-outs can be unreliable, and that vibration/control responses may shift as bits age. These effects complicate simple cross-bit comparisons (Rafezi and Hassani, 2020, 2021). As a result, fleet-level trendlines may hide, or even reverse, the underlying within-bit wear signal if differences in drilling conditions are not controlled.

The study tests the hypothesis that blasthole penetration rates decrease as drill-bit wear increases. We aim to achieve this by narrowly focusing on the within-bit relationship between ROP and cumulative wear across automated blasthole drill rigs. This narrow framing is intended to isolate the wear signal from rock type and site differences. Broader multivariate performance modelling incorporating weight on bit, torque, rotation speed and air variables relevant to mechanical specific energy is purposely reserved for future work.

METHOD

Data and scope

This study uses a data set of MWD records from rotary blasthole rigs from iron ore mines in the Hamersley Basin, Western Australia. The data set, summarised in Table 1, spans approximately 800 000 blastholes drilled only by automated Pit Viper drill rigs using tricone bits of 251 mm diameter. MWD data were recorded at 25 mm intervals down each blasthole, which averaged ~12 m depth. The data set represents drilling across all major rock types in the Hamersley Group, including Brockman and Marra Mamba banded iron formations, channel iron deposits, shales, dolomites and igneous intrusions.

TABLE 1

Data set overview.

Item	Value
Total blastholes	800 938
Total metres drilled	10 011 000
Number of drill bits	4054
Drill rig type	Pit Viper (automated)
Number of drill rigs	30

MWD data were appended to drill bit records and flagged with block model variables including modelled stratigraphy and chemistry, which were subsequently used to determine basic lithology types. The flagging process sourced variables from those blocks through which the majority of each drilling interval intersected.

Statistical approach

To isolate the effect of bit wear on penetration rate, we use a within-bit fixed-effects design that treats each bit as its own baseline and compares it only to itself as metres accumulate. This removes time-invariant differences between bits, rigs and mines. The model captures the change in ROP attributable to wear rather than cross-bit or cross-site noise. Because wear does not progress linearly, the model includes a penalised spline on cumulative bit metres; this allows the trend to flex naturally (early wear-in, steadier mid-life, and any late-life changes) while the penalty limits over-fitting so the curve reflects system behaviour rather than short-term fluctuations. The lithology variable sourced from block models is included to absorb differences between major rock types. This reduces the risk that geological transitions are misread as changes in bit condition. Finally, to check that longer-lasting bits do not skew the fleet-level curve by contributing more late-life data, we applied inverse-probability censoring weights based on the probability that an observation is present at a given cumulative wear; weights were normalised, inspected for extremes, and used to re-estimate the spline. The weighted and unweighted curves were nearly identical, indicating that survivorship does not materially affect the estimated wear response over the 0–6000 m range.

Mine-level comparison

While the spline-based wear curve is appropriate at the fleet scale, individual mines often have narrower wear ranges and fewer high-life bits, making spline fits unstable at a site level. For this reason, each mine's behaviour was summarised using a linear within-bit fixed-effects slope, which captures whether ROP within that site tends to rise, flatten or decline with wear. This provides a simple, comparable indicator of wear response across operations while retaining the noise-reduction advantages of the within-bit approach.

RESULTS

Results are presented progressively, moving from raw bit survivorship and per-bit variability to controlled fleet and mine-level wear responses.

Bit survival curve

Figure 1 shows a decrease in the proportion of bits remaining as cumulative bit metres increase. Approximately 20 per cent of bits expire by 1000 m, with the percentage reaching approximately 50 per cent at 2500 m. Beyond this point, the curve tapers gradually, with around 10 per cent of bits remaining beyond 6000 m.

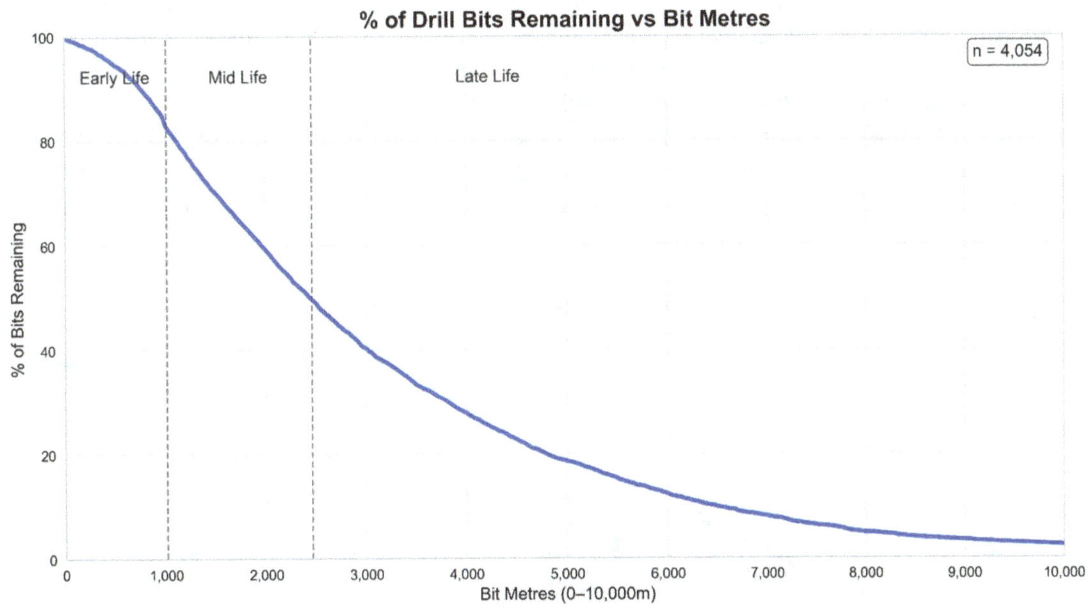

FIG 1 – Bit survival curve across the fleet. Percentage of drill bits remaining versus cumulative bit metres for 4054 tricone bits (0–10 000 m). Dashed lines mark indicative early/mid/late-life intervals used elsewhere in the paper.

Per-bit Pearson diagnostic

Pearson correlation provides a simple, per-bit measure of the linear association between penetration rate and accumulated bit wear. When applied across the full data set of ~4000 drill bits in Figure 2, the correlations spread out in both directions, with most sitting close to zero and the rest scattered across moderate positive and negative values. This demonstrates that at the individual bit scale, there is variability between ROP increases, decreases and an absence of linear trends.

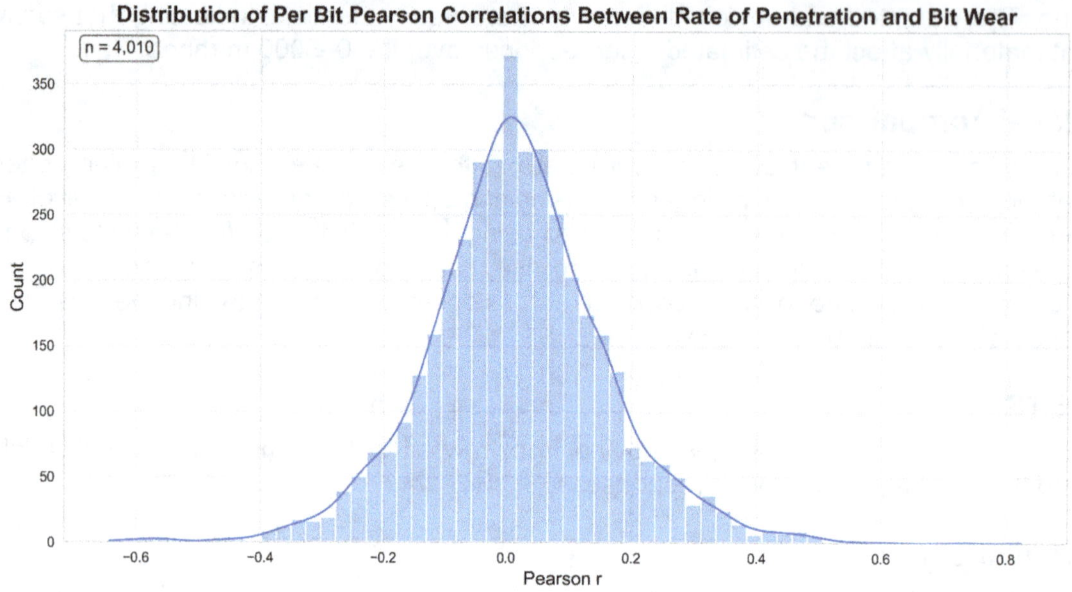

FIG 2 – Distribution of per-bit Pearson correlations between ROP and bit wear (drill bits = 4010).

Fleet-level wear curve

The within-bit fixed-effects spline in Figure 3 normalises mine site and drilled lithology variation to recover the fleet-level ROP versus bit wear relationship. Units on the y axis represent original ROP values in metres per minute. The curve contradicts the original hypothesis that penetration rates decline with bit wear. Instead, the curve indicates that within-bit ROP increases rapidly over the first ~1000 m, rising from near zero to ~ 0.10 units of mean-centred ROP change. Through the mid-life interval (~1000–2500 m), the rate of increase moderates but remains positive, with the curve reaching ~ 0.12. Beyond ~2500 m and into late bit life, the upward trend continues with a gradual, near-linear slope, reaching ~ 0.16 by 6000 m.

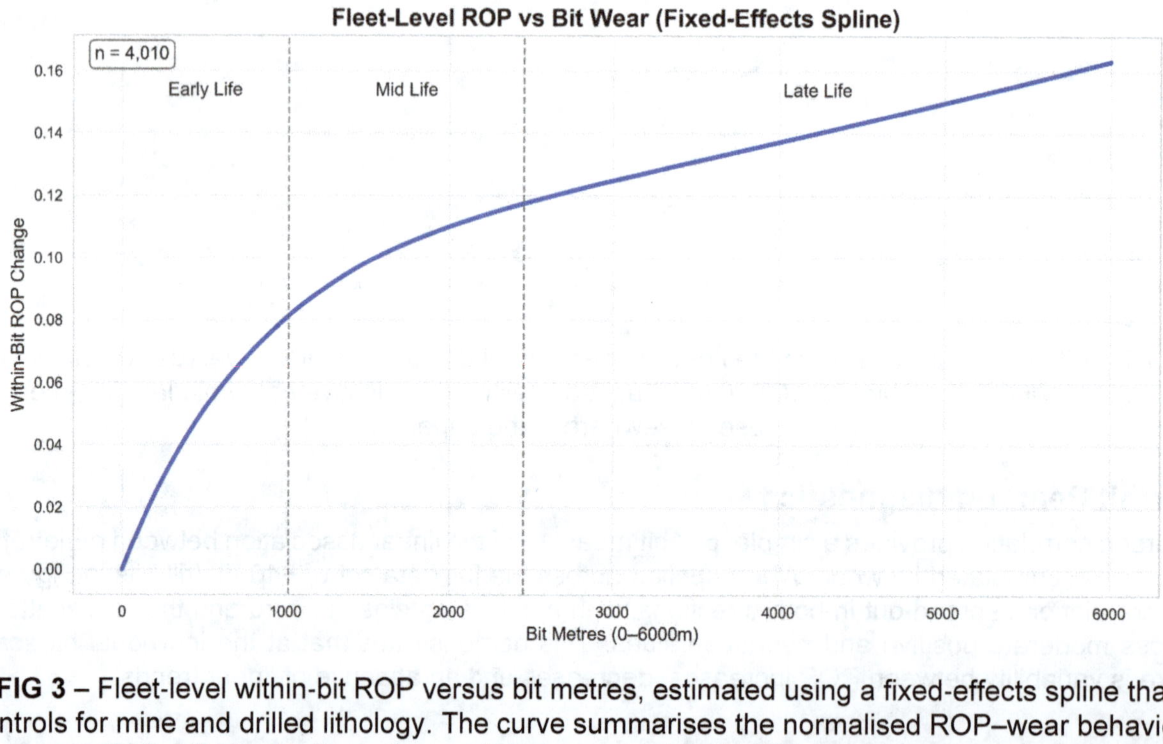

FIG 3 – Fleet-level within-bit ROP versus bit metres, estimated using a fixed-effects spline that controls for mine and drilled lithology. The curve summarises the normalised ROP–wear behaviour across 4010 drill bits. Units on the y axis represent original ROP values in metres per minute.

This result reflects an average within-bit response under automated drilling control, and does not imply that unlimited wear leads to ever-increasing penetration rates.

Stratified fixed effect slopes by bit lifespan

Figure 4 shows the fixed-effects slopes stratified into short, medium, and long-life bit populations. Across all groups, the slope remains positive, confirming that within-bit ROP consistently increases with accumulated wear. However, the magnitude of the effect varies substantially with bit lifespan. Short-life bits show only a weak positive trend (slope ~0.023), medium-life bits show a clearer increase (~0.044), while long-life bits exhibit the strongest positive trend (~0.086), more than three times the slope observed in short-life bits. This stratification demonstrates that bits which survive longer tend to exhibit more pronounced ROP gains as wear progresses.

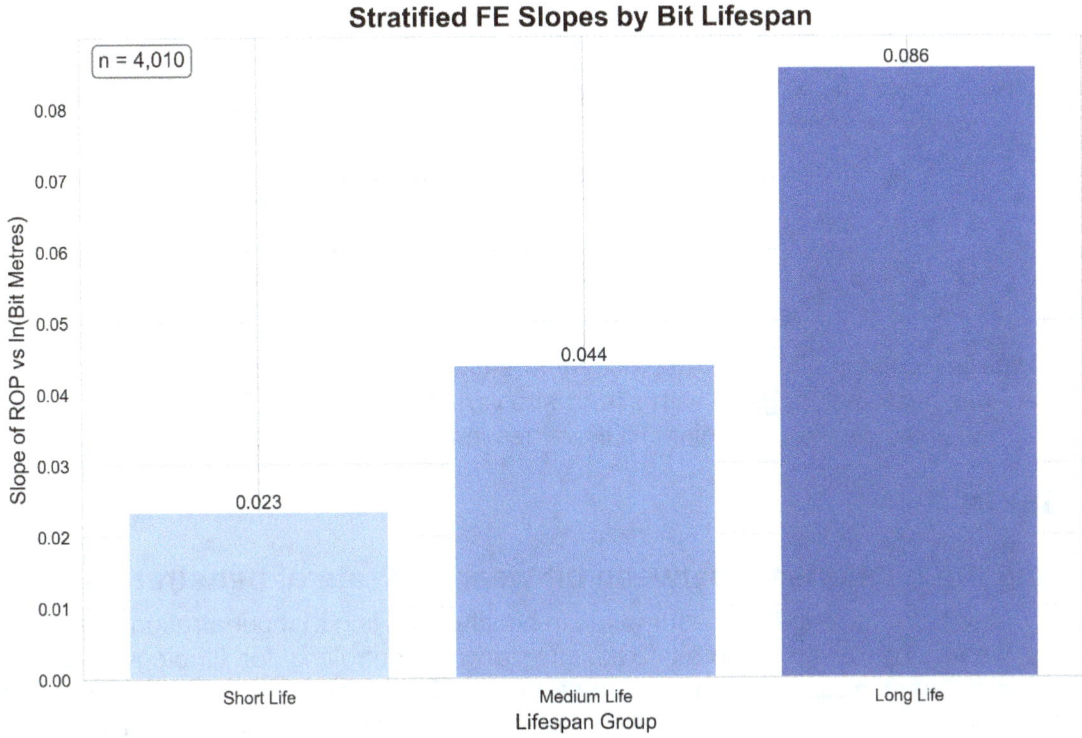

FIG 4 – Within-bit fixed effects slopes by bit lifespan: the positive wear–ROP slope strengthens from short to long-life groups (drill bits = 4010).

Mine-level within-bit ROP versus bit wear (linear fixed-effect slopes)

Figure 5 presents the linear fixed-effects wear slopes for each mine in this study, showing how within-bit ROP trends differ from site to site. While the magnitude differs between mines, the overall direction of the trend is consistent with the fleet-level result: most sites show a steady increase in normalised ROP as wear accumulates. Several mines display strong positive trajectories, with modelled ROP gains exceeding 0.20 units by 6000 m of wear, whereas others show more modest increases or early-life flattening.

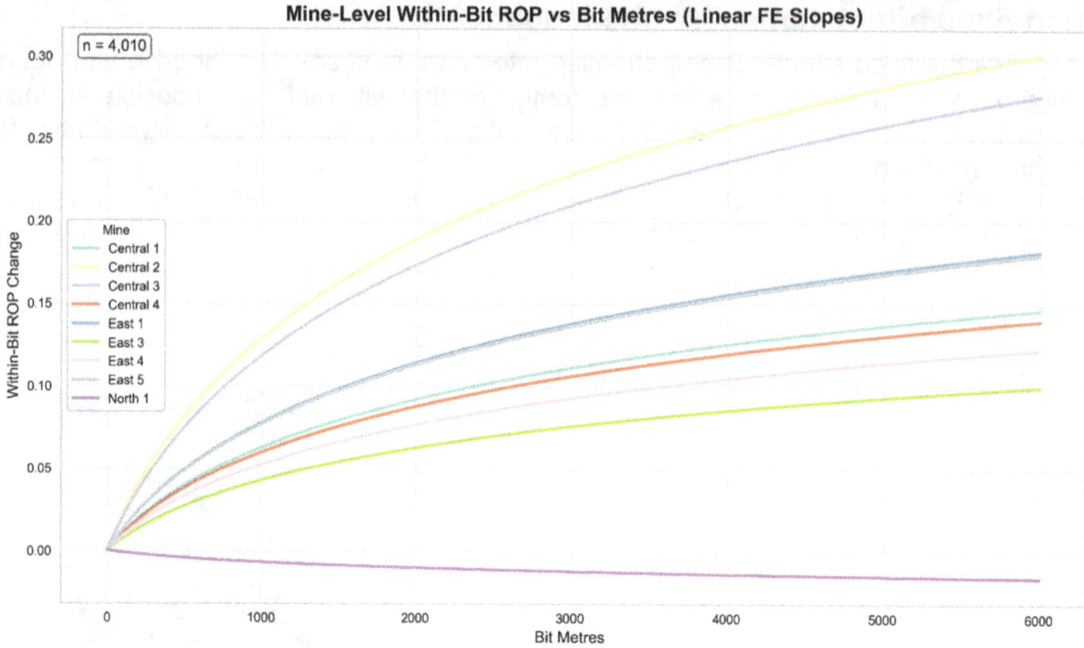

FIG 5 – Mine-level within-bit ROP versus bit metres (linear within-bit slopes), showing generally positive wear-related ROP gains with site-to-site variation in magnitude. Units on the y axis represent original ROP values in metres per minute.

DISCUSSION

Rethinking the relationship between bit wear and rate of penetration

This study aimed to test a long-held assumption in blasthole drilling that penetration rate declines as the drill bit wears. By using within-bit fixed effects and controlling for lithology and mine site differences, a different picture emerged. Across the 0–6000 m bit wear range, penetration rates tended to increase, not decrease, within individual bits. The strongest rise occurred early in life, moderating through mid-life and continuing gradually into late life. This trend persisted across most mines and was also reflected in aggregated fleet-level behaviour. These results suggest that, in automated Pilbara drilling systems, bit wear does not produce the steady performance deterioration historically expected, and may in fact produce conditions that enhance penetration.

Why the statistical approach matters

MWD data are influenced by a range of operational, mechanical and geological variables. Raw cross-bit comparisons blur these variables together, often hiding the true behaviour occurring within individual bits. The fixed-effects design – treating each bit as its own baseline – has allowed us to isolate the change in ROP caused by wear alone, independent of how the bit performed relative to others. Restricting the analysis to the common wear range (0–6000 m) avoided bias from rare high-wear events and potential data errors, and survivorship checks showed minimal distortion. This approach gives a clearer view of the wear signal itself and helps explain why simple fleet averages or trendlines have historically produced contradictory interpretations.

Field observations

Field observations of new and expired drill bits attempted to determine factors that may explain the trends found in this study. While data should be treated as anecdotal only, it was found that worn bits experience a reduction in bit diameter from 251 mm (new) to approximately 235 mm (expired). These measurements can be variable depending on the degree and type of wear – and will also be highly specific to bit type and design – but the observation suggests that hole diameter decreases as a bit wears. This is potentially a significant cause of penetration rate changes with bit wear.

FIG 6 – Example of new tricone drill bit versus worn/expired bit with bit diameter measurements.

Why might worn bits drill faster?

Although this study does not attempt to resolve the full mechanical explanation, several mechanisms could plausibly contribute to increasing ROP with bit wear:

- Improved cuttings removal may occur as the hole narrows. We know from field observations that the diameter of the bit decreases with bit wear, suggesting the hole diameter also shrinks. With compressor airflow essentially constant, the reduced annulus area potentially increases uphole air velocity. This could improve cuttings removal, reduce bottom-of-hole accumulation, and decrease the amount of regrinding the bit must perform. Cleaner bottom conditions would logically cause higher rates of penetration.
- Loosening of tricone bearings during mid-life ('wearing in') can reduce internal friction and allow the cones to rotate more freely, improving cutting efficiency and contributing to higher ROP even as the cutters themselves dull.
- Automated rigs adjust pulldown, torque thresholds, and rotational speed to maintain target drilling energy. As bit geometry changes, the rig may deliver additional energy that offsets or surpasses cutter dulling. This may be confirmed with further studies incorporating all available MWD parameters.
- Some bits may drill more efficiently after early wear smooths overly aggressive teeth ('wearing in the teeth'), reducing vibration or torque spikes and potentially improving penetration efficiency.
- Chip size changes may occur with bit wear. Duller bits would logically produce finer or more uniform cuttings, which can be lifted more effectively by the airflow, reducing chip-bed formation.

These mechanisms should be treated as working hypotheses until validated with a more complete multi-parameter model.

Implications for MWD interpretation and operational decisions

One of the most important implications of this study is that wear-related acceleration could plausibly mimic changing rock types. Without correcting for bit wear, ROP-based geological modelling risks falsely identifying soft zones or obscuring genuine stratigraphic changes. Detrending ROP for wear, using fleet-level or site-specific curves, offers a cleaner pathway for integrating MWD into geological models.

Operationally, the results suggest that bit-change decisions based solely on perceived ROP decline are unreliable under automation, because wear does not generally reduce penetration rate. Instead, change-out decisions should focus on dimensional limits, bit-life risk, and in-field observations.

CONCLUSIONS

This study reframes bit wear from a nuisance variable into a quantifiable, systematic component of MWD signals that must be addressed before geological interpretation. By isolating within-bit behaviour using fixed effects, the analysis shows that ROP generally increases with cumulative wear, contradicting the long-held expectation of performance decline. This trend is consistent across most mine sites in the study, however it does not imply that ROP will increase indefinitely with bit wear.

The findings help explain why MWD cross-sections often show unexpected acceleration or deceleration zones that do not align with geological boundaries, creating confusion for geologists trying to interpret the data. These anomalies may reflect bit-life effects rather than geology alone. For instance, parts of an MWD profile that appear to be 'soft ground' may in fact reflect worn-bit behaviour or automated control responses, not lithological change.

Several system-level mechanisms may contribute to this behaviour, including improved cuttings removal as the hole narrows, control-system compensation, stabilisation of bit geometry, finer chip production, and the possibility that loosening tricone bearings mid-life reduces rotational resistance, allowing cones to turn more freely. These mechanisms require further study and should be viewed as hypotheses rather than confirmed causes.

The results emphasise that raw ROP should not be used directly for geological boundary modelling or operational decision-making without correcting for bit wear. Further work should integrate weight on bit, torque, rotation speed and air parameters into a full multivariate model of drilling performance. Such future work will allow more detailed understanding and provide practical tools for detrending MWD data, supporting higher-confidence geological interpretation.

ACKNOWLEDGEMENTS

This research was supported by a PhD research agreement between BHP and the University of Western Australia. The authors would like to acknowledge drilling expertise provided by the BHP WAIO drill control team, drill and blast leadership and frontline operators at BHP that was integral in understanding trends in blasthole drilling methods. They would also like to acknowledge data assistance by Cameron Leslie, Darryl Werth and Ryan Gair from BHP.

REFERENCES

Arnold, J R, 2023. Application of Measurement While Drilling Data for Mine Blast Optimization Utilizing Machine Learning Techniques with Iron Ore Mine Data, MSc thesis, Virginia Tech.

Bourgoyne, A T Jr and Young, F S Jr, 1974. A multiple regression approach to optimal drilling and abnormal pressure detection, *SPE Journal*, 14(4):371–384.

Goldstein, D M, Aldrich, C and O'Connor, L, 2024. A review of orebody knowledge enhancement using machine learning on open-pit mine measure-while-drilling data, *Machine Learning and Knowledge Extraction*, 6(2):1343–1360.

Goldstein, D M, Aldrich, C, Shao, Q and O'Connor, L, 2025. Unlocking subsurface geology: A case study with measure-while-drilling data and machine learning, *Minerals*, 15(3):241.

Karlson, L, Jessell, M, Keep, M, Pirot, G, Lindsay, M and Minniakhmetov, I, 2023. Identification of dykes using measure while drill data, AEGC Extended Abstracts.

Navarro, J, Seidl, T, Hartlieb, P, Sanchidrián, J A, Segarra, P, Couceiro, P, Schimek, P and Godoy, C, 2021. Blastability and ore grade assessment from drill monitoring for open pit applications, *Rock Mechanics and Rock Engineering*, 54:2665–2685.

Rafezi, H and Hassani, F, 2020. Practicality of tricone bit wear condition assessment, *Experimental Results*, 1:e35:1–6.

Rafezi, H and Hassani, F, 2021. Drilling signals analysis for tricone bit condition monitoring, *International Journal of Mining Science and Technology*, 31(2):187–195.

Rai, P, Schunnesson, H, Lindqvist, P-A and Kumar, U, 2014. An overview on measurement-while-drilling technique and its scope in excavation industry, *Journal of The Institution of Engineers (India):Series D*, 95(2):173–181.

Sobhi, I, Dobbi, A and Hachana, O, 2022. Prediction and analysis of penetration rate in drilling operation using deterministic and metaheuristic optimization methods, *Journal of Petroleum Exploration and Production Technology*, 12:1341–1352.

Teale, R, 1965. The concept of specific energy in rock drilling, *International Journal of Rock Mechanics and Mining Sciences*, 2(1):57–73.

Interpretable XGBoost modelling of blast-induced rock fragmentation and airblast using SHAP for transparent open pit blast design

O Saubi[1], R S Jamisola Jr[2], R S Suglo[3] and O Matsebe[4]

1. PhD candidate, Botswana International University of Science and Technology (BIUST), Palapye, Botswana. Email: so13000604@biust.ac.bw
2. Associate professor, BIUST, Palapye, Botswana. Email: jamisolar@biust.ac.bw
3. Professor, BIUST, Palapye, Botswana. Email: suglor@biust.ac.bw
4. Doctor, BIUST, Palapye, Botswana. Email: matsebeo@biust.ac.bw

INTRODUCTION

Open pit blast design must deliver safe, compliant outcomes (eg controlled airblast) while meeting production targets, such as desired fragmentation for loading, crushing, and downstream throughput. In practice, blast response is nonlinear and site dependent because design parameters (burden, spacing, stemming, charge distribution and timing) interact with rock mass behaviour and structural conditions. This motivates the use of machine learning (ML) approaches for prediction; however, operational adoption is often limited when models behave as black boxes and cannot justify why a prediction changes under design modifications. This work develops interpretable extreme gradient boosting (XGBoost) regression models for two key blast outcomes: airblast level (dBL) and fragmentation median size (X_{50}, mm). Explainability is provided using SHapley Additive exPLanations (SHAP), enabling both global identification of dominant drivers and local diagnosis of individual blast predictions (SHAP is an explainable AI technique that uses Shapley values to quantify how each feature contributes to a prediction relative to a baseline, enabling interpretation of both overall model behaviour and individual predictions).

DATA AND METHODS

A data set of 120 preprocessed production blasts was compiled from the Jwaneng Diamond Mine open pit operation in Botswana. Inputs represent routinely recorded blast design and rock descriptors, including distance to the monitoring point, charge per delay, spacing, burden, stemming length, hole depth, hole diameter, powder factor, blastability index and rock factor. Separate XGBoost regressors were trained for airblast and X_{50}. Hyperparameters were optimised using grid search with three-fold cross-validation, and generalisation was assessed on a held-out test set using coefficient of determination (R2), root mean square error (RMSE), mean absolute error (MAE), variance accounted for (VAF), and mean absolute percentage error (MAPE) (Table 1). To support transparent engineering interpretation, SHAP was applied at two levels: (i) global summary (beeswarm) plots that rank features by mean absolute contribution and show whether high or low feature values drive predictions, and (ii) local explanations that decompose a single prediction into feature-wise contributions for outlier investigation and corrective action. The modelling and explainability workflow follows established gradient-boosting and SHAP methodologies (Chen and Guestrin 2016; Lundberg and Lee 2017).

TABLE 1

Test-set performance of the optimised XGBoost models.

Output	R2	RMSE	MAE	VAF (%)	MAPE (%)
Airblast (dBL)	0.86	4.23	3.49	86.18	3.26
Fragmentation X_{50} (mm)	0.81	4.31	3.52	83.19	3.41

Model performance indicates strong predictive skill for airblast and practical accuracy for fragmentation, given the higher inherent variability of X_{50} in heterogeneous rock masses (Table 1).

RESULTS AND DISCUSSION

Figure 1 presents cross-validated parity plots, which show stable performance across folds and limited bias. The airblast model exhibits tight clustering around the 1:1 line, supporting confidence for compliance-related screening. The X_{50} model shows moderate dispersion consistent with fragmentation variability yet remains sufficiently accurate for design guidance and post-blast learning.

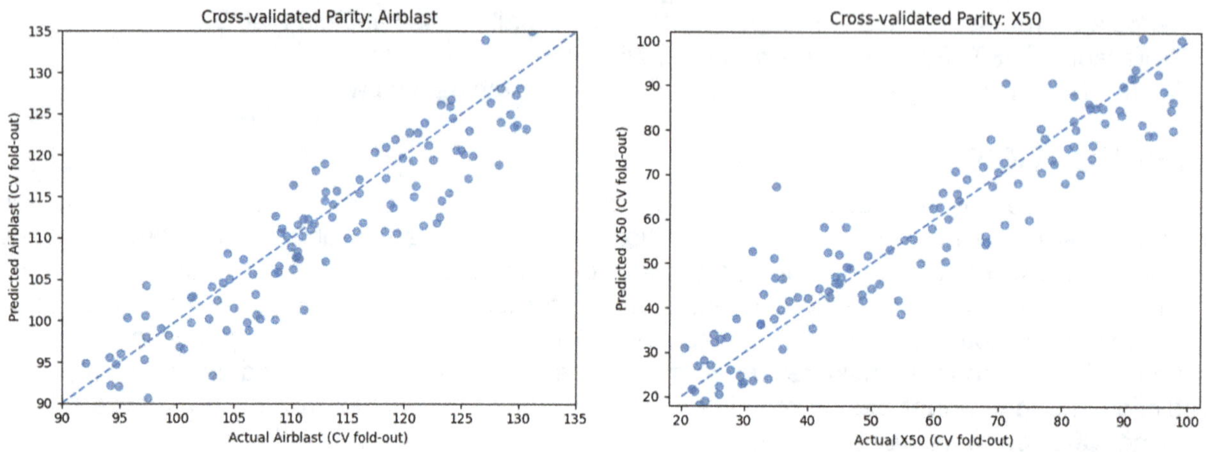

(a) (b)

FIG 1 – Cross-validated parity plots for model stability across folds. (a) Airblast parity; (b) X_{50} parity.

In Figure 2 for airblast, distance is the dominant contributor: larger distances reduce predicted airblast, while shorter distances increase it. Charge per delay and spacing are the next most influential parameters, reflecting the effects of instantaneous energy release and geometric confinement on airblast generation, while rock descriptors (blastability index and rock factor) show a secondary but consistent influence. For fragmentation, the most influential variables are charge per delay, blastability index, powder factor and spacing, where higher explosive energy per unit rock tends to reduce X_{50} and increased spacing generally trends toward coarser fragmentation if not compensated by energy distribution. Hole diameter remains an important contributor, reinforcing the need to co-optimise diameter with energy distribution and timing rather than treating it as an independent parameter.

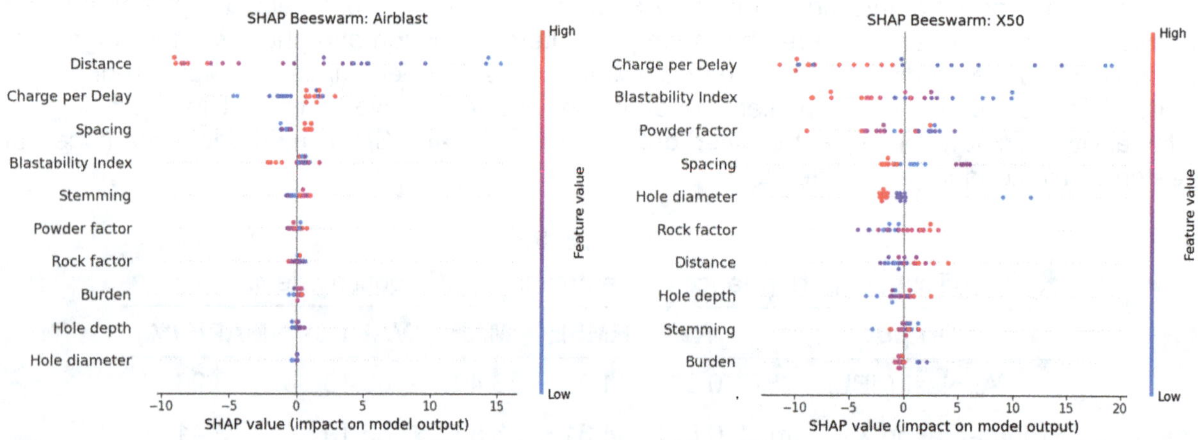

(a) (b)

FIG 2 – Global SHAP beeswarm plots showing feature influence for (a) airblast SHAP summary; and (b) X_{50} SHAP summary fragmentation.

Figure 3 shows SHAP waterfall plots illustrating how individual blast design parameters cumulatively contribute from the baseline to the final predicted airblast and fragmentation (X_{50}) outcomes for a representative blast, highlighting the dominant drivers responsible for deviations from the mean response.

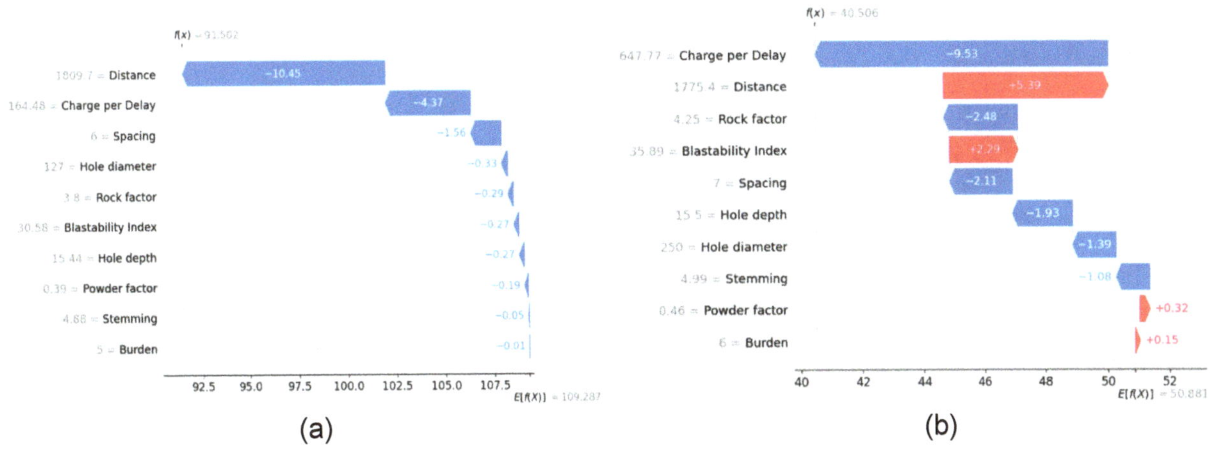

FIG 3 – Local SHAP waterfall plots showing feature influence for (a) airblast waterfall; and (b) X_{50} waterfall fragmentation.

PRACTICAL IMPLICATIONS

The combined accuracy and explainability support routine engineering use cases. First, the airblast model can screen proposed designs for elevated risk at specific monitoring distances, supporting proactive compliance management and improved community outcomes. Second, the X_{50} model can guide fragmentation adjustments by linking predicted changes to interpretable levers (energy per delay, powder factor and geometry), enabling targeted trials rather than broad trial-and-error. Local SHAP explanations provide a structured method for investigating anomalous blasts by identifying the small set of variables that shifted a prediction away from fleet-level trends, supporting corrective action such as stemming adjustments, spacing reviews, or energy redistribution. Operational implementation is the next phase of the work; however, the results identify the parameters most suitable for targeted field trials and design refinement.

CONCLUSIONS

Interpretable gradient-boosted tree models provide a practical pathway to embed ML into day-to-day blast planning while preserving engineering accountability. Using production blast data, XGBoost achieved an R2 of 0.86 for airblast and 0.81 for fragmentation X_{50}, with low percentage errors. SHAP explanations quantified the dominant drivers for each outcome and clarified how incremental changes are expected to shift predicted results. The approach supports transparent, evidence-based decision-making and can be extended to inverse design and multi-objective trade-off analysis for simultaneous control of blast impacts.

ACKNOWLEDGEMENTS

The authors would like to thank Jwaneng Diamond Mine for supplying the data for this research and Debswana Diamond Company for funding it.

REFERENCES

Chen, T and Guestrin, C, 2016. XGBoost: a scalable tree boosting system, in *Proceedings of the 22nd ACM SIGKDD International Conference on Knowledge Discovery and Data Mining*, pp 785–794 (ACM: New York).

Lundberg, S M and Lee, S-I, 2017. A unified approach to interpreting model predictions, in *Proceedings of the 31st International Conference on Neural Information Processing Systems*, pp 4765–4774 (Curran Associates: New York).

Optimised machine learning blast design generator with vibration constraints in open pit mine operation

R Supriatna[1] and S Raval[2]

1. The School of Minerals and Energy Resources Engineering, University of NSW, Sydney NSW 2052. Email: r.supriatna@unsw.edu.au
2. MAusIMM, The School of Minerals and Energy Resources Engineering, University of NSW, Sydney NSW 2052. Email: simit@unsw.edu.au

ABSTRACT

Effective bench planning is an essential parameter to achieve productivity, safety, and environmental compliance in the early stage of open pit mine planning. Every mine site establishes its own design criteria based on operational priorities. While safety is universally critical, some sites place more detailed focus on vibration control due to proximity with critical infrastructure, sites, or residential areas. While many engineering software packages now provide automation to generate a bench plan, they often lack integrated consideration for the potential vibration generated from the blast.

Peak Particle Velocity (PPV) is a primary metric used to evaluate the impact of blasting activities on nearby receptor sites Key blast design parameters such as burden, spacing, hole depth, and diameter were used to train regression models capable of accurately estimating PPV levels. Building on this predictive model, a multi-objective optimisation algorithm is built to recommend blast designs that minimise PPV while considering spatial and operational constraints. A custom scoring function was constructed using weighted factors, including proximity to vibration receptors, ore and waste classification, and total blast tonnage. Polygon clusters were generated using hierarchical clustering with shape control, and parameter optimisation was performed using simulated annealing to explore high-performing blast configurations under real-world constraints.

Results demonstrate that integrating optimised machine learning with spatial analytics and metaheuristic search algorithms enables more consistent compliance with vibration limits, while also improving fragmentation and diggability. The framework provides a scalable decision-support tool for engineers to evaluate bench plans that are both operationally effective and contextually aware of vibration impact in mine operations.

INTRODUCTION

Blasting is a critical operation in open pit mining, which significantly influences overall mine productivity through fragmentation. However, it has potential to generate significant environmental and social impacts if not managed properly, particularly in regard to ground vibration. Among various metrics, Peak Particle Velocity (PPV) is widely accepted as the standard measure for evaluating the impact of blasting vibrations on nearby structures, ecosystems, and culturally sensitive areas.

In regions where mining operations interface with residential areas and important significant infrastructure, the challenge becomes more complex. Although safety remains a universal priority in blast design, some mine sites must also prioritise vibration control as a key operational constraint. These site-specific sensitivities demand more adaptive and spatially aware design approaches.

Modern engineering software has made significant effort in automating bench design and blast planning. For example Orica (2024) BlastIQ, Maptek (2024) BlastLogic, and Dassault Systemes' (2024) GEOVIA products. The JKTech (2024) Technical Bulletin introduces the Mining Polygons Generator Tool, a digital solution that automates the creation of blast and excavation polygons from block models with emphasis in ore control. Most commercial tools focus on geometric, geotechnical, or production-driven parameters, and rarely incorporate spatial heritage constraints or localised vibration sensitivity into their algorithms. This creates a technical gap for operations situated near sensitive heritage sites, where conventional blast design may miss the trigger exceedance of site-specific PPV thresholds.

LITERATURE REVIEW

Over the past four decades, extensive study has been devoted to optimising open pit mine design and production scheduling through mathematical programming and heuristic algorithms. Early foundational work by Gershon (1983) and Ramazan (2001) laid the groundwork for applying large-scale linear and tree-based scheduling models to enhance long-term productivity and resource utilisation. Mixed Integer Linear Programming (MILP) is also commonly used to improve precision and scalability, particularly in large operations (Askari-Nasab *et al,* 2010; Pourrahimian, Askari-Nasab and Tannant, 2012).

Efforts to improve computational efficiency have also led to the use of aggregation techniques (Weintraub *et al,* 2008; Zipkin, 1980) and the development of hierarchical frameworks that balance model fidelity with solution speed (Newman and Kuchta, 2007). These optimisation models have been extended to complex mining contexts such as oil sands (Ben-Awuah and Askari-Nasab, 2011) and underground block caving (Pourrahimian, Askari-Nasab and Tannant, 2012). This shows the versatility of the hierarchical framework to resource operation such as open pit mining.

A significant stream of literature has focused on machine learning and heuristic methods such as genetic algorithms (GA) and clustering techniques implementation for mine planning and grade control. Samanta, Bandopadhyay and Ganguli (2005) applied GA for optimising grade control in bauxite, while Zhang (2006) combined GA with topological sorting to optimise open pit sequences. Further work by Tabesh and Askari-Nasab (2011) and Eivazy and Askari-Nasab (2012) demonstrated the application of clustering (including hierarchical and tabu search-based methods) for block aggregation and short-term planning. These methods primarily target grade control, dig limit optimisation, or equipment productivity, rather than environmental or socio-cultural considerations.

Clustering and pattern recognition techniques have been used in other domains, with algorithms such as improved particle swarm optimisers (Feng *et al,* 2010), normalised Minkowski metrics (Pham, Dimov and Nguyen, 2006), and noise-resistant variants of K-means (Barca and Rumantir, 2007). These have yet to be widely adopted in blasting or vibration-sensitive planning in mining. A gap persists in the integration of machine learning and spatial optimisation for vibration analysis, particularly in the context of sensitive heritage zones. Most existing models assume uniform risk tolerance across the mine environment and are not equipped to adapt to localised constraints such as PPV thresholds near culturally protected areas. The current study addresses this limitation by combining PPV-focused supervised learning with spatial clustering and optimisation to generate heritage-aware blast polygons.

A range of commercial software platforms have been developed to assist with blast design, drill and blast management, and short-term mine planning. Tools such as Orica (2024) BlastIQ, Maptek (2024) BlastLogic, and Dassault Systèmes' (2024) GEOVIA Minex and Surpac offer capabilities for integrating geological models, designing drill patterns, tracking blast performance, and supporting production reporting. These systems have significantly advanced the automation and digitisation of bench-level planning, with features such as automated burden and spacing calculations, fragmentation prediction, and tie-up visualisation.

However, despite these technological advancements, most of these platforms are primarily designed with a focus on geotechnical efficiency, fragmentation quality, and production throughput. They do not natively account for site-specific socio-environmental constraints, such as proximity to heritage zones or localised PPV risk management, which are increasingly important in regions with stringent cultural heritage obligations. For instance, while GEOVIA includes some environmental modules and BlastIQ can integrate sensor feedback, the blast optimisation algorithms remain largely production driven.

In response to these limitations, JKTech (2024) introduced the Mining Polygons Generator Tool, which enables automated delineation of excavation and blast polygons directly from block models. This tool improves consistency in blast planning by reducing manual interpretation; however, it also does not integrate heritage avoidance buffers, vibration prediction models, or AI-driven parameter tuning based on empirical PPV data. Thus, while commercial platforms support efficient execution, there remains a significant opportunity to augment them with machine learning and spatial optimisation algorithms that actively account for cultural and environmental sensitivities in blast design.

Supervised learning model

Random Forest Regressor

The Random Forest Regressor (RFR) is an ensemble-based supervised learning algorithm that constructs multiple decision trees during training and outputs the mean prediction of all individual trees. Each tree is trained on a random subset of the data set and a random subset of features, which reduces overfitting and improves generalisation. This ensemble approach allows the model to capture nonlinear relationships and complex feature interactions, making it highly effective for regression tasks with noisy or heterogeneous data — such as predicting PPV, where multiple interdependent blast parameters contribute to the outcome.

In this study, the RFR model was applied to learn the relationship between design parameters (burden, spacing, hole diameter, and depth) and the resulting PPV values. Its key advantage lies in interpretability — the model provides feature importance scores that highlight which parameters most influence PPV. This property is particularly useful in blast design optimisation, where engineers must understand the relative effects of each variable on vibration intensity. The RFR also demonstrated robustness to outliers and variable correlations, making it a reliable baseline model for vibration prediction in heritage-sensitive mining environments.

XGBoost (Extreme Gradient Boosting) XGBoost, or Extreme Gradient Boosting, is an advanced ensemble learning algorithm based on gradient-boosted decision trees. Unlike Random Forest, which builds trees independently, XGBoost constructs trees sequentially, where each new tree corrects the errors of the previous ones. The model optimises a differentiable loss function using gradient descent, with built-in regularisation terms to prevent overfitting. This structure makes XGBoost exceptionally efficient and powerful for structured tabular data — especially when relationships between variables are nonlinear and include subtle interactions, as is the case with blast vibration modelling.

In this study, XGBoost was used to enhance predictive accuracy for PPV estimation by exploiting its ability to handle large data sets and fine-tune model complexity through hyperparameters such as learning rate, tree depth, and regularisation coefficients. The algorithm's capacity for parallel computation and feature weighting provided fast convergence and strong generalisation on unseen blast records. Moreover, XGBoost's explainability via SHAP (Shapley Additive Explanations) allowed detailed interpretation of each feature's contribution to PPV, offering valuable insights for decision-making in vibration-sensitive blasting operations.

Multilayer Perceptron (MLP) neural network

The Multilayer Perceptron (MLP) is a type of feed forward artificial neural network that learns complex, nonlinear relationships by passing input data through multiple layers of interconnected neurons. Each neuron applies a weighted transformation followed by a nonlinear activation function, enabling the model to capture higher-order feature interactions that simpler models may overlook. MLPs are especially suitable for regression problems where the relationship between inputs and outputs is highly nonlinear and multidimensional, such as the interaction between blast geometry, explosive energy, and ground vibration response.

In this study, the MLP was implemented to test the capacity of deep learning approaches in predicting PPV from empirical blast parameters. The model was trained using backpropagation and stochastic gradient descent, with hyperparameters such as the number of hidden layers, neurons per layer, learning rate, and an activation function optimised for best performance. Although neural networks generally require more data and careful regularisation, the MLP's ability to generalise nonlinear dynamics between design variables and vibration outcomes made it a valuable component of the model comparison. Its results provided a benchmark for evaluating how well deep learning can complement traditional ensemble methods in blast design automation.

Support Vector Regression (SVR)

Support Vector Regression (SVR) is an extension of the Support Vector Machine (SVM) algorithm for regression tasks. It operates by finding a hyperplane that best fits the data within a defined margin of tolerance (epsilon), rather than minimising the absolute error of predictions. By using kernel functions such as the radial basis function (RBF), SVR can efficiently model nonlinear relationships between input features and the target variable. This makes it well-suited for applications like PPV prediction, where the interaction between design parameters and vibration response is highly nonlinear and site dependent.

In this study, SVR was used as a comparative algorithm to evaluate performance against ensemble and neural network models. Its strength lies in its robustness to high-dimensional feature spaces and its ability to perform well even with smaller data sets, a common constraint in mine vibration records. By tuning the regularisation parameter (C), kernel type, and epsilon value, the SVR model achieved smooth and stable PPV predictions without overfitting. Although computationally more intensive than tree-based models for large data sets, SVR provided a reliable benchmark for understanding how classical kernel-based learning techniques handle vibration modelling in complex mining environments as discussed in Table 1.

TABLE 1

Supervised learning model in comparison.

Random Forrest Regressor	XG-Boost	Multilayer Perceptron	Support Vector Regression
Predicts PPV by averaging many simple decision trees, so it can learn messy, non-linear relationship in blast data	**Predicts PPV by building a decision trees one after another – each new tree will fixes mistakes of the last**	**Predicts PPV by sending inputs through several "hidden" layers**	**Predicts PPV by fitting a best fit line with tolerance band with smaller datasets**
● Why it works? It trains different random samples, then **averages** them to reduce overfitting and handles data noices	● It uses **gradient boosting** with control to avoid overfitting	● It uses **gradient boosting** with control to avoid overfitting	● Find a function that keeps most points within margin
● What it learns? It reports **feature importance**	● Can handle large, tabular datasets, and **tricky nonlinear interactions** between inputs	● Great for **highly nonlinear patterns** across many inputs that simpler model might miss	● Strong with **limited data** and many features
● Why using it? It's a strong, dependable baseline to **simplify correlated inputs**	● Show how each input pushes a PPV up or down	● Needs more data and **careful regularization**	● Smooth, **stable predictions,** although can be **slower** than the other 3 models

● Why it works? ● What it learns? ● Why using it?

PPV control

A PPV result depends on distance, charge per delay (Q^d), timing, blast geometry, and geological structure. Those parameters are non-linear and interaction-rich mapping. SVR with RBF kernel is purpose-built for this given its suitability for flexible functions without over-parameterising. Given the tabular, moderate-size blast data set and the need for auditable uncertainty at monitoring point, this study adopts SVR (RBF kernel) wrapped with Conformal Prediction as our primary baseline.

Study to develop PPV predictors from operational blast data using a wide panel of supervised learners and wraps them with Conformal Prediction to produce statistically valid uncertainty bands can address the practical need to quantify exceedance risk at regulated limits (Feliciano, 2025). Empirically, kernel methods (notably SVR) and strong ensembles achieve top accuracy, while SHAP analyses reaffirm physics-consistent drivers such as distance and charge, thus more credible for regulator-facing reporting (Feliciano, 2025).

This study diverges by moving from prediction to prescription through integrating PPV modelling into a full bench-planning pipeline with blast polygon generation (hierarchical clustering with shape controls), multi-objective optimisation (PPV risk, ore/waste, tonnage, access). This framework adds the spatial geometry, operational constraints, and decision-analytic layer needed to produce deployable, CAD-ready blast plans.

METHODOLOGY

PPV

PPV in blasting is dominated by the scaled-distance relationship: vibration decreases with greater distance to the monitoring point or receptor (R) and increases with higher maximum charge per delay (Qd/MIC). These two variables form the backbone of empirical PPV laws, so the first design instinct is to maximise stand-off to sensitive sites and keep the per-delay mass as low as practicable. However, this is not always viable due to ore recovery constraints or geological structure.

Figure 1 explains how a few design choices control how much energy hits the ground and where it goes. Put it simply, If you fire fewer holes at the same time(use longer delays between holes/rows), PPV goes down. Using smaller hole diameters, lower-density explosive, or decked charges cuts the mass per delay, also lowering PPV near sensitive areas. Good stemming traps more energy in the ground (PPV up but air blast down). Finally, adjust burden/spacing, pattern orientation, and firing direction to give the rock more relief and to fire away from receptors. Those approaches can be adopted to reduce vibration at the receptor. However, site limitation, such as drill availability, drill size, or even charging capacity may limit a sites ability to implement these measures as indicated by the flowchart in Figure 1 where the production schedule has options to be flexible or not flexible due to a firm target set during the pre-schedule or pre-planning.

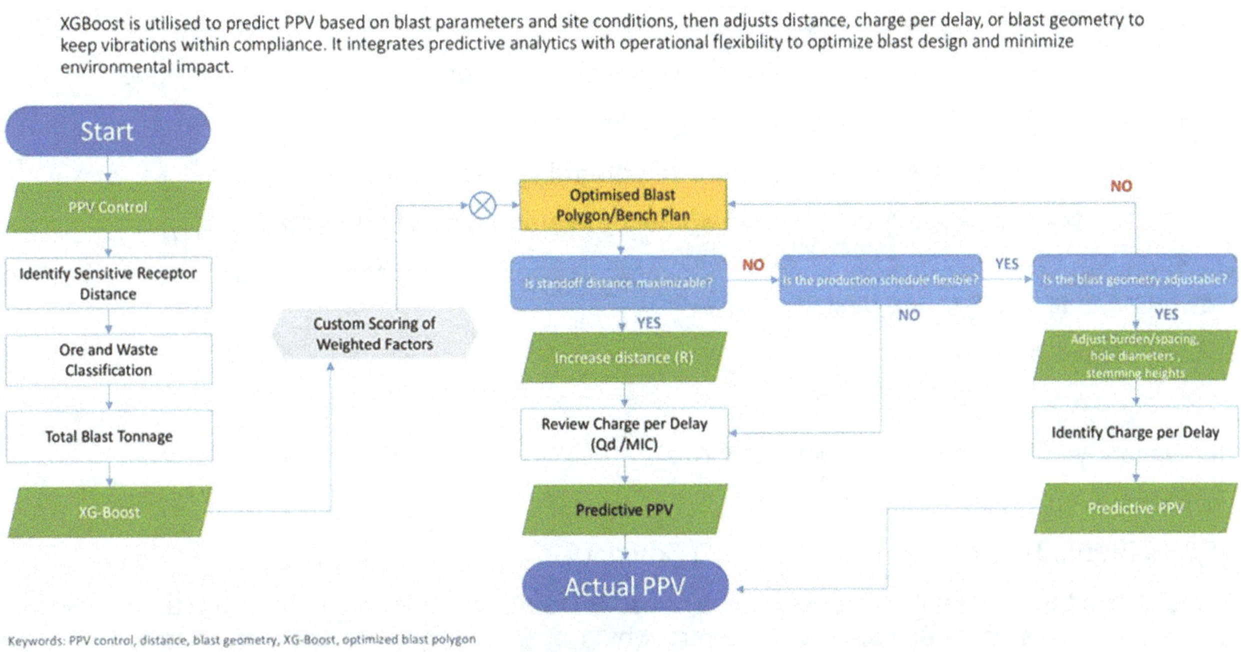

FIG 1 – XG boost flow chart to generate custom scoring and predictive PPV.

Custom scoring of weighted factors

The core contribution of this paper is a custom multi-objective scoring function that converts blast design parameter into a single optimisation target while maintaining safety limit (PPV value between 25 mm/s to 50 mm/s, to maintain vibration impact) and operability constraints (blast radius to sensitive receptor).

For a candidate blast design (x), the score is:

[J(x)=w_v S_v(x)+w_r S_r(x)+w_m S_m(x)+w_t S_t(x)+w_o S_o(x)]

where lower (J(x)) indicates an optimum design parameter

(S_v): vibration compliance (mm/s)

(S_r): receptor sensitivity/proximity penalty

(S_m): material-priority term (ore/waste)

(S_t): tonnage alignment rating

(S_o): operational practicality rating (polygon shape)

Weights satisfy (\sum w_i=1).

In order to enforce regulatory compliance, vibration exceedance is penalised nonlinearly by using the below formula:

[S_v(x)=\left[\max\left(0,\frac{\widehat{PPV}(x)-PPV_{lim}}{PPV_{lim}}\right)\right]^2]

This approach will discourage designs near or above threshold. Receptor-sensitive zones are handled through (S_r) using distance and rating multiplier. Thus, giving higher penalty for residential/critical infrastructure receptors. The custom score is then minimised using simulated annealing, producing ranked feasible designs that explicitly reflect site priorities, whether it is polygon size to achieve production target, bench turnover time, or distance to sensitive receptor.

Simulated annealing optimisation

Optimisation is performed to seek the best feasible design vector for each blast polygon cluster (c), [x=[B,S,H,D,Q_d,\ldots] \in \Omega_c] where (B)=burden, (S)=spacing, (H)=hole depth, (D)=diameter, (Q_d)=charge per delay, and (\Omega_c) is the feasible set defined by geometry, equipment, and safety constraints.

The objective is the weighted score:

[\min_{x \in \Omega_c} J(x)=\sum_i w_i \tilde S_i(x),\quad \sum_i w_i=1; w_i\ge0]

with normalised penalty components (\tilde S_i(x)\in[0,1]).

At iteration (k), a neighbour (x') is generated by perturbing 1–3 decision variables:

[x'=\text{Repair}!\left(x + \delta_k \odot \eta\right),\quad \eta_j\sim U(-1,1)]

where (\delta_k) is step size (reduced as temperature cools), and Repair(.) projects values back to allowed bounds and discrete design sets.

Define (\Delta J = J(x')-J(x)).

Acceptance rule:

If (\Delta J \le 0): accept (x') (improving move)

If (\Delta J > 0): accept with probability

[P(\text{accept})=\exp!\left(-\frac{\Delta J}{T_k}\right)]

To make this formulation deployable on-site constraint setting is imperative. Constrain handling will enable the algorithm to eliminate parameters which are not align with the site specification. For example distance to receptor, bench boundary (which will govern by approval boundary and ultimate pit design), and drill limitations (hole diameter and hole depth).

Constraint handling

To maintain an objective weighting factor, two sets of constraints are applied:

1. Hard constraints (safety distance, max hole depth, bench boundary, drill limits): candidate rejected or repaired before scoring.
2. Soft preferences (shape boundary, tonnage alignment): included in (J(x)) via normalised.

In simulated annealing, each design candidate is checked against constraints before it can compete. Hard constraints will be strictly enforced by rejecting invalid candidates or repairing them (for example, clamping variables back to allowed ranges), while soft constraints (like tonnage closeness or polygon shape preference) are converted into penalty terms added to the weighting score.

RESULTS

Blast polygon optimisation

The blast polygon is determined using a script. First, the feasible search space is defined as the intersection of the ultimate pit design and the clean-up area prepared by production, so the algorithm never proposes polygons outside the area which has been cleaned or prepared for drilling. Within that space, hard constraints are enforced: the polygon must stay fully inside the feasible boundary, remain contiguous, and satisfy the production target of 120 000 t (to satisfy minimum four days of production mining of typical EX3600 digger).

The feasible area is discretised into small cells, and each cell is assigned estimated tonnage from area, bench height, density, and recovery factor. A weighted objective is then evaluated for candidate polygons: ($J = w_{ton} \cdot |T-T_{target}|/T_{target} + w_{shape} \cdot (1-\text{compactness})$), where compactness is treated as a soft shape-quality constraint. Simulated annealing is used to explore the solution space through controlled add/remove cell moves which then allows broad exploration early and gradual convergence as confidence decreases. Any shape that violates hard constraints is rejected immediately, thus ensuring only executable polygon shapes are considered.

This approach produces a blast polygon that is both executable and optimised for production targets. The hard constraints guarantee compliance with ultimate pit design, preparation status (area clean up), and tonnage requirements, while the soft compactness term improves shape practicality for drilling and blasting implementation. Such as allowing a staggering pattern to improve blast throw and material movement and also allowing for minimum linear intersection between neighbouring polygons to minimise bridge drilling and post blast operation. The output includes the selected polygon geometry and traceable metrics (estimated tonnage, tonnage error, shape continuity, and final objective score) as shown in Figure 2.

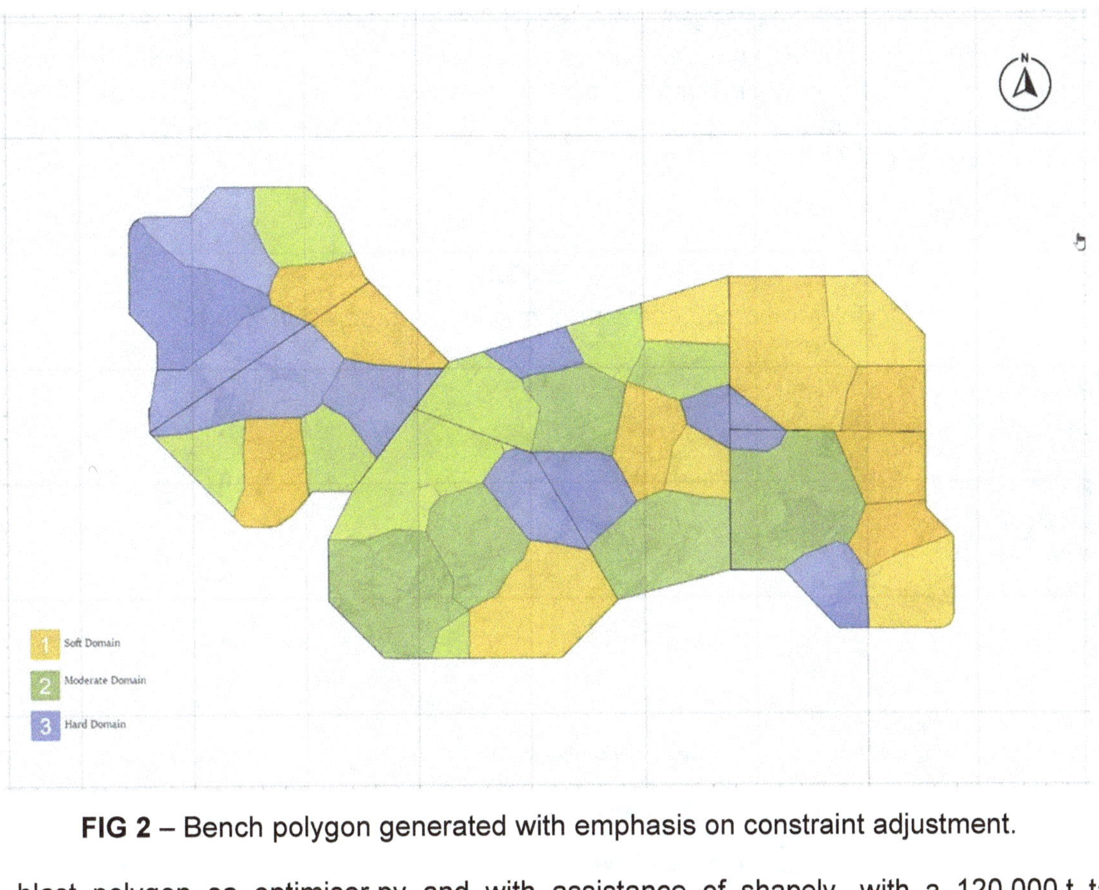

FIG 2 – Bench polygon generated with emphasis on constraint adjustment.

Using blast_polygon_sa_optimiser.py and with assistance of shapely, with a 120 000 t target, ±3 per cent tonnage tolerance, and weighted objective terms of 0.70 (tonnage alignment) and 0.30 (shape compactness), the simulated annealing workflow generated an optimised blast polygon (blast_polygon_opt.geojson) that satisfies all hard constraints (inside ultimate pit design, inside clean-up area, and contiguous geometry). The selected solution demonstrates an operationally executable and align with production target as below.

Design parameters and PPV performance

The baseline variability is a useful signal for model learning, but this also required a careful split strategy to avoid leakage and overfitting. Leakage means that the model overestimates the result, while overfitting means that the model underestimates the result that could be optimum but discard it. This split strategy is done through iterating a few design parameters based on iterative scenarios. To ensure applicability, the design variables were bounded using an empirical approach. Burden and spacing were constrained to 3.5–8.0 m to represent hard, medium, and soft geology domains (with lower values typically applied in harder ground and higher values in softer ground). Bench height was limited to 4–10 m, consistent with common open pit production bench configurations. Hole diameter was restricted to 127–250 mm to match typical production drill-rig capability. These limits were implemented as hard constraints in the optimisation so that all generated designs remain practical.

Optimisation was executed at blast-polygon cluster level with hard bounds of (B,S \in [3.5,8.0]) m (domain-conditioned), (H_b \in [4,00]) m, (D \in [127 250]) mm, and minimum receptor distance (d \ge 300) m. The weighted objective (\min J(x)=\sum_i w_i \tilde S_i(x)) was evaluated only for feasible candidates, ensuring that all recommended designs satisfied geometry, drill specification, and PPV limit before scoring. This produced executable output. Using the constraint sensitive simulated annealing framework, the demonstration run soft domain, the best feasible solution was (B=7.10) m, (S=7.20) m, (H_b=8.00) m, (D=165) mm, and (Q_d=179.59) kg/delay, with total score (J=0.1323). The predicted PPV was (7.08) mm/s (below the 10 mm/s limit), giving zero exceedance penalty ((\tilde S_v=0)). Component scores were (\tilde S_r=0.6071), (\tilde S_m=0), (\tilde S_t=0.0002), and (\tilde S_o=0.1080), while estimated tonnage was 119 996.26 t (target 120 000 t).

In the medium domain, the best feasible solution was (B=5.00) m, (S=5.70) m, (H_b=8.00) m, (D=127) mm, and (Q_d=218.55) kg/delay, with total score (J=0.1323). The predicted PPV was (9.28) mm/s against a 10 mm/s limit, yielding a zero-exceedance penalty ((\tilde S_v=0)). Component scores were (\tilde S_r=0.6071), (\tilde S_m=0), (\tilde S_t=0.0005), and (\tilde S_o=0.1080), while estimated tonnage was 120 010.62 t (target 120 000 t).

The results are shown in Table 2.

TABLE 2

Model iteration outcome through weighted objective (J(x)).

Hard Constraint

Non-negotiable, accept or rejecting assumption

- Burden and Spacing range (3.5m – 8m)
- Bench Height (4m or 8 m)
- Drill rig specification (127-250mm)
- Distance to receptor >= 300m
- 10mm/s PPV

Soft Constraint

Preference based and ranking feasible options through weighted penalties

- Shape compactness
- Tonnage alignment
- Polygon contiguity

Design Parameters	Soft	Moderate	Hard
Burden (m)	7.1	5.0	3.7
Spacing (m)	7.2	5.70	4.9
Bench Height (m)	8	8	8
Diameter (mm)	165	127	165
Charge per Delay (kg)	179.59	218.55	80.06
Weighted Objective	0.1323	0.1323	0.1323
Estimate tonnage (ton)	119,996.26	120,010.62	187,334
Predicted PPV (mm/s) Hard limit of 10mm/s	7.08	9.28	4.9

Design parameters and PPV performance

In this iteration, the objective scores ($J(x)$) are similar because several penalty terms were intentionally held constant (same receptor distance/class, same ore/waste match, similar tonnage alignment, and identical shape compactness), and all three designs stayed below the PPV limit, yielding ($S_v=0$). In real operations, these inputs vary across benches and polygons. For example, the receptor distance changes, the polygon compactness changes due to clean up availability and bench turnover, tonnage alignment also changes to follow mine plan target, and PPV may approach or exceed limits, so each component ($\tilde S_i(x)$) changes, and the weighted sum ($J(x)$) differentiates designs more obviously. Adding richer site data (variable distances, distinct polygons, and actual production targets) naturally will produce distinct objective scores, which then impact the overall design parameters and vibration prediction outcomes.

Fragmentation and diggability factor

Fragmentation was evaluated using the Kuz–Ram approach to translate the optimised blast parameters into an expected size distribution after the blast. For each domain, the *in situ* volume per hole was estimated as ($V = B^* S ^*H_b$), and the powder factor as ($PF = Q_e/V$) (with (Q_e) taken as effective explosive mass per hole or per delay). The Kuznetsov mean fragment size was then computed in the standard form ($x_{50} = A,(V/Q_e)^{0.8}(115/RWS)^{0.19}$), where ($A$) is the site specific rock factor and (RWS) is the relative weight strength of the explosive. The size distribution was described using the Rosin–Rammler relationship ($P(<x)=1-\exp[-(x/x_c)^n]$), with (x-c) tied to (x_{50}) and (n) reflecting pattern regularity and timing uniformity. This formulation is intentionally site adaptive: key inputs such as the rock factor (A) and explosive strength (RWS) are not fixed constants but are calibrated from local geology data and explosive product data. That means the same Kuz–Ram structure can be implemented to different lithologies, explosives, and operational practices without changing the model logic. This flexibility makes the approach robust since it remains valid across sites. Using this framework, the design parameter outcome was utilised as input to predict the fragmentation size.

Based on the indicative strand data (clay, massive rock, density 2.49 g/cc, ($E=12$) GPa), the Kuz–Ram rock factor was calculated as ($A=3.98$) using the standard formulation ($A=0.06(RMD + RDI + HF)$), with ($RMD=50$) for massive rock, ($RDI=12.25$) from density, and ($HF=4.0$) from Young's modulus. Adopting an ANFO baseline for explosive strength (($RWS=100$)) yields a constant explosive term ($(115/RWS)^{19/20}=1.142$).

Applying the updated Kuz–Ram equation ($X_m = A K^{-0.8} Q^{1/6}(115/RWS)^{19/20}$), with ($K=Q/V$) and ($V=B ^*S^*H_b$), produced the following mean fragment sizes: medium domain ($X_m=$ 11.54), hard domain ($X_m=$ 15.54 mm), and soft domain ($X_m=$20.85 mm. These results indicate that in the medium domain design, the fragmentation tends to be more finest. While the soft domain design might produces the coarsest fragmentation due to its larger burden and spacing and lower specific charge.

While diggability is directly influenced by fragmentation distribution, oversize percentage, and muck pile looseness, all of which determine how the digger or excavator load the blasted material. In general, finer fragmentation improves bucket fill and reduces refusal, while coarser fragmentation increases digging resistance and may require secondary breakage. The Kuz–Ram results therefore provide a practical proxy for assessing likely diggability across domains, although they do not explicitly model equipment response or loading cycle performance. Sometimes if the material is too fine, the digger will expect some challenge as the material is too loose to be dug productively.

It is important to note that the optimum fragmentation target for a specific power shovel or digger is not defined in this study. Equipment specific fragmentation thresholds depend on bucket geometry and operational practices, such as operator skills. Thus should be derived from site production data and shovel and operator performance monitoring. As such, while the current framework ensures vibration compliance and provides relative fragmentation comparisons, it does not prescribe the exact fragment size distribution required for peak shovel efficiency.

Integrating diggability optimised fragmentation targets (eg target (P_{80}), oversize fraction limits, or shovel specific size windows) is identified as a future study focus. The study might Incorporate these equipment-based targets as an additional soft constraint or weighted objective would allow the optimisation framework to balance vibration, production, and excavation performance in a fully integrated decision-support system.

Sensitivity and trade-off analysis

Sensitivity and trade-off analysis was conducted to evaluate how optimisation outcomes respond to changes in objective weighting and constraint settings. The analysis emphasise on the weighted scoring function, where tonnage alignment and polygon shape compactness are balanced through **(w_{ton}) and (w_{shape}), with (w_{ton}+w_{shape}=1).** Multiple optimisation runs were performed across representative weight scenarios while also maintaining the same hard constraints: polygon inclusion within ultimate pit and clean-up area, contiguity, and target tonnage of 120 000 t within tolerance. This set-up isolates the effect of preference weighting on solution behaviour without changing executability rules. When tonnage weight was increased (eg (w_{ton}=0.8), (w_{shape}=0.2)), the optimiser consistently produced solutions with lower tonnage deviation from the target, but with reduced geometric compactness. These polygons tended to follow feasible boundary availability more aggressively to meet production target tonnage. The result of this is a slightly less regular shapes. In contrast, when shape weight was increased (eg (w_{ton}=0.5), (w_{shape}=0.5)), the optimiser favoured more compact and operationally cleaner polygons, with a modest increase in tonnage deviation (still within hard tolerance). This confirms the expected behavioural trade-off between production precision and geometric quality. A balanced weighting ((w_{ton}=0.7), (w_{shape}=0.3)) is proven to provide the most stable compromise for operational use. Under this weighting, tonnage outcome remained close to the 120 000 t target while maintaining acceptable compactness for practical drill pattern implementation. The sensitivity results indicate that the proposed framework is robust and flexible for different planning priorities. Where production delivery is critical, higher tonnage weighting can be applied; where execution simplicity and pattern quality are emphasised, higher shape weighting may be preferred. Because the hard constraints remain fixed, this tuning can be performed safely without compromising spatial compliance, target control, or most importantly vibration constrain. The analysis supports the use of weighted multi-objective optimisation as a transparent decision-support mechanism for blast polygon planning.

CONCLUSIONS

This study developed an integrated machine-learning optimisation framework for vibration-constrained blast planning in open pit mining, combining PPV prediction with spatial and operational optimisation in one decision pipeline. A supervised PPV model was used to estimate vibration response from controllable blast variables, and this predictor was embedded in simulated annealing to search feasible designs that minimise risk while preserving production target. The approach moves beyond prediction-only workflows by converting model outputs into prescriptive, executable design recommendations.

The script documented the spatial optimisation workflow for blast polygon generation, where the feasible search space was defined by the intersection of ultimate pit design and clean-up area. The polygon solver enforced hard constraints (inside approved boundaries, contiguity, and 120 000 t target within tolerance) and applied a soft shape objective using compactness. Documenting the design-parameter optimisation workflow, where burden, spacing, bench height, hole diameter, charge per delay, and receptor distance were optimised under domain-based and equipment constraints using the same simulated annealing logic to perform vibration result prediction.

The decision engine used weighted multi-objective scoring, (J(x)=\sum_i w_i \tilde S_i(x)), with penalties for vibration compliance, receptor sensitivity, material priority, tonnage alignment, and operational practicality, plus a practical rating form (\text{Rating}=100(1-J)). The optimum feasible solutions showed strong vibration compliance across geology domains: medium domain ((B=5.00, S=5.70 {m}, H_b=8.00 {m}, D=127 {mm}, Q_d=218.55 kg/delay})) with predicted PPV = 9.28 mm/s; soft domain ((B=7.10 {m}, S=7.20 {m}, H_b=8.00 {m}, D=200 {mm}, Q_d=179.59 kg/delay})) with predicted PPV=7.08 mm/s; and hard domain ((B=3.7 {m}, S=4.9 {m}, H_b=8.00 {m}, D=165 {mm}, Q_d=80.06 kg/delay})) with predicted PPV=4.90 mm/s. These outcomes were achieved with low objective scores ((J \approximately 0.1323), rating (\approximately 86.8/100)) and zero exceedance penalty in all three domains.

The core decision engine was the custom weighted scoring function, (J(x)=\sum_i w_i \tilde{S}_i(x)), with normalised penalties for vibration compliance, receptor sensitivity, material priority, tonnage alignment, and operational practicality. This was paired with a practical rating transform, (\text{Rating}=100(1-J)), to make outcomes easier for engineering interpretation and ranking. Hard constraints were enforced by reject/repair rules before scoring, while soft constraints were represented as weighted penalties, ensuring that optimisation remained both within vibration acceptance limit and production tonnage target.

Results and sensitivity analysis showed that the framework consistently produced feasible designs with compliant predicted PPV and near-target tonnage across hard, medium, and soft domains. Trade-off testing confirmed expected behaviour: higher tonnage weight improved production alignment, while higher shape weight improved geometric compactness and field practicality. A balanced setting ((w_{ton}=0.7; w_{shape}=0.3)) provided the most stable operating compromise, supporting the framework as a deployable decision-support tool for vibration-aware drill and blast planning.

ACKNOWLEDGEMENT

The author sincerely acknowledges Luke P (Senior analyst at Microsoft) for valuable mentoring in Python, Shapely, and script development. Appreciation is also extended to Leonard Rowe (Principal Drill and Blast, Newmont) for his practical feedback and support in assessing the executability of the recommended blast design parameters.

REFERENCES

Askari-Nasab, H, Frimpong, S, Awuah-Offei, K and Szymanski, J, 2010. Modelling open pit dynamics using discrete simulation, *International Journal of Mining, Reclamation and Environment*, 24(2):111–135.

Barca, C and Rumantir, G, 2007. Robust clustering methods for noisy data sets using modified K-means algorithms, *Pattern Recognition Letters*, 28(8):984–991.

Ben-Awuah, E and Askari-Nasab, H, 2011. Incorporating stockpile reclamation and material handling into oil sands mine planning optimization, *International Journal of Surface Mining, Reclamation and Environment*, 25(2):144–162.

Dassault Systèmes, 2024. *GEOVIA Mining Solutions,* Paris: Dassault Systèmes.

Eivazy, H and Askari-Nasab, H, 2012. A mixed integer linear programming model for short-term open pit mine production scheduling, *Mining Technology*, 121(2):97–108.

Feliciano, J, 2025. Predictive modelling of blast-induced ground vibration using supervised machine learning and conformal prediction, *International Journal of Rock Mechanics and Mining Sciences*, 182:1–15.

Feng, Y, Teng, G, Wang, A and Yao, Y, 2010. Chaotic inertia weight in particle swarm optimization, *International Journal of Computational Intelligence Research*, 6(2):145–152.

Gershon, M E, 1983. Mine scheduling optimization with mixed integer programming, *Mining Engineering*, 35(4):351–354.

JKTech, 2024. *Mining Polygons Generator Tool Technical Bulletin,* Brisbane: JKTech Pty Ltd.

Maptek, 2024. *BlastLogic Drill and Blast Management System,* Adelaide: Maptek Pty Ltd.

Newman, A and Kuchta, M, 2007. Using aggregation to optimize long-term production planning at an underground mine, *European Journal of Operational Research*, 176(2):1205–1218.

Orica, 2024. *BlastIQ Intelligent Blasting Platform,* Melbourne: Orica Mining Services.

Pham, D T, Dimov, S S and Nguyen, C D, 2006. Selection of K in K-means clustering, in *Proceedings of the Institution of Mechanical Engineers, Part C: Journal of Mechanical Engineering Science*, 220(1):103–119.

Pourrahimian, Y, Askari-Nasab, H and Tannant, D, 2012. A multi-step approach for block-cave production scheduling optimization, *International Journal of Mining Science and Technology*, 22(5):739–750.

Ramazan, S, 2001. Backflushing in block scheduling to improve the economics of open pit mining operations, *International Journal of Surface Mining, Reclamation and Environment*, 15(1):5–14.

Samanta, B, Bandopadhyay, S and Ganguli, R, 2005. Data mining applications in the mining industry, *International Journal of Mining, Reclamation and Environment*, 19(3):244–260.

Tabesh, M and Askari-Nasab, H, 2011. Two-stage clustering algorithm for block aggregation in mining applications, *Mining Technology*, 120(3):158–169.

Weintraub, A, Epstien, R, Morales, R, Serón, J and Traverso, P, 2008. Operational research models and the management of mining operations, *Annals of Operations Research*, 159(1):65–86.

Zhang, W, 2006. Optimization of open pit mine production scheduling using genetic algorithms and topological sorting, *Mining Science and Technology*, 16(4):405–412.

Zipkin, P H, 1980. Bounds for aggregating multi-location production models, *Operations Research*, 28(2):456–469.

Open Pit

People, capability and operational systems

Advanced mine planning for mineral sands

T S Baillie[1] and P Federici[2]

1. Technical Lead Open Pit Mining, AMC Consultants, Melbourne Vic 3000.
   Email: tbaillie@amcconsultants.com
2. Principal Mining Engineer, AMC Consultants Pty Ltd, Melbourne Vic 3000.
   Email: pfederici@amcconsultants.com

ABSTRACT

The mining of mineral sands presents a unique set of challenges that distinguish it from hard rock open pit operations. Unlike conventional truck-and-shovel mining, mineral sands operations are characterised by continuous mining methods, complex interactions between excavation, tailings deposition, progressive rehabilitation, and highly constrained spatial environments. These factors demand a fully integrated mine planning and scheduling approach that accounts for operational flexibility, backfill sequencing, and final landform design from the outset.

This paper examines the application of mine planning software tools – originally developed for other mining contexts – to model and manage these complexities in mineral sands operations. Using a series of case studies, the paper demonstrates how common mine scheduling, landform simulation, and environmental analysis tools can be adapted and combined to represent mining progression, tailings and overburden placement, and rehabilitation in a dynamic and internally consistent manner.

The result is a fully integrated, end-to-end workflow that captures the spatial and temporal interdependencies inherent in mineral sands operations. This approach not only enhances technical accuracy and operational realism but also improves communication between engineering, environmental, and closure teams.

Fully integrated mineral sands mine planning is achievable through innovative application of modern tools. Practitioners can move beyond siloed processes to a holistic approach that connects design, scheduling, cost estimation, and rehabilitation into a single framework. This not only reduces planning time but also positions operations for sustainable and efficient mine performance.

INTRODUCTION

Hard rock open pit operations are typically mined using flexible truck-shovel systems and generally unconstrained waste rock placement. This decoupling of mining and placement activities allow the ex-pit mining schedule to be completed first, with dump sequencing and rehabilitation modelled as a post-processed scenario from the mining schedule.

Mineral sands operations, however, often rely on continuous mining units such as dredges, dozer traps or bucket wheel excavators, and require progressive backfilling to maintain short haul/tailing distances and meet environmental and rehabilitation requirements. These conditions necessitate that tailings deposition, drying times, containment bund construction, and vertical sequencing dependencies are considered simultaneously when modelling the mining sequence.

Through this interdependent relationship, it can be surmised that any adjustment to the mining sequence will have a direct impact on the backfill sequence. Conversely, adjustments to the inputs related to backfill (such as tailings density) can materially impact mining rates and plant throughputs. The omission of any one of these components during the scheduling phase has the potential to compromise operational practicality and result in an unachievable mining schedule.

TERMINOLOGY

Throughout the mining industry similar terms are often used which may have different meanings based on the operation or context in which they are used. For the purposes of this paper to remove ambiguity, the following terms have been defined:

- Mining path – In a continuous mining method this represents the economic mining limit defined through an optimisation of the resource model or similar analysis.

- Mining blocks – Scheduling blocks within the mining path which have been appropriately designed to consider geotechnical, operational, and geological factors.
- Dump blocks – Blocks designed within the void left in a continuous mining operation. Typically, these are linked to the mining blocks to reflect operational constraints and can receive either tailings or overburden (or a combination of the two).
- Mining schedule/sequence – In the context of this paper the mining schedule refers specifically to the ex-pit mining operations and doesn't considering downstream activities like backfill.
- Backfill schedule/sequence – The sequence in which dump blocks (both tailings and overburden) are backfilled including material quality constraints that may exist.

The following is a list of general definitions relevant to this paper:

- Wet Concentrator Plant (WCP) – The primary separation facility in a mineral sands operation. The WCP receives feed and uses gravity-based processes to separate heavy mineral concentrate (HMC) from lighter sand fractions (tailings).
- Mining Unit Plant (MUP) – Used in dry mining operations, is a mobile or semi-mobile unit, positioned near the active mining location to minimise haulage distances. The MUP screens and pumps run-of-mine ore and pumps it as a slurry to the WCP.
- Deswik.Sched/'Sched' – The Deswik Gantt chart scheduling tool. Used to generate resource (mining equipment and activities) based, tactical level schedules modelling all key aspects of the mining process.
- Deswik.LHS/'LHS' – The Deswik landform and haulage simulation module. Scenarios are post-processed based on mining physicals generated in Deswik.Sched with outputs including haulage distances, energy consumption, equipment hours and dump decision-making according to input targets and constraints.
- Deswik.EnviroTools/'EnviroTools' – Deswik.EnviroTools is an integrated set of tools for landform engineering scenario analysis of surface water catchment, tailing, and waste flows.
- Overflows – A modelled haul path that does not have a defined, accessible or available dumping destination.

MINING BLOCK PREPARATION

The preparation of appropriately sized mining blocks within the defined economic mining limit is the foundation of any mining schedule, from open pit hard rock mining to stratigraphic style deposits such as those exploited in mineral sands operations. Mining blocks should consider the impacts of ore loss and dilution, geotechnical design parameters, and operational constraints applicable to the mining method proposed. An approach is outlined in the following case study.

Case study 1 – dredging operation

Case Study 1 is a dredging operation and details the process of generating detailed mining blocks for use in further scheduling.

For continuous mining operations it is useful to talk about the economic mining limits in terms of a mining path as opposed to the open pit shapes produced through a conventional Lerchs-Grossman (LG) or Pseudoflow optimisation. For this reason, these traditional approaches to open pit optimisation are more often used as an indicator of starting points and direction when determining an appropriate mining path for a mineral sands operation.

A common approach is to generate dummy drill holes across a grid covering the extents of the resource model. Block values within the resource model cells are then aggregated downhole to determine the point of maximum value (Figure 1). These points are then used to generate a base-of-ore surface to identify the economic limits of the mining path.

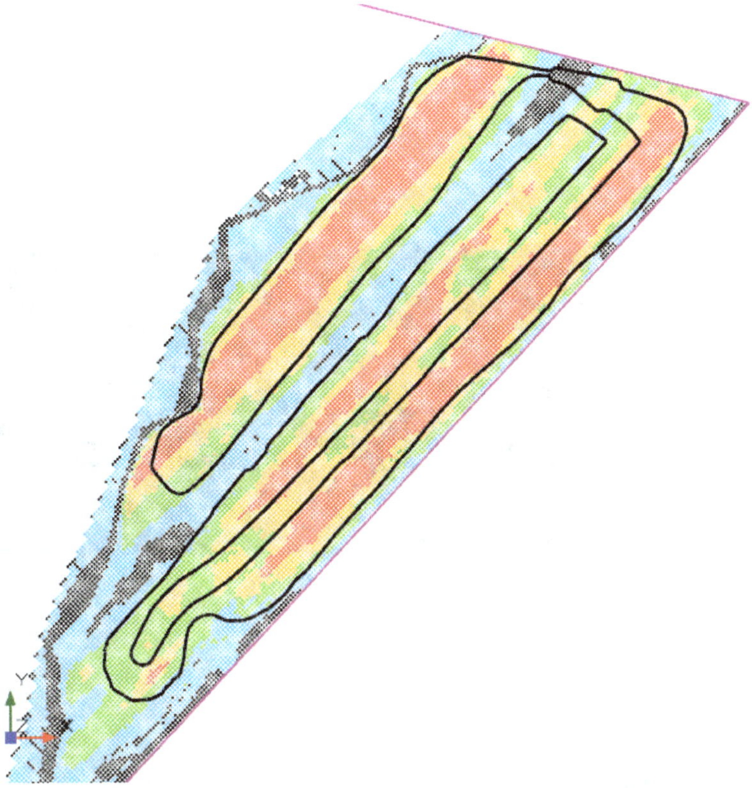

FIG 1 – Example of a heat map showing drill holes by value.

The practical execution of mining within these limits is then assessed according to the proposed mining method, in this case dredging. In a dredging operation it may not be possible to completely recover all mineralisation above cut-off grade within the economic mining limits due to the operational constraints of the dredge. The proposed dredge path is often designed manually, using tools developed in Microsoft Excel (or similar). This process is generally iterative and relies on the experience of the practitioner to maximise value while producing a mining path that is practical.

The mining path is then partitioned into mining blocks. The mining blocks are required to be appropriately sized to reflect the time periods to be scheduled (eg days, weeks, months etc) and the productivity of the resources deployed to them.

A common approach to creating mining blocks is using the 'cookie-cutter' method. Using this approach the mining path is segmented into blocks based on an orthogonal grid of a specified dimension. One of the key limitations of this approach is that by cutting the blocks orthogonally it can be difficult to honour mining direction accurately in a continuous mining operation. Additionally, the boundaries between mining blocks are cut vertically, ignoring both the operational constraints of the proposed mining method, and the geotechnical slope conditions applicable to the *in situ* material.

Using underground tunnelling tools in Deswik.CAD, AMC developed a novel approach that quickly and easily allowed users to create mining blocks directionally along the centreline of the mining path. Geotechnical slope inputs were also able to be defined between weathering surfaces.

This approach reduced manual intervention, improved spatial accuracy, and enhanced the ability to model progressive mining paths aligned with dredge or dozer trap systems. By modelling the underlying commands using Deswik Process Maps the process became repeatable and auditable.

An overview of the process is shown below:

- Step 1 – Layer structure was initialised with all relevant information. This included the topographic starting surface, mining path boundaries, and surfaces used to delineate material types with differing geotechnical characteristics (eg weathering surfaces).
- Step 2 – Mining block dimensions were defined using the 'Tunnel Sections' tool in Deswik.CAD (Figure 2). It was important that the defined width aligned with the maximum width of the mining path.

- Step 3 – The spacing between mining blocks was defined. The output of this step was a longitudinal centreline spanning the length of the mining path with points spaced according to the input spacing distance.

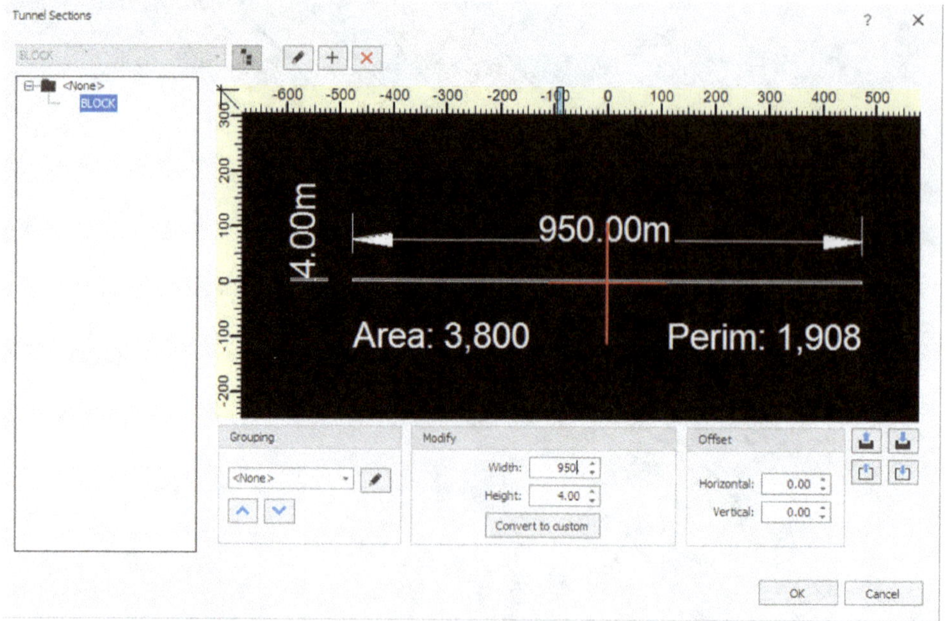

FIG 2 – Tunnel section input dialog window.

This spacing should be appropriate to the proposed scheduling period length. A small spacing can create excessive scheduling tasks inconsistent with the selected mining equipment and schedule accuracy requirements. Conversely, a large spacing can result in mining block grades being smoothed, not adequately capturing the variability of the feed grade.

- Step 4 – The tunnel section tool was applied based on the provided inputs to create 2D mining block polygons. These mining blocks were coded with a sequence attribute based on the mining path direction (Figure 3).
- Step 5 – Geotechnical design parameters were defined based on material types and zones. In this example, the top-of-ore surface was mapped to delineate different material types and different slope angles for defined geotechnical zones.
- Step 6 – A Process Map was set-up and run using the provided input information to progressively generate 3D mining block solids. Each mining block was cut in sequence against the depletion surface (eg topography, existing mining areas, material types) to create practical 3D mining block solids that honoured the mining direction and geotechnical design parameters (Figure 4).

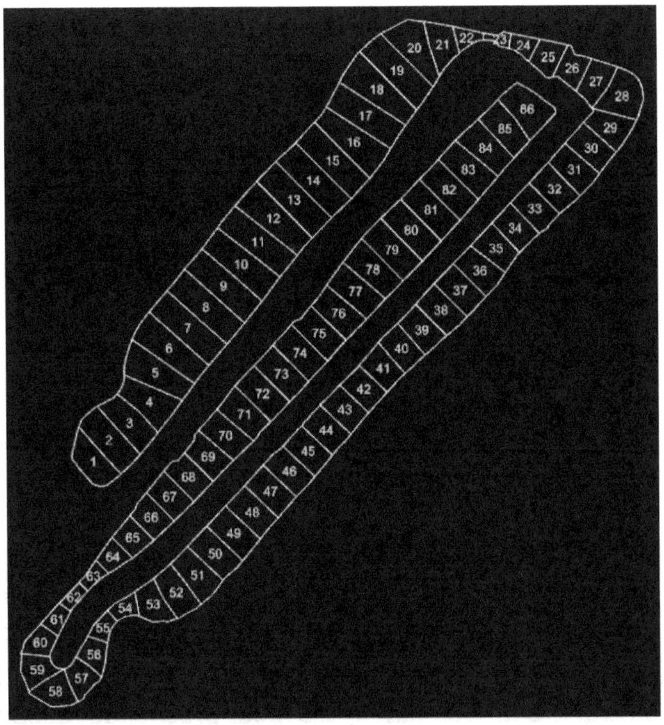

FIG 3 – Mining block polygons with sequence numbers.

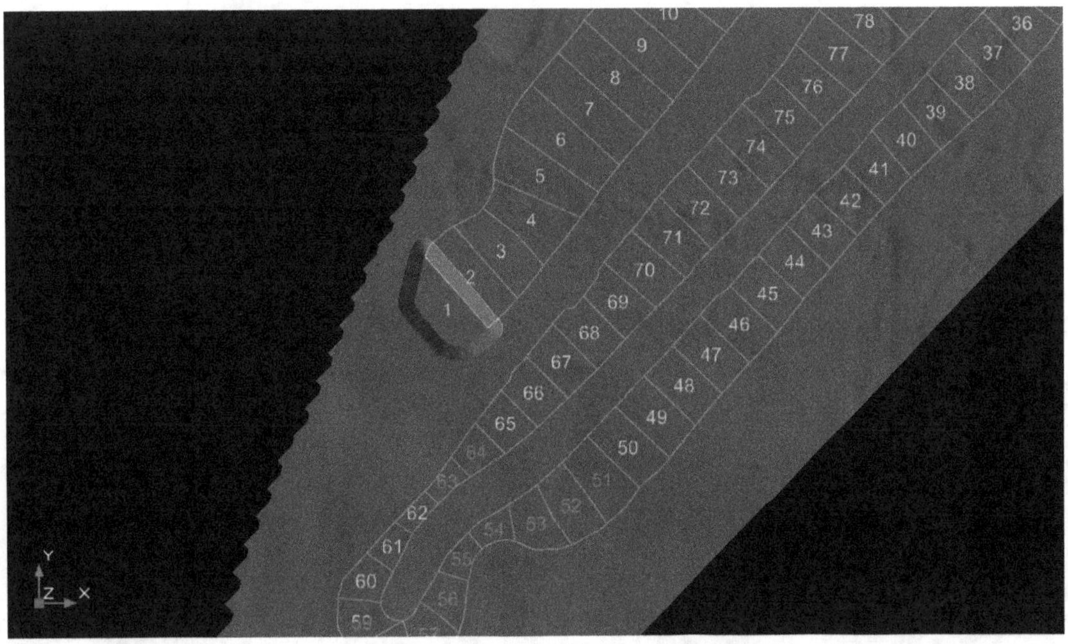

FIG 4 – Mining block solid progress plot after first block has been depleted.

Using Deswik Process Maps the approach developed by AMC is highly automated and requires minimal user intervention in the set-up and creation of mining blocks. This ensures that outputs are not only valid but importantly standardised, minimising the likelihood of user error and creating an audit trail of the process.

DYNAMIC BACKFILL MODELLING

In hard rock open pit operations, mining and backfill activities are typically able to be decoupled and assessed independently. For continuous mining operations, however, backfill modelling is one of the key integration considerations when completing a mine schedule.

If we consider the previous example of a dredging operation, the backfill and mining sequence defines the overall pond size which is generally a constraint on the mining schedule. If the pond becomes too large, the ability to discharge tailings from the WCP can become difficult due to the limitations of float lines, booster pumps etc. Conversely, if the pond is too small there may not be enough space for key infrastructure and dredge mobility could become restricted.

In a dry-mining setting, increased haul distances between active mining areas and backfill locations can have a large impact on project economics. Environmental constraints around maximum allowable disturbance zones are often also applicable and need to be considered.

Therefore, it is critical to consider backfill in the context of any proposed mining sequence in a continuous mining operation. Material changes to the mining sequence will typically result in an update of the backfill sequence which can be an iterative process if not able to be modelled dynamically.

A common approach to backfill modelling is to use spreadsheets employing heuristic methods to define a backfill sequence based on an input mining schedule and then design these dump blocks accordingly. This approach can be time-consuming and often very iterative.

The approach developed by AMC using Deswik.LHS enables a dynamic, visual, and repeatable process for simulating backfill placement dependencies, haulage routes, and sequencing. Progressive tailings and overburden placement can now be modelled simultaneously with mining progression, facilitating real-time evaluation of landform evolution and rehabilitation strategies.

Deswik.LHS is conventionally used to model truck haulage and waste rock placement scenarios in a hard rock open pit environment. Scenarios are post-processed based on the ex-pit mining schedule. The tool however has a high level of adaptability that allows for the modelling of non-trucked 'haulage' for example – tailings deposition.

AMC has developed a process where LHS can be used to accurately, and in real-time, model the entire mining cycle in a mineral sands operation discussed in the following case study.

Case study 2 – dry mining operation

Case Study 2 is a dry mining operation using a combination of a excavator and truck mining method coupled with a semi-mobile MUP for feed delivery to the WCP.

An overview of how the active mining and backfill areas were sequenced is shown in Figure 5.

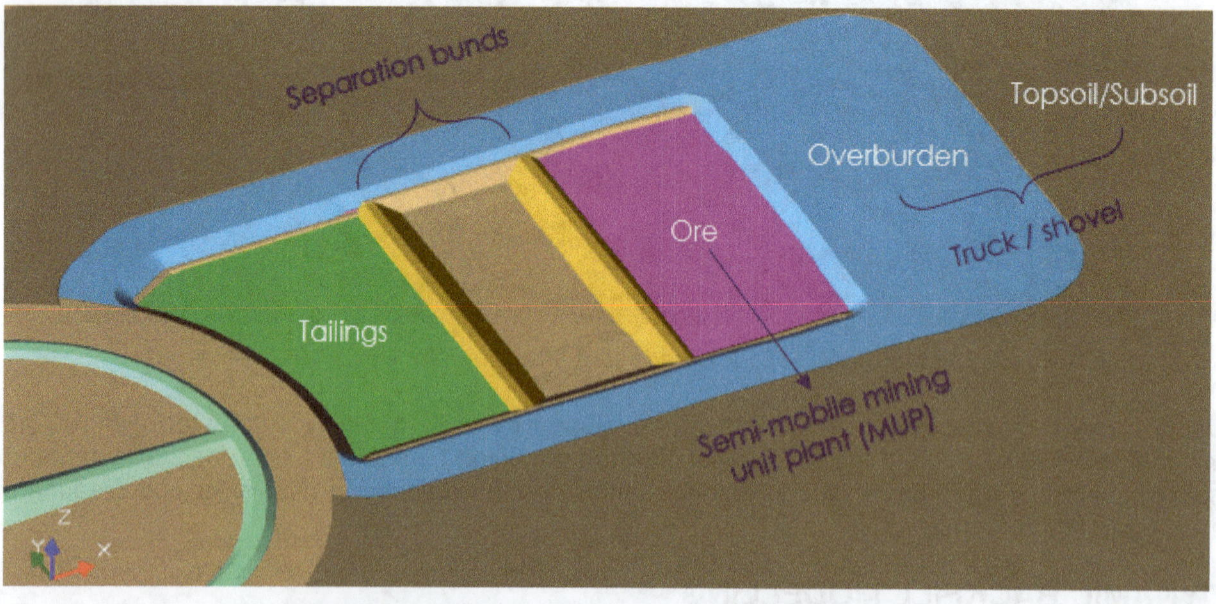

FIG 5 – Schematic of active mining area layout.

The overburden and tailings placement comprised of a combination of both backfill placed directly into the void created by mining ('on-path') and overburden stockpiles external to the mining path ('off-path'). To meet environmental requirements and minimise rehandle, a priority was placed on direct placement of tailings and overburden into on-path locations if available.

Overburden was first removed through conventional mining methods employing the use of excavators and trucks. Based on the overburden quality this material was sent to the following locations (in order of priority):

- Priority 1 – used to construct the embankment of the external tailings storage facility (TSF). This facility was used in the early-schedule periods when no on-path tailing was available.
- Priority 2 – used in the construction of separation bunds to delineate the tailings cells.
- Priority 3 – placed on top of filled tailings cells in an on-path overburden landform.
- Priority 4 – overburden placed in a temporary stockpile and reclaimed once an on-path destination is available.

Ideally tailings produced from the WCP was pumped as a slurry directly back into the mining void however in early-schedule periods where this was not possible it was directed to the external TSF.

Additional constraints were used to control the size of the active mining area to ensure there was sufficient time for settlement and dewatering of tailings cells prior to dumping of on-path overburden above. These constraints were modelled using the dump dependency tool within LHS.

The next step modelled the haulage routes for both overburden and tailings. Two data sets were defined:

- Surface hauls: overburden blocks connected to a surface haulage network and overburden dumping destinations (embankment bunds, on-path landform, stockpiles).
- Tailings hauls: feed (ore) blocks connected to a dummy network modelling the flow of tailings to the TSF and tailings cells.

When setting up the LHS scenario, tailings hauls were modelled as 'cast' hauls to ensure no trucking information was calculated. An overview of the two haulage networks is shown in Figure 6.

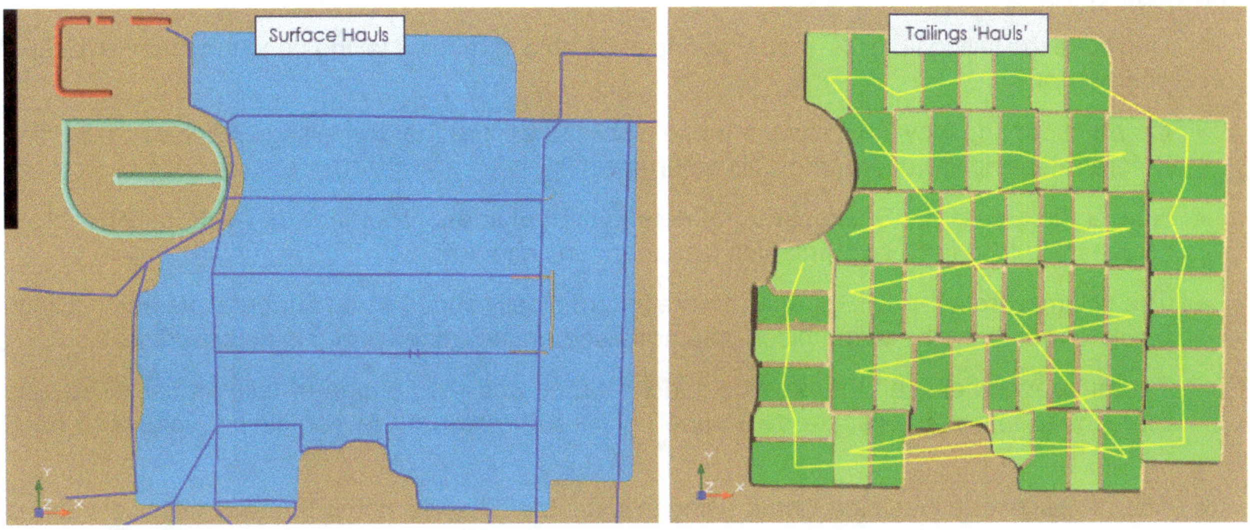

FIG 6 – Overview of overburden (left) and tailings (right) haulage networks.

To ensure that mining voids were released for backfill whilst honouring the size constraints of the active working area, the mining sequence was controlled through dependency rules between the tailings blocks and the on-path overburden blocks.

When using a typical iterative approach, the only way to control the distance between the active mining and backfill areas would be to design the backfill blocks specifically based on the quantities reporting from the mining schedule by period. Based on the variable nature of tailings produced by period, the backfill blocks would need to be adjusted according to the block geometry and the desired fill height. This would then release the dependency between the tailings blocks and the overburden blocks above for backfilling. Apart from the time taken to design these blocks, it can be observed that any change to the mining schedule, or adjustment to the underlying assumptions (eg tailings density), would require a full redesign of backfill blocks.

The dynamic AMC approach was to initially design all tailings blocks to the same elevation (eg the pre-mining surface) and run an LHS scenario that minimised the distance between the active mining area and available backfill locations.

This ensures that the active mining area size constraint was honoured, but introduced a secondary issue. Where tailings blocks were not filled to the pre-mining surface, the dependency between them and the overburden blocks above was not released and this material reported to overflow (Figure 7).

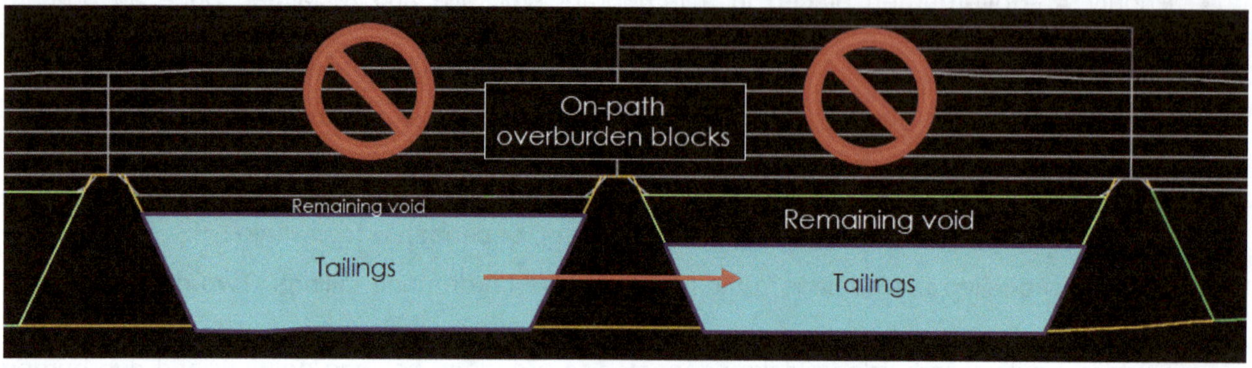

FIG 7 – Example first pass backfill sequence before releasing overburden blocks (with vertical exaggeration).

In LHS a dump block can receive material of multiple types (eg overburden and tailings) however allowing this created a situation where potential tailings cell capacity was instead filled with overburden, resulting in either tailings being sent to the TSF, or overflow in tailings hauls.

The solution was to:

- Duplicate the tailings blocks, mapping one to be able to receive tailings and the other to receive overburden.
- Create a dependency link between these blocks to ensure that the tailings block was filled first before releasing the overburden block for backfill.
- Run a first pass LHS scenario to determine volumes reporting to each tailings block (processing time for a scenario was less than ten seconds).
- The volumes from this LHS scenario were automatically coded as an attribute into each tailings block to represent the target block volume based on the underlying mining schedule.
- The difference between the physical block volume and the volume of tailings backfilled was then coded into the duplicate overburden block to represent the 'remaining' volume in each cell.
- The LHS scenario was then re-run using this volume attribute rather than the physical volume of each backfill block to generate the final backfill sequence representative of the underlying mining schedule.

By adopting this approach tailings blocks were filled to the required level before moving to the next cell as the mining front progressed. The remaining void was then backfilled with overburden. Once the blocks reach the design capacity (pre-mining surface), the dependency between the on-path overburden blocks was released and overburden dumping continued to produce the final landform (Figure 8).

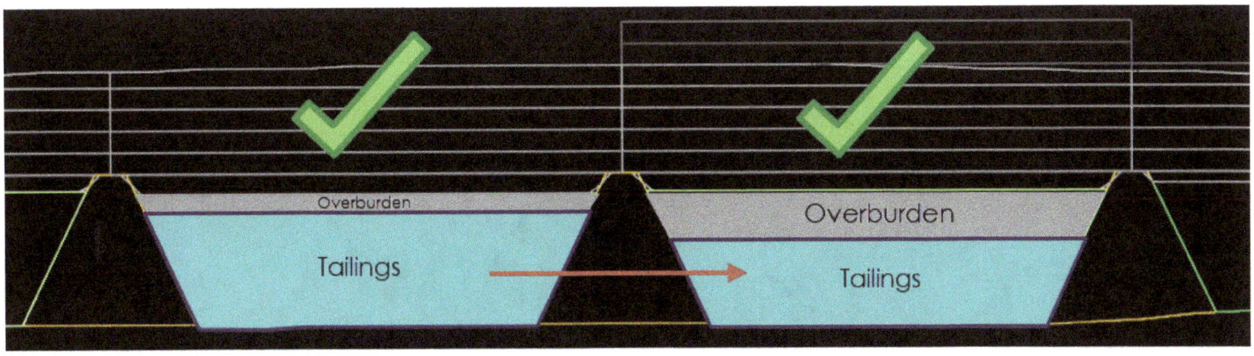

FIG 8 – Final backfill sequence showing combined deposition of tailings and overburden.

This method was efficient and provided the user instant visual feedback through the animation of the tailings and overburden placement sequence. Additionally, it was repeatable and allowed any changes to the mining schedule to be dynamically updated without the need for external tools or iterative post-processing.

The same approach has been used successfully to model backfill across a number of other mineral sands projects, including dredging operations. The flexibility of the LHS tool allows for a range of placement strategies to be modelled, from minimising haulage distances, to ensuring that final landform shapes are completed as soon as possible to expedite progressive rehabilitation activities.

REHABILITATION DESIGN

Progressive rehabilitation of the final landform in mineral sands operations is often a project requirement as part of environmental permitting agreements. The Deswik.EnviroTools module has features which assist in both the design and scheduling of progressive rehabilitation activities.

The reshape landform tool is an efficient and easy way to take an as-dumped landform shape and generate a final landform surface. Using this tool users can define rehabilitation slope angles by landform zone using polygons, as well as by elevation if generating geomorphic slopes.

The tool can be used to develop cut-and-fill volumes which can be imported into a scheduling package and incorporated into a master schedule or haulage scenario. Drainage can also be assessed using the tool, from which culverts or diversion channels can be designed to manage surface water.

An integrated approach to rehabilitation design is discussed in the following case study.

Case study 3 – mineral sands landform design

The Case Study 3 project was situated in an environmentally sensitive setting and hydrology was a key consideration in the permitting process. Additionally, the rehabilitated final landform design was required to be provided based on the mine plan submitted.

AMC completed mine planning to inform the approvals process using Deswik.Sched with a final as-dumped landform shape generated from an LHS Scenario. The final landform rehabilitation slopes were used as a design input, as well as a maximum landform footprint constraint.

Using the reshape landform tool, AMC generated a final landform based on the provided rehabilitated slope parameters, and cut-and-fill volumes which could be scheduled accordingly (Figure 9).

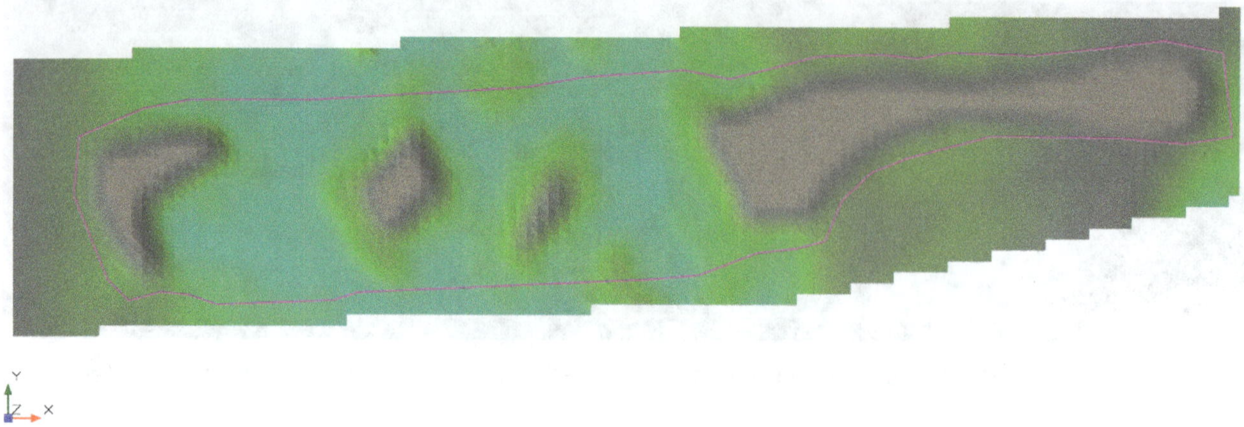

FIG 9 – Final reshaped rehabilitation landform design.

Based on the final landform, an integrated dozer push model was used to determine a trade-off between the use of a track dozer for rehabilitation activities compared with conventional excavator and truck methods. Track dozer productivity curves were applied and assessed based on the push distance between cut-and-fill areas. Where these distances exceeded the maximum productive push distance for the track dozer, movements were assigned to the truck/shovel resources and costed accordingly.

Catchment assessment tools were used to evaluate the impact of surface run-off due to the influence of the final landform. Specified rain events, such as a 1:100 year event, were modelled to determine catchment zones and identify areas where run-off needed to be managed. This tool used the surface topography as an input, and generated drainage lines, identifying discrete catchment zones and fill points, as well as catchment volumes based on the input rainfall data. An overview of the baseline rainfall assessment depicting catchment areas and volumes relevant to the final landform design is shown in Figure 10.

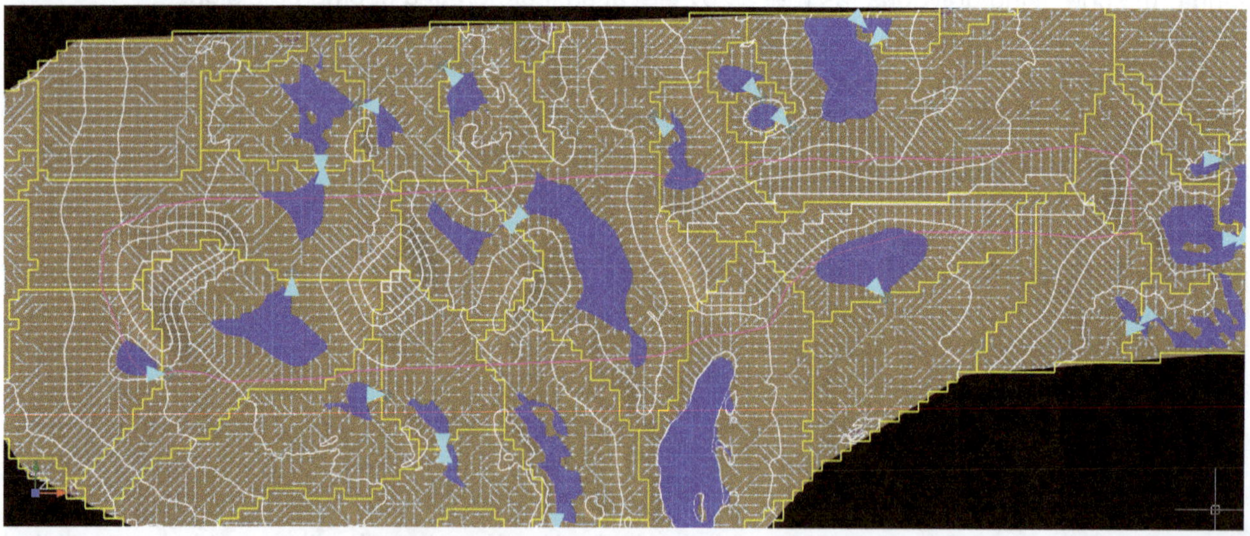

FIG 10 – Baseline rainfall catchment volume assessment on the final landform design (catchment volumes shown in blue).

The assessment identified that run-off to the central catchment zone needed to be redirected to prevent potential contamination of the floodplain in this area. The tool allowed the user to design multiple water structures such as drains, diversion bunds and culverts, each of which could be customised based on specific design requirements. In this instance it was decided that drains at the toe of each of the landforms directly adjacent to the catchment zone was most appropriate. These drains were designed with a fall towards the southern catchment zone and the catchment assessment re-run, resulting in the redirection of surface water as shown in the final assessment (Figure 11).

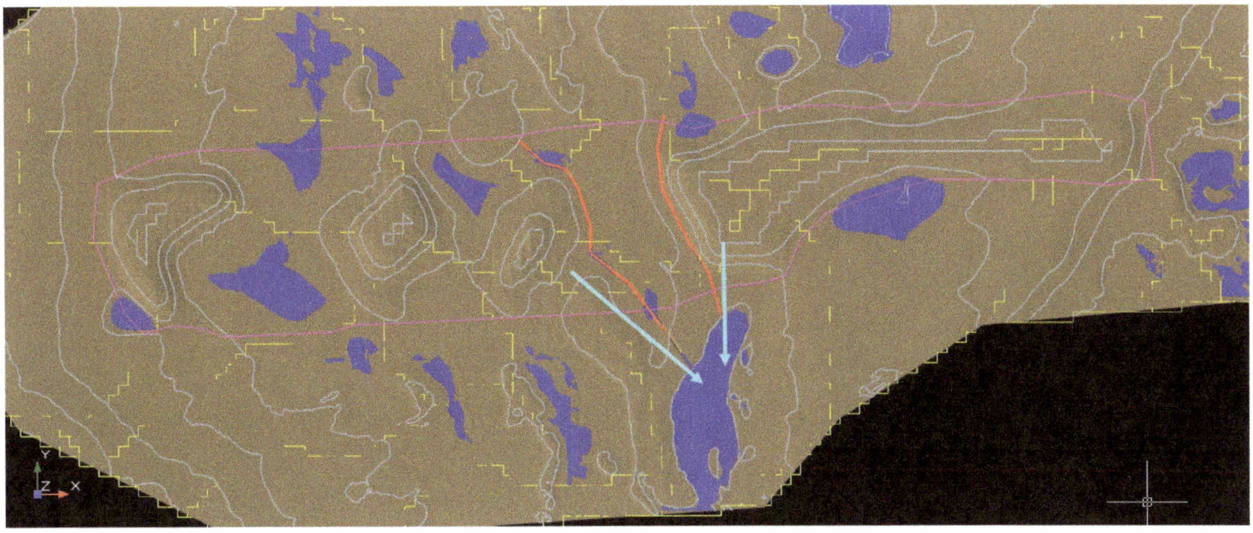

FIG 11 – Final rainfall catchment volume assessment with the introduction of drainage structures (red).

In this case study, the integration of final landform design, construction methodology, and surface water assessments allowed the user to develop alternative rehabilitation designs and approaches quickly.

CONCLUSIONS

Novel problems require novel solutions and while this paper focuses specifically on the use of the Deswik suite of tools to solve these problems, solutions are not inherently limited to one piece of mining software.

There are tools that the mining industry use daily that may have a range of applications beyond their intended use and this paper describes a few of these applications. The ability to find ways to adapt these tools is a skill that still lies solely in the hands of the practitioner and cannot yet be replaced by artificial means.

Case Study 1 has shown that by applying tools originally designed for application in underground mining operations, users are able to generate practical mining blocks through a process that is highly automated and auditable.

Case Study 2 addresses the shortfalls in traditional backfill modelling by using landform and haulage simulation tools to create dynamic backfill models in real-time. This approach has been successfully applied across a range of mineral sands projects and mining methods, significantly reducing processing time when assessing a range of scenarios.

Case Study 3 shows how the introduction of environmentally focused tools into conventional mine planning can assist operations in their permitting and approvals process and provide a greater and timely level of feedback when assessing the potential implications of landform design compared to traditional approaches.

These processes have been developed internally by AMC over many years and continue to be developed as software capabilities improve. By adopting these methods into traditional mine planning in a mineral sands environment, operators and project owners can maximise asset success, ultimately resulting in the generation of more realistic and achievable mine plans.

ACKNOWLEDGEMENTS

The authors acknowledge AMC Consultants Pty Ltd for supporting this publication.

ACKNOWLEDGEMENTS

A geotechnical reconciliation journey implementing a risk-based framework utilising unmanned aerial vehicle (UAV) photogrammetry-based mapping

R C N Botha[1], C J McDiarmid[2], J Dube[3] and M Kostadinovski[4]

1. Senior Geotechnical Engineer (Reconciliation), BHP WAIO, Newman WA 6753. Email: robert.botha@bhp.com
2. Geotechnical Engineer, MineGeo Tech Pty Ltd, Perth WA 6000. Email: chrismcdiarmid@minegeotech.com.au
3. Principal Geotechnical Engineer, formerly BHP WAIO, Perth WA 6000. Currently Superintendent – Operations Technical Support. Email: james.dube@oktedi.com
4. Superintendent Geotechnical Engineering, BHP WAIO, Perth WA 6000. Email: michael.kostadinovski@bhp.com

ABSTRACT

Geotechnical reconciliation is a critical practice in open pit mining, serving as an execution foundation for ensuring operational pit slope stability. This is accomplished through the adoption of the Plan-Do-Check-Act (PDCA) strategy, wherein open pit geotechnical design assumptions are rigorously validated against actual as-mined conditions, and necessary design adjustments are made to mitigate slope stability risks and enhance operational efficiency. Within this framework, the structural geology plays a pivotal role in determining the complexity of the encountered field conditions.

BHP WAIO operations in the Eastern Pilbara region comprise three major hubs, that are geographically widespread consisting of over 50 concurrently mined cutbacks and pits between 100 m to over 300 m final depth. The geotechnical reconciliation framework is an enabler to plan, schedule and execute this process since it is physically impractical to perform detailed mapping of all these exposed walls at the scale of the WAIO operations.

A geotechnical reconciliation framework based on the main input parameters such as risk-based pit or pushback design residual risk rating and structural framework complexity provides the underpinning framework to determine the type of face mapping to be utilised. Various levels of face mapping are outlined commensurate with the level of geotechnical detail required. This reconciliation framework forms the basis for the planned face mapping schedule utilising UAV photogrammetry techniques of mined temporary and final pit walls.

This paper will also outline the routine escalation process based on either structural or stratigraphic non-compliance outcomes. A Context-Purpose-Outcome-Resources-Timing-Safety (CPORTS) format is utilised to enable this geotechnical mapping is incorporated into the short-term tactical geology model updates to improve the main structural and stratigraphic model confidence for stability assessment.

Since implementing this reconciliation mapping framework, a number of significant economic and/or safety benefits have been realised based on timely, proactive decisions. Case histories related to these improvements are presented using 2D and 3D slope stability modelling software to evaluate stability.

INTRODUCTION

The geotechnical slope stability design process within West Australian Iron Ore (WAIO) relies on using a 3D geology model as the basis for the geotechnical stability model and evaluation of the stability of any pushback or pit. This geology model forms the basis for the geotechnical stability evaluation at the strategic level time frame, which typically occurs during the initial greenfields phase of any new pushback or pit evaluation. This geology model has in many respects been regarded by the Design Geotechnical Engineer as comprising a suitably appropriate and elevated confidence model where the model is taken as 'gospel' and the geotechnical design completely underpinned by this geology model.

Geotechnical reconciliation can be considered to be part of the continuous improvement cycle model utilising the Plan-Do-Check-Act (PDCA) principles in implementing, measuring, acting on identified changes to the geology model and design input parameters. Reconciliation plays an important role in closing the design loop: assessing performance, enabling the identification and action plan of key learnings, and identifying deficiencies or new efficiencies in the process (Armstrong, 2022).

In 2021, following one of the improvement projects to assess the suitability of the geotechnical design process with new, improved and evolving BHP-wide corporate geotechnical standards, the following was recognised:

- The initial geology model produced by the Strategic Modelling team as part of the GeoScience Group (Planning, Technical and Environment Department), classified the geology model as not a 'structural model' since the model did not represent the level of structural complexity that was called for by the updated WAIO Geotechnical standards.
- There was not a routine, embedded and consistent process for face mapping processes where stratigraphic, structural and lithological data is consistently being fed back by involved stakeholder field-based teams into the iterative model update(s). In this respect, the particular geology model is termed the Short-Term Geology Model (STGM) and/or Short-Term Production Model (STPM), which is the same above model that is continuously reviewed and updated in the Tactical Modelling time frame (ie within the 3 month to 2-year timeline). The above model terms are for this paper to be regarded as interchangeable and to be consistent, the referenced model term used will be the STGM.

This had the potential to impact the BHP WAIO operations comprising two main geographic mining hubs, namely the Eastern and Central/Northern Pilbara hubs, respectively. In the Eastern Pilbara Hub (EPH) region this comprises three geographical established operational sites namely Whaleback, including Ore Body (OB), OB29-OB30-OB35/Eastern Syncline (ES), Eastern Ridge and Jimblebar mine sites. At any one time, due to the geographical size and scale of the WAIO operations, up to and over 50 active pushbacks within these open pit operations are being mined in predominantly Archean age, deformed, bedded (anisotropic), banded iron formation (BIF) bedrock formation, comprising the Hamersley Group.

Geotechnical slope reconciliation can address these above issues as it forms one of the last field-based controls that reconciles the slope design assumptions with the as-excavated conditions and further seeks to validate the STGM confidence, with the ability to identify and provide iterative structural and stratigraphic inputs to refine and increase the confidence in this model. A WAIO geotechnical reconciliation framework has therefore been devised and implemented to enable and support this reconciliation process. This paper further outlines this geotechnical reconciliation framework, including detailing and expanding on the supporting elements within this process.

REGULATORY AND CORPORATE STANDARDS

The Western Australian, Department of Local Government, Industry Regulation and Safety (LGIRS) does not explicitly define the requirement for geotechnical reconciliation with the recent *Work Health and Safety (Mines) Regulations* (DEMIRS, 2022), encompassing ground control material risk within the requirements for a principal mining hazard management plan (PMHMP). Ground control hazards are covered within the principal mining hazard (PMH) classification. Additional related standards are outlined in the LGIRS ground control for Western Australian mining operations *Code of Practice* and *Guideline* (Department of Minerals and Energy, Western Australia, 2019a, 2019b).

Documentation for BHP WAIO ground control is organised into a hierarchical framework, with links to State and Company governance requirements. This framework is outlined in Figure 1.

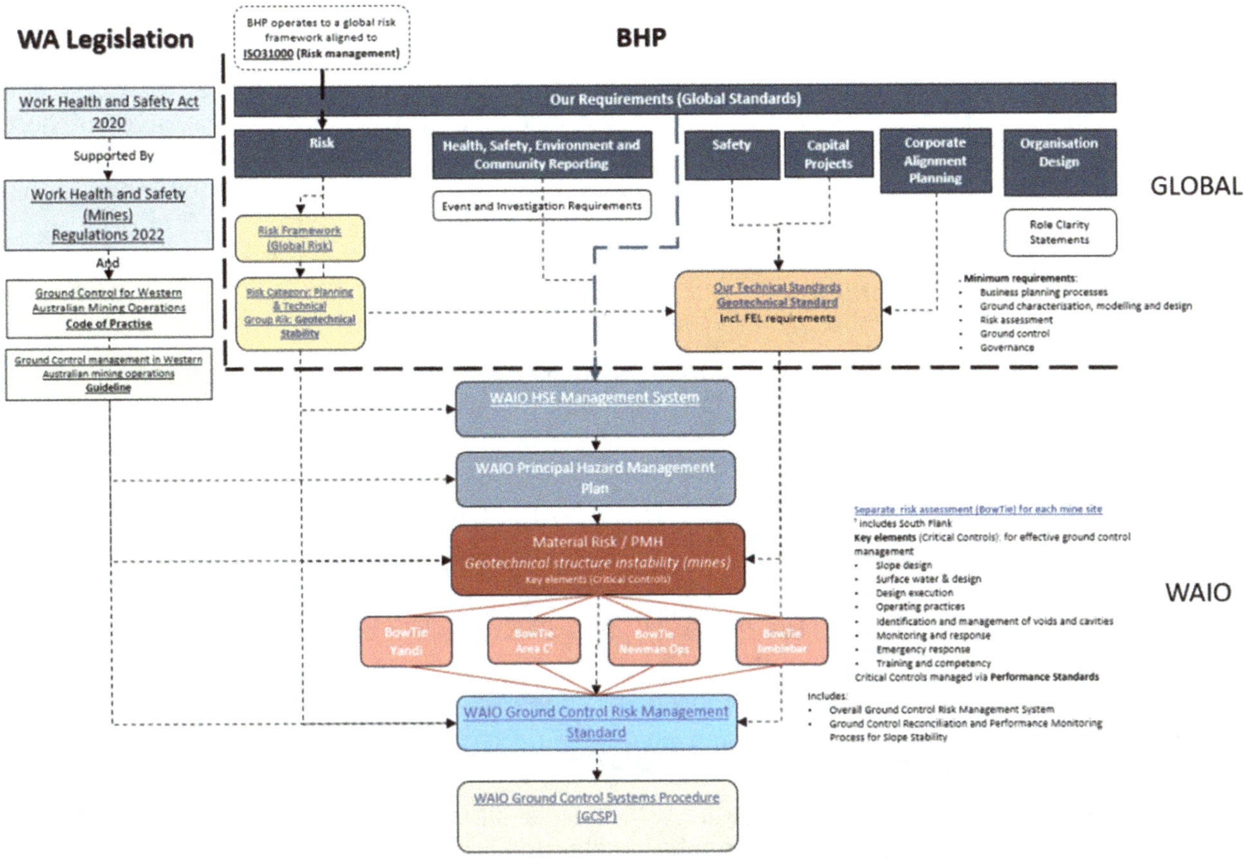

FIG 1 – BHP Corporate and WAIO Ground Control Framework (BHP, 2025).

The WAIO Ground Control Risk Management Standard (BHP, 2025) provides for eight material control measures, of which, the geotechnical reconciliation process is covered under the 'Slope Design' control element. Under this material control element, there is a requirement for a process to be established that, as far as is reasonably practicable, identifies areas of concern and provides appropriate responses to ensure the safety of the operation. The relevant element within this control measure that specifically applies to the reconciliation process is the following:

- 'A process that validates the *in situ* conditions and behaviours, to determine any differences between expected (design) and actual (*in situ*) conditions and behaviours that may materially impact ground control performance'.

Similarly, Our Technical Standards – Geotechnical Standard (BHP, 2021a) engineering objective is to use a risk-based approach to managing geotechnical stability hazards that ensures a safe operating environment while providing operational continuity and optimised business plans. This standard is derived from current worldwide geotechnical best practice and further provides a minimum control framework for managing geotechnical risks and value add to business plans. This standard is based on implementing the geotechnical management (PDCA) process.

GEOLOGY MODEL CONFIDENCE AND ACCURACY

The level of confidence in the geology model is crucial to the geotechnical design process and long-term performance of any slope design. The above WAIO Technical Standard defines the required confidence level for design execution, which for iron ore open pits typically comprised a higher level of model confidence than what was being provided by the geology model delivered as a 'fit-for-purpose' design in the strategic space.

Fundamental to the building of high-quality geology models for geotechnical purposes and for delivering 'fit-for-purpose' geotechnical models, is provision of model accuracy that can be considered to be a measure of the level of confidence. Across both WAIO mining hubs and operations where there could be up to 100 different pushbacks and pits being concurrently mined, the task to proactively support and provide 'fit-for-purpose' geology models is an immense task.

The GeoScience to Geotechnical (G2G) improvement project (BHP, 2021b) recognised a slope reconciliation TARP was required to support the improved impetus for field-based face reconciliation mapping approach and to quantify the model compliance compared to the actual, exposed and mapped conditions. This TARP was a simple and effective way to classify the level of non-compliance and provide a communication process to enable the modellers to understand the current model compliance and incorporate the actual, mapped data into the updated geology model within the STGM update timeline. This TARP is summarised in Figure 2.

Model Compliance Description
Observations Align Well – STRATIGRAPHY within tolerance (±12 m distance measured normal to contact) and STRUCTURE (bedding within ±5° to 10°, folding >50 m wavelength modelled and major fault correctly modelled within ±12 m distance).
Observations Align Moderately – Either STRATIGRAPHY outside of tolerance or STRUCTURE outside of tolerance.
Observations Align Poorly – STRATIGRAPHY and STRUCTURE both outside of tolerance.

FIG 2 – Face mapping model compliance Trigger Action Response Plan (TARP).

UNMANNED AERIAL VEHICLE (UAV) PHOTOGRAMMETRY

The significant advancement in recent years with the use of UAV drones to capture and process open pit wall photogrammetry is now the operating standard across all WAIO operations. The mine site Survey Department is responsible for the execution of all flights to capture high quality, bench scale photogrammetry for use by both the Mine Geology and Geotechnical Engineering teams in reconciliation face mapping.

A DJI Mavic 3E drone with real-time kinematic (RTK) positioning is the main remote piloted aircraft and mode used for collection of the raw photography for face mapping photogrammetry. Typically Survey will either fly the project to capture the pit bench or slope in manual or pre-programmed mode, depending on the accessibility to the pit bench, the scale of the required project (ie should multiple benches be required) and pilot preference for the particular requirements. A minimum of two to three flight passes are required to enable high quality photogrammetry to be produced with in general, photogrammetry flown for every completed interim and final bench mined in any pit. The Survey pilot will aim to capture raw photographic images in single bench application using at least three different flight lines at different heights and gimbal angles to capture all features for accurate model reconstruction (BHP, 2024d).

This means that a continuous photogrammetry record is available of every interim and final wall captured in any open pit mined across WAIO operations.

All processing of the raw photography to create photogrammetry is performed in PIX4D© mapper (PIX4D) software where an.obj file extension is created for import into geotechnical processing software. Both 3DM Analyst© (Adam Technology) and GeologyCore© (Maptek) software platforms are used by Geotechnical Engineering reconciliation team for digitising stratigraphic contacts and structural features, analysis and consolidation of the face mapping for reporting.

There are a number of advantages to using aerial photogrammetry techniques as the basis for geotechnical reconciliation mapping, some of which include:

- No interaction with production active mine areas (surface mining equipment) with flying able to occur from remote, designated safe areas.
- Automated flight path selection based on pre-planned area and planned wall survey coverage.
- Ability to plan and schedule the best time of day and lighting availability to enhance the photogrammetry clarity.

- Enabling significant bench face structural measurements, additional to field-based measurements, to be attained with confidence and accuracy.

GEOTECHNICAL RECONCILIATION FRAMEWORK

The geotechnical reconciliation framework (BHP, 2024c) is a flow chart, detailing a risk-based approach using the structural framework (structural complexity) and material risk rating (residual risk rating or RRR) as the underpinning elements (Figure 3).

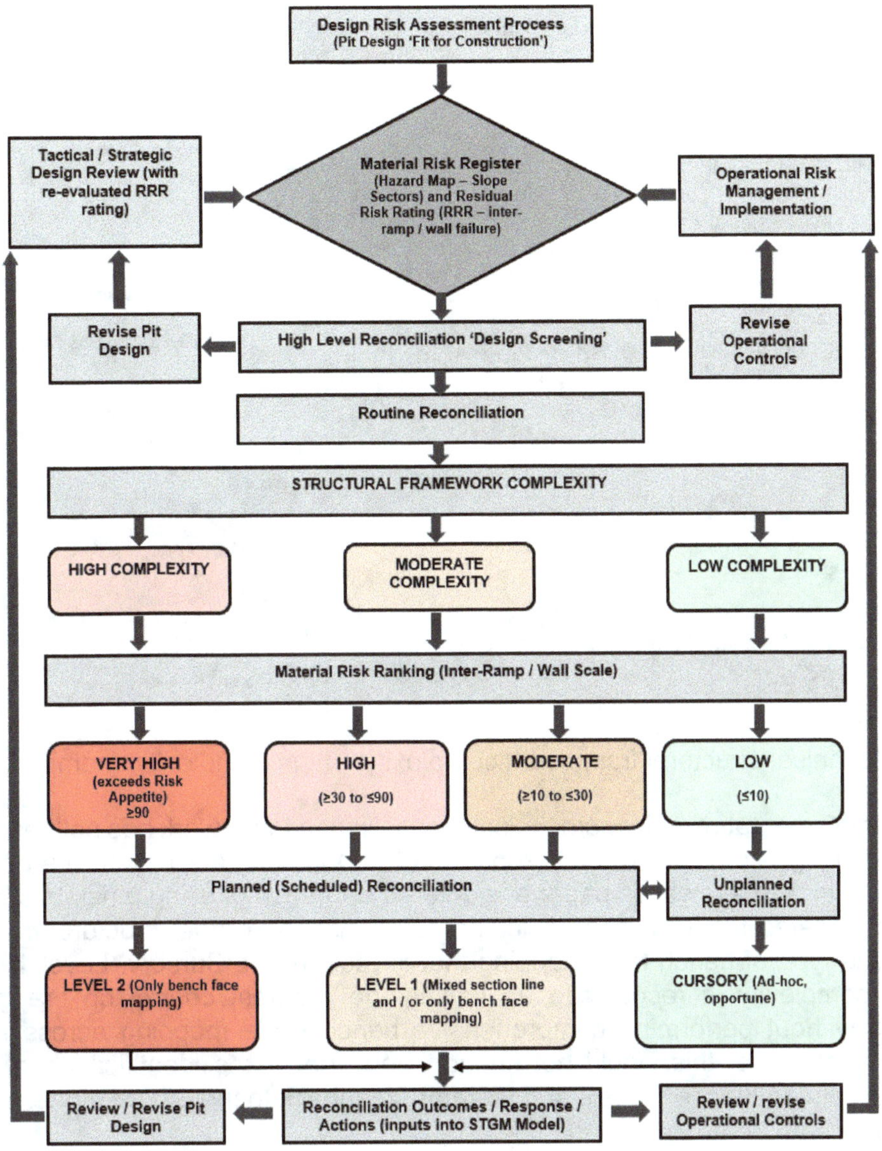

FIG 3 – Geotechnical reconciliation framework.

'Fit-for-Construction' pit designs are released by the Geotechnical Studies and Planning design team in Perth, whose main function is to provide geotechnical detailed designs aligned to the appropriate Design Acceptance Criteria (DAC) and ready for operational execution. The appropriate hazard map, which also defines the design geotechnical domains, forms part of this release which outlines any residual design hazards that may require specific operational controls to be implemented.

Pit risk rankings are continuously assessed through the geotechnical reconciliation reporting and updated to reflect the most recent reconciliation reporting outcomes. Any identified material changes to the potential stability through tolerance non-conformance to stratigraphy and/or structure in the STGM models, will trigger a geotechnical design review to mitigate the elevated risk.

The structural framework complexity is outlined in Figure 4. A hierarchy of structural complexity is defined in this framework comprising various levels of structural complexity from low, to medium and high. Typical cross-section examples from various WAIO pits are illustrated providing a calibration of the described complexity. As described above, the structural complexity category forms one of the underpinning elements in supporting the decision for the type of reconciliation face mapping to be executed (ie bench face mapping, section line(s) or opportunistic, ad hoc local bench face mapping).

HIGH COMPLEXITY

Significant complexity, close, tight to isoclinal, inclined to recumbent, both harmonic and disharmonic folding. Significant overturning and low (listric-shaped) to high angle thrust / faults and internal shearing with stacking / duplication of stratigraphy. Highly variable bedding dip orientation controlled by fold / fault geometry. Regional and / or mine scale fault over-printing structures strike through pit / pushback with varying displacement throw ranging from 10's to 100's of metres.

Modelled fault blocks highly complex shaped in three-dimensions with abundant major faulting extending to depth.

Moderate to thickly developed Tertiary Detrital sequence.

Potential for identification of new faults / splays and high potential for stratigraphic / structural change.

Mainly structural-controlled (bedding and / or fault) failure mechanism(s) expected.

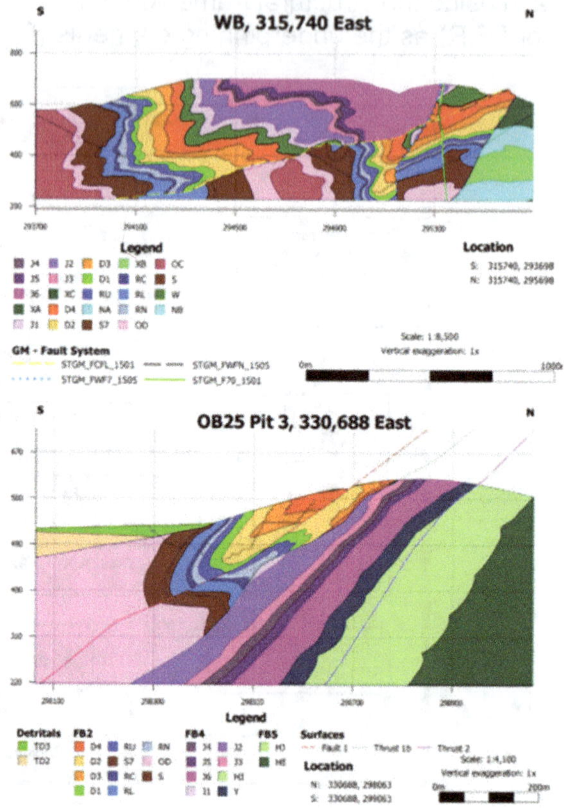

FIG 4 – Geotechnical structural framework complexity – high complexity example (BHP, 2019).

The journey to implementation of a geotechnical reconciliation framework has not been without some of the similar challenges recognised by Dixon *et al* (2011) in comparable iron ore operations in the Pilbara. However, in the case of this paper a two-tiered approach of using a risk-based approach for identification of targeted reconciliation, including categorisation of the structural complexity of the area to define the type of bench face mapping that is required (ie Cursory, Level 1 or 2 mapping). The structural complexity is recognised as having the greatest control on the geology model confidence and without performing comprehensive bench scale mapping across some of these structurally complex walls, this would not support and enable the identification of inherent non-compliance in lower confidence models and therefore feedback to the STGM modellers in improving this confidence.

RECONCILIATION FACE MAPPING

The principal aim of the geotechnical reconciliation mapping across the EPH Pilbara operations is to support the improvement of the structural geological knowledge of these deposits through increased geology model confidence. Across this region, due to the structural and spatial setting of these open pits, the structural complexity can be considered to be predominantly moderate to high, calling for mainly Level 2 type reconciliation mapping or comprehensive bench scale mapping to be performed. This mapping is normally executed as per the pit and/or individual pushback designs.

Precedent experience with the geology model is also utilised where based on this experience, should significant and potentially material design non-compliance be experienced, then this level of mapping is continued to improve the model confidence. Mapping is executed either at a single bench scale (12 m height) or should access not be readily available during mining of this bench the mapping may be delayed until a few concurrent benches (two to three) are exposed.

Discretion is used in implementing Level 1 type mapping and again depending on precedent experience, the geotechnical design domain RRR and the assessed structural complexity, a 'corridor' of mapping defining the chosen section, may be implemented (in this case typically mapping may only be performed within an approximate 50 m distance on either side of the section line).

The current improved geotechnical reconciliation mapping process has been implemented more recently with the previous reconciliation process having been performed on a less systematic and routine basis.

GEOTECHNICAL RECONCILIATION DATA COLLECTION

The principal strategy is to collect, review and provide proactive feedback to the STGM modelling team to enable increased geology model confidence, resulting in a more 'fit-for-purpose' model for structural geotechnical stability confirmation or review. As the geology model confidence is the primary driver for improvement, the geotechnical reconciliation process does not focus on any smaller scale window type or scan-line mapping that primarily focuses on the rock defect and rock mass properties. The focus is to validate at the inter-ramp scale the stratigraphic contacts, lithological types and contacts, overall primary bedding dip orientation structure, fold structure, secondary joint structure and any significant fault structure, whether regarded as a minor or a major structure. This is mainly due to the highly anisotropic rock conditions within the Pilbara BIF-dominant and interbedded stratigraphy, where the main failure mechanism is bedding and/or structural controlled failure, leading to mainly planar or more complex failures with a side or toe release, rock mass component.

Notwithstanding the above, rock mass parameters such as individual lithological and stratigraphic rock material strength, geological strength index (GSI) using the Pilbara modified GSI for heterogenous rock mass, defect (bedding and joint sets) micro-roughness coefficient (JRC), macro-roughness and weathering data is also collected to calibrate against the design parameters.

GEOTECHNICAL RECONCILIATION REPORTING

Geotechnical reconciliation reporting is performed via the SafetyCulture© reporting platform. The report structure is a systematic and structured approach that calibrates all the geotechnical design assumptions against the as-mined, exposed and mapped conditions. Consistent with the primary focus on comparing the current STGM to the as exposed conditions, the available stratigraphic master wireframes and fault block solids are compared to the mapping data to evaluate the model compliance confidence. Use of the STGM geology block model also forms part of the calibration process.

The report structure is based on a TARP 'traffic light' system that evaluates the following reconciliation areas:

- **G2G TARP** – model confidence compliance based on the descriptors in Figure 3.
- **C2D (Compliance to Design) TARP** – classification to inform the achieved bench profile and design berm width (berm reliability index) against the design assumptions.
- **Structural TARP** – classification to inform the structural angle difference (typically bedding), defect parameters and kinematics against the design assumptions.
- **Rock Mass TARP** – classification to inform the GSI, UCS and weathering against the design assumptions.
- **Slope Monitoring TARP** – classification to inform the installed slope monitoring plan and monitoring performance against the planned monitoring installation.

Reporting of the results is based on the completed reconciliation mapped project area and is distributed to the main stakeholders such as Geotechnical Engineering and the Tactical Modelling team in Perth for review and allocation/actioning of follow up actions to the relevant stakeholder group.

GEOTECHNICAL RECONCILIATION ESCALATION OVERVIEW

Various escalation processes are in place to deal with stratigraphic and structural non-compliance highlighted from the geotechnical reconciliation mapping (BHP, 2024c).

STGM non-compliance reporting

Based on the G2G TARP outcomes, an actual level (AL) and potential level (PL) risk-based outcome is awarded to the level of model non-compliance for a classified design hazard category, based on the BHP Corporate Risk Management Global Standard. This is consistent with the geotechnical event reporting guidance for non-compliant model outcomes with a range from AL0, PL2 to AL0, PL3 and PL4 (for moderate and poor model compliance outcomes).

Event management system (EMS) reporting

All classified design hazard events are reported in the BHP SAP Fiori EMS system and the event escalated to Line Management for review and/or investigation actioning. Only PL4 events or greater require a review sign-off by Line Management (Geotechnical Manager).

Context-Purpose-Outcome-Resources-Timing-Safety (CPORTS) format reporting

CPORTS refers to a documentation format that covers the Context (C), Purpose (P), Outcome (O), Resources (R), Timing (T) and Safety (S) aspects of the model non-compliance. This presentation format is provided to the Perth Tactical Modelling team to highlight the main non-compliance issues that have been raised in the geotechnical reconciliation report. It is meant to be a concise, standard format to collate, consolidate and present only the main high-level issues that need to be addressed by the modelling team.

Geotechnical risk and pit design review reporting

Potential material risks in the STGM model non-compliance and lowered confidence are escalated to the Principal Geotechnical Engineer (EPH – Tactical, Perth) for review and further stability analysis.

Depending on the operational urgency, amended risk ranking and potential economic impact to the pit design (and mine plan) of the particular non-compliance, feedback will be provided to the Tactical Modelling team to review and revise the non-compliant stratigraphy/structural wireframes to enable either 2D or 3D stability modelling to be performed.

Geology model non-compliance escalation flow chart

Any geology model non-compliance results in an escalation of the issues that accounts for the materiality of the non-compliance and should more immediate action be required to be implemented to mitigate the risk, then the Tactical Modelling team are provided with feedback around the urgency of the request.

CASE HISTORY

Various case history events have been proactively identified through implementation of the geotechnical reconciliation framework and routine face mapping process. Relevant case studies are presented herein.

Case study event – Ponderosa (Pondy) haul ramp

The Pondy haul ramp is one of two main haul ramps servicing the Whaleback open pit. It is the main waste destination haul ramp routing to the main long-term waste dumps located to the north of the Whaleback Pit. For this reason it is a critical life-of-mine haul ramp which should it be lost or affected by failure, would seriously impact the hauling operations and overall mine plan. The haul ramp traverses the ACS4 Stage 4/5 wall design with the geology comprising a folded sequence of Brockman Iron Formation stratigraphy. The pit design follows the moderate-dipping (±25°), Dales Gorge (D1) unit forming the footwall to the main strong rock (50 MPa to 100 MPa), mineralised and high-grade overlying ore shell comprising mainly D2 to D4 stratigraphy. The D1 unit is partly mineralised comprising strongly anisotropic interbedded Shale – BIF – Chert lithologies comprising varying strength (from 15 MPa to >100 MPa). The mineralised Dales Gorge high-grade ore shell comprises a dense, very high specific gravity ore (approximately 4 g/cm^3).

Figure 5 illustrates a section through the current STGM showing the Pondy Ramp relative to the main Stage 4 wall. The potential failure area occurs under and 'daylighting' below the ramp since the original pit design allowed for the Pondy Ramp infrastructure by the pit wall following the D1 unit to depth, thereby leaving a large 'heavy' wedge (over 300 kilo tonnes (Kt)) of D2 ore underlying the ramp and overlying sheared, anisotropic shale-dominant units (ie D1 and McRae Shale unit RU).

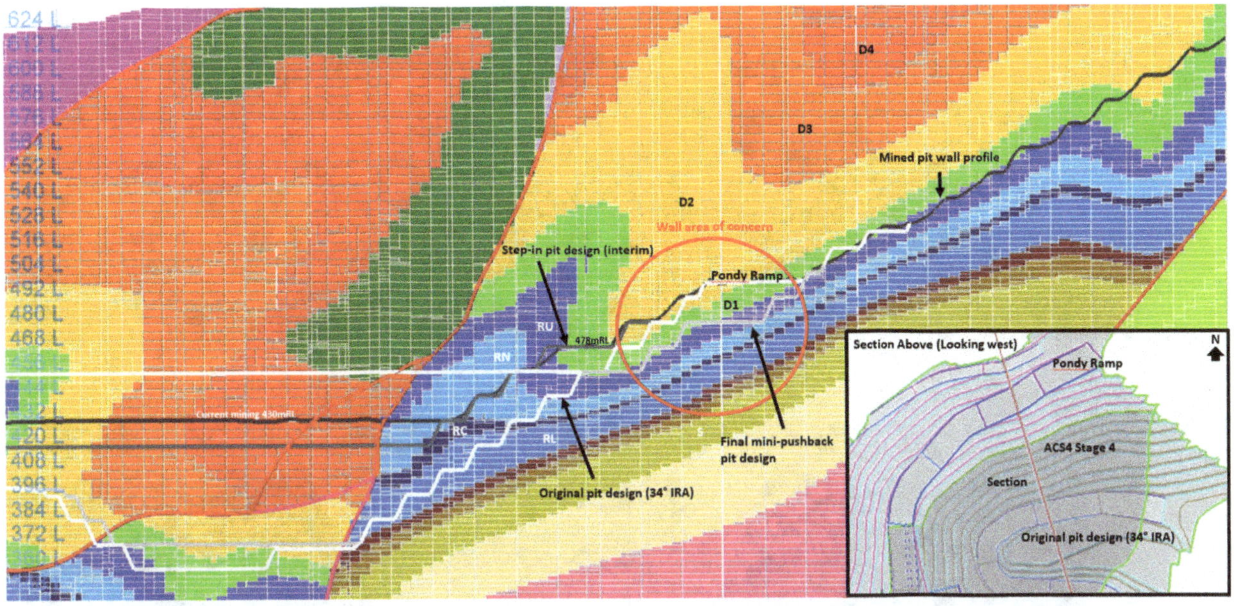

FIG 5 – Current STGM showing cross-section through potential failure area.

The wall was based on a historic design acceptance criteria factor of safety (FoS) ≥ 1.20 (design completed in 2017). However, the resultant and recently updated BHP Geotechnical Standards required a higher level of confidence for this wall with FoS ≥ 1.30 to 1.50.

Reconciliation mapping

Reconciliation mapping of the 478 mRL bench was performed in January 2023 with the potential for planar failure confirmed by the Level 2 comprehensive bench mapping. The bedding dip in the Dales Gorge D2 unit comprised more unfavourably oriented bedding dip than modelled with a dip orientation up to 40°. The Dales Gorge D1 contact comprised sheared, slicken-sided and faulted shale beds (CS5 and CS6 shale beds) with polished and sheared McRae Shale underlying the above unit.

The stability was initially checked with RocScience Inc. Slide 2 (2D analysis), which confirmed that the current wall geometry and 'daylighting' D1 and RN contacts in the following benches would result in a quasi-stable wall. This was significantly below the required design acceptance criteria. Figure 6 indicates the evaluated section view of the potential failure area below the ramp.

To confirm the Slide 2 stability analysis, further stability review was performed using Slide 3 (3D analysis). The results from the 3D review, running various base case (most likely case) and sensitivity scenarios (varying shear strength cases) confirmed the stability concerns with the wall (Figure 7). FoS < 1.20 supported the field-based reconciliation observations with both cases resulting in a FoS of 1.16 (most likely) and 1.08 (bedding sensitivity), respectively.

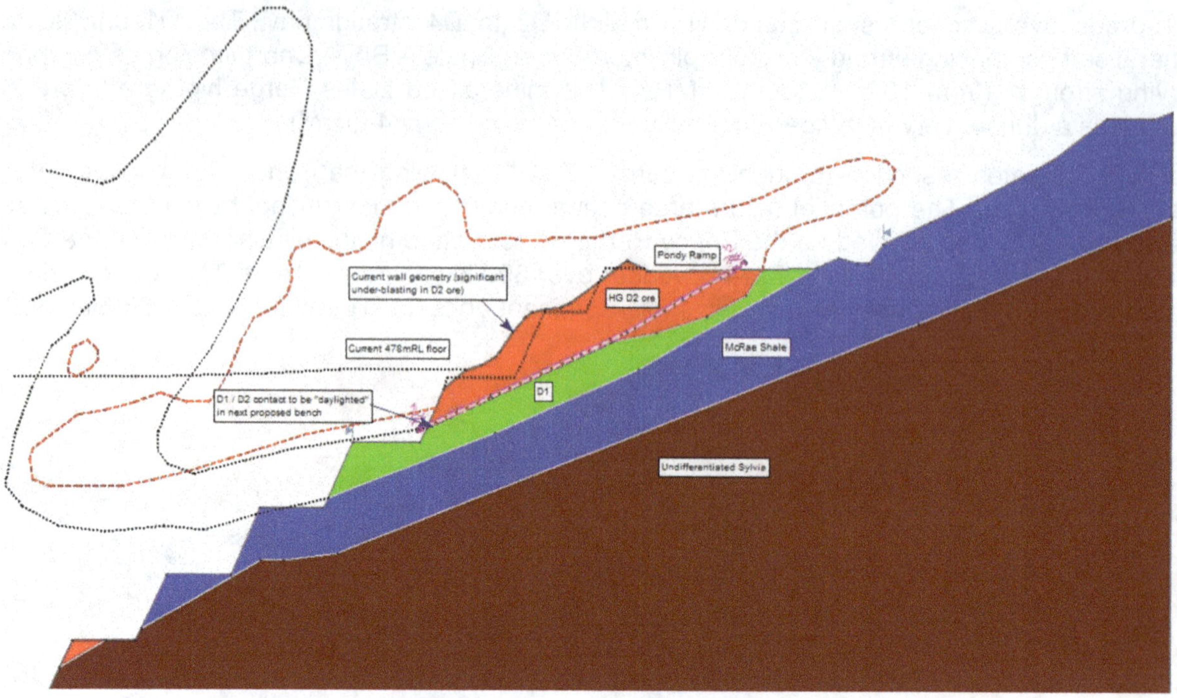

FIG 6 – Slide 2 failure section.

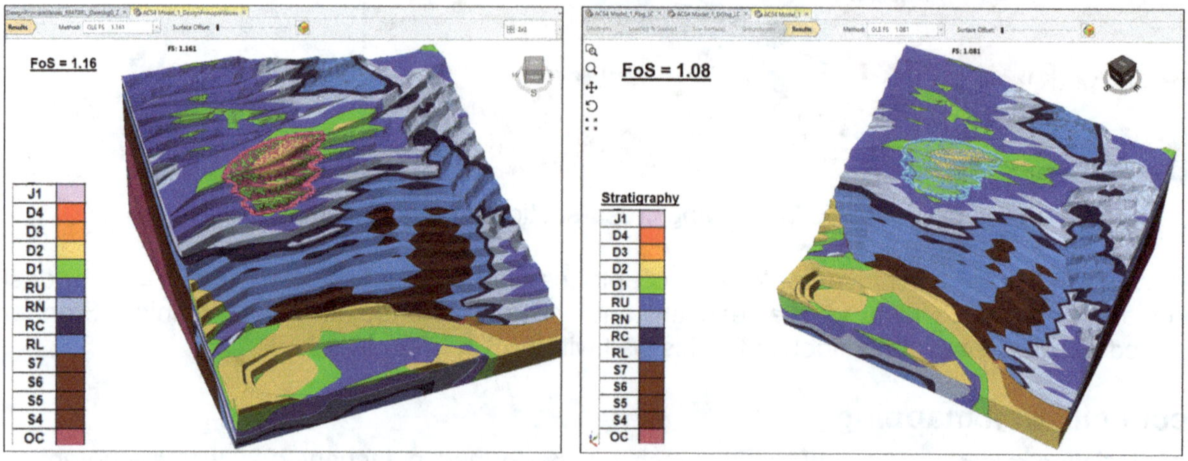

FIG 7 – Slide 3 stability review for base case (left) and sensitivity case (right). Base case comprised bedding shear strength with ϕ_b = 22° (base), ϕ_b = 17° (residual sensitivity) with both comprising 10 kPa cohesion (BHP, 2023a).

Reconciliation remedial outcome

The main options evaluated to mitigate the potential failure risk comprised:

- Step-in interim design was supported by the Production Scheduling Design Engineer with an interim design comprising both wider 478 mRL and 466 mRL berms. This allowed the immediate wall not to be further undercut and allowed sufficient mining width for implementing a 'mini-pushback' behind the Pondy haul ramp at a later stage when a haul ramp to the lower part of the wall was established.
- A subsequent pushback was designed to allow extraction of the high-grade ore underlying the haul ramp and in the mined wall. This recovered an additional 300 Kt with the mining out of the Pondy haul ramp.
- Subsequent re-design of the entire final pushback recovered a further 3.90 million tonnes (Mt) of high-grade ore towards the slope toe (BHP, 2023b), resulting in over 4.2 Mt of additional high-grade ore recovery that was initially identified and triggered by geotechnical reconciliation to safeguard the Pondy haul ramp.

Case study event – H2B1 (Hashimoto) pushback

The Hashimoto pits comprise a series of along strike, adjacent pits (H2A1, H2B1 and H4A1) exposed in mainly very weak to weak, Upper Brockman Iron Formation (Yandicoogina Shale Formation) and Weeli Wolli Formation shale and intrusive dolerite sills. Original geotechnical design analysis of the north walls in these proposed pits utilised residual soil strength (<1 MPa) rock material strength in the Weeli Wolli dolerite lithology based on precedent geotechnical investigation studies in the same members and benchmarked to these pits, resulting in a lowered inter-ramp angle for this north wall.

Reconciliation mapping

Reconciliation mapping based on the existing exposed, mined benches concluded that the Weeli Wolli dolerite was a low strength rock material of 10 MPa, rather than the residual soil strength design assumption used in the original design studies. Secondly, the Yandicoogina Shale member that was modelled as an undifferentiated unit could be differentiated in the field comprising a 'xenolith' zone where thin, irregular, randomly distributed BIF – shale 'xenoliths' were distributed throughout a predominant intrusive dolerite sill with the same strength properties as the main Weeli Wolli dolerite unit. This enabled improved strength parameters and a spatial re-organisation of the STGM to be incorporated into the revised geotechnical stability model.

Reconciliation remedial outcome

The main outcome from this reconciliation-supported review comprised:

- Increased material strength confidence in the stability analyses resulted in the north wall inter-ramp angle to be increased by 10° across all north wall pit designs.
- The revised pit shell designs for two of the pits namely H2A1 and H2B1 allowed for reduced waste stripping of 6 Mt and additional recovery of 126 Kt of high-grade ore (Figure 8).

FIG 8 – Optimised pit design shell for H2B1 north wall pushback.

Case study event – ACS3 South

The ACS3 South haul ramp is a long-term ramp providing access to further pushback development to the west of the main Whaleback Pit. The haul ramp traverses the ACS3 South North Wall design and 'hugs' this north wall with the geology comprising a tightly folded sequence of overlying Brockman Iron Formation stratigraphy juxtaposed and separated by the main, undulating, Central Fault, a significant thrust fault within this open pit (Figure 9). The footwall stratigraphy to this fault comprises the Bee Gorge Member (OC) and Mt. Sylvia Formation (S1 to S7) stratigraphy, which is modelled as a close folded syncline – anticline pair with sub-vertical northern anticline limb.

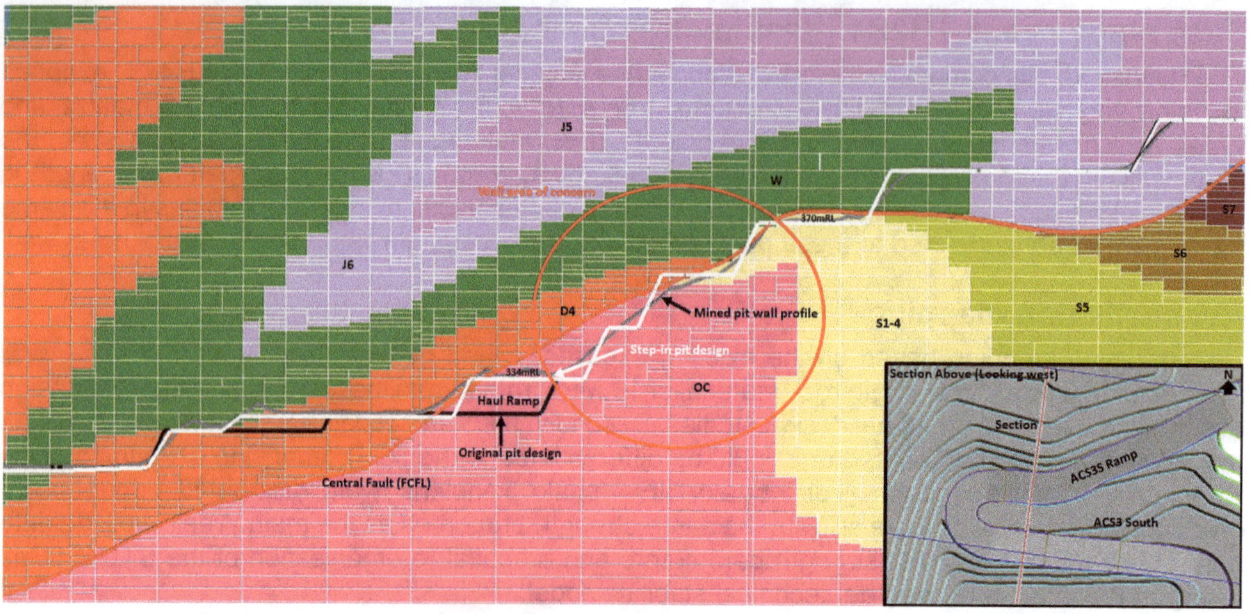

FIG 9 – Historic stability analysis through the north wall showing FoS > 4 (BHP, 2022).

Reconciliation mapping

Reconciliation mapping of the wall below the 370 mRL bench however, confirmed the actual dip orientation of the main Central Fault, including the underlying Bee Gorge sheared and faulted shale stratigraphy was not aligned to the STGM interpretation. The bedding dip comprised a mainly dip slope with dip orientation of ±40° with the Central Fault more unfavourably oriented and also dipping steeper than previously modelled (Figure 10). This obviously had implications to the historic geotechnical design assumptions and stability analyses that implied a rock mass-controlled wall. The changed conditions supported a very high likelihood for mainly planar failure along this structure, confirmed by the Level 2 comprehensive bench mapping performed.

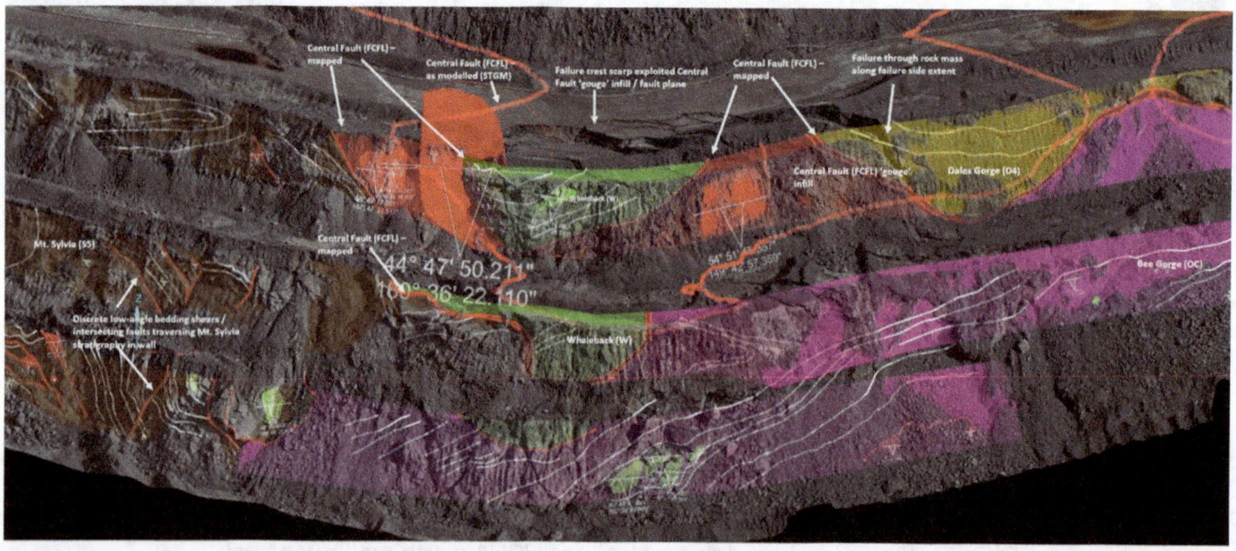

FIG 10 – Photogrammetry-based interpretation showing the Central Fault failure control in January 2024 (BHP, 2024b).

To support the changed conditions and the interpreted quasi-stable wall, further stability review was performed using Slide 3 (3D analysis). The results from the 3D review, running various base case (most likely case) and sensitivity scenarios (lower case) confirmed the stability concerns with the wall (Figure 11). FoS < 1.0 supported the field-based reconciliation observations with both cases resulting in a FoS of 0.76 (most likely) and 0.54 (bedding sensitivity), respectively.

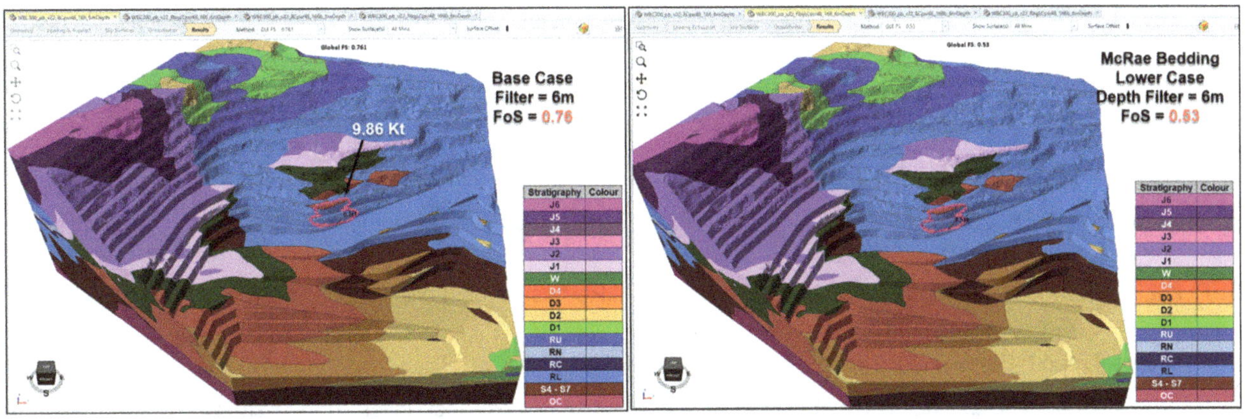

FIG 11 – Slide 3 stability review for base case (left) and sensitivity case (right). Base case comprised bedding shear strength aligned to BHP Geotechnical Engineering Slope Design TPI (BHP, 2024a) with ϕ_b = 18° (base and sensitivity) with cohesion comprising 10 kPa (base) and 0 kPa (sensitivity) BHP (2023c).

Reconciliation remedial outcome

The main options evaluated to mitigate the potential failure risk comprised:

- Initial reconciliation identification of the changed conditions and potential failure occurred in October 2023 with the pit design implementing a 20 m buttress toe step-in on the 334 mRL bench.
- Inter-ramp scale failure was triggered by heavy rainfall events at the end of January 2024 with the failure occurring on the 30 January 2024. The failure was also managed through radar monitoring coverage and closing of the pit prior to the on-set of failure.
- The haul ramp was moved further south to ensure any further wall deterioration was not going to impact the haul ramp. This mitigated mining traffic from traversing under the affected wall. The loss in width in parts of the haul ramp resulted in a tipped waste fill buttress to widen the ramp, to ensure the dual lane continuity along the length of the ramp.

CONCLUSION

The geotechnical reconciliation framework implemented across the EPH Pilbara operations hub, based on both a risk-based and structural complexity decision-making process, including comprehensive photogrammetry-based geotechnical mapping, has been fundamental to proactive and early identification of potential geotechnical failure and design hazards impacting critical mine infrastructure. This process incorporates the required PDCA improvement feedback loop to the Tactical Modeller's thereby allowing for increased comprehensive geological and structural reconciliation data to be incorporated into these iterative short-term models and improve their confidence for geotechnical stability review. Established escalation protocols further support and refine this reconciliation process to ensure improved confidence and model turn-around times. This slope reconciliation framework therefore provides the vital link to the design feedback loop and ensures controls are designed and implemented to safeguard and support efficient mining operations.

ACKNOWLEDGEMENTS

The authors wish to acknowledge the contribution of all BHP colleagues and external Consultants past and present related to historic detailed design reports and reviews.

REFERENCES

Armstrong, J, 2022, 24 Feb. Mine management failure mechanism – a geotechnical perspective, *The AusIMM Bulletin*. Available from: <https://www.ausimm.com/bulletin/bulletin-articles/mine-management-failure-mechanisms--a-geotechnical-perspective/>

BHP, 2019. Guide to the Geology of the Hamersley Iron Ore Province – The Blue Book II, unpublished BHP WAIO internal report, Perth.

BHP, 2021a. Our Technical Standards – Geotechnical Standard v3.0, unpublished BHP WAIO internal report, Perth.

BHP, 2021b. #G2G – GeoScience to Geotech Project, unpublished BHP WAIO internal report, Perth.

BHP, 2022. RPD EPH, 20220127 Memo ACS3 Borefields Design Check, unpublished BHP WAIO internal report, Perth.

BHP, 2023a. RPD EPH, 20230307 WB ACS4 Pondy Ramp Geotechnical Review, unpublished BHP WAIO internal report, Perth.

BHP, 2023b. Newman Operations – WB ACS4 Stage 4 and 5 Re-design, unpublished BHP WAIO internal report, Perth.

BHP, 2023c. RPD EPH, 20231103 wbc300_pb_v22 and Hazard WB096 Geotechnical Review, unpublished BHP WAIO internal report, Perth.

BHP, 2024a. Geotechnical Engineering Slope Design Technical Process Instruction, unpublished BHP WAIO internal report, Perth.

BHP, 2024b. ACS3S North Wall Geotechnical Reconciliation Failure Review, unpublished BHP WAIO internal report, Perth.

BHP, 2024c. WAIO Geotechnical Reconciliation Framework, unpublished BHP WAIO internal report, Perth.

BHP, 2024d. DJI Mavic 3 Enterprise – Face Mapping, unpublished BHP WAIO internal QRG, Perth.

BHP, 2025. WAIO Ground Control Risk Management Standard (STD-PLN-SAF-001) v14.0, unpublished BHP WAIO internal report, Perth.

Department of Energy, Mines, Industry Regulations and Safety (DEMIRS), Western Australia, 2022. *Work Health and Safety Act* 2020. *Work Health and Safety (Mines) Regulations,* Western Australian Government publication, Perth, 699 p.

Department of Minerals and Energy, Western Australia, 2019a. *The Code of Practice – Ground control for Western Australian mining operations,* Western Australian Government publication, Perth.

Department of Minerals and Energy, Western Australia, 2019b. *The Guideline – Ground control in Western Australian mining operations,* Western Australian Government publication, Perth.

Dixon, R A, Johnson, T M, de Graaf, P J H, Wessels, S D N and Venter, J, 2011. Risk Based Geotechnical Slope Reconciliation at Rio Tinto Iron ore, Pilbara Operations, in *Proceedings Slope Stability 2011: International Symposium on Rock Slope Stability in Open Pit Mining and Civil Engineering* (eds: E Eberhardt and D Stead), 14 p (Canadian Rock Mechanics Association (CARMA)).

Risk engineering for the human system – embedding psychosocial safety into open pit mining operations

T Kendall-Davis[1] *and D Head*[2]

1. CEO, Anytime Counselling, Coolamon NSW 2701. Email: tina@anytimecounselling.com.au
2. Principle Consultant and Director, EHS Resolutions, Melbourne Vic 3199.
   Email: darren@ehsresolutions.com.au

EXTENDED ABSTRACT

Open pit mining operations have achieved exceptional capability in identifying, measuring, and controlling physical hazards during the past 30 years. Hazards such as Geotechnical stability, traffic interaction, mobile plant, and energy sources are managed through mature engineering processes, often supported by critical control frameworks and assurance protocols.

Yet despite this technical sophistication, many serious incidents, near misses, and operational disruptions within open pit environments continue to originate not from failures of physical engineering, but from breakdowns within the human system that connects our physical controls. These remain poorly measured, weakly controlled, or normalised over time.

This presentation addresses the increased understanding in psychosocial hazards and reframes the approach to these hazards as risk engineering for the human system: a structured, operational approach to identifying, analysing, and controlling human-centred hazards with the same rigour applied to physical hazards. Rather than treating mental health as an individual well-being issue; workplaces need to shift towards understanding that poor mental health (well-being) is impacted by and the outcome of organisational and individual behaviour, decision-making, fatigue accumulation, communication quality, and cultural drift in the workplace. Each of these are system variables that directly influence operational safety, workforce stability, and production continuity in mining environments. The significance of this being that poor mental health is believed to be a precursor condition to Major Unwanted Events (MUEs) which include fatalities.

There are, of course, issues outside of the workplace that also impact mental health, but oftentimes, these are beyond the scope of workplace control.

Recent industry data highlights the urgency of this shift. Research across Australian mining operations indicates persistently high levels of burnout, low workforce confidence in leadership capability to prevent stress-related harm and delayed organisational responses to psychosocial risk indicators. In open pit contexts, which are characterised by extended rosters, large contractor workforces, high supervision ratios, and sustained production pressure, these conditions create latent hazard states that are rarely captured within traditional safety systems. While physical hazards are systematically engineered out, behavioural and cultural risk pathways are often tolerated as 'part of the job' until material harm has already occurred.

On a mid-sized open pit operation, haul truck operators were routinely extending cognitive effort beyond safe thresholds during the final hours of shift. Fatigue was recognised but culturally reframed as 'part of getting the job done'. No formal effective controls existed beyond self-reporting. Following identification of fatigue as a psychosocial Material Contributing Factor (MCF), and a deep causal pathway analysis, objective fatigue triggers were identified, with subsequent controls put in place. These included increased supervisor intervention thresholds, and re-structured end-of-shift handover protocols being implemented. Within three months, level of near-miss reporting increased and fatigue-related operational errors were subsequently reduced.

The authors propose a structured, multi-step methodology that embeds psychosocial risk management into existing operational risk frameworks, rather than treating it as a independent, parallel or advisory function. The process begins with identifying Similar Work Groups (SWGs) based on operational reality. As an example, these would include haulage operators, drill and blast crews, supervisors, maintenance teams, and contractor interfaces – ensuring psychosocial hazards are analysed within the specific work environment profiles, decision environments, and production constraints in which they arise.

Each SWG is subsequently workshopped to identify mental-health-related MCF's with possible MUE correlated. In open pit operations, these MCF's commonly include fatigue-related loss of attention during haulage, supervision and communication breakdown across large spans of control, conflict escalation between crews under production pressure, psychological injury resulting in extended absence, and workforce withdrawal through presenteeism or attrition. These conditions are treated not as well-being outcomes, but as measurable consequences within safety and productivity.

Each identified MCF is formally risk-assessed and modelled using bow-tie analysis, enabling causal and consequence pathways to be examined with clarity. This methodology is used to surface early human-system indicators such as cumulative fatigue, bullying and harassment, isolation anxiety, degraded handovers, inconsistent supervisory decision-making, and the gradual normalisation of unsafe behavioural shortcuts that frequently precede serious incidents in the workplace. These indicators are positioned as legitimate precursors and control points, not informal observations that need to be addressed.

Autonomous thermal inspection for early detection of haul truck tyre injuries in iron ore open pit mining

A Sacca[1] and A Lautenschlager[2]

1. Global Head of Sales, Pitcrew AI, Newcastle NSW 2294. Email: anthony.sacca@pitcrew.ai
2. Hardware Engineering Manager, Pitcrew AI, Newcastle NSW 2294.
   Email: alex.lautenschlager@pitcrew.ai

ABSTRACT

Haul truck tyres are a major operating cost and reliability constraint in iron ore open pit mining. Tyre life is driven by many interacting factors including payload, road condition, and tyre construction; no single measurement captures them in combination. Tyre Pressure and Temperature Monitoring Systems (TPMS) provide useful general health indicators but rely on static thresholds and offer limited insight into internal injuries that progress to tyre fires or full vehicle loss.

This paper presents mid-trial outcomes from an Autonomous Inspection System (AIS) deployed at a Pilbara iron ore mine for a six-month trial. Results reported here cover the first three months, January to March 2026; the trial is ongoing at the time of writing. The platform uses solar-powered side-of-road skids equipped with FLIR thermal imaging and computer vision to scan ultra-class haul truck tyres without slowing the fleet. A continuously trained model identifies tread injuries; a static Trigger Action Response Plan (TARP) matrix translates injury dimensions and temperatures into operator-actionable classifications. The system has been in continuous operation across mining sites in Australia, North America, and South America since February 2020, with approximately five million haul truck scans logged to date.

Across the three months reported here, the site's third-party tyre management contractor recorded 36 early life component failures. The AIS auto-detected 28 of these (78 per cent), with three further failures identified manually from system imagery. Average lead time from first detection to removal was 8.8 days; eight detections provided 14 days of lead time or more, enabling intervention within the planned fortnightly maintenance window.

A rock cut tread separation on a rear-axle position progressed through HIGH and EXTREME TARP classifications over a sustained multi-day alerting window, and was removed in a vented separation condition with nine days of lead time from first detection. The case illustrates the prediction-to-intervention workflow the system enables.

The paper presents the failure-mode taxonomy targeted by the system, the validation chain against independent post-removal reports, observed limitations, and continuous-improvement work including front-axle damage detection released in April 2026.

INTRODUCTION

Haul truck tyres are among the largest contributors to operating cost and unscheduled production loss in open pit iron ore mining. Off the road tyres fitted to 200 t-plus payload haul trucks accumulate thousands of operating hours under demanding payload and surface conditions. Failures range from gradual wear to catastrophic events; the most severe are tyre fires originating from internally progressing sealed separations.

Tyre life is multi-factorial. Component degradation is governed by interacting drivers including payload and load distribution, vehicle speed and braking, haul road condition, ambient temperature, tyre construction, and operator behaviour. No single sensor captures these drivers in combination.

TPMS provides absolute pressure and temperature trending and is widely deployed across iron ore haulage fleets, but it is configured around static thresholds and does not identify localised tread injuries or provide visual evidence of their progression. Sealed separations can develop and grow internally with limited TPMS signature until terminal failure approaches.

The methodology described in this paper addresses a complementary monitoring need: continuous, autonomous, image-based inspection of every haul truck tyre as it passes a designated point on the haul route, with no truck stop or modification required. The objective is to identify progressing tread injuries early enough for intervention to be planned within scheduled maintenance windows.

The Autonomous Inspection System (AIS) discussed here is not a new platform. It has been in continuous operation across heavy-haulage mining sites in Australia, North America, and South America since February 2020, with cumulative scan counts exceeding six million across all vehicle classes and approximately five million haul truck transits. Long-term deployments include multiple Pilbara iron ore operations and large open cut copper operations elsewhere; ongoing deployments at a second major Pilbara iron ore producer have logged over 480 000 vehicle scans across haul fleet and ancillary equipment.

What this paper contributes is the first structured detection-versus-removal validation captured against an independently produced monthly failure-report data set at this granularity. The detection methodology, hardware, and TARP framework are common to all current deployments; what is new is the quantified record of performance reported below.

SYSTEM ARCHITECTURE AND DETECTION FRAMEWORK

Hardware platform

The AIS is a solar-powered, rapid-deployment skid installed at the side of an active haul road. Each skid carries calibrated FLIR thermal cameras, visual cameras, and a local compute and communications stack. As a vehicle passes at normal operating speed, the cameras capture thermal and visual imagery of the tread and outer sidewall of every tyre on the visible side. No modification to the truck or its tyres is required. Figure 1 shows a representative deployment.

FIG 1 – AIS skid deployed adjacent to an active haul road.

The deployment is fully autonomous: no haul road shutdown, no truck stop, and no driver interaction. The configuration was selected to avoid operational disruption and enable rapid mobilisation on-sites.

Detection model

A continuously trained computer vision model processes each scan in near real time, identifying tread injury thermal signatures, primarily separations and foreign object penetrations. Confirmed injury data returned from post-removal investigations is fed back into the training set, so the model improves as deployments accumulate. The detection model is distinct from the alerting framework described below.

Figure 2 illustrates the platform's central proposition. The left panel shows a thermal scan in which the AIS auto-detected an injury at Position 4; the right panel shows the same tyre after removal. The injury was clearly resolved in the thermal channel and classified on transit, but on the recovered tyre it was not visible to the naked eye on a walk-around inspection. Thermal imaging combined with continuously trained computer vision identifies injuries that visual inspection alone does not.

FIG 2 – Left: thermal scan with AIS-detected injury at Position 4 (bounding box). Right: tyre after removal, confirmed as a rock cut tread separation not visible on a walk-around inspection.

TARP framework

A static Trigger Action Response Plan (TARP) matrix translates each detected injury into an operator-actionable classification of LOW, MEDIUM, HIGH, or EXTREME based on the maximum value across injury height, injury width, and injury temperature for that scan. The matrix used in the trial is shown in Table 1.

TABLE 1

TARP threshold matrix used during the trial.

TARP level	Injury temperature	Injury height	Injury width	Operator action
Low	>20°C	>0 mm	>0 mm	Monitor on subsequent scans
Medium	>60°C	>500 mm	>350 mm	Schedule for inspection
High	>75°C	>1000 mm	>500 mm	Prioritise for inspection
Extreme	>80°C	>1500 mm	>650 mm	Immediate action per site procedures

The TARP matrix is intentionally simple, deterministic, and auditable. Thresholds are configured at site commissioning and are not modified by the detection model. This separation of concerns is deliberate: site operations teams set their alert tolerances based on local fleet, road, and tyre conditions, while the detection model is free to improve continuously without altering the alerting logic. The dashboard reports the highest TARP classification triggered by any single parameter for a given tyre position on a given scan; the matrix does not currently support conditional triggering across multiple parameters simultaneously.

Coverage envelope and limits

By design, the cameras image tread and outer sidewall surfaces from the scanning lane. Injuries that develop on the inboard sidewall or that are shadowed by other tyres may not be detected. This is an acknowledged geometric limitation of the current installation and is addressed in the limitations and continuous-improvement sections later in the paper.

FAILURE-MODE TAXONOMY TARGETED BY THE SYSTEM

The data set reported here comprises 36 tyre component failures recorded across the haul fleet over the first three months of the trial. Classifying each failure by its post-removal investigation reason produces the distribution shown in Figure 3.

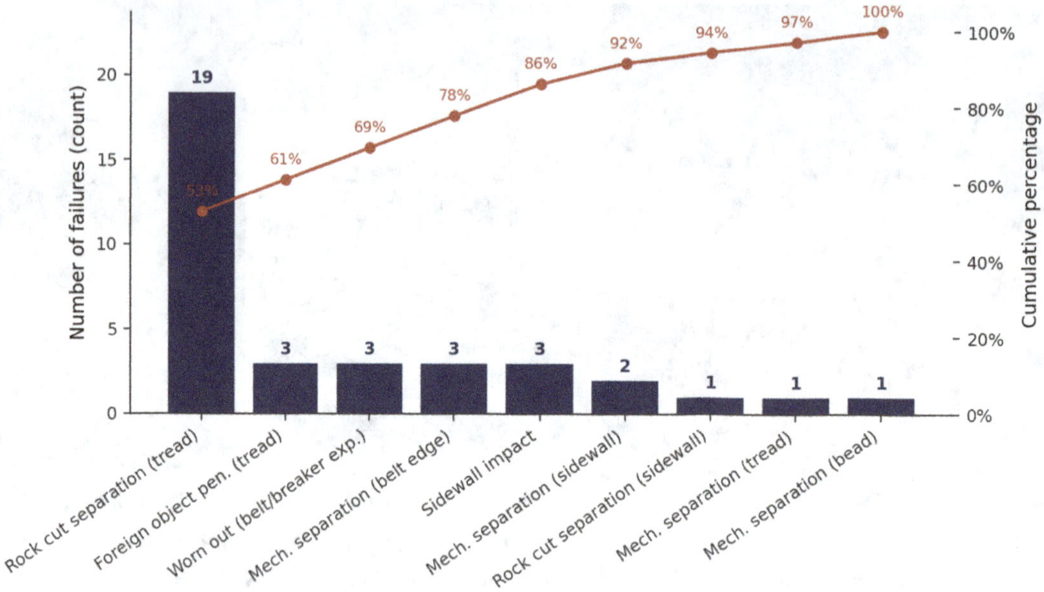

FIG 3 – Pareto chart of tyre failure modes recorded across the first three months of the trial (n = 36 failures).

The dominant failure mode, accounting for 53 per cent of all failures in the trial period, is rock cut separation of the tread. This is an internally progressing failure mode characterised by an external rock cut that initiates progressive separation between belt plies and the surrounding rubber.

Rock cut tread separations can present as either sealed, where the defect remains contained beneath the tread surface, or vented, where the defect has breached the tread surface. Both presentations require tyre removal. Sealed separations carry the higher fire risk because the contained defect retains heat. Terminology for the vented condition varies between operations: it is referred to as a puffer at some sites, while at others the term puffer is reserved for an actively smoking tyre. Early identification of both presentations is the primary value proposition the AIS is engineered around.

The surface signature of a rock cut separation on a removed tyre is often subtle. Figure 4 shows a close-up after AIS detection: the white painted arear marks where post-removal investigation confirmed an internally progressing separation. The visible markings are not, in isolation, sufficient to identify the failure on a walk-around inspection.

FIG 4 – Close-up of a removed tyre tread with the painted location of an AIS-detected rock cut separation. The visual signature is subtle and may not be identified on a walk-around inspection.

A further 17 per cent of failures were tread injuries of other types: foreign object penetration (8 per cent), worn tyres with the belt or breaker exposed (8 per cent), and tread mechanical separations (3 per cent). All sit within the system's detection envelope as they present on the visible tread surface.

The remaining 30 per cent comprise sidewall injuries (impacts, rock cuts, and mechanical separations) and bead-area mechanical separations. Sidewall injuries on the scanning side are detectable; injuries on the inboard sidewall and in the bead lie outside line of sight and represent the bounded limit of detection capability.

Approximately 70 per cent of all failures in the trial period are tread injuries within the system's target population, dominated by rock cut tread separation.

TRIAL METHODOLOGY AND VALIDATION

The trial commenced on 20 January 2026 at a Pilbara iron ore mine and is scheduled to run for six months. Results presented here cover the first three months (January to March 2026); the trial is ongoing at the time of writing. Four AIS skids were deployed at locations selected for repeated daily coverage of the production haul fleet. Trucks transit each scanning zone multiple times per shift. Across the reporting period the system recorded approximately 61 000 scans of which approximately 43 000 were dump trucks (averaging 605 per day).

Each scan is logged with thermal and visual imagery, computed injury metrics, and the resulting TARP classification.

Validation relies on independent post-removal investigation. Every tyre removed from the haul fleet is inspected by the site's third-party tyre management contractor, who issues a weekly Tyre Failure Report classifying each removal by failure mode. AIS detection records are then triangulated against these reports to assess whether each failure was identified, with what lead time, and at what TARP classification. This evaluates detection performance against independently produced ground truth rather than against system-internal classifications. For failures not auto-flagged, system imagery for the affected truck and position over the preceding weeks was reviewed manually to determine whether the injury had been visible but not detected, or was outside the detection envelope.

RESULTS

Detection performance

Of the 36 component failures recorded during the reporting period, the AIS auto-detected 28 (78 per cent) prior to tyre removal. Three further failures were identified manually from system imagery: two predated full system commissioning, and one was a thermal anomaly that the model did not classify as an injury. Five failures were not detected, all involving injuries on the inboard sidewall, in the bead, or on a worn surface, outside the detection envelope. Trial performance metrics are summarised in Table 2.

TABLE 2

Performance summary for the first three months of the six month trial, January to March 2026.

Metric	Value
Reporting period	3 months of 6 month trial
AIS skids deployed	4
Total vehicle scans	61 069
Dump truck scans	42 582
Average scans per day	605
Total component failures	36
Auto-detected by AIS	28 (78 per cent)
Manually identified from imagery	3 (8 per cent)
Not detected	5 (14 per cent)
Average lead time, auto-detected	8.8 days
Lead time range	1 to 18 days
Detections with lead time of 14 days or more	8 (29 per cent)
Detections with lead time of 8 days or more	17 (61 per cent)

Lead time distribution

Lead times for auto-detected failures are shown in Figure 5. The average was 8.8 days, with a range of one to 18 days. Eight detections (29 per cent) gave 14 days or more, falling within the planned fortnightly maintenance window. A further nine (32 per cent) gave eight to 13 days, supporting opportunistic intervention but typically requiring a non-scheduled stand-down. Eleven (39 per cent) gave one to seven days, predominantly in cases where the failure was rapidly progressing.

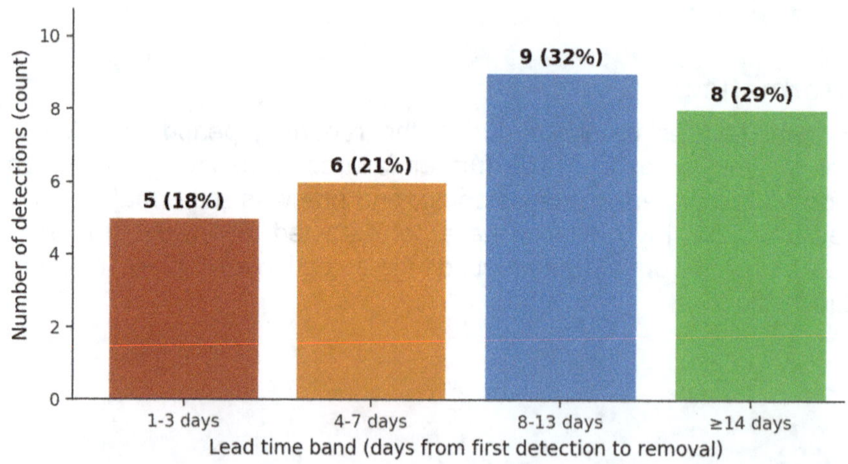

FIG 5 – Distribution of detection lead times for auto-detected failures (n = 28).

The widespread reflects the variability of internally progressing tyre failures: detectability windows vary by an order of magnitude between cases. The system detects injuries as early as they are thermally or geometrically resolvable; its operational value at any site depends on the underlying mix of failure modes and rates of progression.

CASE STUDY – ROCK CUT TREAD SEPARATION

This section describes the prediction-to-intervention workflow for a single high-consequence event during the trial. The vehicle is anonymised as Truck A; the affected position is rear axle Position 3.

The failure mode was confirmed post-removal as a rock cut tread separation that had progressed to a vented condition at the point of removal.

Detection and escalation timeline

The injury was first detected on 9 March 2026 at HIGH TARP, with injury temperature 51.05°C and dimensions of 528 × 527 mm; the detection was raised on the dashboard at that point. The next scan, on 10 March, escalated to EXTREME TARP with injury temperature 58.05°C (the peak observed during the case) and dimensions of 795 × 722 mm. Over 11 to 17 March, the system recorded HIGH and EXTREME classifications on every transit; injury temperatures remained above 48°C and dimensions oscillated between roughly 500 and 750 mm as the separation propagated. The escalation profile is shown in Figure 6 and the per-scan record in Table 3.

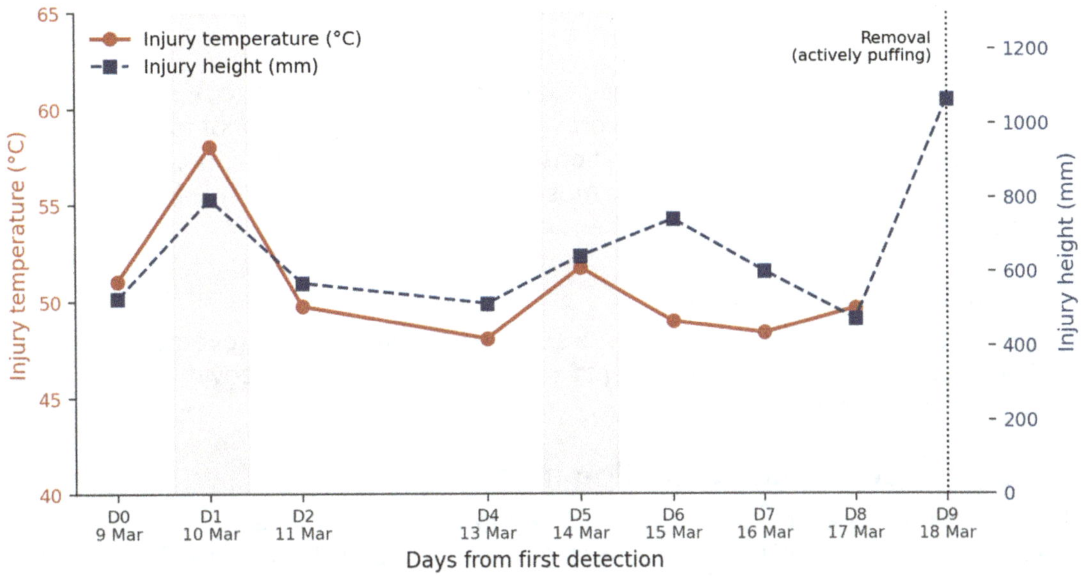

FIG 6 – Truck A injury progression: injury temperature (left axis) and injury height (right axis) across the nine-day detection window. EXTREME TARP scans are shaded.

TABLE 3

Truck A scan progression: per-scan TARP classifications and injury metrics from first detection to removal.

Date	TARP classification	Injury temperature (°C)	Injury height (mm)	Injury width (mm)
9 Mar 2026	HIGH (first detection)	51.05	527.8	526.7
10 Mar 2026	EXTREME (escalation, peak)	58.05	795.0	722.3
11 Mar 2026	HIGH	49.75	570.1	593.9
13 Mar 2026	HIGH	48.05	514.4	637.0
14 Mar 2026	EXTREME	51.75	641.7	750.0
15 Mar 2026	HIGH	48.95	740.8	573.6
16 Mar 2026	HIGH	48.35	601.3	587.6
17 Mar 2026	HIGH	49.65	472.8	533.0
18 Mar 2026	Removal (vented separation)	n/a	1064.5	712.4

Outcome

The tyre was removed on 18 March 2026. The post-removal inspection confirmed a vented separation. Final injury dimensions were 1064 by 712 mm. Lead time from first detection to removal was nine days.

This case illustrates each element of the intended prediction-to-intervention workflow:

- Early autonomous detection at HIGH TARP on the first scan to resolve the thermal and dimensional signature.
- Sustained alerting over multiple days, providing time for inspection planning and confirmation.
- Operator action to remove the tyre while the failure was still containable.
- Independent post-removal validation confirming the failure mode the system was designed to intercept.

The case demonstrates the system's capacity to autonomously detect a rock cut tread separation, classify it through the TARP framework, and provide multi-day lead time prior to removal. The same detection workflow applies to sealed separations, where early identification carries the additional benefit of intercepting the failure presentation most closely associated with tyre fire risk. The targeted sealed-separation detection model is described below.

CONTINUOUS IMPROVEMENT

The detection model and hardware platform are subject to continuous improvement informed by trial data and field operations. Two improvements released in the period covering and immediately following the trial are described here.

Targeted sealed-separation detection model

A targeted detection model focused on sealed separations was released during the trial period. Sealed separations are the failure type most closely associated with tyre fire. The targeted model produces fewer alerts than the general-purpose tread injury model, but each alert is more strongly correlated with the safety-critical failure mode, reducing operator triage burden. The general-purpose model continues to operate alongside it.

Front-axle damage detection (April 2026)

In April 2026, damage detection was extended to front-axle tyres (Positions 1 and 2). Damage detection had previously been limited to rear-axle positions (3 to 6) although temperature monitoring covered all six. The release extends damage-detection coverage to the full tyre complement.

The initial release returns absolute and differential temperature measurements; injury size estimation for front positions is in development. Temperature-based TARP classifications operate at the same thresholds as the rear positions. The front-axle model is an extension of the rear-position sealed-separation model and represents a capability uplift over the previous generic tread damage detection model used on the rear positions for the preceding six years.

DISCUSSION

Limitations

The system does not detect injuries on the inboard sidewall, in the bead, or in any location obscured by the truck body from the scanning lane. This is a geometric limit of the deployment topology, observed in five trial cases (14 per cent of failures). The choice to scan from the side of the haul road, with no truck stop, is deliberate; a bounded sidewall blind spot is the trade-off for continuous unobtrusive coverage of every transit.

Predictive confidence is lower under previously unseen operating conditions; the model improves as deployments expose it to a broader range of injury signatures, ambient regimes, and tyre constructions. The system's near-term value is best characterised as early detection and prioritisation of progressing injuries rather than deterministic prediction of remaining tyre life.

Operational relevance

Tyre life remains multi-factorial. The AIS does not address payload, road, route, or driver behaviour drivers in isolation. It is intended as one element of a broader haulage reliability strategy, providing an early-warning signal for progressing internal injuries that complements existing TPMS, scheduled inspection, road maintenance, and fleet management practices.

As noted in the introduction, the platform is in operation across multiple Pilbara iron ore producers and across global hard rock open pit operations including large open cut copper. The failure-mode mix and lead-time figures will vary by site, fleet, road condition, and tyre product, but the detection methodology and TARP framework are common to all deployments. The mid-trial results reported here are the first published quantification of detection performance against an independent post-removal data set; the numbers should be read as a representative single-site benchmark rather than a global average.

CONCLUSIONS

Three months into a six-month trial of an Autonomous Inspection System at a Pilbara iron ore mine, the AIS has demonstrated:

- Autonomous detection of 78 per cent of all tyre component failures recorded by the site's independent tyre management contractor during the reporting period, with three further failures identified manually from system imagery.
- An average lead time of 8.8 days from first detection to removal, with eight detections (29 per cent) providing 14 days of lead time or more, sufficient to intercept the failure within a planned fortnightly maintenance window.
- Continuous autonomous detection of a rock cut tread separation, classified through the TARP framework with nine days of lead time from first detection to removal in a vented condition.
- Detection coverage matched to the dominant failure modes, with rock cut tread separation (53 per cent of failures) targeted directly and a documented limit on inboard sidewall and bead-area injuries (14 per cent, not detected).

Tyre life is multi-factorial; effective fleet management combines proactive injury detection with payload, road, route, and operator management. Within that strategy, autonomous side-of-road thermal inspection provides a measurable, operator-actionable early-warning signal for the failure mode of greatest safety and production consequence. The platform has been in continuous operation since 2020 across Australian, North American, and South American mines; this paper is the first quantification of its detection performance against an independently produced failure data set and reports the first three months of a six-month trial that is ongoing at the time of writing. Continuous improvements including the targeted sealed-separation model and the front-axle extension demonstrate that the underlying capability continues to develop in step with field experience.

ACKNOWLEDGEMENTS

The authors thank the operations and tyre management teams at the host iron ore operation for site access, trial collaboration, and the independent post-removal investigation reports against which detection performance was measured. The engineering, software, and product teams at Pitcrew AI are acknowledged for their work on the platform.

ACKNOWLEDGEMENTS

Driving excellence in surface mining operations – a case study on Nutrien's Aurora phosphate mine

N Smit[1], J Pierce[2], R Jahan[3] and A Premachandra[4]

1. Global Director, Operations Transformation, Hatch Ltd, Houston TX 77024, USA. Email: nico.smit@hatch.com
2. Director, Mining-Phosphate, Nutrien Ltd, Aurora NC 27806, USA. Email: jeremy.pierce@nutrien.com
3. Senior Analyst, Hatch Ltd, Toronto ON L5K 2R7, Canada. Email: rounaq.jahan@hatch.com
4. Senior Analyst, Hatch Ltd, Toronto ON L5K 2R7, Canada. Email: abhishek.premachandra@hatch.com

ABSTRACT

Operational-excellence initiatives in mining seek to increase production and reduce operating costs through low-capital, high-impact improvements that can be rapidly deployed across the value chain. These initiatives typically focus on bottleneck elimination, improved equipment performance, and stronger operational decision-making. This paper presents a case study of an operational transformation at Nutrien's Aurora phosphate mine, demonstrating how efficiency gains were achieved across key surface-mining domains, including ore handling, equipment reliability, mine planning, and digital automation.

The engagement aimed to strengthen Nutrien's operational capabilities by identifying performance gaps and implementing targeted improvements. A two-month diagnostic phase identified 23 improvement opportunities across the mines, beneficiation plant, and site-wide asset management, of which ten were prioritised for implementation. Beginning in March 2025, a series of quick-win initiatives were launched to deliver rapid impact and support Nutrien's broader operational-excellence journey.

Key improvements focused on reducing non-productive time and stabilising ore flow. Measures included implementing hot-seating during shift changeovers, optimising refuelling practices, repositioning excavators to minimise swing times, and improving coordination between mobile equipment. Reliability of critical ore-handling infrastructure was enhanced through improved spare-equipment strategies, real-time slurry-density monitoring, and process simplification. Slurry handling was further stabilised by consolidating feed tanks and optimising pumping systems to increase slurry density, reduce water usage, and improve downstream performance.

Digital tools were deployed to improve downtime tracking, data integrity, and real-time operational visibility, enabling proactive stockpile management and maintenance planning while sustaining production targets. Overall, the case study demonstrates that sustained performance improvement is driven by accurate root-cause identification, integration of real-time insights into daily operations, and strong workforce ownership aligned with technical and digital solutions.

INTRODUCTION

Operational excellence in surface mining is essential to achieving production targets, controlling costs, and meeting rising expectations for safety, sustainability, and digital innovation. This case study highlights an operational excellence transformation initiative conducted by Nutrien in partnership with Hatch Ltd. at Nutrien's phosphate operations in Aurora, North Carolina – the largest integrated phosphate mine and chemical plant in the world. Beginning in January 2025, the transformation project focused on rapidly increasing ore production through efficiency improvements across mine planning, equipment effectiveness, equipment maintenance, and digital operations by implementing low-cost, quick-win initiatives with minimal capital investment.

Context

Nutrien Aurora's phosphate operation uses strip mining to extract ore from 30 foot thick phosphate layers buried under 70–150 feet of overburden. The mine operates with three dragline excavators that create stockpiles for downstream processing. Mobile ore slurrying tanks are positioned near active mining areas and are relocated as mining progresses. From ore stockpiles adjacent to the draglines, 40 ton haul trucks transport ore to excavators positioned around the slurrying tanks, where ore is loaded onto conveyors for screening before being fed into the tanks.

The operation utilises two slurry tanks that pump ore slurry to the beneficiation plant via booster pumps through a 13.45 km long piping network. The battery limit for this paper extends from the mine to the point of entry at the beneficiation plant. The scale and complexity of the operation demand precise coordination of people, equipment, and processes to maximise productivity and minimise downtime. A schematic showing the key process steps in the end-to-end operation is shown in Figure 1.

FIG 1 – Block flow diagram of phosphate mining operation at Nutrien Aurora.

PURPOSE FOR ENGAGEMENT

The purpose of this engagement was to strengthen Nutrien Aurora's operational capabilities by identifying key challenges and implementing practical, site-specific solutions to improve process efficiency and production performance.

In recent years, Nutrien Aurora has undertaken a review of operating practices across the site with the objective of strengthening production resilience, improving operational efficiency, and positioning the operation for sustainable long-term performance. As part of this effort, Nutrien Aurora identified opportunities to enhance integration across its value chain, with the goal of maintaining consistent production and operational flexibility as ore grade and operating requirements evolved. The project also progressed into new mining areas, prompting a broader assessment of layout, material flow, and work practices to ensure continued efficiency and reliability across the operation.

In parallel, Nutrien Aurora prioritised reinforcing organisational capability and operational discipline to support continuous improvement and knowledge continuity. To advance these objectives, the company established an Operations Reliability Task Force and subsequently engaged Hatch Ltd. to integrate with the task force and identify and execute initiatives to increase production across the mines and beneficiation plant through a coordinated 'one-team' approach.

TRANSFORMATION APPROACH

Starting in January 2025, a project-wide transformation was initiated with the primary objective of rapidly increasing concentrate production from the beneficiation plant. It was determined that improvement areas needed to be systematically identified and mapped into discrete initiatives, followed by the development of a clear execution strategy to guide the transformation effort.

The transformation programme was structured into two phases: Diagnostic and Implementation.

Diagnostic phase

The diagnostic phase, which ran for two months, was conducted in collaboration with Nutrien's Reliability Task Force. This phase focused on identifying operational challenges through a comprehensive, data-driven analysis of key value drivers. The assessment included deep dives into:

- Process efficiency.
- Plant bottlenecks.
- Equipment utilisation.
- Workforce capability.
- Production and maintenance planning.

- Equipment reliability and availability.
- Other critical performance areas.

The diagnostic resulted in the identification of 23 improvement initiatives related to the mines, the mill, and the Asset and Maintenance Management (AMM), of which ten were prioritised for implementation.

Implementation phase

The implementation phase began in March 2025 with quick-win initiatives focused on AMM, particularly reducing conveyor downtime through operator-driven preventive maintenance. Mine improvement initiatives focused on increasing efficiency in ore-handling practices and boosting the volume and consistency of ore delivered to the mill, ensuring a stable supply that could support operational changes in the beneficiation plant.

Throughout implementation, the emphasis was on:

- On-site coaching to build workforce capability.
- Introducing Short Interval Control (SIC) tools to enhance process monitoring and enable timely operator intervention.
- Establishing clear leadership accountability and sponsorship, with Nutrien personnel accountable as initiative owners under a 'one-team' approach.
- Embedding a culture of ownership and continuous improvement at site-leadership levels.
- Executing initiatives with urgency to ensure rapid time-to-value.

KEY CHALLENGES AND INTERVENTIONS

Ore handling optimisation

A notable source of idle time was identified during shift changeovers and refuelling activities, amounting to nearly two hours of lost production time per day. A cost–benefit analysis demonstrated that increasing operator shift times to maintain production during shift changeovers, together with deploying an additional fuelling crew, would materially reduce downtime. The projected gains in productivity and tonnage were shown to far exceed the marginal costs associated with extending crew schedules by 30 mins and contracting an additional fuelling crew to increase fuelling capacity.

Accordingly, a hot-seating method was implemented during shift changes, enabling incoming operators to assume control of equipment without shutdown. This was achieved by staggering day and night shifts to create an overlap period during which shift handovers and pre-shift inspections occurred on operating equipment. It was also identified that the existing fuelling schedule contributed to avoidable downtime, and that refuelling equipment during the handover window – while units remained operational – would be a more efficient approach. In addition, the equipment refuelling order was revised to prioritise production-critical equipment. An additional fuelling crew was contracted to refuel equipment around the slurry tanks and stockpile simultaneously, minimising overall fuelling time.

Figure 2a shows total daily pre-shift downtime from January to June 2025, with the initiative implemented on 14 February. Immediately following implementation, average pre-shift downtime declines noticeably, and the spread of daily values tightens.

Figure 2b compares the distribution of pre-shift downtime before and after initiative implementation. Prior to implementation, average downtime was higher and more variable, with total variability averaging 0.48 hrs – as shown by the height of the distribution band. After implementation, average downtime decreased by approximately 0.55 hrs, while variability declined to 0.26 hrs, representing a reduction of 0.22 hrs in variability. Together, these results demonstrate that the initiative delivered both downtime reduction and consistency.

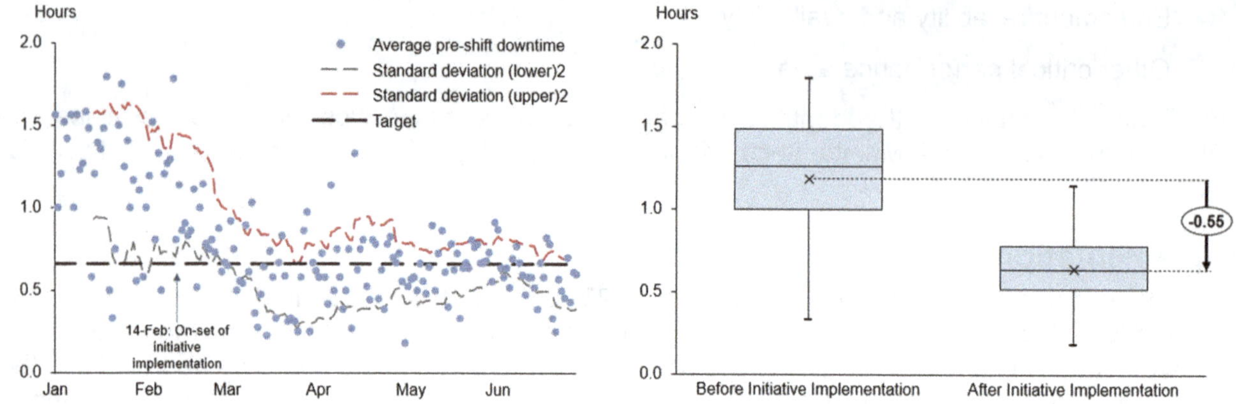

(a) (b)

FIG 2 – Impact of pre-shift downtime initiative.

Additionally, automated triggers were implemented on PowerBI to notify leaders when the fuelling and pre-shift downtime started to deviate from set limits. Fuelling audits and root cause analysis were initiated when deviation persisted for long periods. This intervention of combining hot-seating with an optimised refuelling process, reduced idle time and increased effective operating hours, unlocking approximately one additional hour of production per day.

Another challenge identified was that the excavators loading from truck dumps and unloading into screen feed chutes exhibited inefficient swing motions between pick-up and drop-off, resulting in extended cycle times. This inefficiency stemmed from the lack of prior planning for excavator positioning relative to surrounding work areas. Analysis showed that a 90° swing angle between pick-up and drop-off was the most efficient configuration. Achieving this required planned positioning of excavators relative to the dump ground and feed chutes. Accordingly, excavator parking pads around the dump ground were reconstructed to enable the optimal swing angle. Figure 3 shows an overhead view of the equipment set-up around the slurrying tank before and after the screens and excavators were reoriented.

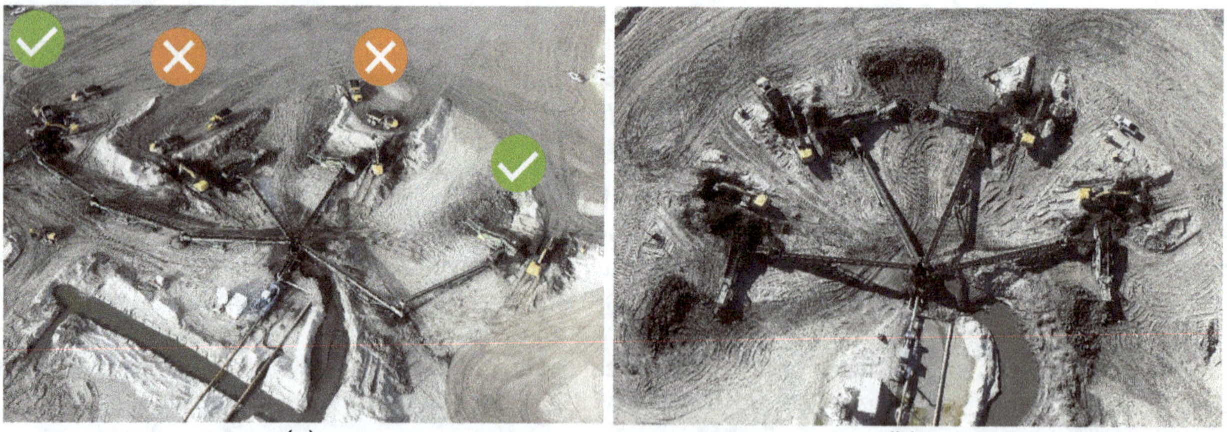

(a) (b)

FIG 3 – (a) Before initiative implementation: Excavators 2 and 3 demonstrate inefficient swing motion. (b) After initiative implementation: All excavators demonstrate optimised swing motion.

Furthermore, hauling trucks offloading to these excavators lacked visibility into which excavator had low working volume. To address this, operator-controlled flashing lights were installed at each excavator parking pad to indicate ore requirements – reducing truck idle time. Implementing optimal swing rotation and signalling systems increased hauled volume by approximately 22 tph.

Equipment reliability

Conveyor reliability was identified as a critical challenge, as the mobile conveyors feeding the slurry tank frequently failed, disrupting production. To address this, a spare conveyor was positioned at the slurry-tank location, enabling replacement of a failed unit within 15 mins. In parallel, the maintenance team prioritised repairs to ensure a consistent supply of operational spares.

Slurry density in the tank was monitored as a key indicator of conveyor performance. A drop in density signalled a reduction in ore volume within the tank, indicating a backlog on the upstream conveyor. With support from Nutrien's on-site instrumentation team, a real-time slurry-density monitoring dashboard was developed and integrated into the operators' human–machine interface (HMI) screens. Operators were trained to monitor slurry density, coordinate the diversion of production to the spare screen, and raise work orders for repair of the failed screen. This approach minimised downtime and ensured continuity of ore processing.

Additionally, slurry production, which was previously managed through two feed tanks, was consolidated into a single tank to minimise density fluctuations. Operating with two pumping systems required additional water to maintain flow rates, resulting in a lower-density slurry (approximately 33 per cent). This meant that a high volume of water was being pumped to the beneficiation plant, frequently straining downstream dewatering equipment. Transitioning to a single pumping system immediately reduced the water load to the beneficiation plant and eliminated dewatering bottlenecks.

In parallel, a pumping-system model was developed using the Applied Flow Technology Fathom software (AFT Fathom) to determine the maximum allowable slurry density under existing pressure and motor-power constraints. This analysis supported increasing slurry density to approximately 40–45 per cent, further reducing water throughput. The combined effect improved recovery and increased uptime at the downstream beneficiation plant.

Furthermore, Nutrien Aurora invested in a custom-made ore-feeding system with the capacity to supply the slurry pumps at 55 kilotonnes per day (ktpd), compared with the existing capacity of 40 ktpd. Subsequent upgrades to the pumping system were required to match this increased capacity, and the AFT Fathom model was used to perform an options analysis to identify the system upgrades required to enable production at 55 ktpd.

As part of this analysis, an operating window was defined to describe the combinations of flow rate and slurry density achievable with the existing pumping system. These were then compared against the flow rate and slurry-density requirements needed to achieve the 55 ktpd production target.

Seven upgrade scenarios were evaluated using AFT Fathom and assessed based on whether the resulting flow rate and density combinations fell within the target operating window. The estimated cost of each scenario was also evaluated and used as an additional basis for comparison. The key results from the option analysis performed in ATF Fathom, as well as the estimated cost of each option, are presented in Table 1.

Assessed upgrade options included installing higher-power motors, switching to smaller-diameter piping, adding booster pumps, and reducing impeller size. These options were compared by assessing the expected improvements in maximum flow rate and slurry density against their respective capital costs.

These options were plotted to identify the most optimal pumping-system configuration that met the 55 ktpd throughput requirement while maintaining slurry density nearing 50 per cent.

The green region in Figure 4 represents the operable envelope of the pumping system in terms of slurry density and flow rate. Options 1B, 2, 4A, and 4B (highlighted in gold) were identified as capable of operating within the target region.

TABLE 1

Description and results of the options analysis. NB: Units are presented in US customary units, as the project was executed using US standards and cannot be reliably converted to metric units.

Option	Scenario description	Base case	Modelled scenario	Flow rate (gal/min)	Density (%)	Potential tonnage (kt)	Cost ($/kt-mile-yr)
Base case	Existing operation	N/A	N/A	14 576	33%	30	22.21
1A	Increase motor capacity	1500 hp	2000 hp	14 426	50%	53	16.16
1B	Increase motor capacity + additional pump(s)	1500 hp, 6 pumps	2000 hp, 8 pumps	16 523	50%	60	18.85
2	Decrease impeller size + additional pump(s)	45'', 6 pumps	42'', 8 pumps	15 345	50%	56	16.52
3A	Increase pipe diameter	20''	22''	16 744	40%	45	20.44
3B	Increase pipe diameter	20''	24''	No Solution			
4A	**Increase motor capacity + pipe diameter**	**1500 hp, 20''**	**2000 hp, 22''**	**16 631**	**50%**	**60**	**18.37**
4B	Increase motor capacity + pipe diameter	1500 hp, 20''	2000 hp, 24''	20 332	50%	74	17.20

gal – gallons; min – minute; kt – kilotonnes; yr – year.

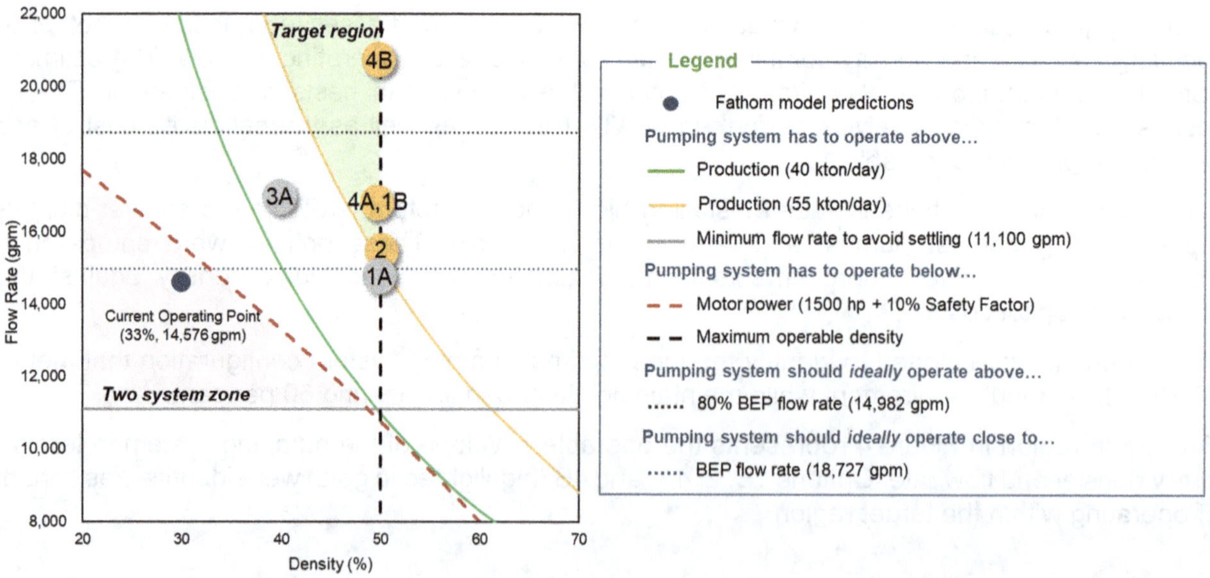

FIG 4 – Results of options analysis on the pumping system operating window.

While Option 4B offered the highest potential production rate, the use of larger-diameter piping increased the minimum flow rate required to prevent slurry settling, thereby reducing operational flexibility during downstream upset conditions. Option 2 was not selected, as the smaller impeller size would have required additional motors to be installed as mining advanced in order to compensate for reduced power. Between the remaining options, 4A and 1B, Option 4A was selected due to its lower capital cost and more favourable operating flexibility.

It was determined that implementing Option 4A would enable pumping at slurry densities of up to 50 per cent and increase the maximum achievable flow rate by approximately 15 per cent. Combined, these upgrades were estimated to reduce pumping costs from $22.21 per kilotonnes-mile-year to $18.37 per kilotonnes-mile-year.

Mine planning

Nutrien Aurora's strip-mining method required close coordination between draglines that supplied ore to the stockpile and downstream mobile equipment that continuously consumed material from the stockpile. During periods of overburden removal, draglines did not add ore to the stockpile; however, stockpile drawdown continued at a steady rate to support downstream operations. As a result, stockpile volumes fluctuated continuously. These fluctuations were further amplified by concurrent initiatives to increase mining productivity and higher ore-pumping rates to the beneficiation plant.

Under normal operating conditions, with all three draglines in service, Nutrien Aurora was generally able to maintain a healthy stockpile volume. However, during scheduled one-month dragline turnaround periods, available ore supply was reduced and stockpile levels declined to the point that ore-pumping rates to the beneficiation plant had to be curtailed.

To address this, options to maintain adequate stockpile levels during dragline turnarounds were evaluated. A scenario analysis was conducted to determine the dragline utilisation (operating hours per month) and productivity (tonnes per hour) required to sustain a defined minimum stockpile volume (Figure 5). Benchmarking of previous turnaround periods showed that acceptable stockpile levels could be maintained with two operating draglines, provided utilisation and productivity were closely monitored and managed.

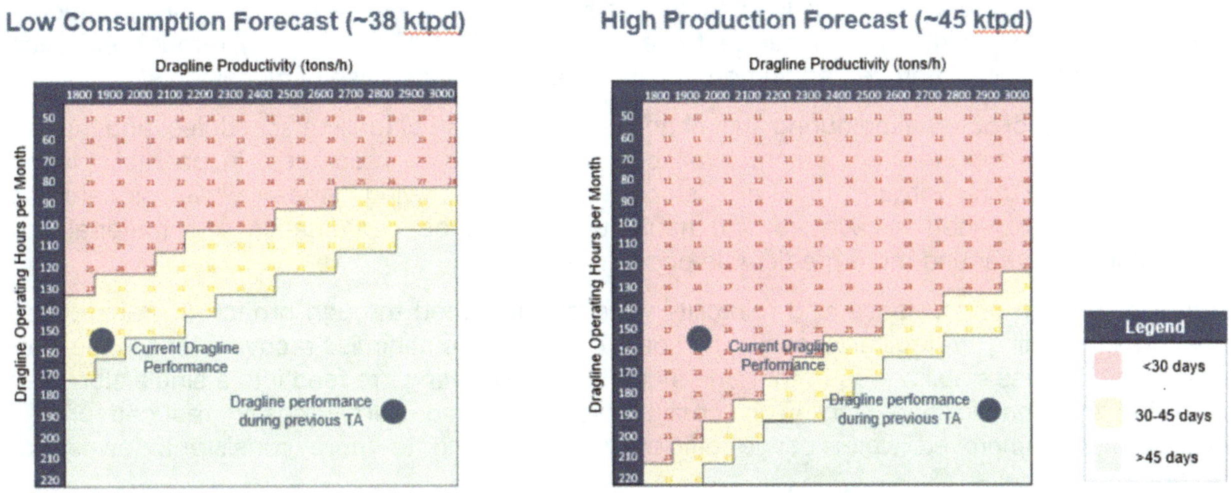

FIG 5 – Results of the dragline utilisation and productivity scenario analysis.

In Figure 5, the red, yellow, and green regions represent the range of days required for the stockpile to reach the minimum acceptable threshold. The red region corresponds to less than 30 days (minimum dragline turnaround duration), the yellow region to 30–45 days (average turnaround duration), and the green region to greater than 45 days (maximum turnaround duration).

Historical performance during turnaround periods, when one of the three draglines was out of service, generally fell within the green region. However, periods of reduced performance were observed when operating conditions shifted into the yellow and red regions in both the low and high stockpile consumption scenarios. Based on this analysis, a new operating target was proposed to maintain dragline productivity and utilisation within the green region.

Building on these insights, enhanced planning and monitoring processes were implemented to improve stockpile-level forecasting and better align operations with maintenance schedules. LiDAR drone surveys were used to regularly measure stockpile volumes. Dragline productivity was redefined to explicitly incorporate the stockpile volume changes in addition to the tonnes per hour mined.

By continuously tracking dragline productivity and utilisation and comparing actual performance against the scenario-analysis results, early indicators of under-performance could be identified. When productivity or utilisation fell below established thresholds, corrective actions were initiated through short-interval control before stockpile levels dropped below the minimum acceptable limit. These proactive measures improved operational resilience and flexibility, allowing draglines to be taken offline for maintenance without adversely affecting production targets.

Digital automation

Prior to the transformation, downtime tracking relied on manual Excel spreadsheets, which posed risks of data loss and limited analytical capability. To address this, the operation migrated to a secure, server-based database with pre-configured fields for standardised data entry. A structured process was developed to ensure that root causes of failures were logged, and entries linked to corresponding work orders. Superintendents were held accountable for verifying that supervisors' entries met these requirements.

This system enabled real-time reporting, improved data integrity, and supported the development of Power BI dashboards for operational insights. Close coordination between Information Technology (IT) team and operations during the transition ensured smooth adoption and comprehensive user training.

Furthermore, mobile surveying applications were introduced for equipment operators to log equipment health data. Maintenance teams used these survey results to assess equipment condition and plan proactive maintenance activities, reducing unplanned downtime and improving reliability.

RESULTS AND IMPACT

The interventions implemented at Nutrien Aurora's mining area delivered significant productivity gains, unlocking an additional 9.7 per cent over the base 33 kt of ore per day through enhanced operational efficiency, improved equipment reliability, and expanded use of digital tools.

The introduction of hot-seating during shift changes, optimisation of fuelling schedules, and targeted operator training improved equipment availability and workforce readiness. In parallel, planned excavator positioning around the ore stockpile reduced swing times, while signal-based communication between excavators and haul trucks improved coordination and material flow, resulting in smoother and more predictable operations.

Conveyor system performance and reliability were strengthened through proactive preparedness measures, including strategically positioning spare equipment to minimise recovery time in the event of failures. Process stability was further enhanced by consolidating ore feed into a single slurry tank supplying the beneficiation plant. This configuration stabilised slurry density, reduced process variability, and improved water usage efficiency, contributing to more consistent downstream performance.

Digital enablement and preventative mine planning played a critical role in sustaining these gains. Enhanced downtime tracking improved data accuracy and transparency, enabling more reliable performance analysis and forecasting of dragline excavator maintenance requirements. These insights supported the refinement of stockpiling and maintenance strategies, further reducing operational risk and improving system resilience.

Overall, these combined initiatives elevated operational performance, increased resilience, and strengthened the reliability of the mine-side value chain.

LESSONS LEARNED

- **Diagnosing the root cause:** Combining data analysis with on-site observations helps uncover low-hanging opportunities for improvement and ensures that interventions address the actual problem rather than symptoms.
- **Data-driven decisions:** Real-time data systems are essential for identifying bottlenecks and enabling proactive interventions. However, data alone is not enough – findings must be

embedded into workflows that trigger action rather than passive monitoring. Real-time insights also support maintenance planning and outage management forecasting.

- **On-site ownership matters:** Sustainable improvements require combining technical solutions and digital initiatives with workforce engagement. Ensuring operator ownership and buy-in throughout the transformation process is critical for success.

CONCLUSIONS

Driving operational excellence in surface mining requires a holistic approach that addresses people, processes, and technology. Combining data analysis with on-site observations helps identify low-hanging opportunities for improvement. The operational excellence transformation at Nutrien Aurora's phosphate mining demonstrates how targeted interventions across multiple domains can deliver measurable gains in productivity and reliability through cost-effective solutions. These lessons provide a practical roadmap for other mining operations seeking to enhance performance in an increasingly complex and demanding environment.

ACKNOWLEDGEMENTS

The authors wish to thank the teams at Nutrien Aurora and Hatch Ltd. for their collaboration and support throughout this operational excellence initiative. We acknowledge the contributions of the Reliability Task Force, site leadership, and the operations and maintenance teams whose insights, data, and on-site engagement were essential to this work.

Open Pit

Risk management

Board assurance of critical controls effectiveness

A C Atkins[1]

1. PhD candidate; Corporate Advisor, Alex Atkins and Associates Pty Ltd, Perth WA 6056.
   Email: alexandra.c.atkins@postgrad.curtin.edu.au

ABSTRACT

There is a gap in generic, sector-agnostic governance frameworks – created mainly by lawyers and accountants, mainly for lawyers and accountants who dominate the C-suite, on Boards, and amongst those commercially advising, auditing and regulating listed companies.

Underground mining is a high hazard industry requiring subject matter expert (SME) assurance on critical risks and critical controls effectiveness to prevent fatalities and project underperformance or failure. The benefit of this assurance is improved due diligence, which facilitates fatality prevention and also improved operational performance. The cost of this gap in mining governance frameworks is potentially existential – to companies, Directors and Officers and to frontline workers – and must be resolved.

BACKGROUND AND INTRODUCTION

The complexity and uncertainty of mining makes it inherently dangerous

The mining industry operates in an environment where risk is omnipresent – mainly because mines are a dynamic, complex system with multi-variate interdependencies (see Figure 1). Australia's mining industry had a fatality rate of 2.1 fatalities per 100 000 workers in 2021, which is comparatively low compared to the US's 14.2 fatality rate, yet mining was ranked fifth highest for Australian worker fatalities in 2023 (Newnam and Dee, 2026).

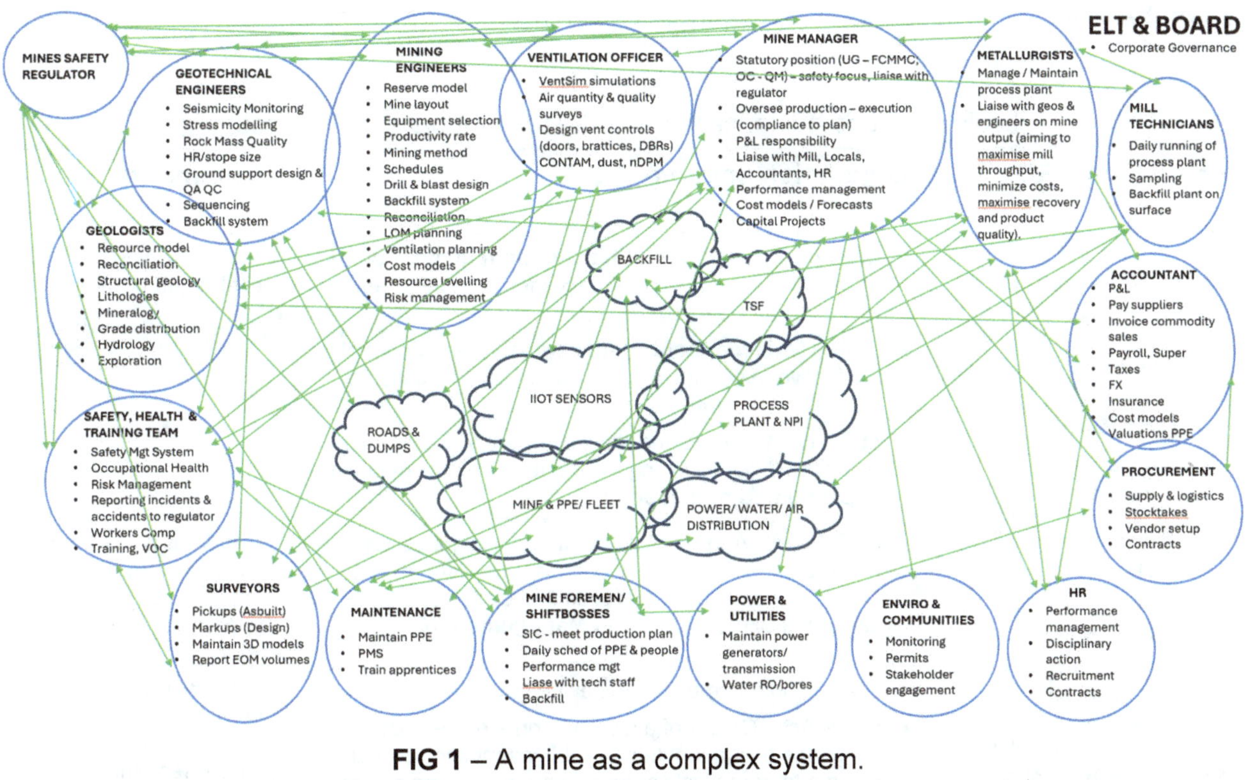

FIG 1 – A mine as a complex system.

The most frequently recurring multiple fatality event among International Council of Mining and Metals (ICMM) members since 2015 is 'fall of ground' (Deakin, 2024) and according to the Minerals Council of Australia (MCA) 2026 Fatality Prevention Research Project Report 'rockfalls were the leading cause of death in Australian mining'.

Mining, particularly underground, exposes companies to a range of technical and operational risks, including: uncertainty of orebody knowledge (ie confidence for executability of Life-of-mine Plan (LOMP)/ore reserves is never a given); rock mass response to mining (ie potential for seismicity and/or falls of ground/pit slips); mine ventilation (ie provide a safe, breathable, non-explosive atmosphere with regulated temperature in all working areas); equipment interactions in confined spaces; and general worker safety in a 'hostile' environment – all of which can have significant safety implications as well as commercial consequences due to constraints on mine schedule flexibility exacerbated by bottlenecks of human capabilities, equipment and financial resources.

Governance frameworks are sector-agnostic and generic

While current board assurance frameworks – often designed, directly or indirectly, by large global accounting and consulting firms – focus heavily on financial and legal risks, they frequently neglect the nuances of technical and operational risk (Atkins and Ritchie, 2019), particularly 'critical controls effectiveness', which is vital to the success of mining operations and increasingly pertinent when it comes to compliance with Australian legislation.

A survey of Australian Institute of Company Director (AICD) members in 2019 found only 3 per cent had science and technology expertise and one in five respondents said there was a short-term financial focus mainly due to institutional investors being focused on the next dividend (Kehoe, 2019). This is despite the ASX Corporate Governance Principles and Recommendations requiring boards to 'recognise and manage risk' with risk committee members having the 'necessary technical knowledge and a sufficient understanding of the industry' (ASX Limited, 2019).

Directors' duties include the duty to act with care and diligence, to act in good faith in the best interests of the company and to ensure there is appropriate continuous disclosure to the public (Baxt, 2012). There are multiple sections of the Australian Corporations Act (see Table 1) which highlight duties of directors when it comes to due diligence, risk management, checking the effectiveness of controls and continuous disclosure – which are generally well complied with for commercial aspects but appear to be underdone, if at all, for the technical and operational risks behind high consequence, low probability events in high hazard industries like mining.

TABLE 1

Legislation and guidance – verify critical controls effectiveness.

Regulation/CRM Guidance	Legislation section	Comments
Work Health and Safety (Mines) Regulations 2022 (WA)	• r.34 and r.36 – **Principal hazard management plans** must include controls and processes to monitor and review control effectiveness. • r.37 – **Review of control measures** when: a control is ineffective; a change occurs and/or after a notifiable incident. • r.638–r.641 – Audit and review of Safety Management System (SMS), including **verification that controls are implemented and effective**.	Explicit duty to check control effectiveness.
Mining and Quarrying Safety and Health Act 1999 (Qld) Mining and Quarrying Safety and Health Regulation 2017 Resources Safety and Health Legislation Amendment Act 2024 (introduces high-reliability organisation principles).	• s.42 (MQSH Act) – Risk management requires controls that are effective. The law defines '**due diligence**' to include specific responsibilities that closely mirror the key steps of a **Critical Controls Management (CCM)** framework. • s.43 – Controls must be reviewed and revised. • s.47A Act – Officer of a Corporation – officer **due diligence** obligation: Legally obliges mining company officers to proactively manage and assure the company's safety and risk controls, effectively embedding **critical controls management** into **directors' and executives' core responsibilities**. For board members, fulfilling this duty requires oversight of critical risk and control identification, control implementation, resource allocation, information flow on control performance, and	Mandates a systematic critical controls management (CCM) approach. Regulations provide practical requirements for SHMS, risk management processes, and principal hazard management.

Regulation/CRM Guidance	Legislation section	Comments
	verification of control effectiveness (ie a rigorous critical controls assurance program). • Part 11, Reg. 89–91 – Safety and Health Management System (SHMS): performance standards; audit and review requirements.	
Work Health and Safety (Mines and Petroleum Sites) Regulation 2022 (NSW)	• cl.14–17 – Management of risks and mandatory review of control measures. • cl.15 – Controls must be reviewed: if ineffective; before changes; and/or after incidents. • cl.18–22 – Safety Management System, including: performance standards; audits; regular reviews. • cl.27–30 – Principal hazard management plans and principal control plans, requiring: defined controls; review and update of effectiveness.	Explicit duty to check control effectiveness.
Occupational Health and Safety Regulations 2017 (Vic) – Chapter 5.3 (Mines)	• r.5.3.7–5.3.10 – Hazard identification and risk control. • r.5.3.11 – Review of risk control measures when: controls fail; incidents occur; conditions change. • r.5.3.22–5.3.25 – Safety Management System: audit; review; continuous improvement.	Explicit duty to check control effectiveness.
Work Health and Safety (Mines) Regulations 2020 (SA) WHS Act 2012 (SA)	• r.613–r.616 – Risk management for mining operations. • r.617 – Review of control measures. • r.621–r.624 – Safety Management System, including: monitoring; audit; review.	Explicit duty to check control effectiveness.
Australian Corporations Act 2001 (C'wlth)	• **Section 180** – Director duty of care and diligence – put in place or monitor effective risk-management and control processes. ASIC has made clear that today's 'duty of care' encompasses actively monitoring non-financial risks (like safety, environmental, cyber) alongside traditional financial and strategic risks. • **Section 189** – Reliance on information or advice – boards to obtain assurance (eg internal/external audit reports, expert reviews) on the effectiveness of controls while still requiring a critical, questioning approach. • **Section 299A** – Operating and Financial Review (OFR) – annual Directors' Report – disclose significant risks facing the business and how they are managed. Reporting on material risks – such as safety hazards, environmental risks, or cyber threats – and describe the risk controls or mitigation strategies in place. • **Section 674** – operates in conjunction with **ASX Listing Rule 3.1** – continuous disclosure obligations require listed companies to promptly disclose material information and risks to the market. Disclosure requirements include timely reporting of significant risk events or control failures. • **ASIC Information Sheet 183** – directors' obligations to maintain adequate risk management and control frameworks. • **Regulatory Guide 247** – directors of listed companies to discuss material business risks and how they are managed or mitigated in the Operating and Financial Review (OFR) section of their annual directors' report. • **ASX Corporate Governance Council's Principles and Recommendations (4th Edition, 2019) – recognise and manage risk (Principle 7)** – Board should annually review to satisfy itself company is operating within the risk appetite set by the board. Boards should have processes for independent, objective assurance on the adequacy of the company's risk management and	Boards must ensure proper internal control mechanisms and oversight to manage uncertainties and protect the company and stakeholders. Principles: highlight importance of clear disclosure – eg informing shareholders how the board receives assurance on risk and controls and what material risks the company faces. If a company's governance practices diverge from these recommendations, it must publicly explain why. International governance guidance (IIA, COSO, OECD) implies that **board assurance over critical controls requires layered assurance across the Three Lines**, with internal audit providing independent confirmation that management-identified critical controls are

Regulation/CRM Guidance	Legislation section	Comments
	controls (whether via an internal audit department, specialist risk reviews, or external assurance services). • **ASIC's 2019 Corporate Governance Taskforce report on Director and Officer Oversight of Non-Financial Risk** (in the financial services sector) – review and challenge the company's risk management framework, risk appetite, and major risk controls, ensure material risks (including emerging non-financial risks) are being monitored and mitigated, and confirm that management is promptly escalating significant incidents or control failures to the board. • ASIC's focus on non-financial risk in recent years – covering areas such as work health and safety, environmental compliance, cyber-security, and conduct risk – reinforces that boards must treat risk management as a fundamental part of their duty of care, just as they do for financial performance.	effectively designed and operating, beyond management self-reporting.
ICMM CCM Guideline (2015a, 2015b, 2026)	• **Risk knowledge and hazard identification:** Board/executives must identify and understand material hazards and critical control measures for major risks. • **Identifying 'Material Unwanted Events' (MUEs):** Recognising high-consequence hazards and the controls critical to preventing catastrophic events. • **Implementing and resourcing controls:** Provide sufficient resources, personnel, and systems to implement effective control measures (especially critical controls) to reduce risk to an acceptable level. • **Monitoring and timely response:** Establish robust reporting and monitoring systems to track critical control performance and quickly address any control failures, incidents, or changes in risk levels. • **Governance and integration into SHMS:** Institutionalise critical control management within the Safety and Health Management System (SHMS) and overall corporate risk management processes, ensuring compliance with legal obligations at all levels.	ICMM Language: Material Unwanted Events (MUE's), Bow Tie Analysis (BTA), Critical Controls (CC) ICMM (2026) highlights board responsibilities, which wasn't explained in 2015 guidelines. **Boards must have strategic oversight and governance over the process** (p 10).

Regulatory frameworks such as the Workplace Health and Safety (WHS) Act 2020 (WA) and the WHS (Mines) Regulations 2022 (WA) (and equivalents in other mining states) (see Table 1 and Appendix 1) place an increasing emphasis on the regular assessment of critical risks (ie Principle Mine Hazard Management Plans) and the effectiveness of critical controls through positively demonstrating 'due diligence' (Risk Engineers Society, 2014). The regulations loosely align with ICMM's Critical Control Management Good Practice Guide (ICMM, 2015a) and Implementation Guide (ICMM, 2015b) – recently updated in 2026 with more tangible expectations around Board responsibilities – which advocate for the use of Material Unwanted Events (MUEs) and Bow Tie Analysis (BTA) as methods for identifying, assessing and managing operational and technical risks and to identify critical controls – 'what must go right' (see Figure 2).

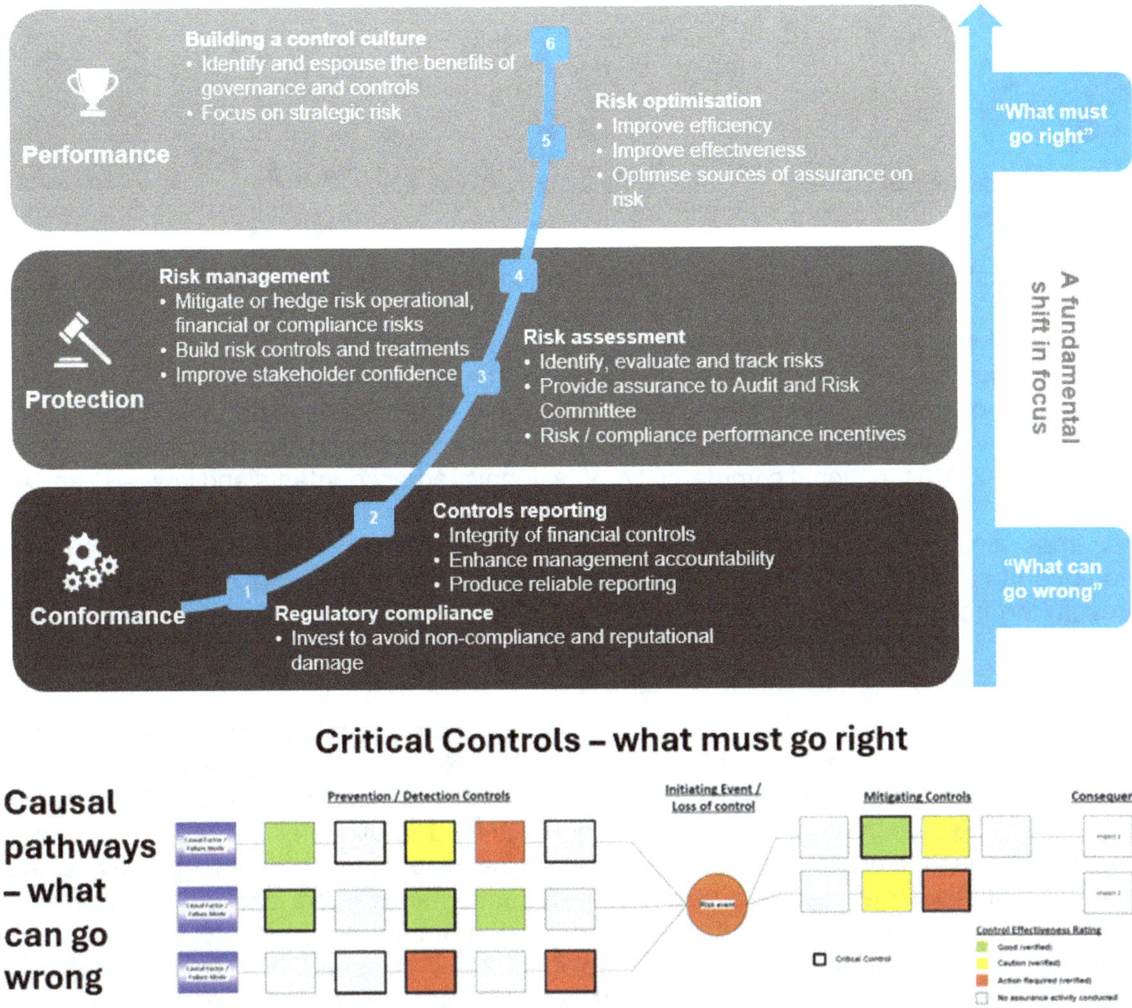

FIG 2 – Material Unwanted Events (MUEs) and Bow Tie Analyses (BTAs) to find Critical Controls (CCs) enabling focus on 'what must go right'. Source: (top) adapted from Deloitte's *The Future of Controls: Becoming a Control Intelligent Organization* (2017); (bottom) adapted from Hassall (2017).

THERE ARE MATERIAL GAPS IN MINING RISK GOVERNANCE

With all these requirements for diligence and competence at the Board level, it is disturbing to see there is currently no universally adopted system that ensures critical risks and critical controls effectiveness are communicated to Boards in any language, let alone one they can understand – commercial. This gap in governance doesn't appear to have been fully identified and explored in the literature.

There are many case studies (Atkins, 2022) (see Figure 3) and academic analyses (Quinlan, 2014) of catastrophic mining accidents which underscore what can happen when there is no mechanism for effective communication between technical an operational teams at the frontline and decision-makers at the Board level, who set the strategy, budgets, organisational structure and it's resultant culture and who signal to markets expected project performance (ie forward looking guidance on costs and production).

FIG 3 – Disaster case studies (ie material unwanted events (MUE's)). Source: Atkins (2022).

Research (Hopkins, 2019) highlights the importance of organisational structure to ensure vital, unfiltered frontline risk information flows up quickly to Boards, but this is only half the problem. The involvement of process safety-competent technical and operational experts at the Board level remains limited with mining industry Boards predominantly comprised of homogeneous, group-think-prone (Hopkins, 2019) functional commercial professionals (eg accountants and lawyers) and former executives with 'P&L line management' careers, rather than technical. While these individuals bring critically important commercial acumen and corporate governance experience, they are often less well-equipped to understand in first principles the technical and operational complexities of mining. As a result, the frontline perspective of technical and operational experts is frequently downplayed (ie as tactical or management, not commercial or strategic) and challenging and supporting management in strategic deliberations is at risk of being underdone.

There are clear lines of accountability from the CEO and Board down to the frontline for commercial matters but missing lines of accountability for major risks, exacerbated by organisational structures which prevent technical professionals from having decision-making authority (Hopkins, 2022) (Quinlan, 2014). Research has found Coronial inquests tend to allocate contributory factors to the frontline, even though they are influenced by systemic issues originating from higher levels and that there is differentiation in exposure to risks between contracted workers and company employees. (Newnam and Dee, 2026). There is a lack of aggregation of information to provide a more holistic assessment of critical controls effectiveness (Deakin, 2024). Wilkinson (2024) finds that reporting on performance of the critical controls is not reported to a senior enough level, even though some of the risks had the potential to have catastrophic consequences, which is concerning considering these risks could be regarded by ASX listed companies as 'material exposures' requiring oversight in accordance with Principle 7 of the ASX Corporate Governance Principles 2019 (ASX Limited, 2019). This is the state of play despite investors – reacting to the Brumadinho tailings dam and Juukan Gorge disasters – being concerned about ethics and protecting their investments, wanting board accountability (Hopkins and Kemp, 2021). The Australian Banking Royal Commission highlighted the increasing importance of addressing non-financial risks in corporate governance (Hayne, 2019) and is yet to be translated into appropriate governance of critical risks in mining.

PHD RESEARCH ON GAPS IN MINING RISK GOVERNANCE

The PhD project, which commenced in July 2025, explores the critical role of board-level assurance in managing technical and operational risks in mining, focusing on the commercial implications of critical risks/MUE's, noting how critical controls influence not only safety but also operational productivity, profitability, and long-term commercial success in the mining industry. Despite the considerable importance of managing technical and operational risks, current Board assurance frameworks in many mining and mining services companies may not be adequately capturing and addressing these risks. The focus of Boards is often disproportionately centred on a broad range of legal compliance requirements, financial performance and matters obviously related to enhancing shareholder value, while technical and operational risks are often relegated to a secondary role, because there is often not an obvious, direct connection to shareholder value.

There are gaps in research on this governance gap, even though it is alluded to by highly regarded researchers (see Table 2). This research seeks to quantify this governance gap, understand why it exists and then make recommendations to bridge the gap by developing a plug-in/add-on framework for existing Board assurance of critical risks' critical controls effectiveness.

TABLE 2
Preliminary literature review themes.

Author(s)	Power structure/ C-suite and Board accountability	Technical/frontline experts voice at top	Systemic issues (higher levels contribution)
Hopkins (2019, 2022)	Full line of accountability – commercial only (financial and legal) (*: a form of abrogated responsibility)	Missing, no authority. Defer to management oversimplifying and giving overconfident assurances	Organisational structure creates culture, information flow. Boards are responsible – but oversight failure
Quinlan (2014)	'Political economy perspective' (*)	Missing Contractor de-risking	Power structures, economic incentives and regulatory regimes shape organisational behaviour
Newnam and Dee (2025)	Priorities and constraints (production > safety)	Frontline blaming Contractor de-risking	Not considered in coronial enquiries
Hopkins and Kemp (2021)	Investors want accountability for disasters		Ethics
Dekker (2017)	Legal blockers to learning (*)	Responsibility shifted to least powerful actors	'Drift Into Failure', bureaucracy as a substitute for accountability
Deakin (2024)		Holistic CCM Gap	CCM not operationalised
Leech (2026)	'Don't Ask/ Don't Tell' (*)	Information asymmetry creates chronic oversight risk	Boards lack systems to know whether 'mission-critical objectives' are on track

Aim of the research is to lift the voices of the frontline and experts

The aim of this research is to find a better way to integrate frontline expert insights into corporate decision-making, by making the information accessible (ie through direct assurance reporting) and understandable to non-mining experts (ie translate into commercial language). This will ensure Directors and Officers make more informed decisions and it will also be a way to ensure the company is, in reality, operating within it's Board approved risk appetite, which should reflect shareholder risk appetite and broader community expectations. Without bridging this governance gap, C-suite, Boards and investors are legally and/or financially exposed when critical risks materialise and critical controls are ineffective. Recent examples include: the Samarco (2015) and Bruhmandino (2019) tailings dam failures, Grosvenor explosion (2024), El Teniente fall of ground (2025) and the Grasberg inrush (2025). To keep the scope manageable, the research is focused on Australian-domiciled, underground metalliferous miners that are small-to-mid cap, but large and stable enough to fund and sustain Critical Risk Management (CRM) and Board-level Critical Controls Effectiveness (CCE) reporting. Outcomes are expected to be broadly generalisable.

PHD RESEARCH HYPOTHESES

Core hypothesis

Boards approve management-set guidance which is potentially over-optimistic and are exposed when things 'go wrong' due to a lack of oversight of critical controls effectiveness.

Background hypotheses (based on experience)

Hypothesis 1 – A mine is like an experiment

Companies start with a hypothesis and fully/partially prove/disprove it as they mine. This eludes to the uncertainty and complexity of a mine as a dynamic system, with multi-variate interdependencies, requiring a plan-do-check-act, first principles based, learning culture to continuously improve performance when things 'go wrong'.

Hypothesis 2 – Bad things happen in mining due to pressure from financial markets

Boards and the C-suite are responsible for setting expectations in the market for future performance (ie 'guidance') – which often has a 'blue-sky' bias, in order to access capital on the best possible terms. Promises made early in a project's life (eg the spruiking stage on the Lassonde Curve) have the potential to set-up unrealistic expectations during operations, putting considerable pressure on the frontline and the experts. By 'blue-sky lying' Boards are inadvertently setting themselves up to be exposed to financial and legal liabilities when 'things go wrong' if they are not checking their critical controls are effective for critical (technical and operational) risks.

This gap in the current board assurance framework leaves C-suite executives and Boards exposed under existing regulatory frameworks such as the Work Health and Safety Act 2020 (WA), the Work Health and Safety (Mines) Regulations 2022 (WA) (and other state's equivalents, see Table 1 and Appendix 1) and the Corporations Act 2001 (Department of the Treasury, 2001). Without a robust process for evaluating and assuring the effectiveness of critical controls, mining companies are also missing opportunities to optimise operational performance and unlock long-term value for stakeholders such as improving productivity and achieving lower cost quartile status (see Figure 4).

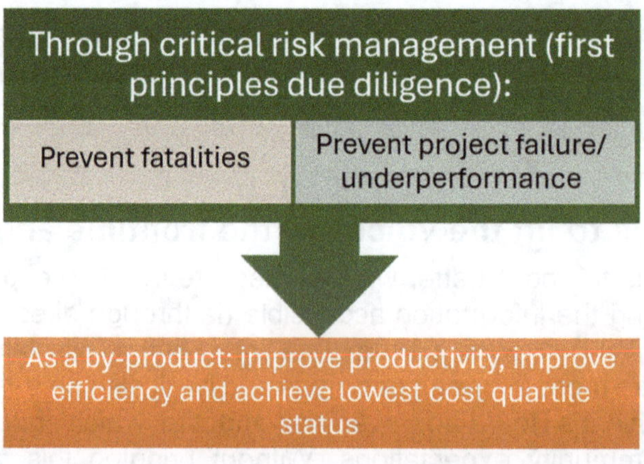

FIG 4 – Outcomes of due diligence/critical risk management.

Critical Control failure is priced after the event, often with a long tail. The aim of this research is to give Boards and investors leading indicators of enterprise value at risk by driving critical control focused accountability from the frontline to the Board, improving both commercial and safety performance.

GOVERNANCE GAPS TO ADDRESS

The research addresses three missing pieces in mining risk governance.

The 'missing vertical'

Three lines board assurance of CC effectiveness for all Critical Risks/MUE's and line of sight into the due diligence process CRM promotes to date doesn't have a place in mining risk governance at the Board level, due to it's perception as being: management, operational, tactical and 'in the weeds' (ie not strategic and not commercially relevant).

It is timely that, along with the MQSH Act s.47A on director and officer responsibilities, that ICMM's latest Critical Control Management (CCM) Guideline's recommended Board level responsibilities are: *'Receive and discuss reports on the organisation-wide implementation of the CCM process and on fatal hazard and critical control 'health' periodically (eg on a quarterly/six monthly basis)'* (ICMM 2026, p 54).

This is the 'missing vertical' (three lines board assurance). Hopkins (2019, 2022) describes it as a missing 'full line of accountability' and Quinlan (2014) describes it as 'political economy perspective', also identified by Leech (2026) as 'Don't Ask/Don't Tell' (see Table 2). This missing vertical keeps the voice of the frontline and technical and operational experts out of boardroom deliberations, where Newnam and Dee (2025) suggest production priorities override safety constraints. Quinlan (2014) suggests this is due to power structures and economic incentives, along with regulatory regimes which have historically not focused on Directors and Officers contributions. This is reinforced by Newnam and Dee's (2025) coronial enquiries research which shows coronial findings tend to blame the frontline actors and fail to consider systemic issues or higher-level contributions.

The 'missing horizontal'

There is poor transparency and accountability across the mining company – contractor interface, when it comes to aligning on what is the best approach for identification and management of critical risks, understanding which controls are critical controls and making their performance criteria specific and auditable, three lines of assurance verifying critical controls effectiveness and escalation when critical controls are ineffective. This interface is the source of many contractor fatalities. The client owns most of the critical controls – think place of work and systems of work, set-up by the mine design and schedule – but the contractor bears the risk. Newnam and Dee (2025), Quinlan (2014) and Dekker (2017) highlight how contractors and the frontline are used as a de-risking mechanism to shift responsibility and blame to the least powerful actors (see Table 2).

ICMM (2026, p 12) state: '*Highly technical hazards may require specialised methodologies to identify critical controls and the involvement of subject-matter experts to verify their effectiveness*'. For contractors, their clients own the majority of the controls for the highly technical hazards (eg fall of ground/seismicity, ventilation/hazardous atmospheres and inrush), so there is a need for improved across interface engagement to obtain alignment on 'what could go wrong' (causal pathways to MUEs) and 'what must go right' (critical controls), not just for fatality prevention but also improved performance.

The 'missing translation'

There is a need for a connective bridge joining the bottom half (ie more technically and operationally focused) and top half (ie more commercially focused) levels of mining organisations to enable conversion of technical and operational language and concepts into commercial language and frameworks, so anyone can understand it. This is needed because traditional governance content creators (ie internal audit and external audit providers) and consumers of governance content (ie Boards) are predominantly lawyers, accountants and former C-suites (who are commercial and usually not very technical, even though they may originate from a technical profession). Without the commercial translation, frontline technical and operational risks may be filtered and curated out of Board packs, leading Directors and Officers to fail to 'join the dots' on material risks and whether the company is operating inside it's risk appetite. Their preference is also for quantifiable metrics, which is a challenge that begs the question: 'how do we aggregate and analyse all industry data on the cost of critical controls failing (ie when MUE's materialise) and weave that into board critical control management assurance packs?'

CONCLUSION

The research being undertaken will: measure the gaps in mining risk governance (ie the missing vertical and missing horizontal); attempt to assess the consequences of the gaps (ie the cost of critical control failure); investigate the reasons for the gaps (eg demographics, lack of commercial translation, missing vertical and horizontal lines of reporting and accountability); identify what good looks like (eg exemplar case study); and recommend how to best inform board deliberations to bridge the gap. On completion, findings will be promulgated across the industry. Until then, the C-suite, Boards, frontline workers and investors are exposed.

CONTRIBUTE TO THE RESEARCH

This research is about lifting the voices of the frontline and mining industry experts into boardroom deliberations, to facilitate better strategic decision-making and fit-for-purpose capital allocation. It is possible to contribute to this PhD research (Curtin University HREC Approval No. 2026–0037) via the anonymous survey (at link: https://curtin.au1.qualtrics.com/jfe/form/SV_2iuf9xPXaeDDpA2) and there is an opt-in for a de-identified interview at the end of the survey (or volunteer interviewees can email: alexandra.c.atkins@postgrad.curtin.edu.au). There are three question blocks for three distinct cohorts: Operators; Governance; Investors (respondents can complete more than one). The survey closes 30 September 2026 and will be analysed by the end of 2026.

ACKNOWLEDGEMENTS

I gratefully acknowledge my PhD supervisors Professor Eric Lilford at Curtin University (who will keep me commercial) and Professor Maureen Hassall at The University of Queensland (who will keep me technically correct). I acknowledge Perenti Limited who, after the privilege of serving 7.5 years on their Board, are collaborating with me on CRM governance, including an 'in-house' deep dive on the contractor interface. I acknowledge the countless dedicated people in the mining industry doing the hard work every day to keep our miners and the community safe and the mining experts – working hard in their roles as the creators and protectors of value – to ensure high professional standards are maintained, protecting the mining industry's reputation as a good investment destination. Last, but not least, I acknowledge my kids for their patience with my obsession (ie Ikigai) with making the mining industry better.

REFERENCES

ASX Limited, 2019. Corporate Governance Principles and Recommendations, 4th edn, ASX Limited. Available from: <https://www.asx.com.au/content/dam/asx/about/corporate-governance-council/cgc-principles-and-recommendations-fourth-edn.pdf>

Atkins, A and Ritchie, M, 2019. Improving board assurance of technical and operational risks in mining, in *MGR 2019 – Proceedings of the First International Conference on Mining Geomechanical Risk* (ed: J Wesseloo), pp 97–110 (Australian Centre for Geomechanics: Perth). https://doi.org/10.36487/ACG_rep/1905_03_Atkins

Atkins, A, 2022. Lessons learnt from six mine disaster case studies to improve audit and assurance and ESG, in *Proceedings of the Minesafe International Conference*, 6 p (The Australasian Institute of Mining and Metallurgy: Melbourne). Available from: <https://www.ausimm.com/publications/conference-proceedings/minesafe-international-conference-2022/lessons-learnt-from-six-mine-disaster-case-studies-to-improve-audit-and-assurance-and-esg/>

Baxt, R, 2012. Duties and responsibilities of Directors and Officers – 20th Edition, Australian Institute of Company Directors.

Deakin, A, 2024. Critical control management – our journey so far and the road ahead, in *Proceedings of the International Mine Health and Safety Conference*, pp 115–128 (The Australasian Institute of Mining and Metallurgy: Melbourne).

Dekker, S, 2017. *Just Culture: Restoring trust and accountability in your organisation* (CRC Press).

Deloitte, 2017. The Future of Controls: Becoming a Control Intelligent Organization, What we do – Services, Deloitte. Available from: <https://www.deloitte.com/global/en/services/consulting-risk/research/becoming-control-intelligent-through-future-of-controls.html>

Department of the Treasury, 2001. Corporations Act, 2001, Attorney-General's Department, Department of the Treasury. Available from: <https://www5.austlii.edu.au/au/legis/cth/consol_act/ca2001172/>

Hassall, M, 2017. UQ R!SK critical controls masterclass (presentation), Sustainable Minerals Institute, The University of Queensland.

Hayne, K, 2019. Royal Commission Report: Misconduct in the Banking, Superannuation and Financial Services Industry. Available from: <https://www.royalcommission.gov.au/banking/final-report>

Hopkins, A and Kemp, D, 2021. *Credibility Crisis: Brumadinho and the Politics of Mining Industry Reform* (Wolters Kluwer).

Hopkins, A, 2019. *Organising for safety: How structure creates culture* (Wolters Kluwer).

Hopkins, A, 2022. *Sacrificing Safety: Lessons for Chief Executives* (Wolters Kluwer).

International Council on Mining and Metals (ICMM), 2015a. *Health and Safety Critical Control Management: Good Practice Guide*, ICMM. Available from: <https://www.icmm.com/website/publications/pdfs/health-and-safety/2015/guidance_ccm-good-practice.pdf?cb=60007>

International Council on Mining and Metals (ICMM), 2015b. *Critical Control Management: Implementation Guide*, ICMM. Available from: <https://www.icmm.com/website/publications/pdfs/health-and-safety/2015/guidance_ccm-implementation.pdf?cb=60006>

International Council on Mining and Metals (ICMM), 2026. *Critical Control Management – Good Practice Guide*, ICMM. Available from: <https://www.icmm.com/en-gb/guidance/health-safety/2026/ccm-good-practice-guide>

Kehoe, J, 2019. Company directors admit boards are too short term, *Australian Financial Review* (18 September 2019). Available from: <https://www.afr.com/technology/company-directors-admit-boards-are-too-short-term-20190917-p52s3t>

Leech, T J, 2026. *Mission-Critical Governance – Focusing on what matters most* (CRC Press).

Minerals Council of Australia (MCA), 2026. Fatality Prevention Research Project Report (confidential to MCA members). Available from: <https://minerals.org.au/fatalityprevention/>

Newnam, S and Dee, A, 2026. Analysing coronial inquests into mining fatalities: A systems thinking approach, *Safety Science*, 193(107023).

Quinlan, M, 2014. *Ten Pathways to Death and Disaster: Learning from Fatal Incidents in Mines and Other High Hazard Workplaces* (The Federation Press).

Risk Engineers Society, 2014. Safety Case Guideline: clarifying the safety case concept to engineer due diligence under the provisions of the model Work Health and Safety Act 2011, Risk Engineers Society. Available from: <https://www.engineersaustralia.org.au/sites/default/files/2025-07/safety-case-guideline.pdf>

Wilkinson, A P, 2024. Preventing unwanted events critical controls – success and failure in implementing the ICMM guidance, in *Proceedings of the International Mine Health and Safety Conference*, pp 197–202 (The Australasian Institute of Mining and Metallurgy: Melbourne).

APPENDIX 1 – AUSTRALIAN LEGISLATION

Jurisdiction	Australian Mining WHS Legislative Framework
New South Wales	*WHS* requirements for mining are regulated through the: • Work Health and Safety Act 2011 (NSW). • Work Health and Safety Regulation 2017 (NSW). • Work Health and Safety (Mines and Petroleum Sites) Act 2013. • Work Health and Safety (Mines and Petroleum Sites) Regulation 2022. More information about NSW's *WHS* regulatory framework for mining is available at the NSW Resources Regulator.
Victoria	*WHS* requirements for mining are regulated through: • Chapter 5.3 (Mines) of the Occupational Health and Safety Regulations 2017 . • Occupational Health and Safety Act 2004. More information about Victoria's *WHS* regulatory framework for mining is available at WorkSafe Victoria.
Queensland	*WHS* requirements for mining are regulated through the: • Mining and Quarrying Safety and Health Act 1999. • Mining and Quarrying Safety and Health Regulation 2017. • Coal Mining Safety and Health Act 1999. • Coal Mining Safety and Health Regulation 2017. More information about Queensland's *WHS* regulatory framework for mining is available at the Queensland Government website.
Western Australia	*WHS* requirements for mining are regulated through the: • Work Health and Safety Act 2020 (WA). • Work Health and Safety (Mines) Regulations 2022. More information on Western Australia's *WHS* regulatory framework for mining is available at the Department of Mines, Industry, Regulation and Safety.
South Australia	*WHS* requirements for mining are regulated through: • Work Health and Safety Act 2012 (SA). • Work Health and Safety Regulations 2012 (SA). • Mines and Works Inspection Act 1920. • Mines and Works Inspection Regulations 2013. More information is available at SafeWork South Australia.
Tasmania	*WHS* requirements for mining are regulated through the: • Work Health and Safety Act 2012 (Tas). • Work Health and Safety Regulations 2022 (Tas). • Mines Work Health and Safety (Supplementary Requirements) Act 2012. • Mines Work Health and Safety (Supplementary Requirements) Regulations 2025. More information about Tasmania's *WHS* regulatory framework for mining is available at WorkSafe Tasmania.
Australian Capital Territory	*WHS* requirements for mining are regulated through the: • Work Health and Safety Act 2011 (ACT). • Work Health and Safety Regulation 2011 (ACT). More information is available at WorkSafe ACT.
Northern Territory	*WHS* requirements for mining are regulated through the: • Work Health and Safety (National Uniform Legislation) Act 2011. • Chapter 10 (Mines) of the Work Health and Safety (National Uniform Legislation) Regulations 2011. More information is available at NT WorkSafe.

Open Pit

Risk, ESG and system-level decision making

Resilient open pits in an ESG-conscious world – a strategy optimisation playbook integrating mining and ESG parameters

T Sharman[1], A Jones[2] and N Barros[3]

1. Mine Sustainability Lead, AMC Consultants Pty Ltd, Melbourne Vic 3000. Email: tsharman@amcconsultants.com
2. Principal Mining Engineer, AMC Consultants Pty Ltd, Melbourne Vic 3000. Email: ajones@amcconsultants.com
3. Mine Sustainability Lead, AMC Consultants Pty Ltd, Melbourne Vic 3000. Email: nbarros@amcconsultants.com

INTRODUCTION

Open pit mining is increasingly executed within binding environmental, social and governance (ESG) constraints that influence pit geometry, sequencing, dewatering regimes, energy systems, and material movement design – not merely cost line-items. Under modern disclosure, permitting expectations and more contemporary sustainability corporate policy, ESG performance permeates the full project life cycle, from evaluation and operations through to closure and post-mining transition.

In conventional optimisation workflows, net present value (NPV) is typically maximised first, with environmental impacts subsequently addressed through cost provisions. By contrast, when considered from the outset, ESG baselines operate as binding constraints that can compress schedules, create exposure to carbon credit liabilities, or limit the extent of disturbance expansion. The implication is fundamental: ESG performance criteria must be integrated into strategic design frameworks, rather than treated as closure contingencies or external cost modifiers applied after optimisation.

AMC's experience across strategy optimisation in open pit to underground transition studies demonstrates that when emissions, water intensity, and disturbed area trajectories are embedded directly within optimisation, materially different life-of-mine strategies emerge. This proposed approach now lends itself to switching the focus to how open pits can remain safe, technically, environmentally and economically viable, and socially acceptable within ESG-bound operating environments, rather than advocating a mining method change.

STRATEGY OPTIMISATION IN AN ESG CONTEXT

Strategy optimisation differs from conventional schedule optimisation by modelling the mining system holistically and reporting multiple metrics simultaneously. Multi-objective evaluation enables explicit trade-offs between financial outcomes and ESG commitments, aligning strategies with corporate goals and regulatory thresholds.

In practical terms, ESG constraints can be implemented in three ways within the optimiser:

1. As binding constraints (eg water abstraction baseline limits).
2. As monetised penalties (eg carbon price above baseline).
3. As secondary objectives (eg minimise disturbed area growth rate).

Scenario-based design incorporates structured uncertainty sets, including:

- Climate severity bins affecting availability.
- Hydrologic states influencing dewatering demand.
- Energy price and grid availability paths.
- Regulatory tightening trajectories.

These scenario inputs feed the optimisation engine to generate trade-off fronts rather than a single 'best case' solution.

TECHNICAL LEVERS FOR OPEN PIT RESILIENCE

Resilience in open pit mining is achieved through the identification and management of controllable design levers that influence both ESG and production outcomes:

- Pushback geometry and sequencing.
- Depressurisation and groundwater control regimes.
- Energy portfolio configuration.
- Integrated waste routing and in-pit storage.
- Bench height, equipment scale, ore control, and selectivity.

These levers interact non-linearly. For example, proactive depressurisation may enable steeper stable slopes, reduce strip ratio and external haul distances while simultaneously lowering pumping energy. Integrated waste routing can shorten haulage cycles, reduce external waste rock storage growth, and pre-condition final landforms – reducing both emissions intensity and closure liability.

ESG PARAMETERS AND PLANNING INTEGRATION

ESG themes must be translated into parameters that pit engineers can control. Typical conversion relationships are outlined in Table 1 and include:

- Diesel consumption per tonne mined.
- Electricity intensity per tonne milled.
- Water consumption per tonne moved.
- Waste storage footprint per volume placed.

TABLE 1

Example ESG conversion parameters embedded in strategy optimisation.

Parameter	Unit	Typical value
Diesel $CO_{2\text{-eq}}$ factor	tCO_2/L	0.0027
Open pit diesel intensity	L/t mined	~ 0.7
Electricity intensity	kWh/t	~ 20
Water intensity (OP mining)	L/t	~ 3–4
Waste storage capacity	m^3/ha	~ 500 000

(Values indicative; site-specific calibration required.)

Energy systems increasingly behave as mine design inputs rather than procurement afterthoughts. Technology pathways such as trolley assist, in-pit crushing and conveying (IPCC), renewables integration, and autonomous fleets alter both emissions intensity and operating risk.

Scenario testing demonstrates that diversified energy portfolios reduce exposure to price shocks while flattening emissions trajectories. When treated as decision variables within optimisation, energy configurations can shift the trade-off front outward – achieving improved ESG outcomes for limited incremental cost.

Embedding such parameters within optimisation allows material movement, haul distance, pit depth, and pumping volumes to translate directly into ESG KPIs. This enables emissions, water intensity, and disturbance trajectories to be optimised jointly with financial metrics.

Social and approval constraints are similarly physicalised through land disturbance ceilings:

- Visual, dust, and noise interface limits.
- Progressive rehabilitation rates.

- Maximum disturbance growth rates.
- Known environmental and social sensitivities.

Governance mechanisms such as emissions baselines and disclosure thresholds convert exceedance into real financial consequences, reinforcing the need for early integration.

SCENARIO DESIGN AND ROBUSTNESS TESTING

Scenario-based optimisation tests strategy robustness across structured uncertainty sets rather than relying on a single commodity price case. Outputs include trade-off fronts across:

- NPV and cash flow.
- Material movement and strip ratio.
- Energy and water intensity.
- Disturbed area evolution.

Robust strategies are those that remain compliant across multiple ESG tightening scenarios while maintaining acceptable value retention. This approach reframes optimisation from maximising peak NPV toward preserving strategic optionality.

CASE STUDIES' INSIGHTS – SMALLER, FASTER, MORE SELECTIVE PITS

Case studies from gold and copper operations demonstrate that resilient strategies often converge toward smaller, faster, more selective pit geometries.

In one operating gold case, the results of a study are illustrated in Figure 1, that mining a smaller pit provided:

- ~3 per cent NPV difference between strategies.
- 13 per cent reduction in Scope 1 $CO_{2\text{-eq}}$.
- 17 per cent reduction in total material movement.

Scenario 1 – NPV (US$ m)

Processing Capacity	US$m	UG Cut off grade						
		1.8	2.1	2.3	2.5	2.7	3.0	3.3
	1.8	575	590	592	571	529	428	325
	2.1	604	635	629	580	530	442	337
	2.3	619	650	634	576	525	449	337
	2.5	611	636	615	558	507	435	336
	2.7	600	611	593	552	501	422	338
	3.0	554	583	554	544	498	411	343
	3.3	534	548	519	533	492	406	351

Scenario 1 – CO_2 emissions

Processing Capacity	CO2 Kt	UG Cut off grade						
		1.8	2.1	2.3	2.5	2.7	3.0	3.3
	1.8	587	585	583	579	577	576	571
	2.1	582	580	578	576	574	573	568
	2.3	580	578	576	574	573	571	567
	2.5	578	576	575	573	572	570	567
	2.7	576	575	573	572	571	569	566
	3.0	574	573	572	571	569	567	565
	3.3	573	572	570	570	569	566	564

Scenario 2 – NPV (US$ m)

Processing Capacity	US$m	UG Cut off grade						
		1.8	2.1	2.3	2.5	2.7	3.0	3.3
	1.8	570	577	603	556	514	421	307
	2.1	594	617	626	572	523	432	314
	2.3	609	627	630	576	525	430	312
	2.5	600	620	629	571	519	427	311
	2.7	604	612	626	561	512	426	311
	3.0	581	585	616	554	508	424	319
	3.3	544	551	599	542	507	432	326

Scenario 2 – CO_2 emissions

Processing Capacity	CO2 Kt	UG COG pit active						
		1.8	2.1	2.3	2.5	2.7	3.0	3.3
	1.8	678	677	673	671	669	664	662
	2.1	673	671	669	667	666	662	660
	2.3	671	669	668	666	664	661	659
	2.5	669	667	666	665	663	660	658
	2.7	667	666	665	664	662	659	657
	3.0	665	664	664	662	661	659	656
	3.3	664	663	663	661	660	658	656

Optimum NPV on heat map

FIG 1 – Heat map outlining NPV and $CO_{2\text{-e}}$.

In a large Cu/Au case:

- 57 per cent reduction in Scope 1 $CO_{2\text{-eq}}$.
- 16 per cent NPV reduction relative to maximum-NPV case.

These results illustrate that modest NPV sacrifice can deliver material reductions in emissions intensity and disturbance exposure. Importantly, optimisation revealed strategies that improved ESG performance without materially compromising value.

Hydrology-aware scheduling reduced pumping peaks and availability losses under wet-climate scenarios. Integrated waste routing reduced external haulage distances and long-term closure reshaping requirements.

A pragmatic geotechnical approach – managing exposure duration and sequencing internal backfilling – enabled steeper but stable slopes in selected sectors, further reducing strip pressures.

CONCLUSION

Building resilience in open pit mining within an ESG-bound operating space requires a shift from post-optimisation mitigation to design-embedded sustainability. By integrating emissions, water, disturbance, and social guardrails as primary optimisation inputs, open pits can remain technically defensible and economically viable under tightening regulatory regimes.

Resilient open pits are not defined by maximum scale but by intelligent configuration: steeper stable slopes, shorter haulage distances, integrated waste management, diversified energy systems, and improved selectivity. Strategy optimisation reveals that ESG constraints do not eliminate open pit viability – they reshape it toward more efficient and robust geometries and operational configuration solutions.

Enhancing dilution and ore loss estimation

J Stoddart[1], C Sprott[2], P Federici[3] and C Reardon[4]

1. Principal Mining Engineer, AMC Consultants Pty Ltd, Melbourne Vic 3000. Email: jstoddart@amcconsultants.com
2. Principal Mining Engineer, AMC Consultants Pty Ltd, Melbourne Vic 3000. Email: csprott@amcconsultants.com
3. Principal Mining Engineer, AMC Consultants Pty Ltd, Melbourne Vic 3000. Email: pfederici@amcconsultants.com
4. Principal Mining Engineer, AMC Consultants Pty Ltd, Melbourne Vic 3000. Email: creardon@amcconsultants.com

ABSTRACT

Controlling waste rock dilution and ore loss is critical to maximising value from a mine. For new projects, bench height analysis and selective mining unit size (implying size of mining equipment) and their impact on dilution and ore loss, and therefore value, is a critical step in optimising a project. For existing mines, accurately accounting for the dilution and ore loss that will occur during the mining process is important for the accurate forecasting of production. Poor reconciliation between forecast production and production reports is often caused by inaccurate estimation of ore loss and dilution, and can lead to material impacts on project value, in some cases resulting in premature mine closure.

For an open pit mine, it is easy to think that ore plus waste creates more ore, but unnecessary waste denies the mine plan significant quantities of lower grade mineralisation that could have been blended into the product stream and used to extend the life of the mine, or can lead to the unnecessary processing of waste that carries negligible value or incurs an economic loss.

In AMC Consultants' (AMC) experience, the estimation of planned dilution and ore loss in both projects and mining operations is often inappropriate. To account for this, global factors are regularly applied as unplanned dilution when reconciliation identifies a variance between the block model estimate and mine production. This approach does not allow for spatial variance in developing the mine or facilitate an understanding of the root causes of poor reconciliation and how they can be addressed.

It is always important that the modelling approach of estimating ore loss and dilution reflects the physical mining process that is proposed or is being undertaken. The modelling methods selected need to reflect the size and geometry of the deposit, distribution of mineralisation through the deposit, the size and type of the mining equipment, bench heights, grade control methods, among many others.

AMC use a range of approaches for modelling ore loss and dilution for both new projects and existing mines, some of which have been developed in-house and tested extensively over many years, including through production reconciliation. There are times when a simple block regularisation process provides an adequate result and times when more sophisticated techniques are required. But AMC very rarely believes that factoring results from a Mineral Resource block model, which takes little to no account of the issues discussed above, is appropriate.

Enhancing dilution and ore loss estimation accuracy allows project developers and operators to better understand the value and opportunity potential of their assets. This paper discusses some of the modelling techniques developed by AMC and their various applications and provides some case studies for mines for which these techniques have been applied.

INTRODUCTION

Dilution and ore loss modelling is fundamental to achieving the planned outcomes from a project. Processing a tonne of barren material not only incurs cost for no value, but when an operation is plant constrained that tonne of waste is displacing valuable material and has a direct effect on the revenue of an operation.

It's often assumed that minimising planned dilution and ore loss must be the best option for a deposit. However, this is not always the case and getting the right mining strategy and dilution modelling approach is fundamental to achieving the best return from a deposit.

Dilution arises from several sources, including orebody geometry, geological uncertainty, blast movement, mining equipment, and mining errors. This paper focuses on planned dilution, which results from practical mining and design constraints. Planned dilution is explicitly introduced during the conversion of Mineral Resources into mineable inventories and is inseparable from ore loss.

Dilution and ore loss is influenced by two main factors: the spatial distribution of the mineralisation; and the operating strategy selected to extract the ore. The decisions made on these factors will be fundamental to what value can be realised from a deposit. For example, a deposit with narrow defined mineralised units may be able to be mined highly selectively on 5 m benches with minimal dilution and ore loss. But this method will limit the potential production rate. Alternatively, this same deposit may be mined in bulk with the mineralised units mined together in 15 m benches. This significantly increases production rates but incurs high levels of dilution and ore loss. Getting dilution modelling right for each case is fundamental to understanding the potential value of a deposit.

Other factors which need to be considered are any physical or visual differences between the mineralised and barren material. If the valuable material is hard and black, and the waste soft and white then minimising dilution should be relatively straightforward, however this is rarely the case.

MODELLING APPROACHES

Dilution modelling can be grouped into three different approaches:

1. Dilution and ore loss factors – These can be applied globally or by rock types or zones. They are a coarse tool and, in many cases, will be selected based on the experience of the team which may not be relevant to the deposit being considered. Generally, AMC avoids this approach for all but preliminary high-level conceptual study or at an operating mine where the factors are well supported by reconciliation and the mineralisation being planned for is demonstrably similar.
2. Regularising to a standard block size (re-blocking) – Aggregating blocks to a defined selective mining unit (SMU) based on the planned digging or drilling equipment, typically in an orthogonal grid. This can be very effective in disseminated mineralisation such as a porphyry deposit with the SMU based on excavation equipment. It can also be effective in shallow dipping deposits with the SMU based upon blast parameters.
3. Consolidating units and adding a dilution skin – This is the most sophisticated approach and is one that AMC often uses. This approach considers the strike orientation, width across strike, and grade of the mineralised units and consolidates them adding internal and external waste to achieve defined minimum mining widths and cut-off grades. This method most accurately reflects the spatial distribution of the mineralisation and the practicalities of mining and is preferable in many situations.

When deciding which approach to use it is vital to consider the practicality of implementing the approach in an open pit and how it can be utilised implementing a cut-off grade (CoG) strategy.

In this paper the authors present case studies on how dilution modelling approaches have been applied specifically looking at:

- Regularisation in a shallow dipping deposit.
- Consolidation using AMC's in-house approach, Drill_Dil.
- Selective consolidation using AMC's in-house approach, Split Drill_dil.
- Consolidation using shape optimiser.

REGULARISATION IN SHALLOW DIPPING DEPOSIT

While regularisation works well in disseminated mineralisation which is to be mined with big equipment, such as a large porphyry copper deposit for example, AMC has found it useful for modelling ore loss and dilution in medium to shallow dipping mineralisation. A key characteristic of these deposits is a much larger apparent horizontal width the deposit is mined across compared to the actual width. The key dilution issue is identifying the waste which will be added as mining enters the mineralisation from the hanging wall and leaves from the footwall. Regularisation allows the inclusion of rock along these hanging wall/footwall boundaries capturing the dilution mined if all the potentially valuable material is treated in an SMU sized block. Then a decision must be made whether to take the mineralisation with the associated waste included in this block and incur dilution or discard it incurring ore loss.

The operational strategy, including the grade control and blasting practices, are key considerations as these define an appropriately sized SMU. For example, if blastholes are to be sampled with a single sample per bench giving a high density of lower quality samples, then the SMU should be based around the spacing and burden using perhaps one to four hole areas for the horizontal X and Y dimensions and the bench or flitch height for the vertical dimension (Z). If reverse circulation (RC) grade control is going to be used giving a lower density of better-quality samples, then a regularisation block X and Y dimensions based on the spacing between holes and rows is a good option where perhaps half the spacing is a good starting point. But this needs to be reviewed considering the mining equipment to be utilised.

A final issue to consider is the minimum size of the sub-cells in the Mineral Resource model. If these are relatively coarse with minimal sub-celling of the parent cell, which may be reasonable for Mineral Resource estimation, they can limit options for modifying the model to account for dilution. Generally, the regularised block should be no less than two times the minimum sub-cell size in the X and Y directions. This is generally less of an issue in the Z direction as the modelling methods often fit exactly to the surface in one direction which is generally Z in shallow dipping deposits.

The sections in Figures 1 and 2 show examples of a regularisation dilution modelling approach applied to a shallow dipping deposit on 2.5 m and 5.0 m excavation heights, where grey blocks less than 0.25 g/t represent dilution. This highlights the potential impact of blast heights on dilution for shallow deposits.

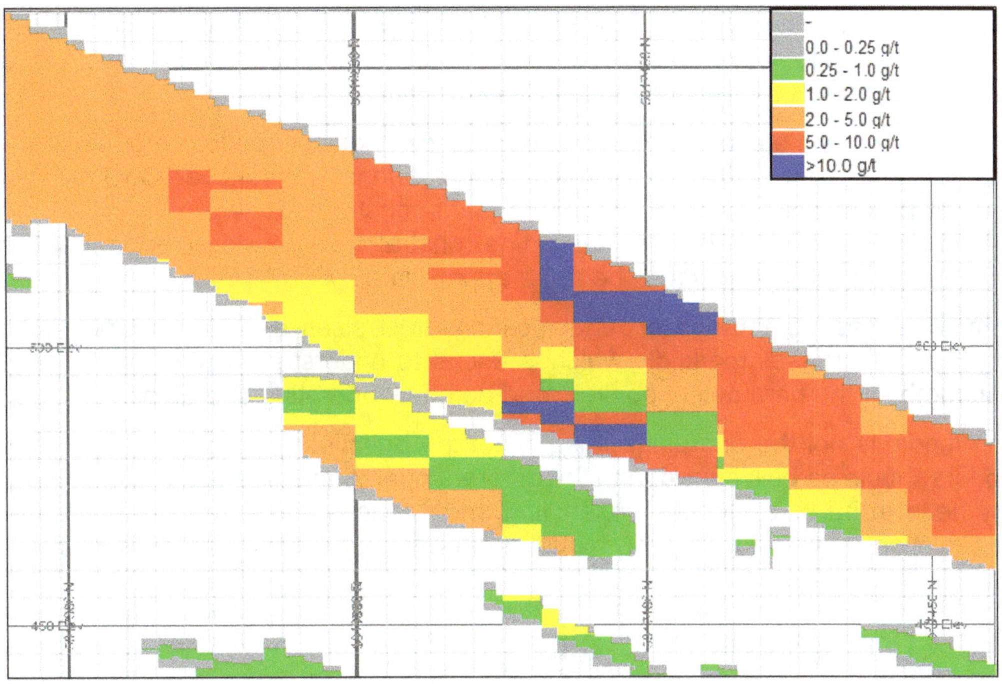

FIG 1 – Shallow dipping deposit 2.5 m bench.

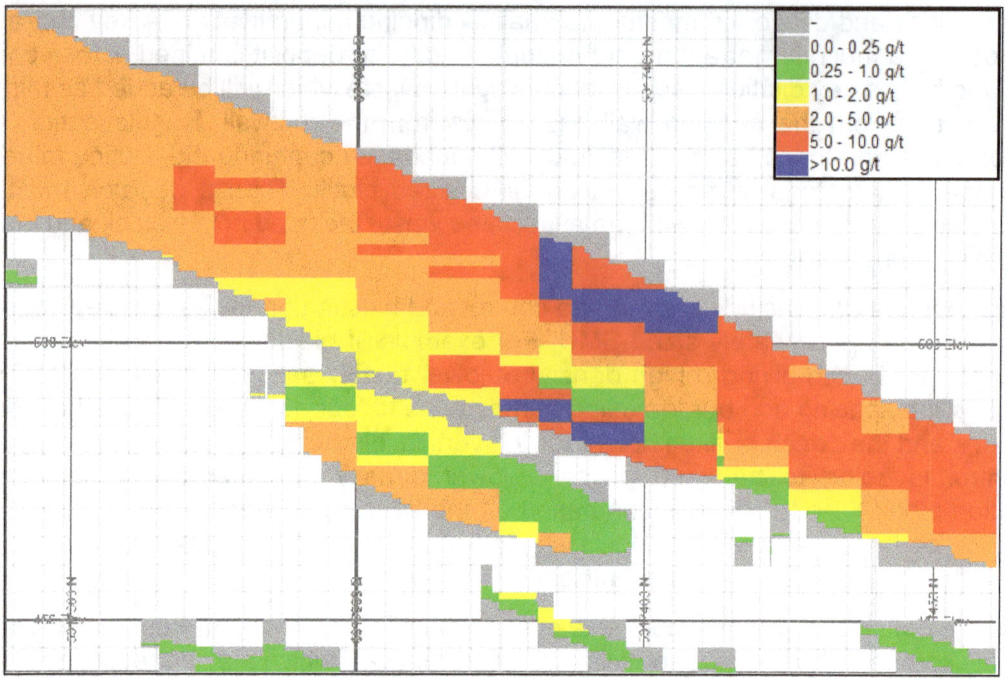

FIG 2 – Shallow dipping deposit 5.0 m bench.

These sections illustrate how this approach can be useful modelling dilution of different thicknesses around mineralisation. The normal approach is to identify any regularised cell which has classified mineralisation in it and estimate the average grade to use in the ore selection process. This is relatively straightforward for continuous variables such as grade, but categorical variables such as Mineral Resource Classification need more careful consideration. It is tempting to use the dominant tonnage, but in a situation where you may have 30 per cent high-grade Measured Mineral Resource diluted with 70 per cent unclassified waste, basing the consolidated Resource classification on tonnage will downgrade that block to unclassified. In this situation AMC recommends using the dominant quantity of metal, which is what the Mineral Resource Classification is expressing confidence in, as the basis for the regularised SMU classification.

But as with all dilution and ore loss modelling approaches, regularised block models provide a basis for making strategic decisions. In this shallow dipping scenario, a breakeven CoG can be applied to decide whether to process a block. But if you're plant constrained there may be an imperative to reduce dilution at the expense of ore loss. Or if mine constrained a very low marginal CoG minimising ore loss at the expense of ore loss may be more appropriate.

AMC recommends testing a range of regularisation SMU sizes as there is often a point where quantity of dilution increases markedly which is useful to understand when determining a dilution modification method, the parameters applied, and potential operating strategies.

The dilution model is not the end point; it is part of the optimisation process. A range of different dilution models should be built considering the different operating strategies which could be applied. These may need to be taken though scheduling and financial analysis allowing for the effect on potential production rates to identify the model and operating strategy which is best suited to the deposit.

CONSOLIDATION USING DRILL_DIL

While regularisation can be an effective modelling approach in some situations, consolidation is a more sophisticated method which allows an SMU to be applied but is not restricted to block model parent cell size or boundaries. The approach considers a row of blocks through the mineralisation and looks for the best method to consolidate them while honouring minimum mining widths, CoG's and adding a dilution skin.

The Drill_Dil modelling approach has been developed by AMC over 30 years. The development was triggered for a project comprising numerous very narrow structures which there would be no way of mining individually and a solution was required to define the optimal mining units.

The approach is based on creating a grid of imitation horizontal drill hole traces perpendicular to the strike of the mineralisation with a drill intercept record for each change in grade. The grid defines two of the dimensions for the SMU, Y and Z. The X dimension is calculated by identifying consolidated block intercepts above a pre-defined CoG. The dimension (length) and grade is analysed to determine which blocks should be consolidated along the intercept increasing the length to achieve the highest grade grouping which adheres to the X minimum length of the SMU and the CoG criteria.

This is a relatively complex process as there can be a situation where an above CoG intercept can have wide below CoG intercepts on one portion, and narrow barren material on the other portion, before a block of above CoG material. If the analysis combined intercepts to achieve the SMU is based solely on the adjacent grade, the Drill_Dil approach will model two intercepts either side of the barren material rather than a potentially better higher grade result which combines the two above CoG intercepts including the barren material between.

The modelling approach can also account for internal waste applying a minimum width to waste (below CoG) as well as mineralisation.

Once the intercepts have been consolidated honouring the width, grade and dilution skin parameters, they are expanded to become strings which are then used to flag the blocks within the YZ range of the hole trace. An average grade for the consolidated block is generated. Figure 3 compares the grade distribution of a plan-section of the original Mineral Resource block model to the diluted model generated using the Drill_Dil approach.

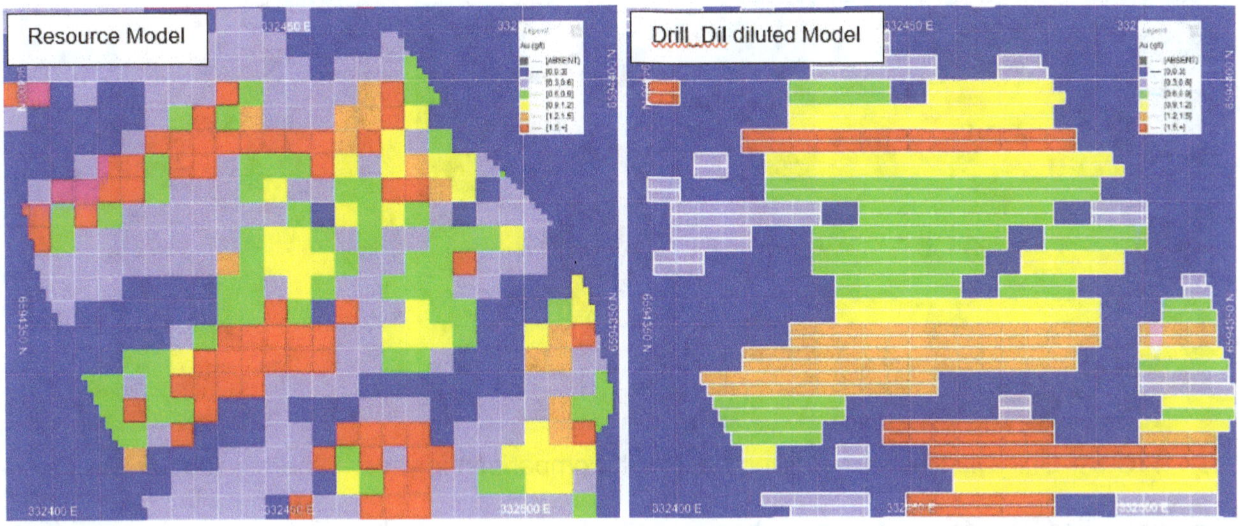

FIG 3 – Comparison of Resource Block Model and Drill_Dil Block Model Plan View Section.

This average grade of the consolidated block can be used in subsequent mine planning or used as a flag to delineate the above CoG mineralisation.

Another example, shown in Figure 4 as vertical cross-sections, illustrates how the Drill_Dil modelling approach produces a more accurate representation of estimated dilution than block regularisation for narrow mineralisation structures. The approach accounts for the spatial grade distribution and mining selectivity by consolidating blocks inclusive of additional mining width where it is required. Conversely in this example, regularisation results in significantly higher estimated dilution and ore loss because the approach does not adequately account for the spatial characteristics of the mineralisation, which would not represent the actual mining approach.

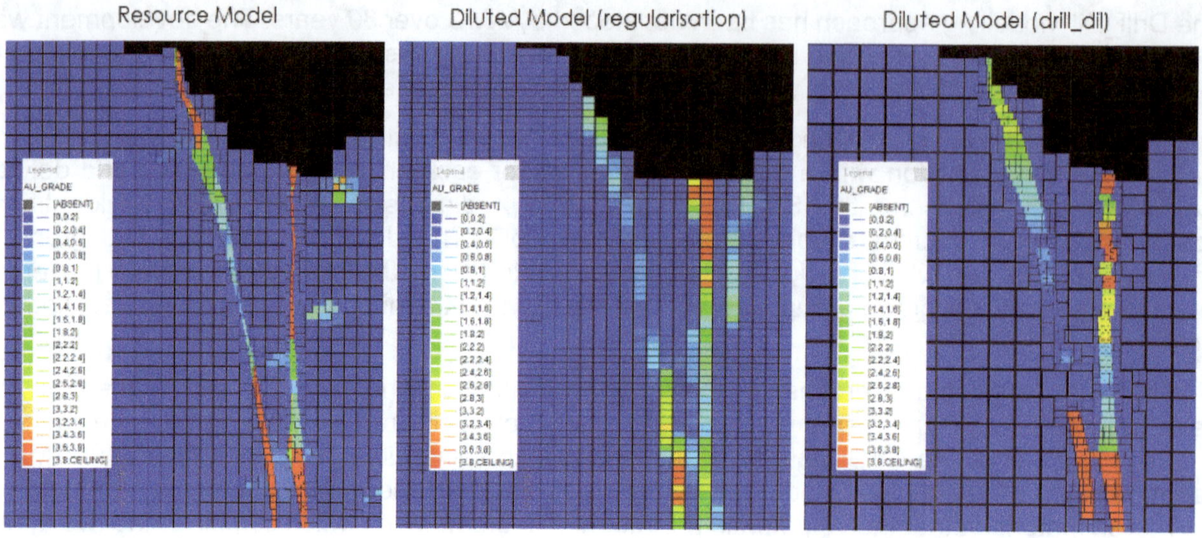

FIG 4 – Drill_Dil dilution modelling approach comparison in narrow mineralisation zones.

The Drill-Dil approach can also be applied to wider deposits. The example in Figure 5 shows a plan view section through a wider mineralised unit, where the Drill_Dil approach identified similar mineralised blocks to block regularisation, but is more effective in estimating edge dilution (skins) and inclusion of internal waste blocks. In this case, a consolidation modelling approach resulted in slightly higher dilution, but a more accurate estimate reflecting the operating strategy.

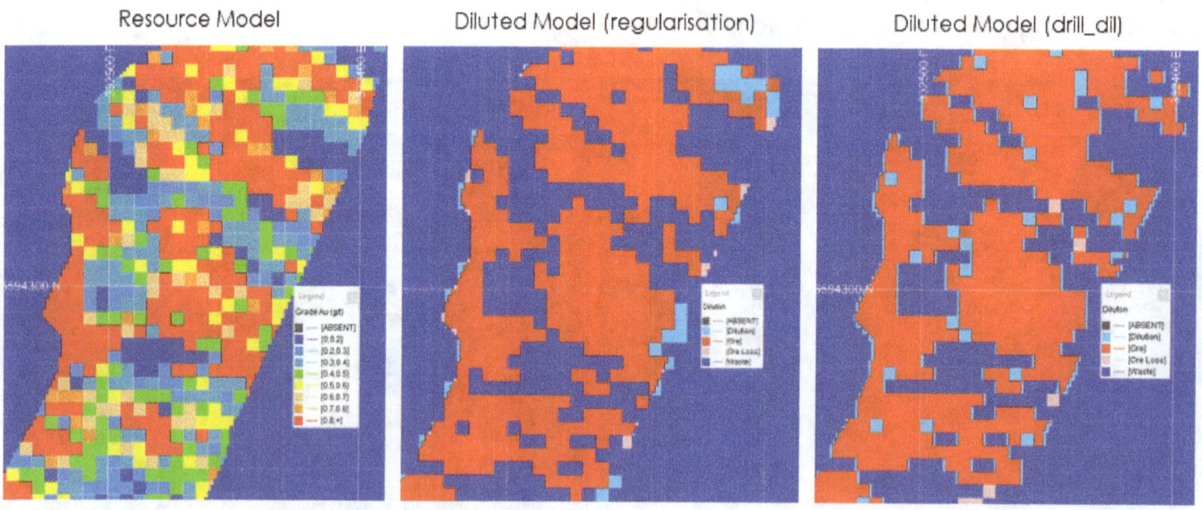

FIG 5 – Drill_Dil dilution modelling approach comparison in wider mineralisation zones.

The application of a consolidation dilution modelling approach can be effective in bulk deposits to consolidate sections by grade allowing different CoG strategies to be tested. The example shown below is for a massive deposit made up of narrow structures with differing grades, where the mineralisation needed to be consolidated to account for dilution and ore loss, but also identify the preferred CoGs for feed grade objectives. In this case, the Drill_Dil modelling approach was applied but could also be undertaken using a stope optimiser approach discussed in the next section.

The approach entailed modelling a relatively high CoG at the point where the mineralisation only just forms a consolidated block according to a specified SMU, then stepping through a range of lower CoG's creating incrementally larger overall blocks. These consolidated blocks were then incorporated into a single block model stamped over the original Mineral Resource block model, illustrated in Figures 6 and 7.

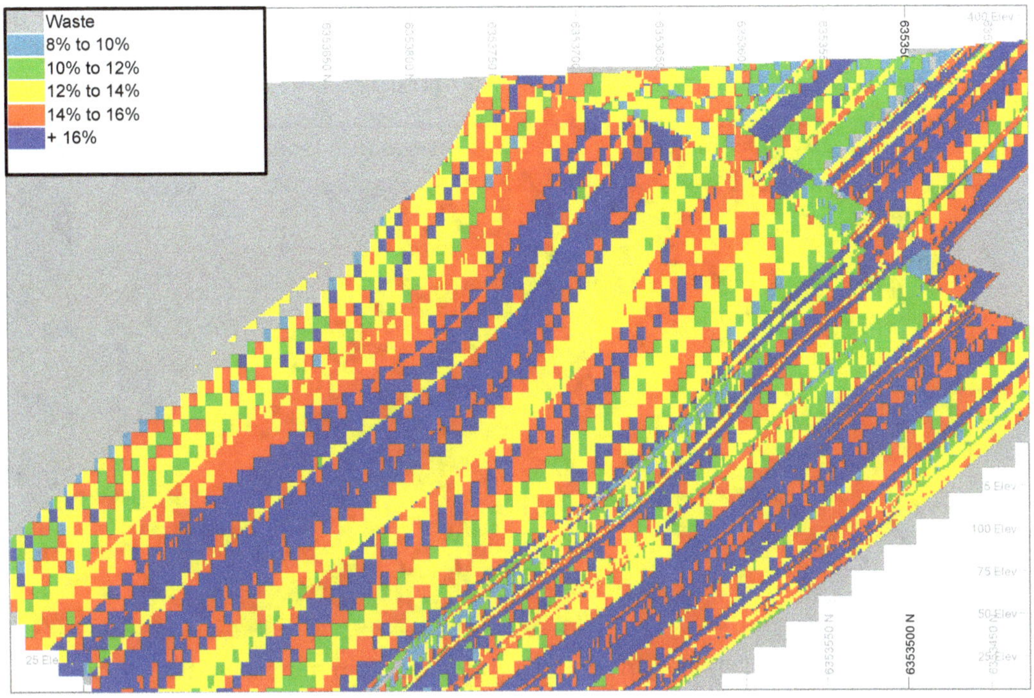

FIG 6 – Bulk deposit mineral resource with multiple mineralisation structures.

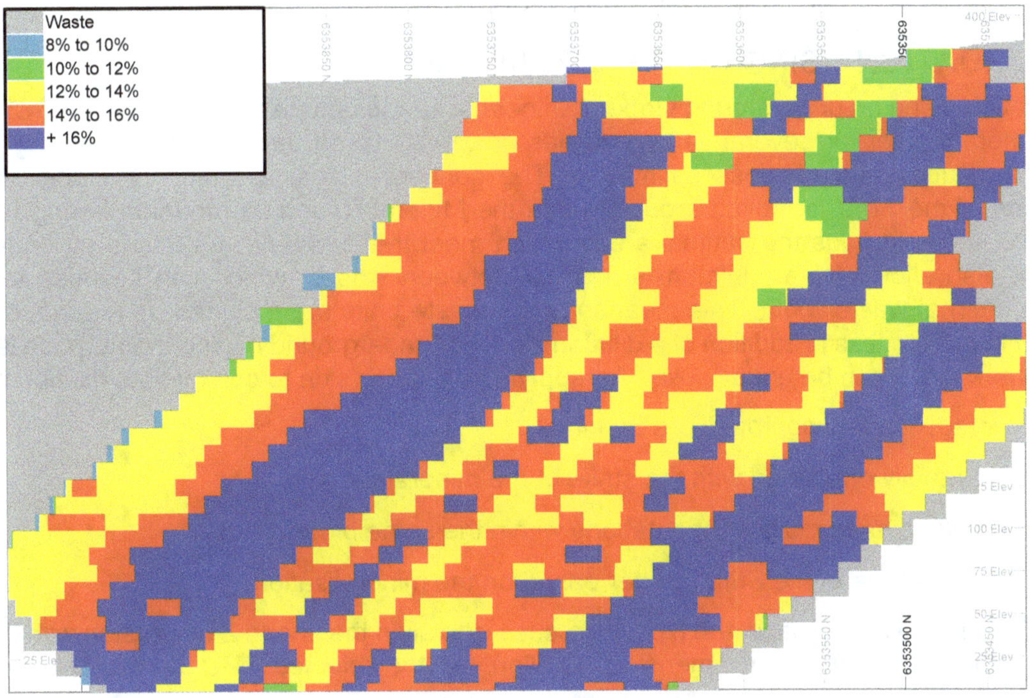

FIG 7 – Drill_Dil dilution modelling approach at various cut-off grades in bulk mineralisation zones.

The diluted block model was used as the basis for a strategic schedule. Decisions whether to direct feed mineralisation to the plant or send to the stockpile were based on the bulked CoG classification shown in Figure 7, with the value of the mineralisation based on the original grade.

This approach enabled the optimal CoG strategy to be identified for the deposit according to the project objectives.

CONSOLIDATION USING SPLIT DRILL_DIL

The Drill_Dill approach described above identifies an overall consolidated block according to a minimum SMU. However, it may not account for the potential selectivity within a consolidated block.

AMC has developed a modified version of Drill_Dil (Split Drill_Dil), which splits the larger units of the consolidated blocks into sub-blocks, still honouring the SMU constraints but enabling the delineation of practical selective blocks at different CoGs within a mineralisation zone (Figure 8).

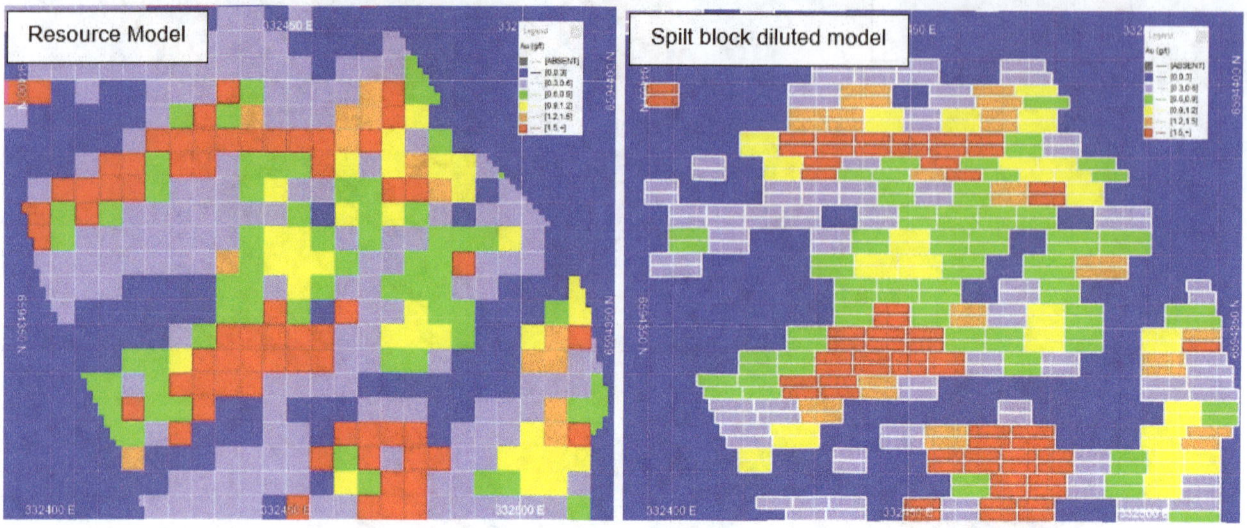

FIG 8 – Split Drill_Dil dilution modelling approach in mineralisation zones.

This approach has proven to be very effective in accurately estimating dilution and ore loss for wider mineralisation zones comprising variable grades.

CONSOLIDATION USING SHAPE OPTIMISER

Consolidation using an underground stope or mineable shape optimiser is a concept AMC has used in an ad hoc manner for several years and has now systemised and added to the range of tools utilised for dilution and ore loss estimation. The concept is that a shape or stope which has dimensions of the SMU is moved around within the Mineral Resource model adhering to defined grade criteria until it flags the optimal areas which meet these criteria. As this is an underground process it also allows for a minimum size pillar between stopes, which can be used to apply a minimum internal waste width. This means narrow waste units are included as internal dilution. A final utility in the process is addition of a dilution skin. For underground mining the purpose is to allow for overbreak, but it can be equally used for applying a dilution skin to consolidate dilution blocks.

Some advantages of this approach are that it:

- Can be readily orientated with the strike of the mineralisation.
- Is possible to have multiple orientations within the same block model.
- Allows the shape of the SMU to alter based on the mineralisation.

Similar to the Drill_Dil approach, the typical process is to align the shape Z dimension with the bench height and then set the X and Y dimensions to align with the selected SMU dimensions.

One of the constraints of the Drill_Dil process is that it works best when the strike of the mineralisation aligns with the model as the drill traces are generated based on the model orientation. As Mineral Resource block models are often created using local grids or virtual grids in rotated models, this is not often an issue. But where there are multiple structures with different strike orientations in a deposit the ability to have multiple dilution orientations and align them parallel to the strike of the mineralised units is very useful. This is shown in Figure 9 where the blue and pink rectangles show the different orientations of the dilution zones.

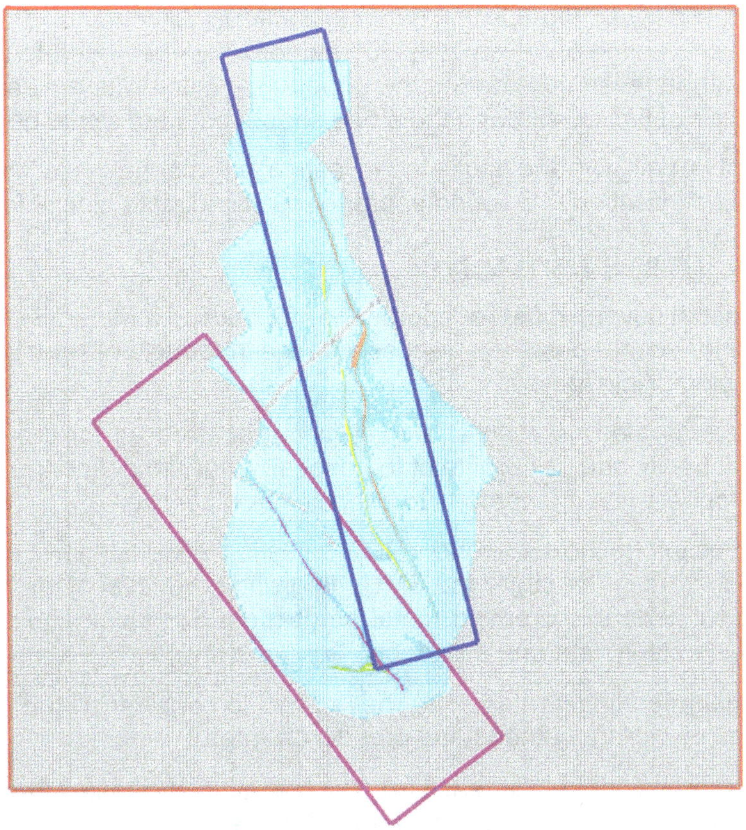

FIG 9 – Dilution orientation zones using stope optimiser.

A stope optimiser can modify the shape being used to align more closely with the mineralisation. While this is vital function for developing stope shapes it should be used with caution in open pit dilution modelling. It must be remembered that the goal is to realistically estimate what dilution and ore loss is likely when mining, and if for example the aim is to define mineralisation before blasting and where possible have discrete waste and ore blasts, then maintaining a vertical dilution shape is probably appropriate. An example of how the stope shape can be adjusted to account for open pit mining is illustrated in Figure 10.

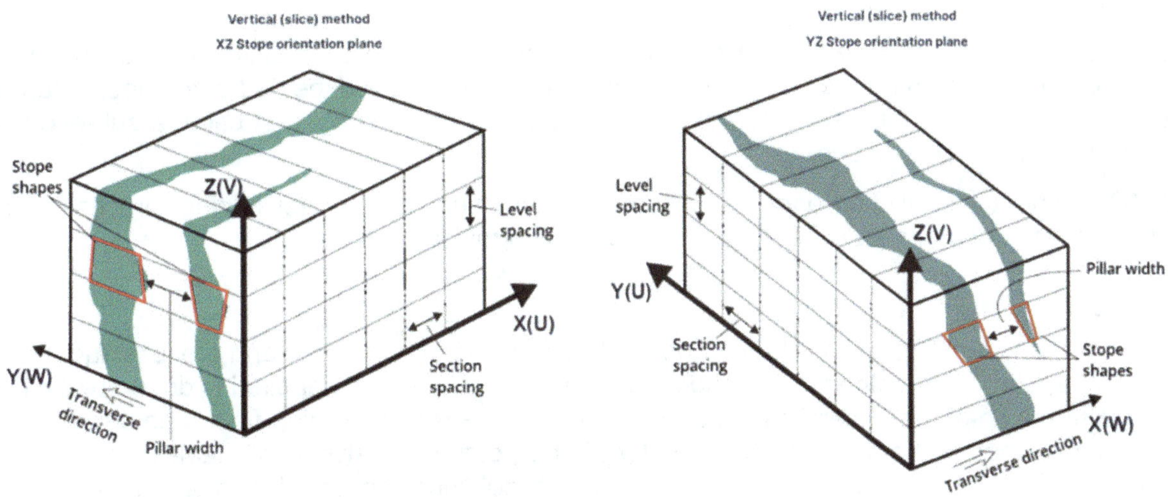

FIG 10 – Stope orientation planes showing primary directions (source: Deswik Help).

If the project is plant constrained, reducing dilution may be the key driver. One of the authors worked at a site where the mineralisation had little visual delineation but the mineralised contacts based on assay samples were quite distinct. The operation was plant constrained and minimising dilution was key to the financial success. So, the site introduced RC grade control, this was in the 1990s and not common practice, which allowed pre-splitting hanging and footwall contacts. This minimised dilution, and analysis showed preventing approximately 300 mm of waste being fed to the plant covered the cost of the pre-split without even considering the opportunity cost. With a clearly defined operation like this, using the stope optimiser allowing the stope shape to adjust would be an excellent option.

AMC comparisons of a Drill_Dil and stope optimisation approaches has shown similar results, confirming that stope optimisation is a sound approach to consolidate blocks for dilution.

SELECTING A DILUTION METHOD

The paper has highlighted several different approaches for diluting a Mineral Resource block model. But which one is right for your deposit? As discussed in the introduction, selection of the appropriate approach is dictated by multiple factors.

Underlying the decision is the operating strategies being considered for the deposit, which must include the financial capacity and risk profile of the deposit owner to define the overall objectives, in conjunction with appropriate mining methods to execute the objective.

The operating strategies inform the subsequent SMU selection influencing two major factors: Dilution and ore loss, and mining rate. The larger the SMU the greater the level of dilution and ore loss, but a higher mining rate will lead to increased metal production. So how is the optimal configuration identified? The following stages are one approach:

- Develop potential operating strategies – These need to consider all the practical options and consider a range of mining methods and production rates.
- Develop SMU dimensions – For each of the operating strategies, an SMU needs to be developed which adheres to the associated mining method, production rate, and associated mining equipment. This could be associated with excavation method, bucket size, bench height, blast row spacing etc.
- Identify appropriate dilution processes and application methods that best represent the mining method and proposed equipment according to the deposit mineralisation characteristics.
- Complete dilution modelling for a range of SMU's and compare the outputs – An initial rejection of SMU's and associated operating strategies is usually possible at this stage.
- Production scheduling and financial modelling – Develop a production schedules and financial models for the range of operating strategies to an appropriate level of detail allowing an informed decision to be made.

Critical to this approach is ensuring that the dilution and ore loss is estimated accurately. An inappropriate estimation method can lead to significant errors in the expected outcomes, potentially leading to the selection of a sub-optimal operating strategy, or in the worst case, a sub-economic decision leading to mine closure.

In AMC's experience, dilution and ore loss is often only considered after mining, processing and production rate decisions have been made, leading to significant loss of value.

CONCLUSION

There are a range of different modelling approaches for estimating dilution and ore loss and a range of ways to apply these methods. The choice of approach and how it's applied is driven by both the spatial distribution of the mineralisation and the operating strategy: Mining fleet, bench height, drill and blast parameters etc. It is vital to consider all the potential operating strategies as they effect mining rate, and for each an appropriate dilution model must be developed as this effects the potential feed grade and tonnage. The decision on which approach to use needs to be supported by appropriate analysis, which may require production scheduling and financial modelling.

The case studies presented demonstrate that enhanced modelling approaches can be applied to accurately estimate dilution and ore loss, allowing project developers and operators to better understand the value and opportunity potential of their assets.

High-resolution 3D models of ground conductivity in a channel iron deposit to de-risk production scheduling

R L van Dam[1], N Tabain[2], A Sanchez-Mora[3] and B Klue[4]

1. Technical Director – Near Surface, Southern Geoscience, Perth WA 6005.
   Email: remke.vandam@sgc.com.au
2. Principal Geophysics, BHP, Perth WA 6000. Email: nathan.tabain@bhp.com
3. Senior Geophysicist, Southern Geoscience, Perth WA 6005.
   Email: alejandro.sanchez@sgc.com.au
4. Senior Principal Geophysicist, Southern Geoscience, Perth WA 6005.
   Email: brodie.klue@sgc.com.au

ABSTRACT

A two-month survey program to provide high-resolution imaging of the near-surface lateral and vertical changes in ground conductivity, related to clay content and moisture content in a channel iron deposit (CID), was undertaken at a mining hub in the Pilbara, Western Australia. The primary aim of the survey was to map high risk ore, including from a materials handling and product quality perspective. A secondary aim was to map deeper channel morphology. The survey was conducted using a time-domain electromagnetic (TEM) system towed behind a light vehicle. Data were acquired in grids using a line spacing of 5 m. Survey areas were prepared as flat surfaces with major obstacles such as windrows and dewatering pipelines removed to maximise data coverage and acquisition efficiency. Post-processing of the data resulted in 3D block models of ground conductivity with smallest voxel cell dimensions of 2.5 × 2.5 × 1 m. When compared against probe drilling at 20 × 20 m spacing, the assessment has shown that higher conductivities correlate reasonably well with elevated moisture and clay, which can be well distinguished from lower conductivities, at a higher vertical and lateral resolution than probe drilling. This improves confidence in the ability of the mine to recover ore from select portions of the high-risk locations, whilst reducing risk of negative impacts to processing and product quality. Further, the two-month program has highlighted that the survey work can be successfully integrated into the operational cycle, with fast-turnaround results to be included in mine planning.

INTRODUCTION

Production scheduling in large open pit mines depends on many factors, including geological, geometallurgical, economical and operational (Maleki *et al*, 2020). To address uncertainty and to optimise the efficiency of various processes in the production chain, new approaches and modern technologies are frequently tested and evaluated (eg Leung, Hill and Melkumyan, 2025).

In the Pilbara region of Western Australia, iron ore is mined from both banded iron formations (BIFs) and channel iron deposits (CIDs). Miocene fluvial CIDs are the infill of meandering palaeochannels and contain a range of textures, from gravelly mudstones to granular rocks and conglomerates. Consolidated fine gravels and interbedded conglomerates and ore contained therein comprise the major source of iron ore export. Clay horizons and non-ore conglomerates are also present (Morris and Ramanaidou, 2007). Some CIDs are underlain by basal clays.

In active CID mining operations, probe drilling is used to assess material characteristics in advance of blasting and recovery. High-risk ore, which is defined as having poor materials handling properties (especially wet clay) or low product quality (high alumina and moisture content), is identified and potentially excluded from the mined envelope. However, at 20 × 20 m (probe) or 50 × 50 m (exploration) spacing and logging intervals of 6 m and 2 m, respectively, the data are relatively low resolution.

To address some of these limitations, a two-month geophysical survey program was undertaken at a mining hub in the Pilbara. The program used a towed TEM system for high-resolution and near-surface imaging of the lateral and vertical changes in ground conductivity, related to variations in clay and moisture content. The primary aim was to map high risk ore at mining bench level (0–12 m below ground surface). A secondary aim was to 'look ahead' at subsequent benches, and map deeper channel morphology. It should be noted that Fe, which is highly resistive, has no direct impact on EC and therefore TEM measurements are not suitable for measuring Fe grade at the required level of detail. An indirect correlation does exist because as alumina, associated with conductive clays and shales, increases, so does silica, leading to a reduction in Fe.

APPROACH

Surveys were conducted using a towed time-domain electromagnetic (tTEM) system manufactured by TEM Company in Denmark (Figure 1). This system consists of a transmitter (Tx) loop and offset receiver (Rx) coil, pulled by a vehicle that contains all electronics. The system uses an alternating low-moment (LM) and high-moment (HM) sequence aimed at achieving a very fast turn-off of the primary field (which allows confident modelling of the near surface conductivity) and a good signal-to-noise ratio (SNR) for greater depths of investigation, respectively (Maurya *et al,* 2023).

FIG 1 – tTEM system in operation during the survey at an open cut pit in the Pilbara.

Real-time kinematic GPS positional data was integrated and used for positional guidance during the survey. The set-up selected for this survey was the 331 configuration, which offers a better SNR than the 221 configuration and better near-surface detail (faster turn-off and earlier first time channels) than the 333 configuration (Table 1).

Although the system has a signal footprint of almost 10 m, the characterisation of small-scale features and mapping of transitions between units was optimised through slow driving speeds (~5 km/hr) and a tight line spacing of 5 m. Survey areas were prepared as flat surfaces with major obstacles such as windrows and dewatering pipelines removed to maximise data coverage and acquisition efficiency (Figure 1). Data were acquired in regular grids with several concentric lines around the edges of each survey area to ensure full coverage where line turns had created gaps. Fifteen individual grids were surveyed for a total of 334 line kilometres.

TABLE 1

Specifications of the tTEM electromagnetic survey system.

Component	Variable	221	331	333
Transmitter (Tx)	Loop size (m)	2 × 2	3 × 3	3 × 3
	Number of loop turns	1	1	3
	Effective area (m^2)	4	9	27
	LM/HM current (A)	3/30	3/30	1/25
	Peak moment (Am^2)	120	270	675
	LM/HM pulse frequency (Hz)	1055/315	1055/315	1055/247
	LM turn-off time (μs)	3	3	5
Receiver (Rx)	LM/HM number of recording gates	10/20	10/20	7/17
	First useable gate (centre time* in μs)	5.6	5.6	12.3
	Last gate (centre time* in μs)	876	876	822

*from start of ramp.

The tTEM data were processed in AGS Workbench software. To improve the SNR of the data, a trapezoidal moving average filter was applied and late time channels with high noise levels or with coupling effects were removed.

To produce electrical conductivity (EC) models of the subsurface, the data were then modelled using a smooth, spatially constrained inversion (SCI) approach with 40 layers and the first and last layer boundaries at 0.5 and 100 m below ground level, respectively. Any data negatives were excluded from the modelling. Multiple standard inversion products were generated for each data set with changes in starting resistivity, filter widths, and use of LM only or both LM and HM channels.

The data post-processing and modelling resulted in full-coverage inversions interpolated into 3D block models of ground conductivity, with smallest (near surface) voxel cell dimensions of 2.5 × 2.5 × 1 m. From these voxels, depth and elevation slices as well as iso-surfaces were extracted.

RESULTS

Results from the geophysical survey were compared against probe drill data. Here, we present results for one of the survey areas, where the geological model identified a meandering channel, and a basal clay layer is situated along the channel/pit margins.

Figure 2 highlights the strong correlation between the elevation of the top of the basal clay surface interpreted and the geological model, based on 50 × 50 m reverse circulation (RC with diamond tail) exploration drilling. The basal clay surface derived from the tTEM data was interpreted by taking the EC iso-surface that best fit the control data, inclusive of the removal of shallow anomalies associated with localised clay pods. Depths of 30 m below ground surface were reliably imaged. Thanks to the significant improvement of resolution compared to the exploration drilling, the narrow thalweg (ie the along-channel line of lowest elevation/greatest depth) of the meandering channel is imaged in detail and without aliasing effects.

Figure 3 presents a comparison between the bench-average EC over a depth range of 0–12 m and smoothed spectral moisture content derived from manual cone samples taken from the 20 × 20 m blast-rig probe drilling. Following filtering and smoothing of the noisy manual blast cone results (McArthur, Jones and Murphy, 2010), the spectral moisture data display a clear trend that aligns well with the trend seen in the EC data from the tTEM survey. One exception is the basal clay along the eastern edge of the pit, which has high electrical conductivities but no noticeable increase in moisture content.

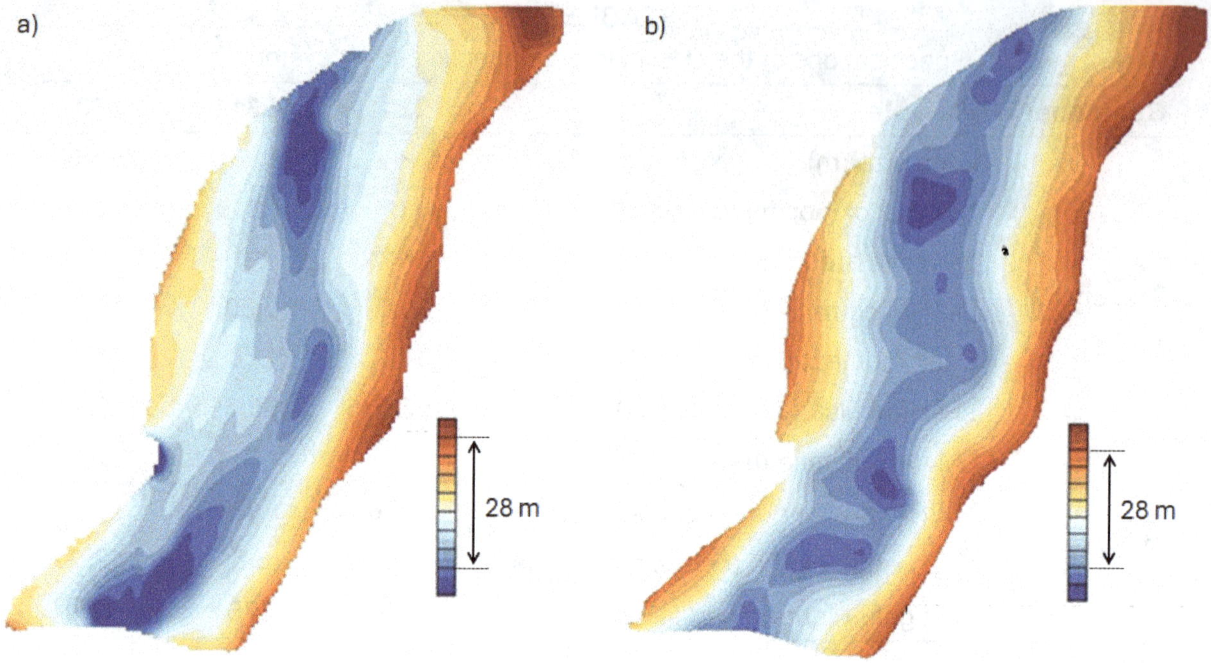

FIG 2 – Relative elevation of top of basal clay interpreted from: (a) the tTEM survey; and (b) the exploration geological model.

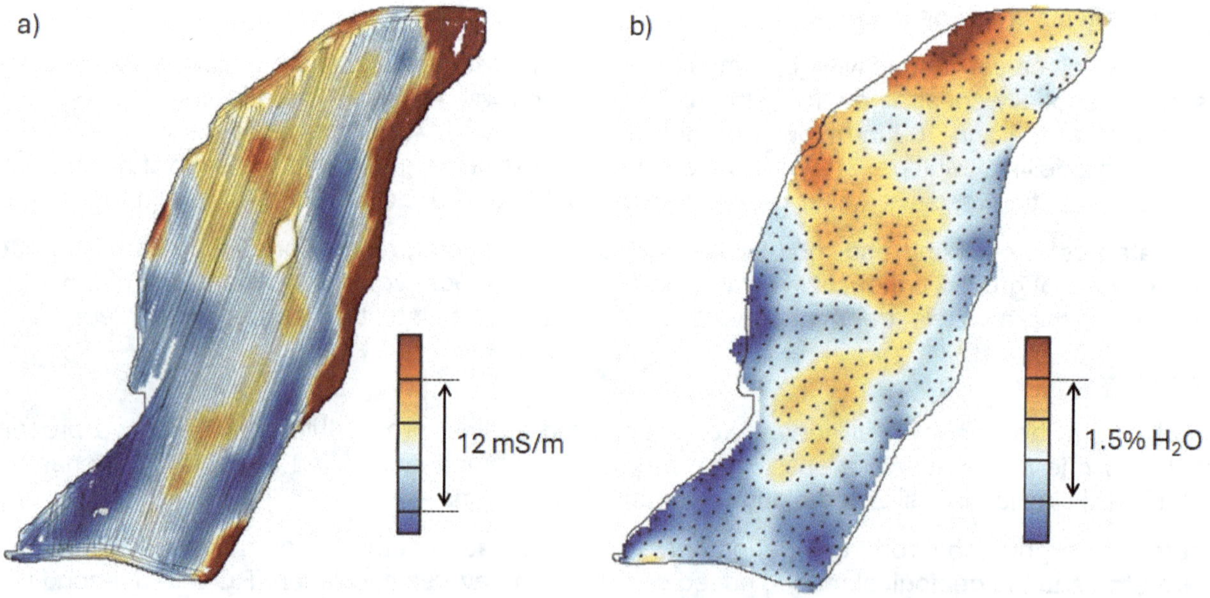

FIG 3 – Comparison between: (a) bench-average (0–12 m) EC from the tTEM survey (survey paths shown as thin grey lines); and (b) spectral moisture content from manual cone samples taken from probe drill holes (sample locations shown as black dots) after filtering and smoothing.

CONCLUSIONS

The assessment has shown that higher conductivity areas identified from the tTEM survey correlate reasonably well with elevated moisture and clay content. These can be well distinguished from lower conductivities which correlate with dry, low alumina content CID. Importantly, the tTEM survey resulted in 3D EC models with a higher vertical and lateral resolution than probe drilling. These outcomes improve confidence in the ability of the mine to recover ore from select portions of high-risk locations, whilst reducing risk of negative impacts to processing and product quality. Further, the two-month program has highlighted that the survey work can be integrated in the operational cycle, with fast-turnaround results to be included in mine planning. However, further work is required to optimise the equipment, acquisition, data processing and integration into other stakeholders' processes.

ACKNOWLEDGEMENTS

We thank Rabea Sedaghat (currently at CSIRO), George Brabec, and Thusitha Nimalsiri for their support with data processing and the survey crew and site staff for the support in safely and efficiently executing this survey.

REFERENCES

Leung, R, Hill, A J and Melkumyan, A, 2025. Automation and Artificial Intelligence Technology in Surface Mining: A Brief Introduction to Open pit Operations in the Pilbara, *IEEE Robotics and Automation Magazine*, 32(3):164–183. https://doi.org/10.1109/MRA.2023.3328457

Maleki, M, Jélvez, E, Emery, X and Morales, N, 2020. Stochastic Open pit Mine Production Scheduling: A Case Study of an Iron Deposit, *Minerals*, 10(7):585. https://doi.org/10.3390/min10070585

Maurya, P K, Foged, N, Madsen, L M and Christiansen, A V, 2023. Comparison of Towed Electromagnetic with Airborne Electromagnetic and Electrical Resistivity Tomography in a Hydrogeophysical Context, *Geophysical Journal International*, 235(1):817–830. https://doi.org/10.1093/gji/ggad276

McArthur, G, Jones, C and Murphy, M, 2010. Blasthole Cone Sampling Experiments for Iron Ore Flitch Mining, in *Proceedings of the Sampling Conference 2010*, 6 p (The Australasian Institute of Mining and Metallurgy: Melbourne).

Morris, R C and Ramanaidou, E R, 2007. Genesis of the channel iron deposits (CID) of the Pilbara region, Western Australia, *Australian Journal of Earth Sciences*, 54(5):733–756. https://doi.org/10.1080/08120090701305251

ACKNOWLEDGEMENTS

We thank [illegible] for [illegible] support with data processing and the survey [illegible] executing this survey.

REFERENCES

[illegible]

Open Pit

Slope stability and geotechnical analysis

Slope stability analysis using Fellenius method with a fractal-enhanced Mohr–Coulomb (FEMC) model

D Kon[1], B K Mawugnon[2], S Ngoie[3] and Y Mbuyu[4]

1. PhD candidate, China University of Mining and Technology, Xuzhou 221000 Jiangsu, China. Email: dinakon25@gmail.com
2. PhD, China Railway Resources Group Survey and Design Co, Ltd. Researcher, Lomé, Togo. Email: gedeons07@yahoo.fr
3. Professor, University of Kolwezi, Kolwezi, Lualaba, Democratic Republic of the Congo. Email: sagesseng@gmail.com
4. Lecturer, University of Kolwezi, Kolwezi, Lualaba, Democratic Republic of the Congo. Email: yadahmbuyu16@gmail.com

ABSTRACT

An accurate slope stability assessment of fractured rock masses remains a major challenge because conventional limit equilibrium methods generally assume a constant shear strength along the failure surface and may therefore overestimate stability in structurally degraded rock. This study presents a fractal-enhanced Mohr–Coulomb (FEMC) model coupled with the classical Fellenius method of slices to explicitly incorporate fracture complexity into slope stability analysis using rock quality designation (RQD) data. In the proposed framework, RQD is transformed into an effective fractal dimension, which is then used to modify the cohesion and friction angle, allowing the factor of safety (F_s) to vary according to rock mass conditions. In simple terms, poor rock quality corresponds to greater fracture complexity, lower effective shear resistance, and reduced slope stability. To illustrate the framework, representative RQD values of 30 per cent, 70 per cent, and 95 per cent, derived from the rock mass quality conditions reported for the Mashamba West deposit in the Democratic Republic of the Congo, were used to generate the corresponding fractal dimensions (D_s) and evaluate their influence on slope stability. The analytical results show that the classical Fellenius solution yields a baseline F_s of 1.55, whereas the FEMC-corrected values are 1.23, 1.46, and 1.57 for RQD values of 30 per cent, 70 per cent, and 95 per cent, respectively. These results indicate a reduction of approximately 21 per cent under poor rock mass conditions and convergence toward the classical solution for competent rocks. Numerical validation using Slide2D produced the same overall trend, with F_s values decreasing from 1.103 to 0.899 for RQD = 30 per cent and increasing slightly to 1.119 for RQD = 95 per cent. Additional finite element verification using PLAXIS further confirmed the sensitivity of slope stability to fractal-dependent strength properties. The results demonstrate that the FEMC–Fellenius framework provides a practical and mechanically meaningful way to link routine rock mass characterisation with slope stability analysis while preserving the simplicity of classical limit equilibrium methods.

INTRODUCTION

Slope stability analysis is a key issue in geotechnical and mining engineering. This is especially true for fractured rock masses, where the failure mechanism of the rock mass is highly dependent on the orientation of the network of discontinuities (or fractures) in the rock mass. The limit equilibrium method remains the most widely used approach for evaluating slope stability because the techniques involved are conceptually straightforward and more computationally efficient than alternative methods. The Fellenius ordinary method of slices is one of the earliest and most widely utilised limit equilibrium methods for estimating slope stability. The safety factor is estimated by establishing an equilibrium between the resisting and driving forces acting on a potential slip surface (Fellenius, 1936). Most classical applications of the Fellenius ordinary method of slices assume that the shear strength of the rock is spatially uniform, which is frequently incorrect in fractured rock slopes with varying joint spacing, roughness, and persistence (Huang, 1983; Doan, 2023). The rock quality designation (RQD) is one of the most widely used empirical indices because of its ease of use and dependence on routine core logging (Deere, 1989; Deere and Deere, 1988; Zhang, 2016; Pells *et al,* 2017). RQD has been used in many rock mass classification systems and to estimate the strength of rock masses; however, its direct use in slope stability analysis is limited. RQD has been used in several studies to determine that fracture spacing and distribution are not random but are scale-invariant and clustered; therefore, they can be modelled using fractal geometry (Mandelbrot *et al,* 1984; Turcotte, 1986, 1997, 2009). In particular, Boadu and Long (1994) established a quantitative correlation between RQD and the fractal dimension of fracture spacing, thereby creating a quantifiable relationship between rock mass quality and fracture complexity.

To date, fractal concepts have been used to quantify joint roughness (Kon *et al,* 2025), fracture density of fractures (Kon, Shu, and Mbako, 2025), and the heterogeneity of rock masses (Barton, 1976; Barton and Choubey, 1977); however, their application to traditional slope stability methods is limited. Current research indicates that incorporating fractal descriptors into constitutive strength models provides a physically meaningful way to represent the degradation in the strength of fractured rock masses (Kon *et al,* 2026). In light of this, the current study outlines a fractal-enhanced Mohr–Coulomb (FEMC) model to be used in conjunction with the Fellenius method so that the factor of safety can be calculated explicitly based on the complexity of fractures, as represented by the RQD. The goal of this study is to provide a means of linking empirical rock mass characterisation with analytical slope stability evaluations.

METHODOLOGY

This study presents a framework for analysing slope stability in fractured rock masses by integrating the classical Fellenius ordinary method of slices and a FEMC strength model (Kon *et al,* 2026). This study explicitly considers the geometric complexity of discontinuities using fractal dimensions derived from RQD measurements. The use of fractal dimensions provides a direct analytical relationship between the factor of safety and spatially variable quality of the rock mass and enhances the physical realism of limit equilibrium analyses for slope stability, while still allowing for practical applications.

Conceptual basis for the FEMC-Fellenius coupling

The conceptual basis of the FEMC–Fellenius model is to enhance the constitutive description of the shear strength within the classical limit equilibrium framework without modifying its fundamental equilibrium assumptions. In this formulation, the Fellenius method continues to provide a global stability analysis through the moment equilibrium of slices, whereas the shear strength parameters are redefined as functions of the fractal dimension. The fractal dimension (D_s) serves as a quantitative descriptor of the discontinuity surface complexity, capturing the influence of roughness, fragmentation, and structural degradation on shear resistance. To enable practical applications, a relationship between the RQD and fractal dimension is established, allowing field-measured rock mass quality to be translated into a geometric parameter suitable for strength modification. This establishes a transfer mechanism between geological observations and mechanical behaviour: slices intersecting highly fractured zones are assigned lower effective shear strength, whereas slices located in more intact rock retain higher resistance. Consequently, the safety factor can reflect spatial variations in rock mass conditions, thereby improving the physical realism of the analysis. Notably, this integration did not alter the fundamental structure of the Fellenius method. Instead, it introduced variability at the constitutive level, thereby preserving the analytical transparency and computational efficiency of the slice method. This balance between simplicity and realism is particularly valuable in open pit mining applications, where rapid yet reliable stability assessments are required for design and operational decision-making.

Derivation of the fractal dimension from RQD

The RQD is a method that is widely used to assess the integrity of a rock mass (Deere and Deere, 1988), and is calculated by determining the percentage of intact drill core pieces greater than 0.1 m (10 cm) extracted over a given length of drill core. In the case of mining operations that use an open pit mining process, the RQD data are typically collected throughout the depth of the open pit mine on a regular basis and therefore allow for routinely recorded data related to the frequency of fractured rock material and the disturbance to the mass of rock. To define the fractal nature of a fracture system geometrically in relation to the RQD, the RQD concept was extended by Boadu and Long (1994) to the use of the fractal point process, which indicates that the spacing distributions of fractures often follow a power-law scaling:

$$n(x) \propto x^{-D_S}\ ;\ (1 < D_S < 2) \tag{1}$$

where:

$n(x)$	denotes the spacing intensity of discontinuities exceeding length x
x	is the fracture spacing scale (m)
D_S	is the fractal dimension describing fracture roughness complexity

Instead of treating $n(x)$ as a probability density function, the model was interpreted as a scaling relationship for the dominant intact core fraction. Under this expectation-based approximation, the proportion of intact segments exceeding the RQD threshold x_0 scales as follows (Boadu and Long, 1994):

$$\mathrm{RQD}=\frac{100}{\mathrm{L}}\int_{x_0}^{\mathrm{L}} x.\mathrm{n}(x)\mathrm{d}x \Rightarrow \mathrm{RQD}=100\left(\frac{x_0}{\mathrm{L}}\right)^{D_S-1} \tag{2}$$

where:

$x_0 = 0.1\mathrm{m}$	is the standard RQD cut-off length
$\mathrm{L}=1\mathrm{m}$	is the reference core run length

Substituting these values yields:

$$\mathrm{RQD} = 100 \times (0.1)^{\mathrm{D_S}-1} \tag{3}$$

Solving for the effective fractal dimension gives:

$$\mathrm{D}_S = 1 + \log_{10}\left(\frac{100}{\mathrm{RQD}}\right) \tag{4}$$

Practically, this equation converts a routine rock quality index (RQD) into a fractal parameter that describes fracture complexity. A lower RQD indicates a more fractured rock mass and, therefore, corresponds to a higher fractal dimension, indicating a more irregular and mechanically weakened discontinuity system. The slice-specific fractal dimension was computed to account for spatial variability as follows:

$$\mathrm{D}_S^{(i)} = 1 + \log_{10}\left(\frac{100}{\mathrm{RQD}^{(i)}}\right) \tag{5}$$

where

$\mathrm{RQD}^{(i)}$	denotes the rock quality value assigned to slice i

This development provides a direct conversion from routine borehole logging data to a quantitative descriptor of the fracture surface complexity. This implies that each slice can be assigned its own fracture-complexity value according to the local rock quality. Consequently, the model can represent spatial variability in the slope instead of assuming that the entire rock mass behaves uniformly.

Brief introduction to FEMC method

The classical Mohr–Coulomb failure criterion assumes constant cohesion c' and friction angle φ'. However, laboratory and field investigations have demonstrated that both parameters increase with joint surface roughness owing to asperity interlocking and roughness-induced dilation (Barton, 1976). The proposed FEMC model incorporates these effects by expressing the shear strength parameters as explicit functions of the fractal dimension.

Cohesion enhancement:

$$c'\left(D_S^{(i)}\right)=c_0+c_1^{\left[k_c\left(D_S-D_0\right)\right]} \tag{6a}$$

Friction angle enhancement:

$$\varphi'\left(D_S^{(i)}\right)=\varphi_0+k_\varphi\left(D_S-D_0\right) \tag{6b}$$

where:

c_0 and φ_0 = baseline parameters for smooth joints

D_0 = reference fractal dimension (typically, D_0 = 1.0)

c_1 = exponential base controlling cohesion growth

k_c and k_φ = material sensitivity coefficients

In simple terms, these equations allow the shear-strength parameters to change with fracture surface conditions rather than remaining constant. Therefore, the model links the rock mass geometry to the mechanical behaviour by adjusting the cohesion and friction angle according to the measured degree of discontinuity complexity.

FEMC model calibration

The baseline parameters, c_0 and φ_0, are obtained from direct shear testing or empirical correlations for planar discontinuities (Barton and Choubey, 1977). The FEMC model was developed to characterise the shear strength of fractured rocks using the coefficients k_c and k_φ, which were determined via least-squares fitting to the experimental shear strength versus analysis of discontinuity surface roughness. When $D_S^{(i)}=D_0$, the FEMC equation is reduced to the conventional Mohr–Coulomb criterion:

$$c_i'=c_0,\ \varphi_i'=\varphi_0 \tag{7}$$

where c_i' is the effective cohesion of slice i (kPa), and φ_i' is the effective friction angle of slice i (degrees). This indicates that the FEMC model is a generalisation of the classical Mohr–Coulomb criterion rather than a separate formulation. When fracture effects are absent or negligible, the proposed model reverts to the conventional strength law.

Unified Fellenius–FEMC factor of safety determination

Slope stability was evaluated using the Fellenius (ordinary) method, in which a trial circular slip surface was discretised into n vertical slices. The resisting shear force at the base of slice i is expressed as:

$$\tau_{res}^{(i)}=c'\left(D_S^{(i)}\right)+\left(\sigma_n^{(i)}-u_i\right)\tan\varphi'\left(D_S^{(i)}\right) \tag{8}$$

where $\sigma_n^{(i)}$ is the normal stress on the slice base (kPa), and u_i is the pore water pressure along the base (kPa). Physically, this equation states that the resistance along the slip surface comes from two sources: the bonding effect represented by cohesion and the frictional resistance mobilised by the normal force. Both contributions are reduced when the rock mass becomes more fractured and structurally degraded. Under the Fellenius assumption, the normal stress is approximated as:

$$\sigma_n^{(i)}=\frac{W_i\cos\alpha_i}{l_i} \tag{9}$$

where:

W_i = Weight of slice i (kN)

α_i	= base inclination angle
l_i	= slice base length (m)

Here, the normal stress is estimated from the slice weight and the inclination of the slice base. This approximation preserves the simplicity of the Fellenius method while still capturing the main mechanical contribution of each slice to slope stability. The global factor of safety is then:

$$\mathrm{F_S} = \frac{\sum_{i=1}^{n}\left[c_i^{'}l_i + (W_i\cos\alpha_i - u_il_i)\tan\varphi_i^{'}\right]}{\sum_{i=1}^{n}(W_i\sin\alpha_i)} \tag{10a}$$

This equation compares the total resisting force with the total driving force along the potential failure surface. If the resisting forces are greater than the driving forces, the slope is stable; otherwise, failure is more likely.

Explicit fractal dependence

Because $c_i^{'}$ and $\varphi_i^{'}$ depend on the fractal dimension, $\mathrm{D}_S^{(i)}$ and $\mathrm{D}_S^{(i)}$ are derived from $\mathrm{RQD}^{(i)}$, and substitution yields a factor of safety explicitly controlled by the spatial rock quality:

$$\mathrm{F_S} = \frac{\sum_{i=1}^{n}\left[c_0 + c_1^{\left[k_c\left(1+\log_{10}\left(\frac{100}{\mathrm{RQD}^{(i)}}\right)-D_0\right)\right]}\cdot l_i + (W_i\cos\alpha_i - u_il_i)\tan\varphi'\varphi_0 + k_\varphi\left(1+\log_{10}\left(\frac{100}{\mathrm{RQD}^{(i)}}\right)-D_0\right)\right]}{\sum_{i=1}^{n}(W_i\sin\alpha_i)} \tag{10b}$$

This is the key step of the proposed framework: the factor of safety becomes directly dependent on rock mass quality through the fractal dimension. Therefore, poorer-quality and more highly fractured zones automatically contribute lower shear resistance and reduce overall slope stability. Thus, zones of poor rock quality (low RQD → higher D_s) reduce stability in a mechanically consistent manner. Dimensional consistency indicates that all calculations were performed for an out-of-plane thickness of one ($b = 1m$). Thus, the contribution to the pore pressure is expressed as u_il_i. If the thickness was not a unit (b > 1), it was expressed as u_il_ib. An enhancement of the proposed platform, the Fellenius–FEMC system, combines both traditional routine RQD measurements to produce fractal characterisation and the original Fellenius equilibrium method, providing a more soundly based and theoretically defensible method for determining slope stability conditions in broken rock masses while preserving the practical applicability of this methodology for engineers and designers.

Fellenius–FEMC model validation

Initially, the slope stability of the open pit rock mass was evaluated using the classical Fellenius method, which provides an initial factor of safety based on spatially uniform shear strength parameters (that is, the same shear strength at all points along the failure surface). The potential circular failure surface was divided into seven vertical 'slices', as illustrated in the corresponding drawing (Figure 1), where the following parameters for each slice were estimated:

1. Weights W_i.
2. Base inclinations α_i.
3. Lengths of the base of the slices Δl_i.

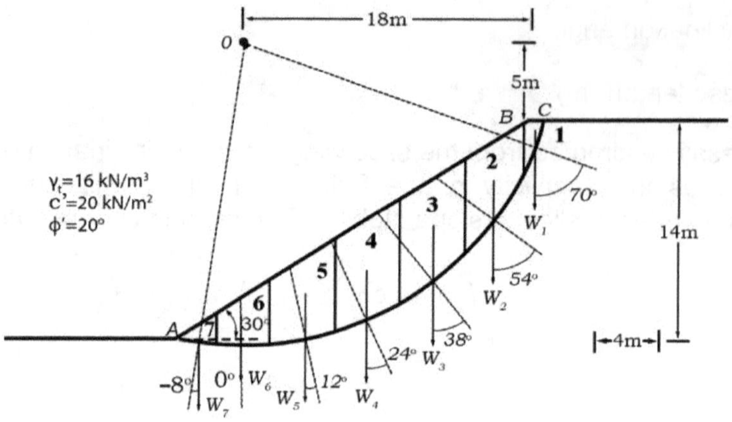

FIG 1 – Slope geometry and slice discretisation used in the classical Fellenius method of slices.

The slice equilibrium components and summations required for the Fellenius formulation are listed in the data set in Figure 1 and Table 1.

TABLE 1
Slice geometry and force components for the Fellenius analysis.

Slice #	W_i(kN/m)	α_i	$\sin\alpha_i$	$\cos\alpha_i$	Δl_i	$W_i \sin\alpha_i$(kN/m)	$W_i \cos\alpha_i$(kN/m)
1	22.4	70	0.94	0.342	2.924	21.1	6.7
2	294.4	54	0.81	0.588	6.803	238.5	173.1
3	435.2	38	0.616	0.788	5.076	268.1	342.94
4	435.2	24	0.407	0.914	4.376	1777.1	397.8
5	390.4	12	0.208	0.978	4.09	81.2	381.8
6	268.8	0	0	1	4	0	268.8
7	66.58	-8	-0.139	0.99	3.232	-9.25	65
$\sum$	-	-	-	-	**30.501**	**776.75**	**1637.04**

The total slip surface length is obtained as follows:

$$\sum_{i=1}^{n} \Delta l_i = 30.501\text{m}$$

where the driving and normal force contributions are computed as follows:

$$\sum_{i=1}^{n} W_i \sin\alpha_i = 776.75 \text{ KN/m}, \sum_{i=1}^{n} W_i \cos\alpha_i = 1637.04 \text{ KN/m}$$

Using the classical Mohr–Coulomb shear strength parameters shown in Figure 1 ($c' = 20\,\text{Kpa}$, $\varphi' = 20^o$, and dry conditions), the Fellenius safety factor was calculated as follows:

$$\text{F}_S^{\text{Classical}} = \frac{c'\sum \Delta l_i + \left(\sum W_i \cos\alpha_i\right)\tan\varphi'}{\sum W_i \sin\alpha_i}$$

Substitution of the calculated values yielded a baseline safety factor of:

$$\text{F}_S^{\text{Classical}} = 1.55$$

This result represents the conventional homogeneous strength stability estimate and serves as a reference for comparison with the proposed fractal-enhanced framework.

Rock mass quality conditions

The data set obtained from the Mashamba West open pit mine (Democratic Republic of Congo) demonstrates the variability in rock quality (and therefore rock mass) owing to the density and complexity of the fractures. The Mashamba West RQD data set indicates considerable differences in terms of rock quality (ie rock mass) because it includes both poorly fractured areas (with RQD < 50) and well-fractured areas (with RQD > 70). This study utilised the following representative RQD values: for RQD = 30 per cent, the examined rock mass was of poor quality and was highly fractured. The rock mass at RQD = 70 per cent had moderate quality and was considered moderately competent. Finally, the examined rock mass at RQD = 95 per cent had very good quality and was considered to have a relatively intact rock mass. These identified RQD values were used to quantify fracture complexity and therefore provide the basis for examining the mechanical influence of fractures on slope stability.

Fractal dimension derived from RQD

The FEMC framework directly links rock mass quality to fracture surface roughness through the fractal dimension computed from the RQD. Using the analytical relationship:

$$\mathrm{D}_S = 1 + \log_{10}\left(\frac{100}{\mathrm{RQD}}\right)$$

The effective fractal dimensions corresponding to the Mashamba West scenarios were obtained as follows:

- RQD = 30 per cent: $D_s = 1.5229$
- RQD = 70 per cent: $D_s = 1.1549$
- RQD = 95 per cent: $D_s = 1.0223$.

The results show an inverse relationship between decreased RQD and increased fractal dimension, such that as the RQD decreases, the complexity of the resulting fractures decreases and the discontinuity within the intact rock increases.

Fellenius based FEMC safety factors calculation

The traditional Fellenius approach assumes that the shear stress parameters remain constant across the entire rock mass (Huang, 1983). However, in the proposed numerical modelling framework (Fellenius–FEMC), the cohesion and friction angle become variable based on the fractal dimension of the rock mass. The FEMC calibration constants were adopted from (Kon *et al*, 2026), as follows: $k_\varphi = 15.32^o$, $k_c = 2.784$, $c_1 = 37\,\mathrm{kPa}$. Using the baseline reference parameters, $c_0 = 20\,\mathrm{kPa}$, $\varphi_0 = 20^o$, and $D_0 = 1.0$, the shear strength parameters were updated according to the FEMC formulation and substituted into the Fellenius stability expression.

$$c'(D_S) = c_0 + c_1^{[-k_c(D_S - D_0)]}$$

$$\varphi'(D_S) = \varphi_0 + k_\varphi(D_S - D_0)$$

Then Fellenius becomes:

$$\mathrm{F}_S(\mathrm{RQD}) = \frac{\left[c'(D_S)\sum \Delta l_i + \left(\sum W_i \cos\alpha_i\right)\tan\varphi'(D_S)\right]}{\sum W_i \sin\alpha_i}$$

Calculated FEMC strength parameters and safety factors: RQD = 30 per cent, 70 per cent, 95 per cent; ($D_S = 1.52288$); $D_S - D_0 = 1.52288 - 1.0 = 0.52288$.

Cohesion:

$$c' = 20 + 37^{-2.784(0.52288)} = 20.0052\,\text{kPa}$$

Friction angle:

$$\varphi' = 20 + 15.32(0.52288) = 11.989^{o}$$

Safety factor:

$$F_S(30) = \frac{\left[30.501(20.0052) + 1637.04\tan(11.989^{o})\right]}{776.75} = 1.233$$

$$F_S(70) = \frac{\left[30.501(20.2107) + 1637.04\tan(17.627^{o})\right]}{776.75} = 1.463$$

$$F_S(95) = \frac{\left[30.501(20.7994) + 1637.04\tan(19.659^{o})\right]}{776.75} = 1.570$$

These findings indicate that incorporating fracture complexity is related to a progressive increase in rock mass instability, as determined by the FEMC, as shown in Table 2. For example, with an RQD of 30 per cent, the safety factor of the slope decreased to approximately 21 per cent from the classical Fellenius baseline ($F_S(30) \approx 1.233$). This indicates that the constant-strength limit equilibrium methods do not adequately account for instability because of the high number of fractures present in this location. As the RQD values improve, the safety factors produced from the FEMC tend toward the classical estimate. For example, at an RQD value of 70 per cent, the slope condition is at the midpoint between ($F_S(70) \approx 1.46$), whereas for competent rock conditions (RQD = 95 per cent), the predicted safety factor slightly exceeds the classical estimate ($F_S(95) \approx 1.57$). Therefore, slope stability assessments based on rock mass degradation using the FEMC-Fellenius method provide an appropriate mechanism for assessing the slope stability of heterogeneous rock masses during open pit quarry operations.

TABLE 2

Factors of safety calculation for classical and FEMC models using an analytical approach.

Method	RQD (%)	D_s	c' (kPa)	φ' (°)	F_s
Classical Fellenius	-	-	20.0	20.00	**1.55**
Fellenius–FEMC	30	1.52	20.005	11.989	**1.23**
Fellenius–FEMC	70	1.15	20.211	17.627	**1.46**
Fellenius–FEMC	95	1.02	20. 799	19.659	**1.57**

Numerical model set-up

To validate the analytical evaluations of the Fellenius and Fellenius–FEMC slope stabilities, as well as the geotechnical procedures, two-dimensional numerical analyses of slope stability were performed using Slide2D software. The numerical models of the slopes had the same slope geometry, material properties, and assumptions regarding the failure surfaces used in the previously described analytical analyses, thereby providing a basis for directly comparing the obtained analytical results with those derived from Slide2D. The slope geometry in Slide 6.0 was defined by a polygonal boundary with vertices at (-4.199, 7.855), (-4.191, 11.151), (11.021, 11.151), (5.572, 15.174), (12.030, 15.174), and (12.030, 7.855) m, forming a 7.319 m-high slope with an approximately 16.23 m-wide base. The model included a planar upper face, a horizontal crest segment, and a toe located at (-4.199, 7.855). The slope material in the plate model was assumed to be homogeneous and to have the same constant shear strength parameters as those used in the analytical methods of assessment for project purposes: c = 20, φ = 20, γ = 20 kN/m^3, and u = 0. Figure 2 presents the results of the numerical slope stability analysis for different rock mass qualities, as indicated by the varying RQD values. In the classical Fellenius model for a highly intact rock mass with RQD = 100 per cent (Figure 2a), the factor of safety (F_s) is approximately 1.103, serving as the baseline for comparison. For a heavily fractured rock mass with RQD = 30 per cent (Figure 2b), the Fellenius–FEMC model predicted a lower F_s of approximately 0.899, indicating reduced and significant instability, as observed in the stress distribution. The scenario with RQD = 70 per cent (Figure 2c), representing a moderately fractured rock mass, resulted in an F_s of 1.049, which was slightly higher than that of the previous case; however, the stability was still affected by the fractures, as indicated by the higher stress levels. In the case of a competent rock mass with RQD = 95 per cent (Figure 2d), F_s increased to 1.119, showing the highest stability, with the stress distribution being more evenly spread across the slope. Figure 2 clearly illustrates how F_s decreases with lower RQD values, highlighting the destabilising effects of fractures and discontinuities. In contrast, higher RQD values represent more stable conditions with fewer failure zones, thus emphasising the importance of fracture characterisation and RQD in slope stability assessments.

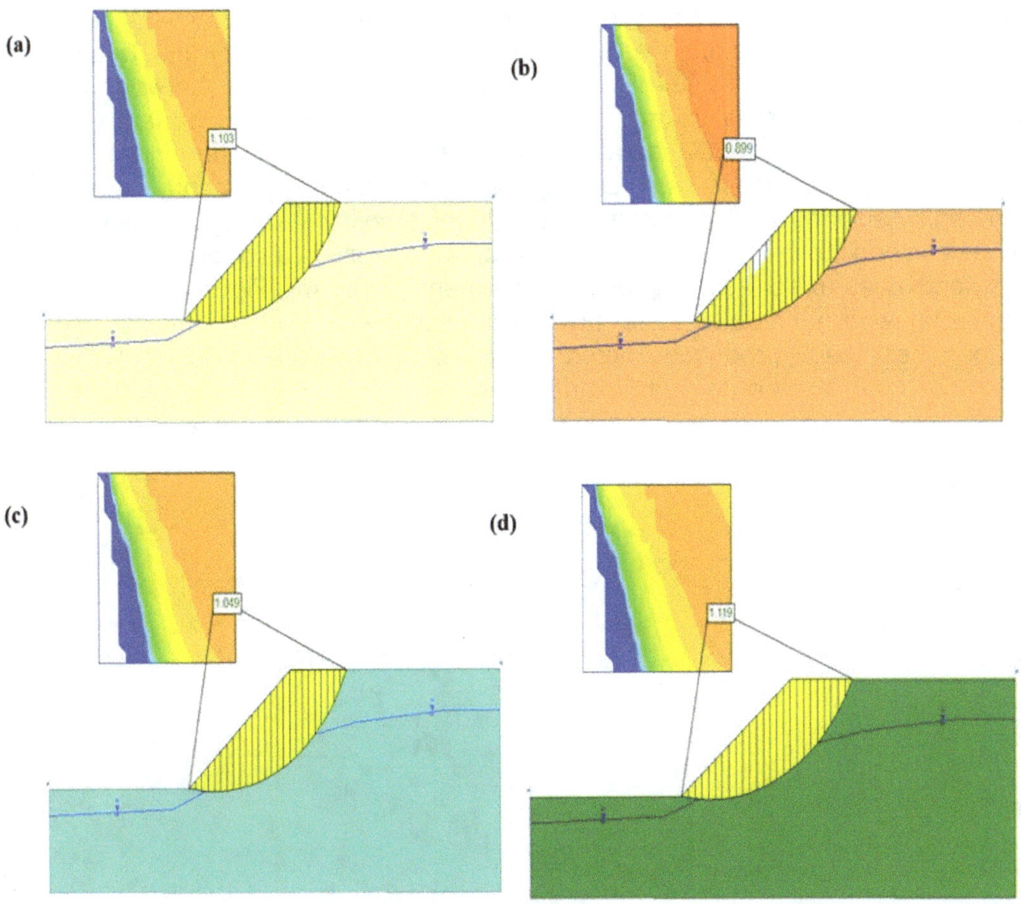

FIG 2 – Numerical evaluation of slope stability using classical Fellenius and Fellenius–FEMC techniques.

Finite element verification using PLAXIS

To further validate the proposed FEMC–Fellenius framework and to address the limitations associated with limit equilibrium methods, numerical simulations were conducted using the finite element method (FEM) implemented in the PLAXIS software. Unlike limit equilibrium approaches, which assume predefined failure surfaces and satisfy overall force or moment equilibrium, the FEM accounts for stress–strain relationships, deformation compatibility, and progressive failure mechanisms within the slope. This provides an independent and more physically rigorous assessment of slope stability, particularly for fractured rock masses, where failure may not strictly follow idealised circular slip surfaces. The 2D slope model is defined by a polygonal boundary with vertices at (0,0), (0,9), (12,9), (20,3), (25,3), and (25,0) (coordinates in metres, origin at the slope toe), forming a 9 m-high slope with a planar face inclined from (12,9) to (20,3) and a 5 m-wide horizontal bench at an elevation of 3 m. The base spans 25 m horizontally, with fixed constraints applied along the bottom edge (y = 0) (see Figure 3). The material properties, boundary conditions, and loading parameters were maintained identical to those used in the previous sections to ensure direct comparability of the results. The influence of the rock mass surface roughness, quantified by the fractal dimension D_s, on slope stability was evaluated through finite element limit analysis in PLAXIS using the φ–c reduction approach. Three scenarios were considered for a representative rock layer:

1. An idealised smooth condition (D_s = 1.0, corresponding to RQD = 100 per cent) (see Figure 3a and Figure 3b).
2. A highly fractured state (D_s = 1.5229, corresponding to RQD = 30 per cent) (see Figure 3c and Figure 3d).
3. A moderately rough condition (D_s = 1.0223, corresponding to RQD = 95 per cent) (see Figure 3e and Figure 3f).

For each case, the material strength parameters were updated using fractal-dependent relationships. The computed factors of safety (F_s) were 2.379, 2.356, and 1.832 for D_s = 1.0, 1.5229, and 1.0223, respectively. Notably, the FoS remained nearly unchanged between the smooth and highly fractured cases (Figure 3a, 3c, and 3e), indicating that the cohesion enhancement associated with asperity interlocking at high D_s partially compensated for the reduction in the friction angle. In contrast, the moderate roughness case, D_s = 1.0223, produced a significantly lower F_s, suggesting a non-monotonic relationship between D_s and global stability, likely governed by the balance between cohesion and friction mobilisation. The deformation results also revealed a strong sensitivity to D_s. The smooth case exhibited negligible displacement (~10^{-17} m) (Figure 3b), whereas the highly fractured case showed nanometer-scale movement (1.022×10^{-9} m) (Figure 3d). By contrast, the moderate roughness case produced micrometer-scale deformation (1.987×10^{-6} m) (Figure 3f), which was more than three orders of magnitude greater than that of the highly fractured case, indicating pronounced shear localisation and strain-softening behaviour near the failure surface.

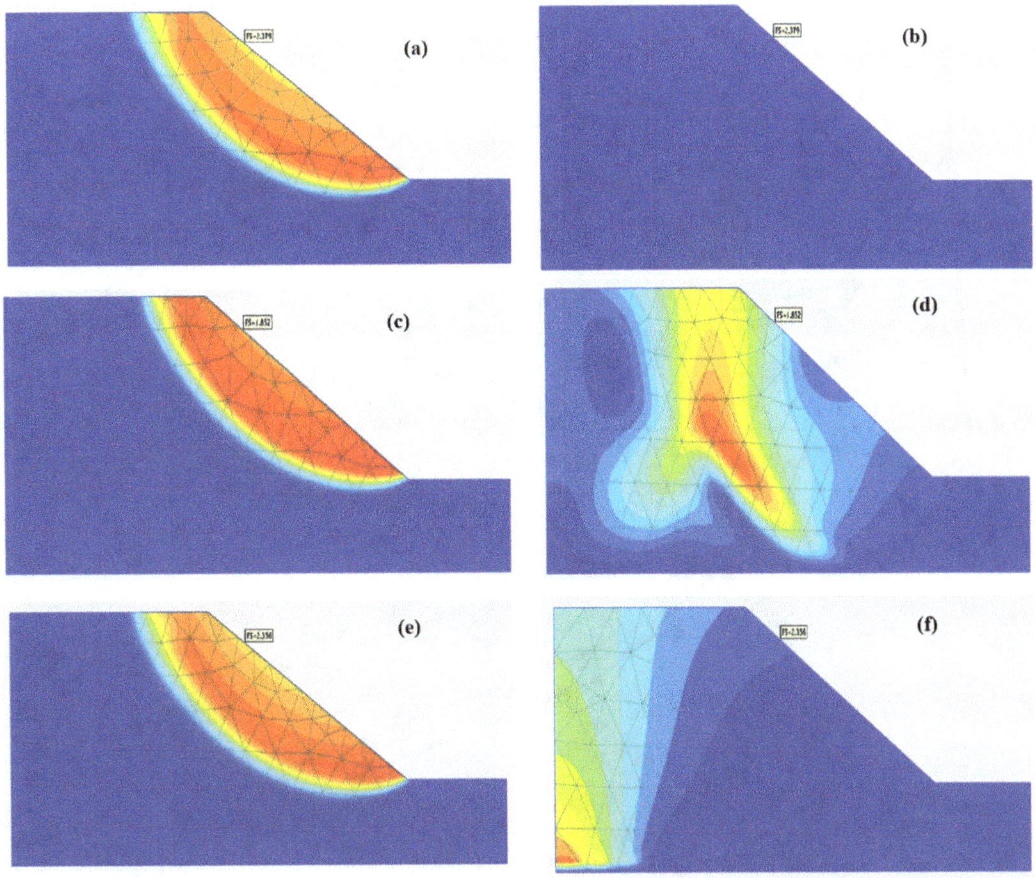

FIG 3 – Influence of fractal dimension D_s on slope stability and deformation characteristics.

These findings suggest that although F_s may appear adequate in highly fractured media because of the apparent gain in cohesion, serviceability, and deformation control remain critical design considerations. Conversely, moderate roughness may trigger more extensive yielding despite a higher nominal strength, underscoring the complex role of surface topology in slope performance.

Sensitivity analysis

The results of the sensitivity analysis depicted in Figure 4 demonstrates a systematic and physically consistent influence of the fractal dimension D_s on the shear strength parameters and slope stability. As the fractal dimension increases, representing greater fracture complexity and rock mass degradation, cohesion (c) shows a relatively limited and nonlinear variation, with slight enhancement at lower D_s, followed by stabilisation, (Figure 4a), suggesting that asperity interlocking provides only marginal compensation for the loss of frictional strength. In contrast, the friction angle (φ) exhibits a pronounced monotonic decrease across all FEM cases (Figure 4b), indicating a progressive loss of frictional resistance due to increased discontinuity roughness and reduced interlocking efficiency. Consequently, the factor of safety (F_s) consistently decreases with increasing D_s, with reductions of approximately 18–22 per cent observed between the lowest and highest fractal dimensions across all cases (Figure 4c). The close agreement among the three FEM cases indicates that the observed trends are robust and not significantly sensitive to minor parameter variations.

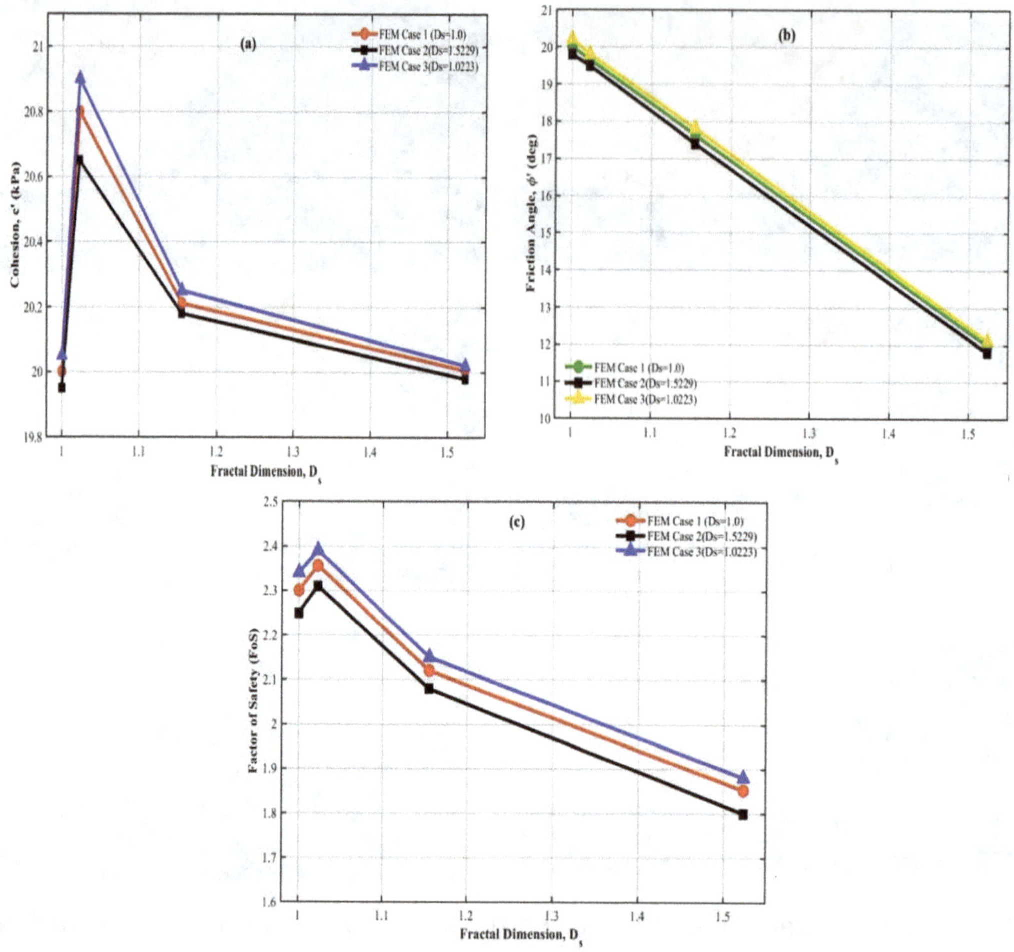

FIG 4 – Sensitivity of shear strength parameters and factor of safety to fractal dimension (D_s) based on FEM analysis: (a) variation of friction angle, (b) variation of cohesion, and (c) variation of factor of safety (F_s).

A direct comparison between the analytically derived safety factors obtained using the traditional Fellenius method and the innovative Fellenius–FEMC framework, as well as the corresponding numerical safety factors calculated using Slide2D, is presented in Table 3. The slope geometries and material properties were kept constant across all comparisons, whereas the rock mass quality was varied using representative RQD values to reflect the effects of fracture complexity on stability. A side-by-side comparison revealed that the analytical and numerical methods were closely related, and by using fractal-based strength degradation in the analyses, the slope stability predictions of the fractured rock masses became more realistic.

TABLE 3

Comparison of analytical (Fellenius/Fellenius–FEMC) and numerical factors of safety under varying rock mass quality.

Method	Approach	RQD (%)	D_s	γ_t (kN/m^3)	c' (kPa)	φ' (°)	F_s
Classical Fellenius	Analytical	–	–	16	20.0	20.0	**1.55**
FEMC–Fellenius	Analytical	30	1.52	16	20.005	11.989	**1.23**
FEMC–Fellenius	Analytical	70	1.15	16	20.211	17.627	**1.46**
FEMC–Fellenius	Analytical	95	1.02	16	20.799	19.659	**1.57**
Classical Fellenius	LEM (Slide2D)	–	–	16	20.0	20.0	**1.103**
FEMC–Fellenius	LEM (Slide2D)	30	1.52	16	20.005	11.989	**0.899**
FEMC–Fellenius	LEM (Slide2D)	70	1.15	16	20.211	17.627	**1.049**
FEMC–Fellenius	LEM (Slide2D)	95	1.02	16	20.799	19.659	**1.119**

DISCUSSION

The results demonstrate that incorporating the FEMC into the classical Fellenius framework produces a consistent and physically meaningful reduction in slope stability in fractured rock masses. Across all analytical, limit equilibrium (LEM), and FEM approaches, the introduction of fracture complexity via the fractal dimension leads to a systematic decrease in the factor of safety (F_s), particularly under poor rock mass conditions, as shown in Figure 5.

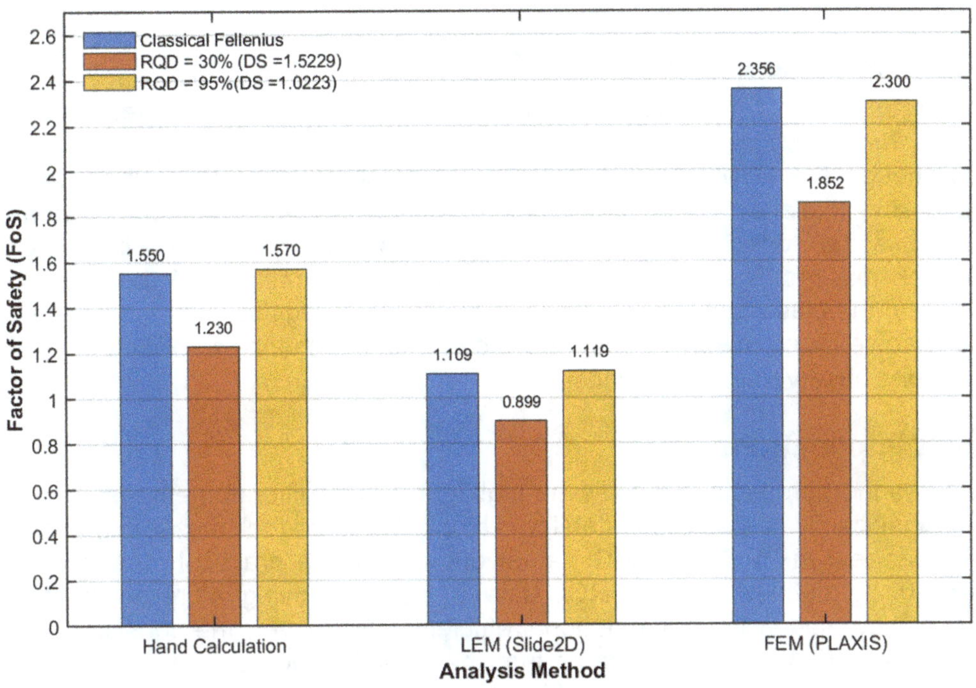

FIG 5 – Factor of safety (F_s) comparison across analysis methods and rock mass quality (RQD).

Comparison of classical and FEMC methods

For the baseline condition (intact or high-quality rock), the analytical Fellenius method yielded an F_s of 1.55, whereas the corresponding LEM (Slide2D) result was 1.103. The FEM (PLAXIS) analysis produced a higher F_s of approximately 2.30, reflecting the ability of continuum-based methods to capture stress redistribution and progressive deformation prior to failure. Despite these differences in absolute magnitude, all methods exhibited consistent trends when fracture effects were introduced. Under highly fractured conditions (RQD = 30 per cent), the analytical FEMC model predicts a reduction in F_s from 1.55 to 1.23, corresponding to a decrease of approximately 20.6 per cent. The LEM results show a comparable reduction from 1.103 to 0.899, representing an 18.5 per cent decrease, indicating a transition to an unstable condition. Similarly, the FEM results decrease from 2.30 to 1.852, corresponding to a 19.5 per cent reduction in stability. These closely aligned reductions across methods confirm that fracture-induced degradation of shear strength is consistently captured by the fractal-enhanced formulation. For high-quality rock (RQD = 95 per cent), the analytical FEMC model shows a slight increase in the FoS to 1.57, representing a +1.3 per cent change relative to the classical Fellenius result. The LEM analysis similarly increases from 1.103 to 1.119 (+1.5 per cent), while the FEM results increase from 2.30 to 2.356 (+2.4 per cent). This indicates that, in relatively intact rock masses, fractal correction does not significantly reduce stability and may even produce marginal increases owing to enhanced interlocking effects associated with surface roughness.

Consistency of trends and methodological implications

Although the absolute F_s values differ between the analytical, LEM, and FEM approaches, the relative trends are remarkably consistent. All methods show (i) a pronounced reduction in stability under low RQD conditions; and (ii) convergence toward or slight enhancement of stability under high RQD conditions. This agreement across fundamentally different modelling frameworks, force equilibrium, limit equilibrium, and continuum mechanics, provides strong validation of the FEMC concept as a robust descriptor of fracture-controlled behaviour. The higher F_s values observed in the FEM simulations are expected, as the method accounts for stress redistribution, strain compatibility, and progressive failure mechanisms that are not explicitly represented in limit equilibrium approaches. However, the fact that the percentage reduction in F_s under fractured conditions remains consistent (approximately 18 per cent–21 per cent across all methods) indicates that the fractal-based modification effectively captures the underlying mechanical impact of fracture complexity.

Mechanical interpretation of fractal effects

The observed reduction in stability under low RQD conditions can be attributed to the degradation of the effective shear strength resulting from increased discontinuity density and reduced frictional resistance. Although the FEMC formulation allows for some cohesion enhancement owing to asperity interlocking, this effect is insufficient to compensate for the loss of frictional strength in highly fractured rock masses. Consequently, the net effect is a significant decrease in F_s. Conversely, for high-quality rock masses, the fractal dimension approaches unity, and the FEMC formulation converges toward the classical Mohr–Coulomb model. In this regime, minor increases in cohesion and the friction angle owing to surface roughness can slightly enhance stability, explaining the small positive deviations observed in all methods.

Engineering implications

From an engineering perspective, the results highlight that traditional methods assuming uniform strength may significantly overestimate stability in highly fractured rock masses, with potential errors exceeding 20 per cent in F_s. The FEMC framework provides a practical means of incorporating fracture effects using routinely available RQD data, thereby improving the reliability of slope stability assessments without sacrificing computational efficiency. Moreover, the consistency between the analytical, LEM, and FEM results suggests that the FEMC approach can be confidently applied across different modelling environments, making it suitable for both preliminary design and advanced numerical analysis in open pit mining and rock slope engineering.

CONCLUSIONS

This study developed a fractal-enhanced Mohr–Coulomb (FEMC) framework integrated with the Fellenius method to incorporate fracture complexity into slope stability analysis using rock quality designation (RQD) as an input parameter. By introducing a fractal dimension to modify shear strength parameters, the proposed approach establishes a direct and practical link between rock mass characterisation and mechanical stability evaluation. The results demonstrate that fracture complexity has a measurable and consistent impact on slope stability. For highly fractured rock masses (RQD = 30 per cent), the factor of safety decreased by approximately 18 per cent–21 per cent across the analytical, limit equilibrium (LEM), and finite element (FEM) approaches, confirming that conventional constant-strength methods can significantly overestimate stability under such conditions. In contrast, for high-quality rock (RQD = 95 per cent), the FEMC model produced F_s values comparable to or slightly higher than classical estimates, indicating that the formulation appropriately converges toward conventional behaviour as the fracture influence diminishes. Importantly, the agreement in trends across analytical (Fellenius), limit equilibrium, and finite element analyses demonstrates the robustness and transferability of the FEMC concept across different modelling frameworks. This consistency supports the physical validity of using fractal descriptors to represent fracture–controlled strength degradation in rock masses. From a practical perspective, the FEMC framework provides a computationally efficient and implementable tool for slope stability assessment, particularly in open pit mining environments, where RQD data are routinely available but detailed fracture mapping may be limited. The method preserves the simplicity of limit equilibrium approaches while enhancing their realism through spatially variable strength representation. Despite these advantages, the current formulation assumes that fracture complexity can be adequately represented by a single fractal dimension and does not explicitly account for anisotropy, progressive failure mechanisms, or highly heterogeneous geological conditions. Future work should focus on extending the model to incorporate directional fracture properties, time-dependent degradation, and validation against field-scale failure cases to further strengthen its applicability.

ACKNOWLEDGEMENTS

All coauthors listed have made valuable contributions to the authorship of this paper. Each author was instrumental in the areas of geotechnical analysis, computational modelling, and interpretation of the results to successfully complete this research.

REFERENCES

Barton, N and Choubey, V, 1977. The Shear Strength of Rock Joints in Theory and Practice, *Rock Mechanics Felsmechanik Mecanique Des Roches*, 10(1–2):1–54. https://doi.org/10.1007/BF01261801

Barton, N, 1976. The Shear Strength of Rock and Rock Joints, *International Journal of Rock Mechanics and Mining Sciences and Geomechanics Abstracts*, 13(9):255–279. https://doi.org/10.1016/0148-9062(76)90003-6

Boadu, F K and Long, L T, 1994. The Fractal Character of Fracture Spacing and RQD, *International Journal of Rock Mechanics and Mining Sciences and Geomechanics Abstracts*, 31(2):127–134. https://doi.org/10.1016/0148-9062(94)92802-9

Deere, D U and Deere, D W, 1988. The Rock Quality Designation (RQD) Index in Practice, in *Rock Classification Systems for Engineering Purposes*, (ASTM International). https://doi.org/10.1520/STP48465S

Deere, D U, 1989. Rock Quality Designation (RQD) after Twenty Years, US Army Corps of Engineers Contract Report GL-89-1, Waterways Experiment Station, Vicksburg, MS 67.

Doan, N S, 2023. Reliability Analysis and Uncertainty Quantification of Clay and Sand Slopes Stability Evaluated by Fellenius and Bishop's Simplified Methods, *International Journal of Geo-Engineering*, 14(1):22. https://doi.org/10.1186/s40703-023-00200-2

Fellenius, W, 1936. Calculation of the Stability of Earth Dams, *Transactions 2nd Congress on Large Dams*, pp 445–462, Washington DC.

Huang, Y H, 1983. *Stability Analysis of Earth Slopes* (Springer US). https://doi.org/10.1007/978-1-4684-6602-7

Kon, Dina, Jisen Shu and Dave Mbako, 2025. The Impact of Vertical Scaling on the Estimation of Fractal Dimension of Slope Failures in Opencast Mine, *Open Journal of Civil Engineering*, 15(02):249–270. https://doi.org/10.4236/ojce.2025.152014

Kon, Dina, Jisen Shu, Liu Han and Sage Ngoie, 2025. Quantifying Failure Surface Roughness and Fractal Characteristics in Open pit Mining: The Case of Mashamba West Mine, *Open Journal of Civil Engineering*, 15(01):91–112. https://doi.org/10.4236/ojce.2025.151006

Kon, Dina, Sage Ngoie, Jisen Shu, Yadah Mbuyu and Dave Mbako, 2026. A Fractal-Enhanced Mohr–Coulomb (FEMC) Model for Strength Prediction in Rough Rock Discontinuities, *Fractal and Fractional*, 10(1):61. https://doi.org/10.3390/fractalfract10010061

Mandelbrot, Benoit, B, Dann, E, Passoja and Alvin, J, Paullay, 1984. Fractal Character of Fracture Surfaces of Metals, *Nature*, 308(5961):721–722. https://doi.org/10.1038/308721a0

Pells, P J, Z T, Bieniawski, S R, Hencher and S E, Pells, 2017. Rock Quality Designation (RQD): Time to Rest in Peace, *Canadian Geotechnical Journal*, 54(6):825–834. https://doi.org/10.1139/cgj-2016-0012

Turcotte, D L, 1997. *Fractals and Chaos in Geology and Geophysics*, 2nd edn (Cambridge University Press).

Turcotte, D L, 1986. Fractals and Fragmentation, *Journal of Geophysical Research: Solid Earth*, 91(B2):1921–1926.

Turcotte, Donald, L, 2009. Fractals in Geology and Geophysics, in *Encyclopedia of Complexity and Systems Science*, (ed: R A Meyers), (Springer New York). https://doi.org/10.1007/978-0-387-30440-3_224

Zhang, Lianyang, 2016. Determination and Applications of Rock Quality Designation (RQD), *Journal of Rock Mechanics and Geotechnical Engineering*, 8(3):389–397. https://doi.org/10.1016/j.jrmge.2015.11.008

Operational decarbonisation in slope stability monitoring – developing and supporting sustainable solutions

D McAllister[1], N Harries[2] and J Wrigley[3]

1. Regional Sales Director, IDS GeoRadar, Perth WA 6000. Email: dean.mcallister@hexagon.com
2. Director APAC, Hexagon, Brisbane Qld 4000. Email: neal.harries@hexagon.com
3. Senior Technical Sales Engineer, Hexagon Geosystems, Perth WA 6000. Email: joshua.wrigley@hexagon.com

ABSTRACT

Operational decarbonisation has become a key focus of large-scale Australian iron ore producers in recent years. As producers continue to target long-term net zero goals, there has been an increase on focus for suppliers to provide solutions that align with these goals and maintain operating effectiveness.

Slope stability radar monitoring requires reliable electrical power to charge batteries on the radar that are utilised in the operation of the radar. In many countries, this has been mains power that is particularly common in mines that use electric shovels. In the iron ore mines of the Pilbara onboard power generation has been required. This has been historically in the form of diesel generators, or bespoke solutions tailored to individual systems to provide green energy. In using diesel gensets, high maintenance requirements often require frequent travel of personnel to remote locations to maintain and repair onboard power solutions, further increasing environmental footprints to maintain real-time operational slope monitoring. However, in the development of ad hoc green solutions, solar systems have been utilised on an as-needed basis and can encounter development difficulties when applied to mass production. Furthermore, solar solutions are often seen as an ancillary power source rather than a single solution.

To achieve the long-term operational needs of producers to meet net zero goals in open pit applications, slope monitoring systems need to see a reduction in their environmental footprint. For this to be achieved, systems need to be manufactured on a scale that is repeatable, reliable, and capable of meeting or exceeding current performance levels. In the case of slope monitoring radars, it is not uncommon for system downtime to have a direct correlation to production, or economical loss due to radar monitoring forming a key piece of the mining risk-management framework. Furthermore, in some instances monitoring outages can have a direct impact to the safety of mine personnel, if the radar system is not available to alert on slope movement.

IDS GeoRadar, in collaboration with key customers in Western Australia, developed a solar-only solution for its IBIS ArcSAR radar systems that that are now produced alongside diesel powered systems at the assembly level. To date, 11 'Green ArcSAR's' have been deployed to iron ore operations in the Pilbara region of Western Australia, utilising the exceptional solar potential of the Pilbara region. This paper discusses the results of long-term deployment performance, availability and benefits related to lower carbon footprint radar monitoring.

As many existing customers already have a need to deploy slope monitoring radar to manage risk associated with pit wall fall-of-ground events, the difference in cost between a diesel-powered solution and a fully solar solution represents an overall cost increase of approximately A$25 000 for additional battery capacity and management systems. In comparison, labour and fuel costs to ensure ongoing diesel generator operation can cost in the order of A$6600 per annum. Over a seven-year service life, this represents a minimum saving of approximately A$52 000 without considering additional overheads typically associated with diesel systems and reduction in net CO^2 emissions.

INTRODUCTION

The mining industry faces increasing demand for operational safety, efficiency, and sustainability. In recent years there has been a push to increase operational decarbonisation to reduce the impact of the resources sector on global warming.

Monitoring slope stability in open pit mining requires precise, continuous measurement of ground displacement to mitigate geotechnical risks. IDS GeoRadar systems employ radar interferometry to provide real-time deformation data with millimetre-scale precision, enabling early detection of instability and supporting operational decision-making. There are often two predominant applications in slope monitoring applications when applied to ground-based radar monitoring systems. In the first instance, diagnostic monitoring is carried out by site-based or remote staff, typically Geotechnical Engineers, to identify slope displacement and review slope performance for potential instability. The second application typically occurs once a potential for instability has been identified and operational controls are required to allow safe mining to continue. During this second phase the radar system becomes a critical control to the ongoing operation, with Geotechnical staff establishing slope alarm parameters and monitoring regimes to identify potential slope failure. It is during this tactical monitoring phase that long-term reliability of the radar system becomes crucial, as any downtime of the radar would typically require a cessation of mining operations to keep personnel out of the line of fire of potential slope failure. This requirement for ongoing reliability in radar monitoring poses the biggest challenge when designing and implementing a solar-powered radar solution to industry.

In previous system deployments, IDS GeoRadar products have primarily been powered by onboard or standalone diesel engines. It should be noted, however, that from early deployments there has been a demonstrated focus on providing sustainable solar solutions that in some cases resulted in fully solar solutions. Through this process of continual improvement, IDS GeoRadar now has 11 fully solar-capable radar systems deployed in the Pilbara region of Western Australia.

REGIONAL BACKGROUND

The Pilbara region of Western Australia is one of the world's most resource-rich areas, hosting extensive iron ore, lithium, and other mineral deposits that underpin global supply chains. Characterised by vast open landscapes, high solar irradiance, and consistent wind patterns, the region offers exceptional conditions for renewable energy deployment with a mean daily solar exposure exceeding 20 MJ/m^2/day, or 6 kWh/m^2 per day (Australian Bureau of Meteorology, 2025), making photovoltaic installations highly viable. The abundance of solar exposure creates opportunities for integrating solar into off-grid energy systems, reducing reliance on diesel generation, and lowering operational emissions. In addition to this, the reduction of complex internal combustion engines and supporting infrastructure in favour of radiant, electrical can be expected to result in fewer mechanical failures. Considering the expanse of mining operations in the Pilbara region, the reduction of maintenance and repairs for mechanical failure can lead to lower operating costs.

However, to accommodate the ongoing operational needs of Pilbara customers, it is necessary to consider the seasonal impact to solar exposure so as not to run into complications with inadequate system design. Despite elevated temperatures and higher solar exposure during the warmer months of November to April, monsoonal weather patterns can present extended periods of lower solar activity. Additionally, lower exposure during the cooler dry season can result in gradational declines of solar performance if not considered in the system design process. Figure 1 (Australian Bureau of Meteorology, 2025) shows comparative solar exposure between dry (May to September) and wet (October to April) seasons. These results show that there can be upwards of a 10 MJ/m^2 difference between wet and dry seasons in the Pilbara region. These seasonal variations need to be factored into design considerations to ensure that system availability is not degraded due to lower solar performance.

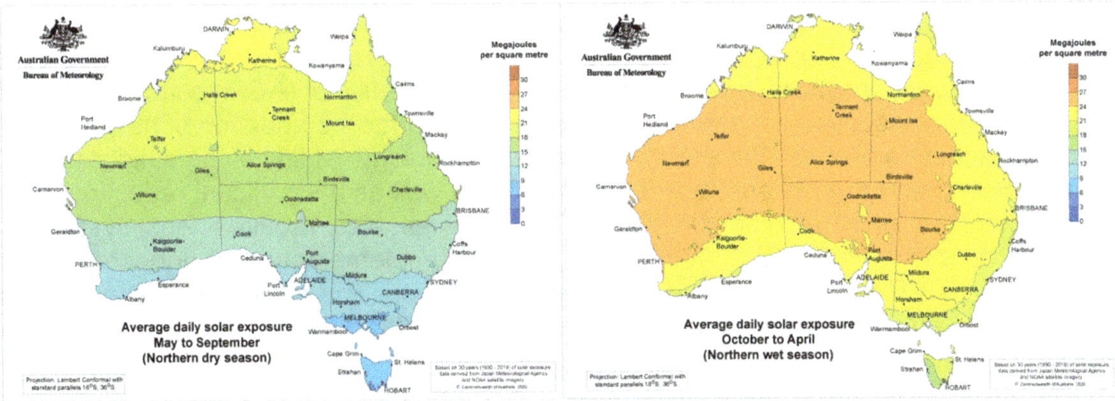

FIG 1 – Seasonal solar exposure maps (Australian Bureau of Meteorology, 2025).

SYSTEM DESIGN

When compared to ground-based dish-style real aperture radar, synthetic aperture radar (SAR) allows for more efficient integration of solar-powered options due to lower power consumption and less interference with solar panels. Whereas real aperture radar (RAR) requires high-powered motors to drive azimuth and elevation movement of the antenna, SAR relies on low current linear motors to move the radar transceiver along a rail or positioner. Additionally, SAR transceivers do not rely on large antennas to provide a beam focal point, thus reducing shade on solar panels. Taking learnings from previous successes with rail-mounted linear scanner systems (McAllister, Harries and Irvine, 2024), a project to design and implement a solar powered option for the flagship ArcSAR (Figure 2) product was conceived by the IDS GeoRadar team to further develop in-house renewable power options.

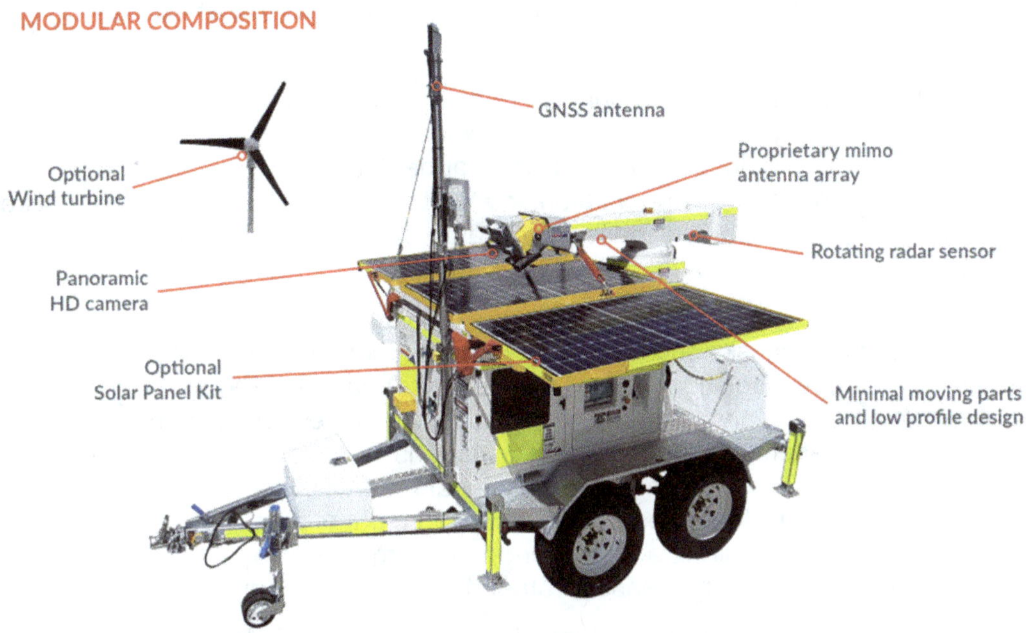

FIG 2 – IBIS ArcSAR.

The decision to develop a fully solar powered ArcSAR system required additional considerations when compared to previous linear SAR solutions, such as those discussed in McAllister, Harries and Irvine (2024). Whereas earlier systems had a low power consumption, ample storage room for multiple batteries and large surfaces for solar panel mounting, the ArcSAR radar is a trailer-based platform with minimal room for adding onboard battery storage and additional solar panels. To develop a fully self-sustaining solar capacity, additional battery storage and solar charging capacity was needed to eliminate the requirement for the onboard diesel generator and fuel storage tank. By removing the fuel tank and generator, additional battery storage could be added that would allow for the additional power consumption of the ArcSAR system.

The additional storage awarded by removing the generator and fuel tank also allowed for further advancements in battery technology. Whereas the diesel ArcSAR variant had been initially designed with onboard chargers configured to support large Absorbed Glass Mat (AGM) batteries, the removal of the generator and associated charge requirements allowed for the adoption of newer-generation lithium iron phosphate ($LiFePO_4$) batteries. This newer technology, while having a higher initial cost, provide a much higher cycle life and power availability, as well as a significant reduction in weight. The higher cycle count when compared to AGM batteries also requires less system battery replacements, as shown in Table 1. Additionally, with the implementation of onboard technology, individual battery health can be monitored via remote connection. This capability was required during development to determine that adequate solar power was generated to provide ongoing monitoring needs.

TABLE 1

Battery specifications: battery discharge = battery capacity/power consumption.

Battery parameters	$LiFePO^4$	AGM
Battery voltage =	24 Volts	24 Volts
Battery capacity =	800 Ahr	740 Ahr
Battery % discharge	90%	60%
Available battery capacity (to discharge level)	720 Ahr	370 Ahr
ArcSAR consumption	5.42A	5.42A
Battery discharge in hours	133 hrs	68.3 hrs
Battery discharge in days	5.55 days	2.84 days

To determine if the available solar exposure and battery storage would be sufficient, total power requirements were assessed across both summer (wet) or winter (dry) seasons. Common off-grid solar failure modes include prolonged low sunlight that depletes batteries; solar array issues such as soiling, shading, damage, poor orientation, or faulty connections, as well as battery faults including imbalance, ageing, trips, and inaccurate charge estimation. Increased load from additional installed hardware also needs to be taken into consideration, such as site communications equipment.

Operational or environmental risks such as vibration, water ingress, corrosion, cable wear, impacts, and human errors (like open isolation or poor maintenance) can also reduce reliability. To mitigate this, several controls were implemented. These include (i) increasing battery autonomy through higher usable capacity batteries, (ii) adequate PV margins to cover seasonal minima, (iii) remote visibility of battery state-of-charge, charge/discharge rates, and alarm states to support proactive intervention, and (iv) defined site procedures for relocation, panel cleaning, and electrical isolation. Where operational criticality requires, a contingency re-charge pathway (eg temporary external AC input) can be retained as a recovery measure rather than the primary energy source.

Using the parameters shown in Table 2, there is a calculated net daily surplus of solar power available across both seasons. These solar requirements are determined by calculating the daily effective hours of solar power for the assessed location and available system solar output from the onboard battery chargers. If the available solar production value is greater than the operating requirements, calculated as 3.12 kWh in the power requirements calculation below, it can be determined that there is sufficient solar power available daily to power the system. With these supporting values and favourable conditions, construction on the 'Green ArcSAR' was commenced at the IDS GeoRadar production facility in Brisbane and onward to a customer site in the Pilbara via Perth.

TABLE 2
Solar power requirements.

Parameter	Value
Daily current draw	24 hrs × 5.42 Amps = 130 Ahr
Daily power consumption	24 volts × 130 Ahr = 3.12 kWh
Solar array max power	1.32 kW
Solar charger max charge current	40 A
Solar charger max power output	2.9 kW (charger rated max)
Expected solar performance	Location specific – as per Figure 1
Effective hrs of sun per day (summer)	PSH = 5.5 hrs
System daily solar production (summer)	1.32 kW × 5.5 hrs = 7.26 kWh
Effective hrs of sun per day (winter)	PSH = 3.13 hrs
System daily solar production (winter)	1.32 kW × 3.13 hrs = 4.13 kWh

In many operations, ground-based radar forms part of a Trigger Action Response Plan (TARP) or equivalent geotechnical risk framework, where defined alarm states drive operational responses (eg exclusion zones, work stoppage, or re-entry). Within this context, the off-grid power system should be treated as a supporting layer that protects the integrity of the critical control: it must sustain continuous monitoring power, provide clear telemetry on energy health, and enable escalation when margins reduce (eg low battery warning) so that the site can enact pre-defined contingencies before monitoring is compromised. While these alarm parameters already exist within the onboard radar hardware, implementation of a solar solution would need to meet the available system uptime of currently available diesel-powered systems to mitigate risks associated with monitoring interruptions, such as unnecessary exclusions, risk to personnel from a loss of monitoring, or a loss to operational production.

RESULTS

To avoid ambiguity, this paper uses 'availability' to mean power availability to the radar (ie the radar remained powered and capable of monitoring) rather than geotechnical monitoring availability inclusive of communications, software, or site operational constraints. During the observed trial window, the solar-only system maintained continuous operation, with outages limited to intentional shutdowns for planned maintenance and controlled run-down testing. This result is a strong indicator of fitness-for-purpose; however, it should not be interpreted as an unconditional guarantee of zero downtime across all sites and operating regimes.

Although average seasonal solar exposure indicates a net daily surplus (Table 2), wet-season operation can be governed by consecutive low-irradiance days rather than monthly averages. A practical stress case for the Pilbara is a multi-day cloud band or monsoonal event that materially reduces effective sun hours below the winter design value (3.13 hrs) and, in some cases, below the daily energy break-even point. Under such conditions the system transitions from 'daily energy balance' to 'autonomy mode,' where battery capacity becomes the controlling factor for continuity of monitoring.

Using the measured daily energy requirement of approximately 3.12 kWh (Table 2) and the demonstrated multi-day discharge window from run-down testing (Table 3), the design intent is to preserve sufficient monitoring continuity through short-to-medium weather events and then recover state-of-charge when irradiation returns. 'Near-miss' events are therefore defined as periods where state-of-charge approaches the site-defined discharge floor (eg a BMS low-voltage trip threshold) without causing a monitoring outage. Capturing these events (time at low State of Charge, minimum SoC reached, and recovery time) is recommended as an operational key performance indicator, because it provides leading indication of PV soiling, battery ageing, atypical load, or seasonal underperformance before an outage occurs. Presently, these values can be monitored remotely via IDS Support staff from Perth and validated during scheduled service visits.

TABLE 3

Testing Results.

Test type	Discharge SoC	Time to discharge	Time to 100% recharge
Dual solar array	25%	7.5 days	3 days
Partial solar array	56.5%	4.5 days	8.5 days
AC genset input	54%	4.6 days	0.4 days

The first Green ArcSAR (Figure 3) was deployed to a customer site during September of 2023 for a six-month trial that concluded in 2024, with the system remaining on-site after the trial was concluded. During this trial, the system delivered uninterrupted power except during intentional shutdowns for maintenance and run down testing. System performance was monitored both locally and remotely to collect data on the health and performance of battery health and longevity, solar and battery performance, and controlled tests. It should be noted that the system was initially deployed with a reduced solar capacity of 1.04 kW, however, was still capable of maintaining uninterrupted operations during the summer trial. This was later upgraded by adding another panel to the onboard array that eliminated the requirement for additional solar arrays to be set-up adjacent to the trailer, further reducing workload to site teams.

FIG 3 – Green ArcSAR.

The system performed as intended, with measured discharge and recharge profiles showing full utilisation of the battery capacity. To validate expected results, a complete run-down test was conducted to measure run down and recharge times and are summarised below in Table 3. To accomplish this, the customer site team isolated solar input to allow system batteries to discharge to various discharge floors, monitored remotely by IDS support. Following this discharge, solar charging was reinstated via either a dual solar array, partial solar array, or external AC generator. The time required to discharge, state of charge, and the time taken to fully recharge to 100 per cent state of charge with each option was measured. This test demonstrated the expected autonomy when batteries were discharged to set capacity. A set of three tests were completed in total, with each test completed with successful results. While using a partial array resulted in a longer total recharge, the results showed that the partial array was still capable of meeting power demands. To mitigate any potential performance impact during cooler months, a fourth panel was added to the initial 1.04 kW array configuration to facilitate a completely mobile solution and reduce potential risks associated with stand-alone arrays at the customer's request. Following installation of the fourth onboard panel, system availability has remained above 98 per cent for the fleet of solar powered systems since 2023. To date, these systems have been exposed to several wet season stress cases, with Figure 4 providing an example of a high average battery percentages despite lower solar yields during the season.

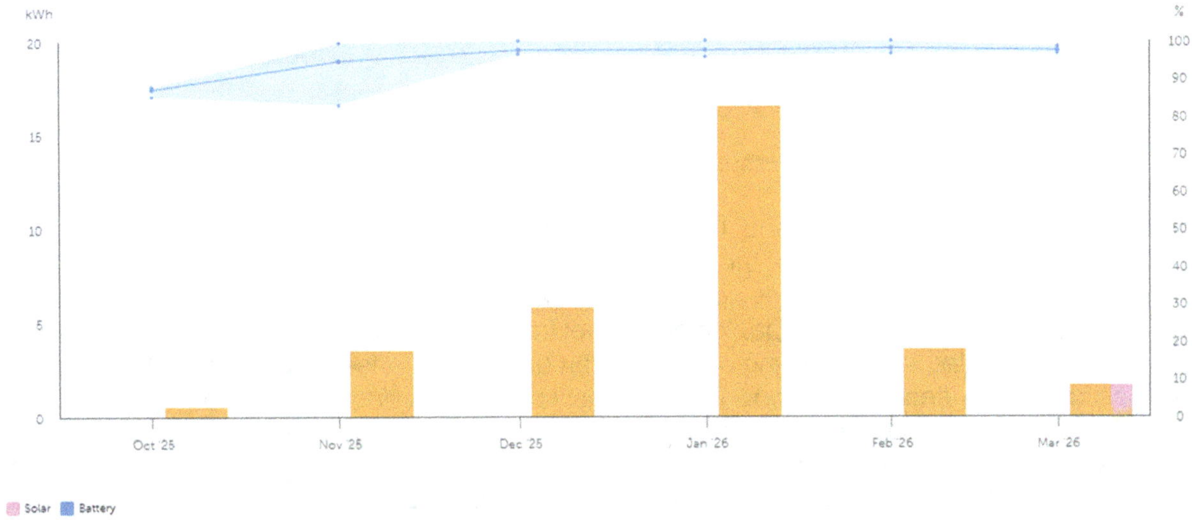

FIG 4 – Wet season yield versus battery SoC example.

To validate the decarbonisation benefits of the Green ArcSAR, it is necessary to measure the initial cost offset of additional solar and battery equipment against the ongoing maintenance costs of the diesel-powered alternative. The additional outlay in net cost to produce a Green ArcSAR is approximately A$25 000 when considering the outlay of additional components, of which the additional battery storage is the greatest percentage. To calculate the ongoing maintenance costs some assumptions must be made, such as expected labour costs associated with travel for refuelling and the cost of fuel itself. Pilbara iron ore mining operations often cover expansive distances when compared to other commodities that offer deeper and narrower pit designs. In this instance, it is common for radars to be located up to a one hour return trip from the nearest maintenance facility or office to refuel a single radar. The business cost for labour hire can be estimated at $180 per hour, with many refuelling schedules occurring on a fortnightly basis at a minimum. Additionally, a diesel-powered ArcSAR can be expected to use approximately 1200 litres per annum of diesel fuel, at an approximate rate of A$1.65 cents per litre. These costs can be calculated over the minimum seven year expected operating life of an ArcSAR system and are summarised in Table 4 and Figure 5.

TABLE 4

Whole-of-life cost comparison (diesel fuel and labour versus solar upgrade premium).

Cost component	Annual cost (AUD)	7-year total (AUD)	Net impact (7 years)
Solar upgrade premium (Green modifications)	–	-$25 000	-$25 000
Diesel operating expenses (labour for refuelling)	$4680	$32 760	+$32 760 (avoided)
Diesel fuel	$1980	$13 860	+$13 860 (avoided)
Net saving (avoided diesel labour + fuel – solar upgrade premium)	**$6660**	**$46 620**	**$21 620**

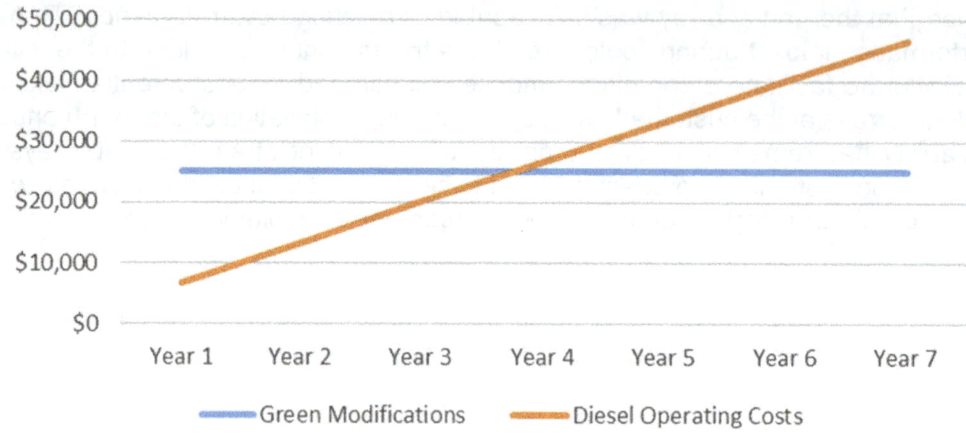

FIG 5 – Green versus diesel operating costs over service life.

When considering only the fuel and labour costs associated with running the onboard diesel generator, there is a net saving over the life of the system of A$21 620, as shown in the Net Saving row in Table 4. Furthermore, this does not consider downtime associated with generator breakdowns or reduced maintenance costs with having personnel travel to site for servicing or repairs. In addition to the values presented above, mining operators can use carbon offsets to further reduce operating expenses related to decarbonisation. The savings calculated using this method were 3.22 t per annum or over the expected operational lifetime of the system of 7 years, a lifetime total of 22.5 t as can be seen in the equation below. Emissions were calculated using energy consumption (EC) and an emission factor (EF) sourced from the Australian National Greenhouse Accounts (NGA) Factors (DCCEEW, 2025).

Carbon offsets

$$E = \frac{Q\ x\ EC\ x\ EF}{1000}$$

Where:

E	is the total emissions released in tonnes CO_2-e
Q	is the quantity of Fuel combusted in kL
EC	is the energy factor of the fuel in GL/kL
EF	is the emission factor for the fuel in kg CO_2-e/GL

$$E = \frac{1.2\ x\ 38.6\ x\ 69.5}{1000}$$

$$E = 3.22\ (annual)\ or\ 22.5\ (lifetime)$$

CONCLUSIONS

The Green ArcSAR trial and subsequent deployments demonstrated that the solar-powered variant can meet key operating requirements, delivering stable battery performance and repeatable recharge capability while maintaining 100 per cent availability during the monitored period. This capability supports ongoing slope monitoring requirements while reducing reliance on diesel generators and associated maintenance and refuelling activities. The findings indicate that a solar-only ArcSAR configuration is a viable pathway to reduce operational emissions in the Pilbara while maintaining operational effectiveness for critical geotechnical monitoring. While the measured period achieved 100 per cent power availability under those conditions, the system is still subject to identifiable failure modes and site-specific stress cases (eg extended low-irradiance wet-season periods, atypical duty cycles, component faults, and operational damage).

Operators may prefer solar over diesel because it reduces the logistics burden of routine refuelling—cutting light-vehicle travel to remote locations, lowering exposure to driving risk, and easing competing demands on-site services teams—while also simplifying maintenance by removing an internal combustion generator and the associated rotating-equipment failures, consumables, and unplanned servicing that can otherwise drive monitoring outages. A solar-only system also offers higher predictability: energy margins can be tracked remotely (eg state-of-charge and charge-current trends), enabling proactive interventions such as panel cleaning or temporary recharge before continuity is threatened. From a whole-of-life perspective, total cost of ownership can favour solar; the results shown above estimate diesel fuel and labour at around $6600 per annum, which exceed the approximate A$25 000 initial outlay over a typical service life of seven years, particularly once indirect costs such as mobilisation, downtime, spares, and administration are included. Finally, choosing solar supports decarbonisation and reporting by eliminating diesel use and reducing emissions tied to monitoring infrastructure, and it improves operational resilience by reducing dependency on consumable fuel delivery when supply interruptions occur due to weather, access restrictions, or competing demand.

ACKNOWLEDGEMENTS

The authors wish to acknowledge the contributions of past and present members of the IDS GeoRadar and customer teams who contributed to the conception, development and testing of the Green ArcSAR.

REFERENCES

Australian Bureau of Meteorology (BoM), 2025. Solar radiation maps – Western Australia [online]. Australian Bureau of Meteorology. Available from: <https://www.bom.gov.au/jsp/awap/solar/index.jsp?colour=colour&time=latest&map=solarave&period=12month&step=0&area=wa> [Accessed 6 January 2026].

Department of Climate Change, Energy, the Environment and Water (DCCEEW) (2025). National Greenhouse Accounts Factors. Australian Government, Canberra. Available from: <https://www.dcceew.gov.au/climate-change/publications/national-greenhouse-accounts-factors> [Accessed: 9 April 2026].

McAllister, D, Harries, N and Irvine, A, 2024. Development of Advanced Green Monitoring Solutions; Providing reduced environmental footprint and delivering improved maintenance outcomes, in *Proceedings of Slope Stability 2024*, Belo Horizonte, Brazil.

Undercover iron ore – an investigation into resource drilling limitations and structural implications of open pit slope stability

S Shahin[1] *and M Nguyen*[2]

1. Specialist Structural Geology, Rio Tinto Iron Ore, Perth WA 6000. Email: sam.shahin@riotinto.com
2. Specialist Structural Geology, Rio Tinto Iron Ore, Perth WA 6000. Email: martin.nguyen@riotinto.com

ABSTRACT

Mining under cover has become common place in Australia, especially with the demand for critical minerals being driven by global development. Interpreting large orebodies in the absence of outcrop and beneath challenging cover sequences is a difficult task, requiring reliable data and orebody knowledge. Rio Tinto takes one such case, an operational open pit with its economical ore underneath Neogene-Quaternary detritus units.

The formation of a structural model is fundamental to slope stability and design. In bulk commodities such as iron ore, studies are often driven by the need to produce a resource model prior to implementation of a structural mode. Said resource models are heavily informed on grade drilling across a set resource grid that is commonly set perpendicular to the strike of the dominating outcrop. Such a decision made in early study can come with potential orientation bias and uncertainty-based risk should drill hole placement and orientation under sample or miss identifying structures pertinent to mine design.

The understanding of The Pit's structural architecture has matured with focused data collection and mapping capable of talking to regional tectonic styles and trends. The resulting interpretation and modelling work have been paramount in preventing slope instability through informed geotechnical design choices and provided pit optimisation solutions to improve economics, reduce mining impact, sustain and protect sensitive sites. Furthermore, the modelled interpretation when verified by in-pit mapping observations highlight the importance of considering regional geological history when implementing data collection strategies.

INTRODUCTION

Terranes buried under cover and without relevant outcrop can present significant geological and operational challenges that extend beyond exploration and evaluation phases of a prospect. At Rio Tinto Iron Ore (RTIO) exists such a case of an active open pit (referred to here on in as 'The Pit') with evaluated mineralisation beneath Neogene-Quaternary detrital sequences to the east. This work presents the current structural characterisation of The Pit and outlines the need to consider data issues inherited from resource drilling campaigns.

The development of an accurate structural model is fundamental to geotechnical design and execution of the mine plan. However, in bulk-commodity operations, early study phases often prioritise resource modelling over structural modelling, with drilling programs commonly designed on grids oriented perpendicular to dominant outcrop trends. While effective for resource estimation, this approach can introduce orientation bias and fail to capture key structures in structurally complex settings, particularly where drill spacing is wide and dominated by unoriented hole orientations.

In 2025, the structural geology team at RTIO planned targeted orientation drilling and pit wall mapping campaigns to investigate and evaluate structures in The Pit's designed south wall. The following work outlines the need to integrate structurally informed data collection methods and interpretations to build robust structural models for engineering objectives.

STRUCTURAL DATA AND TRENDS

Exactly 30 252 holes (excluding blast production holes) have been drilled in the prospect as of the completion of 2025. Of these 30 252 holes, 81.9 per cent of them were drilled near vertical (85–90°

dip angle of completed borehole). From the 18.1 per cent of holes with dip angles below 85° a vast majority were planned with geotechnical or structural purposes.

Some 535 holes were sampled with downhole televiewer orientation data. Of the 535 holes, 208 were Geotech or structural specifically planned hole designs, and the remaining 327 holes were vertical dipping evaluation holes. This further highlight a strong inclination towards evaluation style drill hole campaigns which primarily define the resource in The Pit over geotechnical or structural holes which focus on defining stratigraphy, structures and relevant orientations pertinent to the orebody and final mine design.

Figure 1 visually outlines the distribution of a 75 × 75 m spaced resource grid, orientated N-S (mine grid rotated axis). Represented by red circles are resource evaluation designed hole locations and in green are geotechnical/structural designed holes with blue representing the 2025 structural drilling program aimed at investigating the structural style of the prospect prior to concurrent mining of the area in 2026.

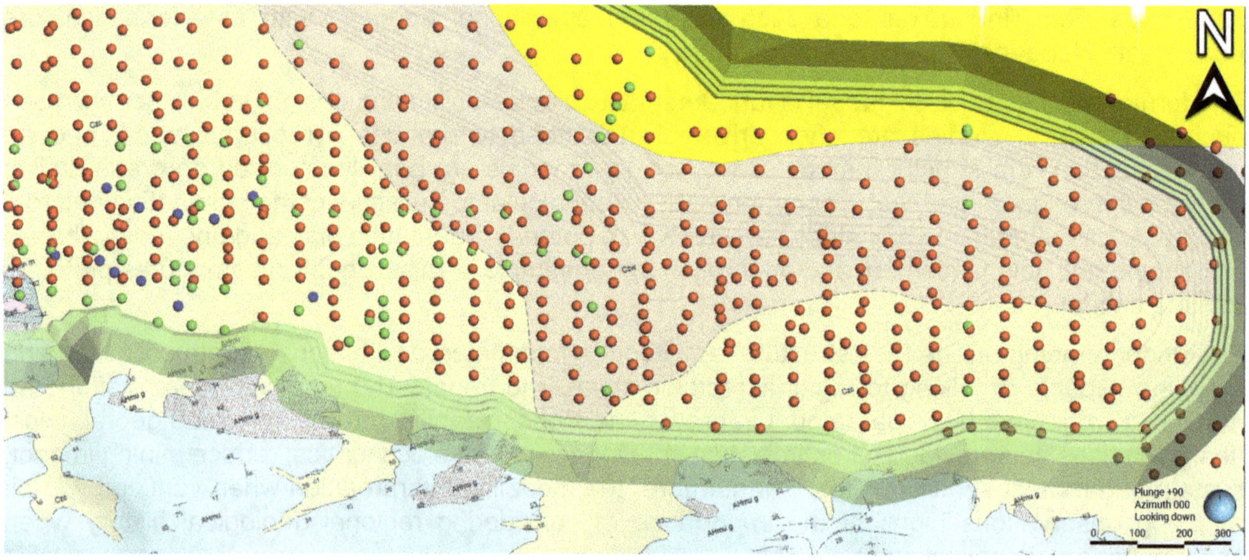

FIG 1 – Location of borehole collars (red – evaluation planned holes, green – geotechnical/structural planned holes, blue – 2025 targeted structural drilling design). Outlined by a green mesh is the final pit crest atop mapped detrital cover sequences.

DATA SAMPLING ISSUES IN TELEVIEWER

In the south wall of the pit design, the televiewer data was re-inspected by the structural geology team at RTIO. The team identified a significant under sampling of steep dipping bedding observable in downhole imagery and diamond core. Under sampling of steep bedding domains was attributed to two primary issues that lead to a recognised bias in data and inadequate characterisation of the orebody.

These two issues were thought to be:

1. Bulk televiewer data interpretation was performed by a wide range of different people with differing levels of training, context and guidance. This led to non-standardised data capture and interpretation which allowed key changes in domains to be missed'.
2. Bedding at similar dip angle to the borehole orientation is difficult to interpret with the available software where planar surfaces are fit to sinusoidal curves.

When comparing the interpreted bedding orientation data from a single hole, it is evident that shallow to moderate dipping beds are well represented by both the original picks (2015) and the revised structural geology picks (2025). However, as represented by Figure 2 steep dipping beds are absent from the original interpretations and would subsequently impact the overall interpretation of the systems architecture and the 3D modelling product relied on by engineering.

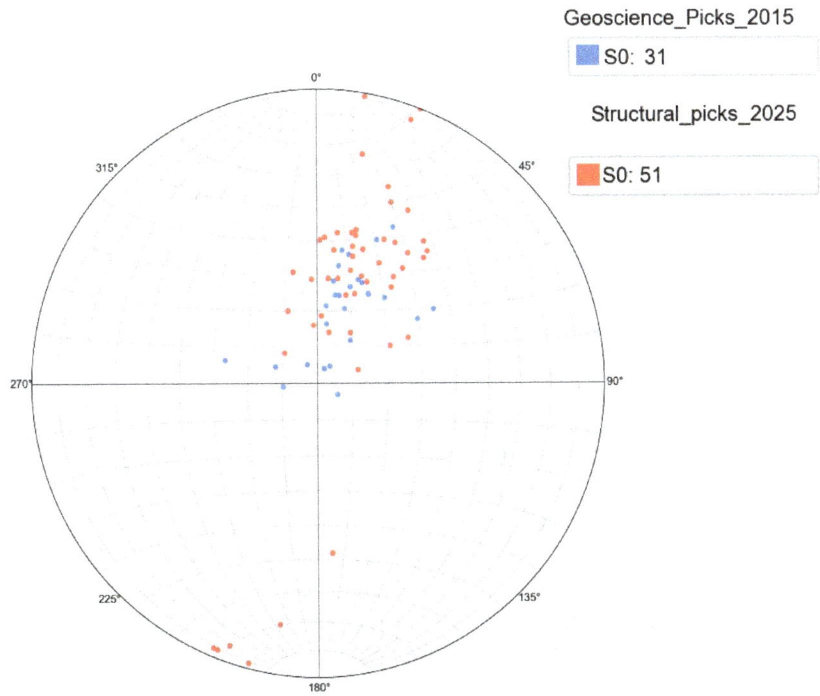

FIG 2 – Stereoplot of orientation data extracted from hole (blue points) and revised structural interpretation (red points).

TELEVIEWER DATA AND TRENDS

Downhole acoustic and optical televiewer across the study area produced a structural data set comprising of 8519 features. The stereoplot of the data (see Figure 3) provides a basis for characterising the geometry of the geology beneath the cover. The data plot indicates a dominant sample set of shallow, north dipping bedforms which indicate a low-angled bedding fabric at depth. The best fit girdle suggests a shallow, east plunging fold axis. The poles to bedding are well spread across the best fit girdle and have a slightly more dominant sample of steeper, north dipping beds as indicated by the large cluster in Figure 3.

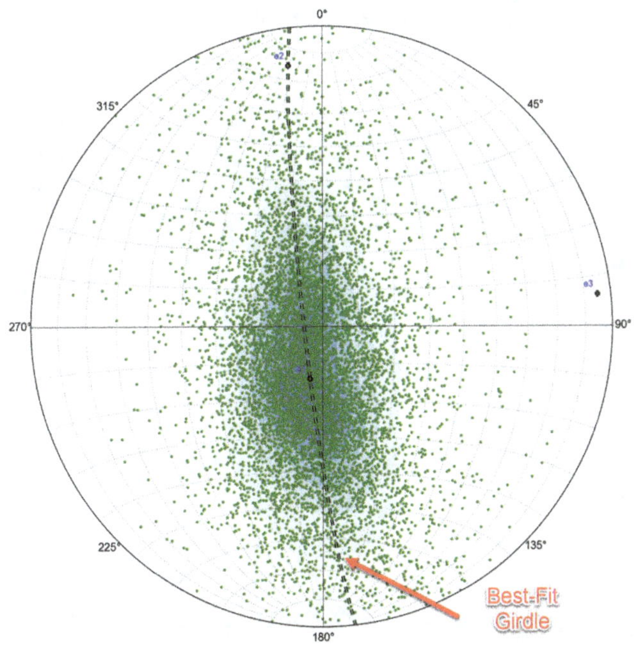

FIG 3 – Stereoplot of all downhole televiewer data (n=8519) with plotted best-fit girdle.

The features of the system as suggested by trends in Figure 3 were confirmed by multi-batter scale fold closures comprising of a shallow, north dipping long limb and a steep south dipping short limb, seen and measured in the newly exposed faces in the east of The Pit. Observations such as brecciation, alteration and other discontinuities indicative of a fault system were also mapped in these faces. These observations and structures were contextualised along section and outlined stratigraphic repetition with the requirement for a thrust fault to separate a plastically deformed hanging wall from a weakly deformed footwall.

RESULTING STRUCTURAL INTERPRETATION

The discrepancies with the televiewer data interpretation revealed how un-orientated drilling and data interpretation have impacted orebody knowledge and model confidence through a data bias.

The structural team undertook reinterpretation of key structural domains with low orientation confidence and reconciliation issues, notably in the east of the unmined pit. The resulting orientation data when brought into a 2D sectional view or 3D model highlighted asymmetrical fold domains and a pronounced, overturned fold nappe system bound by thrust faulting. The asymmetrical fold domains were interpreted to be parasitic folds along the major bounding fold nappe with a short limb closely aligned to the regional bedding trend (shallow dipping to the north) and a long limb (steeply dipping to the north/south). The structural model update triggered a geotechnical design review as per RTIO internal standards. The updated design allowed a more aggressive slope design reduced pit crest extents, steeper inter-ramp angles and improved understating of failure mechanisms present in the system.

CONCLUSION

Challenges will continue to be plentiful when evaluating orebodies under cover. This case study demonstrates a clear need to move away from the reliance on resource grid drilling with vertical or uninformed borehole orientations which can risk misidentifying the structural fabric pertinent to mine design.

Continued work on The Pit takes the form of reconciliation mapping on southern batters of the operation. Identification and contextualisation of fold and fault structures in the pit have provided confidence and positive reconciliation with the updated structural model. Newly excavated batters continue to provide the same opportunity for the structural team to collect data and modify the model where reconciliation is inadequate. Mine design optimisation will continue based on material changes to the structural model and its confidence in contextualising geotechnical design domains.

Does a Factor of Safety of 1.2 mean anything? The hidden drivers of FoS variability in open pit slopes

A Vakili[1] and S Arndt[2]

1. Principal Geotechnical Engineer, Mining One Consultants, Melbourne Vic 3000; Founder, Cavroc Pty Ltd, Melbourne Vic 3000. Email: avakili@miningone.com.au
2. Head of Innovation and Technology, Mining One Consultants, Melbourne Vic 3000; Technology Lead, Cavroc Pty Ltd, Melbourne Vic 3000. Email: sarndt@miningone.com.au

ABSTRACT

The stability of open pit slopes is often described as being controlled primarily by geological conditions and input parameters. A common industry perspective is that, provided input data are robust, the resulting Factor of Safety (FoS) can be considered meaningful and broadly comparable regardless of the numerical method or constitutive model applied. This perspective overlooks a critical reality: the physics and assumptions embedded within analysis software and constitutive formulations can fundamentally alter the outcome, even when geometry and material inputs remain identical.

This paper challenges the prevailing 'it's all about the inputs' view by demonstrating that, even when geometry and primary geotechnical characterisation (eg laboratory testing, rock mass classification, and structural interpretation) are held constant, different analysis tools and constitutive formulations can produce widely different FoS values – in some cases ranging from below 1.0 to above 2.0. Using two real-world case studies, we show how differences in material representation, post-peak behaviour, strength degradation, solver mechanics, and failure detection criteria drive these variations. In several scenarios, the modelling framework exerts greater influence on FoS than plausible variation in geological parameters.

By systematically comparing outcomes across commonly used geomechanical tools, this study highlights the risks of relying solely on software defaults or assuming model equivalence. The findings question commonly accepted interpretations, such as treating FoS = 1.2 as a reliable and transferable indicator of stability. We argue that appropriate model selection is foundational to defensible slope design and present a practical framework for evaluating constitutive suitability and understanding tool limitations in open pit applications.

INTRODUCTION

Factor of Safety (FoS) has long been the central quantitative metric used to assess the stability of open pit slopes. It is embedded in design workflows, regulatory frameworks, and operational decision-making, with commonly adopted design acceptance criteria typically ranging between FoS = 1.2 and 1.3 for inter-ramp and overall slopes (higher for more critical locations like below the ramp etc). In practice, FoS is often treated as an objective and transferable indicator of stability, provided that geological interpretation and material inputs are considered sufficiently robust.

A substantial body of published slope stability literature places primary emphasis on the characterisation of input parameters, such as geological structure, rock mass classification, and laboratory-derived strength properties, while providing limited discussion of the influence of modelling framework, constitutive formulation, or numerical implementation. For example, Hoek and Bray (1981) and Hoek, Carranza-Torres and Corkum (2002) focus extensively on rock mass characterisation and empirical strength estimation using GSI-based approaches, forming the basis for many modern design workflows, yet do not address how different constitutive implementations may influence stability outcomes. Similarly, Wyllie and Mah (2004) present comprehensive guidance on slope stability analysis centred on parameter selection and failure mechanisms, with limited consideration of solver-dependent behaviour. Duncan (1996) highlights the importance of parameter selection and variability in limit equilibrium analysis, but treats the method itself as a stable and reliable framework. More recent studies, such as de Bruyn *et al* (2025), continue this trend by comparing stability outcomes across multiple sites while focusing primarily on geological inputs and disturbance factors, without explicitly examining the role of constitutive models or numerical assumptions. Even in comparative studies of slope stability methods, such as Cheng, Lansivaara and Wei (2007), the emphasis is placed on differences in calculated FoS rather than on how underlying constitutive behaviour or solver mechanics influence those differences. Collectively, these works reflect a prevailing practice in which FoS is interpreted primarily as a function of input data quality, while the influence of modelling framework and constitutive assumptions remains comparatively underexplored. Advances in numerical modelling over the past two decades have increasingly challenged the assumption that FoS is independent of the modelling framework used to compute it. The widespread adoption of continuum-based methods and shear strength reduction (SSR) techniques has shown that FoS is not a unique property of a slope, but an emergent result that depends on how failure is defined, how strength degradation is implemented, and how numerical instability is identified.

This raises a fundamental question: what does it truly mean to declare a slope 'stable' at FoS = 1.2?

This paper addresses that question using two real-world case studies. Rather than advocating a single modelling approach, we demonstrate that FoS is strongly influenced by constitutive representation and numerical framework, and argue that explicit consideration of these factors is essential for defensible slope design.

WHAT DOES 'FACTOR OF SAFETY' ACTUALLY MEAN?

In classical slope stability analysis, the Factor of Safety (FoS) is defined as the ratio of resisting forces (or shear strength) to driving forces acting along a potential failure surface, as typically evaluated using limit equilibrium methods (LEM). Within this framework, FoS represents a kinematic stability margin, where values greater than unity indicate that available shear strength exceeds that required for equilibrium. This definition forms the basis of commonly adopted design thresholds, such as FoS ≈ 1.2 for open pit slopes.

In contrast, in continuum-based analyses using shear strength reduction (SSR), FoS is defined as the factor by which material strength parameters must be reduced to induce numerical instability or failure within the model, following the strength reduction framework described by Dawson, Roth and Drescher (1999). While both approaches aim to quantify a margin of stability, they are conceptually different: LEM FoS is derived from force equilibrium along an assumed failure mechanism, whereas SSR FoS emerges from the interaction between constitutive behaviour, stress redistribution, and the numerical failure criterion (Griffiths and Lane, 1999). As a result, FoS is not a single, objective property of a slope but a method-dependent measure, and values obtained from different analysis approaches are not strictly equivalent and must be interpreted within the context of the modelling framework used to derive them.

Arndt and Vakili (2024) frames this clearly by emphasising that SSR/SRF-based FoS is not simply 'inputs → FoS,' but depends on solver behaviour, definitions of failure, and numerical implementation choices. For example, *implicit* and *explicit* solvers do not even share a common failure trigger: implicit SSR commonly uses non-convergence as the indicator of failure, whereas explicit solvers often rely on displacement/velocity/kinetic energy response to identify instability.

Even when the reduction rule is formally the same (eg reduce c and $\tan\phi$ by SRF), the physical meaning depends on the constitutive model:

- A perfectly plastic Mohr–Coulomb model with constant dilation behaves fundamentally differently under progressive yielding than a strain-softening model.
- Models that include anisotropy, post-peak degradation, or stiffness softening can shift both the *mechanism* and the *SRF at instability.*

This is why 'SSR FoS' is better understood as: the strength reduction factor at which the adopted constitutive model and numerical formulation first satisfy the selected instability or failure criterion.

WHAT DOES IT MEAN TO 'VALIDATE' A SLOPE STABILITY MODEL?

In geotechnical engineering, numerical models are routinely used to support design decisions, yet the concept of model validation is often applied inconsistently. In many slope stability studies, validation is implicitly assumed once a model produces a Factor of Safety (FoS) above an accepted threshold. However, achieving a target FoS does not, in itself, demonstrate that a modelling approach can realistically represent slope behaviour or failure mechanisms.

Model validation, in its strict sense, requires demonstrating that a modelling framework is capable of reproducing observed behaviour under known conditions. For slope stability problems, this extends beyond matching a single FoS value and instead requires the ability to replicate key aspects of real slope response, including deformation patterns, failure mechanisms, spatial extent of damage, and, where applicable, the timing or progression of instability.

Back-analysis as the basis for validation

Back-analysis of documented slope failures remains the most defensible method for validating slope stability models. In this context, validation involves applying a modelling approach to a slope for which the failure mechanism, geometry, deformation history, and geological controls are reasonably well constrained, and demonstrating that the model can reproduce those observations using plausible material parameters.

Critically, this requires more than demonstrating that a model becomes unstable at some point during strength reduction. A validated model should, at a minimum:

- reproduce the observed failure mechanism (eg sliding, toppling, combined modes).
- predict deformation patterns that are consistent with monitoring data or post-failure mapping.
- localise damage in regions consistent with observed failures.
- produce FoS values that are consistent with the observed stability state of the slope (ie stable slopes should yield FoS values greater than unity, whereas slopes that have failed or are actively failing should yield FoS values less than or close to unity).

FoS is a scalar quantity, whereas slope behaviour is inherently spatial and kinematic. Two models may report similar FoS values while predicting fundamentally different deformation mechanisms or failure volumes. Conversely, a model may reproduce realistic deformation patterns while producing a different FoS. Without validation against observed behaviour, FoS alone cannot confirm that a modelling framework captures the true mechanics of failure.

Limited validation of commonly used methods in hard rock slopes

It is important to distinguish between the extensive validation history of slope stability methods in soil and soft-rock engineering, and their application to large-scale hard rock open pit slopes. Limit equilibrium methods combined with Mohr–Coulomb constitutive behaviour have been widely validated for embankments, cut slopes, and natural slopes in soils and weak rock, where failure mechanisms are often relatively shallow, continuous, and governed by low confinement conditions. In these contexts, numerous documented case histories demonstrate that LEM-based FoS values, when interpreted appropriately, can provide realistic and reliable indicators of stability.

However, this validation base does not readily extend to hard rock open pit slopes. Large open pit failures typically involve complex three-dimensional mechanisms, strong confinement effects, stress redistribution during excavation, brittle fracture, progressive damage, and scale effects that are not naturally captured by traditional LEM formulations or by perfectly plastic Mohr–Coulomb representations. In such settings, failure surfaces are rarely predefined or continuous in the classical sense assumed by LEM, and deformation often develops progressively through cracking, yielding, and localisation rather than sudden global collapse.

Continuum-based analyses using linear or bilinear Mohr–Coulomb or linearised Hoek–Brown models, where the nonlinear Hoek–Brown strength envelope is approximated by an equivalent linear Mohr–Coulomb envelope over a selected stress range, face well-recognised limitations when applied to hard rock slopes. While such models can identify regions of yielding or global instability under strength reduction, they often do not reproduce realistic post-peak deformation, progressive failure, or damage localisation consistent with observed hard rock slope behaviour, particularly in situations where confinement-dependent strength and stiffness degradation play a critical role. As a result, although LEM and conventional constitutive models are widely trusted and routinely applied in open pit slope design, their ability to represent real hard rock slope failure mechanisms is often assumed rather than demonstrated. This gap between widespread application and limited published validation highlights the need for explicit back-analysis of hard rock failures when interpreting FoS values derived from these methods.

More advanced formulations, such as strain-softening Hoek–Brown and Itasca's Improved Model for Advanced Strain Softening (IMASS), have been developed to address some of these limitations; however, their application in practice remains less standardised. Strain-softening Hoek–Brown models are implemented in multiple forms across different software platforms, typically requiring selection of softening laws and parameters for which limited published guidance exists, leading to variability in application and results. IMASS similarly represents a significant development in incorporating anisotropy and strain softening, but publicly available methodologies for parameter derivation and calibration remain limited, and published validation through case studies is relatively sparse. In addition, its treatment of anisotropy relies on Mohr–Coulomb-type formulations along weakness planes, inheriting some of the associated limitations. As a result, despite their potential, these models are not yet as widely or consistently applied in open pit slope design practice, and they do not fully resolve the broader challenges of representing confinement-dependent strength, progressive failure, and deformation localisation in hard rock slopes.

Lessons from IUCM validation studies

In contrast, the development of the Improved Unified Constitutive Model (IUCM) placed explicit emphasis on validation through back-analysis of documented case histories. Ford, Lucas and Vakili (2020) presented several open pit case studies in which the IUCM was applied to reproduce observed slope behaviour, including deformation patterns, failure mechanisms, and regions of damage. Importantly, these studies reported both successful replications and cases where the model was unable to accurately reproduce observed behaviour, providing transparency regarding the model's limitations.

However, the authors also noted that many of the case studies were subject to confidentiality constraints, which limited the level of detail that could be published regarding geology, material parameters, structural models, and mine-scale geometry. While sufficient information was provided to demonstrate the general validity of the approach, full disclosure of inputs and assumptions was not always possible, restricting the ability of third parties to independently interrogate or reproduce the analyses.

SOFTWARE AND CONSTITUTIVE MODELS COMPARED

To investigate the extent to which Factor of Safety (FoS) is influenced by modelling framework rather than input data, this study compares results obtained using a subset of widely adopted slope stability software and constitutive models. The tools and models selected reflect current industry practice in open pit slope design, particularly in hard rock mining, and represent the approaches most commonly relied upon for both operational and long-term design decisions.

The objective of this comparison is not to provide an exhaustive review of all available software or material models, but to focus on those that dominate practice and therefore underpin the majority of published FoS-based slope assessments.

Software platforms considered

The following categories of software are included in this study.

Limit equilibrium methods (LEM)

Limit equilibrium methods remain the most widely used class of tools for open pit slope design. In this study, commonly used industry software packages, including Slide2 and Slide3, are considered as representative LEM implementations. These tools compute FoS by satisfying global force and/or moment equilibrium along assumed failure surfaces. While efficient and well established, LEM inherently relies on predefined kinematic assumptions and does not explicitly simulate stress redistribution, deformation, or progressive failure.

Continuum-based numerical tools

Continuum-based numerical modelling tools are widely used to complement or extend LEM analyses, particularly for deep pits, complex geometries, and stress-sensitive problems. The following platforms are considered as representative of common practice:

- **FLAC/FLAC3D** (explicit finite difference method).
- **RS2 and RS3** (implicit finite element method).
- **Abaqus** (implicit and explicit finite element method).

These tools enable stress–strain-based analyses and allow FoS to be derived using the shear strength reduction (SSR) method. As discussed earlier, the interpretation of FoS in these frameworks depends strongly on solver type (implicit versus explicit), numerical controls, and the definition adopted to identify instability or failure.

Constitutive models considered

Across the software platforms listed above, this study focuses on two constitutive modelling approaches that represent the most commonly applied and the most contrasting philosophies in current open pit slope practice.

Mohr–Coulomb (elastic-perfectly plastic)

The Mohr–Coulomb (MC) model remains the most widely used constitutive model in both LEM and continuum-based analyses of open pit slopes. It is characterised by a linear strength envelope, constant stiffness, and elastic-perfectly plastic behaviour. In numerical applications, peak strength parameters are typically derived from laboratory testing or from linearised approximations of nonlinear criteria such as Hoek–Brown.

Despite its simplicity and computational efficiency, the MC model does not inherently account for confinement-dependent strength, progressive damage, stiffness degradation, or post-peak softening. As a result, its application to large-scale hard rock slopes relies heavily on the assumption that peak strength parameters alone provide a sufficient representation of rock mass behaviour.

Inclusion of the Improved Unified Constitutive Model (IUCM)

In addition to the conventional Mohr–Coulomb model, this study includes the **Improved Unified Constitutive Model (IUCM)** (Vakili, 2016). The IUCM was developed to address several recognised limitations of traditional MC-based representations in hard rock slope applications, particularly where confinement dependency, anisotropy, progressive damage, and stiffness degradation influence large-scale behaviour.

The IUCM incorporates a continuous representation of rock mass strength, confinement-dependent post-peak behaviour, evolving dilation, and stiffness softening, and has been implemented within continuum-based numerical frameworks. While its adoption is not yet as widespread as MC-based models, it has been applied in a growing number of complex geomechanical problems and validated against selected underground and open pit case histories.

For the purposes of this study, the IUCM is included as a **contrast case** to conventional modelling approaches, allowing assessment of how a more physics-informed constitutive formulation influences:

- predicted deformation patterns
- failure mechanisms
- derived FoS values under otherwise identical modelling conditions.

CASE STUDY 1 – BACK-ANALYSIS OF A HARD ROCK OPEN PIT SLOPE FAILURE

Case study context

This case study relates to a hard rock open pit operation in Western Australia that experienced a multi-batter slope failure in early 2021. Prior to the failure, the pit had undergone multiple slope design iterations driven by geotechnical stability assessments and independent review. The case has previously been presented at a high level to illustrate the economic implications of advanced numerical modelling; however, the present paper focuses specifically on the geotechnical modelling approaches, assumptions, and interpretation of Factor of Safety (FoS).

To preserve confidentiality, the name of the operation has been removed. All geological, material, and modelling information presented here is derived from a detailed post-failure investigation and numerical back-analysis report.

Pre-failure slope design based on 2D limit equilibrium analysis

Initial slope designs for the hanging wall were assessed using **two-dimensional limit equilibrium methods (LEM)** implemented in **Slide**. These analyses formed the primary basis for early design decisions and independent engineering opinion.

The 2D LEM results indicated marginal stability for the proposed hanging wall geometry, with computed FoS values approaching or falling below commonly adopted acceptance criteria. In response, the slope design was modified by:

- reducing the overall hanging wall slope angle
- introducing a wide geotechnical decoupling berm.

These design changes were implemented explicitly to increase FoS as calculated by the 2D LEM analyses. While effective in meeting FoS targets, the resulting design had a substantial adverse impact on the mine plan and project economics. An example of the original and modified designs derived from 2D LEM is shown in Figure 1.

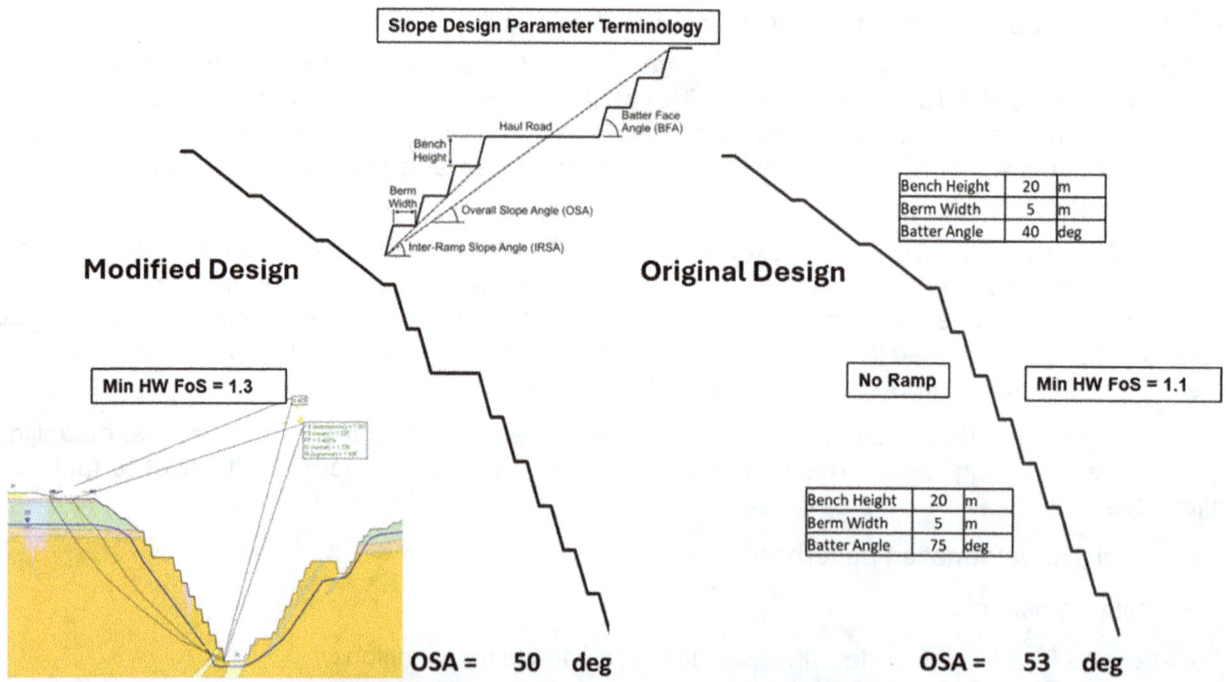

FIG 1 – Original 2D LEM-driven slope design.

Redesign informed by continuum modelling and spatial FoS interpretation

While the 2D LEM approach satisfied conventional FoS criteria after geometric modification, it did not explicitly account for three-dimensional stress redistribution or progressive rock mass behaviour. To investigate these effects, three-dimensional continuum modelling was undertaken.

Following concerns regarding the economic impact of the conservative 2D LEM-based design, three-dimensional continuum modelling was undertaken using **FLAC3D with the IUCM**. These analyses explicitly simulated excavation sequencing, three-dimensional pit geometry, and stress redistribution effects not represented in the 2D LEM framework.

The IUCM analyses indicated that, for the same geometry and material inputs, the hanging wall exhibited a **global FoS exceeding 1.5**, suggesting that the decoupling berm and slope flattening were not required to maintain overall stability. On this basis, the slope was redesigned to remove the geotechnical berm and reinstate steeper overall slope angles.

Importantly, although the global FoS was high, the IUCM results revealed **significant spatial variability in stability**. A localised zone of reduced FoS (approximately **1.2**) was identified on the **north-facing wall**, associated with weaker saprolite and transitional materials and reduced confinement near the crest. This spatial distribution of FoS is illustrated in Figure 2.

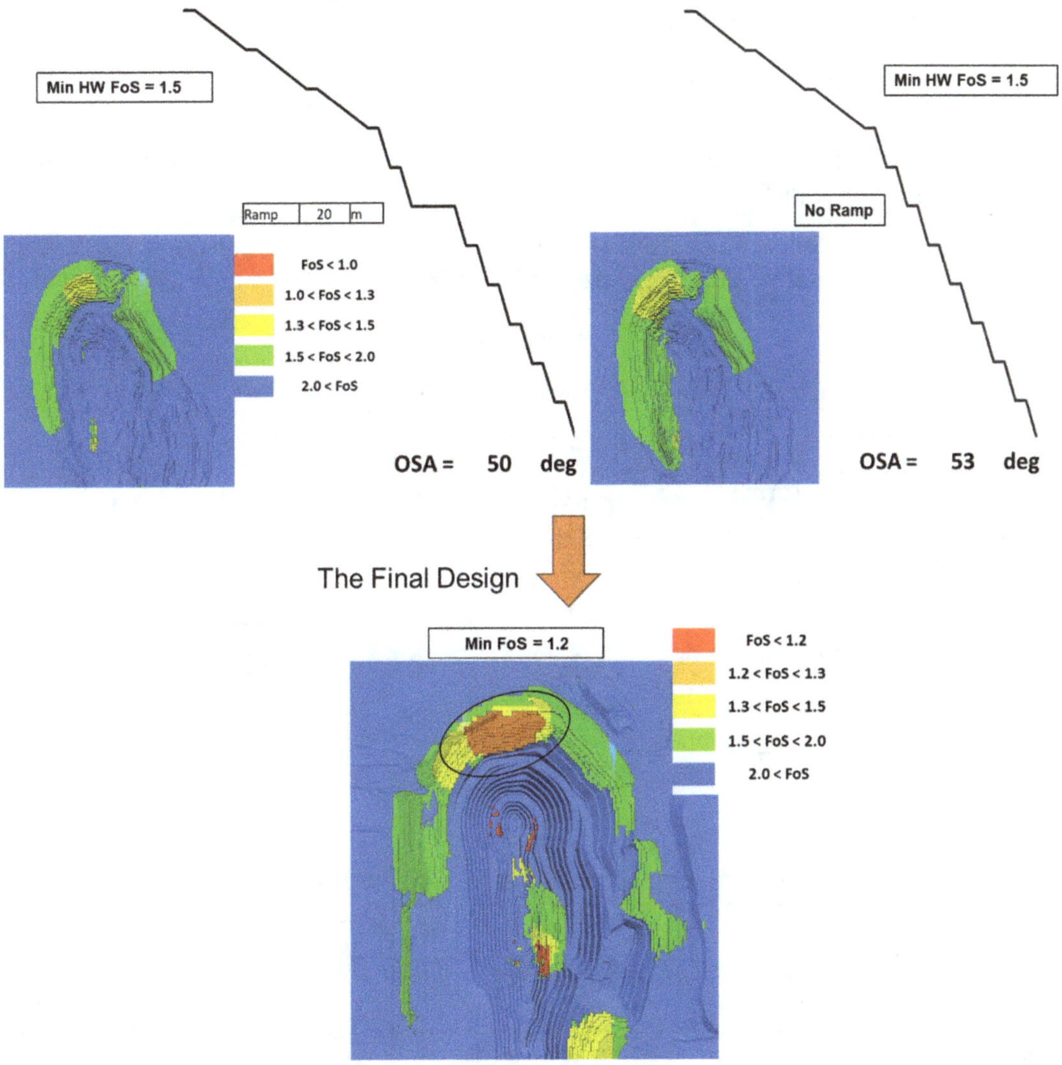

FIG 2 – IUCM-based redesign and spatial FoS distribution.

Post-failure pit geometry used for comparative modelling

In addition to the pre-failure design and back-analysis described in sections 'Pre-failure slope design based on 2D limit equilibrium analysis' and 'Back-analysis as the basis for validation', respectively, a separate modelling exercise was undertaken to directly compare FoS outcomes using different constitutive formulations under controlled conditions. This comparison was performed using a later pit geometry, corresponding to a pit surface developed after the failure event.

Accordingly, the results presented in this subsection do not represent a back-analysis of the historical failure. Rather, they provide a method comparison study, intended to isolate the influence of constitutive model selection on computed FoS for a subsequent pit design stage.

The comparison was conducted using FLAC3D and included two constitutive formulations:

- Mohr–Coulomb (MC), representing standard industry practice.
- The Improved Unified Constitutive Model (IUCM), representing a confinement-dependent rock mass formulation.

Geometry, zoning, groundwater assumptions, boundary conditions, and excavation state were held constant between the two models.

Analysis methods and modelling assumptions

In order to isolate the effect of rock mass behaviour, the modelling strategy intentionally retained the **Mohr–Coulomb formulation for the laterite and saprolite domains** in both simulations. These units behave predominantly as soil-like weathered materials, and Mohr–Coulomb can provide a reasonable first-order representation of their strength behaviour.

The key difference between the two simulations therefore lies in the representation of the **transitional and fresh rock domains**:

- In the baseline model, these units were represented using Mohr–Coulomb with linear strength envelopes as defined in Table 1.
- In the IUCM model, these units were represented using confinement-dependent, anisotropic, and strain-softening behaviour as defined in Table 2.

TABLE 1

Mohr–Coulomb material for Case Study #1.

Domain	Density (t/m^3)	Mass Young's modulus (GPa)	Friction (°)	Cohesion (MPa)	Tensile strength (MPa)
Laterite	1.78	0.25	31	0.085	0.035
Saprolite	2.56	0.25	31	0.085	0.035
Transitional	2.56	1.3	33	0.436	0.005
Fresh rock	2.75	5.7	56	2.0	0.35
Shale	2.75	4.5	42	1.2	0.35

TABLE 2

IUCM material properties for Case Study #1.

Domain	Density (t/m^3)	Intact Young's modulus (GPa)	UCS (MPa)	GSI	miMax	miMin	Anisotropy factor	Residual friction (°)	Residual cohesion (MPa)
Laterite	MC model used with same parameters as Table 1								
Saprolite									
Transitional	2.56	11	12	35	16	8	5.0	38	0
Fresh rock	2.75	11	113	60	16	8	3.25	45	0
Shale	2.75	11	62.5	55	6	3	3.25	45	0

This approach ensures that differences in FoS arise from the constitutive treatment of hard rock, rather than from changes in soil-like material behaviour. Figure 3 shows a comparison of Mohr–Coulomb (MC) and Hoek–Brown (HB) strength envelopes for the transitional, fresh rock, and shale domains, plotted in principal stress space (σ_1-σ_3) for confinement up to 2 MPa. The MC envelopes are derived from the parameters in Table 1, while the HB curves are based on the rock mass properties in Table 2.

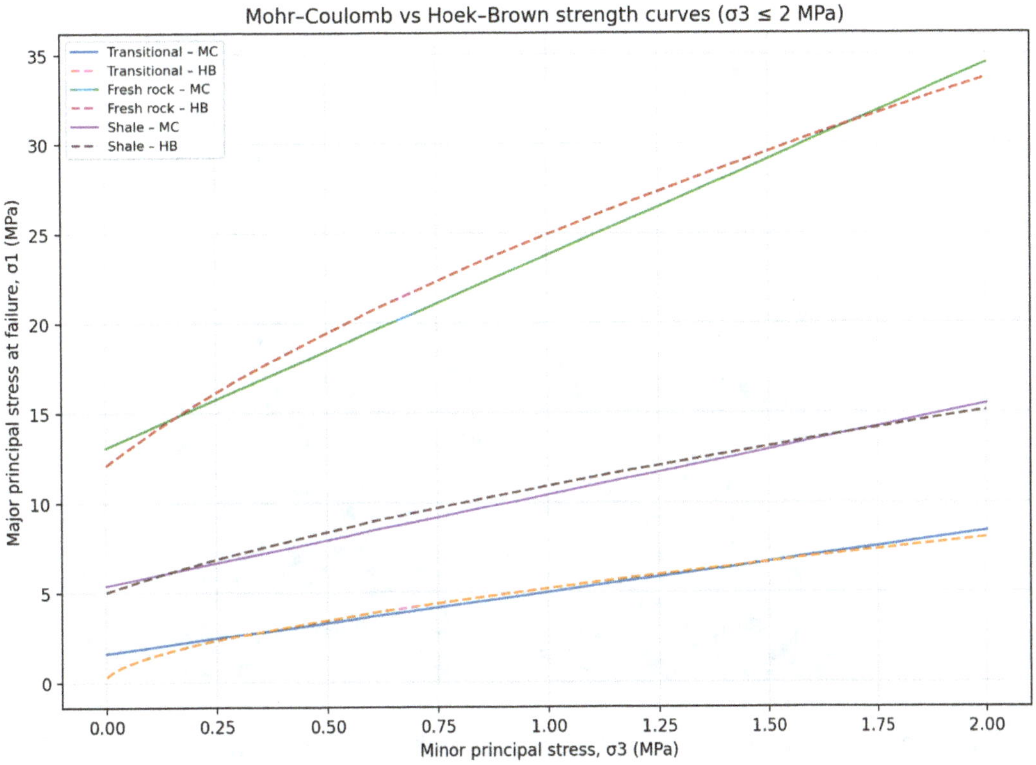

FIG 3 – Comparison of Mohr–Coulomb and Hoek–Brown strength envelopes in σ_1-σ_3 showing differences for Case Study #1.

The IUCM formulation was adopted to assess the influence of confinement-dependent strength and progressive degradation on the predicted stability response. Adopted domains are shown in Figure 4, MC parameters in Table 1 and IUCM parameters are summarised in Table 2, note these properties.

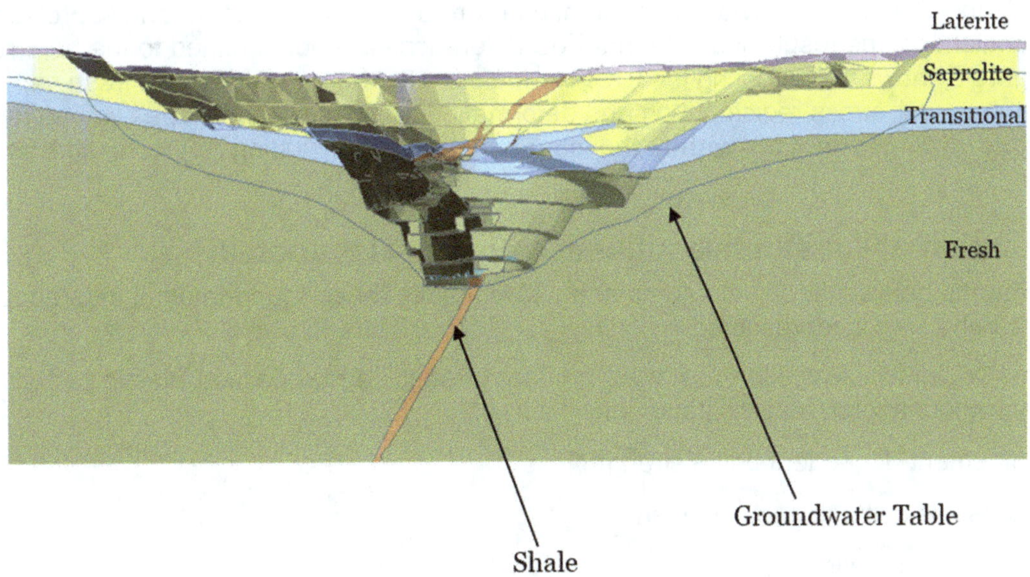

FIG 4 – Material domains for Case Study #1.

Comparison of FoS distributions – Mohr–Coulomb versus IUCM

Figure 5 presents the FoS distributions obtained using Mohr–Coulomb and IUCM for the post-failure pit geometry.

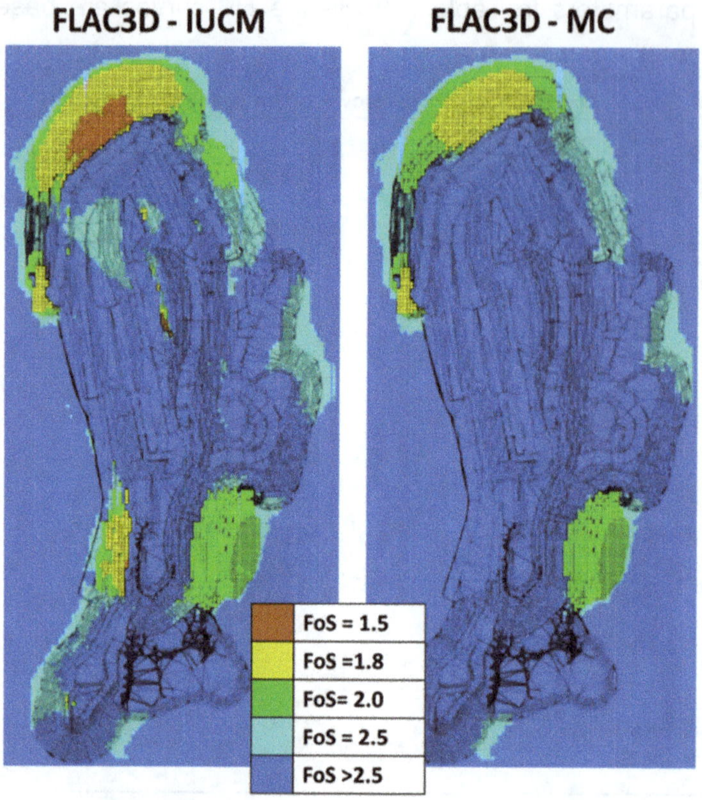

FIG 5 – FoS comparison between Slide3, FLAC3D Mohr–Coulomb (MC), and FLAC3D IUCM.

The IUCM model produced the lowest minimum FoS across all analyses, with a minimum value of approximately FoS = 1.5. In contrast, the Mohr–Coulomb model predicted higher FoS values across most slope segments, including extensive regions with FoS exceeding 2.0.

A key observation is that the Mohr–Coulomb results indicate high stability within the competent rock units, with FoS values remaining elevated across much of the fresh rock slope. In contrast, the IUCM results show meaningful FoS reductions extending into these hard rock domains.

Although Mohr–Coulomb behaviour was intentionally retained for the laterite and saprolite domains in both simulations, the results indicate that FoS differences are not confined to the hard rock units alone. Figure 5 shows that adopting the IUCM formulation for the transitional and fresh rock domains modifies the overall stress redistribution and deformation response of the slope system, resulting in **measurable changes in FoS within the cover sequence**, including saprolite and transported materials.

Interpretation of constitutive influence on global response

The divergence between Mohr–Coulomb and IUCM results reflects fundamental differences in how rock mass behaviour is represented.

The Mohr–Coulomb formulation assumes a linear strength envelope and elastic-perfectly plastic response, without explicit representation of:

- confinement-dependent brittle strength loss
- progressive damage accumulation
- stiffness degradation
- or post-peak strain localisation.

In contrast, the IUCM explicitly incorporates confinement sensitivity and progressive post-peak degradation in rock mass behaviour. As a result, IUCM predicts reduced stability in rock units where stress redistribution and confinement changes influence strength mobilisation.

Importantly, although the soil-like materials were treated identically in both models, changing only the basement rock constitutive behaviour influenced the overall pit response. This indicates that constitutive model selection in hard rock does not merely affect local behaviour, but can alter the global stability field.

Key outcomes from the comparative modelling

The comparative modelling exercise conducted on the post-failure pit geometry demonstrates the following:

- FoS values are strongly dependent on constitutive model selection, even when geometry, material zoning, groundwater conditions, and boundary conditions are unchanged.
- The IUCM model produced the lowest minimum FoS (≈1.5) among the models tested, despite identical treatment of the saprolite and laterite units in both simulations.
- Differences in FoS are not confined to the hard rock domains. Although Mohr–Coulomb behaviour was retained for the soil-like units in both models, adopting IUCM for the transitional and fresh rock altered the stress redistribution and deformation compatibility of the entire slope system, resulting in measurable FoS differences within the overlying cover sequence.
- The most pronounced divergence occurs within competent rock, where the Mohr–Coulomb model predicts consistently higher FoS values relative to IUCM. This suggests that linear elastic–perfectly plastic formulations may overpredict stability in confinement-dependent hard rock environments.
- Constitutive representation of the basement rock mass materially influences not only local rock slope behaviour, but also the global stability field, including the response of weathered and soil-like units.

CASE STUDY 2 – STEEP HARD ROCK SLOPES SUCCESSFULLY MINED UNDER PREDICTED IUCM PERFORMANCE

Case study context

The second case study is drawn from a deep, structurally controlled hard rock open pit mine in South Australia that was successfully mined to completion at a depth exceeding 360 m. The pit included steep overall slope angles and highly persistent, unfavourably oriented discontinuity sets, raising concerns regarding the potential for large-scale instability.

Lucas, Vakili and Hutchison, (2020) describe the geotechnical setting and slope stability challenges of this mine, with particular emphasis on the west wall, which was undercut by a widely spaced but highly persistent joint set dipping into the excavation. In addition to structural controls, concerns were raised regarding the influence of mining-induced stress redistribution due to the depth and three-dimensional confinement of the pit geometry.

Unlike many published case histories where modelling is limited to FoS values, this mine provides an opportunity to compare predicted stability and deformation patterns against extensive monitoring data collected throughout mining.

Early assessments and conventional stability predictions

Prior to adopting advanced three-dimensional modelling, the mine was assessed using multiple conventional stability methods. A summary of these earlier predictions indicates substantial variability in computed stability outcomes:

- Empirical assessments suggested FoS values as low as approximately **0.85**.

- Two-dimensional limit equilibrium (LE) analyses produced FoS values in the range of approximately **0.8 to 1.1**.
- Two-dimensional finite element analyses (RS2) produced FoS values in the range of approximately **1.0 to 1.2**.

These results collectively implied that the slope configuration was either unstable or marginally stable, and that mining to the proposed steep design would involve a high risk of large-scale failure.

IUCM modelling and predicted feasibility of aggressive slope designs

To address uncertainty in the conventional assessments, the mine adopted three-dimensional numerical modelling using FLAC3D. Lucas *et al* (2020) report that multiple phases of 3D modelling were undertaken between 2015 and 2019, focusing on larger-scale stability and deformation mechanisms rather than purely on slip-surface FoS calculations.

The Improved Unified Constitutive Model (IUCM) was selected due to its ability to represent confinement dependency, progressive damage accumulation, and anisotropic rock mass response under higher stress conditions.

The IUCM modelling predicted that the mine could successfully excavate steep walls with an acceptable Factor of Safety against large-scale instability, and that mining-induced stress would not trigger deep-seated failure mechanisms. Based on these results, the mine proceeded with the aggressive steep wall design.

Validation through monitoring and observed deformation behaviour

A key strength of this case study is that the numerical modelling was validated against extensive monitoring data sets. Lucas *et al* (2020) report that the mine employed routine laser scanning and that modelled displacements showed good agreement with observed displacement patterns and magnitudes.

The modelling was able to reproduce:

- the spatial distribution of displacement along the west wall
- the structural influence on movement mechanisms
- the magnitude of observed displacement development over time.

The mine was ultimately excavated to the final pit design, which Lucas *et al* (2020) report as being consistent with modelling predictions indicating FoS values greater than approximately 1.1.

Later comparative study – Slide3 versus IUCM (SlopeX plugin)

Following completion of the mine, a separate study was conducted to compare FoS outcomes between 3D limit equilibrium analysis (Slide3) and IUCM-based numerical modelling implemented using the SlopeX plugin for FLAC3D.

This comparison was performed on a simplified representation of the pit and rock mass domain geometries. Simplification was necessary because Slide3 required reduction of geometric complexity to achieve a workable limit equilibrium model, whereas the FLAC3D-based SlopeX workflow was capable of incorporating more complex surfaces and internal zoning. To ensure comparability, the simplified Slide3 geometry was replicated in FLAC3D and analysed using IUCM. Material domains are shown in Figure 6, model inputs for Slide3 are listed in Table 3 and for IUCM in Table 4.

For the comparative analysis between Slide3 and IUCM in Case Study 2, a generalised Hoek–Brown (HB) formulation was adopted within the Slide3 model. This was done to ensure that the underlying strength representation was as consistent as possible with the IUCM framework, which is fundamentally based on Hoek–Brown-type behaviour. In this context, HB parameters were derived from the same rock mass characterisation used for the IUCM model, allowing a more direct comparison between the two approaches. This represents an intentional departure from the earlier discussion of typical industry practice, where Mohr–Coulomb formulations are more commonly applied. The use of HB in this case was therefore aimed at minimising differences arising purely from strength representation, so that the comparison more clearly reflects differences in modelling framework and failure simulation.

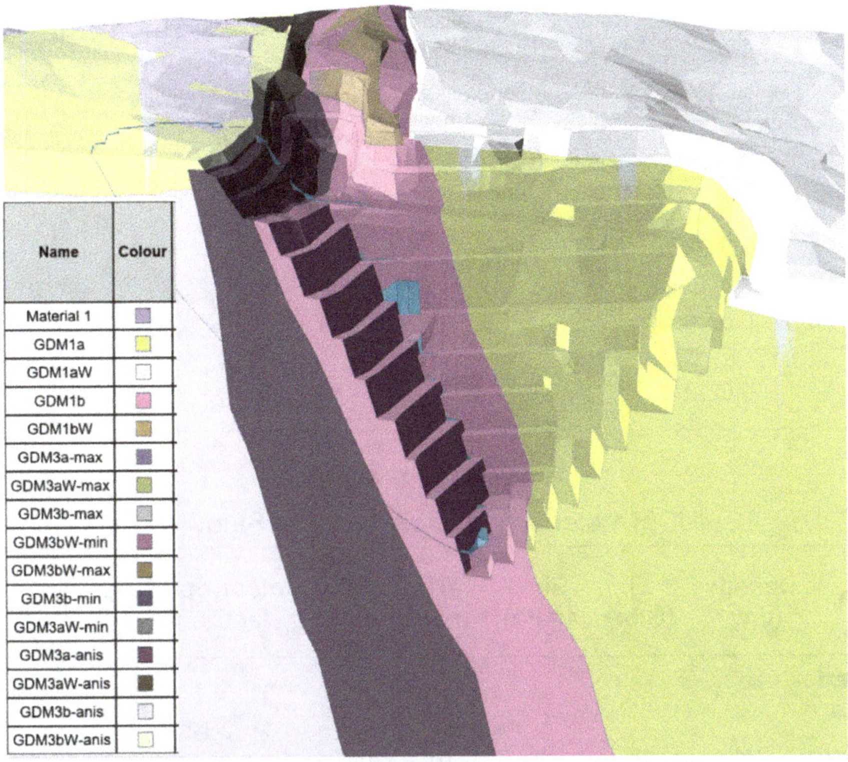

FIG 6 – Material domains for Case Study #2.

TABLE 3

IUCM material properties for Case Study #2.

Name	Colour	Unit weight (MN/m³)	Water surface	Failure criterion	Cohesion (MPa)	Phi (°)	UCS of intact rock (MPa)	Tensile cut-off type	GSI	mi	D
Material 1		0.02	Water Table	Mohr Coulomb	0.001	35					
GDM1a		0.0333	Water Table	Generalised Hoek–Brown			48	None	80	10	0
GDM1aW		0.0333	Water Table	Generalised Hoek–Brown			40	None	70	10	0
GDM1b		0.0284	Water Table	Generalised Hoek–Brown			48	None	69	10	0
GDM1bW		0.0284	Water Table	Generalised Hoek–Brown			40	None	60	10	0
GDM3a-max		0.0274	Water Table	Generalised Hoek–Brown			33	None	55	10	0
GDM3aW-max		0.0274	Water Table	Generalised Hoek–Brown			29	None	50	10	0
GDM3b-max		0.0274	Water Table	Generalised Hoek–Brown			18	None	40	10	0
GDM3bW-min		0.0274	Water Table	Generalised Hoek–Brown			12	None	50	10	0
GDM3bW-max		0.0274	Water Table	Generalised Hoek–Brown			50	None	50	10	0

Name	Colour	Unit weight (MN/m^3)	Water surface	Failure criterion	Cohesion (MPa)	Phi (°)	UCS of intact rock (MPa)	Tensile cut-off type	GSI	mi	D
GDM3b-min		0.0274	Water Table	Generalised Hoek–Brown			12	None	40	5	0
GDM3aW-min		0.0274	Water Table	Generalised Hoek–Brown			19.3	None	50	5	0
GDM3a-anis		0.0274	Water Table	Generalised Anisotropic							
GDM3aW-anis		0.0274	Water Table	Generalised Anisotropic							
GDM3b-anis		0.0274	Water Table	Generalised Anisotropic							
GDM3bW-anis		0.0274	Water Table	Generalised Anisotropic							

TABLE 4

IUCM material properties for Case Study #2.

Domain	Density (t/m^3)	Ei (GPa)	Sigci (MPa)	Mi max	Mi min	Anisotropy factor	Dip (°)	Dip direction (°)	GSI
Weathered domains									
GDM1a-w	3.4	45	40	10	n/a	n/a	n/a	n/a	70
GDM1b-w	2.9	40	40	10	n/a	n/a	n/a	n/a	60
GDM3a-w	2.8	40	29	10	5	1.5	74	103	50
GDM3b-w	2.8	25	18	10	5	1.5	74	103	40
Fresh domains									
GDM1a	3.4	69	48	10	n/a	n/a	n/a	n/a	80
GDM1b	2.9	53	48	10	n/a	n/a	n/a	n/a	70
GDM3a	2.8	53	33	10	5	1.5	74	103	55
GDM3b	2.8	53	18	10	5	1.5	74	103	40

Results of the sensitivity analysis for Massive hard rock is shown in Figures 7 and 8 and for soft and highly jointed rock shown in Figure 9. The study demonstrated that differences between Slide3 and IUCM based model could be significant, particularly under massive rock mass conditions. The results suggested discrepancies in minimum FoS of up to approximately 100 per cent in some scenarios, and substantial differences in predicted failure volume and location.

The Hoek–Brown disturbance factor (D) was applied consistently in both the Slide3 and IUCM (FLAC-based) models, ensuring that equivalent rock mass strength conditions were represented across the comparative analyses.

None of the sensitivity cases presented are intended to represent the design case in terms of absolute FoS values, as the comparative models incorporate geometric and material simplifications, including the treatment of groundwater conditions. However, the dry, anisotropic case with D = 0 most closely reflects the material characterisation adopted in the design model.

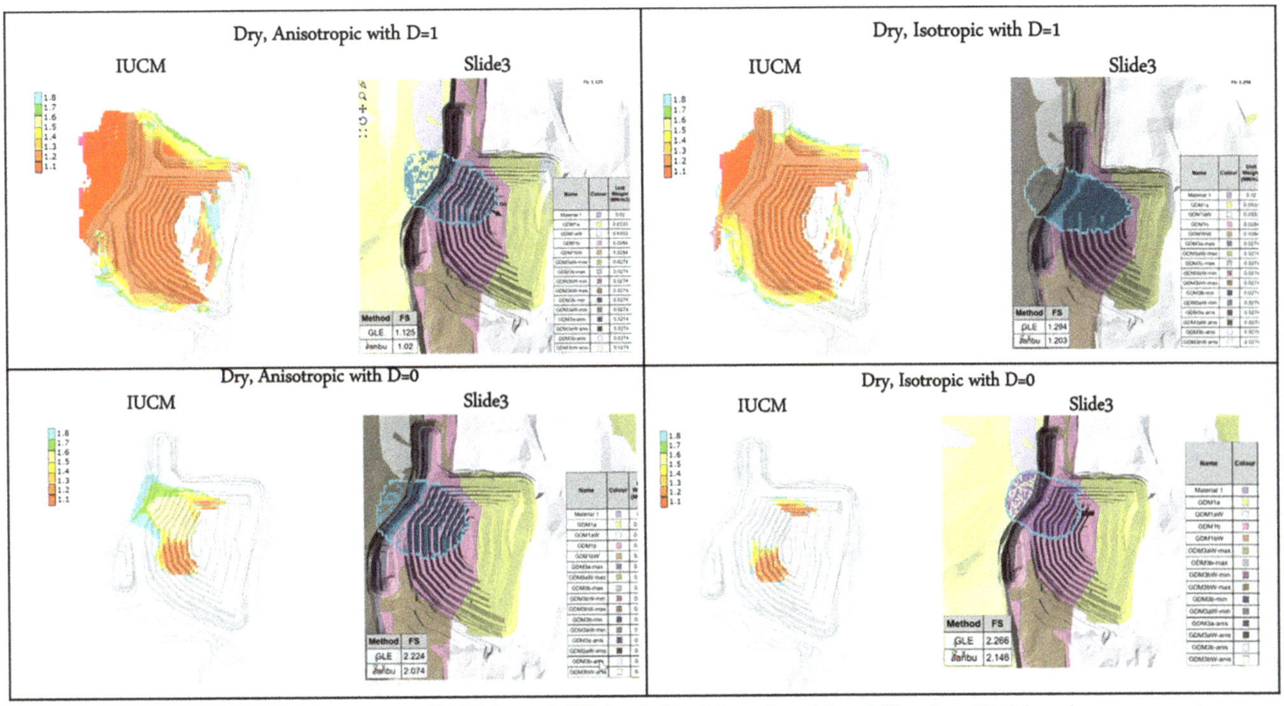

FIG 7 – Comparison between IUCM and Slide 3 for Massive Hard Rock with Varying property changes (Case Study #2 and Dry conditions).

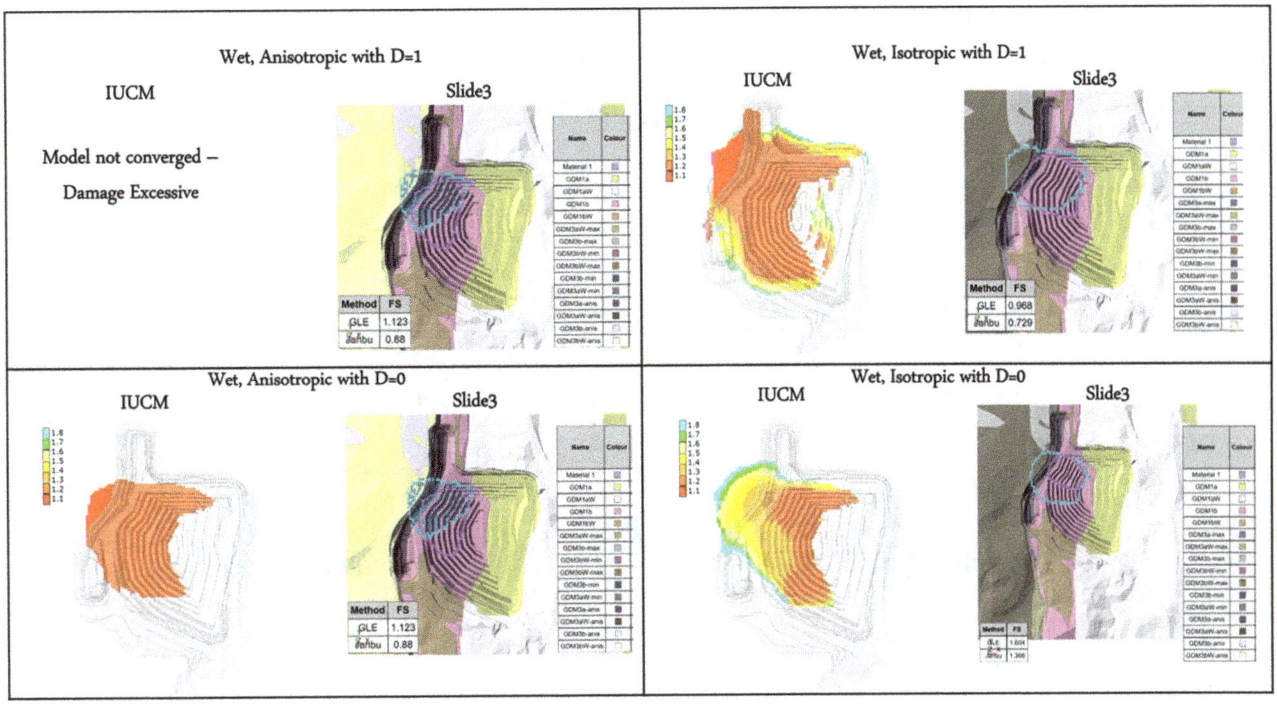

FIG 8 – Comparison between IUCM and Slide 3 for Massive Hard Rock with Varying property changes (Case Study #2 and Wet conditions).

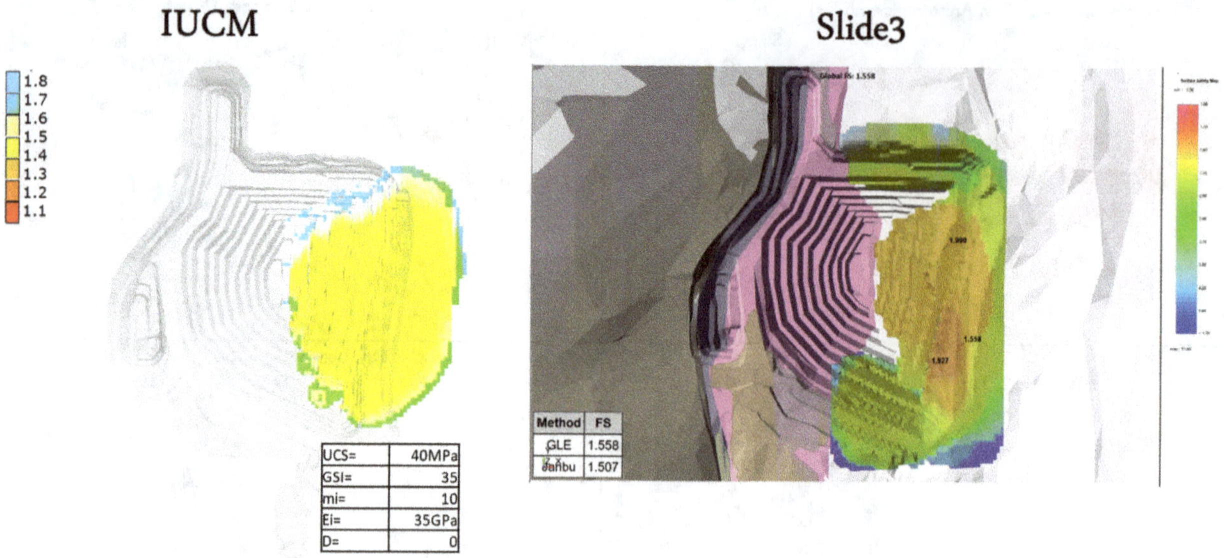

FIG 9 – Comparison between IUCM and Slide 3 for highly jointed and soft rock with varying property changes (Case Study #2 and Dry conditions).

A key finding from this comparative exercise was that agreement between methods improved when a circular or soil-like failure mechanism was expected, while divergence increased when the stability mechanism depended on progressive rock mass damage or confinement-controlled behaviour. A summary of the FoS for varying conditions between the two approaches are in Tables 5 and 6.

TABLE 5

Minimum FoS for multi-batter failure in massive rock mass.

	Dry						Wet					
	Iso			Aniso			Iso			Aniso		
	D=0	D=0.7	D=1	D=0	D=0.7	D=1	D=0	D=0.7	D=1	D=0	D=0.7	D=1
IUCM	1.1	<1	<1	1.1	<1	<1	<1	<1	<1	<1	<1	<1
Slide3D	2.2	1.7	1.3	2.1	1.5	1.1	1.6	1.4	<1	1.5	1.1	1.1
Difference	100%	70%	30%	91%	50%	10%	60%	40%	0%	50%	10%	10%

TABLE 6

Minimum FoS for multi-batter failure in highly jointed and weathered rock mass.

	Dry Iso D=0
IUCM	1.4
Slide3D	1.5
Difference	7%

Recent reanalysis using updated pit shell and stress field

A more recent reanalysis of the pit was undertaken using an updated pit shell and an updated *in situ* stress field interpretation. This modelling included a direct comparison between Mohr–Coulomb and IUCM constitutive representations using consistent geometry and boundary conditions (see Figure 10 for MC parameters and comparison with HB in σ_1-σ_3 space).

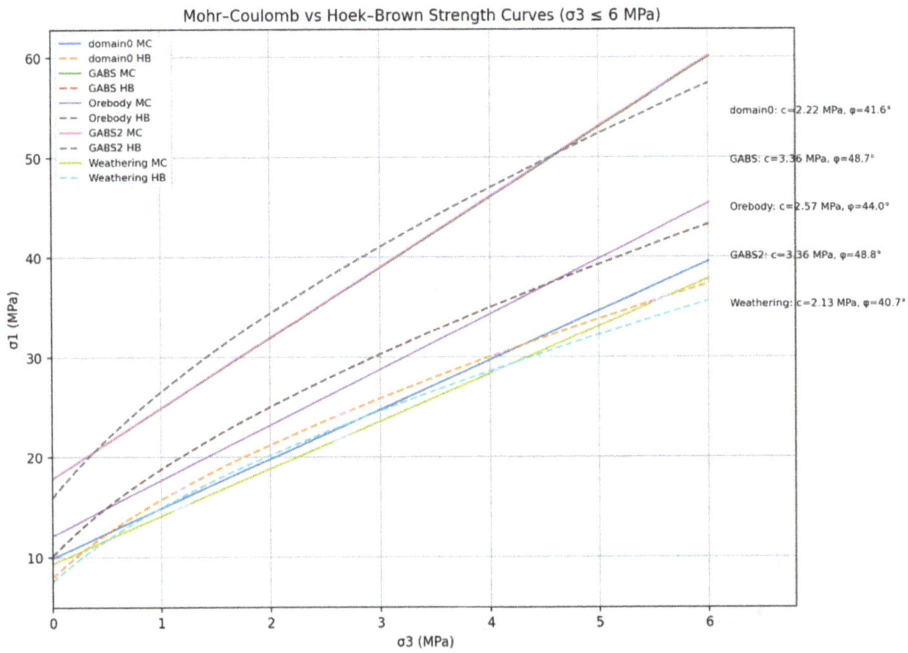

FIG 10 – Comparison of Mohr–Coulomb and Hoek–Brown strength envelopes for Case Study 2 materials in σ_1-σ_3 space.

The results of this updated analysis are presented in Figure 11, which shows that the Mohr–Coulomb formulation produced high FoS values with limited sensitivity in competent rock domains, while IUCM predicted a more differentiated FoS distribution and lower minimum FoS values. Notably, IUCM predicted meaningful stability gradients extending into hard rock regions, whereas Mohr–Coulomb produced predominantly high FoS values across the pit.

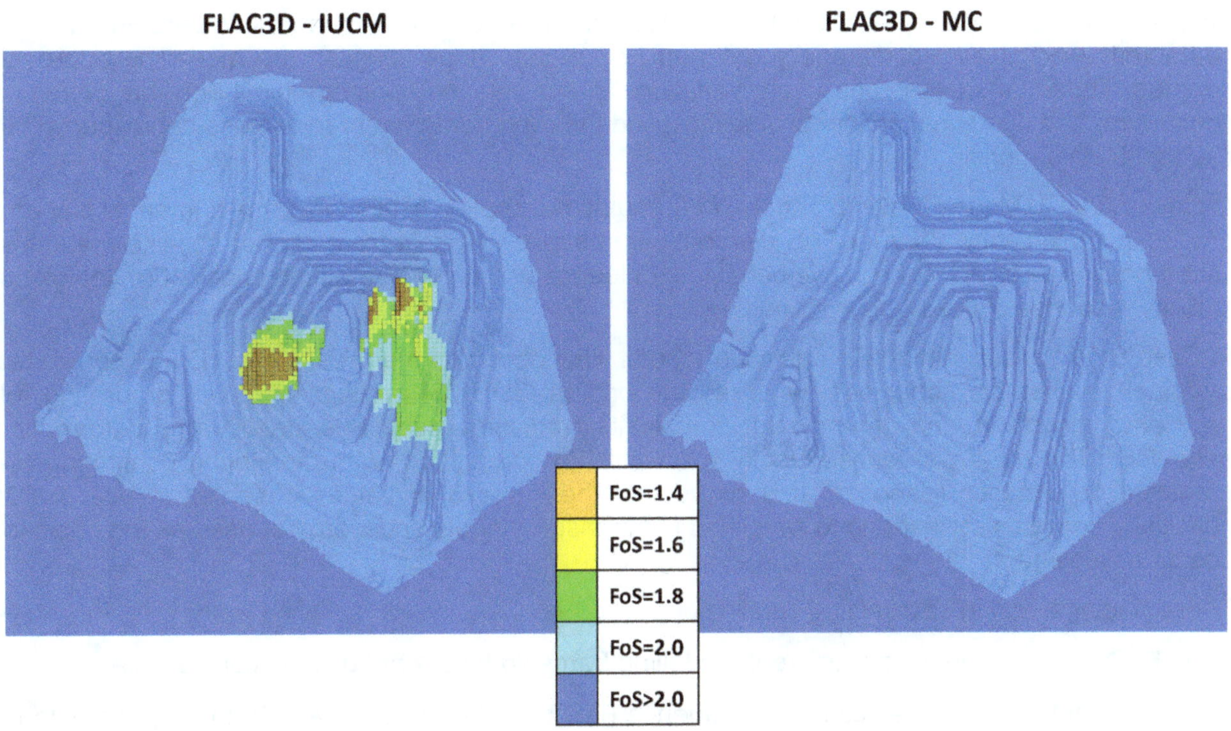

FIG 11 – Updated pit reanalysis comparing FoS distributions obtained using Mohr–Coulomb and IUCM under consistent geometry and updated stress field conditions.

This finding is consistent with Case Study 1 and supports the conclusion that constitutive model selection can materially alter not only local stability predictions, but the entire system response, including cover sequence behaviour and global FoS distribution.

Implications for FoS interpretation

Case study 2 highlights a key contradiction within common slope design practice. Conventional methods (empirical, LE, and 2D FEM) indicated marginal or unstable slope conditions, while the IUCM-based 3D modelling predicted acceptable stability and deformation behaviour. The mine proceeded with aggressive steep slopes and was successfully completed, with laser scanning monitoring confirming that observed deformation mechanisms and magnitudes were consistent with numerical predictions.

The results demonstrate that FoS should not be interpreted as an absolute and transferable measure of stability across modelling frameworks. Instead, the meaning of FoS depends on whether the analysis method can realistically represent:

- confinement-dependent rock strength mobilisation
- stress redistribution during excavation
- structural control and anisotropy
- progressive damage accumulation
- deformation compatibility at slope scale.

This case study therefore provides one of the rare published examples where a confinement-dependent constitutive model, combined with three-dimensional explicit modelling of structures, not only predicted a defensible FoS but also successfully predicted slope deformation trends that were validated through ongoing monitoring.

CONCLUSIONS

This study demonstrates that the Factor of Safety (FoS) is not an intrinsic property of a slope, but a model-dependent outcome governed by the numerical framework and constitutive assumptions used in its calculation.

Across two real-world open pit case studies, identical geometries and comparable material characterisation produced materially different FoS values depending on the modelling approach adopted. These differences arise from variations in analysis methodology (limit equilibrium versus continuum), the numerical definition of failure, and the representation of rock mass behaviour within the constitutive model.

In Case Study 1, Mohr–Coulomb and IUCM formulations applied within the same numerical platform produced significantly different FoS distributions, despite identical treatment of soil-like materials. Differences in hard rock constitutive behaviour influenced stress redistribution and, in turn, the global stability response of the slope system.

In Case Study 2, conventional empirical methods, limit equilibrium analysis, and 2D FEM predicted marginal or unstable conditions, whereas the original 3D IUCM design model-incorporating full geometric, structural, and groundwater complexity-predicted acceptable stability and deformation behaviour. The mine was subsequently excavated to completion, with monitoring confirming the predicted deformation response. It is noted that the lower SRF values observed in later IUCM sensitivity analyses correspond to simplified comparison models and were not used as the basis for design assessment.

These findings lead to several key conclusions:

- FoS values derived from different modelling frameworks are not directly comparable.
- Constitutive model selection can influence FoS to a degree comparable to, or greater than, reasonable parameter variation, even when models are calibrated to the same underlying geotechnical data set. While the derivation of model parameters introduces an additional source of variability, the results presented here demonstrate that differences in constitutive formulation alone can lead to substantial divergence in predicted FoS.
- Validation against observed deformation behaviour and failure mechanisms is essential when interpreting FoS values.

- An FoS value such as 1.2 does not have a universal meaning independent of the modelling assumptions used to derive it.

FoS should therefore be interpreted as a model-conditioned stability indicator rather than an absolute measure of safety. Defensible slope design requires explicit consideration of model-form uncertainty, constitutive representation, solver behaviour, and validation against observed performance, alongside traditional parameter sensitivity analysis. Without this broader perspective, numerical results risk reflecting the assumptions of the modelling approach rather than the true mechanics of the slope.

REFERENCES

Arndt, S and Vakili, A, 2024. Current state and limitations of 3D continuum models for slope stability models using Abaqus with octree meshing and IUCM, presented at Slope Stability Symposium 2024, Nova Lima (Belo Horizonte region), Brazil.

Cheng, Y M, Lansivaara, T and Wei, W B, 2007. Two-dimensional slope stability analysis by limit equilibrium and strength reduction methods, *Computers and Geotechnics*, 34(3):137–150.

Dawson, E M, Roth, W H and Drescher, A, 1999. Slope stability analysis by strength reduction, *Géotechnique*, 49(6):835–840.

de Bruyn, I, Mylvaganam, J, Jung, J, Le Roux, K, Young, D and Casparis, D, 2025. Approach to steepening of slopes in Pilbara iron ore mines, in *SSIM 2025: Fourth International Slope Stability in Mining Conference* (eds: J J Potter and J Wesseloo) (Australian Centre for Geomechanics: Perth). https://doi.org/10.36487/ACG_repo/2535_17

Duncan, J M, 1996. State of the art: Limit equilibrium and finite-element analysis of slopes, *Joumal of Geotechnical Engineering*, 122(7):577–596.

Ford, A, Lucas, D S and Vakili, A, 2020. Validation of the improved unified constitutive model for open pit applications, in *Slope Stability 2020: Proceedings of the 2020 International Symposium on Slope Stability in Open Pit Mining and Civil Engineering* (ed: P M Dight), pp 953–968 (Australian Centre for Geomechanics: Perth). https://doi.org/10.36487/ACG_repo/2025_63

Griffiths, D V and Lane, P A, 1999. Slope stability analysis by finite elements, *Géotechnique*, 49(3):387–403.

Hoek, E and Bray, J W, 1981. *Rock Slope Engineering*, 3rd edn (London: Institution of Mining and Metallurgy).

Hoek, E, Carranza-Torres, C and Corkum, B, 2002. Hoek–Brown failure criterion – 2002 edition, *Proceedings of the 5th North American Rock Mechanics Symposium (NARMS-TAC)*, pp 267–273.

Lucas, D S, Vakili, A and Hutchison, B J, 2020. Geotechnical design and performance of steep open pit slopes in South Australia, in Proceedings of the AusIMM Open Pit Conference 2020 (The Australasian Institute of Mining and Metallurgy: Melbourne).

Vakili, A, 2016. An improved unified constitutive model for rock material and guidelines for its application in numerical modelling, *Computers and Geotechnics*, 80:261–282. https://doi.org/10.1016/j.compgeo.2016.08.020

Wyllie, D C and Mah, C W, 2004. *Rock Slope Engineering: Civil and Mining,* 4th edn (London: Spon Press).

Author index

www.ingramcontent.com/pod-product-compliance
Lightning Source LLC
LaVergne TN
LVHW080545210926
848481LV00024B/2679